Geoenvironmental Disaster Reduction

Series Editor

The International Consortium on Geo-disaster Reduction (ICGdR) decided to create a series of books which could help to reduce the overall risk from natural hazards of different origin in concordance with the main strategy and objectives of ICGdR. These are:

1. Promote geo-disaster reduction for the benefit of human society and the natural environment, and capacity development, including education;
2. Combine international expertise and coordinate their efforts in geo-disaster reduction, thereby resulting in an effective international organization which will act as a partner in various projects; and
3. Promote regional and global, multidisciplinary activity on geo-disaster reduction.

The idea is to draw the attention of authors and research teams to the subject of mitigation of natural risks irrespective the detailed topic of hazards or region of occurrence. The International Consortium on Geo-disaster Reduction in cooperation with Springer Verlag with its broad variety of scientific interests and large international operation is a suitable platform to manage such an important mission. The ICCdR accepts proposals from members as well as non-members. The submitted books will go through the standard editorial process. The ICGdR will help to promote these books. We welcome your proposals.

Matthias Jakob · Scott McDougall ·
Paul Santi
Editors

Advances in Debris-flow Science and Practice

Springer

Editors
Matthias Jakob
BGC Engineering Inc.
Vancouver, BC, Canada

Paul Santi
Colorado School of Mines
Golden, CO, USA

Scott McDougall
Department of Earth and Ocean Sciences
University of British Columbia
Vancouver, BC, Canada

ISSN 2946-5710 ISSN 2946-5729 (electronic)
Geoenvironmental Disaster Reduction
ISBN 978-3-031-48690-6 ISBN 978-3-031-48691-3 (eBook)
https://doi.org/10.1007/978-3-031-48691-3

This Springer imprint is published by the registered company Springer Nature Switzerland AG
The registered company address is: Gewerbestrasse 11, 6330 Cham, Switzerland

Paper in this product is recyclable.

Foreword

Over the past few decades, various types of geo-disasters have frequently occurred around the world, triggered by tropical cyclones, landslides, earthquakes, floods, tsunamis, and volcanic eruptions. These disasters have caused huge damage to infrastructure, the natural environment, and a great loss of human life and property. As disasters are serious threats to human society, the reduction of geo-disasters has become an urgent and critical issue for the protection of human life and economic development. This is why we decided to create a new book series, to support risk reduction and disaster management.

ICGdR and the Chair plan to publish books that are either regionally or process oriented and based on basic and/or applied research, which reflect current disaster events and trends worldwide, and those that are based on systematic research in the selected areas, and which will increase the level of research knowledge. The book series is focused on deepening the theoretical basis, and the synthesis of data, which can lead to proper disaster reduction in the future.

Vít Vilímek
Professor, Faculty of Science, Charles University
Prague, Czech Republic
Treasurer, International Consortium
on Geo-disaster Reduction
Coordinator, ICGdR
Publishing Committee

Fawu Wang
Professor, College of Civil Engineering
Tongji University, China
Professor Emeritus, Shimane University, Japan
Chairholder, UNESCO Chair on Geoenvironmental Disaster Reduction
President, International Consortium on Geo-disaster Reduction

Preface and Dedication

This book was Matthias Jakob's idea and he was the driving force behind it for most of its development. Matthias envisioned it as a follow-up to *Debris-flow Hazards and Related Phenomena*, a book he co-edited with Oldrich Hungr in 2005. Since then, countless new papers on this subject have been published and Matthias felt it was time for an updated summary of the state-of-the-art and state-of-practice.

Tragically, Matthias died in a paragliding accident just before this book was finished. His fingerprints are all over it though, and familiar readers will recognize his unique contributions to the specific chapters he co-authored. Matthias was a passionate geoscientist who helped shape modern geohazard science and practice, and he had a huge influence on many people. His premature death was a profound loss for our community on many levels.

On behalf of all of the contributors to this book, we dedicate it to Matthias.

Vancouver, BC, Canada
Golden, CO, USA
March 2023

Scott McDougall
Paul Santi

The original version of the book has been revised: A correction to this book can be found at https://doi.org/10.1007/978-3-031-48691-3_20

Matthias Jakob (1966–2022)

Acknowledgments

Scott and Paul would like to thank BGC Engineering for maintaining an online hub for our editing. BGC is also purchasing and donating several copies of this book in memory of Matthias.

Contents

Contributors

Simon K. Allen Institute for Environmental Sciences, Climate Change Impacts and Risks in the Anthropocene (C-CIA), University of Geneva, Geneva, Switzerland; Department of Geography, University of Zurich, Zurich, Switzerland

Juan Antonio Ballesteros-Cánovas Institute for Environmental Sciences, Climate Change Impacts and Risks in the Anthropocene (C-CIA), University of Geneva, Geneva, Switzerland; The National Museum of Natural Sciences, Spanish Scientific Research Council, Madrid, Spain

Craig Batchelar Planning Consultant, Auckland, New Zealand

José María Bodoque Mining and Geological Engineering Department, University of Castilla La Mancha, Toledo, Spain

Clarence Edward Choi Department of Civil Engineering, The University of Hong Kong, Hong Kong SAR, China

Velio Coviello Research Institute for Geo-Hydrological Protection, National Research Council (CNR IRPI), Padova, Italy

Vincenzo D'Agostino Department of Land, Environment, Agriculture and Forestry, University of Padova, Padua, Legnaro, Italy

Ashok Dahal Faculty of Geo-Information Science and Earth Observation (ITC), University of Twente, Enschede, AE, Netherlands

Tjalling de Haas Department of Physical Geography, Utrecht University, Utrecht, Netherlands

Tjalling de Haas Faculty of Geosciences, Universiteit Utrecht, Utrecht, The Netherlands

Saskia de Vilder GNS Science, Lower Hutt, New Zealand

Casey Dowling BGC Engineering Inc., Golden, CO, USA

Jeff Farrell Whakatāne District Council, Whakatāne, New Zealand

Joseph E. Gartner BGC Engineering Inc., Golden, CO, USA

David L. George U.S. Geological Survey, Vancouver, WA, USA

Richard Guthrie Stantec, Calgary, AB, Canada

Zheng Han School of Civil Engineering, Central South University, Changsha, Hunan, China

Ken Ho Geotechnical Engineering Office, Civil Engineering and Development Department, Kowloon, Hong Kong

Toshiyuki Horiguchi National Defense Academy, Kanagawa, Japan

Pascal Horton Institute of Geography & Oeschger Centre for Climate Change Research, University of Bern, Bern, Switzerland; Terranum Sàrl, Bussigny, Switzerland

Johannes Hübl Institute of Mountain Risk Engineering, University of Natural Resources and Life Sciences Vienna, Vienna, Austria

Christian Huggel Department of Geography, University of Zurich, Zurich, Switzerland

Marcel Hürlimann Department of Civil and Environmental Engineering, UPC BarcelonaTech, Barcelona, Spain

Akihiko Ikeda Sabo and Landslide Technical Center, Tokyo, Japan

Fumitoshi Imaizumi Faculty of Agriculture, Shizuoka University, Shizuoka, Japan

Richard M. Iverson U.S. Geological Survey, Vancouver, WA, USA

Matthias Jakob BGC Engineering Inc, Vancouver, BC, Canada

Roland Kaitna Institute of Mountain Risk Engineering, University of Natural Resources and Life Sciences, Vienna (BOKU), Vienna, Austria

Chao Kang Key Laboratory of Geological Hazards On Three Gorges Reservoir Area, Ministry of Education, China Three Gorges University, Yichang, Hubei, China

Jason W. Kean U.S. Geological Survey, Golden, CO, USA

Julian S. H. Kwan Civil Engineering and Development Department, Government of the Hong Kong Special Administrative Region, Hong Kong SAR, China

Carlos Lam Geotechnical Engineering Office, Civil Engineering and Development Department, Government of the Hong Kong Special Administrative Region, Hong Kong SAR, China

Carie-Ann Lau BGC Engineering Inc., Vancouver, BC, Canada

Philip LeSueur Geological Survey of Canada, Vancouver, Canada

Haiming Liu Department of Civil and Environmental Engineering, The Hong Kong University of Science and Technology, Hong Kong SAR, China

Frankie Lo Geotechnical Engineering Office, Civil Engineering and Development Department, Kowloon, Hong Kong

Luigi Lombardo Faculty of Geo-Information Science and Earth Observation

(ITC), University of Twente, Enschede, AE, Netherlands

Lorenzo Marchi Research Institute for Geo-Hydrological Protection, National Research Council (CNR IRPI), Padova, Italy

Francesco Marra Department of Geosciences, University of Padova, Padova, Italy; Institute of Atmospheric Sciences and Climate (ISAC), National Research Council of Italy (CNR), Bologna, Italy

Scott W. McCoy Department of Geological Sciences and Engineering, University of Nevada, Reno, USA

Scott McDougall The University of British Columbia, Vancouver, Canada

Martin Mergili Institute of Geography and Regional Sciences, Faculty of Environmental, Regional and Educational Sciences, University of Graz, Graz, Austria

Georg Nagl Institute of Mountain Risk Engineering, University of Natural Resources and Life Sciences Vienna, Vienna, Austria

Charles Wang Wai Ng Department of Civil and Environmental Engineering, The Hong Kong University of Science and Technology, Hong Kong SAR, China

Natalia Nolde Department of Statistics, University of British Columbia, Vancouver, Canada

Nina Oakley California Geological Survey, Sacramento, CA, USA

Marisa C. Palucis Department of Earth Sciences, Dartmouth Collage, Hanover, NH, USA

Guillaume Piton Université Grenoble Alpes, INRAE, CNRS, IRD, Grenoble INP, IGE, Grenoble, France

Nikhil Prakash Colorado School of Mines, Golden, CO, USA

Francis K. Rengers U.S. Geological Survey, Golden, CO, USA

Paul Santi Colorado School of Mines, Golden, CO, USA

Wendy Saunders Toka Tū Ake EQC, Wellington, New Zealand

Manfred Scheikl Alpinfra Engineering, Mattsee, Austria; Canadian Hydrotech Corp., Vancouver, BC, Canada

Sergio A. Sepulveda Simon Fraser University, Burnaby, BC, Canada

Gary Sheridan University of Melbourne, Parkville, VIC, Australia

Markus Stoffel Climate Change Impacts and Risks in the Anthropocene (C-CIA), Institute for Environmental Sciences, University of Geneva, Geneva, Switzerland; Dendrolab.Ch, Department of Earth Sciences, University of Geneva, Geneva, Switzerland; Department F.-A. Forel for Environmental and Aquatic Sciences, University of Geneva, Geneva, Switzerland

Alex Strouth BGC Engineering Inc., Golden, CO, USA

Matthieu Sturzenegger BGC Engineering Inc., Vancouver, BC, Canada

Jürgen Suda Alpinfra Engineering, Vienna, Austria

Joshua Theule The Alpine Environment Research Center (CREALP), Sion, Switzerland

Taro Uchida Faculty of Life and Environmental Sciences, University of Tsukuba, Tsukuba, Japan

James W. Vallance U.S. Geological Survey, Cascades Volcano Observatory, Vancouver, WA, USA

Bastian van den Bout Faculty of Geo-Information Science and Earth Observation (ITC), University of Twente, Enschede, AE, Netherlands

Dario Ventra Department of Earth Sciences, University of Geneva, Geneva, Switzerland

Volker Wichmann Laserdata GmbH, Innsbruck, Austria

Richard Woods GNS Science, Lower Hutt, New Zealand

Sophia Zubrycky BGC Engineering Inc., Vancouver, Canada

Abbreviations

Note: Abbreviations associated with equations or listed in tabular form in chapters are not included. Likewise, the names of software packages are not included. Because of the large number of contained equations, Chaps. 8 and 16 have their own lists of abbreviations and variables.

2D	Two dimensional
3D	Three dimensional
ACR	Alkali-Carbonate Reactions
AEP	Annual Exceedance Probability
AIFR	Annual Individual Fatality Risk
ALARP	As Low as Reasonably Practicable
AMR	Adaptive Mesh Refinement
AMS	Accelerator Mass Spectrometer
AR	Atmospheric Rivers
asl	Above sea level
ASR	Alkali-silica Reactions
BAER	Burned Area Emergency Response
BP	Before Present
C	Centigrade
cm	Centimeters
CNN	Convolutional Neural Networks
CPZ	Coastal Protection Zone
CRE	Cosmic Ray Exposure
CSL	Critical State Line
CSLW	Channel Supply-Limited Watersheds
CSSL	Cost to Save a Statistical Life
DAS	Distributed Acoustic Sensors
DEM	Digital Elevation Model
DF	Debris Flows
dNBR	Differenced Normalized Burn Ratio
DRR	Disaster Risk Reduction
EASM	East Asian Summer Monsoon
ENSO	El Niño-Southern Oscillation
FCNN	Fully Connected Neural Networks
FM, F-M	Frequency-Magnitude

FN, F-N, fN	Graph Showing Frequency of "N or more" Fatalities
FOS, FoS	Factor of Safety
FWUW	Full Supply Unlimited Watersheds
g	Grams
GAM	Generalized Additive Model
GCM	Global climate model
GDP	Gross domestic product
GIS	Geographic information system
GLM	Generalized Linear Model
GLOF	Glacial lake outburst flood
GPD	Generalized Pareto distribution
h, hr	Hour
HWM	High water mark
Hz	Hertz
IDF	Intensity-duration-frequency
IRSL	Infrared stimulated luminescence
IVT	Integrated water vapor transport
k	Hydraulic permeability
Kcv	Compensation index
kg/m^3	Kilograms per cubic meter
km	Kilometers
LIA	Little Ice Age
LS	Landslides
l/s	Liters per second
m	Meters
MAP	Mean Annual Precipitation
MCF	Magnitude-Cumulative Frequency Curves
M-F	Magnitude and Frequency
mm	Millimeters
m/s	Meters per second
NAM	North American monsoon
NCL	Normal Compression Line
NDRF	National Disaster Recovery Framework
NGRI	National Geophysical Research Institute
OSL	Optically Stimulated Luminescence
PDE	Partial Differential Equation
PDI	Probability of Death of an Individual
PIV	Particle Image Velocity
PLL	Probable Life Loss
PPT	Pore Pressure Transducer
PSI	Paleostage Indicator
PTV	Particle Tracking Velocimetry
PVC	Polyvinyl Chloride
Q1, Q2, Q3, Q4	Quarters (3-month intervals) of the Year
QSUW	Quasi Supply-Unlimited Watersheds
RBPA	Risk-based Planning Approach
RCP	Representative Concentration Pathway
RF	Random Forest
RMA	Resource Management Act

ROA	Ranked Order Analysis
s	Seconds
SAM	South American Monsoon
SAR	Synthetic Aperture Radar
SASM	South Asian Summer Monsoon
SDGs	Sustainable Development Goals
SPH	Smooth(ed) Particle Hydrodynamics
STATSGO	State Soil Geographic Database (USA)
SVM	Support Vector Machines
t	Time
TIN	Triangular Irregular Network
TL	Thermoluminescence
TWI	Topographic Wetness Index
UN	United Nations
USGS	United States Geological Survey
VEI	Volcano Explosivity Index
VPD	Vapor Pressure Deficit
WSLW	Watershed Supply-Limited Watersheds
WTP	Willingness to Pay
yr	Year

1 Introduction

Matthias Jakob, Scott McDougall and Paul Santi

1.1 About This Book

We live in truly unprecedented times. Looking at various graphics, be it population growth, greenhouse gas emissions, loss of biodiversity, frequency of heat waves and floods, or even the net worth of the richest individuals, one may begin to think we have entered a new era that overlaps with the Anthropocene: the "Exponocene", defined by the seemingly exponential increase in the frequency of processes and conditions harmful to most members of society. The onslaught of climate-related horror-news is complemented by various wars and extremism of every shade worldwide.

How does this book fit into this apparent mess? It is apolitical, but only to some degree. Environmental destruction and ecosystem annihilation are rampant, as pointed out unemotionally in the latest IPCC (2023) report. All the while, we add some 80 million people to the planet every year who, predominantly seeking urban centres to live and work, also encroach into increasingly marginal environs that include steep slopes and remote valleys. Debris flows that once harmlessly (to humans) descended into uninhabited locations become potentially lethal when spatially coinciding with human habitation and infrastructure (Fig. 1.1).

This, no doubt, is increasingly accentuated by the effects of anthropogenic climate change. Particularly dangerous in this context are rock avalanche-triggered debris flows and lahars that can project their damage potential tens of kilometres beyond their source areas. Glacial retreat, formation of new lakes below steep slopes, permafrost degradation and changes in precipitation patterns are the Damocles sword towering over people living in valleys distant from the sources.

In this book, we take stock of the novelties in science and practice pertaining to debris flows, arguably one of the most frequently occurring and lethal global landslide processes. This broad snapshot of where we stand as a scientific and engineering community is useful for many reasons. Few, if any, of us can physically, let alone mentally, read and absorb the surging waves of

Note: Superscripts are used in this chapter to indicate the original chapters where more detailed descriptions can be found. Please refer to the Table of Contents for corresponding chapter titles and authors.

M. Jakob
BGC Engineering Inc, Vancouver, Canada

S. McDougall (✉)
The University of British Columbia, Vancouver, Canada
e-mail: smcdouga@eoas.ubc.ca

P. Santi
Colorado School of Mines, Golden, USA

M. Jakob et al. (eds.), *Advances in Debris-flow Science and Practice*, Geoenvironmental Disaster Reduction,
https://doi.org/10.1007/978-3-031-48691-3_1

Fig. 1.1 St. Sofia camp, California, USA. Fourteen people were killed on Christmas Day 2003 when a debris flow swept the house off of this garage structure

debris-flow related literature. Nor is it straightforward to filter good from perhaps questionable science, or to extract what may become a useful tool for practitioners.

As editors, we set out with the desire to bring together researchers and practitioners in a wide spectrum of fields to provide an overview of, and importantly an update to, the advances in global research and practice of their respective specialties. We attempted to cover key topics and provide global representation, though we are fully aware of certain centres of debris-flow research excellence that are under-represented. Similarly, we know that significant advances have been made around the world on the topic of field-scale experiments and monitoring in natural channels, but lamentably this book does not include a chapter specifically focused on this area.

This book is aimed at a wide readership. We hope to serve readers from academia, students, practitioners in hazard, risk and mitigation fields, and land use or emergency managers alike. We asked the chapter authors to keep this broad audience in mind, to highlight recent advances and ongoing challenges (as summarized in the next section of this chapter) and to be as practical as their specialty fields allow. Front to back, this book is structured to progress from science to applications, but every chapter contains practical guidance. Most chapters also include cross-references to other chapters where complementary information can be found. While we strived for this sort of integration of the chapters as much as possible, we did not try to stifle any respectful differences of opinion between authors, as constructive scientific debate helps move us all forward.

We are content that we have achieved the principal objectives of this book, though we realize it was already mildly outdated by the time it went to print. This volume is neither the first nor the last of its kind. Some aspects of debris-flow science are reaching maturity, while others may still undergo some revolutionary changes. Irrespective, we believe that we have created a sound foundation for our readers, who can then further pursue their specific interests using the many references that can be found at the end of each chapter.

1.2 Summary of Advances and Challenges

This book highlights a number of key advances that have been made in debris-flow science and practice since the publication of Jakob and Hungr

(2005). At the same time, several ongoing challenges and key research needs are also identified. This section summarizes some of these advances and challenges, in part to encourage readers to dive deeper into specific chapters of this book. The summary is organized into six main themes and superscripts are used to indicate the original chapters where more detailed descriptions can be found. Readers are asked to refer to and cite the original chapters (and other references therein) when using any of this information.

1.2.1 Field and Remote Observations

Observations of real debris flows and their impacts remain fundamental to debris-flow research. With unprecedented advances in field and remote sensing data collection and analysis capabilities over the past two decades, our understanding of debris-flow processes has significantly improved (Fig. 1.2). For example, our ability to identify debris flows as the dominant creek process using watershed and fan morphometric characteristics has improved basic hazard recognition[2]. More detailed observations have also given us a better understanding of avulsion processes and patterns, including the role of channel plugs, as well as a better understanding of the links between debris-flow behaviour and supply-limited versus transport-limited conditions[2]. The use of absolute and relative age dating methods, including dendrogeomorphic approaches, has also advanced[8].

News also spreads faster around the world these days, and recent records of damaging debris-flow events are more complete than ever[3]. With these improving inventories, we are gaining a better appreciation for the global distribution of fatal debris-flow events, with developing countries clearly disproportionately affected[3].

Field and remote observations of watershed-fan systems and real debris-flow events should continue to be a cornerstone of debris-flow research going forward. We still have a lot to learn about the interactions between sediment production, geomorphic characteristics and climatic conditions, in the context of interpreting both paleoenvironmental conditions and ongoing climate change effects[2,10]. Detailed observations are also still needed to improve our ability

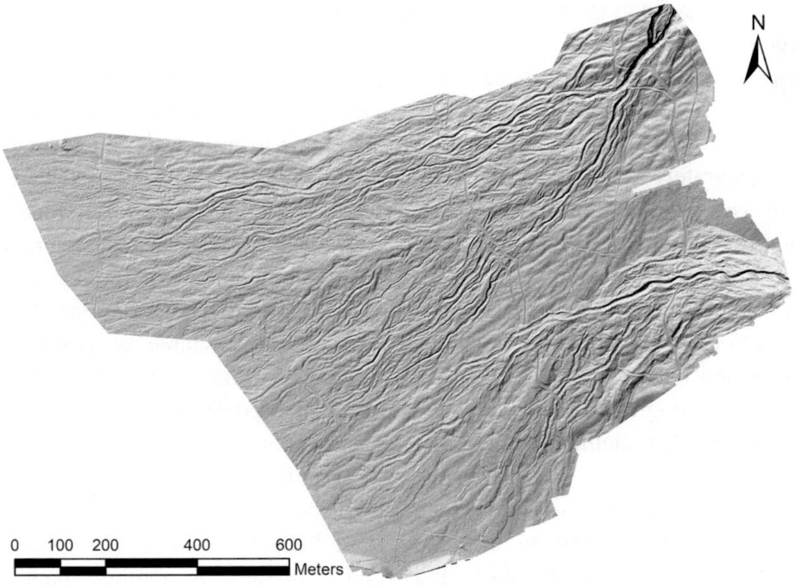

Fig. 1.2 Bare earth, shaded relief image of a very active debris-flow fan system near Lillooet, British Columbia, Canada. High-resolution topographic data collected from low altitude at low speed using a drone-based lidar system (courtesy of Andrew Mitchell)

to predict bulking during debris-flow events, erosion of fans during or between events, and avulsion locations and timing[2]. Continued development and application of robust field sensors could aid some of this work[6]. At the same time, more consistent reporting is needed to distinguish damaging debris-flow events from other hazards and gain a clearer understanding of their consequences worldwide, both in terms of life loss and economic costs[3].

With these important needs in mind, readers are urged to continue to collect field site and case study data, and to share their data with other members of the debris-flow community as much as possible.

1.2.2 Lab Experiments

Alongside field observations, well-controlled flume and centrifuge experiments remain a fundamental tool in debris-flow research. These experiments allow us to isolate specific factors and responses that may not be as readily observable in a natural setting. This capability has made them critically important for investigating underlying physical processes and, in turn, developing and evaluating theoretical and numerical models[4,5,6].

Significant advances have been made over the past two decades in image analysis and instrumentation of flume models, improving our ability to accurately measure parameters such as frontal velocities and impact forces[4]. Our understanding of physical scaling principles has also improved, while at the same time, a broader range of model scales has been used to explore interactions between debris flows and streambeds, trees and barriers[4,15]. Of key interest, some important progress has been made recently on the physical modelling of sediment entrainment[4,6].

Still, more work is needed to investigate realistic boundary conditions and to evaluate complex, multi-phase rheology theories and models, for example, by including boulders in the flow[4,15]. Experiments with varying sediment supply conditions could also provide insights on potential climate change effects[10]. The recent construction of new large-scale flume facilities around the world offers fresh promise for this important line of research.

1.2.3 Theory and Models

Numerical runout models have been a hot topic in research and practice for over two decades now, and dozens of different models have been proposed[5,8,13]. With steadily increasing computational power over the years, the practical viability of increasingly complex, theoretically-rigorous models has significantly improved[5]. Multi-phase models that can simulate pore fluid pressure evolution, even before the onset of motion, are a particularly intriguing advancement and are now widely available[5]. The same trends in computing power have also improved data-driven models, including models designed for use at a regional scale[13]. The increasing availability of high-quality digital elevation models has also been a boon for all types of models[5,13].

Entrainment of material from the path is an important characteristic of debris flows, but it remains a key modelling challenge. While our theoretical understanding of this process has improved and various entrainment models have been proposed, further development and testing are still needed[5,6]. Guidance for practitioners on model setup and input parameter estimation will also need to improve[6]. There may eventually be a practical limit to how deep we need to dive into the micro-scale, but models that simulate realistic three-dimensional interactions between individual grains, including at the interface between moving and stationary material, could become more viable as computational power continues to improve[6]. At the same time, more efficient models are desirable for rapid, multi-basin applications[11].

Simulating the process of grain-size sorting, contributing to important debris-flow macro-scale behaviours including levee/lobe formation and avulsions, is another modelling frontier[5]. The identification of credible debris-flow source locations and volumes is also a key challenge for modellers that needs more attention[5]. As research along these lines continues, practitioners are encouraged to try the latest available numerical models whenever possible, and to provide feedback to the model developers and the broader community on their experiences.

1.2.4 Occurrence and Frequency

Several advances have been made over the past two decades in our understanding of the causes and triggers of debris flows, particularly with respect to meteorology and climatology. This work has improved our knowledge of the effects of long-term versus short-term precipitation and, in turn, our ability to set more reliable rainfall trigger thresholds[7,11,18]. Our awareness of process chains, or cascading events, has also increased, along with broad recognition that anthropogenic climate change will continue to influence debris-flow susceptibility around the world, though the effects will be region- and site-specific[7,10].

Notably, satellite-based monitoring methods have emerged to supplement sometimes spotty ground-based weather gauge networks[7]. Satellite and other remote sensing methods are also increasingly used to fill gaps and improve debris-flow dating and magnitude estimation[8]. At the same time, more rigorous statistical techniques and model ensembles are being applied to such data to decipher debris-flow frequency-magnitude relationships, an important step towards predicting future trends, especially quantitatively[9].

More complete records of debris-flow occurrence are still needed to advance this work[8,10]. Again, further field and remote observations are critical. Watershed-specific studies are one potential path to a better understanding of debris-flow sources, initiation processes, frequencies and the long-term effects of climate change[7,9,10]. Watersheds that already have a relatively long history of monitoring could be a good starting point for this work[10]. Expanded use of satellite-based monitoring of meteorological triggers could also be helpful[7].

In the meantime in practice, these data and knowledge gaps must be acknowledged and judgment must be used, especially when extrapolating far into the future[9]. As consideration of climate change becomes more common in practice, support for the development and refinement of professional practice guidelines will also be needed[10].

1.2.5 Special Types

Two unique types of debris flows were given their own chapters in this book because they warrant special consideration: post-wildfire debris flows and volcanic debris flows (lahars). Both types involve terrain that has been primed for debris flows by other catastrophic processes.

A lot of research has been done over the past two decades to improve our understanding of post-wildfire debris-flow behaviour. Notably, methods for the rapid evaluation of burn severity have significantly advanced[11]. However, much of this work has been focused so far on specific regions with relatively long histories of wildfires. More research is needed in other regions, including areas that are becoming more susceptible to wildfires due to climate change[11]. A better understanding of the response of burned areas to rain-on-snow and rain-plus-snow events is also needed[11].

Our understanding of lahars is comparatively mature, and recent advances have focused more on improvements to hazard mapping, monitoring and communication[12]. Still, the fundamental science continues to evolve. One specific process that merits further research is the proximal bulking and transformation of floods into more sediment-rich lahar phases[12]. Although there are obvious parallels with entrainment processes in non-volcanic floods and debris flows, the volcanic environment involves unique water release and sediment supply conditions that can lead to the formation of relatively large volume events[12].

1.2.6 Hazard and Risk Management

Sound decision-making is the ultimate goal when it comes to reducing losses caused by debris flows, and many advances have been made on this front over the past two decades. At a region-wide scale, more consistent hazard characterization has been made possible by improved data quality and more robust models, which increasingly account for runout in addition to susceptibility[13]. At both a regional and site-specific scale, better guidance for qualitative and semi-quantitative risk assessments exists these days, but arguably the biggest gains have been made in quantitative approaches[14]. With more and more examples of quantitative risk assessments for communities being documented, clearer guidance on input values and a better understanding of uncertainty and risk tolerance have started to emerge[14].

Our approaches to the design of mitigation structures have also improved (Fig. 1.3). There is increasing emphasis on the role of individual mitigation components as parts of larger systems that must be tailored to each site[15]. For example, components designed specifically to catch woody debris or buffer peak discharges have improved overall system functionality[15]. There has also been more emphasis on maintenance and life cycle management recently[17]. At the same time, methods to estimate impact loads have improved, including approaches that account for the flexibility of structures and the use of cushioning materials[16]. Much of this work has been facilitated by lessons learned through better documentation of past failures of mitigation systems[17].

In combination with improved rainfall trigger thresholds and detection algorithms, debris-flow warning systems have also become more advanced with the use of local flow level, trip wire and ground vibration sensors[18]. Like structural mitigation measures, these devices are increasingly used together as components of integrated systems[18].

Land-use planning is a key way to manage, and in some cases outright avoid, debris-flow risks to communities[19]. International frameworks for disaster risk reduction and sustainable development, recently adopted by the United Nations, have provided some high-level motivation and guidance for this work[19]. Case studies

Fig. 1.3 Debris-flow capture system, Colorado Springs, Colorado, USA

that illustrate effective risk-based planning approaches, including public consultation, are also useful guides[19].

There are still many opportunities to improve hazard and risk management practices. Dynamic approaches that adapt with climate and land cover changes may be more appropriate than static maps[13]. More research is also needed to reduce uncertainties in vulnerability estimates[3,14]. Structural mitigation designs should continue to evolve to account for dynamic interactions with debris, and more consistent documentation of actual performance over time is essential for further design improvements[16,17]. The same point is true for early warning systems, especially as new sensor technologies are tested[18]. As functional design of debris-flow mitigation systems matures, more attention could be directed towards aesthetic considerations, which could improve community buy-in[15,19]. Ultimately, support for sound decision-making requires clear communication and the ability to influence policy, which may require many of us to delve out of our comfort zones further into the social science realm.

References

Intergovernmental Panel on Climate Change (IPCC). (2023). *Climate change 2023: AR6 synthesis report*. IPCC.

Jakob, M., & Hungr, O. (2005). *Debris-flow hazards and related phenomena*. Springer.

Debris-Flow Watersheds and Fans: Morphology, Sedimentology and Dynamics

2

Tjalling de Haas, Carie-Ann Lau and Dario Ventra

Abstract

Debris flows typically originate in mountainous watersheds. At the base of these watersheds and where not truncated by a higher order stream or discharging into a body of still water, repeated debris-flow deposition often forms semi-conical debris-flow fans. To mitigate debris-flow hazards, it is important to identify debris flow-prone watersheds and to understand their dynamics. This chapter reviews the morphology, sedimentology and dynamics of debris-flow watersheds and associated fans. Debris flow-generating watersheds are generally smaller and steeper than those dominated by floods and debris floods, and host less well-developed drainage networks. They can be roughly categorized into transport-limited systems (in which each hydroclimatic event producing high discharge triggers debris flows due to an abundance of sediment and high recharge rates) or supply-limited systems (in which debris-flow activity is limited by sediment availability and corresponding slow recharge rates). Debris-flow fans have typical lengths of 0.5–10 km and slopes of 5–15° and develop through spatio-temporal shifts of the locus of deposition through avulsion. Their surface topography and stratigraphy consist of stacked and amalgamated lobe, levee, and channel deposits, with sediment of mud to boulder grade. They provide archives of past flow processes and sediment supply indicative of past climate and environmental change. However, our ability to interpret debris-flow successions is still limited by relatively poor knowledge of how catchment geology and flow properties and composition affect resulting depositional features, from bed scale to fan scale. Recent field-based, experimental, and numerical advances on these topics are starting to fill this knowledge gap.

T. de Haas (✉)
Department of Physical Geography, Utrecht University, Princetonlaan 8a, 3584 CB Utrecht, Netherlands
e-mail: t.dehaas@uu.nl

C. Lau
BGC Engineering Inc., 500-980 Howe Street, Vancouver, BC V6Z 0C8, Canada

D. Ventra
Department of Earth Sciences, University of Geneva, 13 Rue des Maraichers, 1205 Geneva, Switzerland

2.1 Introduction

Debris flows are sediment-gravity flows consisting of highly concentrated mixtures of rock, soil, and water, which rush down mountainsides and into valleys, where they can cause loss of life and property (Dowling & Santi, 2014;

M. Jakob et al. (eds.), *Advances in Debris-flow Science and Practice*, Geoenvironmental Disaster Reduction,
https://doi.org/10.1007/978-3-031-48691-3_2

Iverson, 1997). Furthermore, debris flows are increasingly recognized as one of the most efficient physical processes that transport sediment and erode bedrock in highland settings and at the topographic margins of continental sedimentary basins (Anderson et al., 2015; McCoy, 2015; Stock & Dietrich, 2003, 2006).

Debris flows typically originate in mountainous watersheds by the translation and transformation of open slope landslides into liquified masses (Aaron et al., 2019; Bovis & Dagg., 1988; Campbell, 1974, 1975; Guthrie et al., 2012; Nishiguchi & Uchida, 2022), or by mobilization of accumulated in-channel sediment (Bovis & Dagg., 1988; D'Agostino et al., 2010; Evans et al., 2001; Gregoretti & Fontana, 2007; Kean et al., 2013; Palucis et al., 2021). More rarely, debris flows can also occur by the failure of natural or artificial dams (Chiarle et al., 2007; Clague et al, 1985; Tannant & Skermer, 2013), or by secondary mobilization of unconsolidated volcaniclastic debris (lahars) that incorporate large volumes of water and ice as volcanic edifices collapse (Major et al., 2005; Procter et al., 2021). The unique characteristic of debris-flow watersheds is that, while they are subject to nearly continual sediment-generating processes (e.g., rockfall, rilling, landslides), most of this sediment is exported from the watershed during discrete flow events (Benda et al., 2005; Rickenmann & Jakob, 2015). Figure 2.1 illustrates an example of a typical debris-flow dominated watershed-fan system in which sediment is generated from open-slope rockfall failures

Fig. 2.1 **a** Example of a debris-flow watershed (blue outline) and fan (yellow outline) system at Catiline Creek, British Columbia, Canada. **b** Debris flows are generated in this watershed by rockfall, which accumulates into gullies in the upper watershed. **c** During intense rainfall, the accumulated rockfall debris is transported as a debris flow in a confined channel. **d** Debris flows deposit onto the alluvial fan. Photos by Matthias Jakob

and accumulates in higher-order channels, to be transported by debris flows from the higher-order channels into a principal stream course, and finally deposited onto the debris-flow fan.

While debris flows can occur across all climates and mountain ranges on Earth, they are commonly studied across four main Köppen climate classes: (i) dry (typically described as "arid") (e.g., Al-Farraj & Harvey, 2005; Bull, 1962; Harvey, 1989; Harvey et al., 1999a, 1999b; Hooke, 1967, 1968; Melton, 1965), (ii) temperate (e.g., Chiarle et al., 2007; Crosta & Frattini, 2004; De Scally & Owens, 2004; De Scally et al., 2010; Di et al., 2019; Marchi et al., 1993; Valkanou et al., 2013), (iii) continental (e.g., Clark, 1987; De Scally et al., 2001; Jackson et al., 1987; Wilford et al., 2004), and (iv) tropical (e.g., Arango et al., 2021; Picanço et al., 2016). While debris-flow watershed-fan systems exist also in arctic climates, they are relatively less studied in such settings (e.g., Bernhardt et al., 2017; De Haas et al., 2015a; Tomczyk, 2021; Webb & Fielding, 1999). Aside from local topography, climate is probably the main driver for the frequency and magnitude of sediment-generating and sediment-transport processes, as described further in Sects. 2.2.1 and 2.2.2.

The types of fans linked to watersheds that produce debris flows vary across a spectrum spanning from colluvial cones and aprons dominated by rock-gravity to sediment-gravity flows, typically featuring steep depositional gradients and relatively small volumes (Blikra & Nemec, 1998; Ventra et al., 2013), to larger alluvial fans with lesser gradients, where sediment is transported over larger distances by floods and/or debris flows (Blair & McPherson, 1994; Harvey, 2011). Debris flows may represent a significant component of active processes in both kinds of depositional systems, irrespective of climate and morpho-tectonic setting. Fan and watershed classifications and terminology can be complex and vary among different workers. To discriminate between systems dominated by different geomorphic processes, the classification adopted here emphasizes the importance of dominant geomorphic processes (cf. De Haas et al., 2015a; Tomczyk, 2021):

(1) Flood-dominated systems. These are systems dominated by fluid-gravity flows, i.e., water floods, hyperconcentrated floods, and debris floods. Fans in such systems are commonly referred to as fluvial or alluvial fans (cf. Blair & McPherson, 1994; Church & Jakob, 2020; Harvey, 2005; Ryder, 1971).
(2) Debris-flow dominated systems. These are systems dominated by sediment-gravity flows, i.e., debris flows and mud flows. Irrespective of their radial extent and depositional gradients, the fans aggraded by these systems can be commonly called debris-flow fans or debris fans (Blikra & Nemec, 1998; de Scally et al., 2010).
(3) Colluvial systems. These are systems dominated by rock-gravity and sediment-gravity flows, with dominant activity relating to rockfalls, grain flows, and snow avalanches (in periglacial and alpine settings). Debris flows typically constitute only a relatively minor component of geomorphic processes in such systems. Fans in these systems are also commonly known as colluvial cones or talus cones (Åkerman, 1984; De Haas et al., 2015a; Iturrizaga, 2008; Siewert et al., 2012).

Although each of these systems may be dominated by one type of geomorphic process, other processes may also occur. For example, water floods are not uncommon on many debris-flow dominated systems, while debris-flow deposits are commonly recognized on colluvial cones (Blair & McPherson, 2009; De Haas et al., 2014, 2015a).

Debris-flow fans are depositional landforms formed at the base of steep watersheds or slopes due to a combined loss of channel confinement and topographic gradient, resulting in a decrease in flow velocity and flow depth (Bull, 1964; French, 1987). Debris-flow deposits radiate out from the fan apex, creating a distinctively net aggradational landform with fan-shaped planform. Fans act as temporary sediment storage areas for debris-flow deposits and organic material and are reworked through processes of avulsion, bank erosion, and channel incision, or over longer time scales by secondary processes typically controlled by runoff, eolian deflation and/or pedogenic modification (Blair & McPherson, 1994, 2009; De Haas et al., 2014, 2018a, 2020; Schürch et al., 2011). Ultimately, some of the sediment generated from the debris-flow watershed may be further transported to trunk drainage networks downstream (i.e., larger streams or rivers), while some of the sediment may be preserved within the fan. Debris-flow fans are the product of the debris flows that formed them and are thus intimately linked to the geomorphic history and dynamics of the source watershed. They therefore provide archives of past conditions in the watersheds (D'Arcy et al., 2017; Dietrich & Krautblatter, 2017; Savi et al., 2014).

This chapter details the key characteristics and dynamics of debris-flow watersheds and fans. In Sect. 2.2, we detail the dynamics of supply-limited and transport-limited watersheds (Sect. 2.2.1), elaborate on how changes in watershed dynamics may affect debris-flow activity (Sect. 2.2.2), and discuss how the morphometric characteristics of watersheds can influence their tendency to generate debris flows, how we can use these morphometric characteristics to estimate debris-flow occurrence, and explore the links between watershed and fan morphometry (Sect. 2.2.3). In Sect. 2.3, we detail the primary and secondary surface morphology of debris-flow fans (Sect. 2.3.1), the sedimentology and stratigraphy of debris-flow fans (Sect. 2.3.2), the spatio-temporal evolution of debris-flow fans (Sect. 2.3.3), focusing particularly on avulsion mechanisms (Sect. 2.3.3.1) and on fan incision (Sect. 2.3.3.2), and discuss how debris-flow fans provide archives of past conditions (Sect. 2.3.4). Section 2.4 summarizes how debris-flow watershed and fan characteristics can be used to inform debris-flow hazard management. These sections are followed by an overview of research perspectives (Sect. 2.5).

2.2 Watershed Characteristics and Dynamics

A wide variety of watersheds generate debris flows (Fig. 2.2). Their common characteristics are: (i) the availability of sufficient volumes of sediment for debris-flow generation, (ii) a relatively high probability of hydroclimatic events that exceed the intensity threshold to generate high runoff rates, and (iii) steep slopes that translate into potential energy sufficient to entrain sediment and to sustain debris-flow motion until the watershed outlet at the fan apex. As such, the magnitude (volume) and frequency (M-F) of debris-flow activity is a function of watershed morphometry, material characteristics, and climatic conditions. Sediment may be available and produced from various sources, including primary weathering of bedrock, (shallow) landsliding, soil erosion, and remobilization of unconsolidated sediment sources, for example glacial and volcanic sediments or talus slopes (Bennett et al., 2013, 2014; Berger et al., 2011a; Imaizumi et al., 2017; Jakob et al., 2005; Suwa et al., 2011).

The morphometry of a watershed is generally constant over relatively short time scales (hundreds to thousands of years), whereas the availability and recharge rates of sediment may be more dynamic over time. Therefore, debris-flow watersheds may be subdivided into transport-limited and supply-limited, based on the availability of sediment. We define supply-limited watersheds as those in which the generation of debris flows is constrained by a relatively slow or volumetrically limited production of unconsolidated debris over short time scales (for example, due to weathering-resistant bedrock sources or unfavorable bio-climatic conditions; Jakob et al., 2005). Given the limited availability, available sediment volumes will frequently range below the necessary threshold to trigger debris flows even in the presence of intense hydroclimatic events. Conversely, in transport-limited watersheds sufficient sediment for debris-flow formation is always available, such that debris-flow activity is controlled more directly by the magnitude and frequency of hydroclimatic events.

Fig. 2.2 Examples of debris-flow generating watersheds. **a** Illgraben, Switzerland (McArdell & Sartori, 2021). **b** Symmes Creek, California, USA (Whipple & Dunne, 1992). **c** Gadria, Italy (Comiti et al., 2014). **d** Kamikamihori Creek, Japan (Suwa et al., 2011). **e** Rebaixader, Spain (Hürlimann et al., 2014). **f** Chalk Cliffs, Colorado, USA (Coe et al., 2008). **g** Hanaskogdalen, Svalbard (Bernhardt et al., 2017; De Haas et al., 2015a). Note the range of vegetation types, degree of weathering of bedrock, steepness, and landscape dissection by erosion

The dynamics of these two subsystems may differ substantially. Below, we first describe the dynamics of transport-limited and supply-limited systems in Sect. 2.2.1. Next, in Sect. 2.2.2 we explain how temporal changes in sediment availability may alter the dynamics of debris-flow watersheds and the frequency and magnitude of debris flows. While it is difficult to accurately quantify sediment availability and recharge rates for a given watershed, determining its morphometry from topographic data generally is straightforward. Therefore, morphometric analyses are typically applied to derive first-order estimates of the potential for debris-flow activity in a watershed. The empirical relations between watershed morphometry and debris-flow activity are summarized in Sect. 2.2.3.

2.2.1 Dynamics of Transport-Limited and Supply-Limited Watersheds

Debris-flow watersheds can be transport-limited (also referred to as supply-unlimited) or supply-limited (also referred to as weathering-limited) systems (Bovis & Jakob, 1999; Church & Miles, 1987; Zimmermann et al., 1997). Transport-limited watersheds have a large availability of sediment and rapid rates of recharge from surrounding bedrock or colluvial slopes, such that a hydroclimatic event of sufficient magnitude nearly always triggers a debris flow (Fig. 2.3c, d). In contrast, supply-limited watersheds typically exhibit a lower frequency of events because of slower sediment recharge rates and fewer zones of slope instability (Bovis & Jakob, 1999; Brayshaw & Hassan, 2009) (Fig. 2.3a, b). In such systems, hydroclimatic events that produce enough discharge to trigger a debris flow only do so when sufficient sediment is available from in-channel mobilization or from the transformation of hillslope landslides into debris flows. In such systems exceedance of a hydroclimatic threshold is thus a necessary, but not a sufficient condition for debris-flow occurrence (Jakob et al., 2005).

While simplistic in nature, these two classifications are useful to understand the differences in observed frequencies and magnitudes of debris flows. In transport-limited watersheds, debris-flow frequency and magnitude are

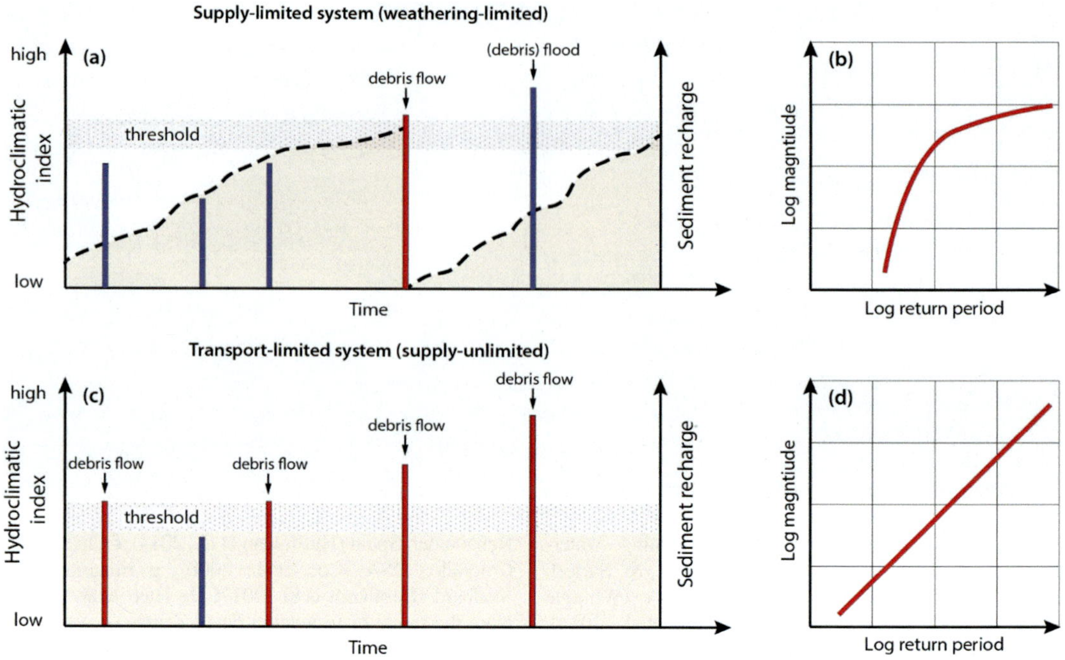

Fig. 2.3 Supply-limited and transport-limited concepts applied to the occurrence of debris flow events and their frequency and magnitude. Bars indicate precipitation, curved rising lines indicate cumulative sediment recharge. Modified after Bovis and Jakob (1999) and Jakob (2022). Note that the magnitude-frequency curve in panel d will eventually level off because of finite hydroclimatic conditions (i.e. there is a limit to maximum rainfall) and hence debris-flow magnitudes cannot be infinitely high

primarily controlled by the frequency and magnitude of hydroclimatic events because sufficient sediment for debris-flow formation is (nearly) always present. In contrast, in supply-limited watersheds, the frequency and magnitude of debris flows is a function of hydroclimatic events and sediment recharge rates, which are dependent on channel recharge rates and hillslope landslide frequency. Therefore, supply-limited watersheds typically exhibit a lower frequency of debris-flow activity than transport-limited watersheds (Bovis & Jakob, 1999). Moreover, maximum debris-flow magnitudes are typically limited by sediment availability in supply-limited watersheds because the sediment reservoir is (nearly) emptied after debris-flow activity (Fig. 2.3b), in contrast to transport-limited watersheds, where debris-flow magnitude generally increases with hydroclimatic magnitude (Fig. 2.3d). A caveat to these observations assumes that increasing hydroclimatic-event magnitudes result in linearly increasing watershed responses, whereas supply-limited watersheds may generate low-frequency, but high-magnitude debris flows when multiple side channel failures contribute to the main channel (Marchi & D'Agostino, 2004).

Little data are available that quantify the relative occurrence of transport-limited and sediment-limited watersheds across mountain ranges and climates worldwide. Bovis and Jakob (1999) studied 33 watersheds in the Coast Mountains region of southwestern British Columbia (Canada) and found 12 transport-limited and 21 supply-limited watersheds. In the same region, Jakob et al. (2005) found that supply-limited watersheds more frequently occur in more competent igneous, meta-sedimentary and meta-volcanic rocks, likely due to the relatively enhanced resistance to weathering of these rocks in temperate humid climates, with consequent low rates of regolith production.

Others have utilized the watershed yield rate to describe the variety in landslide mechanisms, available sediment to trigger in-channel debris flows, and available channel sediment for entrainment (c.f. Hungr et al., 2005; Marchi & D'Agostino, 2004). High channel yield rates occur in talus/scree slopes, volcanic sediments, deeply weathered bedrock, loess deposits, recently burnt areas, channels with high bank erosion, or when the occurrence of large landslides coincides with large floods (Hungr et al., 1984, 2005; Marchi & D'Agostino, 2004; Santi & Morandi, 2013). Low yield rates occur in non-erodible channel beds such as bedrock-controlled channels and consolidated glacial deposits (Hungr et al., 1984, 2005). Channel yield rates vary temporally with sediment recharge rates, weathering rates, and frequency of channel-clearing debris-flow events. While this approach requires more effort to classify watersheds, yield rate can be used to estimate probable debris-flow volumes both at local and large regional scales (Hungr et al., 2005; Marchi & D'Agostino, 2004).

More recently, Jakob (2021) further subdivided debris-flow-prone watersheds into four main types. These are supply-limited and supply-unlimited watersheds and channels. Each of these four types of systems is characterized by a specific debris-flow magnitude-frequency relationship.

1. **Watershed supply-limited**. Watershed supply-limited systems predominantly consist of competent bedrock with very low sediment production rates. These systems typically contain little or no soil, little or no vegetation, and a low drainage density. Once a debris flow occurs in such watersheds, it may take decades for enough sediment to accumulate on hillslopes and in the channel network before another debris flow can occur. In these systems even extreme hydroclimatic events that occur before a threshold recharge is reached will only result in a flood or debris flood.
2. **Channel supply-limited**. These watersheds are defined as those whose channels recharge as described in watershed supply-limited systems. However, the watershed is also covered by soil, and in regions with sufficient precipitation and available soil, also with trees or shrubs. These systems have a moderate drainage density (number of channels within the watershed). Once sufficient volumes and intensities of rain have fallen, small debris flows or debris slides may initiate and travel downhill until they reach a higher order stream where they typically evolve into larger debris flows. This process implies that even if the channel has been depleted of available debris by a preceding debris flow, additional debris flows can occur through momentum and mass transfer from side-slope landslides.
3. **Quasi supply-unlimited**. These watersheds have ample sediment supply in the form of talus slopes, weak and deeply-weathered soils, glacial drift, or artificially introduced sediment (mine tailings or mine waste rock, road fill, garbage dumps) with some apparent or true cohesion. They are typically partially vegetated, have a high drainage density, abundant but cohesive sediment, and fast recharge rates. Debris-flow initiation occurs in the form of side slope failure(s) and/or in-channel sediment mobilization. Such watersheds will readily produce debris flows when a hydroclimatic threshold is exceeded. However, increases in the magnitude of triggering events will not result in infinitely large debris flows due to the apparent or true interparticle cohesion that stabilizes unconsolidated source sediments and inhibits infinite sediment entrainment. This watershed type is characteristic of dormant or extinct volcanoes, denuded dryland/badland terrain, consolidated glacial drift, or areas covered by weak, heavily weathered sedimentary rocks.
4. **Supply-unlimited (transport limited)**. These watersheds have abundant, loose soil cover or heavily weathered, weak bedrock with poor or no cohesion. They have partial or no vegetation cover, very high drainage density and recharge rates, and mostly in-channel debris-flow formation. Debris flows occur every time a hydroclimatic threshold

is exceeded and their magnitude is directly proportional to the intensity and duration of a rainstorm event.

2.2.2 Changes in Watershed Dynamics and Their Effects on Debris-Flow Activity

Changes in the availability and recharge rates of sediment, as well as changes in hydroclimatic triggering conditions, may change watershed dynamics. These changes may range from altered debris-flow frequency and magnitude to transitions from debris-flow dominated to flood-dominated conditions, or vice versa. There is a wide range of processes that may alter sediment availability and recharge rates in a watershed. These include earthquakes (e.g., Domènech et al., 2019; Liu et al., 2021; Shieh et al., 2009; Zhou & Tang, 2014), volcanic outbursts (e.g., Suwa et al., 2009), wildfires (e.g., Cannon, 2001; McGuire et al., 2021; Nymann et al., 2015; Staley et al., 2017; Gartner et al., this volume), glacier retreat and permafrost degradation (e.g., Haeberli et al., 1993; Fisher et al., 2006; Sattler et al., 2011; Yu et al., 2020), climate change (e.g., Hirschberg et al., 2021; Ryder, 1971; Stoffel et al., 2014; Turkington et al., 2016; Stoffel et al., this volume), enhanced landsliding in the watershed (e.g., Chien-Yuan et al., 2007; Imaizumi et al., 2017; Jakob et al., 2020), and a wide range of anthropogenic actions, such as deforestation or reforestation, road construction, mining, drainage modification, and construction of check dams, terraces, or sediment traps (e.g., Huebl & Fiebiger, 2005; Imaizumi et al., 2008; Remaître & Malet, 2010; Roberts & Church, 1986; Yu et al., 2012).

Overall, we identify three main pathways of evolution in watershed dynamics resulting from changes in sediment availability and recharge rate (Fig. 2.4):

1. **Transition from supply-limited to transport-limited conditions**. A large increase in sediment availability may transition a watershed from supply-limited to transport-limited conditions (Fig. 2.4a, b). This will lead to an increase in debris-flow frequency, because each hydroclimatic event above the triggering threshold will generate a debris flow, while previously it might have triggered a (debris) flood, because of a lack of sediment. Similarly, the magnitude of debris flows will no longer be limited by sediment availability, and is therefore expected to increase as well (Fig. 2.3) (Jakob et al., 2021; Zimmermann et al., 1997). Large increases in sediment availability in watersheds may happen near-instantaneously by stochastic geological events, such as earthquakes, wildfires, landsliding, or following anthropogenic interventions, such as mine dumps or logging, or following more gradual environmental change, for example caused by glacial retreat or permafrost degradation. Wildfires can generate temporary transport-limited conditions, as both channel recharge is enhanced following the fire and channel yield is increased due to channel erosion and side slope failures (e.g., DiBiase & Lamb, 2019; Nymann et al., 2015; Santi et al., 2008; Santi & Morandi, 2013).
2. **Supply-limited with an increase in recharge rate**. In supply-limited watersheds an increase in sediment recharge rate will lead to a higher debris-flow frequency (Fig. 2.4c, d). Because of the enhanced sediment recharge rate, the watershed will likely require less time to generate the sediment volumes necessary for debris-flow generation, thereby decreasing the time between individual debris-flow events. Whether flow magnitude also increases will depend on the frequency of hydroclimatic events above the threshold for debris-flow generation. If these hydroclimatic events are frequent, flow magnitudes will not change significantly, because the system is flushed regularly. For larger return periods between triggering hydroclimatic events, more sediment will be able to build up in the watershed than previously, leading to enhanced debris-flow magnitudes. Increases in recharge rate may for example result from enhanced landslide activity,

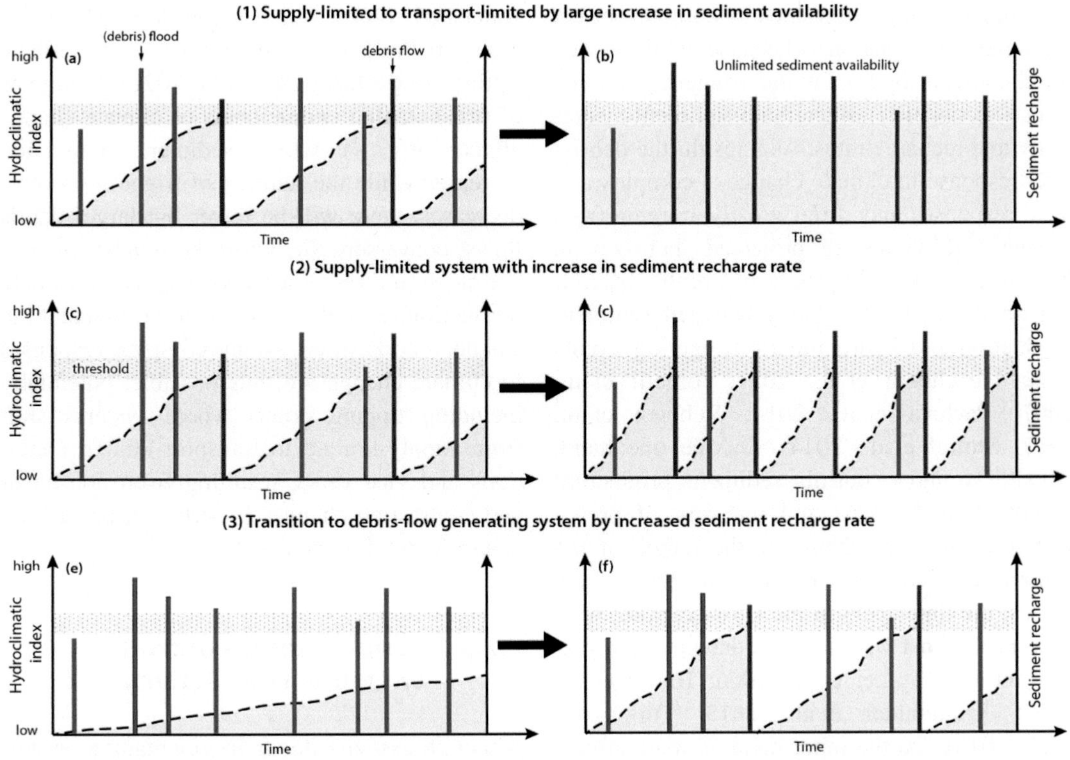

Fig. 2.4 Examples of changes in watersheds dynamics and debris-flow activity as a result of changes in sediment availability and recharge rates. **a**, **b** Transition from supply-limited to transport-limited conditions. **c**, **d** Supply-limited system with an increase in recharge rate. **e**, **f** Transition from flood-dominated to debris-flow dominated

weathering rates, permafrost degradation, and glacial retreat, possibly resulting from climate change.

3. **Transition from flood-dominated to debris-flow dominated**. Watersheds that predominantly generate (debris) floods may transition to systems that predominantly generate debris flows as a result of enhanced sediment availability (Fig. 2.4e, f). Watersheds with morphometric characteristics favoring generation of debris flows may predominantly generate floods under limited sediment availability. In such systems an increase in sediment availability, for example because of an enhanced recharge rate or sudden input of large volumes of sediment by geological events such as earthquakes or wildfires, facilitates the occurrence of debris flows.

A wide range of pathways of watershed evolution are possible, given the many accompanying processes affecting debris-flow watersheds on superimposed time scales. Firstly, the above-mentioned transitions may also occur in reverse, as a result of decrease in sediment supply, such as after paraglacial conditions (Church & Slaymaker, 1989; Ryder, 1971). Moreover, change may vary from rapid (e.g., earthquakes, volcanic outbursts, wildfires) to very gradual (climate change, glacial retreat, permafrost degradation). In many situations where sudden increases in sediment availability occur, watershed dynamics tend to recover to their original state as the excess sediment gradually depletes or vegetation is re-established (e.g., Suwa et al., 2011; Thomas et al., 2021).

Climate change may have a complex effect on watershed dynamics, because while it primarily affects hydroclimatic triggering events, it also secondarily affects both sediment availability and recharge rates. As a result, the debris-flow response to climate change is complex and local responses may differ greatly, ranging from projected decreases to projected increases in magnitude and/or frequency of events, depending on the delicate balance between sediment production and hydroclimatic triggering events (e.g., Hirschberg et al., 2021; Jomelli et al., 2007; Pavlova et al., 2014; Rebetez et al., 1997; Stoffel et al., 2014). On the one hand, climate strongly controls sediment production by affecting the type and intensity of prevalent weathering mechanisms, the length of the snow-free season, the number of freeze–thaw cycles, the time spent in the frost-cracking window, permafrost warming and thawing, and glacier retreat (Fischer et al., 2006; Harris et al., 2009; Krautblatter et al., 2013; Turkington et al., 2016). On the other hand, climate affects hydrological triggering events, mostly rainfall frequency, intensity and duration, the ratio between rainfall and snowfall, and the frequency and duration of snowmelt (Ban et al., 2015, 2018; Giorgi et al., 2016; Huscroft et al., 2003; Jomelli et al., 2004; Rebetez al., 1997; Stoffel et al., 2014). At the same time, climate change could also reduce the mobility or availability of sediment sources if warming increases vegetation cover, especially at high elevations (Eichel et al., 2018). An increase in the frequency and intensity of wildfires could also impact sediment availability for debris flows in fire-affected areas, as a result of reduced canopy interception and evapotranspiration, but especially by releasing sediment anchored to slopes by roots and increasing surface runoff during heavy rainfall (e.g., Cannon & Reneau, 2000; Nyman et al., 2011; Pierce et al., 2004; Tomkins et al., 2008).

This emphasizes that the response of debris-flow activity to climate change depends on complex and as yet poorly constrained interactions between many climatic factors and landscape properties, whose interplay may vary between watershed types (e.g., supply-limited versus transport-limited), mountain ranges, and climate settings (e.g., Jakob & Owen, 2021). Examples of two potential endmembers of climate-change effects are: (i) where sediment production increases while the number of triggering events decreases, there will be fewer but larger debris flows; conversely, (ii) when the number of triggering events increases and outpaces sediment production rates, there will be more frequent but smaller debris flows. Many complex responses to climate change are thus possible, potentially including tipping points where systems shift from supply-limited to transport-limited conditions and vice versa, resulting from non-linear and contrasting changes in sediment production and/or hydroclimatic events.

2.2.3 Morphometric Controls on Debris-Flow Activity

Although extreme floods in mountainous watersheds may induce catastrophic damages and loss of lives, it is generally recognized that the most dangerous watersheds are those prone to debris flows (Bertrand et al., 2013), and especially those with low frequency—large magnitude flows, where previous events are forgotten or outside written documentation. Debris flows achieve maximum discharges that are many times greater than those associated with water floods (Brunkal & Santi, 2017; Kean et al., 2016; Mao et al., 2009; VanDine, 1985). For the same watershed area, debris-flow peak discharges can be 5–40 times higher than those of floods (Hungr et al., 2001). Therefore, identifying debris-flow prone watersheds is of great importance for flood hazard assessment in mountainous regions and is a key step in the prioritization of areas that require detailed hazard and risk studies, or risk mitigation measures (Jakob, 2005). As a first-order estimate, the potential for debris-flow activity in a watershed is often approximated with empirical relationships from morphometric characteristics. Caution should be applied in assuming the empirical classification for watersheds, particularly where regional evidence reports low-frequency, high-magnitude debris flows in otherwise flood-dominated watersheds (e.g., Jakob et al., 2017).

2.2.3.1 Morphometric Dataset of Watersheds and Fans Worldwide

A wide range of morphometric variables (Table 2.1) have been tested for their ability to discriminate debris-flow watersheds from non-debris-flow watersheds (e.g., Arango et al., 2021; Bertrand et al., 2013; Calvache et al., 1997; De Scally & Owens, 2004; De Scally et al., 2010; Jackson et al., 1987; Kostaschuk et al., 1986; Marchi et al., 1993; Rowbotham et al., 2005; Sorriso-Valvo et al., 1998; Stolle et al., 2015; VanDine, 1985; Welsh & Davies, 2011; Wells & Harvey, 1987; Wilford et al., 2004), as well as to predict typical volumes of events in specific watersheds (De Haas & Densmore, 2019; Hungr et al., 1984; Johnson et al., 1991; VanDine, 1985). The wealth of watershed attributes that may affect the dominant process type can be roughly divided into parameters describing watershed relief and steepness (e.g., watershed slope, watershed relief, relief ratio, and Melton index), watershed geometry (e.g., watershed length, area, perimeter, elongation, form, circularity), drainage characteristics (e.g., stream order, stream length, drainage density, drainage intensity), and parameters describing the sediment availability and erodibility of a watershed (e.g., lithology, vegetation cover) (see Table 2.1). One should bear in mind, however, that many of the selected topographic predictors are intercorrelated, which complicates identifying the true controls on debris-flow susceptibility.

In order (i) to identify the morphometric characteristics that define dominant process type in a watershed and (ii) to illustrate the relationships between watersheds and fans, we have compiled morphometric data from watersheds and fans across several continents, mountain ranges, climate zones, and process types. This dataset includes published data from the Himalayas, Ladakh, India (Stolle et al., 2015), the tropical Andes, Columbia (Arango et al., 2021), Spitsbergen, Svalbard (Tomczyk, 2021), British Columbia, Canada (Kostaschuk et al., 1986; Jackson et al., 1987; and newly presented data in this chapter), the southern Carpathians, Romania (Ilinca, 2021), the Southern Alps, New Zealand (De Scally & Owens, 2004; De Scally et al., 2010), the North Cascade Foothills, USA, the European Alps (including Switzerland, Italy, France, and Austria), and the Pyrenees (from multiple authors compiled by Bertrand et al., 2013). The dataset comprises information from colluvial, debris-flow, and flood-dominated (also including debris flood) systems. It contains 231 colluvial systems, 749 debris-flow dominated systems, and 369 flood-dominated systems. Furthermore, we have compiled data on the area, length, relief, relief ratio and Melton index of the watersheds, and slope and area of the associated depositional fans. In total, data were compiled for 1349 systems, although not all information was available for all systems, with data availability ranging from 413 sites for fan area to all 1349 systems for Melton index and process type.

2.2.3.2 Morphometric Characteristics of Debris-Flow Dominated Watersheds

The morphometry of watersheds and fans dominated by debris-flow processes is distinct from those dominated by colluvial processes and floods, although considerable overlap exists (Fig. 2.5; Table 2.2). Overall, debris-flow dominated systems have morphometric characteristics intermediate between colluvium-dominated and flood-dominated systems. From colluvium-dominated to flood-dominated systems, we typically observe an increase in watershed area and watershed relief, while we observe a decrease in Melton ratio and Relief ratio. Fan longitudinal slope decreases from colluvium-dominated and flood-dominated systems. The area of debris-flow dominated and flood-dominated fans in our dataset is roughly similar, while the area of colluvium-dominated fans is much smaller.

Table 2.1 Selection of parameters describing watershed and climatic characteristics, which may affect the dominant watershed process

Parameter	Abbreviation	Equation	Reference
Watershed relief			
Relief	H	$H = H_{max} - H_{min}$	Strahler (1952)
Mean slope	S (°)		
Relief ratio	R_h	$R_h = H/L_b$	Schumm (1956a, 1956b)
Ruggedness number	R_n	$R_n = Rd * H$	Strahler (1964a, 1964b)
Melton index	M	$M = H * A^{0.5}$	Melton (1957)
Watershed geometry			
Length	L_b		Schumm (1956a, 1956b)
Area	A		Schumm (1956a, 1956b)
Perimeter	P		Schumm (1956a, 1956b)
Form factor	F_f	$F_f = A/L_b^{\ 2}$	Horton (1932)
Elongation ratio	R_e	$R_e = (2/L_b) * (A/x)^{0.5}$	Schumm (1956a, 1956b)
Circularity ratio	R_c	$R_c = 4\pi A/P^2$	Miller (1953)
Compactness coefficient	C_c	$C_c = 0.2841 * P/A^{0.5}$	Gravelius (1941)
Texture ratio	R_t	$R_t = N_1/P$	Smith (1950)
Fitness ratio	R_f	$R_f = C_l/P$	Melton (1957)
Wandering ratio	R_w	$R_w = C_l/L_b$	Smart and Surkan (1967)
Drainage characteristics			
Main stream order	S_u	Hierarchical rank	Strahler (1952)
Number of streams	N_u	$N_u = N_1 + N_2 + \ldots + N_n$	Horton (1955)
Stream length	L_u	$L_u = L_1 + L_2 + \ldots + L_n$	Strahler (1964a, 1964b)
Main stream length	C_l	Length of longest streamline	
Stream length ratio	L_{ur}	$L_{ur} = L_u/L_{u+1}$	Strahler (1964a, 1964b)
Bifurcation ratio	R_b	$R_b = N_u/N_{u+1}$	Strahler (1964a, 1964b)
Rho coefficient	ρ	$\rho = L_{ur}/R_b$	Horton (1955)
Stream frequency	F_s	$F_s = N_u/A$	Horton (1955)
Drainage density	D_d	$D_d = L_u/A$	Horton (1932)
Channel maintenance constant	C	$C = 1/D_d$	Schumm (1956a, 1956b)
Drainage intensity	Di	$D_i = F_s/D_d$	Faniran (1968)
Infiltration number	I_f	$I_f = F_s * D_d$	Faniran (1968)
Length of overland flow	L_g	$L_g = A/(2L_u)$	Horton (1955)
Sediment availability/erodibility			
Rock strength	RS	Hierarchical rank	
Soil texture	ST	Hierarchical rank	
Vegetation cover	V_c	%	
Unconsolidated sediment cover	U_c	%	
Seismic intensity	S_I	Hierarchical rank	
Distance from faults	F_d		
Land use	LU	Hierarchical rank	

(continued)

Table 2.1 (continued)

Parameter	Abbreviation	Equation	Reference
Hydroclimate			
Mean annual temperature	*AT*		
Mean annual precipitation	*AP*		
Maximum daily rainfall	*MDR*		
Moisture index	M_I	$M_I = 100 \times (W_S - W_D)/\mathrm{PE}$	Thornthwaite & Mather (1955)
Time in the frost-cracking window	T_c		

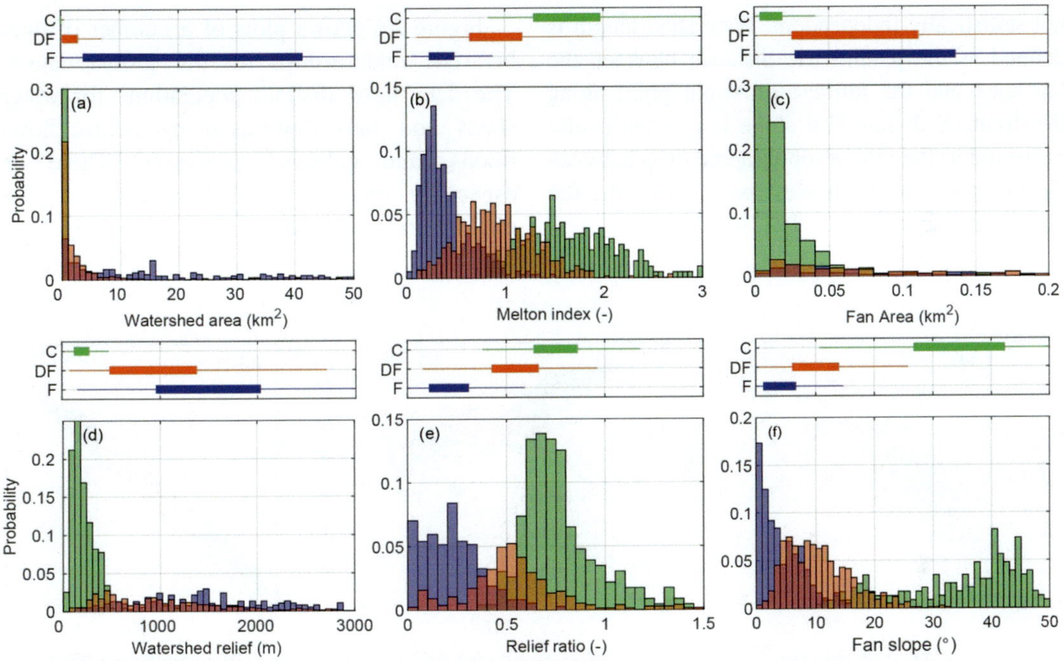

Fig. 2.5 Histograms and box plots of key morphometric variables for colluvial, debris-flow, and flood dominated watersheds and fans. Flood-dominated systems are represented in blue, debris-flow dominated systems in orange, and colluvial systems in green

2.2.3.3 Identifying Debris Flow-Generating Watersheds Based on Morphometry

Many studies have used bivariate predictors to define distinct domains of flood-dominated and debris-flow dominated watersheds (Fig. 2.6) (e.g., Arango et al., 2021; Bertrand et al., 2013; Stolle et al., 2015; Welsh & Davies, 2011; Wilford et al., 2004). The most widely used indicators for determining formative processes are the Melton index, watershed length, and fan slope (e.g., Bertrand et al., 2013; Wilford et al., 2004). The Melton index is a normalized index of the gravitational energy of a watershed, defined as the ratio of watershed relief (difference between the maximal elevation of the watershed and the elevation of the fan apex at the watershed's outlet) by the square root of the watershed area (Melton, 1965) (Table 2.1)—note that watershed relief and area are typically correlated and thus

Table 2.2 Median, 25th percentile, and 75th percentile statistics of morphometric characteristics of colluvium-dominated, debris-flow dominated, and flood-dominated watersheds and fans. Notation: P_{50} (P_{25}–P_{75})

Watershed type	Watershed area (km^2)	Watershed relief (m^2)	Melton ratio	Relief ratio	Fan area (km^2)	Fan slope (°)
Colluvium-dominated	0.01 (0.005–0.04)	179 (124–277)	1.59 (1.3–2.0)	0.73 (0.64–0.86)	0.009 (0.004–0.02)	37.4 (26.7–42.3)
Debris-flow dominated	0.74 (0.22–2.9)	905 (484–1377)	0.88 (0.64–1.18)	0.53 (0.43–0.66)	0.06 (0.03–0.11)	9.8 (6.1–14.0)
Flood-dominated	15.3 (3.8–41.2))	1414 (957–2034)	0.32 (0.23–0.49)	0.21 (0.11–0.31)	0.05 (0.03–0.14)	3.2 (1.3–6.7)

not statistically independent. Watershed length is defined as the planimetric distance between the fan apex and the farthest proximal point along the drainage divide. Fan slope is conventionally measured as the rate of topographic drop between the fan apex and the lowest point along the fan perimeter (Jackson et al., 1987).

Figure 2.6 shows plots of a number of these bivariate relationships for our global dataset. The data show distinct populations for watersheds and fans dominated by debris flows, floods, and colluvial processes. There are especially distinct populations for the plots of Melton index versus fan slope (Fig. 2.6e)

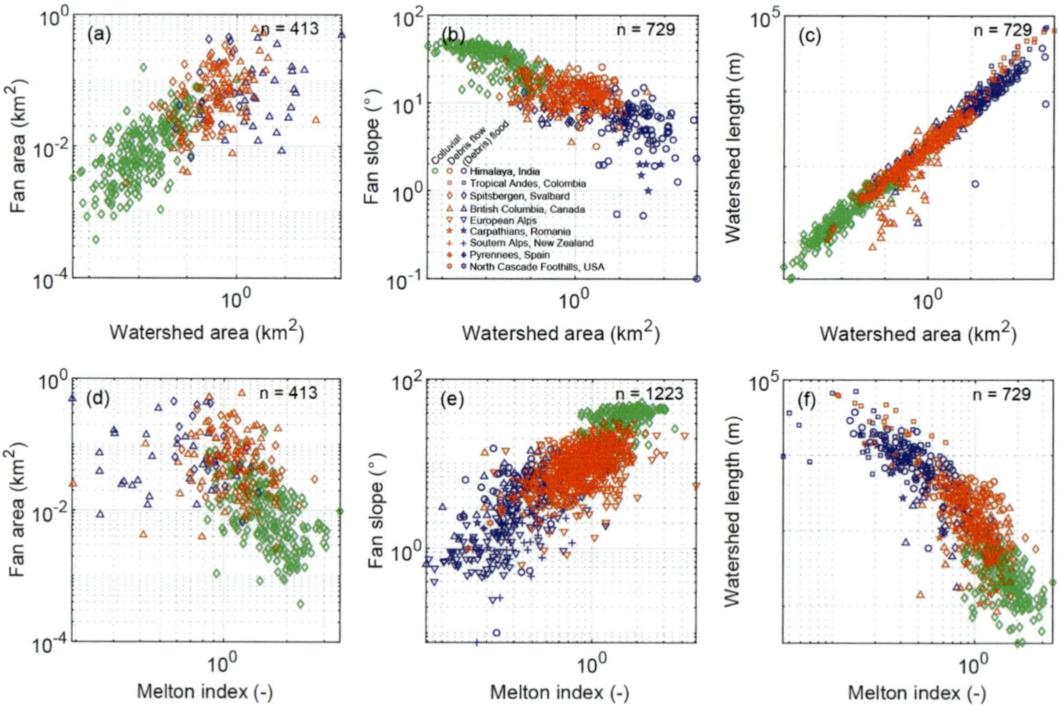

Fig. 2.6 Bivariate combinations of watershed and fan variables showing distinct populations of systems dominated by debris flows, floods, and colluvial processes, but also highlighting that considerable overlap exists

and watersheds length (Fig. 2.6f), in accordance with observations reported in the literature (e.g., Bertrand et al., 2013; Wilford et al., 2004). At the same time, there is also considerable overlap among the edges of these populations, where multiple process types may be equally important. When we break down the relations of Melton index with fan slope and watershed length per region (Fig. 2.7), we observe that these relations are generally useful to discriminate debris flow-dominated from flood-dominated systems. However, the transition between these populations differs between regions and may be either sharp (e.g., Himalaya, India; Carpathians, Romania) or feature substantial overlap (e.g., British Columbia, Canada; European Alps), depending on local characteristics. Only in examples from the Tropical Andes of Colombia there is no clear distinction between debris-flow dominated and flood-dominated systems on a plot of Melton index versus watershed length. However, Arango et al. (2021) show that dominant processes in these systems can be discriminated based on plots of watershed area versus relief ratio and plots of Melton index versus relief ratio.

The morphometric data generally shows that watersheds with high-relief, steep slopes, and a relatively small area are prone to generate debris flows (Bertrand et al., 2013; Crosta & Frattini, 2004). The occurrence of debris flows requires sufficient sediment and the potential of the flow to entrain this sediment, and these conditions are more likely met in systems with high relief and steep slopes. At the same time, larger watersheds have a higher chance that floods will reach the depositional fans, which can be explained by a number of factors: (i) in larger watersheds, debris flows that originate on steeper upper slopes are less likely to reach the fan, as flow deposits can become temporarily stored on lower slopes or in lower-gradient and wider-channel sections within the watersheds (Blair & McPherson, 1994); (ii) larger watersheds are more often host to perennial streams that rework fine sediment continuously by low-discharge floods (Kostachuk et al., 1986); (iii) most importantly, extensive watersheds collect more runoff from overland flow and comprise larger areas where loose regolith can be stored for relatively protracted timespans, which generally leads to the generation of dilute waterflows (Crosta & Frattini, 2004; Imaizumi & Sidle, 2007; Ventra et al., 2013).

The work of Arango et al. (2021) on watersheds in the Tropical Andes of Colombia provides insights into the attributes of drainage networks of watersheds susceptible to debris flows. They show that debris-flow dominated watersheds contain a lower number of streams with lesser stream lengths and stream-length ratios, but a high mean value for main stream length, indicating a young stage of drainage development compared to flood-dominated watersheds. Debris-flow dominated watersheds further show relatively lower values of drainage density, stream frequency, and drainage intensity. They are typically poorly dissected, thus featuring a lower capacity of in-channel sediment storage per unit of drainage area. De Scally et al. (2010) shed light on the effects of watershed lithology and vegetation cover, based on a study of systems in the Southern Alps of New Zealand. They found that a much lower degree of watershed ruggedness or slope is required in debris-flow systems underlain by schist compared to sedimentary rocks. In this region, schist has a lower mechanical strength and breaks down into finer sediments, with both factors favoring the formation of debris flows. They further found that debris-flow dominated watersheds have a lower density of forest cover than their fluvial counterparts. This can be linked to the observation that forests increase the surficial stability of slopes, limiting sediment availability and recharge rates (e.g. Chiaradia et al., 2016; Lu et al., 2001; Wang et al., 2017).

Most of the previous studies (Bertrand et al., 2013; De Scally & Owens, 2004; Jackson et al., 1987; Marchi & Brochot, 2000; Wilford et al., 2004) attempted to provide unique threshold values for morphometric predictors, above or below which watersheds would be dominated by debris flows. However, no clear discontinuity exists between debris-flow dominated and flood-dominated systems in the distribution of their morphometric indicators (Figs. 2.5 and

2.6). Moreover, proposed thresholds are generally only applicable locally since processes may be influenced by other geologic or geomorphic factors not necessarily encompassed by the considered predictors and cannot be extrapolated to other settings (Fig. 2.7). For example, Melton index thresholds reported in the literature above which watersheds would generate debris flows range between 0.3 in the Canadian Rocky Mountains up to 0.75 in the Southern Alps of New Zealand, with many intermediate values reported from other regions (Bovis & Jakob, 1999; de Scally & Owens, 2004; de Scally et al., 2001, 2010; Ilinca, 2021; Jackson et al., 1987; Kostaschuk et al., 1986; Marchi & Brochot, 2000; Marchi et al., 1993; Welsh & Davies, 2011; Wilford et al., 2004). Similarly, the thresholds defined by Wilford et al. (2004) for fans of western British Columbia (Canada), based on Melton index and watershed length, were found to have a low predictive value for the occurrence of debris flows in the Serra da Prata range in Brazil (Picanço et al., 2016).

Variability of sediment supply may explain part of the scatter and overlap in datasets. For example, Crosta and Frattini (2004) observed that systems with morphometric characteristics symptomatic for debris-flow activity (e.g., steep slopes and small watershed area) do not generate them where they are underlain by highly competent bedrock, resulting in slow sediment yield. Conversely, watersheds with gentle terrain gradients that typically generate floods are observed to generate debris flows where they have a high sediment availability. To include these factors, some authors have proposed multivariate models integrating geological controls (Crosta & Frattini, 2004; D'Agostino & Marchi 2001). These studies suggest that multivariate analyses are effective methods to discriminate the processes occurring on the fans, but they are more difficult to implement considering the greater number of indicators involved in natural processes.

Apart from affecting sediment production rates, the bedrock geology of watersheds may guide the preferential generation of water floods versus debris flows by influencing the production of fine sediment fractions (clay to fine silt), which can impart significant control on the rheology of mobile sediment–water mixtures. This has been observed in studies of Late Holocene to recent alluvial-fan records (e.g. Blair, 1999a; De Scally et al., 2010; Franke et al., 2015; Maraga et al., 1998; Mather & Stokes, 2018;

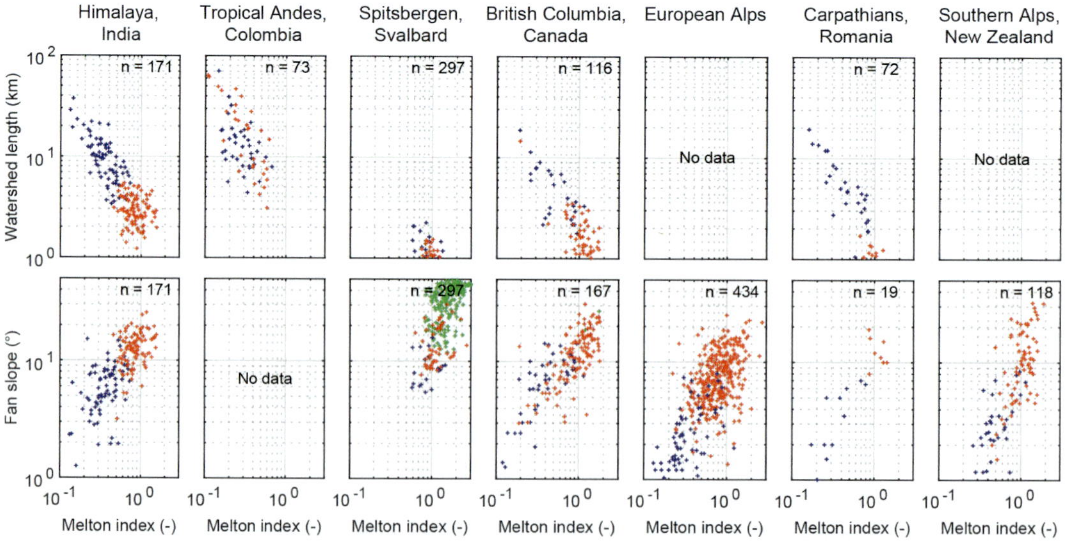

Fig. 2.7 Regional breakdown of the relationships between Melton index and watershed length and Melton index and fan slope for systems dominated by debris flows, floods, and colluvial processes shown in Fig. 2.2. See Fig. 2.6 for color legend

Moscariello et al., 2002), where direct process observations and the accessibility of geomorphically active landscape tracts enabled a link between the type and location of sediment sources to the nature and relative abundance of resulting sedimentary facies. The mobilization of regolith with high volume percentages of mud has been consistently shown to result in preferential formation of debris flows. Bedrock sources consisting dominantly of fine-grained sedimentary and/or intermediate to low-grade metamorphic rocks of fine texture (e.g., schists, mica-schists and phyllites), especially when densely fractured after local tectonic activity (e.g. De Scally et al., 2010; Franke et al., 2015), are more prone to yield mud-rich sediment, thus favoring debris-flow processes (Coussot, 1995; Coussot & Piau, 1994; Major & Pierson, 1992). In contrast, the presence of compositionally pure carbonate rocks or sandstones, as well as intrusive magmatic rocks of acidic, quartz-rich composition, tends to yield relatively large volumes of coarse debris (sandy to gravelly), which is less likely to generate debris flows and thus may result in flood dominance (Major, 2000).

Given the likely absence of a well-defined morphometric discontinuity between debris-flow dominated and flood-dominated systems, we advocate that it would be better to shift from threshold-based approaches to probability-based analyses for more effective hazard assessment and mitigation. Figure 2.8 shows the probability to develop debris-flow dominated, flood-dominated, or colluvium-dominated systems for a combination of Melton index with watershed length and Melton index with fan slope, based on our global data compilation. Such an approach helps to identify combinations of parameter values for which one process is highly likely to be dominant, but also to highlight combinations for which two process types may be equally likely. This approach should lead to more objective and

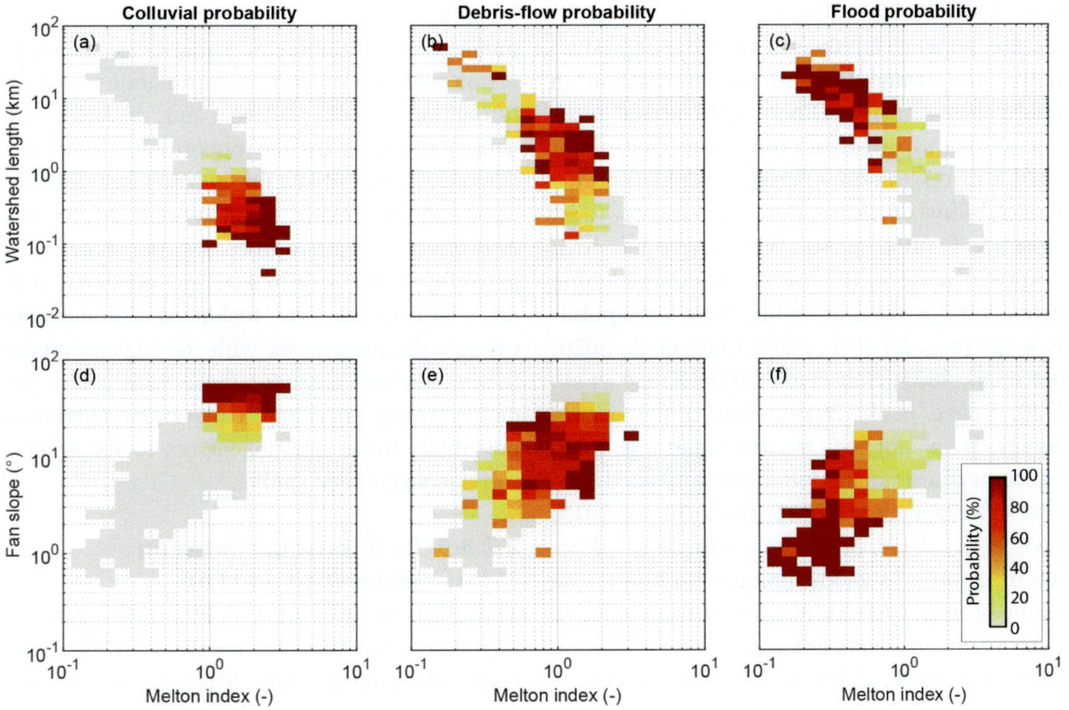

Fig. 2.8 Probability of debris-flow, flood, or colluvial dominance for the systems in our global database, relative to the two other types of systems

realistic decision-making and hazard assessment, especially at regional scales.

In short, debris-flow dominated, flood-dominated, and colluvium-dominated watersheds and fans present generally distinct morphometric characteristics. However, no sharp discontinuity exists based on which such systems may be distinguished, considerable overlap exists, and locally documented thresholds for process generation and/or dominance generally are not suitable for extrapolation to other regions. Therefore, confirming these empirically based classifications with ground-truthing remains a prerequisite approach for accurate process identification from local to regional scale (e.g. Hartley et al., 2005; Ventra et al., 2013).

2.2.3.4 Recent Advances in Identifying Debris Flow-Prone Watersheds

Multivariate association rules and discriminant analyses rank amongst the most popular methods for distinguishing debris flow-prone watersheds (Bovis & Jakob, 1999; Wilford et al., 2004; Welsh & Davies, 2010). The methods are popular for their simplicity, reproducibility, and straightforward implementation in hazard assessments. Other multivariate techniques such as principal component analysis or logistic regression offer further insights into the predictability of debris-flow occurrence (e.g., Crosta & Frattini, 2004). While there has been a rapid increase in the use of machine-learning techniques for creating debris-flow susceptibility maps (e.g., Di et al., 2019; Qing et al., 2020; Qiu et al., 2021; Sun et al., 2021; Xiong et al., 2020; Zhang et al., 2019), only a few recent studies have applied machine-learning techniques to discriminate dominant watershed processes (Arango et al., 2021; Stolle et al., 2015; Wang et al., 2019). Machine learning is a collection of algorithms and techniques used to create computational systems that learn from data to make predictions and inferences (e.g., Bishop, 2006; Hastie et al., 2009; Witten et al., 2011).

Stolle et al. (2015) compared simple discriminant analysis with more sophisticated approaches, including classification tree modelling and Bayesian logistic regression, and found a similar overall success rate compared to the discriminant models. This finding supports observations from data mining that simpler models often have only slightly lower prediction accuracy than more complex models (Witten et al., 2011). Still, more comparisons between simple discriminant analyses and machine-learning approached are needed before definitive conclusions on overall success rates can be drawn.

A disadvantage of more complex machine-learning approaches is that they can be difficult to understand or implement for practical studies on debris-flow susceptibility (Stolle et al., 2015). Yet, the more complex techniques offer several advantages. Bivariate combinations of predictors typically provide high success rates as long as they are tested against observed data, in which case the reported predictive accuracy may be deceptively high (Stolle et al., 2015). With machine learning, datasets can be partitioned into training and test sets. Further advantages are that class membership may be expressed as probabilities and that forecast outcomes may be weighted. Machine learning enables combinations of large numbers of morphometric parameters. In combination with the modern-day availability of high-resolution remote-sensing and hydroclimatic data, this opens up the opportunity to include parameters describing the geology, tectonic setting, and hydroclimatic conditions, which may also affect formative processes, but which have generally been overlooked in previous studies. Furthermore, the technique automatically extracts the parameters with the highest predictive values and excludes those with a low predictive value, providing a first-order ranking of potentially more relevant factors to focus on in subsequent detailed studies.

2.2.3.5 Relationships Between Watershed and Fan

The characteristics of a debris-flow watershed largely control the characteristics of its associated depositional fan. Most commonly, statistically significant relations of watershed area (WA) and Melton index (MI) with fan area (FA) and slope (FS) have been reported across different mountain ranges and climates (e.g., Bertrand et al., 2013; Bull, 1962; Crostra & Frattini, 2004; De Haas et al., 2015a; de Scally et al., 2010; Harvey, 2018; Harvey et al., 1999a; Hooke, 1968; Hooke & Rohrer, 1977; Kostaschuk et al., 1986; Lecce, 1991; Mather et al., 2000; Melton, 1965; Stokes & Gomes, 2020; Tomczyk, 2021). The relationships are power laws of the following forms:

$$FA = aWA^{b} \tag{2.1}$$

$$FA = aMI^{-b} \tag{2.2}$$

$$FS = aWA^{-b} \tag{2.3}$$

$$FS = aMI^{b} \tag{2.4}$$

where a and b are regression constants. For our dataset, extracted from different continents, mountain ranges, climates, and process types, we find statistically significant power-law relations between watershed and fan characteristics (Fig. 2.6; Table 2.3). The correlations between watershed area against fan area and slope are strong. The relation between Melton index and fan slope has a similar strength, while the relation between Melton index and fan area is weaker, although still statistically significant. For a given watershed area or Melton index, fan areas and slopes may vary by over an order of magnitude (Table 2.3; Fig. 2.6), as previously found in many regional studies (e.g., Bertrand et al., 2013; Blair & McPherson, 1994, 2009; Harvey, 1989). Crosta and Frattini (2004) identified larger scatter of data for alpine fans compared to arid-environment fans. They attributed this to a larger variability in accommodation space on the floors of Alpine valleys and to a greater variability in sediment supply in time and space. The latter can result from variations in bedrock erodibility, processes responsible for sediment removal and delivery, and paraglacial effects.

These relationships are independent of systems type (e.g., colluvial, debris flow, and fluvial) and are valid across a wide range of climatic, tectonic, and geologic settings (e.g., de Scally et al., 2010; Hooke, 1968; Kostaschuk et al., 1986; Tomczyk, 2021). The power-law relations between watershed area and fan area found for our global dataset have lower correlation coefficients than those reported from regional studies (Table 2.1), highlighting that those correlations may vary between different regions. Variations in the coefficients in the relationships are typically explained in terms of variations in climatic, geologic, and tectonic setting, as well as other local factors (e.g., Blair & McPherson, 2009; Densmore et al., 2007; Harvey et al., 1999b). Moreover, many watershed and fan features are often dynamic and have a transient nature at geological time scales, including relief, stream piracy, tectonic beheading, sedimentary processes, variable inherited geological discontinuities, bedrock lithology, areal constriction (for example, due to the presence of adjacent fans),

Table 2.3 Overview of power-law relationships between watershed (area, Melton index) and fan (area, slope) properties, from this study and as reported in literature

Relation	Location	Number of systems	a	b	R2	Reference
$FS = aMI^{b}$	Global compilation	1223	11.9	1.21	0.65	This study
$FS = aWA^{b}$	Global compilation	603	11.6	−0.22	0.68	This study
$FA = aMI^{b}$	Global compilation	413	0.03	−1.75	0.31	This study
$FA = aWA^{b}$	Global compilation	413	0.07	0.47	0.60	This study
$FA = aWA^{b}$	Central Alps (Italy)	64	0.15	0.65	0.81	Crosta and Frattini (2004)
$FA = aWA^{b}$	Great basin, Nevada (USA)	106	0.38	0.86	0.96	Harvey (2005)
$FA = aWA^{b}$	Wadi-al-Bih (UAE, Oman)	63	0.18	0.85	0.86	Al-Farraj and Harvey (2005)
$FA = aWA^{b}$	Cabo de Gata (Spain)	20	0.84	0.75	0.92	Harvey (2002)
$FA = aWA^{b}$	North Evoikos Gulf (Greece)	42	0.12	0.70	0.89	Valkanou et al. (2013)

interplay with environments that border the fans, such as fan truncation by valley rivers (Blair & McPherson, 2009), or base-level interference by distally established waterbodies or local tectonics (e.g. González et al., 2021; Harvey, 1987; Stokes & Mather, 2000).

The geomorphological explanation behind the relationships between watersheds and fans of debris-flow systems is that larger and steeper watersheds can generate larger debris-flow volumes (e.g., De Haas & Densmore, 2019), regardless of differences in hydroclimatic triggers, such as precipitation, snow melt, or permafrost thawing (Tomczyk, 2021). Debris-flow runout also scales with flow volume (e.g., Berti & Simoni, 2007; D'Agostino et al., 2010; Griswold & Iverson, 2008; Iverson et al., 1998), and therefore flows of larger volume generate fans with larger areas (e.g. Blair, 2003; Schoch-Baumann et al., 2022; Scott et al., 2001). Fan gradient is generally negatively correlated with watershed area, which is explained by increased water discharge as the drainage area extends (Crosta & Frattini, 2004).

2.3 Fan Characteristics and Dynamics

Debris-flow fans are semi-conical landforms that grow by deposition of sediment from repeated flows issued from a nearby watershed. They are ubiquitous in mountainous regions and, because of their relatively gentle surface slopes, they are commonly occupied for settlement, agriculture, or other human activities. Identification of the dominant process type on a fan is therefore of key importance for hazard assessment (Huebl & Fiebiger, 2005). This requires an understanding of the morphological and stratigraphical characteristics of debris-flow fans, and knowledge of how these differ from fans dominantly formed by floods or colluvial processes. Overviews of debris-flow-fan morphology and sedimentary stratigraphy are presented in Sects. 2.3.1 and 2.3.2, respectively. Table 2.4 summarizes some of the key characteristics for distinguishing debris-flow fans from colluvial and fluvial fans.

Debris-flow fans form and evolve over time mostly through areal shifts in the locus of deposition, which are termed avulsions (De Haas et al., 2018a; Whipple & Dunne, 1992). At any given moment in time, debris-flow deposition typically only affects a minor fraction of the fan surface, centered along and/or at the terminus of the active channel, and this area may shift in position on the fan's surface through avulsion. Debris flows that leave the main channel typically pose the largest threat to settlements and infrastructure on alluvial fans, as mitigation measures such as check dams and retention basins are often applied only to presently active channels (Densmore et al., 2019; Pederson et al., 2015). In addition, shifting of the active channel over time leads to distinct aggradational sectors on the fan surface, each preserving a record of debris-flow activity during a particular time period and over a number of flows, providing archives of flow conditions during that time period. Therefore, it is important to understand the mechanisms of avulsion on debris-flow fans. Section 2.3.3 discusses known patterns of spatio-temporal evolution and avulsion mechanisms.

Debris-flow fans are the products of the debris flows that formed them, and therefore provide archives of past conditions. However, given the generally high energy of these environments and the relatively low frequency of depositional events, the extraction of paleoenvironmental information from these depositional systems remains particularly challenging, especially at high spatio-temporal resolution. Section 2.3.4 summarizes efforts of extracting information of past environmental conditions from debris-flow-fan deposits.

2.3.1 Debris-Flow Fan Morphology

2.3.1.1 Primary Surface Morphology

Debris-flow fans have radial lengths that typically range between 0.5 and 5 km, but can attain exceptional lengths in excess of 10 km (Blair & McPherson, 1994, 2009; Blikra & Nemec, 1998) in rare cases where high debris volumes

Table 2.4 Overview of characteristics used to identify debris-flow fans from colluvial and fluvial fans, partly based on data aggregated for this study and presented in Sect. 2.2.3.2

Characteristic	Colluvial	Debris flow	Fluvial
Radial length	Range: 0.1–1 km	Range: 0.5–10 km	Range: 1–100 km
Fan slope	Average: 35–45° Range: 10–50°	Average: 5–15° Range: 1–30°	Average: 2–7° Range: generally less than 10°
Morphology	Poorly defined channel	Channels, lobes, and levees	Channels, braid-like channel features, terraces, bank erosion or migration
Common grain sizes	Gravel to boulder	Clay to boulder. Biggest clasts can be larger than the expected bankfull depth of stream flow. Coarser grained at lobe fronts and along levees	Sand to boulder
Grain angularity	Angular to subangular	Subrounded to angular	Rounded to subrounded
Sediment morphology	Clast-supported deposits	Matrix supported (clast supported rarely), no clast imbrication, inversely graded deposits	Clast supported, clast imbrication, normally graded deposits
Vegetation	Cleared or damaged trees near fan apex	Near active channel—adventitious roots, impact scarring Outside active channel—buried tree trunks, impact scars above deposit height	Near active channel—adventitious roots, impact scarring, wide breaks in riparian canopy Outside active channel—buried tree trunks, impact scars near deposit height

and a favorable composition enhance the mobility of debris flows (e.g. Blair, 2003; Lorente et al., 2003; Pederson et al., 2015). The slopes of debris-flow fans typically range between 5 and 15°, although in rare cases they may be as low as 1–2° or as high as 30° (Fig. 2.5f). Generally, fan slopes are highest near the fan apex and in the proximal sector, decreasing distally towards the fan toe (Blair & McPherson, 2009). Some debris-flow fans display a distinct change in gradient around one-half to two-thirds of the way down their radial length, commonly referred to as the inflection or intersection point (Hooke, 1967). In comparison, flood-dominated fans generally have a much broader range of radial lengths, ranging from a few kilometers for alluvial fans to more than 100 km in the case of fluvial fans, but featuring lesser gradients ranging from 2 to 7° and rarely over 10° for alluvial fans, down to substantially less than 1° over most of the surface of aggradational fluvial fans (e.g., Fig. 2.5f; Blair & McPherson, 2009; Blikra & Nemec, 1998; Horton & DeCelles, 2001; Wilkinson et al., 2006). Colluvial fans, predominantly formed by rock falls, dry grain flows, and snow avalanches in cold environments, are generally smaller in area and radial length than debris-flow fans, and exhibit steeper slopes close to the angle of repose of coarse granular material, up to 35–45° on the most proximal sectors abutting adjoining topographic relief, and possibly between 10° and 20° on their medial and distal sectors (Fig. 2.5f; Blikra & Nemec, 1998; De Haas et al., 2015a; Tomczyk, 2021).

Morphologically, debris-flow fans tend to attain a lobate, low-relief plan-view shape during constructional phases of the landform, with a concave longitudinal profile of downfan-decreasing gradient (see above) and a generally convex transverse profile (e.g. Blair & McPherson, 1994; Blikra & Nemec, 1998; De Scally & Owens, 2004; Hooke, 1967). The positive topography accreted by debris-flow fans can feature a markedly more prominent relief compared to alluvial fans (Volker et al., 2007); this is due to the tendency of debris flows to construct depositional ridges and lobes of high aspect ratio (thickness versus width), owing to their stabilization by cohesive matrix and by frictional interlocking of very coarse debris fractions (Kim & Lowe, 2004). Aside from possible significant reworking by post-depositional agents (see following Sect. 2.3.1.2), this effect scales up to the entire fan once individual debris-flow deposits stack up into compound, vertically and laterally amalgamated lobes. The overall lobate, convex shape of individual debris-flow fans remains clearly discernible in the case of laterally adjoining alluvial fans ('bajadas'), as recognizable from numerous depositional piedmonts worldwide (e.g. Blair, 1999a; Damanti, 1993; De Scally & Owens, 2004; Ferrill et al., 1996; Hartley et al., 2005; Milana, 2000). However, smaller colluvial fans with more subdued topography can merge laterally, forming aprons in which individual fans are less well distinguishable, maintaining a surface relief of only metric scale (e.g. Blikra & Nemec, 1998; Ventra et al., 2013). At system scale, the aggradational shape of debris-flow fans can be subjected to a high degree of modification when reduced sediment yield from the watershed, changes in dominant processes and/or base-level effects trigger long-term, or large-scale incision, commonly associated with a valleyward shift in the main locus of aggradation (see Sect. 2.3.3.2; Blair & McPherson, 2009; De Haas et al., 2014).

At the mesoscale, the morphology of debris-flow fans is controlled by the dynamics of debris-flow runout and deposition, and generally consists of spatially variable associations of channels, levees, and lobes, depending on whether flows travel on the fan surface in mostly confined or unconfined fashion. In planform, this leads to fans that consist of stacked, vertically and laterally amalgamated, elongate, and straight to sinuous lobes, which, if not buried by subsequent lobes, are connected to (leveed) channels (Figs. 2.9 and 2.10). The lobes and levees of debris flows consist of poorly sorted sediment mixtures ranging from clay to boulder-sized particles (Fig. 2.9) (Beaty, 1963; Blair & McPherson, 1998; Blair, 1999b; Hubert & Filipov, 1989; Kim & Lowe, 2004). On fans with sufficiently steep slopes, boulders with

Fig. 2.9 Lobe and levee deposits on debris-flow fans. **a** Debris-flow lobe deposit on a fan surface in periglacial environment of Svalbard (fan described in detail in Bernhardt et al., 2017). **b** Debris-flow lobe deposit in the Chilean arid Atacama Desert. **c** Stony debris-flow lobe deposit on Pinyon Creek fan, Owens Valley, California, USA (fan described in detail in Whipple and Dunne (1992)). **d** Debris-flow lobe deposit on the forested Ritigraben fan, Switzerland (fan described in Stoffel et al., 2008a, 2008b). **e** Debris-flow channel with clearly-defined lateral levees on Svalbard. This channel is connected to the lobe shown in panel a. **f** Debris-flow channel with lateral levees in the Chilean Atacama Desert. This channel is connected to the lobe shown in panel b. **g** Debris-flow channel with well-defined levees on the Dolomite fan in the Mojave Desert (USA) (fan described in detail in Blair and McPherson et al., 1998). **h** Debris-flow channel with well-defined levees on a forested fan in the Italian Alps

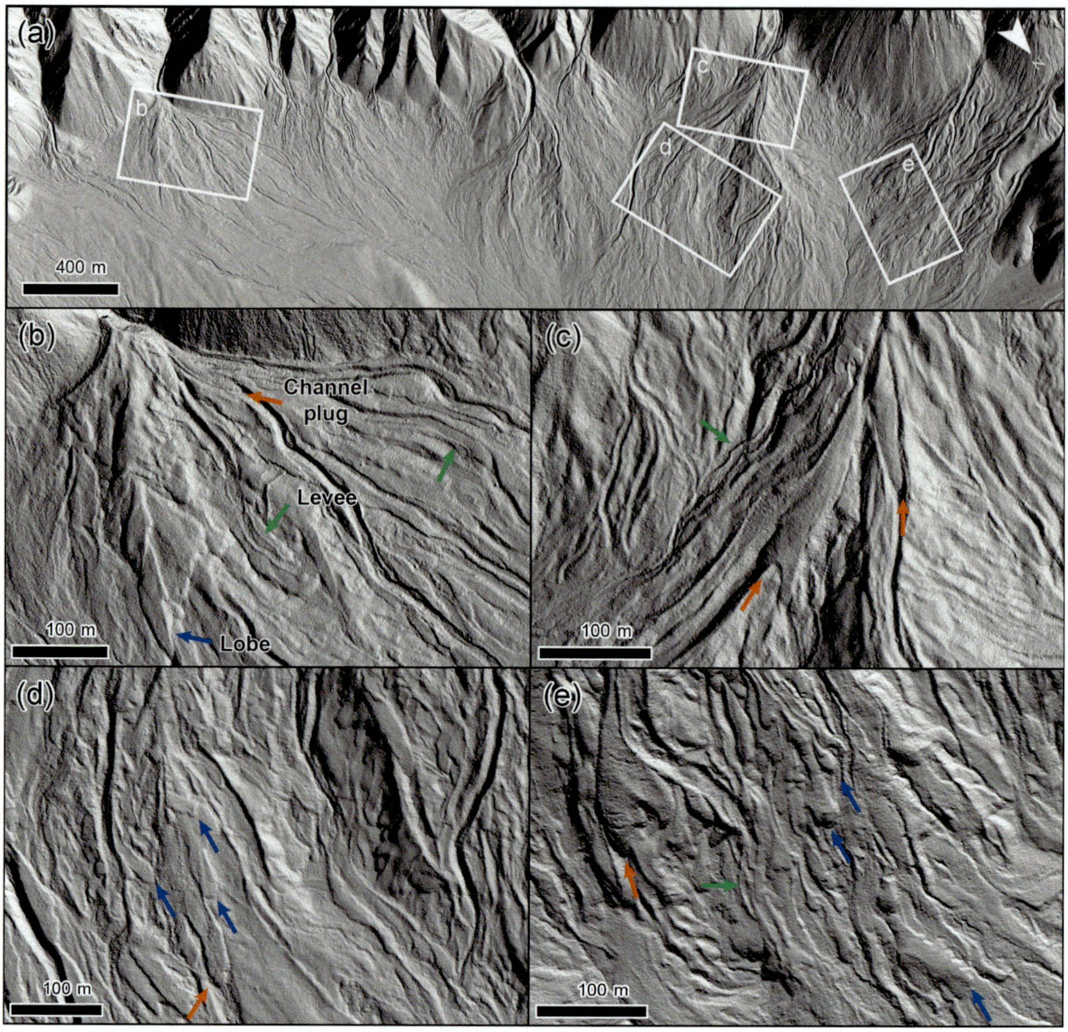

Fig. 2.10 Examples of unmodified overlapping channel, lobe, and levee morphology shown in hillshaded lidar digital elevation models of debris-flow fans in Saline Valley, California, USA. The surfaces of these fans are composed of channels, lobe and levee deposits. Blue arrows point at examples of lobe deposits, green arrows highlight levee deposits, and orange arrows point at channel plugs—lobe deposits that plug a channel and force avulsion upstream of the lobe (see also Fig. 2). See further details in De Haas et al. (2019) and De Haas and Densmore (2019). Fans located around 36°34′33.45″ N, 117°37′53.07″ W

diameters up to several meters can be transported for hundreds of meters by debris flows and can therefore be common on the surfaces of debris-flow fans (e.g. Beaty, 1963, 1989; D'Arcy et al., 2015; Schoch-Baumann et al., 2022), where they may form relict elements of anomalously high relief even after significant reworking or degradation of the surrounding topography.

The typically rugged morphology of debris-flow fans results from the high solid concentration of debris flows. As a result of this high solid concentration, coarse grains are preferentially transported to the front of individual flows and are successively deposited in lateral levees and frontal lobes (De Haas et al., 2015a; Johnson et al., 2012; Major & Iverson, 1999; Sharp, 1942; Suwa, 1988). These levees constitute sharply bounded, downfan-oriented ridges typically 1–2 m wide and 1–4 m tall, coupled in parallel pairs separated by a typically 2–10 m wide channel that occupies the original runout path of the main flowing mass (Fig. 2.9). Lobe deposits typically constitute elongated features 1–100 m wide and 0.1–3 m thick (Blair & McPherson, 2009), and decrease in thickness and increase in width from fan apex to fan toe (De Haas et al., 2019). The largest, often boulder-rich, particles are generally concentrated in the levees and frontal margins of lobe deposits (e.g. Blair & McPherson, 1998; Hubert & Filipov, 1989; Suwa & Okuda, 1983; Ventra et al., 2013). As a result of the distinct mesotopography formed by lobe and levee deposits, local surface relief of debris-flow fans is generally high in comparison to their flood-dominated counterparts, typically ranging between 0.5 and 4 m (Blair & McPherson, 2009; Volker et al., 2007).

The primary topographic texture typical of debris-flow fans, as described above, is more pronounced on proximal to medial sectors, tending to become less evident on distal sectors. Despite sufficient gradients to transport such sediments, the gravel-rich fractions of mobile debris mixtures are more likely to accumulate on proximal fan sectors due to processes such as coarse-clast interlocking at flow margins and the gradual loss of interstitial fluidal phases that support coarse fractions (Blair & McPherson, 1998; Kim & Lowe, 2004; Moscariello et al., 2002). Whereas debris-flow levees and gravel-rich snouts are therefore more commonly accumulated upfan, the farther runout downfan of finer-grained, more fluidal fractions leads to lateral and vertical amalgamation of wider, mud-prone lobes a few centimeters to decimeters thick, forming distal low-relief aprons (De Haas et al., 2014; Matthews et al., 1999; Meyer & Wells, 1997; Whipple & Dunne, 1992). In time, the establishment of this longitudinally bipartite morphology is enhanced by the repeated advection of dilute flows to the distal fan, as secondary runoff keeps reworking the surface.

At the fan scale, surface roughness is much higher on debris-flow fans than on flood-dominated fans. For example, on debris-flow fans in Saline Valley, California, USA (Fig. 2.10), main channels were typically found to be elevated above the surrounding fan surface by 2–5 channel depths, and locally by more than 7 channel depths, in places exceeding 10 m above the overall fan surface (De Haas et al., 2019). In contrast, on flood-dominated fans, channels can be raised above the surrounding surface by a maximum of ~0.5–1 channel depths before superelevation of the main locus of aggradation renders the system locally unstable and avulsion occurs again (Carling et al., 2016; Jerolmack & Mohrig, 2007).

2.3.1.2 Post-Depositional Reworking of Debris-Flow Fan Surfaces

The high return periods of debris flows (often ranging from multiple years to centuries, depending on local settings) and their limited spatial extent relative to the total fan surface combine to render (parts of) fan surfaces susceptible to modification by secondary post-depositional processes (Fig. 2.11) (e.g., Blair & McPherson, 2009; De Haas et al., 2014; Ventra & Clarke, 2018; Walk et al., 2022). Processes of secondary modification include surface runoff or groundwater flow, deflation, bioturbation, weathering and pedogenesis, neotectonics, and a variety of periglacial processes (Blair & McPherson, 1994, 2009; De Haas et al., 2014, 2015a; Harvey, 2011). Post-depositional modification of fan deposits occurs at varying rates and by different processes, mostly depending on the local climatic conditions (André, 1990a; Frankel & Dolan, 2007; Webb & Fielding, 1999). The relative influence of primary versus secondary processes in generating the morphology of debris-flow fans therefore depends on the balance between (i) the characteristic time scales by which primary processes build a fan's morphology and (ii) the rates at which combined secondary processes modify such morphology (De Haas et al., 2014).

Post-depositional modification by secondary processes may result in well-defined morphological and textural differentiation between fan sectors of different ages, whereby sectors that have persisted in a state of depositional inactivity for the longest time will manifest the greatest degree of change (Al Farraj & Harvey, 2000; De Haas et al., 2014; Dickerson et al., 2015; Frankel & Dolan, 2007; Matmon et al., 2006; McFadden et al., 1989a, 1989b; Ritter et al., 1993; Walk et al., 2022; Wells et al., 1987). Hence, in general, the active fan sector is often dominated by debris-flow morphology and texture, while the characteristics of older fan surfaces may have been variably modified or even dominated by secondary processes (e.g., De Haas et al., 2014). On fans where older sectors are modified by post-depositional processes, pronounced morphological and textural differences may therefore be used for relative dating of different sectors of the fan surface (e.g. Frankel & Dolan, 2007; Johnstone et al., 2018; McFadden et al., 1989a, 1989b; Scheinert et al., 2012).

Fan surfaces are exposed to many types of physical and chemical weathering, including salt weathering, frost weathering, and thermal weathering, depending on the climatic setting (Berger & Cooke, 1997; Warke, 2013). Surface weathering breaks down surficial grains, leading to a progressive decrease in overall grain size (Fig. 2.11i) (Berger & Cooke, 1997; De Haas et al., 2014; Walk et al., 2022). Clasts in the fine to medium gravel range (pebbles to cobbles) are susceptible to greater rates of physical disaggregation (e.g. Eppes et al., 2010), favouring the gradual loss of mesoscale relief features and surface roughness, such as debris-flow levees and frontal lobes (e.g. Al-Farraj & Harvey, 2000; Amit et al., 1993; McFadden et al., 1989a, 1989b; Quade, 2001; Scheinert et al., 2012). This characteristic can therefore be used as a typical first-order criterion for the identification of long-abandoned fan sectors. Particle disintegration by weathering in combination with aeolian dust deposition may ultimately lead to the formation of desert pavement on a fan surface (Fig. 2.11g) (Al Farraj & Harvey, 2000; Blair & McPherson, 2009). In debris-flow watersheds that have ephemeral or low-discharge water flow, most fan surface deposits are from larger debris-flow events and are secondarily modified by flood events (Blair & McPherson, 2009). Moreover, direct precipitation on fan surfaces may result in overland flows producing rills, gullies, and coarse sediment lags by selective entrainment of fine sediments from older fan-surface deposits (Fig. 2.11h) (De Haas et al., 2014; Walk et al., 2022). On long-inactive debris-flow fans, rills and gullies may mask the original surface by forming secondary drainage networks (De Haas et al., 2014). Wind may also remobilize the fine sediment fraction from fan surfaces (silt to fine pebble size), and can modify the original morphology (Fig. 2.11a, d) (Al Farraj & Harvey, 2000; Blair & McPherson, 2009; Cook & Pelletier, 2007). Wind may further transport fine sediment to fan surfaces from non-watershed sources, such as adjoining dunefields, sandsheets or playas, thereby blanketing the original surface and masking debris-flow generated morphological features to various degrees (Anderson & Anderson, 1990; Blair & McPherson, 2009; Pelletier, 2007; Pelletier & Cook, 2005; Ventra et al., 2013).

In periglacial environments, permafrost-related processes such as frost heaving, aufeis formation, frequent freeze thaw-cycles and snow avalanches all contribute to post-depositional modification of fan deposits. In these regions, the formation of patterned ground, hummocks and solifluction lobes may modify the surfaces of debris-flow fans (Fig. 2.11b, d, e, f) (Blikra & Nemec, 1998, 2000; Catto, 1993; De Haas et al., 2015a; Legget et al., 1966). Due to their extremely high erosive capacity, where snow avalanches are frequent, they generally rapidly bevel and level the primary morphology and alter the morphological and textural properties of most exposed depositional units at the surface—typically debris-flow levees and lobes (André, 1990b; Blikra & Nemec, 1998; De Haas et al., 2015a). In densely vegetated or forested areas under more temperate, humid climates, the growth of trees is inhibited by frequent debris-flow activity, while trees may grow on inactive parts of the fan surface (Michelini et al., 2017;

Fig. 2.11 Secondary, post-depositional, modification of debris-flow fan surfaces, masking the original depositional morphology. **a** Debris-flow fan surfaces covered by eolian sand, Atacama Desert, Chile. **b** Polygonal ground on top of a debris-flow fan surface on Svalbard (from De Haas et al., 2015a). **c** Hummocks on top of a fluvial fan surface on Svalbard (from De Haas et al., 2015a). **d** Detail of fan shown in panel a (from De Haas et al., 2015c). **e** Detail of polygonal ground on fan in panel b (from De Haas et al., 2015a). **f** Detail of hummocky ground on fan in panel c (from De Haas et al., 2015a). **g** Desert pavement on top of a fan surface in Nevada (USA) (from Blair & McPherson, 2009). **h** Fluvial channels from secondary runoff on a debris-flow fan surface, Atacama Desert, Chile; person for scale is 1.85 m tall (from De Haas et al., 2014). **i** Weathering of a clast on the surface of a debris-flow fan as a result of salt weathering in the Atacama Desert, Chile (from De Haas et al., 2014). Figure modified from Conway et al. (2019)

Mohseni et al., 2017). At the same time, where a fan is incised, the groundwater table is higher and promotes increased riparian vegetation growth compared to the drier, more elevated surfaces farther away from the incised channel (Kellerhals & Church, 1990). Pedogenesis and biological activity represent the principal agents of secondary surface modification in these settings, where the frequency of debris flows and especially the petrographic composition of debris bears a strong influence on the biogeochemical processes and their products. However, even in densely forested piedmonts, debris-flow activity may prove relatively unaffected by the vegetation canopy, with only partial reduction of runout or volume for individual flows (Guthrie et al., 2010; Michelini et al., 2017).

2.3.2 Debris-Flow Fan Sedimentology and Stratigraphy

Debris-flow fans have a unique sedimentology and stratigraphy (Fig. 2.12). This results from the heterogeneous textural composition of debris flows and from their varying, single- to multi-phase state during runout and deposition (Iverson, 1997; Iverson et al., 2010; Pudasaini, 2012), with internal processes controlled concomitantly by fluid–solid interactions and solid–solid interactions within the flowing mass. Debris flows essentially consist of two sedimentological components: (i) a liquefied, mud-rich phase that maintains its fluidal state owing to excess pore (i.e. intergranular) pressures and

(a)
(b)
(c)
(d)
(e)
(f)
(g)
(h)
5 m

Fig. 2.12 Examples of sedimentary facies of debris flows exposed along incised channels on the proximal debris-flow fans (**a**, **f**), and flood-dominated fans for contrast (**g**, **h**). The sections predominantly consist of lobes separated by washed partings and thin beds (arrows). **a** Sedimentary facies of a debris-flow fan in the Chilean Atacama Desert. Flow is from right to left. **b** Facies a few hundred meters upstream of figure panel a, exposing a very large outsized boulder. Flow is from right to left. **c** Debris-flow facies along an incised channel in the Warm Spring fan, Death Valley, USA (see Blair (1999a, 1999b) for a detailed facies description). Note the geological hammer for scale. Flow is from right to left. **d** Sedimentary facies of the Straight Canyon fan in Owens Valley, California, USA (see Hubert and Filipov (1989) for a detailed fan and facies description). Flow is from right to left. **e** Sedimentary facies of a fine-grained debris-flow fan in the Chilean Atacama Desert. Individual debris-flow beds are clear from secondary modification of the bed surfaces. Flow from left to right. **f** Inverse grading in a Miocene debris-flow deposit in the Teruil basin, Spain. Flow is towards the viewer. **g** Flood deposit along the upper valley of the Mendoza River, western Argentina, overlain by a debris-flow deposit. Note the sharp contrast between the underlying fluvial facies (moderate to good sorting, imbrication of well-rounded cobbles and of their sandy-gravelly matrix, finer-upward sorting) and structureless overlying debris-flow deposits with angular clasts dispersed in a muddy matrix. Flow is from left to right. **h** Sedimentary facies of a flood-dominated fan in the Chilean Atacama Desert. The fan is nearby the debris-flow fan shown in panels a-b. Note the narrower textural range, lack of large clasts, and general organization in well-stratified depositional units with clear textural contrasts, in contrast to the debris-flow deposits. Flow is towards the viewer

sustains the coarse fraction of the overall mass through a combination of buoyancy, viscosity, and drag. The liquefied slurry has a finite shear strength because of frictional and especially cohesive grain-to-grain interactions, conferring to the overall flow, at least temporarily, non-Newtonian properties (O'Brien & Julien, 1985, 1988; Phillips & Davies, 1991; Coussot & Piau, 1994); (ii) a coarse sediment fraction (commonly gravel-grade clasts and vegetation macrodebris), which essentially constitutes the solid, particulate load of flows and contributes part of the shear resistance through frictional and interlocking grain-to-grain interactions (Iverson, 2003, 2005; Major & Pierson, 1992; Pierson, 1985, 1986). The characteristics of debris-flow deposits thus depend on the combined properties of these two components, on their degree of separation during overall flow evolution (Pudasaini & Fischer, 2020), and on their dynamic feedbacks. Mechanisms of particle segregation toward the front and lateral margins of the flow result in deposits of coarse-grained levees and frontal snouts that envelop a finer-grained body (De Haas et al., 2015b; Gray, 2018; Gray & Ancey, 2009, 2011; Johnson et al., 2012; Major, 1997; Major & Iverson, 1999; Pierson, 1986; Sharp, 1942; Sharp & Nobles, 1953). These characteristics are reflected in the sedimentology and stratigraphy of debris-flow fans.

The sedimentology of debris-flow deposits is quite variable both laterally and vertically across fan successions, depending on local transport and accumulation processes as levees and lobes are juxtaposed and stacked through time, and on changes in the volume and type of regolith entrained from the substrate at different runout stages. In stratigraphic sections, the signature facies attributes of debris-flow deposits are: (i) internally massive beds with decimetric to metric thickness and (ii) poorly to very poorly sorted texture, from mud-grade matrix to outsized metric boulders, with (iii) a frequent lack of vertical grain-size trends (but basal or bed-scale inverse grading are occasionally recorded, Fig. 2.12f) and (iv) a distinctly matrix-supported fabric in which the coarse fraction does not feature marked grain orientation (e.g. Blair, 1999b; Blikra & Nemec, 1998; Fisher, 1971; Hubert & Filipov, 1989; Johnson, 1984; Nemec & Steel, 1984; Scott, 1971; Sohn et al., 1999; Ventra et al., 2013) (Fig. 2.12).

In general, stacked deposits may appear structureless and amalgamated both vertically and laterally, hampering the recognition of individual depositional units. The frequent absence of basal erosion during the final phases of runout and emplacement of debris flows can also impede recognition of erosional surfaces between deposits. However, bedding surfaces can be highlighted by marginal differences in weathering, weak paleosols, and burned matter in fire-prone areas, but especially by relatively thin, generally centimetric facies with discontinuous lateral distribution, recording secondary reworking of original depositional units by runoff and/or wind during prevalent periods of non-aggradation (Fig. 2.12) (Blair & McPherson, 1994). Individual beds are typically quasi-tabular to slightly irregular, subparallel to the original depositional surface, and continuous over distances up to several hundreds of meters in slope-oriented sections, but sharply tapering or pinching out locally in strike-oriented sections where the lobe margins are preserved. Deposits possess a high degree of internal homogeneity at intermediate (metric to decametric) scale, with minimal variations in facies attributes at a cursory examination. In most instances, gravel and outsized cobbles or boulders are dispersed randomly within a mud-rich matrix (Fig. 2.12a-c). However, they may present some degree of segregation overlying basal bed contacts, where basal shearing of the original flow induces inverse grading in thin, clast-rich flows (Fig. 2.12f; Blair, 1999b; Phillips, 2006; Shultz, 1984), and along bed tops, where clasts may have been supported continuously by the buoyancy and viscosity of the flowing slurry (Fig. 2.12d). In addition, clasts of strongly unequant morphology (discoidal or rod-shaped) may lie more frequently with their long and intermediate axis (*a-b* axis) subhorizontal or oriented subparallel to the bed's depositional dip (Fig. 2.12a), or with their long axis (a-axis) featuring a narrow range of preferential angles with respect to flow margins, having been sheared in transport by the viscous mass (i.e., imbrication) (e.g. Bertran et al., 1997; Enos, 1977; Hubert & Filipov, 1989; Kohlbeck et al., 1994; Major, 1998; Van Steijn & Coutard, 1989; Vallance & Scott, 1997). In strike-oriented sections, debris-flow deposits appear as laterally tapering plano-convex lenses or wedges with facies attributes as mentioned above, but possibly preserving marginal concentrations of coarser debris representing the original emplacement of levees shouldered laterally by the shearing flow. These are recognizable commonly by their limited lateral extent, their clast-supported fabric, by the coarser modal texture of the gravelly fraction compared to that comprised within the main flow body, by the possibly better organized fabric of unequant clasts, and by the locally reduced volume of interstitial matrix (e.g. Blair & McPherson, 1998; Major, 1997, 1998; Ventra et al., 2013).

The most distal sectors of debris flow-dominated fans will commonly host another typical facies, consisting of thin (centimetric to decimetric), massive to faintly stratified sets of strata ranging from sandy and fine-gravelly mud to muddy sand, emplaced by more fluidal, hyperconcentrated to watery tails of flows which manage to expand over lower-gradient surfaces thanks to their lower viscosity and frictional resistance (Blair & McPherson, 1998; De Haas et al., 2014; Matthews et al., 1999; Pierson & Costa, 1987; Sletten & Blikra, 2007; Ventra et al., 2018). Considering that such flow tails can be relatively dilute, tractional transport and deposition can also be significant, leading to associations of sedimentary structures from aggradation under unidirectional, subcritical to transcritical currents, commonly being ripple cross-lamination and plane lamination. These are essentially indistinguishable from the products of shallow, unconfined runoff that will frequently affect fan surfaces between mass-flow events.

Debris-flow dominated fans are also shaped by (debris) flood processes, for example during hydroclimatic events of insufficient intensity to trigger debris flows or when sediment scarcity in the watershed hampers debris-flow formation. On debris-flow dominated fans, deposits of water floods and debris floods (hyperconcentrated flows) are therefore commonly interbedded with debris-flow strata, forming vertical facies sequences and flow-longitudinal facies tracts that originate from the downfan transformation of individual flows and/or from time-phased sedimentation at a place from distinct rheological segments within individual flows (Cronin et al., 1997; Matthews et al., 1999; Pierson, 2005; Pierson & Scott, 1985; Sohn et al., 1999). Debris-flow fans are also affected by runoff carrying lesser sediment volume to the fan, resulting in reworking of the original depositional surface. General criteria to distinguish (debris) flood facies include a higher degree of internal heterogeneity both longitudinally and transversely to the deposits, expressed by a recognizable variety of sedimentary structures and textural segregations, moderate to very good sorting with oriented clast fabrics expressed at multiple scales, common to prevalent clast-supported fabric, and a possible organization into macroforms (e.g. longitudinal and side bars) related to more protracted runoff and depositional stages (Fig. 2.12) (e.g. Blair, 1999b; Costa, 1988; De Haas et al., 2014; Moscariello et al., 2002; Nemec & Steel, 1984).

As mentioned above, debris flows accumulate sediment mainly by lateral and vertical stacking of coarse- and fine-grained levees and lobes that through time will produce a rather disorganized stratigraphy, long recognized as a typical attribute of alluvial- and colluvial-fan successions in piedmont settings. Thick debris-flow successions will commonly feature a dominantly aggradational architecture consisting of juxtaposed, laterally tapering, lens-shaped to tabular beds in which original depositional units (genetic 'beds', sensu Campbell, 1967) may be hard to recognize owing to amalgamation of several debris-flow deposits with similar textural and sedimentological attributes. In strike-oriented stratigraphic sections of limited width, the overall attitude of deposits will be invariably subhorizontal and tabular, but in longitudinal sections the primary bedding can show pronounced valleyward dips up to and exceeding 10°, reflecting the progradational advancement of the inclined fan surface.

The composite aggradational architecture may be disrupted by erosional surfaces at various scales of depth and extent (Fig. 2.12). These are indicative either of local, short-lived erosion by debris flows or runoff along channels, or of protracted degradational phases in the overall system evolution which led to the incision of a longitudinal trench or valley across the fan, commonly signaling long-term morphodynamic disequilibrium related to environmental change (e.g. Decaulne et al., 2007; Harvey, 1987; Viseras & Fernandéz, 1995; White et al., 1996). There have been very few attempts to frame the complex internal architecture of fan successions in terms of hierarchical stratigraphic surfaces or units of genetic significance (e.g. DeCelles et al., 1987), probably owing to poor facies differentiation and organization in the first place. This stands in clear contrast with work on fluvial and eolian successions, for which descriptive and interpretive frameworks of genetic significance are well established (e.g. Brookfield, 1977; Kocurek, 1988; Miall, 1985). The problem has been compounded by the long-standing misinterpretation of fluvial successions at basin margins as originally accumulated by 'alluvial fans' (e.g. Amorosi et al., 1996; Carboni et al., 2006; Meek et al., 2020), a lingering issue of semantic origin, in spite of the clear genetic and process distinction between fluvial and alluvial-fan systems (e.g. Blair & McPherson, 1994; Moscariello, 2005; Ventra & Clarke, 2018). However, recent studies on presently active alluvial fans (e.g. Franke et al., 2015; Hornung et al., 2010; Pederson et al., 2015; Roberts et al., 2003; Schoch-Baumann et al., 2022) emphasize the possibility of defining (i) architectural elements (stratasets at mesoscale, indicative of aggradation from specific process sets) and (ii) system-scale units (stratasets corresponding to periods of protracted aggradation or degradation) towards defining a descriptive framework for fan stratigraphy.

Architectural elements for fan successions, for example, may consist of channel fills, progradational wedges and lobes resulting from areally concentrated deposition by channel-confined processes, mass flows in proximal and medial fan sectors, or they may comprise areally more extensive sheets and tabular elements from wide distribution of sediment by unconfined processes on low-relief medial and distal fan sectors, resulting in amalgamation of thin aggradational units. Thick, system-scale stratasets associated with relatively continuous aggradational phases may be delimited by stratigraphic surfaces of fan-wide extent, recognizable by evidence for pedogenic modification, winnowing and minor aggradation by runoff, or eolian deflation (e.g. Al-Farraj & Harvey, 2000; Andreucci et al., 2014; Franke et al., 2015; Owen et al., 2014; Sanchez-Nuñez et al., 2015;

Steffen et al., 2010), and may offer better potential to extract absolute dates by various methods in order to constrain chronologies of fan evolution. In the case of systems that have been affected mostly by erosion, rather than geomorphic stabilization, evidence in the form of incised longitudinal channels and terraces can be difficult to distinguish from laterally restricted strike-oriented sections, and even hardly noticeable from dip-oriented sections.

In addition, Pederson et al. (2015) demonstrated that the latero-vertical positioning of debris-flow lobes across strike-oriented fan sections can be parameterized by the compensation index (κ_{cv}; Straub et al., 2009; Straub & Pyles, 2012), a statistical metric quantifying the relative tendency of individual depositional units to either spread uniformly over the available substrate (generally by local or nodal avulsion events) or to be stacked within a restricted area. The tendency for different fans to manifest compensational architecture at system scale implies the possibility to define degrees of order even within apparently disorganized clastic successions. Furthermore, the rough correlation between estimated values of κ_{cv} and other sedimentological or spatial properties of debris-flow successions (such as size, textural composition, radial distance of accumulation from the apex) also suggests a possibility to recognize distinct morphodynamic evolutionary patterns in debris-flow dominated systems, with important implications for hazard assessment and prediction (De Haas et al., 2018b; Pederson et al., 2015; Santi et al., 2017; Tsunetaka et al., 2022).

2.3.3 Spatio-Temporal Evolution of Debris-Flow Fans

2.3.3.1 Avulsion Mechanisms and Patterns

Avulsions are critical for the evolution of debris-flow fans and are important for several reasons. Firstly, they control how sediment is distributed across the fan surface and thereby determine the spatio-temporal evolution of debris-flow fans. Secondly, from an applied perspective, they distribute hazards through space and time on debris-flow fans. Despite their importance, avulsions remain an understudied aspect of debris-flow fans and flow hazards. There is a small, but growing set of field-based (e.g., Bernhardt et al., 2017; De Haas et al., 2018a, 2019; Imaizumi et al., 2016; Schaefer et al., 2021; Suwa & Okuda, 1983; Whipple & Dunne, 1992; Zubrycky et al., 2021) and experimental (Adams et al., 2019; De Haas et al., 2016, 2018b; Schumm et al., 1987; Tsunetaka et al., 2022) studies that document avulsion on debris-flow fans. Nevertheless, very few systematic studies of avulsion location, trigger mechanisms, or frequency exist. This may in part be the result of the long return periods of debris flows in many settings, which make direct observations of avulsions rare (de Haas et al., 2018a). In addition, few debris-flow fans are observed worldwide where depositional events have been monitored in space and time (e.g., Imaizumi et al., 2016; Marchi et al., 2002; Suwa & Okuda, 1983; Wasklewicz & Scheinert, 2016) or have been reconstructed (e.g., Dühnforth et al., 2007, 2017; Schürch et al., 2016; Stoffel et al., 2008a) over sufficiently long timespans.

Avulsions on debris-flow fans may result from overtopping of channel banks in sharp outer bends, where superelevation of the debris-flow surface may exceed bank height (Zubrycky et al., 2021), but more commonly result from channel obstruction by debris-flow lobes, which is commonly referred to as channel plugging (De Haas et al., 2018a; Whipple & Dunne, 1992). Recent field and experimental studies have listed several factors that affect debris-flow avulsions, including flow composition and grain-size distribution, longitudinal position on the fan, path obstructions, degree of lateral confinement, channel depth, and magnitude-frequency distributions (de Haas et al., 2016, 2018a, 2018b, 2019; Densmore et al., 2019; Pederson et al., 2015; Schaefer et al., 2021; Tsunetaka et al., 2022; Zubrycky et al., 2021). A complication in understanding avulsion scenarios is the presence of roads or other anthropogenic fan modifications. Road culverts or undersized bridges can become blocked by debris flows, forcing avulsions outside the main channel, or follow the road itself (Millard et al., 2006).

Debris flows stopping inside the channel form plug deposits that can block the channel locally, thereby potentially forcing subsequent surges or flows out of the channel and inducing avulsion (in rare cases channels may also be plugged by individual large boulders or woody debris). The formation of channel plugs is a stochastic process, but appears to depend in part on the sequence of flow volumes, the geometry of the channel, and the composition of the flows (De Haas et al., 2018b, 2019). While channel plugs often occur as a result of lobe deposition in the channel, channel plugging can also occur dynamically where the relatively immobile front of the mass is slowing down while the more mobile materials in the debris flow upstream piles up and diverts because of its higher velocity and momentum.

On debris-flow fans in Saline Valley, California, USA (Figs. 2.10 and 2.14), the thicknesses of lobes that have triggered avulsions roughly equal local channel depths and are on average twice as thick as the local median lobe thickness (De Haas et al., 2019). Thick lobes are thus more likely to induce avulsion, although it is important to realize that smaller lobe deposits can also induce avulsion, especially where a compound plug has developed by deposition from multiple flows or surges.

Based on an analysis of avulsion patterns on 16 debris-flow fans from Japan, USA, Switzerland, France, and Kyrgyzstan, De Haas et al. (2018a) showed that, at the time scale of individual surges or flows, abrupt changes in channel position are driven by the occurrence of channel plugs. Over the course of multiple flows, the active channel gradually shifts toward the topographically lower parts of the fan, allowing fan-scale compensation in response to local surface gradients (De Haas et al., 2016; Pederson et al., 2015), analogue to the topographic control of autogenic dynamics in fluvial systems (e.g. Mohrig et al., 2000; Slingerland & Smith, 2004).

We have identified four common patterns of avulsion on debris-flow fans (Fig. 2.13) (De Haas et al., 2018a, 2018b). Which of these patterns occurs often depends on the frequency and magnitude of debris-flows. When a debris flow forms a channel plug, a number of potential development pathways exist, depending on the magnitudes of subsequent flows. Firstly, when the channel plug is of limited size and the subsequent flow is of considerable magnitude and has high erosive power, the channel plug may be removed and avulsion does not occur (Benda, 1990). Secondly, when the channel plug is effectively blocking the channel, and a subsequent flow is of sufficiently large magnitude and high mobility, it will be forced out of the channel, and may establish a new channel. If the new channel follows a topographically favorable pathway, it may become the new main channel down the fan (scenario 1 in Fig. 2.13 and examples in Fig. 2.14). Alternatively, if the subsequent flow is of minor to medium size, it is likely to deposit behind (upstream of) the channel plug. This may induce a sequence of backstepping depositional events trending toward the fan apex, until a flow of sufficiently large magnitude is able to leave the channel and avulse (scenario 2 in Fig. 2.13 and examples in Fig. 2.14). Large debris flows may overtop a channel and form multiple new pathways down the fan. Afterwards, the topographically most favorable pathway down the fan is enlarged by subsequent debris flows, while the other pathways are progressively sealed by backstepping deposits (scenario 3 in Fig. 2.13). The fourth mechanism that can be common on debris-flow fans, especially on relatively small and steep ones, sees deposition of laterally adjacent debris-flow lobes shifting gradually toward a topographical low (scenario 4 in Fig. 2.13).

Which of these avulsion scenarios materializes depends in part on the frequency and magnitude of events. Based on analysis of field examples (De Haas et al., 2018a) and analogue experiments (De Haas et al., 2016, 2018b), it has been shown that a relatively uniform flow-mass distribution promotes regular avulsion cycles in

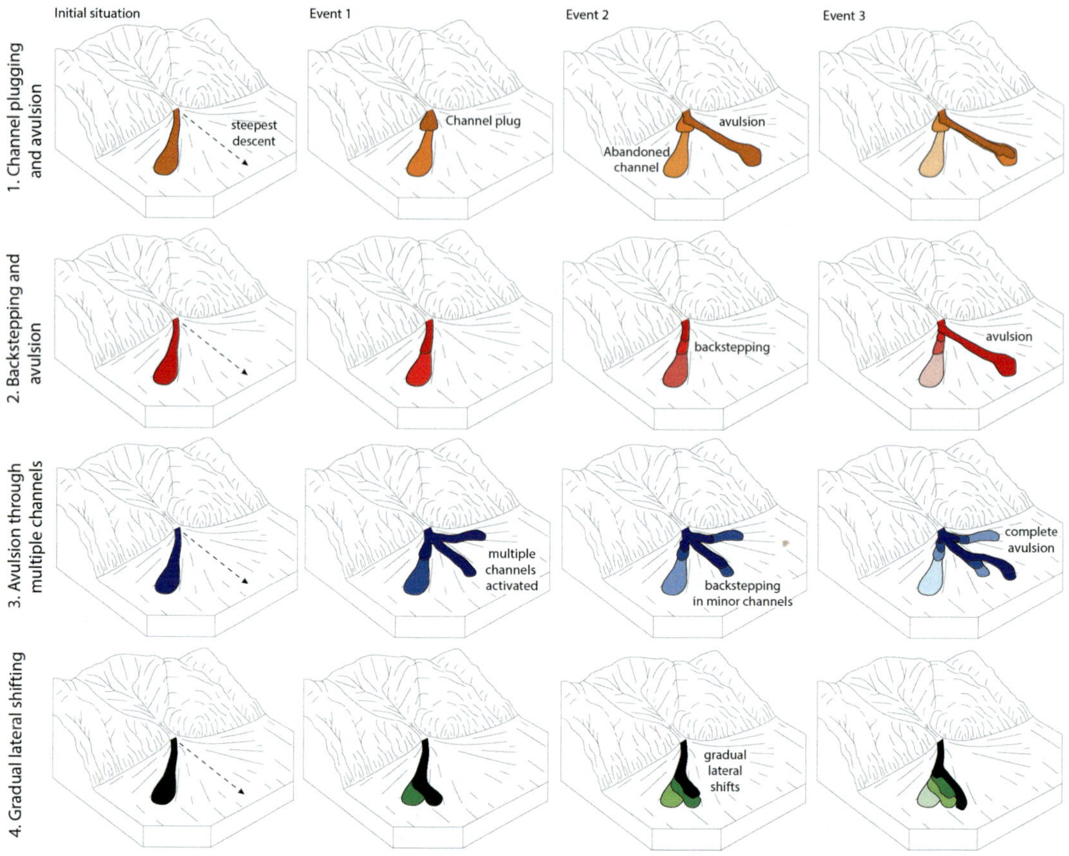

Fig. 2.13 Conceptual diagram illustrating the four most common avulsion scenarios on debris-flow fans. Scenario 1 illustrates how a single debris-flow lobe deposit can block a channel, and during subsequent larger flows the debris flow is forced to leave the main channel due to the limited channel capacity and form a new channel. Scenario 2 illustrates a similar pattern, illustrating that when a channel plug is followed by a number of flows that cannot overtop and form a new channel (typically small to medium volume flows). The lobes backstep toward the fan apex until a sufficiently large flow occurs that can avulse and form a new channel. Scenario 3 illustrates how a large-volume flow can overtop the main channel to occupy and open (or close off) multiple channels. The most topographically favorable channel is subsequently enlarged in subsequent flows, while the other channels are closed off by plug deposition and backstepping. Scenario 4 illustrates how flows can shift laterally toward a topographic depression on the fan. Figure modified and extended from De Haas et al. (2018a)

which avulsion occurs by backstepping deposition (scenario 1). In contrast, a greater range in flow magnitudes promotes avulsion through channel plugging or overtopping by large flows (scenarios 2–3); large events have greater likelihood of overtopping existing channels and establishing new pathways, while sequences of small flows have a higher likelihood to plug the active channel. This suggests that sequences of several small- to medium-sized flows may be critical in forcing avulsion during subsequent larger flow—as observed, for example, on the Kamikamihori Fan in Japan (Suwa & Okuda, 1983). Nevertheless, plugs may form by debris flows with relatively reduced mobility, such as those with mixtures comprising large grain sizes or low water content (De Haas et al., 2015b).

The importance of channel plugging for avulsion on debris-flow fans highlights a major difference between the fluvial avulsion cycle and that observed on debris-flow fans: backfilling and channel plugging need not proceed up the entire fan surface in order to force avulsion on debris-flow fans, as is often the case on fluvial

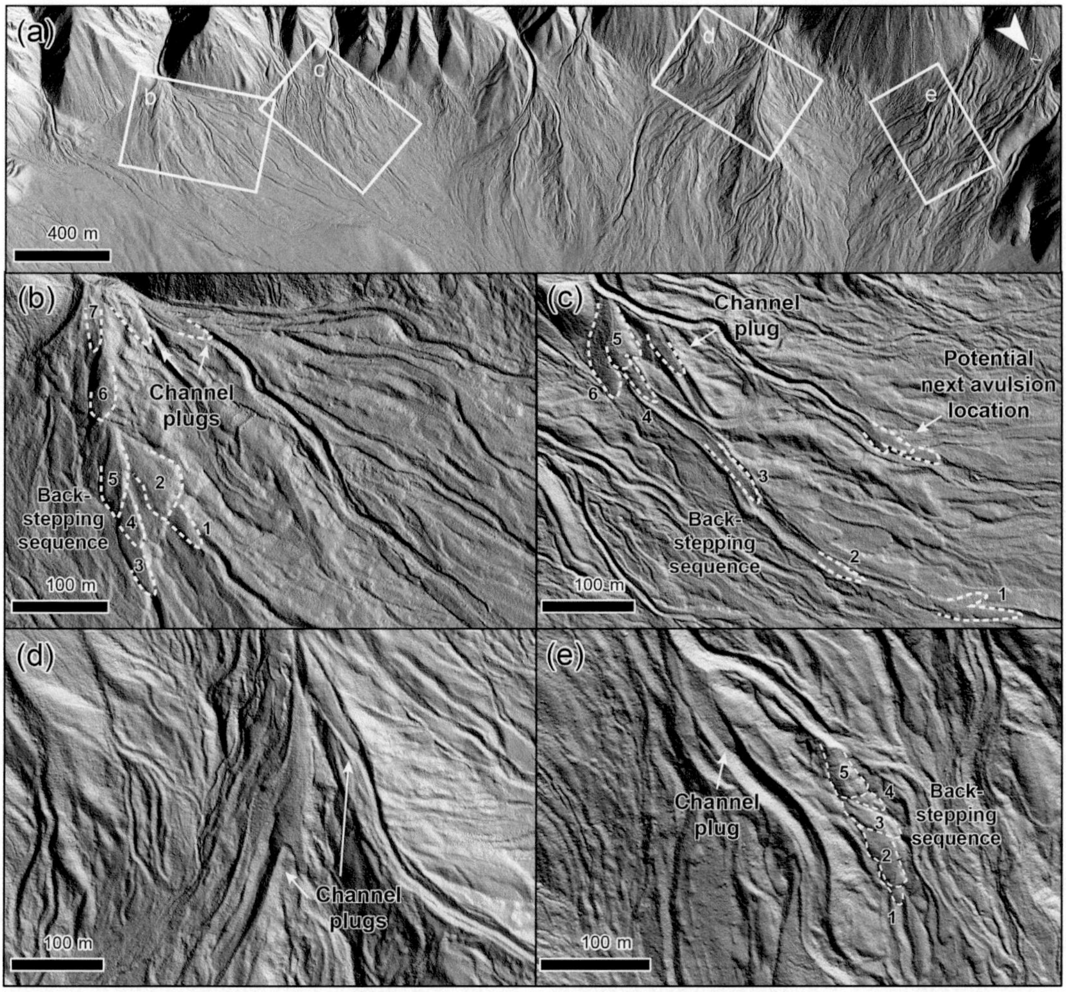

Fig. 2.14 Examples of avulsion by the formation of channel plugs and backstepping sequences (Scenarios 1 and 2 in Fig. 2.11) on a number of pristine fans in Saline Valley, California, USA. On these fans most avulsions are caused by the formation of channel plugs, possible after either an initial backstepping sequence or direct avulsion. See De Haas et al. (2019) for more details on avulsion on these fans. Fans located around 36°34′33.45″ N, 117°37′53.07″ W

fans (Bryant et al., 1995; Clarke et al., 2010; Leenman & Eaton, 2021; Leenman et al., 2022; Reitz & Jerolmack, 2012; Schumm et al., 1987), but can be limited to a key reach, or even a single location. This makes debris-flow avulsions harder to predict compared to their fluvial counterparts and calls for more investigations on likely avulsion locations.

There is limited information on the probability of avulsions. Zubrycky et al. (2021) studied the runout of 146 debris flows in British Columbia, Canada, while Schaefer et al. (2021) studied the runout of 587 debris flows in the Colorado Front Range, USA. Avulsion was common in these events, with 76% of the flows in British Columbia and 33% of the flows in Colorado being avulsive. However, most of these avulsions occupied new flow paths close to their original channel, with less than 10% of the flows in British Columbia reaching more than half-way across the fan relative to their previous flow path (Fig. 2.15). On the Kamikamihori and

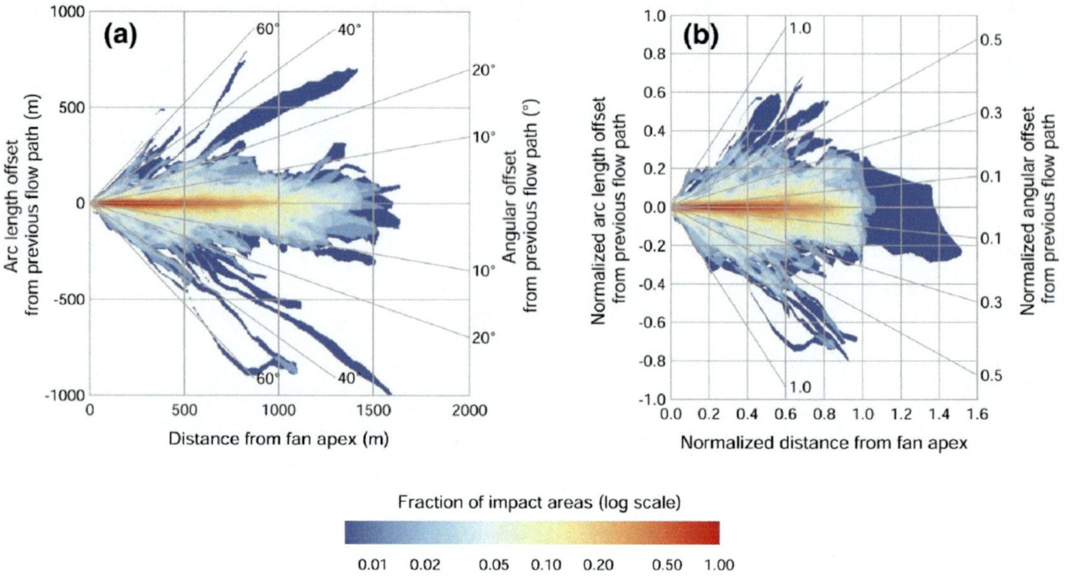

Fig. 2.15 Heat maps showing spatial impacts of debris flows for fans in south-western British Columbia (Zubrycky et al., 2021). The maps were created by summing all impact area plots (146 impact areas across 30 fans) and dividing by the total in the dataset. **a** Un-normalized data. **b** Data normalized to the maximum fan length (maximum straight-line distance from the fan apex to the fan boundary) and maximum fan arc length (maximum arc length intersecting the fan along a circle centered on the fan apex). The coordinates at the right provides lines of **a** equivalent angular offset and **b** normalized angular offset. Figure from Zubrycky et al. (2021)

Ohya fans (Japan), the Chalk Creek Fan (USA), and the Chalance Fan (France), large shifts in channel location occurred approximately every 3–8 flow events (De Haas et al., 2018a).

The likelihood of avulsion has been found to be affected also by flow magnitudes (De Haas et al., 2018a, 2019; Zubrycky et al., 2021). A flow magnitude threshold needs to be exceeded for a flow to leave the channel and to avulse, which means that the probability of avulsion increases with magnitude only until a channel-specific threshold is reached (Zubrycky et al., 2021). This threshold may be lowered by the presence of a plug in the channel, but at the same time, large-magnitude events may also follow the previous flow path, which is a possible moderation effect by their strong erosive ability and the formation of lateral levees. Schaefer et al. (2021) showed that the likelihood of a debris flow to avulse increases with its travel length, because long-runout flows are more likely to encounter obstructions along their path. In addition, debris-flow paths with low lateral confinement, down-fan decreasing slope angles, and distinct changes in slope (knickpoints or inflection points) were found to be relatively prone to avulsion. Furthermore, flows containing boulders were more likely to avulse than their finer-grained counterparts.

Although the point along a channel where a flow loses confinement is a reasonable candidate location for avulsion in a forward analysis, Zubrycky et al. (2021) showed that many avulsions initiate upstream of the fan inflection point. This is the result of channel plugging within the upper fan channel upslope of the inflection. Overall, they found that most avulsions initiated at distances of 20–40% of the maximum fan length from the fan apex and upstream of the fan intersection point (Fig. 2.15). Yet, de Haas et al. (2019) found that the ratio of medium lobe thickness to channel depth on debris-flow fans in Saline Valley remains roughly constant from fan apex to toe.

This suggests an equal likelihood of channel plugging across the fan surface, although the maximum potential distance between the old and new channel locations, and thus the avulsion impact, decreases from fan apex to toe (Densmore et al., 2019).

There are to date no systematic studies to the importance and relative frequency of avulsion in sharp channel bends, prompting superelevation and as a result avulsion in outer bends. This phenomenon has been observed (Zubrycky et al., 2021), and numerical models that simulate runout over three-dimensional terrain can help identify potential avulsion sites at channel bends where superelevation and runup happen (McDougall, 2017a, 2017b).

2.3.3.2 Fan Erosion and Incision

Debris-flow fans may have a deeply incised main channel, typically in the range of several meters and in extreme cases up to tens of meters (Fig. 2.16). These incised channels are common in the upper to medial portions of the fan, although deep incisions may persist across the fan (Blair & McPherson, 2009). In cases where incised streams have shifted the locus of sediment deposition significantly downstream of the inflection point, debris-flow fans can create unique fan landforms termed "telescoping fans", "entrenched fans", or specifically "fan-head entrenchment" when incision is on the proximal fan (Fig. 2.16). Incised channels occur from two mechanisms operating over distinct time scales: erosion during distinct debris-flow events (herein referred to as debris-flow erosion), and gradual incision over longer time periods (herein referred to as fan incision) (see Kang, et al., this volume).

Fan incision over long time periods typically results from several external forces on debris-flow fans including base-level lowering (Harvey, 1987, 2002; Melton, 1965; Ryder, 1971, 1972), tectonic uplift (Blissenbach, 1954; Bull, 1964; Denny, 1967), and climatic shifts resulting in alterations of sediment delivery to the fan and shifts from debris-flow to flood dominance (Sect. 2.3.4; Bull, 1991; Davies & Korup, 2007; Hooke, 1967; Li et al., 2018; Ryder, 1971). At the same time, debris-flow fans can also undergo autogenic cycles of deposition, backfilling, avulsion, and incision, as demonstrated experimentally (de Haas et al., 2016) and observed on frequently active debris-flow fans (Sect. 2.3.3.1; de Haas et al., 2018a).

Debris-flow erosion on time scales of one to a few flows typically occurs through shearing and impact forces on the channel bed and/or liquefaction of the channel bed (Berger et al., 2011b; De Haas et al., 2022; Hungr et al., 2005; Iverson, 2012; Iverson & Ouyang, 2015; Roelofs et al., 2022; Schürch et al., 2011; Stock & Dietrich, 2006; Zheng et al., 2021). The channel can fail progressively through grain collisions and incorporation, or en masse as in the case of liquefaction (de Haas & Woerkom, 2016). Field observations of debris-flow erosion show that the process occurs both during the coarse granular debris-flow front passage (e.g., Berger et al., 2010, 2011b; McCoy et al., 2012) and during later and more fluid after surges (e.g., Berger et al., 2011b; McCoy et al., 2012). Additional observations indicate that debris-flow in-channel erosion generally increases with flow volume, velocity, height, channel slope, and bed wetness (De Haas et al., 2022; Schürch et al., 2011; Theule et al., 2015). Debris-flow erosion can significantly increase debris-flow volumes through entrainment, and the possibility of this process should be considered as input to hazard assessments for debris-flow fans (Frank et al., 2015; Jakob et al., 1997; Suwa & Okuda, 1983).

Extreme cases of debris-flow erosion on fans have recently been reported in several locations (Fig. 2.17, Table 2.5). These case studies are similar in that all watersheds are affected by frequent rockfall processes in addition to debris flows, resulting in transport-limited watersheds and resultantly steep debris-flow fans. The initial debris-flow erosion events occurred during heavy rainfall that generated significant flow volumes, which enhances the erodibility of the flows. For example, the 1995 Hope Creek and Mount Allen debris flows occurred during a moderate-intensity rain-on-snow event (Jakob et al., 1997 and unpublished data). Eye-witness accounts reported that the debris flow surges

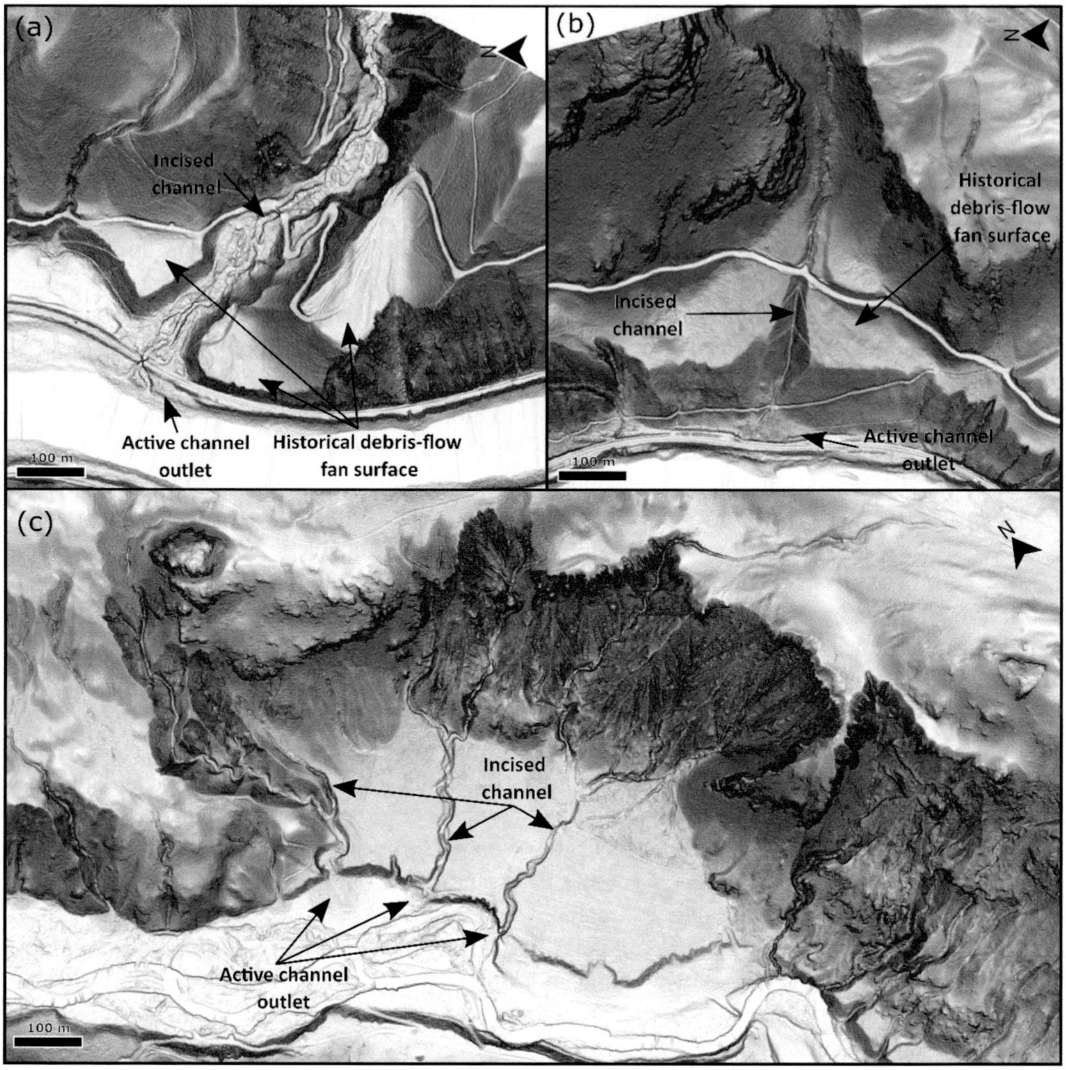

Fig. 2.16 Examples of incised channels on debris-flow fans viewed from slope-shaded lidar digital elevation terrain models. **a** Telescoping debris-flow fan at Pimainus Creek, (British Columbia, Canada) (fan described in detail in Ryder, 1972). **b** Distally incised due to base level lowering on a debris-flow fan at an unnamed creek approximately 15 km north of Pimainus Creek (British Columbia, Canada). **c** Fan-head incision on three debris-flow fans 2 km south of Cache Creek (British Columbia, Canada)

continued at Hope Creek for more than 2 h after the main event (Jakob et al., 1997). The fan erosion during the 2005 Rotalaui/Guttannen debris flow was attributed to the 170 mm of precipitation that fell on the watershed over 48 h (Scheuner et al., 2009). A prolonged rainstorm near Neff and Bear Creeks (approximately 30 km apart) triggered erosive debris flows at both locations.

Frank et al. (2015) observed that debris flows occurring after the main erosive event at Spreitgraben (Switzerland) remained channelized and further eroded and widened the channel. Subsequent debris flows transported sediment to the trunk drainage river downstream, rather than depositing the sediment on the fan itself. Similar processes have been observed at Mount Allen (British Columbia, Canada), where

Fig. 2.17 Examples of observed extreme debris-flow fan erosion resulting from singular or multiple debris flows. **a** Approximately 10 m of debris-flow erosion at Neff Creek (British Columbia, Canada) from one debris flow event. **b** Approximately 30 m of debris-flow erosion at Spreitgraben (Guttanen, Switzerland) from several debris flows beginning in 2009 (fan described in detail in Frank et al., 2015; image provided by Geoprevent)

channelization has more effectively routed to the distal fan, causing several avulsions and damaging railway infrastructure. The significant erosion observed in these case studies reinforces the notion that the potential for debris-flow erosion needs to be accounted for when assessing debris-flow fan hazards, particularly when infrastructure crosses these channels or when settlements are located below the fan inflection point (see Kang et al., this volume).

2.3.4 Debris-Flow Fans as Archives of Past Conditions

Debris-flow fans are the result of deposition of many debris flows over millennia or longer. Changes in hydroclimatic conditions and sediment availability in watersheds affect the frequency and magnitude, as well as the sedimentological characteristics, of debris flows. Therefore, changes in hydroclimatic conditions or sediment availability in watersheds may alter the morphology, stratigraphy and aggradation rates of debris-flow fans. Moreover, periods of sediment deficiency may induce fan incision (Bull, 1991; Dühnforth et al., 2008; Harvey & Renwick, 1987; Li et al., 2018; Ryder, 1971). Debris-flow fans therefore represent sedimentary archives of past flow processes (Dühnforth et al., 2007; D'Arcy et al., 2017), sediment supply (Dietrich & Krautblatter, 2017; Franke et al., 2015), and past climate and environmental change in general (D'Arcy et al., 2017; Dietrich & Krautblatter, 2017; McDonald et al., 2003; Nott et al., 2001; Ryder, 1971, 1972; Savi et al., 2014; Sohn et al., 2007; Thomas, 2004).

Table 2.5 Case studies of extreme debris-flow fan erosion events

Location	Reference	Watershed area (km^2)	Fan gradient (°)	Number of debris flows and dates of events	Estimated debris flow volume (m^3)	Fan erosion depth (m)	Average fan yield rate (m^3/m)
Mount Allen (Yale Subdivision, British Columbia, Canada)	Previously unpublished data	0.8	18	5 1990–2006	Unknown	4–12	N/A
Hope Creek (Hope, British Columbia)	Jakob et al. (1997)	0.9	17	1 1995	50,000	N/A	50
Rotlaui/Guttannen (Guttannen, Switzerland)	Scheuner et al. (2009)	4.1	16	1 2005	500,000	8–12	300
Spretigraben (Guttannen, Switzerland)	Stoffel and Huggel (2012) Tobler et al. (2014); Frank et al. (2015)	4.3	14	Multiple 2009–2013	Total volumes of all events >650,000	Total depth up to 30	N/A
Neff Creek (Birken, British Canada)	Zubrycky et al. (2021); previously unpublished data	3.3	13	1 2015	220,000 ± 30,000	2–14	160
Bear Creek (Seton Portage, British Columbia)	Previously unpublished data	3.4	13	2 2015, 2016	2016 event: 67,000	2–8	N/A

Historic debris-flow activity on fans can be reconstructed from both the surface and stratigraphy. Surficial deposits are often dated through dendrochronology (Bollschweiler & Stoffel, 2010a; Stoffel et al., 2008a, 2008b) and lichenometry (André, 1995; Helsen et al., 2002; Innes, 1983) on decadal to centennial scales, and by optically stimulated luminescence (OSL) (Zhao et al., 2015) or cosmogenic nuclide analysis (Dühnforth et al., 2007; Schürch et al., 2016) for longer time scales. In stratigraphy, debris-flow deposits are often dated through OSL (Chen et al., 2008; Li et al., 2018; Zhao et al., 2017) and ^{14}C (Jakob & Friele, 2010; Chiverrell & Jakob, 2013; Muñoz-Salinas et al., 2020). Avulsive switching of the active channel to different pathways results in the formation of a small number of distinct sectors on the fan surface, each recording debris-flow activity during a specific time period and over a number of flow events (e.g., Dühnforth et al., 2007; Schumm et al., 1987; Schürch et al., 2016). These sectors may capture periods of time spanning as long as tens of thousands of years, depending on the area of the fan, debris-flow frequency and magnitude, and avulsion frequency (e.g., Dühnforth et al., 2007, 2017). However, a common problem in extracting information from colluvial and alluvial fans within a chronological framework is the relatively limited resolution and technical viability of dating methods, such that the selection and application of appropriate techniques has become a field of research unto itself (Schneuwly-Bollschweiler et al., 2013; see Jakob and Nolde, and Ballesteros, et al., this volume).

Multiple authors have coupled inferred changes in debris-flow activity and fan aggradation rates to changing climatic conditions over a wide range of timescales from sedimentary records. This stems from frequent recent observations of debris flows being directly associated, for example, with intense or extreme rainfall and snowmelt events or with specific antecedent meteorologic conditions leading to sediment-charging of watersheds and/or colluvial slope instability (e.g. Barclay et al., 2007; Chen et al., 2013; Chiarle et al., 2007, 2021; Church & Miles, 1987; Clark, 1987; Palladino et al., 2018; Sharp & Nobles, 1953; Van Westen & Daag, 2005). Although, as mentioned above, local topography and other geomorphic factors play a role in determining slope and watershed responses to potential meteorological triggers (e.g. Jomelli et al., 2007; Ventra et al., 2013). Numerous studies have elaborated on linkages between debris flows and climate change at decadal to centennial scales on various continents and geomorphic settings, but especially in Europe, where proxy and historical records of climate conditions are available (e.g. André, 1995; Bollschweiler & Stoffel, 2010b; Innes, 1985, 1997; Jomelli et al., 2004; Kotarba, 1992; Stoffel et al., 2008b; Strunk, 1992). Process reconstructions range from enhanced debris-flow frequency because of increased rainstorm activity over the past decades, to changes in debris-flow activity and grain size and fan aggradation and incision cycles as a result of glacial-interglacial cycles over tens of thousands of years. Examples on the decadal to centennial timescale come from 8 debris-flow fans in the Northern Calcareous Alps (Austria). Here, Dietrich and Krautblatter (2017) found that debris-flow activity has increased by a factor of more than three after 1980, compared to the period 1940–1980. Moreover, the present-day activity also exceeds the Holocene/Late-glacial activity by a factor of 2–3. The authors relate this recently enhanced debris-flow activity to increased rainstorm frequency (Dietrich & Krautblatter, 2017; Kiefer et al., 2021). Over the past decades, debris-flow frequency on fans in the Swiss Alps peaked at the end of the Little Ice Age and in the period 1920–1929, while activity was relatively low in the period 2000–2009 (Bollschweiler & Stoffel, 2010b). Enhanced debris-flow activity in this area is explained by an increase in the frequency of hydroclimatic triggering events, because sediment is not a limiting factor in these watersheds (Stoffel et al., 2008a).

Over periods of multiple tens of thousands of years, Sewell et al. (2015) provide an example from five debris-flow fan complexes in the coastal foothills of Lantau Island, Hong Kong. The authors inferred that over the last 30 ky debris-flow activity was dominated by periodic thick, sediment-rich flows during cool, relatively dry periods. In contrast, during periods of climatic amelioration, especially warmer and wetter conditions associated with intensification of the East Asian Monsoon, the fans were mainly affected by debris floods. They relate this to increased forestation limiting sediment availability and production during warmer and wetter periods. Similarly, along the wet tropical highlands of northeast Queensland (Australia), climatic aridity and reduced plant cover, along with a continuation of discrete high-magnitude rainfall events, were found to cause enhanced slope instability and fan aggradation between 27 and 14 ka (Nott et al., 2001). In the Santa Catalina Mountains, Arizona, USA, Late Pleistocene to early Holocene debris-flow activity was largest because of a relatively high sediment availability that decreased towards the present as a result of a transition to a more arid climate (Youberg et al., 2014).

Fan aggradation and incision history and regional climatic history along the eastern Tibetan plateau margin suggest that aggradation occurred in cold and/or dry climates, while incision took place under warm and wet climates (Li et al., 2018). Cold climates lead to enhanced frost shattering and dry conditions that favour a deterioration of the vegetation canopy, promoting sediment production and yield, and hence aggradation rates on the fans. Along the eastern Sierra Nevada of California, USA, the depth of fan-head incision has been shown to correlate with the pattern and degree of Late Pleistocene watershed erosion by valley glaciers (Dühnforth et al., 2008). Watersheds affected by glacial activity in the past have fans with deeply incised heads, while fans fed by glacially unmodified watersheds are not incised. Dühnforth et al. (2008) attributed this to sediment storage in glaciated watersheds leading to fan starvation, and hence to fan-head incision during deglaciation, while glacially-unmodified watersheds experienced more continuous sediment supply. A similar role for Late Quaternary climatic cyclicity in controlling the frequency and volume of depositional events, mediated by glacial and periglacial agents, has been inferred by Hornung et al. (2010) and Schoch-Baumann et al. (2022) for debris flow-dominated fans of the eastern Swiss Alps, in which 'paraglacial' peaks of sediment yield were linked to phases of climate perturbation that liberated watersheds from glacially produced or stored debris and induced rapid aggradation on the fans. D'Arcy et al. (2017) showed that the last full glacial-interglacial cycle is recorded in the grain-size distribution of debris-flow deposits on fan surfaces in Owens Valley, California, USA. Debris flows were relatively fine-grained during the cooler glacial climate and became systematically coarser-grained as the climate warmed and dried. They attributed this to past changes in rainstorm intensity rather than sediment supply. The warmer and drier climate would have triggered higher-magnitude storms, in turn leading to greater runoff rates entraining coarser sediments. The pre-Quaternary stratigraphic record is harder to decipher, owing to even fewer opportunities to extract climate proxies and chronological information from coarse-clastic successions. Ventra et al. (2018) provided a stratigraphic analysis of alluvial-fan successions from the Miocene of the Spanish interior, relying on interbedding relationships between distal debris-flow-lobe terminations and mudflat to palustrine deposits for which astronomical tuning demonstrated accumulation under astronomically-controlled climate cycles. Distal-fan stratigraphy showed that debris-flow events of largest volume, thus capable of bypassing the fan body, occurred consistently during transitions from relatively arid to humid climate states. This suggests more effective sediment yield from a transport-limited watershed at times of landscape instability, due to increased intensity of seasonal precipitation at the onset of the humid phase, affecting highland slopes poorly stabilized by soils or vegetation during the preceding arid phase (Ventra et al., 2018).

Besides extracting paleoenvironmental information from chronological and stratigraphic analyses of primary debris-flow deposits, it should be noted that a complementary, but often overlooked record of past climatic, hydrologic and biotic conditions is stored within fan strata in the form of facies (associations) that were derived from reworking processes, such as runoff, wind and regolith weathering or pedogenesis (e.g. Blair & McPherson, 1994; Blair et al., 1990; Blikra & Nemec, 1998; De Haas et al., 2014; Kochel & Johnson, 1984; Matmon et al., 2006; Sanchez-Nuñez et al., 2015; Stokes et al., 2007; Ventra et al., 2013). These are more likely to contain geochemical, palynological and sedimentological signatures of specific environmental states than the coarse-clastic facies accumulated by geologically 'instantaneous' aggradational events. The relevance of these secondary deposits is related to the fact that primary fan aggradation takes place by mass flows and floods which are inherently short-lived and infrequent in time, whereas prolonged intervals between main depositional events encompass the overwhelming majority of a fan's history from a temporal perspective. This is an inherent property of the stratigraphic record of most depositional systems (Miall, 2015; Paola et al., 2018), compounded in colluvial and alluvial fans by their characteristically low frequency and high magnitude of sediment-transport processes (Blair & McPherson, 1994; Blikra & Nemec, 1998; Johnson et al., 1991).

2.4 Watershed and Fan Morphology, Sedimentology, and Dynamics from a Hazard Management Perspective

Key steps in debris-flow hazard assessments comprise: (1) debris-flow hazard recognition; (2) estimation of debris-flow probability and intensity (frequency and magnitude); and (3) assessment of debris-flow impact on downstream valued assets, such as communities, infrastructure, and social cultural values (Jakob, 2005, 2021, Strouth et al., this volume). Debris-flow watershed characteristics and hydroclimatic conditions typically set the probability and intensity of debris flows. Debris-flow fans are preferred locations for urban development because they are gently sloping, well drained, and often provide good aquifers. In many mountainous regions fans provide the only readily developable land, despite being prone to debris flows and (debris) floods. Ten percent of the global population lives in mountain areas (Abram et al., 2019), and many mountain settlements are at risk from debris flows (Dowling & Santi, 2014; Fuchs et al., 2013). The recognition and characterization of debris-flow dominated watershed-fan systems is critical for keeping mountain settlements and infrastructure safe from debris flow-related hazards (Borga et al., 2014; Huebl & Fiebiger, 2005; Jakob, 2005; Kellerhals & Church, 1990).

Recognition of the potential for debris-flow hazards is typically based on geomorphic evidence, such as the presence of a debris-flow fan. The morphometry of watersheds and fans can be used for an a priori assessment of debris flow likelihood (Table 2.2; Fig. 2.5, 2.6, and 2.8). Our data generally indicate that debris-flow systems with watersheds with an area <3 km^3, Melton ratio >0.5, and connected to fans with a gradient >5°, have a high debris-flow potential, although numerous exceptions and regional variances exist (Sect. 2.2.3.3). Past debris-flow activity can be identified in the field using the distinct morphology and sedimentology of debris-flow deposits (Sect. 2.3). In particular, well-defined and coarse-grained depositional snouts and levées, scour marks, mudlines, debris-impact scars on trees well above the flood limit, and boulders larger than the bankfull depth of the channel are indicative of debris-flow activity (Fig. 2.9; Table 2.4). In stratigraphy, debris flows are typically recognized by their matrix-supported texture (clast-supported is possible but uncommon), lack of clast imbrication, and possibly inversely graded deposits (Fig. 2.12; Table 2.4).

The hazard potential in a debris-flow system depends on the frequency and magnitude of events. Flow magnitude controls flow velocity, flow depth, peak discharge, inundation area, and thereby the potential consequences of the debris flow, and frequency determines how often life and valued assets are exposed to the hazard (Rickenmann, 1999; Strouth et al., this volume). Frequency is a function of the availability of erodible sediments and the frequency at which hydroclimatic thresholds are exceeded (Bovis & Jakob, 1999). As a result, higher debris-flow frequencies can be expected in transport-limited watersheds compared to supply-limited watersheds (Sects. 2.2.2 and 2.2.3). Flow magnitudes are typically limited by sediment availability in supply-limited watersheds because the sediment reservoir is (nearly) emptied after debris-flow activity, while debris-flow magnitude generally increases with hydroclimatic magnitude in transport-limited watersheds (Fig. 2.3). Debris-flow frequency and magnitudes have been correlated to watershed and fan characteristics (e.g., De Haas & Densmore, 2019; Hungr et al., 1984; Jakob & Bovis, 1996; Jakob et al., 2020; Johnson et al., 1991), and together with local estimates of sediment availability and debris flow triggering thresholds (Sects. 2.2.1 and 2.2.2) they can inform M-F relationship estimates. In the reality of a fast-changing climate, past debris flows may not be appropriate analogues for future events, and a careful examination of shifting watershed dynamics under climate change scenarios should be undertaken (see Stoffel et al., this volume). Exposed stratigraphic sequences can provide information on debris-flow frequency, magnitude, and flow characteristics (Sect. 2.3.2). While post-event re-working of debris-flow deposits (Sect. 2.3.1.2) can appear to impede characterization of past debris flows, knowledge of the rate of these processes can inform minimum ages of surfaces of debris-flow deposits.

Estimating the spatial impact of debris flows on a fan (i.e., the probability of a debris flow with a particular magnitude reaching a certain location) is an important part of hazard mapping, risk assessment, and mitigation design (Zubrycky et al., 2021). However, predicting debris-flow runout can be challenging due to complex flow behaviour. Flows with larger magnitudes have more momentum and tend to spread farther and occupy a larger portion of a fan (Griswold & Iverson, 2008). Topographic controls, such as elevation loss, channel gradient, path curvature, channel confinement, forest cover, and anthropogenic structures influence runout by dissipating energy and promoting deposition over entrainment (De Haas et al., 2022; Fannin & Wise, 2001; Zhang et al., 2013).

Debris flows are prone to avulsion, which are deviations of flow from the active channel. Debris flows that leave the main channel typically pose the largest threat to settlements and infrastructure on alluvial fans, as mitigation measures such as check dams and retention basins are often applied only to the active channel (De Haas et al., 2018a). Although avulsions are a key control on debris-flow impact, little guidance exists for practitioners to predict where and when they occur. Avulsions occur when a channel becomes blocked by debris, forcing subsequent surges to abandon the active channel, or when superelevation exceeds channel bank height (Sect. 2.3.3.1). Numerical models that simulate runout over three-dimensional terrain can help identify potential avulsion sites at channel bends where superelevation and runup happen (McDougall, 2017a, 2017b), but they currently lack the capability to simulate avulsions caused by sporadic channel blocking by plugs of boulders or woody debris unless deliberately specified during model design. Recognizing this limitation, hazard assessors could examine several variables that have been found to increase the tendency for debris flows to avulse, including decreasing or relative slope angle minima along an expected path, distinct changes in slope (knickpoints), decreasing lateral confinement, length, large changes in elevation, boulders present in the debris flows, and path obstructions (Schaefer et al., 2021; Zubrycky et al., 2021; Sect. 2.3.3.1). Recent research has also illustrated the potential sequence of avulsion mechanisms through channel plugging and backstepping (Fig. 2.13; De Haas et al., 2016, 2018a, 2018b, 2019). The likelihood of plug formation and backstepping sequences in part depends on flow frequency and magnitude (De Haas et al., 2018b). Abandoned channels, identified in part from their morphological characteristics (Sect. 2.3.1.1), can be expected to become preferred pathways on a debris-flow fan following a future avulsion, but new pathways are also frequently occupied (Kellerhalls & Church, 1990; De Haas et al., 2018a). While these observations provide some guidelines on when and where to expect debris-flow avulsions, predictions remain associated with large uncertainties and site-specific judgement should be applied.

Debris-flow fans are sometimes erroneously characterized as an exclusively aggradational landform, even though erosion is generally observed during debris-flow events and in stratigraphic sections. Improper recognition of debris-flow erosion could lead to disastrous consequences if erosion undermines building foundations or exposes buried linear infrastructure, such as the case at Spreitgraben, Switzerland (Tobler et al., 2014). Debris-flow erosion can also significantly bulk event volume and exacerbate downslope deposition and avulsion if not accurately characterized, as demonstrated in several cases in British Columbia, Canada (Sect. 2.3.3.2). Similar to the modelling limitations for debris-flow avulsion, debris-flow erosion and bulking is generally not (correctly) analyzed by modelling software packages. Preliminary research by Lau (2017) showed that fan slope, watershed area, and watershed relief may be a first-order guide for identifying highly erosive debris-flow fans. More work is required to address debris-flow fan erosion for practical implementation in hazard management (Jakob, 2021).

2.5 Research Perspectives and Outlook

In this section, we summarize some of the key research priorities that will help to develop a more complete understanding of the dynamics of debris-flow watersheds and fans for the benefit of science and hazard assessments.

2.5.1 Watershed Dynamics

Debris-flow watersheds can be classified into supply-limited and transport-limited systems (see Sect. 2.2). Especially in supply-limited debris-flow watersheds, sediment recharge rates exert a strong control on debris-flow frequency and magnitude. Yet, we have a limited understanding and few direct measurements of sediment recharge rates and how these depend on local environmental conditions. Enhanced understanding of the interaction between sediment production and hydroclimatic and terrain variables may lead to better estimates of the occurrence of debris flows and of their magnitude (Bonneau et al., 2022). Adding urgency, ongoing climate change will alter sediment availability and recharge rates in many debris-flow watersheds through variation in hydroclimatic conditions. To adapt to the impacts of climate change on debris flow hazards, we need to enhance our understanding of the processes that control sediment recharge rates across the wide range of debris-flow watersheds worldwide.

Critically, we lack numerical models and analogue models that can simulate the dynamics of debris-flow watersheds. Such tools would enable to define and isolate specific boundary conditions, determining their effects on watershed dynamics and on debris-flow activity. A promising numerical model is SedCas, which has been originally developed and calibrated for the Illgraben watershed in Switzerland (Bennett et al., 2014; Hirschberg et al., 2021). This model conceptualizes a debris-flow watershed as a spatially lumped cascade of connected reservoirs representing hillslope and channel storages, wherein sediment goes through cycles of storage and subsequent export by debris flows and floods. The hydrological component is based on the linear reservoir concept and includes all relevant hydrological processes that lead to runoff formation in alpine watersheds, including precipitation, snow accumulation, snowmelt, evapotranspiration, and soil water storage. In the current model, sediment is supplied stochastically by slope failures derived from a probability function based on measurements from the Illgraben, which may be triggered by frost-weathering, rainfall or randomly. However, there are many different types of watersheds with alternative processes dominating sediment production. To include this variability, and to make this model more widely available for studying the interplay between sediment production and hydroclimatic events and the effects of changes therein on debris-flow activity in watersheds, these processes need to be included. Such a model would represent a key tool for assessing the effects of changes in sediment availability and hydroclimatic conditions in debris-flow watersheds, for example because of climate change, wildfires, and earthquakes.

Schumm et al. (1987) created large analogue models of watershed-fan systems in the Rainfall Erosion Facility (REF) at Colorado State University. In these experiments, they applied rainfall to a watershed of 15 × 9.2 m to trigger geomorphic activity and sediment transport, forming a depositional fan at the watershed outlet over the course of multiple events. In agreement with our current understanding of differences between debris-flow and flood-dominated watersheds (Arango et al., 2021; Bertrand et al., 2013; Wilford et al., 2004), a decrease in watershed area and length and an increase in watershed relief led to a transition from a flood-dominated to a debris-flow dominated system. Schumm et al. (1987) only performed one experiment with a debris-flow dominated system, the analysis of which remained limited to qualitative fan evolution and was only of marginal relevance in the scope of their work. Yet, this experiment testifies to the feasibility of creating analogue models of debris-flow dominated watershed-fan systems. The large scale of the REF inhibits the reproducibility of such experiments in most laboratories, but later landscape- and watershed-evolution experiments have shown that watershed evolution mimics nature even in experimental set-ups smaller than 1 m^2 in area (Babault et al., 2005, 2007; Bigi et al., 2006; Bonnett & Crave, 2003; 2006; Graveleau et al., 2015; Hasbargen and Paolo, 2000; 2003), justifying the possible set-up of watershed-fan analogue experiments.

2.5.2 Identifying Debris-Flow Watersheds

Although debris-flow-generating watersheds are distinct from watersheds dominated by floods and colluvial processes, no clear distinguishing characteristics exist between these types of watersheds. There is considerable morphometric overlap, and locally ascertained threshold values of specific morphometric or hydroclimatic variables that generally cannot be extrapolated to other areas (Figs. 2.5, 2.6 and 2.7). Therefore, for more objective and realistic decision-making and hazard assessment, it would be more productive to shift research attention from threshold-based approaches to probability-based approaches. The recent advances in machine-learning workflows may facilitate this transition (Arango et al., 2021; Wang et al., 2019).

Secondly, while debris-flow activity is a function of watershed morphometry and material characteristics, identification of debris-flow watersheds is often carried out based on morphometric characteristics alone. To improve the identification of debris-flow watersheds and possible application of identified thresholds to different areas, we should also account for material characteristics (i.e. grain-size distribution, mineralogical and petrographic composition) and sediment availability. For example, this could be accomplished by including lithology, sediment contributing area, and forest cover, and including hydroclimatic conditions that may affect sediment recharge rates, such as exposure time of bedrock or regolith to the frost-cracking window. The wealth of freely available global remote-sensing and climate data (Muñoz-Sabater et al., 2021) could facilitate the consideration of such parameters. We further observe that morphometric studies would benefit from relating the recurrence interval of debris flows to the observed morphometry. For example, testing whether morphometric differences control higher or lower frequency and magnitude of debris flows, or shifts in frequency and processes over time.

2.5.3 Spatio-Temporal Evolution of Debris-Flow Fans and Avulsion

Despite the importance of avulsion for the spatio-temporal evolution of debris-flow fans and for the prediction of potential hazard to communities, the process remains poorly studied and understood in this kind of depositional system. In particular, future research should focus on: (1) improving our understanding of avulsion frequencies and their controls; (2) identifying spatial probabilities of avulsion on fans (for example, identify whether hotspots exist on a fan surface where avulsion is most likely); (3) identifying conditions leading to avulsion and fan morphological configurations that increase the likelihood of avulsions; (4) increasing our overall understanding of how debris-flow fans evolve in time and space in different climatic and geomorphic settings.

Our current understanding of the spatio-temporal evolution of debris-flow fans and avulsions comes mainly from field monitoring and reconstructions, and from a limited number of analogue experiments. Yet, there are no studies based on numerical modelling. At flow scale, numerical models that simulate debris-flow runout over three-dimensional terrain can help identify potential avulsion sites at the outer bends of channels, where superelevation and runup occur (McDougall, 2017a, 2017b), but these currently lack the capability for automatic terrain modification to simulate avulsions caused by the key process of channel blocking by in-channel deposition. Moreover, these models are computationally expensive, making it challenging to model the evolution of debris-flow fans over the course of hundreds to thousands of events. At system scale, developing a numerical model that simulates debris-flow-fan evolution over the course of numerous events would enable us to greatly enhance our understanding of the spatio-temporal evolution of these fans. Such a model would allow systematic exploration of the effects of boundary conditions and changes therein, such as the frequency and magnitude of debris flows, on the evolution, morphology, and stratigraphy of debris-flow fans. Thereby, they could also be used to anticipate the effects of climate change, wildfires, and earthquakes on the evolution and dynamics of debris-flow fans.

2.5.4 Stratigraphy of Debris-Flow Fans

The process sedimentology of debris flows and bed-scale facies analysis of related deposits have long been researched, and have benefited from advances in understanding the mechanics of these gravity-driven phenomena. However, to interpret the stratigraphic archives of debris-flow fans and to improve our ability to decipher paleoenvironmental information, we need to develop comprehensive 3D architectural models of debris-flow fans at system scale, ideally capturing systems across a range of geologies and climates. Presently such models are largely lacking, in contrast to those for fluvially-dominated fans and fan deltas. A key limitation is that the small size of outcrops prevents fan-wide stratigraphic correlations necessary to extract 3D architectural models, although the problem has recently been partly overcome by the application of ground-penetrating radar and electrical resistivity tomography to both alluvial and colluvial fans, enabling imaging of stratigraphic and architectural patterns across presently active systems even in the absence of suitable sections through their deposits (Franke et al., 2015; Siewert et al., 2012). As mentioned above, a promising conceptual tool for extracting debris-flow magnitude-frequency data from fan stratigraphy may be analysis in terms of compensational index (κ_{cv}), as demonstrated by application to debris-flow fans in the field (Pederson et al., 2015; Santi et al., 2017) and in analogue models (De Haas et al., 2018b). The approach suggests that characterization of debris-flow dominated successions in terms of quantifiable parameters holds promise to unravel their possible variability within individual systems (for example, proximal-to-distal trends downfan, or vertically through stratigraphy) or among different systems, a possible advance in the analysis of deposits that have long remained cryptic because of their apparent stratal homogeneity. Considerable advances should also derive from systematic sets of analogue experiments in which downscaled debris-flow systems are observed to evolve morphologically under controlled boundary conditions. Early results show the feasibility and merit of the experimental approach (e.g. De Haas et al., 2015a, 2015b, 2015c, 2018b) in addressing specific questions mostly concerning debris-flow dynamics, but future steps should involve a broader range of set-ups and boundary conditions in order to extract conceptual guidelines to the analysis of stratigraphic successions.

The improving ability to interpret debris-flow fans should be applied to discern possible systematic patterns and differences in the spatial and temporal dynamics of their evolution across different settings. Debris flows occur in a wide range of morphoclimatic settings. Alluvial fans and colluvial aprons are the two depositional systems in which debris flows are more commonly recorded, as well as those more likely to be encroached by anthropic settlements or infrastructure. Considering significant differences in the catchment types and morphodynamics between these two kinds of fans, hazard prediction should be considerably improved by identifying possible short- and long-term differences in debris-flow histories between the two settings, as well as how those could respond to expected environmental change. In this regard, variable dynamics should be expected across a spectrum of climatic zones, as changes in weathering intensity and precipitation patterns in debris-flow-prone watersheds should trigger varying responses from (semi)arid to temperate to periglacial/arctic regions.

2.5.5 Fan Erosion and Incision

Recent examples of severe debris-flow erosion and resultant volume bulking (Sect. 2.3.3) have highlighted the lack of tools to identify debris-flow fans and watersheds that may generate these processes. Extreme debris-flow erosion can damage buried infrastructure such as pipelines, waterlines, or fibre optic cables and increase debris volume towards the distal portion of the fan. Practitioners often simplistically view debris-flow fans as largely depositional landforms and may be placing buried infrastructure at unquantified risk (Jakob, 2021). Preliminary research by Lau (2017) showed that fan slope, watershed area, and watershed relief may be a first-order guide for identifying highly erosive debris-flow fans. Additional empirical and experimental research may confirm these morphometrics and/or additional processes that can be used to identify these fans prior to extreme erosion occurrence.

Field observations suggest that the deepest channel incisions occur near the fan apex. We are not aware of experimental studies that monitor the changes in channel erosion with distance along debris-flow fans, and whether the observed incision at the fan apex is due to increased erosion. Anecdotal observations of "firehose effects" from prolonged post-wildfire debris floods and debris flows have been postulated to generate significant fan erosion during the November 2021 atmospheric river in Southern British Columbia, Canada, resulting in billion-dollar damages to highway and railway infrastructure (Lau et al., 2021). Multi-temporal topographic surveys, in-channel basal shear stress observations, or repeat channel surveys could identify temporal and spatial patterns in debris-flow erosion along debris-flow fans. Results from these observations could be used to generate debris-flow erosion spatio-temporal models similar to those recently developed for debris-flow fan avulsions (e.g., Zubrycky et al., 2021).

Similar to the limitations for modelling debris-flow avulsions, some debris-flow modelling software does not allow for terrain modification and volumetric bulking from fan erosion. In addition, modern debris-flow models that do incorporate erosion processes are often not robustly tested and validated, because of a lack of field observations (e.g., Baggio et al., 2021; Frank et al., 2015; Pudasaini & Krautblatter, 2021). In addition, it is tricky in many models to include the formation of channel plugs for modelling potential avulsions.

2.6 Summary

Debris flows typically originate in mountainous watersheds. They form semi-conical fans that prograde onto topographically lower terrain, with apex at the watershed's outlet, mainly as a result of a loss in gradient and flow confinement that lead to a decrease in flow velocity and depth.

Debris-flow activity is a function of both watershed morphometry, material characteristics, and hydroclimatic conditions. Debris-flow watersheds generate sufficient runoff during intense hydroclimatic events, and they have sufficiently steep slopes to entrain and transport sediment and to sustain debris-flow motion throughout the watershed and transfer it to an external basin. Sediments may be available and produced from various sources, including weathering of bedrock, (shallow) landsliding, soil erosion, and erosion of a multitude of unconsolidated sediment sources, including glacial and volcanic successions or talus slopes.

Debris-flow watersheds can be transport-limited (supply-unlimited) or supply-limited (weathering-limited). Because of the comparatively higher availability of sediment in transport-limited watersheds, each hydroclimatic event producing enough runoff to trigger a debris flow will do so. Conversely, in supply-limited watersheds debris-flow activity is limited by sediment availability and recharge rates, and therefore hydroclimatic events that produce enough discharge to trigger a debris flow only do so when sufficient sediment is available. As a result, changes in sediment availability or in the recharge rates of watersheds may significantly alter the frequency and magnitude of debris flows, or even the dominant processes (e.g., debris flow versus floods).

Watersheds and fans dominated by debris flows present distinct morphometric properties from those of watersheds and fans dominated by other colluvial processes and by (debris) floods, although there is no sharp discontinuity between process types, and considerable overlap exists. Debris-flow watersheds are typically smaller and steeper than those dominated by floods, and have less well-developed and more immature drainage patterns with lower capacity of channel storage per unit of drainage area. Therefore, the morphometric characteristics of watersheds are often used to determine their potential for generating debris flows. Common indicators are combinations of Melton index (watershed relief divided by the square root of watershed area), watershed length, and fan slope. However, there are no distinct thresholds between debris-flow dominated, flood-dominated and colluvially-dominated systems, and locally significant thresholds are generally not suitable for extrapolation to other areas. We therefore advocate a methodological shift from threshold-based approaches to probability-based approaches in order to characterize geomorphic areas potentially liable to debris-flow activity.

Debris-flows fans have radial lengths roughly ranging between 0.5 and 10 km and slopes roughly ranging between 5° and 15°. Their surfaces typically consist of stacked, vertically and laterally amalgamated, elongate, straight to sinuous lobes, which, if not buried by subsequent lobes, are connected to (leveed) channels. These lobes and levee systems consist of poorly sorted sediments ranging from clay to boulder-sized particles. Large boulders with diameters up to several meters may be common on the surfaces of debris-flow fans. The often high return periods of debris flows and their limited spatial extent relative to the total fan surface make inactive parts of fan surfaces susceptible to considerable modification by secondary (post-depositional) processes. These include surface or groundwater flow, wind, bioturbation, neotectonics, weathering and pedogenesis, and a variety of periglacial processes. In stratigraphy, debris-flow-fan deposits are poorly stratified and internally massive, typically consisting of gravel and outsized boulders dispersed and generally randomly oriented within a finer matrix, but may present segregations or alignments of coarse material at flow tops and/or margins.

On debris-flow fans, the main locus of deposition shifts over time through avulsion. Avulsions on debris-flow fans are often induced by debris-flow deposits blocking a channel (termed channel plugs), or less commonly by flows overtopping the channel in correspondence of outer bends. During individual flows or flow surges, deposition of channel plugs forces subsequent flows to avulse into alternative paths. Plug deposition is a stochastic process, but it appears to depend in part on the sequence of flow volumes, the geometry of the channel, and the composition of the flows. The occurrence of channel plugs makes prediction of debris-flow avulsion difficult. However, over time scales of multiple events, the average locus of debris-flow deposition shifts toward the topographically lower parts of a fan, highlighting the importance of topographic compensation for the evolution of debris-flow fans.

Changes in the hydroclimatic conditions and sediment availability of watersheds affect the frequency and magnitude of debris flows, as well as their sedimentological characteristics. Debris-flow fans therefore may provide archives of past flow processes, sediment supply, and sedimentary signals indicative of past climate and environmental change, which can be reconstructed from both their surface features and from their stratigraphy. More research focus is needed, however, in fine-tuning our ability to extract paleoenvironmental information from these coarse-clastic, mass-flow-dominated successions. This is especially important in present times of profound climate change, as the assessment of future debris-flow hazard at different time scales is dependent also on our ability to decipher the sedimentary record of debris-flow fans.

References

Aaron, J., McDougall, S., & Jordan, P. (2019). Dynamic analysis of the 2012 Johnsons Landing landslide at Kootenay Lake, British Columbia: The importance of undrained flow potential. *Canadian Geotechnical Journal, 57*(8), 1172–1182. https://doi.org/10.1139/cgj-2018-0623

Abram, N., Gattuso, J.-P., Prakash, A., Cheng, L., Chidichimo, M. P., Crate, S., Enomoto, H., Garschagen, M., Gruber, N., Harper, S., Holland, E., Kudela, R. M., Rice, J., Steffen, K., & von Schuckmann, K. (2019). Framing and context of the report. In H.-O. Pörtner, D.C. Roberts, & V. Masson-Delmotte (Eds.), *IPCC Special Report on the Ocean and Cryosphere in a Changing Climate*.

Adams, K., Wasklewicz, T., De Haas, T., Lecce, S., & Gares, P. (2019). Reproducibility of debris-flow fan physical modeling experiments. In J. Kean, J. Coe, P. Santi, & B. Guillen (Eds.), *Debris-flow hazards mitigation: Mechanics, prediction, and assessment* (pp. 637–644). Millpress.

Åkerman, H. J. (1984). Notes on talus morphology and processes in Spitsbergen. *Geografiska Annaler: Series A, Physical Geography, 66*(4), 267–284.

Al-Farraj, A., & Harvey, A. M. (2000). Desert pavement characteristics on wadi terrace and alluvial fan surfaces: Wadi Al-Bih, U.A.E. and Oman. *Geomorphology, 35*(34), 279–297.

Al-Farraj, A., & Harvey, A. M. (2005). Morphometry and depositional style of Late Pleistocene alluvial fans: Wadi Al-Bih, northern UAE and Oman. *Geological Society, London, Special Publications, 251*(1), 85.

Amit, R., Gerson, R., & Yaalon, D. H. (1993). Stages and rate of the gravel shattering process by salts in desert Reg soils. *Geoderma, 57*, 295–324.

Amorosi, A., Farina, M., Severi, P., Preti, D., Caporale, L., & Di Dio, G. (1996). Genetically related alluvial deposits across active fault zones: An example of alluvial fan-terrace correlation from the upper Quaternary of the southern Po Basin, Italy. *Sedimentary Geology, 102*, 275–295.

Anderson, S. P., & Anderson, R. S. (1990). Debris-flow benches: Dune-contact deposits record paleo-sand dune positions in north Panamint Valley, Inyo County, California. *Geology, 18*(6), 524–527.

Anderson, S. W., Anderson, S. P., & Anderson, R. S. (2015). Exhumation by debris flows in the 2013 Colorado front range storm. *Geology, 43*(5), 391–394.

André, M.-F. (1990a). Geomorphic impact of spring avalanches in northwest Spitsbergen (79 N). *Permafrost and Periglacial Processes, 1*(2), 97–110.

André, M.-F. (1990b). Frequency of debris flows and slush avalanches in Spitsbergen: A tentative evaluation from lichenometry. *Polish Polar Research, 11*, 345–363.

André, M.-F. (1995). Holocene climate fluctuations and geomorphic impact on extreme events in Svalbard, Geografiska. *Annaler, 77A*, 241–250.

Andreucci, S., Panzeri, L., Martini, I. P., Maspero, F., Martini, M., & Pascucci, V. (2014). Evolution and architecture of a west mediterranean upper pleistocene to Holocene coastal apron-fan system. *Sedimentology, 61*, 333–361.

Arango, M. I., Aristizábal, E., & Gómez, F. (2021). Morphometrical analysis of torrential flows-prone catchments in tropical and mountainous terrain of the Colombian Andes by machine learning techniques. *Natural Hazards, 105*(1), 983–1012.

Babault, J., Bonnet, S., Crave, A., & Van Den Driessche, J. (2005). Influence of piedmont sedimentation on erosion dynamics of an uplifting landscape: An experimental approach. *Geology, 33*(4), 301–304.

Babault, J., Bonnet, S., Van Den Driessche, J., & Crave, A. (2007). High elevation of low-relief surfaces in mountain belts: Does it equate to post-orogenic surface uplift? *Terra Nova, 19*(4), 272–277.

Baggio, T., Mergili, M., & D'Agostino, V. (2021). Advances in the simulation of debris flow erosion: The case study of the Rio Gere (Italy) event of the 4th August 2017. *Geomorphology, 381*, 107664.

Ban, N., Schmidli, J., & Schär, C. (2015). Heavy precipitation in a changing climate: Does short-term summer precipitation increase faster? *Geophysical Research Letters, 42*, 1165–1172.

Ban, N., Rajczak, J., Schmidli, J., & Schär, C. (2018). Analysis of Alpine precipitation extremes using generalized extreme value theory in convection-resolving climate simulations. *Climate Dynamics, 55*, 61–75.

Barclay, J., Alexander, J., & Susnik, J. (2007). Rainfall-induced lahars in the Belham Valley, Montserrat, West Indies. *Journal of the Geological Society of London, 164*, 815–827.

Beaty, C. B. (1963). Origin of alluvial fans, white mountains, California and Nevada. *Annals of the Association of American Geographers, 53*(4), 516–535.

Beaty, C. B. (1989). Great big boulders I have known. *Geology, 17*, 349–352.

Benda, L. (1990). The influence of debris flows on channels and valley floors in the Oregon Coast Range, U.S.A. *Earth Surface Processes and Landforms, 15*, 457–466.

Benda, L., Hassan, M. A., Church, M., & May, C. L. (2005). Geomorphology of steepland headwaters: The transition from hillslopes to channels 1. *JAWRA Journal of the American Water Resources Association, 41*(4), 835–851.

Bennett, G. L., Molnar, P., McArdell, B. W., Schlunegger, F., & Burlando, P. (2013). Patterns and controls of sediment production, transfer and yield in the Illgraben. *Geomorphology, 188*, 68–82.

Bennett, G. L., Molnar, P., McArdell, B. W., & Burlando, P. (2014). A probabilistic sediment cascade model of sediment transfer in the Illgraben. *Water Resources Research, 50*(2), 1225–1244.

Berger, I. A., & Cooke, R. U. (1997). The origin and distribution of salts on alluvial fans in the Atacama Desert, northern Chile. *Earth Surface Processes and Landforms, 22*(6), 581–600.

Berger, C., Mcardell, B. W., Fritschi, B., & Schlunegger, F. (2010). A novel method for measuring the timing of bed erosion during debris flows and floods. *Water Resources Research, 46*, 1–7.

Berger, C., McArdell, B. W., & Schlunegger, F. (2011a). Sediment transfer patterns at the Illgraben catchment, Switzerland: Implications for the time scales of debris flow activities. *Geomorphology, 125*(3), 421–432.

Berger, C., Mcardell, B. W., & Schlunegger, F. (2011b). Direct measurement of channel erosion by debris flows, Illgraben, Switzerland. *Journal of Geophysical Research, 116*, 1–18.

Bernhardt, H., Reiss, D., Hiesinger, H., Hauber, E., & Johnsson, A. (2017). Debris flow recurrence periods and multi-temporal observations of colluvial fan evolution in central Spitsbergen (Svalbard). *Geomorphology, 296*, 132–141.

Berti, M., & Simoni, A. (2007). Prediction of debris flow inundation areas using empirical mobility relationships. *Geomorphology, 90*(1–2), 144–161.

Bertran, P., Bernard, H., Texier, J.-P., & Van Steijn, H. (1997). Fabric characteristics of subaerial slope deposits. *Sedimentology, 44*, 1–16.

Bertrand, M., Liébault, F., & Piégay, H. (2013). Debris-flow susceptibility of upland catchments. *Natural Hazards, 67*(2), 497–511.

Bigi, A., Hasbargen, L. E., Montanari, A., & Paola, C. (2006). Knickpoints and hillslope failures: interactions in a steady state experimental landscape. In S.D. Willett, N. Hovius, M.T. Brandon, & D.M. Fisher (Eds.), *Tectonics, Climate, and Landscape Evolution: Penrose Conference Series Special Paper* (pp. 295–307). Geological Society of America.

Bishop, C. M. (2006). *Pattern recognition and machine learning*. Springer.

Blair, T. C., & McPherson, J. G. (1994). Alluvial fans and their natural distinction from rivers based on morphology, hydraulic processes, sedimentary processes, and facies. *Journal of Sedimentary Research, A64*, 451–490.

Blair, T. C., & McPherson, J. G. (1998). Recent debris-flow processes and resultant form and facies of the Dolomite alluvial fan, Owens Valley, California. *Journal of Sedimentary Research, 68*(5), 800–818.

Blair, T. C. (1999a). Cause of dominance by sheetflood vs. debris-flow processes on two adjoining alluvial fans, Death Valley, California. *Sedimentology, 46*, 1015–1028.

Blair, T. C. (1999b). Sedimentology of the debris-flow-dominated Warm Spring Canyon alluvial fan, Death Valley, California. *Sedimentology, 46*(5), 941–965.

Blair, T. C. (2003). Features and origin of the giant Cucomungo Canyon alluvial fan, Eureka Valley, California. *Special Paper of the Geological Society of America*, 105–126

Blair, T. C., & McPherson, J. G. (2009). Alluvial fan forms and processes. In *Geomorphology of desert environments* (2nd Ed., pp. 413, 467). Springer Science+ Business Media.

Blair, T. C., Clark, J. S., & Wells, S. G. (1990). Quaternary continental stratigraphy, landscape evolution, and application to archeology: Jarilla piedmont and Tularosa graben floor, White Sands Missile Range, New Mexico. *Geological Society of America Bulletin, 102*, 749–749.

Blikra, L. H., & Nemec, W. (1998). Postglacial colluvium in western Norway: Depositional processes, facies and palaeoclimatic record. *Sedimentology, 45*(5), 909–960.

Blikra, L. H., & Nemec, W. (2000). Post-glacial colluvium in western Norway: Depositional processes, facies and palaeocolimatic record. Reply. *Sedimentology, 47*, 1058–1068.

Blissenbach, E. (1954). Geology of alluvial fans in semiarid regions. *Geological Society of America Bulletin, 65*, 175–190.

Bollschweiler, M., & Stoffel, M. (2010a). Tree rings and debris flows: Recent developments, future directions. *Progress in Physical Geography, 34*(5), 625–645.

Bollschweiler, M., & Stoffel, M. (2010b). Changes and trends in debris-flow frequency since AD 1850: Results from the Swiss Alps. *The Holocene, 20*(6), 907–916.

Bonneau, D., Hutchinson, D. J., McDougall, S., Difrancesco, P. M., & Evans, T. (2022). Debris-flow channel headwater dynamics: Examining channel recharge cycles with terrestrial laser scanning. *Frontiers in Earth Science*, 1204.

Bonnet, S., & Crave, A. (2003). Landscape response to climate change; insights from experimental modeling and implications for tectonic versus climatic uplift of topography. *Geology, 31*(2), 123–126.

Bonnet, S., & Crave, A. (2006). Macroscale dynamics of experimental landscapes. In S. J. H. Buiter & G. Schreurs (Eds.), *Analogue and numerical modelling of crustal-scale processes* (pp. 327–339). Special Publications.

Borga, M., Stoffel, M., Marchi, L., Marra, F., & Jakob, M. (2014). Hydrogeomorphic response to extreme rainfall in headwater systems: Flash floods and debris flows. *Journal of Hydrology, 518B*, 194–205.

Bovis, M., & Dagg, B. (1988). A model for debris accumulation and mobilization in steep mountain streams. *Hydrological Sciences Journal, 33*(6), 589–604. https://doi.org/10.1080/02626668809491292

Bovis, M. J., & Jakob, M. (1999). The role of debris supply conditions in predicting debris flow activity. *Earth Surface Processes and Landforms, 24*(11), 1039–1054.

Brayshaw, D., & Hassan, M. (2009). Debris flow initiation and sediment recharge in gullies. *Geomorphology, 109*, 122–131.

Brookfield, M. E. (1977). The origin of bounding surfaces in ancient aeolian sandstones. *Sedimentology, 24*, 303–332.

Brunkal, H., & Santi, P. (2017). Consideration of the validity of debris-flow bulking factors. *Environmental & Engineering Geoscience, 23*(4), 291–298.

Bryant, M., Falk, P., & Paola, C. (1995). Experimental study of avulsion frequency and rate of deposition. *Geology, 23*(4), 365–368.

Bull, W. B. (1962). Relations of alluvial fan size and slope to drainage basin size and lithology in western Fresno County, California. *U.S. Geological Survey Professional Paper*, 450-B.

Bull, W. B. (1964). Alluvial fans and near-surface subsidence in Western Fresno County California. *Geological Survey Professional Paper*, 437-A

Bull, W. B. (1991). *Geomorphic responses to climate change* (p. 326). Oxford Unvi. Press.

Calvache, M. L., Viseras, C., & Ferndez, J. (1997). Controls on fan development—evidence from fan morphometry and sedimentology; Sierra Nevada SE Spain. *Geomorphology, 21*(1), 69–84.

Campbell, C. V. (1967). Lamina, laminasetit, bed and bedset. *Sedimentology, 8*, 7–26.

Campbell, R. H. (1974). Debris flows originating from soil slips during rainstorms in southern California. *Quarterly Journal of Engineering Geology, 7*, 339–349.

Campbell, R. H. (1975). Soil slips, debris flows and rainstorms in the Santa Monica Mountain and vicinity, southern California. *U.S. Geological Survey Professional Paper*, 851, 51 pp.

Cannon, S. H. (2001). Debris-flow generation from recently burned watersheds. *Environmental & Engineering Geoscience, 7*(4), 321–341.

Cannon, S. H., & Reneau, S. L. (2000). Conditions for generation of fire-related debris flows, Capulin Canyon, New Mexico. *Earth Surface Processes and Landforms, 25*, 1103–1121.

Carboni, S., Palomba, M., Vacca, A., & Carboni, G. (2006). Paleosols provide sedimentation, relative age, and clastic information about the alluvial fan of the River Tirso (central-western Sardinia, Italy). *Quaternary International, 156–157*, 79–86.

Carling, P., Gupta, N., Atkinson, P., & He, H. Q. (2016). Criticality in the planform behavior of the Ganges River meanders. *Geology, 44*(10), 859–862.

Catto, N. R. (1993). Morphology and development of an alluvial fan in a permafrost region, Aklavik Range, Canada. *Geografiska Annaler: Series a, Physical Geography, 75*(3), 83–93.

Chen, J., Dai, F., & Yao, X. (2008). Holocene debris-flow deposits and their implications on the climate in the upper Jinsha River valley, China. *Geomorphology, 93*(3–4), 493–500.

Chen, J. C., Dan, C. D., & Huang, W. S. (2013). Characteristics of rainfall triggering of debris flows in the Cenyulan watershed, Taiwan. *Natural Hazards and Earth Systems Science, 13*, 1015–1023.

Chiaradia, E. A., Vergani, C., & Bischetti, G. B. (2016). Evaluation of the effects of three European forest types on slope stability by field and probabilistic analyses and their implications for forest management. *Forest Ecology and Management, 370*, 114–129.

Chiarle, M., Iannotti, S., Mortara, G., & Deline, P. (2007). Recent debris flow occurrences associated with glaciers in the Alps. *Global and Planetary Change, 56*, 123–136. https://doi.org/10.1016/j.gloplacha.2006.07.003

Chiarle, M., Geertsema, M., Mortara, G., & Clague, J. C. (2021). Relations between climate change and mass movement: Pespectives from the Canadian Cordillera and the European Alps. *Global and Planetary Change, 202*, 103499.

Chien-Yuan, C., Fan-Chieh, Y., Sheng-Chi, L., & Kei-Wai, C. (2007). Discussion of landslide self-organized criticality and the initiation of debris flow. *Earth Surface Processes and Landforms, 32*(2), 197–209.

Chiverrell, R., & Jakob, M. (2013). Radiocarbon dating: alluvial fan/debris cone evolution and hazards. In *Dating torrential processes on fans and cones* (pp. 265–282). Springer.

Church, M., & Miles, M. J. (1987). Meteorological antecedents to debris flow in southwestern British

Columbia; some case studies. In J.E. Costa & G.F. Wieckzorek (Eds.), *Debris flows/avalanches*. Geological Society of America, Reviews in Engineering Geology (Vol. 7, pp. 63–79).

Church, M., & Slaymaker, O. (1989). Disequilibrium of Holocene sediment yield in glaciated British Columbia. *Nature, 337*, 452–454.

Church, M., & Jakob, M. (2020). What is a debris flood? *Water Resources Research, 56*(8). https://doi.org/10.1029/2020WR027144

Clark, G. M. (1987). Debris slide and debris flow historical events in the Appalachians south of the glacial border. In J.E. Costa & G.F. Wieckzorek (Eds.), *Debris flows/avalanches*. Geological Society of America, Reviews in Engineering Geology (Vol. 7, pp. 125–138).

Clague, J. J., Evans, S. G., & Blown, I. G. (1985). A debris flow triggered by the breaching of a moraine-dammed lake, Klattasine Creek, British Columbia. *Canadian Journal of Earth Science, 22*, 1492–1502.

Clarke, L., Quine, T. A., & Nicholas, A. (2010). An experimental investigation of autogenic behaviour during alluvial fan evolution. *Geomorphology, 115*(3), 278–285.

Coe, J. A., Kinner, D. A., & Godt, J. W. (2008). Initiation conditions for debris flows generated by runoff at Chalk Cliffs, central Colorado. *Geomorphology, 96*(3–4), 270–297.

Comiti, F., Marchi, L., Macconi, P., Arattano, M., Bertoldi, G., Borga, M., Theule, J., et al. (2014). A new monitoring station for debris flows in the European Alps: First observations in the Gadria basin. *Natural Hazards, 73*(3), 1175–1198.

Conway, S. J., de Haas, T., & Harrison, T. N. (2019). Martian gullies: A comprehensive review of observations, mechanisms and insights from Earth analogues. *Geological Society, London, Special Publications, 467*(1), 7–66.

Cook, J. P., & Pelletier, J. D. (2007). Relief threshold for eolian sand transport on alluvial fans. *Journal of Geophysical Research: Earth Surface, 112*(F2).

Costa, J. E. (1988). Rheologic, geomorphic, and sedimentologic differentiation of water floods, hyperconcentrated flows, and debris flows. In V. R. Baker & P. C. Patton (Eds.), *Flood geomorphology* (pp. 113–122). Wiley.

Coussot, P. (1995). Structural similarity and transition from newtonian to non-newtonian behavior for clay-water suspensions. *Physical Review Letters, 74*, 3971–3974.

Coussot, P., & Piau, J.-M. (1994). On the behavior of fine mud suspensions. *Rheologica Acta, 33*, 175–184.

Cronin, S. J., Neall, V. E., Lecointre, J. A., & Palmer, A. S. (1997). Changes in Whangaehu River lahar characteristics during the 1995 eruption sequence Ruapehu Volcano, New Zealand. *Journal of Volcanology and Geothermal Research, 76*, 47–61.

Crosta, G. B., & Frattini, P. (2004). Controls on modern alluvial fan processes in the central Alps, northern Italy. *Earth Surface Processes and Landforms, 29*(3), 267–293. https://doi.org/10.1002/esp.1009

D'Agostino, V., Cesca, M., & Marchi, L. (2010). Field and laboratory investigations of runout distances of debris flows in the Dolomites (Eastern Italian Alps). *Geomorphology, 115*, 294–304.

Damanti, J. F. (1993). Geomorphic and structural controls on facies patterns and sediment composition inn a modern foreland basin. In M. Marzo & C. Puigdefabregas (Eds.), *Alluvial sedimentation* (pp. 17, 221–233). IAS Special Publication.

D'Arcy, M., Boluda, D. C. R., Whittaker, A. C., & Carpineti, A. (2015). Dating alluvial fan surfaces in Owens Valley, California, using weathering fractures in boulders. *Earth Surface Processes and Landforms, 40*, 487–501.

D'Arcy, M., Roda-Boluda, D. C., & Whittaker, A. C. (2017). Glacial-interglacial climate changes recorded by debris flow fan deposits, Owens Valley, California. *Quaternary Science Reviews, 169*, 288–311.

Davies, T. R. H., & Korup, O. (2007). Persistent alluvial fanhead trenching resulting from large, infrequent sediment inputs. *Earth Surface Processes and Landforms, 32*, 725–742.

Decaulne, A., Sæmundsson, T., Jónsson, H. P., & Sandberg, O. (2007). Changes in deposition on a colluvial fan during the upper Holocene in the Tindastóll Mountain, Skagafjörður district, North Iceland: Preliminary results. *Geografiska Annaler, Series A, 89*, 51–63.

DeCelles, P. G., Gray, M. B., Ridgway, K. D., Cole, R. B., Pivnik, D. A., Pequera, N., & Srivastava, P. (1987). Controls on synorogenic alluvia-fan architecture, Beartooth Conglomerate (Palaeocene), Wyoming and Montana. *Sedimentology, 38*, 567–590.

De Haas, T., Ventra, D., Carbonneau, P. E., & Kleinhans, M. G. (2014). Debris-flow dominance of alluvial fans masked by runoff reworking and weathering. *Geomorphology, 217*, 165–181.

De Haas, T., Kleinhans, M. G., Carbonneau, P. E., Rubensdotter, L., & Hauber, E. (2015a). Surface morphology of fans in the high-Arctic periglacial environment of Svalbard: Controls and processes. *Earth Science Reviews, 146*, 163–182.

De Haas, T., Braat, L., Leuven, J. R., Lokhorst, I. R. & Kleinhans, M. G. (2015b). Effects of debris flow composition on runout, depositional mechanisms, and deposit morphology in laboratory experiments. *Journal of Geophysical Research – Earth Surface, 120*(9), 1949–1972.

De Haas, T., Ventra, D., Hauber, E., Conway, S. J., & Kleinhans, M. G. (2015c). Sedimentological analyses of Martian gullies: The subsurface as the key to the surface. *Icarus, 258*, 92–108.

De Haas, T., van den Berg, W., Braat, L., & Kleinhans, M. G. (2016). Autogenic avulsion, channelization and backfilling dynamics of debris-flow fans. *Sedimentology, 63*(6), 1596–1619.

De Haas, T., & van Woerkom, T. (2016). Bed scour by debris flows: Experimental investigation of effects of debris-flow composition. *Earth Surface Processes and Landforms, 41*(13), 1951–1966.

De Haas, T., Densmore, A. L., Stoffel, M., Suwa, H., Imaizumi, F., Ballesteros-Cánovas, J., & Wasklewicz, T. (2018a). Avulsions and the spatio-temporal evolution of debris-flow fans. *Earth-Science Reviews, 177*, 53–75.

De Haas, T., Kruijt, A., & Densmore, A. L. (2018b). Effects of debris-flow magnitude–frequency distribution on avulsions and fan development. *Earth Surface Processes and Landforms, 43*(13), 2779–2793.

De Haas, T., Densmore, A. L., den Hond, T., & Cox, N. J. (2019). Fan-surface evidence for debris-flow avulsion controls and probabilities, Saline Valley, California. *Journal of Geophysical Research - Earth Surface, 124*(5), 1118–1138.

De Haas, T., & Densmore, A. L. (2019). Debris-flow volume quantile prediction from catchment morphometry. *Geology, 47*(8), 791–794.

De Haas, T., Nijland, W., De Jong, S. M., & McArdell, B. W. (2020). How memory effects, check dams, and channel geometry control erosion and deposition by debris flows. *Scientific Reports, 10*(1), 1–8.

De Haas, T., McArdell, B. W., Nijland, W., Åberg, A. S., Hirschberg, J., & Huguenin, P. (2022). Flow and bed conditions jointly control debris-flow erosion and bulking. *Geophysical Research Letters*, e2021GL097611.

Denny, C. S. (1967). Fans and Piedomonts. *American Journal of Science, 265*, 81–105.

De Scally, F., Slaymaker, O., & Owens, I. (2001). Morphometric controls and basin response in the Cascade Mountains. *Geografiska Annaler: Series a, Physical Geography, 83A*, 117–130.

De Scally, F., & Owens, I. (2004). Morphometric controls and geomorphic responses on fans in the southern Alps New Zealand. *Earth Surface Processes and Landforms, 29*(3), 311–322.

De Scally, F. A., Owens, I. F., & Louis, J. (2010). Controls on fan depositional processes in the schist ranges of the Southern Alps, New Zealand, and implications for debris-flow hazard assessment. *Geomorphology, 122*(1–2), 99–116. https://doi.org/10.1016/j.geomorph.2010.06.002

Densmore, A. L., Allen, P. A., & Simpson, G. (2007). Development and response of a coupled catchment fan system under changing tectonic and climatic forcing. *Journal of Geophysical Research: Earth Surface, 112*(F1). https://doi.org/10.1029/2006JF000474

Densmore, A. L., de Haas, T., McArdell, B., & Schürch, P. (2019). Making sense of avulsions on debris-flow fans. In J. Kean, J. Coe, P. Santi, & B. Guillen (Eds.), *Debris-flow hazards mitigation: Mechanics, prediction, and assessment* (pp. 637–644). Millpress.

Di, B., Zhang, H., Liu, Y., Li, J., Chen, N., Stamatopoulos, C. A., Zhan, Y., et al. (2019). Assessing susceptibility of debris flow in southwest China using gradient boosting machine. *Scientific Reports, 9*(1), 1–12.

Dickerson, R. P., Bierman, P. R., & Cocks, G. (2015). Alluvial fan surfaces and an age-related stability for cultural resource preservation: Nevada Test and Training Range, Nellis Air Force Base, Nevada, USA. *Journal of Archaeological Science, 2*, 551–568.

Dietrich, A., & Krautblatter, M. (2017). Evidence for enhanced debris-flow activity in the Northern Calcareous Alps since the 1980s (Plansee, Austria). *Geomorphology, 287*, 144–158.

DiBiase, R. A., & Lamb, M. (2019). Dry sediment loading of headwater channels fuels post-wildfire debris flows in bedrock landscapes. *Geology, 48*, 189–193.

Domènech, G., Fan, X., Scaringi, G., van Asch, T. W., Xu, Q., Huang, R., & Hales, T. C. (2019). Modelling the role of material depletion, grain coarsening and revegetation in debris flow occurrences after the 2008 Wenchuan earthquake. *Engineering Geology, 250*, 34–44.

Dowling, C. A., & Santi, P. M. (2014). Debris flows and their toll on human life: A global analysis of debris-flow fatalities from 1950 to 2011. *Natural Hazards, 71*(1), 203–227.

Dühnforth, M., Densmore, A. L., Ivy-Ochs, S., Allen, P. A., & Kubik, P. W. (2007). Timing and patterns of debris flow deposition on Shepherd and Symmes creek fans, Owens Valley, California, deduced from cosmogenic 10Be. *Journal of Geophysical Research: Earth Surface, 112*, F03S15.

Dühnforth, M., Densmore, A. L., Ivy-Ochs, S., & Allen, P. A. (2008). Controls on sediment evacuation from glacially modified and unmodified catchments in the eastern Sierra Nevada, California. *Earth Surface Processes and Landforms, 33*(10), 1602–1613.

Dühnforth, M., Densmore, A. L., Ivy-Ochs, S., Allen, P., & Kubik, P. W. (2017). Early to late pleistocene history of debris-flow fan evolution in western Death Valley (California) using cosmogenic 10Be and 26Al. *Geomorphology, 281*, 53–65.

Eichel, J., Draebing, D., & Meyer, N. (2018). From active to stable: Paraglacial transition of Alpine lateral moraine slopes. *Land Degradation & Development, 29*(11), 4158–4172.

Enos, P. (1977). Flow regimes in debris flow. *Sedimentology, 24*, 133–142.

Eppes, M. C., McFadden, L. D., Wegmann, K. W., & Scuderi, L. A. (2010). Cracks in desert pavement rocks: Further insights into mechanical weathering by direction insolation. *Geomorphology, 123*, 97–108.

Evans, S. G., Hungr, O., & Clague, J. J. (2001). Dyanmics of the 1984 rock avalanche and associated distal debris flow on Mount Cayley, British Columbia, Canada: Implications for landslide hazard assessment on dissected volcanoes. *Engineering Geology, 61*, 29–51.

Faniran, A. (1968). The index of drainage intensity—a provisional new drainage factor. *Australian Journal of Science, 21*, 328–330.

Fannin, R. J., & Wise, M. P. (2001). An empirical-statistical model for debris flow travel distance. *Canadian Geotechnical Journal, 38*(5), 982–994.

Ferrill, D. A., Stamatakos, J. A., Jones, S. M., Rahe, B., McKague, H. L., Martin, R. H., & Morris, A. P. (1996). Quaternary slip history of the Bare Mountain fault (Nevada) from the morphology and distribution of alluvial fan deposits. *Geology, 24*, 559–562.

Fisher, R. V. (1971). Features of coarse-grained, high-concentration fluids and their deposits. *Journal of Sedimentary Petrology, 41*, 916–927.

Fischer, L., Kääb, A., Huggel, C., & Noetzli, J. (2006). Geology, glacier retreat and permafrost degradation as controlling factors of slope instabilities in a high-mountain rock wall: The Monte Rosa east face. *Natural Hazards and Earth System Sciences, 6*(5), 761–772.

Frank, F., McArdell, B. W., Huggel, C., & Vieli, A. (2015). The importance of entrainment and bulking on debris flow runout modeling: Examples from the Swiss Alps. *Natural Hazards Earth Systems Science, 15*, 2569–2583. https://doi.org/10.5194/nhess-15-2569-2015

Franke, D., Hornung, J., & Hinderer, M. (2015). A combined study of radar facies, lithofacies and three-dimensional architecture of an alpine alluvial fan (Illgraben Fan, Switzerland). *Sedimentology, 62*, 57–86.

Frankel, K. L., & Dolan, J. F. (2007). Characterizing arid region alluvial fan surface roughness with airborne laser swath mapping digital topographic data. *Journal of Geophysical Research, 112*(F2), F02025.

French, R. H. (1987). *Hydraulic processes on alluvial fans*. Elsevier Science Publishers.

Fuchs, S., Keiler, M., Sokratov, S., & Shnyparkov, A. (2013). Spatiotemporal dynamics: The need for an innovative approach in mountain hazard risk management. *Natural Hazards, 68*(3), 1217–1241.

Giorgi, F., Torma, C., Coppola, E., Ban, N., Schär, C., & Somot, S. (2016). Enhanced summer convective rainfall at Alpine high elevations in response to climate warming. *Nature Geoscience, 9*(8), 584–589.

González, C., Kelley, M., Marvin, M. C., López-Castañeda, N., Dorn, R. I., & Schmeeckle, M. (2021). Regional piedmont incision during base-level rise in the northeastern Sonoran Desert, Arizona, USA. *Physical Geography* (in press). https://doi.org/10.1080/02723646.2021.1934964

Gravelius, H. (1941). *Flusskunde*. Goschen'sche Verlagshandlung, Berlin.

Graveleau, F., Strak, V., Dominguez, S., Malavieille, J., Chatton, M., Manighetti, I., & Petit, C. (2015). Experimental modelling of tectonics–erosion–sedimentation interactions in compressional, extensional, and strike–slip settings. *Geomorphology, 244*, 146–168.

Gray, J. M. N. T., & Ancey, C. (2009). Particle size-segregation, recirculation, and deposition at coarse particle rich flow fronts. *Journal of Fluid Mechanics, 629*, 387–423.

Gray, J. M. N. T., & Ancey, C. (2011). Multi-component particle-size segregation in shallow granular avalanches. *Journal of Fluid Mechanics, 678*, 535–588.

Gray, J. M. N. T. (2018). Particle segregation in dense granular flows. *Annual Review of Fluid Mechanics, 50*, 407–433.

Gregoretti, C., & Fontana, G. (2007). The triggering of debris flow due to channel-bed failure in some alpine headwater basins of the Dolomites: Analyses of critical runoff. *Hydrological Processes, 22*(13), 2248–2263. https://doi.org/10.1002/hyp.6821

Griswold, J. P., & Iverson, R. M. (2008). Mobility statistics and automated hazard mapping for debris flows and rock avalanches. US Department of the Interior, US Geological Survey.

Guthrie, R. H., Hockin, A., Colquhoun, L., Nagy, T., Evans, S. G., & Ayles, C. (2010). An examination of controls on debris flow mobility: Evidence from coastal British Columbia. *Geomorphology, 114*, 601–613.

Guthrie, R. H., Friele, P., Allstadt, K., Roberts, N., Evans, S. G., Delaney, K. B., Roche, D., Clague, J. J., & Jakob, M. (2012). The 6 August 2010 Mount Meager rock slide-debris flow, Coast Mountains, British Columbia: Characteristics, dynamics, and implications for hazard and risk assessment. *Natural Hazards and Earth Systems Science, 12*, 1277–1294. https://doi.org/10.5194/nhess-12-1277-2012

Haeberli, W., Guodong, C., Gorbunov, A. P., & Harris, S. A. (1993). Mountain permafrost and climatic change. *Permafrost and Periglacial Processes, 4*(2), 165–174.

Harris, C., Arenson, L. U., Christiansen, H. H., Etzelmüller, B., Frauenfelder, R., Gruber, S., Haeberli, W., Hauck, C., Hölzle, M., Humlum, O., Isaksen, K., Kääb, A., Kern-Lütschg, M. A., Lehning, M., Matsuoka, N., Murton, J. B., Nötzli, J., Phillips, M., Ross, N., Seppälä, M., Springman, S. M., & Mühll, D. V. (2009). Permafrost and climate in Europe: Monitoring and modelling thermal, geomorphological and geotechnical responses. *Earth-Science Reviews, 92*(3), 117–171.

Hartley, A. J., Mather, A. E., Jolley, E., & Turner, P. (2005). Climatic controls on alluvial-fan activity, Coastal Cordillera, northern Chile. In A.M. Harvey, A.E., Mather, & M. Stokes (Eds.), *Alluvial fans: Geomorphology, sedimentology, dynamics* (pp. 251, 95–115). Geological Society of London, Special Publication.

Harvey, A. M. (1987). Patterns of quaternary aggradational and dissectional landform development in the Almeria region, southeast Spain: A dry-region, tectonically active landscape. *Die Erde, 118*, 193–215.

Harvey, A. M. (1989). The occurrence and role of arid zone alluvial fans. In D. S. G. Thomas (Ed.), *Arid zone geomorphology* (pp. 136–158). Belhaven.

Harvey, A. M., Wigand, P. E., & Wells, S. G. (1999a). Response of alluvial fan systems to the late Pleistocene to Holocene climatic transition: Contrasts between the margins of pluvial Lakes Lahontan and

Mojave, Nevada and California, USA. *CATENA, 36*, 255–281.
Harvey, A. M., & Renwick, W. (1987). Holocene alluvial fan and terrace formation in the Bowland Fells, northwest England. *Earth Surface Processes and Landforms, 12*, 249–257.
Harvey, A. M., Silva, P. G., Mather, A. E., Goy, J. L., Stokes, M., & Zazo, C. (1999b). The impact of Quaternary sea-level and climatic change on coastal alluvial fans in the Cabo de Gata ranges, southeast Spain. *Geomorphology, 28*(1), 1–22. https://doi.org/10.1016/S0169-555X(98)00100-7
Harvey, A. M. (2002). The role of base-level change in the dissection of alluvial fans: Case studies from southeast Spain and Nevada. *Geomorphology, 45*(1), 67–87.
Harvey, A. M. (2005). Differential effects of base-level, tectonic setting and climatic change on quaternary alluvial fans in the northern Great Basin, Nevada, USA. *Geological Society, London, Special Publications, 251*(1), 117–131. https://doi.org/10.1144/gsl.Sp.2005.251.01.09
Harvey, A. (2011). Dryland alluvial fans. In *Arid zone geomorphology: Process, form change drylands* (3rd Ed., pp. 333–371).
Harvey, A. M. (2018). *Alluvial fans, reference module in earth systems and environmental sciences*. Elsevier.
Hasbargen, L. E., & Paola, C. (2000). Landscape instability in an experimental drainage basin. *Geology, 28*(12), 1067–1070.
Hasbargen, L. E., & Paola, C. (2003). How predictable is local erosion rate in erosional landscapes? In P.R. Wilcock & R.M. Iverson (Eds.), *Prediction in geomorphology. Geophysical monograph* (pp. 231–240). American Geophysical Union, Washington, D.C.
Hastie, T., Tibshirani, R., & Friedman, J. (2009). *The elements of statistical learning: Data mining, inference, and prediction*. Springer.
Helsen, M. M., Koop, P. J. M., & Van Steijn, H. (2002). Magnitude–frequency relationship for debris flows on the fan of the Chalance torrent, Valgaudemar (French Alps). *Earth Surface Processes and Landforms, 27*(12), 1299–1307.
Hirschberg, J., Fatichi, S., Bennett, G. L., McArdell, B. W., Peleg, N., Lane, S. N., Schlunegger, F., & Molnar, P. (2021). Climate change impacts on sediment yield and debris-flow activity in an alpine catchment. *Journal of Geophysical Research: Earth Surface, 126*(1), e2020JF005739.
Hooke, R. L. (1967). Processes on arid-region alluvial fans. *The Journal of Geology, 75*(4), 438–460.
Hooke, R. L. (1968). Steady-state relationships of arid-region alluvial fans in closed basins. *American Journal of Science, 266*, 609–629.
Hooke, R. L., & Rohrer, W. L. (1977). Relative erodibility of source area rock types, as determined from second-order variations in alluvial-fan size. *Geological Society of America Bulletin, 88*, 117–182.
Hornung, J., Pflanz, D., Hechler, A., Beer, A., Hinderer, M., Maisch, M., & Bieg, U. (2010). 3-D architecture, depositional patterns and climate triggered sediment fluxes of an alpine alluvial fan (Samedan, Switzerland). *Geomorphology, 115*, 202–214.
Horton, R. E. (1932). Drainage-basin characteristics. *Eos Transactions of the American Geophysical Union*. https://doi.org/10.1029/TR013i001p00350/full
Horton, R. F. (1955). Erosional development of streams and their drainage basins, hydrophysical approach to quantitive morphology. *Journal of the Japanese Forest Society, 37*(2), 79–82.
Horton, B., & DeCelles, P. (2001). Modern and ancient fluvial megafans in the foreland basin system of the central Andes, southern Bolivia: Implications for drainage network evolution in fold-thrust belts. *Basin Research, 13*(1), 43–63.
Hubert, J. F., & Filipov, A. J. (1989). Debris-flow deposits in alluvial fans on the west flank of the White Mountains, Owens Valley, California, USA. *Sedimentary Geology, 61*(3–4), 177–205.
Huebl, J., & Fiebiger, G. (2005). Debris-flow mitigation measures. In *Debris-flow hazards and related phenomena* (pp. 445–487). Springer.
Hungr, O., Morgan, G. C., & Kellerhals, R. (1984). Quantitative analysis of debris torrent hazards for design of remedial measures. *Canadian Geotechnical Journal, 21*, 663–677.
Hungr, O., Evans, S. G., Bovis, M. J., & Hutchinson, J. N. (2001). A review of the classification of landslides of the flow type. *Environmental & Engineering Geoscience, 7*(3), 221–238.
Hungr, O., McDougall, S., & Bovis, M. (2005). Entrainment of material by debris flows. In M. Jakob & O. Hungr (Eds.), *Debris-flow hazards and related phenomena*. Springer Praxis.
Hürlimann, M., Abancó, C., Moya, J., & Vilajosana, I. (2014). Results and experiences gathered at the Rebaixader debris-flow monitoring site, Central Pyrenees, Spain. *Landslides, 11*(6), 939–953.
Huscroft, C. A., Lipovsky, P., Bond, J. D., Emond, D. S., & Lewis, L. L. (2003). Permafrost and landslide activity: Case studies from southwestern Yukon Territory. *Yukon Exploration and Geology*, 107–119.
Ilinca, V. (2021). Using morphometrics to distinguish between debris flow, debris flood and flood (Southern Carpathians, Romania). *CATENA, 197*, 104982.
Imaizumi, F., & Sidle, R. C. (2007). Linkage of sediment supply and transport processes in Miyagawa Dam catchment, Japan. *Journal of Geophysical Research, 112*, F03012. https://doi.org/10.1029/2006JF000495
Imaizumi, F., Sidle, R. C., & Kamei, R. (2008). Effects of forest harvesting on the occurrence of landslides and debris flows in steep terrain of central Japan. *Earth Surface Processes and Landforms, 33*, 827–840.
Imaizumi, F., Trappmann, D., Matsuoka, N., Tsuchiya, S., Ohsaka, O., & Stoffel, M. (2016). Biographical sketch of a giant: Deciphering recent debris-flow

dynamics from the Ohya landslide body (Japanese Alps). *Geomorphology, 272*, 102–114.

Imaizumi, F., Hayakawa, Y. S., Hotta, N., Tsunetaka, H., Ohsaka, O., & Tsuchiya, S. (2017). Relationship between the accumulation of sediment storage and debris-flow characteristics in a debris-flow initiation zone, Ohya landslide body, Japan. *Natural Hazards and Earth System Sciences, 17*(11), 1923–1938.

Innes, J. L. (1983). Lichenometric dating of debris-flow deposits in the Scottish Highlands. *Earth Surface Processes and Landforms, 8*(6), 579–588.

Innes, J. L. (1985). Magnitude-frequency relations of debris flows in northwest Europe. *Geografiska Annaler A, 67*, 23–32.

Innes, J. L. (1997). Historical debris-flow activity and climate in Scotland. In J.A. Matthews, D. Brunsden, B. Frenzel, & B. Gläser (Eds.), *Rapid mass movement as a source of climatic evidence* (pp. 19, 233–240). Paläoklimaroschung.

Iturrizaga, L. (2008). Paraglacial landform assemblages in the Hindukush and Karakoram Mountains. *Geomorphology, 95*(1–2), 27–47.

Iverson, R. M. (1997). The physics of debris flows. *Reviews of Geophysics, 35*(3), 245–296.

Iverson, R. M. (2003). The debris-flow rheology myth. In D. Rickenmann & C. L. Chen (Eds.), *Debris-flow hazards mitigation: Mechanics, prediction, and assessment* (pp. 303–314). Millpress.

Iverson, R. M. (2005). Debris-flow mechanics. In H. Jakob & O. Hungr (Eds.), *Debris-flow hazards and related phenomena* (pp. 105–134). Springer.

Iverson, R. M. (2012). Elementary theory of bed-sediment entrainment by debris flows. *Journal of Geophysical Research, 117*. https://doi.org/10.1029/2011JF002189

Iverson, R. M., & Ouyang, C. (2015). Entrainment of bed material by earth-surface mass flows: Review and reformulation of depth-integrated theory. *Reviews of Geophysics, 53*(1), 27–58.

Iverson, R. M., Schilling, S. P., & Vallance, J. W. (1998). Objective delineation of lahar-inundation hazard zones. *Geological Society of America Bulletin, 110*(8), 972–984.

Iverson, R. M., Logan, M., LaHusen, R. G., & Berti, M. (2010). The perfect debris flow? Aggregated results from 28 large-scale experiments. *Journal of Geophysical Research, 115*, F03005. https://doi.org/10.1029/2009JF001514

Jackson, L., Kostaschuk, R., & MacDonald, G. (1987). Identification of debris flow hazard on alluvial fans in the Canadian Rocky Mountains. In J.E. Costa & G.F. Wieckzorek (Eds.), *Debris flows/avalanches*: Geological Society of America, Reviews in Engineering Geology (Vol. 7, pp. 115–124).

Jakob, M., & Bovis, M. J. (1996). Morphometric and geotechnical controls of debris flow activity, southern Coast Mountains, British Columbia, Canada. *Zeitschrift fur geomorphologie, 104*, 13–26.

Jakob, M., Hungr, O., & Thomson, B. (1997). Two debris flows of anomalously high magnitude. In C.-I. Chen (Ed.), *Debris-flow hazards mitigation: Mechanics, prediction and assessment* (pp. 382–394). American Society of Civil Engineers.

Jakob, M., Bovis, M., & Oden, M. (2005). The significance of channel recharge rates for estimating debris-flow magnitude and frequency. *Earth Surface Processes and Landforms: THe Journal of the British Geomorphological Research Group, 30*(6), 755–766.

Jakob, M. (2005). Debris-flow hazard analysis. In *Debris-flow hazards and related phenomena* (pp. 412–443). Springer.

Jakob, M., & Friele, P. (2010). Frequency and magnitude of debris flows on Cheekye River, British Columbia. *Geomorphology, 114*(3), 382–395.

Jakob, M., Weatherley, H., Bale, S., Perkins, A., & MacDonald, B. (2017). A multi-faceted debris-flood hazard assessment for Cougar Creek, Alberta, Canada. *Hydrology, 4*(1).

Jakob, M., Mark, E., McDougall, S., Friele, P., Lau, C.-A., & Bale, S. (2020). Regional debris-flow and debris-flood frequency-magnitude relationships. *Earth Surface Processes and Landforms, 45*, 2954–2964.

Jakob, M. (2021). Debris-flow hazard assessments—A practitioner's view. *Environmental and Engineering Geoscience, 27*(2), 153–166. https://doi.org/10.2113/EEG-D-20-00110

Jakob, M., & Owen, T. (2021). Projected effects of climate change on shallow landslides, North Shore Mountains, Vancouver, Canada. *Geomorphology, 393*, 107921.

Jakob, M. (2022). Landslides in a changing climate. In *Landslide hazards, risks, and disasters* (pp. 505–579). Elsevier.

Jerolmack, D. J., & Mohrig, D. (2007). Conditions for branching in depositional rivers. *Geology, 35*(5), 463–466.

Johnson, A. M. (1984). Debris flow. In D. Brunsden & D. B. Prior (Eds.), *Slope instability* (pp. 257–361). Wiley.

Johnson, C., et al. (2012). Grain-size segregation and levee formation in geophysical mass flows. *Journal of Geophysical Research: Earth Surface,* (2003–2012) *117*(F1).

Johnson, P. A., McCuen, R. H., & Hromadga, T. (1991). Magnitude and frequency of debris flows. *Journal of Hydrology, 123*, 68–82.

Johnstone, S. A., Hudson, A. M., Nicovich, S., Ruleman, C. A., Sare, R. M., & Thompson, R. A. (2018). Establishing chronologies for alluvial-fan sequences with analysis of high-resolution topographic data: San Luis Valley, Colorado, USA. *Geosphere, 14*, 2487–2504.

Jomelli, V., Pech, V. P., Chochillon, C., & Brunstein, D. (2004). Geomorphic variations of debris flows and recent climatic change in the French Alps. *Climatic Change, 64*, 77–102.

Jomelli, V., Brunstein, D., Grancher, D., & Pech, P. (2007). Is the response of hill slope debris flows to recent climate change univocal? A case study in the Massif des Ecrins (French Alps). *Climatic Change, 85*, 119–137.

Kean, J. W., McCoy, S. W., Tucker, G. E., Staley, D. M., & Coe, J. A. (2013). Runoff-generated debris flows: Observations and modeling of surge initiation, magnitude, and frequency. *Journal of Geophysical Research Earth Surface, 118*(4), 2190–2207. https://doi.org/10.1002/jgrf.20148

Kean, J., McGuire, L., Rengers, F., Smith, J., & Staley, D. (2016). Amplification of postwildfire peak flow by debris. *Geophysical Research Letters, 43*, 8545–8553.

Kellerhals, R., & Church, M. (1990). Hazard management on fans, with examples from British Columbia. In A. H. Rachocki & M. Church (Eds.), *Alluvial fans: A field approach* (pp. 335–354). Wiley.

Kiefer, C., Oswald, P., Moernaut, J., Fabbri, S. C., Mayr, C., Strasser, M., & Krautblatter, M. (2021). A 4,000 year debris-flow record based on amphibious investigations of fan delta activity in Plansee (Austria, Eastern Alps). *Earth Surface Dynamics Discussions*, 1–27.

Kim, B. C., & Lowe, D. R. (2004). Depositional processes of the gravelly debris flow deposits, South Dolomite alluvial fan, Owens Valley, California. *Geosciences Journal, 8*(2), 153–170.

Kochel, R. C., & Johnson, R. A. (1984). Geomorphology and sedimentology of humid-temperate alluvial fans, central Virginia. In E.H. Koster & R.J. Steel (Eds.), *Sedimentology of gravels and conglomerates* (pp. 10, 109–122). Canadian Society of Petroleum Geologists Memoir.

Kocurek, G. (1988). First-order and super bounding surfaces in eolian sequences—Bounding surgfaces revisited. *Sedimentary Geology, 56*, 193–206.

Kohlbeck, F., Mojica, J., & Scheidegger, A. E. (1994). Clast orientation of the 1985 lahars of the Nevado del Ruiz, Colombia, and implications for depositional processes. *Sedimentary Geology, 88*, 175–183.

Kostaschuk, R. A., Macdonald, G. M., & Putnam, P. E. (1986). Depositional process and alluvial fan-drainage basin morphometric relationships near banff, Alberta, Canada. *Earth Surface Processes and Landforms, 11*(5), 471–484.

Kotarba, A. (1992). High-energy geomorphic events in the Polish Tatra Mountains. *Geografiska Annaler A, 74*, 123–131.

Krautblatter, M., Funk, D., & Günzel, F. K. (2013). Why permafrost rocks become unstable: A rock–ice-mechanical model in time and space. *Earth Surface Processes and Landforms, 38*(8), 876–887.

Lau, C.-A. (2017). *Channel scour on temperate alluvial fans in British Columbia*. Simon Fraser University M.Sc. Thesis, 200 pages.

Lau, C.-A., Jakob, M., Baumgard, A., Rios, A., Brayshaw, D., Brugman, M., Millard, T., & Giblin, L. (2021). Hydrogeomorphic effects of the November 2021 atmospheric rivers on infrastructure in southwestern British Columbia. *Innovative Geoscience for Tomorrow - Proceedings of Geohazards 8, Canadian Geotechnical Society*.

Lecce, S. A. (1991). Influence of lithologic erodibility on alluvial fan area, western White Mountains, California and Nevada. *Earth Surface Processes and Landforms, 16*, 11–18.

Leenman, A., & Eaton, B. (2021). Mechanisms for avulsion on alluvial fans: Insights from high-frequency topographic data. *Earth Surface Processes and Landforms, 46*(6), 1111–1127.

Leenman, A. S., Eaton, B. C., & MacKenzie, L. G. (2022). Floods on alluvial fans: implications for reworking rates, morphology and fan hazards. *Journal of Geophysical Research: Earth Surface, 127*(2), e2021JF006367.

Legget, R. F., Brown, R. E., & Johnston, G. H. (1966). Alluvial fan formation near Aklavik, Northwest Territories, Canada. *Geological Society of America Bulletin, 77*(1), 15–30.

Li, Y., Armitage, S. J., Stevens, T., & Meng, X. (2018). Alluvial fan aggradation/incision history of the eastern Tibetan plateau margin and implications for debris flow/debris-charged flood hazard. *Geomorphology, 318*, 203–216.

Liu, S., Hu, K., Zhang, Q., Zhang, S., Hu, X., & Tang, D. (2021). Quantitative analysis of the effects of an earthquake on rainfall thresholds for triggering debris-flow events. *Frontiers in Earth Science, 9*, 417.

Lorente, A., Begueria, S., Bathurst, J. C., & Garcia-Ruiz, J. M. (2003). Debris flow characteristics and relationships in the central Spanish Pyrenees. *Natural Hazards and Earth System Science, 3*, 683–692.

Lu, S.-Y., Cheng, J. D., & Brooks, K. N. (2001). Managing forests for watershed protection in Taiwan. *Forest Ecology and Management, 143*, 77–85.

Major, J. J. (1997). Depositional processes in large-scale debris-flow experiments. *Journal of Geology, 105*, 345–366.

Major, J. J. (1998). Pebble orientation on large, experimental debris-flow deposits. *Sedimentary Geology, 117*, 151–164.

Major, J. J. (2000). Gravity-driven consolidation of granular slurries—Implications for debris-flow deposition and deposit characteristics. *Journal of Sedimentary Research, 70*, 64–83.

Major, J. J., & Pierson, T. C. (1992). Debris flow rheology: Experimental analysis of fine-grained slurries. *Water Resources Research, 28*, 841–857.

Major, J. J., & Iverson, R. M. (1999). Debris-flow deposition: Effects of pore-fluid pressure and friction concentrated at flow margins. *Geological Society of America Bulletin, 111*(10), 1424–1434.

Major, J. J., Pierson, T. C., & Scott, K. M. (2005). Debris flows at Mount St. Helens, Washington, USA. In *Debris-flow hazards and related phenomena* (pp. 685–731). Springer.

Mao, L., Cavalli, M., Comiti, F., Marchi, L., Lenzi, M., & Arattano, M. (2009). Sediment transfer processes in two Alpine catchments of contrasting

morphological settings. *Journal of Hydrology, 364*(1–2), 88–98.
Marchi, L., Pasuto, A., & Tecca, P. (1993). Flow processes on alluvial fans in the Eastern Italian Alps. *Zeitschrift Für Geomorphologie, 37*(4), 447–458.
Marchi, L., & Brochot, S. (2000). Les cônes de déjection torrentiels dans les Alpes françaises. Morphométrie et processus de transport solide torrentiel. *Revue De Geographie Alpine, 88*, 23–38.
Marchi, L., Arattano, M., & Deganutti, A. M. (2002). Ten years of debris-flow monitoring in the Moscardo Torrent (Italian Alps). *Geomorphology, 46*, 1–17.
Marchi, L., & D'Agostino, V. (2004). Estimation of debris-flow magnitude in the Eastern Italian Alps. *Earth Surface Processes and Landforms, 29*(2), 207–220.
Maraga, F., Marchi, L., Mortara, G., & Moscariello, A. (1998). Colate detritiche torrentizie: Aspetti granulometrici e influenza sul territorio. *Memorie Della Societá Geologica Italiana, 53*, 75–96.
Mather, A. E., Harvey, A. M., & Stokes, M. (2000). Quantifying longterm catchment changes of alluvial fan systems. *Geological Society of America Bulletin, 112*, 1825–1833.
Mather, A. E., & Stokes, M. (2018). Bedrock structural control on catchment-scale connectivity and alluvial fan processes, High Atlas Mountains, Morocco. In D. Ventra & L.E. Clarke (Eds.), *Geology and geomorphology of alluvial and fluvial fans: Terrestrial and planetary perspectives* (pp. 440, 103–128). Geological Society of London, Special Publication.
Matmon, A., Nichols, K., & Finkel, R. (2006). Isotopic insights into smoothening of abandoned fan surfaces, southern California. *Quaternary Research, 66*, 109–118.
Matthews, J. A., Shakesby, R. A., McEwen, L. J., Berrisford, M. S., Owen, G., & Bevan, P. (1999). Alpine debris-flow in Leirdalen, Joutnheimen, Norway, with particular reference to distal fans, intermediate-type deposits, and flow types. *Arctic, Antarctic, and Alpine Research, 31*, 421–435.
McDougall, S. (2017a). 2014 Canadian geotechnical colloquium: Landslide runout analysis—Current practice and challenges. *Canadian Geotechnical Journal, 54*(5), 605–620.
McFadden, L. D., Ritter, J. B., & Wells, S. G. (1989a). Use of multiparameter relative-age methods for age estimation and correlation of alluvial fan surfaces on a desert piedmont, eastern Mojave Desert, California. *Quaternary Research, 32*, 276–290.
Meek, S. R., Carrapa, B., & DeCelles, P. G. (2020). Recognizing allogenic controls on the stratigraphic architecture of ancient alluvial fans in the western US. *Frontiers in Earth Science*. https://doi.org/10.3389/feart.2020.00215
Melton, M. A. (1957). An analysis of the relations among elements of climate, surface properties, and geomorphology; office of naval research technical report no. 11." Office of naval research, Technical Report: 99.
Melton, M. A. (1965). The geomorphic and paleoclimatic significance of alluvial deposits in southern Arizona. *Journal of Geology, 73*, 1–38.
McArdell, B. W., & Sartori, M. (2021). The illgraben torrent system. In *Landscapes and landforms of Switzerland* (pp. 367–378). Springer.
McCoy, S. W., Kean, J. W., Coe, J. A., Tucker, G. E., Staley, D. M., & Wasklewicz, T. A. (2012). Sediment entrainment by debris flows: In situ measurements from the headwaters of a steep catchment. *Journal of Geophysical Research, 117*, F03016. https://doi.org/10.1029/2011JF002278
McCoy, S. W. (2015). Research focus: Infrequent, large-magnitude debris flows are important agents of landscape change. *Geology, 43*(5), 463–464.
McDonald, E. V., McFadden, L. D., & Wells, S. G. (2003). Regional response of alluvial fans to the Pleistocene-Holocene climatic transition, Mojave Desert, California. In Y. Enzel, S. G. Wells, & N. Lancaster (Eds.), *Paleoenvironments and paleohydrology of the Mojave and Southern Great Basin Deserts* (pp. 189–205). Geological Society of America.
McDougall, S. (2017). 2014 Canadian geotechnical colloquium: Landslide runout analysis—Current practice and challenges. *Canadian Geotechnical Journal, 54*, 605–620.
McGuire, L. A., Rengers, F. K., Oakley, N., Kean, J. W., Staley, D. M., Tang, H., Youberg, A. M., et al. (2021). Time since burning and rainfall characteristics impact post-fire debris-flow initiation and magnitude. *Environmental & Engineering Geoscience, 27*(1), 43–56.
McFadden, L. D., Ritter, J. B., & Wells, S. G. (1989b). Use of multiparameter relative-age methods for age estimation and correlation of alluvial fan surfaces on a desert piedmont, eastern Mojave Desert, California. *Quaternary Research, 32*(3), 276–290.
Meyer, G. A., & Wells, S. G. (1997). Fire-related sedimentation events on alluvial fans, Yellowstone National Park, U.S.A. *Journal of Sedimentary Research, 67*, 776–791.
Miall, A. D. (1985). Architectural-element analysis: A new method of facies analysis applied to fluvial deposits. *Earth-Science Reviews, 22*, 261–308.
Miall, A. D. (2015). Updating uniformitarianism: Stratigraphy as just a set of 'frozen accidents'. In D.G. Smith, R.J. Bailey, P.M. Burgess, & A.J. Fraser (Eds.), *Geological society of London* (pp. 404, 11–36). Special Publication.
Michelini, T., Bettella, F., & D'Agostino, V. (2017). Field investigations of the interaction between debris flows and forest vegetation in two Alpine fans. *Geomorphology, 279*, 150–164.
Milana, J. P. (2000). Characterization of alluvial bajada facies distribution using TM imagery. *Sedimentology, 47*, 741–760.
Millard, T., Wilford, D. J., & Oden, M. E. (2006). Coastal fan destabilization and forest management. Forest Research Technical Report, Coast Forest

Region. Forest Service of British Columbia Technical Report 034.
Miller, V. C. (1953). *A quantitative geomorphic study of drainage basin characteristics in the ...-Victor C.* Miller-Google Books.
Mohrig, D., Heller, P. L., Paola, C., & Lyons, W. J. (2000). Interpreting avulsion process from ancient alluvial sequences: Guadalope-Matarranya (northern Spain) and Wasatch Formation (western Colorado). *Geological Society of America Bulletin, 112*, 1787–1803.
Mohseni, N., Hosseinzadeh, S. R., Sepehr, A., Golzarian, M. R., & Shabani, F. (2017). Variations in spatial patterns of soil-vegetation properties and the emergence of multiple resilience thresholds within different debris flow fan positions. *Geomorphology, 290*, 365–375.
Moscariello, A. (2005). Exploration potential of the mature Southern North Sea basin margins: Some unconventional plays based on alluvial fan and fluvial fan sedimentation models. In A.G. Doré & B.A. Vining (Eds.), *Petroleum Geology: North-West Europe and Global Perspectives – Proceedings of the 6th Petroleum Geology Conference* (pp. 595–605). Geological Society of London.
Moscariello, A., Marchi, L., Maraga, F., & Mortara, G. (2002). Alluvial fans in the Italian Alps: Sedimentary facies and processes. In I.P. Martini, V.R. Baker, & G. Garzón (Eds.), *Flood and megaflood processes and deposits: Recent and ancient examples* (pp. 32, 141–166). IAS Special Publication.
Muñoz-Sabater, J., Dutra, E., Agustí-Panareda, A., Albergel, C., Arduini, G., Balsamo, G., Thépaut, J. N., et al., (2021). ERA5-Land: A state-of-the-art global reanalysis dataset for land applications. *Earth System Science Data Discussions*, 1–50.
Muñoz-Salinas, E., Castillo, M., Franco-Ramos, O., Arce, J. L., Sanderson, D. C. W., & Cresswell, A. C. (2020). Assessing paraglacial processes at Nexpayantla Gorge (Popocatépetl volcano, Central Mexico) using OSL and 14C. *Earth Surface Processes and Landforms, 45*(11), 2450–2462.
Nemec, W., & Steel, R. J. (1984). Alluvial and coastal conglomerates: Their significant features and some comments on gravelly mass-flow deposits. In E.H. Koster & R.J. Steel (Eds.), *Sedimentology of gravel and conglomerates* (pp. 10, 1–31). Canadian Society of Petroleum Geologists Memoir.
Nishiguchi, Y., & Uchida, T. (2022). Long-runout-landslide-induced debris flow: The role of fine sediment deposition processes in debris flow propagation. *Journal of Geophysical Research: Earth Surface, 127*, e202JF006452.
Nott, J. F., Thomas, M. F., & Price, D. M. (2001). Alluvial fans, landslides and late quaternary climatic change in the wet tropics of northeast Queensland. *Australian Journal of Earth Sciences, 48*, 875–882.
Nyman, P., Sheridan, G. J., Smith, H. G., & Lane, P. N. (2011). Evidence of debris flow occurrence after wildfire in upland catchments of south-east Australia. *Geomorphology, 125*(3), 383–401.
Nymann, P., Smith, H. G., Sherwin, C. B., Langhans, C., Lane, P. N. J., & Sheridan, G. J. (2015). Predicting sediment delivery from debris flows after wildfire. *Geomorphology, 250*, 173–186.
O'Brien, J. S., & Julien, P.Y. (1985). Physical properties and mechanics of hyperconcentrated sediment flows. In D.S. Bowles (Ed.), *Delineation of Landslide, Flash Flood, and Debris Flow Hazards in Utah. Proceedings of Specialty Conference*, Utah State University, 1984 (pp. 260–279).
O'Brien, J. S., & Julien, P. Y. (1988). Laboratory analysis of mudflow properties. *Journal of Hydraulic Engineering, 114*, 877–887.
Owen, L. A., Clemmens, S. J., Finkel, R. C., & Gray, H. (2014). Late Quaternary alluvial fans at the eastern end of the San Bernardino Mountains, southern California. *Quaternary Science Reviews, 87*, 114–134.
Palladino, M. R., Viero, A., Turconi, L., Brunetti, M. T., Peruccacci, S., Melillo, M., Luino, F., Deganutti, A. M., & Guzzetti, F. (2018). Rainfall thresholds for the activation of shallow landslides in the Italian Alps: The role of environmental conditioning factors. *Geomorphology, 303*, 53–67.
Paola, C., Ganti, V., Mohrig, D., Runkel, A. C., & Straub, K. M. (2018). Time not our time: Physical controls on the preservation and measurement of geologic time. *Annual Review of Earth and Planetary Sciences, 46*, 409–438.
Palucis, M. C., Ulizio, T. P., & Lamb, M. P. (2021). Debris flow initiation from ravel-filled channel bed failure following wildfire in a bedrock landscape with limited sediment supply. *GSA Bulletin, 133*, 2079–2096.
Pavlova, I., Jomelli, V., Brunstein, D., Grancher, D., Martin, E., & Déqué, M. (2014). Debris flow activity related to recent climate conditions in the French Alps: A regional investigation. *Geomorphology, 219*, 248–259.
Pederson, C. A., Santi, P. M., & Pyles, D. R. (2015). Relating the compensational stacking of debris-flow fans to characteristics of their underlying stratigraphy: Implications for geologic hazard assessment and mitigation. *Geomorphology, 248*, 47–56.
Pelletier, J. D., & Cook, J. P. (2005). Deposition of playa windblown dust over geologic time scales. *Geology, 33*(11), 909–912.
Pelletier, J. D. (2007). Cantor set model of eolian dust deposits on desert alluvial fan terraces. *Geology, 35*(5), 439–442.
Phillips, C. J., & Davies, T. R. H. (1991). Determining rheological parameters of debris flow material. *Geomorphology, 4*, 101–110.
Phillips, E. (2006). Micromorphology of a debris flow deposit: Evidence of basal shearing, hydrofracturing, liquefaction and rotational deformation during

emplacement. *Quaternary Science Reviews, 25*, 720–738.

Picanço, J. L., Tanaka, H. S., Mesquita, M. J., Costa, V. V., Luiz, E. F. O., Lopes, A. B. B., Piment, V. et al. (2016). Debris flow hazard zonation in Serra da Prata Range, Paraná State, Brazil: Watershed morphometric Constrain. Landslide and engineered slopes. Experience, theory and practice. In *Proceedings of the 12th International Symposium on Landslides*.

Pierce, J. L., Meyer, G. A., & Timothy Jull, A. J. (2004). Fire-induced erosion and millennial-scale climate change in northern ponderosa pine forests. *Nature, 432*(7013), 87–90.

Pierson, T. C. (1985). Effects of slurry composition on debris flow dynamics, Rudd Canyon, Utah. In D.S. Bowles (Ed.), *Delineation of landslide, flash flood, and debris flow hazards in Utah. Proceedings of Specialty Conference Utah State University*, 1984 (pp. 132–152).

Pierson, T. C. (1986). Flow behavior of channelized debris flows, Mount St. Helens, Washington. In A.D. Abrahams (Ed.), *Hillslope processes* (pp. 269–296). Allen & Unwin, Hempstead.

Pierson, T. C. (2005). Hyperconcentrated flow. In M. Jakob & O. Hungr (Eds.), *Debris-flow hazards and related phenomena* (pp. 159–202). Springer.

Pierson, T. C., & Scott, K. M. (1985). Downstream dilution of a lahar: Transition from debris flow to hyperconcentrated streamflow. *Water Resources Research, 21*, 1511–1524.

Pierson, T. C., & Costa, J. E. (1987). A rheologicla classification of subaerial sediment and water flows. In J.E. Costa & G.F. Wieczorek (Eds.), *Debris flows/avalanches*, Geological Society of America Reviews in Engineering Geology (pp. 7, 1–12).

Procter, J., Zernack, A., Mead, S., Morgan, M., & Cronin, S. (2021). A review of lahars; past deposits, historic events and present-day simulations from Mt. Ruapehu and Mt. Taranaki, New Zealand. *New Zealand Journal of Geology and Geophysics, 64*(2–3), 479–503.

Pudasaini, S. P. (2012). A general two-phase debris flow model. *Journal of Geophysical Research, 117*, F03010.

Pudasaini, S. P., & Fischer, J.-T. (2020). A mechanical model for phase separation in debris flow. *International Journal of Multiphase Flow, 129*, 103292.

Pudasaini, S. P., & Krautblatter, M. (2021). The mechanics of landslide mobility with erosion. *Nature Communications, 12*(1), 1–15.

Qing, F., Zhao, Y., Meng, X., Su, X., Qi, T., & Yue, D. (2020). Application of machine learning to debris flow susceptibility mapping along the China-Pakistan Karakoram highway. *Remote Sensing, 12*(18), 2933.

Qiu, C., Su, L., Zou, Q., & Geng, X. (2021). A hybrid machine-learning model to map glacier-related debris flow susceptibility along Gyirong Zangbo watershed under the changing climate. *Science of the Total Environment*, 151752.

Quade, J. (2001). Desert pavements and associated rock varnish in the Mojave Desert: How old can they be? *Geology, 29*, 855–858.

Remaître, A., & Malet, J. P. (2010). The effectiveness of torrent check dams to control channel instability: example of debris-flow events in clay shales. Check dams, morphological adjustements and erosion control in torrential streams, edited by: Garcia, CC and Lenzi, MA, Nova Science Publishers Inc., New York, 211–237.

Reitz, M. D., & Jerolmack, D. J. (2012). Experimental alluvial fan evolution: Channel dynamics, slope controls, and shoreline growth. *Journal of Geophysical Research. Earth Surface, 117*(F2), F02021.

Rickenmann, D. (1999). Empirical relationships for debris flows. *Natural Hazards, 19*(1), 47–77.

Rickenmann, D., & Jakob, M. (2015). Erosion and sediment flux in mountain watersheds. *The High Mountain Cryosphere: Environmental Changes and Human Risks*, 166–183.

Ritter, J., Miller, J., Enzel, Y., Howes, S., Nadon, G., Grubb, M., Hoover, K., Olsen, T., Reneau, S., Sack, D., Summa, C., Taylor, I., Touysinhthiphonexay, K., Yodis, E., Schneider, N., Ritter, D., & Wells, S. (1993). Quaternary evolution of Cedar Creek alluvial fan, Montana. *Geomorphology, 8*(4), 287–304.

Roberts, R. G., & Church, M. (1986). The sediment budget in severely disturbed watersheds, Queen Charlotte Ranges, British Columbia. *Canadian Journal of Forestry Research, 16*, 1092–1106.

Rebetez, M., Lugon, R., & Baeriswyl, P.-A. (1997). Climatic change and debris flows in high mountain regions: The case study of the Ritigraben torrent (Swiss Alps). *Climate Change, 36*, 371–389.

Roberts, M. C., Niller, H.-P., & Helmstetter, N. (2003). Sedimentary architecture and radar facies of a fan delta, Cypress Creek, West Vancouver, British Columbia. In C.S. Bristow & H.M. Jol (Eds.), *Ground penetrating radar in sediments* (pp. 211, 111–126). Geological Society of London Special Publication.

Roelofs, L., Colucci, P., & de Haas, T. (2022). How debris-flow composition affects bed erosion quantity and mechanisms: An experimental assessment. *Earth Surface Processes and Landforms, 47*(8), 2151–2169.

Ryder, J. (1971). Some aspects of the morphometry of paraglacial alluvial fans in South-central British Columbia. *Canadian Journal of Earth Sciences, 8*, 1252–1264.

Ryder, J. (1972). The stratigraphy and morphology of para-glacial alluvial fans in South-central British Columbia. *Canadian Journal of Earth Sciences, 8*, 279–298.

Rowbotham, D., De Scally, F., & Louis, J. (2005). The identification of debris torrent basins using morphometric measures derived within a GIS. *Geografiska Annaler: Series A, 87*(4), 527–537.

Sanchez-Nuñez, J. M., Macias, J. L., Saucedo, R., Zamorano, J. J., Novelo, D., Mendoza, M. E., & Torres-Hernández, J. R. (2015). Geomorphology, internal structure and evolution of alluvial fans at Motozintla, Chiapas, Mexico. *Geomorphology, 230*, 1–12.

Santi, P., deWolfe, V. G., Higgins, J. D., Cannon, S., & Gartner, J. (2008). Sources of debris flow material in burned areas. *Geomorphology, 96*, 310–321.

Santi, P., & Morandi (2013). Comparison of debris-flow volumes from burned and unburned areas. *Landslides, 10*, 757–769. https://doi.org/10.1007/s10346-012-0354-4

Santi, P. M., Pyles, D. R., & Pederson, C. A. (2017). Debris flow avulsion. *International Journal of Erosion Control Engineering, 10*, 67–73.

Sattler, K., Keiler, M., Zischg, A., & Schrott, L. (2011). On the connection between debris flow activity and permafrost degradation: A case study from the Schnalstal, South Tyrolean Alps, Italy. *Permafrost and Periglacial Processes, 22*(3), 254–265.

Savi, S., Norton, K. P., Picotti, V., Akçar, N., Delunel, R., Brardinoni, F., Schlunegger, F., et al. (2014). Quantifying sediment supply at the end of the last glaciation: Dynamic reconstruction of an alpine debris-flow fan. *Bulletin, 126*(5–6), 773–790.

Schaefer, L. N., Santi, P. M., & Duron, T. C. (2021). Debris flow behavior during the September 2013 rainstorm event in the Colorado Front Range, USA. *Landslides, 18*(5), 1585–1595.

Scheuner, T., Keusen, H. R., Mcardell, B. W., & Huggel, C. (2009). Murgangmodellierung mit dynamisch-physikalischem und GIS-basiertem Fliessmodell. *Wasser Energie Luft, 101*, 15–21.

Scheinert, C., Wasklewicz, T., & Staley, D. (2012). Alluvial fan dynamics – Revisiting the field. *Geography Compass, 6*, 752–775.

Schneuwly-Bollschweiler, M., Stoffel, M., & Rudolf-Miklau, F. (2013). *Dating torrential processes on fans and cones* (p. 456). Springer Science.

Schoch-Baumann, A., Blöthe, J. H., Munack, H., Hornung, J., Codilean, A. T., Fülöp, R. H., Wilcken, K., & Schrott, L. (2022). Postglacial outsize fan formation in the Upper Rhone valley, Switzerland – Gradual or catastrophic? *Earth Surface Processes and Landforms, 47*, 1032–1053.

Shultz, A. W. (1984). Subaerial debris-flow deposition in the Upper Paleozoic Cutler Formation, western Colorado. *Journal of Sedimentary Petrology, 54*, 759–772.

Schumm, S. A. (1956a). *Evolution of drainage systems and slopes in badlands at Perth Amboy*. Geol Soc Am Bull.

Schumm, S. A. (1956b). Evolution of drainage systems and slopes in badlands at Perth Amboy, New Jersey. *Bulletin of the Geological Society of America, 67*(5), 597–646.

Schumm, S., Mosley, M., & Weaver, W. (1987). *Experimental fluvial geomorphology*. Wiley.

Schürch, P., Densmore, A. L., Rosser, N. J., & McArdell, B. W. (2011). Dynamic controls on erosion and deposition on debris-flow fans. *Geology, 39*(9), 827–830.

Schürch, P., Densmore, A. L., Ivy-Ochs, S., Rosser, N. J., Kober, F., Schlunegger, F., McArdell, B., & Alfimov, V. (2016). Quantitative reconstruction of Late Holocene surface evolution on an alpine debris-flow fan. *Geomorphology, 275*, 46–57.

Scott, K. M. (1971). Origin and sedimentology of 1969 debris flows near Glendora, California. *U.S. Geological Survey Professional Paper* 422-M (pp. 242–247).

Scott, K. M., Maciás, J. L., Naranjo, J. A., Rodríguez, S., & McGeehin, J. P. (2001). Catastrophic debris flows transformed from landslides in volcanic terrains: Mobility, hazard assessment, and mitigation strategies. *U.S. Geological Survey Professional Paper* 1630 (59 pp.).

Sewell, R. J., Parry, S., Millis, S. W., Wang, N., Rieser, U., & DeWitt, R. (2015). Dating of debris flow fan complexes from Lantau Island, Hong Kong, China: The potential relationship between landslide activity and climate change. *Geomorphology, 248*, 205–227.

Sharp, R. P. (1942). Mudflow levees. *Journal of Geomorphology, 5*, 222–227.

Sharp, R. P., & Nobles, L. H. (1953). Mudflow of 1941 at Wrightwood, southern California. *Geological Society of America Bulletin, 64*, 547–560.

Shieh, C. L., Chen, Y. S., Tsai, Y. J., & Wu, J. H. (2009). Variability in rainfall threshold for debris flow after the Chi-Chi earthquake in central Taiwan, China. *International Journal of Sediment Research, 24*(2), 177–188.

Siewert, M. B., Krautblatter, M., Christiansen, H. H., & Eckerstorfer, M. (2012). Arctic rockwall retreat rates estimated using laboratory-calibrated ERT measurements of talus cones in Longyeardalen, Svalbard. *Earth Surface Processes and Landforms, 37*(14), 1542–1555.

Sletten, K., & Blikra, L. H. (2007). Holocene colluvial (debris-flow and water-flow) processes in eastern Norway: Stratigraphy, chronology and palaeoenvironmental implications. *Journal of Quaternary Science, 22*, 619–635.

Slingerland, R., & Smith, N. D. (2004). River avulsions and their deposits. *Annual Reviews of Earth and Planetary Sciences, 32*, 257–285.

Smart, J. S., & Surkan, A. J. (1967). The relation between mainstream length and area in drainage basins. *Water Resources Research, 3*(4), 963–974.

Smith, K. G. (1950). Standards for grading texture of erosional topography. *American Journal of Science, 248*(9), 655–668.

Sohn, Y. K., Rhee, C. W., & Kim, B. C. (1999). Debris flow and hyperconcentrated flood-flow deposits in an alluvial fan, northwestern part of the Cretaceous Yongdong Basin, central Korea. *Journal of Geology, 107*, 111–132.

Sohn, M. F., Mahan, S. A., Knott, J. R., & Bowman, D. D. (2007). Luminescence ages for alluvial-fan deposits in Southern Death Valley: Implications for climate-driven sedimentation along a tectonically active mountain front. *Quaternary International, 166*, 49–60.

Sorriso-Valvo, M., Antronico, L., & Pera, E. L. (1998). Controls on modern fan morphology in Calabria Southern Italy. *Geomorphology, 24*(2–3), 169–187.

Staley, D. M., Negri, J. A., Kean, J. W., Laber, J. L., Tillery, A. C., & Youberg, A. M. (2017). Prediction of spatially explicit rainfall intensity–duration thresholds for post-fire debris-flow generation in the western United States. *Geomorphology, 278*, 149–162.

Steffen, D., Schlunegger, F., & Preusser, F. (2010). Late Pleistocene fans and terraces in the Majes Valley, southern Peru, and their relation to climatic variations. *International Journal of Earth Science, 99*, 1975–1989.

Stock, J., & Dietrich, W. E. (2003). Valley incision by debris flows: Evidence of a topographic signature. *Water Resources Research, 39*(4).

Stock, J. D., & Dietrich, W. E. (2006). Erosion of steepland valleys by debris flows. *Geological Society of America Bulletin, 118*(9–10), 1125–1148.

Stoffel, M., Conus, D., Grichting, M. A., Lièvre, I., & Maître, G. (2008a). Unraveling the patterns of late holocene debris-flow activity on a cone in the Swiss Alps: Chronology, environment and implications for the future. *Global and Planetary Change, 60*(3), 222–234.

Stoffel, M., Bollschweiler, M., Leutwiler, A., & Aeby, P. (2008b). Tree-ring reconstruction of debris-flow events leading to overbank sedimentation on the Illgraben Cone (Valais Alps, Switzerland). *The Open Geology Journal, 2*, 18–29.

Stoffel, M., & Huggel, C. (2012). Effects of climate change on mass movements in mountain environments. *Progress in Physical Geography, 36*(3), 421–439. https://doi.org/10.1177/0309133312441010

Stoffel, M., Mendlik, T., Schneuwly-Bollschweiler, M., & Gobiet, A. (2014). Possible impacts of climate change on debris-flow activity in the Swiss Alps. *Climatic Change, 122*(1), 141–155.

Stokes, M., & Mather, A. E. (2000). Response of Plio-Pleistocene alluvial systems to tectonically induced base-level changes, Vera Basin, SE Spain. *Journal of the Geological Society of London, 157*, 303–316.

Stokes, M., Nash, D. J., & Harvey, A. M. (2007). Calcrete 'fossilisation' of alluvial fans in SE Spain: The roles of groundwater, pedogenic processes and fan dynamics in calcrete development. *Geomorphology, 85*, 63–84.

Stokes, M., & Gomes, A. (2020). Alluvial fans on volcanic islands: Amorphometric perspective (São Vicente, Cape Verde). *Geomorphology*, 107356. https://doi.org/10.1016/j.geomorph.2020.107356

Stolle, A., Langer, M., Blöthe, J. H., & Korup, O. (2015). On predicting debris flows in arid mountain belts. *Global and Planetary Change, 126*, 1–13.

Strahler, A. N. (1952). Hypsometric (area-altitude) analysis of erosional topography. *Bulletin Geological Society of America, 63*(11), 1117–1142.

Strahler, A. N. (1964a). Part I.I Quantitative geomorphology of drainage basins and channel networks. In *Handbook of applied hydrology* (pp. 4–39). McGraw-Hill.

Strahler, A. N. (1964b). Quantitative geomorphology of drainage basin and channel networks. In *Handbook of applied hydrology*. https://ci.nii.ac.jp/naid/10021.22978.9/

Straub, K. M., Paola, C., Mohrig, D., Wolinsky, M., & George, R. (2009). Compensational stacking of channelized sedimentary deposits. *Journal of Sedimentary Research, 79*, 673–688.

Straub, K. M., & Pyles, D. R. (2012). Quantifying the hierarchical organization of compensation in submarine fans using surface statistics. *Journal of Sedimentary Research, 82*, 889–898.

Strunk, H. (1992). Reconstructing debris flow frequency in the southern Alps back to AD 1500 using dendrogeomorphological analysis. *Erosion, Debris Flows and Environment in Mountain Regions. IAHS Publication, 209*, 299–306.

Sun, J., Qin, S., Qiao, S., Chen, Y., Su, G., Cheng, Q., Guo, X., et al. (2021). Exploring the impact of introducing a physical model into statistical methods on the evaluation of regional scale debris flow susceptibility. *Natural Hazards, 106*(1), 881–912.

Suwa, H. (1988). Focusing mechanism of large boulders to a debris-flow front. *Transactions of the Japanese Geomorphological Union, 9*, 151–178.

Suwa, H., & Okuda, S. (1983). Deposition of debris flows on a fan surface, Mt. Yakedake, Japan. *Zeitschrift fur Geomorphologie NF Supplementband, 46*, 79Ű-101.

Suwa, H., Okano, K., & Kanno, T. (2009). Behavior of debris flows monitored on test slopes of Kamikamihorizawa Creek, Mount Yakedake, Japan. *International Journal of Erosion Control Engineering, 2*(2), 33–45.

Suwa, H., Okano, K., & Kanno, T. (2011). Forty years of debris flow monitoring at Kamikamihorizawa Creek, Mount Yakedake, Japan. In *5th International Conference on Debris-Flow Hazards Mitigation: Mechanics, Prediction and Assessment*. Casa Editrice UniversitaLa Sapienza, Roma (pp. 605–613).

Tannant, D. D., & Skermer, N. (2013). Mud and debris flows and associated earth dam failures in the Okanagan region of British Columbia. *Canadian Geotechnical Journal, 50*(8), 820–833.

Theule, J., Frédéric, L., & Dominique, L. (2015). Spatial variability of channel erosion by debris-flows (field observations in the French Alps). In G. Lollino, M. Arattano, M. Rinaldi, O. Giustolisi, J.-C. Marechal, & G.E. Grant (Eds.), *Engineering geology for society and territory*. Springer Praxis.

Thomas, M. F. (2004). Landscape sensitivity to rapid environmental change—A quaternary perspective

with examples from tropical areas. *CATENA, 55*, 107–124.
Thomas, M. A., Rengers, F. K., Kean, J. W., McGuire, L. A., Staley, D. M., Barnhart, K. R., & Ebel, B. A. (2021). Postwildfire soil-hydraulic recovery and the persistence of debris flow hazards. *Journal of Geophysical Research: Earth Surface*. https://doi.org/10.1029/2021JF006091
Thornthwaite, C. W., & Mather, J. R. (1955). *The water balance. Publications in climatology*. Laboratory of Climatology, Drexel Institute of Technology.
Tomczyk, A. M. (2021). Morphometry and morphology of fan-shaped landforms in the high-Arctic settings of central Spitsbergen, Svalbard. *Geomorphology, 392*, 107899.
Tobler, D., Kull, I., Jacquemart, M., & Haehlen, N. (2014). Landslide science for a safer geoenvironment: Volume 3: Targeted landslides. In K. Sassa, P. Canuti, & Y. Yin (Eds.), *Landslide science for a safer geoenvironment: Volume 3: Targeted landslides* (pp. 25–30). Springer.
Tomkins, K. M., Humphreys, G. S., Gero, A. F., Shakesby, R. A., Doerr, S. H., Wallbrink, P. J., & Blake, W. H. (2008). Postwildfire hydrological response in an El Niño-Southern Oscillation-dominated environment. *Journal of Geophysical Research Earth Surface, 113*. https://doi.org/10.1029/2007jf00853,F02023
Tsunetaka, H., Hotta, N., Sakai, Y., & Wasklewicz, T. (2022). Effect of debris-flow sediment grain-size distribution on fan morphology. *Earth Surface Dynamics, 10*, 775–796.
Turkington, T., Remaître, A., Ettema, J., Hussin, H., & van Westen, C. (2016). Assessing debris flow activity in a changing climate. *Climatic Change, 137*(1), 293–305.
Valkanou, K., Karymbalis, E., Papanastassiou, D., Gaki-Papanastassiou, K., & Giles, P. (2013). Analysis of relationships among coastal alluvial fans and their contributing catchments in North Evoikos Gulf (Central Greece). *Bulletin of the Geological Society of Greece, 47*(1). https://doi.org/10.12681/bgsg.10961
Vallance, J. W., & Scott, K. M. (1997). The Osceola mudflow from mount reainier: Sedimentology and hazard implications of a huge clay-rich debris flow. *Geological Society of America Bulletin, 109*(2), 143–163.
VanDine, D. (1985). Debris flows and debris torrents in the southern Canadian Cordillera. *Canadian Geotechnical Journal, 22*(1), 44–68.
Van Steijn, H., & Coutard, J. P. (1989). Laboratory experiments with small debris flows: Physical properties related to sedimentary characteristics. *Earth Surface Processes and Landforms, 14*, 587–596.
Van Westen, J. C., & Daag, A. S. (2005). Analysing the relation between rainfall and lahar activity at Mount Pinatubo, Philippines. *Earth Surface Processes and Landforms, 30*, 1663–1674.
Ventra, D., & Clarke, L. E. (2018). Geology and geomorphology of alluvial and fluvial fans: Current progress and research perspectives. Geological SocietyIn D. Ventra & L. E. Clarke (Eds.), *Geology and geomorphology of alluvial and fluvial fans: Terrestrial and planetary perspectives* (Vol. 440, pp. 1–22). London.
Ventra, D., Chong Díaz, G., & De Boer, P. L. (2013). Colluvial sedimentation in a hyperarid setting (Atacama Desert, northern Chile): Geomorphic controls and stratigraphic facies variability. *Sedimentology, 60*, 1257–1290.
Ventra, D., Abels, H. A., Hilgen, F. J., & DeBoer, P. L. (2018). Orbital-climate control of mass-flow sedimentation in a Miocene alluvial-fan succession (Teruel Basin, Spain). In D. Ventra & L.E. Clarke (Eds.), *Geology and geomorphology of alluvial and fluvial fans: Terrestrial and planetary perspectives* (pp. 440, 129–157). Geological Society of London, Special Publication.
Viseras, C., & Fernández, J. (1995). The role of erosion and deposition in the construction of alluvial fan sequences in the Guadix Formation (SE Spain). *Geologie En Mijnbouw, 74*, 21–33.
Volker, H. X., Wasklewicz, T. A., & Ellis, M. A. (2007). A topographic fingerprint to distinguish alluvial fan formative processes. *Geomorphology, 88*(1–2), 34–45.
Walk, J., Bartz, M., Stauch, G., Binnie, A., Brückner, H., & Lehmkuhl, F. (2022). Weathering under coastal hyperaridity–Late Quaternary development of spectral, textural, and gravelometric alluvial fan surface characteristics. *Quaternary Science Reviews, 277*, 107339.
Wang, S., Meng, X., Chen, G., Guo, P., Xiong, M., & Zeng, R. (2017). Effects of vegetation on debris flow mitigation: A case study from Gansu Province, China. *Geomorphology, 282*, 64–73.
Wang, N., Cheng, W., Zhao, M., Liu, Q., & Wang, J. (2019). Identification of the debris flow process types within catchments of Beijing Mountainous Area. *Water, 11*(4), 638.
Warke, P. (2013). Weathering in arid regions. *Treatise on Geomorphology (Elsevier), 4*, 197–227.
Wasklewicz, T., & Scheinert, C. (2016). Development and maintenance of a telescoping debris flow fan in response to human-induced fan surface channelization, Chalk Creek Valley natural debris flow laboratory, Colorado, USA. *Geomorphology, 252*, 51–65.
Webb, J., & Fielding, C. R. (1999). Debris flow and sheetflood fans of the northern Prince Charles Mountains, East Antarctica. In A.J. Miller & A. Gupta (Eds.), *Varieties of fluvial form* (pp. 317–341). Wiley.
Wells, S. G., & Harvey, A. M. (1987). Sedimentologic and geomorphic variations in storm-generated alluvial fans, Howgill Fells, northwest England. *Geological Society of America Bulletin, 98*(2), 182–198.

Wells, S. G., McFadden, L. D., & Dohrenwend, J. C. (1987). Influence of late Quaternary climatic changes on geomorphic and pedogenic processes on a desert piedmont, Eastern Mojave Desert, California. *Quaternary Research, 27*(2), 130–146.

Welsh, A., & Davies, T. (2011). Identification of alluvial fans susceptible to debris-flow hazards. *Landslides, 8*(2), 183–194.

Whipple, K. X., & Dunne, T. (1992). The influence of debris-flow rheology on fan morphology, Owens Valley, California. *Geological Society of America Bulletin, 104*(7), 887–900.

White, K., Drake, N., Millington, A., & Stokes, S. (1996). Constraining the timing of alluvial fan response to Late Quaternary climatic changes, southern Tunisia. *Geomorphology, 17*, 295–304.

Wilford, D. J., Sakals, M. E., Innes, J. L., Sidle, R. C., & Bergerud, W. A. (2004). Recognition of debris flow, debris flood and flood hazard through watershed morphometrics. *Landslides, 1*(1), 61–66.

Wilkinson, J. M., Marshall, L. G., & Lundberg, J. G. (2006). River behavior on megafans and potential influences on diversification and distribution of aquatic organisms. *Journal of South American Earth Sciences, 21*(1), 151–172.

Witten, I. H., Frank, E., & Hall, M. A. (2011). *Data mining* (3rd Ed., 629 pp.). Morgan Kaufman.

Xiong, K., Adhikari, B. R., Stamatopoulos, C. A., Zhan, Y., Wu, S., Dong, Z., & Di, B. (2020). Comparison of different machine learning methods for debris flow susceptibility mapping: A case study in the Sichuan Province, China. *Remote Sensing, 12*(2), 295.

Youberg, A. M., Webb, R. H., Fenton, C. R., & Pearthree, P. A. (2014). Latest Pleistocene-Holocene debris flow activity, Santa Catalina Mountains, Arizona; implications for modern debris-flow hazards under a changing climate. *Geomorphology, 219*, 87–102.

Yu, G. A., Huang, H. Q., Wang, Z., Brierley, G., & Zhang, K. (2012). Rehabilitation of a debris-flow prone mountain stream in southwestern China-Strategies, effects and implications. *Journal of Hydrology, 414*, 231–243.

Yu, G. A., Yao, W., Huang, H. Q., & Liu, Z. (2020). Debris flows originating in the mountain cryosphere under a changing climate: A review. *Progress in Physical Geography: Earth and Environment*, 0309133320961705.

Zhang, S., Zhang, L.-M., Chen, H.-X., Yuan, Q., & Pan, H. (2013). Changes in runout distances of debris flows over time in the Wenchuan earthquake zone. *Journal of Mountain Science, 10*(2), 281–292.

Zhang, Y., Ge, T., Tian, W., & Liou, Y. A. (2019). Debris flow susceptibility mapping using machine-learning techniques in Shigatse area, China. *Remote Sensing, 11*(23), 2801.

Zhao, Q., Thomsen, K. J., Murray, A. S., Wei, M., & Song, B. (2017). Single-grain quartz OSL dating of debris flow deposits from Men Tou Gou, south west Beijing, China. *Quaternary Geochronology, 41*, 62–69.

Zhao, Q., Thomsen, K. J., Murray, A. S., Wei, M., Pan, B., Song, B., Chen, H., et al. (2015). Testing the use of OSL from quartz grains for dating debris flows in Miyun, northeast Beijing, China. *Quaternary Geochronology, 30*, 320–327.

Zheng, H., Shi, Z., Yu, S., Fan, X., Hanley, K. J., & Feng, S. (2021). Erosion mechanisms of debris flow on the sediment bed. *Water Resources Research, 57*(12), e2021WR030707.

Zhou, W., & Tang, C. (2014). Rainfall thresholds for debris flow initiation in the Wenchuan earthquake-stricken area, southwestern China. *Landslides, 11*(5), 877–887.

Zimmermann, M., Mani, P., & Romang, H. (1997). Magnitude-frequency aspects of alpine debris flows. *Eclogae Geologicae Helvetiae, 90*(3), 415–420.

Zubrycky, S., Mitchell, A., McDougall, S., Strouth, A., Clague, J. J., & Menounos, B. (2021). Exploring new methods to analyse spatial impact distributions on debris-flow fans using data from south-western British Columbia. *Earth Surface Processes and Landforms, 46*(12), 2395–2413.

Fatalities from Debris Flows: Worldwide Distribution and Trends

3

Nikhil Prakash, Paul Santi, Alex Strouth, Sergio A. Sepulveda and Casey Dowling

Abstract

Debris flows are the extremely rapid (often channelized) movement of unconsolidated and saturated slope forming materials. Once triggered, these mass movements can gain enough momentum to destroy life and infrastructure exposed to their travel path. Exceptional historic debris flows have even killed thousands of people in one single event (for example, 15,000 deaths due to debris flows in 1999 in Venezuela). However, not enough studies have addressed the societal risk of global debris-flow occurrences in isolation, as they are mostly considered/combined together with all other types of landslides (sliding or rotational failure, rockfalls, etc.). In one study, (Dowling and Santi, Natural Hazards 71:203–227, 2013) compiled a database of 213 fatal debris flows that occurred from 1950 to 2011, recording 77,779 fatalities (1275 fatalities/year). However, no similar global database was created after this study, which is essential to understand the evolution of fatal debris flow occurrence over the last decade. To help address this gap, we created a new database of all fatal debris-flow events in 2019 that recorded 72 events causing 1628 fatalities. The debris flows accounted for ~46% of the total landslide fatalities of 2019.

3.1 Introduction

Debris flows are viscous mixtures of water, loose soil, rocks, and sometimes organic matter traveling down-slope in channelized paths (Highland & Bobrowsky, 2008). These mass movements initiate on steep slopes (gradient > 15°) and are triggered directly by climatic factors like intense rainfall/snowmelt or by secondary events like earthquakes, volcanic activity, and landslides that disturb soil previously saturated by rainfall/snowmelt (Jakob, 2005; Wieczorek & Glade, 2005). Upon initiation, a debris flow entrains sediment to increase its volume while traveling long distances at rapid to extremely rapid velocities (McCoy et al., 2012; Kang et al., this volume). The momentum of a catastrophic flow can cause fatalities and heavy destruction to infrastructure exposed to its travel path. Generally, the affected population does not have sufficient time to respond

N. Prakash · P. Santi (✉)
Colorado School of Mines, 1500 Illinois Street, Golden, CO 80401, USA
e-mail: psanti@mines.edu

A. Strouth · C. Dowling
BGC Engineering Inc., 600 12th Street, Golden, CO 80401, USA

S. A. Sepulveda
Simon Fraser University, 8888 University Dr W, Burnaby, BC V5A 1S6, Canada

M. Jakob et al. (eds.), *Advances in Debris-flow Science and Practice*, Geoenvironmental Disaster Reduction,
https://doi.org/10.1007/978-3-031-48691-3_3

and escape from the upcoming impact. Deaths have been recorded at inundation depths as low as 0.8 m from traumatic injuries, asphyxiation, and drowning (Bulmer et al., 2005; Pollock & Wartman, 2020). The high likelihood of damage to the head and airways can lead to near-instantaneous deaths, making rescue operations ineffective.

Debris flows often co-occur with other geohazards like flooding, cloudbursts, tropical cyclones, wildfires, earthquakes, dam breaches, or volcanic activity. For example, a tropical cyclone can trigger a large number of debris flows along with other types of landslides and flooding. A landslide can also dam a steep stream and be breached, causing a debris flow phase and eventually dilution into a debris flood (Church & Jakob, 2020). Such a multi-hazard situation amplifies the damage caused by a single hazard in time and space relative to each constituent hazard occurring separately (Zhu et al., 2021).

Local communities typically do not sufficiently understand the risk from debris flows, and they often continue to live in hazardous locations (either legally or illegally), like narrow valleys or debris fans. The driving factors for such behavior can be attributed to many things, including a general lack of knowledge, increase in population, loyalty to ancestral history and land ownership, or even the availability of water and arable land, among many others (Jakob, 2005; Santi et al., 2010). As a result, a catastrophic debris flow can result in many fatalities per event, with more than 77,000 cumulative deaths reported between 1951 and 2011 (Dowling & Santi, 2013). For example, the extraordinary rainstorm-induced debris flows of 1999 in Venezuela were reported to kill 15,000 people and destroy (either partially or entirely) 88,000 buildings. More than 2 million people were displaced after this incident, with the total economic losses estimated at around $2 billion. Similarly, high fatalities have been reported for recent catastrophic events like the 1985 event in Nevado del Ruiz, Colombia (23,000 killed), the 2010 Gansu Mudslide, China (1,471 killed), and the 2013 Kedarnath Mudslides, India (6,000 killed).

Because debris flows are often part of a cascading series of hazards, isolating their impacts requires a clarification of terminology. This chapter refers to a "slide" as a mass movement of one hillslope face that is subjected to shear failure along a well-defined rupture surface. On the other hand, we have used the term "debris flow" to identify all events driven by flow type movement during their transport and deposition phase. A translational or rotational slide on steep slopes or even a rock fall can evolve into a debris flow (Fustos-Toribio et al., 2021; Geertsema et al., 2022), but our focus is on the runout of the event, not on the mechanics of the triggering landslide in the source area, hence, we would still refer to this type of event as a "debris flow." This includes channelized flows, debris avalanches (open slope), earth flows, lahars, and even tailing dam failures.

It is common for the non-scientific media to colloquially report debris flows using synonyms such as "landslides", "flooding", "mudflows", "mudslides", "flow slides", and similar terms. Scientific and technical literature correctly identifies debris flows but mainly investigates the physical characteristics and underlying process rather than fatalities. Major landslide fatality databases, an essential input for subsequent risk assessment studies, compile different landslide types (i.e., slides, debris flows, rock falls, etc.) into a single database (e.g., Froude & Petley, 2018). As a result, we do not clearly understand the societal risk (and vulnerability) associated with the individual components. Long-term records of fatal debris flows can be used to investigate and quantify the risk to human lives in terms of annualized probability of fatality. In this chapter, we compare the risk to human lives from debris flows and landslides using existing global long-term databases (Dowling & Santi, 2013; Froude & Petley, 2018), as well as data from online news aggregator websites. We further compare societal risk from just debris flows to risk from all landslides. In subsequent sections, we further discuss two interesting case studies from 2019 before delivering our concluding remarks.

3.2 Debris-Flow Fatality Distribution

A high-quality database of historical events is needed for hazard assessment and to quantify risk estimates (Froude & Petley, 2018). However, no systematic and global protocols are defined for communicating/recording deaths caused by debris flows. Apart from some exceptional events, the majority of such deaths are reported only by local/regional media coverage, which is weak in rural areas or developing countries where these events occur. These deaths are often linked with inconsistent terminologies like floods, mudslides, landslides, etc., making it difficult to search them by just using relevant keywords. Immediately after a fatal event, news articles attempt to report the estimated or confirmed death, missing, and injured counts. Over time, the number of deaths increases as additional bodies are recovered during rescue operations or the seriously injured victims die. However, depending on the sustained media interest in the event, the updated casualty numbers may or may not be reported again in follow-up articles. Moreover, the reported death counts from debris flows are lumped together with other co-occurring geohazards like earthquakes, extreme rainfall, cyclones, or even other slides. All the above factors combined challenge the compilation of any global debris-flow fatality database.

3.2.1 Fatalities from 1951 to 2011

The first detailed study focusing exclusively on debris flow fatalities was done by Dowling and Santi (2013), who compiled a global database of debris-flow fatalities (DF_{LT}) for a span of 62 years (1950–2011), recording 77,779 deaths in 213 unique events (Table 3.1, Fig. 3.1). They collected information from various sources, including academic publications, newspaper articles, government documents, aid organization filings, personal correspondence, and open surveys. This database did not add the count of missing persons to the fatality numbers and relied on official sources for death tolls whenever available.

For their study period of 62 years, we notice an unequal distribution of debris-flow fatalities in terms of geographical spread and fatality count. Figure 3.1c shows that the fatality count histogram is skewed towards lower values, indicating that the majority of events have a low fatality count, with few exceptional high fatality events. For example, the Nevado del Ruiz lahars in Colombia (13th November 1985) and Vargas debris flows in Venezuela (14–16th December 1999) killed 23,000 and 19,000 people, respectively, accounting for more than 50% of the total recorded fatalities. Because of such extreme outliers, the authors chose to use median values to describe the central tendency of debris-flow causalities.

Table 3.1 Summary of global landslide and debris flow fatality database used in this study. Please note that the database abbreviated as "LS" contains fatalities from all types of landslides, including slides, debris flows, falls, etc. However, the database abbreviated as "DF" contains fatalities only from debris flows

Study	Abbr	Time period	Landslide type	Number events	Number fatalities
Froude and Petley (2018)	LS_{LT}	2004–2017	All LS type	5490	64,218
Dowling and Santi (2013)	DF_{LT}	1950–2011	DF only	213	77,779
"The Landslide Blog" Petley (2020)	LS_{2019}	2019	All LS type	534	3513
This study	DF_{2019}	2019	DF only	72	1628

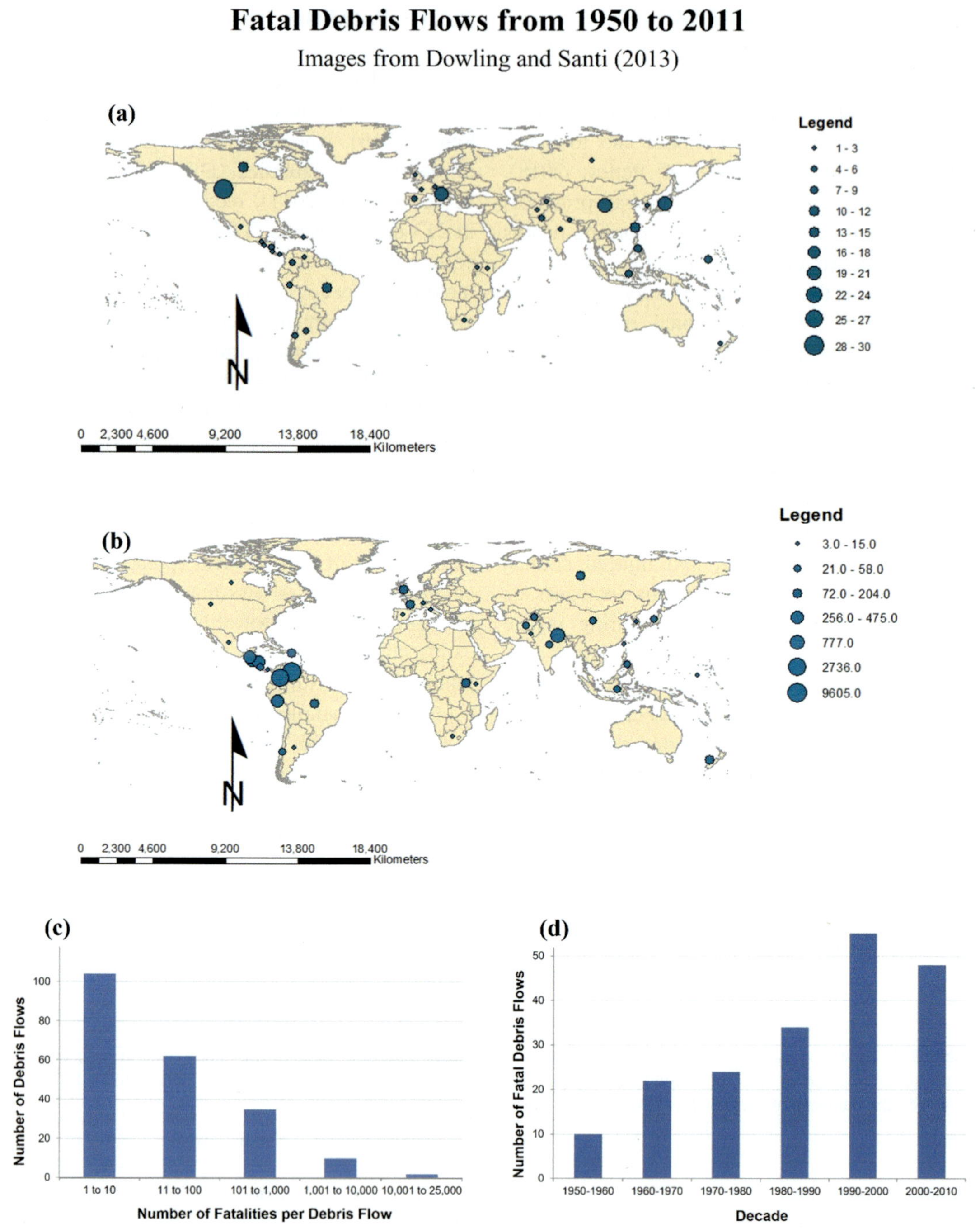

Fig. 3.1 Images reproduced from Dowling and Santi (2013). **a** Countrywise distribution of fatal debris-flow events. **b** Countrywise distribution of median debris-flow fatality count. **c** Histogram of fatalities per debris flow event. **d** Bar graph showing the number of recorded fatal debris flows in the database by decade

Fatal debris-flow activity can be observed in the continents of North & South America, Europe, parts of Africa, Asia, and the islands of Oceania (Fig. 3.1a and b). The highest median debris-flow fatality counts per event were in developing countries of South America and Asia. On the other hand, advanced nations like the United States, Italy, and Japan had low median debris-flow fatality counts even though their records have a high number of events. The authors attributed the high event count with low fatalities to better reporting and abundant debris-flow research in these countries. Figure 3.1d shows the year-on-year increase of recorded fatal debris flows in every decade (except 2000–2010). The authors further analyzed this graph to conclude: (i) The records of fatal debris flows in developing countries are increasing every decade, which indicates an improved reporting and documentation of smaller events. (ii) A decrease in fatal debris flows recorded in advanced countries indicates an improvement in debris-flow mitigation and hazard management.

The authors also found that extreme precipitation events, including tropical cyclones and monsoon rainfall, were the most common trigger and were associated with 67% of the debris flows in the database, causing a median fatality count of 9 per event. On the other hand, less common triggers, like earthquakes and dam bursts, were associated with a median fatality count of more than 500 per event.

3.2.2 Fatalities in 2019

The DF_{LT} database records events only until 2011, and no similar database was created for a global scale that adequately captures small events in developing countries. Hence, to help understand the evolution of fatal debris flow occurrence over the last decade, we created a new database for debris-flow deaths recorded in one recent year, 2019 (Table 3.1).

Extracting information about past debris flows and their impact on human life and infrastructure is still challenging. We created this database in a two-step approach. In the first step, we identified potential/probable events by running automated scripts programmed to scan specific keywords on ReliefWeb (a popular information source on global disasters) and records from the Landslide Blog, which is included as LS_{2019}, the landslide fatality database of 2019 (Petley, 2020). We did not assume to find "debris flows" directly, so we included search terms like "landslide", "mudslide", "debris", "flow", "flood", "death", "kill", "dead", "died", "fatal", etc. to generate the preliminary list of potential/probable debris flows that would require closer inspection.

In the second step, we manually retrieved information about each potential/probable debris flow by searching for information in relevant scientific publications, disaster reports, specialized blogs, news articles, or social media (Twitter, Facebook, and YouTube). These online contents have two components that help in identifying a debris-flow event: textual description and associated images (or videos). The description of an event such as "*flood brought a large amount of soil, stone and even wood in the form of logs*" (Rumagit, 2019), "*heavy rain caused the landslide, sending mud, fallen trees and debris*" (News, 2019), or "*dam break*" are some examples that can indicate a possible debris flow. However, the visual media associated with the event description were the most critical and convincing evidence to identify a debris flow (from other slides). These media fell under two broad categories: (i) aerial/satellite photographs or photographs/video taken from a distant and high vantage point, or (ii) photographs/video taken from very close proximity.

For the first category of media, which are taken from a distance or aerial/spaceborne cameras, we tried to assess whether the event was a debris flow by evaluating the images for characteristics typical of debris-flow events. The characteristics and reasoning included:

- Attempting to identify evidence of multiple travel paths (which can be curved as well). A debris flow funnels into existing channels/valleys, or can even create new paths—often breaking and merging during the travel.
- Attempting to identify a high length-to-width ratio. Long runouts are the most prominent feature of debris flows.
- Attempting to identify a long runout event, where source area and deposit area might not be visible in the same frame (when the

coverage of the photograph is not wide enough).
- Attempting to identify when the width of the source area is not equal to the width of the deposition area. The deposition area has a distributary pattern created by shifts in the active channel.

On the other hand, photographs in the second category were taken from very close proximity, capturing just one small view of the event. Depending on the visible features, interpreting these photographs can be difficult to distinguish between debris flows and other slides. Here are some key characteristics that helped us identify a debris flow:

- A large quantity of both fine and coarse-grained sediments, and possibly large boulders, logs, trees, or even cars.
- Erosional downcutting of travel channels. The transported sediments are deposited on the sides of the travel paths, forming coarse levees on one or both sides.
- Boulders covered with soil, indicating signs of recent movement.
- Flat and spread out deposits.
- Typical to observe post-event water drainage, as a debris flow is a mixture containing more than 50% water.

Figure 3.2 shows a few example photographs from these online sources, along with our interpretation of the event type (debris flows vs. slides). However, care was given to avoid news articles using stock photographs of unrelated landslides as representative photographs to accompany their post.

The above observations are non-exhaustive, but serve as a starting point for identifying

Fig. 3.2 Example of images from online articles reporting landslide events in 2019, with our interpretation of the type of event labeled on each image. **a** West Manggarai, Indonesia—7th March 2019. **b** Shuicheng, China—23rd July 2019. **c** Nganjuk, East Java, Indonesia—14th February 2019. **d** Jayapura, Indonesia—16th March 2019. **e** Bududa, Uganda—3rd December 2019. **f** Risaralda Department, Colombia—13th October 2019

debris flows and were used in this study to distinguish between debris flows and other landslides. Once a debris-flow event was identified, we recorded the most updated casualty numbers in our database. Similar to the Dowling and Santi (2013) approach, deaths from multi-hazard situations or disasters involving a population of debris flows were broken down into individual events. The combined fatalities were still recorded if the available information did not support such separation.

Following the data collection steps listed above, we recorded 1628 fatalities in 72 unique debris flow events in our database (DF_{2019}—Table 3.1). This accounts for 48% of all landslide fatalities recorded in 2019 (Table 3.1). The worldwide distribution of events is shown in Fig. 3.3. Similar to the past debris-flow occurrence trends, the fatality distribution was skewed towards lower values, with only 12% of events with more than 50 deaths (Fig. 3.4). The Chimanimani and Chipinge districts of Zimbabwe were the worst hit, with 344 deaths caused by debris flows triggered by cyclone Idai in March 2019. Likewise, the islands of Indonesia witnessed ten different fatal debris-flow events throughout 2019, causing 239 deaths. Brazil (261 fatalities in 3 events), India (255 fatalities in 7 events), and China (166 fatalities in 4 events) were also among the countries highest affected by debris flows in 2019.

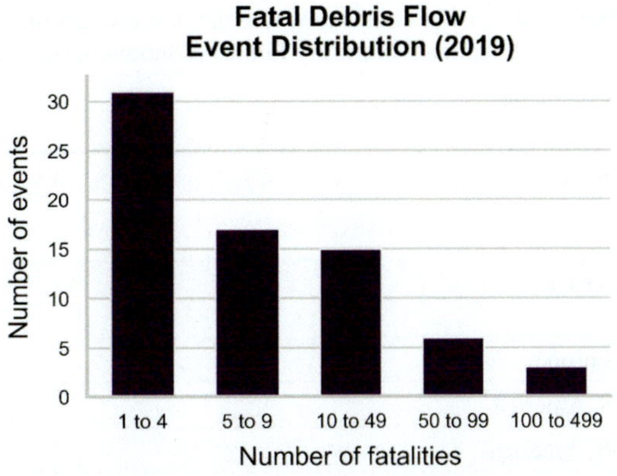

Fig. 3.4 Distribution of fatalities per debris-flow event recorded in 2019

Similar to the past trends, debris flows still disproportionately affect developing nations (Table 3.2). Countries in Asia, Africa, and South

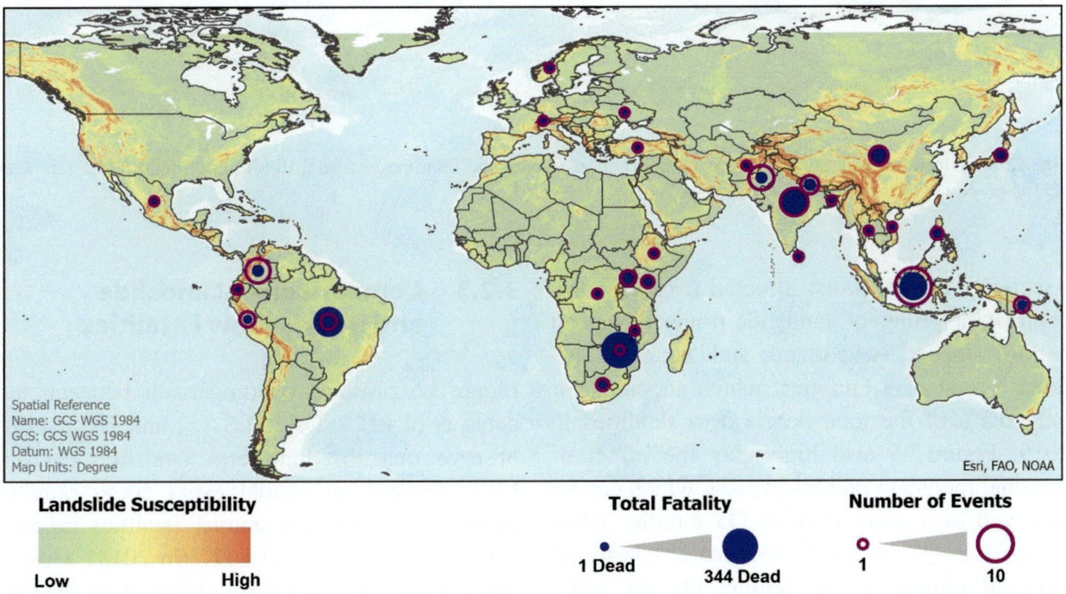

Fig. 3.3 Distribution of fatal debris-flow events, by country, recorded in 2019. The size of the filled circle represents the cumulative fatalities, and the size of the empty circle represents the number of events recorded in our database. The background map shows the global landslide susceptibility from Stanley and Kirschbaum (2017)

Table 3.2 List of countries where debris-flow fatalities were recorded in 2019 (DF_{2019}). The "advanced economies" (as per International Monetary Fund's economic groupings) are identified in bold together with the "*Adv.*" Tag

Continents	Country where debris flow fatality was recorded in 2019
Asia	**Japan** (*Adv.*), India, Sri Lanka, Philippines, Afghanistan, Thailand, Vietnam, Turkey, Pakistan, Bangladesh, Nepal, China, Indonesia
Africa	Malawi, South Africa, Ethiopia, Democratic Republic of Congo, Uganda, Kenya, Zimbabwe
Europe	**Norway** (*Adv.*), **Switzerland** (*Adv.*), Ukraine
S. America	Ecuador, Peru, Colombia, Brazil
N. America	Mexico
Oceania	Papua New Guinea

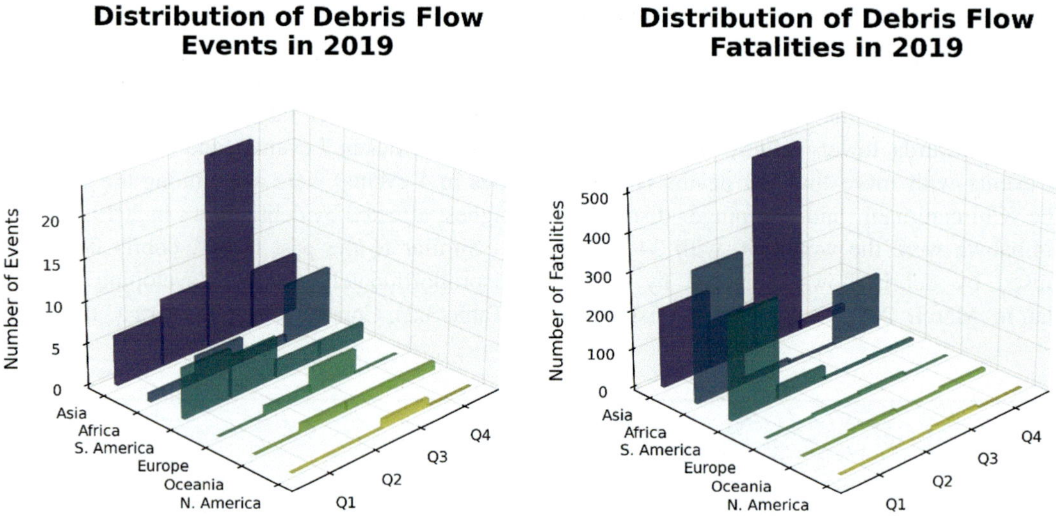

Fig. 3.5 Distribution and annual timing of debris flow events and fatalities across different continents. (Q1: Jan–Mar | Q2: Apr–Jun | Q3: Jul–Sept | Q4: Oct–Dec)

America were the worst affected (Fig. 3.5). By contrast, a group of landslide-prone advanced nations (Japan, Switzerland, and Norway) suffered a total of 8 fatalities, which accounts for only 0.5% of the total debris flow fatalities in 2019. Figure 3.5 also highlights the effect of seasonal monsoon rainfall in Asia, which can be observed as a sharp peak in Q3 months. Also, the peak in Q1 for South America is due to the seasonal rainfall in the region. On the other hand, the Q1 high fatality in Africa was caused by the debris flows triggered by the March 2019 Cyclone Idai in Mozambique.

3.2.3 Comparison of Landslide and Debris-Flow Fatalities

Figure 3.6 presents a comparison between the datasets of LS_{LT}, DF_{LT}, LS_{2019}, and DF_{2019}. We observe that the long-term landslide dataset, LS_{LT}, showed an annual event count ranging from 250 to 496 and annual fatalities ranging from 2253 (in 2016) to 8677 (in 2013). On the other hand, the long-term debris-flow dataset, DF_{LT}, records an annual event count ranging from 0 to a maximum of 15 events per year (in

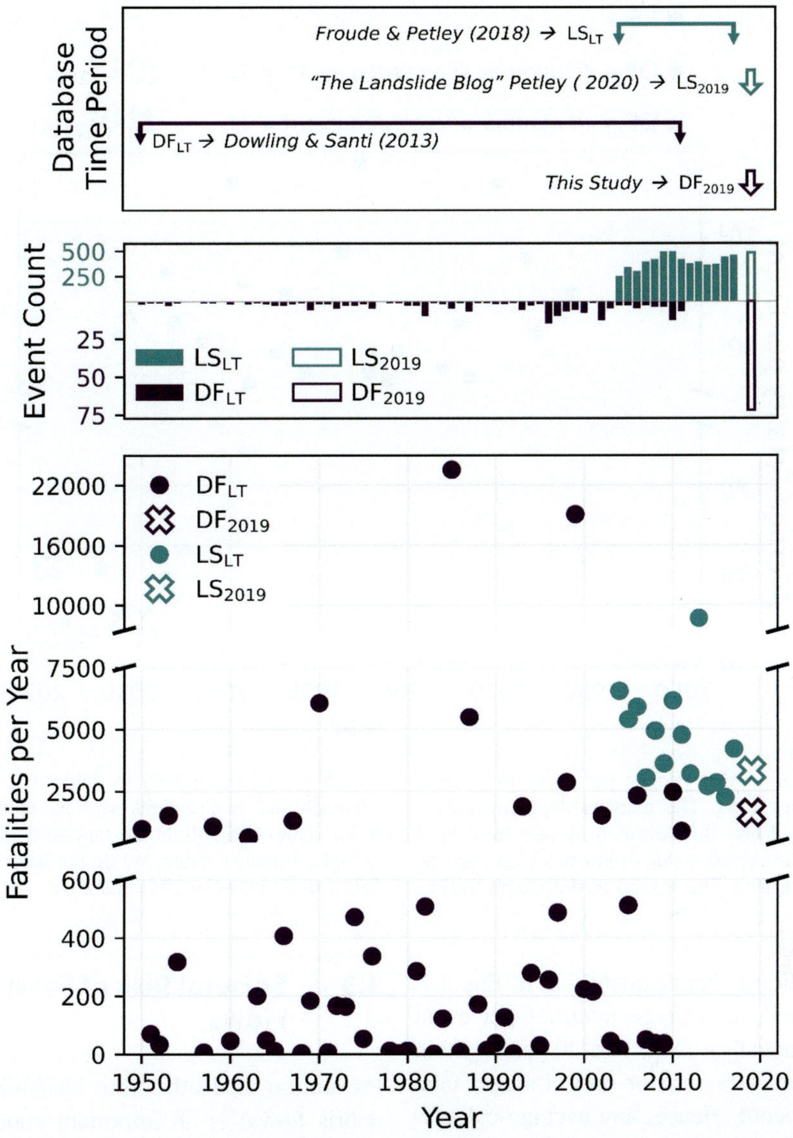

Fig. 3.6 Comparison between landslide and debris-flow fatality database. The legend of this graph follows the abbreviations defined in Table 3.1. The timeline (top) illustrates the observation period of all the databases. For debris flows, we do not have fatality data between 2011 and 2019. The fatalities per year graph (bottom) has been split into three sections to highlight the variability across the lowest to highest fatality range

1996). However, the recorded annual fatalities had a large variation for all years when debris flows were recorded—ranging from only one death per year to a maximum of 23,553 deaths (in 1985). The low count of one death per year is likely a consequence of lack of reporting, while the 23,000+ deaths in 1985 were dominated by the Nevado del Ruiz lahar. On the other hand, the dataset created in our study (DF_{2019}) records a much higher debris flow count per year (with 72 debris flows in 2019), indicating that DF_{LT} could be under-representing the true scale of this disaster type.

From the above-discussed historical records, we can observe that debris flows have a substantial number of fatalities for fewer events.

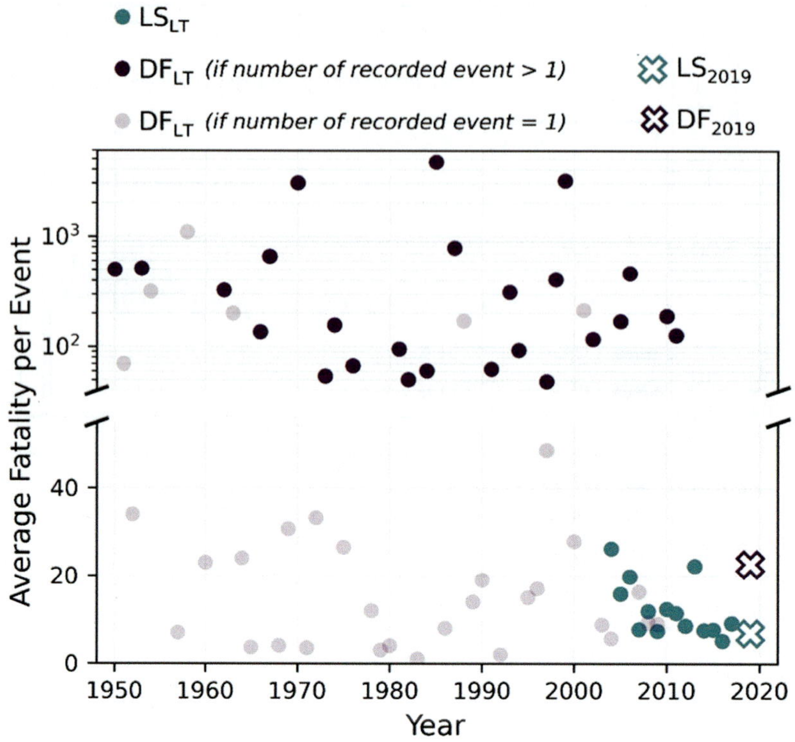

Fig. 3.7 Average annual fatality per event from landslides and debris flows. The markers depicting only a single recorded debris flow event in a year have been faded out—as the average value of just one record has no statistical significance. The legend of this graph follows the abbreviations defined in Table 3.1. The graph has been split into two sections (with the upper range plotted in log scale) to highlight the variability across the lowest to highest fatality range. We do not have fatality data for debris flows between 2011 and 2019

This observation is better quantified in Fig. 3.7, where we track the average fatalities per event (calculated annually). DF_{LT} had 20 years within its time period with zero or only a single fatal debris-flow record. Hence, any average calculations in such cases are not statistically significant and have been faded in Fig. 3.7. From all the valid records, we can observe that debris flows, on average, can kill more humans per event compared to corresponding landslides. However, the gap between average fatality per event of DF_{LT} and DF_{2019} is significantly large, suggesting that DF_{LT} might have underrepresented lower fatality debris-flow events.

3.3 Societal Risk of Fatal Debris Flows

Prevention of deaths from landslides (including debris flows) is an important concern for civil protection authorities operating in territories with a moderate to high likelihood of landslide occurrences. The FN criteria are one tool used to measure the societal risk (i.e., the cumulative risk to a group of people) from a hazard (landslides, floods, traffic accidents, aircraft crashes, dam failures, industrial mishaps, etc.). An FN curve is generally created to display the frequency of "N or more" fatalities per year on a log–log scale. However, it can be customized to track other consequences like infrastructure damage and economic losses. Many studies have adopted this tool to quantify historical landslide fatalities and to display risk estimates (see Strouth et al., this volume). Here, we briefly introduce the concept to compare the global risk of landslides and debris-flow fatalities in subsequent paragraphs.

We plot the FN curves for the global fatality datasets identified in Table 3.1 to quantitatively compare the societal risk of fatal debris flows and landslides (Fig. 3.8). In global databases, the low fatality events are likely under-reported (Dowling & Santi, 2013; Froude & Petley, 2018), making any observations to the left of 10^1 in X-axis unreliable. However, we believe that the high fatality events receive enough coverage from scientific and mainstream media, and hence, have a good chance of being recorded in the fatality datasets. We observe that the frequency of fatal events (intercept on Y-axis) for landslides is higher than debris flows, particularly for smaller magnitude events. However, gradually the frequency of occurrence of debris flows matches with the landslides for the high fatality events—indicating that they are the major contributor to all high fatality events in the LS datasets (landslide databases also contains debris-flow events).

We also observe that the FN curves show significantly lower risk from DF_{LT} (1950–2011)

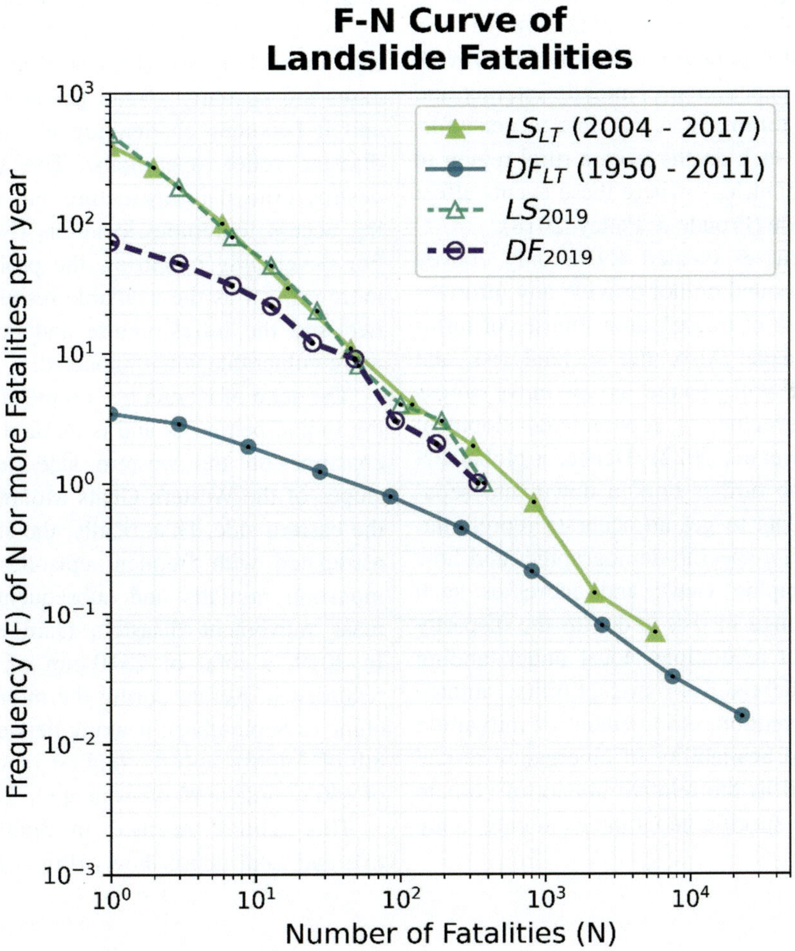

Fig. 3.8 FN curve plotted for global historical landslide and debris flow fatalities datasets in Table 3.1

when compared to DF_{2019} (in 2019). This can be due to lower records of fatal events in previous decades, as DF_{LT} has ~210 events for 50+ years, while DF_{2019} has ~70 events recorded in a single year. The possible explanation is a combination of: (i) 2019 saw increased occurrences of landslides and debris flows. The frequency of extreme events (including debris flows) is rising and is expected to increase further in our warming world (Affairs, 2020; Gariano & Guzzetti, 2016; Gaupp, 2020; Jakob, 2022; Rahmstorf & Coumou, 2011). (ii) Vulnerable populations are continuing to be driven into higher hazard locations due to population growth, migrations, economic inequality, internal discrimination, and other societal factors. Combined with poor engineering practices on unstable slopes, the toll on the exposed population will continue to grow (Garcia-Delgado et al., 2022). Or, (iii) The reporting of debris-flow fatalities has improved, and with the penetration of mobile internet and social media forums, it is possible to get information about such deaths from a rural region in developing countries—where these events affect the most people (Froude & Petley, 2018).

The FN curves created above only convey risk to life-loss and do not provide any information on the risk of major/minor injuries or infrastructure damage. Also, the societal risk and risk tolerance will change as we move across different geographies or economic/development statuses (Sim et al., 2022). Hence, a global FN curve must be further broken down into separate components to get an accurate representation of risk to a specific demographic, and how each demographic views and addresses such risk. This change in risk is due to the disparity in government regulations, local understanding of landslide/debris flow susceptibility, anthropogenic intervention, enforcement of mitigation measures, and availability of emergency rescue teams, as well as the relative impact of climate change in the specific geographies among many others.

3.4 Case Study of Selected Events from 2019

The following section provides a detailed discussion on selected case studies of deadly and destructive debris flows that occurred in 2019. We focus on discussing the conditioning factors that led to the disaster and the resulting human and economic losses.

3.4.1 The 2019 Landslides and Debris Flows in Kerala (India)

We selected this case as it demonstrated a commonly observed cascading hazard disaster situation where an extreme rainfall event is followed by multiple debris flows, landslides, and flooding. In such cases, the casualties and critical cases are spatially spread across large regions, and it becomes challenging to organize quick disaster relief operations. The debris could destroy critical infrastructure and roads, blocking access to remote locations (NDRF, 2019). For developing countries, the problem is even more critical as the available resources are limited and the infrastructure and population are more vulnerable when impacted.

The state of Kerala lies on the southern tip of the Indian peninsula and is marked by a 590 km coastline on the western side and the steep slopes of the Western Ghats Mountain range on the eastern side. Historically, the state has been associated with frequent episodes of extreme monsoon rainfalls and subsequent floods and mass movement disasters (Jain et al., 2021). In 2019, a total of 2310 mm of rainfall was recorded in Kerala during the monsoon months (June to September), peaking between the 4th to 8th of August with the highest recorded rainfall of 140 mm/day (Wadhawan et al., 2020).

This rainfall resulted in multiple destructive and fatal debris flows (Fig. 3.9a and b). In

Fig. 3.9 Debris flows in Kerala (India) after the extreme monsoon rainfall of August 2019. **a** and **b** shows the post-event photographs of two major events at Pothukal town (Malappuram) and Puthumala village (Wayanad), respectively. Images are reproduced from Aswanth (2019). **c** A synoptic view of the debris flow observed from cloud-free Sentinel-2 satellite imagery

the first three days, "only" 80 landslides were identified (Benjamin, 2022). However, this is a substantially underestimated count, which is evident from post-event cloud-free satellite imagery (Fig. 3.9c). In later studies, Jain et al. (2021) mapped 756 landslides in Kerala for all of 2019. The events resulted in severe infrastructure damage, destroying many houses, roads, bridges, and tea estates, disrupting economic activities in the region for many days.

In response to the loss of infrastructure and human lives, nine teams from the National Disaster Recovery Framework (NDRF) were dispatched to different locations. They faced resistance from heavy rainfall and road obstruction (due to debris) and started their relief operations on 9th August, retrieving 61 bodies (NDRF, 2019). Local assistance was provided by state police, fire and rescue services, civilian volunteers, and a team from the National

Geophysical Research Institute (NGRI) (Sabith, 2019). In subsequent news articles, the number of reported fatalities from landslides increased to 120 (Benjamin, 2022). In this multi-hazard situation, all the news articles and reports combined the fatality counts to a single number. Apart from two prominent debris flows, it is not possible to individually determine the death toll of all the individual smaller events (including other slides).

In 2018, Kerala had already suffered from a similar disaster scenario, where the monsoon rainfall triggered more than 1000 landslides and debris flows, causing 155 fatalities (Premlet, 2019). The high fatality of landslides and debris flows were linked to avoidable factors like modifications of slopes not based on proper geotechnical analysis and design, mismanagement of water drainage, unplanned residential developments within active floodplains, and lack of early warning schemes (Wadhawan et al., 2020). However, these frequent disasters should be treated as a learning phase to implement sufficient mitigation measures to prevent a similar disaster in the future (Ali & George, 2022).

3.4.2 Brumadinho Mine Tailings Dam Collapse (Brazil)

The catastrophic Brumadinho mine tailings dam collapse of 25th January 2019 occurred at the “Corrego do Feijao” Iron ore mine, which was operated by the mining giant Vale (Fig. 3.10a). The entire face of the dam ruptured in less than 10 s, releasing 9.7 million m^3 of tailings (75% of the stored tailings) in less than 5 min, flowing downhill at the speed of 66 km/h (Petley, 2019; Robertson et al., 2019). The mudflow had a 10 m high wave which traveled 10 km (Fig. 3.10b and c), killing 259 people and damaging office buildings, houses, roads, and farmlands in its travel path (Rotta et al., 2020).

Tailings deposition occurred until July 2016. The dam was fully saturated, and the ultimate failure was the result of flow (static) liquefaction within the brittle materials of the dam (Robertson et al., 2019; Rotta et al., 2020).

This tailing flow event is an example of a man-made disaster. The operating company, which also operates 133 iron ore dams in the region, was involved in another recent dam failure disaster (120 km away in Mariana, Brazil) in 2015 that caused 19 deaths. The Corrego do Feijao mine tailings dam showed episodes of subsidence in 2018 (Rotta et al., 2020). It lacked adequate internal drainage systems to reduce excess water in the dam (CIAEA, 2020; Robertson et al., 2019). One day before the failure, a borehole of 100 mm in diameter and 70 m deep was drilled to install piezometers, which could have disturbed the tailing layers, however, experts debate the effect of this drilling activity on the ultimate failure (CIAEA, 2020).

Apart from the toll on human lives and economic losses, the mudflow spread toxic mining by-products in the environment and contaminated the Paraopeba River, which is a critical source of water supply for the downstream towns. It destroyed multiple buildings (including mine offices, houses, the Nova Estancia hotel), farms, roads, and railway infrastructure including a bridge (Darlington et al., 2019; Rotta et al., 2020). Overall, the operating company had to pay $7 billion in compensation for disaster victims and for “socio-economic” and “socio-environmental” reparations, funding projects to repair the surrounding environment, including a massive clean-up of the Paraopeba River (BBC, 2021).

3.5 Concluding Remarks

Debris flows are a geohazard that caused 1628 deaths in 2019. In a few extreme historic cases, debris flows have even caused thousands of fatalities in a single event. They show an unequal distribution in fatality counts, with the developing countries in Asia, South America, and Africa suffering the most severe impacts. In this chapter, we created a new debris-flow fatality database for events reported in 2019 (DF_{2019}) and subsequently compared it with a similar published database (DF_{LT}) containing records from 1950 to 2011. A 33% increase is noted in

Fig. 3.10 The tailing dam collapse of "Corrego do Feijao" mine, Brumadinho (Brazil) in January 2019. **a** A photograph showing the aftermath of the mudflow event near the source with the collapsed dam (source) in view. **b** and **c** show the pre- and post-event satellite images indicating the area affected by the debris. Images modified from Mendonça (2019) and Rotta el al. (2020)

2019 over the long-term average for the reference period (which may at least be partially attributable to more complete reporting). It also showed that the divide in debris-flow fatalities between advanced and developing countries increased in 2019 compared to the 1950 to 2011 period. This indicates that allocating proper resources to scientific studies and mitigation measures can reduce the impact on human lives. As shown in both case studies, many debris-flow deaths are preventable if we take lessons from past events and implement preventive actions. Global reporting of death tolls has also increased in recent years, primarily due to the increased reach of mobile connectivity, the internet, and social media platforms (Froude & Petley, 2018). We see more death reports in DF_{2019} from African countries.

We found that debris flows are more likely to cause high fatality events compared to other landslide types on a global scale. This is attributable to the long runout distance of debris flows and their poor predictability. The societal risk for both these disasters is high, indicating a need to take appropriate mitigation measures to decrease risk. To improve our understanding of the debris-flow impacts, future research and regulatory improvements should address the following points:

- No globally accepted reporting channels exist for debris-flow fatalities. The life-loss vulnerability relationships are not consistent, and each proposed relationship is based on a very small dataset. We need to know how many people were in the impact zones (at varying intensity levels) and the number of fatalities. But we typically only record the number of fatalities, making it difficult to estimate the probability of life loss from a given impact (see Strouth et al., this volume).
- In multi-hazard situations, the debris-flow fatalities are often combined with other co-occurring hazards, which makes it difficult to isolate the contributions from debris flows alone.
- Information on economic losses is even more difficult to obtain. Indirect economic losses (like business interruption) can be much more important than direct losses and need to be estimated. These losses are necessary to include to tip the scale in benefit–cost assessments that support proactive mitigation measures.

References

Affairs, U. N. O. for D. (2020). *Human Cost of Disasters*. UN. https://doi.org/10.18356/79b92774-en

Ali, S., & George, A. (2022). Fostering disaster mitigation through community participation- case of Kochi residents following the Kerala floods of 2018 and 2019. *Natural Hazards, 111*(1), 389–410. https://doi.org/10.1007/s11069-021-05058-0

Aswanth, S. M. (2019). In Photos: Landslides in Kerala wipe out entire villages. *TheWire*. https://thewire.in/environment/kerala-rains-landslides-flood

BBC. (2021). *Vale dam disaster: $7bn compensation for disaster victims—BBC News*. https://www.bbc.com/news/business-55924743

Benjamin, H. S. (2022). Frequency of landslides rising in Kerala due to land-use patterns, climate change. *OnManorama*. https://www.onmanorama.com/news/kerala/2021/10/22/landslides-frequency-increase-kerala-climate-change-land-use-pattern.html

Bulmer, M., Bissell, R., Petley, D., & Bond, M. C. (2005). *A study of injuries sustained in 10th January 2005 La Conchita Landslide Disaster, California, USA*. Landslide Observatory, University of Maryland, Baltimore County, USA, International Landslide Centre, University of Durham, UK.

Church, M., & Jakob, M. (2020). What Is a debris flood? *Water Resources Research, 56*, e2020WR027144.https://doi.org/10.1029/2020WR027144

CIAEA. (2020). *Executive summary of the independent investigation report failure of Dam 1 of the Córrego do Feijão Mine – Brumadinho, MG*. Comitê Independente de Assessoramento Extraordináriode Apuração - CIAEA.

Darlington, S., Glanz, J., Andreoni, M., Bloch, M., Pecanha, S., Singhvi, A., & Griggs, T. (2019). A tidal wave of mud. *The New York Times*. https://www.nytimes.com/interactive/2019/02/09/world/americas/brazil-dam-collapse.html

Dowling, C. A., & Santi, P. M. (2013). Debris flows and their toll on human life: A global analysis of debris-flow fatalities from 1950 to 2011. *Natural Hazards, 71*(1), 203–227. https://doi.org/10.1007/s11069-013-0907-4

Froude, M. J., & Petley, D. N. (2018). Global fatal landslide occurrence from 2004 to 2016. *Natural Hazards and Earth System Sciences, 18*(8), 2161–2181. https://doi.org/10.5194/nhess-18-2161-2018

Fustos-Toribio, I. J., Morales-Vargas, B., Somos-Valenzuela, M., Moreno-Yaeger, P., Muñoz-Ramirez, R., Rodriguez Araneda, I., & Chen, N. (2021). Debris flow event on Osorno volcano, Chile, during summer 2017: New interpretations for chain processes in the southern Andes. *Natural Hazards and Earth System Sciences, 21*(10), 3015–3029. https://doi.org/10.5194/nhess-21-3015-2021

Garcia-Delgado, H., Petley, D. N., Bermúdez, M. A., & Sepúlveda, S. A. (2022). Fatal landslides in Colombia (from historical times to 2020) and their socio-economic impacts. *Landslides, 19*(7), 1689–1716. https://doi.org/10.1007/s10346-022-01870-2

Gariano, S. L., & Guzzetti, F. (2016). Landslides in a changing climate. *Earth-Science Reviews, 162*, 227–252. https://doi.org/10.1016/j.earscirev.2016.08.011

Gaupp, F. (2020). Extreme events in a globalized food system. *One Earth, 2*(6), 518–521.

Geertsema, M., Menounos, B., Bullard, G., Carrivick, J. L., Clague, J. J., Dai, C., Donati, D., Ekstrom, G., Jackson, J. M., Lynett, P., Pichierri, M., Pon, A., Shugar, D. H., Stead, D., Del Bel Belluz, J., Friele, P.,

Giesbrecht, I., Heathfield, D., Millard, T., Sharp, M. A., et al. (2022). The 28 November 2020 landslide, Tsunami, and outburst flood – A hazard cascade associated with rapid deglaciation at Elliot Creek, British Columbia, Canada. *Geophysical Research Letters, 49*(6). https://doi.org/10.1029/2021GL096716

Highland, L. M., & Bobrowsky, P. (2008). The landslide handbook—A guide to understanding landslides. *US Geological Survey*. https://doi.org/10.3133/cir1325

Jain, N., Martha, T. R., Khanna, K., Roy, P., & Kumar, K. V. (2021). Major landslides in Kerala, India, during 2018–2020 period: An analysis using rainfall data and debris flow model. *Landslides, 18*(11), 3629–3645. https://doi.org/10.1007/s10346-021-01746-x

Jakob, M. (2005). Debris-flow hazard analysis. In *Debris-flow hazards and related phenomena* (pp. 411–443). Springer. https://doi.org/10.1007/3-540-27129-5_17

Jakob, M. (2022). Landslides in a changing climate. In *Landslide hazards, risks, and disasters* (pp. 505–579). Elsevier. https://doi.org/10.1016/B978-0-12-818464-6.00003-2

McCoy, S. W., Kean, J. W., Coe, J. A., Tucker, G. E., Staley, D. M., & Wasklewicz, T. A. (2012). Sediment entrainment by debris flows: In situ measurements from the headwaters of a steep catchment. *Journal of Geophysical Research: Earth Surface, 117*(F3), n/a–n/a. https://doi.org/10.1029/2011jf002278

Mendonça, V. (2019). *Catástrofe socioambiental provocada pelo rompimento de barragem da mineradora Vale em Brumadinho (MG)*. https://en.wikipedia.org/wiki/Brumadinho_dam_disaster#/media/File:Brumadinho,_Minas_Gerais_(47021723582).jpg

NDRF. (2019). *Landslide in Kerala - 2019 | NDRF - National Disaster Response Force*. http://ndrf.gov.in/operations/landslide-kerala-2019

News, F. (2019). *Peru – Deadly Landslide in Junín Region*. https://floodlist.com/america/peru-landslide-chanchamayo-junin-region-october-2019

Petley, D. (2019). *The speed of the Brumadinho tailings dam landslide—The Landslide Blog—AGU Blogosphere*. https://blogs.agu.org/landslideblog/2019/02/04/brumadinho-tailings-dam-landslide/

Petley, D. (2020). A map of 2019 fatal landslides. In *The Landslide Blog*. https://blogs.agu.org/landslideblog/2020/05/13/2019-fatal-landslides-2/

Pollock, W., & Wartman, J. (2020). Human vulnerability to landslides. *GeoHealth, 4*(10), e2020GH000287.

Premlet, B. (2019). Landslide 18 Kerala - A Report. In B. Premlet (Ed.), *Proceedings of the Symposium on Landslides in Kerala 2018*. Phasor Books.

Rahmstorf, S., & Coumou, D. (2011). Increase of extreme events in a warming world. *Proceedings of the National Academy of Sciences, 108*(44), 17905–17909. https://doi.org/10.1073/pnas.1101766108

Robertson, P. K., Melo, L. de, Williams, D. J., & Wilson, G. W. (2019). *Report of the Expert Panel on the Technical Causes of the Failure of Feijão Dam I*. Comitê Independende de Assessoramento Extraordináriode Apuração - CIAEA. https://bdrb1investigationstacc.z15.web.core.windows.net/assets/Feijao-Dam-I-Expert-Panel-Report-ENG.pdf

Rotta, L. H. S., Alcântara, E., Park, E., Negri, R. G., Lin, Y. N., Bernardo, N., Mendes, T. S. G., & Filho, C. R. S. (2020). The 2019 Brumadinho tailings dam collapse: Possible cause and impacts of the worst human and environmental disaster in Brazil. *International Journal of Applied Earth Observation and Geoinformation, 90*, 102119. https://doi.org/10.1016/j.jag.2020.102119

Rumagit, A. (2019). *Jalan raya Sentani-Kemiri putus akibat diterjang banjir*. https://www.antaranews.com/berita/811117/jalan-raya-sentani-kemiri-putus-akibat-diterjang-banjir

Sabith, M. (2019). Ground report: Entire families have perished in the Kerala floods and landslides. *TheWire*. https://thewire.in/environment/kerala-floods-landslides-rainfall

Santi, P. M., Hewitt, K., VanDine, D. F., & Cruz, E. B. (2010). Debris-flow impact, vulnerability, and response. *Natural Hazards, 56*(1), 371–402. https://doi.org/10.1007/s11069-010-9576-8

Sim, K. B., Lee, M. L., & Wong, S. Y. (2022). A review of landslide acceptable risk and tolerable risk. *Geoenvironmental Disasters, 9*(1). https://doi.org/10.1186/s40677-022-00205-6

Stanley, T., & Kirschbaum, D. B. (2017). A heuristic approach to global landslide susceptibility mapping. *Natural Hazards, 87*(1), 145–164. https://doi.org/10.1007/s11069-017-2757-y

Strouth, A., & McDougall, S. (2021). Historical landslide fatalities in British Columbia, Canada: Trends and implications for risk management. *Frontiers in Earth Science, 9*, 606854. https://doi.org/10.3389/feart.2021.606854

Wadhawan, S. K., Singh, B., & Ramesh, M. V. (2020). Causative factors of landslides 2019: Case study in Malappuram and Wayanad districts of Kerala, India. *Landslides, 17*(11), 2689–2697. https://doi.org/10.1007/s10346-020-01520-5

Wieczorek, G. F., & Glade, T. (2005). Climatic factors influencing occurrence of debris flows. In *Debris-flow hazards and related phenomena* (pp. 325–362). Springer. https://doi.org/10.1007/3-540-27129-5_14

Zhu, L., He, S., Qin, H., He, W., Zhang, H., Zhang, Y., Jian, J., Li, J., & Su, P. (2021). Analyzing the multi-hazard chain induced by a debris flow in Xiaojinchuan River, Sichuan, China. *Engineering Geology, 293*, 106280. https://doi.org/10.1016/j.enggeo.2021.106280

Flume Modeling of Debris Flows

4

Clarence Edward Choi, Charles Wang Wai Ng
and Haiming Liu

Abstract

Debris flows are unique compared to other types of landslides (e.g., rock avalanches; dry granular flows) because their dynamics are governed by both the soil grains and interstitial fluid. On one hand, debris flows may exhibit some features of dry granular flows, such as particle size segregation. On the other hand, contact and collisional grain stresses are strongly influenced by the viscosity and excess pressure of the interstitial fluid. Furthermore, debris flows travel over complex erodible boundaries and have been reported to reach volumes of up to 10^9 m^3 and travel over 100 km. Evidently, the scale and dynamics of debris flows are complex and difficult to reproduce. One of the most common approaches to investigate the dynamics of debris flows is by conducting scaled laboratory experiments, which provide idealized and high-quality evidence to evaluate theoretical and numerical models. This chapter provides an overview of the fundamental principles, scaling considerations, commonly adopted setups (e.g., flumes and geotechnical centrifuge), and the state-of-the-art in instrumentation and image analysis for modeling debris flows.

C. E. Choi (✉)
Department of Civil Engineering, The University of Hong Kong, Hong Kong SAR, China
e-mail: cechoi@hku.hk

C. W. W. Ng · H. Liu
Department of Civil and Environmental Engineering, The Hong Kong University of Science and Technology, Hong Kong SAR, China
e-mail: cecwwng@ust.hk

H. Liu
e-mail: haimingliu@ust.hk

4.1 Introduction

Physical modeling of debris flows is crucial for revealing their underlying mechanisms, as the spatiotemporal unpredictability of debris flows means that their mechanisms are difficult to observe in the field. However, the objective of physical modeling is not to mimic natural debris flows in their entirety; instead, idealized models are developed to simplify the complex nature of debris flows, thereby enabling their key features to be examined in a reproducible and systematic manner. Reliable high-quality data can then be used to calibrate theoretical and numerical models (Ng, 2014).

It is common to see theoretical and numerical models being directly evaluated using field data. However, one should be cautious when a "good" match is achieved between calculated and simulated results with field data given the complexity and heterogeneity of materials and complexity of the conditions in natural environments. As

M. Jakob et al. (eds.), *Advances in Debris-flow Science and Practice*, Geoenvironmental Disaster Reduction,
https://doi.org/10.1007/978-3-031-48691-3_4

such, it is crucial to obtain evidence from physical experiments that omit competing effects that are not of interest by adopting well-controlled material composition, and boundary and initial conditions, as this facilitates the development of well-characterized and simplified benchmarks to evaluate theoretical and numerical models. Ultimately, any high-fidelity theoretical or numerical model should be able to replicate at least the simplest cases at any scale.

Many experimental studies have been published in the literature in recent decades to investigate the initiation (Iverson, 2000), transportation and entrainment (Iverson et al., 2011; Song & Choi, 2021), deposition (Major, 1996), and impact mechanisms of debris flows (Ng et al., 2021a; Vicari et al., 2022). For example, Song et al. (2017) described the physical modeling of debris flows with different solid fractions impacting rigid barriers in the geotechnical centrifuge, and Ng et al. (2019) reported novel geotechnical centrifuge modeling techniques to investigate the effects of dynamic fragmentation on the impact dynamics on a rigid barrier. These studies have used various types of physical models, such as steady and unsteady flumes of various scales, and geotechnical centrifuges.

In this chapter, we describe the fundamental aspects of scaling that must be considered when modeling debris-flow processes, the working principles and applications of commonly used model types, and the state-of-the-art instrumentation and image analysis techniques that are needed to obtain meaningful experimental data.

4.2 Scaling

Any reliable physical model must be subjected to careful scaling considerations. Otherwise, engineers and scientists cannot apply the experimental outcomes to the field scale. In this section, we briefly discuss key aspects of scaling related to quasi-static and flow processes, and how these aspects are used to correlate phenomena at different scales and to guide the design of experiments.

4.2.1 Quasi-Static Processes

Soil is a stress- and path-dependent material, and thus the initial state of a slope-forming soil before failure influences the mobility of a debris flow that may occur because of failure (Iverson et al., 2000). Figure 4.1 illustrates the important effect of stress state on the shear and volumetric response of a soil slope, based on the critical state soil framework (Schofield & Wroth, 1968). NCL is the normal compression line, corresponding to the virgin compression of soil. CSL is the critical state line, where plastic shearing continues indefinitely without changes in volume or effective stresses, i.e., $\frac{\partial p'}{\partial \varepsilon_q} = \frac{\partial q}{\partial \varepsilon_q} = \frac{\partial v}{\partial \varepsilon_q} = 0$ (p', q, v, and ε_q represent mean effective stress, deviator stress, specific volume and shear strain, respectively) (Wood, 1991). Soil sample A, which is initially located below the CSL, dilates and thus generates negative pore-water pressures when it is sheared under a relatively low mean effective stress, p' (e.g., at laboratory scale). In contrast, soil sample B, which has the same initial void ratio as soil sample A but is initially located above the CSL, contracts and thus generates positive pore-water pressures when it is sheared under relatively high mean effective stress (e.g., at field scale). This demonstrates that soil motion rates at an incorrect initial state in the laboratory condition are decreased by negative pore-water pressures (Iverson et al., 2000). Thus, to give

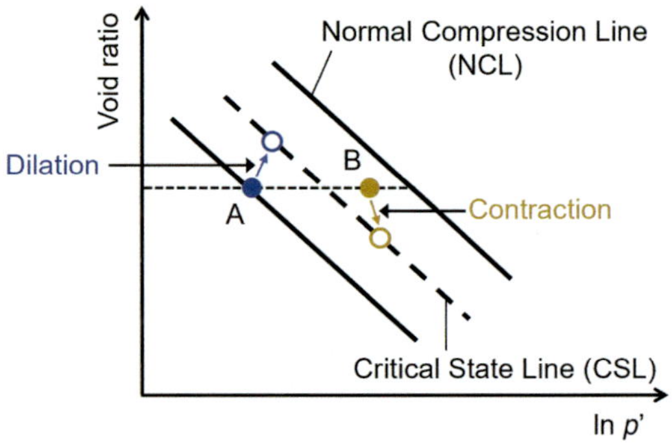

Fig. 4.1 Distinct responses of two soil samples of the same density sheared under different confining stresses (Ng, 2014)

the correct response, a physical model must be constructed to faithfully represent the initial state of a soil slope in the field.

4.2.2 Flow Processes

It is well established that debris flows are scale-dependent (Iverson, 1997), and this is most easily demonstrated by using the following dimensional analysis equation to describe the dynamics of an idealized fluid-filled landslide (Iverson, 2015; Iverson et al., 2004):

$$\frac{\overline{u}}{\sqrt{gH}} = f\left(\frac{H}{L}, \frac{\sigma}{\rho g H}, \frac{t}{\sqrt{\left(\frac{L}{g}\right)}}, \theta, \phi, m, \frac{\rho_s}{\rho_o}, \frac{\rho_f}{\rho_o}, \frac{c_f}{\rho_o g H}, \frac{E_f}{\rho_o g H}, \frac{E}{\rho_o g H}, \frac{(L/g)^{1/2}}{\mu_f H^2/kE}, \frac{\rho_o H (gL)^{1/2}}{\mu_f}\right) \tag{4.1}$$

where $\frac{\overline{u}}{\sqrt{gH}}$ represents the Froude number, $\overline{u}$ is the flow velocity, H is the flow depth, L is the flow length, σ is the internal stress, ρ is the bulk density, g is the gravitational acceleration, t is time, θ is the channel inclination, ϕ is the basal Coulomb friction angle, ρ_s is the solid density, ρ_f is the fluid density, ρ_o is the initial bulk density, c_f is the cohesive shear strength of the fluid phase, E_f is the elastic bulk modulus of the fluid phase, E is the elastic bulk modulus of the flow, μ_f is the viscosity of the fluid phase, k is the hydraulic permeability, and m is the solid volume fraction. Among the dimensionless numbers shown in Eq. 4.1, the following two numbers play key roles in the scaling of debris flow experiments:

$$N_{\mathrm{Re}} = \frac{\rho_o (gL)^{1/2}}{\mu_f / H} \tag{4.2}$$

$$N_{\mathrm{P}} = \frac{(L/g)^{1/2}}{\mu_f H^2 / kE} \tag{4.3}$$

where N_{Re} is the Reynolds number, which is the ratio of flow inertial force and viscous shear resistance; and N_{P} is the pore pressure diffusion number, which is the timescale ratio between the motion of debris flow and slope-normal diffusion of excess pore-fluid pressure. Based on the assumption that ρ_o, L, μ_f, k, and E are constant for debris flows with the same composition but different flow depths, Fig. 4.2 shows that N_{Re} is proportional to H, whereas N_{P} is proportional

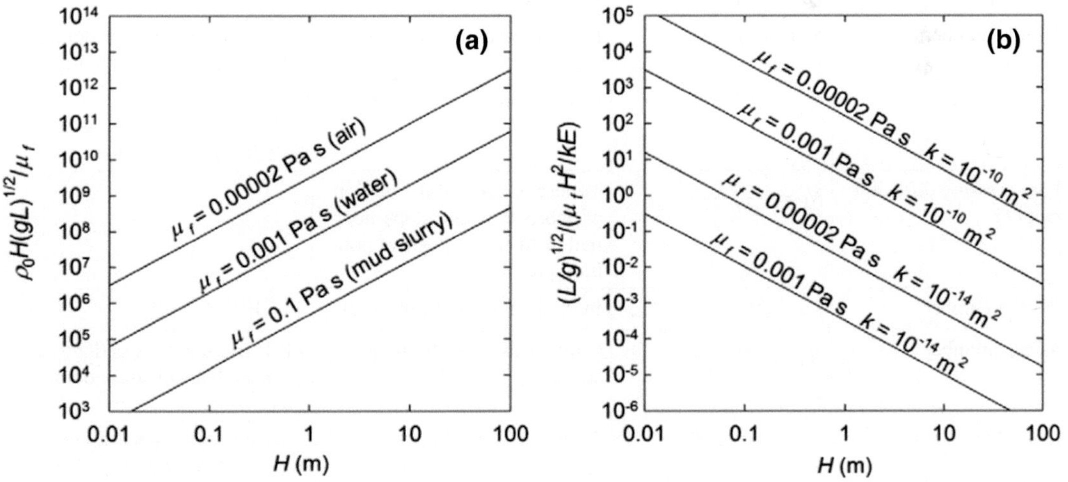

Fig. 4.2 Graphs of the Reynolds number (**a**) and pore pressure diffusion number (**b**) as functions of landslide or debris-flow size (described by the flow depth H) for various typical interstitial fluid viscosities of air, water, and mud slurry and for various typical hydraulic permeabilities of clean sand and poorly sorted debris with a muddy matrix. These graphs were generated by using the following quantities and fixed values: a length-to-scale ratio (H/L) = 100 (typical of many mass movements), a compressive stiffness (E) = 1 × 10^7 Pa (typical of loose granular soils), and a bulk density (ρ_0) = 1,500 kg/m^3 (typical of loosely packed granular soils) (Iverson, 2015; Iverson et al., 2004)

to $1/H^2$. Thus, if H is decreased by a factor of n times, such as by converting from field scale to laboratory scale, N_{Re} decreases by a factor of n and N_P increases by a factor of n^2. Consequently, the viscous shear resistance increases by a factor of n, whereas the pore-pressure diffusion decreases by a factor of n^2. Therefore, to maximize the geomorphological relevance of debris flow experiments to field events, such experiments must be conducted at the largest feasible scales. A summary of key dimensionless numbers is given in Table 4.1.

4.3 Experimental Models

4.3.1 Flume Models

Chow (1959) defined a flume as a channel with a regular geometrical cross-section that is supported on or above the surface of the ground to transport fluids or material mixtures. Flume modeling was adopted from hydraulic engineering to study debris-flow processes in the early 1900s (Gilbert & Murphy, 1914), and the flume models reported in the literature have ranged in size from those that fit on laboratory benchtops (MiDi, 2004) to those that have been constructed on natural hillsides (Iverson et al., 2010; Paik et al., 2012; Won et al., 2016). Although scaling considerations (Sect. 4.2.2) mean that the largest possible flume models should be used to closely replicate the dynamics of field-scale debris flows, laboratory-scale flume models still have utility for the simulation of landslide processes provided that the results are interpreted within the constraints of the range of dynamics that can be achieved. In this section, we describe the various types and scales of flumes used to conduct debris-flow research, and we examine their advantages, limitations, and range of applications.

Figure 4.3 shows a comparison of the ranges of N_{Re} and N_P for debris flows modeled by laboratory-scale and field-scale flume models and geotechnical centrifuge models that have been reported in the literature. The ranges of N_{Re} and

Table 4.1 Key dimensionless numbers in debris flow research

	Formulation	Description	Threshold value
Froude number	$Fr = \frac{\bar{u}}{\sqrt{gH}}$	Inertial force/gravitational force	<1: gravitational force dominates >1: inertial force dominates
Reynolds number	$N_{Re} = \frac{\rho_o H (gL)^{1/2}}{\mu_f}$	Inertial force/viscous force	<500: laminar flow 500–12,500: transitional flow >12,500: turbulent flow
Pore pressure diffusion number	$N_P = \frac{(L/g)^{1/2}}{\mu_f H^2/kE}$	Timescale ratio of the motion of debris flow and slope-normal diffusion of positive pore-fluid pressure	
Inertial number	$I = \frac{\dot{\gamma}\delta}{\sqrt{P/\rho_0}}$	Inertial stress/confining stress	1×10^{-5}–1×10^{-1}(a)
Savage number	$N_S = \frac{\rho_s}{\rho_s - \rho_f} \frac{\dot{\gamma}^2 \delta^2}{gH}$	Collisional stress/frictional stress	<0.1: frictional stress dominates >0.1: collisional stress dominates
Bagnold number	$N_B = \frac{v_s}{1 - v_s} \frac{\rho_s \dot{\gamma} \delta^2}{\mu_f}$	Collisional stress/viscous stress	<40: viscous stress dominates >450: collisional stress dominates
Friction number	$N_{fric} = \frac{v_s}{1 - v_s} \frac{(\rho_s - \rho_f) gH}{\dot{\gamma} \mu_f}$	Frictional stress/viscous stress	<100: viscous stress dominates >100: frictional stress dominates

(a) Typical range reported for natural debris flows (Jerolmack & Daniels, 2019)

N_P for typical field debris flows are based on the assumption of a typical flow velocity of 10 m/s, field flow depths from 0.2 to 4.0 m (Hong et al., 2015), and interstitial fluid viscosities from 0.001 to 0.1 Pa s (Iverson, 2015). In addition, the length-to-scale ratio, compressive stiffness, and bulk density used are the same as those adopted in Fig. 4.2.

4.3.1.1 Laboratory-Scale Flume Models

No clear definition exists of what constitutes a "small-scale" or a "large-scale" flume model. Moreover, even the largest flume models in the world can only simulate an initial debris volume on the order of 1×10^2 m^3, whereas the largest subaerial debris volumes can be as high as 1 ×

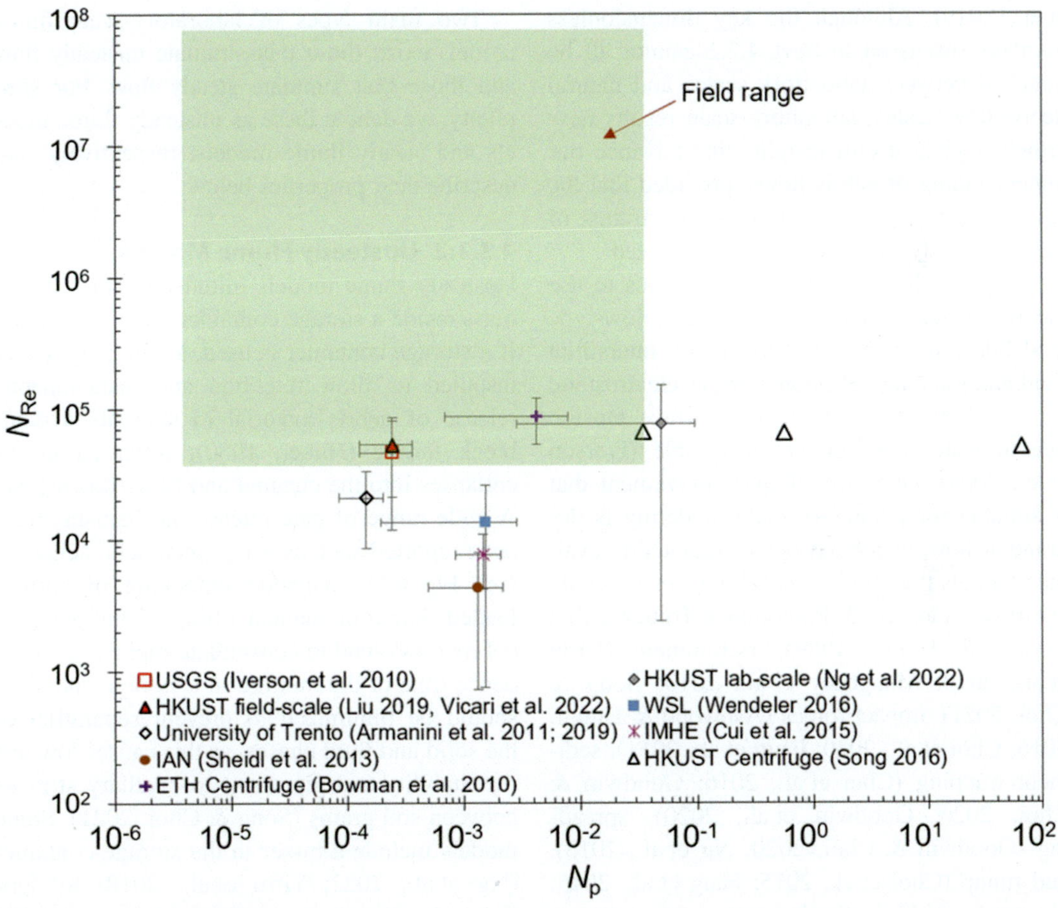

Fig. 4.3 Range of Reynolds numbers (N_{Re}) and pore pressure diffusion numbers (N_P) of debris flows generated by flume models at various institutions worldwide: a field-scale flume model at the United States Geological Survey (USGS) (Iverson et al., 2010); laboratory-scale and field-scale flume models at the Hong Kong University of Science and Technology (HKUST) (Liu, 2019; Ng et al., 2022; Vicari et al., 2021); a laboratory-scale flume model at the Swiss Federal Research Institute (WSL) (Wendeler & Volkwein, 2015); a laboratory-scale flume model at the University of Trento (Italy) (Armanini et al., 2011, 2019); a laboratory-scale flume model at the Institute of Mountain Hazards and Environment (IMHE), Chinese Academy of Sciences (Cui et al., 2015); a laboratory-scale flume model at the Institute of Mountain Risk Engineering (IAN) at the University of Natural Resources and Life Sciences (Austria) (Scheidl et al., 2013); a geotechnical beam-centrifuge model at the HKUST (Hong Kong) (Song, 2016); and a geotechnical drum-centrifuge model at the Swiss Federal Institute of Technology (ETH) (Bowman et al., 2010)

10^9 m^3 (Takahashi, 2007). As such, laboratory-scale flume models are regarded as those that can be installed in a conventional laboratory.

Laboratory-scale flume model tests are cost- and time-effective to perform, and have been used to investigate various debris-flow processes, such as runout (de Haas et al., 2015; Jing et al., 2018), entrainment (de Haas & van Woerkom, 2016; Li et al., 2018, 2020), and flow-barrier interaction (Ng et al., 2021a; Song et al., 2021; Wendeler & Volkwein, 2015; Zhou et al., 2019). Although the key dimensionless numbers discussed in Sect. 4.2.2 cannot all be matched between laboratory scales and natural debris-flow scales, laboratory-scale results nevertheless yield useful insights that advance our understanding of debris flows, provided that the results are interpreted within the constraints of the range of dynamics that can be achieved.

For example, it may be advantageous to use laboratory-scale models to simulate flow-type landslides, in which the effects of interstitial fluid and granular cohesion, such as electrostatic forces, chemical bonding, and surface tension due to water menisci, are negligible (Iverson et al., 2004). One type of mass movement that is suitable for laboratory-scale modeling is dry granular flow, which has been simulated to evaluate models that predict mobility (Iverson et al., 2004; Ng et al., 2013; Pudasaini & Hutter, 2007; Savage & Hutter, 1989), entrainment (Farin et al., 2014; Mangeney et al., 2010; Song & Choi, 2021), impact forces (Ashwood & Hungr, 2016; Choi et al., 2020; Faug et al., 2011), sediment trapping (Choi et al., 2016; Goodwin & Choi, 2020; Goodwin et al., 2020), spreading (Goodwin & Choi, 2020; Ng et al., 2013), and runup (Choi et al., 2015; Faug et al., 2015; Ng et al., 2017a). Furthermore, although dry granular flows are regarded as geomorphologically irrelevant to natural debris flows (Iverson, 2015), dry granular flow experiments serve as idealized benchmarks for the evaluation of theories and numerical models. Any sufficiently robust theory or numerical model should be able to replicate the dynamics of the simplest flow cases, such as a frictional dry granular flow at the laboratory scale.

Another type of flow that is amenable to laboratory-scale simulation is mud flow with a large concentration of suspended particles (i.e., silt and clay). For example, Jing et al. (2018) used a laboratory-scale model to develop scaling principles of mud flows and to study their deposition morphology. Laboratory-scale flume models have also been used to investigate the rheological behaviors of mud flows (Ancey & Cochard, 2009; Chambon et al., 2014; Coussot, 1995).

Two main types of laboratory-scale flume models exist: those that simulate unsteady flow and those that simulate steady flow. For simplicity, we denote these as unsteady flume models and steady flume models, respectively, and describe their properties below.

4.3.1.2 Unsteady Flume Models

Unsteady flume models initially retain a debris mass inside a storage container, tank, or hopper. If a storage container is used, a gate is typically installed to allow retention and instantaneous release of debris material to simulate a dam-break failure (Huber, 1980). Released debris collapses into the channel and flows downslope. A wide range of gate release mechanisms have been reported such as using pneumatic actuators (see Fig. 4.4), magnetic unlocking of spring-loaded doors, or manual lifting of the gate. As debris flows tend to consolidate rapidly, the residence time of the mixture in a storage container should be minimized to prevent separation of the solid and fluid phases, as this causes low test repeatability and gives rise to capillary stresses between soil grains (Song & Choi, 2021). Some models include a mixer in the storage container (Ng et al., 2022; Yifru et al., 2018) to continuously disturb the solid–fluid mixtures and thereby hinder consolidation.

Among the different modes of initiation, Iverson (2015) noted that debris flows in the field can be initiated from a static sloping ground in response to subtle and progressive changes in pore water pressures, which lead to force imbalances and slope instability, culminating in landslides (Iverson, 2000). In contrast, a dam-break model instantaneously generates a

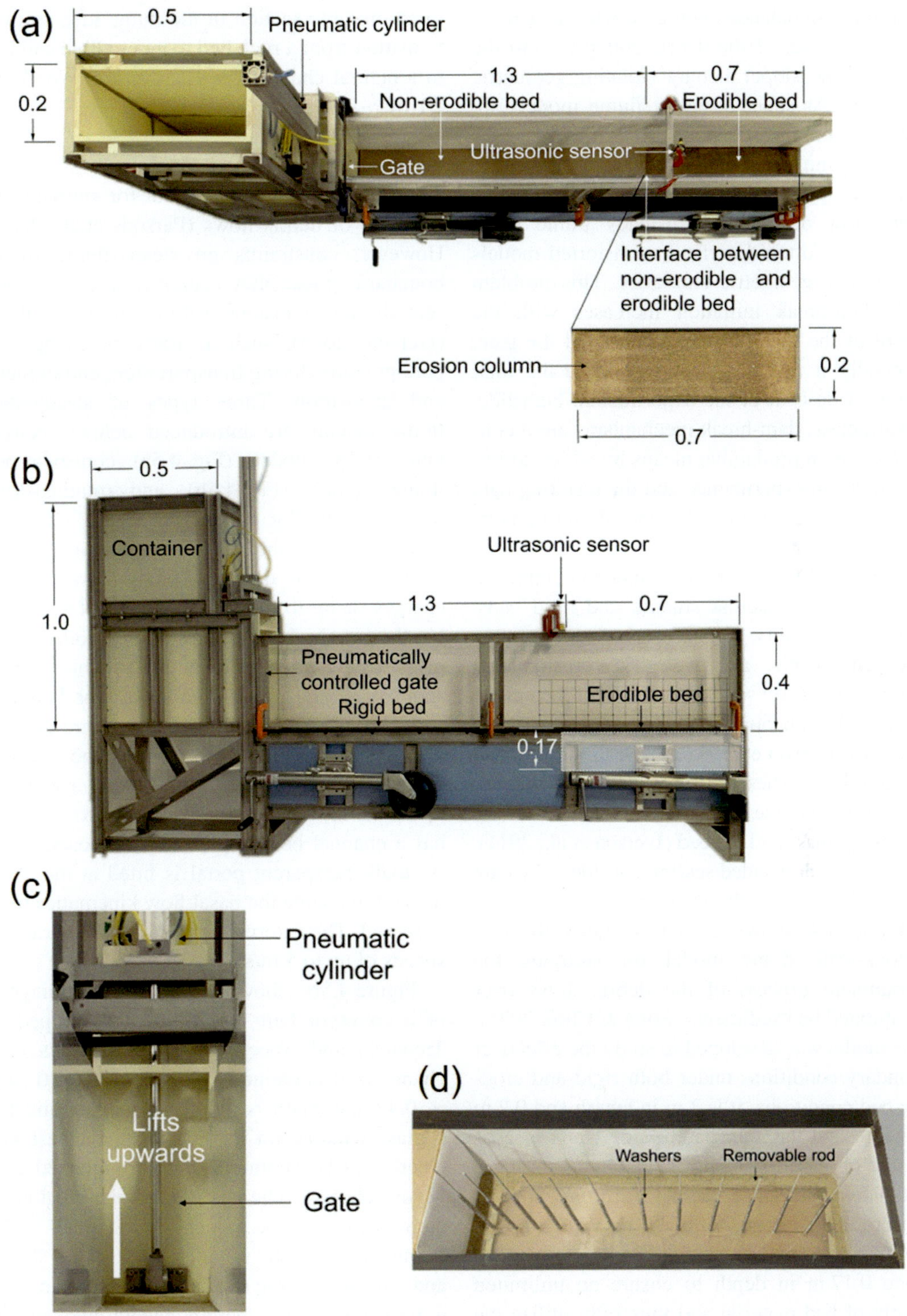

Fig. 4.4 A laboratory-scale flume model for studying entrainment (Song & Choi, 2021). **a** Top view of the flume; **b** side view of the flume; **c** front view of the pneumatically controlled gate; **d** top view of the erodible section, with the erosion columns devoid of soil. All dimensions are in meters. The pneumatically controlled gate lifts vertically to simulate dam-break initiation

large force imbalance upon gate release, and the downslope face of the debris collapses into the channel downslope, thereby violating geometric similarity. As such, unsteady flume models are generally only suitable for studying processes following initiation and cannot model the entire landslide process from slope instability to debris-flow deposition. Unsteady flume models are thus often described as distorted models that violate geometric similarity. This problem with dam-break initiation increases with the height of the material retained behind the gate, especially if this height is similar to the total height of descent of the experimental landslide. Nevertheless, dam-break mechanisms are a convenient and reproducible means by which to initiate flows in experiments, and the resulting data can be used to evaluate theoretical and numerical models.

Most unsteady flume models replicate flows that move across smooth and rigid beds. However, in reality, soil beds are rough, deformable, entrainable, and wet. The erosion and entrainment of wet soil beds are two of the unique features that enable debris flows to travel long distances (Iverson et al., 2011). Scaling must also be considered when modeling flow–bed interactions, especially the scaling of the thickness of the overriding mass and the bed (Iverson et al., 2011). However, such detailed scaling considerations are beyond the scope of this chapter.

Figure 4.4 shows a typical unsteady laboratory-scale flume model for studying the entrainment process of the debris flows over unsaturated bed sediments (Song & Choi, 2021). This model was developed to study the effects of boundary conditions under both rigid and erodible bed conditions. It is 2 m in length and 0.2 m in width and has transparent side walls to enable the observation of flow kinematics during a test. The channel bed has a 1.3-m long, smooth non-erodible section, followed by a 0.7-m long erodible section. The erodible bed was designed to be 0.17 m in depth to ensure an unlimited supply of bed material and thus fully utilize the erosion potential of the debris material in the storage container. The transition from a rigid to an erodible section in the flume idealizes the transition from a rock bed to an erodible soil bed in a natural channel (de Haas & van Woerkom, 2016).

4.3.1.3 Steady Flume Models

Steady flume models are ideal for studying the rheology of debris flows (Parsons et al., 2001). However, constraints on steady-flume model boundaries mean they cannot be used to represent the key transient features of debris flows (Iverson, 2015), such as transient changes in pore pressure during transportation, entrainment, and deposition. Three types of steady-state flume models are introduced below: recirculating flume models (Fig. 4.5a), conveyor belt flume models (Fig. 4.5b), and rotating-drum flume models (Fig. 4.6).

Figure 4.5a shows a recirculating flume model that can develop steady flows at discharges of up to 20 L/s (Armanini et al., 2005; Larcher et al., 2007). This flume model has a glass-walled open channel that enables image acquisition and an external conveyor belt that recirculates a solid–fluid debris mixture upslope after it has reached the bottom of the channel. The channel has a length of 6 m and a width of up to 0.4 m, can be tilted from 0° to 25°, and has a channel bed roughened with coarse sand. A small transparent portal is fitted at the channel bed to enable the basal flow kinematics to be observed. The external conveyor belt operates at speeds of up to 5 m/s.

Figure 4.5b shows a schematic diagram of a conveyor belt flume model developed by Trewhela and Ancey (2021). It consists of a channelized container that is 1.41 m × 0.14 m × 0.42 m (length × width × height) and has a glass window on its side to enable test kinematics to be captured, and a roughened conveyor-belt base. The latter is a typical feature of conveyor belt flume models (Marks et al., 2017; Perng et al., 2006; van der Vaart et al., 2018), and it serves to drag debris to counterbalance the gravitational stress, thereby generating a steady-state flow. This model can simulate flow depths of up to 0.15 m.

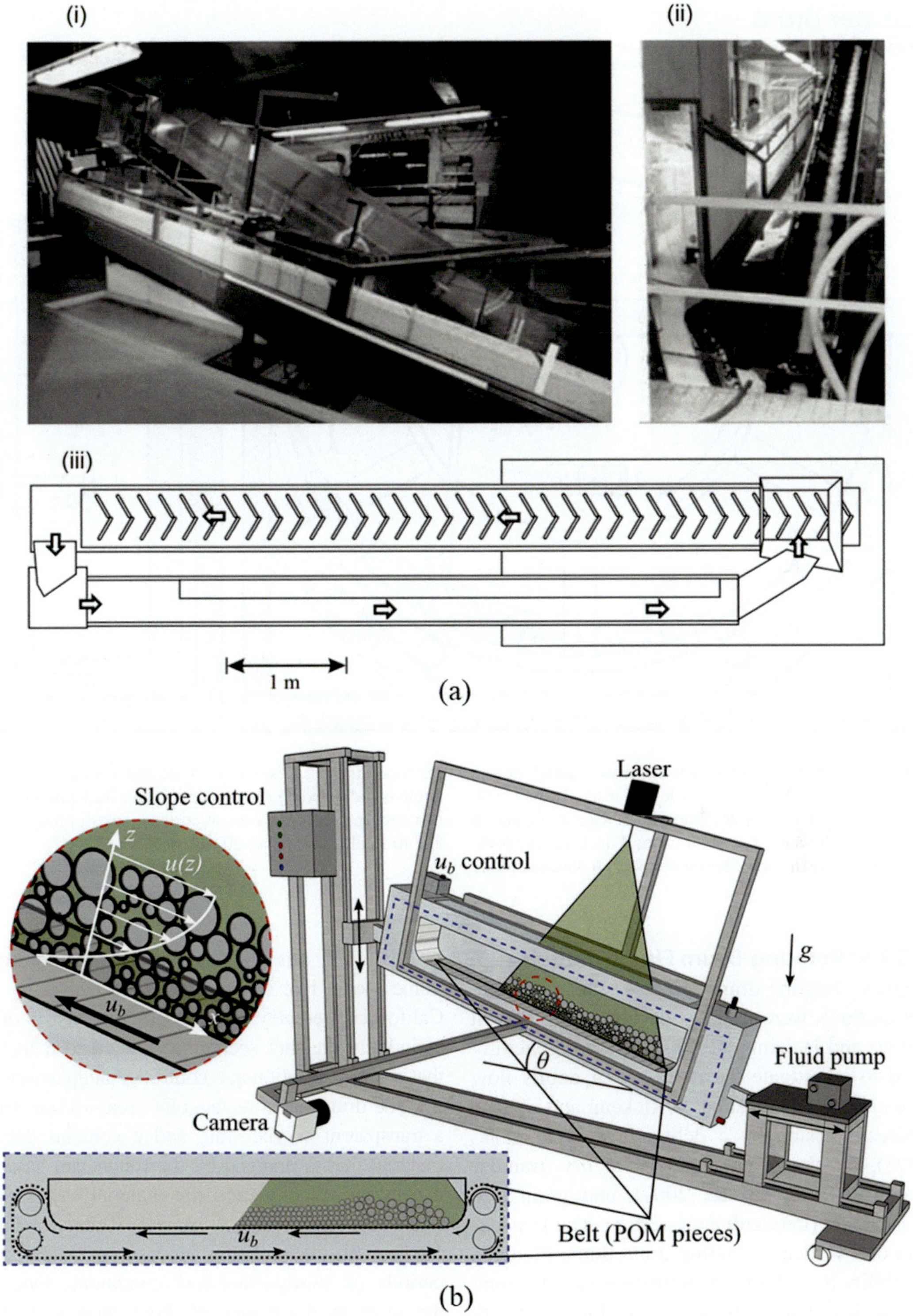

Fig. 4.5 Steady flumes: **a** recirculating flume—(i) side view; (ii) downstream view; and (iii) plane view (Armanini et al., 2005); **b** conveyor flume (Trewhela & Ancey, 2021)

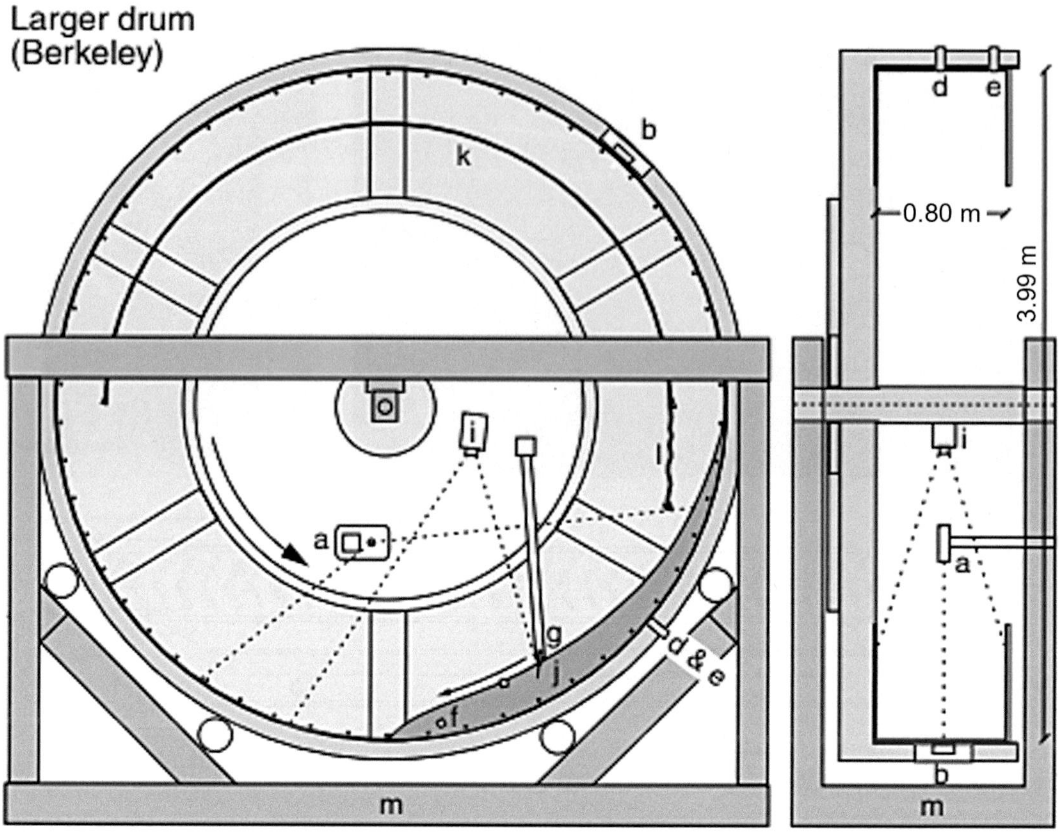

Fig. 4.6 A typical rotating-drum flume model at the University of California, Berkeley: **a** laser; **b** load cell; **c** shear cell; **d** pore-water pressure sensor at center; **e** pore-water pressure sensor at side; **f** in-flow temperature loggers; **g** in-flow thermocouples; **h** thermocouple for recording the laboratory temperature; **i** camera (view angle is indicated); **j** gravel–ice mixture flow lobe; **k** roof (for drop protection); **l** brush/curtain (splash protection); and **m** steel frame (Hsu, 2010)

4.3.1.4 Rotating-Drum Flume Models

Vertical rotating-drum models that simulate debris flows were initially developed by Brown (1992) and Huizinga (1993). They are generally used to investigate the rheology of debris flow (Huizinga, 1993; Kaitna & Rickenmann, 2007), bedrock erosion by a debris flow (Hsu et al., 2008), the flow characteristics of dry granular material (Chou & Lee, 2009), and grain size segregation (Leonardi et al., 2015). The key feature of a vertically rotating drum flume model is its ability to sustain a recirculating flow for long timescales at a controllable rotational velocity (Hsu, 2010).

Figure 4.6 shows a typical rotating-drum flume model that is installed at the University of California, Berkeley (Hsu, 2010). It consists of a cylindrical channel section connected to a motor that drives its motion at a controlled angular velocity. The drum has a rectangular cross-section with a transparent acrylic wall, and it contains debris material that is dragged by the roughened base at a constant velocity until the material is released under the influence of gravity (Cabrera et al., 2015). This recirculating process generates long periods of steady-state flow conditions. One of the inherent limitations of these models is that debris material becomes attached to the roughened base, and it subsequently drops from the base or is trapped on the inner roof of the drum, which results in a gradual decrease in the debris volume during a test (Cabrera et al., 2015; Leonardi et al., 2015).

Six typical flow regimes have been identified based on the operational conditions (e.g. rotational speed; filling level, which is the maximum depth of initial mass over the diameter of the drum; and flow-bed interface friction). These regimes are slipping, slumping, rolling, cascading, cataracting, and centrifuging (Henein et al., 1983), and are depicted in Fig. 4.7. The rolling regime was reported to be able to replicate the dynamics of dense granular flows (MiDi, 2004), which are flows with persistent force chains between particles. The other flow regimes may be less relevant for studying debris flow processes.

The various flow regimes that evolve under different operational conditions (rotational speeds, filling levels, and Froude numbers) in a specific type of flowing material can be characterized in a bed behavior diagram (Henein et al., 1983; Mellmann, 2001). Figure 4.8 shows a bed behavior diagram for a dry sand flow in a rotating-drum flume model (Henein, 1981; Henein et al., 1983). Detailed descriptions of the transitions between various regimes based on mathematical models and experimental measurements have been provided by Henein et al. (1983) and Mellmann (2001).

4.3.1.5 Field-Scale Flume Models

As mentioned in Sect. 4.2.2, the largest feasible facilities should be used to model debris flows. The term "field-scale" is used instead of "large-scale" because the former better describes models that do not fit inside conventional laboratories, whereas the latter is an ambiguous descriptor of natural events. Field-scale flume models have been used for studying the onset of slope failure (Iverson, 2000; Iverson

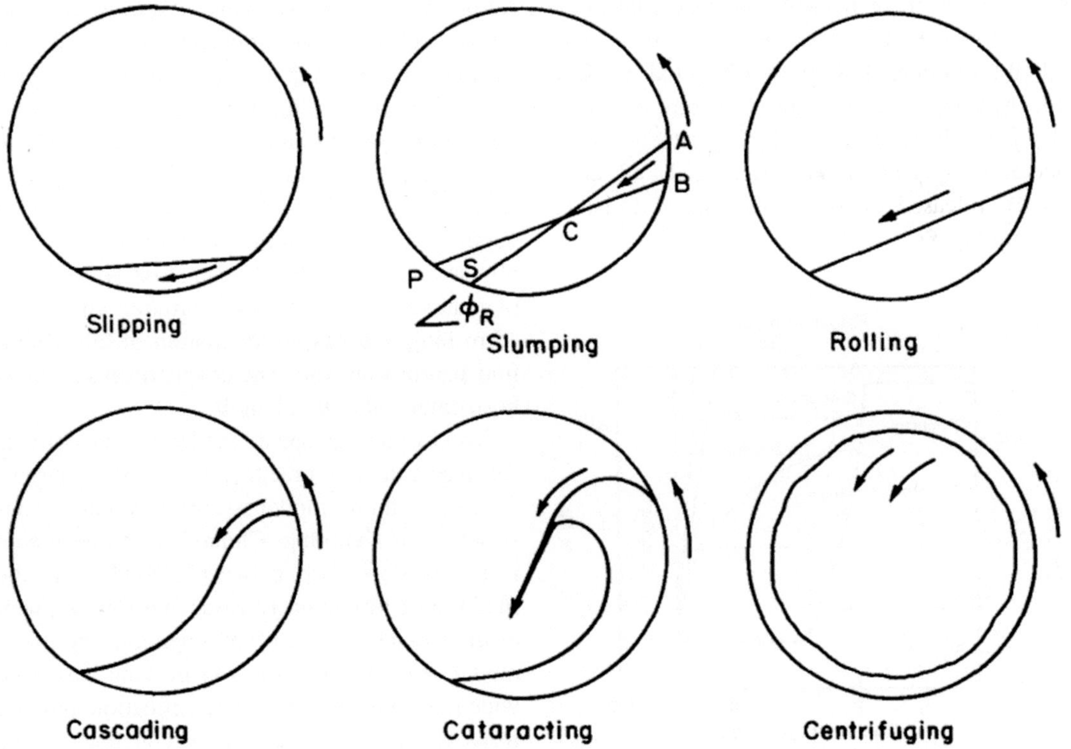

Fig. 4.7 Schematic diagram showing different regimes of transverse bed motion in a vertically rotating-drum flume model (Henein et al., 1983)

et al., 2000) or analyzing debris-flow dynamics, impact on structures, and runout (Bugnion et al., 2012; Iverson et al., 2010, 2016; Ng et al., 2021a; Paik et al., 2012; Vicari et al. 2021; Won et al., 2016).

Field-scale flume models operate like unsteady laboratory-scale flume models, with a storage container at the upslope end of an inclined channel that retains the debris material, and they typically feature a gate to allow the retention and release of debris material. However, gates for field-scale flume models are subject to higher soil pressures than laboratory-scale flume models, and thus the former require more rigorous design than the latter. To this end, field-scale flume model gates are often reinforced and equipped with seals around their edges to prevent water from leaking and accelerating the consolidation of debris material inside the storage container. In addition, when the gate in a field-scale flume model is released, the debris material can flow down either synthetic or natural channels. The field-scale flume models described in Iverson et al. (2010), Tan et al. (2018), and Ng et al. (2021a, 2021b) released debris onto concrete and steel channels. In contrast, the experiments conducted by Bugnion et al. (2012) and Paik et al. (2012) released debris down natural hillsides and gullies.

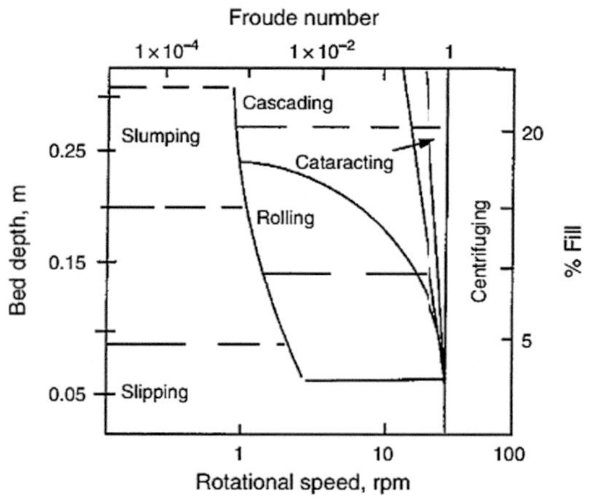

Fig. 4.8 A bed behavior diagram for a typical sand (limestone) (Henein, 1981)

Akin to laboratory-scale flume models, the boundary conditions of field-scale flume models can be varied along a channel bed. For example, the United States Geological Survey flume model initially had a smooth, broom-finished concrete surface; this was subsequently roughened by anchoring specially fabricated rough concrete tiles to better model natural conditions (Iverson, 2020). Compared with the distance of debris runout on the smooth bed, the distance of debris runout on the roughened bed was much greater because of the size-segregation of grains, which led to the formation of coarse-grained lateral levees and constrained the lateral spreading of debris flows (Iverson, 2020).

Through a research collaboration between The Hong Kong University of Science and Technology (HKUST) and The University of Hong Kong, funded by the Research Grants Council of HKSAR, a 28-m-long field-scale flume model was developed and built at the Kadoorie Centre of The University of Hong Kong (see Fig. 4.9a). This flume is used to model debris flows impacting on rigid and flexible barriers and, more recently, has also focused on replicating debris flows on an erodible bed impacting on barriers (Ng et al., 2021a; Vicari et al. 2021). The steel flume has a uniform cross-section with a width of 2 m and a depth of 1 m. The storage container inclines at 30°, is 5 m in length, and can store up to 10 m^3 of material. The main transportation channel is 15 m long and has an inclination of 20° with a transparent side wall. The downstream section is horizontal and 8 m in length.

To further enhance modeling capabilities for research and practice, the world's largest field-scale flume model is currently under construction in Kunming, China. The flume model is jointly developed between HKUST and the Institute of Mountain Hazards and Environment of the Chinese Academy of Sciences. The flume model (Fig. 4.9b) has a 144-m long and 6-m wide main channel with an inclination ranging from 16° to 32°. The storage container at the upstream end of the flume can store up to 500 m^3 of material.

Fig. 4.9 **a** The 28 m-long field-scale flume in Hong Kong, China. **b** Design drawing of 172 m-long flume in Kunming, China

4.3.2 Geotechnical Centrifuge Models

Geotechnical centrifuge models generate increased gravitational acceleration to produce identical self-weight stresses to their corresponding prototypes, thereby enabling unique experimental evidence to be gathered on complex problems. In recent years, such models have become increasingly popular for studying mass movements (Bowman et al., 2010; Bryant et al., 2015; Ng et al., 2017b; Song et al., 2017, 2018). In particular, it has been shown that the flow regimes achieved in geotechnical centrifuge models better represent those of field-scale debris flows than the flow regimes achieved in other flow models (Bowman et al., 2010).

The two main types of geotechnical centrifuge models are beam-centrifuge models and drum-centrifuge models: the former have a box container and the latter have a circular channel container. Figure 4.10 shows examples of a geotechnical beam-centrifuge model and a geotechnical drum-centrifuge model at HKUST. The geotechnical beam-centrifuge model (Fig. 4.10a) (Ng, 2014) has a diameter of 8.4 m and is equipped with two swinging platforms. The platforms can accommodate a model size of 1.5 m × 1.5 m × 1.0 m and can be subjected to a centrifugal acceleration of up to 150 g bearing a payload of up to 400 g-ton. The geotechnical drum-centrifuge model (Fig. 4.10b) (Lu & Zhang, 2020) has a circular channel with an inner diameter, height, and depth of 2.2 m, 0.7 m, and 0.3 m,

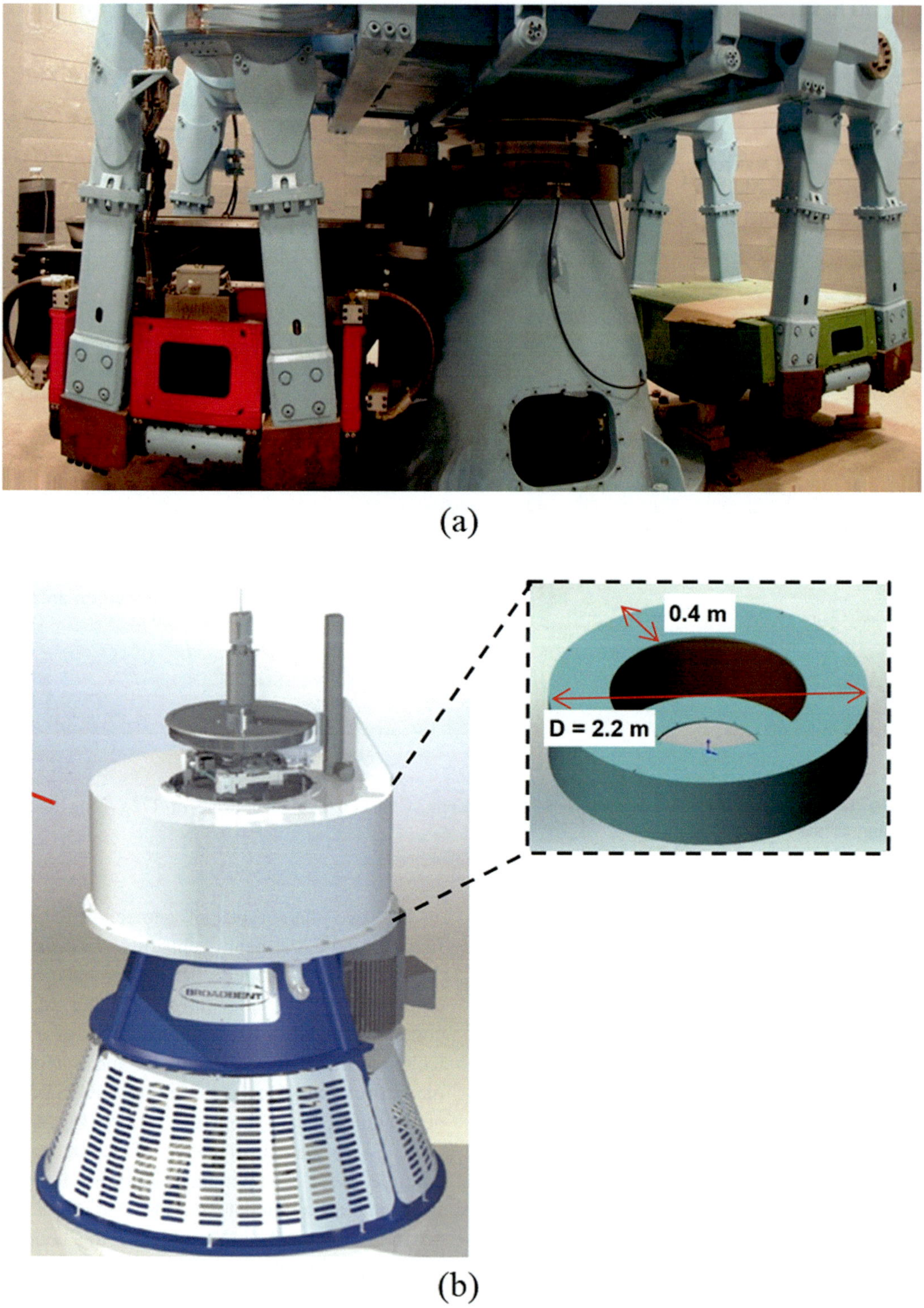

Fig. 4.10 The two geotechnical centrifuges at HKUST: **a** the geotechnical beam-centrifuge, which has a diameter of 8.4 m, a maximum *g*-level of 150 g, and a maximum payload of 400 g-ton (Ng, 2014); **b** the geotechnical drum-centrifuge, which is under installation and has an inner diameter of 2.2 m, a model channel depth of 0.4 m, and a maximum payload of 870 g-ton (Lu & Zhang, 2020)

respectively. The channel can be accelerated up to 440 g bearing a payload of up to 870 g-ton.

Some key advantages of geotechnical beam-centrifuges are that the model preparation, instrumentation and in-flight manipulation are simpler than those in a drum-centrifuge. Moreover, it is easier to preserve the deposited profile of the debris after spinning down, thereby enabling close examination of the composition and morphology of the debris material. However, it is more ideal to use a drum-centrifuge to investigate flow mobility as it has a long flow path. It is also easier to inject debris material from outside of a drum-centrifuge model to prevent in-flight consolidation of the debris material (Bowman et al., 2010).

The first attempt at modeling debris flows in a centrifuge was conducted by Bowman et al. (2010), who studied entrainment in a geotechnical drum-centrifuge model at ETH. Subsequently, Gue (2012) used a geotechnical drum-centrifuge model to investigate submarine debris flows and develop relevant scaling laws. Ng et al. (2016) pioneered a geotechnical beam-centrifuge model to simulate the impact of granular flow against a flexible barrier. Moreover, Song et al. (2017, 2018) used the geotechnical beam-centrifuge to determine the mechanisms by which debris flows affect rigid and flexible barriers in HKUST.

4.3.2.1 Working Principles

Figure 4.11 shows a schematic of the acceleration field acting on a single particle of debris flow in a geotechnical beam-centrifuge model as a function of its distance from the axis r, its velocity u, and any other external force (Cabrera et al., 2020). The rotation axis is defined as $\mathcal{O}$ and the angular rotation vector is defined as $\boldsymbol{\omega}$. The centrifugal acceleration field for a centrifuge test, including the Earth's gravitational acceleration $\boldsymbol{g}$, can be expressed as follows:

$$\boldsymbol{a} = \boldsymbol{a}_{\mathrm{cf}} + \boldsymbol{a}_{\mathrm{co}} + \boldsymbol{g} \tag{4.4}$$

where $\boldsymbol{a}_{\mathrm{cf}}$ is the centrifugal acceleration and $\boldsymbol{a}_{\mathrm{co}}$ is the Coriolis acceleration. The centrifugal acceleration $\boldsymbol{a}_{\mathrm{cf}}$ acts radially outward from the rotation axis, and is expressed as follows:

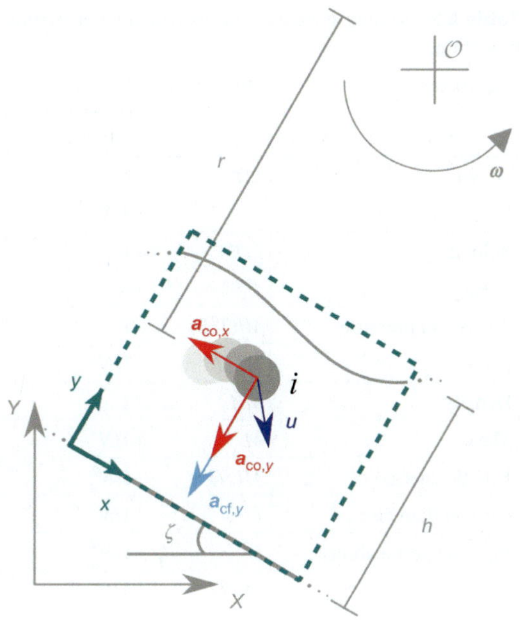

Fig. 4.11 Centrifugal acceleration field components $\boldsymbol{a}_{\mathrm{cf}}$ and $\boldsymbol{a}_{\mathrm{co}}$ on a single particle i with local velocity $\boldsymbol{u}_{\mathrm{i}}$ in the global coordinate system (X, Y), with particle i rotating around $\mathcal{O}$ at a distance r_{i} and at an angular velocity $\boldsymbol{\omega}$; ζ is the slope angle (Cabrera et al., 2020)

$$\boldsymbol{a}_{\mathrm{cf}} = \boldsymbol{\omega}^2 \boldsymbol{r} \tag{4.5}$$

The virtual Coriolis acceleration $\boldsymbol{a}_{\mathrm{co}}$ also deflects the debris flow material, and is defined as follows:

$$\boldsymbol{a}_{\mathrm{co}} = 2\boldsymbol{v} \times \boldsymbol{\omega} \tag{4.6}$$

The Earth's gravitational acceleration $\boldsymbol{g}$ cannot be ignored, as it acts orthogonally to $\boldsymbol{a}_{\mathrm{cf}}$. However, with sufficiently large values of $\boldsymbol{a}_{\mathrm{cf}}$, the effect of gravity becomes secondary. In centrifuge modeling, $\boldsymbol{a}_{\mathrm{cf}}$ is usually related via a scaling factor N to the radius R of the centrifuge model, in the form of $\boldsymbol{a}_{\mathrm{cf}} = N\boldsymbol{g} = \boldsymbol{\omega}^2\boldsymbol{R}$. Therefore, a field-scale problem can be downscaled in a geotechnical centrifuge model by a factor of N while also rotating the model up to a centrifugal acceleration proportional to N.

Typical scaling laws have been well established for geotechnical centrifuge modeling

Table 4.2 Scaling laws used in geotechnical centrifuge modeling

Parameter	Dimension	Scaling law (model/prototype)
Gravity	L/T^2	N
Linear dimension	L	$1/N$
Velocity	L/T	1
Shear rate	$1/T$	N
Stress and pressure	$M/(T^2L)$	1
Force	ML/T^2	$1/N^2$
Density	M/L^3	1
Mass	M	$1/N^3$
Kinetic energy	ML^2/T^2	$1/N^3$
Particle diameter	L	$1/N$
Intrinsic permeability	L^2	$1/N^2$
Pore fluid viscosity	$M/(TL^3)$	$1/N$
Flexible barrier stiffness	M/T^2	$1/N$
Bending moment	ML^2/T^2	$1/N^3$

(Ng, 2014; Song et al., 2018), and a summary is given in Table 4.2.

Examination of the scaling laws reveals that a geotechnical centrifuge model has the advantages of enabling the modeling of large flow depths, particle sizes, and flow volumes. As the gravitational acceleration in such models increases by a factor of N, the length scale dimension decreases by a factor of N. As such, if we abide by scaling laws for a geotechnical centrifuge model of scaling the flow depth by a factor of N, then much deeper flows can be simulated in the model. According to the scale of linear dimensions, the volume scales by a factor of N^3, which is important because field observations have suggested that the runout distance of landslides increases as their initial volume increases (Heim, 1932; Lucas et al., 2014). The energy in a geotechnical centrifuge model also scales by a factor of N^3, and thus, such models are ideal for evaluating energy-based approaches for the design of barriers (Sun & Law, 2015) for resisting debris flows. Following the work by Bowman et al. (2010), Ng et al. (2017a, 2017b) exploited the advantages of a geotechnical centrifuge model to study debris flow impacts on barriers. In such modeling, both the barrier and flow are subjected to the same gravitational field, and thus, the same set of scaling principles can be adopted to scale an impact problem (Ng et al., 2016). Moreover, geotechnical beam-centrifuge models are usually used to model flow-structure interactions, whereas geotechnical drum-centrifuge models are more widely used in modeling flow mobility.

4.3.2.2 Modeling Considerations for Debris Flows

One of the challenges in studying debris flows in a geotechnical centrifuge model is to match the inertial timescale (which scales by a factor of $1/N$) with the diffusional timescale (which scales by a factor of $1/N^2$); these can be quantified by the shear rate and soil permeability, respectively (Bowman et al., 2010). According to the scaling laws shown in Table 4.2, the shear rate scales by a factor of N, and the intrinsic permeability of soil scales by a factor of $1/N^2$. Bowman et al. (2010) reported that the enhanced gravitational field accelerated the timescale for pore pressure diffusion indicated by the Hagen–Poiseuille equation, where the soil hydraulic permeability k is proportional to $\rho_f g R_p^2/\mu_f$ (where ρ_f is the density of the fluid, g is the gravitational acceleration, R_p represents the characteristic particle size for soils, and μ_f is the dynamic viscosity of fluid). Consequently, Bowman et al. (2010) developed two approaches for matching the inertial and diffusional timescales in geotechnical centrifuge modeling: increasing the viscosity of the pore fluid or decreasing the particle size.

In the first approach, the pore fluid is changed from water to a mixture of water and glycerine to increase the fluid viscosity by a factor of N, which increases the period required for diffusion-consolidation by a factor of N^2 and increases inertia by N. Therefore, the overall scales of inertia and diffusion are the same as those of the prototype that uses water. In the second approach, the particle size is scaled by a factor of $N^{0.5}$ at the same pore fluid viscosity. Thus, based on the Hagen–Poiseuille equation, the intrinsic permeability k is scaled down by a factor of N, meaning that both the inertial and diffusional timescales match at 1/N of the prototype timescale. Song et al. (2017) developed another approach that involves reducing both the particle size and the pore fluid viscosity by a factor of N to fully match the dimensionless group of numbers (the Froude number, Savage number, Bagnold number, and pore pressure diffusion number shown in Table 4.1) with prototype conditions.

Furthermore, Bowman et al. (2010) reported that the debris material consolidated in-flight and became unable to flow. Thus, modeling techniques were required to inject the material in-flight from outside of the geotechnical centrifuge model or to continuously mix the debris material in-flight. As such, Song et al. (2017) developed a storage container with a built-in mixer to mix the debris slurry in-flight before releasing it into the channel.

Another modeling consideration for debris flows is the Coriolis effect. Equation 4.9 indicates that at a given angular velocity, the Coriolis acceleration increases proportionately with the flow velocity. Although the Coriolis effect is negligible in traditional quasi-static geotechnical problems, it is non-negligible in dynamic processes such as debris flows (Bryant et al., 2015; Cabrera et al., 2020). Therefore, the results of geotechnical centrifuge model tests of debris flows must be carefully interpreted with reference to the Coriolis acceleration.

Bryant et al. (2015) conducted a series of geotechnical centrifuge model tests to investigate the role of the Coriolis effect on the runout behavior of dry granular flows (Fig. 4.12). They observed that when the Coriolis acceleration acted on a slope, it increased the effective gravitational force of the flow, thereby reducing the runout of the granular flow compared to tests performed with the same geometry under 1 g conditions. In contrast, when the additional Coriolis force acted away from a slope, the material had a reduced effective gravitational force, and a longer runout of the granular flow was observed. The Coriolis effect was also magnified in high-velocity simulations. Wang (2020) carried out a series of geotechnical centrifuge model tests on the impact of granular flow against barriers with different stiffnesses. The centrifuge tests were back-analyzed using discrete element method simulations. Computed results showed that without considering the Coriolis effect, the computed peak impact force was larger than the measured force by up to two times. Consistent results were obtained between physical centrifuge tests and numerical simulations only by considering the Coriolis effect.

Cabrera et al. (2020) performed a series of discrete element method simulations to model steady-state granular flows with a Bagnold-like velocity profile (Silbert et al., 2001) inside a rotating frame of reference. In these simulations, the Coriolis acceleration acted to compress the flow. The results showed that when the Coriolis acceleration was considered, both the g-level, N, and the distance from the rotation center to granular flow, R, significantly affected the flow dynamics. In particular, a large R was beneficial because it reduced the spurious effects of the non-uniform gravitational field created by the curvature of the centrifugal acceleration field and by the components of Coriolis acceleration. Therefore, to minimize the effects of the non-uniform gravitational field, they proposed that granular flow experiments should always maintain a rotation radius that is more than 50 times the flow depth (i.e., $R/H > 50$). In sum, the Coriolis acceleration reduces the flow velocity proportionally to Iu_{co}/u^*, where I is the inertial number, u_{co} is the Coriolis reference velocity, and u^* is a flow characteristic velocity:

$$u_{co} = \frac{\omega H^2}{d} \tag{4.7}$$

$$u^* = \sqrt{\omega^2 RH\phi \cos\theta} \tag{4.8}$$

4.4 Instrumentation

Experimental tests are only as useful as the data that can be meaningfully measured. Typical instrumentation and imaging techniques used for modeling landslides and debris flows are introduced in this section.

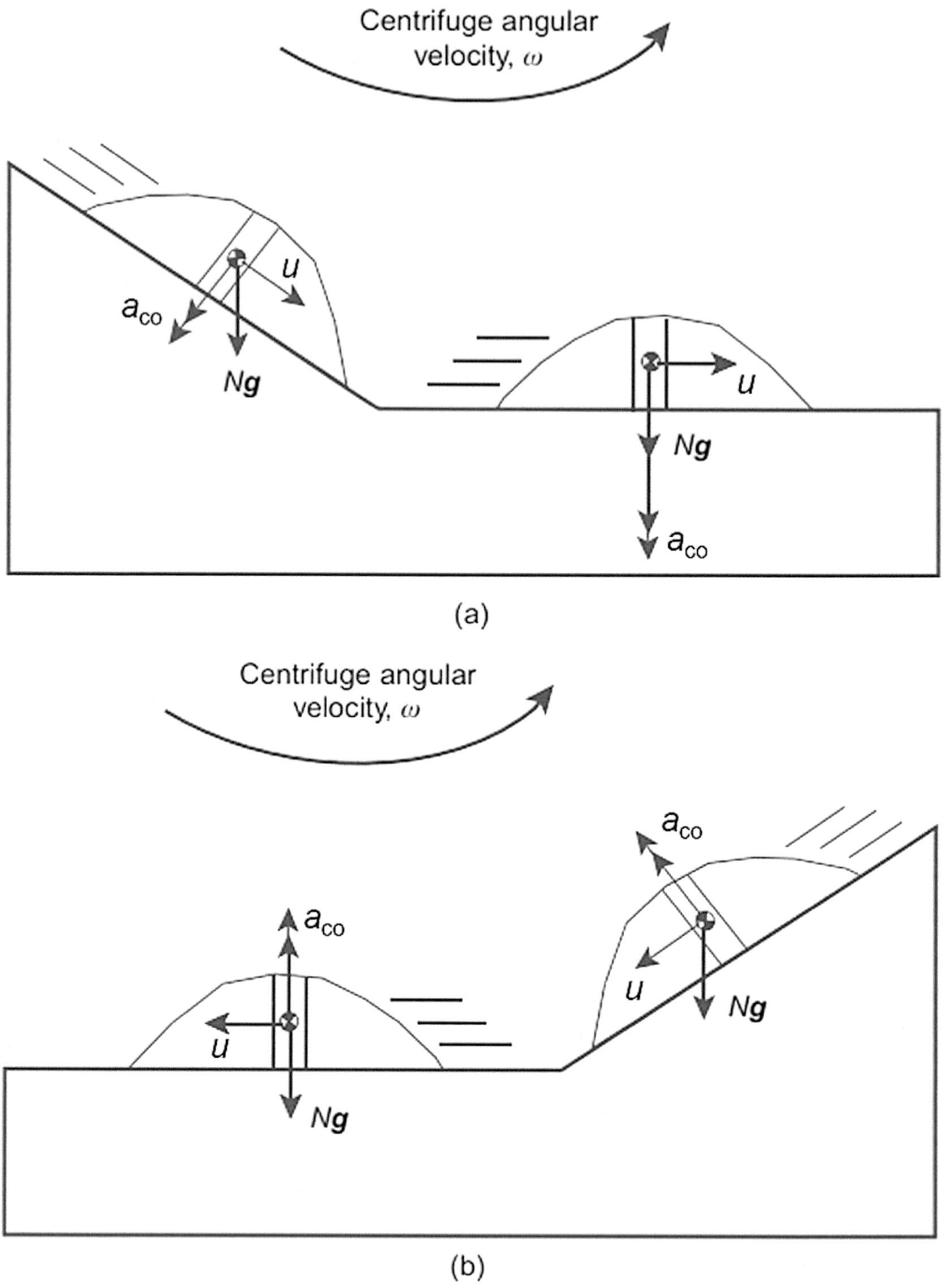

Fig. 4.12 Definition of the Coriolis effect in a centrifuge with **a** acceleration due to the Coriolis force ($\boldsymbol{a}_{co}$) acting on the slope and horizontal base and **b** acceleration due to the Coriolis force acting away from the slope and horizontal base. The normal force due to gravity (*N**g***) and the velocity of the landslide (***u***) are defined with respect to the center of gravity of a slice of the landslide mass (Bryant et al., 2015)

4.4.1 Flow-Bed Interfaces

To investigate the dynamics of debris flows, it is essential to obtain measurements of their basal normal and shear stresses, as well as their pore pressures. As such, force plates are often installed at the base of a channel (Iverson et al., 2010; Ng et al., 2021b). Figure 4.13 shows a typical force plate design. The plate is a circular polyvinyl chloride (PVC) plate (Fig. 4.13a) with a diameter of 0.32 m (with PVC being used to minimize inertial effects), which is supported by a triaxial load cell (Liu, 2019; Vicari et al. 2021) fixed to a stainless-steel frame (Fig. 4.13a and b). Conventionally, force plate designs use uniaxial load cells (McArdell et al., 2007). However, multiple uniaxial load cells are required to capture stresses in the shear and normal directions, so triaxial load cells are used to reduce the number of load cells and moving components. Triaxial load cells measure forces along three axes, such that the normal and shear stresses induced by a debris flow on a force plate can be simultaneously measured at the same point. Each triaxial load cell contains three orthogonal flexures, to each of which are attached four strain gauges. When a load is applied to an axis, the flexure corresponding to that axis deforms, and thus, the resistance changes. This change serves as a signal, which is calibrated and can then be correlated with the loads on all three axes.

The pore-fluid pressure is measured using piezoresistive pressure transducers with their diaphragms separated from the passing debris by a reservoir of water (Berger et al., 2010; Iverson et al., 2010; Ng et al., 2021b) or hydraulic oil (Kaitna et al., 2016; Nagl et al.,

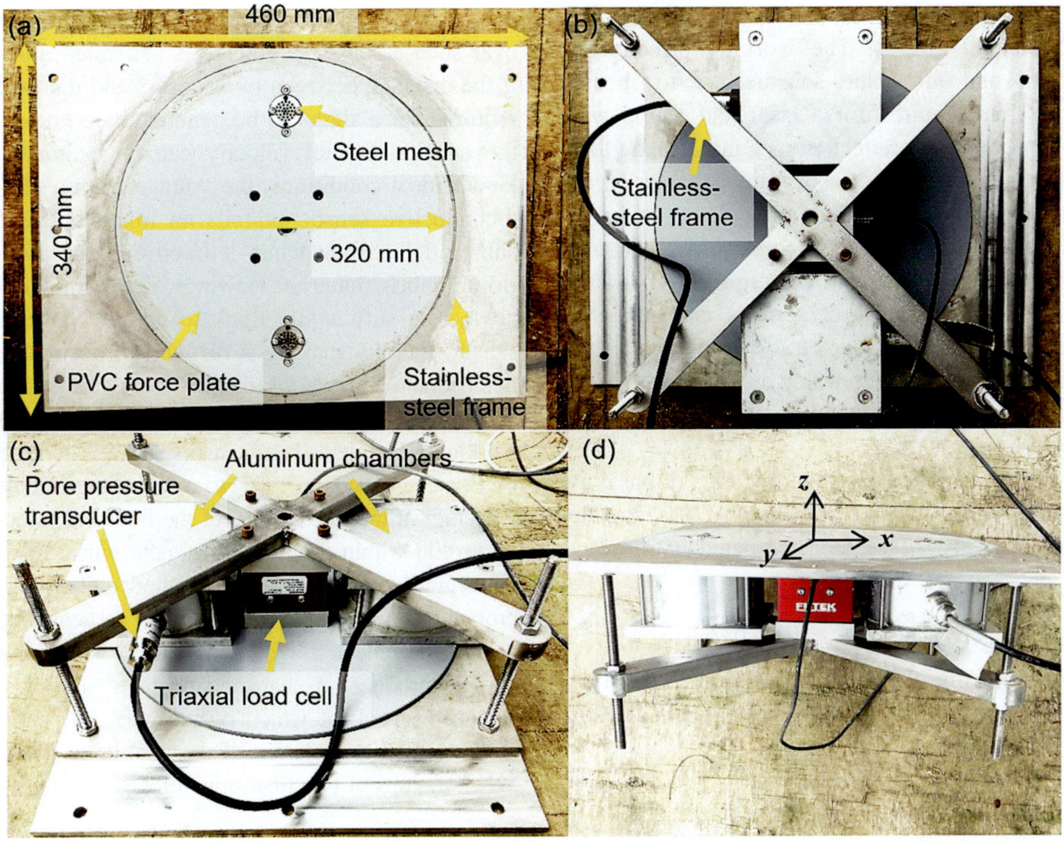

Fig. 4.13 Details of a typical force plate: **a** top view; **b** bottom view; **c** bottom side view; **d** top side view (Liu, 2019)

2020). Figure 4.13c shows the use of cylindrical chambers with an inner diameter of 0.10 m and a height of 0.05 m that house pore pressure transducers (PPTs). The opening at the top of each chamber is connected to the force plate and covered with a steel mesh to filter out fines to prevent them from entering the chamber (Fig. 4.13a). Before each test, the chambers are filled with water. The surfaces upstream and downstream of the PPT openings on the force plate should be as smooth as possible to ensure that accurate dynamic pore pressures are obtained. Moreover, the edges of the force plate should be sealed so that flow cannot uplift the force plate, as this would lead to erroneous basal stress measurements. The seals should also be soft enough to allow the force plate to deform to enable shear measurements to be obtained.

Laser and ultrasonic sensors are generally installed above the center of a force plate to measure the flow depth of the overriding debris (Iverson et al., 2010; Nagl et al., 2020; Ng et al., 2021b). The working principles of laser and ultrasonic sensors are to measure the time required for a laser and sound wave, respectively, to reflect from a target back to the transmitter. Laser sensors are capable of capturing details of a fluctuating flow surface and have a much shorter response time than ultrasonic sensors because light travels faster than sound. However, laser sensors are less effective when measuring transparent or translucent targets, such as water or muddy flows, and ultrasonic sensors thus provide more reliable measurements.

By identifying the time histories of basal normal stress $\sigma(t)$ and flow height $H(t)$, the flow bulk density $\rho_0(t)$ can be estimated as follows:

$$\rho_o(t) = \frac{\sigma(t)}{gH(t)\cos\theta} \tag{4.9}$$

Additionally, the volumetric solid fraction $\phi(t)$ can be calculated as follows (Nagl et al., 2020):

$$\phi(t) = \frac{\rho_o(t) - \rho_w}{\rho_s - \rho_w} \tag{4.10}$$

4.4.2 Frontal Velocity

An array of sensors may be installed along the flow direction to deduce the frontal debris velocity. For example, in the flume tests reported by Baselt et al. (2021), an array of pressure sensors and ultrasonic sensors were deployed along the flow direction (Fig. 4.14). Although these sensors were for measuring basal pressure and flow height, by tracking the moments at which the sensors generate readings, the kinematics of the flow could also be approximated (Liu, 2019). A similar example can be seen in Take et al. (2004), where a set of pore pressure transducers were deployed within a slope to monitor the pore pressure evolution. The readings of these transducers could be back-analyzed to reveal the pore pressure contour. Photoconductive sensors (a type of light sensor) embedded into the channel bed at fixed intervals can also be used to estimate the frontal velocity of a flow (Choi et al., 2015). When granular material travels over such a sensor, a voltage is generated; thus, if the distance between two sensors and the time required for a signal to be generated are known, the average frontal velocity can be estimated. Under ideal conditions, the voltage from a photoconductive sensor exhibits an immediate and substantial change when it is covered by overriding debris material. However, debris may not always be sufficiently thick to entirely cover a sensor, thereby making it difficult to accurately determine the arrival of the flow front at a sensor.

The fiber-optic probing method has a similar working principle as laser sensors and was adopted by Boateng and Barr (1997) to measure the particle velocities at the free surface of the flow and velocity profile within the flow in a geotechnical rotating-drum flume model. In this method, a fiber-optic probe transmits light to the target and subsequently detects the reflected light signal. The passage of a particle produces a characteristic quasi-sinusoidal signal, the peak of which occurs when the particle is located over the tip of the fiber-optic probe. When the tips of two fiber-optic probes are aligned

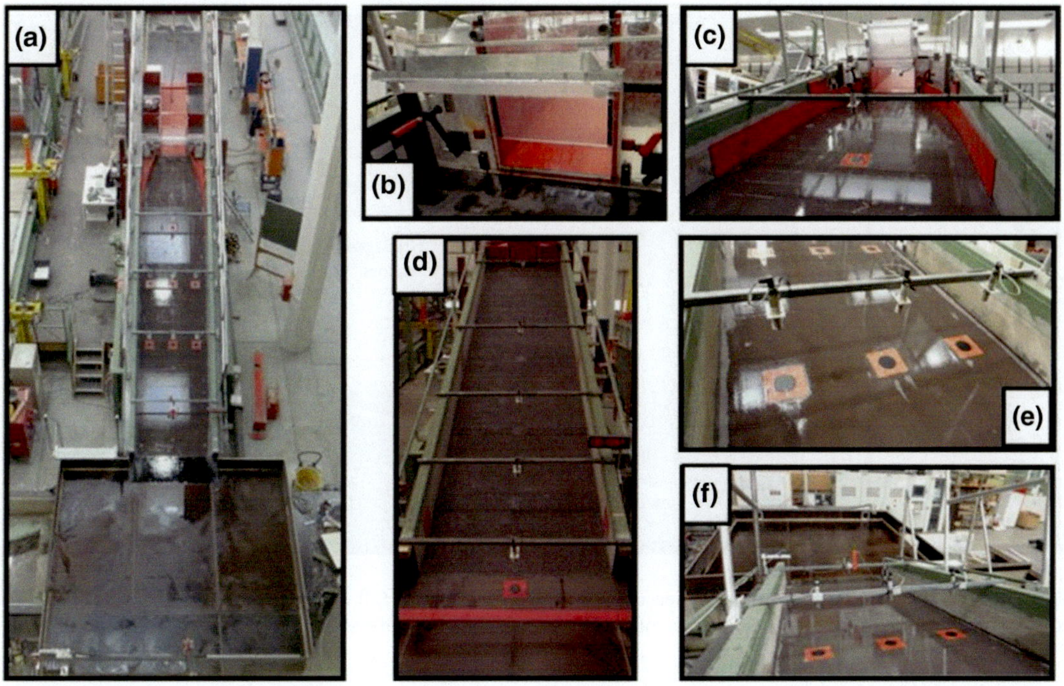

Fig. 4.14 Indirect tracking of flow kinematics using arrays of sensors (the black dots the inside red boxes on the slope; image from (Baselt et al., 2021). **a** The full view of the laboratory flume. **b** Storage container with flap system. **c** The up view of the flume close to the release area with installed deflection walls. **d** Flume with a rough surface. **e** Location of ultrasonic transducers and pressure sensors. **f** The end of the flume and the runout plane

with the streamwise direction of the flow, the passage of a single particle produces an identical signal in each probe but with a delay in the downstream signal equal to the time required for the particle to travel the distance between the probes.

4.4.3 Impact Force

Estimating the design impact forces exerted by debris flows on obstacles and barriers is an increasingly important engineering challenge (Goodwin & Choi, 2021; Ng et al., 2021a; Sun & Law, 2015). To model debris-flow impact on rigid barriers, piezoresistive load cells are fixed to force plates and a reaction frame to measure impact forces (Liu, 2019). Additionally, miniature load cells and PPTs can be embedded on a load-bearing surface to measure local forces and pore pressures (Song et al., 2021). As shown in Fig. 4.15, PPTs are covered with a steel mesh with square openings with a nominal length of 0.4 mm. The capillary tension between the water and the openings of the steel mesh prevents the outflow of the water inside the chamber. Another consideration in the design of load-measuring systems is to ensure that the force plate does not rotate or bend during impact. Furthermore, the force plate should not be in contact with the channel, as this would create additional frictional resistance. Goodwin and Choi (2021) reported scaling considerations for the design of rigid load-measuring systems, which include the relative mass of the grains and the load-measuring system, the stiffness of the load-measuring system, and the grain impact velocity. In the design of flexible barriers, cable forces are usually measured by using load cells to estimate the total impact forces. For example, Fig. 4.16 shows a flexible barrier model used in the geotechnical centrifuge flume model developed by Ng et al. (2016). Load cells are installed along

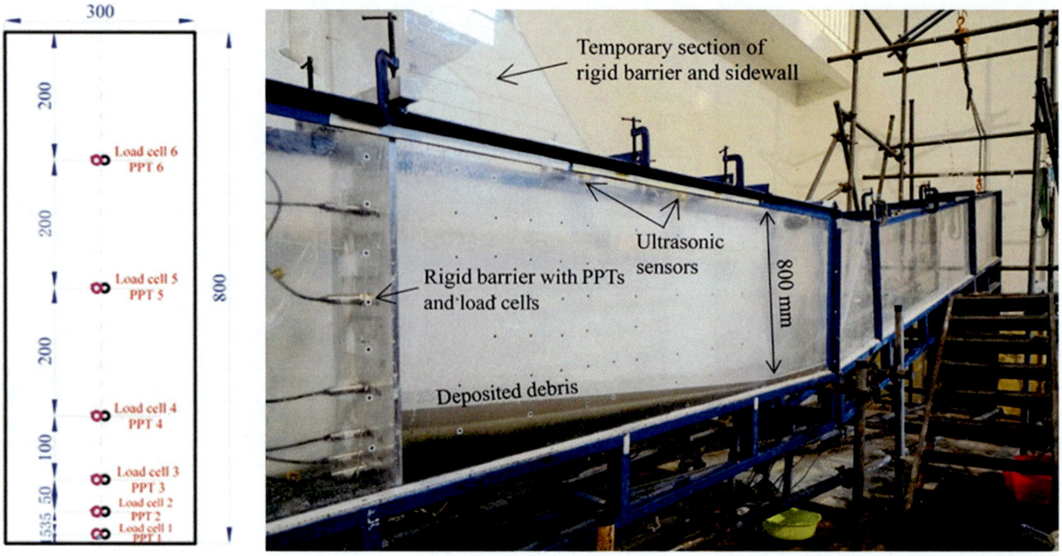

Fig. 4.15 Local load and pore pressure measurements on a rigid barrier (all dimensions in mm; Song et al., 2021): **a** schematic diagram of the load-bearing face of the barrier; **b** photograph of the flume setup from behind the barrier

each horizontal cable of the flexible barrier to measure axial forces, and laser sensors are used to measure the displacement of each cable.

4.4.4 Soil Bed Erosion

When modeling soil bed erosion, it is important to measure the erosion depth and rate. However, the erosion depth is difficult to measure because it is often masked by deposition. Berger et al. (2010) addressed this by developing an erosion sensor from a stack of 50-mm-tall aluminum cylinders buried in an erodible soil bed (Fig. 4.17). The height of each cylinder was selected to represent a characteristic grain size. Each cylinder had a resistor, which was connected to cylinders below and above using thin wires. When a cylinder was eroded from the soil bed by an overriding debris flow, the thin wires were pulled out, thereby breaking the circuit and causing a change in resistance. This erosion sensor could measure both the rate of erosion and the maximum erosion depth. A similar simplified concept was later used by Song and Choi (2021) and (Vicari et al. 2021) to measure the maximum erosion depth with erosion columns.

4.4.5 Image Analysis

Image analysis is the most adopted approach to deduce the velocity field of experimental debris flows. Image analysis techniques can be characterized as those that track foreign objects, texture, or overall kinematics.

The addition of tracers that can move with a flow is a commonly used method for capturing the flow kinematics. Such methods were pioneered by fluid researchers (Post and Walsum 1993), and include dye injection or smoke injection into a flow. The use of tracers has since been borrowed and modified by geotechnical and geophysical researchers. For example, in the two-dimensional soil collapse experiments reported by Bui et al. (2008), the deformation of soil was captured by preparing a grid consisting of dyed particles. By tracking the deformation of the grids, the deformation of soil could be back-calculated. Furthermore, in an experiment on snow failure (Gaume et al., 2018), a set of black markers were inserted into a snowpack. By tracking the displacement of these markers, the snowpack deformation was obtained.

Particle image velocimetry (PIV) (White et al., 2003) is one of the most commonly used

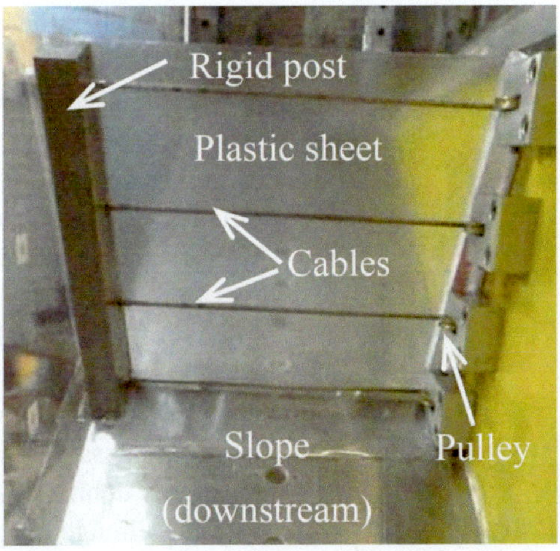

(a)

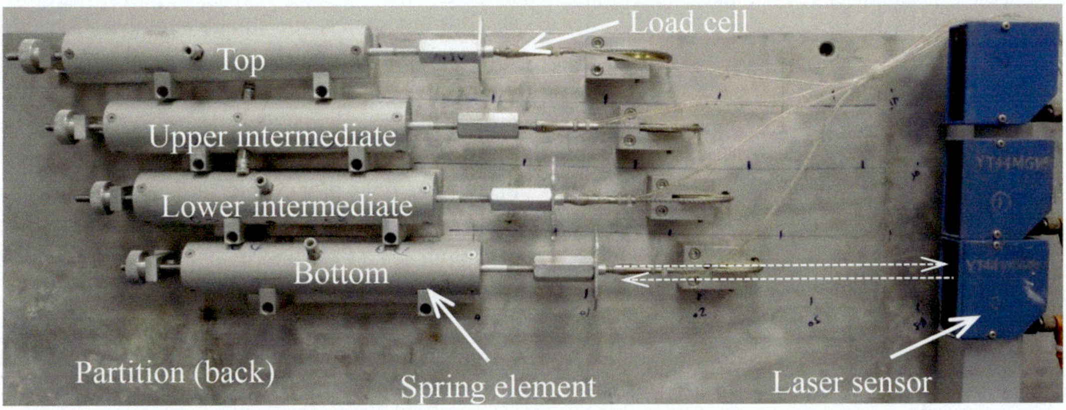

(b)

Fig. 4.16 A flexible barrier model used in geotechnical centrifuge modeling (Ng et al., 2016): **a** a rigid post, a plastic sheet, and four horizontal cables in front of the partition; **b** spring elements and instruments mounted on the back of the partition

methods to visualize the deformation of geomaterials and capture their flow kinematics (Song & Choi, 2021). When carrying out PIV analysis, video frames are divided into a grid of patches (Fig. 4.18a). The displacement of each patch between two consecutive frames can be detected and calculated by comparing the texture of the patches between the two frames. Consequently, the displacement field of the material is obtained. Another method often compared with PIV is particle tracking velocimetry (PTV) (Maas et al., 1993). Instead of comparing textures of patches (Fig. 4.18a), PTV tracks the velocity field of tracer particles (Fig. 4.18b). This means that PTV requires the addition of an adequate amount of tracer particles with sufficient color contrast against the background particles. PIV usually does not require any tracer particles for granular flows (Pudasaini et al., 2005), but if the experimental materials being examined by PIV do not have clearly distinguishable textures (e.g., clay), then the velocity fields may not be detectable. Aside from estimating the deformation and kinematics of flow material, PIV and PTV can also be used to estimate the granular temperature, which

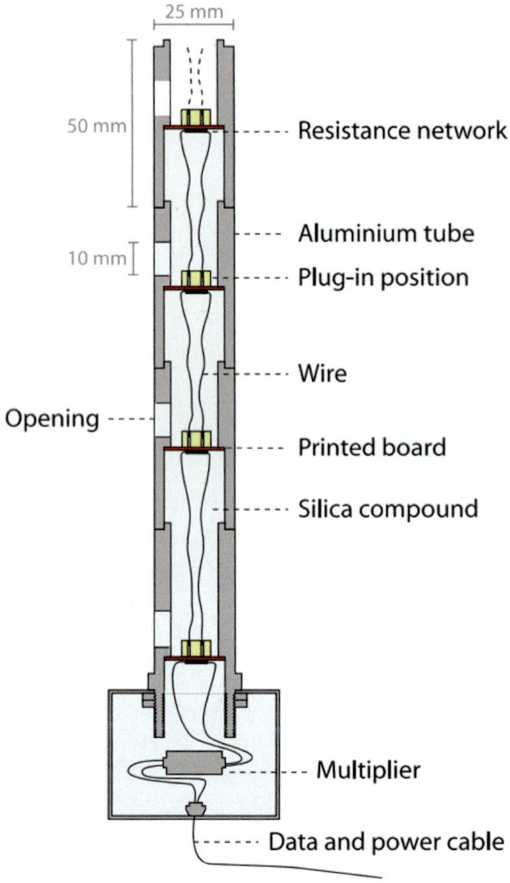

Fig. 4.17 Cross-section of the bottom end of an erosion sensor column. Each 50-mm tall cylindrical-machined aluminum element contained a printed circuit board with a resistor, all of the elements were connected with thin wires, and the entire column was filled with silicone to insulate the sensors from moisture (Berger et al., 2010)

is a key parameter for evaluating kinetic theories for granular flows (Taylor-Noonan et al., 2021). Reasonably accurate granular temperature can be obtained by using newly-proposed algorithms by Taylor-Noonan et al. (2021) together with a proper frame rate as recommended by Gollin et al. (2017). However, such approaches are still limited to coarse grains and the granular temperatures at the side wall may not be entirely representative of the temperatures in the flow body.

Textures can also be visualized by tracking the distortion of laser sheets projected onto a material's surface and then performing back-calculations to determine the material's topography (Fig. 4.19; Iverson et al., 2004). However, the field of view of this method is limited by the size of the laser device.

When dealing with rapid flows, it may be more important to focus on the overall flow kinematics than the internal field of deformation. Consequently, tracking the overall flow boundaries may be the key task. Flow boundary tracking is often equivalent to the machine vision process of segmenting pictures into backgrounds and flow bodies. Moreover, the image segmentation process required for visualizing geomaterials or debris flows involves both field-scale satellite image segmentation (Barlow et al., 2006) and laboratory-scale image segmentation (Luong et al., 2019). Although these images are at different scales, the image segmentation process is the same.

Segmentation is essentially a digital image processing-based classification task: each pixel of an image is classified as either a background pixel or a flow pixel. One popular method that is used to classify pixels is thresholding, which distinguishes a flow and its background based on color or color-related properties (Luong et al., 2019). However, simple thresholding methods struggle to process complex images. For example, when applied to an image with a complex background comprising both shadows and reflected lights (Fig. 4.20a), a classical thresholding method typically yielded noisy results (Fig. 4.20b and c). By comparison, our recent observations (Liang et al., 2023) suggested that a deep learning-empowered image segmentation method may outperform the classical method and yield a clean and accurate result (Fig. 4.20d).

4.5 Relationships Among Physical Modeling, Theoretical Modeling, and Field Monitoring

Physical models serve as a bridge between theoretical models and field observations. Figure 4.21 shows the inter-relationships among physical modeling, theoretical modeling and full-scale field monitoring/testing. These three

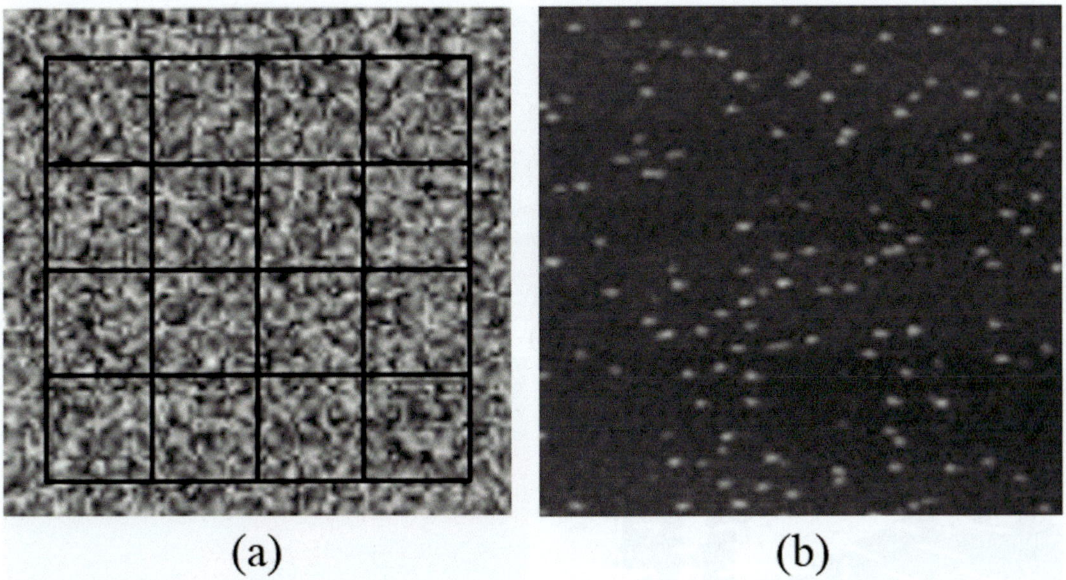

Fig. 4.18 **a** PIV can be used to track condensed textures of each patch (White et al., 2003); **b** PTV tracks scattered tracer particles (Maas et al., 1993)

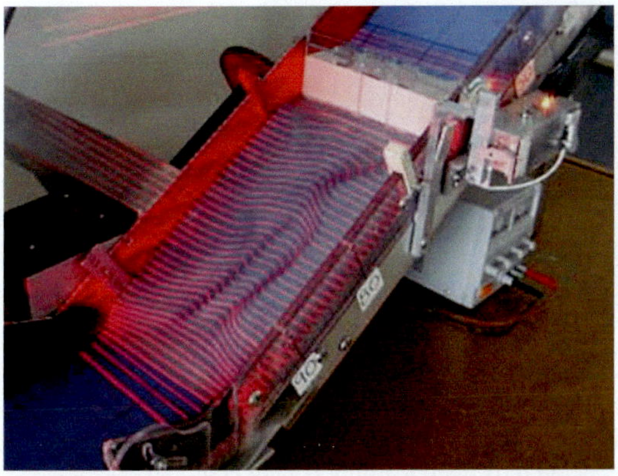

Fig. 4.19 Flume bed topography illuminated by superposed laser contours (Iverson et al., 2004)

methods are complementary to each other because there is no perfect method for solving debris-flow problems in terms of quality, reliability, time, and cost. These three methods each have their own advantages and disadvantages. Full-scale monitoring/testing may provide direct insight into the complexity of debris flows and can be used to evaluate proposed theory. However, field events are unique and difficult to compare, have poor spatiotemporal predictability, and are difficult to interpret. Physical modelling can produce experimental evidence in a systematic, cost and time effective manner. However, careful scaling considerations are required to ensure model-scale phenomena have relevance to prototype conditions. Numerical modelling enables detailed back-analysis and parametric studies to be carried out. However, numerical models require careful validation of the constitutive and rheological models used. More importantly, input parameters and model assumptions need to be carefully selected.

The principal applications of physical modeling include the following aspects: (i) modeling of prototypes such as debris-flow initiation, transportation, runout, and impact on barriers; (ii) investigating new phenomena and mechanisms; (iii) carrying out parametric studies; and (iv) calibrating theoretical and numerical methods.

Although physical modeling and field monitoring can produce data, the validation of theoretical models using field data requires caution due to potential compensating errors (Fig. 4.22). For instance, field data is subjected to many uncertainties because the actual ground conditions, flow compositions, soil

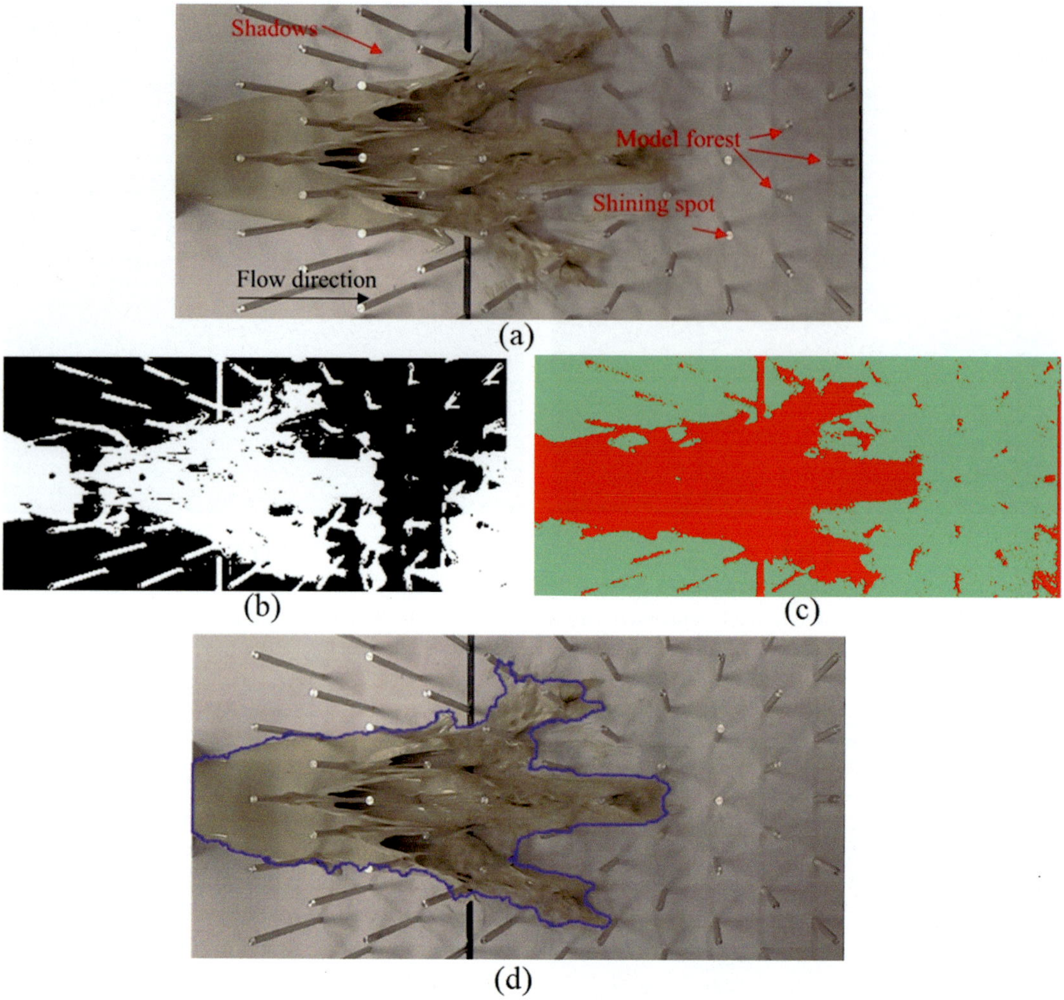

Fig. 4.20 **a** An example image of flow-structure interaction; **b** the noisy results of flow segmentation created by automated thresholding via ImageJ (Schindelin et al., 2012); **c** the noisy results of flow segmentation created by trainable Weka segmentation (Arganda-Carreras et al., 2017); **d** image segmentation results (the purple outline) generated by deep learning (Liang et al., 2023)

homogeneity, degree of saturation, number of debris surges, and boundary conditions are generally unknown. Any computed results that "match" with observed and measured field data may be fortuitous, resulting from compensating errors. In contrast, a mismatch between computed and measured data does not necessarily imply that the field measurements or numerical predictions are incorrect. As illustrated in Fig. 4.23, a missing link exists in the procedure of calibrating numerical models and their input parameters using field data. Prior to validating numerical models and calibrating model parameters against field data/case histories, a vital intermediate step of using physical model tests is desirable and necessary to provide "known" initial and boundary conditions and flow parameters for numerical modelers.

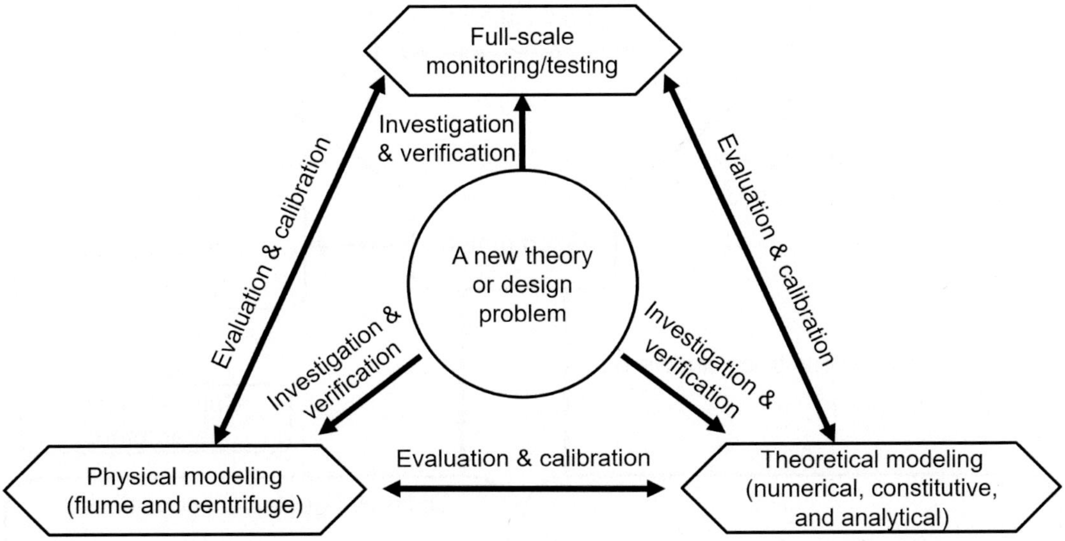

Fig. 4.21 Relationships among physical modeling, theoretical modeling and full-scale monitoring/testing (modified from Ng, 2014)

4.6 Concluding Remarks

This chapter reviews the fundamental scaling considerations, various types of commonly used physical models, and state-of-the-art instrumentation and imaging methods that have been used to model debris flows. The complex scale-dependent behavior of debris flows means that experiments conducted using field-scale flume models or geotechnical centrifuge models have the greatest geomorphological relevance to field flows. Nevertheless, laboratory-scale models are useful for revealing the rheological behavior of debris flow and validating theoretical and numerical models, provided that their results are interpreted within the constraints of the range of dynamics that such models can achieve.

Well-controlled flume and geotechnical centrifuge tests provide idealized and high-quality physical data to evaluate theoretical and numerical models. It is recommended that any numerical model should be validated by well-controlled model tests (i.e., flume or centrifuge tests) rather than field data prior to predicting actual performance in the field. This is because field data often involves many more uncertainties than those in experiments.

The past two decades have witnessed a great advancement in physical modelling of debris flow. The advancement in the scaling principles (Iverson, 2015) has substantially increased the recognition of the need for field-scale facilities to model debris flows, leading to the development of field-scale facilities around the world (Ng et al., 2021a; Paik et al., 2012). Increasing research has been carried out using various scales of physical modelling to study complex process such as flow-forest interaction (Liang et al., 2023), flow impact on flexible barriers (Ng et al., 2021a), flow-bed-barrier interaction (Vicari et al., 2022), and the entrainment of post-fire debris flow (Ng et al., 2023). Moreover, improved image analysis also makes the measurement of complex boundary of debris flow possible and improves the interpretation of physical modelling results (Liang et al., 2023). Despite the advancements that have been made, great challenges still need to be solved in terms of modelling realistic boundary conditions and the inclusion of boulders in debris flows. Lastly, many debris-flow theories are highly simplified and have yet to be evaluated. As such, further advancement in physical modeling is crucial to help advance our understanding of debris flows,

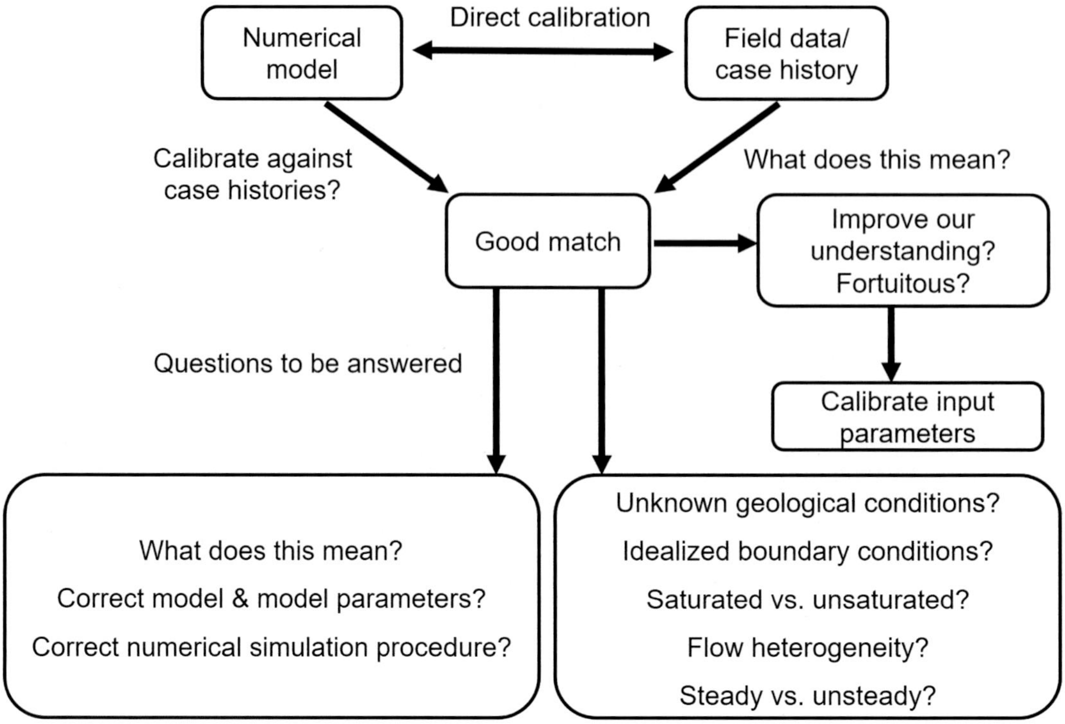

Fig. 4.22 Concept of compensating errors (modified from Ng, 2014)

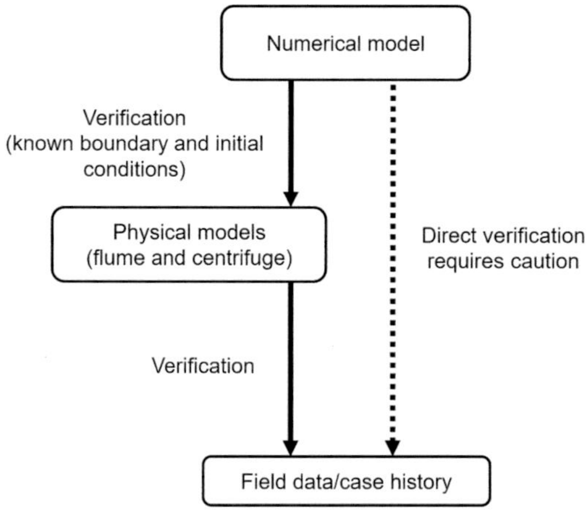

Fig. 4.23 A missing link for calibrating numerical models and model parameters with field data/case histories (modified from Ng, 2014)

narrow the gap between theory and reality, and provide effective engineering solutions to safeguard the communities.

Acknowledgements The authors are grateful for the generous financial sponsorship of the Research Grants Council of Hong Kong (Early Career Scheme 27205320; General Research Fund Grants 16209717; GRF 16212618; 16210219; and 17206622; Areas of Excellence Scheme AoE/E-603/18).

References

Ancey, C., & Cochard, S. (2009). The dam-break problem for Herschel-Bulkley viscoplastic fluids down steep flumes. *Journal of Non-Newtonian Fluid Mechanics, 158*, 18–35. https://doi.org/10.1016/j.jnnfm.2008.08.008

Arganda-Carreras, I., Kaynig, V., Rueden, C., Eliceiri, K. W., Schindelin, J., Cardona, A., & Seung, H. S. (2017). Trainable Weka segmentation: A machine learning tool for microscopy pixel classification. *Bioinformatics, 33*, 2424–2426. https://doi.org/10.1093/bioinformatics/btx180

Armanini, A., Capart, H., Fraccarollo, L., & Larcher, M. (2005). Rheological stratification in experimental free-surface flows of granular-liquid mixtures. *Journal of Fluid Mechanics, 532*, 269–319. https://doi.org/10.1017/s0022112005004283

Armanini, A., Larcher, M., & Odorizzi, M. (2011). Dynamic impact of debris flow front against a vertical wall. *Italian Journal of Engineering Geology and Environment, 2011*, 1041–1049. https://doi.org/10.4408/IJEGE.2011-03.B-113

Armanini, A., Rossi, G., & Larcher, M. (2019). Dynamic impact of a water and sediments surge against a rigid wall. *Journal of Hydraulic Research, 58*, 1–12. https://doi.org/10.4408/IJEGE.2011-03.B-113

Ashwood, W., & Hungr, O. (2016). Estimating total resisting force in flexible barrier impacted by a granular avalanche using physical and numerical modeling. *Canadian Geotechnical Journal, 53*, 1700–1717. https://doi.org/10.1139/cgj-2015-0481

Barlow, J., Franklin, S., & Martin, Y. (2006). High spatial resolution satellite imagery, DEM derivatives, and image segmentation for the detection of mass wasting processes. *Photogrammetric Engineering and Remote Sensing, 72*, 687–692. https://doi.org/10.14358/pers.72.6.687

Baselt, I., de Oliveira, G. Q., Fischer, J. T., & Pudasaini, S. P. (2021). Evolution of stony debris flows in laboratory experiments. *Geomorphology, 372*. https://doi.org/10.1016/j.geomorph.2020.107431

Berger, C., McArdell, B. W., Fritschi, B., & Schlunegger, F. (2010). A novel method for measuring the timing of bed erosion during debris flows and floods. *Water Resources Research, 46*. https://doi.org/10.1029/2009wr007993

Boateng, A. A., & Barr, P. V. (1997). Granular flow behaviour in the transverse plane of a partially filled rotating cylinder. *Journal of Fluid Mechanics, 330*, 233–249. https://doi.org/10.1017/s0022112096003680

Bowman, E. T., Laue, J., Imre, B., & Springman, S. M. (2010). Experimental modelling of debris flow behaviour using a geotechnical centrifuge. *Canadian Geotechnical Journal, 47*, 742–762. https://doi.org/10.1139/t09-141

Brown, S. M. (1992). *The vertically rotating flume for use as a rheometer*. University of Missouri-Rolla.

Bryant, S. K., Take, W. A., Bowman, E. T., & Millen, M. D. L. (2015). Physical and numerical modelling of dry granular flows under Coriolis conditions. *Géotechnique, 65*, 188–200. https://doi.org/10.1680/geot.13.P.208

Bugnion, L., McArdell, B. W., Bartelt, P., & Wendeler, C. (2012). Measurements of hillslope debris flow impact pressure on obstacles. *Landslides, 9*, 179–187. https://doi.org/10.1007/s10346-011-0294-4

Bui, H. H., Fukagawa, R., Sako, K., & Ohno, S. (2008). Lagrangian meshfree particles method (SPH) for large deformation and failure flows of geomaterial using elastic–plastic soil constitutive model. *International Journal for Numerical Analytical Methods in Geomechanics, 32*, 1537–1570. https://doi.org/10.1002/nag.688

Cabrera, M. A., Gollin, D., Kaitna, R., & Wu, W. (2015). Viscous effects on granular mixtures in a rotating drum. In W. Wu (Ed.), *Recent advances in modeling landslides and debris flows* (pp. 57–71). Springer.

Cabrera, M. A., Leonardi, A., & Peng, C. (2020). Granular flow simulation in a centrifugal acceleration field. *Géotechnique, 70*, 894–905. https://doi.org/10.1680/jgeot.18.P.260

Chambon, G., Ghemmour, A., & Naaim, M. (2014). Experimental investigation of viscoplastic free-surface flows in a steady uniform regime. *Journal of Fluid Mechanics, 754*, 332–364. https://doi.org/10.1017/jfm.2014.378

Choi, C. E., Ng, C. W. W., Song, D., Kwan, J. S. H., Shiu, H. Y. K., Ho, K. K. S., & Koo, R. C. H. (2015). Flume investigation of landslide debris-resisting baffles. *Canadian Geotechnical Journal, 52*, 255–255. https://doi.org/10.1139/cgj-2014-0521

Choi, C. E., Goodwin, G. R., Ng, C. W. W., Cheung, D. K. H., Kwan, J. S. H., & Pun, W. K. (2016). Coarse granular flow interaction with slit-structures. *Géotechnique Letters, 6*(4), 1–8. https://doi.org/10.1680/jgele.16.00103

Choi, C. E., Ng, C. W. W., Liu, H., & Wang, Y. (2020). Interaction between dry granular flow and rigid barrier with basal clearance: Analytical and physical modelling. *Canadian Geotechnical Journal, 57*, 236–245. https://doi.org/10.1139/cgj-2018-0622

Chou, H. T., & Lee, C. F. (2009). Cross-sectional and axial flow characteristics of dry granular material in rotating drums. *Granular Matter, 11*, 13–32. https://doi.org/10.1007/s10035-008-0118-y

Chow, V. T. (1959). *Open-channel hydraulics*. McGraw Hill Book Company INC.

Coussot, P. (1995). Structural similarity and transition from newtonian to non-newtonian behavior for clay-water suspensions. *Physical Review Letters, 74*, 3971–3974. https://doi.org/10.1103/PhysRevLett.74.3971

Cui, P., Zeng, C., & Lei, Y. (2015). Experimental analysis on the impact force of viscous debris flow. *Earth Surface Processes and Landforms, 40*, 1644–1655. https://doi.org/10.1002/esp.3744

de Haas, T., Braat, L., Leuven, J., Lokhorst, I. R., & Kleinhans, M. G. (2015). Effects of debris flow composition on runout, depositional mechanisms, and deposit morphology in laboratory experiments. *Journal of Geophysical Research-Earth Surface, 120*, 1949–1972. https://doi.org/10.1002/2015jf003525

de Haas, T., & van Woerkom, T. (2016). Bed scour by debris flows: Experimental investigation of effects of debris-flow composition. *Earth Surface*

Processes and Landforms, 41, 1951–1966. https://doi.org/10.1002/esp.3963

Farin, M., Mangeney, A., & Roche, O. (2014). Fundamental changes of granular flow dynamics, deposition, and erosion processes at high slope angles: Insights from laboratory experiments. *Journal of Geophysical Research: Earth Surface, 119*, 504–532. https://doi.org/10.1002/2013jf002750

Faug, T., Caccamo, P., & Chanut, B. (2011). Equation for the force experienced by a wall overflowed by a granular avalanche: Experimental verification. *Physical Review E, 84*. https://doi.org/10.1103/PhysRevE.84.051301

Faug, T., Childs, P., Wyburn, E., & Einav, I. (2015). Standing jumps in shallow granular flows down smooth inclines. *Physics of Fluids, 27*. https://doi.org/10.1063/1.4927447

Gaume, J., Gast, T., Teran, J., van Herwijnen, A., & Jiang, C. (2018). Dynamic anticrack propagation in snow. *Nature Communications, 9*. https://doi.org/10.1038/s41467-018-05181-w

Gilbert, G. K., & Murphy, E. C. (1914). *The transportation of debris by running water*. US Government Printing Office.

Gollin, D., Brevis, W., Bowman, E. T., & Shepley, P. (2017). Performance of PIV and PTV for granular flow measurements. *Granular Matter, 19*. https://doi.org/10.1007/s10035-017-0730-9

Goodwin, G. R., & Choi, C. E. (2020). Slit structures: Fundamental mechanisms of mechanical trapping of granular flows. *Computers and Geotechnics, 119*, 103376. https://doi.org/10.1016/j.compgeo.2019.103376

Goodwin, G. R., Choi, C. E., & Yune, C. Y. (2020). Towards rational use of baffle arrays on sloped and horizontal terrain for filtering boulders. *Canadian Geotechnical Journal, 58*(10), 1571–1589. https://doi.org/10.1139/cgj-2020-0363

Goodwin, G. R., & Choi, C. E. (2021). Translational inertial effects and scaling considerations for coarse granular flows impacting landslide-resisting barriers. *Journal of Geotechnical and Geoenvironmental Engineering, 147*. https://doi.org/10.1061/(asce)gt.1943-5606.0002661

Gue, C. S. (2012). *Submarine landslide flows simulation through centrifuge modelling*. University of Cambridge.

Heim, A. (1932). *Bergsturz und Menschenleben*. Fretz and Wasmuth.

Henein, H. (1981). *Bed behaviour in rotary cylinders with applications to rotary kilns*. Ph.D. Thesis, University of British Columbia, Canada.

Henein, H., Brimacombe, J. K., & Watkinson, A. P. (1983). Experimental study of transverse bed motion in rotary kilns. *Metallurgical Transactions B, 14*, 191–205. https://doi.org/10.1007/BF02661016

Hong, Y., Wang, J. P., Li, D. Q., Cao, Z. J., Ng, C. W. W., Cui, P. (2015). Statistical and probabilistic analyses of impact pressure and discharge of debris flow from 139 events during 1961 and 2000 at Jiangjia Ravine, China. *Engineering Geology, 187*, 122–134. https://doi.org/10.1016/j.enggeo.2014.12.011

Hsu, L., Dietrich, W. E., & Sklar, L. S. (2008). Experimental study of bedrock erosion by granular flows. *Journal of Geophysical Research: Earth Surface, 113*(F2). https://doi.org/10.1029/2007JF000778

Hsu, L. (2010). *Bedrock erosion by granular flow*. Ph.D. Thesis, University of California, Berkeley, USA.

Huber, A. (1980). Schwallwellen in seen als Folge von Felsstürzen. VAW-Mitteilung 180, Vischer. ETH Zurich, Zurich (in German), Switzeland.

Huizinga, R. J. (1993). *An analysis of the two-dimensional flow in a vertically rotating flume*. University of Missouri-Rolla.

Iverson, R. M. (1997). The physics of debris flows. *Reviews of Geophysics, 35*, 245–296. https://doi.org/10.1029/97rg00426

Iverson, R. M. (2000). Landslide triggering by rain infiltration. *Water Resources Research, 36*, 1897–1910. https://doi.org/10.1029/2000wr900090

Iverson, R. M. (2015). Scaling and design of landslide and debris-flow experiments. *Geomorphology, 244*, 9–20. https://doi.org/10.1016/j.geomorph.2015.02.033

Iverson, R. M. (2020). Landslide disparities, flume discoveries, and oso despair. *Perspectives of Earth and Space Scientists, 1*, e2019CN000117. https://doi.org/10.1029/2019CN000117

Iverson, R. M., Reid, M. E., Iverson, N. R., LaHusen, R. G., Logan, M., Mann, J. E., & Brien, D. L. (2000). Acute sensitivity of landslide rates to initial soil porosity. *Science, 290*, 513–516. https://doi.org/10.1126/science.290.5491.513

Iverson, R. M., Logan, M., & Denlinger, R. P. (2004). Granular avalanches across irregular three-dimensional terrain: 2. Experimental tests. *Journal of Geophysical Research: Earth Surface, 109*. https://doi.org/10.1029/2003JF000085

Iverson, R. M., Logan, M., LaHusen, R. G., & Berti, M. (2010). The perfect debris flow? Aggregated results from 28 large-scale experiments. *Journal of Geophysical Research: Earth Surface, 115*. https://doi.org/10.1029/2009jf001514

Iverson, R. M., Reid, M. E., Logan, M., LaHusen, R. G., Godt, J. W., & Griswold, J. P. (2011). Positive feedback and momentum growth during debris-flow entrainment of wet bed sediment. *Nature Geoscience, 4*, 116–121. https://doi.org/10.1038/ngeo1040

Iverson, R. M., George, D. L., & Logan, M. (2016). Debris flow runup on vertical barriers and adverse slopes. *Journal of Geophysical Research: Earth Surface, 121*, 2333–2357. https://doi.org/10.1002/2016jf003933

Jerolmack, D. J., & Daniels, K. E. (2019). Viewing Earth's surface as a soft-matter landscape. *Nature Reviews Physics, 1*, 716–730. https://doi.org/10.1038/s42254-019-0111-x

Jing, L., Kwok, C. Y., Leung, Y. F., Zhang, Z., & Dai, L. (2018). Runout scaling and deposit morphology of rapid mudflows. *Journal of Geophysical Research: Earth Surface, 123*, 2004–2023. https://doi.org/10.1029/2018jf004667

Kaitna, R., & Rickenmann, D. (2007). A new experimental facility for laboratory debris flow investigation. *Journal of Hydraulic Research, 45*(6), 797–810. https://doi.org/10.1080/00221686.2007.9521817

Kaitna, R., Palucis, M. C., Yohannes, B., Hill, K. M., & Dietrich, W. E. (2016). Effects of coarse grain size distribution and fine particle content on pore fluid pressure and shear behavior in experimental debris flows. *Journal of Geophysical Research: Earth Surface, 121*, 415–441. https://doi.org/10.1002/2015jf003725

Larcher, M., Fraccarollo, L., Armanini, A., & Capart, H. (2007). Set of measurement data from flume experiments on steady uniform debris flows. *Journal of Hydraulic Research, 45*, 59–71. https://doi.org/10.1080/00221686.2007.9521833

Leonardi, A., Cabrera, M., Wittel, F. K., Kaitna, R., Mendoza, M., Wu, W., & Herrmann, H. J. (2015). Granular-front formation in free-surface flow of concentrated suspensions. *Physical Review E, 92*. https://doi.org/10.1103/PhysRevE.92.052204

Li, P., Hu, K. H., & Wang, X. K. (2018). Debris flow entrainment rates in non-uniform channels with convex and concave slopes. *Journal of Hydraulic Research, 56*, 156–167. https://doi.org/10.1080/00221686.2017.1313321

Li, P., Wang, J. D., Hu, K. H., & Wang, F. (2020). Experimental study of debris-flow entrainment over stepped-gradient beds incorporating bed sediment porosity. *Engineering Geology, 274*. https://doi.org/10.1016/j.enggeo.2020.105708

Liang, Z. Y., Choi, C. E., Zhao, Y., Jiang, Y., & Choo, J. (2023). Revealing the role of forests in the mobility of geophysical flows. *Computers and Geotechnics* (In press).

Liu, H. (2019). *Impact mechanisms of debris flow against multiple rigid barriers with basal clearance*. Ph.D. Thesis, Hong Kong University of Science and Technology, Hong Kong, China.

Lu, W., & Zhang, L. (2020). Modelling of overtopping and erosion of embankment dams on a drum centrifuge. In *The International Conference on Embankment Dams* (pp. 201–205). Springer

Lucas, A., Mangeney, A., & Ampuero, J. P. (2014). Frictional velocity-weakening in landslides on earth and on other planetary bodies. *Nature Communications, 5*. https://doi.org/10.1038/ncomms4417

Luong, T. H., Baker, J. L., & Einav, I. (2019). Spread-out and slow-down of granular flows through model forests. *Granular Matter, 22*. https://doi.org/10.1007/s10035-019-0980-9

Maas, H., Gruen, A., & Papantoniou, D. (1993). Particle tracking velocimetry in three-dimensional flows. *Experiments in Fluids, 15*, 133–146.

Major, J. J. (1996). *Experimental studies of deposition by debris flows: Process, characteristics of deposits, and effects of pore-fluid pressure*. Ph.D. Thesis, University of Washington, USA.

Mangeney, A., Roche, O., Hungr, O., Mangold, N., Faccanoni, G., & Lucas, A. (2010). Erosion and mobility in granular collapse over sloping beds. *Journal of Geophysical Research: Earth Surface, 115*. https://doi.org/10.1029/2009jf001462

Marks, B., Eriksen, J. A., Dumazer, G., Sandnes, B., & Maloy, K. J. (2017). Size segregation of intruders in perpetual granular avalanches. *Journal of Fluid Mechanics, 825*, 502–514. https://doi.org/10.1017/jfm.2017.419

McArdell, B. W., Bartelt, P., & Kowalski, J. (2007). Field observations of basal forces and fluid pore pressure in a debris flow. *Geophysical Research Letters, 34*. https://doi.org/10.1029/2006gl029183

Mellmann, J. (2001). The transverse motion of solids in rotating cylinders - forms of motion and transition behavior. *Powder Technology, 118*, 251–270. https://doi.org/10.1016/s0032-5910(00)00402-2

MiDi, G. D. R. (2004). On dense granular flows. *European Physical Journal E, 14*, 341–365. https://doi.org/10.1140/epje/i2003-10153-0

Nagl, G., Huebl, J., & Kaitna, R. (2020). Velocity profiles and basal stresses in natural debris flows. *Earth Surface Processes and Landforms, 45*, 1764–1776. https://doi.org/10.1002/esp.4844

Ng, C. W. W. (2014). The state-of-the-art centrifuge modelling of geotechnical problems at HKUST. *Journal of Zhejiang University-Science A, 15*, 1–21. https://doi.org/10.1631/jzus.A1300217

Ng, C. W. W., Choi, C. E., & Law, R. P. H. (2013). Longitudinal spreading of granular flow in trapezoidal channels. *Geomorphology, 194*, 84–93. https://doi.org/10.1016/j.geomorph.2013.04.016

Ng, C. W. W., Song, D., Choi, C. E., Koo, R. C. H., & Kwan, J. S. H. (2016). A novel flexible barrier for landslide impact in centrifuge. *Géotechnique Letters, 6*. https://doi.org/10.1680/jgele.16.00048

Ng, C. W. W., Choi, C. E., Liu, L. H. D., Wang, Y., Song, D., & Yang, N. (2017a). Influence of particle size on the mechanism of dry granular run-up on a rigid barrier. *Géotechnique Letters, 7*(1), 79–89. https://doi.org/10.1680/jgele.16.00159

Ng, C. W. W., Song, D., Choi, C. E., Liu, L. H. D., Kwan, J. S. H., Koo, R. C. H., & Pun, W. K. (2017b). Impact mechanisms of granular and viscous flows on rigid and flexible barriers. *Canadian Geotechnical Journal, 54*, 188–206. https://doi.org/10.1139/cgj-2016-0128

Ng, C. W. W., Choi, C. E., Cheung, D. K. H., & Cui, Y. (2019). Effects of dynamic fragmentation on the impact force exerted on rigid barrier: Centrifuge modelling. *Canadian Geotechnical Journal, 56*, 1215–1224. https://doi.org/10.1139/cgj-2018-0092

Ng, C. W. W., Choi, C. E., Liu, H., Poudyal, S., & Kwan, J. S. H. (2021a). Design recommendations for single and dual debris flow barriers with and without basal clearance. In K. Sassa, M. Mikoš, S. Sassa, P. T. Bobrowsky, K. Takara, & K. Dang (Eds.), *Understanding and reducing landslide disaster risk*. Sendai Landslide Partnerships and Kyoto Landslide Commitment (Vol. 1, pp. 33–53). Springer International Publishing.

Ng, C. W. W., Liu, H., Choi, C. E., Kwan, J. S. H., & Pun, W. K. (2021b). Impact dynamics of boulder-enriched debris flow on a rigid barrier. *Journal of Geotechnical and Geoenvironmental Engineering, 147*. https://doi.org/10.1061/(asce)gt.1943-5606.0002485

Ng, C. W. W., Majeed, U., & Choi, C. E. (2022). Effects of solid fraction of saturated granular flows on overflow and landing mechanisms of rigid barriers. *Géotechnique*, 1–41. https://doi.org/10.1680/jgeot.21.00170

Ng, C. W. W., Zheng, M., Liu, H., & Poudyal, S. (2023). Effects of bed hydrophobicity on post-fire debris flow entrainment and momentum growth. *Journal of Geophysical Research: Earth Surface* (In press).

Paik, J., Son, S., Kim, T., & Kim, S. (2012). A real-scale field experiment of debris flow for investigating its deposition and entrainment. AGU Fall Meeting Abstracts, pp EP53A-1010

Parsons, J. D., Whipple, K. X., & Simoni, A. (2001). Experimental study of the grain-flow, fluid-mud transition in debris flows. *Journal of Geology, 109*, 427–447. https://doi.org/10.1086/320798

Perng, A. T. H., Capart, H., & Chou, H. T. (2006). Granular configurations, motions, and correlations in slow uniform flows driven by an inclined conveyor belt. *Granular Matter, 8*, 5–17. https://doi.org/10.1007/s10035-005-0213-2

Post, F. H., & Walsum, T. (1993). Fluid flow visualization. In *Focus on scientific visualization* (pp. 1–40). Springer.

Pudasaini, S. P., Hsiau, S. S., Wang, Y., & Hutter, K. (2005). Velocity measurements in dry granular avalanches using particle image velocimetry technique and comparison with theoretical predictions. *Physics of Fluids, 17*, 093301. https://doi.org/10.1063/1.2007487

Pudasaini, S. P., & Hutter, K. (2007). *Avalanche dynamics: Dynamics of rapid flows of dense granular avalanches*. Springer Science & Business Media.

Savage, S. B., & Hutter, K. (1989). The motion of a finite mass of granular material down a rough incline. *Journal of Fluid Mechanics, 199*, 177–215. https://doi.org/10.1017/s0022112089000340

Scheidl, C., Chiari, M., Kaitna, R., Müllegger, M., Krawtschuk, A., Zimmermann, T., & Proske, D. (2013). Analysing debris-flow impact models based on a small scale modelling approach. *Surveys in Geophysics, 34*, 121–140. https://doi.org/10.1007/s10712-012-9199-6

Schindelin, J., Arganda-Carreras, I., Frise, E., Kaynig, V., Longair, M., Pietzsch, T., Preibisch, S., Rueden, C., Saalfeld, S., Schmid, B., Tinevez, J. Y., White, D. J., Hartenstein, V., Eliceiri, K., Tomancak, P., & Cardona, A. (2012). Fiji: An open-source platform for biological-image analysis. *Nature Methods, 9*, 676–682. https://doi.org/10.1038/nmeth.2019

Schofield, A. N., & Wroth, P. (1968). *Critical state soil mechanics*. McGraw-Hill.

Silbert, L. E., Ertas, D., Grest, G. S., Halsey, T. C., Levine, D., & Plimpton, S. J. (2001). Granular flow down an inclined plane: Bagnold scaling and rheology. *Physical Review E, 64*. https://doi.org/10.1103/PhysRevE.64.051302

Song, D. (2016). *Mechanisms of debris flow impact on rigid and flexible barriers*. Hong Kong University of Science and Technology.

Song, D., Ng, C. W. W., Choi, C. E., Zhou, G. G. D., Kwan, J. S. H., & Koo, R. C. H. (2017). Influence of debris flow solid fraction on rigid barrier impact. *Canadian Geotechnical Journal, 54*, 1421–1434. https://doi.org/10.1139/cgj-2016-0502

Song, D., Choi, C. E., Ng, C. W. W., & Zhou, G. G. D. (2018). Geophysical flows impacting a flexible barrier: Effects of solid-fluid interaction. *Landslides, 15*, 99–110. https://doi.org/10.1007/s10346-017-0856-1

Song, D., Chen, X. Q., Zhou, G. G. D., Lu, X. Q., Cheng, G. W., & Chen, Q. (2021). Impact dynamics of debris flow against rigid obstacle in laboratory experiments. *Engineering Geology, 291*. https://doi.org/10.1016/j.enggeo.2021.106211

Song, P. J., & Choi, C. E. (2021). Revealing the importance of capillary and collisional stresses on soil bed erosion induced by debris flows. *Journal of Geophysical Research: Earth Surface, 126*. https://doi.org/10.1029/2020jf005930

Sun, H. W., & Law, R. P. H. (2015). A preliminary study on impact of landslide debris on flexible barriers. Technical Note No TN 1/2012, Hong Kong SAR, China, Geotechnical Engineering Office, Civil Engineering and Development Department, The HKSAR Government.

Takahashi, T. (2007). *Debris flow: Mechanics, prediction and countermeasures*. Taylor & Francis.

Take, W. A., Bolton, M. D., Wong, P. C. P., & Yeung, F. J. (2004). Evaluation of landslide triggering mechanisms in model fill slopes. *Landslides, 1*, 173–184. https://doi.org/10.1007/s10346-004-0025-1

Taylor-Noonan, A. M., Gollin, D., Bowman, E. T., & Take, W. A. (2021). The influence of image analysis methodology on the calculation of granular temperature for granular flows. *Granular Matter, 23*. https://doi.org/10.1007/s10035-021-01153-y

Tan, D. Y., Yin, J. H., Feng, W. Q., Qin, J. Q., & Zhu, Z. H. (2018). Large-scale physical modelling study of a flexible barrier under the impact of granular flows. *Natural Hazards and Earth System Sciences, 18*, 2625–2640. https://doi.org/10.5194/nhess-18-2625-2018

Trewhela, T., & Ancey, C. (2021). A conveyor belt experimental setup to study the internal dynamics of granular avalanches. *Experiments in Fluids, 62*. https://doi.org/10.1007/s00348-021-03299-0

van der Vaart, K., Thornton, A. R., Johnson, C. G., Weinhart, T., Jing, L., Gajjar, P., Gray, J., & Ancey, C. (2018). Breaking size-segregation waves and mobility feedback in dense granular avalanches. *Granular Matter, 20*. https://doi.org/10.1007/s10035-018-0818-x

Vicari, H., Ng, C. W. W., Nordal, S., Thakur, V., De Silva, W. A. R. K., Liu, H., & Choi, C. E. (2022). The effects of upstream flexible barrier on the debris flow entrainment and impact dynamics on a terminal barrier. *Canadian Geotechnical Journal, 59*, 1007–1019. https://doi.org/10.1139/cgj-2021-0119

Wendeler, C., & Volkwein, A. (2015). Laboratory tests for the optimization of mesh size for flexible debris-flow barriers. *Natural Hazards and Earth System Sciences, 15*, 2597–2604. https://doi.org/10.5194/nhess-15-2597-2015

White, D. J., Take, W. A., & Bolton, M. D. (2003). Soil deformation measurement using particle image velocimetry (PIV) and photogrammetry. *Géotechnique, 53*, 619–631. https://doi.org/10.1680/geot.53.7.619.37383

Wang, C. (2020). *Effects of barrier stiffness on the impact mechanisms of debris flow with different particle sizes: Centrifuge and numerical modelling*. Ph.D. Thesis, Hong Kong University of Science and Technology, Hong Kong, China

Won, S., Lee, S. W., Paik, J., Yune, C.-Y., & Kim, G. (2016). Analysis of erosion in debris flow experiment using terrestrial LiDAR. *Journal of the Korean Society of Surveying, Geodesy, Photogrammetry and Cartography, 34*, 309–317. https://doi.org/10.7848/ksgpc.2016.34.3.309

Wood, D. M. (1991). *Soil behaviour and critical state soil mechanics*. Cambridge University Press.

Yifru, A. L., Laache, E., Norem, H., Nordal, S., & Thakur, V. J. H. T. (2018). Laboratory investigation of performance of a screen type debris-flow countermeasure. *HKIE Transactions, 25*, 129–144. https://doi.org/10.1080/1023697X.2018.1462104

Zhou, G. G. D., Li, S., Song, D., Choi, C. E., & Chen, X. Q. (2019). Depositional mechanisms and morphology of debris flow: Physical modelling. *Landslides, 16*, 315–332. https://doi.org/10.1007/s10346-018-1095-9

Numerical Modeling of Debris Flows: A Conceptual Assessment

5

Richard M. Iverson and David L. George

Abstract

Real-world hazard evaluation poses many challenges for the development and application of numerical models of debris flows. In this chapter we provide a conceptual overview of physically based, depth-averaged models designed to simulate debris-flow motion across three-dimensional terrain. When judiciously formulated and applied, these models can provide useful information about anticipated depths, speeds, and extents of debris-flow inundation as well as debris interactions with structures such as levees and dams. Depth-averaged debris-flow models can differ significantly from one another, however. Some of the greatest differences result from simulation of one-phase versus two-phase flow, use of parsimonious versus information-intensive initial and boundary conditions, use of tuning coefficients versus physically measureable parameters, application of dissimilar numerical solution techniques, and variations in computational speed and model accessibility. This overview first addresses these and related attributes of depth-averaged debris-flow models. It then describes model testing and application to hazard evaluation, with a focus on our own model, D-Claw. The overview concludes with a discussion of outstanding challenges for development of improved debris-flow models and suggestions for prospective model users.

R. M. Iverson (✉) · D. L. George
U.S. Geological Survey, 1300 SE Cardinal Ct., Vancouver, WA 98683, USA
e-mail: riverson@usgs.gov

D. L. George
e-mail: dgeorge@usgs.gov

5.1 Introduction

This chapter assesses the state of practical, physically based debris-flow modeling. By "practical" we mean modeling that addresses real-world questions about the anticipated speeds, depths, and extents of inundation caused by debris flows moving across three-dimensional terrain, including terrain that has been modified by construction of levees, roads, buildings, and dams. By "physically based" we mean modeling that obeys physical conservation laws as closely as possible and that minimizes use of adjustable coefficients. The chapter does not include a comprehensive literature review or detailed descriptions of individual debris-flow models, which number in the dozens (e.g., Trujillo-Vela et al., 2022). Instead, it focuses on concepts and principles that are relevant across a spectrum of models and on features that distinguish models from one another.

M. Jakob et al. (eds.), *Advances in Debris-flow Science and Practice*, Geoenvironmental Disaster Reduction,
https://doi.org/10.1007/978-3-031-48691-3_5

The chief goal of the chapter is to provide a nonmathematical summary of information that may be useful to scientists and engineers who are not model developers themselves but are knowledgeable about debris flows and wish to use numerical models to simulate them. Such model users are effectively consumers of products created or marketed by others. The chapter aims to equip such model users with a healthy skepticism that characterizes consumers who are well-informed.

We are model developers ourselves, but we are also model skeptics. Our skepticism comes from studying numerous models, from recognizing our own limitations in model development, and from humility born of long experience with the natural complexities of debris flows. Unlike geological phenomena that are hidden by the veils of deep Earth or deep time, debris flows occur routinely and are readily observed. It consequently can be easy to see when predictions of debris-flow models are wrong. However, some prediction errors result from fundamental model shortcomings, whereas others result from uncertainties about debris properties, volumes, or initial states. Thus, some types of erroneous predictions should increase the skepticism of model users, whereas others should boost users' confidence by quantifying the effects of uncertainties that are inherently part of the natural world (e.g., Barnhart et al., 2021b).

Our overview first explains how the diversity of depth-averaged debris-flow models reflects differences in model conceptualization, formulation, idealization, parameterization, testability, and accessibility. It then describes examples of model testing and application to hazard evaluation, with a focus on our own model, D-Claw. Model testing entails comparing model predictions with detailed measurements made in a laboratory setting, and also with less-detailed but highly relevant field measurements of phenomena such as flow speeds, depths, runup heights, and inundation patterns. The overview concludes with a discussion of outstanding challenges for development of improved debris-flow models and with suggestions for prospective model users.

5.2 Physical Characteristics of Debris Flows

The design of useful debris-flow models should account for the fact that debris flows encompass a range of Earth-surface flow phenomena intermediate in character between rock avalanches and sediment-laden water floods. Like rock avalanches, debris flows typically begin on slopes steeper than 25° and have definite starting and ending points in space and time, but like water floods, debris flows can travel great distances across low-relief terrain—more than 100 km in the case of Earth's largest volcanic debris flows (Vallance & Iverson, 2015). The observation that debris flows share some attributes with both rock avalanches and water floods has influenced the development of many debris-flow models, some of which are rooted in rock-avalanche modeling, and some of which are rooted in flood-wave modeling (McDougall, 2017; Rickenmann et al., 2006; Trujillo-Vela et al., 2022).

Debris flows differ from either rock avalanches or water floods, however, and these differences have important implications for model formulation and application. Most conspicuously, debris flows contain roughly equal volumes of liquid water and solid grains. Whereas momentum transfer and energy dissipation are dominated by solid grain forces in rock avalanches and by fluid dynamical forces in water floods, mechanical interactions between grains and fluid have large effects on the physics of debris flows. Moreover, solid–fluid interactions evolve as debris flows transition from static initial states to rapidly flowing states and then transition again to form nearly rigid deposits. The prominent role that solid–fluid interactions play in causing these transitions and mediating flow resistance is a key feature of debris flows (Iverson, 1997, 2003a).

Another key feature of debris flows is the presence of grains with diverse shapes, sizes, and compositions (Hungr et al., 2001, 2014; Varnes, 1978). This trait distinguishes debris flows from phenomena investigated in laboratory or theoretical studies that focus on flows of identical solid spheres immersed in liquid (e.g., Berzi et al., 2010). Such investigations can have great scientific value, but they have limited relevance for practical modeling of debris flows in which grains commonly range from clay-sized particles to boulders exceeding 1 m in diameter. In some cases debris flows also carry a significant freight of woody debris, including large logs (Fig. 5.1).

The diversity of grains in debris flows permits a great variety of grain arrangements, which affect debris bulk density and dilatancy (i.e., shear-induced volumetric dilation/contraction of the granular fraction of deforming debris). Debris bulk density can also vary in response to changes in debris agitation, grain-size segregation, debris dilution during mixing with water bodies, and erosion or sedimentation. Consequently, from a modeling perspective, debris mixtures effectively behave as compressible materials with bulk densities that commonly vary from about 1600 to 2400 kg/m^3—even though the individual solid and liquid constituents of debris are nearly incompressible (Iverson, 1997; Iverson et al., 2010).

Changes in debris bulk density are generally accompanied by changes in intergranular fluid pressure that affects intergranular friction and bulk flow resistance (George & Iverson, 2014; Iverson, 1997; Iverson & George, 2014; Iverson et al., 2010). The diversity of grain sizes in debris flows enhances this effect because it reduces the hydraulic permeability of debris by increasing solid–fluid drag forces that result from fluid moving with respect to grains on a local scale (Iverson, 1997). On a continuum scale encompassing thousands of grains and their adjacent fluid, high intergranular fluid pressures that develop during contractive debris deformation can lead to partial or complete debris liquefaction. The presence of more than several weight percent silt- and clay-sized grains helps perpetuate liquefaction and thereby increases debris-flow mobility (Iverson et al., 2010). In contrast, the presence of silt- and clay-sized sediment can reduce the mobility of miniature, laboratory-scale debris flows because it provides some cohesive yield strength. Mobility reduction caused by cohesive strength is a scale-dependent phenomenon, however, and it typically has little importance in natural debris flows with volumes ranging from thousands to billions of cubic meters (Iverson, 2003a, 2015).

Conspicuous grain-size segregation occurs in many debris flows, and it can significantly affect macroscopic flow dynamics and the formation of deposits (Iverson et al., 2010; Major & Iverson, 1999). One consequence of size segregation is development of coarse-grained, high-friction surge fronts that impede the motion of finer-grained trailing debris in which liquefaction is persistent (Iverson et al., 2010; Sharp & Nobles, 1953). Typically, this liquefied sediment-rich debris is trailed by more-dilute debris that in turn is trailed by muddy water (Pierson,

Fig. 5.1 Boulders, finer sediment, and wood fragments deposited by a rain-triggered debris flow that descended the Eliot Branch and Middle Fork Hood River, Oregon, USA, on 7 November 2006. The debris flow began roughly 8 km distant and 1 km higher in elevation than this deposition site just upstream of the confluence of the Eliot Branch and Middle Fork Hood River. It was sourced primarily from landslides on lateral moraines of Eliot Glacier on snow-clad Mount Hood in the background. Photo courtesy of Darryl Lloyd

1986). Another consequence of grain-size segregation is the development of coarse-grained lateral levees that commonly form where debris flows begin to escape or overtop constricted channels (Johnson et al., 2012). These natural levees can channelize ensuing flow and thereby increase runout distance by focusing debris-flow momentum downstream (Fig. 5.2). Alternatively, levees can be breached or overtopped by successive debris-flow surges, leading to flow avulsion and redirection of downstream inundation (de Haas et al., 2018). Thus, development of multiple surge fronts and levees during individual debris-flow events can strongly influence the distribution of associated hazards.

Entrainment of debris along flow paths is another phenomenon that influences debris-flow dynamics and hazards. Entrainment is typically concentrated in the steep, narrow, upper reaches of debris-flow paths, particularly in areas where channels slope at angles greater than about 10° (Hungr et al., 2005). Entrainment can occur through scour of bed material, especially if the bed material is wet and loose enough to weaken as a result of excess pore-fluid pressure generated when it is suddenly overrun (Iverson et al., 2011; McCoy et al, 2012; Reid et al., 2011). It can also occur by undermining of steep channel banks, resulting in sloughing of debris onto the surface of passing debris flows. In some cases the majority of a debris flow's final volume results from entrainment that occurs along its path, greatly increasing the scope of impacts in debris-flow runout zones (Breien et al., 2008; Hungr et al., 2005; Pierson et al., 1990).

Finally, an important feature of most debris flows is that they are "shallow," meaning that their typical thicknesses are much less than their typical planimetric dimensions—commonly by a factor of 10 or more (Fig. 5.3). Most river floods, storm surges, and tsunamis inundating coastlines also satisfy this shallowness criterion. Debris-flow modeling consequently has exploited a large body of mathematical and computational knowledge developed to simulate shallow-water flow phenomena (e.g., Berger et al., 2011; Vreugdenhil, 1994). Nevertheless, as summarized in the preceding paragraphs, debris flows have many distinctive attributes, and simple adjustments of flow resistance coefficients in shallow-water flow models are unlikely to yield debris-flow models that reliably predict quantities of practical importance such as flow speeds, depths, and inundation patterns.

Fig. 5.2 Boulder-rich lateral levee deposited by the same debris flow that emplaced the deposits pictured in Fig. 5.1. The levee formed approximately 7 km downstream from the site shown in Fig. 5.1, and it focused debris-flow momentum into the channel of the Middle Fork Hood River, visible on the right side of the photo. Largest boulders in the levee were >2 m in diameter

5.3 Depth-Averaged Conservation Equations

The mathematical core of nearly all practical, physically based debris-flow models consists of partial differential equations that represent depth-averaged continuum-mechanical mass and momentum conservation laws for flows that are "shallow" as illustrated in Fig. 5.3 (e.g., Iverson & George, 2014). The same is true of most physically based models of high-speed landslides that don't qualify as debris flows (e.g., McDougall, 2017). However, a potential source of confusion is that we categorize depth-averaged models for motion across three-dimensional terrain as 2-D models, whereas McDougall (2017) categorized them as 3-D models. We reserve the term "3-D" for models that do not use depth averaging.

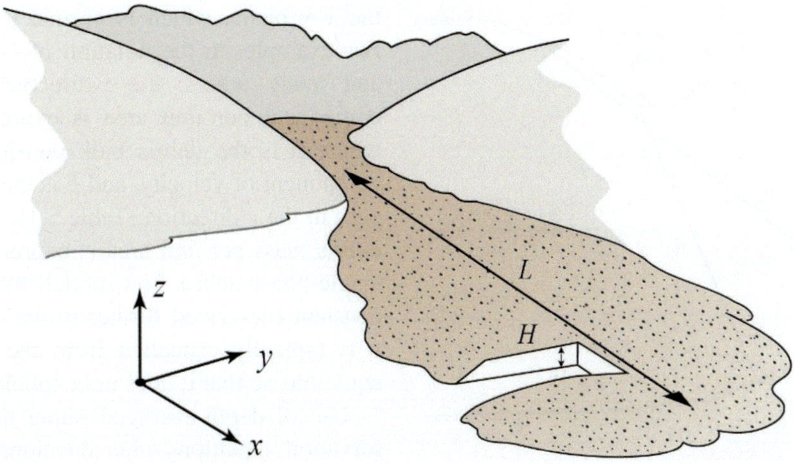

Fig. 5.3 Debris flows satisfy a "shallowness" criterion if their characteristic thickness H in the z direction is much less than their characteristic extent L in the x and y directions, as illustrated in this schematic (adapted from Iverson, 2005)

Whereas 3-D models calculate evolving momentum fluxes in three coordinate directions, most depth-averaged models assume that momentum fluxes normal to the bed can be neglected—implying that basal normal stresses balance the static weight of overlying debris. Iverson (2005) provided a quantitative assessment of this assumption and of ways that it can be relaxed. Although the assumption of static basal normal stress might seem highly restrictive, experience with depth-averaged models of diverse phenomena ranging from tsunamis and floods to granular avalanches has shown that these models commonly yield suitable practical predictions even for flows in which the assumption is locally or transiently violated (e.g., George, 2010; Gray et al., 2003; LeVeque et al., 2011).

Depth-averaged debris-flow models also generally assume that depth-dependent variations in downstream momentum are negligible. As in shallow-water flow models, corrections for the effects of these variations can be made by introducing momentum-distribution coefficients (e.g., Vreugdenhil, 1994). However, values of the distribution coefficients depend on the shape of flow velocity and density profiles, which are poorly constrained for debris flows. Most depth-averaged debris-flow models consequently exclude the coefficients as an unwarranted complication. They instead assume that downstream debris-flow momentum can be approximated as uniform at all depths.

Depth-averaged conservation equations are derived by integrating 3-D conservation equations through the debris-flow thickness in the coordinate direction we denote with z in Fig. 5.3 (Iverson, 2005). However, the direction of z is not the same in every model. Terrain-fitted curvilinear coordinates with z rotated so that it is everywhere normal to the bed are the most rigorous coordinates from a mathematical standpoint, and they facilitate accounting for the effect of centripetal accelerations on basal stresses (Gray et al., 1999; Pudasaini et al., 2005; Savage & Hutter, 1991). However, curvilinear coordinates can be difficult to employ in numerical simulations of flow across rugged 3-D terrain, in part because adjacent computational cells can have sharp differences in bed-normal z directions, potentially resulting in conflicting definitions of flow depths (Fig. 5.4). Moreover, if the bed geometry changes significantly owing to erosion or deposition, curvilinear coordinates fitted to the original terrain lose at least some of their relevance. Use of Earth-centered Cartesian coordinates with z uniformly vertical is simpler, but it requires care in calculation of momentum

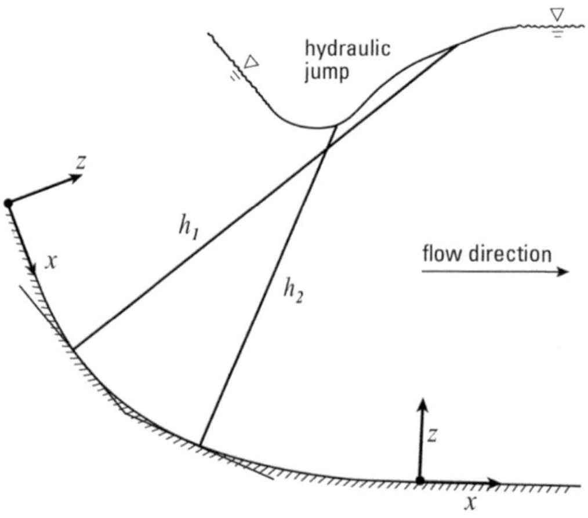

Fig. 5.4 Schematic vertical cross section illustrating a debris flow moving across variably sloping, discretized bed topography with terrain-fitted coordinates. As the flow passes through a hydraulic jump, the flow thickness h in the z direction is incorrectly represented with respect to discretized bed topography, because the flow depth h_1 (upstream) is larger than h_2 (downstream). Use of Earth-centered Cartesian coordinates with z vertical everywhere eliminates this issue

fluxes and basal stresses, because the orientation of sloping beds is not normal to a vertical z coordinate (Denlinger & Iverson, 2004; Iverson, 2005; Iverson & George, 2019a).

In all depth-averaged debris-flow models the fundamental conserved quantities (mass and linear momentum) are expressed per unit area in the x–y plane, which is normal to z (Fig. 5.3). For example, in the notation of Iverson (2005) and many others, the x-direction momentum component per unit area is expressed as ρuh, where ρ is the debris bulk density, u is the x component of velocity, and h is the debris thickness in the z direction (Table 5.1). Similarly, the debris mass per unit area is expressed as ρh. In single-phase debris-flow models that assume ρ is constant (described further in the next section), ρ is typically cancelled from the conservation equations so that it does not explicitly appear.

Use of depth-averaged rather than 3-D conservation equations aids development of efficient numerical solution techniques, in part because free-surface and basal boundary conditions are embedded in the equations during the depth-integration procedure (Iverson, 2005; Iverson & Ouyang, 2015). This feature eliminates the need to separately resolve the evolving boundary positions as computational solutions proceed. However, during depth integration it can be challenging to properly incorporate the effects of a basal boundary where erosion or deposition occurs because fluxes of mass and momentum across an irregular, evolving surface are involved. The problem can be addressed rigorously by adopting a two-layer perspective in which the debris flow constitutes the upper layer and the bed constitutes the lower layer, and then

Table 5.1 Definitions of variables commonly used in some depth-averaged debris-flow models

Typical symbols	Definitions
Independent variables	
x, y	Planimetric spatial coordinates
z	Vertical or slope-normal coordinate
t	Time coordinate
Dependent variables that evolve as functions of x, y ant t	
u, v	Flow velocity components in x and y directions[a]
h	Flow thickness in z direction[a]
ρ	Debris bulk density[b]
m	Debris solid volume fraction[b]
p_{bed}	Basal pore-fluid pressure

[a]Two-phase models typically include separate values of u, v, and h for each phase
[b]Use of both ρ and m as variables can be redundant because $\rho = \rho_s m + \rho_f(1 - m)$ applies and the solid grain density ρ_s and intergranular fluid density ρ_f are generally treated as constants

enforcing mass and momentum conservation in each layer as well as in the two-layer system as a whole (Iverson & Ouyang, 2015). Han et al. (2015) implemented a simplified form of this approach (cf. Iverson, 2012), but to our knowledge a more complete implementation has not been utilized in model applications to practical hazard assessment.

Finally, we note that depth-averaged numerical models are most likely to be sound if details of the depth-integration procedure have been presented in publications that can be scrutinized. In our experience, model formulations that skip steps in the depth-integration procedure can lead to spurious equations with hidden assumptions or even to equations that violate physical conservation laws. No aspect of debris-flow modeling is more important than satisfying physical conservation laws, which provide the strongest block in the foundation of debris-flow science.

5.4 One-Phase Versus Two-Phase Models

The simplest depth-averaged models idealize debris flows as one-phase materials with fixed bulk densities. These models include three conservation equations, one expressing mass conservation and two expressing conservation of orthogonal components of either bed-parallel or horizontal linear momentum (e.g., McDougall & Hungr, 2004; Rickenmann et al., 2006). The three equations generally differ from the standard shallow-water equations only in their definitions of flow resistance and longitudinal stress coefficients. Their similarity to the shallow-water equations has significant advantages from a mathematical and computational standpoint because the shallow-water equations have a long history of theoretical analysis, numerical solution, and practical application (e.g., Berger et al., 2011; Vreugdenhil, 1994). However, one-phase debris-flow models omit a central feature of debris-flow behavior: natural evolution of flow resistance that results from evolving interactions of solid and fluid constituents during evolution of debris velocity and bulk density. This evolution indicates that flow resistance is an emergent property that coevolves with flow dynamics (Iverson, 2003a), implying that it is inappropriate to specify evolving flow resistance a priori.

Models that treat debris flows as two-phase materials with evolving bulk densities, solid–fluid interactions, and flow resistance typically include four to six conservation equations. One of these equations expresses conservation of solid mass and another expresses conservation of either fluid mass or conservation of mass for the two-phase mixture as a whole. The other conservation equations express either orthogonal momentum components of separate but interacting solid and fluid constituents (e.g., Bouchut et al., 2016; Meyrat et al., 2022; Pitman & Le, 2005; Pudasaini, 2012) or orthogonal momentum components of a two-phase mixture in which the solid and fluid volume fractions evolve as the solid and fluid constituents interact (Iverson & George, 2014; Kowalski & McElwaine, 2013). The conceptual underpinnings of these two approaches are similar but differ in one important respect. An approach that considers distinct solid and fluid phases allows for the possibility of complete separation of the two phases into bodies composed entirely of fluid or entirely of solid grains, whereas an approach that considers a two-phase mixture assumes that some amalgamation of grains and fluid is always present—albeit with grain or fluid concentrations that might become very small in some circumstances.

A fundamental issue in the formulation of all two-phase debris-flow models concerns definition of the fluid phase. A rationale provided by Iverson (1997) is that the fluid phase includes small grains that can remain suspended by purely hydrodynamic forces—without need of grain interaction forces—for the duration of a debris flow. These suspended small particles increase the effective fluid viscosity relative to the viscosity of pure water, and in sufficiently high concentrations they can confer some fluid-phase yield strength. Typically, these small particles are clay-sized and silt-sized (<0.0625 mm), whereas larger particles are treated as part of the granular phase. Muddy water that drains from the margins of fresh debris-flow deposits provides evidence that silt- and clay-sized particles can remain in suspension even after debris-flow motion ceases (e.g., Logan et al., 2007).

The use of more than three conservation equations in two-phase and two-phase mixture models makes their mathematical and computational structure more complicated than that of one-phase models. However, the use of additional conservation equations simplifies two-phase and two-phase mixture models from a physical standpoint because it reduces the breadth of assumptions used in deriving the equations and it sharpens the definitions of model parameters. The total number of differential equations in any debris-flow model must of course equal the total number of dependent variables in the model. A basic tenet of mathematical physics, applicable in debris-flow modeling, is that predicting the values of increasing numbers of dependent variables enables increasingly diverse and stringent model tests. Models that can successfully predict the values of the largest number of dependent variables simultaneously are the most likely to be physically correct (Iverson, 2003b).

5.5 Ancillary Equations, Variables, Parameters, and Tuning Coefficients

Depth-averaged debris-flow models generally include ancillary equations that supplement the core conservation equations. For example, our D-Claw model includes a differential equation describing evolution of basal pore-fluid pressure, which is derived and solved in conjunction with the conservation equations but does not express a physical conservation law (George & Iverson, 2014; Iverson & George, 2014, 2016). Other differential equations with similar relationships to conservation laws have been derived to describe phenomena such as grain-size segregation (Gray & Kokelaar, 2010) or granular temperature (a measure of kinetic energy associated with random grain motions in agitated debris). Such ancillary equations account for evolution of dependent variables that is fully coupled to evolution of the basic dependent variables that appear in depth-averaged conservation laws (Table 5.1). A model's use of ancillary differential equations accounts for the fact that relationships among quantities that serve as dependent variables in the equations may evolve in a variety of ways, contingent on the evolution of all other dependent variables in the model and on the auxiliary conditions they satisfy (see Sect. 5.6).

The ancillary equations most commonly used in debris-flow models are not differential equations; instead, they are algebraic equations that prescribe relationships that do not evolve. Unlike use of a differential equation, use of an algebraic formula relating two or more quantities implies knowledge of an immutable relationship between the quantities, whereas use of a differential equation implies less omniscience. In debris-flow models algebraic formulas are typically used to specify various kinds of basal or internal flow resistance, solid–fluid drag forces, lateral stress transfer, basal erosion or sedimentation rates, or other energy-dissipating phenomena that are not easily characterized by using continuum conservation laws alone. In some cases authors identify these ancillary formulas as "rheologies" or even "laws," but such labels are generally inappropriate. True rheological formulas—sometimes called constitutive equations—must satisfy strict mathematical and physical criteria and be cast in three-dimensional vector/tensor forms (e.g., Malvern, 1969). Few ancillary formulas used in depth-averaged debris-flow models satisfy these criteria.

Ancillary equations are commonly the place where most parameters and tuning coefficients are introduced in debris-flow models. Here we use "parameter" to describe model quantities that have a clear physical basis and are measurable outside the context of any model (Table 5.2). For example, debris friction angles, which are measurable in geotechnical tests and estimable from sediment angles of repose, qualify as physical parameters. The same is true of debris hydraulic permeabilities, which can be measured

Table 5.2 Definitions of parameters and tuning coefficients commonly used in some depth-averaged debris-flow models

Typical symbols	Definitions
Physical parameters	
ϕ_{int}	Internal friction angle of granular debris
ϕ_{bed}	Basal friction angle of granular debris in contact with bed
ρ_s	Mass density of solid grains
ρ_f	Mass density of intergranular fluid
μ	Viscosity of intergranular fluid
k_0	Debris hydraulic permeability in initial state
α_0	Debris compressibility in initial state
m_{crit}	Debris critical-state solid volume fraction in initial state
Tuning coefficients	
A	Coefficient of basal flow resistance independent of flow velocity
B	Coefficient of basal flow resistance proportional to flow velocity
C	Coefficient of basal flow resistance proportional to flow velocity squared

in permeameter tests or slug tests, and of the viscosity of muddy intergranular water, which is readily measured using a standard benchtop viscometer.

In contrast to our definition of parameters, we define model tuning coefficients as quantities that are used to fit model results to observations of macroscopic debris-flow behavior but that lack clear links to independent physical measurements outside the context of any model. For example, some debris-flow models such as the well-known FLO-2D model use basal flow-resistance formulas similar to $R = A + Bv + Cv^2 + ...$, where R is the total flow resistance, v is the magnitude of flow velocity, and A, B, and C are tuning coefficients (Table 5.2). An equation of this form can provide a power-series representation of almost any functional dependence of R on v. Owing to this flexibility, tuning the values of A, B, and C may yield appealing fits of model output to observations of debris-flow behavior. There is nothing fundamentally wrong with such a model tuning procedure, provided that its scientific basis and interpretation are not misrepresented. Some publications have stated that a coefficient analogous to A represents a combination of cohesive and frictional resistance, a coefficient proportional to B represents viscous resistance, and a coefficient proportional to C represents resistance due to turbulence and/or gain inertia during shearing (e.g., O'Brien et al., 1993). If these descriptions were scientifically valid, then values of A, B, and C could be readily interpreted and could be determined from experiments or estimated from field data that are wholly independent of any model. When A, B, and C are used as tuning coefficients, however, that is not the case.

From a practical rather than scientific perspective, why does the difference between physical parameters and tuning coefficients matter? An important reason is that appropriate values of physical parameters are commonly constrained by a great body of accumulated knowledge that is independent of specific models but is potentially useful in all models. For example, few experienced debris-flow investigators would estimate that applicable values of the Coulomb friction angle ϕ of the granular fraction of debris lie outside the range $20° \leq \phi \leq 50°$. Such a range constraint is applicable regardless of whether the debris of interest is part of a small alpine debris torrent or a great volcanic lahar. Indeed, applicability from one location to another—and from one model to another—is a great advantage that physical parameters hold over tuning coefficients. If all debris-flow models used physically defined parameters rather than tuning coefficients, then debris-flow science and model-based hazard evaluation could advance more rapidly because comparisons between predictions of different models could be more direct (e.g., Barnhart et al., 2021b).

5.6 Initial and Boundary Conditions

Numerical solution of the system of partial differential equations in depth-averaged debris-flow models requires specification of appropriate initial and boundary conditions, collectively called auxiliary conditions. These conditions must satisfy physical constraints and must be ascertainable from a priori knowledge. Additionally, they must be posed in a manner consistent with the mathematical properties of the conservation equations themselves, which generally can be classified as hyperbolic. Mathematical properties of hyperbolic partial differential equations are beyond the scope of this chapter, but an overarching property is that auxiliary conditions appropriate for various uses of the equations can differ greatly.

The boundary conditions for some debris-flow models can be easily and unequivocally specified because the domain boundaries can be placed arbitrarily far from any location where debris might originate or travel. At these distant boundary locations, values of all dependent variables can be set equal to zero for the duration of a simulation (Fig. 5.5). Use of this parsimonious boundary condition is standard when our own model, D-Claw, is applied to real-world problems (Barnhart et al., 2021a; George et al., 2017, 2022; Iverson & George, 2016), but not all debris-flow models are structured or implemented in this way. For example, some models require information-intensive specification of the depths and velocities (i.e., hydrographs) of incoming flows that enter upstream domain boundaries. Such differences in the type and scope of a priori information requirements reflect fundamental differences in the epistemological framework of the models, complicating the task of comparing the models' predictive value (Barnhart et al., 2021b).

If the goal is merely to fit model results to the outcome of a past debris-flow event, models

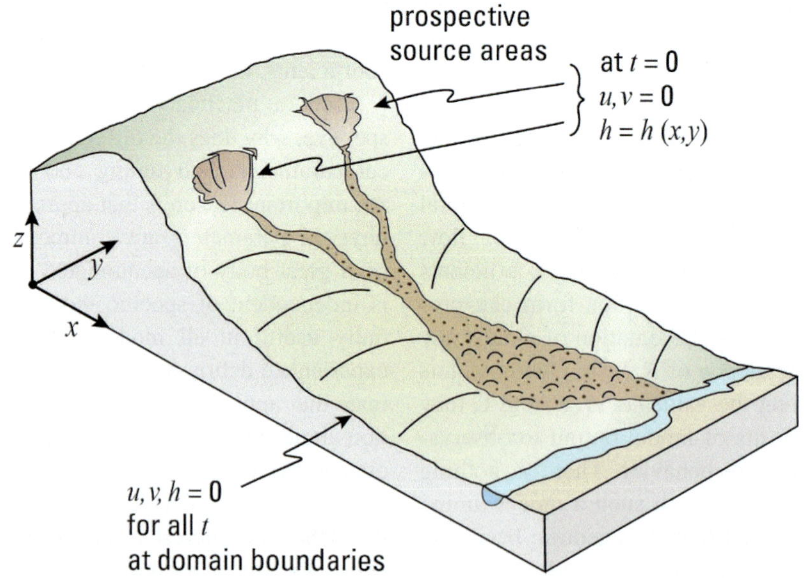

Fig. 5.5 Schematic block diagram illustrating debris-flow paths that lie entirely within the boundaries of a rectangular computational domain. The dependent variables u, v, h, and independent variables x, y, z, and t are defined in Table 5.1. The initial and boundary values of u, v, and h, as implemented in D-Claw, are shown in the figure

that use freely adjustable, information-intensive boundary information have a competitive advantage over those that use fixed, parsimonious boundary information because adjustable boundary information can be tailored to produce the desired model output. On the other hand, if the goal is to predict the behavior of future debris flows, the use of freely adjustable boundary information diminishes a model's value because it requires some omniscience by the model user who makes the adjustments. Parsimonious boundary conditions are consistent with unambiguous physical constraints and don't require such adjustments.

Initial conditions used in depth-averaged debris-flow models can be just as diverse as boundary conditions. However, regardless of the type of initial condition used, the volume of potentially mobilized debris has an overarching importance for anticipating debris-flow runout (Griswold & Iverson, 2008; Iverson et al., 1998; Reid et al., 2016) and simulating debris-flow motion (Barnhart et al., 2021b). Any model application that assumes or calculates a seriously flawed debris-flow volume cannot be expected to yield good predictions of debris-flow speeds, depths, and inundation patterns. Estimation of future flow volumes and source locations is consequently an essential step in specifying initial conditions for all debris-flow models that are used to make predictions—and it is a step in which considerable uncertainties may be involved.

On the other hand, some physical certainties can be exploited in specifying initial conditions and flow volumes for debris-flow models. For example, the initial state of debris-flow material is unquestionably a state of static mechanical equilibrium with balanced driving and resisting forces. Static debris is initially poised on slopes or in channels and is ready to move under the influence of a physical perturbation such as rainfall that increases pore-water pressures in slopes and/or the discharge of water in channels. A small perturbation of such a statically balanced state is the initial condition typically used in D-Claw, for example (Fig. 5.6a). This initial condition is parsimonious because the statically balanced initial state satisfies unambiguous physical constraints.

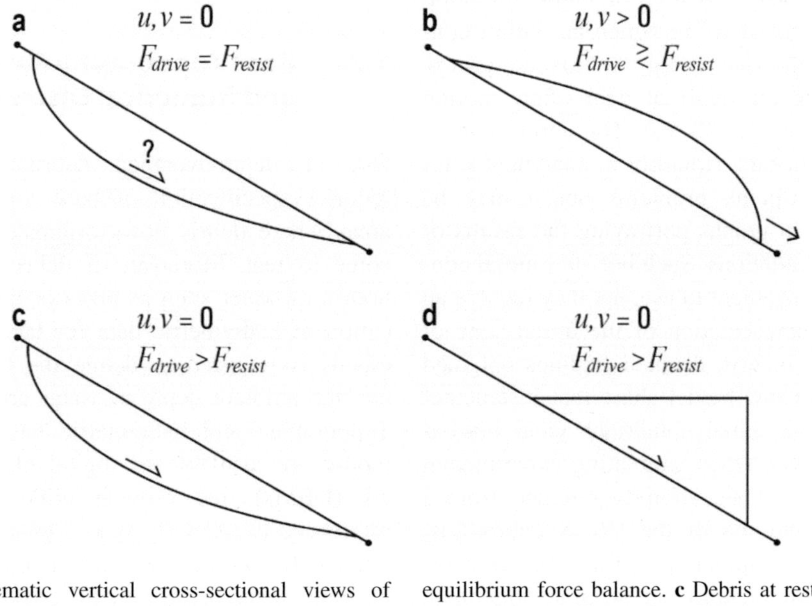

Fig. 5.6 Schematic vertical cross-sectional views of initial conditions used in various debris-flow models. **a** Debris at rest satisfies a static equilibrium force balance with driving forces (F_{drive}) equal to resisting forces (F_{resist}). **b** Debris in motion may or may not satisfy an equilibrium force balance. **c** Debris at rest does not satisfy a static equilibrium force balance, but motion is prohibited until a computer command is issued. **d** Debris at rest is constrained by an imaginary vertical dam that vanishes when a computer command is issued

Some other debris-flow models use information-intensive initial conditions in which a priori flow depths and velocities must be specified (Fig. 5.6b). This type of initial condition is commonly used in flood-wave modeling, but river floods differ fundamentally from debris flows because most rivers flow continually. Moreover, the flows of many rivers are gauged, providing data to constrain the flow depths and velocities that precede a flood. By contrast, specifying flow depths and velocities in advance of a debris flow requires considerable guesswork.

Still other debris-flow models use various forms of dam-break initial conditions. In these cases modelers specify a spatial distribution of static debris in prospective debris-flow source areas, but they do not require the debris to satisfy an initial state of static mechanical equilibrium in which gravitational driving forces are balanced by resisting forces (Fig. 5.6c). Instead, the simulated debris is artificially held in place by withholding a computer command. Issuing a command instantaneously places the debris in a far-from-equilibrium state in which driving forces greatly exceed resisting forces, mimicking the effect of an idealized, instantaneous dam break (Fig. 5.6d). This type of initial condition produces a "hot start" in which an abrupt conversion of potential energy to kinetic energy occurs during the onset of debris-flow motion (Iverson & George, 2019b). The abrupt energy conversion appears explicitly in analytical solutions of dam-break problems but it may be less obvious in graphs portraying the results of numerical simulations. Such hot-start initial conditions are convenient to use, but they involve an unphysical representation of the initial state of static debris. Indeed, detailed findings obtained using our D-Claw model show that instantaneous dam-break initial conditions yield inaccurate results even when simulating experimental debris flows that are abruptly released from a headgate at the top of the USGS debris-flow flume. Accurate simulations of these experimental debris flows require an accurate kinematic representation of the headgate opening style and rate (George & Iverson, 2014).

From a practical rather than scientific perspective, why does the difference between physical and unphysical auxiliary conditions matter? Perhaps the most important reason is that physical auxiliary conditions can be constrained by direct a priori observations of static debris on a landscape, whereas unphysical conditions cannot. Physical law requires that static mechanical equilibrium is satisfied by debris poised on slopes or in channels prior to the onset of debris-flow motion. Moreover, the volume and distribution of static debris that may be susceptible to motion can be inferred from topographic, geologic, and soils maps, supplemented by field observations, subsurface investigations, and geomechanical analysis. Such detailed information is not available in every circumstance, but obtaining such information is at least plausible. Accurately specifying future disequilibrium states of debris (i.e., states that do not satisfy mechanical equilibrium) is a more formidable task. Indeed, modern understanding of nonlinear dynamical systems indicates that it is fundamentally impossible to specify future disequilibrium states, because debris flows are nonlinear phenomena that may evolve in diverse ways contingent on initial conditions.

5.7 Base Topography Resolution and Numerical Discretization

Use of depth-averaged debris-flow models requires specification of base topography for areas where debris flows originate, travel, and come to rest. Moreover, if debris flows enter bodies of water such as lakes or rivers, specification of bathymetric data for lakebeds or riverbeds is required to define the basal surface and the ambient depth of water above the bed. Topographic and bathymetric base data commonly are available as digital elevation models (DEMs) that provide gridded elevations referenced to regularly spaced planimetric coordinates. If such data are unavailable, DEMs can be constructed by digitizing analog data such as topographic contours, although this type of DEM construction can be labor-intensive.

The resolution of DEM data is an important consideration in debris-flow modeling. A coarse DEM with 10 or 20 m horizontal resolution may be satisfactory for modeling great lahars (volcanic debris flows) involving hundreds of millions of cubic meters of debris and inundating 100 km^2 or more of low-relief terrain. On the other hand, DEMs with sub-meter resolution (typically acquired using Lidar, structure-from-motion digital photography, or some analogous technology) may be necessary for modeling debris flows that involve only thousands of cubic meters of material descending mountain-front channels and alluvial fans. In other cases, hybrid resolutions may be warranted, because relatively coarse DEMs may be satisfactory for some parts of debris-flow paths but high-resolution DEMs may be required to portray accurately some high-value/high-consequence features such as buildings, levees, or dams in debris-flow runout zones. However, not all numerical methods are well-suited for solving depth-averaged conservation equations for flow across topographic surfaces portrayed with variable resolutions (see Sect. 5.8 that follows).

Use of a computational grid resolution that corresponds to the maximum available DEM resolution throughout any domain might seem desirable, but in many cases it isn't practical because it comes at the expense of greatly increased computation time. For 2-D depth-averaged models, the number of computational grid cells in the domain scales with the square of the lineal spatial resolution. However, computation time typically scales with the cube of the lineal resolution, because fine spatial discretization requires commensurately fine temporal discretization in order to satisfy the Courant- Friedrichs–Lewy condition and thereby permit numerical accuracy and stability. Thus, doubling the lineal topographic resolution by replacing one square grid cell with four square grid cells generally implies that the computation time will increase about eightfold. Moreover, a large-scale computation requiring 1 day of runtime on a 10 m grid would typically require about 1000 days of runtime on a 1 m grid. Considerable care is consequently required in selecting the numerical grid resolution for a given problem. Furthermore, the optimal grid resolution generally varies throughout the course of a simulation, implying that fixed or static computational grids are suboptimal.

The incorporation of DEM data into a model's numerical discretization can be accomplished in a variety of ways. The simplest approach is to utilize fixed computational grids that align with gridded DEM data. However, this approach has disadvantages such as the requirement of a uniform resolution throughout a rectangular domain within which a debris flow typically inundates only a small fraction of the grid cells (e.g., Fig. 5.5). An alternative is to construct non-uniform computational meshes that conform to topographic features through interpolation of the gridded DEM data. The meshes may be unstructured (for example, triangular irregular networks (TINs)) or based on terrain-fitted curvilinear coordinates. While these approaches can provide improvements in terms of computational efficiency or numerical accuracy in some cases, irregular features in natural terrain, particularly in mountainous regions where most debris flows occur, present challenges for mesh generation or the definition of suitable curvilinear coordinate systems. Moreover, these approaches may require the modeler to anticipate the likely debris-flow path and extent.

Because DEM base data provide elevations at discrete points, a model's numerical discretization (or the generation of derived topographic data sets described above) entails implicit assumptions regarding the smoothness of the underlying surface represented by the raw data. Some modeling approaches rely on continuity or smoothness to achieve numerical accuracy or numerical stability. However, real terrain almost invariably contains irregularities and discontinuities. Such features can include not only natural cliffs but also the vertical faces of some structures such as buildings and dams. In many cases debris-flow interaction with these structures is a prime motivation for performing hazard assessments, and imprecise model representation of their geometries is undesirable.

In our own model, D-Claw, the treatment of DEM data is influenced by the class of numerical methods that we utilize to solve the governing equations (see Sect. 5.8 that follows). It also results in several properties that we consider advantageous for debris-flow modeling. First, the raw DEM data (from one or more possibly overlapping and non-aligned DEM data sets) are used without preprocessing or modification, eliminating the need to interpolate topographic data for debris-flow paths that may be difficult to define in advance. Second, the numerical discretizations (i.e., the computational grids) are not restricted or constrained by the alignment or resolution of the DEM data, although the available DEM resolution(s) may influence accuracy or a modeler's selection of computational grid parameters. Finally, smooth regions as well as topographic discontinuities are represented in the numerical discretization to the degree possible given the raw DEM data. A sharp vertical discontinuity such as a vertical wall at a known location within otherwise coarse DEM data is well-represented by utilizing an intersecting DEM data subset for the structure or feature. Exploiting this representation requires use of numerical methods that are accurate and robust in the presence of such variable DEM resolutions.

In D-Claw we utilize DEM data by employing the following multistep methodology executed during the computation runtime. Initially, the discrete elevation data from a single or multiple DEMs are used to define a piecewise continuous basal surface that intersects the points defined by the most highly prioritized DEM (usually the highest resolution DEM) in any region within the computational domain. The average elevation value of this globally defined surface within a computational grid cell determines a unique elevation value for each cell in the numerical discretization. Therefore, the dynamically derived topographic surface used for the numerical discretization can be interpreted as a piecewise constant surface with step discontinuities at computational cell boundaries. This methodology confers several properties that are advantageous for D-Claw's numerical solution schemes. First, it preserves volume integrals of the debris lying above the basal surfaces defined by the DEM(s) and derived piecewise constant computational topography for any resolution and alignment of the computational grids, allowing conservation of mass and momentum in the presence of evolving computational grids (see Sect. 5.8 for a description of adaptive mesh refinement). Second, elevation gradients in the DEM data that may represent smooth regions or underlying discontinuities in terrain are retained to the degree possible in the piecewise constant computational grids. Finally, step discontinuities arising from vertical surfaces that are well-defined by intersecting DEMs can be retained in the computational topography.

5.8 Numerical Solution Techniques

The mathematical core of physically based, depth-averaged debris-flow models is a system of nonlinear hyperbolic partial differential equations (PDEs). This class of PDEs, which arise in diverse problems involving wave propagation, presents unique challenges for obtaining accurate numerical solutions, in part because the equations can have non-unique, discontinuous solutions. The discontinuous solutions apply to (but are not limited to) shock waves, a term borrowed from gas dynamics but also used to describe analogous features such as hydraulic jumps in debris flows (as illustrated in Fig. 5.4 and in Sect. 5.9 below). Standard numerical methods such as classical finite-difference methods are poorly suited for these problems because they may produce physically spurious solutions, numerical dispersion, or numerical instabilities. These shortcomings have motivated the development of specialized classes of numerical solution techniques, including the shock-capturing, finite-volume methods provided by the Clawpack open-source software project (LeVeque, 2002, www.clawpack.org).

The numerical methods implemented in Clawpack utilize fixed (Eulerian) coordinate systems and analytical wave-propagation solutions to Riemann problems, a term that describes mathematical problems involving discontinuous steps in values of variables at computational grid-cell interfaces. Originally developed for solving shock-wave problems (Godunov, 1959), the Riemann wave-propagation methodology also provides a logical means for the accurate representation of discontinuous steps in elevation data that may exist when modeling flows across topography represented by DEMs (George, 2008, 2010). The representation of discontinuous steps is particularly relevant for resolving debris-flow fronts moving across steep terrain or impacting vertical structures (See Sect. 5.10 below).

Our D-Claw model uses numerical solution techniques built within the Clawpack framework and on top of the GeoClaw subpackage developed to simulate depth-averaged flows across variable topography (George & LeVeque, 2006; Berger et al., 2011; LeVeque et al., 2011). D-Claw also provides additional numerical strategies aimed at properly resolving balanced steady states (Bouchut, 2004; LeVeque, 2002). Because debris accelerations are absent in such states, the governing conservation equations reduce to steady force balances, which are simple in principle but difficult to satisfy in practice because minuscule numerical imprecision can upset a balanced state, particularly if thousands of numerical iterations are involved. Such balanced steady states exist, for example, in static bodies of water overlying variable bathymetry (George, 2008) and in static masses of sediment resisting downslope gravitational forces prior to the onset of debris-flow motion (George & Iverson, 2014; George et al., 2022; Iverson & George, 2016). Thus, maintaining well-balanced steady states plays a particularly important role when statically balanced initial debris conditions are considered (see Sect. 5.6 above).

Additional numerical challenges exist because many hyperbolic problems, including those that arise in depth-averaged modeling of debris flows, feature a broad spectrum of spatial scales that vary in time, location, and extent. This breadth of scales motivated the development of a computational strategy known as adaptive mesh refinement (AMR) (Berger and Oliger 1984), which utilizes evolving patches of nested computational grids with varying cell sizes based on dynamic features of the numerical solution. Various implementations of AMR exist, but the aim of all AMR techniques is to provide optimal grid resolutions for attaining numerical accuracy and computational efficiency throughout the duration of a simulation.

Standard AMR techniques developed for general hyperbolic conservation equations are not well-suited for modeling depth-averaged shallow flows moving across variable topography, because the techniques cannot simultaneously preserve balanced steady states (such as static states) and conserve mass, momentum, and energy. Specialized AMR techniques were developed to overcome this problem in the context of tsunami modeling (George & LeVeque, 2006; LeVeque et al., 2011). The techniques were later extended to modeling of general depth-averaged flows over topography such as overland floods (Berger et al., 2011; George, 2010) and debris flows (George & Iverson, 2014).

Use of the specialized AMR algorithms that preserve balanced states also enables D-Claw to use evolving, high-resolution grids to dynamically resolve advancing flow fronts moving through complex terrain, while retaining coarse grids where no evolution of variables occurs (e.g., as in those parts of the computational domain where flow is absent). This approach makes possible the use of modern sub-meter-resolution DEM data only where it is needed within a large domain. By doing so, use of AMR in D-Claw dramatically affects the simulation results where there are important yet small-scale topographic features at isolated locations (for example, dams, levees, buildings, and so on), which require fine grid resolutions that would be prohibitive computationally if applied to the entire domain.

An alternative open-source software package, r.avaflow (Mergili et al., 2017), solves the two-phase depth-averaged debris-flow equations of Pudasaini (2012). Like D-Claw, r.avaflow employs Eulerian numerical methods developed specifically for hyperbolic systems of differential equations. However, it utilizes total variation diminishing non-oscillatory central differencing (TVD-NOC) numerical solution schemes (Nessyahu & Tadmor, 1990), which have been successfully applied to mass flow problems more generally (see Mergili et al., 2017, for examples). Unlike D-Claw, r.avaflow solves equations in local terrain-fitted coordinates. This approach confers some advantages in terms of physical fidelity, but it can present practical challenges for modeling flows across complex, real-world topography. For instance, variables expressed in local terrain-fitted coordinates and DEMs expressed in spatially uniform Cartesian coordinate systems are not always compatible. In general, a transformation of variables computed in terrain-fitted coordinates to uniform Cartesian DEM coordinates is necessary, and it must be approximated (Mergili et al., 2017).

Another approach for tackling the difficulty of accurately modeling shallow flows across rugged topography is implemented in the model SHALTOP (Bouchut et al., 2003; Lucas et al., 2007), which solves model equations in global Cartesian coordinates but corrects the equations in order to satisfy the shallowness assumption applied in the local bed-normal direction. This approach can capture bed-normal accelerations due to terrain curvature (Peruzzetto et al., 2021), but it still requires that bed-normal variables be transformed to a Cartesian reference frame as represented by a standard DEM. Like D-Claw, SHALTOP utilizes well-balanced finite-volume numerical schemes for solving hyperbolic systems of differential equations (Bouchut, 2004).

Other Eulerian debris-flow models include TITAN2D (Pitman et al., 2003), which solves single-phase depth-averaged equations, and RAMMS (Christen et al., 2010), which was originally developed for modeling snow avalanches but utilizes a single-phase Voellmy flow resistance formula for modeling debris flows. These models employ finite-volume solution schemes broadly similar to those used in D-Claw.

A more distinctive method for solving the hyperbolic system of PDEs in depth-averaged debris-flow models is the meshless Lagrangian method known as smooth-particle hydrodynamics (SPH) (McDougall & Hungr, 2004.). As opposed to Eulerian methods, Lagrangian methods formulate the model PDEs in a moving frame of reference that translates with the flow. In applications of SPH to debris flows, the continuum mechanical representation of debris-flow material is effectively replaced by a collection of interacting, columnar "particles" that span the flow thickness as they move downstream (McDougall & Hungr, 2004). Moreover, SPH removes the effects of discontinuities in solutions of the governing PDEs by using a smoothing formula to average particle-interaction mechanics among groups of neighboring particles.

In some respects the SPH methodology codifies a modeling philosophy that is the opposite of that of D-Claw. Whereas SPH produces smoothed results, the shock-capturing AMR methodology of D-Claw aims to preserve the effects of discontinuities and the details of abrupt flow-dynamics features such as hydraulic jumps and flow interactions with vertical barriers. Each of these methodologies—as well as other numerical methodologies—may have advantages in different circumstances. However, we are unaware of any application of SPH methodology to two-phase debris flows in which the debris solid volume fraction, bulk density and apparent rheology naturally coevolve as flow dynamics evolve.

5.9 Computer Source Code Verification and Accessibility

A problematic aspect of computational modeling of complex natural phenomena such as debris flows is that appealing graphical results may be generated even if the underlying model is inadequate from a scientific perspective. (Few scientists would deny that Hollywood movie animators are adept at generated numerical output that superficially resembles disastrous natural phenomena, ranging from tornadoes to tsunamis to asteroid impacts, but few animators would claim that their model outputs have scientific validity merely because they look good.) The value of scientific numerical models cannot be judged solely on the basis of the appearance of their output. Instead, the models should be thoroughly documented, tested, and available for scrutiny, and a key component of such models is the computer source code.

Just as thorough derivations of debris-flow model equations should be freely available for critical inspection and validation by prospective model users, computer source codes that are used to solve the equations and generate output should also be freely available. No scientific computation should involve use of proprietary computer source code that is inaccessible for independent verification. Source code errors or inadequacies are commonly revealed only through scrutiny of the code by diverse users who pursue diverse applications, and issues may never be revealed if the source code remains in a black box.

Trujillo-Vela et al. (2022) provide an extensive table summarizing source code accessibility as well as some other properties of many debris-flow models, landslide models, and granular-avalanche models. The table contains some errors, but it is nevertheless useful. We have avoided presenting a similar table here, in part because it would largely reproduce that of Trujillo-Vela et al. (2022), and in part because it is difficult to ensure that the content of such tables is accurate. Many models are presented and utilized without full disclosure of all of their salient details, making it challenging for others to evaluate the models thoroughly or characterize them accurately.

In the twentieth century scientific computer source codes were commonly printed out in full and made available for inspection, but full printouts of recently developed source codes typically aren't available. Modern computer source codes can be long and complex, and they can evolve with time as improvements and additions are made. Consequently, developers of some scientific codes have adopted practices established in computer science and software engineering. These practices include the use of version control systems (e.g., subversion, git), open-source licensing, and freely available source-code repositories. Our model D-Claw has been developed following these practices, and its components are publicly hosted on github.com (see Mandli et al., 2016, www.github.com/clawpack, and www.github.com/geoflows/D-Claw).

LeVeque (2013) suggested that all aspects of computational procedures use to generate scientific output, including post-processing programs used to generate graphical representations of output, ought to be fully documented and freely accessible for scrutiny. Similar views have been expressed by others in the computational sciences (e.g., Barnes, 2010; Peng, 2011; Stodden et al., 2016). It is difficult to argue against this full-disclosure protocol if reproducibility and verifiability of model output is the objective.

5.10 Model Testing

Systematic testing of the predictions of debris-flow models should precede use of the models for hazard assessment, but conclusive model testing can be hampered by the effects of uncertainties about initial flow volumes, auxiliary conditions, or appropriate values of parameters or tuning coefficients. Nonetheless, experimental data and well-documented historical debris flows afford opportunities for model testing that is more rigorous than is generally possible for models of geological processes that occur in deep Earth or deep time.

Depth-averaged debris-flow models can potentially predict a wide range of quantities, each of which can be used for model testing. The most rigorous tests involve testing of model predictions against several quantities measured simultaneously. All depth-averaged models of debris-flow motion across three-dimensional terrain predict values of flow depth $h(x, y, t)$ and two flow velocity components, $u(x, y, t)$ and $v(x, y, t)$ (which yield two flow momentum components, $\rho uh(x, y, t)$ and $\rho vh(x, y, t)$), as described in Sect. 5.3 above. Therefore, direct tests of model predictions can be made when direct measurements of these evolving quantities at various locations are available, and these tests can be expanded if there are simultaneous measurements of additional quantities such as basal shear stresses, total normal stresses and pore-fluid pressures. Reproducible measurements of these co-evolving quantities generally can be made only in idealized experimental settings (cf. George & Iverson, 2014; Iverson et al., 2010), but some analogous measurements have been made in carefully instrumented cross sections constructed at field sites (McArdell et al., 2007; McCoy et al., 2010). The spatial distributions and thicknesses of natural debris-flow deposits also afford opportunities for model tests, although predictions of these quantities are strongly influenced by assumptions about initial conditions—especially concerning initial flow volumes, velocities, and depths (Barnhart et al., 2021b). In some circumstances the seismic signatures of large natural debris flows can be used for testing model predictions of temporally varying momentum exchanges between debris flows and their beds (Moretti et al., 2015). Runup heights measured where natural debris flows encounter obstacles in their paths can be used in a similar manner, particularly if independent constraints on flow velocities exist (e.g., Allstadt, 2013).

Here we illustrate some tests of our own debris-flow model, D-Claw, against the results of large-scale laboratory experiments and a ~50×10^6 m^3 debris flow that occurred in 2010 at Mount Meager, British Columbia, Canada. Each type of test serves a particular purpose for evaluating model simulations of debris-flow onset, flow dynamics, or extent of downstream inundation. No single test is definitive, but confidence in any model grows if it performs well in a variety of tests.

5.10.1 Debris-Flow Mobilization from Landslides

Mobilization of debris flows from landslides provides a model-testing benchmark problem because it involves a precisely defined initial condition (static mechanical equilibrium of a wet sediment mass) and because the physical properties of static debris are readily measured. Many experiments on debris-flow mobilization from landslides have been performed at the U.S. Geological Survey (USGS) debris-flow flume (Iverson et al., 1997, 2000; Logan et al., 2007; Reid et al., 1997). Figures 5.7 and 5.8 illustrate the results of D-Claw simulations of one of these experiments, conducted on 17 June 1998. The experiment was described in detail by Iverson et al. (2000) and documented in video recordings that are viewable online (Logan et al., 2007). In the experiment a liquefying landslide was triggered by slowly adding groundwater to a 6 m^3 static prism of loosely packed, moist debris resting on the 31° concrete flume bed and held in place by a rigid retaining wall. When basal pore pressures became large enough, the slope began to fail (at $t = 0$ in Fig. 5.7). Then, over the course of about 1 s, the failing debris liquefied in response to porosity contraction and attendant pore-pressure increase. The liquefied debris flowed over the retaining wall and then down a ramp and onto the flume bed, as illustrated in the last few frames of Fig. 5.7 (George & Iverson, 2014).

The accuracy of the modeled behavior depicted in Fig. 5.7 can be tested quantitatively by comparing it with sensor data collected during the experiment. Data from two ground-surface extensometers and three pore-pressure sensors buried at different depths are shown in Fig. 5.8 along with two sets of model predictions for a 2 s period that spans the time of slope

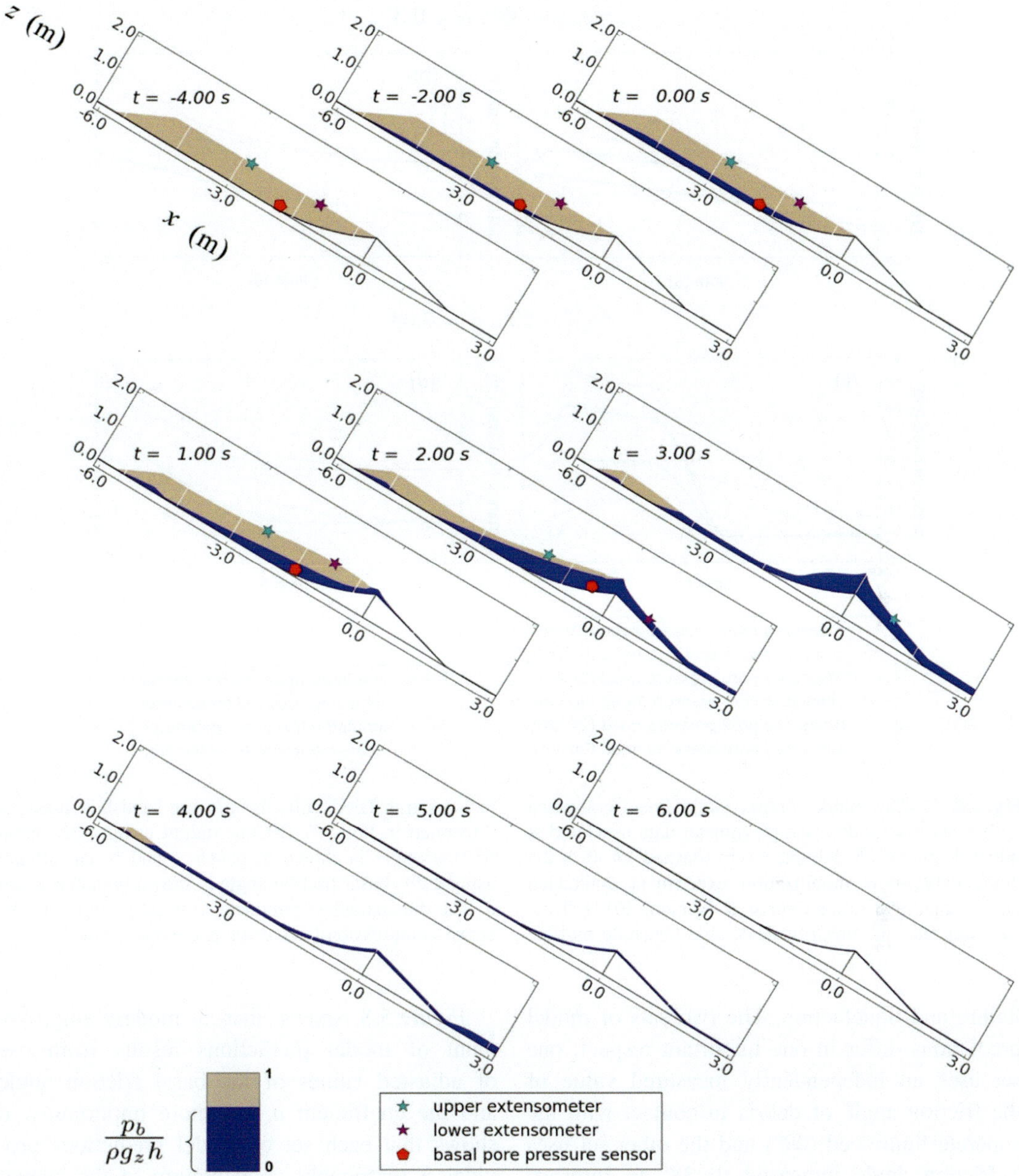

Fig. 5.7 Time sequence of vertical cross-sectional views of a D-Claw simulation of a liquefying landslide at the USGS debris-flow flume on 17 June 1998 (after George & Iverson, 2014). The relative height of blue shading within the brown-shaded landslide mass depicts the basal pore-water pressure p_b as a fraction of the total lithostatic basal normal stress, $\rho g_z h$. A condition of complete liquefaction is indicated by $p_b/\rho g_z h = 1$ and full blue shading. Symbols show the positions of surface extensometers and the basal pore-pressure sensor used to test model predictions, as illustrated in Fig. 5.8. The sensors are motionless until $t = 0$, when slope failure commences. They subsequently move with the sediment and pass over the retaining wall and ramp centered at $x = 0$ m. The shape of the curved basal surface of the sediment mass just upslope from the bed-normal retaining wall is defined by data from a stack of tiltmeters that revealed the location of the basal slip surface (Iverson et al., 2000)

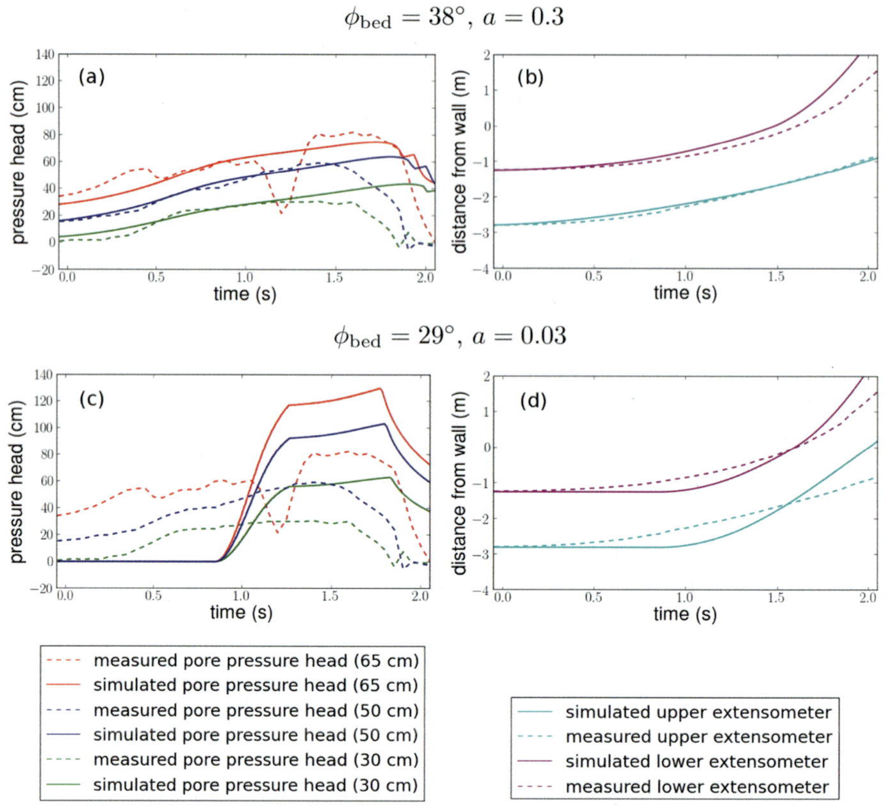

Fig. 5.8 D-Claw model output (solid lines) compared with extensometer data and piezometer data measured at initial depths of 65, 50 and 30 cm (dashed lines) in the flume debris-flow mobilization experiment conducted on 17 June 1998 (after George & Iverson, 2014). Time t = 0 indicates the onset of detectable landslide motion. Sensors translated with the moving landslide mass as illustrated in Fig. 5.7. D-Claw output using a 38° basal friction angle is shown in panels a and b, and output using a 29° basal friction angle is shown in panels c and d. The dimensionless parameter a is proportional to the debris compressibility (Iverson & George, 2014)

failure and liquefaction. The two sets of model predictions differ in one important respect: one set uses an independently measured value of the friction angle of debris in contact with the concrete flume bed (29°) and the other set uses a friction angle increased to 38° to improve the fit to the experimental data (Fig. 5.8). The increased friction angle helps account for the influence of flume sidewall friction that was not explicitly represented in the depth-averaged numerical simulations (George & Iverson, 2014). The two sets of predictions also use differing values of a dimensionless coefficient a, which is proportional to the debris compressibility and is not as well-constrained as other debris properties (Iverson & George, 2014).

Figure 5.8 shows that a modest improvement of model predictions results from use of adjusted values of the basal friction angle and the coefficient a, but more importantly, it shows that each set of model predictions provides a reasonably good picture of the coevolution of several dependent variables as slope failure occurred and debris liquefaction ensued. Consequently, we infer that the underlying physics represented by the model is approximately correct. If the values of fewer dependent variables were predicted simultaneously, then model tests would be weaker and inferences about model validity would be more equivocal. It is desirable whenever possible to test debris-flow models against such multivariate data.

5.10.2 Debris-Flow Runup

Runup of debris flows on barriers or adverse slopes can provide model-testing opportunities because runup occurs commonly and maximum runup heights can be measured with confidence. However, in most field settings, debris-flow dynamics prior to the onset of runup are poorly constrained. In this section we compare D-Claw simulations of runup with observations obtained in experiments conducted with 10 m^3 debris flows at the USGS debris-flow flume. An advantage of the experiments is that flow speeds and depths prior to runup were precisely measured, eliminating a source of uncertainty in model testing (Iverson et al., 2016).

Runup that occurs when a debris flow encounters a vertical barrier is most significant when the incoming flow is supercritical and an abrupt conversion of kinetic energy to potential energy generates a shock (i.e., hydraulic jump). Runup of subcritical debris flows is less significant because subcritical flows merely decelerate and accumulate as "backwater" deposits when they encounter a barrier. Figure 5.9 compares snapshots of D-Claw simulations with time-stamped video frames that depict a supercritical 10 m^3 debris flow running up approximately 2 m on a rigid vertical wall. Details of the debris properties and parameter values used in the simulations were presented by Iverson et al. (2016). The flow generating the runup had multiple surges, the largest of which is visible in Fig. 5.9 exiting the flume mouth at $t = 10.12$ s. Both constructive and destructive interference of successive debris-flow surges strongly influenced runup behavior in this experiment and in the model simulations by transiently increasing or decreasing the runup height (Iverson et al., 2016). Analogous interactions of surges can be difficult to decipher from runup deposits in the field.

Runup can also occur when debris flows encounter adverse slopes and travel some distance up their faces. Model tests against data from debris-flow flume experiments addressed this scenario by using a 30-degree adverse slope with a width of 3 m, so that some debris spilled off the edges of the slope as runup occurred when a 10 m^3 debris flow discharged from the mouth of the 2-m wide USGS debris-flow flume (Fig. 5.10). Details of the slope configuration and bed roughness were described by Iverson et al. (2016).

As in the vertical barrier case, runup on the adverse slope was influenced by arrival of successive surges of flow and by the development of a shock. However, shock development on the 30-degree slope was preceded by a smooth upslope flux of momentum, as illustrated in the upper frames of Fig. 5.10. The D-Claw model accurately simulated the transition from this smooth momentum flux to a shock-interrupted flux. It also accurately simulated the maximum runup height, which developed as the peak shock collapsed and shunted some debris both upslope and downslope (final frames in Fig. 5.10).

Successful testing of the D-Claw model (or any other debris-flow model) under the controlled, reproducible conditions described above can bolster confidence in the model. It can also prompt a next step in testing: application to historical real-world debris flows that are more complicated than experimental debris flows but are reasonably well-constrained by observations and data.

5.10.3 Mount Meager Debris-Flow Dynamics

One of the world's largest debris flows in recent decades began as a landslide on the south flank of Mount Meager volcano in British Columbia, Canada on 6 August 2010. Most of the ~50×10^6 m^3 landslide mobilized as a debris flow that descended the steep, tortuous valley of Capricorn Creek before it reached a confluence with Meager Creek and ran up to height of 270 m on an adjacent mountainside (Fig. 5.11). Much of the debris then traveled downstream along the Meager Creek drainage and onto the broad floodplain of the Lillooet River, but some

Fig. 5.9 Time sequence of oblique 3-D perspective views of a D-Claw simulation (left column) and corresponding video frame captures (right column) of a vertical barrier runup experiment at the USGS debris-flow flume on 23 June 1994. Measured incoming flow properties were used to initiate the numerical simulation. Timestamp in each frame of model results and video recordings refers to the time since the flume headgate began to open. An animated version of this figure is available as Movie S5 in the Supporting Information of Iverson et al. (2016)

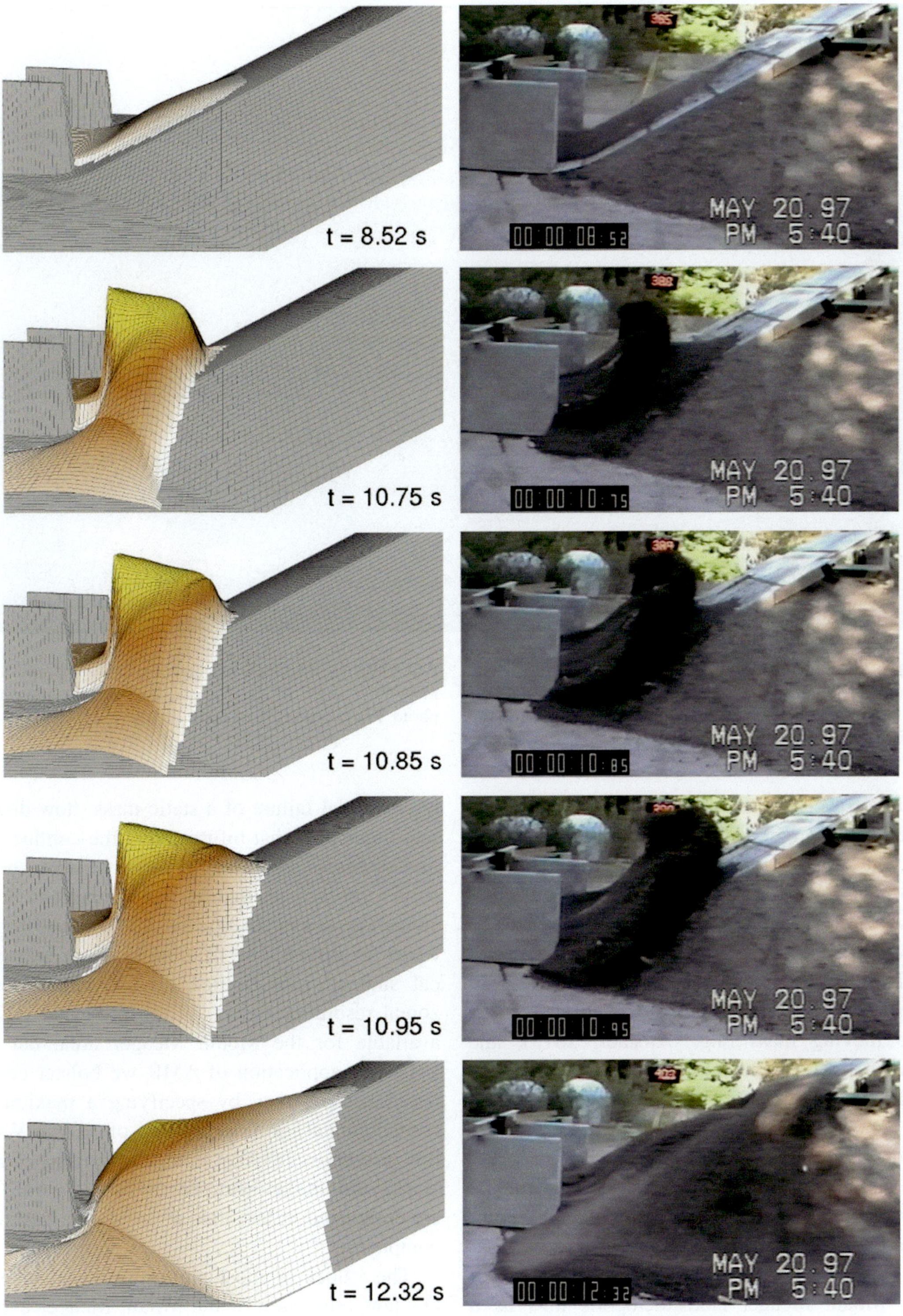

Fig. 5.10 Time sequence of oblique 3-D perspective views of a D-Claw simulation (left column) and corresponding video frame captures (right column) of an adverse-slope runup experiment at the USGS debris-flow flume on 20 May 1997. Measured incoming flow properties were used to initiate the simulation. Timestamp in each frame of model results and video recordings refers to the time since the flume headgate began to open. An animated version of this figure is available as Movie S6 in the Supporting Information of Iverson et al. (2016)

Fig. 5.11 Oblique aerial photographs of the site where the 2010 Mount Meager debris flow ran up 270 m vertically on the mountainside at the confluence of Capricorn and Meager Creeks, located at 50.606°N/123.430°W. Red arrow in the small inset photo shows the approximate look angle of the camera used to take the large photo. Photos courtesy of Dave Steers

debris traveled more than 3 km upstream along the Meager Creek valley.

The 2010 Mount Meager landslide/debris flow was thoroughly examined in several studies, beginning with that of Guthrie et al. (2012). Some of the studies inferred the detailed chronology of the event through analysis of the long-period seismic energy radiated by the moving mass (e.g., Allstadt, 2013), and some involved modeling the event numerically (e.g., McDougall, 2017; Moretti et al., 2015). Knowledge gained from these studies and detailed field work by Roberti et al. (2017, 2018) makes the Mount Meager landslide/debris flow better understood than most events of comparable scale.

Here we describe a D-Claw simulation of the Mount Meager landslide/debris flow as it interacted with complex topography. The event provides a multifaceted test of model capabilities because it involved a combination of gravitational failure of a static mass, flow down a tortuous path that bifurcated at the confluence of Capricorn and Meager Creeks (with some debris traveling upstream), flow runup, and flow spreading and deposition on the broad Lillooet River floodplain. The resolution of our numerical simulation is limited by the coarse horizontal resolution (20 m) of the pre-event DEM available for the Mount Meager area, but in D-Claw's application of AMR we bolster computational accuracy by specifying a maximum computational grid resolution of 10 m. With this resolution, D-Claw simulations of the entire slope failure/landslide/debris-flow sequence are completed in less than an hour on a desktop computer.

The values of most parameters used in our D-Claw simulation of the Mount Meager event are the same as values discussed in detail by Iverson and George (2016) for our best-fit simulation of a large, liquefying landslide that occurred near Oso, Washington, USA in 2014. However, significant differences exist in the values of the sediment basal friction angle (42° for Mount Meager versus 36° for Oso) and the initial sediment hydraulic permeability (5×10^{-10} m^2 for Mount Meager versus 1×10^{-8} m^2 for Oso). The larger basal friction angle for Mount Meager is required to balance forces within the initially static rock/sediment mass, which occupied a slope much steeper than that at Oso. The smaller hydraulic permeability for Mount Meager reflects the presence of fine-grained sediment derived from hydrothermally altered volcanic rock, which causes the modeled Mount Meager debris to remain liquefied longer than the debris at Oso. Thus, for our Mount Meager simulation neither the friction angle nor permeability values were measured directly, but their assigned values are consistent with physical evidence.

As in the Oso case, onset of slope failure in our Mount Meager simulation is triggered by imposing gradually rising basal pore pressures that perturb the balanced static state. The geometry of the failing mass at Mount Meager is specified by using a method detailed in prior publications on the Oso event (Iverson & George, 2016; Iverson et al., 2015). In brief, the method involves creation of a smooth basal failure surface beneath the pre-event topography so as to replicate the observed ~50×10^6 m^3 landslide volume and match the observed perimeter of the landslide source area.

We stress that the simulation of the Mount Meager event we present here is not comprehensive because it does not consider diverse possible scenarios for the onset of motion and does not consider many of the geological complexities that were recognized after the event occurred (cf. Guthrie et al., 2012; Moretti et al., 2015; Roberti et al., 2017, 2018). Precise accounting for *ex post facto* observations is not our intention. Instead, our goal is testing of a model designed to be useful for prediction, and predictions never benefit from the clarity offered by post-event hindsight. Thus, rather than trying to mimic the precise geological character of the Mount Meager debris-flow source area materials, geometry, and failure sequence, we use an approach that might be described as "intentionally naïve" in order to test how our model performs when using a limited amount of a priori information.

A map view of the overall behavior of our Mount Meager simulation (Fig. 5.12) illustrates the debris flow's rapid initial acceleration and descent of Capricorn Creek, where the flow reached inferred speeds as high as ~90 m/s (Allstadt, 2013; Guthrie et al., 2012). Approximately 140 s after the onset of slope failure, the modeled flow front reaches the T-shaped confluence of Capricorn Creek with Meager Creek, and runup on the adjacent mountainside quickly ensues (Fig. 5.13). This timing is consistent with the timing inferred from seismological evidence (Allstadt, 2013). Simulated runup behavior includes development of a shock (i.e., hydraulic jump), and it resembles the runup behavior observed in flume experiments with debris flows encountering an adverse slope (Fig. 5.10).

During and following runup, the simulated debris flow bifurcates and travels upstream as well as downstream in the valley of Meager Creek (Figs. 5.12 and 5.13). The runup process influences this bifurcation because a reversal of flow momentum occurs as the debris running up the mountainside collapses back into the Meager Creek valley, where it collides with debris still arriving from upstream via Capricorn Creek. The details of such a collapse-and-collide process cannot be inferred with confidence from field observations of deposits, but the simulation is consistent with field evidence of the maximum 270 m runup height, and it provides insight to the possible sequence of events during the flow bifurcation.

Following debris-flow bifurcation, the simulated inundation of the Lillooet River floodplain occurs relatively slowly as the advancing debris flow loses speed and eventually stops (Fig. 5.12). This part of the simulation appears adequate, but it inspires the lowest degree of confidence because nuances of topographic steering and frictional energy dissipation play an increasingly prominent role as the modeled

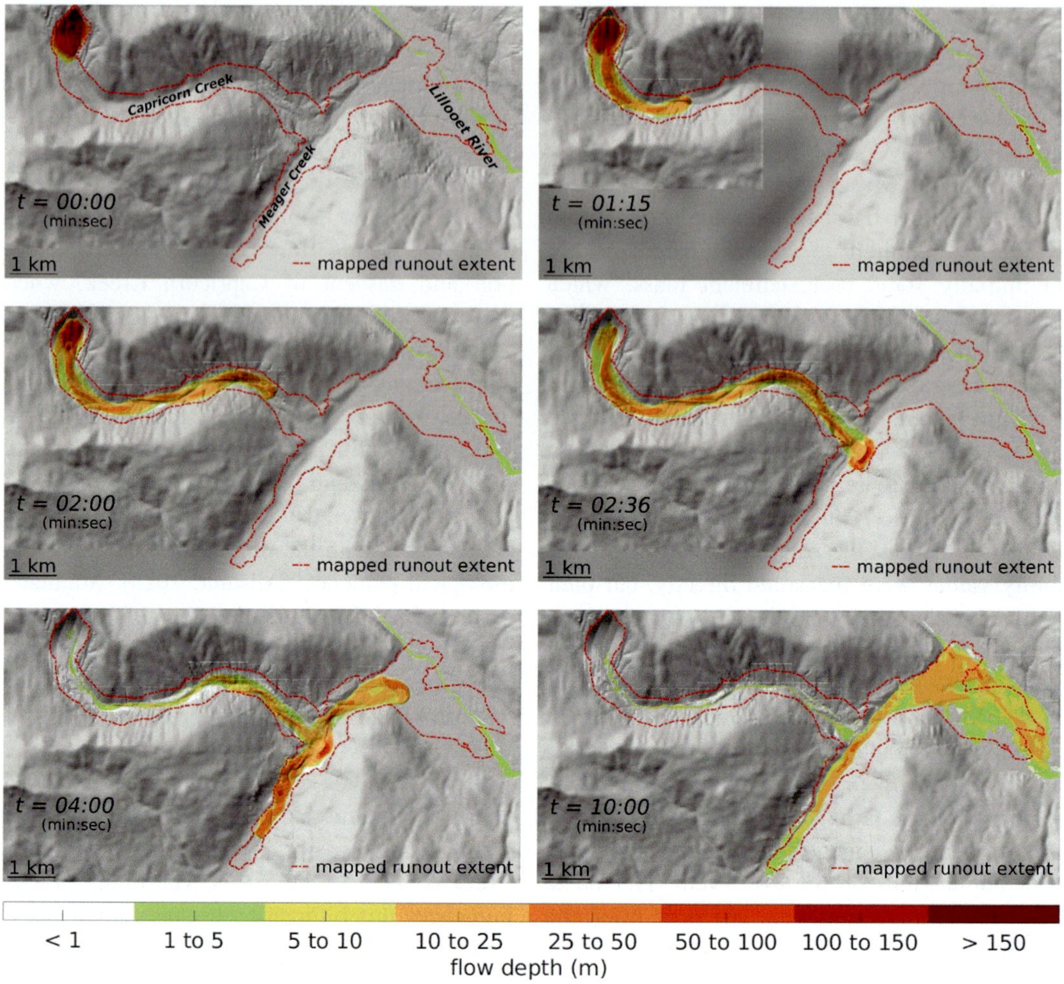

Fig. 5.12 Time sequence of shaded-relief maps of a D-Claw simulation of the 2010 Mount Meager landslide/debris flow, illustrating the distribution of debris thicknesses at selected times (t) after the onset of motion. The first frame ($t=0$) shows the initial, static landslide mass, and the final frame ($t=10$ min) shows the distribution of static deposits at the conclusion of the simulation. The dashed red line shows mapped flow path boundaries inferred from satellite imagery. Shaded relief appears blurry where D-Claw's use of AMR affords very low resolution

debris-flow slows and momentum fluxes play a diminishing role. Whereas conservation of momentum and formulation of numerical methods to ensure that momentum is conserved are very solidly grounded in physical law and mathematical analysis, specification of frictional forces that cause flow stoppage is an inexact science. DEM representation of topographic subtleties that influence the margins of modeled deposits is similarly inexact.

We note that prediction of the extent of debris-flow runout paths has great practical importance but also that matching the observed distribution of downstream debris-flow deposits provides an insufficient test of D-Claw or any other debris-flow model. Better tests are possible if, in addition to observational constraints on runout extent, there are observational constraints on flow dynamics and the distributions of statically balanced material in debris-flow source

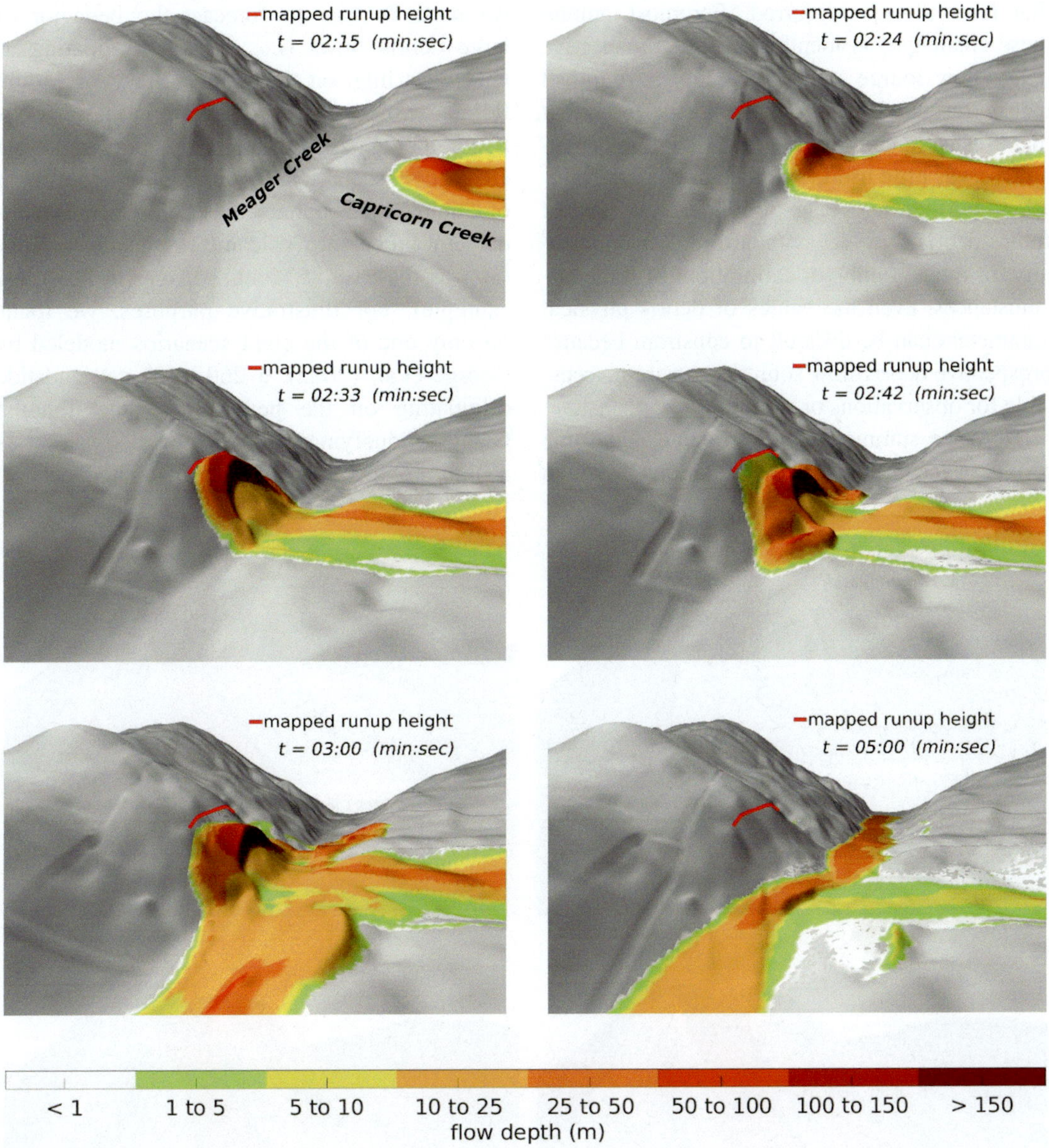

Fig. 5.13 Time sequence of oblique shaded relief images of a D-Claw simulation of the Mount Meager debris flow running up to a vertical height of 270 m, shown by a red marker line on the mountainside adjacent to the T confluence of Capricorn and Meager Creeks. Times (t) shown in each frame refer to the time elapsed since the onset of landslide motion at $t=0$

regions. The presence of these constraints makes the Mount Meager case well-suited for testing of numerical models, irrespective of the coarse resolution of the pre-event DEM for the area.

5.11 Model Application to Hazard Forecasting

Application of numerical models to debris-flow hazard forecasting poses several challenges that don't arise when modeling debris flows that have already occurred. Foremost among these challenges is identification of prospective debris-flow source areas and volumes. A model that represents debris-flow physics perfectly but that uses an inappropriate source location or volume serves little useful purpose in hazard evaluation. Furthermore, use of models to forecast future debris-flow behavior affords no opportunity for model tuning or calibration. In some circumstances, even the values of debris physical parameters can be difficult to constrain because prospective debris-flow source areas are inaccessible for observations or sampling.

Here we summarize the rationale and some results of recent numerical simulations in which we used D-Claw to forecast the behavior of large debris flows (i.e., lahars) originating as landslides high on the western flank of Mount Rainier volcano in Washington State, USA (George et al., 2022). The study addressed eight alternative scenarios for initiation and runout of "unheralded" lahars that could occur without any precursory volcanic or seismic activity (as was the case at Mount Meager in 2010, for example). For illustrative purposes, we focus on only one of the eight scenarios modeled by George et al. (2022): a 260×10^6 m^3 landslide originating on the headwall of the Tahoma Glacier, liquefying, and moving downslope as a high-mobility lahar (Fig. 5.14). This scenario

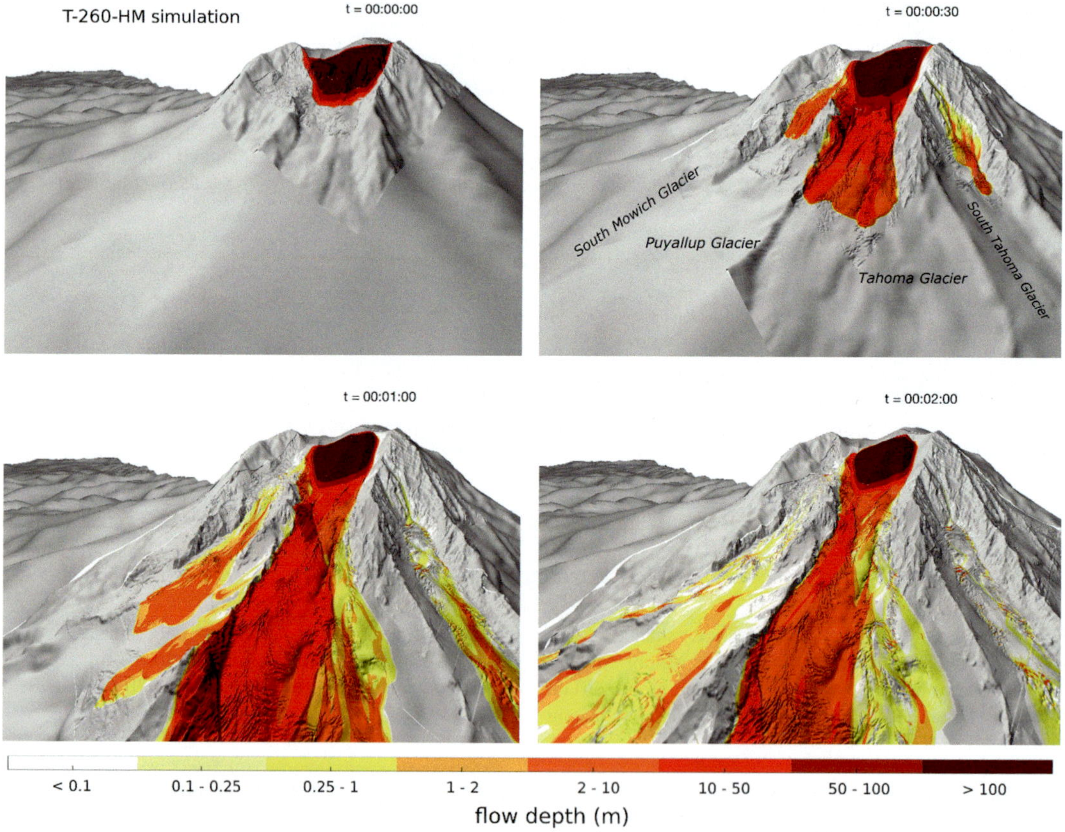

Fig. 5.14 Time sequence of oblique shaded-relief images of Mount Rainier's western flank showing thicknesses of the simulated T-260-HM landslide and lahar at selected times during the first 2 min of motion (from George et al., 2022). Imagery appears blurry where lahar material is absent because D-Claw's adaptive mesh refinement employs very coarse resolution in those areas. Time (t), indicated in hours:minutes:seconds, refers to the time elapsed since landslide motion commences. An animated version of this figure is linked to the report of George et al. (2022)

is named T-260-HM to distinguish it from the other seven scenarios modeled by George et al. (2022).

George et al. (2022) used 260×10^6 m^3 as the maximum credible volume of an unheralded lahar at Mount Rainier because it matches the estimated volume of the largest documented prehistoric lahar that has occurred there without evidence of accompanying eruptive activity (i.e., the Electron Mudflow, Scott et al., 1995). The study used the Tahoma Glacier headwall as a probable lahar source area because it has been identified as the part of Mount Rainier most prone to slope failure due to its steep slopes and abundance of weak, hydrothermally altered rock (Reid et al., 2001). Within the headwall region, George et al. (2022) constructed a landslide basal slip surface so that the source area conformed with natural topographic boundaries and enclosed the desired 260×10^6 m^3 of material, including glacial ice (cf. Iverson & George, 2016; Iverson et al., 2015). As in our Mount Meager simulation, motion of material within the source area was instigated by gradually raising basal pore-fluid pressure until slope failure began in the weakest area. Failure then propagated to adjacent areas by momentum exchange.

The Mount Rainier D-Claw simulations of George et al. (2022) distinguished high-mobility lahars from low-mobility events by employing different values of the debris hydraulic permeability, k (cf. Iverson & George, 2016). Simulations with a low debris permeability ($k = 10^{-11}$ m^2) were used to mimic the behavior of clay-rich, high-mobility lahars in which liquefaction can persist for hours, whereas simulations with a higher permeability ($k = 10^{-9}$ m^2) were used to simulate low-mobility events in which liquefaction persists for only several minutes. Such low-mobility events exhibit behavior similar to that of some volcanic debris avalanches. Values of key parameters that were held constant in all simulations include those of the basal friction angle, set at 38°, and of the initial difference between the solid volume fraction and quasi-static critical-state volume fraction, set at −0.02. A detailed discussion of all parameter values was provided by George et al. (2022).

The subtlety of glacial drainage divides high on the western flank of Mount Rainier causes the simulated T-260-HM lahar to diverge into multiple drainages before funneling into the canyons of the Puyallup and Nisqually Rivers (Figs. 5.14 and 5.15). This flow divergence illustrates the difficulty of anticipating the paths of future lahars in advance, and it highlights the consequent need for large computational domains. Map views of simulation results show that within 8 min of landslide onset, the fronts of the divergent lahar branches descending the Puyallup and Nisqually River valleys travel more than 15 km from their source (Fig. 5.15). Additional map views and animations provided by George et al. (2022) use a larger domain to show the continuation of the simulated lahar paths beyond the margins of Fig. 5.15.

Map views of modeled lahar propagation can be very useful, but for hazard assessment it is important to provide additional information about simulated lahar behavior. For example, Fig. 5.16 shows D-Claw time-series output of lahar depths, speeds, and discharges computed for a location on the Nisqually River floodplain near the community of Ashford, Washington (denoted by a crosshair symbol and the label "4" in Fig. 5.15). At this location the simulated flow front arrives approximately 20 min after landslide onset, followed quickly by peak flow depths close to 14 m and peak flow speeds close to 10 m/s. These values underscore the practical implications of the modeled lahar dynamics in a way that may not be evident from inundation maps alone.

Approximately 20 km downstream from Ashford, the flow of the modeled T-260-HM lahar in the Nisqually River valley is blocked by Alder Dam, a 100-m high concrete structure that impounds Alder Lake reservoir. The reservoir capacity of roughly 300×10^6 m^3 is similar to the volume of the simulated landside and lahar shown in Fig. 5.14. Less than half of the volume of the simulated lahar reaches the head of Alder Lake, but upon entering the lake it mixes with

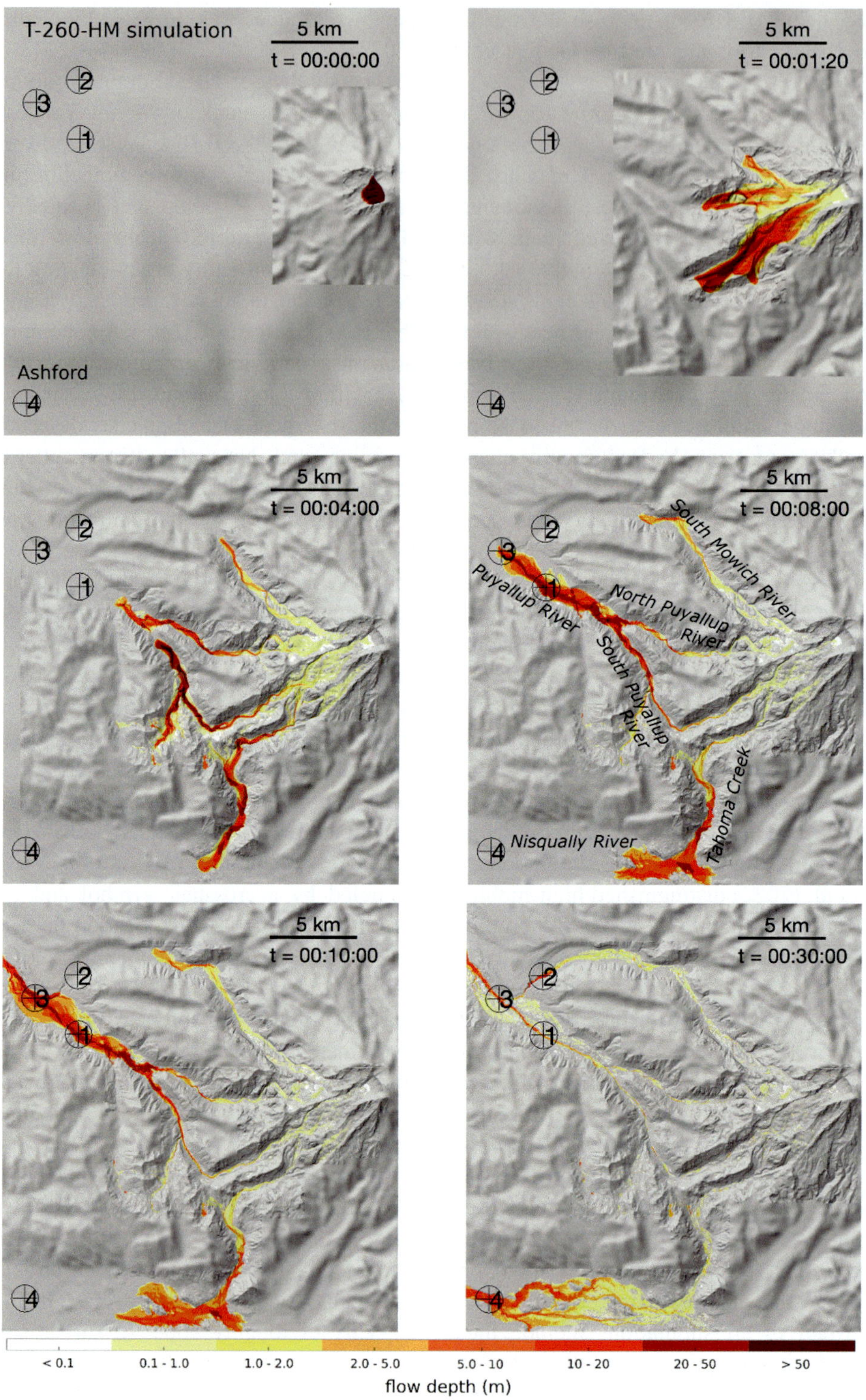

Fig. 5.15 Time sequence of shaded relief maps showing thicknesses of the simulated T-260-HM lahar at selected times during the first 30 min of motion (from George et al., 2022). Imagery appears blurry where lahar material is absent because D-Claw's adaptive mesh refinement employs very coarse resolution in those areas. Circled crosshairs with numbers indicate locations for computation of time-series output similar to that shown in Fig. 5.16. Time (t), in hours:minutes:seconds, refers to the time elapsed since the onset of landslide/lahar motion

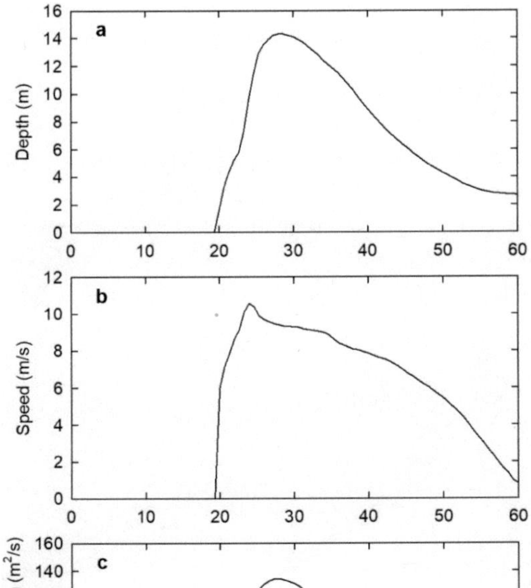

Fig. 5.16 Graphs of D-Claw time-series output showing the evolution of lahar depths, speeds, and discharges at gauge site 4, located on the Nisqually River floodplain near Ashford, Washington (see Fig. 5.15 for location at the circled crosshair labeled 4). Discharge is given per unit valley width perpendicular to the local flow direction (after George et al., 2022)

and displaces sufficient lake water to cause dam overtopping (George et al., 2022). Figure 5.17 illustrates how D-Claw simulations portray the mixing and overtopping processes for a condition in which the lake's initial water level is 4 m below the dam crest elevation. An important finding of the simulations is that the effects of dynamic wave propagation caused by lahar entry into the shallow headwaters of the reservoir are small in comparison to the effects of volumetric displacement of lake water (George et al., 2022).

The scenario-based computation of lahar hazards downstream of Mount Rainer by George et al. (2022) may someday be supplanted by Monte Carlo simulations that begin by considering a much broader spectrum of possible lahar source areas, volumes, and mobilities. Such an approach is indeed feasible for simulations of relatively small debris flows traveling within relatively restricted domains. However, for high-resolution simulations of flow across domains that span many hundreds of square kilometers of rugged terrain, the computation time necessary for a Monte Carlo approach remains very substantial. The lahar simulation illustrated in Figs. 5.14, 5.15, 5.16 and 5.17 used Lidar-derived DEM's with a horizontal resolution as fine as 1 m in some high-consequence areas and used a computational grid resolution refinable to 4 m in all areas. With this high resolution and a computational domain spanning roughly 2500 km^2 (most of which is not shown in Fig. 5.15), some of the individual simulations reported by George et al. (2022) required more than one month of computer run time. This practical limitation will likely decrease as desktop computers become faster and more model users gain access to supercomputers. In the meantime, this limitation highlights the importance of making judicious choices about maximum resolutions used in numerical simulations, as explained in more detail Sect. 5.7 above.

5.12 Remaining Challenges and Concluding Advice

Numerical modeling of debris flows has advanced greatly during the past two decades, in part due to increased computational power and availability of DEMs, and in part due to increased sophistication of model equations, numerical solution methods, and model testing. As a result, depth-averaged models of debris flows moving across three-dimensional terrain have reached a state of fruition that enables their useful application to hazard forecasting. Nonetheless, significant challenges remain before debris-flow modeling reaches full maturity and models can be used to forecast hazards with great confidence.

The first and biggest challenge in using numerical models for hazard assessment involves identification of prospective debris-flow

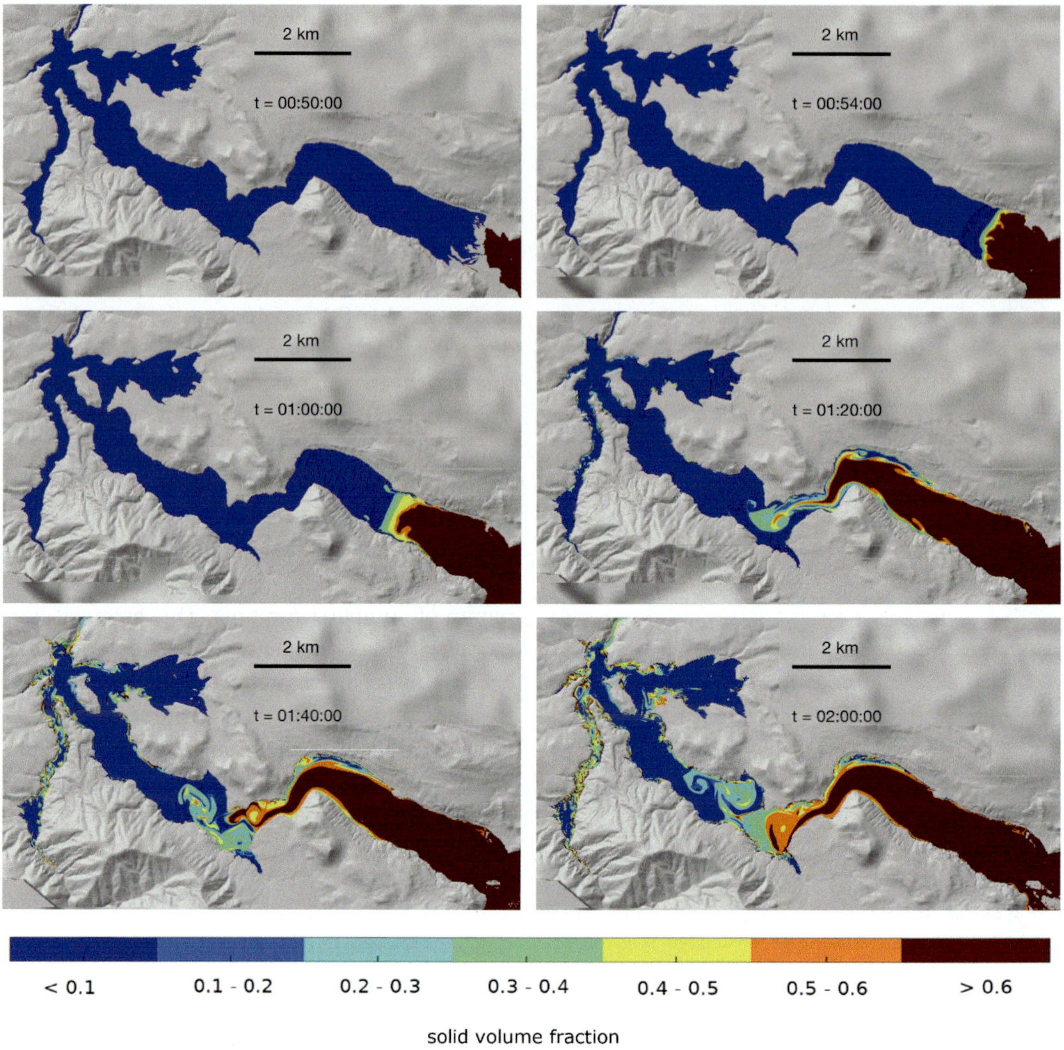

Fig. 5.17 Time sequence of shaded relief maps showing evolution of solid volume fractions (that is, volumetric sediment concentrations) at selected times as the simulated T-260-HM lahar enters Alder Lake reservoir and mixes with lake water, shown in dark blue (from George et al., 2022). Time (t) is the time elapsed since the onset of landslide/lahar motion, indicated in hours:minutes:seconds. An animated version of the figure is linked to the report of George et al. (2022)

source areas and volumes. Relatively simple, statistically based hazard forecasting models such as Laharz address this problem by considering all plausible source areas and a wide range of flow volumes within a study area (Griswold & Iverson, 2008; Iverson et al., 1998; Reid et al., 2016). However, the computational resources needed to apply Laharz are small in comparison to those required to apply a physically based numerical model such as D-Claw. Thus, in the foreseeable future, physically based numerical modeling of debris flows will likely focus on relatively limited ranges of possible flow volumes and source areas. This limitation underscores the importance of selecting volumes and source areas based on an a priori assessment of topography and the spatial distribution of debris that can be mobilized. The structure of some numerical models facilitates this process by explicitly considering distributions of potentially unstable, static debris, whereas other models require information about debris that is assumed to be inherently unstable or already in motion.

A second outstanding challenge, which is related to the first, entails accounting for the effects of debris entrainment or deposition along flow paths. A large body of evidence indicates that entrainment can greatly increase debris-flow volumes and the scope of downstream hazards (e.g., Hungr et al., 2005; Pierson et al., 1990; Reid et al., 2016). Insertion of entrainment terms in depth-averaged debris-flow models is trivial—although appropriate forms of the terms must satisfy mass and momentum conservation constraints identified by Iverson and Ouyang (2015). On the other hand, any entrainment term necessarily includes at least one parameter that cannot be constrained by conservation laws, and accurate computation of entrainment rates that evolve in time and space consequently remains a frontier problem. Large-scale experiments have revealed some key features of the entrainment process (Iverson et al., 2011; Reid et al., 2011), and various models of the process have been suggested (Han et al., 2015), but to our knowledge none have been tested sufficiently to use with great confidence in hazard evaluation.

Another outstanding challenge involves simulating the interrelated processes of grain-size segregation, lateral levee formation, and flow-path avulsion. Grain-size segregation is pervasive in debris flows and commonly leads to formation of boulder-rich lateral levees that help channelize flow downstream (e.g., Fig. 5.3). On the other hand, breaching of levees or overtopping of channels by ensuing debris-flow surges can lead to avulsion and redirection of flow, especially on the divergent topographic surfaces of alluvial fans. Consequently, from a hazards perspective, improved models for forecasting areas of debris-flow inundation should simulate the interrelated processes of grain-size segregation, lateral levee formation, and possible flow-path avulsion. Understanding of the pertinent processes is improving (e.g., Johnson et al., 2012; de Haas et al., 2018), but rigorous inclusion of these processes in physically based models is in its earliest stages.

As we stated in our Introduction, we are model developers ourselves, but we are also model skeptics. All mathematical models of dissipative physical processes such as debris flows are inexact, and all involve tradeoffs between mathematical simplicity and physical realism. In reference to model formulation, Einstein purportedly said that "Everything should be made as simple as possible, but not simpler." From a practical perspective, debris-flow models aren't useful if they're so simple that they omit or misrepresent essential physical phenomena that influence debris-flow hazards. Alternatively, models aren't useful if they have so many complications that clear links between causes and effects of their predictions aren't discernible.

Finally, we note that debris-flow models, like debris-flow scientists, provide a benefit to society only insofar as they're transparent and truthful. Prospective model users should ask questions of models and their developers. Ask about the origin of the model equations—are thorough derivations and explanations of the equations published, or are the equations merely stated? Ask about the model parameters—how are their values determined, and how do they differ for debris flows with various settings and compositions? Ask about the model's computer source code—is it freely accessible and available for scrutiny? If a model developer makes a bold claim, perhaps that a debris-flow model "simulates entrainment," ask, "How does it simulate entrainment, and does the method of simulation satisfy physical conservation laws and pass empirical tests?" Finally, it may be appropriate to ask the ultimate question, "If I lived near the path of a simulated future debris flow, would I bet my life on the model's prediction?".

Acknowledgements We thank Charles Cannon for his contributions to the Mount Rainier lahar modeling study summarized in Sect. 5.11 of this paper, and we thank the many USGS colleagues who participated in the debris-flow flume experiments considered in Sects. 5.10.1 and 5.10.2. Oldrich Hungr originally directed our attention to the Mount Meager landslide/lahar event considered in Sect. 5.10.3, and Richard Guthrie and John Clague shared information from their study of the event. Our paper was improved as a result of review comments provided by Katherine Barnhart, Scott McDougall, Paul Santi, Matthias Jakob, Larry Mastin, and Christopher Magirl. Any use of trade, firm, or product names is for descriptive purposes only and does not imply endorsement by the U.S. Government.

References

Allstadt, K. (2013). Extracting source characteristics and dynamics of the August 2010 Mount Meager landslide from broadband seismograms. *Journal of Geophysical Research: Earth Surface, 118*(3), 1472–1490.

Barnes, N. (2010). Publish your computer code: It is good enough. *Nature, 467*, 753.

Barnhart, K. R., Jones, R. P., George, D. L., Coe, J. A., & Staley D. M. (2021a). Preliminary assessment of the wave generating potential from landslides at Barry arm, prince William sound, Alaska. U.S. Geol Sur Open-File Report 2021-1071.

Barnhart, K. R., Jones, R. P., George, D. L., McArdell, B. W., Rengers, F. K., Staley, D. M., & Kean, J. W. (2021b) Multi-model comparison of computed debris flow runout for the 9 January 2018 Montecito, California post-wildfire event. *Journal of Geophysical Research: Earth Surface, 126*, e2021JF006245. https://doi.org/10.1029/2021JF006245

Berger, M. J., George, D. L., LeVeque, R. J., & Mandli, K. T. (2011). The GeoClaw software for depth-averaged flows with adaptive refinement. *Advances in Water Resources, 34*(9), 1195–1206. https://doi.org/10.1016/j.advwatres.2011.02.016

Berger, M. J., & Oliger, J. (1984). Adaptive mesh refinement for hyperbolic differential equations. *Journal of Computational Physics, 53*, 484–512.

Berzi, D., Jenkins, J. T., & Larcher, M. (2010). Debris flows: Recent advances in experiments and modeling. *Advances in Geophysics, 52*, 103–138.

Bouchut, F. (2004). *Nonlinear stability of finite volume methods for hyperbolic conservation laws and well-balanced schemes for sources*. Birkhauser Verlag.

Bouchut, F., Fernandez-Nieto, E. D., Mangeney, A., & Narbona-Reina, G. (2016). A two-phase two-layer model for fluidized granular flows with dilatancy effects. *Journal of Fluid Mechanics, 801*, 166–221.

Bouchut, F., Mangeney-Castelnau, A., Perthame, B., & Vilotte, J. P. (2003). A new model for Saint Venant and Savage-Hutter type for gravity driven shallow water flows. *Comptes Rendus De L'académie Des Sciences—Series I, 336*, 531–536.

Breien, H., De Blasio, F. V., Elverhøi, A., & Høeg, K. (2008). Erosion and morphology of a debris flow caused by a glacial lake outburst flood, Western Norway. *Landslides, 5*, 271–280.

Christen, M., Kowalski, J., & Bartelt, B. (2010). RAMMS: Numerical simulation of dense snow avalanches in three-dimensional terrain. *Cold Regions Science and Technology, 63*, 1–14.

de Haas, T., Densmore, A. L., Stoffel, M., Suwa, H., Imaizumi, F., Ballesteros-Cánovas, J. A., & Wasklewicz, T. (2018). Avulsions and the spatio-temporal evolution of debris-flow fans. *Earth Science Reviews, 117*, 53–75. https://doi.org/10.1016/j.earscirev.2017.11.007

Denlinger, R. P., & Iverson, R. M. (2004). Granular avalanches across irregular three-dimensional terrain: 1. Theory and computation. *Journal of Geophysical Research: Earth Surface, 109*(F01014). https://doi.org/10.1029/2003JF000085

George, D. L. (2008). Augmented Riemann solvers for the shallow water equations over variable topography with steady states and inundation. *Journal of Computational Physics, 227*(6), 3089–3113.

George, D. L. (2010). Adaptive finite volume methods with well-balanced Riemann solvers for modeling floods in rugged terrain: Application to the Malpasset dam-break flood (France, 1959). *International Journal for Numerical Methods in Fluids, 66*(8), 1000–1018.

George, D. L., & Iverson, R. M. (2014). A depth-averaged debris-flow model that includes the effects of evolving dilatancy. II. Numerical predictions and experimental tests. *Proceedings of the Royal Society A: Mathematical, Physical and Engineering Sciences, 470*. https://doi.org/10.1098/rspa.2013.0820

George, D. L., Iverson, R. M., & Cannon, C. M. (2017). New methodology for computing tsunami generation by subaerial landslides: Application to the 2015 Tyndall Glacier landslide, Alaska. *Geophysical Research Letters, 44*(14), 7276–7284.

George, D. L., Iverson, R. M., & Cannon, C. M. (2022). Modeling the dynamics of lahars that originate as landslides on the west side of Mount Rainier, Washington. U.S. Geol Survey Open-file Report 2021–1118, Reston, Virginia. https://doi.org/10.3133/ofr20211118

George, D. L., & LeVeque, R. J. (2006). Finite volume methods and adaptive refinement for global tsunami propagation and inundation. *Science of Tsunami Hazards, 24*(5), 319–328.

Godunov, S. K. (1959). A finite difference method for the computation of discontinuous solutions of the equations of fluid dynamics. *Journal of Matematičeskij Sbornik, 47*, 271–306.

Gray, J. M. N. T., & Kokelaar, B. P. (2010). Large particle segregation, transport and accumulation in granular free-surface flows. *Journal of Fluid Mechanics, 652*, 105–137.

Gray, J. M. N. T., Tai, Y.-C., & Noelle, S. (2003). Shock waves, dead-zones and particle-free regions in rapid granular free surface flows. *Journal of Fluid Mechanics, 491*, 161–181.

Gray, J. M. N. T., Wieland, M., & Hutter, K. (1999). Gravity driven free surface flow of granular avalanches over complex basal topography. *Proceedings of the Royal Society of London. Series A, 455*, 1841–1874.

Griswold, J. P., & Iverson, R. M. (2008). Mobility statistics and automated hazard mapping for debris flows and rock avalanches (59 p.). U.S. Geol Sur Sci Invest Rep 2007-5276.

Guthrie, R. H., Friele, P., Allstadt, K., Roberts, N., Evans, S. G., Delaney, K. B., Roche, D., Clague, J. J., & Jakob, M. (2012). The 6 August 2010 Mount Meager rock slide-debris flow, Coast Mountains, British Columbia: Characteristics, dynamics, and implications for hazard and risk assessment. *Natural Hazards and Earth System Sciences, 12*(5), 1277–1294.

Han, Z., Chen, G., Li, Y., Tang, C., Xu, L., He, Y., Huang, X., & Wang, W. (2015). Numerical simulation of debris-flow behavior incorporating a dynamic method for estimating the entrainment. *Engineering Geology, 190*, 52–64.

Hungr, O., Evan, S. G., Bovis, M. J., & Hutchinson, J. N. (2001). A review of the classification of landslides of the flow type. *Environmental and Engineering Geoscience, 7*(3), 221–238.

Hungr, O., Leroueil, S., & Picarelli, L. (2014). The Varnes classification of landslide types, an update. *Landslides, 11*(2), 167–194.

Hungr, O., McDougall, S., & Bovis, M. (2005). Entrainment of material by debris flows. In M. Jakob & O. Hungr (Eds.), *Debris flow hazards and related phenomena* (pp. 135–158). Springer-Praxis.

Iverson, R. M. (1997). The physics of debris flows. *Reviews of Geophysics, 35*, 245–296.

Iverson, R. M. (2003a). The debris-flow rheology myth. In D. Rickenmann & C. L. Chen (Eds.), *Debris-flow hazards mitigation: Mechanics, prediction, and assessment, 1* (pp. 303–314). Millpress.

Iverson, R. M. (2003b). How should mathematical models of geomorphic processes be judged? In P. R. Wilcock & R. M. Iverson (Eds.), *Prediction in geomorphology, geophysical Monograph* (Vol. 135, pp. 83–94). Amer Geophys Union.

Iverson, R. M. (2005). Debris-flow mechanics. In M. Jakob & O. Hungr (Eds.), *Debris flow hazards and related phenomena* (pp. 105–134). Springer-Praxis.

Iverson, R. M. (2012). Elementary theory of bed-sediment entrainment by debris flows and avalanches. *Journal of Geophysical Research: Earth Surface, 117*(F03006). https://doi.org/10.1029/2011JF002189

Iverson, R. M. (2015). Scaling and design of landslide and debris-flow experiments. *Geomorphology, 244*, 9–20. https://doi.org/10.1016/j.geomorph.2015.02.033

Iverson, R. M., & George, D. L. (2014). A depth-averaged debris-flow model that includes the effects of evolving dilatancy I. Physical basis. *Proceedings of the Royal Society A: Mathematical, Physical and Engineering Sciences, 470*, 20130819. https://doi.org/10.1098/rspa.2013.0819

Iverson, R. M., & George, D. L. (2016). Modeling landslide liquefaction, mobility bifurcation and the dynamics of the 2014 Oso disaster. *Geotechnique, 66*, 175–187. https://doi.org/10.1680/jgeot.15.LM.004

Iverson, R. M., & George, D. L. (2019a). Basal stress equations for granular debris masses on smooth or discretized slopes. *Journal of Geophysical Research: Earth Surface, 124*, 1464–1484. https://doi.org/10.1029/2018JF004802

Iverson, R. M., & George, D. L. (2019b). Valid debris-flow models must avoid hot starts. In J. W. Kean, J. A. Coe, P. M. Santi, & B. K. Guillen (Eds.), *Debris-flow hazards mitigation: Mechanics, monitoring, modeling, and assessment, Proceedings of the Seventh International Conference on Debris-Flow Hazards Mitigation Monitoring* (Vol. 28, pp. 25–32). Association of Environmental & Engineering Geologists, Special Publication.

Iverson, R. M., George, D. L., & Logan, M. (2016). Debris flow runup on vertical barriers and adverse slopes. *Journal of Geophysical Research: Earth Surface, 121*, 2333–2357. https://doi.org/10.1002/2016JF003933

Iverson, R. M., George, D. L., Allstadt, K., Reid, M. E., Collins, B. D., Vallance, J. W., Schilling, S. P., Godt, J. W., Cannon, C. M., Magirl, C. S., Baum, R. L., Coe, J. A., Schulz, W. H., & Bower, J. B. (2015). Landslide mobility and hazards: Implications of the 2014 Oso disaster. *Earth and Planetary Science Letters, 412*, 197–208. https://doi.org/10.1016/j.epsl.2014.12.020

Iverson, R. M., Logan, M., LaHusen, R. G., & Berti, M. (2010). The perfect debris flow? Aggregated results from 28 large-scale experiments. *Journal of Geophysical Research: Earth Surface, 115*(F03005). https://doi.org/10.1029/2009JF001514

Iverson, R. M., & Ouyang, C. (2015). Entrainment of bed material by Earth-surface mass flows: Review and reformulation of depth-integrated theory. *Reviews of Geophysics, 53*, 27–58. https://doi.org/10.1002/2013RG000447

Iverson, R. M., Reid, M. E., Iverson, N. R., LaHusen, R. G., Logan, M., Mann, J. E., & Brien, D. L. (2000). Acute sensitivity of landslide rates to initial soil porosity. *Science, 290*, 513–516.

Iverson, R. M., Reid, M. E., & LaHusen, R. G. (1997). Debris-Flow Mobilization from Landslides. *Annual Review of Earth and Planetary Sciences, 25*, 85–138.

Iverson, R. M., Reid, M. E., Logan, M., LaHusen, R. G., Godt, J. W., & Griswold, J. G. (2011). Positive feedback and momentum growth during debris-flow entrainment of wet bed sediment. *Nature Geoscience, 4*(2), 116–121. https://doi.org/10.1038/NGEO1040

Iverson, R. M., Schilling, S. P., & Vallance, J. W. (1998). Objective delineation of lahar-inundation hazard zones. *Geological Society of America Bulletin, 110*, 972–984.

Johnson, C. G., Kokelaar, B. P., Iverson, R. M., Logan, M., LaHusen, R. G., & Gray, J. M. N. T. (2012). Grain-size segregation and levee formation in geophysical mass flows. *Journal of Geophysical Research, 117*. https://doi.org/10.1029/2011JF002185

Kowalski, J., & McElwaine, J. N. (2013). Shallow two-component gravity-driven flows with vertical variation. *Journal of Fluid Mechanics, 714*, 434–462. https://doi.org/10.1017/jfm.2012.489

LeVeque RJ (2002) Finite volume methods for hyperbolic problems. Cambridge University Press.

LeVeque, R. J. (2013). Top ten reasons to not share your code (and why you should anyway). *SIAM News, 46*(3), 15.

LeVeque, R. J., George, D. L., & Berger, M. J. (2011). Tsunami modeling with adaptively refined finite volume methods. *Acta Numerica, 20*, 211–289. https://doi.org/10.1017/S0962492911000043

Logan, M., Iverson, R. M., & Obryk, M. K. (2007, revised 2018). Video documentation of experiments at the USGS debris-flow flume, 1992–2017. U.S. Geol Sur Open-file Report 2007-1315. http://pubs.usgs.gov/of/2007/1315/

Lucas, A., Mangeney, A., Bouchut, F., Bristeau, M-O., & Mege, D. (2007). Benchmarking exercises for granular flows. In K. Ho & V. Li (Eds.), *The 2007 International forum on landslide disaster management: Proceedings of the 2007 international forum on landslide disaster management*, Hong Kong, 10–12 December 2007 (pp. 967–986). Geotechnical Division, Hong Kong Institution of Engineers.

Major, J. J., & Iverson, R. M. (1999). Debris-flow deposition—effects of pore-fluid pressure and friction concentrated at flow margins. *Geological Society of America Bulletin, 111*, 1424–1434.

Malvern, L. E. (1969). *Introduction to the mechanics of a continuous medium*. Prentice-Hall.

Mandli, K. T., Ahmadia, A. J., Berger, M. J., Calhoun, D., George, D. L., Hadimichael, Y., Ketcheson, D. I., Lemoine, G. I., & LeVeque, R. J. (2016). Clawpack: Building an open-source ecosystem for solving hyperbolic PDEs. *PeerJ Computer Science, 2*, e68.

McArdell, B. W., Bartlet, P., & Kowalski, J. (2007). Field observations of basal forces and fluid pore pressure in a debris flow. *Geophysical Research Letters, 34*, L07406. https://doi.org/10.1029/2006GL029183

McCoy, S. W., Kean, J. W., Coe, J. A., Staley, D. M., Wasklewicz, T. A., & Tucker, G. E. (2010). Evolution of a natural debris flow: In situ measurements of flow dynamics, video imagery, and terrestrial laser scanning. *Geology, 38*, 735–738.

McCoy, S. W., Kean, J. W., Coe, J. A., Tucker, G. E., Staley, D. M., & Wasklewicz, T. A. (2012). Sediment entrainment by debris flows: In situ measurements from the headwaters of a steep catchment. *Journal of Geophysical Research: Earth Surface, 117*(F03016). https://doi.org/10.1029/2011JF002278

McDougall, S. (2017). 2014 Canadian Geotechnical Colloquium: Landslide runout analysis—current practice and challenges. *Canadian Geotechnical Journal, 54*, 605–620. https://doi.org/10.1139/cgj-2016-0104

McDougall, S., & Hungr, O. (2004). A model for the analysis of rapid landslide motion across three-dimensional terrain. *Canadian Geotechnical Journal, 41*, 1084–1097. https://doi.org/10.1139/T04-052

Mergili, M., Fischer, J.-T., Krenn, J., & Pudasaini, S. P. (2017). R.avaflow v1, an advanced open-source computational framework for the propagation and interaction of two-phase mass flows. *Geoscientific Model Development, 10*, 553–569. https://doi.org/10.5194/gmd-10-553-2017

Meyrat, G., McArdell, B., Ivanova, K., Müller, C., & Bartelt, P. (2022). A dilatant, two-layer debris flow model validated by flow density measurements at the Swiss illgraben test site. *Landslides, 19*, 265–276.

Moretti, L., Allstadt, K., Mangeney, A., Capdeville, Y., Stutzmann, E., & Bouchut, F. (2015). Numerical modeling of the Mount Meager landslide constrained by its force history derived from seismic data. *Journal of Geophysical Research: Solid Earth, 120*, 2579–2599. https://doi.org/10.1002/2014JB011426

Nessyahu, H., & Tadmor, E. (1990). Non-oscillatory central differencing for hyperbolic conservation laws. *Journal of Computational Physics, 87*, 408–463.

O'Brien, J. S., Julien, P. Y., & Fullerton, W. T. (1993). Two-dimensional water flood and mudflow simulation. *Journal of Hydraulic Engineering Division of the American Society of Civil Engineers, 119*, 244–261.

Peng, R. D. (2011). Reproducible research in computational science. *Science, 334*, 1226–1227.

Peruzzetto, M., Mangeney, A., Bouchut, F., Grandjean, G., Levy, C., Thiery, Y., & Lucas, A. (2021). Topography curvature effects in thin-layer models for gravity-driven flows without bed erosion. *Journal of Geophysical Research: Earth Surface, 126*, e2020JF005657.

Pierson, T. C. (1986). Flow behavior of channelized debris flows, Mount St. Helens, Washington. In A. D. Abrahams (Ed.) *Hillslope processes* (pp. 269–296). Allen and Unwin.

Pierson, T. C., Janda, R. J., Thouret, J. C., & Borrero, C. A. (1990). Perturbation and melting of snow and ice by the 13 November 1985 eruption of Nevado del Ruiz, Colombia, and consequent mobilization, flow and deposition of lahars. *Journal of Volcanology and Geothermal Research, 41*, 17–66.

Pitman, E. B., & Le, L. (2005). A two-fluid model for avalanche and debris flows. *Philosophical Transactions of the Royal Society a: Mathematical, Physical and Engineering Sciences, 363*, 1573–1601. https://doi.org/10.1098/rsta.2005.1596

Pitman, E. B., Nichita, C. C., Patra, A. K., Bauer, A., Sheridan, M., & Bursik, M. (2003). Computing granular avalanches and landslides. *Physics of Fluids, 15*, 3638–3646.

Pudasaini, S. P. (2012). A general two-phase debris flow model. *Journal of Geophysical Research: Earth Surface*, *117*(F03010). https://doi.org/10.1029/2011JF002186

Pudasaini, S. P., Wang, Y., & Hutter, K. (2005). Modelling debris flows down general channels. *Natural Hazards and Earth Systems Sciences, 5*, 799–819. https://doi.org/10.5194/nhess-5-799-2005

Reid, M. E., LaHusen, R. G., & Iverson, R. M. (1997). Debris-flow initiation experiments with diverse hydrologic triggers. In C.-L. Chen (Ed.), *Debris-flow hazards mitigation: Mechanics, prediction, and assessment* (pp. 1–11). American Society of Civil Engineers.

Reid, M. E., Iverson, R. M., Logan, M., LaHusen, R. G., Godt, J. W., & Griswold, J. G. (2011). Entrainment of bed sediment by debris flows: Results from large-scale experiments. In R. Genevois, D. L. Hamilton, & A. Prestinizi (Eds.), Fifth international conference on debris-flow hazards mitigation, mechanics, prediction and assessment (pp. 367–374). Casa Editrice Universita La Sapienza.

Reid, M. E., Sisson, T. W., & Brien, D. L. (2001). Volcano collapse promoted by hydrothermal alteration and edifice shape, Mount Rainier, Washington. *Geology, 29*(9), 779–782.

Reid, M. E., Coe, J. A., & Brien, D. L. (2016). Forecasting inundation from debris flows that grow volumetrically during travel, with application to the Oregon Coast Range, USA. *Geomorphology, 273*, 396–4011. https://doi.org/10.1016/j.geomorph.2016.07.039

Rickenmann, D., Laigle, D., McArdell, B. W., & Hübl, J. (2006). Comparison of 2D debris-flow simulation models with field events. *Computers & Geosciences, 10*, 241–264.

Roberti, G., Ward, B., van Wyk de Vries, B., Friele, P., Perotti, L., Clague, J. J., & Giardino, M. (2018). Precursory slope distress prior to the 2010 Mount Meager landslide, British Columbia. *Landslides, 15*, 637–647https://doi.org/10.1007/s10346-017-0901-0

Roberti, G., Friele, P., van Wyk de Vries, B., Ward, B., Clague, J. J., Perotti, L., & Giardino, M. (2017). Rheological evolution of the Mount Meager 2010 debris avalanche, southwestern British Columbia. *Geosphere, 13*(2), 369–390.https://doi.org/10.1130/GES01389.1

Savage, S. B., & Hutter, K. (1991). The dynamics of avalanches of granular materials from initiation to runout, part I, Analysis. *Acta Mechanica, 86*, 201–223.

Scott, K. M., Vallance, J. W., & Pringle, P. T. (1995). Sedimentology, behavior, and hazards of debris flows at Mount Rainier, Washington (56 p.). U.S. Geol Sur Prof Paper 1547.

Sharp, R. P., & Nobles, L. H. (1953). Mudflow of 1941 at Wrightwood, Southern California. *Geological Society of America Bulletin, 64*, 647–660.

Stodden, V., McNutt, M., Bailey, D. H., Deelman, E., Gil, Y., Hanson, B., Heroux, M. A., Ioannidis, J. P. A., & Taufer, M. (2016). Enhancing reproducibility for computational methods. *Science, 354*, 1240–1241.

Trujillo-Vela, M. G., Ramos-Cañón, A. M., Escobar-Vargas, J. A., & Galindo-Torres, S. A. (2022) An overview of debris-flow mathematical modeling. *Earth Science Reviews, 232*, 104135. https://doi.org/10.1016/j.earscirev.2022.104135

Vallance, J. W., & Iverson, R. M. (2015). Lahars and their deposits. In H. Sigurdsson (Ed.), *The encyclopedia of volcanoes* (2nd ed., pp. 649–664). Academic Press. https://doi.org/10.1016/B978-0-12-385938-9.00037-7

Varnes, D. J. (1978). Slope movement types and processes. In R. L. Schuster & R. J. Krizek (Eds.). *Landslides analysis and control* (pp. 11–33). U.S. National Academy of Sciences.

Vreugdenhil, C. B. (1994). *Numerical methods for shallow-water flow*. Kluwer Academic.

6 Sediment Entrainment and Deposition

Chao Kang, Fumitoshi Imaizumi and Joshua Theule

Abstract

Many debris flows have significant amounts of entrainment, resulting in a substantially larger final volume than their initial volume and more complicated mechanics and characteristics of the flow. This chapter reviews some historical cases with substantial entrainment and erosion mechanisms that have been commonly considered in studying debris flow entrainment. The formation of a debris flow along with dominant factors affecting flow characteristics are comprehensively reviewed, and the erosion process is described. The instruments used to monitor erosion depth/rate and factors affecting the rate of erosion are summarized. It is commonly agreed that key factors affecting debris flow entrainment are topography, initial volume, solid concentration and particle size distribution, which are also discussed in this chapter. The entrainment and deposition models are classified based on the erosion mechanisms, and limitations and advantages of these entrainment models are discussed and summarized. In addition, physical modelling equipment is tabulated, and various numerical entrainment runout models are compared and discussed. This chapter provides a review of debris flow entrainment studies in the last two decades and may be of help to researchers studying debris flow entrainment in the future.

C. Kang (✉)
Key Laboratory of Geological Hazards On Three Gorges Reservoir Area, Ministry of Education, China Three Gorges University, Yichang 443002, Hubei, China
e-mail: ckang3@ualberta.ca

F. Imaizumi
Faculty of Agriculture, Shizuoka University, Shizuoka, Japan
e-mail: imaizumi@shizuoka.ac.jp

J. Theule
The Alpine Environment Research Center (CREALP), Sion, Switzerland
e-mail: joshua.theule@crealp.vs.ch

6.1 Entrainment and Deposition Overview

In many debris-flow events, flow channels are typically covered by surficial deposits, sometimes several meters thick, of loose granular material. A rapidly moving debris flow could mobilize these materials, which could change the volume of the debris flow significantly during the flow process. The inclusion of material through erosion of the flow channel is called entrainment (Iverson, 1997; Iverson & Ouyang, 2015; Larsen et al., 2010). Many debris flows have significant amounts of entrainment, resulting in a substantially larger final volume than their initial volume (see Table 6.1). The entrained material can

M. Jakob et al. (eds.), *Advances in Debris-flow Science and Practice*, Geoenvironmental Disaster Reduction,
https://doi.org/10.1007/978-3-031-48691-3_6

Table 6.1 Examples of debris flows with substantial entrainment

Case	Initial volume (m^3)	Entrained volume (m^3)	Average slope angle of entrainment zone	Final volume (m^3)	Entrainment ratio (%)	Entrainment material	References
Nomash River, Canada, 1999 Rock slide-debris avalanche	3.0×10^5	3.0×10^5	20°–45°	6.0×10^6	50	Till-derived colluvium	Hungr and Evans (2004)
Tsingshan, Hong Kong 1990 Debris flow	4.0×10^2	1.9×10^4	23°–35°	2.0×10^4	98	Colluvium, residual soil	Lo and Chau (2003)
Faucon, French,2003 Debris flow	8.5×10^3	3.65×10^4	3°–50°	4.5×10^3	81	Tills, kame terraces and moraines	Remaître et al. (2009)
Fjærland, Norway, 2004 Debris flow	2.5×10^4	2.2×10^5	4°–60°	2.4×10^5	90	Quaternary till and older debris-flow material	Breien et al. (2008)
Wenjiagou, China, 2010 Debris flow	2.9×10^6	3.1×10^6	12°–32°	6.0×10^6	50	Colluvium	Tang et al. (2012); Kang et al. (2021)
Guanling, China, 2010 Flowslide	1.2×10^6	4.7×10^5	25° - 35°	1.7×10^6	28%	Quaternary sediments	Xing et al. (2014)
Dolomites, Italy, 1997 Debris flows	6.0×10^2	5.4×10^3	7°–30°	6.0×10^3	90	Postglacial sediments	Berti et al. (1999)

also change the mechanics and characteristics of the flow, thus making prediction and back analysis of debris flows much more difficult (Iverson et al., 2004). The rate of this process is affected by the shear stress applied to the bed sediments, internal friction between the particles, velocity of the flowing debris, slope angle of the channel, cohesion of the bed material, pore water pressure generated by the shearing of fine-grained materials in the bed, and characteristic particle size (Iverson et al., 2011; Kang & Chan, 2017; Medina et al., 2008; Takahashi, 2014).

One of the mechanisms causing material entrainment in debris flows is bed destabilization and erosion. Destabilization of bed material is the result of drag forces acting at the base of the flow channel, which may be aided by strength loss due to rapid undrained loading, impact loading and liquefaction of the saturated channel fill (Hutchinson & Bhandari, 1971; Sassa, 1985). Another important mechanism of material entrainment results from instability of stream banks undercut by bed erosion. It is important to consider that steep stream and gully channels are often being actively incised. Thus, their banks may exist in a state of marginal equilibrium that is easily disturbed by lowering of the channel bed, such as often occurs during passage of a debris-flow surge. The bank may respond immediately and release a shallow landslide directly into the body of the surge, or may release with a delay, to provide material for incorporation into the next surge (Jakob et al., 2005). Such entrainment mechanisms can change the mobility of the material through changes of the flow volume and its rheological behavior (Quan Luna et al., 2012).

Two mechanisms are typically considered in studying debris flow entrainment: shear failure in macro scale and scouring erosion in micro scale (Fig. 6.1). For shear failure, erosion is considered when the shear stress in the

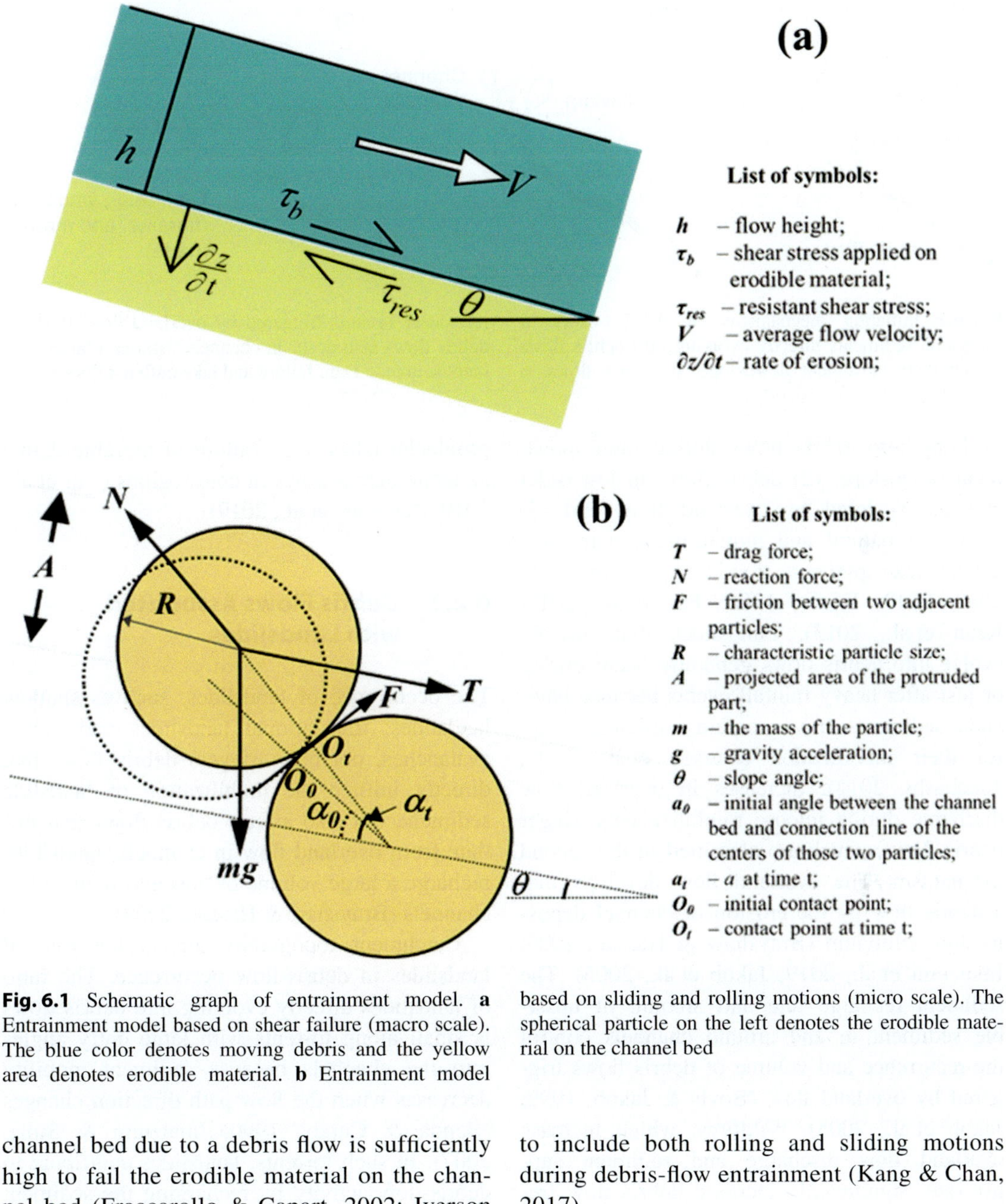

Fig. 6.1 Schematic graph of entrainment model. **a** Entrainment model based on shear failure (macro scale). The blue color denotes moving debris and the yellow area denotes erodible material. **b** Entrainment model based on sliding and rolling motions (micro scale). The spherical particle on the left denotes the erodible material on the channel bed

channel bed due to a debris flow is sufficiently high to fail the erodible material on the channel bed (Fraccarollo & Capart, 2002; Iverson & Ouyang, 2015; Medina et al., 2008). This mechanism is most commonly adopted in erosion model development. A detailed review of the shear failure mechanism and formulating equation was performed by Iverson and Ouyang (2015). At the micro scale, granular material is eroded grain by grain, as observed by McCoy et al. (2012). A progressive scouring entrainment model has been developed to include both rolling and sliding motions during debris-flow entrainment (Kang & Chan, 2017).

6.2 Initiation of Debris Flows

6.2.1 Initiation Mechanisms of Debris Flows

Debris flows can be initiated via three mechanisms (Fig. 6.2): (1) landslide sediments directly

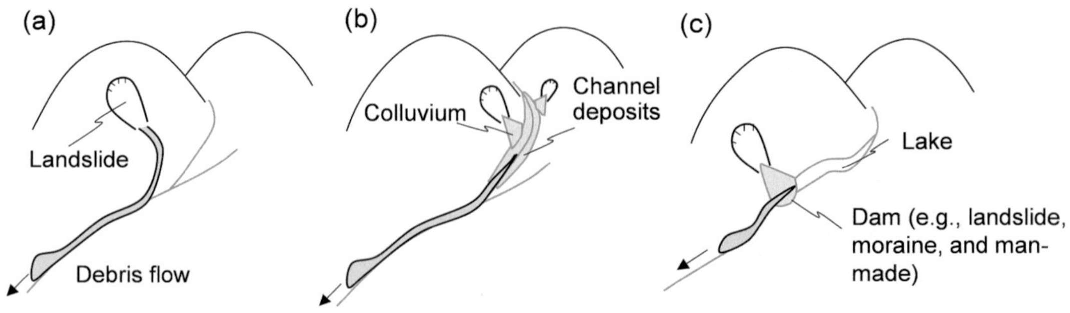

Fig. 6.2 Initiation mechanisms of debris flows. **a** Landslide sediments directly evolving into debris flows during their movement downslope. **b** Debris flows in low-order streams triggered by overland flow. Post-fire debris flows also occur in channels without a large sediment source. **c** Dam failure and lake outburst floods

evolving into debris flows during their movement downslope, (2) debris flows in low-order streams triggered by overland flow, and (3) failure of natural and human-made dams and glacial lake outburst floods (e.g., Cui et al., 2010, 2013; Evans, 1986; Fan et al., 2017; Kean et al., 2013). Landslides that directly evolve into debris flows generally occur during or just after heavy rainfall events because landslide sediments must contain sufficient water for their mobilization (Iverson et al., 1997; Takahashi, 2014). Increases in overland flow discharge during intense rainfall events trigger debris flows, which are classified in the second mechanism. The overland flow develops into a debris flow by the erosion of channel deposits and colluvium (Brayshaw & Hassan, 2009; Imaizumi et al., 2019; Jakob et al., 2005). The sediment recharge rate and amount of unstable sediment in and around channels control the recurrence and volume of debris flows triggered by overland flow (Bovis & Jakob, 1999; Jakob et al., 2005). Wildfires, which increase overland flow discharge and sediment supply rate, significantly increase debris flow frequency triggered by overland flow (Santi & Macaulay, 2021; Staley et al., 2017). The third mechanism, failure of natural and human made dams, occurs both during and after heavy rainfall events (Dunning et al., 2006; Fan et al., 2020). Damming of mountainous rivers by landslide sediment is the common cause of this type of debris flow (Dunning et al., 2006; Fan et al., 2020). Outbursts of subglacial, supraglacial, and proglacial lakes (e.g., failure of moraine dams) generate debris flows in cold regions (Cui et al., 2010; Zaginaev et al., 2019).

6.2.2 Debris Flows Associated with Landslides

The occurrence of landslides, such as shallow landslides, translational landslides and debris avalanches, not only triggers debris flows that directly initiate the mobilization of landslide sediment, but also affects debris flows that initiate from overland flow in channels; landslides recharge a large volume of unstable sediment to channels (Brayshaw & Hassan, 2009).

Catchment topography affects the role of landslides in debris-flow occurrence. The ratio of landslides directly evolving into debris flows is small along torrents with large entry angles into the channels because sediment mobility decreases when the flow path direction changes (Benda & Cundy, 1990; Imaizumi & Sidle, 2007). In such torrents, landslides contribute to debris-flow activity by recharging the sediment (Brayshaw & Hassan, 2009; de Haas et al., this volume). In contrast, a higher ratio of landslides directly evolves into debris flows along torrents with a small entry angle. Debris flows dominate steep channels (e.g., $>15°$), while bedload transport is the predominant transport mechanism in gentler channel reaches (Iverson, 1997; Tillery & Rengers, 2019; VanDine, 1985). Thus, the induction of debris flows by landslides is clearer in steeper channels. Increases in debris-flow frequency are typically associated with increased landslide occurrence in steep terrain because many landslides directly evolve into debris flows (Fan et al., 2017; Imaizumi & Sidle, 2021).

6.2.3 Sediment Recharge

The volume of channel deposits, which are the main source of debris-flow materials mobilized by an increase in overland flow (Fig. 6.3), is controlled by sediment recharge from hillslopes and the evacuation of sediment by debris flows and fluvial processes (Berger et al., 2011a; Imaizumi et al., 2019; Rengers et al., 2020a; Theule et al., 2012; Zhang et al., 2017). Debris flow frequency is generally high in torrents with high sediment supply rates (Jakob et al., 2005; May, 2002).

Many types of sediment supply processes exist other than episodic landslides (e.g., translational landslides and debris avalanches) in debris flow initiation zones (Hürlimann et al., 2012; Jakob et al., 2005). Freeze–thaw cycles, which promote rockfall and dry ravel, are an important part of sediment supply processes in alpine regions (Hürlimann et al., 2012; Imaizumi et al., 2006). Freeze–thaw cycles facilitate rock weathering, producing unstable sediments (Rengers et al., 2020a, 2020b). Freeze–thaw also transports deposits on hillslopes and moraines into channels (Hürlimann et al., 2012). The recharge

Fig. 6.3 Channel debris deposits in the initiation zone of debris flows at the Ohya landslide scar, Japan

rate of the sediment, which is controlled by the frequency of the freeze–thaw cycles, naturally changes with the season (Rengers et al., 2020a, 2020b). As a result, the volume of channel deposits is high in the seasons just after high freeze–thaw activity (Imaizumi et al., 2006). In contrast to landslides, freeze–thaw recharges sediment in debris-flow channels every year. Permafrost warming, active-layer thickening, and movement of rock glaciers can be major sources of sediment in high alpine areas (Stoffel et al., 2014a). The slope aspect is also a factor controlling debris-flow activity (Tillery & Rengers, 2019), since it affects the weathering rate of the rock wall (Rengers et al., 2020a) and the distribution of water (Xu et al., 2013).

The sediment connectivity between the hillslope and the channel also controls the sediment recharge rate. Part of the sediment produced on hillslopes (e.g., landslide sediment) is stored on hillslopes as deposits and talus slopes (Nakamura et al., 1995; Roberts & Church, 1986; Sidle & Ochiai, 2006). Therefore, sediment recharge rates in rivers and streams do not necessarily coincide with overall sediment production. A larger proportion of landslide sediment reaches the channel networks in steeper terrain (Imaizumi & Sidle, 2007), but some morphological conditions, such as glacial cirques, can decouple hillslopes from channels, decreasing the delivery ratio of sediment to channels (Cavalli et al., 2013).

Changes in vegetation coverage sometimes increase the sediment recharge rate and volume of unstable sediment in channels. Wildfire, which promotes dry ravel and bank failure, has some of the most severe impacts on the sediment recharge rate (Gartner et al., this volume; Palucis et al., 2021; Santi & Macaulay, 2021). Hence, the volume of sediment in the channel increases after a wildfire (Palucis et al., 2021). Forest harvesting results in significant changes in vegetation cover, which affects debris flow activity (May, 2002). Decay in tree root strength after forest harvesting facilitates the occurrence of landslides (Sidle, 1992; Vergani et al., 2017), as do poorly designed logging roads whose fill slopes are susceptible to failure if not constructed well and surface runoff managed. These factors can result in high sediment recharge rates (Hatten et al., 2018; Jakob et al., 2005; Rachels et al., 2020); hence, the debris-flow frequency caused by the mass movement of channel deposits can be substantially higher in harvested areas (Imaizumi & Sidle, 2021; Jakob et al., 2005). Ground cover change affects the susceptibility of debris flows not only by changing the sediment recharge rate, but also by changing the hillslope and stream hydrology; for example, by enhancing overland flow and increasing storm runoff (Edeso et al., 1999; McGuire et al., 2018).

6.2.4 Rainfall Thresholds

Assessment of rainfall thresholds in debris flow torrents is important for improving warning systems (Marchi et al., this volume). Rainfall thresholds are also used to understand mechanisms that initiate debris flows. Therefore, rainfall thresholds have been obtained for many debris flow torrents worldwide.

Bovis and Jakob (1999) classified debris flow basins into weathering-limited (supply-limited) basins, which require a long sediment recharge period for the occurrence of a debris flow, and transport-limited basins, which are characterized by debris flow occurrence primarily controlled by hydroclimatic events. Transport-limited basins are characterized by a high frequency of debris flows (Bovis & Jakob, 1999; Schlunegger et al., 2009). A high volume of readily erodible sediment is always present in the basin, so a rainfall threshold can explain the timing of debris flows. The intensity–duration threshold (ID threshold) is a common method used to determine the rainfall threshold (Caine, 1980; Guzzetti et al., 2008; Staley et al., 2012). Short-term rainfall intensity (e.g., 10-min and hourly rainfall intensities) and cumulative rainfall depth are also used in basins with high debris-flow frequency (Imaizumi et al., 2021; Tang et al., 2009).

Many mountainous basins are supply-limited, so the occurrence of debris flows is mostly controlled by the volume of sediment in the channel (Bovis & Jakob, 1999; Jakob et al., 2005; Theule et al., 2015). In this type of basin, the source areas and channels have a smaller amount of loose sediment that can be mobilized (Jakob et al., 2005). Rainfall events exceeding the threshold may not trigger a debris flow immediately after the preceding debris flow if there is not sufficient accumulation of sediment between the rainfall events. Therefore, not only the rainfall magnitude but also the recharge period is important for the timing of debris flows (Theule et al., 2012). Further details on recharge mechanisms are provided in de Haas et al. (this volume).

Jakob (2021) updated the classification of debris flow basins based on the soil depth on hillslopes and frequency-magnitude relationships of debris flows. Watershed supply-limited watersheds (WSLWs) are characterized by little or no soil on hillslopes and a very low sediment production rate. Once a debris flow occurs, it takes decades for enough sediment to be produced for another debris flow. In channel supply-limited watersheds (CSLWs), which are covered by shallow soils, shallow landslides triggered by rainfall travel downhill, typically evolving into debris flows. Quasi supply-unlimited watersheds (QSUWs) and full supply unlimited watersheds (FWUWs) are characterized as watersheds with active sediment supply. Debris flows occur when a hydroclimatic threshold is exceeded. Extreme hydroclimatic events in QSUWs will not result in infinitely large debris flows due to the cohesion of sediment, while debris flow magnitude in FWUWs depends on the intensity and duration of a rainstorm or runoff event.

In some active debris-flow torrents, the rainfall threshold increases when the volume of sediment stored in the basin is larger (Imaizumi et al., 2021; Tsunetaka et al., 2021). This is because the amount of rainfall required to saturate channel deposits increases with a rise in the volume of sediment storage.

The rainfall threshold changes with time and is affected by changes in the hydraulic conditions and sediment supply in a basin. The rainfall threshold after a wildfire is lower than that seen before a wildfire; this can be attributed to a decrease in the infiltration rate on hillslopes resulting in higher runoff (Gartner et al., this volume; McGuire et al., 2018; Rengers et al., 2020b; Staley et al., 2017). Earthquakes also lower the rainfall threshold of debris flows by increasing the sediment supply from landslides (Tang et al., 2009). In areas that have experienced recent forest harvesting, landslides can occur during rainfall events with relatively low return periods (i.e., <10 yr; Imaizumi & Sidle, 2012), likely resulting in a higher frequency of debris flows in harvested areas than in non-harvested areas (Imaizumi & Sidle, 2021). The rainfall threshold also changes owing to the formation of seasonally frozen ground. Debris flows during cold periods, when the infiltration rate decreases due to the formation of ice in the soil matrix, are triggered by lower rainfall intensities and durations compared to warmer periods (Imaizumi et al., 2021).

Climate change is a potential factor that can change rainfall thresholds and debris-flow frequency (Stoffel et al., this volume). Under climate change, increases in the amount of sediment delivered to channels by the thawing of permafrost and extreme rainfall events increase the threat of debris flows on cold mountains (Jakob, 2022; Staffler et al., 2008; Stoffel et al., 2014a, 2014b). At the same time, a stochastic weather generator model by Hirschberg et al. (2021) predicted a future decrease in the frequency and sediment yield of debris flows because of a decrease in sediment supply from hillslopes. The impact of climate change on the rainfall threshold is likely to differ among regions affected by local climate conditions, as well as the type and activity of sediment supply processes.

6.3 Debris Flow Dynamics

Debris flows are typically composed of multiple surges. Discharge rapidly increases at the front of the surges, while it gradually decreases after the peak of surges (Arattano et al., 2012; March et al., 2002). Field monitoring has revealed that a series of mass movements can initiate multiple debris-flow surges (Imaizumi et al., 2019; Kean et al., 2013). Failure of a small in-channel dam formed by the sediment supply and bed load transport develops into a debris-flow surge (Berti et al., 1999; Kean et al., 2013). Debris-flow surges are also triggered by erosion of channel deposits by overland flow, sliding of channel deposits, and sediment supply from channel banks and tributaries (Imaizumi et al., 2016, 2019). A single-source mass can also develop into multiple surges because of the mechanical instability within the flow (Iverson, 1997). In basins where large boulders are supplied into the channel network, the accumulation of boulders has been observed at the front of the surges (March et al., 2002; Okano et al., 2012; Okuda et al., 1977; Suwa, 1988).

Debris flows are saturated, and the pores of the matrix between boulders are filled by interstitial water (Fig. 6.4a). The front of the surges, however, is sometimes unsaturated (Fig. 6.4b; McArdell et al., 2007; McCoy et al., 2013; Okano et al., 2012). Even in the same torrent, the sediment concentration and the presence or appearance of unsaturated flow are different among debris flow events (Hürlimann et al., 2014; Imaizumi et al., 2019; Kean et al., 2013). Okano et al. (2012) reported that the front of a surge was completely saturated when rainfall intensity over a long duration (24 h) was high, whereas unsaturated flow occurred during short-term, intense rainfall events. The sediment supply from the hillslopes and the volume of channel deposits also change the flow characteristics (Gregoretti et al., 2016; Staley et al., 2014; Zhou et al., 2015). Unsaturated flow dominates in the debris-flow initiation zones, where the channel gradient is steep (Imaizumi et al., 2017).

6.4 Entrainment Monitoring

The monitoring of entrainment is essential to measure the variation of elevation of erodible surfaces during debris-flow movement. Measurements commonly take place at specified locations where erosion occurred. However, direct measurement is challenging since the

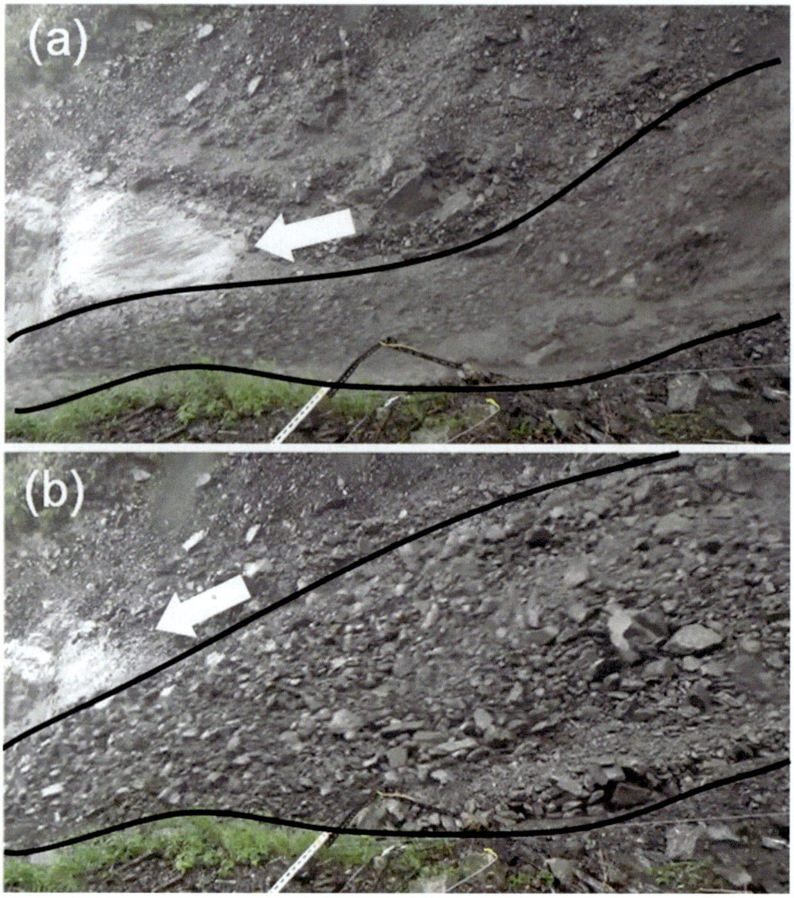

Fig. 6.4 Debris flow observed in the Ohya landslide scar, Japan, on July 13, 2021. **a** Saturated debris flow. **b** Unsaturated debris flow

interface between moving debris and an erodible bed is not easy to distinguish. The interface also moves upward and downward in deposition and erosion processes. Therefore, the total volumes or erosion depths are generally calculated to obtain a lump sum erosion amount. The rate of erosion is mostly indirectly calculated based on measured shear stresses or normal stresses and pore water pressures, along with measured soil properties.

6.4.1 Total Volume/Erosion Depth

Estimates of eroded volume are generally obtained by subtracting the digital elevation model before and after an event (Scheidl et al., 2014). The accuracy of this approach relies heavily on the resolution of the digital elevation models. In addition, if the tail of debris did not move out of the source zone, the overlap of the source and deposition areas makes the calculation of source and deposition volumes very difficult. The technology used to obtain the digital elevation model is the same as that commonly used for debris-flow monitoring (Marchi et al., this volume). Hürlimann et al. (2019) comprehensively reviewed the methods adopted in debris-flow monitoring and discussed the monitoring methods in debris-flow entrainment. Table 6.2 lists the most commonly used sensors related to debris flow entrainment monitoring.

In general, the incision rate of bedrock ranges from 10 to 20 mm/year, making it possible to

Table 6.2 Characteristics of the sensors related to entrainment monitoring of debris flows (adapted from Hürlimann et al., 2019)

Sensor	Parameter	Typical distance between sensor and active channel (m)	Typical cost excluding installation ($ USD)
Time/frequency domain reflectometry (TDR/FDR)	Soil water content	0–100	10^2
Pressure sensor	Pore water pressure	0–100	10^2
Video camera	Flow regime, flow magnitude, surface velocity etc	1–100	10^3
Stage sensor	Flow depth	0	10^3
Geophone	Ground vibration	1–100	10^2
Seismometer	Ground vibration	1–1000	10^3
Infrasound sensor	Air pressure fluctuation	1–1000	10^3
Doppler radar	Surface velocity	1–1000	10^4
Entrainment sensor	Erosion	0	10^3
Load cell	Basal, lateral and impact stress	0	10^3

measure the incision rate using bolts mounted on the channel bed (McCoy, 2012) (Fig. 6.5a). The variation of the exposed bolt is measured at the end of an event, which gives a total eroded depth during an event. However, due to the large normal force applied on the bolt, the bolt mounted on the rock can be bent, so the curvilinear length of the bolt is measured to obtain the total incision depth.

6.4.2 Rate of Erosion

Berger et al. (2010) developed an erosion sensor that can be used to directly measure the variation of the ground surface with time (Fig. 6.5b, c). The erosion sensor columns were first installed at the middle stream of the Illgraben catchment, located in the western part of Switzerland, where deposits on the channel bed were mainly composed of sand and gravel. This innovative development allows the measurement of the rate of erosion in the field while debris is moving. The erosion sensor is primarily composed of a resistor, wire, multiplier, and data and power cable. When erosion occurs, the wires are pulled out and a change of resistance can be measured, which is then interpreted as a change of the channel bed.

McCoy (2012) designed an erosion sensor after Berger et al. (2011b), which can directly measure the variation of the height of bed sediment. The principle of the sensor is to measure the variation of voltage due to the shortening of the resistance chain as a result of erosion. The erosion resolution relies heavily on the height of the resistor unit. For the erosion sensor developed by McCoy (2012), the length of the resistor unit was 5 cm, the resolution of which can be improved by decreasing the size of the resistor unit.

In flume experiments, the eroded depth can also be measured directly using high speed cameras placed perpendicular to the debris-flow channel. (Li et al., 2020; Su et al., 2020; Zhou et al., 2015). The advantage of this approach is that the rate of erosion can be directly measured. However, as the rate of erosion varies transversely, the measured rate of erosion from high-speed cameras should be calibrated before being considered as the average rate of erosion. In addition, boundary effects should also

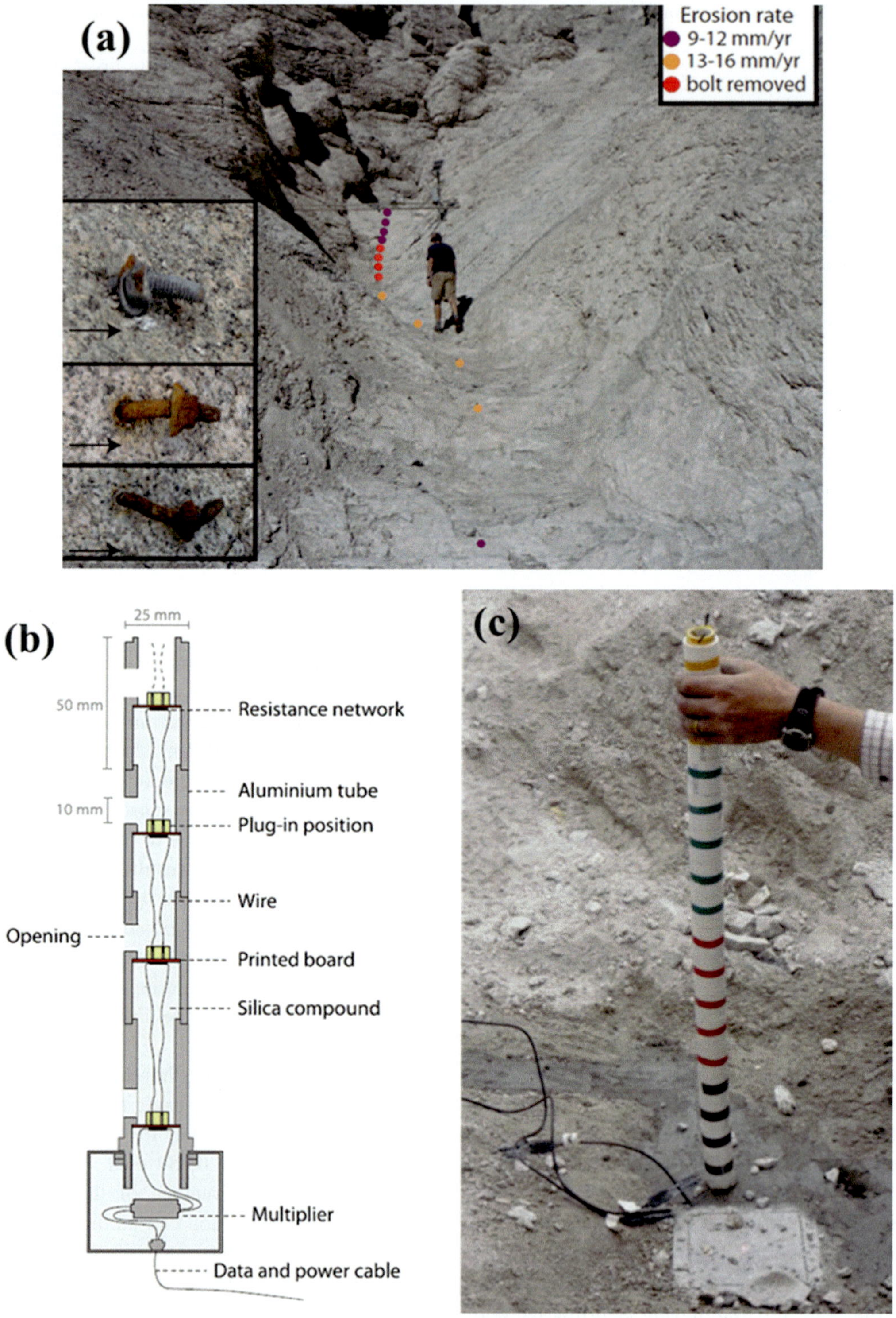

Fig. 6.5 Photographs of erosion sensor. **a** Measurement of eroded depth along the flow channel with close views of steel bolts (adapted from McCoy 2010), **b** Schematic diagram of erosion sensor (adapted from Berger et al., 2010) and **c** Erosion sensors placed on site (adapted from McCoy et al., 2012, 2013)

be considered, as the flow velocity close to the flume sides is not the same as that in the centre of the moving debris.

The rate of erosion can also be indirectly calculated based on pore water pressures, normal forces and moisture contents (McArdell et al., 2007; McCoy, 2012). The ground vibrations generated by moving debris can be measured in time (Kean et al., 2015), and are used to calculate the impact forces applied on the erodible bed. The amount of erodible sediment is then estimated based on the impact force and soil properties, utilizing analytical or empirical approaches. This indirect method has been verified in bedrock erosion and needs further validation for colluvial channels.

6.5 Factors Affecting Entrainment

6.5.1 Topography

It is commonly considered that the rate of erosion is dominated by the slope angle in both the shearing entrainment model and the progressive scouring entrainment model. It is also understood that the rate of erosion is proportional to the tangent of the slope angle in the shearing entrainment model (Gao et al., 2021; Su et al., 2020; Takahashi, 2014; Guo et al., 2022). However, this correlation is only valid in theory, as it is challenging to build a model that solely considers the variation of the slope angle. Mangeney et al. (2010) studied the influence of slope gradient on the erosion efficiency based on laboratory experiments and concluded that erosion is strongly dependent on the slope of the topography. Entrainment can affect the kinematic characteristics of a debris flow when the slope angle is close to a critical angle, which is close to half of the angle of repose of the erodible material. Kang and Chan (2022) studied historical rock avalanches, debris flows, and flowslides using the micro scale entrainment model and tried to correlate the rate of erosion of the debris flow front with the slope angle along the flow path. The results were quite scattered and no clear correlation was identified between slope angle and the rate of erosion. The main reason is that slope angle is not the only factor that may affect the rate of erosion. The contraction and spreading processes, curvature of the flow channel and properties of the erodible material can also contribute to the rate of erosion (Li et al., 2020).

6.5.2 Initial Volume

The initial volume of debris affects the rate of erosion by changing the flow height and initial kinematic characteristics, which then changes the shear and normal stresses applied on the erodible bed (Cui et al., 2013; Hungr et al., 2005; Mangeney et al., 2010; Sovilla et al., 2006). Kang and Chan (2022) considered the initial volume (V_0) of debris in correlating the rate of erosion with the kinematic characteristics of a debris flow and found that the rate of erosion can be correlated to $V_0^{1/3}$ in regression analysis. The equation is described in Sect. 6.6.

6.5.3 Solid Concentration

Egashira et al. (2001a; 2001b) and Papa et al. (2004) proposed a model to calculate the rate of erosion based on the difference between the solid concentrations of the moving debris and the erodible channel bed. It was concluded that, after the solid concentration in the moving debris flow reaches a threshold value, the erosion process is heavily restrained without any further inflow of water to the moving debris. Kang et al. (2021) numerically studied the Wenjiagou debris flow (2010) and found that the unit weight of the moving debris increased to the same unit weight as the source material after the debris traveled 50 s, while the total duration of the debris flow was 300 s. However, continuous runoff from tributaries of the flow channel can temporarily lower the solid concentration of moving debris, which was not considered in the simulation. In conclusion, solid concentration theoretically has an impact on the rate of erosion at the particle scale, as basal shear stress, normal stress and basal friction all change with a variation of solid concentration in moving debris.

6.5.4 Particle Size and Distribution

Theoretically, increasing the particle size of the bed material should result in a larger force required to mobilize the erodible material (Attal et al., 2015; Hsu et al., 2008; Iverson, 1997; Kang et al., 2017; Shields, 1936). Hsu et al. (2008) used a rotational drum to experimentally study the effect of particle size on erosion and concluded that the rate of erosion increases with the particle diameter in the moving debris, as shown in the following equation:

$$-\frac{\partial z}{\partial t} = \frac{K_0 K_1}{\frac{T_0^2}{E_{eff}}(F)} fL \left[\cos(\theta_b) v_s \rho_s D_e^2 \left(\frac{u_s}{h}\right)^w\right]^n \quad (6.1)$$

where K_0 is a constant that relates the bulk inertial normal stress to higher excursions of inertial normal stress, K_1 is a constant that relates rock resistance to incision rate, T_0 (Pa) is the tensile strength of the bedrock, E_{eff} (Pa) is the elastic modulus of the bedrock, F (m) is a function of the fracture spacing of bedrock and size of eroding boulders, f (yr^{-1}) is the frequency of flows over the bedrock per annum, D_e (m) is the effective grain size, u_s (m/s) is the surface velocity, h (m) is the flow height, L (m) is the length of eroding flow, and w and n are empirical exponents (Hsu et al., 2008).

Large particles on the channel bed are difficult to transport. It is also not easy to include particle size distribution in a debris-flow entrainment model since the particle sizes ranges from less than 1 mm to more than 1 m in some cases. Kang et al. (2017) introduced a characteristic particle size in the progressive scouring entrainment model to calculate the rate of erosion. Instead of using a particle size distribution curve directly, the mean particle size (d_{50}) from the particle size distribution curve was used. However, the determination of the characteristic particle size needs further study.

6.5.5 Other Factors

Apart from the factors discussed above, the moisture content of the erodible bed, lithology of granular material, content of fine-grained soil, and kinematic flow characteristics can also contribute to the erosion characteristics of debris. However, very few studies have quantitatively assessed the importance of these factors in debris-flow erosion. It is also possible that these factors may not be dominant components affecting the erosion of debris.

6.6 Entrainment/Deposition Models

Various models exist for estimating the amount and rate of erosion in debris-flow analysis (Iverson & Ouyang, 2015). Models for estimating the amount of material incorporated from channel beds can be roughly categorized into analytical models and empirical models (Table 6.3). In the analytical category, two approaches exist to calculate entrainment: the static approach and the dynamic approach. Newton's law of motion or the force equilibrium is used to calculate the entrainment rate or depth of erosion of the channel base. Shear stress and shear resistance are the most important factors for calculating entrainment (Iverson & Ouyang, 2015). Failure is considered to occur when the static shear stress exceeds the shear strength of the material in the channel bed. The depth at which failure occurs is determined, and the amount of material is calculated and added to the main body of the debris flow. In the dynamic approach, the rate of entrainment is calculated based on the rate of erosion of the material in the channel bed. The rate of erosion is determined based on the shear failure at the surface and the amount of material removed from the surface depending on the flow velocity of the main body of the debris

Table 6.3 Summary of disadvantages and advantages of different approaches to entrainment calculation (adopted from Kang & Chan, 2017)

Approaches	Physics	Complexity	Determination of material parameters	Range of applications	Pore pressure	Application in case histories	References
Static	Shear failure of erodible material based on shear failure	Simple	Conventional geotechnical tests	All types of soil	Y	N/A	Medina et al. (2008)
Dynamic	Shear failure of erodible material based on rate of shear failure	Simple	Conventional geotechnical tests	All types of soil	Y	Font de la Llum event; Cardemeller debris flow; 2003 Faucon debris flow;	Medina et al. (2008); Quan Luna et al. (2012); Iverson and Ouyang (2015); Bouchut et al. (2016); Pudasaini and Krautblatter (2021)
Diffusion process	Difference of solid concentration between flow and erodible bed	Complex	Geotechnical tests and hydraulic tests	Non-cohesive soil	N	Used in flume experiments;	Egashira et al. (2001a);
Empirical approach	Correlate the rate or erosion with average flow velocity and shear stress from runout analysis	Simple	Statistical data	Soil properties generally is not considered	Y	Fjæland debris flow; The 1999 Nomash River landslide;	De Blasio et al. (2011); McDougall and Hungr (2005);
Progressive scouring	Rolling and sliding motion	Simple	Conventional geotechnical tests	Granular material	N	Yigong rock avalanche	Kang et al. (2017); Su et al. (2020)

flow (Bouchut et al., 2016; Medina et al., 2008, Iverson & Ouyang, 2015). It is assumed in this approach that the velocity of newly eroded material is the same as the average velocity of the debris flow (Fraccarollo & Capart, 2002; Medina et al., 2008). The entrainment model proposed by van Asch et al. (2004) is also a one-dimensional dynamic debris-flow entrainment model that accounts for the entrainment concept based on the generation of excess pore water pressure under undrained loading of the in-situ material (Quan Luna et al., 2012). Currently, most entrainment models were developed based on the model proposed by Iverson (2012), which considered the behavior of a slide block descending an erodible slope with the ability to incorporate soil from the bed. However, selecting an appropriate model in entrainment calculations should consider the mechanics of erosion, availability of information about the erodible material, purpose of the calculation, and impact of the potential hazard.

The diffusion process caused by the difference in sediment concentration between the erodible channel bed and the main body of the debris flow is also considered as a possible mechanism of entrainment. Egashira et al. (2001a) proposed the following formula to calculate the erosion rate assuming that the slope of the channel bed is always adjusted to the angle corresponding to the limiting equilibrium conditions:

$$E\Delta x = Ev\Delta t = c_* v\Delta z \tag{6.2}$$

where E is the erosion rate, v is the average velocity of the debris flow, Δx is the distance increment, Δt is the time increment, c_* is the sediment concentration by volume of bed sediment (non-moving layer) and Δz is the erosion depth during Δt. They show that the material in the channel left behind by an unsaturated debris flow will approach the limiting equilibrium slope angle. The geometrical relationship between the initial bed slope and equilibrium slope angle is incorporated into the mass conservation law of the eroded material to obtain the entrainment rate.

In the empirical approach, however, the entrainment rate is empirically related to the flow velocity or shear stress exerted on the erodible bed. De Blasio et al. (2011) suggested a semi-empirical model to calculate entrainment. In the semi-empirical model, the entrainment rate is dependent solely on the tangential component of the weight at the base and on the average velocity of the debris flow, as follows:

$$\frac{dh}{dt} = c|U|^y(\tau_{xy} - \overline{\tau}) \text{ for } \tau_{xy} > \overline{\tau} \tag{6.3}$$

$$\frac{dh}{dt} = 0 \, for \;\; \tau_{xy} < \overline{\tau} \tag{6.4}$$

where dh/dt is the rate of bed erosion (m/s), U is the velocity of the debris flow, $\overline{\tau}$ is the critical shear rate for erosion, and c and y are constants. The critical shear stress was used as the threshold to determine the occurrence of entrainment. This model is promising because it relates debris- flow velocity and basal shear stress to entrainment.

The models mentioned above are all developed at the macro scale. Kang and Chan (2017) proposed a particle-scale progressive scouring entrainment model, in which both rolling and sliding motions of erodible particles were considered. The primary parameters in the model are the characteristic particle size, micro-scale frictional coefficient and internal friction angle. A probability density function is used to incorporate random arrangement of adjacent particles, which further provides a possibility to include a discrete entrainment model in a continuum runout model.

Kang and Chan (2022) also proposed a semi-empirical model and related the rate of erosion with flow height and velocity, based on runout analysis of historical cases using a particle scale entrainment runout model. A best fit equation was found to be:

$$\dot{E}_i \Big/ V_0^{1/3} = 6.4 \times 10^{-4}(H_i + v_i^2/2g)/H_0 - 3.8 \times 10^{-4} \tag{6.5}$$

with an R^2 of 0.80 (Fig. 6.6). The ordinate of the plot, $\dot{E}_t/V_0^{1/3}$ with a unit of *1/s*, refers to the average rate of erosion corresponding to the initial volume, V_0, while the dimensionless abscissa represents a measure of the normal stress based on the height of the debris that generates the shear stress that causes erosion (Kang & Chan, 2022). Once the potential initial debris volume is known, the rate of erosion for an erodible flow channel can be empirically calculated during movement. With additional well-documented historical cases, this correlation can be improved.

6.7 Entrainment/Deposition Physical and Numerical Modelling

Although many characteristic parameters of debris flows, such as surface velocities, runout distances and final volumes, can be observed in field and laboratory tests, the physical erosion process is still not easy to observe nor is the shear stress exerted on the erodible particles. Therefore, different approaches, including numerical modelling and physical modelling, are employed to understand the detailed entrainment process and provide concrete information for model verification.

6.7.1 Physical Modelling

To understand entrainment in debris flows, experiments have been conducted using inclined flumes (Choi et al., this volume), which have different sizes of tanks, and different lengths of the erosional zones and deposition pads (Egashira et al., 2001a; Iverson & Ouyang, 2015; Mangeney et al., 2010). Figure 6.7 shows a schematic diagram of a typical setup used for flume experiments. Table 6.4 summarizes geometric characteristics of some reported flumes.

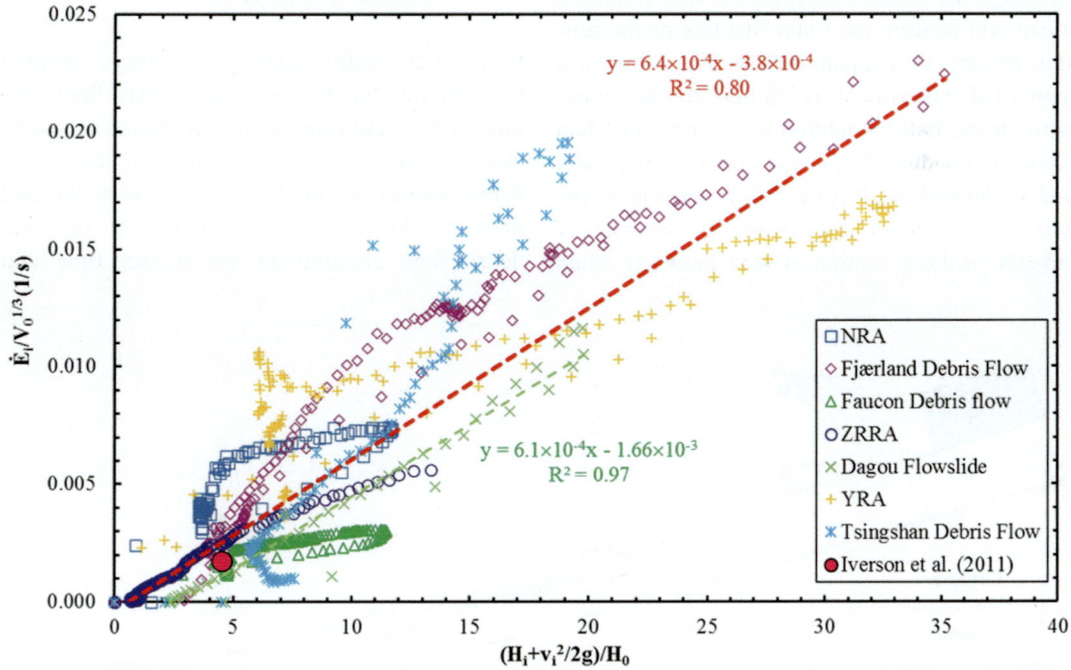

Fig. 6.6 Correlation between normalized kinematic characteristics and the rate of erosion (Kang & Chan, 2022)

6.7.2 Numerical Modelling

There are generally two approaches to numerically study the erosion process: continuum and discontinuum. Cundall (1971) introduced a discrete element method to study rock-mechanics problems, which was then applied to soils by Cundall and Strack (1979). Newton's second law and force–displacement law are used to calculate the motion of the particles. The assembly of particles continues to move until forces are in equilibrium or the mechanical energy has been completely dissipated by friction and/or system damping. As one discrete element method, Particle Flow Code (PFC) can monitor the kinematic characteristics of particles in models (Cundall, 1971). Since spherical particles with a specified radius are used in the model, it can show an idealized situation of physical flume experiments. In this way, other factors influencing the erosion process, which include shape factor of the particles and particle size distribution of the granular flow, can be excluded. Although the model simplifies the real situation, it can still capture the main features of the erosion process in a granular flow. The setup in a numerical experiment is similar to the apparatus in the field or laboratory. Kang and Chan (2018a) conducted a numerical experiment and found that neglecting rolling motion in the entrainment model underestimates the rate of erosion. Rolling motion is first detected when erosion starts and rotational velocities are monitored in the erosion zone, which shows that rolling motion exists in the erodible particles.

The Finite Element Method (FEM), Material Point Method (MPM) and Smoothed Particle Hydrodynamics (SPH) can also be used to study the erosion process of material on the channel bed resulting from moving debris (Han et al., 2020; Luo et al., 2022; Vicari et al., 2022). In these numerical methods, the erosion process is mostly based on the shear failure mechanism (Lee & Jeong, 2022). The material on the channel bed is considered to be eroded when the shear stress overcomes the shear resistance under large deformation, or empirical correlations are used. These methods provide different approaches to capture the erosion characteristics. However, the mechanism considered in these methods still needs verification in the laboratory or field.

6.7.3 Debris Flow Entrainment Runout Analysis

In practice, field mapping is always adopted to estimate the difference in debris flow volume before and after the event, which is generally considered as eroded volume if the tail of debris moves out of the source area at the end of event. However, in the back analysis of a real debris flow, entrainment rate at each time step

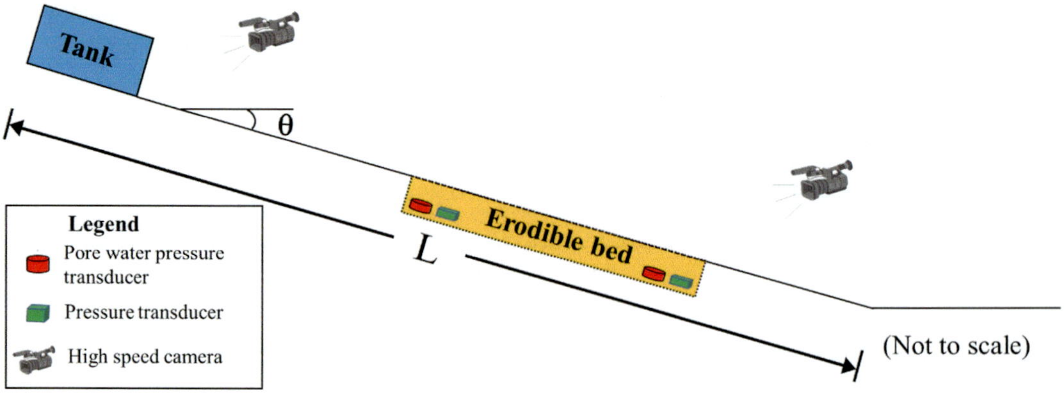

Fig. 6.7 Sketch of a simplified flume experiment setup used in entrainment analysis

Table 6.4 Summary of geometrical characteristics of some reported flume setups used in entrainment studies

Size	Tank volume, V (m^3)	Length, L (m)	Width, W (m)	Slope angle, θ (°)	Erodible material	Collected information	References
Small	N/A	6.0	0.1	12	Sand	Erosion depth; flow depth;	Egashira et al. (2001a) and Papa et al. (2004)
	0.2–0.14	3.0	0.1	0–30	Glass beads	Runout distance; flow height;	Mangeney et al. (2010)
	0.07	8	0.3	18	Gravel and sand	Flow thickness; runout distance	Zheng et al. (2021)
	0.5	6	0.3	10	Gravel and sand	Flow thickness; runout distance; pore water pressure; frontal velocity;	Li et al. (2020)
	N/A	10	0.2	10–19	Glass beads	Flow thickness; flow velocity;	Lee et al. (2019)
Large	10	28	2	20	Gravel and sand	Flow thickness; frontal velocity; normal and share stresses;	Vicari et al. (2021)
	10	80	2	31	Sand and gravel	Runout distance; flow height; frontal speed; deposition thickness; normal stress; pore water pressure;	Iverson et al. (2010)
	12	45	0.7	12	Sand	Total pressure; pore water pressure; deposition thickness; erosion depth;	Zhou et al. (2015)

should be calculated using entrainment models since the rate of erosion is an input parameter in runout analysis. It may be true that an entrainment model cannot be used to describe the erosion process of all types of soil. However, with advancing the understanding of entrainment mechanism, the selection of entrainment model can be improved by considering dominate erosion mechanism as well as soil types.

To calculate entrainment for a real debris flow, a runout model can be used (Pudasaini, 2012). Table 6.5 shows several runout models that are currently available for two-phase flow and single-phase flow simulations. In runout models, the rheological behaviour of the materials are often calculated according to frictional, viscous, and Voellmy models. Some of the models incorporate the rate of erosion through an empirical approach, such as DAN3D (McDougall, 2006) and avaflow (Mergili et al., 2017). However, as discussed in Mergili et al. (2017) and Han et al. (2021), process-based mechanical erosion models (Pudasaini & Fischer, 2016) are better at describing the

Table 6.5 List of selected numerical landslide runout models available or in development (adapted from Kang and Chan (2018b))

Model	Rheology	Entrainment incorporated	Selected references
3dDMM	Frictional and Voellmy	YES	Kwan and Sun (2007)
DAN2D	Frictional, Voellmy, and Bingham	YES	Hungr (1995)
DAN3D	Frictional, Voellmy, and Bingham	YES	McDougall (2006)
FLATModel	Frictional and Voellmy	YES	Medina et al. (2008)
FLO-2D	Quadratic	NO	O'Brien (2006)
Flow-R	Voellmy	NO	Horton et al. (2013)
GeoFlow-SPH	Frictional, Voellmy, and Bingham	YES	Pastor et al. (2009)
D-Claw	Frictional	YES	Iverson and George (2014); Iverson (2012)
MADFLOW	Frictional, Voellmy, and Bingham	YES	Chen and Lee (2000)
MassMov2D	Voellmy and Bignham	YES	Begueria et al. (2009)
PFC	Inter-particle and particle–wall interaction	YES	Kang and Chan (2018b)
RAMMS	Voellmy	YES	Christen et al. (2010)
RASH3D	Frictional, Voellmy, and Quadratic	NO	Pirulli (2005)
r.avalanche	Frictional	NO	Mergili et al. (2012)
r.avaflow	Frictional (solid) and Non-Newtonian (fluid)	YES	Mergili et al. (2017)
Sassa-Wang	Frictional	NO	Wang and Sassa (2002)
SCIDDICA S3-hex	Energy based	YES	D'Ambrosio et al. (2003)
SHALTOP-2D	Frictional	NO	Mangeney-Castelnau et al. (2003)
TITAN2D	Frictional	NO	Pitman et al. (2003)
TOCHNOG	Frictional (elastoplastic model)	YES	Roddeman (2002)
TopFlowDF	Frictional, Voellmy, and Quadratic	YES	Scheidl and Rickenmann (2011); Han et al. (2016)
VolcFlow	Frictional and Voellmy	NO	Kelfoun and Druitt (2005)
Wang	Frictional	YES	Wang et al. (2010); Kang and Chan (2018b)
Massflow	Frictional	YES	Ouyang et al. (2015)

erosion and transportation process. Therefore, other runout models have included the rate of erosion using an analytical approach, such as FLATModel (Medina et al., 2008) and Massflow (Ouyang et al., 2015). Due to the complexity of the rheology of debris material and the size of a typical debris flow, continuum modeling is a practical approach to obtain solutions for realistic debris flow problems.

It is meaningful to compare the performance of entrainment models in studying historical cases, which can provide practitioners some hints as to which ones they should look into for corresponding cases. Shen et al. (2020)

compared three entrainment models when studying the 1990 Tsing Shan debris flow. The models used in the simulation were Linear-Ex, Linear-MC and the model proposed by Iverson and Ouyang (2015). The comparison indicated that the Iverson-Ouyang model produces a much larger but ephemeral entrainment rate, while the Linear-Ex model produces a mild but prolonged entrainment process. The Linear-MC model produces results somewhere between the above two models. Kang and Chan (2018a) also compared the static and dynamic entrainment models with the progressive scouring entrainment model based on numerical experiment results. The comparison indicated that the progressive scouring model produces results closer to the actual measured depth, suggesting that only considering shear failure motion in erosion calculations could result in underestimation of erosion depths. This conclusion has been further verified by Kang and Chan (2018b). Unfortunately, no comprehensive comparison has been carried out for all available entrainment models. Therefore, it is suggested that selecting an appropriate entrainment model should consider erosion mechanisms, availability of case information and expected accuracy, and calculated results should be verified by available measurements and observations.

In entrainment runout analysis, the entrainment runout model is often calibrated based on available observations, such as runout distance, estimated final volume and maximum flow velocities at specific locations calculated based on runup or superelevation (Davies & McSaveney, 1999; Iverson & Ouyang, 2015; Kang et al., 2017; Mergili et al., 2017; Ouyang et al., 2015; Scheidl et al., 2015). However, the calibration process is always challenging due to limited field measurements. Therefore, parameters in entrainment models are empirically determined or based on previous research. Kinematic characteristics, such as velocity and flow height, and erosion characteristics, such as the rate of erosion and erosion depth, are then calculated (Kang & Chan, 2018a). In addition, the average velocity and the variation of the total volume are generally reported, and local rates of erosion are rarely described, except in Kang and Chan (2022).

6.8 Discussion and Conclusions

This chapter reviewed some historical cases with substantial entrainment and erosion mechanisms that were commonly considered in studying debris flow entrainment. The entire post failure process was comprehensively reviewed, and the erosion process was clearly described. The instruments used to monitor erosion depth/rate and factors affecting the rate of erosion were discussed. Entrainment and deposition models were categorized according to different erosion physics. Physical modelling equipment was tabulated and various numerical models were compared and discussed.

There are two mechanisms that are typically considered in the calculation of debris flow entrainment: shear failure in macro scale and scouring erosion in micro scale. For macro scale entrainment models developed based on the shear failure mechanism, curvature effects and 3D erosion have been included. However, the shear failure mechanism is very challenging to capture in the micro scale erosion process. Therefore, a micro scale erosion model will be an important future development. However, there remain two challenges in micro scale entrainment model development. The first is to include 3D particles in the model development, which means that the interactions between particles in the transverse direction should be included. However, the interaction force makes the calculation very complicated, although it is still possible to be solved. The second challenge is to find a characteristic particle size in the entrainment calculation. To solve this problem, it would be helpful to develop an empirical entrainment model based on the calculated erosion characteristics.

There is very limited information available during the erosion process, as the instruments used to measure the variation of erosion depth can be easily destroyed. Comparison of Lidar data before and after an event can provide net erosion amounts, but this approach is limited by the resolution of the data. Installing steel bolts can provide a direct measurement of erosion depth, while using erosion sensors can obtain the rate of erosion. However, the resolution of erosion rate heavily relies on the height of the resistor unit.

For most flume experiments, the erosion depth is the only information available for the verification of erosion calculations. High speed cameras can capture the variation of the interface between the moving debris and the stable channel during erosion. However, identifying the interface between the moving debris and the erodible channel bed is always challenging. Therefore, numerical modelling is mostly adopted to investigate the erosion process of debris flows.

In the last twenty years, more than twenty numerical models were developed with entrainment calculation incorporated. However, in debris flow entrainment runout simulations, the determination of parameters is not well explained for most of the back analyses, as available measures to calibrate the numerical models are always limited (including runout distance, maximum frontal velocity at specified locations and total volume). In addition, none of them can be used to capture the erosion characteristics of all types of soil. Therefore, selecting an appropriate entrainment model is very important. The advancement of computational capacity gives researchers an option to investigate debris flow erosion at the particle scale in the future.

References

Arattano, M., Marchi, L., & Cavalli, M. (2012). Analysis of debris-flow recordings in an instrumented basin: Confirmations and new findings. *Natural Hazards and Earth Systems Sciences, 12*, 679–686. https://doi.org/10.5194/nhess-12-679-2012

Attal, M., Mudd, S. M., Hurst, M. D., Weinman, B., Yoo, K., & Naylor, M. (2015). Impact of change in erosion rate and landscape steepness on hillslope and fluvial sediments grain size in the Feather River basin (Sierra Nevada, California). *Earth Surface Dynamics, 3*(1), 201–222.

Begueria, S., van Asch, Th. W. J., Malet, J. P., & Grondahl, S. (2009). A GIS-based numerical model for simulating the kinematics of mud and debris flows over complex terrain. *Natural Hazards and Earth System Sciences, 9*, 1897–1909. https://doi.org/10.5194/nhess-9-1897-2009

Benda, L. E., & Cundy, T. W. (1990). Predicting deposition of debris flows in mountain channels. *Canadian Geotechnical Journal, 27*, 409–417.

Berger, C., McArdell, B.W., Fritschi, B., & Schlunegger, F., (2010). A novel method for measuring the timing of bed erosion during debris flows and floods. *Water Resources Research*, 46(2).

Berger, C., McArdell, B. W., & Schlunegger, F. (2011a). Sediment transfer patterns at the Illgraben catchment, Switzerland: Implications for the time scales of debris flow activities. *Geomorphology, 125*, 421–432. https://doi.org/10.1016/j.geomorph.2010.10.019

Berger, C., McArdell, B. W., & Schlunegger, F. (2011b). Direct measurement of channel erosion by debris flows, Illgraben, Switzerland. *Journal of Geophysical Research: Earth Surface, 116*(F1). https://doi.org/10.1029/2010JF001722

Berti, M. G., Simoni, A. R., & Tecca, P. R. (1999). Field observations of a debris flow event in the Dolomites. *Geomorphology, 29*, 265–274. https://doi.org/10.1016/S0169-555X(99)00018-5

Bouchut, F., Fernández-Nieto, E. D., Mangeney, A., & Narbona-Reina, G. (2016). A two-phase two-layer model for fluidized granular flows with dilatancy effects. *Journal of Fluid Mechanics, 801*, 166–221.

Bovis, M. J., & Jakob, M. (1999). The roll of debris supply conditions in predicting debris flow activity. *Earth Surface Processes and Landforms, 24*, 1039–1054.

Brayshaw, D., & Hassan, M. A. (2009). Debris flow initiation and sediment recharge in gullies. *Geomorphology, 109*, 122–131.

Breien, H., De Blasio, F. V., Elverhøi, A., & Høeg, K. (2008). Erosion and morphology of a debris flow caused by a glacial lake outburst flood, Western Norway. *Landslides, 5*(3), 271–280.

Caine, N. (1980). The rainfall intensity–duration control of shallow landslides and debris flows. *Geografiska Annaler A, 62*, 23–27.

Cavalli, M., Trevisani, S., Comiti, F., & Marchi, L. (2013). Geomorphometric assessment of spatial sediment connectivity in small Alpine catchments. *Geomorphology, 188*, 31–41.

Chen, H., & Lee, C. F. (2000). Numerical simulation of debris flows. *Canadian Geotechnical Journal, 37*(1), 146–160. https://doi.org/10.1139/t99-089

Christen, M., Kowalski, J., & Bartelt, P. (2010). RAMMS: Numerical simulation of dense snow avalanches in three-dimensional terrain. *Cold Regions Science and Technology, 63*, 1–14. https://doi.org/10.1016/j.coldregions.2010.04.005

Cui, P., Dang, C., Cheng, Z., & Scott, K. M. (2010). Debris flows resulting from glacial-lake outburst floods in Tibet, China. *Physical Geography, 31*, 508–507. https://doi.org/10.2747/0272-3646.31.6.508

Cui, P., Zhou, G. G. D., Zhu, X. H., & Zhang, J. Q. (2013). Scale amplification of natural debris flows caused by cascading landslide dam failures. *Geomorphology, 182*(2013), 173–189.

Cundall, P. A., & Strack, O. D. L. (1979). A discrete numerical model for granular assemblies. *Geotechnique, 29*(1), 47–65.

Cundall, P. A. (1971). A computer model for simulating progressive, large scale movements in blocky rock systems. In: Anonymous Nancy (eds.), *Proceedings of the international symposium on rock mechanics* (vol. 2, pp. 129–36).

D'Ambrosio, D., Di Gregorio, S., & Iovine, G. (2003). Simulating debris flows through a hexagonal cellular automata model: SCIDDICA S3-hex. *Natural Hazards and Earth System Science, 3*(6), 545–559. https://doi.org/10.5194/nhess-3-545-2003

Davies, T. R., & McSaveney, M. J. (1999). Runout of dry granular avalanches. *Canadian Geotechnical Journal, 36*(2), 313–320.

De Blasio, F. V., Breien, H., & Elverhøi, A. (2011). Modelling a cohesive-frictional debris flow: An experimental, theoretical, and field-based study. *Earth Surface Processes and Landforms, 36*(6), 753–766.

Dunning, S. A., Rosser, N. J., Petley, D. N., et al. (2006). Formation and failure of the Tsatichhu landslide dam, Bhutan. *Landslides, 3*, 107–113. https://doi.org/10.1007/s10346-005-0032-x

Edeso, J. M., Merino, A., Gonzalez, M. J., & Marauri, P. (1999). Soil erosion under different harvesting managements in steep forestlands from northern Spain. *Land Degradation and Development, 10*, 79–88.

Egashira, S., Honda, N., & Itoh, T. (2001a). Experimental study on the entrainment of bed material into debris flow. *Physics and Chemistry of the Earth, Part C: Solar, Terrestrial & Planetary Science, 26*(9), 645–650.

Egashira, S., Itoh, T., & Takeuchi, H. (2001b). Transition mechanism of debris flows over rigid bed to over erodible bed. *Physics and Chemistry of the Earth, Part B: Hydrology, Oceans and Atmosphere, 26*, 169–174. https://doi.org/10.1016/S1464-1909(00)00235-5

Evans, S. G. (1986). The maximum discharge of outburst floods caused by the breaching of man-made and natural dams. *Canadian Geotechnical Journal, 23*, 385–387.

Fan, L., Lehmann, P., McArdell, B., & Or, D. (2017). Linking rainfall-induced landslides with debris flows runout patterns towards catchment scale hazard assessment. *Geomorphology, 280*, 1–15.

Fan, X., Dufresne, A., Siva Subramanian, S., Strom, A., Hermanns, R., Tacconi Stefanelli, C., Hewitt, K., Yunus, A. P., Dunning, S., Capra, L., Geertsema, M., Miller, B., Casagli, N., Jansen, J. D., & Xu, Q. (2020). The formation and impact of landslide dams–state of the art. *Earth-Science Reviews, 203*, 103116. https://doi.org/10.1016/j.earscirev.2020.103116

Fraccarollo, L., & Capart, H. (2002). Riemann wave description of erosional dam-break flows. *Journal of Fluid Mechanics, 461*, 183–228.

Gao, L., Zhang, L. M., Chen, H. X., Fei, K., & Hong, Y. (2021). Topography and geology effects on travel distances of natural terrain landslides: Evidence from a large multi-temporal landslide inventory in Hong Kong. *Engineering Geology, 292*, 106266.

Gregoretti, C., Degetto, M., Bernard, M., Crucil, G., Pimazzoni, A., De Vido, G., Berti, M., Simoni, A., & Lanzoni, S. (2016). Runoff of small rocky headwater catchments: Field observations and hydrological modeling. *Water Resources Research, 52*, 8138–8158. https://doi.org/10.1002/2016WR018675

Guo, J., Cui, Y., Xu, W., Yin, Y., Li, Y., & Jin, W. (2022). Numerical investigation of the landslide-debris flow transformation process considering topographic and entrainment effects: A case study. *Landslides, 19*(4), 773–788.

Guzzetti, F., Peruccacci, S., Rossi, M., & Stark, C. P. (2008). The rainfall intensity–duration control of shallow landslides and debris flows: An update. *Landslides, 5*, 3–17. https://doi.org/10.1007/s10346-007-0112-1

Han, Z., Chen, G., Li, Y., Zhang, H., & He, Y. (2016). Elementary analysis on the bed-sediment entrainment by debris flow and its application using the TopFlowDF model. *Geomatics, Natural Hazards and Risk, 7*(2), 764–785.

Han, Z., Su, B., Li, Y., Dou, J., Wang, W., & Zhao, L. (2020). Modeling the progressive entrainment of bed sediment by viscous debris flows using the three-dimensional SC-HBP-SPH method. *Water Research, 182*, 116031.

Han, Z., Ma, Y., Li, Y., Zhang, H., Chen, N., Hu, G., & Chen, G. (2021). Hydrodynamic and topography based cellular automaton model for simulating debris flow run-out extent and entrainment behavior. *Water Research, 193*, 116872.

Hatten, J. A., Segura, C., Bladon, K. D., Hale, V. C., Ice, G. G., & Stednick, J. D. (2018). Effects of contemporary forest harvesting on suspended sediment in the Oregon coast range: Alsea watershed study revisited. *Forest Ecol Manag., 408*, 238–248.

Hirschberg, J., Fatichi, S., Bennett, G. L., McArdell, B. W., Peleg, N., Lane, S. N., Schlunegger, F., & Molnar, P. (2021). Climate change impacts on sediment yield and debris-flow activity in an Alpine catchment. *Journal of Geophysical Research: Earth Surface, 126*, e2020JF005739. https://doi.org/10.1029/2020JF005739

Horton, P., Jaboyedoff, M., Rudaz, B., & Zimmermann, M. (2013). Flow-R, a model for susceptibility mapping of debris flows and other gravitational hazards at a regional scale. *Natural Hazards and Earth System*

Sciences, 13, 869–885. https://doi.org/10.5194/nhess-13-869-2013

Hsu, L., Dietrich, W. E., & Sklar, L. S. (2008). Experimental study of bedrock erosion by granular flows. *Journal of Geophysical Research: Earth Surface, 113*(F2).

Hungr, O. (1995). A model for the runout analysis of rapid flow slides, debris flows, and avalanches. *Canadian Geotechnical Journal, 32*(4), 610–623.

Hungr, O., & Evans, S. G. (2004). Entrainment of debris in rock avalanches: An analysis of a long run-out mechanism. *Geological Society of America Bulletin, 116*(9–10), 1240–1252.

Hungr, O., McDougall, S., & Bovis, M. (2005). Entrainment of material by debris flows. In *Debris-flow hazards and related phenomena*. Springer.

Hürlimann, M., Abancó, C., Moya, J., & Vilajosana, I. (2014). Results and experiences gathered at the Rebaixader debris-flow monitoring site. *Central Pyrenees, Spain, Landslides, 11*, 939–953. https://doi.org/10.1007/s10346-013-0452-y

Hürlimann, M., Coviello, V., Bel, C., Guo, X., Berti, M., Graf, C., Hübl, J., Miyata, S., Smith, J. B., & Yin, H. Y. (2019). Debris-flow monitoring and warning: Review and examples. *Earth-Science Reviews, 199*, 102981.

Hürlimann, M., Abancó, C., & Moya, J. (2012). Rockfalls detached from a lateral moraine during spring season. 2010 and 2011 events observed at the Rebaixader debris-flow monitoring site (Central Pyrenees, Spain). *Landslides, 9*, 385–393. https://doi.org/10.1007/s10346-011-0314-4

Hutchinson, J. N., & Bhandari, R. K. (1971). Undrained loading, a fundamental mechanism of mudflows and other mass movements. *Géotechnique, 21*, 353–358.

Imaizumi, F., & Sidle, R. C. (2007). Linkage of sediment supply and transport processes in Miyagawa Dam catchment, Japan. *Journal of Geophysical Research: Earth Surface, 112*, F3.

Imaizumi, F., & Sidle, R. C. (2012). Effect of forest harvesting on hydrogeomorphic processes in steep terrain of central Japan. *Geomorphology, 169*, 109–122.

Imaizumi, F., & Sidle, R. C. (2021). Effects of terrain on the occurrence of debris flows after forest harvesting. *Geografiska Annaler: Series a, Physical Geography, 103*, 1–14. https://doi.org/10.1080/04353676.2021.1932482

Imaizumi, F., Sidle, R. C., Tsuchiya, S., & Ohsaka, O. (2006). Hydrogeomorphic processes in a steep debris flow initiation zone. *Geophysical Research Letters, 33*, L10404. https://doi.org/10.1029/2006GL026250

Imaizumi, F., Tsuchiya, S., & Ohsaka, O. (2016). Field observations of debris-flow initiation processes on sediment deposits in a previous deep-seated landslide site. *Journal of Mountain Science, 13*, 212–222.

Imaizumi, F., Hayakawa, Y. S., Hotta, N., Tsunetaka, H., Ohsaka, O., & Tsuchiya, S. (2017). Relationship between the accumulation of sediment storage and debris-flow characteristics in a debris-flow initiation zone, Ohya Landslide Body, Japan. *Natural Hazards and Earth Science Systems, 17*, 1923–1938. https://doi.org/10.5194/nhess-17-1923-2017

Imaizumi, F., Masui, T., Yokota, Y., Tsunetaka, H., Hayakawa, Y. S., & Hotta, N. (2019). Initiation and runout characteristics of debris flow surges in Ohya landslide scar, Japan. *Geomorphology, 339*, 58–69.

Imaizumi, F., Ikeda, A., Yamamoto, K., & Ohsaka, O. (2021). Temporal changes in the debris flow threshold under the effects of ground freezing and sediment storage on Mt. Fuji. *Earth Surface Dynamics, 9*, 1381–1398. https://doi.org/10.5194/esurf-9-1381-2021

Iverson, R. M., Reid, M. E., Logan, M., LaHusen, R. G., Godt, J. W., & Griswold, J. P. (2011). Positive feedback and momentum growth during debris-flow entrainment of wet bed sediment. *Nature Geoscience, 4*(2), 116–121.

Iverson, R. M., & George, D. L. (2014). A depth-averaged debris-flow model that includes the effects of evolving dilatancy. I. Physical basis. *Proceedings of the Royal Society of London A: Mathematical, Physical and Engineering Sciences, 2170*, 20130819. https://doi.org/10.1098/rspa.2013.0819

Iverson, R. M., & Ouyang C. J. (2015). Entrainment of bed material by Earth-surface mass flows: Review and reformulation of depth-integrated theory. *Reviews of Geophysics, 53*. https://doi.org/10.1002/2013RG000447

Iverson, R. M., Reid, M. E., & LaHusen R. G. (1997). Debris flow mobilization from landslides. *Annual Review of Earth and Planetary Sciences*.

Iverson, R. M., Logan, M., & Denlinger, R. P. (2004). Granular avalanches across irregular three-dimensional terrain: 2. Experimental tests. *Journal of Geophysical Research: Earth Surface, 109*(F1).

Iverson, R. M., Logan, M., LaHusen, R. G., & Berti, M. (2010). The perfect debris flow? Aggregated results from 28 large-scale experiments. *Journal of Geophysical Research: Earth Surface, 115*(F3). https://doi.org/10.1029/2009JF001514

Iverson, R. M. (1997). The physics of debris flows. *Reviews of Geophysics, 35*, 245–296. https://doi.org/10.1146/annurev.earth.25.1.85

Iverson, R. M. (2012). Elementary theory of bed-sediment entrainment by debris flows and avalanches. *Journal of Geophysical Research: Earth Surface, 117*(F3). https://doi.org/10.1029/2011JF002189

Jakob, M. (2021). Debris-flow hazard assessments: a practitioner's view. *Environmental & Engineering Geoscience, 27*(2), 153–166.

Jakob, M. (2022). Landslides in a changing climate. In *Landslide hazards, risks, and disasters* (pp. 505-579). Elsevier.

Jakob, M., Bovis, M., & Oden, M. (2005). The significance of channel recharge rates for estimating debris-flow magnitude and frequency. *Earth Surface Processes and Landforms, 30*, 755–766.

Kang, C., & Chan, D. (2017). Modelling of entrainment in debris flow analysis for dry granular material.

International Journal of Geomechanics (ASCE), 17(10), 1–20. https://doi.org/10.1061/(ASCE)GM.1943-5622.0000981

Kang, C., & Chan, D. (2018a). Numerical simulation of 2D granular flow entrainment using DEM. *Granular Matter, 20*(1), 13. https://doi.org/10.1007/s10035-017-0782-x

Kang, C., & Chan, D. (2018b). A progressive entrainment runout model for debris flow analysis and its application. *Geomorphology, 323*, 25–40.

Kang, C., Chan, D., Su, F., & Cui, P. (2017). Runout and entrainment analysis of an extremely large rock avalanche—a case study of Yigong, Tibet, China. *Landslides, 14*(1), 123–139.

Kang, C., Ren, D., Gao, X., Han, C., & Wang, Y. (2021). Study of kinematic characteristics of a rock avalanche and subsequent erosion process due to a debris flow in Wenjia gully, Sichuan, China. *Natural Hazards, 106*(1), 937–964.

Kang, C., & Chan, D. (2022). Investigation of erosion characteristics of debris flow based on historical cases. *Engineering Geology, 306*. https://doi.org/10.1016/j.enggeo.2022.106767

Kean, J. W., McCoy, S. W., Tucker, G. E., Staley, D. M., & Coe, J. A. (2013). Runoff-generated debris flows: Observations and modeling of surge initiation, magnitude, and frequency. *Journal of Geophysical Research, 18*, 2190–2207. https://doi.org/10.1002/jgrf.20148

Kean, J. W., Coe, J. A., Coviello, V., Smith, J. B., McCoy, S. W., & Arattano, M. (2015). Estimating rates of debris flow entrainment from ground vibrations. *Geophysical Research Letters, 42*, 6365–6372. https://doi.org/10.1002/2015GL064811

Kelfoun, K., & Druitt, T. H. (2005). Numerical modeling of the emplacement of Socompa rock avalanche, Chile. *Journal of Geophysical Research: Solid Earth, 110*(B12), 2156–2202. https://doi.org/10.1029/2005JB003758

Kwan, J. S. H., & Sun, H. W. (2007). Benchmarking exercise on landslide mobility modelling—runout analyses using 3dDMM. In Ho & Li (eds.), *Proceedings of the 2007 international forum on landslide disaster management* (pp. 945–966). Hong Kong Geotechnical Engineering Office.

Larsen, I. J., Montgomery, D. R., & Korup, O. (2010). Landslide erosion controlled by hillslope material. *Nature Geoscience, 3*(4), 247.

Lee, K., & Jeong, S. (2022). Study on the effect of entrainment on the behavior of debris flows using a 3D Coupled Eulerian-Lagrangian finite element method. *Bulletin of Engineering Geology and the Environment, 81*(4), 1–14.

Lee, K., Kim, Y., Ko, J., & Jeong, S. (2019). A study on the debris flow-induced impact force on check dam with-and without-entrainment. *Computers and Geotechnics, 113*, 103104.

Li, P., Wang, J., Hu, K., & Wang, F. (2020). Experimental study of debris-flow entrainment over stepped-gradient beds incorporating bed sediment porosity. *Engineering Geology, 274*, 105708.

Lo, K. H., & Chau, K. T. (2003). Debris-flow simulations for Tsing Shan in Hong Kong. In *3rd international conference on debris-flow hazards mitigation: Mechanics, prediction, and assessment* (pp. 577–588).

Luo, G., Zhao, Y., Shen, W., & Wu, M. (2022). Dynamics of bouldery debris flow impacting onto rigid barrier by a coupled SPH-DEM-FEM method. *Computers and Geotechnics, 150*, 104936.

Mangeney, A., Roche, O., Hungr, O., Mangold, N., Faccanoni, G., & Lucas, A. (2010). Erosion and mobility in granular collapse over sloping beds. *Journal of Geophysical Research: Earth Surface, 115*(F3).

Mangeney-Castelnau, A., Vilotte, J. P., Bristeau, M. O., Perthame, B., Bouchut, F., Simeoni, C., & Yerneni, S. (2003). Numerical modelling of avalanches based on Saint Venant equations using a kinetic scheme. *Journal of Geophysical Research, 108*(B11), 2527. https://doi.org/10.1029/2002JB002024

March, L., Arattano, M., & Deganutti, A. M. (2002). Ten years of debris-flow monitoring in the Morcardo Torrent (Italian Alps). *Geomorphology, 46*, 1–17.

May, C. L. (2002). Debris flows throughout different forest age classes in the central Oregon coast range. *Journal of the American Water Resources Association, 38*, 1097–1113.

McArdell, B. W., Bartelt, P., & Kowalski, J. (2007). Field observations of basal forces and fluid pore pressure in a debris flow. *Geophysical Research Letters, 34*, L07406. https://doi.org/10.1029/2006GL029183

McCoy, S. W., Tucker, G. E., Kean, J. W., & Coe, J. A. (2013). Field measurement of basal forces generated by erosive debris flows. *Journal of Geophysical Research. Earth Surface, 118*, 589–602. https://doi.org/10.1002/jgrf.20041

McCoy, S. W., Kean, J. W., Coe, J. A., Tucker, G. E., Staley, D. M., & Wasklewicz, T. A. (2012). Sediment entrainment by debris flows: In situ measurements from the headwaters of a steep catchment. *Journal of Geophysical Research: Earth Surface, 117*(F3).

McCoy, S. W. (2012). Controls on erosion and transport of mass by debris flows. PhD Thesis, University of Colorado, USA.

McDougall, S., & Hungr, O. (2005). Dynamic modelling of entrainment in rapid landslides. *Canadian Geotechnical Journal, 42*(5), 1437–1448.

McDougall, S. (2006). A new continuum dynamic model for the analysis of extremely rapid landslide motion across complex 3D terrain. Ph.D. thesis, The University of British Columbia, Vancouver, B.C.

McGuire, L. A., Rengers, F. K., Kean, J. W., Staley, D. M., & Mirus, B. B. (2018). Incorporating spatially heterogeneous infiltration capacity into hydrologic models with applications for simulating post-wildfire debris flow initiation. *Hydrological Processes, 32*, 1173–1187. https://doi.org/10.1002/hyp.11458

Medina, V., Hürlimann, M., & Bateman, A. (2008). Application of FLATModel, a 2D finite volume code, to debris flows in the northeastern part of the Iberian Peninsula. *Landslides, 5*(1), 127–142.

Mergili, M., Schratz, K., Ostermann, A., & Fellin, W. (2012). Physically-based modelling of granular flows with Open Source GIS. *Natural Hazards and Earth System Sciences, 12*, 187–200. https://doi.org/10.5194/nhess-12-187-2012

Mergili, M., Fischer, J. T., Krenn, J., & Pudasaini, S. P. (2017). R. avaflow v1, an advanced open-source computational framework for the propagation and interaction of two-phase mass flows. *Geoscientific Model Development, 10*, 553–569. https://doi.org/10.5194/gmd-10-553-2017

Nakamura, F., Maita, H., & Araya, T. (1995). Sediment routing analyses based on chronological changes in hillslope and riverbed morphologies. *Earth Surface Processes Landforms, 20*, 333–336.

O'Brien, J. D. (2006). FLO-2D user's manual, Version 2006.01. FLOEngineering: Nutrioso.

Okano, K., Suwa, H., & Kanno, T. (2012). Characterization of debris flows by rainstorm condition at a torrent on the Mount Yakedake volcano, Japan. *Geomorphology, 136*, 88–94.

Okuda, S., Suwa, H., Okunishi, K., Nakano, M., & Yokoyama, K. (1977). Synthetic observation on debris flow. *Annual Report of Disaster Prevention Research Institute, 20B-1*, 237–263. (in Japanese with English abstract).

Ouyang, C., He, S., & Xu, Q. (2015). MacCormack-TVD finite difference solution for dam break hydraulics over erodible sediment beds. *Journal of Hydraulic Engineering, 141*(5), 06014026.

Palucis, M. C., Ulizio, T. P., & Lamb, M. P. (2021). Debris flow initiation from ravel-filled channel bed failure following wildfire in a bedrock landscape with limited sediment supply. *GSA Bulletin, 133*(9–10), 2079–2096. https://doi.org/10.1130/B35822.1

Papa, M., Egashira, S., & Itoh, T. (2004). Critical conditions of bed sediment entrainment due to debris flow. *Natural Hazards and Earth System Sciences, 4*(3), 469–474.

Pastor, M., Haddad, B., Sorbino, G., Cuomo, S., & Drempetic, V. (2009). A depth-integrated, coupled SPH model for flow-like landslides and related phenomena. *International Journal for Numerical and Analytical Methods in Geomechanics, 33*, 143–172. https://doi.org/10.1002/nag.705

Pirulli, M. (2005). Numerical modelling of landslide runout: a continuum mechanics approach. Ph.D. thesis, Politechnico di Torino, Torino.

Pitman, E. B., Nichita, C. C., Patra, A., Bauer, A., Sheridan, M., & Bursik, M. (2003). Computing granular avalanches and landslides. *Physics of Fluids, 15*, 3638–3646. https://doi.org/10.1063/1.1614253

Pudasaini, S. P., & Fischer, J. T. (2016). A mechanical erosion model for two-phase mass flows. arXiv preprint arXiv:1610.01806

Pudasaini, S. P., & Krautblatter, M. (2021). The mechanics of landslide mobility with erosion. *Nature Communications, 12*, 6793. https://doi.org/10.1038/s41467-021-26959-5

Pudasaini, S. P. (2012). A general two-phase debris flow model. *Journal of Geophysical Research: Earth Surface, 117*(F3). https://doi.org/10.1029/2011JF002186

Quan Luna, B., Remaitre, A., van Asch, T. W. J., Malet, J. P., & van Westen, C. J. (2012). Analysis of debris flow behavior with a one dimensional runout model incorporating entrainment. *Engineering Geology, 128*, 63–75.

Rachels, A. A., Bladon, K. D., Bywater-Reyes, S., & Hatten, J. A. (2020). Quantifying effects of forest harvesting on sources of suspended sediment to an Oregon Coast Range headwater stream. *Forest Ecology and Management, 466*, 118123.

Remaître, A., Malet, J. P., & Maquaire, O. (2009). Sediment budget and morphology of the 2003 Faucon debris flow (South French Alps): scouring and channel-shaping processes. In Proceedings of the landslide processes: From geomorphological mapping to dynamic modelling (pp 75–80).

Rengers, F. K., McGuire, L. A., Oakley, N. S., Kean, J. W., Staley, D. M., & Tang, H. (2020a). Landslides after wildfire: Initiation, magnitude, and mobility. *Landslides, 17*, 2631–2641. https://doi.org/10.1007/s10346-020-01506-3

Rengers, F. K., Kean, J. W., Reitman, N. G., Smith, J. B., Coe, J. A., & McGuire, L. A. (2020b). The influence of frost weathering on debris flow sediment supply in an alpine Basin. *Journal of Geophysical Research: Earth Surace*. https://doi.org/10.1029/2019JF005369

Roberts, R. G., & Church, M. (1986). The sediment budget in severely disturbed watersheds, Queen Charlotte Ranges, British Columbia. *Canadian Journal of Forest Research, 16*, 1092–1106.

Roddeman, D. G. (2002). TOCHNOG user's manual—a free explicit/implicit FE program. FEAT. Available from www.feat.nl/manuals/user/user.html

Santi, P. M., & Macaulay, B. (2021). Water and sediment supply requirements for post-wildfire debris flows in the Western United States. *Environmental and Engineering Geoscience, 27*(1), 73–85. https://doi.org/10.2113/EEG-D-20-00022

Sassa, K. (1985). The mechanism of debris flows. In *Proceedings of the XI international conference on soil mechanics and foundation engineering, San Francisco* (vol. 1, pp. 1173–1176).

Scheidl, C., McArdell, B. W., & Rickenmann, D. (2015). Debris-flow velocities and superelevation in a curved laboratory channel. *Canadian Geotechnical Journal, 52*(3), 305–317.

Scheidl, C., & Rickenmann, D. (2011). TopFlowDF-a simple GIS based model to simulate debris-flow runout on the fan. In R. Genevois, & P. Hamilton (Eds.), *5th international conference on debris-flow hazards mitigation, mechanics, prediction, and assessment* (pp. 253–262). Casa Editrice Università La Sapienza.

Scheidl, C., Schraml, K., Moser, M., & Hübl, J. (2014). Analysis of material entrainment with an unmanned aerial vehicle (UAV) and simulation of the debris-flow event at the Sattelbach torrent-Austria, 2013. In *EGU general assembly conference abstracts*.

Schlunegger, F., Badoux, A., McArdell, B. W., Gwerder, C., Schnydrig, D., Rieke-Zapp, D., & Molnar, P. (2009). Limits of sediment transfer in an alpine debris-flow catchment, Illgraben, Switzerland. *Quaternary Science Reviews, 28*, 1097–1105. https://doi.org/10.1016/j.quascirev.2008.10.025

Shen, P., Zhang, L., Wong, H. F., Peng, D., Zhou, S., Zhang, S., & Chen, C. (2020). Debris flow enlargement from entrainment: A case study for comparison of three entrainment models. *Engineering Geology, 270*, 105581.

Shields, A. (1936). Application of similarity principles and turbulence research to bed-load movement. California Institute of Technology, Pasadena, CA.

Sidle, R. C. (1992). A theoretical model of the effects of timber harvesting on slope stability. *Water Resources Research, 28*, 1897–1910.

Sidle, R. C., & Ochiai, H. (2006). Landslides: Processes, prediction, and land use. *American Geophysical Union Water Resources Monograph, 18*.

Sovilla, B., Burlando, P., & Bartelt, P. (2006). Field experiments and numerical modeling of mass entrainment in snow avalanches. *Journal of Geophysical Research: Earth Surface, 111*(F3). https://doi.org/10.1029/2005JF000391

Staffler, H., Pollinger, R., Zischg, A., & Mani, P. (2008). Spatial variability and potential impacts of climate change on flood and debris flow hazard zone mapping and implications for risk management. *Natural Hazards and Earth Systems Sciences, 8*, 539–558. https://doi.org/10.5194/nhess-8-539-2008

Staley, D. M., Kean, J. W., Cannon, S. H., Schmidt, K. M., & Laber, J. L. (2012). Objective definition of rainfall intensity–duration thresholds for the initiation of post-fire debris flows in southern California. *Landslides, 10*, 547–562. https://doi.org/10.1007/s10346-012-0341-9

Staley, D. N., Wasklewicz, T. A., & Kean, J. W. (2014). Characterizing the primary material sources and dominant erosional processes for post-fire debris-flow initiation in a headwater basin using multi-temporal terrestrial laser scanning data. *Geomorphology, 214*, 324–338. https://doi.org/10.1016/j.geomorph.2014.02.015

Staley, D. M., Negri, J. A., Kean, J. W., Laber, J. L., Tillery, A. C., & Youberg, A. M. (2017). Prediction of spatially explicit rainfall intensity–duration thresholds for post-fire debris-flow generation in the western United States. *Geomorphology, 278*, 149–162. https://doi.org/10.1016/j.geomorph.2016.10.019

Stoffel, M., Mendlik, T., Schneuwly-Bollschweiler, M., & Gobiet, A. (2014a). Possible impacts of climate change on debris-flow activity in the Swiss Alps. *Climate Change, 122*, 141–155. https://doi.org/10.1007/s10584-013-0993-z

Stoffel, M., Tiranti, D., & Huggel, C. (2014b). Climate change impacts on mass movements—Case studies from the European Alps. *Science of the Total Environment, 493*, 1255–1266. https://doi.org/10.1016/j.scitotenv.2014.02.102

Su, Y., Choi, C. E., Ng, C. W. W., Lam, H. W. K., Wong, L. A., & Lee, C. (2020). New light-weight concrete foam to absorb debris-flow-entrained boulder impact: Large-scale pendulum modelling. *Engineering Geology, 275*, 105724.

Suwa, H. (1988). Focusing mechanism of large boulders to a debris-flow front. *Annual report of the Disaster Prevention Research Institute*, 31B-1, 139–151. (in Japanese with English abstract).

Takahashi, T. (2014). *Debris flow*. CRC Press/Balkema.

Tang, C., van Asch, T. W. J., Chang, M., Chen, G. Q., Zhao, X. H., & Huang, X. C. (2012). Catastrophic debris flows on 13 August 2010 in the Qingping area, southwestern China: The combined effects of a strong earthquake and subsequent rainstorms. *Geomorphology, 139–140*, 559–576.

Tang, C., Zhu, J., Li, W. L., & Liang, J. T. (2009). Rainfall-triggered debris flows following the Wenchuan earthquake. *Bulletin of Engineering Geology and the Environment, 68*, 187–194. https://doi.org/10.1007/s10064-009-0201-6

Theule, J. I., Liébault, F., Laigle, D., Loye, A., & Jaboyedoff, M. (2015). Channel scour and fill by debris flows and bedload transport. *Geomorphology, 243*, 92–105.

Theule, J. I., Liébault, F., Loye, A., Laigle, D., & Jaboyedoff, M. (2012). Sediment budget monitoring of debris-flow and bedload transport in the Manival Torrent, SE France. *Natural Hazards and Earth System Sciences, 12*, 731–749. https://doi.org/10.5194/nhess-12-731-2012

Tillery, A. C., & Rengers, F. K. (2019). Controls on debris-flow initiation on burned and unburned hillslopes during an exceptional rainstorm in southern New Mexico, USA. *Earth Surface Processes and Landforms, 45*, 1051–1066.

Tsunetaka, H., Hotta, N., Imaizumi, F., Hayakawa, Y. S., & Masui, T. (2021). Variation in rainfall patterns triggering debris flow in the initiation zone of the Ichino-sawa torrent, Ohya landslide, Japan. *Geomorphology, 375*, 107529. https://doi.org/10.1016/j.geomorph.2020.107529

van Asch, Th.W.J., Malet, J.P., Remaitre, A., Maquaire, O., (2004). Numerical modelling of the run-out of a muddy debris-flow. The effect of rheology on velocity and deposit thickness along the run-out track. In Lacerda, Ehrlich & Fontoura & Sayao (eds.), *Landslides: Evaluation and Stabilization* (pp. 1433–1438). London: Taylor & Francis.

VanDine, D. F. (1985). Debris flows and debris torrents in the southern Canadian Cordillera. *Canadian Geotechnical Journal, 22*, 44–62.

Vergani, C., Giadrossich, F., Buckley, P., Conedera, M., Pividori, M., Salbitano, F., Rauch, H. S., Lovreglio, R., & Schwarz, M. (2017). Root reinforcement dynamics of European coppice woodlands and their effect on shallow landslides: A review. *Earth Science Reviews, 167*, 88–102.

Vicari, H., Ng, C. W., Nordal, S., Thakur, V., De Silva, W. R. K., Liu, H., & Choi, C. E. (2021). The effects of upstream flexible barrier on the debris flow entrainment and impact dynamics on a terminal barrier. *Canadian Geotechnical Journal* (in press).

Vicari, H., Tran, Q. A., Nordal, S., & Thakur, V. (2022). MPM modelling of debris flow entrainment and interaction with an upstream flexible barrier. *Landslides*, 1–15.

Wang, X. B., Morgenstern, N. R., & Chan, D. H. (2010). A model for geotechnical analysis of flow slides and debris flows. *Canadian Geotechnical Journal, 47*(12), 1401–1414.

Wang, F. W., & Sassa, K. (2002). A modified geotechnical simulation model for the areal prediction of landslide motion. In *Proceedings of the 1st European conference on landslides* (pp. 735–740).

Xing, A. G., Wang, G., Yin, Y. P., Jiang, Y., Wang, G. Z., Yang, S. Y., Dai, D. R., Zhu, Y. Q., & Dai, J. A. (2014). Dynamic analysis and field investigation of a fluidized landslide in Guanling, Guizhou, China. *Engineering Geology, 181*, 1–14.

Xu, W., Yu, W., Jing, S., Zhang, G., & Huang, J. (2013). Debris flow susceptibility assessment by GIS and information value model in a large-scale region, Sichuan Province (China). *Natural Hazards, 65*, 1379–1392. https://doi.org/10.1007/s11069-012-0414-z

Zaginaev, V., Petrakov, D., Erokhin, S., Meleshko, A., Stoffel, M., & Ballesteros-Cánovas, J. A. (2019). Geomorphic control on regional glacier lake outburst flood and debris flow activity over northern Tien Shan. *Global and Planetary Change, 176*, 50–59. https://doi.org/10.1016/j.gloplacha.2019.03.003

Zhang, F., Kang, C., Chan, D., Zhang, X., Pei, X., & Peng, J. (2017). A study of a flowslide with significant entrainment in loess areas in China. *Earth Surface Processes and Landforms, 42*(14), 2295–2305.

Zheng, H., Shi, Z., Hanley, K. J., Peng, M., Guan, S., Feng, S., & Chen, K. (2021). Deposition characteristics of debris flows in a lateral flume considering upstream entrainment. *Geomorphology, 394*, 107960.

Zhou, G. G. D., Cui, P., Tang, J. B., Chen, H. Y., Zou, Q., & Sun, Q. C. (2015). Experimental study on the triggering mechanisms and kinematic properties of large debris flows in Wenjia Gully. *Engineering Geology, 194*(Supplement C), 52–61.

7 Causes and Triggers

Roland Kaitna, Marisa C. Palucis, Francesco Marra and Christian Huggel

Abstract

Debris flows are caused by rapid mobilization of loose debris and sediment by water in steep mountain environments. They occur in a wide range of geological and climatological settings and are mostly triggered by intense rainfall events. Non-rainfall related triggers include volcanic eruptions, earthquakes and glacial-, landslide-, or beaver lake outburst floods, among others. The mechanisms for the formation of debris flows range from channel runoff that leads to erosion of sediment in or along a flow path to slope failures that transform to fluidized landslides or feed the channel for further transfer. The difficulty in reliably predicting (hydro-)meteorological trigger conditions for debris flows is associated with the difficulty in rainfall monitoring, the variations in geomorphological and hydrological susceptibility within a watershed and the inherent complexity of possible initiation mechanisms. For improved hazard analysis and forecasting, as well as assessment of climate change impact, a better mechanistic understanding of initiation processes and the interplay of trigger conditions is needed.

7.1 Introduction

The phenomenon *debris flow* is a geomorphic mass transfer process in mountain regions that can be described as a mixture of sediment and water transiting steep (headwater-) channels as a single surge or as a sequence of surges. A key-feature of debris flows is that the maximum flow depth and the mean velocity can significantly exceed respective values of floods observed in the same channel (Rickenmann & Zimmermann, 1993). Together with high bulk densities and large boulders dispersed in the flow, debris flow

R. Kaitna (✉)
Institute of Mountain Risk Engineering, University of Natural Resources and Life Sciences, Vienna (BOKU), Peter Jordan Strasse 82, 1190 Vienna, Austria
e-mail: roland.kaitna@boku.ac.at

M. C. Palucis
Department of Earth Sciences, Dartmouth Collage, Hanover, NH 03755, USA
e-mail: marisa.c.palucis@dartmouth.edu

F. Marra
Department of Geosciences, University of Padova, Via Gradenigo 6, 35131 Padova, Italy
e-mail: francesco.marra@unipd.it

F. Marra
Institute of Atmospheric Sciences and Climate (ISAC), National Research Council of Italy (CNR), Via Gobetti, 101, 40129 Bologna, Italy

C. Huggel
Department of Geography, University of Zurich, Winterthurerstrasse 190, CH-8057 Zurich, Switzerland
e-mail: christian.huggel@geo.uzh.ch

M. Jakob et al. (eds.), *Advances in Debris-flow Science and Practice*, Geoenvironmental Disaster Reduction,
https://doi.org/10.1007/978-3-031-48691-3_7

deposits modify valleys and endanger human life and infrastructure. Therefore, understanding causes and triggers of such processes is of interest for understanding landscape evolution (e.g., Dietrich et al., 2003), as well as for devising appropriate hazard and risk assessments in alpine regions (Horton et al., and Strouth et al., this volume) and for the design of structural mitigation measures (Piton et al., and Kwan et al., this volume).

Depending on the material composition and scientific discipline, debris flows may be regarded as a fluidized landslide, as granular flows, similar to rock avalanches, but with pores filled with a fluid, or as the end-member of intensive (fluvial) sediment transport. Hence, research on causes and triggers is located at the intersection between geoscience, hydrometeorology, hydraulics, geotechnical engineering, and granular physics, and is thus highly interdisciplinary.

In this chapter we briefly review the definitions and limits of *what is a debris flow* and general concepts for explaining their formation (Sect. 7.1), review rainfall trigger conditions (Sect. 7.2), and investigate other initiation mechanisms (Sect. 7.3). Section 7.4 describes large debris flows that were initiated by non-rainfall related processes, and resulted from flow transformation such as from rock avalanches. Finally, Sect. 7.5 provides a summary and perspectives of changing boundary conditions and research needs.

7.1.1 Characteristics and Terminology

The term debris flow relates to a flow of unsorted debris and water in steep channels (Hungr et al., 2001). There are only a few official definitions for the terminology of debris flows and related phenomena. For example, the Austrian Standards (ONR 24,800, 2009) define debris flows as one of four channel processes that can occur in a creek, having specific quantitative parameters, like a volumetric sediment concentration > 0.4, that separates debris flows from floods, bedload transport and debris floods. The inclination of source areas of hillslope debris flows has been observed at angles between 25° and 45° (Rickenmann, 2016).

The appearance of debris flows ranges from a granular flow with a high concentration of boulders > 1 m in diameter (Fig. 7.1a) to a mudflow of fine sediment and water (Fig. 7.1b). All of these share the occurrence of surges or roll waves, as well as a typical flow geometry: a distinct front that contains the highest fraction of coarse particles followed by a more dilute

(a) (b)

Fig. 7.1 **a** Example of a debris flow containing large clasts at the Gadria monitoring site, Italy; **b** example of a muddy debris flow at the Jiangia gully, PRC (courtesy Dieter Rickenmann)

flow with increased distance from the front (e.g., Pierson, 1986). Bulk densities of debris flows in motion have been recorded in the range of 1800 to 2250 kg/m^3 (e.g., McArdell et al., 2007; McCoy et al., 2010). Field and laboratory studies underline the importance of the two-phase nature of the process for flow resistance and runout patterns (e.g., Iverson et al., 2010; McArdell et al., 2007; Iverson and George, this volume).

In their classification of *flow-like landslides*, Hungr et al., (2014) define debris flows as *very rapid to extremely rapid surging flow of saturated debris*, with *strong entrainment of material and water from the flow path*. They state that a key feature of debris flows is the presence of a confined path, which is typically a first or second order drainage channel in mountain landscapes. Together with segregation processes and the development of surges (Gray, 2017), such confinement yields peak discharges that can exceed tenfold that of extreme flood events and other sediment-laden processes (Hungr et al., 2014; Rickenmann, 2016). Church and Jakob (2020) investigate the phenomena of debris floods and their transition to debris flows. In their definition, debris floods are an intensive form of bedload transport (*"floods during which the entire bed, possibly barring the very largest clasts, becomes mobile…"*). Of interest for the current chapter is that debris flows may originate from debris floods as a result of bulking, i.e. entrainment of sediment. According to Church and Jakob (2020), the process is reversible by dilution of the sediment-fluid mixture. Debris flows lacking coarse particles are often termed mudflows, a highly viscous mixture of water and suspended fine sediment. The transition between debris flows and hyperconcentrated flows is not well-defined. Debris flows composed of fine-grained volcanic tephra, pumice and ash (cf. review by Thouret et al., 2020) are often termed lahars.

7.1.2 Causes and Triggers

The formation of a debris flow is the result of a critical combination of three elements: (1) sufficient sediment, (2) steep terrain, and (3) water (Rickenmann, 2016). In mountain regions these conditions are regularly met in steep head-watersheds, when loose sediment is rapidly mobilized in the course of a rainfall event. Sediment sources include glacial, periglacial and fluvial deposits, scree slopes, soil on the hillslope, and in-channel sediment storage (e.g., Borga et al., 2014; Sklar et al., 2016). Besides precipitation, water may come from outburst floods (Breien et al., 2008; Carey, 2005) or rapid melting of snow and ice (e.g., Decaulne et al., 2005).

Rainfall is the most frequent trigger of debris flows, specifically, intensive, short-duration convective storm events (e.g., Borga et al., 2014). A recent study for the Austrian Alps showed that about 2/3 of past debris flows were triggered by convective rainfall, and 1/4 by long-lasting rainfall events. The remaining were associated with intensive snow melt (Prenner et al., 2019). Extreme precipitation in the course of tropical cyclones and typhoons has the potential to trigger numerous landslides and debris flows over wide regions (e.g. Lin & Jeng, 2000; Chen & Petley, 2005).

The effect of rainfall on a mountain landscape can be two-fold (Fig. 7.2). Firstly, rainfall infiltrates and reduces effective soil shear strength, which eventually initiates landslides on the hillslope at slopes steeper than ~15–35% and/or at the margins of channels (e.g., Take et al., 2004; Sect. 7.3.1). Subsequently, the material either fluidizes (Iverson, 2005) or is transferred to a channel where it causes a temporary blockage of the channel and eventually a sudden release (Rickenmann, 2016; Mikoš et al., 2004). Secondly, rainfall leads to runoff in steep channels and subsequent erosion and bulking of sediment (e.g., Berti & Simoni, 2005; Theule et al., 2015), local channel-bed failure (e.g., Gregoretti & Fontana, 2008), or sudden release of sediment that was temporarily stored during earlier stages of the runoff event or prior floods (e.g., sediment capacitor model of Kean et al., 2013). These initiation mechanisms are associated with the geomorphological characteristics and history of the watershed and progress from

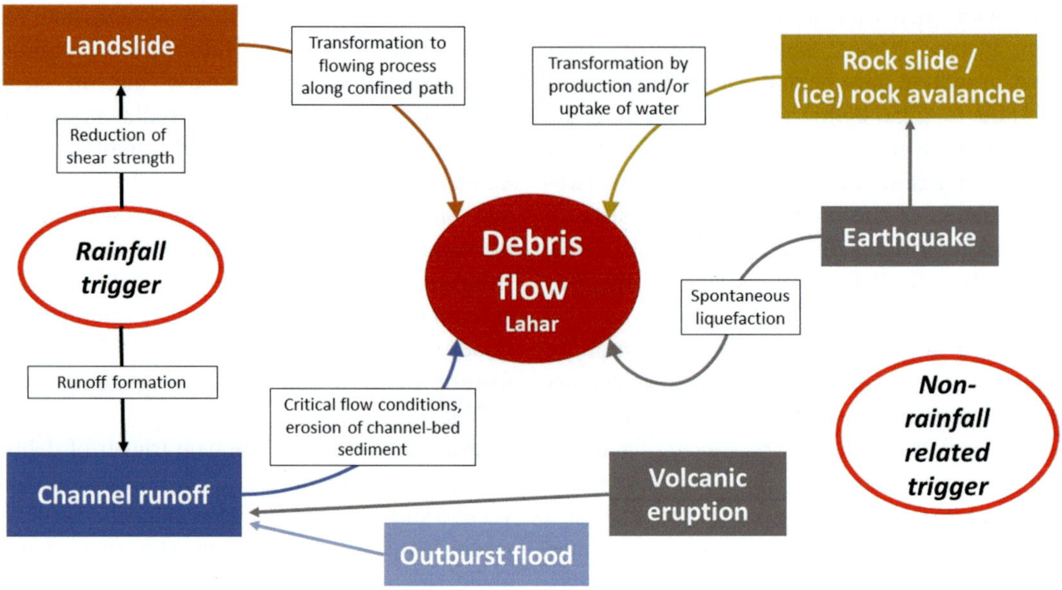

Fig. 7.2 Causes and triggers of debris flows (simplified)

hydro-geomorphic processes at the local scale during a rainfall event. On large temporal and spatial scales, the relative importance and interplay of these initiation mechanisms is currently not understood.

Large, devastating debris flows have been observed in a wide variety of geologic and climatologic settings, with trigger conditions and initiation mechanisms that can differ from those triggered by rainfall (Fig. 7.2). These rare events are often related to geologic/geomorphic activity in the broadest sense and may appear as a cascade of geomorphologic mass-transfer processes. For example, a volcanic eruption can trigger lahars (e.g., Pierson et al., 1990) or an ice-rock avalanche can transform into a debris flow by dissipative melting of snow and ice (e.g., Evans et al., 2009; Huggel et al., 2005; Fischer et al., 2018). These geology/geomorphic-related triggers are selectively described on a case-study basis in Sect. 7.4.

As discussed in Sect. 7.2, it is challenging to connect the initiation of debris flows to a specific amount or intensity of a rainfall event. The main reason for this is that there is a range of geomorphological and hydrological watershed-characteristics that affect the formation of debris flows (e.g. Bogaard & Greco, 2018). Several studies have investigated topographic attributes such as mean slope, relief, and hillslope-channel connectivity (e.g. Carrara et al., 2008; Bertrand et al., 2013; de Haas et al., this volume). Such approaches can define regionally-specific topographic signatures of catchments where mostly fluvial flows occur and catchments where mostly debris flows occur. However, these approaches cannot explain why some catchments sometimes produce fluvial flow and sometimes debris flow, even though a similar rainfall was measured.

To explain temporal and spatial variability in debris-flow occurrence, a conceptual framework was suggested by Kienholz (1995). The idea is that when the combination of basic catchment susceptibility and variable susceptibility are sufficiently high, a rainfall of certain magnitude can trigger a debris flow (Fig. 7.3). Otherwise, an intensive rainfall event may result in fluvial channel runoff. The basic susceptibility does not vary substantially over long time-scales and includes the topographic signature of a watershed, the presence of Quaternary deposits to be eroded, and geologic processes (e.g., uplift rates). This is also discussed in detail in de Haas et al. (this volume). Variable debris-flow

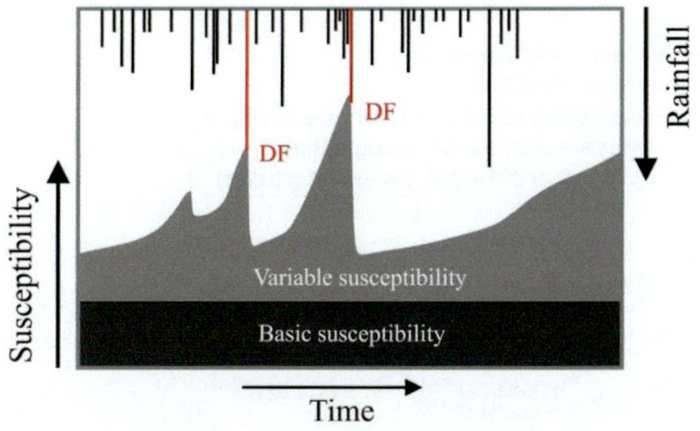

Fig. 7.3 Sketch of the disposition concept for the triggering of debris flows (modified after Zimmermann et al., 1997). Basic susceptibility is constant over geological time-scales while variable susceptibility includes factors like sediment availability, soil moisture or land-use, which may vary at time-scales relevant for hazard assessment. Red and black bars indicate triggering and non-triggering rainfall events, respectively

susceptibility represents factors like channel-sediment refill, antecedent moisture conditions, permafrost- and vegetation conditions, or changes in land use/land cover. In earthquake-prone regions, loose co-seismic landslide debris can dramatically increase the susceptibility to debris flows (Tang et al., 2012). This framework visualizes the importance of the temporal variation of sediment transfer and storage in the catchment (Bennett et al., 2014; Kean et al., 2013) and shows that climate change can have a direct (e.g., water input) and indirect (e.g., sediment production, transfer and storage) effect on debris-flow occurrence (Hirschberg et al., 2021; Stoffel et al., this volume).

7.2 Trigger

7.2.1 Rainfall as the Main Trigger for Debris Flows

The two geomorphological factors needed for debris-flow triggering (i.e., steep slopes and sediment availability) are typically met in steep headwater catchments of a few square kilometers in area (Borga et al., 2014; D'Agostino & Marchi, 2001). The third factor, water, is usually rainfall.

Rainfall amount, event duration and antecedent rainfall may all influence the pore water pressure in steep slopes, increasing the chances of triggering a debris flow (Guzzetti et al., 2008; Hu et al., 2015; Johnson & Sitar, 1990; Prenner et al., 2018, 2019; Sidle & Ochiai, 2006). Lake outbreak floods associated with low-intensity long-lasting rainfall (e.g., in combination with landslides, ice, moraine and beaver-dams ruptures) may also lead to debris-flow initiation. Nevertheless, short-duration peak rain intensity constitutes a key factor: when rain intensity is sufficiently high, it can exceed the soil infiltration capacity causing surface runoff, which eventually leads to the triggering (Church & Miles, 1987; Dunkerley, 2020; Iverson, 2000; Mostbauer et al., 2018).

It is thus not surprising to find that regional-scale debris-flow monitoring and early warning often rely on rainfall thresholds (e.g., Segoni et al., 2018 and references therein; Marchi et al., this volume). Rainfall thresholds are defined as meteorological conditions that, when exceeded in susceptible headwater catchments, are likely to result in debris flow triggering (e.g., Caine, 1980; Guzzetti et al., 2008). Following the typical approach used for landslide early warning, these thresholds often take the form of Intensity-Duration or, equivalently, Depth-Duration thresholds and are expressed as power-law relations between the event duration and average intensity or total rainfall accumulation (e.g., Fig. 7.4). This formulation requires the definition of independent rainfall events (e.g., Melillo et al., 2015) and implicitly accounts for soil moistening during the event and for the rainfall peak that practically causes the triggering. Rainfall thresholds are built either using physical or conceptual triggering models or empirically from observations (Borga et al., 2014; Guzzetti et al., 2007, 2008; Wieczorek & Glade, 2005). Owing to the dominant role of convective rainfall as a trigger of debris flows, rainfall thresholds for debris-flow triggering usually span durations between few minutes and few days, as opposed to the ones derived for landslides which may even exceed monthly

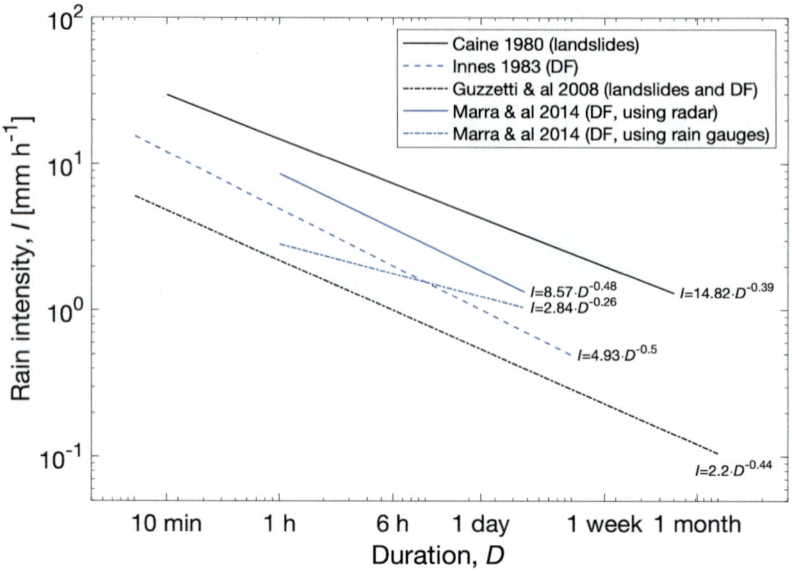

Fig. 7.4 Rainfall intensity–duration threshold for the triggering of landslides and debris flows as reported by: (i) Caine (1980), landslides, based on rain gauge data; (ii) Innes (1983), debris flows, based on rain gauge data; (iii) Guzzetti et al. (2008), landslides and debris flows, based on rain gauge data; (iv) Marra et al. (2014), debris flows based on corrected and gauge-adjusted radar data; (v) Marra et al. (2014), debris flows based on rain gauge data

durations (Borga et al., 2014; Guzzetti et al., 2008; Marra et al., 2016).

It has been found that even for small regions with similar topography, climate and geology, there is often no distinct threshold separating rainfall events leading to debris flows and rainfall events that do not (e.g., Zimmermann et al., 1997). Berti et al. (2012) proposed a Bayesian approach to improve the informative value of rainfall thresholds. This method yields conditional probabilities, i.e. the probability of an event (in our case a debris flow) conditional to another event (i.e., rainfall event of a certain intensity and/or duration). One challenge of this approach is to not only identify triggering rainfall events, but also to measure all other rainfall events in the period of consideration. Additional details on rainfall thresholds and debris-flow hazard assessment can be found in Horton et al. (this volume).

7.2.2 Convective Rainfall in Mountainous Regions

Convective rainfall occurs when abundant moisture is available in the atmosphere and the buoyant instability of air masses causes strong vertical motions. In upward motions, the moisture condenses into large droplets, which eventually fall to the ground at heavy rain intensities. This elementary system of upward and downward motions is termed a "convective cell." Convective cells form within favorable environments, grow to reach the strongest rain rates and then decay, in life cycles that last around 30–60 min. They may extend vertically up to the tropopause and their horizontal scales are of the order of a few square kilometers. Convective storms may consist of numerous convective cells and may thus last much longer than the individual convective cells and cover relatively large areas. For example, the cold fronts typically associated with the transit of low-pressure systems are characterized by numerous convective cells and cover regional scales for many hours. Nonetheless, individual isolated convective cells may form, especially in mountainous areas where the vertical air motions are enhanced by orographic lifting of the air masses and by local heating effects, such as the isolated summer thunderstorms in the Alps (Barry & Chorley, 2010).

Orography physically constrains the motion of air parcels, inducing vertical motions and turbulence, and affects precipitation processes (Bonacina, 1945; Roe, 2005). In general, these vertical motions induce an increase in the precipitation yield on the windward slopes and a decrease in the lee side slopes. Each slope experiences both effects depending the wind direction during storms, but the increasing effect usually dominates. Larger precipitation amounts are indeed observed at higher elevations, the so-called orographic enhancement of precipitation.

Unexpectedly, heavy convective precipitation does not obey this general rule. The intensity of heavy rainfall of short duration, crucial for debris-flows initiation, was found to be lower at higher elevations. This phenomenon, known as 'reverse orographic effect', was observed in several mountain ranges, including the Alps, the Italian Appenines, and the Judean mountains in the southeastern Mediterranean (Allamano et al., 2009; Avanzi et al., 2015; Marra et al., 2021; Mazzoglio et al., 2022). Orography induces turbulence in the air masses, which hinders the vertical motions, redistributing moisture within the convective cells; the precipitation yield of the cells during their lifetime remains roughly similar, but the strongest rain rates convective cells can deliver are generally lower (Formetta et al., 2022; Marra et al., 2021, 2022).

Evidence of intensification of convective precipitation due to climate change is emerging globally (Fowler et al., 2021) and is a concern in mountainous regions such as the Alps (Dallan et al., 2022; Libertino et al., 2019), as it may lead to increased probability of hazards such as flash floods and debris flows. The underlying mechanisms include: an increased water vapor holding capacity of the atmosphere at higher temperatures; temperature-driven changes in the dynamics of the convective cells, which redistribute moisture towards the core of the cells increasing their peak intensity and decreasing their areal extension; and changes in the global circulation, which may alter the frequency of meteorological systems related to debris-flow triggering (Armon et al., 2022; Berg et al., 2013; Lochbiler et al., 2017). These changes will likely influence the future probability of debris flows (Gariano & Guzzetti, 2016 and references therein).

7.2.3 Monitoring Debris-Flow Triggering Rainfall

The spatial and temporal scales of convection are a challenge for our monitoring networks (Hürlimann et al., 2019). Typical rain gauge networks in mountainous areas are too sparse for adequately capturing the scales of convective rainfall. Even in the Alps, where the density is relatively high, a typical distance between two rain gauges is in the order of 10 km, while the typical scales of convective cells are smaller than 10 km (Panziera et al., 2011; Smith et al., 2003). Additionally, rain gauges are usually located in places with easy access, such as valley floors, while the high elevations where debris flows occur (Stoffel et al., 2011) are not well sampled.

The use of rain gauges (usually a few kilometers away from the triggering locations; Nikolopoulos et al., 2014) to estimate the triggering rainfall (typically with sharp spatial variability and peaking over the triggering locations; Destro et al., 2017; Marra et al., 2014) causes a systematic underestimation of the rainfall amounts, which turns into an underestimation of the thresholds used for early warning in operational systems such as the rainfall thresholds described above, with a consequent increase of the false alarms (Marra et al., 2016; Fig. 7.4). These issues cannot be reduced with interpolation of the rain gauge data (Nikolopoulos et al.,

2015a, 2015b), so alternatives for monitoring rainfall are required. Rain gauge networks with daily temporal resolution are sometimes denser than high-resolution networks, and have longer records (Leonarduzzi & Molnar, 2020). Due to the sharp temporal variability of convective processes, however, using daily data for estimating debris-flow triggering rainfall may lead to systematic underestimations as well (Gariano et al., 2020; Marra, 2019).

Remotely sensed precipitation estimates, such as the ones provided by weather radars and satellite-borne sensors, could alleviate the sampling problems of rain gauges (Michaelides et al., 2009). Weather radars are particularly relevant, because they can provide rainfall estimates at the regional scale with spatial resolutions on the order of a kilometer and temporal resolutions of about ten minutes, which are optimal for capturing convective cells (Borga et al., 2022). For example, it was shown that, with respect to rain gauges, weather radars could improve the detection of heavy rainfall at hourly scales by almost six-fold (Lengfeld et al., 2020). Nonetheless, weather radars also have limitations. Accuracy of the radar-based rainfall estimates is often low, especially in mountainous regions, due to echo signals caused by the mountains, blockages of the radar beam, and mixed phase precipitation (Germann et al., 2022). Additionally, attenuation of the radar beam is frequently observed in the heavy convective rainfall typical of debris flows, with important decreases in the estimation accuracy. Dedicated correction procedures are necessary to quantitatively use weather radar estimates of debris flow triggering rainfall (Marra et al., 2014). Operational systems based on weather radar rainfall estimates exist, but are still not common (e.g., Abanco et al., 2016; Piciullo et al., 2018; Postance et al., 2018; Tiranti et al., 2014).

Several satellite-based rainfall products have quasi-global coverage and can reach spatial resolutions of a few kilometers and temporal resolutions of about an hour (Kidd & Levizzani, 2022). In addition, satellites cover ungauged regions for which no rain gauge or radar coverage is available (Kidd et al., 2017). Indeed, satellite applications are gaining momentum in the monitoring of landslides (Kirschbaum & Stanley, 2018; Kirschbaum et al., 2012; Rossi et al., 2017), but not much is yet available for the specific case of debris flows, probably because of the noise in satellite estimates at the fine spatiotemporal scales typical of debris-flow triggering. A preliminary study focusing on debris flows in the Alps showed ~30% deviations of the warning levels estimated from satellite data with respect to the ones obtained from rain gauges (Nikolopoulos et al., 2017), but no other study specific on debris flows was conducted since.

7.3 Initiation Mechanisms

7.3.1 Landslides

While debris flows can initiate in a variety of steep topographic settings, perhaps the most studied initiation mechanism is mobilization from landslides (Gabet & Mudd, 2006; Iverson et al., 1997; Johnson, 1984). The transition from a landslide to a debris flow can be sudden or gradational and the two types of events are differentiated by their failure style (Selby, 1993). Landslide motion tends to be more rigid; the failure occurs along a defined plane or surface (where deformation is localized), material within the failure can move as a discrete unit, and the relict structure of the hillslope can be preserved (Bierman & Montgomery, 2014; Iverson et al., 1997). Debris flows, in contrast, have more fluid-like deformation (Nagl et al., 2020), where both solid and fluid forces strongly influence downslope motion (Iverson, 1997). Observations from the field, laboratory experiments, and theory suggest that debris flows transition from landslides when there is widespread failure within the landslide mass, high pore pressures develop due to saturation/liquefaction/undrained loading of the mass, and/or there is conversion of translational energy to vibrational energy (e.g., Ellen & Fleming, 1987; Iverson, 1997; Iverson & LaHusen, 1989; Iverson et al., 1997).

7.3.1.1 Theory

A common framework for assessing when debris flows initiate from landslides involves the Coulomb failure criterion (Iverson et al., 1997). This was first described by Coulomb (1973) and has been used extensively in soil and grain-flow mechanics ever since (e.g., Iverson & Denlinger, 2001). It has been shown through field and experimental data that failure of a soil mass occurs when

$$\tau = \sigma\prime \tan\phi + c \quad (7.1)$$

where τ is the mean shear stress acting on the failure surface, σ' is the effective normal stress acting on the failure surface, ϕ is the friction angle of the soil (due to both surface friction within the soil and clast interlocking), and c is the apparent cohesion (Lambe & Whitman, 1991). Cohesion varies depending on the system and can be due to forces between clay particles within the soil (Ikari & Kopf, 2011), (bio) cementation (Cui et al., 2017), roots and vegetation (Zhang et al., 2010), and even ice (Arenson & Springman, 2005). The effective normal stress, σ', is defined as

$$\sigma' = \sigma - p \quad (7.2)$$

where σ is the total stress that would be measured normal to the failure surface and p is the pore water pressure (Terzaghi, 1936). It should be noted that Eqs. 7.1 and 7.2 are simplifications, and do not account for stress and pore pressure fields (i.e., the spatial distribution of stress and pore pressure in the x, y, and z directions). A simple way to estimate the pore-pressure distribution in a soil (i.e., $p(x,y,z)$) is to relate it to the distribution of hydraulic head (i.e., $h(x,y,z)$), which drives groundwater and subsurface water flow (Bear, 1988). Pore pressure is related to the hydraulic head via

$$p = \rho g(h + z) \quad (7.3)$$

where ρ is the density of water, g is gravity, and z is oriented downward, in the direction of gravity (and conventionally defined as z=0 at the ground surface). Thus, assessing failure of a soil mass requires one to know, in theory, the flow field through the mass to know the pore pressure distribution.

However, while the Coulomb criterion defines failure along a discrete slip surface, it does not indicate when a hillslope failure will transition into a debris flow. Theory suggests that when there is little to no cohesion, and the vertical pore pressure gradient equals the vertical gradient of the total normal stress, then there is zero effective stress in the soil mass and the soil can flow like a liquid (i.e., liquefaction). These elevated pore pressure distributions can occur due to several reasons in nature, including during transient events like earthquakes and landslides, where changing stresses can disrupt soil structures, flow paths, and cohesive bonds (e.g., Sassa et al., 1996), or during contraction of densely packed soils under shear, where pore-space reduction can occur more quickly than the rate of fluid-pressure dissipation (Goren et al., 2010; Iverson, 2005; Iverson & LaHusen, 1989). Within the soil mechanics community, it has been suggested that debris flow mobilization occurs only in loose soils. These are soils where the *in-situ* bulk densities are less than a "critical state" density (Wood, 1990), and upon contraction (either during soil failure or due to a rock slide, slump or debris avalanche impacting a channel or torrent deposit (Sassa, 2005)) pore pressures are driven up as the pore water cannot dissipate. Geotechnical engineers sometimes refer to this behavior as undrained loading (Terzaghi & Peck, 1948).

Agitation of soil particles and non-cohesive sediment can also lead to enhanced fluidity of a landslide and transformation into a debris flow. Within the granular physics community, the measure of the degree of agitation (or the grain fluctuation kinetic energy) is commonly referred to as the granular temperature and is analogous to molecular kinetic energy that determines the temperature of a gas (Iverson, 1997). Higher granular temperatures correspond to a reduced solids concentration per unit volume and less grain-grain interlocking and they are maintained by the conversion of translation energy (e.g., from a landslide failure or the downslope movement of a debris flow). This energy input is necessary as grain interactions and fluid flow between pores contributes to energy dissipation.

As noted by Iverson et al. (1997), the concept of granular temperature differs in important ways from acoustic fluidization (Melosh, 1987) and dispersive stress (Bagnold, 1954). Acoustic fluidization occurs when elastic waves propagate through a granular medium and cause a reduction of grain-grain contacts and permit flow. Development of granular temperature, however, does not require waves but spreads through conduction (Campbell, 1990). Dispersive stress was a concept put forth by Bagnold (1954, 1966), where he argued that in high sediment concentration flows grain layers must dilate and disperse to move past one another and this dispersion is maintained by grain-grain interactions (i.e., a dispersive stress or pressure). His conceptual and physical models, however, assume a gravity-free environment and a uniform stress field (Iverson & Vallance, 2001), assumptions that are not realistic for natural flows and can be relaxed when assessing granular temperatures.

7.3.1.2 Modeling

Quantitative models for debris-flow mobilization from hillslope failures and landslides generally use the "infinite slope" concept, where Coulomb failure of a soil mass occurs on an inclined plane whose lateral dimensions (i.e., length and width) are much greater than the soil thickness (e.g., Selby, 1993). As such, all physical properties vary only in the direction normal to the slope and all stresses can be calculated from statics (and soil deformation behavior can be ignored). It is also often assumed that sufficient water exists to fully saturate pore spaces. One of the simplest models following the "infinite slope" concept of debris-flow mobilization is Johnson's (1984) one-phase Bingham model. This model assumes that soil fails when shear stresses along a plane exceed a critical yield strength, which is equal to the Coulomb strength defined by Eq. 7.1. This model does not account for changes in physical properties (i.e., soil friction, porosity, pore pressures) as failure proceeds since the solid and fluid phases are not distinctly accounted for (Iverson, 1997). As such, while this model predicts a critical thickness for the initial landslide failure, the slab would need to thicken as failure progresses to keep the mass above the yield strength and allow it to flow, which is rarely observed (Ellen & Fleming, 1987).

Models that account for the two-phase nature of debris flows rely on principles from soil mechanics (Lambe & Whitman, 1991) and granular physics (Campbell, 1990; Savage, 1998). In many of these models, the main variable being adjusted is the pore-water pressure (and the cases under which it becomes elevated above hydrostatic), as partial to full liquefaction is assumed to be a necessary condition for the transformation from landslide to debris flow (Ellen & Fleming, 1987; Iverson et al., 1997). Iverson and Major (1986) developed an analytical model to predict the limiting stable slope for an isotropic soil mass subject to steady, uniform Darcy flow of arbitrary magnitude and direction. They show that minimum slope stability occurs when a component of groundwater flow is oriented vertically upward. Other models focus on the competition between soil drainage (or the rate at which pore pressures dissipate) and soil contraction (or the rate at which pore pressures are generated) as granular materials shear. In soil mechanics, constitutive models for soil failure of loose soils have been developed using the concept of critical state soil mechanics, where at the critical state shear deformation occurs without dilatancy, driving pore pressure up (Casagrande, 1976; Ellen & Fleming, 1987). Two-phase flow models (based on concepts from granular physics and soil mechanics) suggest, however, that debris-flow behavior occurs due to continuously evolving grain rearrangements and pore pressure feedbacks which determine landslide versus debris flow-like behavior (Iverson & George, 2014).

7.3.1.3 Field and Lab Observations

While theory and modeling provide predictive frameworks to understand the transition from landslides to debris flows, observations from the field and/or laboratory experiments are key to test and calibrate these models. For example, after a large rainstorm in the San Francisco Bay Area (USA) in 1982, thousands of shallow landslides occurred and mobilized to different degrees—in some cases the soil masses completely mobilized into debris flows, in others there was partial mobilization and debris flows issued from the failure toe or debris flows did not occur at all (Ellen & Fleming, 1987). Soil analyses suggested that both the style and rate of failure was strongly controlled by whether the soil was contractive (which allowed for complete liquefaction and debris-flow occurrence) or dilative (and slide movement did not produce a debris flow). In contrast, Reid et al. (2003) documented debris-flow initiation from the toe and margin of the large Cleveland Corral landslide complex in the Sierra Nevada Mountains (California, USA) and found that flows were generated due to locally steepened ground (35 to 42°), elevated pore pressures, and dilated soils. The large Oso failure event in Oso, Washington (USA) produced a debris flow at its leading edge when the landslide mass incorporated water and wet floodplain sediment from the North Fork Stillaguamish River (Iverson et al., 2015). A similar observation was made in the Fukaminato River region in Japan, where a landslide-induced debris flow occurred after incorporation of water and fine-sediment, allowing it to travel ~1.2 km downstream (Nishiguchi & Uchida, 2022). Prior work using ring-shear tests in this region suggests that when landslides occur, they occur under undrained loading conditions (Wang et al., 2002). Hürlimann et al. (2015) monitored hillslope failures across Switzerland for over two decades, and based on their field and laboratory observations, they concluded that water content (and to a lesser degree clay content) was the dominant control on whether debris flows formed from landslides. They found that even small increases in water content led to considerably longer runout distances. While differences in soil type, slope, fine-grained sediment, and size of the initial landslide failure differs across these field sites, in almost all cases, as predicted by theory and modeling, partial to full liquefaction of the sediment mass seems to be a necessary ingredient to transform landslides into debris flows (Iverson et al., 2011, 2015; Wang & Sassa, 2003).

As debris flows are episodic in nature, and difficult or dangerous to study in the field, large-scale flumes and artificial hillslopes have been used to supplement field observations (e.g., Iverson, 1997; Iverson et al., 1997; Kaitna et al., 2014, 2016; Palucis et al., 2018; Prancevic et al., 2014; Takahashi, 1978). The United States Geological Survey (USGS) operates a 95 m long and 2 m wide flume, which allows for controlled laboratory experiments on landslides and debris-flow mobilization (Iverson, 1997, 2015). The flume has a slope of 31° and has a retractable retaining wall at its upper end, allowing for the storage and release of soil. To simulate landslide failure and debris-flow runout, two experiments were performed where water was systematically added to pre-characterized non-densified soil until failure occurred. Matric potential, water content, pore pressure, soil-surface displacement, flow depth, and basal stresses were measured pre- and during failure (Iverson et al., 1997). In Experiment 1 water was added as groundwater and in Experiment 2 it was added by sprinkling (i.e., simulating rainfall). Two distinct modes of mobilization were observed, even though soil properties and water contents were similar among the two experiments. Under saturation by groundwater (Mode I), liquefaction was almost spontaneous, pore pressures rose quickly during failure, and a debris flow was instantly generated. During sprinkling (Mode II), failure was slower, involved retrogressive slumping, and debris-flow motion did not occur until the soil mass moved downslope and translational energy could be converted to granular temperature (Iverson et al., 1997). The failure differences between these two modes was observed to be linked to differences in the distribution of elevated pore pressures within the soil mass, which affected the location and characteristics of the failure surface. Specifically, in Mode II, the slip surface was only partially saturated, whereas in Mode I, the entire failure surface was saturated.

Using a smaller flume under a range of slopes (20 to 43°), Prancevic et al. (2018) showed that while landslide failures occurred under near-saturated conditions, increased water levels from those predicted by theory were needed to initiate failure due to boundary stresses from granular contacts. They also showed that the thickness of the failure was proportional to sediment size, and that this grain size dependence supports the observation of increased debris-flow activity after finer-grained sediment is deposited on otherwise coarse channel beds (e.g., Lamb et al., 2011). Kaitna et al. (2016) used a large rotating drum to probe the role of water content, clay content, and grain size distribution on flow mobility; while a unique geometry, the experiments showed the generation of excess pore pressure as debris was sheared for mixtures with wide grain size distributions and abundant fines. These types of experiments are important for understanding how elevated pore pressures are generated from positive or negative dilation at the onset of landslide failure. While experimental data such as these supports field observations that near complete liquefaction of a sediment or soil mass can cause landslides to transform into debris flows, they also point to the importance of grain-grain interactions and granular temperature within dense flows.

7.3.2 Channel Erosion

7.3.2.1 Theory and Flume Studies

A second common class of mechanisms that have been proposed for debris-flow initiation is erosion within stream channels (or regions of steep topographic convergence on hillslopes), where erodible sediment can accumulate. These flows are likely more common in steeper mountain ranges or post-fire landscapes, where hillslope sediment supply may be limited. Perhaps one of the earliest in-channel debris-flow initiation mechanisms was put forth by Takahashi (1978), who proposed that runoff leads to bulking and failure of the channel bed at depth. His model assumes that Coulomb failure of a sediment bed with slope-parallel seepage occurs when surface water causes high enough stresses to exceed Coulomb resistance within a granular mass (Takahashi, 2007). For the case where the overlying shear stress of the water and the resisting stress of the sediment are linearly distributed, the bed angle, θ, where failure will occur can be estimated from

$$\tan\theta = \left\{ \frac{C(\rho_s - \rho)}{C(\rho_s - \rho) + \rho(1 + \frac{h}{a})} \right\} \tan\varphi \tag{7.4}$$

where C is the grain concentration (by volume) of the static sediment bed, ϕ is the internal friction angle of the sediment, h is the surface water flow depth, and a is the distance from the bed surface to the failure plain. Takahashi (1978) showed experimentally that if the failure depth was equal to or greater than 0.5h, then the grains could disperse uniformly through the flow and move as a debris flow. His model thus states that a debris flow will emerge under runoff conditions when

$$\tan\theta \geq \left\{ \frac{C(\rho_s - \rho)}{C(\rho_s - \rho) + 3\rho} \right\} \tan\varphi \tag{7.5}$$

This model neglects the role of soil liquefaction (as well as alternate pore pressure distributions), and as such, predicts failure on hillslopes lower than observed (~25° for most soils versus observed failures between 25 and 45°) (Iverson et al., 1997). However, most channels tend to be less steep than hillslopes, and can concentrate subsurface and surface flow, which adds additional destabilizing forces (Berti & Simoni, 2005; Dunne, 1990).

Prancevic et al. (2014) recast the Takahashi (1978) model in terms of a critical Shields stress for debris flows, τ_c^*,

$$\tau_c^* = (1 - \eta)(\tan\varphi - \tan\theta) - \frac{\rho}{\rho_s - \rho}\tan\theta \tag{7.6}$$

where η is porosity. By setting this model equal to a transport model for initial motion of sediment by fluvial processes (e.g., Lamb et al., 2008), one can theoretically predict the threshold slope or bed angle for initial sediment motion by debris flows for given failure plane and grain-pocket friction angles (Prancevic et al., 2014) (red dashed line, Fig. 7.5).

To test whether a threshold slope exists for in-channel debris flows, Prancevic et al. (2014) experimented in a 5-m-long tilting flume with bed slopes ranging from $\theta = 3$ to 65% (~2 to 33°) and a uniform grain size (median diameter, $D = 1.5$ cm). They found that for slopes less than ~40% (22°) sediment was mobilized by fluvial or river processes; for slopes greater than ~40%, the channel bed failed and initiated a debris flow. To probe whether debris flows could be initiated under high shear stresses ($\tau^* \rightarrow 1$) and slopes less than the threshold slope, Palucis et al. (2018) used a tiltable 12-m-long recirculating flume. They showed that under steady flow conditions, dense granular sheetflow occurred—a hybrid between traditional river transport and debris flows (Fig. 7.5), akin the immature debris flow described by Takahashi (1991). In coarse-bedded channels where debris flows are observed to initiate, shear stresses exceeding the threshold for motion (i.e., $\tau^* > 0.1$) are rare (e.g., Parker et al., 2007). Thus, it is more likely that high shear stresses are generated during flood pulses or due to the input of finer grained sediment which augment buoyancy through increases in fluid density (Lamb et al., 2011; Palucis et al., 2021; Prancevic et al., 2014).

7.3.2.2 Field Observations

While the Takahashi (1978) model provides a quantitative framework for debris-flow initiation within channels and has been tested within laboratory flumes under simplified conditions (i.e., steady flow, uniform grain sizes, planar beds), other mechanisms have been proposed. Possible mechanisms include collapse within the channel bed and adjacent banks that leads to the transformation of clear water flow into a debris flow (Gregoretti & Fontana, 2008; Takahashi, 1978), progressive fluvial entrainment and bulking (Gabet & Bookter, 2008), or channel morphology or spatial variability in flow (e.g., jets and pools, waterfalls) that may induce rapid sediment entrainment and failure (Berti et al., 1999;

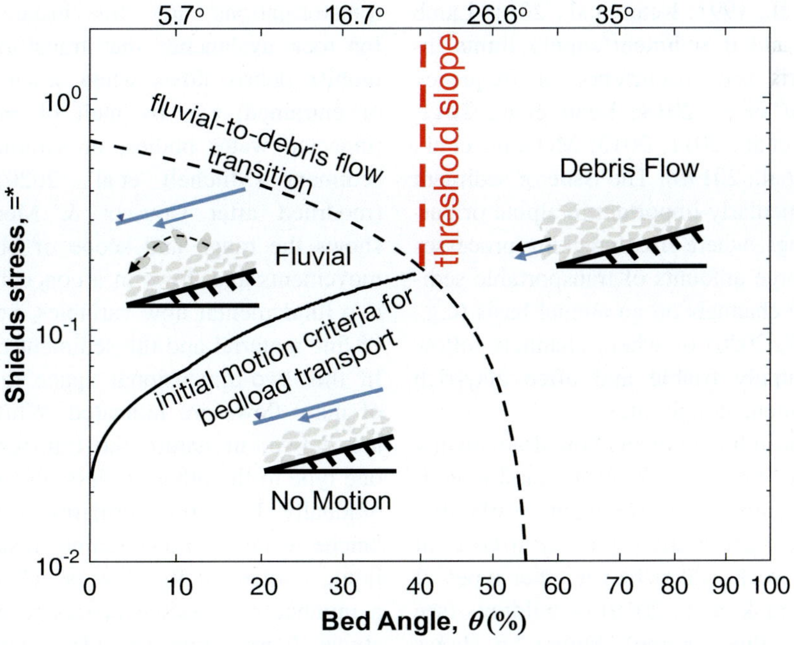

Fig. 7.5 Shields diagram with zones of sediment motion—phase space boundaries are based on models for initial motion criteria for bedload transport and the fluvial-to-debris flow transition (modified from Palucis et al., 2018)

Cannon et al., 2003; Costa, 1984; Kean et al., 2013). The latter mechanism is often referred to as 'the fire-hose effect', in which overland flow becomes concentrated within steep bedrock channels and fluidizes sediment.

The wide range of in-channel debris-flow initiation mechanisms, their sporadic occurrence, and the steep landscapes on which they occur, means that there are limited direct observations of where (e.g., Kean et al., 2013; Staley et al., 2017) and under what precipitation conditions (e.g., Staley et al., 2017; Tang et al., 2019a) in-channel debris flows occur in natural systems (e.g., Berti et al., 2000, 2020; Hirschberg et al., 2021; McCoy et al., 2010; Suwa et al., 2009). Most observations from debris flow monitoring are in higher order channels (e.g., near canyon outlets or downstream of where debris flows initiated) (e.g., Marchi et al., 2002; Hürlimann et al., 2003; McCoy et al., 2010; Suwa et al., 2011); direct observations of debris-flow initiation zones in low order channels are rare (Palucis et al., 2021; Tillery & Rengers, 2020). Therefore, it often remains unclear whether a given flow event initiated from a hillside or in a low-order channel (DiBiase et al., 2017; Florsheim et al., 1991; Kean et al., 2011; Lamb et al., 2011) and if sediment supply limits in-channel debris flow occurrence or frequency (e.g., Bennett et al., 2014; Kean et al., 2011, 2013; Lamb et al., 2011, 2013; McGuire et al., 2017; Tang et al., 2019b). The issue of sediment supply is particularly important in alpine or periglacial settings, where frost cracking processes can deliver large amounts of transportable sediment to steep channels on an annual basis (e.g., Rengers et al., 2020) or where channels follow faults that supply friable and often clay-rich granular sediment at high rates.

Recent research documents how dramatically debris flow activity (i.e., frequency and magnitude) can change when sediment availability is abruptly increased after the occurrence of a rock fall or rock avalanche (or other types of landslides) (Frank et al., 2019) or wildfires (see Gartner et al., this volume). Figure 7.6 shows a systematic analysis of debris-flow activity in six catchments in Switzerland before and after a rock fall or landslide occurred. In all catchments an abrupt increase in debris flow activity was observed although the exact behavior differs from catchment to catchment. In most cases a strong increase is observed during the weeks and months after the event(s), with a leveling off after a certain time. How long the process of increased activity continues is of course highly relevant in terms of hazard prevention and depends on characteristics of the catchment and the sediment input. As analyzed elsewhere (e.g., on volcanoes) the abundance of fine sediments facilitates the initiation of debris flows (see Vallance, this volume).

7.4 Cascading Debris Flow Processes

As outlined in the previous sections, numerous preparatory factors and processes exist that trigger debris flows. Recent events and experiences have fueled the discussion about debris flows resulting from flow transformations from initial slope instabilities with flow characteristics different from debris flows. Common flow transformations are, for instance, observed for rock avalanches that transform into more mobile debris flows when water is produced or entrained, e.g. by melt of snow and ice, impacted water bodies, or saturated entrained sediment (Mitchell et al., 2020). Figure 7.7 (modified after Coussot & Meurnier 1996) shows the range and scope of observed mass movements and flows in a conceptual sketch of two fundamental flow variables, i.e. the content of fine material and the sediment concentration. In this two-dimensional space different types of mass flows are indicated. While solid lines are drawn, in nature the transformation from one type to the other are less obvious and occur gradually. Flow transformation from rock avalanche to debris flows mainly results when the liquid content of the flow is enhanced through a number of possible processes as mentioned above. It may, however, also involve a change of the grain size of the flow, as rock avalanche material comminutes along the flow trajectory

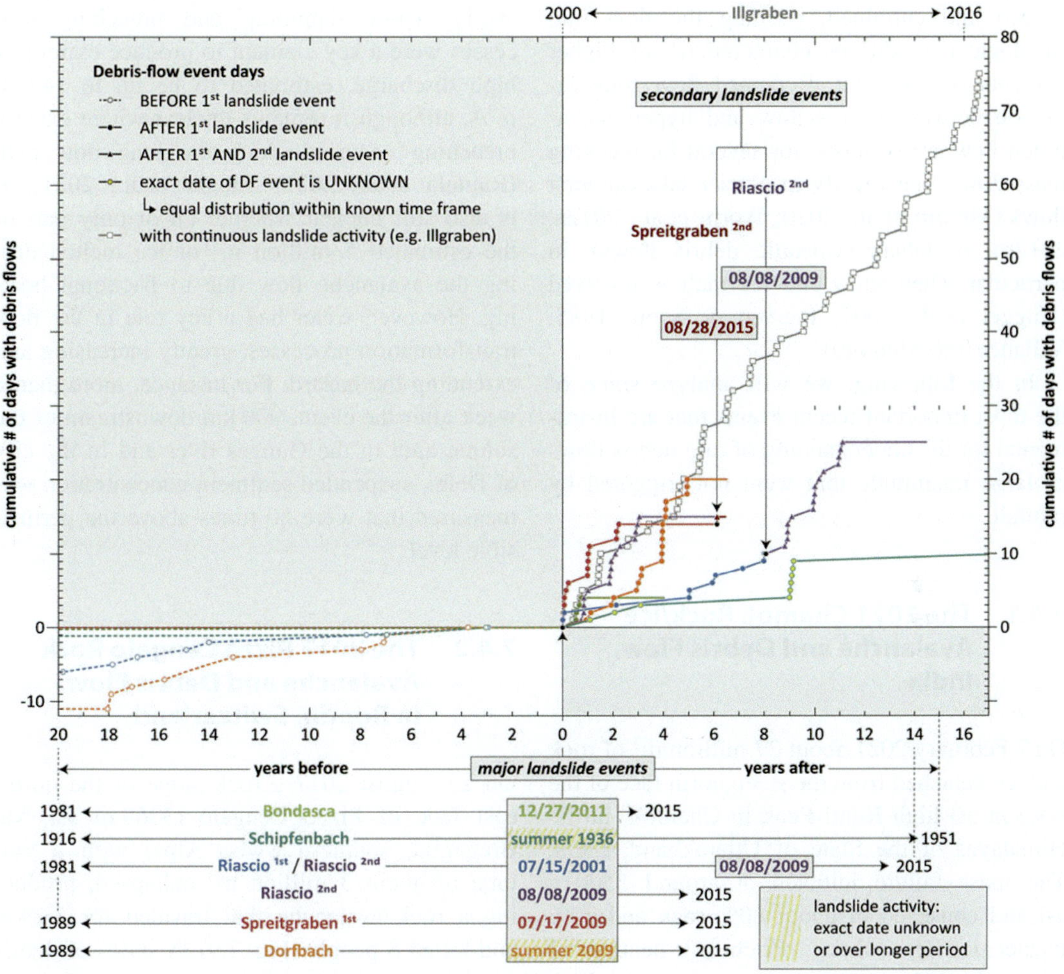

Fig. 7.6 Debris-flow activity (cumulative number [#] of days with debris-flow activity) before and after large landslide events in the five (resp. six) catchments. The day of the landslide event is defined as year = 0 (Frank et al., 2019)

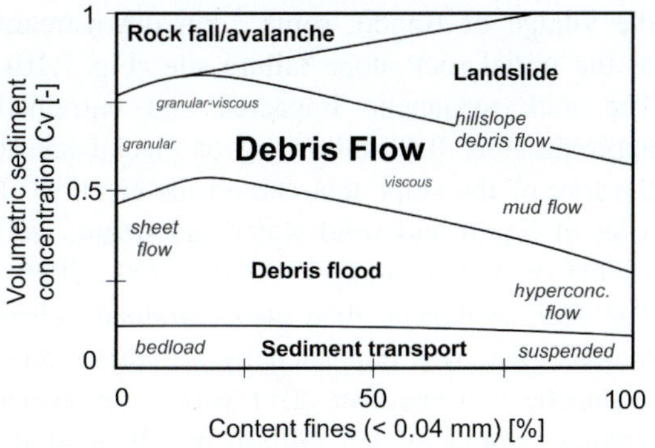

Fig. 7.7 Mass flow types as defined by sediment concentration and content of fine material (modified after Coussot & Meunier, 1996)

due to particle collision and additional entrainment of finer-grained sediment along the flow path.

Furthermore, and equally important, debris flows may also result from flow transformation from more liquid flow types such as hyperconcentrated flows (cf. Church & Jakob, 2020). Section 7.3.2 describes the physical conditions under which erosion occurs in channels as a function of bed slope and acting stresses (Fig. 7.5). In the field, flows with lower sediment concentration than debris flows can transform into debris flows when passing through steeper channel sections where sediment is eroded and entrained, pushing the flow into rheological conditions characteristic of higher contents of sediment. Repeated flow transformation between debris flow and hyperconcentrated flow are common for several far-reaching mass flows, specifically in glacier lake outburst flows (Mergili et al., 2020; Worni et al., 2012a, 2012b), or lahars (volcanic debris flows), in particular when snow and ice melt is involved (Huggel et al., 2007; Pierson & Scott, 1985; Vallance, this volume)).

In the following, we will analyze some of the most important recent events that are instrumental for the understanding of rare debris flows of large magnitude that were not triggered by rainfall.

7.4.1 The 2021 Chamoli Rock/ice Avalanche and Debris Flow, India

On 7 February 2021 about 27 million m^3 of rock and ice detached from the steep north face of the 6063 m asl high Ronti Peak in Chamoli, in the Himalayas of the State of Uttarakhand, India. The mass failure initiated at around 5500 m asl and consisted of about 80% rock and 20% glacier ice by volume. Frictional heating of the ice in the avalanche produced water which then facilitated the transformation to a more fluid flow (Shugar et al., 2021; Schneider et al., 2011). At the confluence with the Rishiganga River, debris consisting of a mixture of ice blocks, rock fragments and material derived from moraines was deposited and blocked the Rishiganga River valley, damming the river flow and forming a lake. A flow with characteristics of a debris flow (based on the sedimentary field record) traveled farther downstream into the Rishiganga River, and joined the Dhauliganga River. This event killed more than 200 people and destroyed important hydropower structures (Fig. 7.8). Field evidence suggests repeated historic debris deposits and damming along a flow path of more than 20 km, with breaching of dams and sudden release of impounded water (Bhardwaj & Sam, 2022; Rautela et al., 2021). These damming and breaching processes were a key element to produce extremely high discharge, estimated to be up to 14,000 m^3/s, although it remains unclear where exactly breaching occurred along the long flow path (Rautela et al., 2021; Shugar et al., 2021). It is also still unclear whether all or only part of the estimated 5 million m^3 of ice melted during the avalanche flow due to frictional heating. However, water had a key role in the flow transformation processes, greatly increasing and extending the hazard. For instance, more than a week after the event, 900 km downstream of the source area in the Ganges river and in the city of Delhi, suspended sediment concentration was measured that were 80 times above the permissible level.

7.4.2 The 2017 Pizzo Cengalo Rock Avalanche and Debris Flow in Bondo, Switzerland

On 23 August 2017 a rock slope of the northeast face of Pizzo Cengalo (3369 m asl, Val Bregaglia, southern Swiss Alps) with a volume of about 3 million m^3 collapsed, producing a rock avalanche that traveled for 3.2 km and killed 8 people (Fig. 7.9). A series of debris flows followed the rock avalanche: the first one only was delayed by only 30 s, and 10 other debris flows occurred within 9.5 h (Walter et al., 2020). Both the rock avalanche and debris flows caused substantial damage to infrastructure and the village of Bondo, some 7 km downstream of the initial rock slope failure site (Fig. 7.10). The rock avalanche impacted and entrained approximately 0.6 million m^3 of glacial ice at the foot of the slope that caused the ejection of a jet of liquid and solid water and steam, well visible on video footages (Walter et al., 2020). The rock avalanche then also entrained debris material that was deposited by an earlier rock avalanche in December 2011 with a somewhat smaller volume of 1.5 million m^3 (Baer et al., 2017). The debris flows that formed immediately (30-s delay) after the rock avalanche and during the following hours originated from the

Fig. 7.8 The Chamoli debris flood on 7 February 2021 reaching settlements and hydropower infrastructure in the Dhauliganga river valley, producing extremely high discharges. The flood rushes down the river bed and shows a mixed appearance resulting from water, sediment and rock involved in the flow with repeated splashes (picture: Carbon Brief)

rock avalanche deposit area; two more debris flows were observed two days after the rock avalanche, all in absence of rainfall. Walter et al. (2020) suggest that the debris flows were triggered by dynamic loading on saturated sediments, causing excess pore pressures maintained over hours and days, and thus enabling debris-flow initiation. This coupling of quasi-immediate to delayed debris flows and rock avalanche has rarely been observed and the exact processes involved are still not well understood.

7.4.3 Flow Transformations in Volcanic Environments

Volcanoes are the origin of many well-known, often devastating debris-flow events and disasters. The term lahar is commonly used to describe debris flows in volcanic environments but is often also applied to hyperconcentrated flow types. Discussion of volcanic mass flow types and their transformation can be found in Pierson and Scott (1985), Scott et al. (2001), Thouret et al. (2020) and Vallance (this volume). Lahars often have exceptional high magnitude and runout, in some cases with peak discharge up to several 10,000 m^3/s, volumes of 100 million m^3 or more (see Jakob, 2005), and travel exceeding 100 km (Huggel et al., 2007; Thouret et al., 2020; Worni et al., 2012a). Primary lahars are triggered by volcanic eruptions while secondary lahars are triggered by non-eruptive processes such as heavy rainfall events causing the initiation of a debris flow by mobilizing volcanic sediment on the slopes of a volcano. Ice-clad active volcanoes pose particular hazards because eruptive processes can melt snow and ice and produce particularly large and far-reaching lahars (Huggel et al., 2007; Pierson et al., 1990; Tuffen, 2011). In this respect, one of the most severe and best studied lahar

Fig. 7.9 Transition zone from the 2017 rock avalanche deposits to the debris flow trajectory. On the rock avalanche deposits incised debris flow channels are visible (photo: swisstopo)

disasters worldwide was triggered by the 1985 eruption of ice-capped Nevado del Ruíz volcano in Colombia which claimed the lives of over 20,000 people in the city of Armero (Pierson et al., 1990). The paradox of the Ruíz disaster is that a relatively small eruption led to one of the world's largest volcanic and debris-flow disasters. Pyroclastic density currents generated large amounts of melt water as hot debris mechanically mixed and entrained snow and ice from the volcano's ice cap. Dilute flows rapidly transformed into debris flows in the river channels draining the volcano by entrainment of sediment (Huggel et al., 2007). A peak discharge of 48,000 m^3/s was reached 9 km downstream of the source, and total flow volume was estimated at 90 million m^3 (Pierson et al., 1990).

Figure 7.11 shows the various ways in which flow transformations can result in lahars, distinguishing between eruption and non-eruption induced flows. Intermediate mass flow types than can transform to lahars include lava flows, pyroclastic flows (density currents) or ice and rock / debris avalanches. The production of meltwater through heat transfer and liquefaction play a key role for lahar generation on ice-capped volcanoes.

For flow transformations to, or from, debris flows several factors are decisive: (i) variability in the flow-path topography, (ii) availability of liquid water, and (iii) availability of sediment. Variability in topography influences sediment erosion and deposition and flow velocity; simply speaking, flows entrain sediment on steeper slopes and deposit on shallower slopes (Iverson et al., 2011). Depending on topography and sediment availability flows may repeatedly transform between debris flows to debris floods, such as those observed from glacier lake outburst flows along long-extending flow paths (Mergili et al., 2020; Worni et al., 2012b). Channel geometry variations can modulate flow transformation processes, e.g. from confined to unconfined flows.

Fig. 7.10 Debris-flow runout zone in the village of Bondo. Several buildings and infrastructure were flooded and heavily damaged. Photograph shows heavy machines removing the deposited sediment to create retention space for later debris flows (photo: swisstopo)

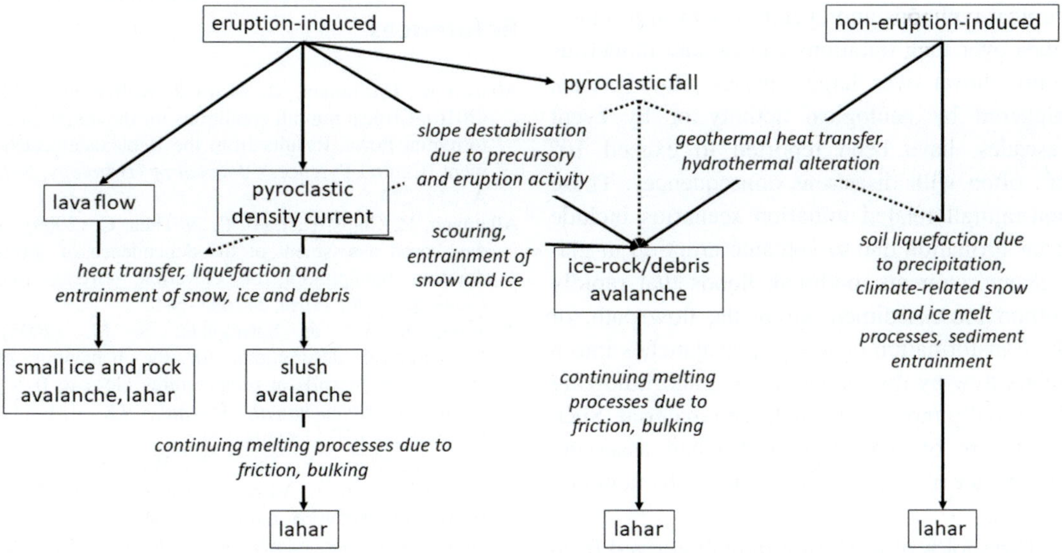

Fig. 7.11 A generic scheme of triggering and flow transformation processes resulting in lahars (volcanic debris flows) on ice-capped volcanoes. Dotted arrows indicate volcanic deposits that facilitate mobilization in the subsequent onset of flow (modified after Huggel et al., 2007)

Availability of water is not only crucial for initiating debris flows but also for controlling flow transformation. Several events over the past years (including those described above) have raised questions about the sources of water required to produce the observed flow types, and in particular transformation to highly mobile flows that largely extend the area of impact and destruction. Observations indicate that the role of entrainment of water-saturated sediment as a factor for transformation to debris flows (or in rare cases possibly even for initiating debris flows, cf. Pizzo Cengalo case) has been underestimated. More research in this field is critical because of the large implications it has for hazard analysis.

7.5 Summary and Perspectives

The causes and triggers of debris flows can be as diverse as the composition and event magnitude of the process itself. Debris flows occurring in mountain regions on a regular basis typically show event volumes between 10^3 and 10^5 m3 and are triggered by rainfall—dominantly convective storms—but also by long-lasting precipitation and snow-melt. Precipitation during tropical typhoons and cyclones with high intensities over long durations can initiate numerous debris flows over large regions. Rare events, triggered by geological activity or by event cascades, have been reported to exceed 10^6 m^3, often with disastrous consequences. These non-rainfall related initiation scenarios include lahar formation due to volcanic eruption in glacial environments, outbreak floods that rapidly entrain loose sediment along the flow path, or the transformation of ice-rock avalanches into a debris flow by dissipative heating and fluidizing the initially more granular frozen material. Such events are of special interest when assessing the impact of climate change in high mountain regions.

Thresholds of triggering rainfall derived from monitoring networks are systematically underestimated, due to the small spatial and temporal scales of convective cells. This underestimation propagates to the rainfall thresholds often used for regional early warning. Weather radars could improve our ability to detect heavy rainfall, but extensive work is needed to obtain quantitatively reliable rainfall estimates over a region of interest.

Debris flows are typically initiated by a reduction of shear resistance of loose material at the hillslope or along the channel (landslide) or by channel runoff that leads to progressive erosion or failure of the channel-bed material (channel erosion). The relative importance of these mechanisms and their interplay during a rainfall event are not yet well understood.

Similarly, dynamics and size distribution of sediment that is produced at the hillslope and delivered to and stored within the channel between debris-flow events, as well as the dependency of the debris-flow initiation on past events, needs further attention for an improved and more reliable hazard prediction on the local as well as on the regional scale.

Acknowledgements The authors would like to acknowledge Dieter Rickenmann for proof-reading and feedback, and Samuel Bissig for support on figure drawing.

References

Abanco, C., Hurlimann, M., Moya, J., & Berenguer, M. (2016). Critical rainfall conditions for the initiation of torrential flows. Results from the Rebaixader catchment (Central Pyrenees). *Journal of Hydrology, 541*, A, 218–229.

Allamano, P., Claps, P., Laio, F., & Thea, C. (2009). A data-based assessment of the dependence of short-duration precipitation on elevation. *Physics and Chemistry of the Earth, 34*, 635–641.

Arenson, L. U., & Springman, S. M. (2005). Mathematical descriptions for the behaviour of ice-rich frozen soils at temperatures close to 0 °C. *Canadian Geotechnical Journal, 42*, 431–442. https://doi.org/10.1139/t04-109

Armon, M., Marra, F., Enzel, Y., Rostkier-Edelstein, D., Garfinkel, C. I., Adam, O., Dayan, U., & Morin, E. (2022). Reduced rainfall in future heavy precipitation events related to contracted rain area despite increased rain rate. *Earth's Future, 9*, e2021EF002397.

Avanzi, F., De Michele, C., Gabriele, S., Ghezzi, A., & Rosso, R. (2015). Orographic signature on extreme precipitation of short durations. *Journal of Hydrometeorology, 16*, 278–294.

Baer, P., Huggel, C., McArdell, B. W., & Frank, F. (2017). Changing debris flow activity after sudden sediment input: A case study from the Swiss Alps. *Geology Today, 33*, 216–223. https://doi.org/10.1111/gto.12211

Bagnold, R. A. (1954). Experiments on a gravity-free dispersion of large solid spheres in a Newtonian fluid under shear. In: *Proceedings of the royal society of London A: Mathematical, physical and engineering sciences* (pp. 49–63). The Royal Society.

Bagnold, R. (1966). An approach to the sediment transport problem. *General Physics Geological Survey*.

Barry, R. G., & Chorley, R. J. (2010). *Atmosphere, weather and climate*. Taylor & Francis. ISBN10: 0-415-46569-9.

Bear, J. (1988). *Dynamics of fluids in porous media*. Courier Corporation.

Bennett, G. L., Molnar, P., McArdell, B. W., & Burlando, P. (2014). A probabilistic sediment cascade model of sediment transfer in the Illgraben. *Water Resources Research, 50*, 1225–1244. https://doi.org/10.1002/2013WR013806

Berg, P., Moseley, C., & Haerter, J. O. (2013). Strong increase in convective precipitation in response to higher temperatures. *Nature Geoscience, 6*(3), 181–185.

Berti, M., & Simoni, A. (2005). Experimental evidences and numerical modelling of debris flow initiated by channel runoff. *Landslides, 2*, 171–182. https://doi.org/10.1007/s10346-005-0062-4

Berti, M., Genevois, R., Simoni, A., & Tecca, P. R. (1999). Field observations of a debris flow event in the Dolomites. *Geomorphology, 29*, 265–274.

Berti, M., Genevois, R., LaHusen, R., et al. (2000). Debris flow monitoring in the Acquabona watershed on the Dolomites (Italian Alps). *Physics and Chemistry of the Earth, Part b: Hydrology, Oceans and Atmosphere, 25*, 707–715.

Berti, M., Martina, M., Franceschini, S., et al. (2012). Probabilistic rainfall thresholds for landslide occurrence using a Bayesian approach. *Journal of Geophysical Research: Earth Surface, 117*, F04006. https://doi.org/10.1029/2012JF002367

Berti, M., Bernard, M., Gregoretti, C., & Simoni, A. (2020). Physical interpretation of rainfall thresholds for runoff-generated debris flows. *Journal of Geophysical Research: Earth Surface, 125*, e2019JF005513.

Bertrand, M., Liébault, F., & Piégay, H. (2013). Debris-flow susceptibility of upland catchments. *Natural Hazards, 67*, 497–511. https://doi.org/10.1007/s11069-013-0575-4

Bhardwaj, A., & Sam, L. (2022). Reconstruction and characterisation of past and the most recent slope failure events at the 2021 rock-ice avalanche site in Chamoli, Indian Himalaya. *Remote Sensing, 14*, 949. https://doi.org/10.3390/rs14040949

Bierman, P., & Montgomery, D. (2014). *Key concepts in geomorphology*. Freeman.

Bogaard, T., & Greco, R. (2018). Invited perspectives: Hydrological perspectives on precipitation intensity-duration thresholds for landslide initiation: Proposing hydro-meteorological thresholds. *Natural Hazards and Earth System Sciences, 18*, 31–39. https://doi.org/10.5194/nhess-18-31-2018

Bonacina, L. C. W. (1945). Orographic rainfall and its place in the hydrology of the globe. *Quarterly Journal of the Royal Meteorological Society, 71*, 41–55.

Borga, M., Stoffel, M., Marchi, L., et al. (2014). Hydrogeomorphic response to extreme rainfall in headwater systems: Flash floods and debris flows. *Journal of Hydrology, 518*, 194–205. https://doi.org/10.1016/j.jhydrol.2014.05.022

Borga, M., Marra, F., & Gabella, M. (2022). Rainfall estimation by weather radar. In R. Morbidelli (Ed.), *Rainfall. Modeling, measurement and applications* (pp. 109–134). Elsevier.

Breien, H., Blasio, F. V. D., Elverhøi, A., & Høeg, K. (2008). Erosion and morphology of a debris flow caused by a glacial lake outburst flood, Western Norway. *Landslides, 5*, 271–280. https://doi.org/10.1007/s10346-008-0118-3

Caine, N. (1980). The rainfall intensity-duration control of shallow landslides and debris flows. *Geografiska annaler: Series A, physical geography, 62*, 23–27.

Campbell, C. S. (1990). Rapid granular flows. *Annual Review of Fluid Mechanics, 22*, 57–90.

Cannon, S., Gartner, J., Parrett, C., & Parise, M. (2003). *Wildfire-related debris-flow generation through episodic progressive sediment-bulking processes, western USA* (pp. 71–82). Mechanics, Prediction, and Assessment Millpress, Rotterdam.

Carey, M. (2005). Living and dying with glaciers: People's historical vulnerability to avalanches and outburst floods in Peru. *Global and Planetary Change, 47*, 122–134. https://doi.org/10.1016/j.gloplacha.2004.10.007

Carrara, A., Crosta, G., & Frattini, P. (2008). Comparing models of debris-flow susceptibility in the alpine environment. *Geomorphology, 94*, 353–378. https://doi.org/10.1016/j.geomorph.2006.10.033

Casagrande, A. (1976). *Liquefaction and cyclic deformation of sands-a critical review*. Harvard Soil Mechanics Series, Harvard University.

Chen, H., Petley, D. N. (2005). The impact of landslides and debris flows triggered by Typhoon mindulle in Taiwan. *Quarterly Journal of Engineering Geology and Hydrogeology, 38*(3), 301–304. https://doi.org/10.1144/1470-9236/04-077

Church, M., & Jakob, M. (2020). What is a debris flood? *Water Resources Research, 56*, e2020WR027144. https://doi.org/10.1029/2020WR027144

Church, M., & Miles, M. J. (1987). Meteorological antecedents to debris flow in southwestern British Columbia; Some case studies. *Reviews in Engineering Geology, 45*, 63–79

Costa, J. E. (1984). Physical geomorphology of debris flows. In J. E. Costa & P. J. Fleisher (Eds.), *Developments and applications of geomorphology* (pp. 268–317). Springer.

Coulomb, C. A. (1973). Essai sur une application des regles de maximis et minimis a quelques problemes de statique relatifs a l'architecture (essay on maximums and minimums of rules to some static problems relating to architecture).

Coussot, P., Meunier, M. (1996). Recognition, classification and mechanical description of debris flows. *Earth-Science Reviews, 40*, 209–227.

Cui, M.-J., Zheng, J.-J., Zhang, R.-J., et al. (2017). Influence of cementation level on the strength behaviour of bio-cemented sand. *Acta Geotechnica, 12*, 971–986. https://doi.org/10.1007/s11440-017-0574-9

D'Agostino, V., & Marchi, L. (2001). Debris flow magnitude in the Eastern Italian Alps: Data collection and analysis. *Physics and Chemistry of the Earth, 26*(9), 657–663.

Dallan, E., Borga, M., Zaramella, M., & Marra, F. (2022). Enhanced summer convection explains observed trends in extreme subdaily precipitation in the Eastern Italian Alps. *Geophysical Research Letters* (in press).

Decaulne, A., Saemundsson, P., & Petursson, O. (2005). Debris flow triggered by rapid snowmelt: A case study in the glei. Arhjalli area, northwestern Iceland. *Geografiska Annaler: Series a, Physical Geography, 87*, 487–500. https://doi.org/10.1111/j.0435-3676.2005.00273.x

Destro, E., Marra, F., Nikolopoulos, E. I., Zoccatelli, D., Creutin, J. D., & Borga, M. (2017). Spatial estimation of debris flows-triggering rainfall and its dependence on rainfall return period. *Geomorphology, 278*, 269–279.

DiBiase, R. A., Lamb, M. P., Ganti, V., & Booth, A. M. (2017). Slope, grain size, and roughness controls on dry sediment transport and storage on steep hillslopes. *Journal of Geophysical Research: Earth Surface, 122*, 941–960.

Dietrich, W. E., Bellugi, D. G., Sklar, L. S., et al. (2003). Geomorphic transport laws for predicting landscape form and dynamics. *Prediction in Geomorphology, 135*, 103–132. https://doi.org/10.1029/135GM09

Dunkerley, D. (2020). Rainfall intensity in geomorphology: Challenges and opportunities. *Progress in Physical Geography*, *45*(4), 488–513.

Dunne, T. (1990). Hydrology, mechanics, and geomorphic implications of erosion by subsurface flow. In C. G. Higgins & D. R. Coates (eds.), *Groundwater geomorphology: The role of subsurface water in earth-surface processes and landforms* (pp. 1–28). Geologic Society of America.

Ellen, S. D., & Fleming, R. W. (1987). Mobilization of debris flows from soil slips, San Francisco Bay region, California. In *Debris flows/avalanches: Process, recognition, and mitigation* (pp. 31–40). Geological Society of America.

Evans, S. G., Bishop, N. F., & Smoll, L. F., et al. (2009). A re-examination of the mechanism and human impact of catastrophic mass flows originating on Nevado Huascarán, Cordillera Blanca, Peru in 1962 and 1970. *Engineering Geology, 108*, 96–118. https://doi.org/10.1016/j.enggeo.2009.06.020

Fischer, J. T., Kaitna, R., Heil, K., & Reiweger, I. (2018). The heat of the flow: Thermal equilibrium in gravitational mass flows. *Geophysical Research Letters, 45*, 11,219–11,226. https://doi.org/10.1029/2018GL079585

Florsheim, J. L., Keller, E. A., & Best, D. W. (1991). Fluvial sediment transport in response to moderate storm flows following chaparral wildfire, Ventura County, southern California. *Geological Society of America Bulletin, 103*, 504–511.

Formetta, G., Marra, F., Dallan, E., Zaramella, M., & Borga, M. (2022). Differential orographic impact on sub-hourly, hourly, and daily extreme precipitation. *Advances in Water Resources, 149*, 104085.

Fowler, H. J., Lenderink, G., Prein, A. F., Westra, S., Allan, R. P., Ban, N., Barbero, R., Berg, P., Blenkinshop, S., Do, H. X., Guerreiro, S., Haerter, J. O., Kendon, E. J., Lewis, E., Schaer, C., Sharma, A., Villarini, G., Wasko, C., & Zhang, X. (2021). Anthropogenic intensification of short-duration rainfall extremes. *Nature Reviews Earth & Environment, 2*, 107–122.

Frank, F., Huggel, C., McArdell, B. W., & Vieli, A. (2019). Landslides and increased debris-flow activity: A systematic comparison of six catchments in Switzerland. *Earth Surface Processes and Landforms, 44*, 699–712. https://doi.org/10.1002/esp.4524

Gabet, E. J., & Bookter, A. (2008). A morphometric analysis of gullies scoured by post-fire progressively bulked debris flows in southwest Montana, USA. *Geomorphology, 96*, 298–309.

Gabet, E. J., & Mudd, S. M. (2006). The mobilization of debris flows from shallow landslides. *Geomorphology, 74*, 207–218.

Gariano, S. L., & Guzzetti, F. (2016). Landslides in a changing climate. *Earth-Science Reviews, 162*(2016), 227–252.

Gariano, S. L., Melillo, M., Peruccacci, S., & Brunetti, M. T. (2020). How much does the rainfall temporal resolution affect rainfall thresholds for landslide triggering? *Natural Hazards, 100*, 655–670.

Germann, U., Boscacci, M., Clementi, L., Gabella, M., Hering, A., Sartori, M., Sideris, I., & Calpini, B. (2022). Weather radar in complex orography. *Remote Sensing, 14*(3), 503.

Goren, L., Aharonov, E., Sparks, D., & Toussaint, R. (2010). Pore pressure evolution in deforming granular material: A general formulation and the infinitely stiff approximation. *Journal of Geophysical Research: Solid Earth*, 115. https://doi.org/10.1029/2009JB007191

Gray, J. M. N. T. (2017). Particle segregation in dense granular flows. *Annual Review of Fluid Mechanics*, 50. https://doi.org/10.1146/annurev-fluid-122316-045201

Gregoretti, C., & Fontana, G. D. (2008). The triggering of debris flow due to channel-bed failure in some alpine headwater basins of the Dolomites: Analyses of critical runoff. *Hydrological Processes, 22*, 2248–2263. https://doi.org/10.1002/hyp.6821

Guzzetti, F., Peruccacci, S., Rossi, M., & Stark, C. P. (2007). Rainfall thresholds for the initiation of landslides in central and southern Europe. *Meteorology and Atmospheric Physics, 98*, 239–267. https://doi.org/10.1007/s00703-007-0262-7

Guzzetti, F., Peruccacci, S., Rossi, M., & Stark, C. P. (2008). The rainfall intensity-duration control of shallow landslides and debris flows: An update. *Landslides, 5*, 3–17.

Hirschberg, J., Fatichi, S., & Bennett, G. L., et al. (2021). Climate change impacts on sediment yield and debris-flow activity in an alpine catchment. *Journal of Geophysical Research: Earth Surface, 126*, e2020JF005739. https://doi.org/10.1029/2020JF005739

Hu, W., Xi, Q., Wang, G. H., van Asch, T. W. J., & Hicher, P.-Y. (2015). Sensitivity of the initiation of debris flow to initial soil moisture. *Landslides, 12*, 1139–1145.

Huggel, C., Ceballos, J. L., Pulgarin, B., et al. (2007). Review and reassessment of hazards owing to volcano-glacier interactions in Colombia. *Annals of Glaciology, 45*, 128–136.

Huggel, C., Zgraggen-Oswald, S., & Haeberli, W. et al. (2005). The 2002 rock/ice avalanche at Kolka/Karmadon, Russian Caucasus: Assessment of extraordinary avalanche formation and mobility, and application of QuickBird satellite imagery. *Natural Hazards and Earth System Science, 5*,173–187. https://doi.org/1684-9981/nhess/2005-5-173

Hungr, O., Evans, S., Bovis, M., & Hutchinson, J. (2001). A review of the classification of landslides of the flow type. *Environmental & Engineering Geoscience, 7*, 221–238.

Hungr, O., Leroueil, S., & Picarelli, L. (2014). The Varnes classification of landslide types, an update. *Landslides, 11*, 167–194. https://doi.org/10.1007/s10346-013-0436-y

Hürlimann, M., Rickenmann, D., & Graf, C. (2003). Field and monitoring data of debris-flow events in the Swiss Alps. *Canadian Geotechnical Journal, 40*, 161–175.

Hürlimann, M., McArdell, B. W., & Rickli, C. (2015). Field and laboratory analysis of the runout characteristics of hillslope debris flows in Switzerland. *Geomorphology, 232*, 20–32. https://doi.org/10.1016/j.geomorph.2014.11.030

Hürlimann, M., Coviello, V., Bel, C., Guo, X., Berti, M., Graf, C., Hubl, J., Miyata, S., Smith, J. B., & Yin, H.-Y. (2019). Debris-flow monitoring and warning: Review and examples. *Earth-Science Reviews, 199*, 102981.

Ikari, M. J., & Kopf, A. J. (2011). Cohesive strength of clay-rich sediment. *Geophysical Research Letters*, 38. https://doi.org/10.1029/2011GL047918

Iverson, R. M. (1997). The physics of debris flows. *Reviews of Geophysics, 35*, 245–296. https://doi.org/10.1029/97RG00426

Iverson, R. M. (2000). Landslide triggering by rain infiltration. *Water Resources Research, 36*(7), 1897–1910.

Iverson, R. M. (2005). Regulation of landslide motion by dilatancy and pore pressure feedback. *Journal of Geophysical Research: Earth Surface, 110*, 16. https://doi.org/10.1029/2004JF000268

Iverson, R. M. (2015). Scaling and design of landslide and debris-flow experiments. *Geomorphology, 244*, 9–20. https://doi.org/10.1016/j.geomorph.2015.02.033

Iverson, R. M., & Denlinger, R. P. (2001). Flow of variably fluidized granular masses across three-dimensional terrain: 1. Coulomb mixture theory. *Journal of Geophysical Research: Solid Earth, 106*, 537–552. https://doi.org/10.1029/2000JB900329

Iverson, R. M., & LaHusen, R. G. (1989). Dynamic pore-pressure fluctuations in rapidly shearing granular materials. *Science, 246*, 796–799. https://doi.org/10.1126/science.246.4931.796

Iverson, R. M., & Major, J. J. (1986). Groundwater seepage vectors and the potential for hillslope failure and debris flow mobilization. *Water Resources Research, 22*, 1543–1548. https://doi.org/10.1029/WR022i011p01543

Iverson, R. M., & Vallance, J. W. (2001). New views of granular mass flows. *Geology, 29*, 115–118. https://doi.org/10.1130/0091-7613(2001)029<0115:NVOGMF>2.0.CO;2

Iverson, R. M., Reid, M. E., & LaHusen, R. G. (1997). Debris-flow mobilization from landslides. *Annual Review of Earth and Planetary Sciences, 25*, 85–138. https://doi.org/10.1146/annurev.earth.25.1.85

Iverson, R. M., Reid, M. E., Logan, M., et al. (2011). Positive feedback and momentum growth during debris-flow entrainment of wet bed sediment. *Nature Geoscience, 4*, 116–121. https://doi.org/10.1038/ngeo1040

Iverson, R. M., George, D. L., Allstadt, K., et al. (2015). Landslide mobility and hazards: Implications of the 2014 Oso disaster. *Earth and Planetary Science Letters, 412*, 197–208. https://doi.org/10.1016/j.epsl.2014.12.020

Iverson, R. M., & George, D. L. (2014). A depth-averaged debris-flow model that includes the effects of evolving dilatancy. I. Physical basis. *Proceedings of the Royal Society A: Mathematical, Physical and Engineering Sciences, 470*, 20130819. https://doi.org/10.1098/rspa.2013.0819

Iverson, R. M., Logan, M., LaHusen, R. G., & Berti, M. (2010). The perfect debris flow? Aggregated results from 28 large-scale experiments. *Journal of Geophysical Research: Earth Surface, 115*, 29. https://doi.org/10.1029/2009JF001514

Johnson, A. M. (1984). Debris flows. In D. Brunsden & D. B. Prior (Eds.), *Slope instability* (pp. 257–361). John Wiley.

Johnson, K. A., & Sitar, N. (1990). Hydrologic conditions leading to debris-flow initiation. *Canadian Geotechnical Journal, 27*, 789–801.

Jakob, M. (2005). A size classification for debris flows. *Engineering Geology, 79*, 151–161. https://doi.org/10.1016/j.enggeo.2005.01.006

Kaitna, R., Dietrich, W. E., & Hsu, L. (2014). Surface slopes, velocity profiles and fluid pressure in coarse-grained debris flows saturated with water and mud. *Journal of Fluid Mechanics, 741*, 377–403. https://doi.org/10.1017/jfm.2013.675

Kaitna, R., Palucis, M. C., & Yohannes, B., et al. (2016) Effects of coarse grain size distribution and fine particle content on pore fluid pressure and shear behavior in experimental debris flows. *Journal of Geophysical Research: Earth Surface, 121*, 2015JF003725. https://doi.org/10.1002/2015JF003725

Kean, J. W., McCoy, S. W., Tucker, G. E., et al. (2013). Runoff-generated debris flows: Observations and modeling of surge initiation, magnitude, and frequency. *Journal of Geophysical Research: Earth Surface, 118*, 2190–2207. https://doi.org/10.1002/jgrf.20148

Kean, J. W., Staley, D. M., & Cannon, S. H. (2011). In situ measurements of post-fire debris flows in southern California: Comparisons of the timing and magnitude of 24 debris-flow events with rainfall and soil moisture conditions. *Journal of Geophysical Research: Earth Surface*, 116.

Kidd, C., Becker, A., Huffman, G. J., Muller, C. L., Joe, P., Skofronick-Jackson, G., & Kirschbaum, D. B. (2017). So, how much of the earth's surface is covered by rain gauges? *Bulletin of the American Meteorological Society, 98*(1), 69–78.

Kidd, C., & Levizzani, V. (2022). Satellite rainfall estimation. In Morbidelli, R. (ed.), *Rainfall. Modeling, measurement and applications* (pp. 135–170). Elsevier.

Kienholz, H. (1995). Gefahrenbeurteilung und-bewertung-auf dem Weg zu einem Gesamtkonzept. *Schweizerische Zeitschrift Für Forstwesen, 146*, 701–725.

Kirschbaum, D. B., & Stanley, T. (2018). Satellite-based assessment of rainfall-triggered landslide hazard for situational awareness. *Earth's Future, 6*(3), 505–523.

Kirschbaum, D. B., Adler, R., Hong, Y., Kumar, S., Peters-Lidard, C., & Lerner-Lam, A. (2012). Advances in landslide nowcasting: Evaluation of a global and regional modeling approach. *Environment and Earth Science, 66*, 1683–1696.

Lamb, M. P., Dietrich, W. E., & Venditti, J. G. (2008). Is the critical Shields stress for incipient sediment motion dependent on channel-bed slope? *Journal of Geophysical Research, 113*, 1–20. https://doi.org/10.1029/2007JF000831

Lamb, M. P., Levina, M., DiBiase, R. A., & Fuller, B. M. (2013). Sediment storage by vegetation in steep bedrock landscapes: Theory, experiments, and implications for postfire sediment yield. *Journal of Geophysical Research: Earth Surface, 118*, 1147–1160.

Lamb, M. P., Scheingross, J. S., & Amidon, W. H., et al. (2011). A model for fire-induced sediment yield by dry ravel in steep landscapes. *Journal of Geophysical Research: Earth Surface*, 116.

Lambe, T. W., & Whitman, R. V. (1991). *Soil mechanics*. John Wiley & Sons.

Lengfeld, K., Kirstetter, P. E., Fowler, H. J., Yu, J., Becker, A., Fleming, Z., & Gourley, J. J. (2020). Use of radar data for characterizing extreme precipitation at fine scales and short durations. *Environmental Research Letters, 15*, 085003.

Leonarduzzi, E., & Molnar, P. (2020). Deriving rainfall thresholds for landsliding at the regional scale: Daily and hourly resolutions, normalisation, and antecedent rainfall. *Natural Hazards and Earth Systems Sciences, 20*, 2905–2919.

Libertino, A., Ganora, D., & Claps, P. (2019). Evidence for increasing rainfall extremes remains elusive at large spatial scales: The case of Italy. *Geophysical Research Letters, 46*, 7437–7446.

Lin, M., & Jeng, F. (2000). Characteristics of hazards induced by extremely heavy rainfall in Central Taiwan Typhoon Herb. *Engineering Geology, 58*, 191–207.

Lochbihler, K., Lenderink, G., & Siebesma, A. P. (2017). The spatial extent of rainfall events and its relation to precipitation scaling. *Geophysical Research Letters, 44*, 8629–8636.

Marchi, L., Arattano, M., & Deganutti, A. M. (2002). Ten years of debris-flow monitoring in the Moscardo Torrent (Italian Alps). *Geomorphology, 46*, 1–17.

Marra, F. (2019). Rainfall thresholds for landslide occurrence: Systematic underestimation using coarse temporal resolution data. *Natural Hazards, 95*(3), 883–890.

Marra, F., Nikolopoulos, E. I., Creutin, J. D., & Borga, M. (2014). Radar rainfall estimation for the identification of debris-flow occurrence thresholds. *Journal of Hydrology, 519*, 1607–1619.

Marra, F., Nikolopoulos, E. I., Creutin, J. D., & Borga, M. (2016). Space–time organization of debris flows-triggering rainfall and its effect on the identification of the rainfall threshold relationship. *Journal of Hydrology, 541*, 246–255.

Marra, F., Armon, M., Borga, M, & Morin, E. (2021). Orographic effect on extreme precipitation statistics peaks at hourly time scales. *Geophysical Research Letters*, *48*, e2020GL091498.

Marra, F., Armon, M., & Morin, E. (2022). Coastal and orographic effects on extreme precipitation revealed by weather radar observations. *Hydrology and Earth System Sciences* (in press).

Mazzoglio, P., Butera, I., Alvioli, M., & Claps, P. (2022). The role of morphology on the spatial distribution of short-duration rainfall extremes in Italy. *Hydrology and Earth System Sciences* (in press).

McArdell, B. W., Bartelt, P., & Kowalski, J. (2007). Field observations of basal forces and fluid pore pressure in a debris flow. *Geophysical Research Letters, 34*, 4. https://doi.org/10.1029/2006GL029183

McCoy, S. W., Kean, J. W., Coe, J. A., et al. (2010). Evolution of a natural debris flow: In situ measurements of flow dynamics, video imagery, and terrestrial laser scanning. *Geology, 38*, 735–738. https://doi.org/10.1130/G30928.1

McGuire, L. A., Rengers, F. K., Kean, J. W., & Staley, D. M. (2017). Debris flow initiation by runoff in a recently burned basin: Is grain-by-grain sediment bulking or en masse failure to blame? *Geophysical Research Letters, 44*, 7310–7319.

Melillo, M., Brunetti, M. T., & Peruccacci, S. et al. (2015). An algorithm for the objective reconstruction of rainfall events responsible for landslides. *Landslides, 12*, 311–320. https://doi.org/10.1007/s10346-014-0471-3

Melosh, H. J. (1987). The mechanics of large rock avalanches. In J. E. Costa & G. F. Wieczorek (eds.), *Debris flows/avalanches*. Geological Society of America. ISBN 978-0-8137-5807-7. https://doi.org/10.1130/REG7-p41

Mergili, M., Pudasaini, S. P., Emmer, A., et al. (2020). Reconstruction of the 1941 GLOF process chain at Lake Palcacocha (Cordillera Blanca, Peru). *Hydrology and Earth System Sciences, 24*, 93–114. https://doi.org/10.5194/hess-24-93-2020

Michaelides, S., Levizzani, V., Anagnostou, E., Bauer, P., Kasparis, T., & Lane, J. E. (2009). Precipitation: Measurement, remote sensing, climatology and modeling. *Atmospheric Research, 94*(4), 512–533.

Mikoš, M., Četina, M., & Brilly, M. (2004). Hydrologic conditions responsible for triggering the Stože landslide, Slovenia. *Engineering Geology, 73*, 193–213. https://doi.org/10.1016/j.enggeo.2004.01.011

Mitchell, A., McDougall, S., Aaron, J., & Brideau, M. A. (2020). Rock avalanche-generated sediment mass flows: Definitions and hazard. *Frontiers in Earth Science, 8*. https://doi.org/10.3389/feart.2020.543937

Mostbauer, K., Kaitna, R., Prenner, D., & Hrachowitz, M. (2018). The temporally varying roles of rainfall, snowmelt and soil moisture for debris flow initiation in a snow-dominated system. *Hydrology and Earth System Sciences, 22*, 3493–3513.

Nagl, G., Hübl, J., & Kaitna, R. (2020). Velocity profiles and basal stresses in natural debris flows. *Earth Surface Processes and Landforms, 45*, 1764–1776. https://doi.org/10.1002/esp.4844

Nikolopoulos, E. I., Crema, S., Marchi, L., Marra, F., Guzzetti, F., & Borga, M. (2014). Impact of uncertainty in rainfall estimation on the identification of rainfall thresholds for debris flow occurrence. *Geomorphology, 221*, 286–297.

Nikolopoulos, E. I., Borga, M., Marra, F., Crema, S., & Marchi, L. (2015a). Debris flows in the Eastern Italian Alps: Seasonality and atmospheric circulation patterns. *Natural Hazards and Earth Systems Sciences, 15*, 647–656.

Nikolopoulos, E. I., Borga, M., Creutin, J. D., & Marra, F. (2015b). Estimation of debris flow triggering rainfall: Influence of rain gauge density and interpolation methods. *Geomorphology, 243*, 40–50.

Nikolopoulos, E. I., Destro, E., Maggioni, V., Marra, F., & Borga, M. (2017). Satellite-rainfall estimates for debris flow prediction: An evaluation based on rainfall accumulation—duration thresholds. *Journal of Hydrometeorology, 18*, 2207–2214.

Nishiguchi, Y., & Uchida, T. (2022). Long-runout-landslide-induced debris flow: The role of fine sediment deposition processes in debris flow propagation. *Journal of Geophysical Research: Earth Surface, 127*, e2021JF006452. https://doi.org/10.1029/2021JF006452

Palucis, M. C., Ulizio, T., Fuller, B., & Lamb, M. P. (2018). Intense granular sheetflow in steep streams. *Geophysical Research Letters, 45*, 5509–5517.

Palucis, M. C., Ulizio, T. P., & Lamb, M. P. (2021). Debris flow initiation from ravel-filled channel bed failure following wildfire in a bedrock landscape with limited sediment supply. *GSA Bulletin, 133*, 2079–2096. https://doi.org/10.1130/B35822.1

Panziera, L., Germann, U., Gabella, M., & Mandapaka, P. V. (2011). Nora—Nowcasting of orographic rainfall by means of analogues. *Quarterly Journal Royal Meteorological Society, 137*, 2106–2123.

Parker, G., Wilcock, P. R., & Paola, C. et al. (2007). Physical basis for quasi-universal relations describing bankfull hydraulic geometry of single-thread gravel bed rivers. *Journal of Geophysical Research, 112*. https://doi.org/10.1029/2006JF000549

Piciullo, L., Calvello, M., & Cepeda, J. M. (2018). Territorial early warning systems for rainfall-induced landslides. *Earth-Science Reviews, 179*, 228–247.

Pierson, T. C., & Scott, K. M. (1985). Downstream dilution of a lahar: Transition from debris flow to hyperconcentrated streamflow. *Water Resources Research, 21*, 1511–1524. https://doi.org/10.1029/WR021i010p01511

Pierson, T. C., Janda, R. J., Thouret, J. C., & Borrero, C. A. (1990). Perturbation and melting of snow and ice by the 13 November 1985 eruption of Nevado del Ruiz, Colombia, and consequent mobilization, flow and deposition of lahars. *Journal of Volcanology and Geothermal Research, 41*, 17–66.

Pierson, T. (1986). Flow behavior of channelized debris flows, Mount St. Helens, Washington. In *Hillslope processes* (pp. 269–296). Allen & Unwin.

Postance, B., Hillier, J., Dijkstra, T., & Dixon, N. (2018). Comparing threshold definition techniques for rainfall-induced landslides: A national assessment using radar rainfall. *Earth Surface Processes and Landforms, 43*(2), 553–560.

Prancevic, J. P., Lamb, M. P., & Fuller, B. M. (2014). Incipient sediment motion across the river to debris-flow transition. *Geology, 42*, 191–194. https://doi.org/10.1130/G34927.1

Prancevic, J. P., Lamb, M. P., Palucis, M. C., & Venditti, J. G. (2018). The role of three-dimensional boundary stresses in limiting the occurrence and size of experimental landslides. *Journal of Geophysical Research: Earth Surface, 123*, 46–65.

Prenner, D., Kaitna, R., Mostbauer, K., & Hrachowitz, M. (2018). The value of using multiple hydrometeorological variables to predict temporal debris flow susceptibility in an Alpine environment. *Water Resources Research, 54*, 6822–6843.

Prenner, D., Hrachowitz, M., & Kaitna, R. (2019). Trigger characteristics of torrential flows from high to low alpine regions in Austria. *Science of the Total Environment, 658*, 958–972. https://doi.org/10.1016/j.scitotenv.2018.12.206

Rautela, P., Kundalia, S. K. S., Joshi, G. C., & Jugran, R. (2021). Sequential damming induced winter season flash flood in Uttarakhand Province of India. *Journal of Environmental & Earth Sciences, 3*, 61–71.

Reid, M., Brien, D., & LaHusen, R., et al. (2003). Debris-flow initiation from large, slow-moving landslides. In D. Rickenmann & C. L. Chen (eds.), Debris-flow hazards mitigation: Mechanics, prediction, and assessment (Vol. 1, pp. 155–166).

Rengers, F. K., Kean, J. W., & Reitman, N. G. et al. (2020). The influence of frost weathering on debris flow sediment supply in an Alpine Basin. *Journal of Geophysical Research: Earth Surface, 125*, e2019JF005369. https://doi.org/10.1029/2019JF005369

Rickenmann, D. (2016). *Methods for the quantitative assessment of channel processes in torrents (steep streams)*. CRC Press.

Rickenmann, D., & Zimmermann, M. (1993). The 1987 debris flows in Switzerland: Documentation and analysis. *Geomorphology, 8*, 175–189.

Roe, G. H. (2005). Orographic precipitation. *Annual Review of Earth and Planetary Sciences, 33*, 645–671.

Rossi, M., Luciani, S., Valigi, D., Brunetti, M. T., Peruccacci, S., & Guzzetti, F. (2017). Statistical approaches for the definition of landslide rainfall thresholds and their uncertainty using rain gauge and satellite data. *Geomorphology, 285*, 16–27.

Sassa, K., Fukuoka, H., Scarascia-Mugnozza, G., & Evans, S. (1996). Earthquake-induced-landslides: Distribution, motion and mechanisms. *Soils and Foundations, 36*, 53–64. https://doi.org/10.3208/sandf.36.Special_53

Sassa, K. (2005). *Mechanism of landslide-triggered debris flows: Liquefaction phenomena due to the undrained loading of torrent deposits. Debris-flow hazards and related phenomena* (pp. 81–104). Springer.

Savage, S. (1998). Analyses of slow high-concentration flows of granular materials. *Journal of Fluid Mechanics, 377*, 1–26.

Schneider, D., Huggel, C., Haeberli, W., & Kaitna, R. (2011). Unraveling driving factors for large rock–ice avalanche mobility. *Earth Surface Processes and Landforms, 36*, 1948–1966.

Scott, K. M., Macías, J. L., & Naranjo, J. A. et al. (2001). *Catastrophic debris flows transformed from landslides in volcanic terrains: Mobility, hazard assessment, and mitigation strategies*. US Geological Survey professional paper 1–59.

Segoni, S., Piciullo, L., & Gariano, S. L. (2018). A review of the recent literature on rainfall thresholds for landslide occurrence. *Landslides, 15*, 1483–1501. https://doi.org/10.1007/s10346-018-0966-4

Selby, M. J. (1993). *Hillslope materials and processes*. Oxford University Press.

Shugar, D. H., Jacquemart, M., Shean, D., et al. (2021). A massive rock and ice avalanche caused the 2021 disaster at Chamoli, Indian Himalaya. *Science, 373*, 300–3006. https://doi.org/10.1126/science.abh4455

Sidle, R. C., & Ochiai, H. (2006). *Landslides: Processes, prediction, and land use, 312*.

Sklar, L. S., Riebe, C. S., Marshall, J. A., et al. (2016). The problem of predicting the size distribution of sediment supplied by hillslopes to rivers. *Geomorphology*. https://doi.org/10.1016/j.geomorph.2016.05.005

Smith, R. B., Jiang, Q., Fearon, M. G., Tabary, P., Dorninger, M., Doyle, J. D., & Benoit, R. (2003). Orographic precipitation and air mass transformation: An Alpine example. *Quarterly Journal Royal Meteorological Society, 129*, 433–454.

Staley, D. M., Negri, J. A., Kean, J. W., et al. (2017). Prediction of spatially explicit rainfall intensity–duration thresholds for post-fire debris-flow generation in the western United States. *Geomorphology, 278*, 149–162. https://doi.org/10.1016/j.geomorph.2016.10.019

Stoffel, M., Bollschweiler, M., & Beniston, M. (2011). Rainfall characteristics for periglacial debris flows in the Swiss Alps: Past incidences—potential future evolutions. *Climatic Change, 105*, 263–280.

Suwa, H., Okano, K., & Kanno, T. (2009). Behavior of debris flows monitored on test slopes of Kamikamihorizawa Creek, Mount Yakedake, Japan. *International Journal of Erosion Control Engineering, 2*, 33–45.

Suwa, H., Okano, K., & Kanno, T. (2011). Forty years of debris flow monitoring at Kamikamihorizawa Creek, Mount Yakedake, Japan. *5th international conference on debris-flow hazards mitigation: Mechanics, prediction and assessment* (pp. 605–613). Casa Editrice UniversitaLa Sapienza.

Takahashi, T. (1978). Mechanical characteristics of debris flow. *Journal of the Hydraulics Division, 104*, 1153–1169.

Takahashi, T. (1991). *Debris flow*. Balkema.

Takahashi, T. (2007). *Debris flow mechanics, prediction and counter measurements*. Taylor and Francis Group.

Take, W., Bolton, M., Wong, P., & Yeung, F. (2004). Evaluation of landslide triggering mechanisms in model fill slopes. *Landslides, 1*, 173–184. https://doi.org/10.1007/s10346-004-0025-1

Tang, C., van Asch, T. W. J., Chang, M., et al. (2012). Catastrophic debris flows on 13 August 2010 in the Qingping area, southwestern China: The combined effects of a strong earthquake and subsequent rainstorms. *Geomorphology, 139–140*, 559–576. https://doi.org/10.1016/j.geomorph.2011.12.021

Tang, H., McGuire, L. A., Rengers, F. K., et al. (2019a). Evolution of debris-flow initiation mechanisms and sediment sources during a sequence of post-wildfire rainstorms. *Journal of Geophysical Research: Earth Surface*. https://doi.org/10.1029/2018JF004837

Tang, H., McGuire, L. A., Rengers, F. K., et al. (2019b). Developing and testing physically based triggering thresholds for runoff-generated debris flows. *Geophysical Research Letters, 46*, 8830–8839.

Terzaghi, K. (1936). The shearing resistance of saturated soils and the angle between the planes of shear. *First International Conference on Soil Mechanics, 1936*, 54–59.

Terzaghi, K., & Peck, R. B. (1948). *Soil mechanics. Engineering practice*. John Wiley and Sons, Inc.

Theule, J. I., Liébault, F., Laigle, D., et al. (2015). Channel scour and fill by debris flows and bedload transport. *Geomorphology, 243*, 92–105. https://doi.org/10.1016/j.geomorph.2015.05.003

Thouret, J.-C., Antoine, S., Magill, C., & Ollier, C. (2020). Lahars and debris flows: Characteristics and impacts. *Earth-Science Reviews, 201*, 103003. https://doi.org/10.1016/j.earscirev.2019.103003

Tillery, A. C., & Rengers, F. K. (2020). Controls on debris-flow initiation on burned and unburned hillslopes during an exceptional rainstorm in southern New Mexico, USA. *Earth Surface Processes and Landforms, 45*, 1051–1066.

Tiranti, D., Cremonini, R., Marco, F., Gaeta, A. R., & Barbero, S. (2014). The DEFENSE (Debris Flows triggered by storms-Nowcasting SystEm): An early warning system for torrential processes by radar storm tracking using a Geographic Information System (GIS). *Computers & Geosciences, 70*, 96–109.

Tuffen, H. (2011). Ice-volcano interactions. In V. P. Singh, P. Singh, & U. K. Haritashya (Eds.), *Encyclopedia of snow, ice and glaciers* (pp. 625–628). Springer.

Walter, F., Amann, F., Kos, A., et al. (2020). Direct observations of a three million cubic meter rock-slope collapse with almost immediate initiation of ensuing debris flows. *Geomorphology, 351*, 106933. https://doi.org/10.1016/j.geomorph.2019.106933

Wang, G., & Sassa, K. (2003). Pore-pressure generation and movement of rainfall-induced landslides: Effects of grain size and fine-particle content. *Engineering Geology, 69*, 109–125. https://doi.org/10.1016/S0013-7952(02)00268-5

Wang, F. W., Sassa, K., & Wang, G. (2002). Mechanism of a long-runout landslide triggered by the August 1998 heavy rainfall in Fukushima Prefecture, Japan. *Engineering Geology, 63*(1–2), 169–185.

Wieczorek, G. F., & Glade, T. (2005). Climatic factors influencing occurrence of debris flows. In M. Jakob & O. Hungr (eds.), Debris-flow hazards and related phenomena, pp. 325–362. Berlin, Heidelberg: Springer.

Wood, D. M. (1990). *Soil behaviour and critical state soil mechanics*. Cambridge University Press.

Worni, R., Huggel, C., Stoffel, M., & Pulgarín, B. (2012a). Challenges of modeling current very large lahars at Nevado del Huila Volcano, Colombia. *Bulletin of Volcanology, 74*, 309–324. https://doi.org/10.1007/s00445-011-0522-8

Worni, R., Stoffel, M., Huggel, C., et al. (2012b). Analysis and dynamic modeling of a moraine failure and glacier lake outburst flood at Ventisquero Negro, Patagonian Andes (Argentina). *Journal of Hydrology, 444–445*, 134–145. https://doi.org/10.1016/j.jhydrol.2012.04.013

Zhang, C.-B., Chen, L.-H., Liu, Y.-P., et al. (2010). Triaxial compression test of soil–root composites to evaluate influence of roots on soil shear strength. *Ecological Engineering, 36*, 19–26. https://doi.org/10.1016/j.ecoleng.2009.09.005

Zimmermann, M., Mani, P., & Gamma, P. (1997). *Murganggefahr und Klimaänderung ein GIS-basierter Ansatz*. Switzerland.

8 Debris Flow Dating and Magnitude Reconstruction

Juan Antonio Ballesteros-Cánovas, Markus Stoffel, Tjalling de Haas and José María Bodoque

Abstract

Debris flows typically occur in mountain regions with high energy relief, complex geology and extreme climate conditions, where they often affect humans as well as their cultural and economic assets. Understanding the frequency and magnitude of debris flows is essential to conduct reliable hazard and risk assessments in the context of climate warming. However, observational records of past and recent debris flows are unusual worldwide and difficult to gather. In this chapter, we review methodological advances (after Jakob, Debris flow hazards and related phenomena, Springer, 2005) on debris flow dating and magnitude estimation. We briefly describe the theoretical basis of classical absolute and relative dating techniques, including radiometric, botanical and historical as well as modern approaches based on remote sensing. We also review different existing strategies to quantify the magnitude of debris flow retrospectively, including empirical equations on different parameters (volume, affected area, peak discharge or frontal flow velocity) and the use of latest-available numerical models. Finally, we discuss the applicability and limitations of the frequency-magnitude relationship in the context of hazard and risk assessments. The approaches provided in this chapter are grounded on field-truthed evidence and contribute to more assertive spatial decision-making and to build more resilient societies in the context of climate warming.

J. A. Ballesteros-Cánovas (✉)
The National Museum of Natural Sciences, Spanish Scientific Research Council, Madrid, Spain
e-mail: juan.ballesteros@mncn.csic.es

M. Stoffel
Climate Change Impacts and Risks in the Anthropocene (C-CIA), Institute for Environmental Sciences, University of Geneva, Boulevard Carl-Vogt 66, 1205 Geneva, Switzerland

M. Stoffel
Dendrolab.Ch, Department of Earth Sciences, University of Geneva, 13 Rue Des Maraichers, 1205 Geneva, Switzerland

M. Stoffel
Department F.-A. Forel for Environmental and Aquatic Sciences, University of Geneva, Boulevard Carl-Vogt 66, 1205 Geneva, Switzerland

T. de Haas
Department of Physical Geography, Utrecht University, Utrecht, Netherlands

J. M. Bodoque
Mining and Geological Engineering Department, University of Castilla La Mancha, Toledo, Spain

Abbreviations

g	Gravity (m/s^2)
r_c	Curvature radii (m)
θ	Superelevation gradient (°)

M. Jakob et al. (eds.), *Advances in Debris-flow Science and Practice*, Geoenvironmental Disaster Reduction,
https://doi.org/10.1007/978-3-031-48691-3_8

α Channel gradient (%)
Δh Superelevation (m)
B Width of the flow (m)
K Viscosity factor
$\varnothing$ Angle of the channel (°);
B Width of the flow (m);
hm Average flow height
γ Weight's unit;
S Channel slope (m/m);
μ Flow viscosity;
K Shape factor;
H Flow thickness (m)
θ Angle of inclination of the adverse slope (°);
$\varnothing$ Effective basal friction angle of debris on the slope;
F_r Froude number;
$\varkappa$ Longitudinal pressure coefficient
h_1 Flow depth (m)
k Relation between exit flow depth and critical depth (Hong et al. 2010).
L Horizontal drop length from spill way of the check dam (in m)
H_c Vertical drop length from spill way to test height (in m)
h_e Vertical drop length from spill way to exit flow depth at spill way (in m)
ρ Fluid density (kg/m^3)
$\overline{h}$ Average flow depth (m)
ξ Lumped coefficient in dilatant grain shearing
C Chézy coefficient
δ Factor related to the average flow depth (see reference)
u* Shear velocity (s/m$^{1/3}$)
C_v Volumetric sediment concentration

8.1 Introduction

Estimating the frequency and magnitude (FM) of debris flows is critical for gaining a better process understanding and conducting reliable hazard and risk assessments (Jakob et al., 2016; Thouret et al., 2020; Jakob and Nolde, this volume). Flow magnitude controls flow velocity, flow depth, peak discharge, inundation area, and thereby the amount of damage and number of fatalities (Dowling & Santi, 2014; Griswold & Iverson, 2008; Rickenmann, 1999), and frequency determines how often life and property are at risk. Furthermore, the FM of debris flows is of crucial importance for land-use and infrastructure planning and understanding landscape evolution (Brardinoni & Hassan, 2006; Mao et al., 2009; McCoy, 2015). Dating and magnitude quantification of past and recent events not only enable us to decipher climate trigger mechanisms (Stoffel, 2010; Stoffel et al., 2014; Schneuwly-Bollschweiler and Stoffel, 2012; Heiser et al., 2022), but also to identify the influence of environmental changes on the debris flow process (Cannon & DeGraff, 2009; de Haas et al., 2018a, 2018b). Long-term records are also the basis for the identification of past, present and future debris flow trajectories, which contributes to determining how they will evolve under climate change scenarios (Jakob et al., 2020; Stoffel et al., 2014; Winter et al., 2010; Stoffel et al., this volume). Importantly, extensive databases of past debris flows provide ground-truthed evidence for the calibration of numerical hydrodynamic models, thereby facilitating the mitigation of impacts to infrastructure and people. However, such long-term records are still uncommon worldwide and are often difficult to gather because debris flows have long return periods (van Steijn, 1996) and frequently occur in mountain regions where observation and monitoring of such events is challenging (Imaizumi et al., 2006). The alternative to this lack of records is the retrospective reconstruction of debris flows that occurred in the past. This approach has become relevant to overcome the lack of baseline data (Savi et al., 2014; Schürch et al., 2016), especially in mountain areas with recent human developments. It relies on the analyses of geomorphic, geological or botanical evidence, as well as historical archives (Bodoque et al., 2015; de Haas & Densmore, 2019; Jakob et al., 2020). The global applicability of near-real-time remote satellite sensors has also contributed to significant methodological advancements in this field over the past few decades.

The applicability of the existing dating techniques is greatly dependent on the nature and requirements of each study, as different techniques offer varying spatio-temporal coverage and precision. For instance, radiometric-based techniques such as radiocarbon (^{14}C), Cosmic Ray Exposure (CRE), Optically Stimulated Luminescence (OSL), and Thermoluminescence (TL) are employed to date debris flow deposits (Dühnforth et al., 2007, 2017; D'Arcy et al., 2015; Schürch et al., 2016). Dendrochronology is also used to determine the absolute date, by identifying the growth response of trees to geomorphic changes caused by debris flows (Jakob, 2010; Stoffel et al., 2016). Radiometric techniques allow absolute dating over thousands of years, although with decadal or interannual precision. By contrast, dendrochronological approaches only cover the past few centuries, but yield data with high temporal precision (annual to seasonal resolution). Lichenometry is typically used to date the absolute (where date control is available) and/or relative age of lichen colonies on mobilized debris and boulders (Tanarro García et al., 2022; Bernhardt et al., 2017; Evans et al., 1999). In recent decades, the increase in the number and quality of satellite sensors have also allowed analysis and recording of debris-flow activity in remote mountain regions with a dating accuracy of often up to a few days (Muñoz-Torrero et al., 2021). Historical archives (i.e. description, old pictures and maps i.e. Google Earth imagery) can also provide quantitative and qualitative information on the occurrence and impact of past debris flows, although they are generally limited to torrents near human settlements or along major infrastructure corridors (Bernhardt et al., 2017). On the other hand, relative dating methods, such as the analyses of rock weathering processes (i.e. Schmidt Hammer), or soil development (Matthews et al., 2020), could be used to supplement absolute dating approaches.

Estimation of the magnitude of debris flows is required to derive FM relationships. In analogy to paleohydraulic techniques, where geological, historical, and botanical records are interpreted to quantify past floods, debris flow magnitudes can be approximated using data retrieved from the above-mentioned records (Wilhelm et al., 2019). Yet, debris flows are much more complex hydrological phenomena, as they represent non-Newtonian flows with solid fractions ranging between 50 and 90% (corresponding to densities of 1,300–2,200 kg/m^3; Takahashi, 1981; Iverson, 2014; Trujillo-Vela et al., 2022). At these sediment concentrations, critical yield strength is generated, so that large particles can be suspended indefinitely in the moving flow (Pierson, 2005). Accordingly, the whole mass will undergo very considerable and approximately continuous deformation, because there is little difference between the velocity of the two phases (water and sediment) (McCoy et al., 2010). Moreover, the flow is governed by drag, weights, particle–particle collisions and electromechanical cohesive forces (in the case of mudflows) (Iverson, 2014). This flow complexity, together with the fact that debris flows often go unnoticed because they occur in mountains where direct measurements are difficult to perform, makes the retrospective estimation of debris flow magnitudes exceedingly difficult and largely reliant on field observations.

In recent years, the application of physically-based hydrodynamic models (Iverson & George, 2014) and the development of new empirical equations have improved estimates of debris flow magnitudes (Gartner et al., 2014; Gartner et al., this volume). The estimation of the debris flow magnitude thereby relies on three parameters: (i) *mobilized volume*, which requires quantification of the mobilized sediment at the initial slope failure, the entrained material along the channel, and the deposited sediments along the fan; (ii) *peak discharge*, generally estimated along the channel using either empirical or hydraulic models given certain conditions at the cross-section such as invariant topography (e.g., bedrock) and critical flow conditions; and (iii) *inundated area*, directly retrieved from field inspection and detailed aerial picture interpretation.

This chapter aims at providing an overview of existing methods for debris flow dating and magnitude estimation, highlighting the most recent developments in these fields over the past 15 or so years. Due to the wide variety of approaches presented in this chapter, we focus on describing the fundamentals and provide a limited suite of application examples. For a deeper understanding of the technical explanation of the dating approaches, the reader is directed to the specific studies on each technique, many of which are cited herein.

8.2 Identifying Debris Flow Occurrence

Field recognition and geomorphic surveys of torrential catchments are the first and most important procedures in identifying suitable evidence of debris flow activity (Brenna et al., 2020; Jakob et al., 2022). An initial assessment should be performed to distinguish between different hydrological processes (i.e. debris flows, hyperconcentrated flows and clear-water floods) that may coexist in the same catchment, either using desktop methods (i.e. aerial pictures; morphometric analyses) (Bertrand et al., 2013; de Haas et al., this volume) or in-field recognitions (Jakob et al., 2020). Debris flow evidence is generally expressed as changes in landforms due to surficial erosion and sedimentation processes (Iverson, 2012; Iverson & Ouyang, 2015), which often produce changes in vegetation pattern (Hoch et al., 2021; Rengers et al., 2016; Fig. 8.1). At the catchment scale, typical characteristics that are indicative of the occurrence of debris flow are:

(i) **Hillslope**: Steep slopes are typically the primary source of mobilized sediments during extreme rainfall events, or temperature increases in case of cryosphere-related landforms. Landslides, erosive landforms and geomorphic changes (i.e. rockglacier) can be traced using historical maps, aerial pictures and recent satellite-based products. These data provide insight into the initial volume of sediments delivered to the channel, which will ultimately depend on the slope-channel connectivity (Steger et al., 2022; Lancaster and Casebeer, 2007).

(ii) **Channel**: Mobilized sediments are typically conveyed along steep channels. Flow energy erodes banks and the channel bed, contributing to the entrainment of sediment, which may leave erosion lines on preexisting levees or channel banks that can serve as paleostage indicators (PSI) (Bodoque et al., 2011). When the valley becomes locally wider, intense debris flow events may overflow the channel banks, resulting in sedimentary landforms such as *lobes* or *chutes*. Mapping these landforms can be used to determine the maximum extent of the affected areas, but also to identify weak cross-sections in topography where further avulsion processes could evolve (de Haas et al., 2018a, b). In the channel, bed material analysis usually yields quantitative data about the sediment transport capacity of streams (Nagl et al., 2020).

(iii) **Fans or debris-flow cones**: Depositions of transported sediments form debris flow fans at the mouths of torrents (Fig. 8.1) where flow energy is significantly reduced. Fans are semi-conical landforms with a relatively low surface slope, although it depends on the bulk rheology of sediments involved (de Haas et al., 2015, 2016). At the fan level, the primary geomorphic features are related to poorly-sorted mixtures of sediment ranging in size from microns to meters (Calhoun & Clague, 2018) and landforms such as *levees*, *lobes* or *snouts* (de Haas et al., this volume). Fans are usually cut by the active main channel, although abandoned scoured tracks or paleochannels, sometimes partially filled, are also frequently recognized. Eventually, in the case of debris flow channels ending in a lake or sea, the application of radar techniques may be needed to identify geomorphic features.

8.3 Debris Flow Dating Techniques

Once debris flow processes have been properly characterized (i.e. Calhoun & Clague, 2018), events can be dated. Debris flow dating can involve both relative and absolute methods.

Fig. 8.1 Characteristic debris flow catchment located in the northern Kyrgyz mountain range (Aksay River catchment). Left: oblique view of the hillslope characterized by the presence of glaciers and lakes. Right: channel and debris flow fan at the bottom valley where paleo channels are easily distinguished. Courtesy: Vitali Zaginaev

Relative dating methods allow for the determination of whether a specific sediment deposit or erosive feature is older or younger than another one. It provides a qualitative sequence of ages (i.e. the order in which events occurred), but without defining specific dates. This approach can be used to understand the evolution and dynamics of debris flow cones (de Haas et al., 2018a, b; Dühnforth et al., 2007), but it cannot be used to determine the frequency-magnitude relationship required for quantitative hazard or risk assessments (Schürch et al., 2016). On the other hand, absolute dating methods refer to techniques that may—ideally—provide the calendar ages of the analyzed events. Absolute dating allows relating the occurrence of debris flows with temporally-dependent processes (i.e. climate and environmental factors) (Ballesteros-Cánovas & Stoffel, 2017), as well as quantification of the return period of events by defining the frequency-magnitude relationship (Helsen et al., 2002; Stoffel, 2010). The application of one approach or another is highly dependent on the objective of the study and the nature of evidence available in the field. Relative dating techniques may be employed in conjunction with absolute dating methods where appropriate, but their utility depends on the objective of the study.

8.3.1 Remote Sensing Approach

Remote sensing approaches may include aerial photo interpretation and the advanced analyses of satellite imagery products, and are commonly conducted initially as desktop-based studies. These approaches have gained great popularity, especially in remote mountain areas. Analyses of aerial photographs have traditionally been an indispensable tool for the determination of areas that have been affected by debris flows. The earliest aerial pictures date back to the 1930s, although only few countries can rely on a continuous record over the past 90 years. The spatial resolution of the first aerial pictures barely allowed for an identification of larger geomorphic features coupled with debris flows. Jakob et al. (2022) underlined that having an error lower than 5–10 cm in the GIS delineation of geomorphic features may require a picture scale of 1:40,000 or larger. Later improvements of optical sensors in drones, airborne vehicles or satellites and the use of platform analyses such as Google Earth Engine have exploited the possibilities of such records (Ghamisi et al., 2019). Thus, multi-temporal digital aerial imagery has become a cost-effective approach to determine the source of sediments involved in debris flows and landslides and to delineate their runout

(Imaizumi et al., 2019; Muñoz-Torrero et al., 2021, Fig. 8.2). However, the time span between available imagery can become an issue in forested regions because rapid regrowth of vegetation can mask evidence. In coniferous forest, bracketing event ages is facilitated by identifying deciduous pioneer vegetation with lighter coloration during the growth season or lack of foliage during the dormant season. Usage of drones has increased exponentially, especially to obtain imagery of affected areas during post-event field recognitions (Adams et al., 2016; de Haas et al., 2020, 2022). Still, the ad hoc character of the use of drone imagery hinders a chronological comparison unless systematic flights are conducted in the area of interest (Bremer & Sass, 2012; de Haas et al., 2022).

The large array of satellite imagery products (for simplicity, here also considered airborne-based products) also enables temporal comparison, as they are systematically delivering imagery throughout their operational lifetime. Satellite imagery sensors allow the acquisition of multispectral and hyperspectral imagery with different spatial and temporal resolution (Woodget et al., 2015; Cao et al., 2021). The application of advanced analyses based on optical imagery (e.g., Landsat, Sentinel, Planet Labs), synthetic aperture radar (SAR) data or LiDAR datasets (Bull et al., 2010; Frodella et al., 2017) has improved the quantification of geomorphic changes at the hillslope and fan levels by identifying topography, vegetation cover, color or textural changes in materials (D'Arcy et al., 2018; Zhang & Matsushima, 2018). The combination of different imagery products and new automatic recognition processes based on machine learning has been successful in covering larger regions (Lary et al., 2016), particularly in detecting sediments sources and creating susceptibility maps (Mwaniki et al., 2017). The capacity to redirect satellites to specific regions is also a promising research avenue for real time debris-flow disaster management and forensic analyses.

8.3.2 Radiometric Dating Techniques

Radiometric dating techniques are employed to date different debris-flow deposits, generally recognized by trenches excavated into fan deposits (Fig. 8.3). This absolute dating

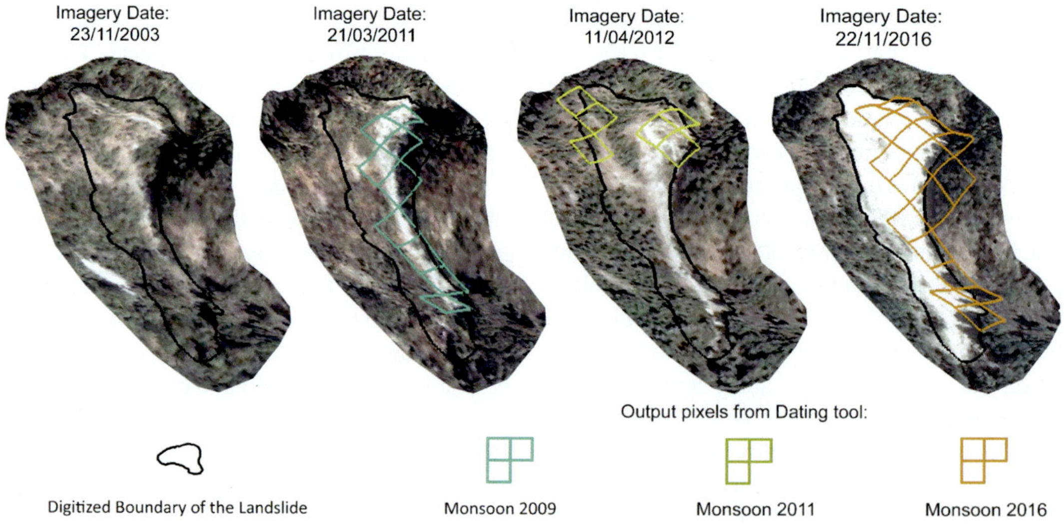

Fig. 8.2 Example of multi-temporal digital aerial imagery analyses (automatic tool) to determine the occurrence of landslides in the source area of catchments that results in debris flows downvalley (Muñoz-Torrero et al., 2021). Squares indicate cloud-free pixels on which the automatic tool identified vegetation changes related to landslides

Fig. 8.3 Example of debris-flood stratigraphy and units from a cutbank at Cougar Creek Creek and Three Sisters Creek, Canmore, Alberta (from Church & Jakob, 2020, with permission)

technique relies on the rate of radioactive isotope decay. By comparing the presence of a radioactive isotope in a sample to its abundance on Earth, one can estimate the elapsed time in absolute terms, as the decay rate is intrinsic to each isotope. Since radioactive components normally have a slow decay rate, this dating method can be used to date organic material or rocks embedded in debris-flow deposits or in paleosols underlying or overlying debris-flow sediments on a scale ranging from decades to many millennia. This timeframe and accuracy depend on the radioactive isotope being analyzed and if accelerator mass spectrometer (AMS) techniques are being used.

Although radiometric dating has been used to determine the timing of exposure of sediment sources in some cases, it is more commonly used to date sediments on fans or in lakes (Kiefer et al., 2021). It means that sediment records must be carefully recognized and interpreted. Sediment characteristics are highly dependent on process type, but are generally composed of a heterogeneous mixture of poorly graded sediment, including gravels, cobble, or boulders (Iverson et al., 2010; Jeong, 2010). Debris flows can also include coarse-grained sand or silt and clay (mudflows) or volcanic material (lahars). During field inspection, the debris-flow fan should be trenched or excavated where land ownership allows and the surface has not been artificially altered. In the excavations, the different units related to different events are differentiated by the texture, color, structure, and lithology of the sediments (Fig. 8.3). Generally, the limits between two events are characterized by the presence of buried soil, or break points between inversely graded soil units. Differentiating events in the sedimentary record may be challenging, as sediment processes may involve the act of local erosive processes (Iverson et al., 1997). Thus, unless the trench covers the complete cross-fan section, this approach tends to miss debris-flow events, and therefore may underestimate the return period.

Radiocarbon dating: this is the most prevalent method (i.e. Macklin et al., 1992; Johnstone et al., 2006; Lancaster and Casebeer, 2007), and is based on analyses of the radioactive ^{14}C isotope contained in organic samples (i.e. vegetation remains, charcoal or organics) that can be found among the various overlaying debris flow deposits. Typically, the temporal coverage of the radiocarbon method expands up to 50,000 years before present (BP). Radiocarbon dating requires the calibration of obtained ages with precise curves because of secular variations in radiocarbon production in the atmosphere (Benito et al., 2022). Thus, getting precisely calibrated ages over the last 300 years is challenging. Moreover, nuclear weapon testing in the 1950s and 1960s doubled the levels of artificial ^{14}C in the atmosphere, inducing a sharp peak in 1963 known as the "bomb curve". The artificial delivery of ^{14}C during this period can be used as age control for younger deposits (Chiverrell & Jakob, 2013). ^{14}C dating requires multiple samples that need to be collected and stored properly, preferably under dry and dark conditions. Generally, just a few grams of organic matter are needed when a conventional method is applied to measure the ^{14}C decay rate. Smaller samples are allowed, if an accelerator mass spectrometry (AMS) method is used (Reimer et al., 2013). These limitations should be considered to define the sampling strategy. Care needs to be taken in interpreting the bounding ages from paleosols vs. the actual dates of organics found within debris-flow units. Also, the age of wood found within a debris-flow unit does not necessarily indicate the age of the unit itself. For example, should the piece of wood be the inner ring of a 200-year old ring, the debris-flow date may be the radiocarbon date minus the 200-years of the tree age.

Luminescence dating: The luminescence approach relies on the accumulation of electrons in crystalline minerals, usually quartz and feldspar, due to their ionizing radiation from cosmic radiation and radioactive isotopes contained in sediments (Lian & Roberts, 2006). Radiation induced by electrons can be measured and depends on sediment burial duration (Lian & Roberts, 2006). Electrons can be stimulated by heat (Thermoluminescence, TL), optical (Optically Stimulated Luminescence, OSL) or even infrared light (Infrared Stimulated Luminescence, IRSL). OSL is the most common technique, normally applied to sediments containing quartz minerals. Similar to flood sediments (Benito et al., 2022), OSL becomes an important dating technique in cases where debris flow deposits lack organic material, or are older than 40,000 years or younger than 200 years (out of the suitable range of ^{14}C) (i.e. Fuchs & Owen, 2008; Zhao et al., 2015; Muñoz-Salinas et al., 2020). To assure the reliability of dating, sample collection and processing must comply to stringent protocols (Liritzis et al., 2013).

Other radiometric techniques include the **Caesium-137** (^{137}Cs) isotope, which is used to determine the age of sediments over the last century and date debris flow events (i.e. Irmler et al., 2006). The isotope ^{137}Cs has an anthropogenic origin derived from the nuclear weapon testing that started in the 1950s to produces two characteristic peaks: an initial signal in 1954 and a maximum in 1963 (Détriché et al., 2010). These signals are used to calibrate sediment age. By contrast, analyses of **Berilium-10** (^{10}Be), an isotope produced on Earth by the interaction of energetic cosmic rays, allow dating of the exposure of ancient rock surfaces (above 1ka years). This technique is typically used to date paleodeposits and to explain landscape heritages or ancient events (i.e. Dühnforth et al., 2007; Patton et al., 2018).

8.3.3 Biological Dating Techniques

Biological dating techniques rely on information gathered from living organisms such as plants or lichens. This approach is based on the autotrophic conditions forcing organisms to get the energy they need to survive directly from their living environments Thus, these organisms are silent witnesses of environmental and geomorphic changes (Ballesteros-Cánovas & Stoffel, 2017; Stoffel & Corona, 2014) and adjust their growth conditions accordingly. Analyses of the annual growth rates of these organisms allow an absolute dating of landforms created by debris flows.

Tree-ring records: Dendrochronology is the science of dating growth patterns in trees and shrubs species living in zones where rings are formed annually. Dendrochronology has been applied to a large array of disciplines, including archaeology, climatology or ecology (Cook and Kairiukstis, 2013). Tree rings have also been utilized to retrospectively date the activity of geomorphic processes such as debris flows (i.e. dendrogeomorphology). The interaction between the material transported in a debris flow events and the existing vegetation, mostly on fans or along channels, will leave different types of dendrogeomorphic evidence in trees (e.g. damage on tree stems, buried or tilted stems or exposed roots) (Ballesteros-Cánovas et al., 2015; Bollschweiler & Stoffel, 2010). These mechanical disturbances can be identified in tree-ring records, allowing the dating of past debris flow events with annual or even seasonal precision. Dendrogeomorphic evidence induced by debris flows include: (i) *Stem wounds (or scars):* Scars are the most common evidence of debris flow activity caused by the impact of boulders on tree stems (Fig. 8.4a and b). When such damages occur, trees react upon this disturbance by compartmentalizing the wounded surface to avoid pathogen infections (Fig. 8.4c). As a result, trees generate characteristic cells including chaotic callus tissues and smaller cells that are easily recognizable in the tree-ring record (Fig. 8.4d) as well as—in the case of some conifers—tangential rows of traumatic resin ducts (TRD); (ii) *Tilted trees*: as a result of the sudden and unidirectional pressure exerted on tree stems by debris flow material, trees can lean towards the channel or in the flow direction after debris flow impact. Trees that are tilted will form reaction wood to regain the vertical axis, and this reaction is characterized in the growth-ring record by changes in the eccentric of growth rings as well as wood properties (i. e. colour); (iii) *Buried trees*: sediment deposited around trees by a debris flow will prevent water and/or nutrient supply to reach the roots, hence decreasing growth rates. In some cases, trees (especially younger individuals) may lose their crown or branches, which will result in a sudden growth suppression as well; (iv) *Exposed roots*: erosive processes along the channel or levees can expose roots, and the latter will react to denudation with conspicuous wood anatomical changes; (v) *Surviving trees and new colonization:* debris flows can uproot and remove trees in heavily affected areas. Survivor trees will benefit from less competition, more light, nutrients and water availability, thus experiencing an increase in their growth rates. Moreover, new plants often germinate on bare surfaces providing evidence of the minimum age of landforms.

Dendrogeomorphic dating involves the following methodological steps: (i) recognition and record of botanical evidence, generally at the level of fans and/or in and along channels, (ii) sampling of trees using increment bores (eventually hand or machine saws); (iii) preparation of samples in the laboratory; (iv) measurement of tree-ring widths and its synchronization for cross-dating purposes; (v) identification of growth

Fig. 8.4 Examples of botanical evidence (i.e. scars on trees) affected by debris flows. **a** cross-section sample of a dead tree scarred by past debris flow, showing how the tree tried to close the wound. **b** Sampling procedure of a living scarred tree using an increment borer. **c** Core sample extracted from the edge of a wound, where it is recognized the growth disturbance related to debris flow processes (assumed it produced the scar on the tree) in the tree ring records. d) anatomical signatures in the tree-ring record in the form of chaotic callus tissues and tangential rows of traumatic resin ducts (anatomical pictures are from Bollschweiler et al., 2008)

Fig. 8.5 Different generations of lichens growing on a debris flow deposit containing material from different events. Note that the combination of the dating of disturbance in trees and the assessment of lichen thalli can improve the interpretation of deposits. Photo: Juan A. Ballesteros, Aksay cone, Kyrgyzstan

anomalies associated to each type of dendrogeomorphic evidence. If the ultimate objective is to estimate debris flow magnitude, one must carefully record tree positions aiming to determine the areas that have been affected by each event as well as the height of scars in order to estimate flow discharge with numerical models or empirical equations (Ballesteros-Cánovas et al., 2015).

Lichenometry: Lichenometry is a biological dating technique using growth rates of lichens to assess the age of exposed rocks (Armstrong, 2016). It assumes that the age of older lichens growing on boulder surfaces can be used to determine the exposure time of that rock. Typically, lichenometry has been applied to estimate the minimum age of monuments, but it has also been used to establish absolute dating of geomorphic landforms such as debris flows (Alcalá-Reygosa et al., 2018; Armstrong, 2016; Bernhardt et al., 2017; Helsen et al., 2002; Fig. 8.5). The initial assumption of lichenometry is that growth of lichens depends on the species being characterized and environmental factors (Innes, 1985). The age of lichens is then compared with reference age-growth curves, specific to each species and site. These curves are based on the ages of individual lichens growing on a given object that is known to be of a certain age (normally anthropogenic surfaces such as monuments, buildings or cemetery headstones). The approach is especially useful for the analysis of debris flow deposits in the absence of trees (or of older trees) with a temporal coverage of up to 1,500 years depending on the lichen species (Rosenwinkel et al., 2015).

Rhizocarpon sp. is the most trustworthy (and commonly used) lichen species since it is adaptable to a variety of environments. It can survive under severe climatic conditions and is easily recognizable. The main steps in lichenometric assessments include: (i) measuring lichens with a millimetric feeler or by photographing them both on boulders to be analyzed and on anthropogenic surfaces for establishing the lichens reference chronology; (ii) applying statistical procedures to determine size distribution (i.e. using generalized extreme value GEV); (iii) building the reference chronology (growth-age curves) using known ages, and (iv) determining absolute ages of ensured lichens using the established reference chronology. Lichenometry has been successfully applied for the dating of debris flow deposits in a number of environments (André, 1990; Bernhardt et al., 2017; Helsen et al., 2002; Jomelli, 2013). Nonetheless, the approach remains highly sensitive to changes in environmental conditions (e.g. wildfires) or pollution, which can induce significant uncertainty into dating procedure (Rosenwinkel et al., 2015).

8.3.4 Historical Records

The historical record refers to a set of documentary archives (manuscripts, newspaper libraries, ecclesiastical documents, administrative records, letters of varying natures), oral records (testimonials of people and direct observation from land-use planners/managers), audio-visual records (old photographs and maps or modern social media posts) and epigraphic marks (physical signs left by past events) associated to unrecorded past debris flow events that eventually impacted infrastructures or inhabitants (Marchi & Cavalli, 2007). Overall, these records are constrained to the past few decades and only exist for a few torrents, which are often located close to urban areas. The historical record can provide quantitative and qualitative data, with a high space–time resolution. Quantitative data can include the volume mobilized, a rough estimation of peak discharges or even areas affected. Qualitative data includes descriptions of the impact and the relative magnitude (e.g. this event is the largest ever seen by locals).

This approach is based on hermeneutic research, which requires a critical analysis of data sources and their analysis in the historical context (Marchi and Rosella-Tecca, 2006). Archival data is beneficial to understand debris flow occurrence at regional and catchment scales. Regional analyses can provide a broad perspective and helps to identify morphometric parameters of susceptible basins, while analyses at the catchment level is relevant for hazard assessment and to understand the relationship between specific events and sediment supply processes (Brenna et al., 2020). Other examples have shown how historical debris flow databases can be used to decipher climatic triggers (i.e. rainfall thresholds and temperature; Pavlova et al., 2014; Schlögl et al., 2021; Witt, 2005).

There are a number of considerations that must be taken into account when using historical records of past debris flows, notwithstanding their great value (Marchi and Rosella-Tecca, 2006). The sustained growth of elements-at-risk (e.g. road infrastructures, paths, houses, etc.) could affect the homogeneity of historical records, because events that did not cause damage in the past will tend to occur unnoticed. The availability of data sources has the same impact on historical database quality. Moreover, it may be challenging to differentiate between event types based just on historical records. On the other hand, the homogeneity of records may be affected by different sediment-supply conditions within or between catchments. Thus, debris flows in catchments with an unlimited debris supply are deemed normal processes, and only extreme events are recorded. By contrast, events that occur in catchments with limited debris supply are seen as unusual, regardless of their magnitude, and will therefore more often be recorded.

8.4 Debris Flow Magnitude Estimates

Knowing the magnitude of debris-flows is essential for any rational hazard and risk assessment (Ballesteros-Cánovas et al., 2016; Jakob, 2021). Debris flow magnitude relates to quantitative or qualitative data regarding the volume of sediments mobilized (m^3), peak discharge (m^3/s) or inundated area (m^2) (Jakob & Hungr, 2005). These parameters can be estimated directly by applying physical-based equations and resolving hydrodynamic models with step-backwater schemes (Jakob, 2021), or indirectly, by means of established empirical relationships

(Rickenmann, 1999). Selection of one approach over another is highly influenced by the primary objective of the analyses, as well as data availability and terrain inaccessibility.

8.4.1 Estimating Debris Flow Volume

Debris flow volume is defined as the amount of material transported from headwaters to fans by debris flows. During debris flow events, the transported material (water and sediments) can range in volume from a few cubic meters to several million cubic meters. Quantifying these values is essential to physically dimension hydraulic structures aimed at retaining sediment (e.g., check dams, retention basins). In the few torrents where measurement stations are in operation volume may be directly measured, typically by integrating discharge over time (Schimmel et al., 2022; Schlunegger et al., 2009), but typically volume can only be reconstructed from past deposits. Estimation of debris flow volume depends on three processes: (i) initiation volumes related to failure(s) (V_i) at hillslopes (i.e. landsliding); (ii) volumes entrained along channel(s) (V_e) from the channel floor and banks; and (iii) volumes of sediment deposited (V_d) along the channel and fan (i.e. levees, chutes, lobes, avulsions phenomena) (Jakob, 2005). Mathematically, it can be expressed as (Eq. 8.1):

$$V_t = \sum V_i + \sum V_e - \sum V_d \qquad (8.1)$$

where V_t is the total volume reaching the apex of the fan. In complex medium-to-large catchments, it may be quite challenging to estimate these values in detail, whereas in small, accessible catchments, it may be feasible. The volume entrained and deposited along the fan should be considered into this equation if the assessment is done downstream of the fan. Moreover, this volume estimate only provides an idea of a volume should the debris flow occur in the present time. Recharge over time needs to be considered (Jakob, 2005), especially in supply-limited catchments, as well as the possibility of multiple point source failures that are particularly prevalent during earthquakes or extreme rainstorms.

8.4.2 Initiation Volumes at Hillslopes

As a result of various geomorphic processes, such as landslides, rock avalanches, rock glacier advances, wet snow avalanches, or the outburst of artificial or natural dammed-lakes, large sediment volumes can be mobilized at the hillslope and incorporated into the channel (Pastorello et al., 2020). However, the volume of sediments reaching the channel will depend on the connectivity between the slope processes and the channel (Tarolli and Della Fontana, 2009; Germain et al., 2020; Torresani et al., 2021), as this is an important catchment attribute driving debris flow activity. Remote sensing approaches are highly valuable for predicting which geomorphic processes are involved at sediment sources, as well as for quantifying the initially mobilized volume.

If scars on hillslopes are not hidden by a dense tree canopy, geomorphic changes related to landsliding or other processes can be traced spatially using available chronosequence archives of old and more recent aerial photos (Muñoz-Torrero et al., 2021; 2022). Regarding the above, optical sensors used on aerial pictures cannot provide relevant information, thus making the use of LiDAR sensors essential (Blasone et al., 2014; Bremer & Sass, 2012; Dietrich & Krautblatter, 2019). Recently, the use of Interferometric Synthetic Aperture Radar (inSAR) sensors has opened the possibility for large-scale studies of slope deformation under forested terrains, especially using longer-wavelength (L/P-band) sensors (Cao et al., 2021; Frodella et al., 2017). When using topographic data retrieved by photogrammetry, LiDAR or inSAR sensors, the volume can be determined by subtracting DEM values (Difference of DEMs or DoD). Still, these estimates may need to be checked in the field and depth of slope failures measured at regular time intervals to be sure they are correct.

8.4.3 Entrainment of Sediment Along Channels

The entrainment of sediment by debris flows can radically change channel morphology, thereby increasing flow volume by orders of magnitude (Iverson et al., 2011). There are two major processes that can lead to the material of entrainment along channels, namely: (i) channel bed erosion as a result of drag forces; and (ii) undercutting and instability of stream banks (generally as a consequence of bed erosion) (Iverson et al., 2011). The yield rate approach is the most straightforward way to calculate channel entrainment (V_e), which assumes a homogeneous erosion depth (D_i, in m) along a given path width (B_i, in m) and channel reach (L_i, in m). It can be mathematically expressed as (Eq. 8.2):

$$V_e = \sum_{i=1}^{n} D_i x B_i x L_i, \quad (8.2)$$

where i is the number of reaches into which the stream has been discretized.

This approach, however, can produce significant inaccuracies depending on the complexity of the channel, the number of channel reaches and the nature of the substrate (especially in erodible-base channels). It also requires detailed field observations along the channel(s), which are not always easy to obtain. In addition, the approach assumes invariant rheology and disregards the potential of material remobilization during debris flows. In recent years, the assessment of channel entrainment has been facilitated by photogrammetric and structure from motion (SfM) techniques (de Haas et al., 2020, 2022; Ellett et al., 2019), the versatility of drones, and the use of high-resolution topographic techniques such as laser scanning (Schürch et al., 2011), as well as the use of proxies such as timing of root exposure to reconstruct channel topography and to determine the widening and erosion retreat (Malik and Matzja, 2008; Ballesteros-Cánovas et al., 2017; Franco-Ramos et al., 2022).

8.4.4 Estimation of Deposited Volumes

Estimating the volume of deposited sediment can be addressed based on (i) a detailed mapping of the surficial deposits—either based on field recognition (Stoffel, 2010) or remote sensing techniques (de Haas & Densmore, 2019; Imaizumi et al., 2016b); and (ii) quantification of deposit depths. Once fans have been mapped, deposit depths can be estimated by developing pits tests and trenches. This approach is necessary when studying past debris flows, but it is time-consuming and labor-intensive, as it requires an observation period that is long enough to determine an absolute date. In forested areas, the number of observations could be increased by using buried trees that can record the time of a given debris flow and sediment depth (Friedman et al., 2005). However, for recent events the most accurate approach for estimating the volume of deposits is to conduct regular topographic surveys (Lato, 2020).

8.4.5 Estimating Peak Discharge

Peak discharge along channels can be determined at a given cross-section by solving the discharge equation (Eq. 8.3):

$$Q = \overline{v} x A \quad (8.3)$$

,where $\overline{v}$ is the frontal flow velocity and A the affected cross-sectional area. In addition, numerical-based models along the entire channel system or empirical relationships can be employed (see Sects. 3.4 and 3.5, respectively). Estimation of the cross-sectional area depends directly on the evidence left by past debris flows from which the affected area can be estimated, while the velocity needs to be indirectly estimated by recommended equations.

Estimating the affected cross-sectional area. Applying the discharge equation requires approximate stability of the channel's cross-section geometry. If this requirement can be satisfied, the existence of fresh high watermarks (HWM, e.g., erosion lines, fresh-thin deposits or mudlines), or paleostage indicators (PSI, such as scars on trees) is required (Bodoque et al., 2015; Benito et al., 2022). Generally, HWM are unequivocal evidence of the maximum flow stage, but they are perishable and only physically present for a few years at most after an event. Erosion lines may persist longer in the field, but they are difficult to date. PSIs are evidence of the flood stage but not necessarily the maximum stage. However, by including multiple observation locations and multiple scars, this bias can be reasonably minimized (<15%) (Ballesteros et al., 2011; Gottesfeld, 1996). Scars on trees can be retrospectively dated, enabling for the reconstruction of past events (Tichavský et al., 2022; Cato & Goforth, 2021; Rodriguez-Morata et al., 2019). Pictures and video-records, if available, can be employed to reconstruct peak discharge and estimate hydrographs and potentially sediment concentrations (Coviello et al., 2021).

Estimating flow velocity. Back-calculation of the flow velocity of debris flows is highly challenging and depends on indirect equations such as flow superelevation, run-up, or channel characteristics. Tables 8.1 and 8.2 show the main equations used to estimate debris flow velocity along with the required parameters.

The *superelevation equation* relies on the deformation properties of the free surface of a fluid due to the centrifugal force experienced in channel bends. It is expressed as the difference in flow height between the inner (lower) and outer (higher) curves and has been applied largely in debris flow studies (Iverson et al., 1994; Prochaska et al., 2008a, 2008b; Wang et al., 2021; Scheidl et al., 2019; Fig. 8.6). This approach is highly sensitive to the definition of channel width and curve radius, which may lead to uncertainties in the estimation of flow velocities of up to 30% (Iverson et al., 1994). The superelevation equation was originally developed for Newtonian fluids, and applying the method to non-Newtonian debris flows may result in large under- or overestimations of flow velocity (e.g., Costa, 1984; Procter, 2012). Therefore, many studies have used a correction factor k for debris flows (e.g., Bulmer et al., 2002; Han et al., 2015; Hungr et al., 1984; Scheidl et al., 2019). This correction factor is generally expressed as a dimensionless number that is related to the vertical velocity distribution and corrects for variations in material properties, viscosity and vertical sorting in the flow (Procter, 2012; Scheidl et al., 2015, 2019). Studies indicate that a wide range of possible correction factors is possible (Procter, 2012). For example, Bulmer et al. (2002) found correction

Table 8.1 List of equations to determine debris flow velocity. Other equations based on flow depth are listed in Rickenmann (1999)

Equations	Type	Author
$v = (gr_c\cos\theta\tan\alpha)^{0.5}$	Superelevation	Chow (1959)
$v = \sqrt[2]{\frac{r_c g}{k}\frac{\Delta h}{B}}$	Superelevation	Hungr et al. (1984)
$v = \left[\frac{r_c\Delta h(g\cos\varnothing)^{0.4}}{4.4Bhm^{0.6}}\right]^{1.19}$	Superelevation	Scheidl et al., (2015)
$v = (2g\Delta h)^{0.5}$	Runup (Δh= runup against an obstacle)	Chow (1959)
$v = 1.21g\Delta h$	Runup	Wigmosta (1983)
$v = (\gamma S/K\mu)H^2$	Runup	Hungr et al., (1984)
$\frac{H}{\vartheta^2/g} = \frac{\left(1+\frac{\varkappa}{2}\frac{cos^3\theta}{Fr^2}\right)^2}{1+\frac{\tan\varnothing}{\tan\theta}}$	Runup (momentum-jump model)	Iverson et al., (2016)
$\frac{H}{h_m} = 1 + \frac{v^2}{2gh_m}$	Runup (point mass model)	Rickenmann et al., (2019)
$v = 0.6\frac{L}{k}\sqrt{\frac{g}{2(Hc+h_e)}}$	Critical flow in spillways	Takahashi et al. (2019)

Table 8.2 Overview of debris flow velocity equations primarily based on mean flow depth and channel slope

Equation	Type	Author
$v = \frac{1}{3}\rho g \overline{h}^2 \frac{S}{\mu}$	Newtonian laminar flow	Rickenmann (1999)
$v = \frac{2}{3}\xi \overline{h}^{1.5} S$	Dilatant grain shearing	Takahashi (1991)
$v = \frac{1}{n}\overline{h}^{2/3} S^{0.5}$	Newtonian turbulent flow: Manning–Strickler equation	Mizuyama (1988)
$v = C\overline{h}^{0.5} S^{0.5}$	Newtonian turbulent flow: Chézy equation	Rickenmann (1990)
$v = 2.17\overline{h}^{1.5} S^{0.5}$	Empirical equations	Hungr et al. (1984)
$v = 18.5\overline{h}^{2/3} S^{0.5}$		Yang (1985)
$v = 28.5\overline{h}^{2/3} S^{0.5}$		Du et al. (1987)
$v = C_1\overline{h}^{0.3} S^{0.5}$		Koch (1998)
$v = \delta\overline{h}^{2/3} S^{1/5}$		Lo (2000)
$v = 0.35\overline{h}^2 S + 5.36$		Prochaska et al., (2008a, b)
$v = 2.379u^* S^{-0.446} C_v^{\;0.011\ln(S)+0.391}$		Wang et al. (2018)

Fig. 8.6 Example of differences in flow height (Δh) between in both banks (right—RB - and left bank - LB) after an extreme debris flow event occurred in September 2022 in Ordesa National Park (Pyrenees, Spain) (Photo: Juan A. Ballesteros-Cánovas)

Fig. 8.7 Example of a critical section in a debris-flow torrent where erosive sigmoidal lines are identified along the levees. Arroyo Cabrera, Spain (Bodoque et al., 2011)

factors ranging from 2 to 10, VanDine (1996) found values between 1 and 5, and Hunger et al. (1984) found a best-fit value of 2.5.

The *run-up equation* (also known as the velocity head equation) assumes that kinetic energy is converted into potential energy when debris flows impact obstacles perpendicularly. At experimental sites, this method has been shown to overestimate flow velocity (Iverson et al., 1994). Recently, these concepts have been applied to investigate run-up on flows against vertical barriers and sloping obstacles, providing reliable results if prior knowledge of the flow dynamics is available (Iverson et al., 2016). Yet, Rickenmann et al. (2019) observed that run-up is highly dependent on the flow sediment type and volume, suggesting that it is challenging to apply a universal equation to varied flow conditions. If a critical-depth approach is employed, flow velocity estimation uncertainties could be substantially minimized (Bodoque et al., 2011; Fig. 8.7). This approach requires evidence on cross sections where critical flow occurs (Froude number = 1) due to local drops, narrow stream constrictions, or check dam spillways (Takahashi et al., 2019).

The velocity equations described above rely on superelevation and runup. Therefore, they cannot be applied to debris flows in a straight channel (Chen et al., 2007). To estimate debris flow velocity in straight channels a number of studies have developed equations based on mean flow depth and channel slope (Table 8.2). Detailed overviews of these equations can be found in Rickenmann (1999), Prochaska et al. (2008b) and Wang et al. (2018).

8.4.6 Estimating the Affected Area

Evaluation of the area affected by debris flows is crucial for both hazard mapping and the design of countermeasures. At the fan level, avulsion phenomena are primarily responsible for the spatial and temporal distribution of sediments. Recognition of the affected area of debris flows can be achieved by combining remote sensing with field inspections and analyzing the spatial distribution of affected trees. Imaizumi et al. (2016b) used a multi-approach (i.e. chronosequence of aerial pictures, high-resolution topographic data (Zubrycky et al., 2021), and dendrogeomorphic evidence (Schraml et al., 2013)) to identify the affected area at the fan level, and discussed their weaknesses and strengths. They found mismatches between methods, although their combination improves the assessment of areas affected by individual events. Higher resolution can be achieved using high-resolution topographic data and DEM differences (Lato et al., 2019), but care should be taken to distinguish geomorphic changes from other local phenomena (Imaizumi et al., 2016b). Zaginaev et al. (2016) employed the spatial representativeness of affected trees on fans to distinguish size-patterns in debris-flow events. However, some limitations may arise on a higher scale because of the nature of data and the limited number of point-type data (Bollschweiler et al., 2008). Schneuwly et al. (2013) provide guidelines for determining the minimum number of trees to sample for capturing the spatiotemporal representativeness of past debris flows. Recently, these approaches have enhanced our understanding of the spatiotemporal patterns of debris flow fan formation and, in particular, on the avulsion formation, revealing that the impact (and consequently the affected area) of large debris flows could depend on the existence of channel plugs shaped by prior smaller debris flows (de Haas et al., 2018a, 2018b; see also de Haas et al., this volume).

8.4.7 Numerical Models

Different numerical debris flow models have been developed to back-estimate flow velocities, where the models are resolved iteratively to reproduce the different HWMs, PSIs, or the affected area (Bodoque et al., 2015). Based on Riemann solvers, finite difference schemes, or finite volume schemes, numerical models are derived from mathematical formalisms describing flow dynamics with different rheology. Most numerically-based models are two-dimensional, although there are a few three-dimensional hydrodynamic models. Two-dimensional models are easier to set up and can simulate runout of large-scale debris flows reasonably well, although they still provide inaccurate predictions of bank erosion and scour during debris flow events. Three-dimensional models can simulate material flow in complex terrain and deliver a more realistic representation of debris flows, but they require a larger amount of computational resources and frequently cannot simulate events on a large scale. A detailed (and up-to-date) review of debris flow models can be found in Trujillo-Vela et al. (2022). Table 8.3 shows a list of available numerical models and for a more comprehensive description, see Iverson and George (this volume).

8.4.8 Empirical Relationships

Empirical equations can be used to represent relationships among flow discharge, flow depth, thickness, affected area and deposit volume. These equations are simple to use, although they may lead to significant errors. Several efforts exist to produce more trustworthy regional empirical equations, although their application is still constrained to the specific climatic, geological and geomorphic settings where they were defined. Table 8.4 synthetizes the main existing empirical relationships between (i) peak discharge with debris flow volume; (ii) volume, watershed area, channel slope and precipitation; and (iii) affected areas. Yet, empirical equations can also misjudge large debris flows, as they often rely on a limited number of (recent) observations, which could relegate the occurrence of large debris flows to "atypical" cases. Moreover, the interdependence of morphometric factors, such as stream slope/length and catchment area, may hide spurious correlations (Jakob, 2005). Nevertheless, empirical relationships can be applied to characterize events with an order-of-magnitude resolution in cases where other detailed assessments are lacking or impossible to perform at a regional scale. Recently, machine learning approaches have improved the ability to predict the frequency and magnitude of debris flows based on a large dataset analysis, which include morphometry characteristics, as well as climatic, soil-type and forest parameters (Huang et al., 2020), which can then be used to produce susceptibility maps (Kadavi et al., 2018).

8.5 Frequency-Magnitude Relationships

Frequency-magnitude (FM) relationships are the most useful tool derived from the reconstruction of past debris flow events (see Jakob & Nolde, this volume). FM relationships allow estimating of the return period (probability of occurrence of a debris flow event of a certain magnitude), which is required for hazard analyses and the design of countermeasures. In Switzerland, for

Table 8.3 List of existing debris flow models, their main characteristics and references

Name	Type	Availability	References
HEC-RAS	Two-dimensional non-Newtonian fluids including single-phase	Non-commercial	https://www.hec.usace.army.mil/software/hec-ras/download.aspx
EDDA 2.0	Two-dimensional non-Newtonian fluids including single-phase	Non-commercial	https://doi.org/10.5281/zenodo.1033377 Shen et al. (2018)
Morpho2DH	Two-dimensional non-Newtonian fluids including single-phase	Non-commercial	https://i-ric.org/en/solvers/morpho2dh
FLO-2D	Two-dimensional non-Newtonian fluids including two-phases	Comercial	https://flo-2d.com/
MIKE 21	Two-dimensional Newtonian fluids including two-phases	Comercial	https://www.mikepoweredbydhi.com/products/mike-21-3
BASEMENT	Two-dimensional Newtonian fluids including two-phases	Non-commercial	https://basement.ethz.ch/
FLOW 3D	Three-dimensional Newtonian fluids including two-phases	Comercial	https://www.flow3d.com/
IBER	Two-dimensional Newtonian fluids including two-phases	Non-commercial	https://www.iberaula.es/
TRENT-2D	Two-dimensional non-Newtonian fluids including two-phases	Commercial	http://www.weezard.eu/index.php
TITAN-2D	Two-dimensional non-Newtonian fluids including two-phases	Non-commercial	https://github.com/TITAN2D/titan2d
Kanako 2D	One and two dimensional Newtonian fluids including two-phases	Comercial	http://hyper-kanako.com/hyper-kanakoenglish/
OpenLISEM	Two-dimensional non-Newtonian fluids including two-phases	Non-commercial	https://blog.utwente.nl/lisem/
Morpho2DH			
D-Claw	Two-dimensional non-Newtonian fluids including two-phases	Non-commercial	https://dlgeorge.github.io/project/dclaw-project
OpenFOAM	Three-dimensional non-Newtonian fluids including two-phases	Non-commercial	https://develop.openfoam.com/Community/avalanche
r.avaflow	Two-dimensional non-Newtonian fluids including one and two-phases	Non-commercial	https://www.landslidemodels.org/r.avaflow/
RAMMS	Two-dimensional non-Newtonian fluids including one phase	Comercial	https://www.slf.ch/de/services-und-produkte/ramms-rapid-mass-movement-simulation.html
Flow-R	Spatially distributed empirical model	Non-commercial	https://www.flow-r.org/ Horton et al. (2013)
DFLOWZ	Based on empirical equations	Non-commercial	Berti and Simoni (2014)
REGOLITH	Empirical and process-based models	Non-commercial	https://www.usgs.gov/software/regolith-fortran-95-program-estimating-soil-mantle-thickness-digital-landscape-landslide
Laharz	Statistically based model	Non-commercial	https://www.usgs.gov/software/laharz

Table 8.4 List of empirical relationships: (i) Eqs. 1–9, peak discharge ($Q_{dp} = m^3/s$) versus volume ($V = m^3$); (ii) Eqs. 10–14, debris flow volume ($V = m^3$) versus watershed area ($S = km^2$), effective runoff volume ($V_{ref} = m^3$), channel slope ($I = \%$) and total rainfall ($VR = m^3$); and (iii) Eqs. 15–19, empirical–statistical methods relating valley cross-sectional area (A), planimetric area (B) and runout (L) inundated by lahars from lahar volume (V). (*) These relations present a large scatter. Readers are directed to the original work (Ma et al., 2013) to limit the upper and lower bounds of estimations

	Equation	Type	References
Peak discharge (m^3/s)	$Qdp = 0.135V^{0.78}$	Bouldery debris flow	Mizuyama et al. (1992)
	$Qdp = 0.019V^{0.79}$	Muddy debris flow	Mizuyama et al. (1992)
	$Qdp = 0.006V^{0.83}$	Volcanic debris flow	Jitousono et al. (1996)
	$Qdp = 0.04V^{0.90}$	Bouldery debris flow	Bovis and Jakob (1999)
	$Qdp = 0.003V^{1.01}$	Volcanic debris flow	Bovis and Jakob (1999)
	$Qdp = 0.1V^{0.83}$	Bouldery debris flow	Rickenmann (1999)
	$Qdp = 0.00014V$	Muddy debris flow—high sediment concentration	Chen and Chuang (2014)
	$Qdp = cV^{5/6}$	c is a constant that ranges between 0.001 and 1, with values of 0.01 typical of muddy flows and 0.1 typical of granular flows	Ikeda et al. (2019)
	$Qdp = 0.2V^{5/6}$	Granular flows	Mitchell et al. (2022)
Debris flow volume (m3)	$V = 1800\ S^{0.98}$(*)	Debris flow on burned areas-USA	Ma et al. (2013)
	$V = 4376\ S^{0.73}$(*)	Debris flow on alpine catchments	Ma et al. (2013)
	$V = 337{,}000\ S^{1.4}$(*)	Debris flow on tropical catchments	Ma et al. (2013)
	$V = 1.25V_{ref}\ I^{1.5}$	Debris flow and debris floods	Rickenmann and Koschni (2010)
	$V = 10^{(0.019I\_1.55)VR0.877}$	Debris floods with channel gradients from approximately 2%–24%	Jakob et al. (2022)
Affected area and runout (m^2/m)	$A = 0.05V^{2/3}$ and $B = 200V^{2/3}$	Lahars	Iverson et al. (1998)
	$B = 6.2V^{2/3}$	Granular debris flows	Crosta and Negro (2003)
	$A = 0.1V^{2/3}$, $B = 20V^{2/3}$	Non-volcanic debris flows	Griswold and Iverson (2008)
	$A = 0.14V^{0,52}$ and $B = 6{,}9V^{0,83}$	Granular debris flows	Berti and Simoni (2007)
	$L \cong 65.3\varnothing^{-\frac{1}{2}}V^{\frac{1}{3}}$	Fans with uniform runout patterns and aperture angle ranging from 13° to 51°	Rickenmann and Scheidl (2013)
	$\alpha = 0.88\beta$	Being α and β the runout angle and fanhead angle, respectively, measured from the point on the stream located in the mid-altitude between the apex of the fan and the catchment divide	Prochaska et al. (2008a, b)

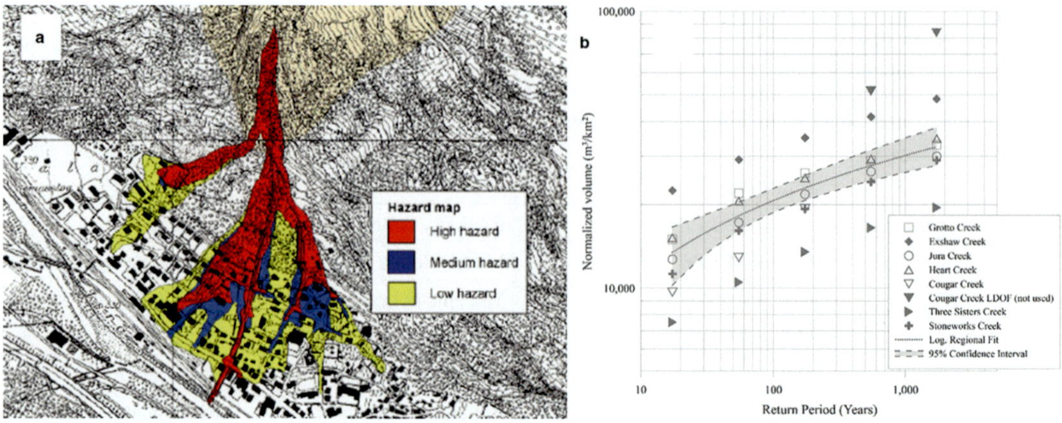

Fig. 8.8 **a** Example of debris-flow hazard zonation in Switzerland. **b** Example of regional frequency-magnitude curve built for Bow Valley debris floods normalized by fan area (from: Jakob et al., 2020, with permission).

instance, hazard maps are required for inhabited areas and form the basis of land-use planning, risk assessments and the design of technical or nature-based countermeasures. Hazard maps are based on information of past events, process modeling and an explanatory report; they show the expected extent of debris flows (in terms of their intensity) for three probabilities of occurrence (typically given with 30, 100 and 300 years, corresponding to 1, 3 and 10 generations of residents) (Fig. 8.8a). In Canada, the return period ranges include 1000, 3000 and sometimes even 10,000 years. In the Swiss case, the combination of intensities with variable event probabilities will result in five distinct danger levels indicated with colors as in Fig. 8.8. Zones for which a significant hazard (red) is defined, people are considered at risk both inside and outside buildings. Here, the rapid destruction of buildings is to be expected or events are of smaller intensity but occur with high probability. The red area is excluded from any future development. In areas considered to be at medium hazard (blue), persons are mostly exposed to the hazard outside of buildings. Some damage to buildings is expected, but their complete destruction is unlikely if certain construction standards are present. The blue area is an area in which severe damage can be avoided by taking appropriate precautionary measures on individual building level. In the third zone, defined by low hazard (yellow), persons are considered relatively safe. Minor damage to buildings or obstructions can occur. The yellow area is essentially a warning area and rarely shown on debris-flow hazard maps (unlike flooding or snow avalanche hazard maps). Similar schemes are also recommended in other European countries (EXCIMAP, 2007).

In palaeohydrology (Wilhelm et al., 2019), FM relationships have been established by fitting a statistical distribution to a censored dataset of reconstructed events at the catchment or regional scales. Reconstructed debris flow datasets can be fitted using the extreme-value theory, specifically peak over threshold (POT) and Generalized Pareto Distributions (GPD) functions (Jakob et al., 2017). Bayesian approaches can be an appropriate technique for combining expert opinion with available datasets and related uncertainties to infer FM relationships as well, however with a certain prediction limit (Nolde & Joe, 2013; Jakob & Nolde, this volume). Statistical methods to establish FM relationships are not to be interpreted as precise. First, reconstructed series are far from being perfect datasets, and require proper assessment and management of uncertainties associated with debris-flow reconstructions. Second, torrential catchments may produce events with different rheology (i.e. rock avalanches, debris flows, hypercontracted floods, clear-water floods), which are generally difficult to distinguish in reconstructed series. Lack of this information may violate the statistical assumption which states that random variables are independent and identically distributed, when fitting a distribution function. Third, often FM statistical methods routinely assume stationarity, which could be wrong for debris-flow processes, since these events could respond not only to thresholds (Bardou & Jaboyedoff, 2008) but also to climate change. For the abovementioned reasons, the applicability of FM relationships should be performed with caution (i.e. Heiser et al., 2022) and validated by an expert approach, especially when applied at catchment scale and for debris-flow hazard assessment (Jakob & Nolde, this volume).

Several studies have focused on addressing these limitations at regional and catchment scales over the past few decades. At the *regional scale*, for instance, Hungr et al. (2008) generated FM regional curves along the west coast of British Columbia by plotting the frequency against the magnitude on a logarithmic scale that accounted for magnitude categories, thereby avoiding effects of data censoring. In another study at a regional scale, Liu et al. (2008) used a large debris flow database in China to infer FM relationships by applying power-law probability distributions and using the catchment area as a surrogate of the frequency of events. More recently, Jakob et al. (2020) used a multiproxy-based reconstruction to retrieve regional FM by fitting logarithmic curves and normalizing by morphometry of fans (Fig. 8.8b). At the *catchment scale*, Bardou and Jaboyedoff (2008) and Bennett et al. (2014) reported that the FM relationship derived from observed and historical debris flow data from the Alps followed a heavy-tailed pattern. This was corroborated by Liu et al. (2008) who used more than 40 years of observations to fit exponential functions for generating FM relationships. However, they also showed the difficulty of providing a unique function in catchments where different types of events could occur. Consideration of two distinct populations of data according to the different initiation processes (i.e. rainfall-generated landsliding causing smaller debris flows and large debris flows as consequence of rock avalanches) could be a solution to avoid issues related to the heterogeneity of data (Jakob & Friele, 2010). Gao et al. (2019) used the power law equation to relate the frequency and magnitude of debris flow events in different catchments in China and they also employed a linear model to investigate the links between the regression coefficients of the equation and precipitation records and drainage area. Recent observations in experimental flumes by de Haas et al. (2018b) revealed that changes in the FM curves, could also condition the occurrence of avulsion phenomena and the development of fans, which have important consequences for landscape evolution.

8.5.1 Limitations and Prospects

Driven by climate change, the possible increase in debris-flow risk is often a direct consequence of a human-caused increase of exposed elements-at-risk on fans (Stoffel et al., this volume). Consequently, in populated mountain regions, the development of appropriate disaster risk reduction policies has become increasingly urgent. Land-use planning, based on robust baseline historical data, is the most efficient way for minimizing economic losses and avoiding new settlements in highly exposed places. The methods reviewed herein provide an overview of how to collect baseline data by reconstructing the spatio-temporal activity of debris flows based on field-truthed evidence. Yet, when employing these approaches, several limitations and uncertainties still arise, which should be considered to avoid misinterpretation in risk assessment.

It is almost impossible to reconstruct the entire debris-flow activity at a specific location over a certain period of time. Even when applying a multiproxy (multi-ensemble) approach, dating of debris flows is generally biased toward high frequency-low magnitude events. This implies that debris-flow data should be considered as censored, which complicates the applicability of classical frequency analyses. On the one hand, the occurrence of large debris flows may hinder the recognition of field-based evidence of past debris flows. This is the specific case of botanical dating methods, since the forest could be partially destroyed following the occurrence of a debris flow (Ballesteros-Cánovas et al., 2015). Due to the limited formation of deposition features or enduring evidence, debris flows that are strictly constrained to the channel may also go unnoticed. Each dating technique has its own limitations, scope, temporal coverage, and accuracy, and sometimes it is challenging to integrate different data sources.

Reconstructing the magnitude of debris flows has its own limitations. It is challenging to assume channel bedrock stability or to consider pre-event topographic condition at the level of a fan or a channel. To reconstruct the topography before the event, it is feasible to use preceding topographic data, old maps, aerial pictures, photographs, or field evidence. Yet, differences in accuracy, particularly with respect to elevation, could add significant uncertainties into the final outcomes. Moreover, approaches used to model previous debris flows may face some issues. Thus, debris flows are complex phenomena that could evolve from Newtonian toward non-Newtonian phases, involving sophisticated numerical models that require large sets of a-priori unknown parameters (Trujillo-Vela et al., 2022). To the contrary, the application of simplified empirical equations or model approaches could result in misleading magnitude estimates (Jakob, 2005). The need to apply one approach or another will depend on the ultimate objective of the study, as well as its financial and time limitations, which are generally determined by experts or end users.

The statistical analysis to create frequency-magnitude relationships requires independent and homogeneity assumptions to be fulfilled, which are often violated in debris-flow analyses. According to Jakob et al. (2013), the independent assumption is only warranted in transport-limited catchments. However, de Haas et al. (2018a, 2018b) have recently demonstrated that the magnitude of debris flows, when evaluated at the fan level, could be influenced by previous events. The homogeneity assumption is also difficult to meet because debris flows can be caused by many different preconditions and triggering factors. This is particularly the case in permafrost and transport-unlimited alpine catchments (Bardou & Jaboyedoff, 2008; Lugon and Stoffel, 2010).

Despite these limitations, the most efficient and accurate approach to performing risk assessment is to gather baseline debris-flow data based on field-truthed evidence. In particular, absolute dating techniques provide a time scale ranging from thousand to centuries, which is an appropriate timeframe for most engineering projects to be undertaken in mountain regions (Jakob, 2021). Requesting high return periods for any assessment could be considered quite *naïve*, given the limitations of the FM curves when extrapolating beyond the observation period and in view of the uncertainties that arise with climate change (Jakob et al., 2020). Further, lumped-based models or the application of hydrological/hydraulic models forced with climate variables (either observed, interpolated or from ensembles of global/regional models) require, in any case, a reliable calibration against truth-evidence that can only come from the historical record or from the reconstruction of past events (Jakob et al., 2022). Also, these approaches tend to oversimplify processes that occur at the interface between the hillslope and the channel. Instead, different international and national policies and strategies dealing with disaster risk reduction recommend that a robust assessment based on the observed and reconstructed impacts of larger events is a more prudent and rational approach for developing further adaptation strategies in highly vulnerable mountain regions (i.e. EXCIMAP, 2007). This will be even more relevant in developing countries, where the impacts of socio-economic development and climate change could be significantly greater than their land-use planning capacity, which relies on limited-baseline information, thereby causing long-lasting anomalies and trigger future risk situations (Allen et al., 2018).

Detailed analyses of past debris-flow activity will always be necessary at the local level not only to minimize future impacts but also to contribute to the scientific advancement of the discipline, by allowing hypothesis testing and discarding of previously established theories. Nevertheless, in the coming decades, remote sensing approaches will become more relevant, since the quality and quantity of available remote sensors will improve, allowing near-real-time monitoring of debris-flow activity worldwide. These data sources will undoubtedly contribute to enriching regional archives with the potential to systematically cover more than half a century (Landsat 1 was launched in 1972). When combined with techniques for machine learning, remote sensing, and in-situ assessments, these will improve definition of debris flow hazard at regional and local scales.

8.6 Concluding Remarks

Since the publication of the first edition of the book *Debris-Flow Hazard and Related Phenomena* (Jakob & Hungr, 2005), several advances have been made in the reconstruction of debris-flow events. Field-based assessments are required to characterize hazard processes and to distinguish between different flow processes that could co-occur in mountain catchments. In the absence of historical and observed records, dendrogeomorphic approaches are highly valuable as they can help improve the spatiotemporal resolution of debris-flow activity on forested fans and cover events that occurred over the past centuries. Radiometric dating techniques are essential to date sediment, especially in the case of ancient events or in environmental settings where botanical evidence is not available. Approaches based on remote sensing are advantageous when studying debris-flow activity over the past few years or in remote areas. The application of these approaches is greatly dependent on catchment characteristics, although the optimum spatial and temporal resolution is obtained by combining multiple methods (Imaizumi et al., 2016a, 2016b). On the other hand, over the past few decades, a variety of empirical equations have been proposed, intending to estimate debris-flow magnitude. Similarly, the increased availability of easily configurable numerical hydrodynamic models and the availability of reasonable resolved topography data make the use of retrospective modelling procedures against physical evidence highly recommended, particularly if the ultimate objective is to perform hazard and risk assessments. In this sense, the approaches provided here are grounded in truth-evidence and will contribute to assertive spatial decision-making aimed at building a more resilient society to foreseeable challenges.

Acknowledgements This chapter is dedicated to the memory of Matthias Jakob, who contributed significantly to methods of debris flow reconstruction and largely inspired the development of this discipline to solve practical issues to the society. The authors gratefully acknowledge Matthias Jakob, Scott McDougall and Paul Santi for reviewing this chapter. Juan A. Ballesteros-Cánovas received funding from the INOVA-RISK (2020-T1/AMB-19913) and EXTreeM (PID2021-124573OA-I00) projects.

References

Adams, M. S., Fromm, R., & Lechner, V. (2016). High-resolution debris flow volume mapping with unmanned aerial systems (UAS) and photogrammetric techniques. International Archives of the Photogrammetry, Remote Sensing & Spatial Information Sciences, 41.

Alcalá-Reygosa, J., Palacios, D., Schimmelpfennig, I., Vázquez-Selem, L., García-Sancho, L., Franco-Ramos et al., (2018). Dating late Holocene lava flows in Pico de Orizaba (Mexico) by means of in situ-produced cosmogenic 36Cl, lichenometry and dendrochronology. *Quaternary Geochronology*, 47, 93-106.

Allen, S. K., Ballesteros-Canovas, J., Randhawa, S. S., Singha, A. K., Huggel, C., & Stoffel, M. (2018). Translating the concept of climate risk into an assessment framework to inform adaptation planning: Insights from a pilot study of flood risk in Himachal Pradesh, Northern India. *Environmental Science & Policy, 87*, 1–10.

André, M. F. (1990). Frequency of debris flows and slush avalanches in Spitsbergen: a tentative evaluation from lichenometry. Polish Polar Research, 11(3–4).

Armstrong, R. A. (2016). Lichenometric dating (Lichenometry) and the biology of the lichen genus

Rhizocarpon: Challenges and future directions. *Geografiska Annaler: Series a, Physical Geography, 98*(3), 183–206.

Ballesteros Canovas, J. A., Eguibar, M., Bodoque, J. M., Díez-Herrero, A., Stoffel, M., & Gutiérrez-Pérez, I. (2011). Estimating flash flood discharge in an ungauged mountain catchment with 2D hydraulic models and dendrogeomorphic palaeostage indicators. *Hydrological Processes, 25*(6), 970–979.

Ballesteros-Cánovas, J. A., Stoffel, M., St George, S., & Hirschboeck, K. (2015). A review of flood records from tree rings. *Progress in Physical Geography, 39*(6), 794–816.

Ballesteros-Cánovas, J. B., Stoffel, M., Corona, C., Schraml, K., Gobiet et al., (2016). Debris flow risk analysis in a managed torrent based on a stochastic life-cycle performance. *Science of the Total Environment*, 557, 142-153.

Ballesteros-Cánovas, J.A. and Stoffel, M. (2017). Biologic Dating Techniques. The International Encyclopedia of Geography. Richardon, D et al., (eds) John Wiley & Sons, DOI: https://doi.org/10.1002/9781118786352.wbieg0806

Ballesteros Canovas, J. A., Stoffel, M., Martín-Duque, J. F., Corona, C., Lucía, A., Bodoque, J. M., & Montgomery, D. R. (2017). Gully evolution and geomorphic adjustments of badlands to reforestation. *Scientific Reports, 7*(1), 1–8.

Bardou, E., & Jaboyedoff, M. (2008). Debris flows as a factor of hillslope evolution controlled by a continuous or a pulse process? *Geological Society, London, Special Publications, 296*(1), 63–78.

Bennett G, Molnar P, McArdell B, Burlando P.(2014). A probabilistic sediment cascade model of sediment transfer in the Illgraben. *Water Resources Research* 50(2): 1225– 1244.

Benito G, Ballesteros-Cánovas JA, Díez-Herrero A (in review, 2022). Palaeoflood hydrology: Reconstructing rare events and extreme flood discharges. In book: *Hydro-Meteorological Hazards, Risks, and DisastersEdition: Hazards and Disasters Series Chapter*: 3 Publisher: Elsevier

Bernhardt, H., Reiss, D., Hiesinger, H., Hauber, E., & Johnsson, A. (2017). Debris flow recurrence periods and multi-temporal observations of colluvial fan evolution in central Spitsbergen (Svalbard). *Geomorphology, 296*, 132–141.

Berti, M., & Simoni, A. (2007). Prediction of debris flow inundation areas using empirical mobility relationships. *Geomorphology, 90*(1–2), 144–161.

Berti, M., & Simoni, A. (2014). DFLOWZ: A free program to evaluate the area potentially inundated by a debris flow. *Computers & Geosciences, 67*, 14–23.

Bertrand, M., Liébault, F., & Piégay, H. (2013). Debris flow susceptibility of upland catchments. *Natural Hazards, 67*(2), 497–511.

Blasone, G., Cavalli, M., Marchi, L., & Cazorzi, F. (2014). Monitoring sediment source areas in a debris flow catchment using terrestrial laser scanning. *CATENA, 123*, 23–36.

Bodoque, J. M., Díez-Herrero, A., Eguibar, M. A., Benito, G., Ruiz-Villanueva, V., & Ballesteros-Cánovas, J. A. (2015). Challenges in paleoflood hydrology applied to risk analysis in mountainous watersheds–A review. *Journal of Hydrology, 529*, 449–467.

Bodoque, J. M., Eguibar, M. A., Díez-Herrero, A., Gutiérrez-Pérez, I., & Ruíz-Villanueva, V. (2011). Can the discharge of a hyperconcentrated flow be estimated from paleoflood evidence?. *Water Resources Research*, 47(12).

Bollschweiler, M., & Stoffel, M. (2010). Tree rings and debris flows: Recent developments, future directions. *Progress in Physical Geography, 34*(5), 625–645.

Bollschweiler, M., Stoffel, M., & Schneuwly, D. M. (2008). Dynamics in debris flow activity on a forested cone—a case study using different dendroecological approaches. *CATENA, 72*(1), 67–78.

Bovis, M. J., & Jakob, M. (1999). The role of debris supply conditions in predicting debris flow activity. *Earth Surface Processes and Landforms, 24*(11), 1039–1054.

Brardinoni, F., & Hassan, M. A. (2006). Glacial erosion, evolution of river long profiles, and the organization of process domains in mountain drainage basins of coastal British Columbia. *Journal of Geophysical Research: Earth Surface, 111*, F01013.

Bremer, M., & Sass, O. (2012). Combining airborne and terrestrial laser scanning for quantifying erosion and deposition by a debris flow event. *Geomorphology, 138*(1), 49–60.

Brenna, A., Surian, N., Ghinassi, M., & Marchi, L. (2020). Sediment–water flows in mountain streams: Recognition and classification based on field evidence. *Geomorphology, 371*, 107413.

Bull, J. M., Miller, H., Gravley, D. M., Costello, D., Hikuroa, D. C. H., & Dix, J. K. (2010). Assessing debris flows using LIDAR differencing: 18 May 2005 Matata event New Zealand. *Geomorphology, 124*(1–2), 75–84.

Bulmer, M. H., Barnouin-Jha, O. S., Peitersen, M. N., & Bourke, M. (2002). An empirical approach to studying debris flows: Implications for planetary modeling studies. *Journal of Geophysical Research: Planets, 107*(E5), 9–1.

Calhoun, N. C., & Clague, J. J. (2018). Distinguishing between debris flows and hyperconcentrated flows: An example from the eastern Swiss Alps. *Earth Surface Processes and Landforms, 43*(6), 1280–1294.

Cannon, S. H., & DeGraff, J. (2009). The increasing wildfire and post-fire debris flow threat in western USA, and implications for consequences of climate change. In: *Landslides–disaster risk reduction* Springer, Berlin, Heidelberg. pp. 177–190.

Cao, C., Zhang, W., Chen, J., Shan, B., Song, S., & Zhan, J. (2021). Quantitative estimation of debris flow source materials by integrating multi-source data: A case study. *Engineering Geology, 291*, 106222.

Cato, K., & Goforth, B. (2021). Alluvial Fan Alteration due to debris flow deposition, incision, and channel migration at forest falls California. *Environmental & Engineering Geoscience, 27*(1), 29–41.
Chen, N. S., Yue, Z. Q., Cui, P., & Li, Z. L. (2007). A rational method for estimating maximum discharge of a landslide-induced debris flow: A case study from southwestern China. Geomorphology, 84(1–2), 44–58.
Chen, J. C., & Chuang, M. R. (2014). Discharge of landslide-induced debris flows: case studies of Typhoon Morakot in southern Taiwan. *Natural Hazards and Earth System Sciences, 14*(7), 1719–1730.
Chiverrell, R., & Jakob, M. (2013). Radiocarbon dating: Alluvial fan/debris cone evolution and hazards. *Dating Torrential Processes on Fans and Cones* (pp. 265–282). Springer.
Chow, V.T., 1959, Open-channel hydraulics: New York, McGraw-Hill Book Co., 680 p.
Church, M., & Jakob, M. (2020). What is a debris flood?. *Water resources research, 56*(8), e2020WR027144.
Cook, E. R., & Kairiukstis, L. A. (Eds.). (2013). Methods of dendrochronology: applications in the environmental sciences. Springer Science & Business Media.
Costa, J. E. (1984). Physical geomorphology of debris flows. In Developments and applications of geomorphology (pp. 268–317). Springer, Berlin, Heidelberg.
Coviello, V., Theule, J. I., Crema, S., Arattano, M., Comiti, F., Cavalli, M., ... & Marchi, L. (2021). Combining instrumental monitoring and high-resolution topography for estimating sediment yield in a debris flow catchment. *Environmental and Engineering Geoscience*, 27(1), 95–111.
Crosta, G. B., & Dal Negro, P. (2003). Observations and modelling of soil slip-debris flow initiation processes in pyroclastic deposits: The Sarno 1998 event. *Natural Hazards and Earth System Sciences, 3*(1/2), 53–69.
De Haas, T., Kleinhans, M. G., Carbonneau, P. E., Rubensdotter, L., & Hauber, E. (2015). Surface morphology of fans in the high-Arctic periglacial environment of Svalbard: Controls and processes. *Earth-Science Reviews, 146*, 163–182.
De Haas, T., Densmore, A. L., Stoffel, M., Suwa, H., Imaizumi, F., Ballesteros-Cánovas, J. A., & Wasklewicz, T. (2018a). Avulsions and the spatio-temporal evolution of debris flow fans. *Earth-Science Reviews, 177*, 53–75.
De Haas, T., Kruijt, A., & Densmore, A. L. (2018b). Effects of debris-flow magnitude–frequency distribution on avulsions and fan development. *Earth Surface Processes and Landforms, 43*(13), 2779–2793.
De Haas, T., van den Berg, W., Braat, L., & Kleinhans, M. G. (2016). Autogenic avulsion, channelization and backfilling dynamics of debris-flow fans. *Sedimentology, 63*(6), 1596–1619.
De Haas, T., & Densmore, A. L. (2019). Debris flow volume quantile prediction from catchment morphometry. *Geology, 47*(8), 791–794.
De Haas, T., Nijland, W., De Jong, S. M., & McArdell, B. W. (2020). How memory effects, check dams, and channel geometry control erosion and deposition by debris flows. *Scientific Reports, 10*(1), 1–8.
de Haas, T., McArdell, B. W., Nijland, W., Åberg, A. S., Hirschberg, J., & Huguenin, P. (2022). Flow and Bed Conditions jointly control Debris-Flow Erosion and Bulking. Geophysical Research Letters, e2021GL097611.
Détriché, S., Rodrigues, S., Macaire, J. J., Bonté, P., Bréhéret, J. G., Bakyono, J. P., & Jugé, P. (2010). Caesium-137 in sandy sediments of the River Loire (France): Assessment of an alluvial island evolving over the last 50 years. *Geomorphology, 115*(1–2), 11–22.
Dietrich, A., & Krautblatter, M. (2019). Deciphering controls for debris-flow erosion derived from a LiDAR-recorded extreme event and a calibrated numerical model (Roßbichelbach, Germany). *Earth Surface Processes and Landforms, 44*(6), 1346–1361.
Dowling, C. A., & Santi, P. M. (2014). Debris flows and their toll on human life: A global analysis of debris flow fatalities from 1950 to 2011. *Natural Hazards, 71*(1), 203–227.
Du, R. H., Kang, Z., Chen, X. Q., & Zhu, P. (1987). *A comprehensive investigation and control planning for debris flow in the Xiaojiang River Basin of Yunnan Province* (p. 33). Sichuan Science and Technology Press.
Dühnforth, M., Densmore, A. L., Ivy-Ochs, S., Allen, P. A., & Kubik, P. W. (2007). Timing and patterns of debris flow deposition on Shepherd and Symmes creek fans, Owens Valley, California, deduced from cosmogenic 10Be. Journal of Geophysical Research: Earth Surface, 112, F03S15.
Dühnforth, M., Densmore, A. L., Ivy-Ochs, S., Allen, P., & Kubik, P. W. (2017). Early to Late Pleistocene history of debris flow fan evolution in western Death Valley (California) using cosmogenic 10Be and 26Al. *Geomorphology, 281*, 53–65.
D'Arcy, M., Roda Boluda, D. C., Whittaker, A. C., & Carpineti, A. (2015). Dating alluvial fan surfaces in Owens Valley, California, using weathering fractures in boulders. *Earth Surface Processes and Landforms, 40*(4), 487–501.
D'Arcy, M., Mason, P. J., Roda-Boluda, D. C., Whittaker, A. C., Lewis, J. M., & Najorka, J. (2018). Alluvial fan surface ages recorded by Landsat-8 imagery in Owens Valley, California. *Remote Sensing of Environment, 216*, 401–414.
Ellett, N. G., Pierce, J. L., & Glenn, N. F. (2019). Partitioned by process: Measuring post-fire debris-flow and rill erosion with Structure from Motion photogrammetry. *Earth Surface Processes and Landforms, 44*(15), 3128–3146.
Evans, D. J. A., Archer, S., & Wilson, D. J. H. (1999). A comparison of the lichenometric and Schmidt hammer dating techniques based on data from the proglacial areas of some Icelandic glaciers. *Quaternary Science Reviews, 18*(1), 13–41.

EXCIMAP (2007) Handbook on good practices for flood mapping in Europe. European exchange circle on flood mapping. EU-JRC. pp 60

Franco-Ramos, O., Ballesteros-Cánovas, J. A., Terrazas, T., Stoffel, M., Vázquez-Selem, L., & Cerano-Paredes, J. (2022). Reconstruction of gully erosion based on exposed tree roots in a recent landform of Paricutin Volcano Mexico. *Earth Surface Processes and Landforms, 47*(3), 742–755.

Friedman, J. M., Vincent, K. R., & Shafroth, P. B. (2005). Dating floodplain sediments using tree-ring response to burial. *Earth Surface Processes and Landforms, 30*(9), 1077–1091.

Frodella, W., Salvatici, T., Pazzi, V., Morelli, S., & Fanti, R. (2017). GB-InSAR monitoring of slope deformations in a mountainous area affected by debris flow events. *Natural Hazards and Earth System Sciences, 17*(10), 1779–1793.

Fuchs, M., & Owen, L. A. (2008). Luminescence dating of glacial and associated sediments: Review, recommendations and future directions. *Boreas, 37*, 636–659.

Gao, Y. C., Chen, N. S., Hu, G. S., & Deng, M. F. (2019). Magnitude-frequency relationship of debris flows in the Jiangjia Gully China. *Journal of Mountain Science, 16*(6), 1289–1299.

Gartner, J. E., Cannon, S. H., & Santi, P. M. (2014). Empirical models for predicting volumes of sediment deposited by debris flows and sediment-laden floods in the Transverse Ranges of southern California. *Engineering Geology, 176*, 45–56.

Germain, D., Gavrilă, I. G., Elizbarashvili, M., & Pop, O. T. (2020). Multidisciplinary approach to sediment connectivity between debris flow channel network and the Dolra River, Mazeri Valley, Southern Caucasus Georgia. *Geomorphology, 371*, 107455.

Ghamisi, P., Rasti, B., Yokoya, N., Wang, Q., Hofle, B., Bruzzone, L., ... & Benediktsson, J. A. (2019). Multisource and multitemporal data fusion in remote sensing: A comprehensive review of the state of the art. *IEEE Geoscience and Remote Sensing Magazine, 7*(1), 6–39.

Gottesfeld, A. S. (1996). British Columbia flood scars: Maximum flood-stage indicators. *Geomorphology, 14*(4), 319–325.

Griswold, J. P., & Iverson, R. M. (2008). *Mobility statistics and automated hazard mapping for debris flows and rock avalanches*. US Department of the Interior, US Geological Survey.

Han, Z., Chen, G., Li, Y., Wang, W., & Zhang, H. (2015). Exploring the velocity distribution of debris flows: An iteration algorithm based approach for complex cross-sections. *Geomorphology, 241*, 72–82.

Heiser, M., Schlögl, M., Scheidl, C., & Fuchs, S. (2022). Comment on "Hydrometeorological Triggers of Periglacial Debris Flows in the Zermatt Valley (Switzerland) Since 1864" by Michelle Schneuwly-Bollschweiler and Markus Stoffel. *Journal of Geophysical Research: Earth Surface*, 127(3), e2021JF006562.

Helsen, M. M., Koop, P. J. M., & Van Steijn, H. (2002). Magnitude–frequency relationship for debris flows on the fan of the Chalance torrent, Valgaudemar (French Alps). *Earth Surface Processes and Landforms, 27*(12), 1299–1307.

Hoch, O. J., McGuire, L. A., Youberg, A. M., & Rengers, F. K. (2021). Hydrogeomorphic recovery and temporal changes in rainfall thresholds for debris flows following wildfire. *Journal of Geophysical Research: Earth Surface*, 126(12), e2021JF006374

Horton, P., Jaboyedoff, M., Rudaz, B. E. A., & Zimmermann, M. (2013). Flow-R, a model for susceptibility mapping of debris flows and other gravitational hazards at a regional scale. *Natural Hazards and Earth System Sciences, 13*(4), 869–885.

Huang, J., Hales, T. C., Huang, R., Ju, N., Li, Q., & Huang, Y. (2020). A hybrid machine-learning model to estimate potential debris flow volumes. *Geomorphology, 367*, 107333.

Hungr, O., McDougall, S., Wise, M., & Cullen, M. (2008). Magnitude–frequency relationships of debris flows and debris avalanches in relation to slope relief. *Geomorphology, 96*(3–4), 355–365.

Hungr, O., Morgan, G. C., & Kellerhals, R. (1984). Quantitative analysis of debris torrent hazards for design of remedial measures. *Canadian Geotechnical Journal, 21*(4), 663–677.

Ikeda, A., Mizuyama, T. and Itoh, T. (2019) Study of prediction methods of debris flow peak discharge. 7th International Conference on Debris flow Hazards Mitigation, Golden, USA

Imaizumi, F., Sidle, R. C., Tsuchiya, S., & Ohsaka, O. (2006). Hydrogeomorphic processes in a steep debris flow initiation zone. *Geophysical Research Letters*, 33(10).

Imaizumi, F., Tsuchiya, S., & Ohsaka, O. (2016a). Field observations of debris flow initiation processes on sediment deposits in a previous deep-seated landslide site. *Journal of Mountain Science, 13*(2), 213–222.

Imaizumi, F., Trappmann, D., Matsuoka, N., Tsuchiya, S., Ohsaka, O., & Stoffel, M. (2016b). Biographical sketch of a giant: Deciphering recent debris flow dynamics from the Ohya landslide body (Japanese Alps). *Geomorphology, 272*, 102–114.

Imaizumi, F., Masui, T., Yokota, Y., Tsunetaka, H., Hayakawa, Y. S., & Hotta, N. (2019). Initiation and runout characteristics of debris flow surges in Ohya landslide scar, Japan. *Geomorphology, 339*, 58–69.

Innes, J. L. (1985). Lichenometric dating of debris-flow deposits on alpine colluvial fans in southwest Norway. *Earth Surface Processes and Landforms, 10*(5), 519–524.

Irmler, R., Daut, G., & Mäusbacher, R. (2006). A debris flow calendar derived from sediments of lake Lago di Braies (N. Italy). *Geomorphology*, *77*(1–2), 69–78.

Iverson, R. M. (2014). Debris flows: Behaviour and hazard assessment. *Geology Today, 30*(1), 15–20.

Iverson, R. M. (2012). Elementary theory of bed-sediment entrainment by debris flows and avalanches. *Journal of Geophysical Research: Earth Surface*, 117(F3)

Iverson, R. M., & George, D. L. (2014). A depth-averaged debris flow model that includes the effects of evolving dilatancy. I. Physical basis. *Proceedings of the Royal Society A: Mathematical, Physical and Engineering Sciences,* 470(2170), 20130819.

Iverson, R. M., & Ouyang, C. (2015). Entrainment of bed material by Earth-surface mass flows: Review and reformulation of depth-integrated theory. *Reviews of Geophysics, 53*(1), 27–58.

Iverson, R. M., LaHusen, R. G., Major, J. J., & Zimmerman, C. L. (1994). Debris flow against obstacles and bends: Dynamics and deposits. *Eos, Transactions, American Geophysical Union, 75*, 274.

Iverson, R. M., Reid, M. E., & LaHusen, R. G. (1997). Debris flow mobilization from landslides. *Annual Review of Earth and Planetary Sciences, 25*(1), 85–138.

Iverson, R. M., Schilling, S. P., & Vallance, J. W. (1998). Objective delineation of lahar-inundation hazard zones. *Geological Society of America Bulletin, 110*(8), 972–984.

Iverson, R. M., Reid, M. E., Logan, M., LaHusen, R. G., Godt, J. W., & Griswold, J. P. (2011). Positive feedback and momentum growth during debris flow entrainment of wet bed sediment. *Nature Geoscience, 4*(2), 116–121.

Iverson, R. M., Logan, M., LaHusen, R. G., & Berti, M. (2010). The perfect debris flow? Aggregated results from 28 large-scale experiments. *Journal of Geophysical Research: Earth Surface*, 115(F3).

Iverson, R. M., George, D. L., & Logan, M. (2016). Debris flow runup on vertical barriers and adverse slopes. *Journal of Geophysical Research: Earth Surface, 121*(12), 2333–2357.

Jakob, M. (2005). Debris flow hazard analysis. *Debris flow hazards and related phenomena* (pp. 411–443). Springer.

Jakob, M. (2010). State of the art in debris flow research: the role of dendrochronology. *Tree Rings and Natural Hazards,* 183–192.

Jakob, M. (2021). Debris flow hazard assessments: A practitioner's view. *Environmental & Engineering Geoscience, 27*(2), 153–166.

Jakob, M., & Hungr, O. (2005). *Debris flow hazards and related phenomena* (Vol. 739). Springer.

Jakob, M., & Friele, P. (2010). Frequency and magnitude of debris flows on Cheekye River British Columbia. *Geomorphology, 114*(3), 382–395.

Jakob, M., McDougall, S., Weatherly, H., & Ripley, N. (2013). Debris flow simulations on cheekye river, british columbia. *Landslides, 10*(6), 685–699.

Jakob, M., Holm, K., & McDougall, S. (2016). Debris flow risk assessment. In: Oxford Research Encyclopedia of Natural Hazard Science.

Jakob, M., Weatherly, H., Bale, S., Perkins, A., & MacDonald, B. (2017). A multifaceted debris flood hazard assessment for Cougar Creek, Alberta, Canada. *Hydrology, 4*, 4–7.

Jakob, M., Mark, E., McDougall, S., Friele, P., Lau, C. A., & Bale, S. (2020). Regional debris-flow and debris-flood frequency–magnitude relationships. *Earth Surface Processes and Landforms, 45*(12), 2954–2964.

Jakob, M., Davidson, S., Bullard, G., Busslinger, M., Collier-Pandya, B., Grover, P., & Lau, C. A. (2022). Debris-Flood Hazard Assessments in Steep Streams. *Water Resources Research, 58*(4), e2021WR030907.

Jeong, S. W. (2010). Grain size dependent rheology on the mobility of debris flows. *Geosciences Journal, 14*(4), 359–369.

Jitousono, T., Shimokawa, E., & Tsuchiya, S. (1996). Debris flow following the 1984 eruption with pyroclastic flows in Merapi volcano, Indonesia. *Journal of the Japan Society of Erosion Control Engineering, 48*(Special), 109–116.

Johnstone, E., Macklin, M. G., & Lewin, J. (2006). The development and application of a database of radiocarbon-dated Holocene fluvial deposits in Great Britain. *CATENA, 66*(1–2), 14–23.

Jomelli, V. (2013). Lichenometric dating of debris avalanche deposits with an example from the French Alps. *Dating Torrential Processes on Fans and Cones* (pp. 211–224). Springer.

Kadavi, P. R., Lee, C. W., & Lee, S. (2018). Application of ensemble-based machine learning models to landslide susceptibility mapping. *Remote Sensing, 10*(8), 1252.

Kiefer, C., Oswald, P., Moernaut, J., Fabbri, S. C., Mayr, C., Strasser, M., & Krautblatter, M. (2021). A 4000-year debris flow record based on amphibious investigations of fan delta activity in Plansee (Austria, Eastern Alps). *Earth Surface Dynamics, 9*(6), 1481–1503.

Koch, T. (1998). Testing various constitutive equations for debris flow modelling. IAHS Publications-Series of Proceedings and Reports- Intern Assoc Hydrological Sciences, 248, 249–258.

Lancaster, S. T., & Casebeer, N. E. (2007). Sediment storage and evacuation in headwater valleys at the transition between debris flow and fluvial processes. *Geology,* 35(11), 1027–1030.

Lary, D. J., Alavi, A. H., Gandomi, A. H., & Walker, A. L. (2016). Machine learning in geosciences and remote sensing. *Geoscience Frontiers, 7*(1), 3–10.

Lato, M. J. (2020). Canadian Geotechnical Colloquium: Three-dimensional remote sensing, four-dimensional analysis and visualization in geotechnical engineering – State of the art and outlook. *Canadian Geotechnical Journal, 58*, 1065–1076.

Lato, M. J., Anderson, S., & Porter, M. J. (2019). Reducing landslide risk using airborne lidar scanning data. *Journal of Geotechnical and Geoenvironmental Engineering, 145*(9), 06019004.

Lian, O. B., & Roberts, R. G. (2006). Dating the quaternary: Progress in luminescence dating of sediments. *Quaternary Science Reviews, 25*(19–20), 2449–2468.

Liritzis, I., Singhvi, A. K., Feathers, J. K., Wagner, G. A., Kadereit, A., Zacharias, N., & Li, S. H. (2013). *Luminescence dating in archaeology, anthropology, and geoarchaeology: An overview* (pp. 2826–2885). Springer.

Liu, J. J., Li, Y., Su, P. C., & Cheng, Z. L. (2008). Magnitude–frequency relations in debris flow. *Environmental Geology, 55*(6), 1345–1354.

Lo, D. O. K. (2000). Review of natural terrain landslide debris-resisting barrier design. *Geotechnical Engineering Office, Civil Engineering Department.*

Lugon, R., & Stoffel, M. (2010). Rock-glacier dynamics and magnitude–frequency relations of debris flows in a high-elevation watershed: Ritigraben Swiss Alps. *Global and Planetary Change, 73*, 202–210.

Ma, C., Hu, K., & Tian, M. (2013). Comparison of debris flow volume and activity under different formation conditions. *Natural Hazards, 67*(2), 261–273.

Macklin, M. G., Rumsby, B. T., & Heap, T. (1992). Flood alluviation and entrenchment: Holocene valley-floor development and transformation in the British uplands. *Geological Society of America Bulletin, 104*(6), 631–643.

Malik, I., & Matyja, M. (2008). Bank erosion history of a mountain stream determined by means of anatomical changes in exposed tree roots over the last 100 years (Bílá Opava River—Czech Republic). *Geomorphology, 98*(1–2), 126–142.

Mao, L., Cavalli, M., Comiti, F., Marchi, L., Lenzi, M. A., & Arattano, M. (2009). Sediment transfer processes in two Alpine catchments of contrasting morphological settings. *Journal of Hydrology, 364*(1–2), 88–98.

Marchi, L., & Cavalli, M. (2007). Procedures for the documentation of historical debris flows: Application to the Chieppena Torrent (Italian Alps). *Environmental Management, 40*(3), 493–503.

Marchi, L., & Tecca, P. R. (2006). Some observations on the use of data from historical documents in debris flow studies. *Natural Hazards, 38*(3), 301–320.

Matthews, J. A., McEwen, L. J., Owen, G., & Los, S. (2020). Holocene alluvial fan evolution, Schmidt-hammer exposure-age dating and paraglacial debris floods in the SE Jostedalsbreen region, southern Norway. *Boreas, 49*(4), 886–902.

McCoy, S. W., Kean, J. W., Coe, J. A., Staley, D. M., Wasklewick, T. A., & Tucker, G. E. (2010). Evolution of a natural debris flow: In situ measurements of flow dynamics, video imagery, and terrestrial laser scanning. *Geology, 38*(8), 735–738.

McCoy, S. W. (2015). Research Focus: Infrequent, large-magnitude debris flows are important agents of landscape change. *Geology, 43*(5), 463–464.

Mitchell, A., Zubrycky, S., McDougall, S., Aaron, J., Jacquemart, M., Hübl, J., ... & Graf, C. (2022). Variable hydrograph inputs for a numerical debris flow runout model. *Natural Hazards and Earth System Sciences*, *22*(5), 1627–1654.

Mizuyama, T., Kobashi, S., & Ou, G. (1992). Prediction of debris flow peak discharge. In: Proc. Int. Symp. INTERPRAEVENT, Bern, Switzerland, Bd. 4, pp. 99–108.

Muñoz-Salinas, E., Castillo, M., Franco-Ramos, O., Arce, J. L., Sanderson, D. C. W., & Cresswell, A. C. (2020). Assessing paraglacial processes at Nexpayantla Gorge (Popocatépetl volcano, Central Mexico) using OSL and 14C. *Earth Surface Processes and Landforms, 45*(11), 2450–2462.

Muñoz-Torrero Manchado, A., Allen, S., Ballesteros-Canovas, J. A., Dhakal, A., Dhital, M. R., & Stoffel, M. (2021). Three decades of landslide activity in western Nepal: New insights into trends and climate drivers. *Landslides, 18*(6), 2001–2015.

Muñoz-Torrero Manchado, A., Ballesteros-Cánovas, J. A., Allen, S., & Stoffel, M. (2022). Deforestation controls landslide susceptibility in far-Western Nepal. *CATENA, 219*, 106627.

Mwaniki, M. W., Kuria, D. N., Boitt, M. K., & Ngigi, T. G. (2017). Image enhancements of Landsat 8 (OLI) and SAR data for preliminary landslide identification and mapping applied to the central region of Kenya. *Geomorphology, 282*, 162–175.

Mizuyama, T. (1988). Technical standard for the measures against debris flow. Technical memorandum of PWRI, 2632, 48.

Nagl, G., Hübl, J., & Kaitna, R. (2020). Velocity profiles and basal stresses in natural debris flows. *Earth Surface Processes and Landforms*, 45(8), 1764–1776.

Nolde, N., & Joe, H. (2013). A Bayesian extreme value analysis of debris flows. *Water Resources Research, 49*(10), 7009–7022.

Pastorello, R., D'Agostino, V., & Hürlimann, M. (2020). Debris flow triggering characterization through a comparative analysis among different mountain catchments. *CATENA, 186*, 104348.

Patton, A. I., Rathburn, S. L., Bilderback, E. L., & Lukens, C. E. (2018). Patterns of debris flow initiation and periglacial sediment sourcing in the Colorado Front Range. *Earth Surface Processes and Landforms, 43*(15), 2998–3008.

Pavlova, I., Jomelli, V., Brunstein, D., Grancher, D., Martin, E., & Déqué, M. (2014). Debris flow activity related to recent climate conditions in the French Alps: A regional investigation. *Geomorphology, 219*, 248–259.

Pierson, T. C. (2005). Hyperconcentrated flow; transitional process between water flow and debris flow. In M. Jakob & O. Hungr (Eds.), *Debris flow Hazards and Related Phenomena* (pp. 159–202). Springer.

Prochaska, A. B., Santi, P. M., Higgins, J. D., & Cannon, S. H. (2008a). Debris flow runout predictions based on the average channel slope (ACS). *Engineering Geology, 98*(1–2), 29–40.

Prochaska, A. B., Santi, P. M., Higgins, J. D., & Cannon, S. H. (2008b). A study of methods to estimate debris flow velocity. *Landslides, 5*(4), 431–444.

Procter, C. (2012). Debris flow dynamics: A flume study of velocity and superelevation (Doctoral dissertation, Durham University).

Reimer PJ, Bard E, Bayliss A, Beck JW, Blackwell PG, Bronk Ramsey C, Buck CE et al., (2013). IntCal13 and Marine13 radiocarbon age calibration curves 0–50,000 years cal BP. *Radiocarbon* 55(4):1869–1887.

Rengers, F. K., McGuire, L. A., Coe, J. A., Kean, J. W., Baum, R. L., Staley, D. M., & Godt, J. W. (2016). The influence of vegetation on debris flow initiation during extreme rainfall in the northern Colorado Front Range. *Geology, 44*(10), 823–826.

Rickenmann, D. (1990). Debris flows 1987 in Switzerland: modelling and fluvial sediment transport. Hydrology in Mountainous Regions, 371–378.

Rickenmann, D. (1999). Empirical relationships for debris flows. Natural hazards, 19(1), 47–77.

Rickenmann, D., & Koschni, A. (2010). Sediment loads due to fluvial transport and debris flows during the 2005 flood events in Switzerland. *Hydrological Processes: An International Journal, 24*(8), 993–1007.

Rickenmann, D., Scheidl, C. (2013). Debris flow Runout and Deposition on the Fan. In: Schneuwly-Bollschweiler, M., Stoffel, M., Rudolf-Miklau, F. (eds) Dating Torrential Processes on Fans and Cones. Advances in Global Change Research, vol 47. Springer, Dordrecht. https://doi.org/10.1007/978-94-007-4336-6_5

Rickenmann, D., Karrer, T., McArdell, B., & Scheidl, C. (2019). Small scale debris flow experiments on run-up height. Association of Environmental and Engineering Geologists; special publication 28.

Rodríguez-Morata, C., Villacorta, S., Stoffel, M., & Ballesteros-Cánovas, J. A. (2019). Assessing strategies to mitigate debris flow risk in Abancay province, south-central Peruvian Andes. *Geomorphology, 342*, 127–139.

Rosenwinkel, S., Korup, O., Landgraf, A., & Dzhumabaeva, A. (2015). Limits to lichenometry. *Quaternary Science Reviews, 129*, 229–238.

Savi, S., Norton, K. P., Picotti, V., Akçar, N., Delunel, R., Brardinoni, F., ... & Schlunegger, F. (2014). Quantifying sediment supply at the end of the last glaciation: Dynamic reconstruction of an alpine debris flow fan. *Bulletin*, 126(5–6), 773–790

Scheidl, C., McArdell, B. W., & Rickenmann, D. (2015). Debris flow velocities and superelevation in a curved laboratory channel. *Canadian Geotechnical Journal, 52*(3), 305–317.

Scheidl, C., McArdell, B., Nagl, G., & Rickenmann, D. (2019). Debris flow behavior in super-and subcritical conditions. Association of Environmental and Engineering Geologists; special publication 28.

Schimmel, A., Coviello, V., & Comiti, F. (2022). Debris flow velocity and volume estimations based on seismic data. *Natural Hazards and Earth System Sciences, 22*(6), 1955–1968.

Schlögl, M., Fuchs, S., Scheidl, C., & Heiser, M. (2021). Trends in torrential flooding in the Austrian Alps: A combination of climate change, exposure dynamics, and mitigation measures. *Climate Risk Management, 32*, 100294.

Schlunegger, F., Badoux, A., McArdell, B. W., Gwerder, C., Schnydrig, D., Rieke-Zapp, D., & Molnar, P. (2009). Limits of sediment transfer in an alpine debris flow catchment, Illgraben. *Switzerland. Quaternary Science Reviews, 28*(11–12), 1097–1105.

Schneuwly, M., Corona, C., & Stoffel, M. (2013). How to improve dating quality and reduce noise in tree-ring based debris flow reconstructions. *Quaternary Geochronology, 18*, 110–118.

Schneuwly-Bollschweiler, M., & Stoffel, M. (2012). Hydrometeorological triggers of periglacial debris flows in the Zermatt valley (Switzerland) since 1864. *Journal of Geophysical Research: Earth Surface*, 117(F2).

Schraml, K., Kogelnig, B., Scheidl, C., Stoffel, M., & Kaitna, R. (2013). Estimation of debris flood magnitudes based on dendrogeomorphic data and semi-empirical relationships. *Geomorphology, 201*, 80–85.

Schulte, L., Wetter, O., Wilhelm, B., Peña, J. C., Amann, B., Wirth, S. B., ... & Gómez-Bolea, A. (2019). Integration of multi-archive datasets for the development of a four-dimensional paleoflood model of alpine catchments. *Global and Planetary Change, 180*, 66–88.

Schürch, P., Densmore, A. L., Ivy-Ochs, S., Rosser, N. J., Kober, F., Schlunegger, F., ... & Alfimov, V. (2016). Quantitative reconstruction of late Holocene surface evolution on an alpine debris flow fan. *Geomorphology, 275*, 46–57.

Schürch, P., Densmore, A. L., Rosser, N. J., Lim, M., & McArdell, B. W. (2011). Detection of surface change in complex topography using terrestrial laser scanning: Application to the Illgraben debris-flow channel. *Earth Surface Processes and Landforms, 36*(14), 1847–1859.

Shen, P., Zhang, L., Chen, H., & Fan, R. (2018). EDDA 2.0: integrated simulation of debris flow initiation and dynamics considering two initiation mechanisms. *Geoscientific Model Development*, 11(7), 2841–2856.

Steger, S., Scorpio, V., Comiti, F., & Cavalli, M. (2022) Data-driven modelling of joint debris flow release susceptibility and connectivity. *Earth Surface Processes and Landforms*.

Stoffel, M. (2010). Magnitude-frequency relationships of debris flows – A case study based on field surveys and tree-ring records. *Geomorphology, 116*, 67–76.

Stoffel, M., Mendlik, T., Schneuwly-Bollschweiler, M., & Gobiet, A. (2014). Possible impacts of climate change on debris flow activity in the Swiss Alps. *Climatic Change, 122*, 141–155.

Stoffel, M., Ballesteros Cánovas, J. A., Corona, C., & Trappmann, D. G. (2016). Dendrogeomorphology. *International Encyclopedia of Geography: People, the Earth, Environment and Technology: People, the Earth, Environment and Technology*, 1–9.

Stoffel, M., & Corona, C. (2014). Dendroecological dating of geomorphic disturbance in trees. *Tree-Ring Research, 70*(1), 3–20.

Takahashi, T. (1981). Debris flow. *Annual Review of Fluid Mechanics, 13*(1), 57–77.

Takahashi, T. (1991). Debris flow. IAHR monograph series. Rotterdam: Balkema. 165 pp.

Takahashi, Y., Fujimura, N., Akita, H., & Mizuno, M. (2019). Dynamic characteristics of extreme superelevation of debris flows observed by laser profile scanners in Sakura-jima volcano, Japan. *Association of Environmental and Engineering Geologists; special publication 28*.

Tanarro García, L. M., Palacios Estremera, D., Fernández Fernández, J. M., de Pablo, A., Oliva, M., Rodríguez Mena, M., ... & Keddadouche, K. (2022). Origins of the divergent evolution of mountain glaciers during deglaciation: Hofsdalur cirques, Northern Iceland. *Ene, 8*, 32.

Tarolli, P., & Dalla Fontana, G. (2009). Hillslope-to-valley transition morphology: New opportunities from high resolution DTMs. *Geomorphology, 113*(1–2), 47–56.

Thouret, J. C., Antoine, S., Magill, C., & Ollier, C. (2020). Lahars and debris flows: Characteristics and impacts. *Earth-Science Reviews, 201*, 103003.

Tichavský, R., Fabiánová, A., Koutroulis, A., & Spálovský, V. (2022). Occasional but severe: Past debris flows and snow avalanches in the Helmos Mts.(Greece) reconstructed from tree-ring records. *Science of The Total Environment*, 157759.

Torresani, L., d'Agostino, V., & Piton, G. (2021). Deciphering sediment Connectivity Index and erosion pattern in a debris flow catchment. In 14th INTERPRAEVENT Congress: Natural hazards in a changing world (pp. 303–311).

Trujillo-Vela, M. G., Ramos-Cañón, A. M., Escobar-Vargas, J. A., & Galindo-Torres, S. A. (2022). An overview of debris flow mathematical modelling. *Earth-Science Reviews*, 104135.

van Steijn, H. (1996). Debris flow magnitude—frequency relationships for mountainous regions of Central and Northwest Europe. *Geomorphology, 15*(3–4), 259–273.

VanDine, D. F. (1996). Debris flow control structures for forest engineering. Res. Br., BC Min. For., Victoria, BC, Work. Pap, 8, 1996.

Wang, T., Chen, J., Chen, X., You, Y., & Cheng, N. (2018). Application of incomplete similarity theory to the estimation of the mean velocity of debris flows. *Landslides, 15*(10), 2083–2091.

Wang, Z., You, Y., Zhang, G., Feng, T., Liu, J., Lv, X., & Wang, D. (2021). Superelevation analysis of the debris flow curve in Xiedi gully, China. *Bulletin of Engineering Geology and the Environment, 80*(2), 967–978.

Wigmosta, M. S. (1983). *Rheology and flow dynamics of the Toutle debris flows from Mt. St. Helens* (Doctoral dissertation, University of Washington).

Wilhelm, B., Ballesteros Cánovas, J. A., Macdonald, N., Toonen, W. H., Baker, V., Barriendos, M., ... & Wetter, O. (2019). Interpreting historical, botanical, and geological evidence to aid preparations for future floods. *Wiley Interdisciplinary Reviews: Water, 6*(1), e1318.

Winter, M. G., Dent, J., Macgregor, F., Dempsey, P., Motion, A., & Shackman, L. (2010). Debris flow, rainfall and climate change in Scotland. *Quarterly Journal of Engineering Geology and Hydrogeology, 43*(4), 429–446.

Witt, A. C. (2005). A brief history of debris flow occurrence in the French Broad River Watershed, western North Carolina. *The North Carolina Geographer, 13*, 59–82.

Woodget, A. S., Carbonneau, P. E., Visser, F., & Maddock, I. P. (2015). Quantifying submerged fluvial topography using hyperspatial resolution UAS imagery and structure from motion photogrammetry. *Earth Surface Processes and Landforms, 40*(1), 47–64.

Yang, Z. N. (1985). Preliminary study on the flowing velocity of viscous debris flows due to intense rainfall. Research of debris flows. Proceeding of the Lanzhou Institute of Glaciology and Cryopedology, Chinese Academy of Sciences, (4).

Zaginaev, V., Ballesteros-Cánovas, J. A., Erokhin, S., Matov, E., Petrakov, D., & Stoffel, M. (2016). Reconstruction of glacial lake outburst floods in northern Tien Shan: Implications for hazard assessment. *Geomorphology, 269*, 75–84.

Zhang, N., & Matsushima, T. (2018). Numerical investigation of debris materials prior to debris flow hazards using satellite images. *Geomorphology, 308*, 54–63.

Zhao, Q., Thomsen, K. J., Murray, A. S., Wei, M., Pan, B., Song, B., ... & Chen, H. (2015). Testing the use of OSL from quartz grains for dating debris flows in Miyun, northeast Beijing, China. *Quaternary Geochronology, 30*, 320-327.

Zubrycky, S., Mitchell, A., McDougall, S., Strouth, A., Clague, J. J., & Menounos, B. (2021). Exploring new methods to analyse spatial impact distributions on debris-flow fans using data from south-western British Columbia. *Earth Surface Processes and Landforms, 46*(12), 2395–2413.

9 Statistical Techniques for Debris-Flow Frequency–Magnitude Analyses

Matthias Jakob and Natalia Nolde

Abstract

Debris flows are a curse where they interfere with the built environment because of their poor predictability, their long runout and high impact forces. This has prompted a vast body of research and hundreds of thousands of structures have been erected worldwide to cope with the corresponding debris-flow risk. Debris-flow science has bloomed over the past decades in response to urban encroachment in steep terrain. In most jurisdictions, a design event is chosen based on a specific return period or annual exceedance probability. In other cases, quantitative risk assessments for loss of life and/or economic losses guide the type and scale of structure (see Strouth et al., this volume). Sometimes, the decision is based purely on cost–benefit analysis amalgamating the other two techniques. Finally, decisions are made based on a finite budget to be spent to achieve maximum risk reduction. Irrespective of which method guides the mitigation efforts, an understanding of debris-flow frequency-magnitude relationships is valuable, and failure to incorporate such relationships has ripple effects for the risk assessment and design of mitigation works. In this chapter, we present various statistical techniques at the disposal of the practitioner, discuss their respective benefits and describe three case studies. As for most predictive geophysical phenomena, we argue for model ensembles rather than reliance on a single approach. Uniting expert judgment with empirical/statistical methods appears the winning bet for deciphering how often and how big debris flows may become.

9.1 Introduction

Debris flows are amongst the most intensively studied landslide types, with over one hundred papers written on various aspects of debris-flow research every year, dedicated journal special issues and books like this one. The wide interest in debris flows can be explained by their frequent occurrence in all mountain ranges that receive at least occasional rainfall, their long runout distance and their potential for widespread destruction, loss of life and infrastructure disruption.

Since Jakob and Hungr (2005), much progress has been made in various aspects of debris-flow research (e.g., Jakob, 2020), as summarized in this volume. One critical aspect,

M. Jakob
BGC Engineering Inc, Vancouver, BC, Canada

N. Nolde (✉)
Department of Statistics, University of British Columbia, Vancouver, Canada
e-mail: natalia@stat.ubc.ca

M. Jakob et al. (eds.), *Advances in Debris-flow Science and Practice*, Geoenvironmental Disaster Reduction,
https://doi.org/10.1007/978-3-031-48691-3_9

indeed the very backbone of any debris-flow risk assessment, is analysis of how often and how large debris flows may become. The frequency-magnitude (F-M) relationship is typically quantified via the concept of the return period and return level, in which the "level" is equivalent with the magnitude (debris volume or peak discharge). The return period of a given magnitude is defined as the expected waiting time until this magnitude is next exceeded. The N-year return level is the magnitude for which the expected waiting time of exceedance is N years. The return level plot summarizes the frequency-magnitude relationship by plotting a range of return periods against the estimates of the associated return levels. Estimation of the return level curve for debris flow volumes is notoriously difficult because, unlike for gauged streams for which these curves are usually developed, data are scarce and often imprecise.

Practicing geomorphologists, geologists and geological or geotechnical engineers, as well as government agencies who are tasked with assessing and mitigating debris flow hazards and risks, are well aware of the difficulties to reliably determine their annual frequency, volume, peak discharge, runout distance and area covered by debris. Based on observed or reconstructed debris flow dates, temporal predictions are made for future occurrence. Frequency-magnitude data also serve as input parameters to numerical models, which in turn are used to create debris-flow intensity maps. Higher probability debris flows (given similar magnitudes) not only result in a higher hazard, but also in a higher risk, as return period is multiplied with measures of consequences to arrive at numerical values for total risk (see Strouth et al., this volume). Therefore, the accurate prediction of future events is at least as important as the establishment of an accurate record for the past from which future inferences are made. This chapter highlights a specific aspect of debris-flow hazard analysis, which is the estimation of the debris flow frequency-magnitude relationship.

Every hazard assessment requires a reconstruction of events into the past, recognizing that with climate change, extrapolations from the past should, as a minimum, be scrutinized and perhaps revised. Commonly applicable rules as to the time window to be used to construct debris-flow frequencies do not exist, but regulatory guidance or legislation worldwide mandate between several tens up to 10,000-year return periods. For example, in British Columbia, Canada (B.C.), the current guidance to Ministry of Transportation approving officers is that a 10,000-year return period (0.0001 annual exceedance probability or AEP) be considered for all life-threatening landslide processes (MoTI, 2009).

Frequency estimates over several hundred to thousands of years require application of one or more absolute dating methods, such as dendrochronology, radiocarbon dating or varve chronologies. With appropriate methods and suitable datable material within the fan stratigraphy, it is often possible to decipher a record of debris flows. This chronology, however, is censored in the sense that small events (often less than size class 3 or 4, Jakob, 2005) have been eroded or obliterated by subsequent events, or have moved through the existing channel and into higher order streams without depositing. This results in little or no physical and datable evidence of such events. In that sense, debris-flow records differ fundamentally from streamflow time series that provide continuous and regularly measured records. Larger debris flows may be easier to detect though they are not immune to erosion, which challenges their exact volumetric reconstruction. Associating volumes with the respective dates of debris-flow occurrence requires numerous trenches or boreholes to reconstruct and interpolate debris-flow thickness of events of known date.

Once a reasonable documentation of events with date and volume has been achieved, the return level curve that signifies the relationship between return periods and return levels of debris flow volumes can be estimated. There exist different approaches for return level curve estimation, which range from simple curve fitting to the empirically determined return periods of the historical volume measurements, to fitting a probability distribution (model) to historical event volumes and deriving the return level curve from the fitted model. Parametric or semi-parametric modelling allows us to extrapolate the fitted return level curve to return period values beyond those observed in the data record. The different methods are reviewed in this chapter.

The principal weakness of analyzing extremes in the debris-flow record is that it fails to account for hydro-climatic or geomorphic mechanisms that produced the observed extremes. As pointed out by Klemes (2000) for extreme flood frequency analysis, the application of traditional frequency methods is based on sometimes arbitrary postulates of preconceived probabilistic mechanisms. The results obtained by such methods are therefore not an accurate prediction of the future, but rather reflect results obtained under the assumption that the probabilistic model that provided a good description of the historical data will remain valid for future events.

Frequency analyses, including those used for debris flows, rely on the premise of stationarity over time and that they underlie an ergodic (independence from initial conditions) stochastic process. Both assumptions can be questioned. For one, extrapolation into high return periods that are a multiple of the initial record length is a daring proposition in the absence of information on how climatic or geomorphic watershed conditions may have changed. Source material depletion, vegetation changes, wildfire suppression, changes in the frequency and/or magnitude of hydroclimatic events and the occurrence of cataclysmic events, such as large landslides in the watershed, can all alter the stationarity assumption at different temporal scales. Ergodicity, in turn, demands that the geophysical process (in this case debris flows) can be viewed as an infinite number of equally likely stochastic events. This assumption can be challenged when physical realities are examined: For example, a formerly active volcano, now dormant or extinct, may have created geomorphic conditions and resulting processes that cannot be repeated indefinitely over time, such as the supply of easily erodible materials or the propensity of generating debris avalanches or rock avalanches. Similarly, the past ice age (or Little Ice Age) may have provided a finite amount of erodible materials, which, once depleted or stabilized through vegetation, can no longer be recruited by contemporary geomorphic processes (Church and Slaymaker, 1989). Therefore, the probability of occurrence in any given year from the observations should only be extrapolated with caution.

Following a discussion of these analytical methods in more depth, we use case studies to examine the difficulties in determining these frequency-magnitude relationships for three debris-flow prone watersheds in southwestern B.C. and one in Whatcom County in Washington State, USA.

9.2 Analytical Approaches

Analytical approaches rely on historical data to estimate the frequency-magnitude relationship of debris flows. In general, for these estimates to be reliable, one requires a sufficient number of observations and confidence as to the accuracy. For instance, for an extreme value analysis using maximum likelihood estimation, one requires in the order of 100 data points. This will create parameter estimates that are sufficiently robust and reliable, which is the case, for example, for many streamflow data that are continuously being gathered.

Given the sparsity of direct debris-flow observations or historical reconstruction via absolute dating methods, the practitioner is often left with perhaps 10 to 20 data points. While analytical approaches discussed in this chapter provide guidance for risk management decisions, their results need to be applied with caution. This realization calls for the use of model ensembles, defined by using several analytical methods, much like in weather forecasts, and complement the models with professional judgment. This does not imply simply averaging the results from each approach, but at least weighing them by the confidence inherent in each method.

9.2.1 Magnitude Cumulative Frequency Analysis (MCF) and Rank Order Analysis (ROA)

Seismology has been the precursor to the use of regional magnitude-cumulative frequency curves (MCF), (Gutenberg & Richter, 1954). This method is appealing because of its ease of use. The general process is as follows.

Let $\{(t_1, V_1), (t_2, V_2), \dots, (t_n, V_n)\}$ denote the inventory of debris-flow volumes, where V_i is the event volume recorded at time t_i. Let T denote the time interval (in years) over which historical data are available.

Rank the debris flow volumes in the sample from largest to smallest: $V_{(1)} \geq \dots \geq V_{(n)}$, where observation with rank k, $V_{(k)}$, corresponds to the k-th largest value. Hungr et al. (2007) recommend binning debris- flow volume data into categories to avoid the effects of data censoring, particularly for small flows that have been overridden and covered by subsequent flows or may not show up in the fan stratigraphy. The incremental debris-flow frequency of rank k is calculated as 1/T and the MCF is then simply the cumulative incremental frequencies:

$$F_k = \sum_{i=1}^{k} 1/T = k/T, \tag{9.1}$$

where F_k is the annual debris-flow frequency of greater than the *k-th* largest volume. The MCF curve is then produced by plotting F_k against $V_{(k)}$. While very easy to apply, the correct use of MCF implies that either all events need to be known, volumes be combined in reasonable volume classes, or the data set is stratified into volume classes where confidence exists that all such events have been included. Furthermore, the selection of different plotting methods (cumulative versus non-cumulative, linear and logarithmic binning, different bin sizes and choice of trendlines for extrapolations) can all bias the results (Brardioni & Church, 2004).

ROA is very similar to MCF. The data are ranked, typically annual extremes or values over a threshold, in increasing order of magnitude from the smallest $m=1$ to the largest $m=n$ and a cumulative probability P_m is associated to the m-th smallest value in the sample. Second, one fits a curve through the points of the ranked values of the historical data against the cumulative probability by some fitting procedure. Third, one interpolates or extrapolates from the graph so that the return period of the extreme value of interest is estimated. For events with identical volumes (typically due to rounding), different ranks are assigned, which requires using the original non-rounded data point. A plotting position formula is used to determine the cumulative probability of a given event with many plotting position formulae available in the literature. One of the most commonly used plotting positions is the Weibull formula (Weibull, 1939):

$$\mathrm{P_m} = m/T + 1, \mathrm{m} = 1, \dots, \mathrm{n} \tag{9.2}$$

where P_m is the cumulative probability, m is the rank of individual observations and T as above is the number of years in the data record. The return period associated with the event having the m-th largest volume is equal to $RP_m = 1/(1-P_m)$.

Makkonen (2006) and Makkonen et al. (2013) argued that Weibull's (1939) plotting formula uniquely provides the non-exceedance probability of the r-th smallest value in order-ranked N observations and that it should therefore be used in plotting.

This analysis is identical to the MCF analysis with the difference that individual frequencies are not summed. P_m therefore provides the simple annual frequency of the m-th smallest volume, while F_i in Eq. 9.1 is the annual frequency of greater than V_i. In reality, however, because of the small incremental changes from cumulating frequencies, there is very little difference between the results obtained from using either MCF or ROA. The values are then plotted on probability paper or simply on a log–log scale. Once plotted, the exceedance probability can be acquired graphically or by fitting a function to the point distribution, as was done by Jakob and Friele

(2009) for their Cheekeye River data set. The obvious problem with this approach is the choice of the curve to be fitted. The practitioner may find that linear, power and log fits all approximate the observed data reasonably well over the observed range, but their extrapolation over multiples of the observation range may result in fundamentally different results for the chosen curve. Without additional information on which of the fitted functions best matches the understanding of geomorphic constraints in a watershed, this exercise may produce large errors.

9.2.2 Fan-Area/Volume F-M Relationships

Establishing frequency-magnitude relationships on a regional scale is challenging. For example, pipelines or other buried or above-ground linear infrastructure often cross mountainous terrain. Oil and gas pipeline owners have a particularly low tolerance to pipeline rupture due to the substantial clean-up costs, environmental fines and very poor publicity, especially in an age in which fossil fuel industries are vilified and investments increasingly divested from that sector (Rifkin, 2019). Moreover, they wish to know the chance of pipeline impact for a range of return periods; in other words, a frequency-magnitude analysis is needed for tens if not hundreds of fans that their pipeline may cross. However, detailed dendrogeomorphic studies, radiocarbon dating, or even detailed mapping of previous deposits are very time and cost intensive and typically not practical or funded.

Jakob et al. (2020) provided a simple, yet effective regional method to estimate frequency-magnitude relationships (Fig. 9.1). Using well-researched debris-flow frequency-magnitude relationships from nine studies in Southwest BC, Canada, Jakob et al. (2020) were able to show that fan area or fan volume can predict these relationships with quantifiable error. The underlying rationale is that the fan, unless truncated by a higher order stream, or obfuscated by a rapidly aggrading floodplain, reflects the sum of all fan-forming events. This method is particularly helpful in areas with late Pleistocene glaciation, where the fan area or volume integrates all debris-flow events that occurred since deglaciation. However, this method is not particularly applicable if a high degree of precision is warranted.

If enough individual frequency-magnitude relationships are known for a particular region

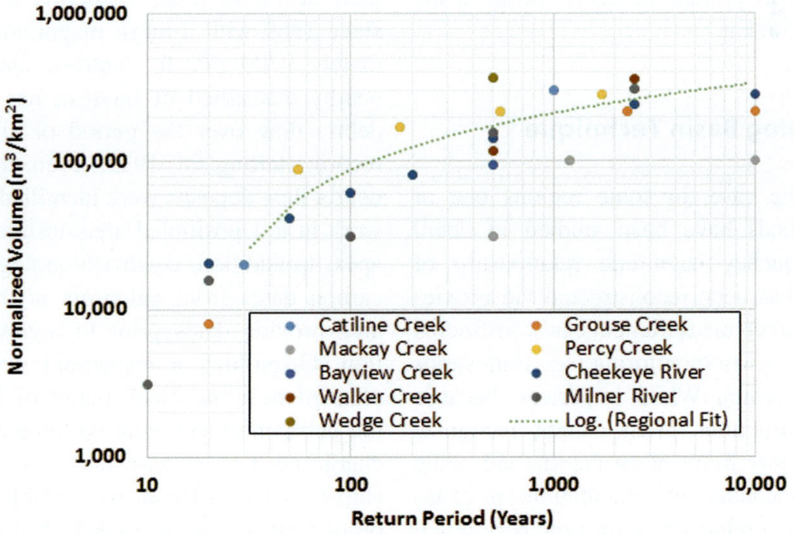

Fig. 9.1 Regional frequency-magnitude curve for southwestern B.C. debris flows normalized by fan area (after Jakob et al., 2020)

of study, a similar relationship could be developed in the form of:

$$V_S = A_f[a\ln(\mathrm{T}) - \mathrm{b}] \quad (9.3)$$

where T is the event return period which is entered manually for the return periods of interest, Vs is the normalized total sediment volume in m^3/km^2, A_f is the fan area, and a and b are coefficients. The values of the coefficients depend entirely on the data and are estimated by least squares analysis. The equation has predictive value only for events/locations that would fall within the distribution of the data set and have similar ancillary properties.

Fan areas are typically simple to measure in Google Earth, digital terrain models from lidar or from orthorectified aerial photographs. However, caveats must be identified, such as abnormally large fans for their respective watersheds that may be attributable to previous large landslides that have now been overprinted by debris flows. In such cases, the regional method above may result in overly conservative results (predicting volumes higher than the "true" debris-flow volumes). Alternatively, river erosion at the toe of a fan or aggrading floodplains that mask some of the fan area may result in a fan area that is small in relation to the frequency-magnitude of debris flows in the upstream catchment.

9.2.3 Analog Basin Technique

It is often the case, in some regions one or more watersheds have been studied in detail and the frequency-magnitude relationship of debris flows has been reconstructed via proxies or through direct measurements in constructed debris basins, or via monitoring (see Ballesteros et al., this volume). While imperfect, because proxy reconstructions often cannot recognize the smaller, more frequent events that are being obliterated. Secondly instrumentations in channels only span a relatively short time period that limits the statistical robustness of any models derived from only a few years of data. Where decades of data exist, the frequency-magnitude relationship of one basin can be transferred to another if the characteristics that explain magnitude and frequency are understood. For example, Bovis and Jakob (1999) provided multiple regression equations to estimate debris flow frequencies and magnitudes for weathering-limited and transport-limited basins in southwestern BC. These equations allow the estimates of mean magnitude for any basin in their study area.

The basic logic is then that once a detailed study has identified a credible F-M relationship for that specific watershed, the appropriate regression equation can be used as a "transfer function" to calculate the mean magnitude for the well-studied watershed and the watershed of interest. The factor difference between the well-studied watershed and the one of interest could then be applied to adjust the F-M curve. The following example illustrates these points.

Charles Creek, located some 30 km north of Vancouver, B.C. has a 53-year historical record of debris flows. Most of these events were directly observed, and their volume measured in a debris basin constructed in 1983, after debris flows in 1981 had killed several people due to bridge destruction and motorists plunging to their death. In total, 18 debris flows occurred since 1969 with a mean magnitude of approximately 9700 m^3. In contrast, Daisy Creek, a nearby watershed of interest, has no observed debris flow over the period of air photograph record starting in 1932, even though recent debris-flow deposits were identified during fieldwork in a depositional area upstream of the fan apex. In addition, dendrochronologic and radiocarbon dates from paleosols in the deposition area provide dates prior to highway construction. Given that a residential development is planned near the creek outlet of Daisy Creek, the debris-flow potential, volume and peak discharge need to be estimated. The F-M relationship for Charles Creek was established via rank order analysis using Eq. 9.1. The data are then plotted as per Fig. 9.2.

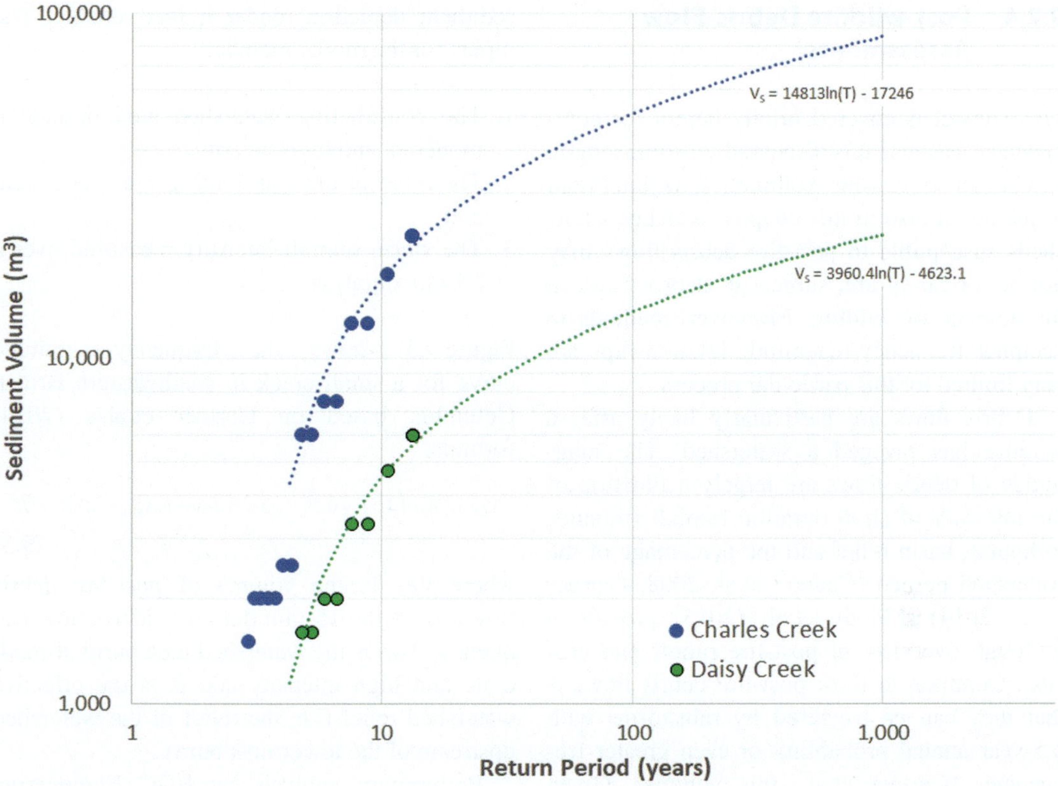

Fig. 9.2 Charles creek and daisy creek frequency-magnitude data with transfer function applied to daisy creek. Logarithmic fit and extrapolation to 10,000 years

The second step is to use the mean magnitude equation from Bovis and Jakob (1999) in the same general region in the form of:

$$Log\overline{V} = 0.48 + 2Z_T + 0.1A\% \tag{9.4}$$

where $\overline{V}$ is the mean volume, Z_T is the watershed relief and $A\%$ is the percentage of the area that actively contributes sediment to the channel system. In the case of Charles Creek, this yields an approximate mean volume of 11,200 m^3. This compares favorably with the approximately 9,700 m^3 of the arithmetic mean from observed debris-flow volumes. Two possibilities for this mild discrepancy arise: One is that given the relatively short record of 53 years, larger debris flows likely occurred in the deeper past with larger volumes than in the past 53 years. The second possibility is that some of the larger (> 20,000 m^3) debris flows overspilled the barrier at unknown volumes. Irrespective, assuming that a long-term mean magnitude of approximately 11,000 m^3 is correct, Eq. 9.4 is then used to calculate the respective volume for Daisy Creek, which was found to be approximately 3000 m^3. Hence, the factor difference between the two watersheds is approximately 4, which was then used to divide all Charles Creek data to arrive at a new F-M relationship for Daisy Creek.

This method is quick and relies on F-M relationships determined for nearby watersheds. It is sensitive to the percentage sediment-contributing watershed area, which is not easy to estimate given that, in extreme storms, hillsides may fail that were not originally added to the sediment-contributing area. Despite such limitations, this method provides an additional tool to construct F-M curve families.

9.2.4 Post-wildfire Debris-Flow Assessments

This subject is covered briefly herein for completeness because it is discussed, in some length, in Gartner et al. (this volume). It is important to include it also in this chapter as many watersheds susceptible to post-fire debris flows may not be, or rarely are, subject to debris flows in the absence of wildfire. Moreover, methods to decipher frequency-magnitude relationships are very limited for this particular process.

Debris flows are particularly likely after a wildfire has ravaged a watershed. The magnitude of debris flows are largely a function of the intensity of short-duration rainfall (minutes to hours), basin relief and the percentage of the watershed burned (Cannon et al. 2008; Gartner et al., 2014). Moody et al. (2013) provide a thorough overview of post-fire runoff and erosion. Common to most post-fire debris flows is that they can be triggered by rainstorms with 0.5-year annual probability or even greater frequencies (Gartner et al., this volume). Given that the worst effects of post-fire debris flows last for at least 2 years (see Gartner et al., this volume), it is likely that this rainfall threshold will be exceeded over that time period, resulting in a high likelihood of debris flow. The key then is how large the debris flow may become, as this influences runout distance and area inundated, which in turn affect the impact and consequences on alluvial fans, as well as any buried infrastructure along the debris-flow transport zone.

Empirical models for predicting post-wildfire debris-flow volumes (e.g., Cannon et al., 2010; Gartner et al., 2014) can be used to assess hazards posed by debris flows following wildfires. The Gartner et al. (2014) model is used by the United States Geological Survey (USGS) for emergency assessments of post-wildfire debris flow hazards (USGS, n.d.). The model is applicable for up to two years following the wildfire, after which plant re-growth and/or source area sediment depletion render it less reliable. The inputs for the model include:

1. The contributing watershed area burned at moderate and high severity,
2. The relief of the contributing watershed area, and
3. The storm rainfall intensity measured over a 15-min duration.

Figure 9.3 shows the frequency-magnitude curve for a small creek in southwestern British Columbia based on Gartner et al.'s (2014) formula:

$$V_{DF} = Exp\left(4.22 + 0.39 \cdot \sqrt{I_{15}} + 0.36 \cdot lnA_{MH} + 0.13 \cdot \sqrt{R}\right) \tag{9.5}$$

where V_{DF} is the volume of post-fire debris flows, I_{15} is the 15-min duration debris-flow frequency, A_{MH} is the watershed area burnt at moderate and high intensity and R is the effective watershed relief (i.e. the relief of the watershed upstream of the lowermost burn).

Preliminary analysis by BGC Engineering has shown that the Gartner et al. (2014) model substantially overestimates debris-flow volumes by a factor of 2 to 4, though much more work is required to confirm this adjustment factor. Irrespective, Intensity–Duration–Frequency (IDF) are readily available for many stations worldwide and can be artificially-generated from regional or global reanalysis precipitation datasets such as ERA5-Land (Sabater, 2021). This, and the knowledge of A_{MH}, and R, both of which are measurable, allows an estimation of debris-flow volumes.

While this is a quick and elegant method to produce an F-M relationship for the specific watershed in question, some confirmation via additional models is recommended. For example, the fan-area or fan-volume empirical relationship outlined above should be developed, as it clearly includes post-fire debris flows in this pre-historic record. In doing so, however, one needs to realize that the F-M relationship shown in Fig. 9.3 is

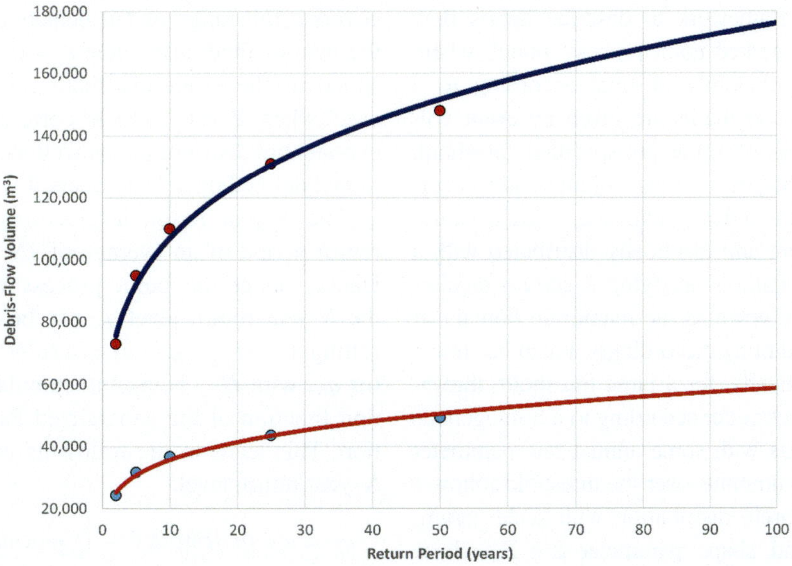

Fig. 9.3 Frequency-debris-flow magnitude curve for Lytton Creek using the Gartner et al. (2014) model. The upper curve shows the unadjusted model, the lower is adjusted by a factor of three which appears to be more applicable to south-central BC

specific to the next two years; it is not the long-term average. To calculate the average return period of the respective volumes, one needs to multiply the average fire frequency and divide it by the duration for which the Gartner et al. (2014) volume model is valid. Thus, the long-term debris-flow return period is calculated as:

$$T_{DF} = T_F \cdot \frac{1}{2} \tag{9.6}$$

where T_{DF} is the return period of debris flows and T_F is the return period of wildfires.

This correction presumes that wildfire return period does not change and that there are no sediment-supply limitations. In a rapidly warming world, wildfire frequency is increasing in many regions (Gartner et al. and Stoffel et al., this volume), which will decrease T_F. This, in turn, may affect sediment availability, especially in feeder gullies and the main stream channel because more frequent fires and subsequent debris flows will shrink the time available for debris recharge. Hence, while debris flow frequency may increase, their respective volumes may decrease depending on the intensity of channel recharge processes and channel-flanking landforms.

9.2.5 Extreme Value Statistics

As mentioned in the Introduction to this chapter, hazard assessment for debris flows in Canada requires estimation of return levels that correspond to return periods of up to 10,000 years. This is a very difficult statistical problem for any hydrological process, but especially so for debris flows owing to highly limited data records both due to relative rarity of the phenomenon and data collection challenges. As a consequence, estimation in such situations has to rely on some form of model-based extrapolation beyond the available data range. An area of statistics that deals with this type of question is known as Extreme Value Analysis (EVA), with statistical models motivated by asymptotic results coming from extreme value theory, a branch of probability theory that studies behaviour of extreme values. Classical textbooks on statistics of extremes include Coles (2001) and Beirlant et al. (2004).

A natural framework to describe debris flow data is via a marked point process model, where events occur randomly in time according to a point process and marks are given by event volumes in excess of some pre-specified threshold. We assume that debris-flow volumes and occurrence times are independent, and volume values are independent and identically distributed with a common distribution satisfying a certain regularity condition (known as the maximum domain of attraction condition). Accordingly, it can be shown that, asymptotically, for a large threshold, threshold exceedances occur according to a homogenous Poisson process with some annual rate parameter $\lambda > 0$. Excess amounts over the threshold follow a generalized Pareto distribution with scale parameter $\tilde{\sigma} > 0$ and shape parameter $\xi \in \mathbb{R}$, whose cumulative distribution function has the form:

$$H_{\tilde{\sigma},\xi}(x) = \begin{cases} 1-(1+\xi x/\tilde{\sigma})^{-1/\xi}, & \xi \neq 0 \\ 1-e^{-x/\tilde{\sigma}}, & \xi = 0 \end{cases}, \quad x \in \mathcal{D}_\xi := \begin{cases} 0 \leq x < \infty, & \xi \geq 0 \\ 0 \leq x \leq -1/\xi, & \xi < 0 \end{cases}. \tag{9.7}$$

The method based on this model is known as peaks-over-threshold (POT) method. It should be noted that this asymptotic model holds more generally when events are not independent over time as long as there is no clustering effect among extreme values, which mathematically corresponds to a stochastic process with extremal index of one.

Given the nature of debris flow observations, the threshold choice and interpretation need to be handled differently here than in the case of regularly spaced time series data. Since small debris flow events are unlikely to be detected and included in the historical record, it is natural to interpret the threshold as a detection threshold, rather than a suitably large threshold, so that all available historical observations can be included in the modelling. In this situation, the asymptotic argument for the distribution of excesses over the threshold may not be immediately applicable, so the generalized Pareto distribution should be assessed to ensure that it provides a reasonable fit to the historical data. The generalized Pareto family of distributions covers a full range of tail behaviours, including the heavy-tailed case when $\xi > 0$, which corresponds to the Pareto distribution, the light-tailed case when $\xi = 0$, which corresponds to the exponential distribution, as well as a short-tailed model defined on a finite interval when $\xi < 0$.

The N-year return level corresponds to the return period of an average of λN observations. Hence, under the point process model above, the N-year return level q_N can be computed by setting $1 - H_{\tilde{\sigma},\xi}(q_N - u) = 1/(\lambda N)$ and solving for q_N, with $H_{\tilde{\sigma},\xi}$ being the cumulative distribution function of the generalized Pareto distribution. This leads to the following expression the N-year return level:

$$q_N = u + (\tilde{\sigma}/\xi)[(\lambda N)^\xi - 1], \text{provided } N > 1/\lambda. \tag{9.8}$$

A number of methods can be used to fit parameters of the generalized Pareto distribution for threshold excesses (marks) in the Poisson point process model above. These include the method of moments and the method of (generalized) probability-weighted moments (Hosking and Wallis (1987), Diebolt et al. (2007)), maximum likelihood estimation (Davison and Smith (1990)) and Bayesian inference (e.g., Stephenson (2016)).

The idea behind the method of moments is to equate model-based moments with their empirical counterparts. For example, for a random variable X with a generalized Pareto distribution with parameters $\tilde{\sigma}$ and ξ, the mean (first moment) and variance (second central moment) have the form:

$$\mathbb{E}(X) = \frac{\tilde{\sigma}}{1-\xi}, \ \mathrm{Var}(X) = \frac{\tilde{\sigma}^2}{(1-\xi)^2(1-2\xi)}, \text{provided } \xi < 1/2. \tag{9.9}$$

Given a random sample $x_1, \ldots, x_n$, the empirical estimates of the mean and variance are given by:

$$\bar{x} = \sum_{i=1}^n x_i, s_X^2 = \frac{1}{1-n} \sum_{i=1}^n (x_i - \bar{x}). \tag{9.10}$$

Setting $\mathbb{E}(X) = \bar{x}$ and Var(X)= s_X^2 and solving for parameters $\tilde{\sigma}$ and ξ gives the following method of moments estimates for parameters of the generalized Pareto distribution:

$$\tilde{\sigma}_{MoM} = \frac{\overline{x}}{2}(1 + \frac{\overline{x}^2}{s_X^2}), \hat{\xi}_{MoM} = \frac{1}{2}(1 - \frac{\overline{x}^2}{s_X^2}). \quad (9.11)$$

Other moments-based methods use the same approach but different versions of moments such as probability-weighted moments. One advantage of such generalizations is to extend the range of admissible shape parameter values. As seen above, variance for a generalized Pareto random variable is defined only if $\xi < 1/2$, which would be restrictive if the actual distribution has a heavier tail. In such cases, the method-of-moments estimates will be biased downward and lead to underestimation of return levels.

The maximum likelihood estimates are obtained by maximizing the likelihood function. For the point process model introduced above, the likelihood function based on the sample of debris flow volumes $\{x_1, \ldots, x_n\}$ is given by:

$$L(\tilde{\sigma}, \xi, \lambda; x_1, \ldots, x_n) = e^{-\lambda n_y} \prod_{i=1}^{n} \lambda \, \tilde{\sigma}^{-1} (1 + \xi \frac{x_i - u}{\tilde{\sigma}})^{-1/\xi - 1}, \quad (9.12)$$

where n_y is the number of years spanned by the data and u is the (detection) threshold.

Bayesian inference offers a natural paradigm to incorporate additional information in the analysis through the use of an informative prior. This is useful in general as extra information is likely to lead to reduction in estimation uncertainty, but especially so in situations where only short data records are available as is the case for debris flow risk analyses. Relying only on historical volume data results in confidence bounds for return levels which are too wide to be of practical relevance. The use of a non-informative prior distribution for the model parameters will lead to parameter estimates which are similar to those obtained under maximum likelihood estimation. For example, let $\widehat{\boldsymbol{\Theta}}$ denote the estimator of model parameters with density function $\pi_{\widehat{\boldsymbol{\Theta}}}$. This density is specified without reference to the data and is referred to as the prior density. It quantifies initial uncertainty about the vector of model parameters. Letting $\mathbf{X} = (X_1, \ldots, X_n)$ denote the data vector, the likelihood function is the conditional density of $\mathbf{X}$ given the model parameters $\widehat{\boldsymbol{\Theta}}$, $f_{\mathbf{X}|\widehat{\boldsymbol{\Theta}}}$. The Bayes' theorem can be used to compute the conditional density of model parameter estimator $\widehat{\boldsymbol{\Theta}}$ given data $\mathbf{X}$:

$$\pi_{\widehat{\boldsymbol{\Theta}}|\mathbf{X}}(\boldsymbol{\theta}|\mathbf{x}) = f_{\mathbf{X}|\widehat{\boldsymbol{\Theta}}}(\mathbf{x}|\boldsymbol{\theta})\pi_{\widehat{\boldsymbol{\Theta}}}(\boldsymbol{\theta})/f_{\mathbf{X}}(\mathbf{x}) \quad (9.13)$$

where

$$f_{\mathbf{X}}(\mathbf{x}) = \int f_{\mathbf{X}|\widehat{\boldsymbol{\Theta}}}(\mathbf{x}|\boldsymbol{\theta}^*)\pi_{\widehat{\boldsymbol{\Theta}}}(\boldsymbol{\theta}^*)d\boldsymbol{\theta}^* \quad (9.14)$$

Density $\pi_{\widehat{\boldsymbol{\Theta}}|\mathbf{X}}$ is called the posterior density. The posterior distribution characterizes the stochastic behaviour of the estimator of model parameters based on information from both the prior distribution and historical data. The mean of the posterior distribution is usually used as the point estimate for the vector of model parameters, while the posterior density itself provides a complete description of the uncertainty about parameter values.

An important question in the Bayesian approach is how to construct an informative prior distribution. Ribatet et al. (2006) propose the use of regional information with application to flood frequency analysis. However, applying such a method to the debris flow analysis would require a database of debris flow events at different locations. As an alternative, Nolde and Joe (2013) developed a methodology in which the prior distribution is elicited using expert judgement in the context of debris flow frequency-magnitude analysis. They found a substantial reduction in uncertainty associated with return level estimation compared to maximum likelihood estimation based on the historical data alone.

The relative merits of various estimation methods for model parameters depend on a number of factors including the tail behaviour of the underlying distribution and sample size. The moments-based methods impose an implicit assumption on the existence of moments, which in turn restricts the range of possible estimates for the shape parameter ξ. In situations where the underlying distribution is not too heavy-tailed, the method of moments is shown to outperform other methods in small samples in terms of the bias and mean squared error of the estimator; Madsen et al. (1997) show this for $0 < \xi < 0.3$. For heavier-tailed data, these methods will lead to negative bias in the estimator of the shape parameter and consequent underestimation of return levels (Nolde and Joe (2013)). On the other hand, the maximum likelihood estimator tends to overestimate the shape parameter and return levels especially in small samples. To improve performance of the maximum likelihood method in smaller samples, it was suggested to include a penalty on the likelihood function which effectively restricts the range of possible parameter values; see Coles and Dixon (1999) and Martins and Stedinger (2001) for proposals of such penalized/generalized likelihood methods. A Bayesian approach with an informative prior similarly leads to an improved performance by balancing information contained in the data with that from the prior.

9.2.6 Effects of Climate Change on Debris-Flow Frequency-Magnitude Relationships

Most field-based and empirical frequency-magnitude relationships implicitly assume that past debris flow activity is a proxy for the future. This may or may not be reasonable, but it entirely depends on the time frame considered to reconstruct such F-M relationships and the specific watershed in question. For example, the method proposed by Jakob et al. (2020), in which fan area or fan volume allows reconstruction of frequency-magnitude relationships, encompasses the time since the fan has developed (in previously glaciated terrain). The Holocene era not only underwent considerable climatic fluctuation, but more importantly, included a period in which geomorphic activity was vastly enhanced due to the abundance of unconsolidated and unvegetated sediment found in the forefield of retreating glaciers (the paraglacial cycle as defined by Church and Ryder, 1972). Therefore, the formation of alluvial fans (even those not truncated by receiving rivers) was equally non-stationary. The practicing geoscientist or geotechnical engineer must then consider how much this non-stationarity affects the outcome of their analysis. For example, a radiocarbon dated surface unit of 7500-year BP overlying 3 more debris flows could indicate that this particular portion of the fan witnessed debris flows every 2500 years (frequency of 0.0004 or 0.04%). An alternative explanation is the effect of the paraglacial cycle, whereby a portion of the fan could be a paleo-landform, hence no longer subject to contemporary processes. The repercussions of this finding can be profound in that new development on a paleo-landform may be permissible. In cases where the fan sector is believed to be affected by debris flows with a 0.04% chance of annual occurrence, a group risk analysis may prove that development cannot proceed without substantial mitigation efforts.

As tempting as it may be, broad statements on how extreme weather may cause bigger and more frequent debris flows are often misplaced. Jakob (2021) attempted to link different debris supply limitation types to likely responses in a changing climate (Fig. 9.4). While it is relatively easy to create conceptual sketches connecting watershed supply limitations to climate-change responses, quantifying those is a rather different story and requires a deep understanding of both watershed type and activity, and also a firm grasp on the likely changes in precipitation intensity, distribution, changing snow regime and ecological externalities (beetle infestation-related tree mortality, wildfire impacts). Doing that with a reasonable degree of confidence is challenging to the extreme. Early work has attempted to do so at a regional scale (e.g., Jakob and Owen, 2021), but at a watershed scale, it is still fraught with broad uncertainty.

9.3 Case Studies

9.3.1 Capricorn Creek

Capricorn Creek drains the southwestern slopes of Mount Meager in the upper Lillooet River valley. It has a long history of debris flows. Jordan (1994) noted that several large debris flows had

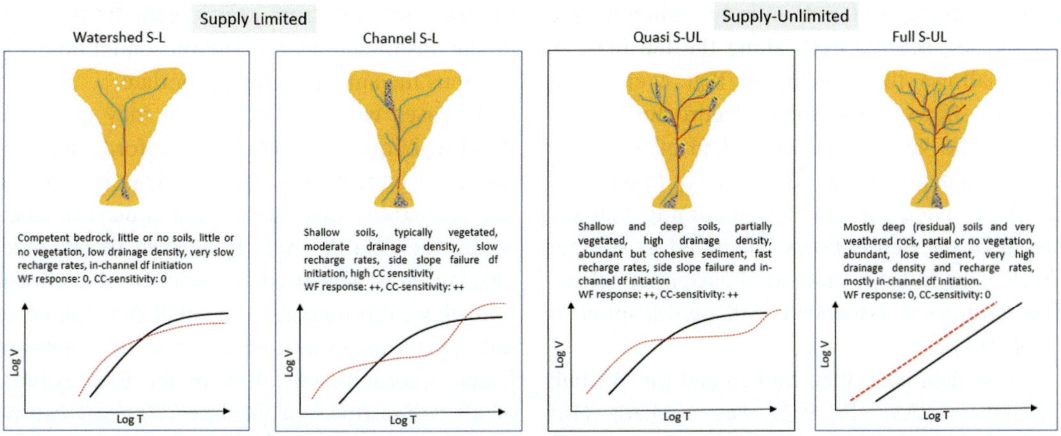

Fig. 9.4 Watershed supply differentiation for debris-flow prone basins. The top row shows conceptual watersheds with varying drainage density and sediment sources. The lower row shows conceptually the frequency-magnitude relationships and expected climate change changes where T stands for return period and V for volume. The black line conceptually indicates the long-term averaged frequency-magnitude curve in absence of climate change (and ignoring paraglacial effects) while the red dotted line signifies the expected conceptual response to climate change. Local variations are clearly expected

descended Capricorn Creek in the past 25 years. Approximately 2500 m upstream from the confluence with Meager Creek, two large landslide scars extend approximately 700 m along the east ridge of Mount Meager. Croft (1983) suggested that this landslide with an estimated volume of 10 Mm^3 descended the creek in the 1920s. However, extensive dendrochronologic studies carried out on the fan, summarized below, do not support this date.

A large event on Capricorn Creek was also dated to 5250 yrs BP (Evans, pers. comm.), which created a deposit that built the modern fan. However, the total volume of this event and the process that generated it are unknown. Furthermore, inclusion in the time series would create extreme bias because undoubtedly there were many more debris flows between 1841 AD and 5250 BP for which there is no more stratigraphic evidence.

Coarse unstratified and matrix-supported deposits on the fan of Capricorn Creek indicate the presence of debris flows and debris floods. Some 25 trees were cored to determine debris flow frequency on this creek (Jakob, 1996). Twelve trees were found where impact scar tissue could likely be attributed to debris flow impact and those were dated by counting tree rings of living trees back to the traumatic resin ducts that herald an impact scar. The record extends 170 years from 1841 to 2010. The debris-flow return period on Capricorn Creek is thus 13 years, which equates to an annual probability of 8%.

Without detailed test trenching, a volumetric assessment of each of these flows is not possible. Even if funds allowed such an investigation, a large portion of the total debris volume would likely have ended up in Meager Creek and been eroded by fluvial processes. A direct assessment of past debris flow volumes was only possible for the 1998, 2009 and 2010 events that were observed directly by Bovis and Jakob (2000), Friele (2009) and Friele (2010).

However, the number of trees affected and the height of the impact scars above the current fan surface allowed a rough approximation of likely debris flow magnitude. Events of up to Class 3 (10^3 to 10^4 m^3) (Jakob, 2005) are unlikely to distribute over the fan and impact trees on the fan margin. Single tree impacts were assigned with a smaller magnitude class than dates that were replicated several times in widely-spaced trees (Jakob, 1996). Even the large 1998 and 2009 events (Class 6 and Class

5, respectively) did not entirely obliterate the fan and left scarred trees along the fan margin, which allows the conclusion that even Class 5 debris flows leave some dendrochronologic record. The volumes of individual events were thus estimated based on the height of impact and the likely width of the flow assuming that the runout of the debris flows continued to Meager Creek. This indirect method of assigning debris-flow volumes is associated with significant error (Table 9.1).

These data were then ranked and the Weibull plotting position (Eq. 9.2) was applied. This method may be appropriate when one can be reasonably confident that most of the events that have occurred have been recorded, and that it spans a long period of observations. A plot of the data shown in Table 9.1 yields a frequency-magnitude graph for Capricorn Creek that indicates that the last three events had respective return periods of (rounded) 170 years (2010), 90 years (1998) and 60 years (2009). The problem with the method lies in the inherent assumption that the largest event (the August 6th 2010 debris flow) is associated with the return period for the time spectrum considered. In this particular case, such an assumption cannot be made and thus the analysis is inappropriate.

In cases where a large event may dominate the distribution, a Pareto (or Bradford) distribution may be used for analysis. This is a power law probability distribution that is used in cases where the data are heavily skewed to one side. Originally it was used to describe the allocation of wealth owned by a small percentage of individuals in society. In the case of Capricorn Creek, approximately 90% of the total volume of all debris flows in the last 170 years can be attributed to the August 2010 debris flow, and thus the method appears well suited.

However, even this assessment is limited with interpretive caveats. The last three events can be attributed to in situ collapses and static liquefaction (1998 and 2009) events and likely undrained loading (2010 event, Guthrie et al., 2011). Most other events that fall into Class 4 or low Class 5 are likely triggered by comparatively small failures along the channel that, through channel scour, evolve into debris flows large enough to travel all the way to Meager

Table 9.1 Summary of debris flow frequency and magnitude data at Capricorn Creek (modified from Nolde & Joe, 2013)

Date	R (replications)	$V_{min} \times 10^5$ [m^3]	$V_{max} \times 10^5$ [m^3]	$V_{mean} \times 10^5$ [m^3]
2010	obs			485
2009	obs	3	7	5
1998	obs	12	15	13.5
1991/92	3	0.3	1	0.65
1988/89	1	0.3	1	0.65
1972/73	4	1	2	1.5
1958/59	5	1	3	2.0
1944/45	3	1	2	1.5
1933/34	2	1	2	1.5
1903/03	1	200	400	300
1873/74	1	3	7	5
1850/51	1	12	15	13.5
1841/42	3	0.3	1	0.65

Note A debris flow in 1669/70 was discarded from the analysis as it is likely that there were many others between 1670 and 1841 but those were likely obliterated by subsequent events leading to fan aggradation and therefore the inclusion of this event would skew the average return period

Creek, but not large enough to erase the sporadic population of trees on the fan. Therefore, the population of debris flows on Capricorn Creek fan consists at least of two separate triggering mechanisms and source area origins. A very similar case was made by Jakob and Friele (2010) for debris flows from Cheekeye River (see Strouth et al., this volume). Separation of the two data populations, however, would yield even less data on which to make reasonably reliable conclusions.

If the GPD analysis is accepted as a reasonable approximation for return periods of debris flows on Capricorn Creek, an event of similar magnitude as the August 2010 event would have a probability of approximately 7% to occur in the next 100 years, or a probability of approximately 50% in the next 1000 years.

Any curve fitted to the data may not be extrapolated indiscriminately because the geomechanical prerequisites for even larger failures need to exist. For example, a large flank collapse of the Meager Mountain face that could yield a Class 8 (10^8–10^9 m^3) rock avalanche and debris flow (e.g., Friele & Clague, 2004) is conceivable, but this may exhaust the volume of future events, so Class 9 events are much less likely to occur on Capricorn Creek. Furthermore, a strong paraglacial signal may be influencing the frequency-magnitude relationship during the past two centuries. During this time, the Capricorn Glacier retreated considerably and also lost most of its mass. This process evacuated the ice from the valley, leaving oversteepened and debuttressed slopes in inherently unstable material. The sequence of increasing magnitudes in the last 12 years may have pointed towards a culmination in the geomorphic response of the Capricorn basin to glacial retreat. Such a trend would violate the assumptions of data stationarity and independence that underlie statistical treatments of this problem. Finally, the last eruption of Mount Meager some 2500 years ago created an environment favourable for rapid recharge of talus aprons and likely steepened the unstable volcanic edifice. These factors, in combination or individually, likely violate assumptions of independence and stationarity and the frequency estimates reported here need to be evaluated with caution.

We revisit the extreme value analysis of debris-flow volumes at Capricorn Creek in Nolde and Joe (2013). The dataset studied in Nolde and Joe (2013) uses the volume estimate for the 2010 event as 3×10^7 m^3 as opposed to the higher precision updated estimate of 4.85×10^7 m^3. The updated dataset is plotted in Fig. 9.5. Refitting the Poisson point process

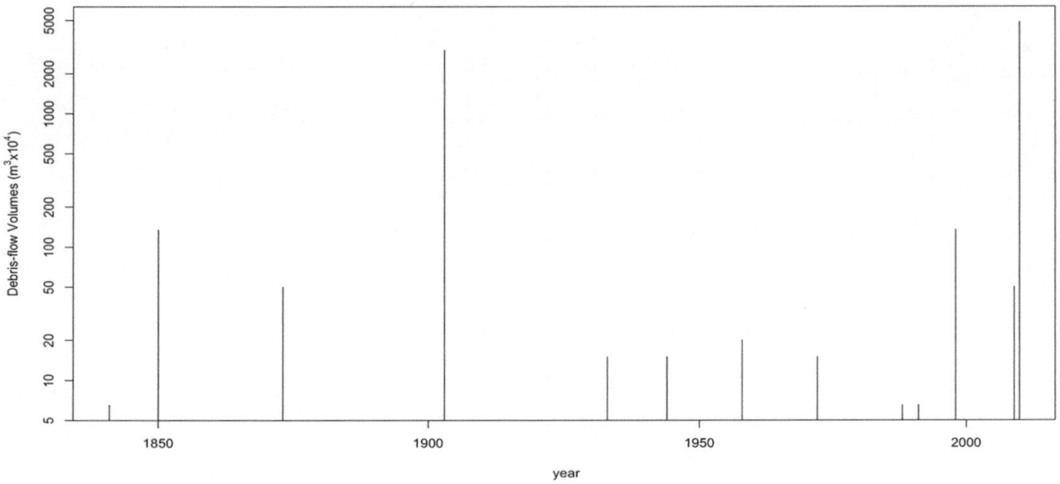

Fig. 9.5 Time series of debris flow volumes at Capricorn Creek (modified from Nolde & Joe, 2013)

model with threshold of 3×10^4 m^3 as in the previous analysis, we find that the maximum likelihood estimate of the shape parameter ξ is 1.92 with standard error of 0.76. The shape parameter estimate is positive and fairly large in magnitude, indicating a very heavy tail for the underlying distribution of debris flow volumes. The fit is influenced by the two largest events in 1903 and 2010, whose magnitudes are several orders larger than any other event in the sample.

To reduce estimation uncertainty, we perform a Bayesian extreme value analysis using an informative prior elicited using expert judgement, as proposed in Nolde and Joe (2013). For the purpose of this illustration, we use the same prior specification as given in Nolde and Joe (2013). Figure 9.6 shows return level plots estimated using maximum likelihood (dashed curves) and median of the posterior distribution based on the Bayesian inference with an informative prior (solid curves). The return level plot is split into two parts for return periods up to 70 years (left panel) and from 70 years up to 750 years (right panel) to improve visualization. As expected, incorporation of additional information in the form of expert judgement leads to reduction in estimation uncertainty, as evidenced by a narrow funnel formed by the solid blue curves corresponding to the posterior 95% confidence bounds relative to the dashed blue curves indicating the approximate 95% confidence bounds for the maximum likelihood estimates. It is also not surprising that the confidence regions become very wide for return periods extending beyond the available data range, reflecting uncertainty when extrapolating into the tail region with no data. It should be noted that the prior distribution uses the effective sample size of 14 observations (this is determined through uncertainty bounds specified by the expert), which is on par with the historical record, and so both likelihood for the observed events and expert-based prior play about equal roles in determining the posterior distribution.

The frequency-magnitude curve (Fig. 9.6) based on the Bayesian inference is markedly lower than the curve based on the maximum likelihood estimation. This can be attributed to the prior distribution conveying the expert's belief in less severe debris flow events, partly due to the consequences of the extremely large 2010 event. The likelihood function attempts to capture the observed events and the resulting return level curve captures these historical records fairly well, with the exception of the 1903 event, which falls outside the 95%

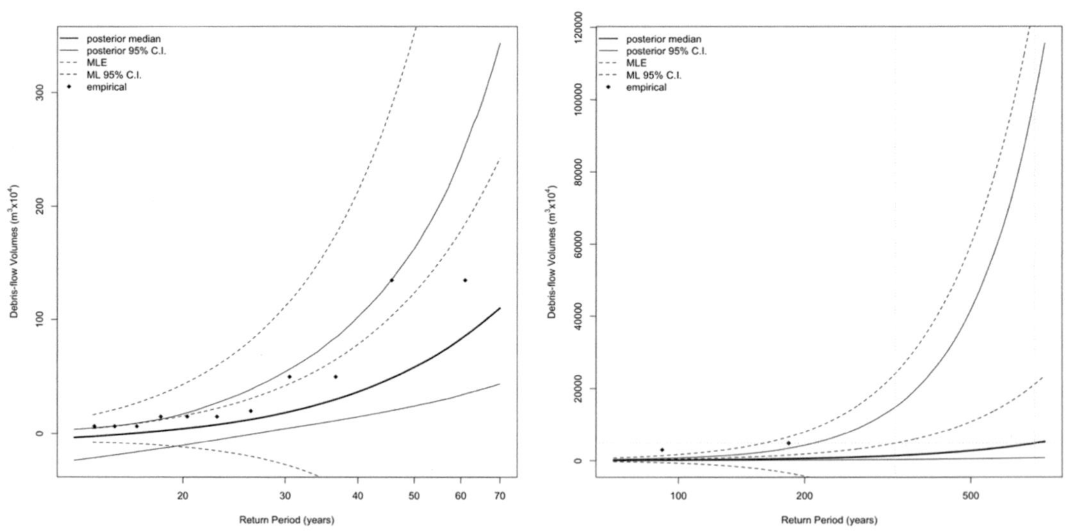

Fig. 9.6 Frequency-magnitude curves for debris flow volumes at Capricorn Creek (modified from Nolde & Joe, 2013). MLE stands for maximum likelihood estimation; C.I. abbreviates point-wise confidence intervals

confidence region. The posterior distribution, on the other hand, balances between the historical events and the expert's belief on the severity and associated frequency of future events.

One can use return level curves to obtain estimates of return periods for events of specified magnitude. For example, the 2010 debris flow has a return period of 330 years under the maximum likelihood estimation, while the posterior distribution suggests a much longer return period of 710 years.

9.3.2 Cheekeye River

Cheekeye River fan is the best studied fan in Canada with numerous consulting and academic studies having focused on the geomorphic development of the fan and the associated hazards and risks (summarized in Jakob and Friele, 2010). It also features in the chapter on debris flow risk (Strouth et al., this volume).

This example underscores a key issue that data are never complete and, depending on the methods chosen to date debris flows, the data are typically heavily skewed towards the more recent events. Statistically, this introduces a significant problem. Fitting, for example, a generalized Pareto distribution (GPD) to the data set and then analyzing them to determine a frequency-magnitude graph, the annual observations have to be specified. In case of Cheekye River, for example, 12 debris flows have been documented over the course of 11,600 years. This implies an annual observation frequency of approximately 0.0001. This value heavily influences the outcome and is an artefact of the number of recorded events. The true number of events is much higher, but unknown.

The second problem is that of extrapolation. In case of Cheekye River, assuming that all data are homogenous (i.e. all debris flows are caused by the same processes), the outcome will be different. For example, if the three large events had not been documented because of insufficient resources, an extrapolation of the other data would have resulted in a 10,000-year event of approximately 7 million m^3 (Fig. 9.7). Equally,

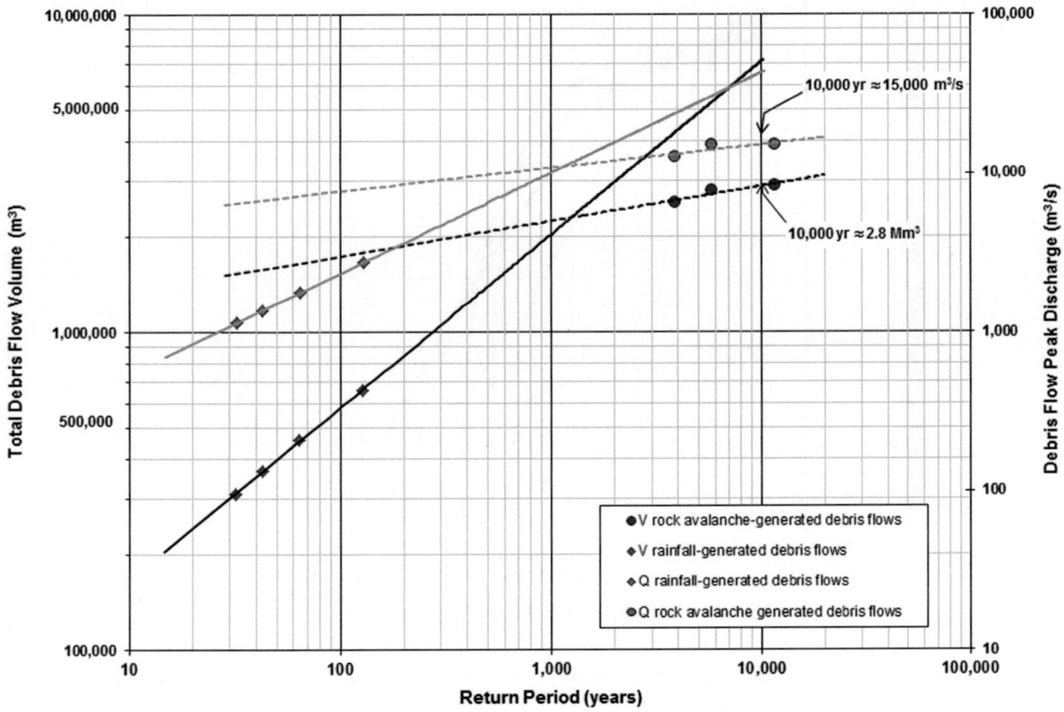

Fig. 9.7 Frequency-volume and frequency-peak discharge relationship at Cheekye river

if only the largest events had been reconstructed, for example, if there were no trees to attempt dendro-chronology, extrapolation towards the lower return periods would yield exaggerated volumes for those return periods. For example, the 100-year return period volume would be approximately 1.8 million m^3 as opposed to 600,000 m^3, which is the result one obtains from an analysis of the stratified dataset (Fig. 9.7).

This example illustrates that the method of extrapolating data must consider different sample populations and examine the reasonableness in extrapolation. Failure to do so could result in highly erroneous results.

9.3.3 Charles Creek

Charles Creek drains a 1.8 km^2 watershed with approximately 30% of the upper drainage consisting of steep, unvegetated cliffs. The remainder of the watershed is characterized by colluvial veneers, talus slopes and minor deposits of glacial till covered by mature forest of fir, hemlock and cedar. The watershed is described more fully by Hungr and Wilson (2023).

Using data from Hungr and Wilson (2023), the cumulative frequency function was plotted and extrapolated using linear, logarithmic and power law fits. Furthermore, the GDP was superimposed (solid line on Fig. 9.8). The objective is to estimate the volume of a 500-year event to determine if the current debris basin is adequately sized. The cumulative frequency function only allows a direct assessment to the probability derived from the record length (i.e. 0.02). The linear, log and power fits yield 500-year (F = 0.002) volume estimates of (rounded) 22,000, 30,000 and 50,000 m^3. The GPD fit yields a total volume of 82,000 m^3 with a 95% confidence interval (not shown on Fig. 9.8) of 15,000 and 230,000 m^3. Any practitioner daring

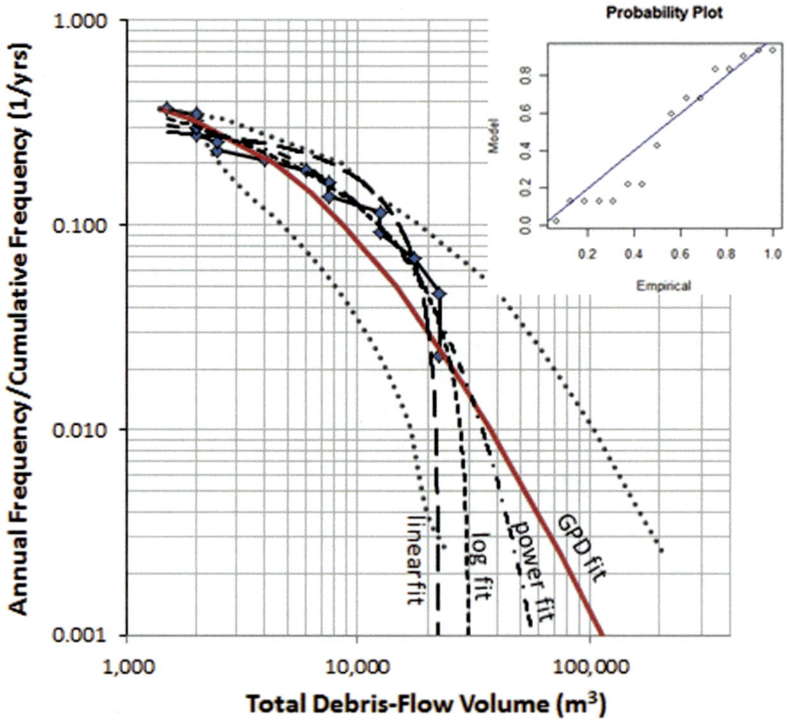

Fig. 9.8 Charles Creek CMF and GDP fits. Linear, logarithmic and power functions were fitted to the CMF. The dotted lines show the upper and lower 95% confidence interval for the GDP fit. Note that the frequency-magnitude relationship is plotted purposely different as, for example, in Fig. 9.7, to demonstrate a different plotting technique that is preferred by some practitioners

to present such a large range may jeopardize the relationship with his/her client. The question, therefore, is which fit is judged to be most credible.

Without an understanding of the watershed geomorphology and the debris-flow triggering mechanisms, such a decision cannot be made. The Charles Creek channel system is recharged frequently by rock fall that has accumulated a talus apron extending between 800 and 1200 m elevation. The high frequency of debris flows indicates that, once a critical hydroclimatic threshold is exceeded, debris flows occur. However, debris-flow volume and severity of the hydroclimatic event may not be related but cannot be proven because of the absence of high elevation weather stations nearby. It appears that the debris-flow volume is limited by the available debris, which in turn is a function of recharge rate and the time since the last debris flow. Very large accumulations of debris are unlikely given that even low frequency (say a 1 in 3 to 5-year storm) may trigger a debris flow that removes available sediment. Therefore, one is led to believe that the GDP and power fits are unlikely candidates because the watershed is unlikely to produce the large events they predict. The linear fit may be discarded on the notion that it asymptotically follows the approximate volume of the largest measured events (~22,500 in 1969 and 2006). It is very unlikely (0.7% chance) that two 500-year events would occur within a 43-year time period. These considerations leave the log and power fits. Following the premise of supply limitations, the log fit appears most promising, and a practitioner may choose 30,000 m^3 as the design volume for the 500-year return period debris flow at this site.

Even though this result is derived from some justifiable reasoning, it neglects the potential of an extreme event, such as an earthquake-triggered rock slide that may evolve into a debris flow. To test the hypothesis that such an event will not occur (at least not with an annual probability of 0.2%), it may be advisable for the professional to investigate questionable rock slopes in the field and perhaps analyse the rock mass kinematics. If a rock slide of several tens of thousands of cubic metres appears feasible, additional analysis may be advisable to ascertain that sufficient water can be recruited at impact with the talus slope and/or on the downslope path to fully transform the rock slide into a debris flow. Alternatively, the practitioner may wish to analyse the residual risk of a 30,000 m^3 retaining structure being overtopped by variable volumes.

9.4 Discussion

This chapter highlights one of the most difficult aspects in debris-flow hazard- and risk assessments: the establishment of a reliable frequency-magnitude relationship on which to base empirical or numerical modelling and subsequent risk calculations. A profound error in the establishment of the frequency-magnitude relationship will result in erroneous model results (see Iverson and George, this volume) and, consequently, errors of the life-loss and/or economic risk estimates and lead to under- or overdesign of mitigation measures, both with undesirable outcomes (see Strouth et al., this volume).

The examples showcased herein demonstrate that there is likely not one single best method to decipher the frequency-magnitude relationships on fans, especially since the dating techniques (Ballesteros et al., this volume) or procedures to identify past debris-flow volumes depend greatly on available budgets, morphologic, climatic and vegetation conditions, as well as the skill and experience of the study team.

The application of traditional frequency analysis that consists of fitting mathematical probability distribution models to recorded events and extrapolates the distributions' tails to very low exceedance probabilities is questionable. It is even more uncertain than in the analysis of, for example, annual stream flow measurements, because volume data are typically extremely sparse, often truncated due to evidence obliteration, and often suffer non-homogeneity in the debris-flow triggering mechanisms. The individual debris-flow triggering mechanics and supply limitations may heavily influence the outcome of the analysis if they are known, which, more often than not, will preclude combining observations from multiple watersheds to produce a regional debris-flow frequency-magnitude curve, as is customary, for example, for snow avalanches.

In the case of individual fan studies, the practitioner is often limited in scope by the financial resources that have been made available to solve the problem (i.e. to create a debris-flow hazard or debris-flow intensity map). This typically precludes application and exhaustion of available methods, such as dendrochronology and test trenching with radiometric dating of organic materials, that allow volumetric and frequency analysis. More often than not, this data limitation leaves the practitioner with an incomplete data set on which to base zoning decisions, with important repercussions on human safety and land value.

An additional complication is introduced through strong data censoring, with small events being reconstructed and clustered in the recent past. Such data truncation influences the tail of any statistical distribution to which data could be fitted, which is likely to result in significant error in the best estimate. Application of, for example, generalized Pareto distributions, requires a designation of annual probabilities that hinge on the number of observations over the time period examined. The lack of data continuity in the deeper past, however, will skew the results. This can be overcome by using a peak-over-threshold approach in which only large events are considered, if a quasi-complete record of those events can be established). Neglect of smaller events can often be justified in practice because larger events determine the hazard and risk on fans.

Probability density functions are meant to be reasonable approximations of the physical distribution. Because of non-stationarity, non-ergodicity and non-homogeneity, many of these distributions are inappropriate for use in discontinuous and fragmentary data sets. More suited are distributions that do not require such underlying statistical assumptions. Still, the outcome from such distributions need to be scrutinized because of the geomorphic and geologic idiosyncrasies inherent in many watersheds analysed. Such anomalies are to be identified by the investigator and used in interpreting the results from extreme value statistics.

In the case of Capricorn Creek, the suggested frequency-magnitude graph may not be extrapolated indiscriminately because the geo-mechanical prerequisites for even larger failures need to exist. A large flank collapse of the Meager Mountain face is conceivable that could yield a Class 8 (10^8–10^9 m^3) rock avalanche and debris flow (e.g. Friele & Clague, 2004), but this may exhaust the volume of future events; Class 9 events are thus much less likely to occur on Capricorn Creek. Furthermore, a strong paraglacial signal may be influencing the frequency-magnitude relationship during the past two centuries. During this time, the Capricorn Glacier retreated considerably and lost most of its mass. This process evacuated the ice from the valley, leaving oversteepened and debuttressed slopes in inherently unstable material. The sequence of increasing magnitudes in the last several years may have pointed towards a culmination in the geomorphic response of the Capricorn basin to glacial retreat and perhaps deterioration of mountain permafrost that could lead to a loss of ice cohesion. Such a trend would violate the assumptions of data stationarity and independence that underlie statistical treatments of this problem. Finally, the last eruption of Mount Meager, some 2500 years ago, has created an environment favourable for rapid recharge of talus aprons and has likely steepened the unstable volcanic edifice.

At Charles Creek, the only example with a reasonably accurate continuous record over the past 40 years, extrapolation of the record to a design-return period is limited by the short period for which data are available, and the choice of the standard trendline fitting to the cumulative frequency distribution will strongly influence the outcome.

These factors, in combination or individually, may violate assumptions of data homogeneity, independence and stationarity, and the frequency estimates reported here need to be evaluated with caution. The authors caution against extrapolations far into the future from limited data records solely because of regulatory demands (e.g. MoTI, 2009). Attempting such extrapolations without a firm physical basis (Higman et al., 2011) may result in over- or underdesign of mitigative works, putting people and infrastructure unnecessarily at risk or unnecessarily sterilizing habitable land. The authors propose that, instead, the practitioner clearly outlines the limitation of an analysis, or in absence of a reasonable data set, refuses to specify the volume of a hypothetical debris flow of an extreme (for example, 10,000 year) return period event.

9.5 Conclusion

Geomorphologists, engineering geologists and geotechnical engineers often need to work with discontinuous debris-flow magnitude data (mostly volumes) and equally sporadic dates of past debris flows. Moreover, the data, once painstakingly deciphered from various proxies or empiricism, are typically associated with significant, yet poorly quantifiable, errors. Even in the case of post-fire debris flows, a subject studied for two decades in meticulous detail, reliable prediction of debris-flow volumes is still fraught with difficulty, especially in areas outside of the southwestern United States, where most of the research on post-fire debris flows originates.

Detailed debris-flow frequency analysis of individual fans forms the backbone of hazard assessments and provides input to quantitative risk assessments. A poorly resolved frequency-magnitude relationship translates into error-prone numerical modeling (see Iverson and George, this volume) and an equally poorly constrained risk assessment (see Strouth et al., this volume) on which decisions for mitigation may be based. In short, mitigation may be substantially over-or under-designed if the original F-M relationship is wrong.

Various statistical procedures are available to construct debris-flow frequency-magnitude relationships. Those are extrapolated to return periods that either lie outside the observation range or are interpolated to specific return periods mandated by regulations. This chapter highlights common and uncommon techniques and discusses their advantages and pitfalls. We conclude that no single technique fits all cases. Rather, each watershed and fan requires careful consideration of what techniques should be applied to what kind of data. Blind extrapolation of fitted curves is likely to result in very large confidence bands that require reconciliation with the need for hazard zonation and design of mitigative works. An honest appraisal of debris-flow volumes associated with specific return periods, apart from careful selection of the statistical technique, must include judgment and common sense linked to an understanding of the watershed's geomorphology, sediment supply limitations, hydroclimatic conditions, and debris-flow triggering mechanisms. This should be a consolation to those who fear their jobs will be replaced by artificial intelligence, machine learning and ultimately geo-robots.

This paper examined cases where data on debris flow age and their respective volumes had been gathered. An attempt was made to establish a relationship between the frequency of corresponding debris-flow magnitudes. Cumulative magnitude frequency functions are used to determine the potential magnitude of a given event frequency. Such curves are particularly useful for regional datasets, where many observations can be gathered and potentially stratified by geological idiosyncrasies. In most cases, however, a regional dataset has not been established and may be cost-prohibitive. Therefore, age and volume data gathered on the fan of interest will need to be used for frequency-magnitude analysis.

Ultimately, no mathematical proof is available to ascertain the validity of the methods discussed herein to debris flows and more in-depth research is needed to verify their applications. Until then, the practitioner may need to combine intuition, judgment, common sense and statistical finesse to arrive at a credible outcome in debris-flow frequency analysis.

References

Beirlant, J., Goegebeur, Y., Segers, J., & Teugels, J. (2004). *Statistics of extremes. Theory and applications*. Wiley Series in Probability; Statistics.

Bovis, M. J., & Jakob, M. (1999). The role of debris supply conditions in predicting debris flow activity. *Earth Surface Processes and Landforms, 24*(11), 1039–1054.

Bovis, M. J., & Jakob, M. (2000). The July 30, 1998 debris flow and landslide dam at Capricorn Creek, Mount Meager Volcanic complex, British Columbia. *Canadian Journal of Earth Sciences, 37*(10), 1321–1334.

Brardioni, F., & Church, M. (2004). Representing the magnitude-frequency relation: Capilano river basin. *Earth Surface Processes and Landforms., 29*, 115–124.

Cannon, S. H., Gartner, J. E., Wilson, R. C., Bowers, J. C., & Laber, J. L. (2008). Storm rainfall conditions for floods and debris flows from recently burned areas in southwestern Colorado and southern California. *Geomorphology, 96*(3–4), 250–269. https://doi.org/10.1016/j.geomorph.2007.03.019

Cannon, S. H., Gartner, J. E., Rupert, M. G., Michael, J. A., Rea, A. H., Parrett, C. (2010). Predicting the probability and volume of postwildfire debris flows in the intermountain western United States. *Geological Society of America Bulletin, 122*(1–2), 127–144. https://doi.org/10.1130/B26459.1

Church, M., & Ryder, J. M. (1972). Paraglacial sedimentation: A consideration of fluvial processes conditioned by glaciation. *Geological Society of America Bulletin, 83*(10), 3059–3072.

Church, M., & Slaymaker, O. (1989). Disequilibrium of holocene sediment yield in glaciated British Columbia. *Nature, 337*, 452–454.

Coles, S. G., & Dixon, M. J. (1999). Likelihood-based inference for extreme value models. *Extremes, 2*, 5–23.

Coles, S. G. (2001). *An introduction to statistical modelling of extreme values*. Springer Series in Statistics.

Croft, S. A. S. (1983). *Stability assessment of the Capricorn Creek valley, British Columbia*. BASc thesis, University of British Columbia.

Davison, A. C., & Smith, R. L. (1990). Models for exceedances over high thresholds (with discussion). *Journal of the Royal Statistical Society Series B, 52*, 393–442.

Diebolt, J., Guillou, A., & Rached, I. (2007). Approximation of the distribution of excesses through a generalized probability-weighted moments method. *Journal of Statistical Planning and Inference, 137*, 841–857.

Friele, P., & Clague, J. J. (2004). Large holocene landslides from Pylon peak, Southwestern British Columbia. *Canadian Journal of Earth Science, 41*, 165–182.

Friele, P. (2009). *Capricorn Creek debris flow hazard & risk management review*. Report prepared by Cordilleran Geoscience for Ministry of Forests, November 13, 2009.

Friele, P. (2010). *The august 6, 2010 Capricorn Creek landslide, Meager Creek valley, southwestern British Columbia: Description, emergency response, infrastructure damage and future hazards*. Report prepared by Cordilleran Geoscience for BC Ministry of Environment. August 8, 2010.

Gartner, J. E., Cannon, S. H., Santi, P. M. (2014). Empirical models for predicting volumes of sediment deposited by debris flows and sediment-laden floods in the transverse ranges of southern California. *Engineering Geology, 176*, 45–56.

Gutenberg, B., & Richter, C. F. (1954). *Seismicity of the earth* (2nd ed.). Princeton University Press.

Guthrie, R. H., Friele, P., Allstadt, K., Robertos, N., Evans, S. G., Delaney, K., Roche, D., Clague, J. J., & Jakob, M. (2011). *The august 6, 2010 Mount Meager rock avalanche-debris flow, Coast Mountains, British Columbia: Characteristics, dynamics, and implications for hazard and risk assessment*. Submitted to Natural Hazards and Earth System Science.

Higman, S., Martin, S., Gustafson, R., & Card, L. (2011). *Geological hazard and risk assessment of Testalinden Creek near Oliver, B.C.* Paper No. 142. Geohazards 2011.

Hosking, J. R. M., & Wallis, J. R. (1987). Parameter and quantile estimation for the generalized pareto distribution. *Technometrics, 29*, 339–349.

Hungr, O., McDougall, S., Wise, M., & Cullen, M. (2007). Magnitude-frequency relationships of debris flows and debris avalanches in relation to slope relief. *Geomorphology*. https://doi.org/10.1016/j.geomorph.2007.03.020

Hungr, O., & Wilson, P. (2023). The Charles Creek case study. In S. Leroueil, & L. Picarelli (Eds.), *Understanding landslides through case studies*. Taylor and Francis.

Jakob, M. (1996). *Morphometric and geotechnical controls of debris-flow frequency and magnitude in southwestern British Columbia*. Unpublished Ph.D. thesis. Geography Department. The University of British Columbia.

Jakob, M. (2005). A size classification for debris flows. *Engineering Geology., 79*, 151–161.

Jakob, M. (2021). Debris-flow hazard assessments: A practitioner's view. *Environmental & Engineering Geoscience, 27*(2), 153–166.

Jakob, M., & Hungr, O. (2005). *Debris-flow hazards and related Phenomena*. Berlin, Heidelberg: Springer.

Jakob, M., & Friele, A. P. (2010). Landslide hazards and risks from volcanic debris flows at Mount Garibaldi, British Columbia. *Geomorphology, 114*, 382–395.

Jakob, M., Friele, P., Mark, E., McDougall, S., Friele, P., Lau, C.-A., & Bale, S. (2020). Regional debris-flow and debris-flood frequency-magnitude curves. *Earth Surface Processes and Landforms, 45*(12), 2954–2964.

Jakob, M., & Owen, T. (2021). Projected effects of climate change on shallow landslides, North Shore Mountains, Vancouver, Canada. *Geomorphology, 393*, 107921.

Jakob, M., & Weatherly, H. (2005). *Debris flow hazard and risk assessment. Jones Creek, Washington*. In O. Hungr, R. Fell, R. Couture, & E. Eberhardt (Eds.), *Landslide risk management. Proceedings* (pp. 533–542).

Jordan, P. (1994). *Debris flows in the southern Coast Mountains, British Columbia: Dynamic behaviour and physical properties*. Unpublished PhD thesis, University of British Columbia.

Klemes, V. (2000). *Common sense and other heresies. Selected papers on hydrology and water resources engineering* (p. 378). Canadian Water Resources Association.

Madsen, H., Rasmussen, P. F., & Rosbjerg, D. (1997). Comparison of annual maximum series and partial duration series methods for modeling extreme hydrologic events. *Water Resources Research, 33*, 747–757.

Makkonen, L. (2006). Plotting positions in extreme value analysis. *Journal of Applied Meteorology and Climatology, 45*, 334–340.

Makkonen, L., Pajari, M., & Tikanmäki, M. (2013). Discussion on Plotting positions for fitting distributions and extreme value analysis. *Canadian Journal of Civil Engineering*. https://doi.org/10.1139/cjce-2013-0227

Martins, E. S., & Stedinger, J. R. (2001). Generalized maximum likelihood pareto-poisson estimators for partial duration series. *Water Resources Research, 37*, 2551–2557.

McClung, D. (1999). The encounter probability for mountain slope hazards. *Canadian Geotechnical Journal, 36*(6), 1195–1196.

Ministry of Transportation and Infrastructure. (2009). Non-referenced bulletin to approving officers: Subdivision preliminary layout review–Natural Hazard Risk.

Moody, J. A., Shakesby, R. A., Robichaud, P. R., Cannon, S. H., & Martin, D. A. (2013). Current research issues related to post-wildfire runoff and erosion processes. *Earth-Science Reviews, 122*, 10–37.

Muñoz-Sabater, J., Dutra, E., Agustí-Panareda, A., Albergel, C., Arduini, G., Balsamo, G., Boussetta, S., Choulga, M., Harrigan, S., Hersbach, H., Martens, B., Miralles, D. G., Piles, M., Rodríguez-Fernández, N. J., Zsoter, E., Buontempo, C., & Thépaut, J.-N. (2021). ERA5-Land: A state-of-the-art global reanalysis dataset for land applications. *Earth system science data, 13*, 4349–4383. https://doi.org/10.5194/essd-13-4349-2021.

Nolde, N., & Joe, H. (2013). A Bayesian extreme value analysis of Debris flows. *Water Resources Research, 49*, 7009–7022.

Ribatet, M., Sauquet, E., Gresillon, J. M., & Ouarda, T. B. M. J. (2006). A regional Bayesian POT model for flood frequency analysis. *Stochastic Environmental Research and Risk Assessment, 21*, 327–339.

Rifkin, J. (2019). *The green new deal: Why the fossil fuel civilization will collapse by 2028, and the bold economic plan to save life on earth*. St. Martin's Press.

Stephenson, A. (2016). Bayesian inference for extreme value modelling. In *Extreme Value Modeling and Risk Analysis: Methods and Applications* (pp. 257–80). Chapman & Hall/CRC.

Weibull, W. (1939). A statistical theory of strength of materials. *Proceedings of the Royal Swedish Institute for Engineering Research, 151*, 45.

Climate Change Effects on Debris Flows

10

Markus Stoffel, Simon K. Allen, Juan Antonio Ballesteros-Cánovas, Matthias Jakob and Nina Oakley

Abstract

In mountain areas with human infrastructure, debris flows are a curse. They are abundant and dreaded as they arrive, usually without warning, and damage or destroy much of what they encounter along their path. It has been widely shown that climate is changing and that the past few decades of global warming are almost entirely attributable to human greenhouse gas emissions. There are, unfortunately, very few signs that emissions will be reduced as much as necessary to avoid further warming. In fact, the current trajectory projects between 3.3 and 5.7 °C warming by the end of the century, with amplified warming rates at high latitudes and high altitudes. Recent geopolitical events have led to a weaponization of energy, which in turn has led to another surge in fossil fuel production. Surely, changes in debris-flow activity may pale in comparison to the myriad of other catastrophic changes that would be unleashed commensurate with such unprecedented warming in human history. Nonetheless, debris flows (including their volcanic expression as lahars), especially as part of complex hazard chains in high mountains, have the potential for vast destruction of human habitat and above-ground and buried infrastructure. In this chapter, we provide a review of recent advances in climate change effects on debris flows. We demonstrate that, for some watersheds, an increase in frequency may result in a decrease in magnitude and hence runout distance. This, in turn, could result in a net risk reduction. In other cases, however, higher rainfall intensities and magnitudes will unleash more vigorous debris-flow activity. We also highlight the secondary effects of warming through tree mortality by insect infestations and drought stress, as well as

M. Stoffel (✉) · S. K. Allen · J. A. Ballesteros-Cánovas
Institute for Environmental Sciences, Climate Change Impacts and Risks in the Anthropocene (C-CIA), University of Geneva, Geneva, Switzerland
e-mail: Markus.Stoffel@unige.ch

M. Stoffel
Department of Earth Sciences, University of Geneva, Geneva, Switzerland

Department F.-A. Forel for Environmental and Aquatic Sciences, University of Geneva, Geneva, Switzerland

S. K. Allen
Department of Geography, University of Zurich, Zurich, Switzerland

J. A. Ballesteros-Cánovas
The National Museum of Natural Sciences, Spanish Scientific Research Council, Madrid, Spain

M. Jakob
BGC Engineering Inc, Vancouver, BC, Canada

N. Oakley
California Geological Survey, Sacramento, CA, USA

M. Jakob et al. (eds.), *Advances in Debris-flow Science and Practice*, Geoenvironmental Disaster Reduction,
https://doi.org/10.1007/978-3-031-48691-3_10

the consequences of wildfires on debris-flow occurrence, frequency, and magnitude. Similarly, we emphasize the importance of glacial retreat and permafrost degradation that destabilize steep, high mountainous terrain worldwide. Importantly, we stress that quantifying debris-flow changes in a warming world is highly complex and subject to much uncertainty.

10.1 Introduction

Global surface temperature today is 1.1 °C higher than during the reference period 1850–1900, and each of the last four decades has been successively warmer than any decade that preceded it since 1850 (IPCC, 2021). Some of the warming that occurred since the peak of the Little Ice Age around 1850–1890 was natural, however, warming of the last decades is unambiguously attributed to anthropogenic greenhouse gas emissions (IPCC, 2021). Warming is now at a rate that is unprecedented in at least the last 2000 years (Hantemirov et al., 2022), with pronounced, obvious, widespread, and rapid changes now occurring in the atmosphere, cryosphere, and biosphere (Hock et al., 2019).

At current rates of CO_2 and other greenhouse gas emissions, global surface temperatures will continue to increase rapidly. In the most optimistic scenario assuming deep reductions in greenhouse gas emissions, warming will continue until at least mid-twenty-first century and halt at +1 to 1.8 °C by the end of the twenty-first century compared to 1850–1900 (IPCC, 2021). By contrast, if humanity should fail to take any notable action to reduce greenhouse gas emissions, global surface temperatures will very likely reach anywhere between 3.3 and 5.7 °C above the 1850–1900 reference period. Noteworthy, and to put these figures into context, the last time global surface temperature was sustained at or above 2.5 °C higher than 1850–1900 was over 3 million years ago, when there were no humans on Earth (IPCC, 2021).

In mountainous environments, where debris flows occur, climate change is even more pronounced (Dimri et al., 2021; Gobiet et al., 2014; Hock et al., 2019; Pepin et al., 2022; Stoffel, 2022), with current levels of warming already well beyond +2 °C in many places. This observation is in line with climate models projecting enhanced climate warming to persist in mountain (*high altitude*) and polar (*high latitude*) environments at rates that are usually at least twice as large as the global mean (England et al., 2021; Gaglioti et al., 2022). In addition to changes in mean conditions, observed and projected changes in extremes are and will be larger in frequency and intensity with every additional increment of global warming (IPCC, 2021).

The ongoing warming has a suite of consequences, including effects on the climate system and climate variables, but also on the overall susceptibility of catchments to produce mass-movement activity in general (Borga et al., 2014; Stoffel & Huggel, 2012; Stoffel et al., 2014a; Zheng et al., 2021a) or debris flows in particular (Stoffel et al., 2014b; Beniston et al., 2018; Prenner et al., 2019; Hirschberg et al., 2021). In this chapter, we disentangle the role of ongoing climate change on: (i) external, climatic variables (or inputs), (ii) the susceptibility within the system to produce debris flows, and (iii) possible reactions of debris-flow catchments to these changes in terms of the frequency and magnitude of individual events. The chapter is hence organized around three domains in which effects of climate change will likely be felt. As shown in Fig. 10.1, the chapter focuses on observed and anticipated changes in climatic input parameters or variables, addresses likely effects of climate change on the basic or variable susceptibility of catchments to produce debris flows, and looks at the likely reactions or consequences that these changes might have on process activity.

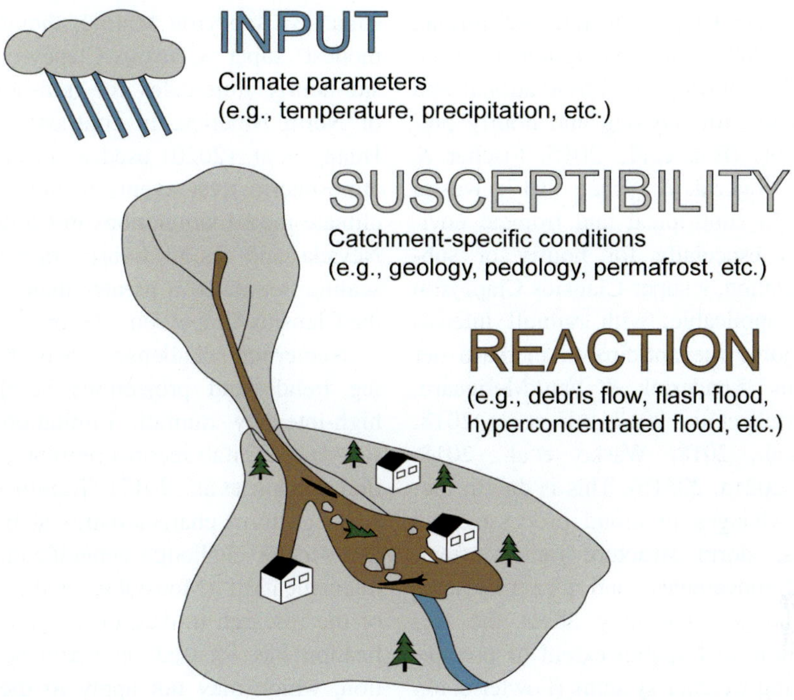

Fig. 10.1 Climatic inputs, susceptibility, and reactions of a debris-flow system as used in this chapter (*Graphic* Sophia Zubrycky, Markus Stoffel)

10.2 Input: Effects of Climate Change on Precipitation

Climatic factors are an important subject for a better understanding of hydrologic response of soils and of how global climate change can influence debris-flow activity. In addition, climatic factors are essential for developing debris-flow warning systems. While independent of climate change, climate factors will in many instances signal more frequent threshold exceedances. Climatic factors have extreme spatial and temporal variability. Rapid infiltration of prolonged intense rainfall, causing soil saturation and a transient increase in pore-water pressure, is generally believed to be the mechanism by which most shallow landslides, and more specifically debris flows, are generated during rainstorms (Iverson, 2000).

10.2.1 Heavy Precipitation in a Warmer Atmosphere—An Overview

A warming climate is expected to lead to an intensification of the hydrological cycle. Global climate models usually predict an increase in global mean precipitation by 2–3% for every degree Celsius of warming (Held & Soden, 2006), a phenomenon that can be explained by global energy constraints (Trenberth et al., 2009). Extreme precipitation does not follow the same scaling, as global constraints do not apply in that case. By contrast, the capacity of the atmosphere to hold moisture increases at a Clausius-Clapeyron rate of about 6–7% per degree Celsius warming. During extreme precipitation events, one would expect that in a warmer atmosphere, such events will scale with the Clausius-Clapeyron rate (O'Gorman & Schneider, 2009). This assumption has been supported by a multitude of observational and numerical studies for daylong and hourly precipitation events (Ban et al., 2015; Fischer & Knutti, 2016; Rajczak & Schär, 2017; Westra et al., 2014). In subtropical and tropical environments and especially for hourly or sub-hourly precipitation, a super-Clausius Clapeyron rate may be applicable, with rainfall intensities up to double the anticipated increase per degree Celsius (Lenderink & Van Meijgaard, 2008; Lenderink et al., 2017; Ali et al., 2018; Guerreiro et al., 2018; Wasko et al., 2018; Fowler et al., 2021a, 2021b). This is due to factors such as changes in cloud processes and characteristics, storm structure (areal extent, duration, and movement), and planetary-scale circulation changes that may affect the frequency, duration, and spatial extent of precipitation-producing weather systems (Fowler et al., 2021a and references therein, Oakley, 2021).

As a result, it should be expected that these changes will lead to increased risks of flooding and landsliding, including debris-flow activity, in a warmer climate. A modeling experiment only accounting for thermodynamic changes associated with a warming climate analyzed hourly precipitation at a 4-km resolution. It showed that a 300–450% increase is expected for an exceedance probability of 99.95th percentile hourly precipitation for the Pacific Northwest in the winter season, and a 100–300% increase is expected for the southwestern US in the late-twenty-first century, as compared to a 2000–2013 baseline (Liu et al., 2017; Prein et al., 2017). For the summer season, the greatest increases are predicted for British Columbia (~200–300%) or the Desert Southwest (~250–450+%), with other areas experiencing <100% increase. Using the same model output, Moustakis et al. (2021) found that a historical 20-year, 1-h rainfall event in the control period will become a 7-year rainfall event, on average, for ~75% of model grid points over the US. In agreement with Prein et al. (2017), Cannon and Innocenti (2019) found projections in hourly extremes to be consistent with the Clausius-Clapeyron scaling, though they noted modest super Clausius-Clapeyron scaling at sub-hourly timescales for mid-latitude regions of North America. In contrast, for California, Huang et al. (2020) used a selection of strong atmospheric river events in downscaled global climate model simulations in California's Sierra Nevada and found hourly rainfall intensities scaling $\sim$6 to 16% greater than expected from the Clausius-Clapeyron relation.

Numerous challenges persist in evaluating trends and projections in short-duration, high-intensity rainfall. Limitations in historic sub-daily rainfall records present a major limitation (Zhang et al., 2017). Terrain influence and varying storm characteristics at both local and large-scales challenge generalization of rainfall intensification (Moustakis et al., 2020). Much of the research to-date on precipitation intensification has focused on warm season convection, which may not apply to the cool season sub-hourly rainfall extremes, for example, that can drive post-fire debris-flow activity in western North America (Fowler et al., 2021a, 2021b; Oakley, 2021). Additionally, it remains difficult to assess these changes effectively in western North America due to the lack of existing datasets with sufficient spatial and temporal resolution (Zhang et al., 2017). To assess debris-flow hazards requires model output that is at high enough spatial and temporal scales (e.g. a few km and $\leq$1 h) to be meaningful (Oakley, 2021). However, with advances in computing power, several datasets have emerged recently with sufficient spatial and temporal resolution that account for both circulation and thermodynamic changes such that the science community can begin to address changes in precipitation driving debris-flow hazards (e.g. Gensini et al., 2022).

Climate models also project a shift in seasonal precipitation patterns in the European Alps with a drying of future summers and a wetting of future winters (Brönnimann et al., 2018; Gobiet et al., 2014). At lower elevations, the higher zero-degree isotherm is expected to lead to a decrease of the snow precipitation ratio in winter (Berghuijs et al., 2014) as well as more frequent rain-on-snow episodes (Beniston &

Stoffel, 2016; Morán-Tejeda et al., 2016), with more and more direct runoff (Fehlmann et al., 2019). There is also indication that exceptional weather situations like the unusual warm and wet January 2018, with repeated rain-on-snow events causing landslides and debris flows, may become more frequent in the European Alps by the end of the century (Stoffel & Corona, 2018). Whereas Turkington et al. (2016) expect only limited changes in debris-flow activity for two regions in France and Italy, Stoffel et al., (2014a, 2014b) expect a shift of the debris-flow activity away from summer to spring and fall. Whereas the increase in debris-flow activity in spring can be ascribed to a thinning snow wedge due to rising snow levels, and the reduced ability of snow to absorb rainfall, a second spike in the debris-flow season will likely be observed in fall (Stoffel & Beniston, 2006). As a result, debris flows, even at higher elevations, will likely occur over most months of the year and will no longer be limited to summer and early fall. Decreasing precipitation frequency in future summers will facilitate sediment accumulation in the channels, thereby enhancing debris-flow probability later in the season (the fall spike; Stoffel et al., 2005a, 2005b, 2008; van den Heuvel et al., 2016).

10.2.2 Effects of Climate Change on Tropical Cyclones

Tropical cyclones (referred to as hurricanes, typhoons, or cyclones depending on location; Fig. 10.2) are among the most devastating weather systems on Earth. They affect coastal regions or islands directly, but also affect inland regions as humid air masses are lifted to generate or enhance storms. Possible changes in tropical cyclone numbers, intensity, and genesis and decay locations all are relevant for changes in landslide activities. With copious rainfall in a relatively short period of time (usually within a couple of days), they often generate a large number of landslides when they impact mountainous regions.

Chand et al. (2022) used a reconstructed long-term proxy of annual tropical cyclone numbers together with high-resolution climate model experiments. They showed statistically robust

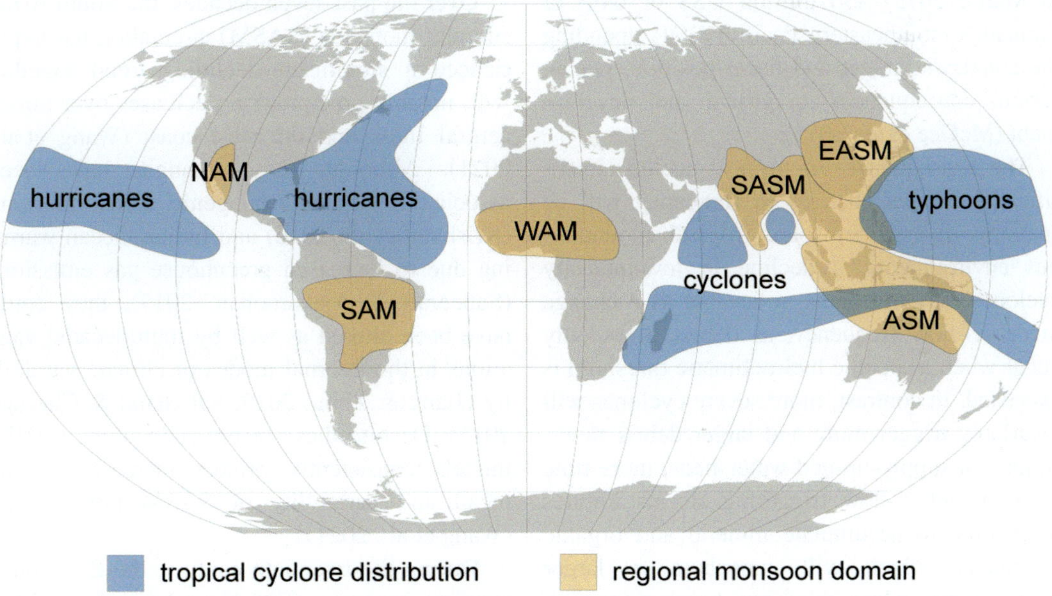

Fig. 10.2 Estimated distribution of regional monsoon domains and tropical cyclones; for abbreviations see text (*Graphic* Sophia Zubrycky, Nina Oakley; based on information from the UK MetOffice and IPCC, 2021)

declining trends in the annual number of tropical cyclones at global and regional scales during the twentieth century. The projected declining trends are consistent with the twentieth century weakening of the Hadley and Walker circulations, which make conditions for tropical cyclone formation less favourable. However, a declining trend in the annual frequency of tropical cyclones does not imply their weakening. Instead, a number of studies have shown that tropical cyclones are likely to become more intense with warming, and their moisture capacity is also likely to increase substantially (Knutson et al., 2020). Equally concerning are the poleward trends of tropical cyclone tracks, which are evident in the global historical data in both the northern and the southern hemispheres, with poleward migration rates of 53 and 62 km/decade, respectively (Kossin et al., 2014). The trends in each hemisphere result in a global-average migration of tropical cyclone activity poleward at a rate of about one degree of latitude per decade (Kossin et al., 2014). The devastating effects of extreme tropical cyclones have been amply demonstrated (Quesada-Román et al., 2019). For example, the total economic cost of the severe Tropical Cyclone *Pam* in March 2015) 450 million US$ or 64% of Vanuatu's (southeastern Pacific) GDP, upending the country's budget and fiscal position, posing serious consequences for growth and development (McGee et al., 2015).

The consequences of tropical cyclone intensification versus changes in frequency will be spatially diverse. In some tropical mountainous environments, a decline in low-intensity cyclones may not be as noticeable as a change in debris-flow frequency, as debris flows only occur when a specific hydroclimatic threshold is exceeded. In contrast, more severe cyclones will invariably trigger more and larger debris flows. In fact, in supply-limited watersheds, more time between debris-flow triggering storms implies more time to accumulate mineral and organic sediments, which will contribute to larger debris flows when they are being triggered. This effect will be particularly tragic in some tropical islands (for example, in the Caribbean) where a multitude of other environmental concerns coincide. Increased population pressures, continued deforestation, rising sea levels and, in some cases, volcanic activity, conspire such that organized and civilized life in some tropical regions may become unsustainable.

10.2.3 Anticipated Changes of Precipitation in Monsoon Regions

Monsoon is commonly defined as the seasonal reversal in prevailing winds (AMS, 2022) and is a response to the large-scale land–ocean thermal contrast that occurs in various regions of Earth. More than two-thirds of the world's population live in regions (Fig. 10.2) where monsoonal precipitation accounts for a great portion of annual rainfall (Wang et al., 2021). Within these regions, the monsoon is a clear and well-established driver of shallow landsliding and erosion of hillslopes (Bookhagen et al., 2005; Kirschbaum et al., 2015; Petley, 2010), and as such, changes in monsoon precipitation will fundamentally alter debris flow activity (Kirschbaum et al., 2020).

Over the past seven decades, the South Asian summer monsoon (SASM) circulation has experienced a significant declining trend together with notable precipitation decreases over north-central India and the west coast (Wang et al., 2021). Although these changes have been associated with anthropogenic aerosol forcing (Krishnan et al., 2016) and Indian Ocean warming due to increased greenhouse gas emissions (Sabeerali & Ajayamohan, 2017), they could have been altered as well by multidecadal variations in the internal modes of climate variability (Jiang & Zhou, 2019; Salzmann & Cherian, 2015). Despite these recent reductions, CMIP6 models consistently project increases in the mean and variability of SASM precipitation (Wang et al., 2021).

During the twentieth century, the East Asian summer monsoon (EASM) exhibited considerable multidecadal variability with a weakened circulation and a 'south flood–north drought' pattern since the late 1970s (Ding et al., 2009; Zhou et al., 2009). This pattern has been predominantly attributed to internal variability (Ha et al., 2020a; Li & Wang, 2018). CMIP6 models indicate that EASM precipitation will increase at 4.7% $°C^{-1}$ (Ha et al., 2020b). Also, EASM duration is projected to lengthen by about 25 days due to earlier onset and delayed retreat (Ha et al., 2020b).

West Africa rainfall totals in the Sahel have been increasing since the 1980s (Wang et al., 2021) owing to positive trends in mean rainfall intensity (Sarr et al., 2013), total rainfall (Sanogo et al., 2015), and the frequency of intense mesoscale convective systems (Taylor et al., 2017). Several West African countries have experienced trends toward a wetter late season and delayed cessation of the rains (Lodoun et al., 2013). All the above changes are qualitatively consistent with the CMIP5 response to greenhouse gas emissions (Marvel et al., 2020). Preliminary results from CMIP6 confirm that the Sahel will become wetter, except for the west coast, and the rainy season will extend later. In East Africa, increases have been observed in the boreal autumn short-duration rains (Cattani et al., 2018) and negative trends in the spring long-duration rains (Maidment et al., 2015). CMIP6 models project little agreement on how East African rainfall will change, with some tendency for enhanced rainfall during the short rains and a curtailed long-rains season (Cook & Vizy, 2013; Han et al., 2019).

Observations show an increasing trend in mean and extreme rainfall during the Australian monsoon, especially over northwestern Australia and since the early 1970s (Dey et al., 2019). The CMIP6 models project insignificant changes in the Australian summer monsoon (Moise et al., 2020).

Observed long-term twentieth century rainfall trends of the North American Monsoon (NAM) are either negative or null, but trends vary substantially within the region (Pascale et al., 2019). Recent observations and modeling studies show an increase in the magnitude of extreme events in NAM and Central American rainfall under anthropogenic global warming (Aguilar et al., 2005; Luong et al., 2017). Climate models suggest—although with limited confidence—an early-to-late redistribution of the mean NAM precipitation with no overall reduction (Cook & Seager, 2013; Seth et al., 2013), and a more substantial reduction for Central American precipitation (Colorado-Ruiz et al., 2018). Uncertainties also persist regarding the future of NAM moisture surges and the track of the upper-level inverted troughs, which are key synoptic processes controlling convective activity (Wang et al., 2021).

The South American monsoon (SAM) has exhibited a significant positive precipitation trend between the 1950s and 1990s, especially in southeast South America, and has been related to interdecadal variability (Grimm & Saboia, 2015) and increasing greenhouse gas emissions (Gonzalez et al., 2014; Vera & Diaz, 2015). The dry season has been lengthened and become drier, especially over southern Amazonia, with significant impacts on vegetation and moisture transport to the SAM core region (Fu et al., 2013). The models project little change in annual mean precipitation, with drier winter/spring and increased peak monsoon rainfall. These findings are consistent with El Niño impacts (Grimm, 2011) showing a delay and shortening of the monsoon season, but intensification in its peak (Seth et al., 2013), and prolonged dry spells between the rainy events (Christensen et al., 2013).

Recent evidence suggests that the attribution of specific monsoon rainfall episodes to hydrogeomorphic processes is complex, as these patterns operate worldwide and they are interdependent (Su et al., 2022). For instance, the coupling of weak EASM and ENSO seems to explain extreme flood events in China (Wei et al., 2014). In mountain catchments, abnormal monsoon years control the sediment fluxes related to the occurrence of hydrogeomorphic processes in the high-elevation Himalayas (Bookhagen et al., 2005). At the same time, the projected changing of monsoon season and timing could enhance the occurrence of hydrometeorological triggering of lake outburst and debris flow disasters in the Himalaya (Allen et al., 2016), as seen during the Kedarnath disaster (India) in June 2013, or recently in the disaster in Uttarakhand (India) in February 2021 (Thayyen et al., 2022). Likewise, the NAM regime has also been linked to intense rainfall events resulting in flash floods and debris flows in western north America, especially in post wildfire mountain catchments (Moody & Martin, 2001; Underwood & Schultz, 2004).

10.2.4 Effects of Climate Change on Atmospheric Rivers

Atmospheric rivers (AR) are long, narrow corridors of high water–vapor transport in the atmosphere and one of the major drivers of moisture redistribution from the tropics to the mid-latitudes (Ralph et al., 2018). In some cases, landfalling ARs result in major flood events in western North America (Corringham et al., 2019; Ralph et al., 2006; Sharma & Déry, 2020) and Europe (Ballesteros-Cánovas et al., 2019; Benito et al., 2021). ARs have been associated with landslide and debris-flow events in western North America (Biasutti et al., 2016; Cordeira et al., 2019; Jakob & Owen, 2021; Oakley et al., 2017, 2018) and western Europe (Ramos et al., 2015).

The strength of atmospheric rivers is often described by Integrated Water Vapor Transport (IVT), a quantity that considers the vertical profile of both water vapor content and wind speed (Rutz et al., 2014). IVT is projected to intensify in a warming climate due to the increased vapor-holding capacity of the atmosphere as temperatures rise (e.g. Espinoza et al., 2018; Gershunov et al., 2019; Payne et al., 2020; O'Brien et al., 2022). Following the Representative Concentration Pathway 8.5 (RCP 8.5), Espinoza et al. (2018) found that AR strength is projected to increase by 25% globally, with ARs also becoming 25% wider and longer by late twenty-first century (2073–2096) as compared to an historic period (1979–2002). While studies are in agreement on a projected increase in the frequency of AR conditions (e.g. Espinoza et al., 2018; Gershunov et al., 2019; Payne et al., 2020; O'Brien et al., 2022), there is some uncertainty as to whether the number of AR events will increase or decrease. Globally, Espinoza et al. (2018) noted a 10% decrease in AR events, but a 50% increase in timesteps with AR conditions by late twenty-first century (2073–2096). This increase in AR timesteps despite a decrease in the number of AR events is associated with the longer duration and increased size of the individual AR events. In contrast, O'Brien et al. (2022) note a projected increase in AR events. In the historic period (1981–2010), they identified a global median AR event count of approximately 5–20 ARs at any given time, depending on AR detection method and climate model assessed. Most AR detection algorithms display an increase of approximately two ARs in future projections (2070–2099), with variations depending on detection algorithm or climate model assessed. It is important to consider that results of AR characteristics assessed in climate model projections are sensitive to the AR detection algorithm applied as well as climate model assessed, thus results may vary across studies (O'Brien et al., 2022).

The mid-latitude regions where ARs are commonplace are projected to experience an increasing number of dry (non-precipitation) days by mid-to-late century (Polade et al., 2014). While some of these regions are anticipated to become drier (e.g. the Mediterranean), in other regions that experience ARs, precipitation is expected to increase (e.g. Pacific Northwest and northern Europe) or have little change (e.g. California, Polade et al., 2014). In the western US, ARs are projected to make a greater contribution to annual precipitation totals by 10–25%, coupled with a decrease in precipitation from non-AR storm events. An increase in extreme precipitation associated with ARs is also projected for the western US (Gershunov et al., 2019). Though there is high confidence thermodynamic response in ARs result in increased IVT, there is uncertainty about how this will translate into precipitation changes due to changes in dynamic processes in the atmosphere. Circulation changes and changes in precipitation efficiency could affect how much IVT is realized as precipitation (Payne et al., 2020). Nevertheless, the projected increases in AR intensity and size portend an increase in flood and debris-flow impacts in areas that experience ARs (e.g. Corringham et al., 2022; Curry et al., 2019; Jakob & Owen, 2021).

10.2.5 Effects of Climate Change on Droughts and Wildfires

Climate change has resulted in increased temperatures, more frequent heat extremes, prolonged dry periods, changes in precipitation patterns, increased drought risk, and, therefore, also increased wildfire activity in the Mediterranean (Ruffault et al., 2020) and western North America (Medellín-Azuara et al., 2018; USGCRP, 2018; Zhang et al., 2019; Dalton & Fleishman, 2021). Numerous studies have documented an increase in wildfire activity (area burned, number of incidents, see Fig. 10.3) in western North America over the last 50+years (Westerling et al., 2006; Flannigan et al., 2009; Dennison et al., 2014; Abatzoglou & Williams, 2016; Hanes et al., 2018; Haugo et al., 2019; Coogan et al., 2021), with several studies noting the key role of anthropogenic climate change in driving these increases (Abatzoglou & Williams, 2016; Littell et al., 2009; Williams et al., 2019).

In the summer season in the western US, warming temperatures have increased vapor pressure deficit (VPD), which correlates strongly with low fuel moisture, a condition favorable for wildfire activity (Abatzoglou & Williams, 2016; Williams et al., 2019). More frequent and prolonged droughts weaken trees or other vegetation, making them more susceptible to pests such as pine bark beetles or sudden oak death that can result in large stands of dead trees that can influence wildfire characteristics (Littell et al., 2010; Metz et al., 2017; Wayman & Safford, 2021). In some areas, delayed onset of precipitation in the autumn season as well as drier, warmer conditions in the spring season and earlier spring snowmelt have increased the length of the fire season (Goss et al., 2020; Westerling et al., 2006). Warmer temperatures

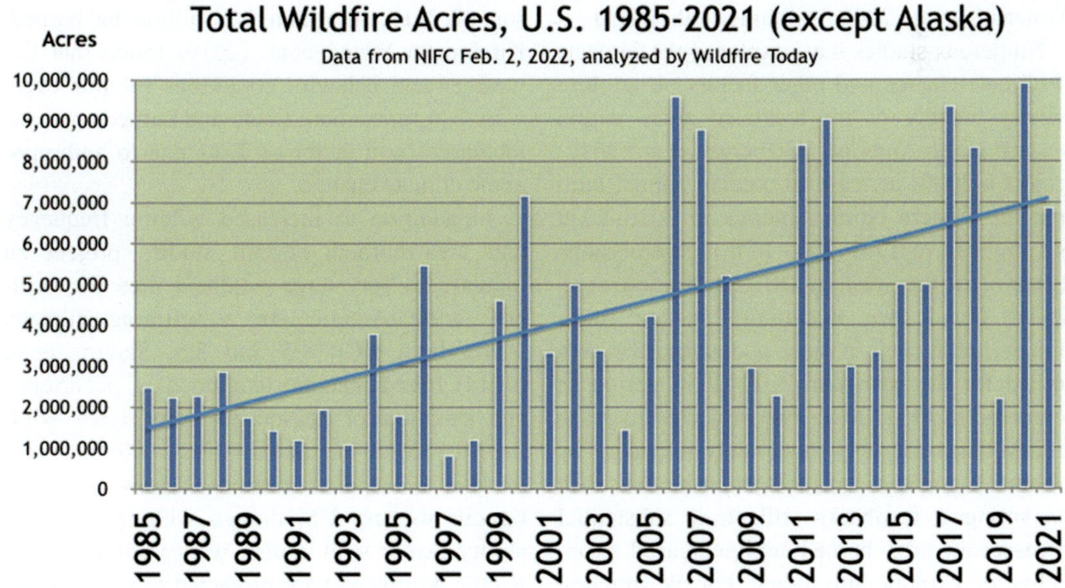

Fig. 10.3 Annual surface burned (in acres) in the United States (dark blue bars) and long term trend (clear blue line), 1985–2021. *Source* Wildfire Today (https://wildfiretoday.com/2021/11/21/acres-burned-in-lower-49-states-this-year-is-more-than-average/)

and increased warm season VPD have allowed wildfires to advance upslope to higher elevations in the western US over the 1984–2017 period (Alizadeh et al., 2021), a situation of particular concern for post-fire debris flow hazards. In California, the delayed onset of autumn precipitation allows the wildfire season to further overlap with the Santa Ana and Diablo wind seasons (Goss et al., 2020; Guzman-Morales et al., 2016; Smith et al., 2018; Swain, 2021).

Two additional factors have contributed to the observed increase in wildfire activity. First, over a century of fire suppression and fire exclusion has led to a buildup of fuels and increased the potential for high severity fires (Haugo et al., 2019; Higuera et al., 2015). Second, expanding populations living and recreating in the wildland urban interface has resulted in an increased number of and potential for human-started wildfires (Balch et al., 2017; Keeley & Syphard, 2018; Radeloff et al., 2018). These changes have also expanded the geographical and seasonal extent of wildfire (Balch et al., 2017). The primary drivers of increased fire activity and observed trends in wildfire activity are spatially heterogeneous across western North America (e.g. Kitzberger et al., 2017; Littell et al., 2010; Westerling et al., 2006; Williams et al., 2019).

Numerous studies have explored the impacts of climate change and other factors on wildfire activity broadly or for localized areas within western North America. Kitzberger et al. (2017) project a 200% increase in median annual burn area for western North America in 2010–2039 as compared to 1961–2004 in a high-emissions scenario. In the western US, Abatzoglou et al. (2021) found that, assuming constant fuels, climate-fire models project a doubling of area burned for the near-term 2021–2050 period as compared to 1991–2020. The projected increase in area burned is only moderately impacted by eventual constraints on fuels, even models with the strongest feedbacks still show substantial increases in area burned and increased probability of extreme fire years. For Washington state, Littell et al. (2010) used a high-emission scenario and project that area burned will double to triple by the 2080s as compared to an historic baseline (1970–1999). Over the next 50–100 years, area burned and fire frequency are projected to increase substantially in Oregon, a 200% increase in area burned is projected in association with 2 °C of warming (Dalton & Fleishman, 2021; McKenzie et al., 2004). In California, Westerling (2018) suggests that the mean annual area burned statewide for the 2070–2099 period is projected to increase by 77% following RCP 8.5 as compared to 1961–1990. The greatest increases were projected for the forested areas of the Sierra Nevada, with mean annual area burned doubling to quadrupling by the end of the century. In Canada, Wotton et al. (2010) used two GCMs to assess changes in wildfire occurrence by the end of the 21st Century as compared to an historic baseline (1975–1995 for Canadian Climate Centre and 1975–1990 for Hadley Centre). They found an increase of fire occurrence of 74 and 144% in the two models, averaged across the country, with an increase of 21 and 190% for British Columbia specifically. In an attribution study for British Columbia's extreme wildfire season of 2017, in which 1.2 million ha burned, Kirchmeier-Young et al. (2019) found that fire weather and behavior conditions for this event were 2–4 times more likely and burned area was increased by a factor of 7–11 due to anthropogenic climate change.

In addition to increased wildfire frequency and area burned, several studies project an increase in very large wildland fires for western North America in a warming climate. Using both RCP 4.5 and 8.5, Stavros et al. (2014) found that the likelihood of occurrence and duration of season for very large wildfires (>20,000 ha) are projected to increase in the twenty-first century across the contiguous western US, though with spatial variability. Areas with cooler, wetter climates (e.g. Pacific Northwest) are projected to have greater increases than areas with hot, dry climates. Similarly, Barbero et al. (2015) used the RCP 8.5 and found very large fires (defined as top 5–10% in area burned) will increase by 200–400% for 2041–2070 as compared to a 1971–2000 baseline. In California, Westerling (2018) predicts that "extreme" fires (>100,000 ha) will occur 50% more often by 2100 than in a baseline historic period (1961–1990).

Burn severity is a key variable in the assessment of post-fire debris flow hazards (Gartner et al., 2015; Staley et al., 2016), though burn severity trends over time are challenging to evaluate due to limitations in data archives (Kolden et al., 2015; Picotte et al., 2020). Research for western North America suggest widespread reduction in burn severity in a warming climate, with a few locations in California, the western Great Basin and Columbia Plateau experiencing slight increases (Guindon et al., 2021; Parks et al., 2016). However, this trend toward decreasing burn severity may have limited impact on future post-fire debris flow hazards. In an experiment to assess climate change and frequency of post-fire debris flow hazards in Southern California, Kean and Staley (2021) found post-fire debris flow activity to be more sensitive to changes in precipitation intensity than changes in burn severity or fire frequency.

10.3 System Susceptibility: Effects of Climate Change on Wildfires, Sediment Production, Availability, and Transfer

The potential of catchments to produce debris flows is often assessed with a susceptibility analysis. The approach distinguishes between the general (or basic) and the variable susceptibility of a system to produce any given mass movement process. In general terms, the basic susceptibility will influence the extent of natural hazard processes, whereas the variable susceptibility will control the frequency of a process.

In this approach, illustrated in Fig. 10.4, the basic susceptibility is represented by parameters that remain fairly "constant" over longer periods of time, these are, for instance, relief, geology or vegetation cover. Yet, individual parameters may well exhibit a longer-term development trend in a certain direction, even more so in the context of climate change. By way of example, the increase of the lower limit of permafrost may result in an increase of the surfaces susceptible to yield debris-flow material and thus increase the basic susceptibility of a system to erosion and mass movement. Likewise, recent droughts have resulted in considerable forest dieback and

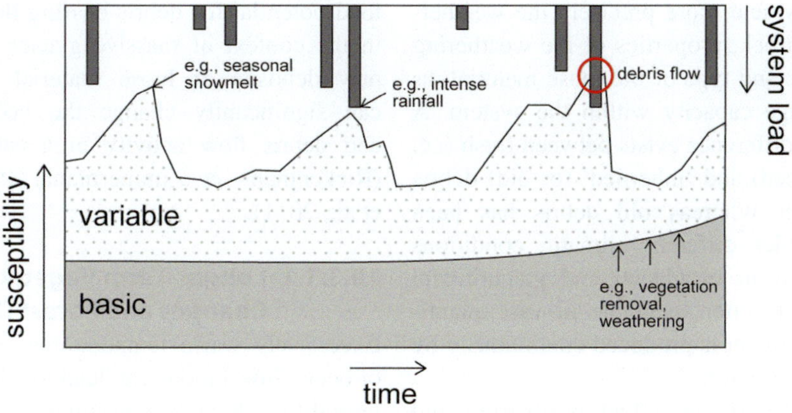

Fig. 10.4 Concept of basic and variable debris-flow susceptibility with system load. When the degree of susceptibility and the system load are high, a debris flow is triggered (*Graphic* Markus Stoffel, Sophia Zubrycky; modified after Zimmermann et al., 1997)

beetle infestations, thereby influencing the vegetation cover.

By contrast, the variable susceptibility is characterized by parameters that are temporally variable within a given system. They are typically time-transient, partly controlled by the season and time of day, such as the meteorological situation, soil water balance, and/or vegetation conditions.

When combined, the susceptibility approach describes the readiness of a system to produce a debris flow. Provided that the current susceptibility is high, only a low system load is required to trigger a process. On the contrary, if current susceptibility is low, a correspondingly higher system load is required. Yet, for an event to occur, a triggering process is still needed, and a debris flow will occur as soon as the system is loaded above a given threshold. This trigger can be, for example, heavy precipitation or intensive snowmelt.

10.3.1 Basic Susceptibility: Geology, Relief, Vegetation Cover

Relief is a key factor influencing the susceptibility of a catchment to produce debris flows as it defines the surfaces from which material can enter the debris-flow system. Another factor that has a strong effect on the basic susceptibility is geology, and more precisely the weatherability of bedrock, properties of the weathering products, size and type of the loose material, as well as storage capacity within the system. A fundamental difference exists between fresh (i.e. newly produced) and 'inherited' (or old) debris (Stiny, 1910). Whereas old debris has been deposited under different climatic conditions (e.g. in the form of glacial and glaciofluvial gravels) and is often available in vast quantities, fresh sediment is produced continuously by weathering processes.

Climate change can affect weathering and thus the production of fresh sediment. The greater the weathering susceptibility of the bedrock, the stronger is the effect that climate change has on its production. This means that in supply-limited watersheds, characterised as those consisting largely of competent bedrock with very low sediment production rates, it may take decades for enough sediment to be produced on hillslopes and the channel system, for another debris flow large enough to travel to the fan to be possible (Bovis & Jakob, 1999; de Haas et al., 2018; de Haas et al., this volume). This also means that even an extreme rainstorm occurring prior to a threshold recharge being reached will only result in a flood, or perhaps debris flood (Borga et al., 2014; Church & Jakob, 2020) on the fan reaches.

By contrast, in quasi supply-unlimited watersheds, characterized as watersheds with ample debris supply in the form of talus slopes, weak and deeply weathered soils, till, glaciofluvial sediments or artificially introduced sediment (mine tailings or mine waste rock, road fill and/or garbage dumps) with small apparent or true cohesion, debris flows will readily occur whenever a hydroclimatic threshold is exceeded. Here, more frequent and/or intense rainfall can increase erosion in the stream and on gully slopes, thus increasing material input into the system. Likewise, climate change may also affect the stability of inherited debris. Precipitation patterns and snowmelt may affect the stability of loose material. As a result, large amounts of material can become unstable and break loose as debris flows or increase the bedload potential for debris-bearing floods. Finally, in the context of massive glacier retreat, large new deposits of loose material arise, which can significantly change the bedload budget and debris flow activity in a catchment area (Rickenmann & Zimmermann, 1993; Seinova et al., 2011).

10.3.1.1 Longer-Term Vegetation Changes and Forest Dieback

Excessively warm temperatures, droughts and/or beetle infestations can lead to widespread tree mortality, which can negatively affect surface water hydrology. For example, Cullingham et al. (2011) point out that the leading edge of the mountain pine beetle epidemic in Canada has reached western Alberta and jack pines are now under beetle attack, which will likely be exacerbated by further warming. Bark beetle outbreaks driven by climate change may also result in trajectories beyond the historical resilience boundaries of some forest ecosystems, causing irreversible ecosystem regime shifts, Bentz et al. (2010) show that for two species of bark beetles, climate projections suggest regimes conducive to population success. They also predict a shift of beetle infestations towards higher latitudes and elevations in the coming decades and centuries. Accompanying droughts can affect tree health and population vigor, further contributing to mortality. Tree mortality associated with beetle infestations can lead to increases in forest fire extent, frequency, and intensity, although those interactions are complex and not well researched (Jenkins et al., 2008). Whether or not accompanied by wildfire changes, enhanced tree mortality will affect slope stability through loss of root strength, decreases or annulation of canopy interception and evaporation, as well as reduced or eliminated evapotranspiration. Similarly, numerous studies have examined the long-term changes in vegetation type and spacing (e.g. Van der Maarel, 1988; Winkler et al., 2021).

10.3.1.2 Hydrological Changes Following Wildfires

Many wildfire studies focus on the dramatic geomorphic and hydrological effects immediately after a wildfire. However, few examine long-term effects (defined as >5 years), which are relevant to regions dependent on water from forested areas. Forest fires can cause profound changes to the hydrology of a catchment that can lead to increased frequency and magnitude of debris flows (Cannon & Gartner, 2005; Moody et al., 2013; Gartner et al., this volume). Changes to the hydrologic response of a catchment to rainfall have been attributed to the intense heat of wildfire removing vegetation and exposing bare ground to rainfall (Ebel et al., 2012; Moody & Martin, 2001), introducing ash to the soil (Balfour & Woods, 2007) and generating water-repellant soils (Doerr et al., 2000). These changes to the hydrological characteristics of a catchment by forest fire result in increased runoff, which erodes creek channels to form hazardous debris flows and debris floods (Cannon & Gartner, 2005; Moody & Martin, 2001). A recent study also demonstrated that wildfire has advanced upslope over the last few decades in the western United States, thereby increasing the area exposed to potential post-fire debris-flow hazards.

Debris flows following forest fires are typically greater in magnitude, with larger volumes of sediment deposited and higher peak discharge than those in pre-fire conditions (Santi & Morandi, 2013). In catchments burned by the 2003 Lost Creek Fire in Alberta, significantly higher mean annual water yield and mean peak flows were measured in comparison to nearby, unburned catchments (Mahat et al., 2016; Silins et al., 2009). In the northern Rocky Mountains of the United States, numerous post-fire debris flows have been triggered by storms with 1- to 2-year return periods, whereas similar magnitude storms did not initiate debris flows in adjacent unburned areas (Cannon & Gartner, 2005).

The period of increased debris-flow activity following fire is most pronounced within the first 1–5 years after the fire (Cannon & Gartner, 2005; DeGraff et al., 2015), after which vegetation begins to re-establish on hillslopes. A second period of post-fire debris-flow activity is conceivable about 2–10 years after a fire. This is attributable to the coincidence of progressive root decay and concordant loss of root strength akin to clearcut logging and a sizable rainfall or rain-on-snow event (Degraff et al., 2015; Klock & Helvey, 1976; Rengers et al., 2020; Sidle, 1991, 2005).

10.3.1.3 Cryospheric Change and Susceptibility

The sensitive mountain cryosphere is changing rapidly in response to climate warming, with implications for erosion and mass movement processes (Hock et al., 2019). Debris flows in cryospheric environments can originate from steep moraines, talus slopes at the foot of eroding rockwalls, destabilised rock glacier tongues, and fluvioglacial deposits within steep stream channels (Evans & Delaney, 2015). Peri- and

Fig. 10.5 Debris flow triggered by the outburst of a glacial lake, dammed behind the steep and unstable lateral moraine of the Chorabari glacier, Northern India, leading to destruction and hundreds of fatalities in the town of Kedarnath (*Photo* D. Grossman, used with permission)

para-glacial zones are particularly susceptible for debris-flow initiation owing to the abundant supply of loose, unconsolidated material, coupled with steep topography, melting of snow and ice, glacial lakes (Fig. 10.5) and heavy convective or monsoon precipitation that occurs in many mountain regions (Allen et al., 2016; Chiarle et al., 2007; Evans & Clague, 1994; Jomelli et al., 2007; Stoffel et al., 2011).

The enhanced susceptibility of the paraglacial environment was illustrated in the Swiss Alps in 1987, when around 600 debris flows were triggered by heavy precipitation. More than 50% of these events initiated from zones that had deglaciated within the previous 150 years (Rickenmann & Zimmermann, 1993; Zimmermann & Haeberli, 1992). Similarly, studies in southern Russia have demonstrated enhanced debris-flow activity associated with recent rapid deglaciation and exposure of morainic material (Fig. 10.6), with source areas often characterised by thermokarst features which can become oversaturated with meltwater or surface runoff (Lugon & Stoffel, 2010; Seinova et al., 2011; Stokes et al., 2006). Thawing permafrost in non-consolidated material leads to a loss of cohesion and increase in pore water pressure and slumping where massive ground ice bodies disappear (Harris, 2005). Because re-vegetation of deglaciated terrain is slow, slopes can remain susceptible to erosion over extensive time periods of several decades or more (Kääb et al., 2005). While most attention in periglacial environments is given to permafrost warming and degradation, cold permafrost bodies in very high or shady locations can act as a barrier to groundwater percolation, leading to local saturation in the overlying non-frozen material (Zimmermann & Haeberli, 1992).

In peri- and paraglacial environments, a distinction is made between sediment reservoirs that are *active* and those that are considered *relict* (Sattler, 2014). Active reservoirs, such as talus slopes, are continuously replenished

Fig. 10.6 Debris flow originating from glacial moraines above the town of Tyrnyauz, southern Russia, July 2000. The most probable trigger of the debris flow was an outburst from an englacial cavity (*Photo* A. Aleinikov)

by weathering and mass movements, but typically produce relatively low magnitude events, because the volume of material is *supply-limited.* Joint density, tectonic wear by uplift and folding, and dip direction all influence rates of sediment production from rock walls, and certain lithologies (e.g. sedimentary rocks) are more susceptible to frost weathering than crystalline rocks (André, 2003). Because reservoirs such as talus slopes must regenerate before a subsequent event can occur, associated frequencies can be low, and the effects of climate change can be complex to disentangle. Frost-weathering is strongly dependent on elevation and aspect, with enhanced rates of rockfall activity and sediment production observed on shaded slopes, and in areas of permafrost (Morel et al., 2015; Sass, 2005; Stoffel et al., 2005a, 2005b). On the one hand, increasing temperatures might reduce the number of freezing days and associated frost-weathering in the long term (e.g. Jomelli et al., 2004, 2007), while on the other hand, permafrost degradation is leading to enhanced rockfall activity from a deepening active layer (Gruber & Haeberli, 2007). By comparison, relict debris reservoirs, such as glacial moraines from the Little Ice Age (LIA), are no longer replenished by active processes, and have formed in response to long term climate change. Related debris flows from these reservoirs can attain very large magnitudes, as an abundant amount of sediment is available for entrainment (*supply-unlimited*), and events may reoccur at high frequencies until the large supply of sediment is eventually exhausted (Sattler, 2014).

10.3.2 Variable Susceptibility

The variable susceptibility includes several parameters and conditions that are changing over time, such as snowpack thickness, soil moisture, or vegetation condition. These parameters are in part controlled by season or time of day.

In cryospheric environments, a strong seasonality to debris-flow activity has been identified, with events occurring more frequently in summer and autumn in the European Alps (Bollschweiler & Stoffel, 2010; Schneuwly-Bollschweiler & Stoffel, 2012; Stoffel et al., 2011), and during summer in the Caucasus (Petrov et al., 2017). This is logical because triggering from heavy convective precipitation or melt-related processes is more likely during these warmer months. Furthermore, sediments are also less likely to be frozen and hence, the availability of material for erosion is greater. Particularly in permafrost zones, a close association can be observed between debris-flow activity and thickness of the permafrost active layer. As a consequence of gradual top-down thawing of the near-surface material over the warm months, slopes become most prone to instability by late summer or autumn. However, critical precipitation thresholds required to trigger an event may actually be lower earlier in the summer, when the active layer is shallower and already saturated from spring snowmelt (Schneuwly-Bollschweiler & Stoffel, 2012).

In high mountain environments, characterized by large local relief and presence of snow, ice, and permafrost, debris flows can be distinct events, but are also often associated with far-reaching cascading process chains, the frequency of which is expected to increase under a warmer climate (Hock et al., 2019). In fact, some of the most devastating disasters in high mountain regions have involved such process chains (Mani et al., 2023), where an immediate or subsequent transition to a debris flow played a significant role in enhancing downstream reach and losses (Allen et al., 2016; Huggel et al., 2005; Lliboutry et al., 1977; Shugar et al., 2021; Mani et al., 2023).

A classic and well-studied process chain scenario involves a mass movement of ice or rock into a glacial lake, leading to an overtopping wave, erosion of the moraine dam and transition to a debris flow (e.g. Worni et al., 2014). The widespread retreat of mountain glaciers since the Little Ice Age has resulted in the formation of lakes trapped behind proglacial moraines, some of which may be spectacularly large with volumes of up to 100 million m^3, and depths exceeding 200 m (Cook & Quincey, 2015). Once initiated, glacial lake outburst floods (GLOFs) tend to entrain large amounts of sediment, with the potential to transport massive boulders, as erosive forces are far greater than typical meteorological floods would produce (Cook et al., 2018). Depending on downstream channel characteristics and erodibility, GLOFs have been observed to undergo phase transformations from debris flows to hyperconcentrated flows, and back again (Worni et al., 2012). With new glacial lakes forming in increasingly close proximity to steep, destabilising mountain flanks (Haeberli et al., 2017; Schwanghart et al., 2016; Zaginaev et al., 2019), the potential for such process chains involving GLOFs are projected to increase dramatically over the next decades (Haeberli et al., 2016; Majeed et al., 2021; Petrakov et al., 2020; Zheng et al., 2021a, 2021b).

Examples from the Alps (Mergili et al., 2020), British Columbia (Guthrie et al., 2012), Caucasus (Huggel et al., 2005), Himalaya (Shugar et al., 2021), and elsewhere, have also shown that ice and/or rock avalanches can initiate devastating debris flows even in the absence of a glacial lake or other obvious liquid water source (Fig. 10.7). The catastrophic mass movement from the 5500 m Ronti Peak, in Chamoli district of Uttarakhand, India, February 7, 2021, initiated as a 27 million m^3 avalanche of mostly rock (80%) and glacial ice (20%) (Bhardwaj & Sam, 2022). However, melting of this ice from frictional heating as the flow descended over a local elevation range of >3000 m transformed the rock and ice avalanche into a highly mobile debris flow and subsequent flood that resulted in more than 200 fatalities (Shugar et al., 2021). Similarly, on August 2017, an estimated 2 million m^3 rock avalanche detached from the northeast face of Pizzo Cengalo, a 3369 m high summit located in the Bondasca Valley, Switzerland (Mani et al., 2023). The rock avalanche entrained about 0.6 million m^3 of glacier ice, which, in addition to water stored within the sediment from an overrun earlier avalanche

Fig. 10.7 The 2010 rock avalanche and debris flow on the west flank of Mount Meager, British Columbia. The initial bedrock failure mobilised additional volcanic material below, and enhanced with meltwater from snow and ice, transformed into a high-velocity debris flow that travelled 7 km downstream and ran up 250 m on the opposing slope. Deposition from the debris flow created a transient blockage of Meager Creek, forming a 1.5 km long lake (*Photo* T. Spurgeon)

deposit, was sufficient to transform the avalanche into a mobile debris flow that reached the village of Bondo (Mergili et al., 2020; Walter et al., 2020). Over the subsequent days, additional debris flows were observed to initiate out of the freshly deposited, saturated rock avalanche deposit, adding to the inundation in Bondo.

While individual process chains such as described above might be difficult to conclusively attribute to climate change (Mani et al., 2023; Schauwecker et al., 2019), the combination of warming permafrost and continued glacial retreat is increasing the likelihood and potential magnitude of such events.

10.4 Consequences: Effects of Climate Change on Debris-Flow Occurrence and Size

Each debris flow-prone basin has its own characteristics which may change in response to changing climate. On a regional scale, debris-flow hazard can be characterized once the key hydroclimatic variables responsible for such change have been isolated and the concordant climate change models queried.

A study by Jakob and Owen (2021) provided some guidance on how such study could be executed. The authors used a substantial database

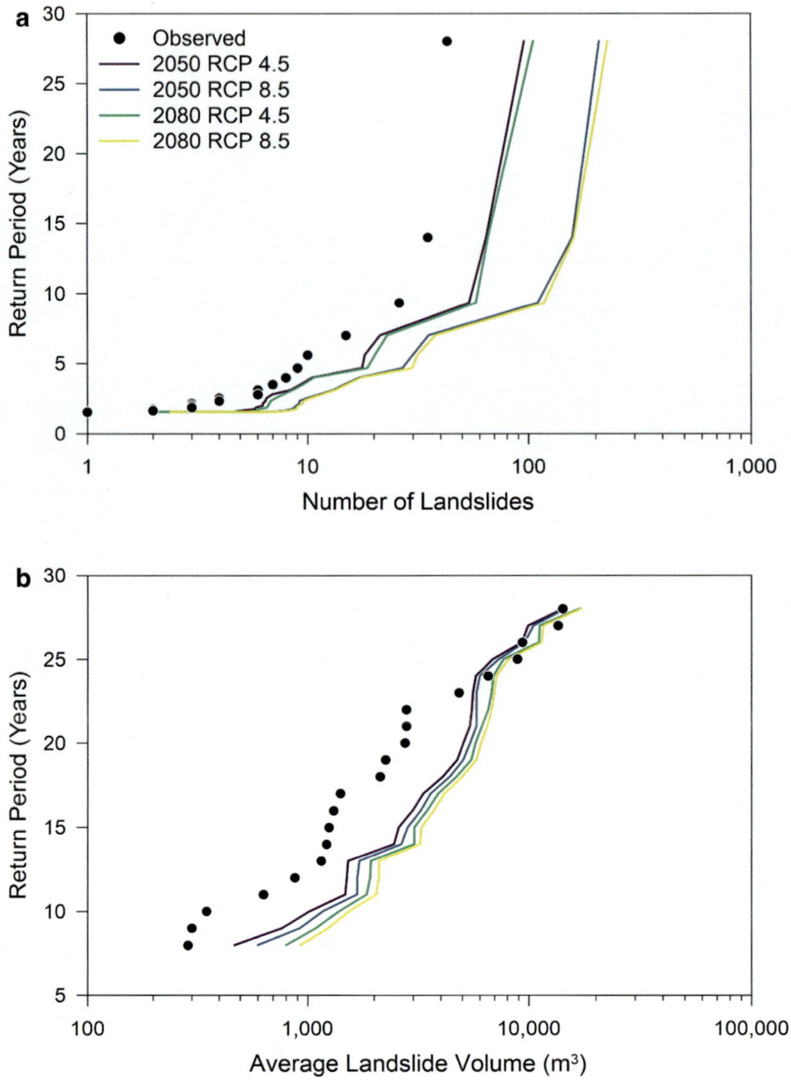

Fig. 10.8 Changes in the frequency (**a**) and magnitude (**b**) relationships of landslides under different RCP scenarios and for the 2050s and 2080s time horizons (data from Jakob & Owen, 2021)

of occurrences (dates, location and type) of shallow landslides and debris flows in British Columbia, Canada and created a model explaining which combination of hydroclimatic variables are responsible for debris-flow occurrence, namely, a combination of antecedent moisture conditions (about a 3–5 week window of substantial fall rains) and spikes of high intensity rainfall (2–6 h intensities). Based on these variables, statistical models were created to explain both debris-flow occurrence (number of debris flows per storm) (Fig. 10.8a) and debris-flow volumes (Fig. 10.8b) for the study area. Their findings showed that the frequency of debris flows may increase by as much as 300% by the 2080s, while accompanied by an increase in mean magnitude of some 30%. The much lower response in magnitude can be explained, slopes on which shallow landslides occur that evolve into debris flows are of finite length and can only increase volumes by multiple point source failures (typically debris avalanches) per watershed, or a more vigorous widening of debris avalanche paths and concordant entrainment. These findings pose a major challenge for debris-flow management. Higher sediment input rates to receiving streams may change their fluvial characteristics from single-threat to braiding, foul water resources by the introduction of fine sediments, and decrease fish habitat (e.g. if fine gravels acting as spawning habitat are destroyed directly by landslides or indirectly by fine-sediment coverage that does not allow fish fry to survive).

Regarding the nature of basins, an increase in temperature and rainfall will likely increase the weathering rates and thus recharge rates in the channel in supply-limited watersheds. This will result in a slight increase in debris-flow frequency, accompanied by a decrease in debris-flow magnitude given the predicted increase in the frequency and magnitude of extreme precipitation events (Prein et al., 2017). A higher frequency of channel-cleaning debris flows then results in a decrease in their respective volumes and peak discharges.

Channel supply-limited watersheds are characterized as those whose channels recharge as described in the previous section. However, the watershed is also covered by soil, and in regions with sufficient precipitation and available soil, also with trees or shrubs. Once sufficient volumes and intensities of rain have fallen, shallow landslides (debris avalanches and debris flows) occur and travel downhill until they reach a higher order stream, where they typically evolve into debris flows. This process implies that even when the channel has been depleted of available debris by a debris flow, additional debris flows can occur in absence of channel debris, through momentum and mass transfer of side-slope landslides.

In terms of climate change response, an increase in both antecedent moisture conditions and rainfall intensity will likely lead to more frequent, but smaller debris flows, much as for water-supply limited basins. However, once a critical hydroclimatic threshold is exceeded, debris flows can occur through single or multiple side-slope failures that evolve into debris flows when entering the main channel. Therefore, rare hydroclimatic events triggering multiple side slope failures may result in larger debris flows than observed in the past. A channel-supply limited watershed could, at least temporarily, turn into quasi-supply-unlimited watersheds if a stand-replacing wildfire were to remove protective vegetation, providing abundant sediment supply until vegetation has re-established and/or created hydrophobic soils.

In fully supply-unlimited watersheds, debris flows will occur every time a hydroclimatic threshold is exceeded, and their magnitude depends on the intensity and duration of a rainstorm or runoff event. The consequences of climate change in such watersheds would be manifest by an increase in the magnitude and frequency of debris flows (Stoffel & Beniston, 2006; Stoffel et al., 2011). The only limitation to debris flow magnitude is hence a meteorological one.

In line with the general model of inputs, susceptibility, and reactions in a debris-flow system (Fig. 10.1), possible responses to consideration of climate change effects can be queried as follows:

1. What are the observed and projected changes in extreme precipitation for the study area? Is there a clear increasing trend in the frequency and/or magnitude of extreme rainfall? Are climate projections and/or ensembles in general agreement with respect to the direction and magnitude of the trend? If significant increases in hydroclimatic extremes can be expected, consider integrating those in the rainfall-runoff models that are input to sediment transport equations.
2. Does the watershed in question have a history of extensive timber harvesting, recent wildfires and/or beetle infestations, and what is the scientific consensus for changes in these processes in the study area? If these processes are predicted to occur or intensify, consider integration in rainfall-runoff models, build in uncertainty in sediment transport considerations or integrate additional uncertainty in the design of mitigation measures.

3. Are there areas underlain by permafrost in the watershed? If so, are they bedrock or in the form of unconsolidated, potentially ice-rich debris? Would degradation of permafrost decrease ice cohesion and lead to the liberation of hitherto ice-cemented sediments? Could permafrost changes lead to large rock slope failures that could impact the streamflow system (i.e. through damming or higher sedimentation rates)?
4. Are there glaciers in the upper watershed and how could they affect runoff as they shrink and the glacial forefield is increasingly colonized by vegetation? Could glacial shrinkage lead to the formation of pro-glacial or supra-glacial lakes that could catastrophically drain through various mechanisms? If so, how could these risks be addressed through monitoring (i.e. lake levels) and warning systems, or through engineered structures? The impacts of climate change on steep creek processes are multi-faceted.

Broadly, one can differentiate between temperature and precipitation-related changes. The former will affect various secondary processes such as the thermal state of permafrost, changes in soil moisture, wildfire frequency, beetle infestations and glacial changes. In terms of precipitation, variations in the frequency-magnitude (F-M) characteristics of high intensity rainfall, proportionality between snowmelt and rain, changes in rainfall timing and storm duration can all affect debris flow F-M relationships.

10.5 Consequences: Debris-Flow Risk Assessments in a Changing Climate

This section describes a framework to integrate climate change effects into risk assessment, organized according to effects of climate change on each parameter contributing to risk. It provides a rationale for identifying the key effects of climate change through an examination of the factors that describe risk. This does not imply that a quantitative risk assessment is always warranted, but it helps to disaggregate those components of total risk that are affected by climate change effects. This section is supported by some key examples below.

For example, imagine a debris-flow prone creek discharging onto an adjacent floodplain. As the creek in question is prone to debris flows, the primary consequences are loss of life and economic impacts, as well as loss of some critical infrastructure. Equation 10.1 highlights the factors that need to be considered in a formal risk assessment (see Strouth et al., this volume).

For quantitative risk assessments, the common risk equation is:

$$R_j = \sum_{i=1}^{n} h_i * S_{i,j} * T_{i,j} * E_j * V_{i,j} \qquad (10.1)$$

Changes in hydroclimate conditions will change the debris-flow probability, h_i, the spatial probability of impact, $S_{i,j}$ (via changes in magnitude), and the vulnerability of the elements at risk, $V_{i,j}$ (via corresponding changes in intensity). The temporal probability of impact, $T_{i,j}$ and the exposure of the elements at risk, E_j, are not affected by climate change. This is exemplified by Fig. 10.9 for a hypothetical debris flow-prone system. To qualify or quantify the effects of climate change, the different components of risk must be examined separately. For example, if the debris flow in question is sediment supply-limited, it may produce more debris flows and h_i increases; however, their magnitude (volume and peak discharge) may decrease, hence reducing the spatial probability ($S_{i,j}$). A reduction in magnitude also implies lesser impact forces and a reduction of vulnerability ($V_{i,j}$).

Some debris flows (e.g. those whose sediment sources are is not fully depleted after an event) will become smaller and thus less intense if they occur more frequently (largely because the sediment in the source zone requires time-dependent recharge). A decrease in debris-flow size will (in most cases) be accompanied by a decrease in runout length and area inundated, which affects the spatial probability. Hence, despite an increase in landslide probability, the overall risk may not increase, or may even

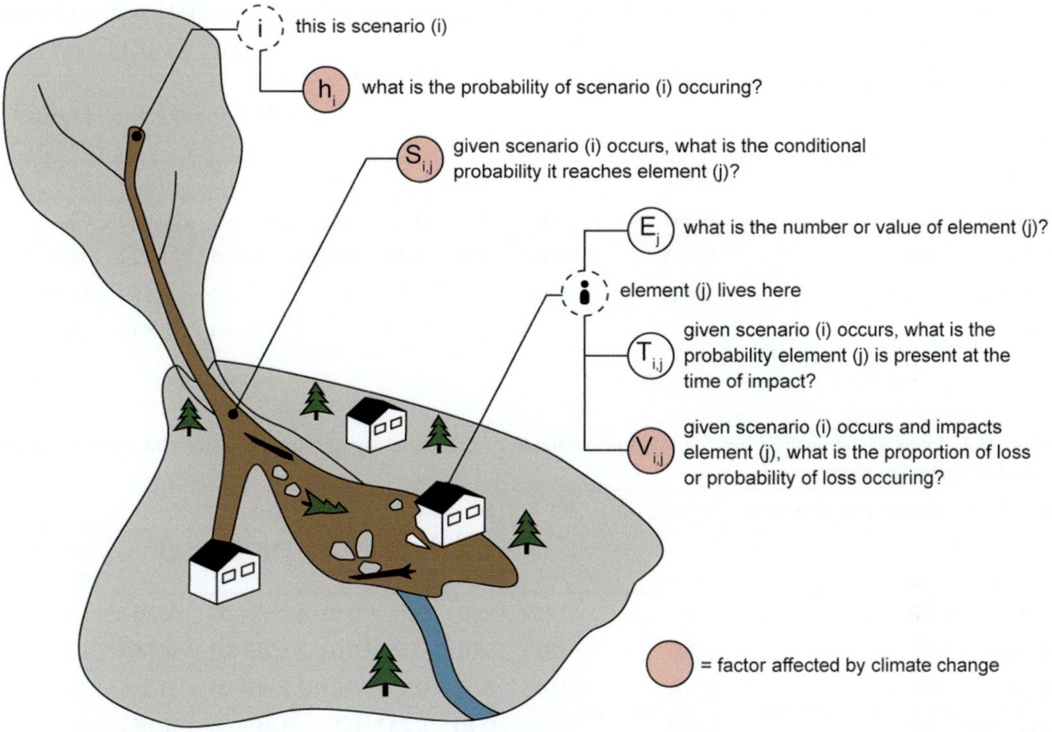

Fig. 10.9 Example of a debris flow-prone watershed with the various components of the risk equation that can be influenced by climate change (*Graphic* Sophia Zubrycky)

decrease. With a decrease in landslide volume and runout, the intensity decreases too, which reduces the vulnerability of a person inside or outside of a building. Hence, the overall risk is further reduced.

The following example explains how to include climate change in a debris-flow risk assessment, thus building on Strouth et al. (this volume). The assignment revolves around a debris-flow hazard assessment for new residential development on the fringes of an alluvial fan, where four new single-family residences and two duplexes are proposed.

Suggested Analysis:

- The practitioner determines if the watershed is watershed supply-limited, watershed supply-unlimited, channel supply-limited or channel supply-unlimited.
- In this example, the watershed is found to be channel supply-limited. This implies that, with climate change (higher precipitation intensities and higher frequency of extreme storms), the frequency of debris flows may increase, however, the magnitude of individual events may decrease due to supply limitations.
- The practitioner then establishes a frequency-magnitude (F-M) curve based on historical data and current conditions and then determines the 'likely' hydroclimatic threshold for debris-flow initiation from historical data.
- The F-M curve is updated to incorporate projected climate change, including the use of appropriate equations to estimate post-fire debris-flow hazard for at least three scenarios (recognizing that post-fire debris flows are more likely to occur in the future in the study area due to increasing temperatures and more frequent periods of drought).
- A numerical runout model of climate-change-adjusted debris flows is developed for various return periods and avulsion scenarios.

Table 10.1 Input to risk assessment for fictitious example of a channel supply-limited watershed based on historical frequency-magnitude data

Size class	Return period (years)	Volume (m^3)	h	S	T	V	Risk		
							PDI	RP	Micro-morts
Small	16	3,000	0.053	0.05	0.75	0.05	9.8E−05	10,159	98
Medium	100	10,000	0.007	0.1	0.75	0.1	5.0E−05	20,000	50
Large	300	17,000	0.002	0.25	0.75	0.2	8.7E−05	11,429	87.5
Very large	1,000	25,000	0.001	0.5	0.75	0.3	1.1E−04	8,889	112.5
Total							**3.5E−04**	**2,870**	**236**

Table 10.2 Input to risk assessment for fictitious example of a channel supply-limited watershed based on future frequency-magnitude data

Size class	Return period (years)	Volume (m^3)	h	S	T	V	Risk		
							PDI	RP	Micro-morts
Small	3	600	0.263	0.01	0.75	0.02	1.6E−05	60,952	16
Medium	20	2,000	0.033	0.01	0.75	0.03	1.2E−05	84,000	12
Large	60	3,400	0.012	0.05	0.75	0.07	2.9E−05	34,286	29
Very large	200	5,000	0.005	0.10	0.75	0.1	3.8E−05	26,667	38
Total							*9.5E−05*	**10,529**	**57**

Notes
1. h is hazard probability
2. S is spatial probability (chance of the debris flow reaching the element at risk)
3. T is temporal probability (chance of element at risk being present during the debris flow)
4. V is vulnerability (chance of a person dying, or chance of a specific loss occurring)
5. PDI is the probability of death of an individual (1/years)
6. RP is return period of the PDI (years)
7. Micromorts are one millions of the chance of someone dying. 1 million micromorts = death is certain

- The climate-change adjusted numerical model results are integrated into the risk assessment.

The risk assessment for the probability of death of an individual (PDI) is based on four different volume classes (first column in Tables 10.1 and 10.2). The values entered in these two tables are fictitious but close to a typical risk assessment outcome.

The frequency and magnitude relationship for this example is plotted in Fig. 10.10a, b. Further examples are provided in Fig. 10.10c, d for supply-unlimited conditions, for comparison.

Figure 10.10 demonstrates how life-loss risk for individuals could change through the effects of climate change compared to historical conditions. In this case, it could be demonstrated that, while the proposed development is currently not approvable, it might become approvable at some later time due to a decreasing debris-flow risk.

The opposite is true in the case of Fig. 10.10c–d, where debris flows become more frequent and larger, and hence result in a higher encounter probability and higher vulnerability due to increased impact forces. In those cases, the development may become un-approvable.

These examples illustrate the complexity inherent in including climate change considerations for debris flows and the considerable effort included in deciphering it. This degree of effort is warranted whenever a larger number of

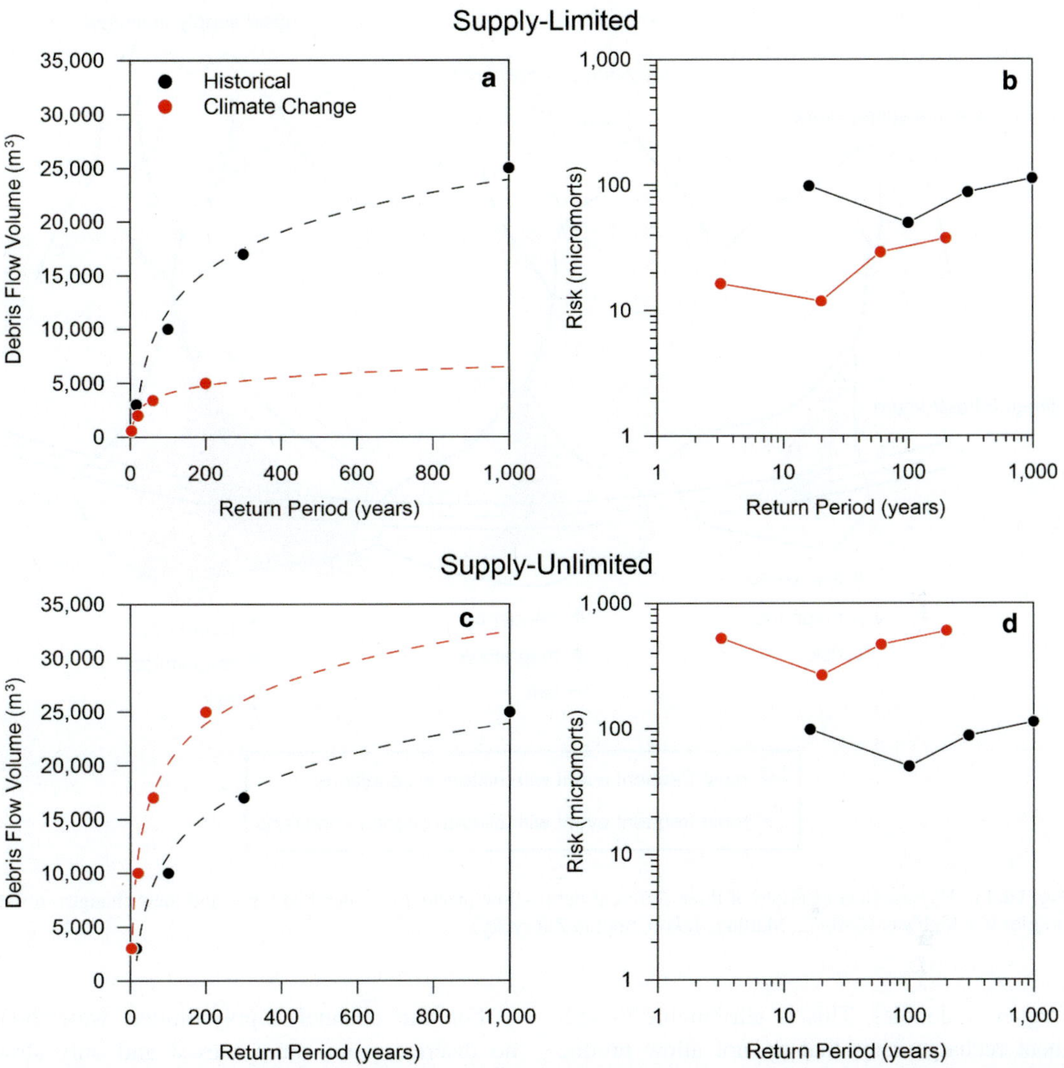

Fig. 10.10 Summary of frequency-magnitude relationship and risk assessment results for fictitious home on a debris-flow fan for historical and climate change (year 2100) conditions. Graphs A and B show historical (black) and climate change (red) risk for a supply-limited basin, C and D show historical and climate change risk for a supply-unlimited basin

people are in harm's way, and will allow realistic quantification of risk and decisions on risk avoidance or appropriate engineering measures to reduce risk to tolerable levels.

The examples provided in Fig. 10.10 were for urban developments. However, the same principle applies to a road or other linear structure. Figure 10.11 shows a hypothetical example of a debris-flow prioritization exercise along a highway. Three watershed types are shown and respective changes in debris-flow frequency-magnitude relationships exemplified graphically. We assume that the most frequent event reaching the highway is used for prioritization of mitigation measures. The logic behind such a decision is that the most frequent event causing damage is often the one that governs total risk. For the channel supply-limited watershed, the risk may remain constant, because the area inundated by a debris flow halves while the

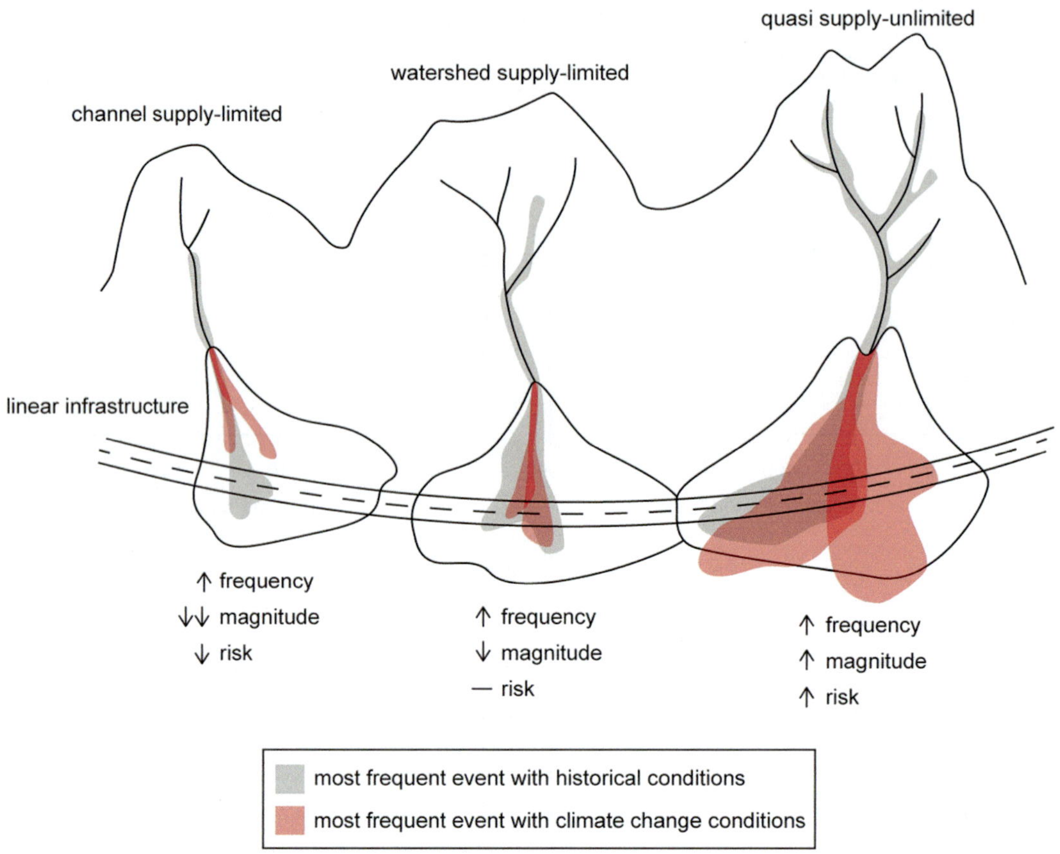

Fig. 10.11 Hypothetical example of three different debris-flow producing watershed types and their changes in risk profiles to a highway (*Graphic* Matthias Jakob, Sophia Zubrycky)

frequency doubles. This is attributable to sediment recharge rates that do not allow production of debris flows large enough to cover the area previously inundated. In this watershed type, larger debris flows can only occur if several debris avalanches or debris slides occur simultaneously.

For the quasi-unlimited watershed type, abundant debris exists in the form of numerous deep sediment sources, such as vast talus slopes, glacial moraines or other Quaternary sediments, or highly erodible rock. In this case, the effect of more frequent and higher intensity rainfall will likely result directly in more frequent and larger debris flows. This implies a greater area of the highway potentially impacted. Hence, the total risk will increase substantially.

For the channel supply-limited watershed, no discrete point sources exist and only slow accumulation of debris by rock fall and ravel can recharge the channel system. An increase in hydroclimatic threshold exceedance for debris flows will invariably mean smaller debris flows because of reduced recharge time. In this case, it may imply that debris flows of a lesser magnitude may not reach the highway at all, which in turn substantially reduces risk.

In summary, umbrella statements that climate change, in a specific region, or along a major infrastructure, increases risk are likely rarely warranted unless watershed-specific studies have been conducted. Such studies entail understanding watershed type and deciphering the time since the last debris flow using absolute dating methods (e.g. Ballesteros et al., this volume). This is followed by establishing common hydroclimatic thresholds for debris-flow initiation (often a combination of antecedent moisture conditions and rainfall intensities) and querying downscaled climate models with respect to how the key hydroclimatic variables will change. This may be a cumbersome exercise, but it will allow for a science-based justifiable design of mitigation measures.

10.6 Sources of Uncertainty

Uncertainty is defined as the state, even a partial state, of deficiency of information related to the understanding or knowledge of an event, and its consequence or likelihood. Uncertainty can be subdivided into three categories: natural (aleatoric), knowledge-source (epistemic), and analysis uncertainty, as described in the following sub-sections. Uncertainties in predicting the effects of climate change on debris-flow assessments are substantial. It is the responsibility of the practitioner to describe or quantify such uncertainties in their assessments:

Natural Uncertainty is inherent to a system due to natural variability or randomness (e.g. the difference in causes and triggers of debris flows being attributable to changes in rainfall, snowmelt or other changes in hydroclimatic variables). It cannot be reduced, due to random variability such as precipitation rates, volumes, changes in phreatic levels and pore water pressures in different soil units.

Knowledge-Source Uncertainty arises from limited information or understanding about a model or process, limits on the accumulated experience of the assessor, and limitations of the underlying data (Aven & Renn, 2015). It can be reduced, for example, through detailed instrumentation over extended time periods, although the benefit in doing so varies based on the level of effort. Another knowledge-source uncertainty is related to the order of the climate change effect. Uncertainty propagates from first order to higher order effects of climate change (i.e. the more indirect the global warming effect) and should be analyzed separately. This propagation is also valid for errors that may spread through different orders of climate change effects. This emphasizes the importance of careful review of all inputs early, as mistakes will affect all follow-up analyses and interpretations.

Limited observational data produces uncertainties, particularly in variables that are not spatially coherent (e.g. convective precipitation, Schauwecker et al., 2021) or are heavily affected by topography. Enhancing observational capacity can reduce this source of uncertainty. *Future atmospheric greenhouse gas concentrations* are unknowable but must be estimated by climate models. The climate change analysis should be based on an assumption that global carbon emissions, and thus atmospheric concentrations of greenhouse gases, will continue to rise along their current trajectory. Global–scale efforts to mitigate emissions and/or remove carbon from the atmosphere would alter that trajectory, as would the future release of carbon from other sources (e.g. methane currently sequestered in permafrost).

Analysis Uncertainty arises, for example, when the debris-flow triggering mechanisms are not fully understood. For example, individual side slope failures may be the dominant failure mechanism and lead to debris flows of a certain magnitude range for a spectrum of return periods. However, at high return periods (several hundreds or even millions of years), the failure mechanism may be associated with large rock slope failures or outburst floods that can result in much higher peak flows and transported debris volumes. Further uncertainties exist in the extrapolation of frequency well beyond the observed record length and the interpretation of confidence bounds as derived from statistical software packages.

Global Climate Models (GCMs) operate on a relatively coarse scale, and thus many physical processes that operate at scales finer than the GCM's grid (e.g. convection, cloud physics) must be parameterized. These parameterizations can contain substantial uncertainty, particularly in environments that have limited empirical data underpinning scientific understanding of the process. Climate models are designed to project changes in atmospheric variables over time. Each model scenario presents one of many possible realizations. Disagreement among models highlights the effects of uncertainty. Model skill/performance can be evaluated against available historical observations. It is assumed that models that most faithfully reconstruct historical conditions will more accurately project future climate states. Uncertainty arises in that current (and possibly future) climate change has no historical analog.

The coarse spatial and temporal scale of GCMs and gridded datasets may not adequately capture all physical processes in hydrological modelling, and thus their frequency, magnitude or duration may be underestimated. For example, high rainfall rates covering a square km or occurring in under an hour may not be adequately captured compared to other processes like snowmelt that occurs over several weeks and exhibits coherence over larger areas. Dynamical downscaling (e.g. using weather forecasting models) enhances the spatial and temporal resolution of GCM output and resolves finer scale physical processes. It can thus improve the estimation of extreme hydrometeorological conditions, but this advantage comes at a high computational cost. Statistical downscaling improves spatial and temporal resolution, and is less computationally intense, but does not integrate the physical processes that could not be resolved at the GCM scale, thus propagating uncertainty. Such uncertainty can be evaluated by comparing the results from members of an ensemble of hydrological models driven by various downscaled GCM scenarios.

Uncertainty (confidence) in climate projections increases (decreases) as bias derived from incomplete knowledge of physical processes propagates through time and through the calculation of second or third order variables. Longer lead times and higher order variables will have the greatest uncertainty. Such uncertainty can be evaluated and perhaps corrected for by analyzing output from multiple numerical model runs and through calibration and validation against direct observations of the variable.

10.7 Conclusions

The current climate change trajectory paints a dire picture for earth hazards, exacerbated by continued ecosystem degradation, overpopulation and political strife and war. Debris flows are a relatively small component of the profound changes that we can expect in every ecosystem and human-built infrastructures. Nonetheless, hundreds of thousands of alluvial fans prone to debris flows exist worldwide with at least some form of development. Debris flows occur when the terrain is steep enough, sufficient sediment is available and climate conditions allow for meteorological triggers. Those conditions are dramatically changing in many environments due to anthropogenic climate change.

At least as important as the direct changes in climate and weather, are the indirect changes. Glacial retreat, permafrost degradation, trees being increasingly stressed by drought or insects, as well as vast, frequent and intense wildfires and deforestation, all conspire towards a potentially lethal cocktail. However, practicing geoscientists and engineers must acknowledge that the risk (i.e. the combination of hazard and consequence) will have different trajectories in various mountain ranges on Earth. Sediment supply limitations, and, in some instances, moisture limitations, may in fact decrease debris-flow risk regionally, or at least for specific watersheds. In other regions, especially those still blanketed by rapidly retreating glaciers and alpine permafrost, debris-flow activity will increase until vegetation has encroached into the glacial forefields and begins to stabilize talus slopes. Especially of concern are glacial collapses and breaches of pro-glacial lakes, which can unleash hazard chains with consequences reaching tens of kilometers downstream. While direct destruction and flooding are most often highlighted in media reporting and scientific investigations, trends towards higher debris-flow activity in mountainous regions can lastingly change riverine planform by introducing vast amounts of sediment. This, in turn, can change flood hazard and also negatively affect aquatic habitat for many fish species.

Much more research is needed in terms of climate change effects on debris-flow activity. Watersheds with a long (decadal) history of monitoring can serve as prototypes to detect and quantify the effects of climate change, and lessons from such archetype watersheds can be applied to similar ones elsewhere. Experimental designs in which lab-scale watersheds are constructed with various levels of sediment supply conditions can yield useful insights in geomorphic response because changes in moisture regime (both antecedent and intensities) can be controlled while holding other factors constant.

Forestry, mining, agriculture, road building and other human activities in high mountain areas are all conducted either without or with only rudimentary consideration of climate change. At this time, such watershed management is anachronistic and must be discontinued. Instead, complex watershed response models need to be created in which the cumulative effects of the various stressors need to be considered in unison and their effects on the changes in shallow (and deep-seated) landslide and debris-flow activity quantified. If this is not done, catastrophic losses are unavoidable. In short, much higher risk management standards need to be applied than are customary today. As competing financial interests may limit the motivation to attain high standards, government will need to be involved to a much higher degree by creating and strictly enforcing guidelines based on best- and evolving science.

The globe is at the cusp of profound and increasingly irreversible (at a human time scale) environmental catastrophe. Cataclysmic events such as hazard chains in densely developed mountain valleys will become more common, as well as rainfall events that trigger hundreds or thousands of debris flows in a larger region. This predictable outcome requires imagination and creativity from academics and practitioners alike to meet this challenge with the appropriate tools to quantify changes and invent, improve or amend techniques to avoid catastrophic losses. We hope that this chapter highlights some of those challenges and acts as an impetus for a new generation of debris flow researchers and practitioners to invest themselves in deciphering the impacts of climate change on debris-flow activity worldwide.

Acknowledgements This chapter is dedicated to our colleague and friend Matthias Jakob who tragically died on October 17, 2022 in a paragliding accident. Matthias contributed very actively to this chapter. The authors warmly acknowledge Sophia Zubrycky for her tremendous help with illustrations.

References

Abatzoglou, J. T., & Williams, A. S. P. (2016). Impact of anthropogenic climate change on wildfire across western US forests. *Proceedings of the National Academy of Sciences, 113*, 11770–11775.

Abatzoglou, J. T., et al. (2021). Projected increases in western US forest fire despite growing fuel constraints. Communications Earth & Environment, 2, 227. https://doi.org/10.1038/s43247-021-00299-0

Aguilar, E., et al. (2005). Changes in precipitation and temperature extremes in Central America and northern South America, 1961–2003. *Journal of Geophysical Research, 110*, D23107. https://doi.org/10.1029/2005JD006119

Ali, H., Fowler, H. J., & Mishra, V. (2018). Global observational evidence of strong linkage between dew point temperature and precipitation extremes. *Geophysical Research Letters, 45*, 320–330.

Alizadeh, M. R., Abatzoglou, J. T., Luce, C. H., Adamowski, J. F., Farid, A., & Sadegh, M. (2021). Warming enabled upslope advance in western US forest fires. *Proceedings of the National Academy of Sciences, 118*(22), e2009717118. https://doi.org/10.1073/pnas.2009717118

Allen, S. K., Rastner, P., Arora, M., Huggel, C., & Stoffel, M. (2016). Lake outburst and debris flow disaster at Kedarnath, June 2013: Hydrometeorological triggering, and topographic predisposition. *Landslides, 13*(6), 1479–1491. https://doi.org/10.1007/s10346-015-0584-3

AMS. (2022). Glossary of meteorology. https://glossary.ametsoc.org/wiki/Monsoon

André, M. F. (2003). Do periglacial landscapes evolve under periglacial conditions? *Geomorphology, 52*(1), 149–164. https://doi.org/10.1016/S0169-555X(02)00255-6

Aven, T., & Renn, O. (2015). An evaluation of the treatment of risk and uncertainties in the IPCC reports on climate change. *Risk Analysis, 35*(4), 701–712.

Balch, J. K., Bradley, B. A., Abatzoglou, J. T., Nagy, R. C., Fusco, E. J., & Mahood, A. L. (2017). Human-started wildfires expand the fire niche across the United States. *Proceedings of the National Academy of Sciences, 114*, 2946–2951.

Balfour, V., & Woods, S. W. (2007). Does wildfire ash block soil pores? A micromorphological analysis of burned soils. In: *AGU fall meeting abstracts* (pp. H43F-1695).

Ballesteros-Cánovas, J. A., et al. (2019). On the extraordinary winter flood episode over the North Atlantic Basin in 1936. *Annals of the New York Academy of Sciences, 1436*(1), 206–216.

Ban, N., Schmidli, J., & Schär, C. (2015). Heavy precipitation in a changing climate: Does short-term summer precipitation increase faster? *Geophysical Research Letters, 42*(4), 1165–1172. https://doi.org/10.1002/2014GL062588

Barbero, R., Abatzoglou, J. T., Larkin, N. K., Kolden, C. A., & Stocks, B. (2015). Climate change presents increased potential for very large fires in the contiguous United States. *International Journal of Wildland Fire, 24*, 892–899.

Beniston, M., & Stoffel, M. (2016). Rain-on-snow events, floods and climate change in the Alps: Events may increase with warming up to 4 °C and decrease thereafter. *Science of the Total Environment, 571*, 228–236.

Beniston, M., et al. (2018). The European mountain cryosphere: A review of its current state, trends and future challenges. *The Cryosphere, 12*, 759–794.

Benito, G., Castillo, O., Ballesteros-Cánovas, J. A., Machado, M., & Barriendos, M. (2021). Enhanced flood hazard assessment beyond decadal climate cycles based on centennial historical data (Duero basin, Spain). *Hydrology and Earth System Sciences, 25*(12), 6107–6132.

Bentz, B. J., et al. (2010). Climate change and bark beetles of the western United States and Canada: Direct and indirect effects. *BioScience, 60*(8), 602–613.

Berghuijs, W., Woods, R., & Hrachowitz, M. A. (2014). Precipitation shift from snow towards rain leads to a decrease in streamflow. *Nature Climate Change, 4*, 583–586. https://doi.org/10.1038/nclimate2246

Bhardwaj, A., & Sam, L. (2022). Reconstruction and characterisation of past and the most recent slope failure events at the 2021 rock-ice avalanche site in Chamoli, Indian Himalaya. *Remote Sensing, 14*(4), 949. https://doi.org/10.3390/RS14040949

Biasutti, M., Seager, R., & Kirschbaum, D. B. (2016). Landslides in West Coast metropolitan areas: The role of extreme weather events. *Weather and Climate Extremes, 14*, 67–79.

Bollschweiler, M., & Stoffel, M. (2010). Changes and trends in debris-flow frequency since 1850—results from eight torrents in the Zermatt valley. *Holocene, 20*, 907–916.

Bookhagen, B., Thiede, R. C., & Strecker, M. R. (2005). Abnormal monsoon years and their control on erosion and sediment flux in the high, arid northwest Himalaya. *Earth and Planetary Science Letters, 231*(1–2), 131–146. https://doi.org/10.1016/J.EPSL.2004.11.014

Borga, M., Stoffel, M., Marchi, L., Marra, F., & Jakob, M. (2014). Hydrogeomorphic response to extreme rainfall in headwater systems: Flash floods and debris flows. *Journal of Hydrology, 518*, 194–205.

Bovis, M. J., & Jakob, M. (1999). The role of debris supply conditions in predicting debris flow activity. *Earth Surface Processes and Landforms, 24*(11), 1039–1054.

Brönnimann, S., Rajczak, J., Fischer, E. M., Raible, C. C., Rohrer, M., & Schär, C. (2018). Changing seasonality of moderate and extreme precipitation events in the Alps. *Natural Hazards and Earth System Sciences, 18*(7), 2047–2056.

Cannon, S. H., & Gartner, J. E. (2005). Wildfire-related debris flow from a hazards perspective. In *Debris-flow hazards and related phenomena* (pp. 363–385). Springer.

Cannon, A. J., & Innocenti, S. (2019). Projected intensification of sub-daily and daily rainfall extremes in convection-permitting climate model simulations over North America: Implications for future intensity–duration–frequency curves. *Natural Hazards and Earth System Sciences, 19*, 421–440. https://doi.org/10.5194/nhess-19-421-2019

Cattani, E., Merino, A., Guijarro, J. A., & Levizzani, V. (2018). East Africa rainfall trends and variability 1983–2015 using three long-term satellite products. *Remote Sensing, 10*, 931. https://doi.org/10.3390/rs10060931

Chand, S. S., et al. (2022). Declining tropical cyclone frequency under global warming. *Nature Climate Change, 12*, 655–661. https://doi.org/10.1038/s41558-022-01388-4

Chiarle, M., Iannotti, S., Mortara, G., & Deline, P. (2007). Recent debris flow occurrences associated with glaciers in the Alps. *Global and Planetary Change, 56*, 123–136.

Christensen, J. H., et al. (2013). Climate phenomena and their relevance for future regional climate change. In Stocker, T. F., et al. (Eds.), *Climate change 2013: The physical science basis* (pp. 1217–1308). Cambridge University Press.

Church, M., & Jakob, M. (2020). What is a debris flood? *Water Resources Research, 56*(8), e2020WR027144.

Colorado-Ruiz, G., Cavazos, T., Salinas, J. A., De Grau, P., & Ayala, R. (2018). Climate change projections from Coupled Model Intercomparison Project phase 5 multi-model weighted ensembles for Mexico, the North American monsoon, and the mid-summer drought region. *International Journal of Climatology, 38*, 5699–5716. https://doi.org/10.1002/joc.5773

Coogan, S. C., et al. (2021). Fifty years of wildland fire science in Canada. *Canadian Journal of Forest Research, 51*(2), 283–302. https://doi.org/10.1139/cjfr-2020-0314

Cook, B. I., & Seager, R. (2013). The response of the North American monsoon to increased greenhouse gas forcing. *Journal of Geophysical Research: Atmospheres, 118*, 1690–1699. https://doi.org/10.1002/JGRD.50111

Cook, K. H., & Vizy, E. K. (2013). Projected changes in East African rainy seasons. *Journal of Climate, 26*, 5931–5948. https://doi.org/10.1175/JCLI-D-12-00455.1

Cook, K. L., Andermann, C., Gimbert, F., Adhikari, B. R., & Hovius, N. (2018). Glacial lake outburst floods as drivers of fluvial erosion in the Himalaya. *Science, 362*(6410), 53–57. https://doi.org/10.1126/science.aat4981

Cook, S. J., & Quincey, D. J. (2015). Estimating the volume of Alpine glacial lakes. *Earth Surface Dynamics, 3*(4), 559–575. https://doi.org/10.5194/esurf-3-559-2015

Cordeira, J. M., Stock, J., Dettinger, M. D., Young, A. M., Kalansky, J. F., & Ralph, F. M. (2019). A 142-year climatology of northern California landslides and atmospheric rivers. *Bulletin of the American Meteorological Society, 100*(8), 1499–1509.

Corringham, T. W., Ralph, F. M., Gershunov, A., Cayan, D. R., & Talbot, C. A. (2019). Atmospheric rivers drive flood damages in the western United States. *Science Advances, 5*(12), eaax4631.

Corringham, T. W., Mcarthy, J., Shulgina, T., Gershunov, A., Caya, D. R., & Ralph, F. M. (2022). Climate change contributions to future atmospheric river flood damages in the western United States. *Science and Reports, 12*, 13747.

Cullingham, C. I., Cooke, J. E., Dang, S., Davis, C. S., Cooke, B. J., & Coltman, D. W. (2011). Mountain pine beetle host-range expansion threatens the boreal forest. *Molecular Ecology, 20*(10), 2157–2171.

Curry, C. L., Islam, S. U., Zwiers, F. W., & Déry, S. J. (2019). Atmospheric rivers increase future flood risk in Western Canada's largest Pacific river. *Geophysical Research Letters, 46*(3), 1651–1661.

Dalton, M. M., & Fleishman, E. (Eds.). (2021). Fifth Oregon climate assessment. Oregon Climate Change Research Institute, Oregon State University, Corvallis, Oregon. https://blogs.oregonstate.edu/occri/oregon-climate-assessments/

DeGraff, J. V., Cannon, S. H., & Gartner, J. E. (2015). The timing of susceptibility to post-fire debris flows in the Western United States. *Environmental & Engineering Geoscience, 21*(4), 277–292.

de Haas, T., Densmore, A., Stoffel, M., Suwa, H., Imaizumi, F., Ballesteros Canovas, J. A., & Wasklewicz, T. (2018). Avulsions and the spatio-temporal evolution of debris-flow fans. *Earth-Science Reviews, 177*, 53–75.

Dennison, P. E., Brewer, S. C., Arnold, J. D., & Moritz, M. A. (2014). Large wildfire trends in the western United States, 1984–2011. *Geophysical Research Letters, 41*, 2928–2933. https://doi.org/10.1002/2014GL059576

Dey, R., Lewis, S. C., Arblaster, J. M., & Abram, N. J. (2019). A review of past and projected changes in Australia's rainfall. *Wiley Interdisciplinary Reviews: Climate Change, 10*, e577. https://doi.org/10.1002/wcc.577

Dimri, A. P., Allen, S., Huggel, C., Mal, S., Ballesteros-Cánovas, J. A., Rohrer, M., Shukla, A., Tiwari, P., Maharana, P., Bolch, T., Thayyen, R. J., Stoffel, M., & Pandey, A. (2021). Climate change, cryosphere and impacts in the Indian Himalayan Region. *Current Science, 120*, 774–790.

Ding, Y., Sun, Y., Wang, Z., Zhu, Y., & Song, Y. (2009). Interdecadal variation of the summer precipitation in China and its association with decreasing Asian summer monsoon Part II: Possible causes. *International Journal of Climatology, 29*, 1926–1944. https://doi.org/10.1002/joc.1759

Doerr, S. H., Shakesby, R. A., & Walsh, R. (2000). Soil water repellency: Its causes, characteristics and hydro-geomorphological significance. *Earth-Science Reviews, 51*(1–4), 33–65.

Ebel, B. A., Moody, J. A., & Martin, D. A. (2012). Hydrologic conditions controlling runoff generation immediately after wildfire. *Water Resources Research, 48*(3), W03529. https://doi.org/10.1029/2011WR011470

England, M. R., Eisenman, I., Lutsko, N. J., & Wagner, T. J. (2021). The recent emergence of Arctic Amplification. *Geophysical Research Letters, 48*(15), e2021GL094086.

Espinoza, V., Waliser, D. E., Guan, B., Lavers, D. A., & Ralph, F. M. (2018). Global analysis of climate change projection effects on atmospheric rivers. *Geophysical Research Letters, 45*, 4299–4308. https://doi.org/10.1029/2017GL076968

Evans, S. G., & Clague, J. J. (1994). Recent climatic change and catastrophic geomorphic processes in mountain environments. *Geomorphology, 10*, 107–128.

Evans, S. G., & Delaney, K. B. (2015). Catastrophic mass flows in the mountain glacial environment. In W. Haeberli & C. Whiteman (Eds.), *Snow and ice-related hazards, risks, and disasters* (pp. 568–606). Elsevier.

Fehlmann, M., Gascón, E., Rohrer, M., Schwarb, M., & Stoffel, M. (2019). Improving medium-range forecasts of rain-on-snow events in pre-alpine areas. *Water Resources Research, 55*, 7638–7661.

Fischer, E. M., & Knutti, R. (2016). Observed heavy precipitation increase confirms theory and early models. *Nature Climate Change, 6*(11), 986–991. https://doi.org/10.1038/nclimate3110

Flannigan, M. D., Krawchuk, M. A., de Groot, W. J., Wotton, M. B., & Gowman, L. M. (2009). Implications of changing climate for global wildland fire. *International Journal of Wildland Fire, 18*(5), 483–507. https://doi.org/10.1071/WF08187

Fowler, H. J., Wasko, C., & Prein, A. F. (2021a). Intensification of short-duration rainfall extremes and implications for flood risk: Current state of the art and future directions. *Philosophical Transactions of the Royal Society A, 379*(2195), 20190541.

Fowler, H. J., et al. (2021b). Anthropogenic intensification of short-duration rainfall extremes. *Nature Reviews Earth & Environment, 2*(2), 107–122.

Fu, R., et al. (2013). Increased dry-season length over southern Amazonia in recent decades and its implication for future climate projection. *Proceedings of the National Academy of Sciences of the United States of America, 110*, 18110–18115. https://doi.org/10.1073/pnas.1302584110

Gaglioti, B. V., Mann, D. H., & Wiles, G. C. (2022). Ecosystems at glacier margins can serve as climate-change laboratories. *Geophysical Research Letters, 49*, e2022GL098574. https://doi.org/10.1029/2022GL098574

Gartner, J. E., Santi, P. M., & Cannon, S. H. (2015). Predicting locations of post-fire debris-flow erosion in the San Gabriel Mountains of southern California. *Natural Hazards and Earth System Sciences, 77*, 1305–1321. https://doi.org/10.1007/s11069-015-1656-3

Gensini, V. A., Haberlie, A. M., & Ashley, W. S. (2022). Convection-permitting simulations of historical and possible future climate over the contiguous United States. *Climate Dynamics*https://doi.org/10.1007/s00382-022-06306-0

Gershunov, A., Shulgina, T., Clemesha, R. E. S., et al. (2019). Precipitation regime change in Western North America: The role of Atmospheric Rivers. *Science and Reports, 9*, 9944. https://doi.org/10.1038/s41598-019-46169-w

Gobiet, A., Kotlarski, S., Beniston, M., Heinrich, G., Rajczak, J., & Stoffel, M. (2014). 21st century climate change in the European Alps—a review. *Science of the Total Environment, 493*, 1138–1151.

Gonzalez, P. L., Polvani, M. L. M., Seager, R., & Correa, G. J. (2014). Stratospheric ozone depletion: A key driver of recent precipitation trends in south eastern South America. *Climate Dynamics, 42*, 1775–1792. https://doi.org/10.1007/s00382-013-1777-x

Goss, M., et al. (2020). Climate change is increasing the likelihood of extreme autumn wildfire conditions across California. *Environmental Research Letters, 15*(9), 094016.

Grimm, A. M. (2011). Interannual climate variability in South America: Impacts on seasonal precipitation, extreme events and possible effects of climate change. *Stochastic Environmental Research and Risk Assessment, 25*, 537–554. https://doi.org/10.1007/s00477-010-0420-1

Grimm, A. M., & Saboia, J. P. J. (2015). Interdecadal variability of the South American precipitation in the monsoon season. *Journal of Climate, 28*, 755–775. https://doi.org/10.1175/JCLI-D-14-00046.1

Gruber, S., & Haeberli, W. (2007). Permafrost in steep bedrock slopes and its temperature-related destabilization following climate change. *Journal of Geophysical Research: Earth Surface, 112*, F02S18. https://doi.org/10.1029/2006JF000547

Guerreiro, S. B., et al. (2018). Detection of continental-scale intensification of hourly rainfall extremes. *Nature Climate Change, 8*(9), 803–807.

Guindon, L., et al. (2021). Trends in wildfire burn severity across Canada, 1985 to 2015. *Canadian Journal of Forest Research, 51*(9), 1230–1244.

Guthrie, R. H., et al. (2012). The 6 August 2010 Mount Meager rock slide-debris flow, Coast Mountains, British Columbia: Characteristics, dynamics, and implications for hazard and risk assessment. *Natural Hazards and Earth System Sciences, 12*, 1277–1294. https://doi.org/10.5194/nhess-12-1277-2012

Guzman-Morales, J., Gershunov, A., Theiss, J., Li, H., & Cayan, D. (2016). Santa Ana Winds of Southern California: Their climatology, extremes, and behavior spanning six and a half decades. *Geophysical Research Letters, 43*, 2827–2834. https://doi.org/10.1002/2016GL067887

Ha, K. J., Kim, B. H., Chung, E. S., Chan, J. C. L., & Chang, C. P. (2020a). Major factors of global and regional monsoon rainfall changes: Natural versus anthropogenic forcing. *Environmental Research Letters, 15*, 034055. https://doi.org/10.1088/1748-9326/ab7767

Ha, K. J., Moon, S., Timmermann, A., & Kim, D. (2020b). Future changes of summer monsoon characteristics and evaporative demand over Asia in CMIP6 simulations. *Geophysical Research Letters, 47*, e2020GL087492. https://doi.org/10.1029/2020GL087492

Haeberli, W., Buetler, M., Huggel, C., Lehmann Friedli, T., Schaub, Y., & Schleiss, A. J. (2016). New lakes in deglaciating high-mountain regions—opportunities and risks. *Climate Change, 139*, 201–214.

Haeberli, W., Schaub, Y., & Huggel, C. (2017). Increasing risks related to landslides from degrading permafrost into new lakes in de-glaciating mountain ranges. *Geomorphology, 293*, 405–417.

Han, F. K., Cook, H., & Vizy, E. K. (2019). Changes in intense rainfall events and dry periods across Africa in the twenty-first century. *Climate Dynamics, 53*, 2757–2777. https://doi.org/10.1007/s00382-019-04653-z

Hanes, C., Wang, X., Jain, P., Parisien, M., Little, J., & Flannigan, M. (2018). Fire-regime changes in Canada over the last half century. *Canadian Journal of Forest Research, 49*(3), 256–269. https://doi.org/10.1139/cjfr-2018-0293

Hantemirov, R. M., et al. (2022). Current Siberian heating is unprec-edented during the last seven millennia.

Nature Communications, 13, 4968. https://doi.org/10.1038/s41467-022-32629-x

Harris, C. (2005). Climate change, mountain permafrost degradation and geotechnical hazard. In U. M. Huber, H. K. M. Bugmann, & M. A. Reasoner (Eds.), *Global change and mountain regions. An overview of current knowledge* (pp. 215–224). Springer.

Haugo, R. D., et al. (2019). The missing fire: Quantifying human exclusion of wildfire in Pacific Northwest forests, USA. *Ecosphere, 10*, e02702. https://doi.org/10.1002/ecs2.2702

Held, I. M., & Soden, B. J. (2006). Robust responses of the hydrological cycle to global warming. *Journal of Climate, 19*(21), 5686–5699. https://doi.org/10.1175/JCLI3990.1

Higuera, P. E., Abatzoglou, J. T., Littell, J. S., & Morgan, P. (2015). The changing strength and nature of fire-climate relationships in the northern Rocky Mountains, USA, 1902–2008. *PLoS ONE, 10*, e0127563. https://doi.org/10.1371/journal.pone.0127563

Hirschberg, J., et al. (2021). Climate change impacts on sediment yield and debris-flow activity in an alpine catchment. *Journal of Geophysical Research: Earth Surface, 126*(1), e2020JF005739.

Hock, R., et al. (2019). High mountain areas. In H. O. Pörtner, et al. (Eds.), *IPCC special report on the ocean and cryosphere in a changing climate. IPCC, 2019* (pp. 131–202). Cambridge University Press. https://doi.org/10.1017/9781009157964

Huang, X., Swain, D. L., & Hall, A. D. (2020). Future precipitation increase from very high resolution ensemble downscaling of extreme atmospheric river storms in California. *Science Advances, 6*(29), eaba1323.

Huggel, C., et al. (2005). The 2002 rock/ice avalanche at Kolka/Karmadon, Russian Caucasus: Assessment of extraordinary avalanche formation and mobility and application of QuickBird satellite imagery. *Natural Hazards and Earth System Sciences, 5*, 173–187.

IPCC. (2021). *Climate change 2021: The physical science basis. Contribution of working Group I to the sixth assessment report of the intergovernmental panel on climate change*. Cambridge University Press. https://doi.org/10.1017/9781009157896.

Iverson, R. M. (2000). Landslide triggering by rain infiltration. *Water Resources Research, 36*(7), 1897–1910.

Jakob, M., & Owen, T. (2021). Projected effects of climate change on shallow landslides, North Shore Mountains, Vancouver, Canada. *Geomorphology, 393*, 107921.

Jenkins, M. J., Hebertson, E., Page, W., & Jorgensen, C. A. (2008). Bark beetles, fuels, fires and implications for forest management in the Intermountain West. *Food Policy, 254*(1), 16–34.

Jiang, J., & Zhou, T. (2019). Global monsoon responses to decadal sea surface temperature variations during the twentieth century: Evidence from AGCM simulations. *Journal of Climate, 32*, 7675–7695. https://doi.org/10.1175/JCLI-D-18-0890.1

Jomelli, V., Brunstein, D., Grancher, D., & Pech, P. (2007). Is the response of hill slope debris flows to recent climate change univocal? A case study in the Massif des Ecrins (French Alps). *Climate Change, 85*, 119–137. https://doi.org/10.1007/s10584-006-9209-0

Jomelli, V., Pech, V. P., Chochillon, C., & Brunstein, D. (2004). Geomorphic variations of debris flows and recent climatic change in the French Alps. *Climate Change, 64*(1/2), 77–102. https://doi.org/10.1023/B:CLIM.0000024700.35154.44

Kääb, A., Reynolds, J. M., & Haeberli, W. (2005). Glacier and permafrost hazards in high mountains. In U. M. Huber, H. K. M. Bugmann, & M. A. Reasoner (Eds.), *Global change and mountain regions. An overview of current knowledge* (pp. 225–234). Springer.

Kean, J. W., & Staley, D. M. (2021). Forecasting the frequency and magnitude of post-fire debris flows across southern California. *Earth Future, 9*, e2020EF001735. https://doi.org/10.1029/2020EF001735

Keeley, J. E., & Syphard, A. D. (2018). Historical patterns of wildfire ignition sources in California ecosystems. *International Journal of Wildland Fire, 27*, 781–799. https://doi.org/10.1071/WF18026

Kirschbaum, D., Stanley, T., & Zhou, Y. (2015). Spatial and temporal analysis of a global landslide catalog. *Geomorphology, 249*, 4–15. https://doi.org/10.1016/j.geomorph.2015.03.016

Kirschbaum, D., Kapnick, S. B., Stanley, T., & Pascale, S. (2020). Changes in extreme precipitation and landslides over high mountain Asia. *Geophysical Research Letters, 47*(4), e2019GL085347. https://doi.org/10.1029/2019GL085347

Kirchmeier-Young, M. C., Gillett, N. P., Zwiers, F. W., Cannon, A. J., & Anslow, F. S. (2019). Attribution of the influence of human-induced climate change on an extreme fire season. *Earth Future, 7*, 2–10. https://doi.org/10.1029/2018EF001050

Kitzberger, T., Falk, D. A., Westerling, A. L., & Swetnam, T. W. (2017). Direct and indirect climate controls predict heterogeneous early-mid 21st century wildfire burned area across western and boreal North America. *PLoS ONE, 12*, e0188486. https://doi.org/10.1371/journal.pone.0188486

Klock, G. O., & Helvey, J. D. (1976). Soil-water trends following wildfire on the Entiat Experimental Forest [Washington]. In *Proceedings Tall Timbers Fire Ecology Conference*.

Knutson, T., et al. (2020). Tropical cyclones and climate change assessment: Part II: Projected response to anthropogenic warming. *Bulletin of the American Meteorological Society, 101*(3), E303–E322. https://doi.org/10.1175/BAMS-D-18-0194.1

Kolden, C. A., Smith, A., & Abatzoglou, J. (2015). Limitations and utilisation of monitoring trends in burn severity products for assessing wildfire severity

in the USA. *International Journal of Wildland Fire, 24*, 1023–1028.
Kossin, J. P., Emanuel, K. A., & Vecchi, G. A. (2014). The poleward migration of the location of tropical cyclone maximum intensity. *Nature, 509*(7500), 349–352.
Krishnan, R., et al. (2016). Deciphering the desiccation trend of the South Asian monsoon hydroclimate in a warming world. *Climate Dynamics, 47*, 1007–1027. https://doi.org/10.1007/s00382-015-2886-5
Lenderink, G., & Van Meijgaard, E. (2008). Increase in hourly precipitation extremes beyond expectations from temperature changes. *Nature Geoscience, 1*(8), 511–514. https://doi.org/10.1038/ngeo262
Lenderink, G., Barbero, R., Loriaux, J. M., & Fowler, H. J. (2017). Super-Clausius–Clapeyron scaling of extreme hourly convective precipitation and its relation to large-scale atmospheric conditions. *Journal of Climate, 30*(15), 6037–6052.
Li, J., & Wang, B. (2018). Origins of the decadal predictability of East Asian land summer monsoon rainfall. *Journal of Climate, 31*, 6229–6243. https://doi.org/10.1175/JCLI-D-17-0790.1
Littell, J. S., McKenzie, D., Peterson, D. L., & Westerling, A. L. (2009). Climate and wildfire area burned in western US ecoprovinces, 1916–2003. *Ecological Applications, 19*(4), 1003–1021.
Littell, J. S., et al. (2010). Forest ecosystems, disturbance, and climatic change in Washington State, USA. *Climate Change, 102*, 129–158. https://doi.org/10.1007/s10584-010-9858-x
Lliboutry, L., Morales, A. B., Pautre, A., & Schneider, B. (1977). Glaciological problems set by the control of dangerous lakes in Cordillera Blanca, Peru. I. Historic failure of morainic dams, their causes and prevention. *Journal of Glaciology, 18*, 239–254.
Liu, C., et al. (2017). Continental-scale convection-permitting modeling of the current and future climate of North America. *Climate Dynamics, 49*(1), 71–95.
Lodoun, T., Giannini, A., Traoreì, P. S., Someì, L., Sanon, M., Vaksmann, M., & Rasolodimby, J. M. (2013). Changes in seasonal descriptors of precipitation in Burkina Faso associated with late 20th century drought and recovery in West Africa. *Environment and Behaviour, 5*, 96–108. https://doi.org/10.1016/j.envdev.2012.11.010
Lugon, R., & Stoffel, M. (2010). Rock-glacier dynamics and magnitude–frequency relations of debris flows in a high-elevation watershed: Ritigraben, Swiss Alps. *Global and Planetary Change, 73*, 202–210.
Luong, T. M., Castro, C. L., Chang, H. I., Lahmers, T., Adams, D. K., & Ochoa-Moya, C. A. (2017). The more extreme nature of North American monsoon precipitation in the southwestern U.S. as revealed by a historical climatology of simulated severe weather events. *Journal of Applied Meteorology and Climatology, 56*, 2509–2529. https://doi.org/10.1175/JAMC-D-16-0358.1
Mahat, V., Silins, U., & Anderson, A. (2016). Effects of wildfire on the catchment hydrology in southwest Alberta. *CATENA, 147*, 51–60.
Maidment, R. I., Allan, R. P., & Black, E. (2015). Recent observed and simulated changes in precipitation over Africa. *Geophysical Research Letters, 42*, 8155–8164. https://doi.org/10.1002/2015GL065765
Majeed, U., Rashid, I., Sattar, A., Allen, S., Stoffel, M., Nüsser, M., & Schmidt, S. (2021). Recession of Gya glacier, proglacial lake evolution, and the 2014 glacial outburst flood in the Trans-Himalayan region of Ladakh, India. *Science of the Total Environment, 756*, 144008.
Mani, P., Allen, S., Evans, S. G., Kargel, J. S., Mergili, M., Petrakov, D., & Stoffel, M. (2023). Geomorphic process chains in high-mountain regions—A review and classification approach for natural hazards assessment. *Reviews of Geophysics, 61*, e2022RG000791. https://doi.org/10.1029/2022RG000791
Marvel, K., Biasutti, M., & Bonfils, C. (2020). Fingerprints of external forcing agents on Sahel rainfall: Aerosols, greenhouse gases, and model-observation discrepancies. *Environmental Research Letters, 15*, 084023. https://doi.org/10.1088/1748-9326/ab858e
McGee, J. S., Phelan, L., & Holbrook, N. J. (2015). *Cyclone Pam: The Pacific needs a climate insurance scheme*. The Conversation Media Group Ltd. https://theconversation.com/cyclone-pam-the-pacific-needs-aclimate-insurance-scheme-38894
McKenzie, D., Gedalof, Z., Peterson, D. L., & Mote, P. (2004). Climatic change, wildfire, and conservation. *Conservation Biology, 18*, 890–902.
Medellín-Azuara, J., et al. (2018). *Economic and environmental implications of California crop and livestock, adaptation to climate change*. California Natural Resources Agency. (University of California, Davis and University of California, Merced. Publication number: CCCA4-CNRA-2018-018).
Mergili, M., Jaboyedoff, M., Pullarello, J., & Pudasaini, S. P. (2020). Back calculation of the 2017 Piz Cengalo-Bondo landslide cascade with r.avaflow: What we can do and what we can learn. *Natural Hazards and Earth System Sciences, 20*(2), 505–520. https://doi.org/10.5194/nhess-20-505-2020
Metz, M. R., Varner, J. M., Simler, A. B., Frangioso, K. M., & Rizzo, D. M. (2017). Implications of sudden oak death for wildland fire management. *Forest Phytophthoras, 7*(1), 30–44. https://doi.org/10.5399/osu/fp.7.1.4027
Moise, A., Smith, I., Brown, J. R., Colman, R., & Narsey, S. (2020). Observed and projected intra-seasonal variability of Australian monsoon rainfall. *International Journal of Climatology, 40*, 2310–2327. https://doi.org/10.1002/joc.6334
Moody, J. A., & Martin, D. A. (2001). Initial hydrologic and geomorphic response following a wildfire in the

Colorado Front Range. *Earth Surface Processes and Landforms, 26*(10), 1049–1070.

Moody, J. A., Shakesby, R. A., Robichaud, P. R., Cannon, S. H., & Martin, D. A. (2013). Current research issues related to post-wildfire runoff and erosion processes. *Earth-Science Reviews, 122*, 10–37.

Morán-Tejeda, E., López-Moreno, I., Stoffel, M., & Beniston, M. (2016). Rain-on-snow events in Switzerland: Recent observations and projections for the 21st century. *Climate Research, 71*, 111–125.

Morel, P., Trappmann, D., Corona, C., & Stoffel, M. (2015). Defining sample size and sampling strategy for dendrogeomorphic rockfall reconstructions. *Geomorphology, 236*, 79–89.

Moustakis, Y., Onof, C. J., & Paschalis, A. (2020). Atmospheric convection, dynamics and topography shape the scaling pattern of hourly rainfall extremes with temperature globally. *Communications Earth and Environment, 1*, 11.https://doi.org/10.1038/s43247-020-0003-0

Moustakis, Y., Papalexiou, S. M., Onof, C. J., & Paschalis, A. (2021). Seasonality, intensity, and duration of rainfall extremes change in a warmer climate. *Earth Future, 9*, e2020EF001824. https://doi.org/10.1029/2020EF001824

Oakley, N. S. (2021). A warming climate adds complexity to post-fire hydrologic hazard planning. *Earth Future, 9*, e2021EF002149. https://doi.org/10.1029/2021EF002149

Oakley, N. S., Lancaster, J. T., Hatchett, B. J., Stock, J., Ralph, F. M., Roj, S., & Lukashov, S. (2018). A 22-year climatology of cool season hourly precipitation thresholds conducive to shallow landslides in California. *Earth Interaction, 22*(14), 1–35.

Oakley, N. S., Lancaster, J. T., Kaplan, M. L., & Ralph, F. M. (2017). Synoptic conditions associated with cool season post-fire debris flows in the Transverse Ranges of southern California. *Natural Hazards and Earth System Sciences, 88*(1), 327–354.

O'Brien, T. A., et al. (2022). Increases in future AR count and size: Overview of the ARTMIP Tier 2 CMIP5/6 experiment. *Journal of Geophysical Research: Atmospheres, 127*, e2021JD036013.

O'Gorman, P., & Schneider, T. (2009). Scaling of precipitation extremes over a wide range of climates simulated with an idealized GCM. *Journal of Climate, 22*, 5676–5685. https://doi.org/10.1175/2009JCLI2701.1

Parks, S. A., Miller, C., Abatzoglou, J. T., Holsinger, L. M., Parisien, M. A., & Dobrowski, S. Z. (2016). How will climate change affect wildland fire severity in the western US? *Environmental Research Letters, 11*(3), 035002.

Pascale, S., Carvalho, L. M., Adams, D. K., Castro, C. L., & Cavalcanti, I. F. A. (2019). Current and future variations of the monsoons of the Americas in a warming climate. *Current Climate Change Reports, 5*, 125–144. https://doi.org/10.1007/s40641-019-00135-w

Payne, A. E., et al. (2020). Responses and impacts of atmospheric rivers to climate change. *Nature Reviews Earth & Environment, 1*, 143–157.

Pepin, N. C., et al. (2022). Climate changes and their elevational patterns in the mountains of the world. *Reviews of Geophysics, 60*, e2020RG000730.

Petley, D. (2010). On the impact of climate change and population growth on the occurrence of fatal landslides in South, East and SE Asia. *Quarterly Journal of Engineering Geology and Hydrogeology, 43*, 487–496.

Petrakov, D., Chernomorets, S., Viskhadzhieva, K., Dokukin, M., Savernyuk, E., Petrov, M., Erokhin, S., Tutubalina, O., Glazyrin, G., Shpuntova, A., & Stoffel, M. (2020). Putting the poorly documented 1998 GLOF disaster in Shakhimardan River valley (Alay Range, Kyrgyzstan/Uzbekistan) into perspective. *Science of the Total Environment, 724*, 138287.

Petrov, V., Chernomorets, S., Budarina, O., Savernyuk, E., & Leontyeva, T. (2017). Debris flow hazards for mountain regions of Russia: Regional features and key events. *Natural Hazards and Earth System Sciences, 88*, 199–235. https://doi.org/10.1007/s11069-017-2841-3

Picotte, J. J., et al. (2020). Changes to the monitoring trends in burn severity program mapping production procedures and data products. *Fire Ecology, 16*(1), 1–12.

Polade, S. D., Pierce, D. W., Cayan, D. R., Gershunov, A., & Dettinger, M. D. (2014). The key role of dry days in changing regional climate and precipitation regimes. *Science and Reports, 4*(1), 4364. https://doi.org/10.1038/srep04364

Prein, A. F., Rasmussen, R. M., Ikeda, K., Liu, C., Clark, M. P., & Holland, G. J. (2017). The future intensification of hourly precipitation extremes. *Nature Climate Change, 7*(1), 48–52.

Prenner, D., Hrachowitz, M., & Kaitna, R. (2019). Trigger characteristics of torrential flows from high to low alpine regions in Austria. *Science of the Total Environment, 658*, 958–972.

Quesada-Román, A., Fallas-López, B., Hernández-Espinoza, K., Stoffel, M., & Ballesteros-Cánovas, J. A. (2019). Relationships between earthquakes, hurricanes and landslides in Costa Rica. *Landslides, 16*, 1539–1550.

Radeloff, V. C., et al. (2018). Rapid growth of the US wildland-urban interface raises wildfire risk. *Proceedings of the National Academy of Sciences, 115*(13), 3314–3319.

Rajczak, J., & Schär, C. (2017). Projections of future precipitation extremes over Europe: A multimodel assessment of climate simulations. *Journal of Geophysical Research: Atmospheres, 122*(20), 10773–10810. https://doi.org/10.1002/2017JD027176

Ralph, F. M., et al. (2006). Flooding on California's Russian River: Role of atmospheric rivers. *Geophysical Research Letters, 33*, L13801. https://doi.org/10.1029/2006GL026689

Ralph, F. M., Dettinger, M. D., Cairns, M. M., Galarneau, T. J., & Eylander, J. (2018). Defining "atmospheric river": How the Glossary of Meteorology helped resolve a debate. *Bulletin of the American Meteorological Society, 99*(4), 837–839.

Ramos, A. M., Trigo, R. M., Liberato, M. L. R., & Tomé, R. (2015). Daily precipitation extreme events in the Iberian Peninsula and its association with atmospheric rivers. *Journal of Hydrometeorology, 16*(2), 579–597.

Rengers, F. K., et al. (2020). Landslides after wildfire: Initiation, magnitude, and mobility. *Landslides, 17*, 2631–2641. https://doi.org/10.1007/s10346-020-01506-3

Rickenmann, D., & Zimmermann, M. (1993). The 1987 debris flows in Switzerland: Documentation and analysis. *Geomorphology, 8*, 175–189.

Ruffault, J., et al. (2020). Increased likelihood of heat-induced large wildfires in the Mediterranean Basin. *Science and Reports, 10*(1), 13790. https://doi.org/10.1038/s41598-020-70069-z

Rutz, J. J., Steenburgh, W. J., & Ralph, F. M. (2014). Climatological characteristics of atmospheric rivers and their inland penetration over the Western United States. *Monthly Weather Review, 142*(2), 905–921.

Sabeerali, C., & Ajayamohan, R. (2017). On the shortening of Indian summer monsoon season in a warming scenario. *Climate Dynamics, 50*, 1609–1624. https://doi.org/10.1007/s00382-017-3709-7

Salzmann, M., & Cherian, R. (2015). On the enhancement of the Indian summer monsoon drying by Pacific multidecadal variability during the latter half of the twentieth century. *Journal of Geophysical Research: Atmospheres, 120*, 9103–9118. https://doi.org/10.1002/2015JD023313

Sankey, J. B., Kreitler, J., Hawbaker, T. J., McVay, J. L., Miller, M. E., Mueller, E. R., Vaillant, N. M., Lowe, S. E., Sankey, T. T. (2017). Climate, wildfire, and erosion ensemble foretells more sediment in western USA watersheds, *Geophysical Research Letters,, 44*, 8884–8892. https://doi.org/10.1002/2017GL073979

Sanogo, S. A., Fink, H., Omotosho, J. A., Ba, A., Redl, R., & Ermert, V. (2015). Spatio-temporal characteristics of the recent rainfall recovery in West Africa. *International Journal of Climatology, 35*, 4589–4605. https://doi.org/10.1002/joc.4309

Santi, P. M., & Morandi, L. (2013). Comparison of debris-flow volumes from burned and unburned areas. *Landslides, 10*, 757–769.

Sarr, M., Am, Z., Seidou, O., Bryant, C. R., & Gachon, P. (2013). Recent trends in selected extreme precipitation indices in Senegal–a changepoint approach. *Journal of Hydrology, 505*, 326–334. https://doi.org/10.1016/j.jhydrol.2013.09.032

Sass, O. (2005). Rock moisture measurements: Techniques, results, and implications for weathering. *Earth Surface Processes and Landforms, 30*, 359–374.

Sattler, K. (2014). *Periglacial preconditioning of debris flows in the Southern Alps, New Zealand*. Springer Theses. https://doi.org/10.1007/978-3-319-35074-5

Schauwecker, S., Gascón, E., Park, S., Ruiz-Villanueva, V., Schwarb, M., Sempere-Torres, D., Stoffel, M., Vitolo, C., & Rohrer, M. (2019). Anticipating cascading effects of extreme precipitation with pathway schemes—three case studies from Europe. *Environment International, 127*, 291–304.

Schauwecker, S., Schwarb, M., Rohrer, M., & Stoffel, M. (2021). Operational prediction of heavy precipitation with increased lead times—challenges and implications for risk preparedness in mountain environments. *Weather and Climate Extremes, 34*, 100372.

Schneuwly-Bollschweiler, M., & Stoffel, M. (2012). Hydrometeorological triggers of periglacial debris flows—a reconstruction dating back to 1864. *Journal of Geophysical Research: Earth Surface, 117*, F02033.

Schwanghart, W., Worni, R., Huggel, C., Stoffel, M., & Korup, O. (2016). Uncertainties at the water-energy nexus—the case of Himalayan hydropower and glacial lake outburst floods. *Environmental Research Letters, 11*, 074005.

Seinova, I. B., Andreev, B., Krylenko, N., Chernomorets, S. S., & Lomonosov, M. V. (2011). Regional short-term forecast of debris flow initiation for glaciated high mountain zone of the Caucasus. *Italian Journal of Engineering Geology and Environment*, 1003–1011. https://doi.org/10.4408/IJEGE.2011-03.B-109

Seth, A., Rauscher, S. A., Biasutti, M., Giannini, A., Carmargo, S. J., & Rojas, M. (2013). CMIP5 projected changes in the annual cycle of precipitation in monsoon regions. *Journal of Climate, 26*, 7328–7351. https://doi.org/10.1175/JCLI-D-12-00726.1

Sharma, A. R., & Déry, S. J. (2020). Linking atmospheric rivers to annual and extreme river runoff in British Columbia and South-eastern Alaska. *Journal of Hydrometeorology, 21*(11), 2457–2472. https://doi.org/10.1175/JHM-D-19-0281.s1

Shugar, D. H., et al. (2021). A massive rock and ice avalanche caused the 2021 disaster at Chamoli, Indian Himalaya. *Science, 373*(6552), 300–306.

Sidle, R. C. (1991). A conceptual model of changes in root cohesion in response to vegetation management. *Journal of Environmental Quality, 20*(1), 43–52.

Sidle, R. C. (2005). Influence of forest harvesting activities on debris avalanches and flows. In *Debris-flow hazards and related phenomena* (pp. 387–409). Springer.

Silins, U., Stone, M., Emelko, M. B., & Bladon, K. D. (2009). Sediment production following severe wildfire and post-fire salvage logging in the Rocky Mountain headwaters of the Oldman River Basin, Alberta. *CATENA, 79*(3), 189–197.

Smith, C., Hatchett, B. J., & Kaplan, M. (2018). A surface observation based climatology of diablo-like winds in California's wine country and western sierra Nevada. *Fire, 1*(2), 25. https://doi.org/10.3390/fire1020025

Staley, D. M., Negri, J. A., Kean, J. W., Laber, J. M., Tillery, A. C., & Youberg, A. M. (2016). *Updated logistic regression equations for the calculation of post-fire debris-flow likelihood in the western United States* (p. 13). US Department of the Interior, US Geological Survey.

Stavros, E. N., Abatzoglou, J. T., McKenzie, D., & Larkin, N. K. (2014). Regional projections of the likelihood of very large wildland fires under a changing climate in the contiguous western United States. *Climate Change, 126*, 455–468.

Stiny, J. (1910). *Die Muren: Versuch einer Monographie mit besonderer berücksichtigung der verhältnisse in Den Tiroler Alpen*. Wagner.

Stoffel, M. (2022). Blowing hot and cold: Glacier microclimate can help understand impacts of climate change on forests communities. *Geophysical Research Letters, 49*, e2022GL100883. https://doi.org/10.1029/2022GL098574

Stoffel, M., & Beniston, M. (2006). On the incidence of debris flows from the early Little Ice Age to a future greenhouse climate: A case study from the Swiss Alps. *Geophysical Research Letters, 33*, L16404.

Stoffel, M., & Corona, C. (2018). Future winters glimpsed in the Alps. *Nature Geoscience, 11*, 458–460.

Stoffel, M., & Huggel, C. (2012). Effects of climate change on mass movements in mountain environments. *Progress in Physical Geography, 36*, 421–439.

Stoffel, M., Bollschweiler, M., & Beniston, M. (2011). Rainfall characteristics for periglacial debris flows in the Swiss Alps: Past incidences-potential future evolutions. *Climate Change, 105*, 263–280.

Stoffel, M., et al. (2005a). 400 years of debris flow activity and triggering weather conditions: Ritigraben, Valais, Switzerland. *Arctic, Antarctic, and Alpine Research, 37*(3), 387–395.

Stoffel, M., Lièvre, I., Monbaron, M., & Perret, S. (2005b). Seasonal timing of rockfall activity on a forested slope at Täschgufer (Valais, Swiss Alps)—a dendrochronological approach. *Zeitschrift Für Geomorphologie, 49*(1), 89–106.

Stoffel, M., Conus, D., Grichting, M. A., Lièvre, I., & Maître, G. (2008). Unraveling the patterns of late Holocene debris-flow activity on a cone in the Swiss Alps: Chronology, environment and implications for the future. *Global and Planetary Change, 60*, 222–234.

Stoffel, M., Tiranti, D., & Huggel, C. (2014a). Climate change impacts on mass movements—case studies from the European Alps. *Science of the Total Environment, 493*, 1255–1266.

Stoffel, M., Mendlik, T., Schneuwly-Bollschweiler, M., & Gobiet, A. (2014b). Possible impacts of climate change on debris-flow activity in the Swiss Alps. *Climate Change, 122*, 141–155.

Stokes, C. R., Gurney, S. D., Shahgedanova, M., & Popovnin, V. (2006). Late-20th-century changes in glacier extent in the Caucasus Mountains, Russia/Georgia. *Journal of Glaciology, 52*(176), 99–109. https://doi.org/10.3189/172756506781828827

Su, T., et al. (2022). Mechanisms responsible for interdecadal variation of the thermodynamic component of East Asian summer monsoon moisture transport. *International Journal of Climatology, 42*(9), 4925–4939. https://doi.org/10.1002/joc.7512

Swain, D. L. (2021). A shorter, sharper rainy season amplifies California wildfire risk. *Geophysical Research Letters, 48*, e2021GL092843. https://doi.org/10.1029/2021GL092843

Taylor, C. M., et al. (2017). Frequency of extreme Sahelian storms tripled since 1982 in satellite observations. *Nature, 544*, 475–478. https://doi.org/10.1038/nature22069

Thayyen, R. J., et al. (2022). Hanging glacier avalanche (Raunthigad–Rishiganga) and debris flow disaster on 7 February 2021, Uttarakhand, India: A preliminary assessment. *Natural Hazards and Earth System Sciences, 114*, 1939–1966. https://doi.org/10.1007/s11069-022-05454-0

Trenberth, K. E., Fasullo, J. T., & Kiehl, J. (2009). Earth's global energy budget. *Bulletin of the American Meteorological Society, 90*(3), 311–323. https://doi.org/10.1175/2008BAMS2634.1

Turkington, T., Remaître, A., Ettema, J., Hussin, H., & van Westen, C. (2016). Assessing debris flow activity in a changing climate. *Climate Change, 137*(1), 293–305.

Underwood, S. J., & Schultz, M. D. (2004). Patterns of cloud-to-ground lightning and convective rainfall associated with post wildfire flash floods and debris flows in complex terrain of the western United States. *Journal of Hydrometeorology, 5*(5), 989–1003.

USGCRP. (2018). *Impacts, risks, and adaptation in the United States: Fourth national climate assessment, Volume II* (1515 pp.). [D. R. Reidmiller, C. W. Avery, D. R. Easterling, K. E. Kunkel, K. L. M. Lewis, T. K. Maycock, & B. C. Stewart (Eds.)]. U.S. Global Change Research Program, Washington, DC, USA. https://doi.org/10.7930/NCA4.2018

van den Heuvel, F., Goyette, S., Rahman, K., & Stoffel, M. (2016). Circulation patterns related to debris-flow triggering in the Zermatt valley in current and future climates. *Geomorphology, 272*, 127–136.

van der Maarel, E. (1988). Vegetation dynamics: Patterns in time and space. In J. Miles, W. Schmidt, & E. van der Maarel (Eds.), *Temporal and spatial patterns of vegetation dynamics*. Advances in vegetation science 9. Springer. https://doi.org/10.1007/978-94-009-2275-4_1

Vera, C. S., & Diaz, L. (2015). Anthropogenic influence on summer precipitation trends over South America in CMIP5 models. *International Journal of Climatology, 35*, 3172–3177. https://doi.org/10.1002/joc.4153

Walter, F., et al. (2020). Direct observations of a three million cubic meter rock-slope collapse with almost immediate initiation of ensuing debris flows. *Geomorphology, 351*, 106933. https://doi.org/10.1016/j.geomorph.2019.106933

Wang, B., et al. (2021). Monsoons Climate Change Assessment. *Bulletin of the American Meteorological Society, 102*(1), E1–E19.

Wasko, C., Lu, W. T., & Mehrotra, R. (2018). Relationship of extreme precipitation, dry-bulb temperature, and dew point temperature across Australia. *Environmental Research Letters, 13*(7), 074031.

Wayman, R. B., & Safford, H. D. (2021). Recent bark beetle outbreaks influence wildfire severity in mixed-conifer forests of the Sierra Nevada, California, USA. *Ecological Applications, 31*(3), e02287. https://doi.org/10.1002/eap.2287

Wei, W., Zhang, R., Wen, M., Rong, X., & Li, T. (2014). Impact of Indian summer monsoon on the South Asian High and its influence on summer rainfall over China. *Climate Dynamics, 43*(5), 1257–1269.

Westerling, A. L. (2018). *Wildfire simulations for California's fourth climate change assessment: Projecting changes in extreme wildfire events with a warming climate*. California's Fourth Climate Change Assessment, California Energy Commission. Publication Number: CCCA4-CEC-2018-014.

Westerling, A. L., Hidalgo, H. G., Cayan, D. R., & Swetnam, T. W. (2006). Warming and earlier spring increase western US forest wildfire activity. *Science, 313*, 940–943.

Westra, S., et al. (2014). Future changes to the intensity and frequency of short-duration extreme rainfall. *Reviews of Geophysics, 52*(3), 522–555. https://doi.org/10.1002/2014RG000464

Williams, A. P., et al. (2019). Observed impacts of anthropogenic climate change on wildfire in California. *Earth Future, 7*, 892–910. https://doi.org/10.1029/2019EF001210

Winkler, A. J., et al. (2021). Slowdown of the greening trend in natural vegetation with further rise in atmospheric CO_2. *Biogeosci, 18*, 4985–5010. https://doi.org/10.5194/bg-18-4985-2021

Worni, R., Huggel, C., Clague, J. J., Schaub, Y., & Stoffel, M. (2014). Coupling glacial lake impact, dam breach, and flood processes: A modeling perspective. *Geomorphology, 224*, 161–176. https://doi.org/10.1016/j.geomorph.2014.06.031

Worni, R., Stoffel, M., Huggel, C., Volz, C., Casteller, A., & Luckman, B. (2012). Analysis and dynamic modeling of a moraine failure and glacier lake outburst flood at Ventisquero Negro, Patagonian Andes (Argentina). *Journal of Hydrology, 444–445*, 134–145.

Wotton, B. M., Nock, C. A., & Flannigan, M. D. (2010). Forest fire occurrence and climate change in Canada. *International Journal of Wildland Fire, 19*(3), 253–271.

Zaginaev, V., Ballesteros-Cánovas, J. A., Petrakov, D., Erokhin, S., Meleshko, A., & Stoffel, M. (2019). Geomorphic control on regional glacier lake outburst flood and debris flow activity over northern Tien Shan. *Global and Planetary Change, 176*, 50–59.

Zhang, X., Zwiers, F. W., Li, G., Wan, H., & Cannon, A. J. (2017). Complexity in estimating past and future extreme short-duration rainfall. *Nature Geoscience, 10*(4), 255–259.

Zhang, X., et al. (2019). Changes in temperature and precipitation across Canada, Chapter 4. In E. Bush & D. S. Lemmen (Eds.), *Canada's changing climate report* (pp. 112–193). Government of Canada.

Zheng, G., et al. (2021a). Increasing risk of glacial lake outburst floods from future Third Pole deglaciation. *Nature Climate Change, 11*, 411–417. https://doi.org/10.1038/s41558-021-01028-3

Zheng, G., Bao, A., Allen, S. K., Ballesteros-Cánovas, J. A., Jiapaer, G., & Stoffel, M. (2021b). Numerous unreported glacial lake outburst floods in the Third Pole revealed by high-resolution satellite data and geomorphological evidence. *Science Bulletin, 66*, 1270–1273.

Zhou, T., Gong, D., Li, J., & Li, B. (2009). Detecting and understanding the multi-decadal variability of the east Asian summer monsoon—recent progress and state of affairs. *Meteorologische Zeitschrift, 18*, 455–467. https://doi.org/10.1127/0941-2948/2009/0396

Zimmermann, M., & Haeberli, W. (1992). Climatic change and debris flow activity in high mountain areas: A case study in the Swiss Alps. *Catena Supplement, 22*, 49–72.

Zimmermann, M., Mani, P., & Gamma, P. (1997). *Murganggefahr und Klimaänderung – ein GIS-basierter Ansatz* (162 pp.). Vdf Hochschulverlag.

Post-Wildfire Debris Flows

11

Joseph E. Gartner, Jason W. Kean, Francis K. Rengers, Scott W. McCoy, Nina Oakley and Gary Sheridan

Abstract

Post-wildfire debris flows pose severe hazards to communities and infrastructure near and within recently burned mountainous terrain. Intense heat of wildfires changes the runoff characteristics of a watershed by combusting the vegetative canopy, litter, and duff, introducing ash into the soil and creating water repellant soils. Following wildfire, rainfall on bare ground is less able to infiltrate into the fire-altered soils and overland flow is less impeded by vegetation. Rainfall runoff in recently burned areas can erode hillslopes owing to the removal of soil binding organic matter near the soil surface by fire. In channels, loose, dry-ravel deposits composed of sand and gravel are readily entrained by concentrated runoff in channels. Entrainment of soil on hillslopes and in channels bulks up the sediment concentration of the rainfall runoff to generate debris flows capable of transporting boulders and large woody debris. Post-wildfire debris flows can be triggered by rainfall conditions that would typically produce little runoff during unburned conditions. The primary rainfall trigger for post-wildfire debris flows is high intensity rainfall during short duration convective rainstorms or periods of high rainfall intensity embedded within a long-duration frontal storm. Numerous observations of debris flows triggered by storms lasting less than an hour following periods of little to no rainfall indicate that antecedent rainfall is not a requirement for initiation of post-wildfire debris flows. Post-wildfire debris-flow hazard assessment entails estimating probability and magnitude of debris flows in the burned area, estimating debris-flow runout and intensity, and defining rainfall intensity-duration thresholds for debris-flow initiation. In the United States, probability and magnitude is estimated using empirically derived models largely based on data collected in southern California. The models provide maps to identify watersheds

Any use of trade, firm, or product names is for descriptive purposes only and does not imply endorsement by the U.S. Government.

J. E. Gartner (✉)
BGC Engineering Inc., Golden, CO, USA
e-mail: jgartner@bgcengineering.com

J. W. Kean · F. K. Rengers
U.S. Geological Survey, Golden, CO, USA

S. W. McCoy
Department of Geological Sciences and Engineering, University of Nevada, Reno, USA

N. Oakley
California Geological Survey, Sacramento, CA, USA

G. Sheridan
University of Melbourne, Parkville, VIC, Australia

M. Jakob et al. (eds.), *Advances in Debris-flow Science and Practice*, Geoenvironmental Disaster Reduction,
https://doi.org/10.1007/978-3-031-48691-3_11

and drainage paths where post-wildfire hazards are most pronounced. Rainfall intensity-duration thresholds can be incorporated into flood hazard forecasting tools. Currently, work is underway to identify how to best implement debris-flow runout models in burned areas with efficiency and accuracy. Post-wildfire debris flows have been a long-recognized process in the Transverse Ranges of southern California; however, climate change is driving more frequent wildfires to burn more mountainous terrain throughout the western United States and worldwide. As a result, post-wildfire debris flows are becoming a more common threat in areas where they were once infrequent. As the threat of post-wildfire debris flow expands into new areas, evaluating the hazard becomes challenging because the degree to which wildfire increases debris-flow susceptibility varies from region to region. This chapter summarizes the knowledge to date for evaluating post-wildfire debris-flow susceptibility and hazard assessment. We summarize the characteristics of wildfire burn severity, topography, underlying soil and geology, and rainfall conditions that contribute to making a watershed most likely to produce post-wildfire debris flows. Methods for hazard assessment in the United States and other countries are summarized. We highlight knowledge gaps for how post-wildfire debris-flow susceptibility varies throughout the western United States and worldwide and identify research needs to improve hazard assessment methods in different geographies.

11.1 Introduction

Post-wildfire debris flows pose severe hazards to communities and infrastructure located near and within recently burned mountainous terrain. The intense heat of wildfires can greatly change the hydrologic response of a watershed by removing the vegetative canopy, litter, and duff (Moody et al., 2013) and changing the rainfall infiltration characteristics of the soil through the development of water-repellant soils (Debano, 1981; Doerr et al., 2000) and introduction of ash (Ebel et al., 2012; Kinner & Moody, 2010; Rompkins et al., 1990). As a result of these changes, increased water runoff and sediment erosion from hillslopes and channels can transform into debris flows (Cannon et al., 2003a; Kean et al., 2013; McGuire et al., 2017) in response to ordinary rainfall events that would typically produce little runoff from an unburned watershed (Cannon et al., 2008; Staley et al., 2020). This transition from sediment-laden floods to debris flows is particularly concerning owing to much greater discharge of debris flows, which can be over an order of magnitude larger than a comparable water flood at the same point in the drainage network (Kean et al., 2016; VanDine, 1985). Post-wildfire debris flows can contaminate critical water supplies, threaten aquatic species, damage infrastructure, and in the most extreme cases, cause fatalities.

This chapter provides an update to the Cannon and Gartner (2005) chapter in Jakob and Hungr (2005) on wildfire-related debris flows from a hazards perspective, which summarized fire-related debris-flow initiation processes, susceptibility to post-wildfire debris flows, and magnitude of post-wildfire debris-flow response.

Since 2005, post-wildfire debris flows have continued to have severe effects in the western United States (U.S.) and worldwide (e.g., Cannon et al., 2010a; Garcia-Ruiz et al., 2013; Jordan, 2015; Nyman et al., 2011; Wang et al., 2021). A particularly devastating example is the 2018 Montecito, California debris-flow event, which resulted in 23 fatalities, 167 injuries, and damage to 408 homes (Kean et al., 2019; Lancaster et al., 2021). The development of new tools for post-wildfire emergency assessments has enabled rapid assessments of post-wildfire debris-flow hazards (Gartner et al., 2014; Staley et al., 2016) and guidelines for issuing warnings based on forecasted rainfall (Cannon et al., 2008, 2011).

Despite advances in hazard assessment, many challenges persist in assessing post-wildfire debris-flow hazards, and the combination of increased wildfire activity and population growth along the wildland-urban interface has heightened the exposure of communities to debris-flow impacts (Cannon & DeGraff, 2009; Radeloff et al., 2018). Owing to warming temperatures throughout western North America driven by climate change, wildfires are becoming more frequent, widespread, and destructive (e.g., Iglesias et al., 2022; Running, 2006; Westerling et al., 2006; Williams et al., 2019). As a result, wildfires are burning in mountainous terrain where fires have been infrequent or non-existent in the past (e.g., coastal rainforests of the northwestern United States and coastal mountains of British Columbia (B.C.)) (Alizadeh et al., 2021). Very arid areas, such as the Sonoran Desert in the southwestern United States, have also experienced an increase in fire activity due to increased fuels provided by invasive grasses (Fusco et al., 2019). The susceptibility of these watersheds to produce debris flows is less understood than in areas with long histories of debris flows following wildfire (e.g., southern California). Because current post-wildfire debris-flow hazard assessment models have been developed largely based on data collected in southern California (e.g., Gartner et al., 2014; Staley et al., 2017), confidence and accuracy in the results is lower when applying these models to other regions in western North America or worldwide.

This chapter reviews our current understanding of post-wildfire debris-flow susceptibility, initiation mechanisms, rainfall triggers, hazard assessment methods and case histories, and identifies future potential research directions for developing new hazard assessment models for reducing future impacts of post-wildfire debris flows.

11.2 Post-Wildfire Debris-Flow Susceptibility

The susceptibility of a watershed to produce a debris flow following wildfire is dependent on the burn severity in the watershed, terrain, soils, geology (including glaciation history), and rainfall (e.g., Staley et al., 2017). As wildfires have become more frequent and larger in the western United States (Fig. 11.1), observations of post-wildfire debris flows have expanded from southern California to the rest of the western United States and British Columbia, Canada (Fig. 11.2), Australia (e.g., Nyman et al., 2011), Spain (e.g., Garcia-Ruiz et al., 2013), Portugal (e.g., Melo & Zêzere, 2017), and China (e.g., Wang et al., 2021).

Susceptibility to post-wildfire debris flows varies throughout the western United States and the world. Post-wildfire debris flows are typically generated in areas that meet the following criteria:

1. A large portion of the watershed has burned at moderate to high severity,
2. The watershed has confined channels that are continuously steep from the initiation area to the watershed outlet,
3. The watershed has sources of sediment that are readily erodible,
4. High intensity rainfall affects the portions of the watershed where debris flows initiate.

For example, post-wildfire debris flows are more common in the Transverse Ranges of southern California than the Front Range mountains in Colorado (Staley et al., 2017). Moody and Martin (2009b) attribute this difference to California having a more frequent fire regime (<10 years versus 15–30 years in Colorado), slightly higher 2-year 30-min rainfall intensity, and steeper slopes (median slope 0.59 versus 0.31 in Colorado). In southern California, many post-wildfire debris flows are triggered by frontal storms that transport moisture from over the Pacific Ocean to the Transverse Ranges, often in the form of atmospheric rivers. The frontal storms have the potential to produce intense rainfall over an entire burned area and trigger numerous debris flows in multiple watersheds. Fine-scale (mesoscale) features within frontal storms can also produce high intensity rainfall that affects only a portion of the burned area (e.g., the 2018 post-wildfire debris flows in Montecito, California). In contrast, convective rainstorms in the Rocky Mountain ranges in Arizona, New Mexico, Colorado, Idaho, and Montana typically have a smaller spatial footprint and a less-widespread debris-flow response

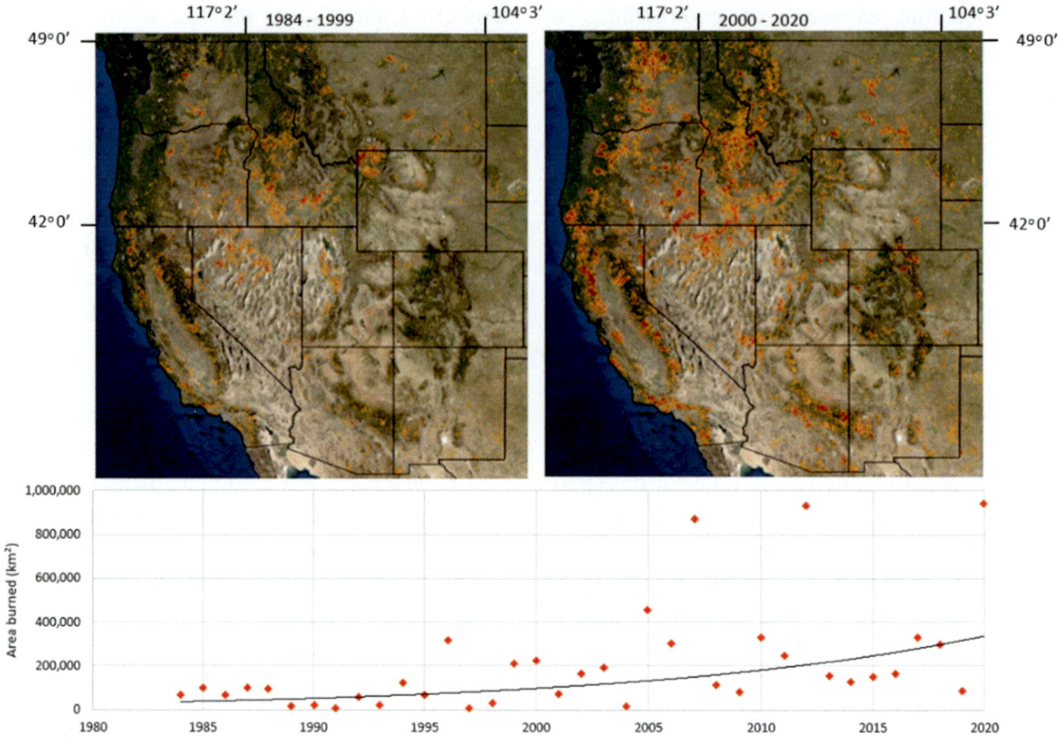

Fig. 11.1 The area burned in the western United States has increased since the 1980s. The images of the western United States show areas burned in red and orange between 1984 and 1999 (upper left) and between 2000 and 2020 (upper right). Data are from the monitoring trends in burn severity database (accessed online at https://mtbs.gov). Burned areas are shown in red and orange

in comparison to the Transverse Ranges of southern California. In northern California, frontal storms have delivered high rainfall amounts and intensities across broad regions that fully impact burned areas; however, few debris flows have been triggered, which is likely due to a lower watershed susceptibility to producing post-wildfire debris flows (East et al., 2021; Perkins et al., 2022; Suter, 2022). Many of the post-wildfire debris flows in the Sierra Nevada Mountains of California have been triggered by warm season convective rainfall (e.g., De Graff et al., 2022; Wagner et al., 2012).

11.2.1 Burn Severity

The intense heat of wildfire is responsible for the changes to the hydrologic response of a watershed to rainfall. Wildfire intensity is a measure of the energy released during the fire and can be approximated by the burn severity, which is a measure of the change in vegetation due to the passage of the wildfire (Keeley, 2009). Burn severity maps can be rapidly generated using remote sensing techniques that compare the difference between the pre- and post-wildfire conditions (Key & Benson, 2006). The analysis provides a map showing areas of low, moderate, and high burn severity, which is based on the differenced normalized burn ratio (dNBR) and verified by field investigation (Keeley, 2009; Key & Benson, 2006; Fig. 11.3). Burn severity maps are critical pieces of information for implementing models that assess the likelihood and magnitude of post-wildfire debris flows in recently burned watersheds.

Soils burned at high temperatures can be characterized by a soil burn severity map, which can provide an indication of how susceptible the

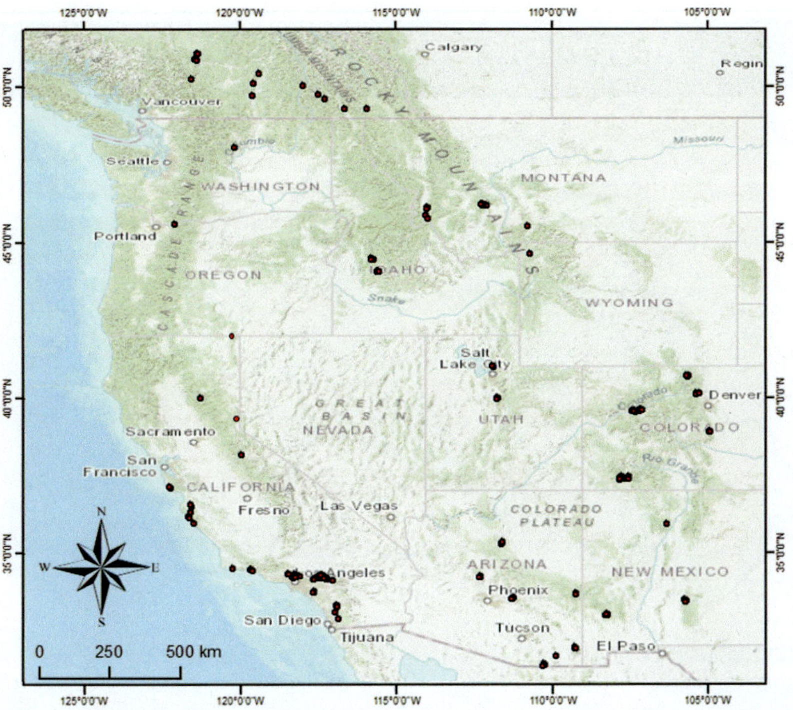

Fig. 11.2 Documented post-wildfire debris flows in the western United States and Canada between 1914 and 2021. Location data were provided from various sources summarized in Gartner et al., (2014), Jordan (2015), Staley et al., (2016); and U.S. Geological Survey event summaries accessed at https://www.usgs.gov/programs/landslide-hazards/science/debris-flow-event-summaries)

soil is to runoff and erosion (Moody & Martin, 2009a; Moody et al., 2008). Spatially continuous hillslopes with high burn severity experience more erosion than slopes that have patchy distributions of high burn severity (Moody & Martin, 2009a). Several studies have identified a statistical relation between the area or proportion of a watershed burned at moderate and high burn severity to the likelihood and magnitude of a post-wildfire debris flow (Cannon et al., 2010a; Gartner et al., 2008, 2014; Staley et al., 2016).

The combustion of the tree canopy, litter, duff, and soil binding root matter and other organics in the soil increases post-wildfire debris-flow susceptibility for various reasons. Removal of vegetation by fire reduces canopy interception of rainfall (Stoof et al., 2012) and increases the amount of bare ground that is susceptible to erosion by direct raindrop impacts (Benavides-Solario & MacDonald, 2005; Larsen et al., 2009; McGuire et al., 2016). Rainsplash can dislodge soil particles and increase sediment concentrations in hillslope runoff. Reductions in forest litter and duff due to vegetation removal can also reduce the overall surface roughness, leading to increased connectivity of runoff and increased runoff velocities immediately after a fire, which in turn acts to increase erosion (Hoch et al., 2021; McGuire et al., 2016; Moody & Kinner, 2006; Tang et al., 2019; Wilson et al., 2020).

Even without the aid of water, severely burnt slopes can become unstable and experience the downslope movement of sediment under dry conditions, commonly called dry ravel. Dry ravel that accumulates in channel beds can increase following fire because sediment that was trapped behind vegetation on hillslopes is free to ravel downslope (Florsheim et al., 1991; Lamb et al., 2011). Large deposits of dry ravel in channels can provide ample sediment for entrainment and bulking of water flows that

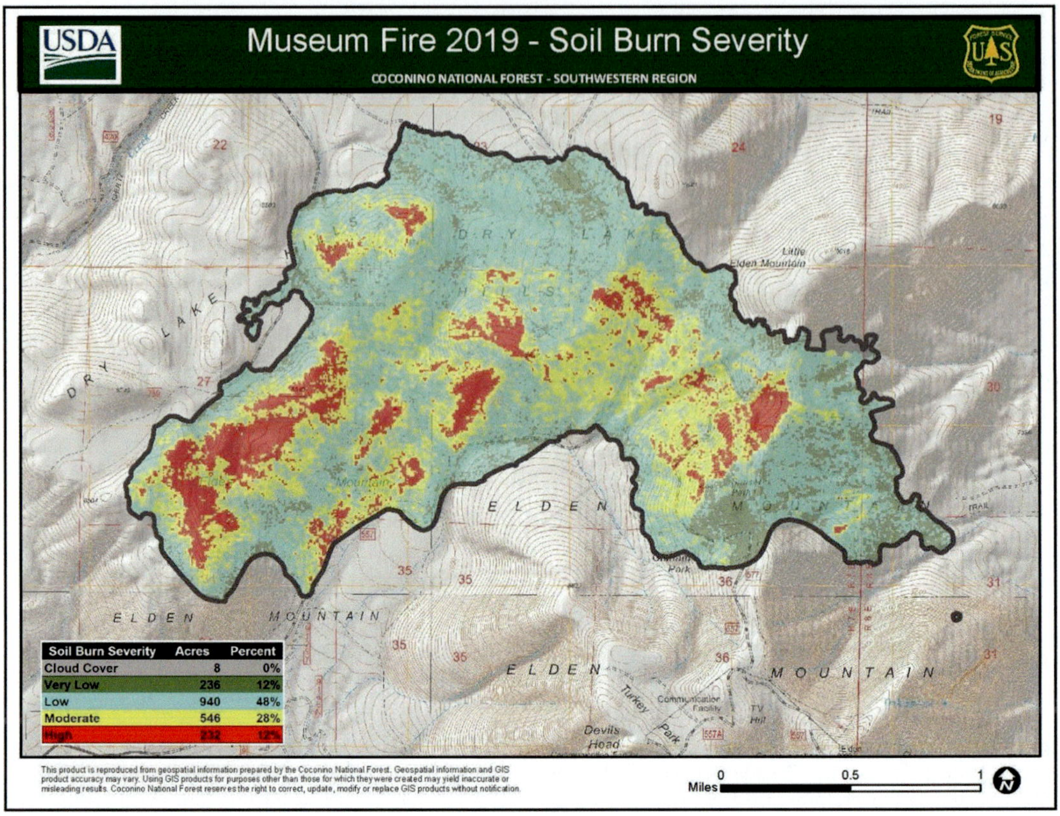

Fig. 11.3 Map of soil burn severity for the 2019 Museum Fire near Flagstaff, Arizona (accessed online at https://www.weather.gov/fgz/MuseumFire2019). Steep slopes burned at moderate and high severity are generally the portions of the burn area most susceptible to debris flows

transition into debris flows (Cannon et al., 2001; DiBiase & Lamb, 2020; Wells, 1987). Dry ravel is particularly common after wildfires in the steep, tectonically active San Gabriel Mountains of southern California (e.g., Fig. 11.4). Dry ravel is less commonly observed in burn areas with slopes at or below the angle of repose. Burn areas without substantial dry ravel deposits can still be susceptible to debris flow.

The rate at which water can infiltrate into soils is frequently reduced after wildfire due to fire-induced hydrophobicity (DeBano, 2000; DeBano et al., 1979), hyper-dry conditions (Moody & Ebel, 2012), and soil-sealing (Larsen et al., 2009). The intense heat of wildfire can vaporize hydrophobic compounds in the litter and duff, which then migrate downwards into the soil profile along a temperature gradient and solidify at cooler horizons to create or enhance a hydrophobic layer a few centimeters below the soil surface (DeBano, 1981; Doerr et al., 2000). Subsequent rainfall that falls on the soil is less able to infiltrate into soil and leads to amplified runoff (Kinner & Moody, 2008). Hillslope runoff erodes rills, the sediment from which converges into channels where bed and bank material can be eroded to supply ample sediment to a debris flow. This eroded sediment can either be progressively bulked into a debris flow (Cannon et al., 2003a) or can be re-deposited in localized low-gradient channel reaches and subsequently re-mobilized during an in-channel failure of the sediment package to initiate a debris flow (Kean et al., 2013; McGuire et al., 2017).

Fig. 11.4 Photograph of dry ravel deposits partially filling the channel by a monitoring station in the 2016 Fish Fire, Los Angeles County, California. The photograph was taken between the fire and the first substantial rainstorm after the fire. These deposits were subsequently entrained by a series of flows during winter rainy season after the fire (Rengers et al., 2021). Photograph by U.S. Geological Survey

11.2.2 Terrain

Watersheds that are prone to producing debris flows during unburned conditions are most susceptible to producing debris flows after wildfire. Debris-flow prone watersheds can be identified based on watershed morphometrics of watershed length and Melton ratio (Melton, 1957; Wilford et al., 2004). Debris flows are typically initiated on slopes steeper than about 27° and travel down channels with slopes 10°–20° prior to depositing material (Hungr et al., 2014; VanDine, 1996). More thorough review of terrain characteristics that are associated with debris-flow initiation are covered elsewhere in this book.

Slope-area thresholds for debris-flow initiation were identified by Cannon et al. (2001) following the 1994 South Canyon Fire in Colorado. Cannon et al. (2001) identified minimum thresholds for debris-flow initiation based on the critical upslope contributing area (A_{cr}, in m^2) and average gradient of the contributing area (θ). The thresholds are defined for debris-flow initiations in the sedimentary rocks of the Maroon Formation with values of θ between 27° and 42°:

$$A_{cr}(\tan\theta)^3 = 200 \quad (11.1)$$

and for debris-flow initiation in older landslide deposits values of θ between 14° and 37°:

$$A_{cr}(\tan\theta)^3 = 300 \quad (11.2)$$

Empirical models for debris-flow probability and magnitude (e.g., Cannon et al., 2010a;

Gartner et al., 2014; Staley et al., 2016) have correlated post-wildfire debris-flow probability and magnitude with variables describing watershed relief, watershed ruggedness and the proportion of burned slopes steeper than 23°. The presence of historical debris-flow deposits is also a good indicator of the potential susceptibility of catchments to debris flows.

11.2.3 Soils and Geology

Underlying geological conditions have not been shown to have a substantial effect on the susceptibility of a burned watershed to produce debris flows and debris floods. Cannon and Gartner (2005) evaluated a database of burned watersheds that both produced and did not produce debris flows and identified that debris flows were triggered in a broad range of physiographies underlain by sedimentary, igneous, and metamorphic rock types. Within this dataset, most of the debris flows initiated by runoff-dominated erosion were underlain by sedimentary and metamorphic rock types (Cannon & Gartner, 2005). In Australia, adjacent observations of burned watersheds with granitic and sedimentary/metamorphic geologies found that debris flows were initiated in the watersheds with sedimentary/metamorphic rock types whereas floods occurred in watersheds with granitic rock where bedrock outcrops limited sediment supply (Sheridan, unpublished data).

Multivariate logistic regression analyses have identified unburned soil properties to be correlated to post-wildfire debris-flow likelihood (Cannon et al., 2010a; Staley et al., 2016). The soil properties evaluated by these analyses were derived from the State Soil Geographic Database (STATSGO) (Schwartz & Alexander, 1995) and the best-performing models included variables for the percent clay content, percent organic matter and liquid limit (Cannon et al., 2010a), and the soil erodibility (or K-factor) (Staley et al., 2018). Moody and Martin (2009b) were unable to find a link between sediment yields and soil erodibility as defined by the STATSGO database. They attribute the lack of correlation to be possibly due to the coarse scale of the STATSGO database, which averages values over a 1-km^2 area.

In southeast Australia, poorly developed soils with little biological activity and soil structure are more likely to produce debris flows than burned areas with well-structured forest soils where high infiltration rates can be maintained despite the presence of fire-induced water repellency (Noske et al., 2016; Sheridan et al., 2015; van der Sant et al., 2018).

Sediment availability has a large influence on the generation of post-wildfire debris flows. In Canada, post-wildfire debris flows are often smaller in magnitude than in the Transverse Ranges of southern California. Post-wildfire debris-flow volume estimates made using the Gartner et al. (2014) model, which is largely based on data collected in southern California, were two to four times greater than volumes measured in southern British Columbia burned areas by Jordan (2015). One possible explanation for the difference in estimated and predicted volumes is the presence of over-consolidated glacial till in many of the burned areas in Canada, which is less erodible than non-glaciated soils in the United States (Jordan, written communication, 2015). Similarly, areas in central and northern California that have notably less available sediment than in the Transverse Ranges in southern California produce markedly smaller debris flows (Neptune et al., 2021) or only produce floods with relatively small sediment yields (East et al., 2021).

11.2.4 Rainfall

Post-wildfire debris flows are most frequently triggered by high-intensity rainfall (Cannon et al., 2008, 2011) that may only last for minutes (e.g., Kean et al., 2011). Convective rainstorms and periods of high intensity rainfall embedded in long-duration frontal rainstorms can trigger post-wildfire debris flows. Rainfall regimes throughout the western United States vary considerably and have been categorized by Moody and Martin (2009b) per seasonal characteristics and rainfall intensity based on the 30-min rainfall intensity of a 2-year return period storm (I_{30}^{2yr}) (Moody & Martin, 2009b). Most of the debris-flow events shown in Fig. 11.2 overlap the Arizona, Pacific, and Plains rainfall regimes and experience rainfall intensity classified as medium (I_{30}^{2yr}= 20–36 mm/h; e.g., California) and high (I_{30}^{2yr}= 36–52 mm/h; e.g., central Colorado and southeast Arizona). In situ measurements of rainfall intensity and debris-flow stage identified debris-flow surges to be best correlated to short-duration (<30-min duration) rainfall intensity with the highest correlation associated with rainfall intensity measured over a 5-min duration.

Rainfall conditions for post-wildfire debris-flow generation do not have to be particularly extreme to result in devastating consequences. For example, storms with recurrence intervals of 2 years or less triggered debris flows following the 2003 Grand Prix and Old fires in southern California that caused 16 fatalities and extensive damage to property and infrastructure (Cannon et al., 2008). Staley et al. (2020) analyzed debris-flow occurrence and rainfall data for areas burned in the southwestern United States (Arizona, California, Colorado, New Mexico, Utah) and found that 77% of debris flows were triggered by storms with a return period of less than 2 years and a mean recurrence interval of approximately 1 year.

Rainfall intensity-duration thresholds have been defined using rainfall data for southern California and Colorado (Cannon et al., 2008, 2011). The thresholds are defined for intensity averaged over durations ranging from 5 min to 24 h and are defined by plotting rainfall intensity duration (the time over which the rainfall intensity was measured) against the rainfall intensity for storms that produced debris flows and storms that did not produce debris flows and identifying a threshold line separating the two sets of data. Subsequent work has focused on short-duration rainfall thresholds for 15-, 30-, and 60-min durations (Raymond et al., 2020; Staley et al., 2017). The definition of rainfall thresholds can be done by visually distinguishing the debris-flow producing storms from storms that did not produce debris flows (e.g., Cannon et al., 2008, 2011) or through empirical methods to objectively delineate rainfall thresholds (e.g., McGuire & Youberg, 2020; Raymond et al., 2020; Staley et al., 2013). Figure 11.5 provides an example of an empirically defined rainfall threshold from Staley et al. (2013). Empirically defined rainfall thresholds show distinct regional differences in susceptibility, which are likely due to differences in geology, topography, and climatology. For example, rainfall thresholds for burn areas are low in the San Gabriel Mountains of southern California (I_{15}= 19 mm/h; Staley et al., 2017), intermediate in the Colorado Front Range (I_{15}= 31 mm/h; Staley et al., 2015), and high in Arizona (I_{15}= 43–66 mm/h; Youberg, 2014; Raymond et al., 2020).

The timing of debris flows within long duration storms is most correlated with rainfall intensities measured over short durations. Following the 2009 Station Fire in southern California, in situ stage sensors and rain gauges identified the following, (1) the timing of debris-flow surge fronts was most correlated with the 5-min intensity duration (Fig. 11.6), and (2) peak rainfall associated with rainfall intensities measured over longer than 15 min occurred after the surge front had passed (Kean et al., 2011).

A storm on 25 December 2003, in the area burned by Grand Prix and Old fires near San Bernardino, California began with low intensity rainfall that gradually increased in intensity until debris flows were reported (Cannon et al., 2008). Similar frontal storms have also produced debris flows in numerous other burned areas in southern California, most notably in Montecito, where reports of debris flows began immediately after intense rainfall began falling (Kean et al., 2019; Oakley et al., 2018).

Antecedent rainfall is not required for generation of a post-wildfire debris flow. Rain gauges in the area burned by the 2002 Coal Seam fire near Glenwood Springs, Colorado, and police dispatch records of a vehicle trapped in sediment indicated that debris flows had been triggered after 5 min of intense rainfall (Cannon et al., 2003b). Santi and MacAulay (2021) tabulated antecedent rainfall amounts for recorded

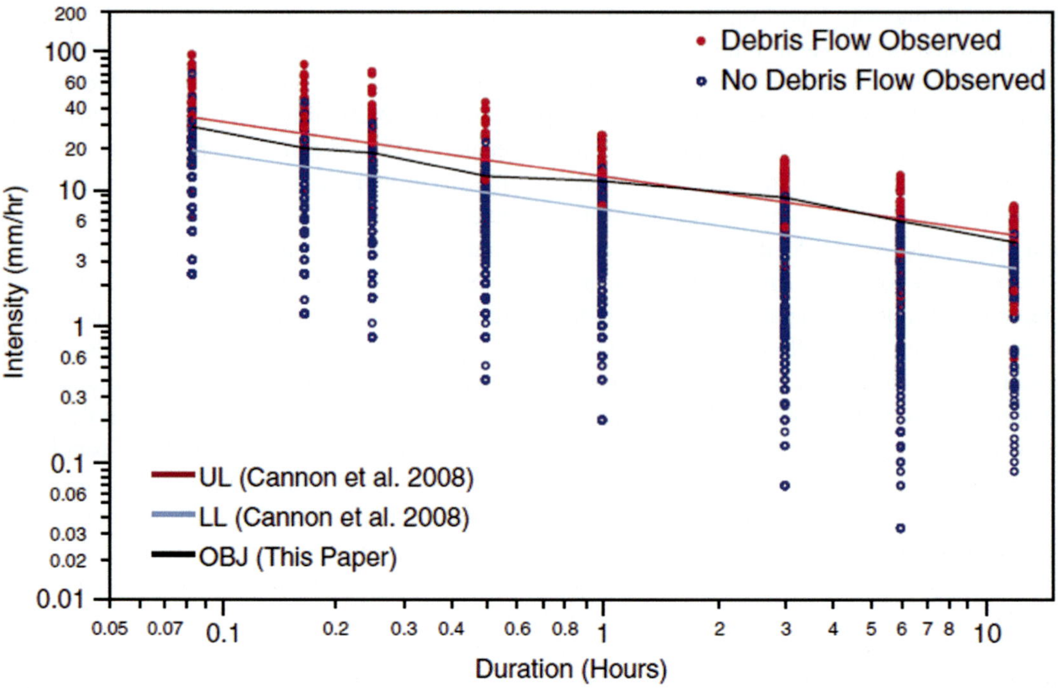

Fig. 11.5 Lower limit (LL), upper limit (UL) and objectively delineated (OBJ) rainfall intensity duration threshold for post-wildfire debris-flow initiation. Figure from Staley et al. (2013)

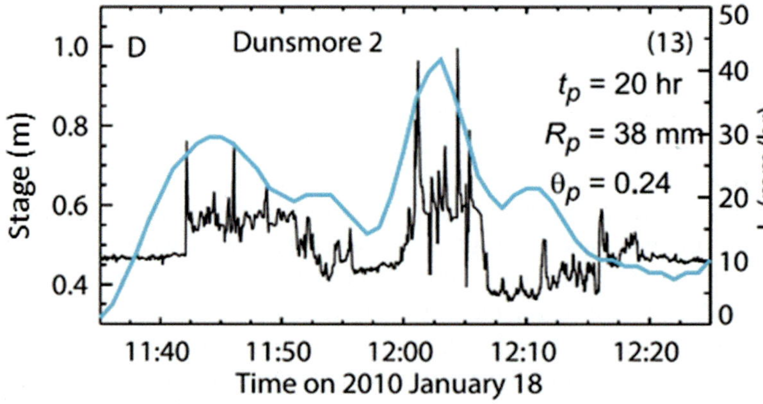

Fig. 11.6 Correlation between debris-flow surge height (stage, in meters as indicted on the left vertical axis and shown as a black line) and the rainfall intensity measured over the 5-min duration (I_5, in mm/h and shown as a blue line on the right vertical axis). The cumulative storm rainfall (R_p), hillslope soil water content (Θ_p) and elapsed time since the beginning of the storm (t_p) are also shown. Figure from Kean et al. (2011)

debris-flow events in the western United States and nearly all debris flows had less than 6 mm of rainfall in the week prior to the debris flow.

Less is known about rainfall-intensity thresholds for debris flows in burn areas in wetter climates, like the Pacific Northwest, or partially recovered burn areas where shallow landslides are sometimes observed in the third or following years after a wildfire (e.g., Rengers et al., 2020). Some events follow the pattern observed in semi-arid areas, whereas others do not. A post-wildfire debris-flow warning system was implemented in the area burned by the 2017 Elephant Hill fire near Cache Creek in British Columbia, Canada, using telemetered rain gauges. Debris flows were reported before the system could issue an alert based on a measured rainfall that exceeded intensity-duration due to the 5-min time lag in the system (Warren Lemky, written communication, 2018). In contrast, the area burned by the 2021 Lytton Creek fire near Lytton, British Columbia, produced debris flows in response to a long duration, "atmospheric river" storm, which only exceeded the thresholds defined by Cannon et al., (2008, 2011) for intensity durations longer than 1 h. These differences could be related to differences in the triggering mechanism (e.g., shallow landslide initiation versus initiation from distributed runoff and erosion) or due to water retention by ash in the soil (e.g., Rompkins et al., 1990). More work would be beneficial to better understand rainfall thresholds in wetter climates and partially recovered burn areas.

The characteristics of post-wildfire debris-flow-triggering rainfall has implications for warning systems and emergency management. Post-wildfire debris flows can be triggered by storms that have short return periods; therefore, after a fire it is very likely that rainfall will occur that has the potential for triggering debris flows. Debris flows may occur shortly after the storm begins, leaving little time to evacuate homes or close roads. Therefore, warning systems are most successful when information is supplied by both telemetered rain gauge and doppler radar data and incorporates high temporal and spatial resolution weather forecasts.

11.3 Initiation Mechanisms

Two general types of debris-flow initiation have been observed in burn areas: landslide-initiated debris flows and runoff-generated debris flows (Cannon et al., 2001; McGuire et al., 2016; Santi & Rengers, 2020) (Fig. 11.7). After wildfire, most observed debris flows (80%) tend to be generated by rainfall runoff (Gartner et al., 2005) and these often occur in the first year following a fire (Cannon et al., 2008; Nyman et al., 2011). In some cases, the rainstorm that helped to extinguish the fire resulted in debris flows (Kean et al., 2019). As described in more detail above, wildfire-induced changes in soil hydrology, surface roughness, and removal of the protective cover provided by organic matter cover increases rainfall runoff and sediment erosion from burned hillslopes. This eroded sediment can either be progressively bulked into a debris flow (Cannon et al., 2003a) or can form sediment dams in low-gradient channel reaches that subsequently fail and initiate a debris flow (Kean et al., 2013; McGuire et al., 2017). As a result of this distributed erosional process, the sediment that forms runoff-generated debris flows is eroded from much of the upstream contributing area that has been burned, including hillslope rill and inter-rill erosion and erosion from the channel network (e.g., Johnson, 1984; McGuire et al., 2016; Meyer et al., 1992; Nyman et al., 2011).

Several studies have quantified the sediment yield in channels scoured by debris flows. A large empirical study incorporating many different debris flows from burned sites showed that the sediment yield from a channel can increase substantially (2.5 m^3/m on average) as debris flows scour bed material (Santi et al., 2008). In steep, tectonically active regions, such as the Transverse Ranges in southern California, investigators have observed a multi-step process: sediment transfer from hillslopes to channels and then debris-flow scour of channelized sediment (Guilinger et al., 2020). For example, during and following a wildfire the vegetation that supports uphill sediment is burned, and lacking support, the uphill sediment ravels downhill infilling channels (DiBiase & Lamb, 2013, 2020; Florsheim et al., 1991; Lamb et al., 2011). This hillslope-derived sediment can account for as much as 90% of the total sediment removed from a basin (Rengers et al., 2021). However, the dry ravel sediment remains in channels until a debris flow scours the channels, ultimately depositing the sediment downstream either as channel infill or as a fan (Rengers et al., 2021).

Fig. 11.7 Left: Rill erosion on hillslopes coalesce into an eroded gully in the Station Fire, California. Right: A small landslide at a channel head and rill erosion on hillslopes in the Mullaly Canyon watershed. Photographs by the U.S. Geological Survey in the 2009 Station Fire burned area, California

The dominant landforms contributing to the sediment budget also change over time in burned watersheds as well as vary regionally. Multiple studies have shown that the first storm events following a wildfire primarily scour material from the channel (e.g., post-wildfire dry ravel and colluvium). Consequently, as that store of sediment becomes exhausted, the relative contribution of rill and inter-rill erosion on hillslopes to the debris-flow sediment budget increases (Guilinger et al., 2020; Tang et al., 2019).

Rilling on hillslopes is common after wildfires, due to a reduction in surface roughness, and the relative influence of rills on sediment budgets varies across many locations. In the Idaho Batholith, rill erosion constituted 25% of the total sediment volume with channel scour contributing 75% (Ellett et al., 2019). In contrast, rill contribution in the Santa Ynez Mountains, California was estimated to be nearly 50% of the total sediment volume (Alessio et al., 2021; Morell et al., 2021).

Landslide-initiated debris flows directly following wildfire are much less common. Directly linking wildfire as the cause of post-wildfire landsliding has been difficult owing to the longer time elapsed between fire and landslides that subsequently mobilize into debris flows. In many cases, landslide-initiated debris flows are triggered several years to decades after a wildfire (e.g., Benda & Dunne, 1997; May & Gresswell, 2003; Meyer et al., 2001; Rice & Foggin, 1971); however, these landslides are frequently generated during large storms that also trigger landslides in unburned catchments. One study was able to link post-wildfire root-decay of burned trees to landslide susceptibility (Jackson & Roering, 2009). A recent study took advantage of a natural experiment, where a single storm affected a mountain range with various states of re-vegetation after wildfire: unburned, 3 years, 5 years, and a patch that burned 10 years prior and reburned 3 years prior (Rengers et al., 2020). They observed that landslide-initiated debris flows were overwhelmingly concentrated in the areas that were burned 3 years prior and that the landslide density in areas burned 5 years prior were like the unburned landslide density. Moreover, areas that were reburned did not have a substantially different landslide density than the area that burned once, 3 years prior to the storm. These observations are useful in linking wildfire activity to observations of increased landsliding. Thomas et al. (2021) describes a mechanistic framework for increased landsliding several years after fire, primarily through increased infiltration and reductions in evapotranspiration, interception, and apparent root cohesion.

11.4 Climate Change Impacts on Wildfire Activity and Precipitation

Climate change driven by anthropogenic activity has disrupted wildfire and precipitation regimes. Increased temperatures, higher frequency of heat extremes, and drought are correlated with decreased fuel moisture, which has resulted in more wildfires and area burned (e.g., Abatzoglou & Williams, 2016; Coogan et al., 2021; Dennison et al., 2014; Flannigan et al., 2009; Hanes et al., 2018; Haugo et al., 2019; Iglesias et al., 2022; Westerling et al., 2006). Rainfall extremes at time scales of several hours to days are intensifying at a rate consistent with the increase in water vapor predicted by the Clausius-Clapeyron (C–C) relation, 7% per degree Celsius of warming (Ali et al., 2018; Fowler et al., 2021a, 2021b; Trenberth et al., 2003). Atmospheric rivers, which are long, narrow, corridors of high water–vapor transport in the atmosphere and the driver of major flood events in western North America (Ralph et al., 2006; Corringham et al., 2019; Sharma & Déry, 2020), are projected to intensify in a warming climate (Curry et al., 2019; Espinoza et al., 2018; Gershunov et al., 2019; Warner et al., 2015). At the hourly to sub-hourly timescales that are most relevant to post-fire debris flows, numerous studies from various locations around the world have indicated that rainfall has intensified at a greater rate than expected following the C–C relation alone (Ali et al., 2018; Fowler et al., 2021a, 2021b; Guerreiro et al., 2018; Lenderink et al., 2017; Wasko et al., 2018).

Collectively, the projections of increased fire occurrence and area burned as well as intensification of short-duration rainfall extremes portends post-fire debris flows of increased frequency, and potentially greater magnitude. Indeed, Kean and Staley (2021) found a projected increase of 110% in the probability of major debris flows in southern California following Representative Concentration Pathway 4.5 (RCP 4.5; Thomson et al., 2011) and scaling rainfall intensity following the C–C relation. Changes in climate leading to a longer wildfire season may also reduce preparation times (Oakley, 2021; Swain, 2021). Post-fire debris-flow hazards may become more common in areas without a history of post-fire flow activity, especially in regions experiencing snowpack decline (Siirila-Woodburn et al., 2021). Stoffel et al. (this volume) provides an in-depth review of how climate change affects wildfire and precipitation regimes.

11.5 Hazard Assessment

Hazard assessment of post-wildfire debris flows consists of (1) estimating the likelihood of a burned watershed to produce a debris flow, (2) estimating the potential magnitude of a debris flow generated by the burned watershed, and (3) mapping the debris-flow intensity (e.g., the depth and velocity of the debris flow) and inundated area.

11.5.1 Estimating Debris-Flow Likelihood

Logistic regression models that estimate the likelihood of a debris flow initiating in a burned watershed have been developed using databases of burned watersheds with records of producing or not producing debris flows (e.g., Cannon et al., 2010a; Staley et al., 2016) and associated watershed morphology, triggering rainfall, burn severity, and soil characteristics. These logistic regression models are periodically updated as more information for debris flows generated from burned areas becomes available. The most current model for estimating post-wildfire debris-flow likelihood developed by the U.S. Geological Survey (USGS; Staley et al., 2016) is:

$$P = \frac{e^x}{(1 + e^x)} \tag{11.3}$$

where:

$$x = -3.63 + (0.41 \times X1 \times R) + (0.67 \times X2 \times R) + (0.7 \times X3 \times R)$$

and:

- $X1$ is the proportion of the upslope area classified as high or moderate soil burn severity and with gradients $\geq 23°$,
- $X2$ is the average dNBR of the upslope area,
- $X3$ is the soil KF-Factor (Schwartz & Alexander, 1995), and
- R is the peak 15-min rainfall accumulation of the design storm (in millimeters).

As described in Staley et al. (2017), the model can also be used to estimate a rainfall threshold for a burned area by rearranging the logistic equation to solve for the rainfall rate that corresponds to debris-flow likelihood of 50% averaged across all of the assessed watersheds. Model-estimated rainfall thresholds are calculated for every USGS debris-flow hazard assessment and close substantial gaps in warning guidance for burn areas not already covered with an empirical rainfall threshold.

Current work to improve predictions of post-wildfire debris-flow likelihood and rainfall thresholds in the United States is focused on filling regional gaps in debris-flow inventories and testing the accuracy of model predictions in recent hazard assessments. The biggest gaps in U.S. data are in wetter climates like northern California, Washington, and Oregon, where post-fire data collection has lagged recent increases in fire activity. Post-wildfire debris flows in this region have been observed to initiate from a variety of mechanisms, including shallow landsliding (e.g., Swanson, 1981; Wondzell & King, 2003; Woodsmith et al., 2004), mass failure of channel material (Calhoun et al., 2022), and distributed runoff and erosion (Wall et al., 2020). In contrast, the Staley et al., (2016, 2017) model was developed and tested with data primarily from burn areas in semi-arid regions where distributed runoff and erosion were the primary debris-flow initiation mechanisms. As such, the accuracy and domain of applicability of this model in northern California, Oregon, and Washington is not fully understood.

Numerous tests of the performance of the Staley et al., (2016, 2017) model for debris-flow likelihood and rainfall thresholds have been performed across the western United States. Three of these locations are highlighted in the event summaries at the end of this chapter (southern California, northern California, and Colorado). Additional tests have been completed in central California (e.g., Neptune et al., 2021) and Arizona (e.g., Raymond et al., 2020) and more are underway in Oregon and Washington. Initial results indicate that, as expected, the model performs best in southern California where it was calibrated (Swanson et al., 2022). The model performs reasonably well in central California but is conservative (i.e., overestimates debris-flow hazard) in northern California and Arizona. Although testing is not complete, the model is not expected to be accurate for burn areas where debris flows initiate by processes other than distributed runoff and erosion (i.e., for burn areas primarily in the wetter climates of the Pacific Northwest). However, the simple model may still be useful for a rapid qualitative assessment of susceptibility because it correctly identifies locations that are both steep and severely burned, factors that likely increase debris-flow susceptibility regardless of the style of initiation.

11.5.2 Estimating Debris-Flow Magnitude

The influence of wildfire to increase the magnitude of debris flows was recognized in the early 1900s in southern California (e.g., Chawner, 1935; Eaton, 1935; Troxell & Peterson, 1937), and models for estimating erosion rates, peak discharges, and sediment yields included factors to account for the occurrence of fire (e.g., Rowe et al., 1949; Scott & Williams, 1978). The models initially developed for sediment yield in southern California were intended to help design sediment retention basins. More recent studies have focused on estimating sediment yields for hazard characterization following wildfire.

Several studies have created models for sediment yield based on site factors. Slope and burn severity are the two parameters used by Pelletier and Orem (2014) to describe sediment yield at a field site in New Mexico:

$$Y_p = aS^bB^c \tag{11.4}$$

where Yp is sediment yield (mm), S is average slope (m/m), B is the average soil burn severity class over the total sub-basin area (using low = 1, moderate = 2, high = 3), $a = 1.53$, $b = 1.6$, and $c = 1.7$. The total volume is obtained by multiplying the sediment yield by the upstream contributing drainage area. A multiple regression model proposed by (Gartner et al., 2014) has been shown to be effective in southern California (Rengers et al., 2021):

$$Ln(V) = 4.22 + \sqrt{i15} + 0.36 \times \text{Ln}(Bmh) + 0.13 \times \sqrt{R} \tag{11.5}$$

variables have dimensional quantities, and V is volume (m^3), $i15$ is the 15-min rainfall intensity (mm/h), Bmh is watershed area burned at moderate and high severity (km^2), and R is the watershed relief (m). This regression model therefore accounts for the rainfall, terrain, and influence of the wildfire. One challenge for these models is that they do not explicitly account for intermittent sediment deposition. Consequently, they are most likely to be valid in basins where the erosional volume is increasing.

In the United States, the USGS provides preliminary post-wildfire debris-flow hazard assessments (Fig. 11.8). The assessment provides estimates for post-wildfire debris-flow likelihood using the logistic regression model developed by Staley et al. (2016), debris-flow volumes using the multiple regression model developed by Gartner et al. (2014), and a combined relative hazard ranking that reflects the probability and volume ratings (Cannon et al., 2010a). The information is calculated for the entire watershed based on conditions at the watershed outlet as well as along the drainage network (Gartner et al., 2011). The hazard assessment models used by the USGS are most applicable to the portions of the watershed where debris flows are entraining and transporting sediment and do not incorporate debris-flow runout models to identify possible inundation areas.

In Australia, agencies responsible for fire, land, water, and emergency management in Victoria developed collaborations with university researchers to develop a model of post-wildfire debris-flow risk called HydroFire that could be applied in a range of agency contexts such as post-wildfire risk assessment and strategic planning. A key requirement was that this model could be operated by the agencies involved, often with limited financial and technical resources, which imposed strong constraints on the model design.

The HydroFire model (Langhans et al., 2016; Nyman et al., 2021) assumes the initiation of post-wildfire debris flows is strongly controlled by *both* the shear stress of surface runoff and the availability of burned, unconsolidated surface soil in convergent upland "headwaters" (Nyman et al., 2013). Headwaters are defined as the area above the channel initiation point, which is assumed to occur (on average) at 0.02 km^2 in these landscapes (Langhans et al., 2016). The landscape is therefore conceptualized as a collection of 0.02-km^2 headwater potential debris-flow initiation points, which can be assigned the necessary parameters for modelling debris-flow

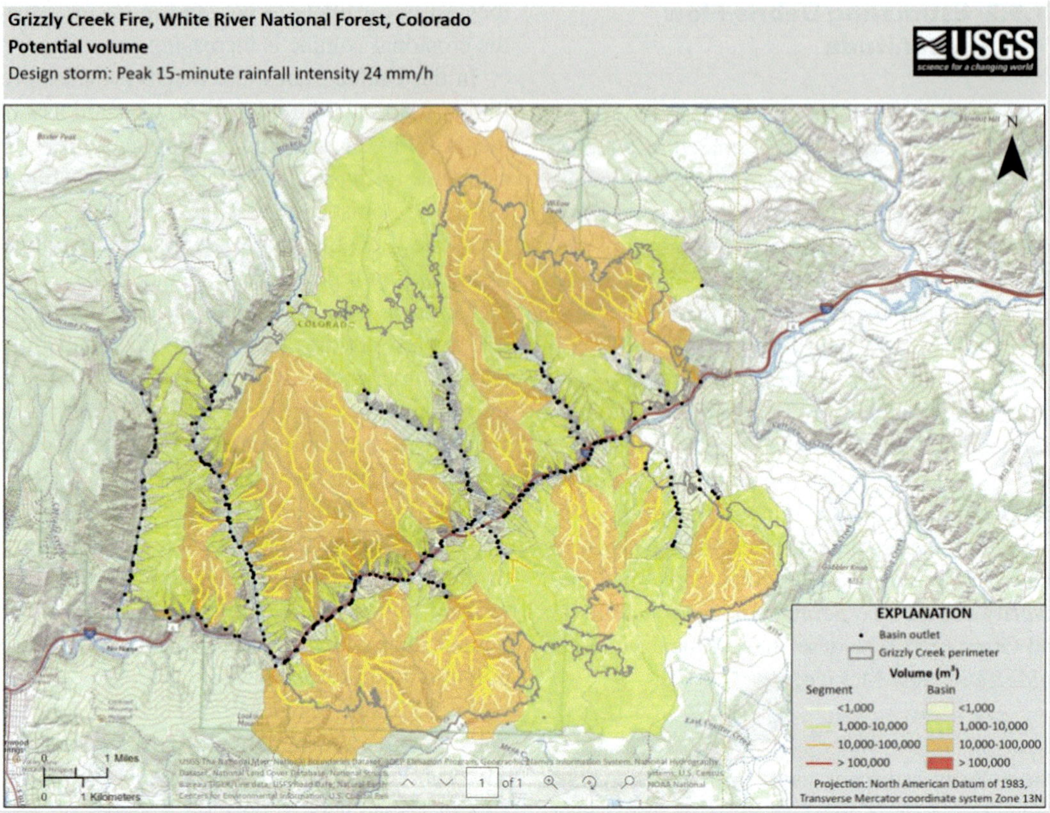

Fig. 11.8 An example of an emergency assessment of post-wildfire debris-flow volumes for the 2020 Grizzly Creek Fire in Colorado

hazard from available spatial data. A compendium of mapped observed debris flows from air photo interpretation, and controlling variables (burn severity, slope, soil type, rainfall intensity), was used to develop an empirical threshold curve for debris-flow initiation (Fig. 11.9). The shear stress for the observed debris-flow events was calculated from the observed rainfall intensity using the process-based rainfall runoff model KINEROS2 (Woolhiser et al., 1990).

The only model input required to operate the HydroFire model is the mapped fire severity (dNBR) from observations, or from a fire spread model, which affects both the runoff generation and the quantity of non-cohesive soil available for transport. The triggering rainfall intensity is the intensity required to generate sufficient streampower to cross the debris-flow initiation threshold (Fig. 11.9).

Estimates of debris-flow runout are used to evaluate post-wildfire risk. In the United States, rapid assessment of risk from debris-flow inundation is primarily based on geomorphic evidence of past flow activity, preexisting flood maps, and expert judgement (e.g., WERT, 2018). Although a variety of debris-flow runout models have been developed (e.g., George & Iverson, 2014; Gorr et al., 2022; McDougall & Hungr, 2004), model estimates of inundation are not yet a routine component of post-wildfire hazard assessment. Cannon et al. (2010b) conducted initial work to model runout paths from select drainages burned in the 2009 Station Fire in southern California. They used an empirical model developed to identify lahar-inundation zones (Iverson et al., 1998) and estimated debris-flow volumes as an initial condition. Similar approaches have since been used in post-event analyses of loss (McCoy et al.,

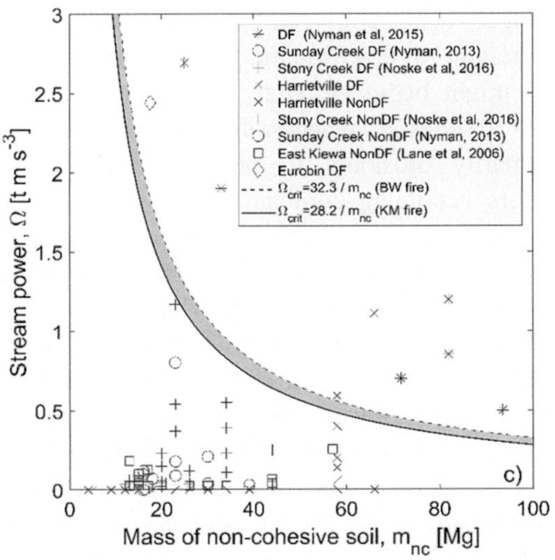

Fig. 11.9 Debris-flow (DF) initiation in the HydroFire model is conceptualized as a threshold-function of the quantity of non-cohesive soil available for transport (which in turn is a function of fire severity), and stream-power at the headwater channel initiation point (assumed to be 0.02 km^2). Each marker represents a monitored intense rainfall event for a headwater with known properties (e.g., burn severity, soil proxy, slope). Red markers indicate observed debris flows, and blue markers indicate events that did not initiate debris flows. Non-cohesive soil and streampower conditions above the threshold curve are assumed to result in debris-flow initiation in the HydroFire model. The Gray shaded area is the envelope between the critical (crit) threshold curves for the Beechworth (BW) and Kilmore-Murrindindi (KW) Fires described in Nyman et al., (2015) (Reproduced from supplemental material in Nyman et al., 2021)

2016) and pre-fire planning using wildfire scenarios (Youberg, 2008).

Despite this progress, the complexity of runout modeling, uncertainty in model parameters, and challenges of the post-wildfire timeline present barriers to establishing operational methods for rapid and accurate post-wildfire debris-flow inundation assessments. In general, runout calculations are computationally more intensive than current predictions of debris-flow likelihood and volume by empirical models. Computational demands are an important consideration in completing hazard assessments before a rainstorm affects the burn area. Not only can time be short between the wildfire and the first rainstorm, but considerable time also would be needed to develop and implement emergency response plans based on model results. The computational challenges are further compounded for megafires, which are becoming increasingly common, and may have thousands of drainages requiring estimates of debris-flow runout.

Recent studies to test predictions of post-wildfire debris-flow runout against field observations are helping to identify a path towards operational predictions of debris-flow inundation (e.g., Barnhart et al., 2021; Bernard et al., 2021; Gibson et al., 2022; Gorr et al., 2022). Studies such as Barnhart et al. (2021) systematically evaluate the effects of uncertainty in model parameters on predictions and identify tradeoffs between model complexity and runtime. Using three different models (RAMMS, FLO2D, and D-CLAW) and data from the 2018 Montecito debris-flow event (Kean et al., 2019), Barnhart et al. (2021) showed that (1) the volume of debris flow was the most important model parameter for matching the extent of inundation, (2) parameters calibrated for an individual drainage worked well on nearby drainages, and (3) calibrated models of intermediate complexity yielded results of comparable accuracy as a computationally more intensive model with better representation of debris-flow physics. Subsequent work by Gorr et al. (2022) using the same dataset showed that a calibrated reduced-complexity model with substantially faster runtimes than the three models evaluated by Barnhart et al. (2021) also provided reasonable estimates of inundation. Further model testing on other datasets of post-wildfire debris-flow inundation would be beneficial to better understand the applicability of model parameters to sites with different topographic and geologic conditions. Presently, very few datasets exist, so additional work to build an appropriate inventory of test cases would be useful.

Another challenge in debris-flow runout modeling is estimating the transition between erosion and deposition of a debris flow. In both burned and unburned watersheds, the cumulative volume eroded in channels tends to increase with distance downstream and then a rapid reversal occurs when deposition dominates

(Santi et al., 2008; Scheip & Wegmann, 2022). Work in the Transverse Ranges of southern California has shown that topographic metrics can be used in a logistic regression model to successfully identify this transition between erosion and deposition in debris-flow paths (Gartner et al., 2015). Some watersheds show a sharp transition between erosion and deposition, such as a 0.3-km^2 basin that burned in the 2019 Woodbury Fire in Arizona (Fig. 11.10). Before and after structure-from-motion elevation data for this study watershed show a sharp transition from erosion (upstream from a waterfall) to deposition (downstream from a waterfall) where 1200 m^3 of sediment were transported from upstream and stored in the low gradient portion of the watershed downstream with an average depth of 0.3 m.

However, many channels show an intermediate transition zone between erosion and deposition, where a local reach may be depositional and the next downstream reach is erosional (Gartner et al., 2015; Brogan et al., 2017; Rengers et al., 2021). In this case, moving from upstream to downstream, a debris-flow channel might begin as purely erosional and move into a series of reaches that alternate between primarily erosional to primarily depositional before becoming dominantly depositional. This transition zone is likely governed by changes in local slope (Cannon, 1989; Fannin & Wise, 2001; Hungr et al., 2005; Hürlimann et al., 2003; Vázquez et al., 2016) and channel width (Cannon, 1989; Fannin & Wise, 2001; Vázquez et al., 2016). Recent work in the San Gabriel Mountains in southern California indicated that this transitional zone was located on channel slopes between 5° and 30°, and once the channel slope decreased to less than 5°, the debris flow became primarily depositional (Rengers et al., 2021). Research has indicated that the primary cause of deposition is related to granular temperature (i.e., the mean square of particle velocity fluctuations) and excess pore-fluid pressure (Iverson, 1997; Major & Iverson, 1999; McCoy et al., 2010; Iverson et al., this volume). Schürch

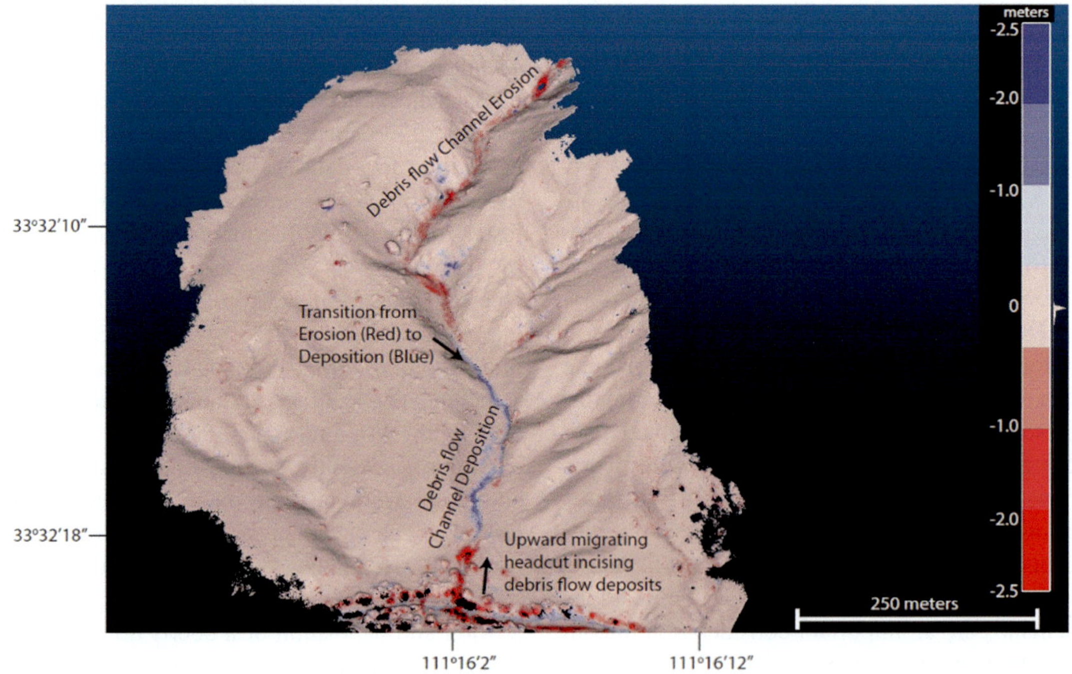

Fig. 11.10 Structure-from-motion vertical change detection map showing erosion (red) and deposition (blue) in a small watershed in the Woodbury Fire in Arizona. Note transition from erosion to deposition. Flow direction is to the bottom of the figure

et al. (2011) shows that erosion tends to increase with flow depth and high basal shear stress, and similar observations have been made in lahars (Vázquez et al., 2016). As material from the granular front is moved into levees, the resulting decrease in height can create positive feedback that further increases deposition (Schürch et al., 2011). In addition, experiments in dry granular flows originating on steep slopes greater than 16°, show that when the flow passes over an erodible bed it terminates in a longer runout than non-erodible beds (Farin et al., 2014). This observation is attributed primarily to a flow propagation phase where the behavior can approach a steady-uniform flow, and where the effective friction is relatively low (Farin et al., 2014).

In many cases debris flows terminate on fans, and the erosion/depositional behavior of the debris flow on the fan can be complex and hard to predict. The primary control of debris-flow avulsion on a fan is the ratio of the debris-flow discharge to the channel depth, both of which change via erosion/deposition during the debris flow itself (Schürch et al., 2011). However, the distribution of debris-flow volume can also affect the rate of avulsions (De Haas et al., 2018; De Haas et al., this volume). Experiments show that when fans experience larger flows, they tend to avulse more frequently De Haas et al., 2018).

11.6 Mitigation

Mitigation to reduce the effects of post-wildfire debris flows is especially challenging because of the typically short time between when the wildfire has been contained and the onset of the rainy season or simply summer convective storms. As such, the period between wildfire containment and debris-flow potential is commonly too short for hazard assessment, mitigation design, and construction of permanent mitigation measures. Post-wildfire debris-flow hazards can also be mitigated by proactive measures that identify the potential hazard prior to wildfire or through rapid response warning and evacuation procedures that are triggered based on forecasted intense rainfall. For example, in the Transverse Ranges of southern California, numerous debris retention basins already exist. Following the 2009 Station Fire near Glendale, California, emergency work to remove sediment from the debris retention basins was done to maximize their effectiveness at capturing sediment from post-wildfire debris flows. As a result of the emergency work to remove sediment from the debris retention basins, only one basin was overtopped by debris-flow sediment, which flowed into the houses located downstream. The overtopped basin had been completely cleaned out prior to the debris flow but could not accommodate the full volume of the flow (Kean et al., 2012).

Many post-wildfire hillslope treatments for reducing erosion have been implemented and studied in burned areas including erosion barriers, aerial seeding, and mulching (Girona-Garcia et al., 2021; Robichaud et al., 2010). Treatments that improve ground cover (e.g., mulching using straw or wood shreds) have been found to be more effective at decreasing erosion than erosion barriers. Wood-based mulches have been implemented on slopes steeper than 60% that were burned at high severity (Robichaud et al., 2010). Re-forestation through aerial seeding and tree planting can reduce the likelihood of landslides in areas with decreased soil cohesion due to the decay of burned tree roots (De Graff, 2018). Mechanical barriers that can be rapidly constructed to capture and store sediment in channels (e.g., ring nets or check dams) have also been implemented following fire with some success (Kane & Jones, 2019). For example, for a watershed burned by the 2002 Missionary Ridge Fire in Colorado and located above the Lemon Reservoir (which is the drinking water supply for the city of Durango, Colorado), combined treatments of straw mulch and check dams resulted in little sediment reaching the reservoir (de Wolfe et al., 2008). Ring nets installed following the 2011 Las Conchas Fire, New Mexico in 2011 provided substantial cost savings for the Santa Clara Pueblo by preventing sediment from affecting the Pueblo's road (Kane & Jones, 2019). Widespread application of mechanical barriers and hillslope treatments throughout an entire burned watershed are oftentimes not feasible for an entire burned area with numerous debris-flow prone watersheds. Therefore, accurate hazard and risk assessment of the burned area using methods described above is critical for optimizing where these treatments are applied.

Post-wildfire debris-flow warning systems based on rainfall thresholds can be used in conjunction with hazard maps to prepare for post-wildfire debris-flow hazards (e.g., NOAA-USGS Debris Flow Task Force, 2005). A benefit of a warning system is that telemetered rain gauges and weather forecasting services are often already in place and can be leveraged for the warning system. Sometimes maintenance of existing debris retention basins to remove previously deposited sediment or installing temporary concrete barriers (e.g., K-Rails or Jersey Barriers) is feasible prior to the onset of debris flows; however, time is often too short to adequately construct physical structures for mitigation. Although warning systems do not protect homes and infrastructure from debris-flow impacts, they can prevent loss of life by guiding evacuations in hazardous areas. Similarly, rainfall thresholds, described earlier in this chapter, form the basis of many operational warning systems and are used in conjunction with forecasted and/or measured rainfall to evaluate if evacuations or closures are warranted (Cannon et al., 2011). Close monitoring of Doppler radar can be done to evaluate if a storm is tracking towards a burned area (Cannon et al., 2011), and measured rainfall can be evaluated to identify if rainfall intensities are approaching threshold conditions.

11.7 Recent Post-Wildfire Debris Flows in the United States, Canada, and Australia

Following are examples of recent wildfires and subsequent post-wildfire runoff and debris flows that illustrate the successes as well as current challenges of the methods reviewed above.

11.7.1 Event Summary: 2017 Thomas Fire, Montecito, Southern California

On 9 January 2018, a narrow cold frontal rain band swept over the Santa Ynez Mountains north of Montecito, California, triggering numerous large debris flows that travelled more than 3 km across an urbanized piedmont plain killing 23 people and damaging or destroying over 400 homes (Kean et al., 2019; Lancaster et al., 2021; Oakley et al., 2018; Fig. 11.11). The steep slopes above Montecito were burned just weeks earlier by the 1,140-km^2 2017 Thomas Fire (Kean et al., 2019), which, at the time, was the largest wildfire in California's history. Additional large debris flows occurred in less populated parts of the extensive burn area (Swanson et al., 2022).

In the short time span between the wildfire and the rainstorm, a large and coordinated interagency effort was expended to plan for post-wildfire flooding and debris flows. This included (1) state and federal post-wildfire emergency response teams that rapidly mapped soil burn severity (an indicator of flood and debris-flow potential) and identified values at risk to flow impacts (e.g., homes, culverts, and bridges), (2) USGS hazard maps identifying the likelihood and potential volume of flow across the Thomas Fire, (3) an early warning system by the National Weather Service that correctly identified the incoming storm 4 days in advance, and (4) a proactive local emergency response community that quickly developed response and evacuation plans. Despite these coordinated efforts, the magnitude of the 9 January 2018 debris-flow event in Montecito was much greater than anticipated, highlighting both challenges and opportunities for improving post-wildfire risk assessments.

The factors that contributed to the event being so destructive include the following: (1) 81% of the very steep hillslopes (median slope of 28°) above Montecito were burned at moderate and high burn severity, (2) the 15-min rainfall intensity (84–99 mm/h) was more than

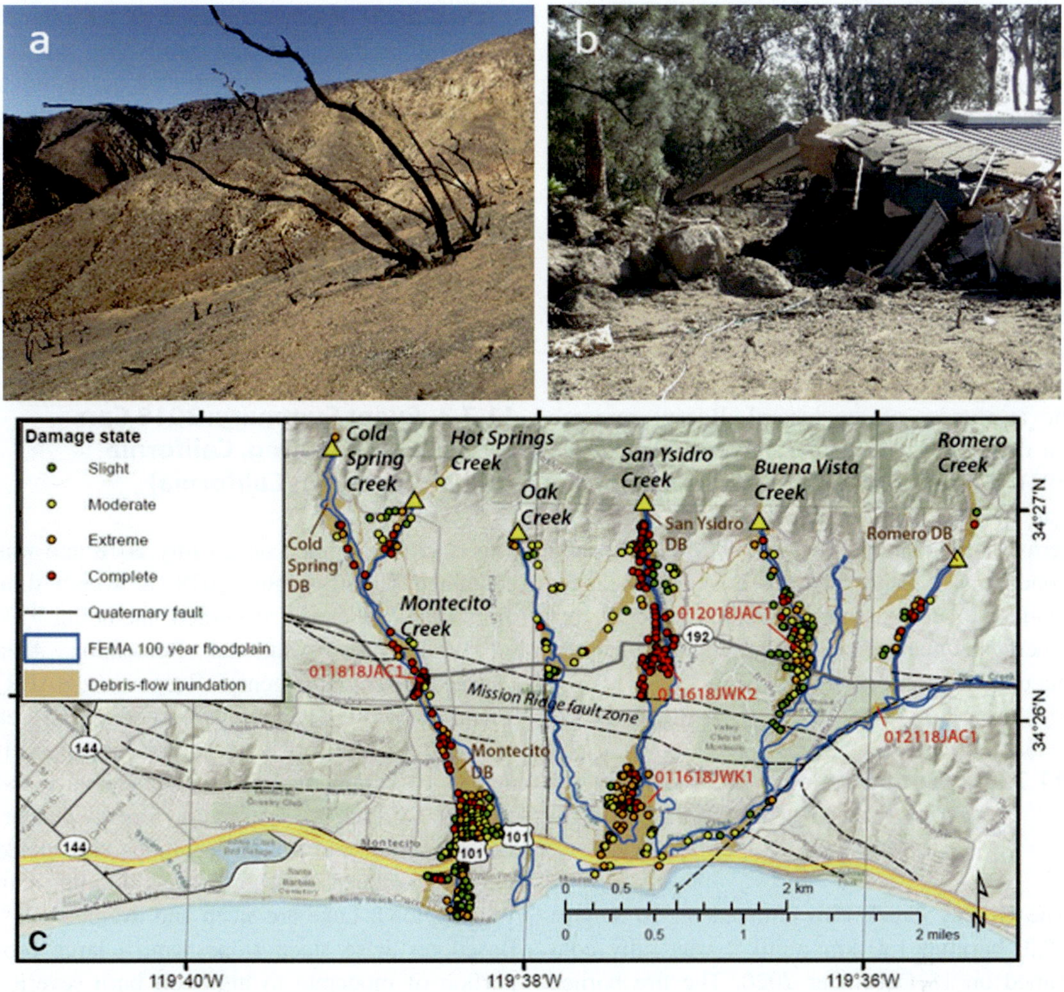

Fig. 11.11 Event Summary: 2017 Thomas Fire, Montecito, California. **a** Burned hillslopes in Hot Springs Creek above Montecito showing the availability of sediment that can be rapidly eroded. **b** Destroyed home near San Ysidro Creek. **c** Main runout paths in the 2018 Montecito debris-flow event, Federal Emergency Management Agency (FEMA) 100-year floodplain, apexes of alluvial fans (yellow triangles), and damaged homes identified by CAL FIRE (California Department of Forestry and Fire Protection) building inspectors (slight, moderate, extreme, and complete damage states) Reproduced from Kean et al. (2019). Photographs by U.S. Geological Survey

three times above the threshold expected for debris-flow initiation, (3) the average size of the six main contributing drainage basins was large (4 km^2), supplying abundant sediment from the hillslopes and channels, (4) large boulders (up to 4 m in diameter) were scoured from the channels and transported through the piedmont, intensifying damage (Morell et al., 2021), (5) a high density of primarily residential construction was located in the flow paths, (6) a short time period between the fire and rainstorm limited capacity to develop and implement emergency response plans and mitigation, and (7) low public awareness intensified the dangers of post-wildfire flooding and debris flow along with a reluctance to heed evacuation orders having just returned home from evacuations associated with the wildfire.

The latter two factors are shared concerns for many burn areas across western North America and highlight an important gap in post-wildfire hazard assessment that remains to be filled. Although the hazard maps for Montecito addressed the potential for debris flows in the source areas, they did not identify where the material would travel and what structures could be affected due to the absence of an operational tool to predict post-wildfire debris-flow inundation. Such information could have been used to develop better evacuation zones and raise public awareness of the hazard. Recent research on debris-flow inundation (e.g., Barnhart et al., 2021; Gorr et al., 2022) is seeking to close that gap by testing runout models against the patterns of inundation observed at Montecito and conducting a user needs assessment of stakeholders to develop an inundation product that emergency managers can readily understand and act upon (Barnhart et al., 2023).

11.7.2 Event Summary: 2020 Grizzly Creek Fire, Glenwood Springs, Colorado

The Grizzly Creek Fire initiated on 10 August 2020 burning 130 km^2 until it was fully contained on 18 December 2020. The fire burned within the narrow Glenwood Canyon, which has an elevation range of 1,750–3,270 m and a median slope of 21°. The burn severity was highest on the steep slopes of the canyon near the Colorado River and decreased in severity as the fire spread uphill and above the canyon rim. Debris-flow hazards were assessed for the burn area using the model described above by Staley et al. (2016).

No significant rainfall events were recorded in the burn area prior to the onset of winter snowfall. However, with the onset of the rainy season in 2021, debris flows were observed following localized convective thunderstorms. With the exception of one storm, all of the rainstorms that triggered debris flows were more intense than the predicted watershed-averaged 15-min rainfall intensity threshold (Fig. 11.12).

Despite the relative accuracy of the rainfall threshold produced by the debris-flow hazard assessment, the small footprint of the summer convective rainstorms triggering debris flows resulted in a localized basin response within the burn area for a single storm event. The debris flows caused extensive damage to Interstate 70 and closed the major transportation corridor numerous times (the longest lasting about two weeks) resulting in a four hour long detour.

11.7.3 Event Summary: 2018 Carr Fire, Redding, California (Northern California)

The Carr Fire ignited on 23 July 2018 and was contained on 30 August 2018, at which time more than 929 km^2 had burned, including 97% of Whiskeytown National Recreation Area (Burned Area Emergency Response [BAER] Team, 2018). The Carr Fire burned in an area of northern California that receives approximately 1500 mm of precipitation annually and supports primarily evergreen forest, chaparral and scrub, and minor deciduous and mixed forest. The mean slopes of basins draining into Whiskeytown Lake are steep and average ~25°. Based on these steep slopes and a large proportion of moderate to high soil burn severity, it was expected that post-wildfire debris flows would be likely and with potentially large volumes (Fig. 11.13; Staley et al., 2017).

The first year after the Carr Fire was unusually wet (Fig. 11.13), with 133% of average rainfall in the study region between 1 October 2018 and 30 September 2019 (National Oceanic and Atmospheric Administration, 2021; East et al., 2021). The first major rain event occurred on 3 October 2018 and had 15-min rainfall intensities that ranged between 35 and 55 mm/h at different gauges across the fire (Fig. 11.13b; East et al., 2021). These intense rains triggered substantial hillslope runoff, rilling, flooding, and sediment erosion but did not trigger debris flows. Moderate to intense rainfall (maximum 15-min intensities ~60 mm/h) continued throughout much of the winter of this first post-wildfire year

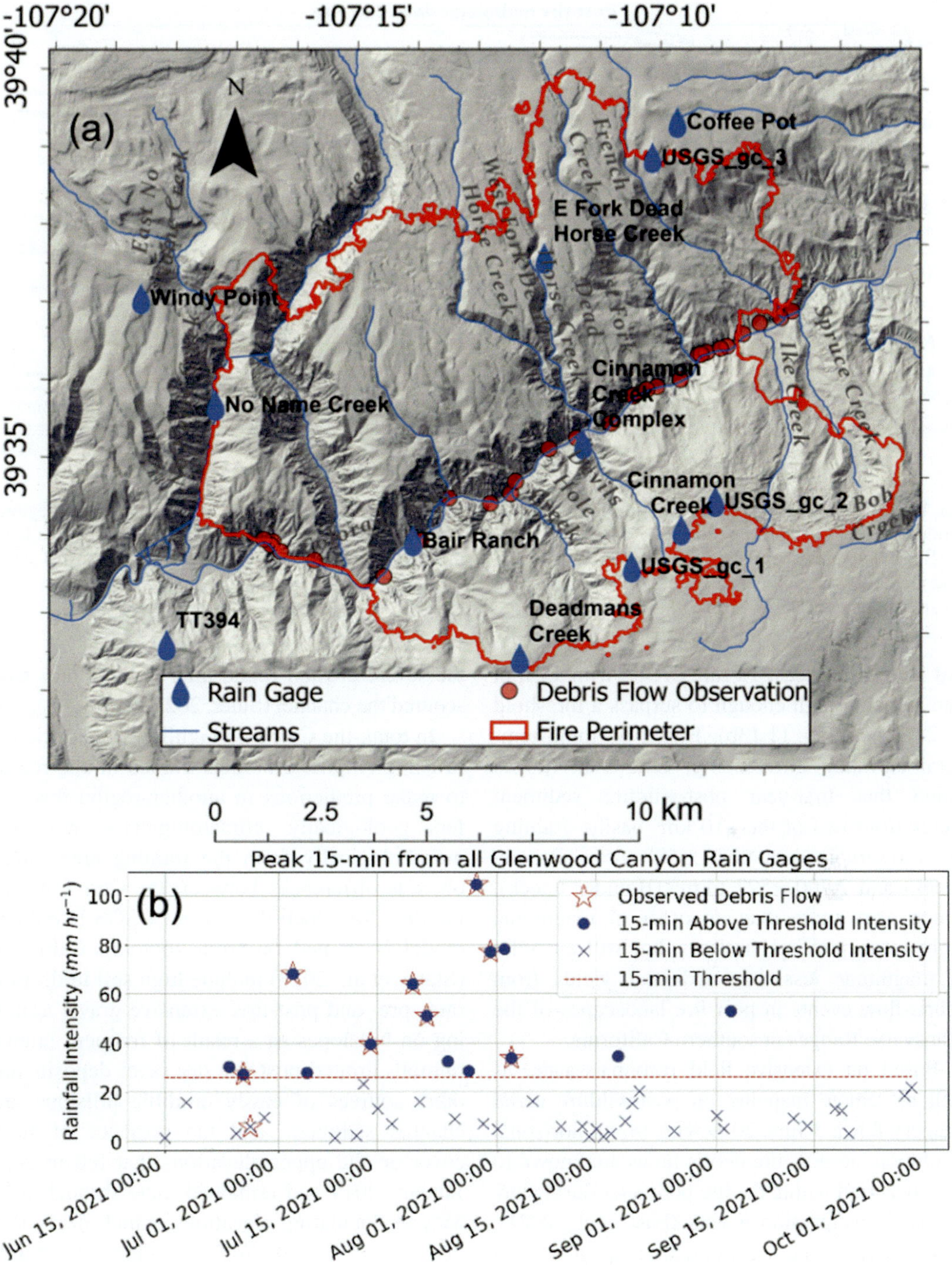

Fig. 11.12 Overview of the 2020 Grizzly Creek Fire, Glenwood Springs, Colorado. **a** Location map showing the burn perimeter, rain gages, and locations of debris-flow observations. **b** A time-series of the peak 15-min rainfall intensities as well as the basin-wide modeled rainfall intensity threshold for 15 min. Rainfall events where debris flows were observed are indicated with a red star

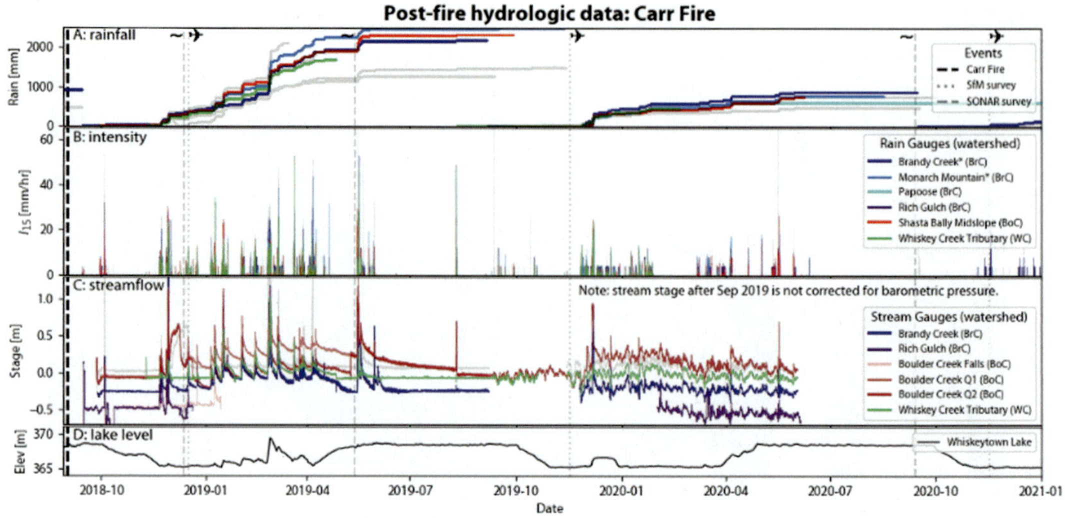

Fig. 11.13 Post-fire hydrologic monitoring data from various locations within the Carr Fire from East et al. (2021). **a** Cumulative rainfall in mm. **b** Rainfall intensity measured over 15-min intervals in mm/h. **c** Approximate stream stage measured using in-channel pressure transducers reported in m. **d** Elevation of Whiskeytown Lake in meters above sea level. Tilda and airplane icons indicate dates of sonar bathymetric surveys and aerial image collection, respectively. Data from East et al. (2021)

and included at least 12 rain events with 15-min rain intensity high enough to surpass a threshold of 24 mm/h (Fig. 11.13b). Based on multi-temporal elevation differencing, East et al. (2021) found that first-year post-wildfire sediment yields from two of the ~10 km^2 basins draining to the reservoir were 4,080 ± 598 t/km^2 (Brandy Creek) and 2,700 ± 527 t/km^2 (Boulder Creek), which were more than an order of magnitude higher than pre-fire baselines, but still an order of magnitude less than sediment yields from debris-flow events in post-fire landscapes of the Transverse Ranges in southern California.

Based on extensive field reconnaissance, as well as remote mapping on post-wildfire aerial imagery flown 3 June 2019 with 14-cm horizonal resolution, no post-fire debris flows are known to have occurred within the fire perimeter during any of these precipitation events (East et al., 2021). Suter (2022) made a continuous map of erosion magnitude in the basins around Whiskeytown Lake that were covered by 3 June 2019 aerial imagery (Fig. 11.14) and found that the majority of basins experienced only low erosion magnitude and almost no basins experienced high erosion magnitude, which typically corresponds to the severe erosion found after debris flows have scoured the channel (Suter, 2022).

In total, the Carr Fire highlights the difficulty of using empirical models trained in one region to make predictions in another region that differs geologically, climatologically and geomorphologically from the training area. Some obvious differences between the Carr Fire and the fires for which the current USGS likelihood model is trained on from southern California (Staley et al., 2017) include high soil infiltration rates pre- and post-fire, extensive gravel armoring on hillslopes as a result of frequent intense rainfall, limited post-fire dry ravel deposits and other sources of easily erodible hillslope and channel sediment, and the presence of snow cover on the upper elevations that fell in early January 2019 and remained there through mid-May at fluctuating elevations, which likely limited erosion from the high elevation contributing areas during these periods. As datasets grow to include examples such as the Carr Fire and as other predictive variables are included into the next generation of empirical hazard models they are expected to become more accurate and flexible.

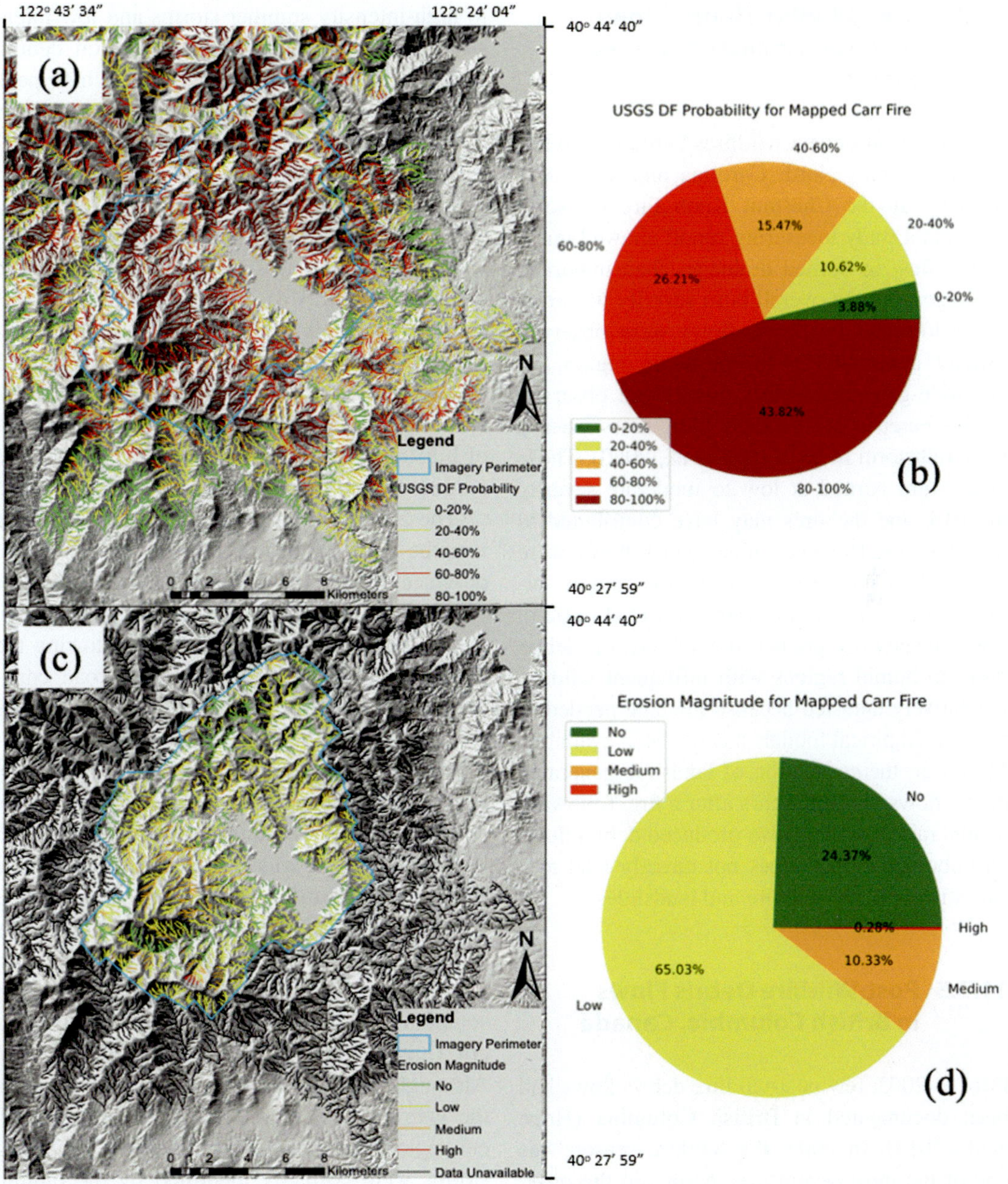

Fig. 11.14 Comparison between USGS debris-flow probability and the remotely mapped erosion magnitude by Suter (2022) for the Carr Fire. The blue perimeter line marks the extent of imagery availability and contains all the data considered here. **a** USGS debris-flow probability map for 24 mm/h rainfall intensity for a 15-min duration. **b** Pie chart of percentages of channels experiencing one of the classes of debris-flow probability. **c** Remotely mapped No, Low, Medium, and High erosion magnitudes. **d** Pie chart of percentages of channels that were given 1 of the 4 scales of erosion magnitude

11.7.4 Post-Wildfire Debris Flows in the Southeastern United States

A series of infrequent wildfires burned steep terrain in western North Carolina and Tennessee in 2016. In the Chestnut Knob Fire in North Carolina a study measuring runoff showed much higher flow in burned drainages than unburned drainages, but the overall burn severity was relatively low and no debris flows were observed (Stiefel et al., 2021). In the same region, 23 landslide-generated debris flows were observed in the Ferebee and Tellico burn areas after a large rainstorm in 2019 (Hill et al., 2020). Those areas were burned at low to moderate severity in 2016, and the fires may have contributed to debris flows. However, nine debris flows were observed in unburned areas in the region following the same storm event (Hill et al., 2020). The capacity to predict the effects of debris flows in humid regions with infrequent wildfire is relatively untested because the data needed to develop empirical models has not been available. Moreover, the correlation of landslide-generated debris flows multiple years after a fire, following a large rain event that also produced debris flows on unburned slopes, does not directly lead to a causative link between fire and landslides.

11.7.5 Post-Wildfire Debris Flows in British Columbia, Canada

Prior to 2003, few post-wildfire debris flows had been documented in British Columbia (Hope et al., 2015). In 2003, the wildfire season was one of the most severe fire seasons in the province's history and post-wildfire debris flows, debris slides, and debris floods were generated in five separate burned areas located throughout the Interior Plateau and southern Rocky Mountain regions (Jordan, 2015; Jordan & Covert, 2009). Debris flows were also reported following two fires in 2007 and a fire in 2010 (Jordan, 2015). The debris flows were triggered by a range of conditions from spring snowmelt to high-intensity summer storms and low intensity fall storms and the most common debris-flow initiation mechanism was due to high peak flow in channels; however, landslide initiations were also observed (Jordan, 2015).

As mentioned previously in this chapter, debris-flow volumes of the 2003, 2007, and 2010 debris flows reported in Jordan (2015) were 2–4 times less than predicted by the Gartner et al. (2014) volume model. The discrepancy is likely related to the glaciation history of burned watersheds in southern British Columbia. Widespread surficial deposits of over-consolidated till may have limited the amount of hillslope and channel sediment that could be eroded (Peter Jordan, Personal communication).

The 2017 wildfire season in British Columbia was the most destructive wildfire season in the province to date and the Elephant Hill Fire near Ashcroft, B.C., burned an estimated 192 km^2. Debris flows within the first year following the fire affected multiple highway and road corridors and resulted in a fatality. The debris flows were triggered by high-intensity convective rainstorms that lasted for a short period (less than about 1 h). Numerous other debris-flow events occurred during subsequent rainstorms; however, these were not as widespread or damaging. The Ministry of Transportation and Infrastructure installed telemetered tipping bucket rain gauges and implemented a warning system based on rainfall thresholds developed for burned areas in southern California and the intermountain western United States (Cannon et al., 2008, 2011). Although the rainfall thresholds used to guide the warning system were developed for different regions, most of the debris-flow triggering storms were correctly identified by the rainfall thresholds. Despite the success of the rainfall thresholds correctly identifying most of the storms that produced debris flows, the warning system was unable to provide advanced warning for debris flows. The warning system had a 5-min lag time between receiving the telemetered rainfall data, processing the data and issuing alerts. Debris flows were reported on roads before the warning system had time to issue alerts. This example highlights the importance of incorporating high resolution forecasts and monitoring of doppler radar into warning systems.

In 2021, a period of hot and dry weather in southern British Columbia led to the ignition and rapid spread of the Lytton Creek Fire. The fire burned nearly the entire town of Lytton, B.C. and burned steep terrain along the TransCanada Highway 1 corridor between Lytton and Spences Bridge, B.C. This transportation corridor includes the White Canyon stretch of the Thompson River, which has interrupted the Canadian National Railway with rockfall hazards (e.g., van Veen et al., 2015). Isolated debris flows occurred in response to convective rainfall following the fire. Subsequently, an atmospheric river in November 2021 drenched southern British Columbia and caused a regional landslide and flooding event that washed out numerous highways, caused fatalities, and inundated vast areas with water. The event was characterized as a long-duration, moderate-intensity rainfall with extraordinary rainfall totals. Debris flows were triggered in the area burned by the Lytton Creek Fire; however, rainfall intensity-duration thresholds identified by Cannon et al., (2008, 2011) were only exceeded for intensities averaged over durations greater than 12 h. This example highlights how low intensity rainfall with high rainfall totals can trigger debris flows in burned watersheds with susceptibility to debris flows.

11.7.6 Post-Wildfire Debris-Flow Hazard Assessment in Australia

In 2003, an intense rainfall event following wildfire triggered post-wildfire debris flows that resulted in the death of a firefighter in the Buckland Valley, in northeast Victoria, Australia (Nyman et al., 2011). Prior to this, few references existed to post-wildfire debris flows in southeast Australia; however, reports of "mud torrents," landslides, and flash floods can be found intermittently in the media in the years after fires. Subsequent research (Nyman et al., 2015) has found that post-wildfire debris flows are a common occurrence in this region, with many similarities to active post-wildfire debris flow regions reported by Cannon et al. (2001) in the western United States. In southeast Australia, post-wildfire debris flows are enabled by highly productive forests that produce extreme fuel loads, steep topography, and a Mediterranean climate that results in high intensity summer bushfires and short-duration high-intensity rainfall.

The runoff-generated debris flows observed in southeast Australia are made possible after fire as a result of three key factors that are similar to those described for western North America: (1) fire-enhanced soil water repellency increases hillslope peak flows during rainfall (van der Sant et al., 2018), (2) soil heating reduces surface soil strength, providing abundant available sediment for transport (Nyman et al., 2013), and (3) the removal of surface roughness elements such as tussock grasses and woody debris, which reduces surface roughness and eliminates sediment trapping (Cawson et al., 2012). When high-intensity convective storms occur in steep convergent headwaters, thin hillslope debris flows are initiated (Langhans et al., 2017) and converge at the channel initiation point, where channel debris-flow processes are initiated.

Post-wildfire debris flows result from increased runoff and peak flows, that in turn have resulted from increased soil water repellency after fire. However, not all soils generate these peak flows; soils associated with more productive forests, with higher levels of soil biological activity, tend to maintain high infiltration rates. Even in the presence of strong water repellency, these soils generate little surface runoff due to stable soil structure and macropores that enable infiltration to by-pass water repellent layers in the soil (Langhans et al., 2016). Recognition of this relation has allowed the identification and mapping of soils most at risk of generating debris flows after fire, using proxies for soil biological activity, such as forest type and measures of climatic aridity (Noske et al., 2016; Nyman et al., 2015; Sheridan et al., 2015; van der Sant et al., 2018).

11.8 Summary and Future Research Directions

Over the previous 20 years, numerous advances have been made to better understand when, where, and how large debris flows can be following wildfire. Refinements to rainfall thresholds for post-wildfire debris-flow initiation have enabled forecasters to better identify storms that could produce hazardous flows. This has improved warnings and evacuation notices for affected communities. New methods for rapidly evaluating burn severity through remote sensing coupled with hazard assessment models have enabled emergency assessments to identify the location of burned watersheds most susceptible to producing post-wildfire debris flows and the magnitude of these events. Efforts to improve post-wildfire hazard assessments have focused on refining existing rainfall thresholds, improving empirical models for estimating debris-flow likelihood and magnitude and incorporating debris-flow runout models into hazard assessments to better portray locations of post-wildfire debris-flow effects.

Climate change is increasing the frequency and magnitude of wildfire and changing precipitation regimes. Post-wildfire debris flows will continue to be problematic in areas where they have been historically problematic and will affect new areas where they have been absent or less common in the past.

As wildfire and associated post-wildfire debris flows expand to new areas, novel challenges and questions arise. For example, little information is available about how a burned area responds to rain that falls on snow or how likely a debris flow is to be triggered by a storm with a mix of rain and snow. Additionally, little is known about how changes to surface runoff hydrology following wildfire affects movement of deep-seated landslides, although some anecdotal evidence of this phenomenon exists in British Columbia and northern California.

The current hazard assessment methods, rainfall thresholds, and research on burn severity, terrain, and geology influences on debris-flow susceptibility have focused on areas with a long history of post-wildfire debris flows (e.g., Arizona, southern California, Colorado, Idaho, Montana, New Mexico). Application of the hazard assessment models and knowledge of post-wildfire debris-flow susceptibility to other areas has not always been successful. In locations such as northern California, Oregon, Washington, and the southeastern United States, estimates of debris-flow likelihood and magnitude by empirical models developed by Staley et al. (2016) and Gartner et al. (2014) have over-predicted the hazard. For this reason, there is high importance in collecting post-wildfire event locations and magnitudes and associated data for triggering rainfall intensities, burn severity characteristics, and soil types in areas where post-wildfire debris flows are less frequent. This information would enable the refinement of existing hazard assessment tools as well as the development of new, region-specific models for hazard assessment.

Rapid assessments of post-wildfire debris-flow hazards would also benefit from improvements to debris-flow runout modeling. Current approaches focus on identifying where post-wildfire debris flows will be initiated, however, the most devastating effects of a post-wildfire debris flows are typically experienced farther downstream in the runout areas. Current modeling approaches are time-consuming and are challenging to implement in multiple basins that might be burned by a large wildfire. Advancements to runout models to allow for rapid implementation at a regional scale and proactive post-wildfire debris-flow runout assessments for areas likely to experience wildfire and subsequent debris flows would benefit susceptible communities.

References

Abatzoglou, J. T., & Williams, A. P. (2016). Impact of anthropogenic climate change on wildfire across western US forests. *Proceedings of the National Academy of Sciences, 113*, 11770–11775. https://doi.org/10.1073/pnas.1607171113

Alizadeh, M. R., Abatzoglou, J. T., Luce, C. H., Adamowski, J. F., Farid, A., & Sadegh, M. (2021).

Warming enabled upslope advance in western US forest fires. *Proceedings of the National Academy of Sciences, 118*(22), e2009717118.

Alessio, P., Dunne, T., & Morell, K. (2021). Post-wildfire generation of debris-flow slurry by rill erosion on colluvial hillslopes. *Journal of Geophysical Research: Earth Surface, 126*(11), e2021JF006108. https://doi.org/10.1029/2021JF006108

Ali, H., Fowler, H. J., & Mishra, V. (2018). Global observational evidence of strong linkage between dew point temperature and precipitation extremes. *Geophysical Research Letters, 45*, 320–330. https://doi.org/10.1029/2018GL080557

Barnhart, K. R., Jones, R. P., George, D. L., McArdell, B. W., Rengers, F. K., Staley, D. M., & Kean, J. W. (2021). Multi-model comparison of computed debris flow runout for the 9 January 2018 Montecito, California post-wildfire event. *Journal of Geophysical Research: Earth Surface, 126*, e2021JF006245. https://doi.org/10.1029/2021JF006245

Barnhart, K. R., Romero, V. Y., & Clifford, K. C. (2023). *User needs assessment for postfire debris flow inundation hazard products*. U.S. Geological Survey Open-File Report 2023-1025

Benavides-Solario, J., & MacDonald, L. H. (2005). Measurement and prediction of post-fire erosion at the hillslope scale, Colorado Front Range. *International Journal of Wildland Fire, 14*, 1–18.

Benda, L., & Dunne, T. (1997). Stochastic forcing of sediment supply to channel networks from landsliding and debris flow. *Water Resources Research, 33*(12), 2849–2863. https://doi.org/10.1029/97WR02388

Bernard, D., Trousil, E., & Santi, P. (2021). Estimation of inundation areas of post-wildfire debris flows in southern California USA. *Engineering Geology, 285*, 105991. https://doi.org/10.1016/j.enggeo.2021.105991

Brogan, D. J., Nelson, P. A., & MacDonald, L. H. (2017). Reconstructing extreme post-wildfire floods: A comparison of convective and mesoscale events. *Earth Surface Processes and Landforms, 42*(15), 2505–2522. https://doi.org/10.1002/esp.4194

Burned Area Emergency Response (BAER) Team. (2018). *Carr fire 2018 burned area emergency response plan* (p. 288). Publication of the Bureau of Land Management (Redding Field Office), National Park Service (Whiskeytown National Recreation Area), and U.S. Department of Agriculture (Shasta-Trinity National Forest). https://www.nps.gov/whis/upload/WHIS-Carr-BAER-Re-dacted-20181004-1.pdf

Calhoun, N., Burns, B., Kean, J., & Rengers, F. (2022). Recent observations of post-fire debris flows in five megafires in the western Cascades, Oregon. *Geological Society of America Abstracts with Programs, 54*, 5. https://doi.org/10.1130/abs/2022AM-382132

Cannon, S. H. (1989). *An evaluation of the travel-distance potential of debris flows* (41 p.). Utah Geological and Mineral Survey, Miscellaneous Publication 89-2.

Cannon, S. H., Kirkham, R. M., & Parise, M. (2001). Wildfire-related debris-flow initiation processes, Storm King Mountain, Colorado. *Geomorphology, 39*, 171–188. https://doi.org/10.1016/S0169-555X(00)00108-2

Cannon, S. H., Gartner, J. E., Holland-Sears, A., Thurston, B. M., & Gleason, J. A. (2003a). Debris-flow response of basins burned by the 2002 Coal Seam and Missionary Ridge fires, Colorado. In D. D. Boyer, P. M. Santi, & W. P. Rogers (Eds.), *Engineering geology in Colorado—contributions, trends, and case histories: AEG Special Publication 15* (31 pp.). Colorado Geological Survey Special Publication 55.

Cannon, S. H., Gartner, J. E., Parrett, C., & Parise, M. (2003b). Wildfire-related debris flow generation through episodic progressive sediment bulking processes, western U.S.A. In D. Rickenmann & C. L. Chen (Eds.), *Debris-flow hazards mitigation—mechanics, prediction, and assessment, Proceedings of the third international conference on debris-flow hazards mitigation, Davos, Switzerland, 10–12 September 2003: A.A. Balkema, Rotterdam* (pp. 71–82).

Cannon, S. H., & Gartner, J. E. (2005). Wildfire-related debris flow from a hazards perspective. In M. Jakob & O. Hungr (Eds.), *Debris-flow hazards and related phenomena* (pp. 363–385). Springer.

Cannon, S. H., Gartner, J. E., Wilson, R. C., Bowers, J. C., & Laber, J. L. (2008). Storm rainfall conditions for floods and debris flows from recently burned areas in southwestern Colorado and southern California. *Geomorphology, 96*(3–4), 250–269. https://doi.org/10.1016/j.geomorph.2007.03.019

Cannon, S. H., & DeGraff, J. V. (2009). The increasing wildfire and post-fire debris-flow threat in western USA, and implications for consequences of climate change. In K. Sassa & P. Canuti (Eds.), *Landslides—disaster risk reduction*. Springer. https://doi.org/10.1007/978-3-540-69970-5_9

Cannon, S. H., Gartner, J. E., Rupert, M. G., Michael, J. A., Rea, A. H., & Parrett, C. (2010a). Predicting the probability and volume of postwildfire debris flows in the intermountain western United States. *Geological Society of America Bulletin, 122*, 127–144. https://doi.org/10.1130/B26459.1

Cannon, S. H., Gartner, J. E., Rupert, M. G., Michael, J. A., Staley, D. M., & Worstell, B. B. (2010b). *Emergency assessment of postfire debris-flow hazards for the 2009 Station Fire, San Gabriel Mountains, southern California* (27 p.). U.S. Geological Survey Open-File Report, 2009-1227.

Cannon, S. H., Boldt, E. M., Kean, J. W., Laber, J., & Staley, D. M. (2011). Rainfall intensity–duration thresholds for postfire debris-flow

emergency-response planning. *Natural Hazards, 59*, 209–236. https://doi.org/10.1007/s11069-011-9747-2
Cawson, J. G., Sheridan, G. J., Smith, H. G., & Lane, P. N. J. (2012). Surface runoff and erosion after prescribed burning and the effect of different fire regimes in forests and shrublands: A review. *International Journal of Wildland Fire, 21*(7), 857–872. https://doi.org/10.1071/WF11160
Chawner, W. D. (1935). Alluvial fan flooding: The montrose, California, flood of 1934. *Geographical Review, 25*(2), 255–263. https://doi.org/10.2307/209600
Coogan, S. C., Daniels, L. D., Boychuk, D., Burton, P. J., Flannigan, M. D., Gauthier, S., & Wotton, B. M. (2021). Fifty years of wildland fire science in Canada. *Canadian Journal of Forest Research, 51*(2), 283–302. https://doi.org/10.1139/cjfr-2020-0314
Corringham, T. W., Ralph, F. M., Gershunov, A., Cayan, D. R., & Talbot, C. A. (2019). Atmospheric rivers drive flood damages in the western United States. *Science Advances, 5*(12).
Curry, C. L., Islam, S. U., Zwiers, F. W., & Déry, S. J. (2019). Atmospheric rivers increase future flood risk in Western Canada's largest Pacific river. *Geophysical Research Letters, 46*(3), 1651–1661.
DeBano, L. F., Rice, R. M., Eugene, C. C., & Conrad, C. (1979). *Soil heating in chaparral fires: Effects on soil properties, plant nutrients, erosion, and runoff. Research Paper PSW-RP-145. Berkeley, CA* (21 p.). U.S. Department of Agriculture, Forest Service, Pacific Southwest Forest and Range Experiment Station.
DeBano, L. F. (1981). Water repellent soils: A state-of-the-art. General Technical Report PSW-GTR-46. Berkeley, CA (21 p.). U.S. Department of Agriculture, Forest Service, Pacific Southwest Forest and Range Experiment Station.
DeBano, L. F. (2000). The role of fire and soil heating on water repellency in wildland environments: A review. *Journal of Hydrology, 231*, 195–206.
De Graff, J. V. (2018). A rationale for effective post-fire debris flow mitigation within forested terrain. *Geoenvironmental Disasters, 5*(7). https://doi.org/10.1186/s40677-018-0099-z
De Graff, J. V., Staley, D. M., Stock, G. M., Takenake, K., Gallegos, A. L., & Neptune, C. K. (2022). Rainfall triggering of post-fire debris flows over a 28-year period near El Portal, California, USA. *Environmental and Engineering Geoscience, 28*(1), 133–145. https://doi.org/10.2113/EEG-D-21-00031
De Haas, T., Kruijt, A., & Densmore, A. L. (2018). Effects of debris-flow magnitude–frequency distribution on avulsions and fan development. *Earth Surface Processes and Landforms, 43*(13), 2779–2793. https://doi.org/10.1002/esp.4432
Dennison, P. E., Brewer, S. C., Arnold, J. D., & Moritz, M. A. (2014). Large wildfire trends in the western United States, 1984–2011. *Geophysical Research Letters, 41*, 2928–2933. https://doi.org/10.1002/2014GL059576
deWolfe, V. G., Santi, P. M., Ey, J., & Gartner, J. E. (2008). Effective mitigation of debris flows at Lemon Dam, La Plata County, Colorado. *Geomorphology, 96*, 366–377. https://doi.org/10.1016/j.geomorph.2007.04.008
DiBiase, R. A., & Lamb, M. P. (2013). Vegetation and wildfire controls on sediment yield in bedrock landscapes. *Geophysical Research Letters, 40*(6), 1093–1097.
DiBiase, R. A., & Lamb, M. P. (2020). Dry sediment loading of headwater channels fuels post-wildfire debris flows in bedrock landscapes. *Geology, 48*(2), 189–193. https://doi.org/10.1130/G46847.1
Doerr, S. H., Shakesby, R. A., & Walsh, R. P. D. (2000). Soil water repellency: Its causes, characteristics and hydro-geomorphological significance. *Earth-Science Reviews, 51*, 33–65.
East, A. E., Logan, J. B., Dartnell, P., Lieber-Kotz, O., Cavagnaro, D. B., McCoy, S. W., & Lindsay, D. N. (2021). Watershed sediment yield following the 2018 Carr Fire, Whiskeytown National Recreation Area, Northern California. *Earth and Space Science, 8*(9). https://doi.org/10.1029/2021EA001828
Eaton, E. C. (1935). Flood and erosion control problems and their solution. *Transactions of the American Society of Civil Engineers, 101*, 1302–1362.
Ebel, B. A., Moody, J. A., & Martin, D. A. (2012). Hydrologic conditions controlling runoff generation immediately after wildfire. *Water Resources Research, 48*. https://doi.org/10.1029/2011WR011470
Ellett, N. G., Pierce, J. L., & Glenn, N. F. (2019). Partitioned by process: Measuring post-fire debris-flow and rill erosion with Structure from Motion photogrammetry. *Earth Surface Processes and Landforms, 44*(15), 3128–3146. https://doi.org/10.1002/esp.4728
Espinoza, V., Waliser, D. E., Guan, B., Lavers, D. A., & Ralph, F. M. (2018). Global analysis of climate change projection effects on atmospheric rivers. *Geophysical Research Letters, 45*, 4299–4308. https://doi.org/10.1029/2017GL076968
Fannin, R. J., & Wise, M. P. (2001). An empirical-statistical model for debris flow travel distance. *Canadian Geotechnical Journal, 38*(5), 982–994. https://doi.org/10.1139/cgj-38-5-982
Farin, M., Mangeney, A., & Roche, O. (2014). Fundamental changes of granular flow dynamics, deposition, and erosion processes at high slope angles: Insights from laboratory experiments. *Journal of Geophysical Research: Earth Surface, 119*(3), 504–532. https://doi.org/10.1002/2013JF002750
Flannigan, M. D., Krawchuk, M. A., de Groot, W. J., Wotton, M. B., & Gowman, L. M. (2009). Implications of changing climate for global wildland fire. *International Journal of Wildland Fire, 18*(5), 483–507. https://doi.org/10.1071/WF08187

Florsheim, J. L., Keller, E. A., & Best, D. W. (1991). Fluvial sediment transport in response to moderate storm flows following chaparral wildfire, Ventura County, southern California. *Geological Society of America Bulletin, 103*, 504–511.

Fowler, H. J., Wasko, C., & Prein, A. F. (2021a). Intensification of short-duration rainfall extremes and implications for flood risk: Current state of the art and future directions. *Philosophical Transactions of the Royal Society A, 379*(2195).

Fowler, H. J., Lenderink, G., Prein, A. F., Westra, S., Allan, R. P., Ban, N., & Zhang, X. (2021b). Anthropogenic intensification of short-duration rainfall extremes. *Nature Reviews Earth & Environment, 2*(2), 107–122.

Fusco, E. J., Finn, J. T., Bach, J. K., & Bradley, B. A. (2019). Invasive grasses increase fire occurrence and frequency across US ecoregions. *Proceedings of the National Academy of Sciences, 116*(47), 23594–23599. https://doi.org/10.1073/pnas.1908253116

Gartner, J. E., Cannon, S. H., Bigio, E. R., Davis, N. K., Parrett, C., Pierce, K. L., Rupert, M. G., Thurston, B. L., Trebesch, M. J., Garcia, S. P., & Rea, A. H. (2005). *Compilation of data relating to the erosive response of 608 recently burned basins in the western United States*. U.S. Geological Survey Open-File Report 2005-1218. https://doi.org/10.3133/ofr20051218

Gartner, J. E., Cannon, S. H., Santi, P. M., & de Wolfe, V. G. (2008). Empirical models to pre-dict the volumes of debris flows generated by recently burned basins in the western U.S. *Geomorphology, 96*(3–4), 339–354. https://doi.org/10.1016/j.geomorph.2007.02.033

Gartner, J. E., Cannon, S. H., & Santi, P. M. (2011). Implementation of post-fire debris-flow hazard assessments along drainage networks, southern California, U.S.A. In R. Genevois, D. L. Hamilton, & A. Prestininzi (Eds.), *Proceedings of the Fifth International Conference on Debris Flow Hazards Mitigation/Mechanics, Prediction, and Assessment Padua, Italy, June 7–11, 2011, Italian Journal of Engineering Geology and Environment–Book: Casa Editrice Universita La Sapienza, Rome* (pp. 855–863). https://doi.org/10.4408/IJEGE.2011-03.B-090

Gartner, J. E., Cannon, S. H., & Santi, P. M. (2014). Empirical models for predicting volumes of sediment deposited by debris flows and sediment-laden floods in the Transverse Ranges of southern California. *Engineering Geology, 176*, 45–56. https://doi.org/10.1016/j.enggeo.2014.04.008

Gartner, J. E., Santi, P. M., & Cannon, S. H. (2015). Predicting locations of post-fire debris-flow erosion in the San Gabriel Mountains of southern California. *Natural Hazards*.https://doi.org/10.1007/s11069-015-1656-3

Garcia-Ruiz, J. M., Arnaez, J., Gomez-Villar, A., Ortigosa, L., & Lana-Renault, N. (2013). Fire-related debris-flows in the Iberian Range, Spain. *Geomorphology, 196*, 221–230.

George, D. L., & Iverson, R. M. (2014). A depth-averaged debris-flow model that includes the effects of evolving dilatancy: II. Numerical Predictions and Experimental Tests. *Proceedings of the Royal Society A, 470*, 20130820. https://doi.org/10.1098/rspa.2013.0820

Gershunov, A., Shulgina, T., & Clemesha, R. E. S. (2019). Precipitation regime change in Western North America: The role of atmospheric rivers. *Scientific Reports, 9*, 9944. https://doi.org/10.1038/s41598-019-46169-w

Gibson, S., Moura, L. Z., Ackerman, C., Ortman, N., Amorim, R., Floyd, I., Eom, M., Creech, C., & Sánchez, A. (2022). Prototype scale evaluation of non-Newtonian algorithms in HEC-RAS: Mud and debris flow case studies of Santa Barbara and Brumadinho. *Geosciences, 12*(3), 134. https://doi.org/10.3390/geosciences12030134

Girona-Garcia, A., Vieira, D. C. S., Silva, J., Fernandez, C., Robichaud, P. R., & Keizer, J. J. (2021). Effectiveness of post-fire soil erosion mitigation treatments: A systematic review and meta-analysis. *Earth Science Reviews*. https://doi.org/10.1016/j.earscirev.2021.103611

Gorr, A. N., McGuire, L. A., Youberg, A. M., & Rengers, F. K. (2022). A progressive flow-routing model for rapid assessment of debris-flow inundation. *Landslides, 19*, 2055–2073. https://doi.org/10.1007/s10346-022-01890-y

Guerreiro, S. B., Fowler, H. J., Barbero, R., Westra, S., Lenderink, G., Blenkinsop, S., & Li, X. F. (2018). Detection of continental-scale intensification of hourly rainfall extremes. *Nature Climate Change, 8*(9), 803–807.

Guilinger, J. J., Gray, A. B., Barth, N. C., & Fong, B. T. (2020). The evolution of sediment sources over a sequence of postfire sediment-laden flows revealed through repeat high-resolution change detection. *Journal of Geophysical Research: Earth Surface, 125*(10), e2020JF005527. https://doi.org/10.1029/2020JF005527

Hanes, C., Wang, X., Jain, P., Parisien, M., Little, J., & Flannigan, M. (2018). Fire-regime changes in Canada over the last half century. *Canadian Journal of Forest Research, 49*(3), 256–269. https://doi.org/10.1139/cjfr-2018-0293

Haugo, R. D., Kellogg, B. S., Cansler, C. A., Kolden, C. A., Kemp, K. B., Robertson, J. C., Metlen, K. L., Vaillant, N. M., & Restaino, C. M. (2019). The missing fire: Quantifying human exclusion of wildfire in Pacific Northwest forests, USA. *Ecosphere, 10*, e02702. https://doi.org/10.1002/ecs2.2702

Hill, J. S., Douglas, T. J., Korte, D. M., Scheip, C. M., Wooten, R. M., & Palmer, J. M. (2020). Debris flows triggered by August 24, 2019 storm in the Nantahala Gorge, western North Carolina; Did the underlying bedrock and the 2016 wildfires increase landslide susceptibility. *Geological Society of*

America Abstracts with Programs, 52(2). https://doi.org/10.1130/abs/2020SE-344730

Hoch, O. J., McGuire, L. A., Youberg, A. M., & Rengers, F. K. (2021). Hydrogeomorphic recovery and temporal changes in rainfall thresholds for debris flows following wildfire. *Journal of Geophysical Research: Earth Surface, 126*(12), e2021JF006374. https://doi.org/10.1029/2021JF006374

Hope, G., Jordan, P., Winkler, R., Giles, T., Curran, M., Soneff, K., & Chapman, B. (2015). Post-wildfire natural hazards risk analysis in British Columbia. Province of B.C., Victoria, B.C. Land Management Handbook 69. https://www.for.gov.bc.ca/hfd/pubs/Docs/Lmh/LMH69.htm

Hungr, O., McDougal, S., & Bovis, M. (2005). Entrainment of material by debris flows. In M. Jakob & O. Hungr (Eds.), *Debris-flow hazards and related phenomena* (pp. 135–158). Springer.

Hungr, O., Leroueil, S., & Picarelli, L. (2014). The Varnes classification of landslide types, an update. *Landslides, 11*, 167–194. https://doi.org/10.1007/s10346-013-0436-y

Hürlimann, M., Rickenmann, D., & Graf, C. (2003). Field and monitoring data of debris-flow events in the Swiss Alps. *Canadian Geotechnical Journal, 40*, 161–175. https://doi.org/10.1139/t02-087

Iglesias, V., Balch, J. K., & Travis, W. R. (2022). U.S. fires became larger, more frequent, and more widespread in the 2000s. *Science Advances, 8*(11), eabc0020. https://doi.org/10.1126/sciadv.abc0020

Iverson, R. M. (1997). The physics of debris flow. *Review of Geophysics, 35*, 245–296. https://doi.org/10.1029/97RG00426

Iverson, R. M., Schilling, S. P., & Vallance, J. W. (1998). Objective delineation of lahar-inundation hazard zones. *Geological Society of America Bulletin, 110*, 972–984. https://doi.org/10.1130/0016-7606(1998)110<0972:ODOLIH>2.3.CO;2

Jackson, M., & Roering, J. J. (2009). Post-fire geomorphic response in steep, forested landscapes: Oregon Coast Range, USA. *Quaternary Science Reviews, 28*, 1131–1146. https://doi.org/10.1016/j.quascirev.2008.05.003

Jakob, M., & Hungr, O. (2005). *Debris-flow hazards and related phenomena* (739 p.). Springer. https://doi.org/10.1007/b138657

Johnson, A. (1984). Debris flow. In: D. P. D. Brunsden & D. B. Prior (Eds.), *Slope instability* (pp. 257–361). Wiley.

Jordan, P., & Covert, S. A. (2009). Debris flows and floods following the 2003 wildfires in southern British Columbia. *Environmental & Engineering Geoscience, 15*, 217–234. https://doi.org/10.2113/GSEEGEOSCI.15.4.217

Jordan, P. (2015). Post-wildfire debris flows in southern British Columbia, Canada. *International Journal of Wildland Fire*. https://doi.org/10.1071/WF14070

Kane, W. F., & Jones, M. A. (2019). Flexible debris-flow nets for post wildfire debris mitigation in the western United States. In J. W. Kean, J. A. Coe, P. M. Santi, & B. K. Guillen (Eds.), *Proceedings of the 7th International Conference on Debris-Flow Hazards Mitigation, Golden, Colorado, USA, June 10–13, 2019*. Association of Environmental and Engineering Geologists Special Publication 28.

Kean, J. W., Staley, D. M., & Cannon, S. H. (2011). In situ measurements of post-fire debris flows in southern California: Comparisons of the timing and magnitude of 24 debris-flow events with rainfall and soil moisture conditions. *Journal of Geophysical Research: Earth Surface, 116*, F04019. https://doi.org/10.1029/2011JF002005

Kean, J. W., Staley, D. M., Leeper, R. J., Schmidt, K. M., & Gartner, J. E. (2012). A low-cost method to measure the timing of postfire flash floods and debris flows relative to rainfall. *Water Resources Research, 48*, 5. https://doi.org/10.1029/2011WR011460

Kean, J. W., McCoy, S. W., Tucker, G. E., Staley, D. M., & Coe, J. A. (2013). Runoff-generated debris flows: Observations and modeling of surge initiation, magnitude, and frequency. *Journal of Geophysical Research: Earth Surface, 118*, 2190–2207. https://doi.org/10.1002/jgrf.20148

Kean, J. W., McGuire, L. A., Rengers, F. K., Smith, J. B., & Staley, D. M. (2016). Amplification of post-wildfire peak flow by debris. *Geophysical Research Letters*, 2016GL069661. https://doi.org/10.1002/2016GL069661

Kean, J. W., Staley, D. M., Lancaster, J. T., Rengers, F. K., Swanson, B. J., Coe, J. A., Hernandez, J. L., Sigman, A. J., Allstadt, K. E., & Lindsay, D. N. (2019). Inundation, flow dynamics, and damage in the 9 January 2018 Montecito debris-flow event, California, USA: Opportunities and challenges for post-wildfire risk assessment. *Geosphere, 15*(4), 1140–1163. https://doi.org/10.1130/GES02048.1

Kean, J. W., & Staley, D. M. (2021). Forecasting the frequency and magnitude of post-fire debris flows across southern California. *Earth's Future, 9*, e2020EF001735. https://doi.org/10.1029/2020EF001735

Keeley, J. E. (2009). Fire intensity, fire severity and burn severity: A brief review and suggested usage. *International Journal of Wildland Fire, 18*, 116–126. https://doi.org/10.1071/WF07049

Key, C. H., & Benson, N. C. (2006). *Landscape assessment (LA) sampling and analysis methods. USDA Forest Service—General Technical Report RMRS-GTR (164 RMRSGTR)* (pp. LA-1–LA-51).

Kinner, D. A., & Moody, J. A. (2008). *Infiltration and runoff measurements on steep burned hillslopes using a rainfall simulator with variable rain intensities* (64 p.). U.S. Geological Survey Scientific Investigations Report 2007-5211.

Kinner, D. A., & Moody, J. A. (2010). Spatial variability of steady-state infiltration into a two-layer soil system on burned hillslopes. *Journal of Hydrology, 381*(3–4), 322–332. https://doi.org/10.1016/j.jhydrol.2009.12.004

Lamb, M. P., Scheingross, J. S., Amidon, W. H., Swanson, E., & Limaye, A. (2011). A model for fire-induced sediment yield by dry ravel in steep landscapes. *Journal of Geophysical Research: Earth Surface, 116*, F03006. https://doi.org/10.1029/2010JF001878

Lancaster, J. T., Swanson, B. J., Lukashov, S. G., Oakley, N. S., Lee, J. B., Spangler, E. R., Hernandez, J. L., Olson, B., DeFrisco, M., Lindsay, D. L., Schwartz, Y. J., McCrea, S. E., Roffers, P. D., & Tran, C. M. (2021). Observations and analyses of the 9 January 2018 debris-flow disaster, Santa Barbara County, California. *Environmental and Engineering Geoscience, 27*(1), 3–27. https://doi.org/10.2113/EEG-D-20-00015

Langhans, C., Smith, H. G., Chong, D. M. O., Nyman, P., Lane, P. N. J., & Sheridan, G. J. (2016). A model for assessing water quality risk in catchments prone to wildfire. *Journal of Hydrology, 534*, 407–426. https://doi.org/10.1016/j.jhydrol.2015.12.048

Langhans, C., Nyman, P., Noske, P. J., van der Sant, R. E., Lane, P. N. J., & Sheridan, G. J. (2017). Post-fire hillslope debris flows: Evidence of a distinct erosion process. *Geomorphology, 295*, 55–75. https://doi.org/10.1016/j.geomorph.2017.06.008

Larsen, I. J., MacDonald, L. H., Brown, E., Rough, D., Welsh, M. J., Pietraszek, J. H., Libohova, Z., Benavides-Solorio, J., & Schaffrath, K. (2009). Causes of post-fire runoff and erosion: Roles of soil water repellency, surface cover, and soil sealing? *Soil Science Society of America Journal, 73*, 1393–1407. https://doi.org/10.2136/sssaj2007.0432

Lenderink, G., Barbero, R., Loriaux, J. M., & Fowler, H. J. (2017). Super-Clausius–Clapeyron scaling of extreme hourly convective precipitation and its relation to large-scale atmospheric conditions. *Journal of Climate, 30*(15), 6037–6052.

Major, J. J., & Iverson, R. M. (1999). Debris-flow deposition: Effects of pore-fluid pressure and friction concentrated at flow margins. *Geological Society of America Bulletin, 11*(10), 1424–1434. https://doi.org/10.1130/0016-7606(1999)111<1424:DFDEOP>2.3.CO;2

May, C. L., & Gresswell, R. E. (2003). Processes and rates of sediment and wood accumulation in headwater streams of the Oregon Coast Range, USA. *Earth Surface Processes and Landforms, 28*(4), 409–424.

McCoy, K., Krasko, V., Santi, P., Kaffine, D., & Rebennack, S. (2016). Minimizing economic impacts from post-fire debris flows in the western United States. *Natural Hazards, 83*, 149–176. https://doi.org/10.1007/s11069-016-2306-0

McCoy, S. W., Kean, J. W., Coe, J. A., Staley, D. M., Wasklewicz, T. A., & Tucker, G. E. (2010). Evolution of a natural debris flow: In situ measurements of flow dynamics, video imagery, and terrestrial laser scanning. *Geology, 38*(8), 735–738. https://doi.org/10.1130/G30928.1

McDougall, S., & Hungr, O. (2004). A model for the analysis of rapid landslide motion across three-dimensional terrain. *Canadian Geotechnical Journal, 41*, 1084–1097. https://doi.org/10.1139/t04-052

McGuire, L. A., Kean, J. W., Staley, D. M., Rengers, F. K., & Wasklewicz, T. A. (2016). Constraining the relative importance of raindrop-and flow-driven sediment transport mechanisms in post-wildfire environments and implications for recovery time scales. *Journal of Geophysical Research: Earth Surface, 121*(11), 2211–2237. https://doi.org/10.1002/2016JF003867

McGuire, L. A., Rengers, F. K., Kean, J. W., & Staley, D. M. (2017). Debris flow initiation by runoff in a recently burned basin: Is grain-by-grain sediment bulking or en masse failure to blame? *Geophysical Research Letters, 44*, 7310–7319. https://doi.org/10.1002/2017GL074243

McGuire, L. A., & Youberg, A. M. (2020). What drives spatial variability in rainfall intensity-duration thresholds for post-wildfire debris flows? Insights from the 2018 Buzzard Fire, NM, USA. *Landslides*. https://doi.org/10.1007/s10346-020-01470-y

Melton, M. A. (1957). *An analysis of the relation among elements of climate, surface properties and geomorphology*. Office of Naval Research, Department of Geology, Columbia University, New York. Technical Report 11. https://doi.org/10.7916/d8-0rmg-j112

Melo, R., & Zêzere, J. L. (2017). Modeling debris flow initiation and run-out in recently burned areas using data-driven methods. *Natural Hazards, 88*, 1373–1407. https://doi.org/10.1007/s11069-017-2921-4

Meyer, G., Wells, S. G., Balling, R., Jr., & Jull, A. J. T. (1992). Response of alluvial systems to fire and climate change in Yellowstone National Park. *Nature, 357*(6374), 147–150.

Meyer, G. A., Pierce, J. L., Wood, S. H., & Jull, A. T. (2001). Fire, storms, and erosional events in the Idaho batholith. *Hydrological Processes, 15*(15), 3025–3038.

Moody, J. A., & Ebel, B. A. (2012). Hyper-dry conditions provide new insights into the cause of extreme floods after wildfire. *CATENA, 93*, 58–63. https://doi.org/10.1016/j.catena.2012.01.006

Moody, J. A., Martin, D. A., Haire, S. L., & Kinner, D. A. (2008). Linking runoff response to burn severity after wildfire. *Hydrological Processes, 22*, 2063–2074. https://doi.org/10.1002/hyp.6806

Moody, J. A., & Martin, D. A. (2009a). Forest fire effects on geomorphic processes. In A. Cerdà & P. R. Robichaud (Eds.), *Fire effects on soils and restoration strategies* (pp. 41–79). Science Publishers.

Moody, J. A., & Martin, D. A. (2009b). Synthesis of sediment yields after wildland fire in different rainfall regimes in the western United States. *International Journal of Wildland Fire, 18*, 96–115. https://doi.org/10.1071/WF07162

Moody, J. A., & Kinner, D. A. (2006). Spatial structures of stream and hillslope drainage networks following gully erosion after wildfire. *Earth Surface Processes and Landforms, 31*, 319–337. https://doi.org/10.1002/esp.1246

Moody, J. A., Shakesby, R. A., Robichaud, P. R., Cannon, S. H., & Martin, D. A. (2013). Current research issues related to post-wildfire runoff and erosion processes. *Earth Science Reviews, 122*, 10–37. https://doi.org/10.1016/j.earscirev.2013.03.004

Morell, K. D., Alessio, P., Dunne, T., & Keller, E. (2021). Sediment recruitment and redistribution in mountain channel networks by post-wildfire debris flows. *Geophysical Research Letters*, e2021GL095549. https://doi.org/10.1029/2021GL095549

NOAA-USGS Debris Flow Task Force. (2005). *NOAA-USGS debris-flow warning system—final report* (47 p.). U.S. Geological Survey Circular 1283. https://doi.org/10.3133/cir1283

National Oceanic and Atmospheric Administration (NOAA). (2021). National Weather Service: Monthly observed precipitation summaries. https://www.cnrfc.noaa.gov/rainfall_data.php

Neptune, C. K., Degraff, J. V., Pluhar, C. J., Lancaster, J. T., & Staley, D. M. (2021). Rainfall thresholds for post-fire debris-flow generation, western Sierra Nevada, CA. *Environmental & Engineering Geoscience, 27*(4), 439–453. https://doi.org/10.2113/EEG-D-21-00039

Noske, P. J., Nyman, P., Lane, P. N. J., & Sheridan, G. J. (2016). Effects of aridity in controlling the magnitude of runoff and erosion after wildfire. *Water Resources Research, 52*(6), 4338–4357. https://doi.org/10.1002/2015WR017611

Nyman, P., Sheridan, G. J., Smith, H. G., & Lane, P. N. J. (2011). Evidence of debris flow occurrence after wildfire in upland catchments of south-east Australia. *Geomorphology, 125*, 383–401. https://doi.org/10.1016/j.geomorph.2010.10.016

Nyman, P., Sheridan, G. J., Moody, J. A., Smith, H. G., Noske, P. J., & Lane, P. N. J. (2013). Sediment availability on burned hillslopes. *Journal of Geophysical Research: Earth Surface, 118*(4), 2451–2467. https://doi.org/10.1002/jgrf.20152

Nyman, P., Smith, H. G., Sherwin, C. B., Langhans, C., Lane, P. N. J., & Sheridan, G. J. (2015). Predicting sediment delivery from debris flows after wildfire. *Geomorphology, 250*, 173–186. https://doi.org/10.1016/j.geomorph.2015.08.023

Nyman, P., Yeates, P., Langhans, C., Noske, P. J., Peleg, N., Schärer, C., Lane, P. N. J., Haydon, S., & Sheridan, G. J. (2021). Probability and consequence of postfire erosion for treatability of water in an unfiltered supply system. *Water Resources Research, 57*(1), 1–17. https://doi.org/10.1029/2019WR026185

Oakley, N. S., Cannon, F., Munroe, R., Lancaster, J. T., Gomberg, D., & Ralph, F. M. (2018). Brief communication: Meteorological and climatological conditions associated with the 9 January 2018 post-fire debris flows in Montecito and Carpinteria, California, USA. *Natural Hazards and Earth System Sciences, 18*(11), 3037–3043. https://doi.org/10.5194/nhess-18-3037-2018

Oakley, N. S. (2021). A warming climate adds complexity to post-fire hydrologic hazard planning. *Earth's Future, 9*, e2021EF002149. https://doi.org/10.1029/2021EF002149

Pelletier, J. D., & Orem, C. A. (2014). How do sediment yields from post-wildfire debris-laden flows depend on terrain slope, soil burn severity class, and drainage basin area? Insights from airborne-lidar change detection. *Earth Surface Processes and Landforms, 39*(13), 1822–1832. https://doi.org/10.1002/esp.3570

Perkins, J. P., Diaz, C., Corbett, S. C., Cerovski-Darriau, C., Stock, J. D., & Prancevic, J. P. (2022). Multi-stage soil-hydraulic recovery and limited ravel accumulations following the 2017 Nuns and Tubbs wildfires in Northern California. *Journal of Geophysical Research: Earth Surface, 127*, e2022JF006591. https://doi.org/10.1029/2022JF006591

Radeloff, V. C., Helmers, D. P., Kramer, H. A., Mockrin, M. H., Alexandre, P. M., Bar-Massada, A., & Stewart, S. I. (2018). Rapid growth of the US wildland-urban interface raises wildfire risk. *Proceedings of the National Academy of Sciences, 115*(13), 3314–3319. https://doi.org/10.1073/pnas.1718850115

Ralph, F. M., Neiman, P. J., Wick, G. A., Gutman, S. I., Dettinger, M. D., Cayan, D. R., & White, A. B. (2006). Flooding on California's Russian River: Role of atmospheric rivers. *Geophysical Research Letters, 33*(13).

Raymond, C. A., McGuire, L. A., Youberg, A. M., Staley, D. M., & Kean, J. W. (2020). Thresholds for post-wildfire debris flows: Insights from the Pinal Fire, Arizona, USA. *Earth Surface Processes and Landforms, 45*(6), 1349–1360. https://doi.org/10.1002/esp.4805

Rengers, F. K., McGuire, L. A., Oakley, N. S., Kean, J. W., Staley, D. M., & Tang, H. (2020). Landslides after wildfire: Initiation, magnitude, and mobility. *Landslides, 17*(11), 2631–2641. https://doi.org/10.1007/s10346-020-01506-3

Rengers, F. K., McGuire, L. A., Kean, J. W., Staley, D. M., Dobre, M., Robichaud, P. R., & Swetnam, T. (2021). Movement of sediment through a burned landscape: Sediment volume observations and model comparisons in the San Gabriel Mountains, California, USA. *Journal of Geophysical Research: Earth Surface, 126*(7), e2020JF006053. https://doi.org/10.1029/2020JF006053

Robichaud, P. R., Ashmun, L. E., & Sims, B. D. (2010). *Post-fire treatment effectiveness for hillslope stabilization. General Technical Report RMRS-GTR-240. Fort Collins, CO* (62 p.). U.S. Department of

Agriculture, Forest Service, Rocky Mountain Research Station.
Rompkins, M. J. M., Prasad, S. N., & Whisler, F. D. (1990). Surface sealing and infiltration. Chapter 5 In: M. G. Anderson, T. P. Burt (Eds.), *Process studies in hillslope hydrology* (pp. 127–172). Wiley.
Rowe, P. B., Countryman, C. M., & Storey, H. C. (1949). *Probable peak discharges and erosion rates from southern California watersheds as influenced by fire* (270 p.). U.S. Department of Agriculture, Forest Service.
Rice, R. M., & Foggin, G. T., III. (1971). Effect high intensity storms on soil slippage on mountainous watersheds in Southern California. *Water Resources Research, 7*(6), 1485–1496.
Running, S. W. (2006). Is global warming causing more, larger wildfires? *Science, 313*, 927–928. https://doi.org/10.1126/science.1130370
Santi, P., & MacAulay, B. (2021). Water and sediment supply requirements for post-wildfire debris flows in the western United States. *Environmental and Engineering Geology, 27*, 73–85. https://doi.org/10.2113/EEG-D-20-00022
Santi, P. M., & Rengers, F. K. (2020). *Wildfire and landscape change*. Elsevier.https://doi.org/10.1016/B978-0-12-818234-5.00017-1
Santi, P. M., deWolfe, V. G., Higgins, J. D., Cannon, S. H., & Gartner, J. E. (2008). Sources of debris flow material in burned areas. *Geomorphology, 96*(3–4), 310–321. https://doi.org/10.1016/j.geomorph.2007.02.022
Scheip, C., & Wegmann, K. (2022). Insights on the growth and mobility of debris flows from repeat high-resolution lidar. *Landslides*, 1–23. https://doi.org/10.1007/s10346-022-01862-2
Schürch, P., Densmore, A. L., Rosser, N. J., & McArdell, B. W. (2011). Dynamic controls on erosion and deposition on debris-flow fans. *Geology, 39*(9), 827–830. https://doi.org/10.1130/G32103.1
Schwartz, G. E., & Alexander, R. B. (1995). *Soils data for the Conterminous United States Derived from the NRCS State Soil Geographic (STATSGO) Data Base*. U.S. Geological Survey Open-File Report 95-449. https://doi.org/10.3133/ofr95449
Scott, K. M., & Williams, R. P. (1978). *Erosion and sediment yield in the Transverse Ranges, southern California* (38 p.). U.S. Geological Survey Professional Paper 1030. https://doi.org/10.3133/pp1030
Sharma, A. R., & Déry, S. J. (2020). Linking atmospheric rivers to annual and extreme river runoff in British Columbia and southeastern Alaska. *Journal of Hydrometeorology, 21*(11), 2457–2472. https://journals.ametsoc.org/view/journals/hydr/21/11/JHM-D-19-0281.1.xml
Sheridan, G. J., Nyman, P., Langhans, C., Cawson, J. G., Noske, P. J., Oono, A., van de Sant, R., & Lane, P. N. J. (2015). Is aridity a high-order control on the hydro-geomorphic response of burned landscapes? *International Journal of Wildland Fire*. https://doi.org/10.1071/WF14079
Siirila-Woodburn, E. R., Rhoades, A. M., Hatchett, B. J., Huning, L. S., Szinai, J., Tague, C., & Kaatz, L. (2021). A low-to-no snow future and its impacts on water resources in the western United States. *Nature Reviews Earth and Environment, 2*(11), 800–819. https://doi.org/10.1038/s43017-021-00219-y
Staley, D. M., Kean, J. W., Cannon, S. H., Schmidt, K. M., & Laber, J. L. (2013). Objective definition of rainfall intensity–duration thresholds for the initiation of post-fire debris flows in southern California. *Landslides, 10*, 547–562. https://doi.org/10.1007/s10346-012-0341-9
Staley, D. M., Gartner, J. E., & Kean, J. W. (2015). Objective definition of rainfall intensity-duration thresholds for post-fire flash floods and debris flows in the area burned by the Waldo Canyon Fire, Colorado, USA. In G. Lollino, et al. (Eds.). *Engineering geology for society and territory* (Vol. 2). https://doi.org/10.1007/978-3-319-09057-3
Staley, D. M., Negri, J. A., Kean, J. W., Laber, J. L., Tillery, A. C., & Youberg, A. M. (2016). *Updated logistic regression equations for the calculation of post-fire debris-flow likelihood in the western United States*. U.S. Geological Survey Open-File Report 2016-1106. https://doi.org/10.3133/ofr20161106
Staley, D. M., Negri, J. A., Kean, J. W., Laber, J. L., Tillery, A. C., & Youberg, A. M. (2017). Prediction of spatially explicit rainfall intensity–duration thresholds for post-fire debris-flow generation in the western United States. *Geomorphology, 278*, 149–162. https://doi.org/10.1016/j.geomorph.2016.10.019
Staley, D. M., Tillery, A. C., Kean, J. W., McGuire, L. A., Pauling, H. E., Rengers, F. K., & Smith, J. B. (2018). Estimating post-fire debris-flow hazards prior to wildfire using a statistical analysis of historical distributions of fire severity from remote sensing data. *International Journal of Wildland Fire, 27*, 595–608. https://doi.org/10.1071/WF17122
Staley, D. M., Kean, J. W., & Rengers, F. K. (2020). The recurrence interval of post-fire debris-flow generating rainfall in the southwestern United States. *Geomorphology, 370*. https://doi.org/10.1016/j.geomorph.2020.107392
Stiefel, L. C., Cooley, S. C., & Johnson, B. G. (2021). Increased colluvial hollow discharge and subsequent recovery after a low intensity wildfire in the Blue Ridge Mountains, USA. *Hydrological Processes, 35*(1), e13971. https://doi.org/10.1002/hyp.13971
Stoof, C. R., Vervoort, R., Iwema, J., Elsen, E., Ferreira, A., & Ritsema, C. (2012). Hydrological response of a small catchment burned by experimental fire. *Hydrology and Earth System Sciences, 16*(2), 267–285. https://doi.org/10.5194/hess-16-267-2012
Suter, I. H. (2022). Mapping post-wildfire erosion across California using very high-resolution multitemporal satellite imagery [M.S., University of Nevada, Reno].

In *ProQuest dissertations and theses*. http://www.proquest.com/pqdtlocal1006038/docview/2693753082/abstract/E003C66C9F37405BPQ/2
Swain, D. L. (2021). A shorter, sharper rainy season amplifies California wildfire risk. *Geophysical Research Letters, 48*, e2021GL092843. https://doi.org/10.1029/2021GL092843
Swanson, F. J. (1981). Fire and geomorphic processes. In: *General technical report, USDA forest service, WO-26, Washington D.C.* (pp. 401–420).
Swanson, B. J., Lukashov, S. G., Schwartz, J. Y., Lindsay, D. N., & Lancaster, J. T. (2022). Assessment of logistic regression model performance and physical controls on January 9, 2018, debris flows, Thomas Fire, California. *Environmental and Engineering Geoscience, 28*(1), 113–131. https://doi.org/10.2113/EEG-D-21-00026
Tang, H., McGuire, L. A., Rengers, F. K., Kean, J. W., Staley, D. M., & Smith, J. B. (2019). Evolution of debris flow initiation mechanisms and sediment sources during a sequence of post-wildfire rainstorms. *Journal of Geophysical Research: Earth Surface*. https://doi.org/10.1029/2018JF004837
Thomas, M. A., Rengers, F. K., Kean, J. W., McGuire, L. A., Staley, D. M., Barnhart, K. R., & Ebel, B. A. (2021). Post-wildfire soil-hydraulic recovery and the persistence of debris flow hazards. *Journal of Geophysical Research: Earth Surface, 126*(6), e2021JF006091. https://doi.org/10.1029/2021JF006091
Thomson, A. M., Calvin, K. V., Smith, S. J., et al. (2011). RCP4.5: A pathway for stabilization of radiative forcing by 2100. *Climatic Change, 109*, 77. https://doi.org/10.1007/s10584-011-0151-4
Trenberth, K. E., Dai, A., Rasmussen, R. M., & Parsons, D. B. (2003). The changing character of precipitation. *Bulletin of the American Meteorological Society, 84*(9), 1205–1218. https://journals.ametsoc.org/view/journals/bams/84/9/bams-84-9-1205.xml
Troxell, H. C., & Peterson, J. Q. (1937). *Flood in la Canada valley* (98 p.). U.S. Geological Survey Water Supply Paper 796-C. https://doi.org/10.3133/wsp796C
van der Sant, R. E., Nyman, P., Noske, P. J., Langhans, C., Lane, P. N. J., & Sheridan, G. J. (2018). Quantifying relations between surface runoff and aridity after wildfire. *Earth Surface Processes and Landforms, 43*(10), 2033–2044. https://doi.org/10.1002/esp.4370
VanDine, D. (1985). Debris flows and debris torrents in the southern Canadian Cordillera. *Canadian Geotechnical Journal, 22*(1), 44–68. https://doi.org/10.1139/t85-006
VanDine, D. F. (1996). *Debris flow control structures for forest engineering*. Research Branch, B.C. Ministry of Forests, Victoria, B.C., Work Paper 08/1996.
van Veen, M., Hutchinson, D. J., Kromer, R., Lato, M., Gauthier, D., & Edwards, T. (2015). *Frequency-magnitude of rockfall events for hazard analysis; a comparison of data from LiDAR scanning with traditional methods of reporting. GeoQuebec. Quebec City, QC, Canada.*
Vázquez, R., Capra, L., & Coviello, V. (2016). Factors controlling erosion/deposition phenomena related to lahars at Volcán de Colima, Mexico. *Natural Hazards and Earth System Sciences, 16*(8), 1881–1895. https://doi.org/10.5194/nhess-16-1881-2016
Wagner, D. L., Lancaster, J. T., & DeRose, M. B. (2012). *The Oak Creek Post fire debris and hyperconcentrated flows of Jusly 12, 2008 Inyo County, California: A geologic investigation* (73 p.). Special Report 225, 1, California Geological Survey.
Wall, S. A., Roering, J. J., & Rengers, F. K. (2020). Runoff-initiated post-fire debris flow Western Cascades, Oregon. *Landslides, 17*, 1649–1661. https://doi.org/10.1007/s10346-020-01376-9
Wang, Y., Hu, X., Wu, L., Ma, G., Yang, Y., & Jing, T. (2021). Evolutionary history of post-fire debris flows in Ren'e Yong valley in Sichuan Province of China. *Landslides*. https://doi.org/10.1007/s10346-022-01867-x
Warner, M. D., Mass, C. F., & Salathe, E. P., Jr. (2015). Changes in winter atmospheric rivers along the North American west coast in CMIP5 climate models. *Journal of Hydrometeorology, 16*(1), 118–128.
Wasko, C., Lu, W. T., & Mehrotra, R. (2018). Relationship of extreme precipitation, dry-bulb temperature, and dew point temperature across Australia. *Environmental Research Letters, 13*(7), 074031.
Wells II, W. G. (1987). The effects of fire on the generation of debris flows in southern California. In J. E. Costa (Ed.), *Debris flow/avalanches: Process, recognition, and mitigation. Reviews in engineering geology, III* (pp. 105–114).
WERT (Watershed Emergency Response Team). (2018). Thomas fire—watershed emergency response team final report, CA-VNC-103156: Sacramento, California, CAL FIRE.
Westerling, A. L., Hidalgo, H. G., Cayan, D. R., & Swetnam, T. W. (2006). Warming and earlier spring increase western US forest wildfire activity. *Science, 313*, 940–943. https://doi.org/10.1126/science.1128834
Williams, A. P., Abatzoglou, J. T., Gershunov, A., Guzman-Morales, J., Bishop, D. A., Balch, J. K., & Lettenmaier, D. P. (2019). Observed impacts of anthropogenic climate change on wildfire in California. *Earth's Future, 7*, 892–910. https://doi.org/10.1029/2019EF001210
Wilford, D. J., Sakals, M. E., Innes, J. L., Sidle, R. C., & Bergerud, W. A. (2004). Recognition of debris flow, debris flood and flood hazard through watershed morphometrics. *Landslides, 1*, 61–66.
Wilson, C., Kampf, S. K., Ryan, S., Covino, T., MacDonald, L. H., & Gleason, H. (2020). Connectivity of post-fire runoff and sediment from nested hillslopes and watersheds. *Hydrological Processes, 35*(1). https://doi.org/10.1002/hyp.13975

Wondzell, S., & King, J. (2003). Postfire erosional processes in the Pacific northwest and Rocky Mountain regions. *Forest Ecology and Management, 178*, 75–87.

Woodsmith, R., Vache, K., McDonnell, J., & Helvey, D. (2004). Entiat experimental forest: Catchment-scale runoff data before and after a 1970 wildfire. *Water Resources Research, 40*, 1–5. https://doi.org/10.1029/2004WR003296

Woolhiser, D. A., Smith, R. E., & Goodrich, D. C. (1990). *KINEROS, A kinematic runoff and erosion model: Documentation and user manual, (1–127)*. U.S. Department of Agriculture, Agricultural Research Service. https://www.tucson.ars.ag.gov/unit/Publications/PDFfiles/703.pdf

Youberg, A. M. (2008). *Delineating post-wildfire debris-flow hazards for pre-fire mitigation, Pine and Strawberry, Arizona, a FEMA 5% initiative study*. Arizona Geological Survey.

Youberg, A. M. (2014). *Modern and ancient debris flows in Arizona*. Unpublished Ph.D. Thesis, University of Arizona, Department of Hydrology and Water Resources, 235 pp.

12 Lahars: Origins, Behavior and Hazards

James W. Vallance

Abstract

Volcanic debris flows that originate at potentially active volcanoes are called lahars. Lahars are like debris flows in non-volcanic terrain but can most notably differ in origin and size. Primary lahars occur during eruptions and may have novel origins such as turbulent mixing of hot rock moving across ice- and snow-clad volcanoes and eruptions through crater lakes. Lahars range in volume to more than a cubic kilometer (10^9 m^3), with the biggest ones caused by huge deep-seated flank collapses of water-saturated edifice rock. Because they can be so voluminous, can have high water contents, and commonly can be clay rich, these lahars can travel tens to even hundreds of kilometers. Long transport causes evolution of flow types from flood flow to hyperconcentrated flow to debris flow. Lahars capable of traveling far downstream are commonly sufficiently liquefied that they drape valley slopes and leave behind thin deposits as they pass downstream. Only in valley bottoms are lahars likely to emplace thick deposits, and even there the deposits are apt to be much thinner than peak flow depths. Flows with long transport change character with time and distance downstream. Deposits, especially those in valley bottoms, can accrete during intervals that represent a significant proportion of the time it takes the flow to pass (typically minutes). The combination of flows changing character and their progressive accretion imposes distinctive characteristics on their deposits such as normal and inverse grading. Historically, lahars have caused thousands of fatalities and destroyed entire towns. Perhaps the most disastrous known lahar occurred in 1985 at Nevado del Ruiz in Colombia and killed more than 23,000 people. Since that disaster, an increasing awareness of lahar hazards has led to efforts to mitigate them. In recent decades, improved land-use decisions, monitoring and communication have improved hazard responses and saved many lives. Lahar hazard maps and development of lahar inundation models have helped planners and people at risk to better understand the nature of the risk owing to lahars.

12.1 Introduction

Lahars can differ in origin, frequency, runout, and size from debris flows in non-volcanic terrain. Lahars can exceed a cubic kilometer in volume and travel more than 300 km from source.

J. W. Vallance (✉)
U.S. Geological Survey, Cascades Volcano Observatory, 1300 SE Cardinal Court, Vancouver, WA 98683, USA
e-mail: jvallance@usgs.gov

M. Jakob et al. (eds.), *Advances in Debris-flow Science and Practice*, Geoenvironmental Disaster Reduction,
https://doi.org/10.1007/978-3-031-48691-3_12

They can be considerably more voluminous than non-volcanic debris flows (Jakob, 2005) owing to abundant loose debris on the steep slopes or on aprons of volcanoes; fine-grained, water-saturated rock within some volcanic edifices; large volumes of water stored in glaciers, crater lakes, and hydrothermal systems; rainfall that washes over denuded slopes after eruptions; and the potential for releasing both water and sediment during and immediately after eruptions.

Lahar origins include the direct and indirect effects of eruptions, as well as those common to any steep terrain. Volumes of hot rock and ash ejected during eruptions can melt snowpack and ice to generate lahars. Volcanic explosions or eruptions through crater lakes sometimes generate floods of water that subsequently incorporate sediment and generate lahars. Eruptions or earthquakes associated with eruptions can cause volcanic edifices to collapse. The resulting landslides sometimes contain sufficient volumes of weakened, hydrothermally altered rock and water that they evolve to form lahars. The largest documented lahars have origins as edifice collapses. After explosive eruptions are over, voluminous deposits of volcanic debris are spread over volcano slopes and aprons. These deposits are loose, unvegetated and easily mobilized to form debris flows during periods of heavy rain.

Because lahars are voluminous and can travel long distances, they show downstream evolutions that are uncommon with non-volcanic debris flows. Many lahars evolve from floods and many remain water rich to their termini (Fig. 12.1). Lahars that move down rivers tend to incorporate river water with distance downstream and evolve over tens of kilometers to less dilute flow types. Volcanic edifice collapses generate debris avalanches that sometimes evolve partially or completely to lahars.

Much of our present knowledge of lahar flow behavior is inferred from study of deposits. Nonetheless, a few key observational and experimental studies have improved understanding of volcanic debris-flow processes. This chapter updates a previous summary of lahars, their origins and behavior (Vallance, 2005). The most notable interim advances in study of lahars includes improvement of numerical models and documentation of substrate erosion and liquefaction processes. The chief purpose of the chapter is to summarize what is known about the causes and behavior of lahars.

After this Introduction, the chapter addresses the following topics: (1) *Discussion of terminology related to lahars* sets the stage for a better understanding of lahars and lahar processes. (2) *Lahar origins* discusses the genesis of primary and secondary lahars as well as those unrelated to eruptions. It also addresses the relationship of volcanic explosivity to lahar volume. (3) *Downstream behavior and deposition* delineates downstream lahar evolution and the complex interaction among erosion, bulking, and deposition when lahars move long distances downstream. (4) *Select case studies* highlights the origins and impacts of several well studied historical and prehistoric lahar events. (5) *Lahar hazards: assessment, modeling, and mitigation* discusses the rationale for lahar-hazard evaluation, summarizes how models developed in the past two decades can help assess lahar hazards, and finally synopsizes the chief mitigation measures that can be taken to address lahar hazards. (6) *Concluding remarks* discusses possible effects of climate change on lahar genesis, highlights key findings of the past two decades, and suggests directions for future research.

12.2 Discussion of Terminology Related to Lahars

Lahar is an Indonesian (Javanese) word describing flowing mixtures of rock debris, ash, and water originating at volcanoes and introduced into the literature by Schmidt (1934) and van Bemmelen (1949). The term is widely used as a synonym for "volcanic mudflow" or "volcanic debris flow" (for example, Crandell, 1971; Fisher & Schmincke, 1984; Pierson & Scott, 1985; Smith, 1986). A definition adopted by the 1988 Geological Society of American Penrose Conference on Volcanic Influences on Terrestrial Sedimentation (Smith & Fritz, 1989) is: *lahar,*

Fig. 12.1 Photographs of lahars illustrating their fluidity. **a** Lahar along Sacobia River, Mount Pinatubo (Philippines). Flow wave is about 1 m high. Photograph by Mike Dolan, Michigan Technological University, 1992. **b** Flow along Whangaehu River, Mount Ruapehu (New Zealand). Concrete Pilar is 4 m tall. Photograph by J.R. Grant, 1995. **c** Lahar at Redoubt Volcano (USA) triggered by melting of snow and ice. Photograph by M. Fulle, 2009, http://www.stromboli.net. **d** Inundation depth of North Fork Toutle River lahar shown by mud coating on trees, Mount St. Helens (USA). Photo by Lynn Topinka, USGS. E. Lahar triggered by release of subglacial water and groundwater, Huila volcano (Colombia). Photograph by B. Pulgarín, Servicio Geológico Colombiano, November 2008

a general term for a rapidly flowing, gravity-driven, mixture of rock debris and water (other than normal streamflow) from a volcano.

In this chapter, the *phase* is the flow type in a lahar *event* at some time and place. Lahar events can include phenomena, such as debris avalanche, flood flow, and pyroclastic flow, not encompassed by the term lahar. A *debris avalanche* is a flowing mixture of debris, rock, and moisture that moves down slope under the influence of gravity. Debris avalanches can contain liquefiable material, but unlike lahars, they lack water saturation and depend on variably lubricated particle interactions for their mobility. Volcanic debris avalanches are commonly the result of deep-seated failures in contrast to those in non-volcanic terrain (Hungr et al., 2014). Pyroclastic flows are hot (several hundred °C) mixtures of ash, rock, and gas that are liquefied by the gas and fine ash mixture.

Lahar behavior is a continuum from muddy flood or debris flood (Church & Jakob, 2020) to bouldery debris flow that varies with time and with distance downstream. The continuum includes debris-, transitional-, or hyperconcentrated-flow, and streamflow phases (Vallance, 2000). Some authors use the terms such as DF (debris-flow) and HF (hyperconcentrated-flow) lahars (Pierson, 2005) or respectively type 1 and type 2 lahars (Major, 2022) to distinguish flow phases. A *debris-flow phase* is one in which the solid and fluid fractions are approximately equal volumetrically and in which the two fractions in a vertical section move downstream approximately in unison. A *streamflow phase* is one in which fine-grained sediment moves in suspension with the fluid (suspended load) and coarse-grained sediment moves along the bed at discrete intervals (bed load). A *transitional flow phase*, called hyperconcentrated flow, has character intermediate between that of debris flow and streamflow. Unlike streamflow, this transitional phase carries very high sediment loads, and unlike the debris-flow phase coarse-grained solids tend to separate vertically from the liquid-and-fine-solids mixture. Although the literature distinguishes hyperconcentrated-flow phase from more dilute and more concentrated phases in terms of solids fraction, transitions are gradational and dependent on other factors like sediment-size distribution and energy of the flow. Thus, flow-phase transitions cannot be precisely defined.

The height above the channel bottom of the flowing lahar is the *stage*. The term stage also refers here to characteristic periods of flow such as the initial rising or waxing stage, the peak-inundation stage, and the final long-duration falling or waning stage.

12.3 Lahar Origins

Lahars may be primary (syneruptive), secondary (posteruptive), or unrelated to eruptions (Table 12.1). Genesis requires (1) sufficient source of water; (2) abundant unconsolidated debris; (3) steep slopes and substantial relief at the source; and (4) a triggering mechanism. Water sources include pore or hydrothermal water, rapidly melted snow and ice, subglacially trapped water, crater or other lake water, and rainfall runoff. Unconsolidated debris derives from pyroclastically fragmented flow and fallout as well as glacial drift, alluvium, colluvium, and soil. Triggering mechanisms owing to volcanic unrest include slope failure, explosions, and earthquakes. The shallow intrusion of hot rock and gases prior to the onset of an eruption drives hydrothermal and pore water toward edifice flanks and the resultant outwardly directed pore-pressure gradient contributes to edifice instability and debris avalanches that sometimes evolve to lahars.

Sudden melting of snow and ice during eruptions by voluminous flows of hot rock and ash that sweep down steep slopes cause water floods that incorporate debris and form lahars. Sudden release of water from crater lakes, other lakes adjacent to volcanoes, and subglacially stored water can also generate debris flows. Lake-breakout-induced debris flows occur during eruptions but can also occur months or years after eruptions.

Temperature and fines content can help to delineate lahar origins. Mixing of hot pyroclastic debris and water results in hot lahars. Such mixtures are necessarily below the boiling point but can contain individual clasts that are hotter. Hot lahars also derive from the release of water heated during volcanic unrest or eruptions. The hot water can derive from subglacial lakes, crater lakes, and hydrothermal systems.

As observed at Mount Rainier (Crandell, 1971), lahars can be divided into those that are rich in silt and clay and those poor in silt and clay. An empirical observation is that fines-rich lahars commonly initiate as flank failures of hydrothermally altered, weakened rock, whereas clay-poor lahars

Table 12.1 Notable examples of lahars and their triggering mechanisms (adapted from Rodolfo, 2000). Data are from Rodolfo (2000) and Siebert et al. (2010) unless otherwise noted. Compare to non-volcanic debris flows in Prakash et al. (this volume)

Triggering mechanism	Volcano, occurrence	Fatalities	VEI[a]	Volumes, Areas	Remarks
Syneruptive (primary)					
Crater lake expulsion	Galunggung, Indonsia, 1822	>400	5	~10^8 m^3	4011 fatalities, at least 10% owing to lahars
	Kelut, Indonesia, 1826	Yes	4		65 villages destroyed
	Kelut, Indonesia, 1848	21	3		11 villages destroyed
	Kelut, Indonesia, 1864	54	2		
	Kelut, Indonesia, 1901	Yes	3		
	Kelut, Indonesia, 1919	5,110	4	4×10^7 m^3 H_2O, 130 km^2	Expelled 4×10^7 m^3 lake. Beginning 1923 tunneled to drain lake to 2×10^6 m^3. No major lahars in 1951, but eruption blocked tunnels, deepened lake to 2.4×10^7 m^3. Drained lake again in 1967.
	Kelut, Indonesia, 1951		3		
	Kelut, Indonesia, 1966	211	4	2×10^7 m^3 H_2O, 45 km^2	
	Ruapehu, New Zealand, 1969, 1975				
	Ruapehu, New Zealand, 1995		4	~10^7 m^3	
Crater lake breakout	Pelée, Martinique, 1902	25	2		
Jökulhlaups	Öraefajökull, Iceland, 1727	4			
	Katla, Iceland, 1918, 1955	4, 1			
	Gjálp, Iceland, 1996				

(continued)

Table 12.1 (continued)

Triggering mechanism	Volcano, occurrence	Fatalities	VEI[a]	Volumes, Areas	Remarks
Pyroclastic flows, hot rock avalanches and hot blasts melting snow and ice	Cotopaxi, Ecuador, 2800 BP Cotopaxi,	Yes	3?	~10^9 m^3	Flowed south and east into Amazon basin, (unpublished data, PA Mothes, ML Hall, and JW Vallance)
	Cotopaxi, Ecuador, 1698	Many	3?		
	Cotopaxi, Ecuador, 1742	Many	3		
	Cotopaxi, Ecuador, 1768	Many	4		
	Cotopaxi, Ecuador, 1877	>400	4		Longest historical runout, L = 230 km, Mothes (1992)
	Nevado del Ruiz, Colombia, 1595	636	4	~10^8 m^3	Thouret et al. (1990)
	Nevado del Ruiz, Colombia, 1845	~1000	3	3×10^8 m^3	Thouret et al. (1990)
	Nevado del Ruiz, Colombia, 1985	23,080	3	~10^8 m^3	Pierson et al. (1990)
	Individual lahars, Ruiz, 1985			Vmax, Vglacier	
	Azufrado-Lagunillas	>21,000		4×10^7, ~9×10^6 m^3	Bulking ~ ×4, Qmax = 5×10^5 m^3/s, Pierson et al. (1990)
	Molinos-Nereidas	>1000		3.2×10^7, 5–8 × 10^6 m^3	Bulking ~ ×5, Qmax = 2×10^5 m^3/s, Pierson et al. (1990)
	Guali	Yes		1.6×10^7, 6–9 × 10^6 m^3	Bulking ~ ×3, Qmax = 2×10^5 m^3/s, Pierson et al. (1990)
	Mount Rainier, USA, 1100 BP		3	1.8 x 10^8 m^3	Iverson et al., (1998)
	Mount St. Helens, USA, 1980		5	3×10^7 m^3	Major et al. (2005a, 2005b)
	Mount St. Helens, USA, 1982		2	1.2×10^7 m^3	Major et al. (2005a, 2005b)
	Redoubt Volcano, USA, 1989				
Expulsion of water from glaciated volcano before or during eruption	Huila, Colombia, 2007	0	0	10^6 m^3	Phreatic eruption, Worni et al. (2012)
	Huila, Colombia, 2007	0	0	5–7 × 10^7 m^3	Phreatic eruption, Worni et al. (2012)
	Huila, Colombia, 2008	<10	3–4	3×10^8 m^3	Phreatomagmatic eruption, Worni et al. (2012)

(continued)

Table 12.1 (continued)

Triggering mechanism	Volcano, occurrence	Fatalities	VEI[a]	Volumes, Areas	Remarks
Lava flows melt ice and ponded water breaks out	Villarica, Chile, 1949	36	2		
	Villarica, Chile, 1963–64	22	2		
	Villarica, Chile, 1971	15	3		
Pyroclastic flows entering streams	Asama, Japan, 1783	~1000	4		Pyroclastic flow dammed river, ephemeral lake breakout 1 h later
Avalanche generated	Mount Rainier, USA, 7200 BP		3	10^8 m^3	Eruption related avalanche liquefied, John et al. (2008)
	Mount Rainier, USA, 5600 BP	Yes	3	4×10^9 m^3	Q = 3–15 × 10^6 m^3/s, avalanche liquefied almost completely to form lahar, Vallance & Scott (1997)
	Mount Rainier, USA, 2500 BP		2	2×10^8 m^3	Incomplete liquefaction of avalanche, Scott et al. (1995)
	Mount Rainier, USA, 1100 BP		2	10^7 to 10^8 m^3	Eruption related avalanche liquefied, Scott et al. (1995)
	Mount Rainier, USA, 500 BP		?	2.5×10^8 m^3	No clear evidence of an eruption, Scott et al. (1995)
	Cotopaxi, Ecuador, ~5000 BP	Yes	3–4	~3×10^9 m^3,	Q = 3–6 × 10^6 m^3/s, L = 330 km, Mothes et al. (1998)
	Mount St. Helens, USA, 1980		5	1.4×10^8 m^3	Avalanche first, lahar 3 h later, Major et al. (2005a, 2005b)
Lava flows melt ice and ponded	Villarica, Chile, 1949	36	2		
Water breaks out	Villarica, Chile, 1963–64	22	2		
	Villarica, Chile, 1971	15	3		
Pyroclastic flows entering streams	Asama, Japan, 1783	~1000	4		Pyroclastic flow dammed river, ephemeral lake breakout 1 h later

(continued)

Table 12.1 (continued)

Triggering mechanism	Volcano, occurrence	Fatalities	VEI[a]	Volumes, Areas	Remarks
Avalanche generated	Mount Rainier, USA, 7200 BP		3	10^8 m^3	Eruption related avalanche which liquefied as it deformed, John et al. (2008)
	Mount Rainier, USA, 5600 BP	Yes	3	4×10^9 m^3,	Q = 3–15 × 10^6 m^3/s, avalanche liquefied almost completely to form lahar Crandell (1971)
	Mount Rainier, USA, 2500 BP		2	2×10^8 m^3	Incomplete liquefaction of avalanche, Scott et al. (1995)
	Mount Rainier, USA, 1100 BP		2	10^7 to 10^8 m^3	Eruption related avalanche which liquefied as it deformed
	Mount Rainier, USA, 500 BP		?	2.5×10^8 m^3	No clear evidence of an eruption, Scott et al. (1995)
	Cotopaxi, Ecuador, ~5000 BP	Yes	3–4	~3×10^9 m^3,	Q = 3–6 × 10^6 m^3/s, L = 330 km, Mothes et al. (1998)
	Mount St. Helens, USA, 1980		5	1.4×10^8 m^3	Avalanche first, lahar 3 h later, Major et al. (2005a, 2005b)
Rain during eruption	Agung, Indonesia, 1963	200	4		Rain during and after eruption
	Pinatubo, Philippines, 1991	~100	6		Typhoon Yunya coeval with climactic eruption of 15 June
Posteruptive (secondary)					
Crater lake failure	Kelut, Indonesia, 1875	Yes			Failure of west crater wall, destroyed 30 villages
	Kelut, Indonesia, 1990	4			Secondary lahar about 1 mo. after eruption
	Ruapehu, New Zealand, 1953	151		2×10^6 m^3,	Q = 850 m^3/s, Cronin et al. (1997)
	Ruapehu, New Zealand, 2007			4.4×10^6 m^3	Bulking ~ ×3, Cronin et al. (1997)
	Parker, Philippines, 1995	310			Caused by illegal gold mining activity?
	Pinatubo, Philippines, 2002	2		1.6×10^8 m^3	~860 mm rainfall in 9 days till breach, bulking ~ ×2

(continued)

Table 12.1 (continued)

Triggering mechanism	Volcano, occurrence	Fatalities	VEI[a]	Volumes, Areas	Remarks
Heavy rain after eruptions	Sakurajima, Japan, post-1955 Irazu,				
	Costa Rica, post-1964				
	Semeru, Indonesia, 1909	221			
	Semeru, Indonesia, 1981	372			
	Mayon, Philippines, 1875	1,500			
	Unzen, Japan, 1990–1995				
	Pinatubo, Philippines, post-1991	24			Many small lahars remobilized 2.5 of 5.5 km^3 of pyroclastic-flow deposits in the 6 years following the eruption
	Sinabung, Indonesia, 2014 to present				Many small lahars
	Merapi, Indonesia, 2010				Small posteruption lahars
	Fuego, Guatemala, 2018				Small posteruption lahars after rain on pyroclastic-deosits
Lake breakouts (volcanically dammed but not crater lakes)	Mount St. Helens, USA, Pine Creek, 2500 BP				Spirit Lake breakout after debris avalanche, Scott (1988)
	Santa Maria, Guatemala, post-1902				
	Pinatubo, Philippines, post-1991				Mapanuepe River was dammed by aggradation to form a lake; that and several smaller lakes with similar origins eventually broke out to form lahars much more voluminous than more common rain-induced ones.
Unrelated to eruptions					
Earthquake-induced avalanche	White Island, New Zealand, 1914	14	n/a		
	Ontake, Japan, 1984		n/a		

(continued)

Table 12.1 (continued)

Triggering mechanism	Volcano, occurrence	Fatalities	VEI[a]	Volumes, Areas	Remarks
Heavy-rain-induced slope failure	Agua, Guatemala, 1541	>1300	n/a		After a period of heavy rains, an earthquake triggered a lahar that destroyed Capital, Cuidad Viejo
	Mombacho, Nicaragua, 1570	400	n/a		Destroyed town of Mombacho
	Mayon, Philippines, 1981	>200	n/a		Typhoon caused lahars, last eruption 1978 destroyed 2
	Casita, Nicaragua, 1998	2,500	n/a		Two towns were destroyed by lahar during Hurricane Mitch, Scott et al. (2005)
Slope failure	Mount Meager, Canada		n/a	5×10^7 m3	13 km runout, Roberti et al. (2017)
Multiple slope failures	Near Huila, Colombia, 1994	~1000	n/a	2×10^8 m3	M6.4 earthquake caused multiple failures that coalesced, Scott et al. (2001)
Glacial outburst floods	Mount Rainier, USA, 1947		n/a	4×10^7 m^3	
	Mount Rainier, USA, 1980s to present		n/a		Small volume lahars occur almost yearly, Scott et al. (1995)

[a] VEI is volcano explosivity index; explained in text

commonly begin with water floods that entrain gravel, sand, and fines. Flood waters leading to the formation of clay-poor lahars have been produced during lake breakout (Pierson, 1999; Pringle & Cameron, 1999), intense rainfall (Rodolfo & Arguden, 1991; Hodgson & Manville, 1999; Lavigne et al., 2000; Lavigne & Thouret, 2003), and from melting of snow and ice by hot pyroclastic flows (Major & Newhall, 1989; Pierson et al., 1990; Scott, 1988; Waythomas et al., 2013).

12.3.1 Lahars Induced by Sudden Water Release

Floods of water move across the easily erodible loose clastic sediments common on the flanks and aprons of volcanoes and readily incorporate that debris to form lahars. Erosion and incorporation of sediment (bulking) is critical to all lahars that begin with floods of water across slopes and aprons of volcanoes. Lahar formation depends on the right mix of readily available sediment and water discharge.

Water floods released at the summit of or on the steep slopes of volcanoes rapidly incorporate sediment to form lahars (Vallance & Iverson, 2015). At volcanoes, lahars induced by sudden water release chiefly occur by five means.

(1) Hot-rock debris usually from pyroclastic flows and surges mix with and rapidly melt glacial ice and firn (Pierson et al., 1990; Waythomas et al., 2013). Such pyroclastic flows come to rest or nearly do to form meltwater. The meltwater then runs off, coalesces, and erodes the pyroclastic debris to form water-rich lahars (Waitt, 1989). The

lahars continue to incorporate volcaniclastic debris, glacial drift, alluvium, and colluvium as they flow downstream so that within a few kilometers to several tens of kilometers they become debris flows (Pierson et al., 1990). Lahars of this type are considered primary.

(2) Eruptions can displace large volumes of crater-lake water that form primary lahars downstream. Crater lakes, caldera lakes, and volcanically debris-dammed lakes can also breach months to years after eruptions to generate secondary lahars (Major, 2022). Such delayed breakouts occur when water levels gradually rise, then overtop and rapidly incise fragile debris dams.

(3) Subglacial eruptions can form subglacial lakes that eventually break out when a section of the ice cap becomes buoyant and releases the trapped water (Maizels, 1997). Although huge eruption-driven outbursts cause sediment-rich, water floods called jökulhlaups (Maizels, 1997), small ones commonly erode loose sediment to form secondary lahars. Small-scale outburst floods, on the other hand, occur as normal glacial processes during periods of ablation (Driedger & Fountain, 1989; Richardson, 1968).

(4) Volumes of groundwater gushing from large fissures or cracks within a volcano's edifice can generate lahars. These lahars are likeliest at glaciated volcanoes. The largest examples of such lahars are directly or indirectly caused by eruptions or unrest (Johnson et al., 2018).

(5) Lahars owing to intense rainfall are common after pyroclastic eruptions deposit abundant loose debris in the form of pyroclastic-flow or fall deposits surrounding vents of volcanoes. Lahars of this type are commonly small in volume but frequent in occurrence during rainy periods. Size and frequency of rain-induced lahars may increase in the months or years following the primary pyroclastic eruption, then slowly decrease as drainage networks and vegetation reestablish themselves (for example, Mount Pinatubo after its 1991 pyroclastic eruption; Scott et al.,1996b).

12.3.2 Edifice- or Flank-Collapse-Induced Lahars

Most slope failures and edifice collapses behave as debris avalanches as they move downslope, but those with sufficient, widely dispersed, pore and hydrothermal water in the pre-collapse rock liquefy as the material deforms during collapse (Vallance & Scott, 1997). Shallow intrusion of magma within an edifice or volcanic unrest are the likeliest triggers of collapse-induced lahars larger than about 0.2 km^3. Smaller collapse-induced lahars may have various triggers, including magmatic or phreatic volcanism, volcanic or tectonic earthquakes, or weakening of steep edifice slopes.

Hydrothermal alteration, especially at glaciated volcanoes, increases the probability and size of edifice-collapse lahars because acid-sulfate leaching weakens the rock so that it more readily disintegrates as it deforms during failure. Collapse-induced lahars are commonly clay-rich because hydrothermal alteration replaces solid rock with clay minerals, which both increases porosity and decreases permeability of the rock (John et al., 2008). The alteration thus traps widely dispersed reservoirs of water within the pre-collapse mass. Water-rich hydrothermally altered rock, unlike fresh rock, readily disintegrates and liquefies as it deforms to generate fines-rich lahars.

At and near some volcanoes, especially tropical ones, tectonic earthquakes cause multiple slope failures that ultimately coalesce and form collapse-induced lahars. Such lahars are unrelated to volcanism but, owing to characteristics of their deposits (steep, commonly clay-rich, hummocky, and voluminous), can be mistaken for lahars generated by major edifice collapse of hydrothermally altered sectors of volcanoes (Scott et al., 2001).

12.3.3 Relationship of Lahar Volume and Eruption Magnitude

Origin is an important determinant in lahar size. Edifice collapse of water saturated slopes cause the largest known lahars, and volumes range up to about 4 km^3 (4×10^9 m^3) (Table 12.1; Fig. 12.2) (Carrasco-Nuñez et al., 1993). Lake-breakout lahar volumes range to several hundred million cubic meters and depend on amount and rate of water release (Table 12.1). Lahars owing to pyroclastic interaction with snow and ice may range in size from tens to a few hundreds of million cubic meters. In contrast, secondary lahars, especially those owing to rainfall on fresh fans of pyroclastic debris, are commonly less than ten million cubic meters but frequent. Their frequency can result in an overwhelming redistribution of pyroclastic debris with cumulative volumes ranging to a cubic kilometer or more.

Primary lahar volume has a somewhat counterintuitive relationship with eruption size and explosivity. Even eruptions modest in volume and not very explosive can generate huge lahars, especially those originating as edifice collapse and hot-rock-ice interaction (Fig. 12.2). Relatively small eruptions can destabilize a volcano's edifice and edifice-collapse lahars are among the most voluminous documented. Thus, for example, a modest VEI 3 eruption of Mount Rainier 5,600 years ago produced a 3.8 km^3 lahar. Volcano explosivity index (VEI) is an exponential measure of the power, explosivity, size, and duration of a volcanic eruption (Fig. 12.2; Newhall & Self, 1982). Very small explosive eruptions (VEI 2-3) of Mount Rainier 1,500 and 1,000 years ago perturbed its icecap to form lahars initiated by hot-rock-and-ice interaction that flowed more than 100 km downstream. Although there is some tendency for large eruptions to produce large lahars, in many cases, especially at ice-clad volcanoes, small to moderate eruptions can produce huge lahars (Fig. 12.2).

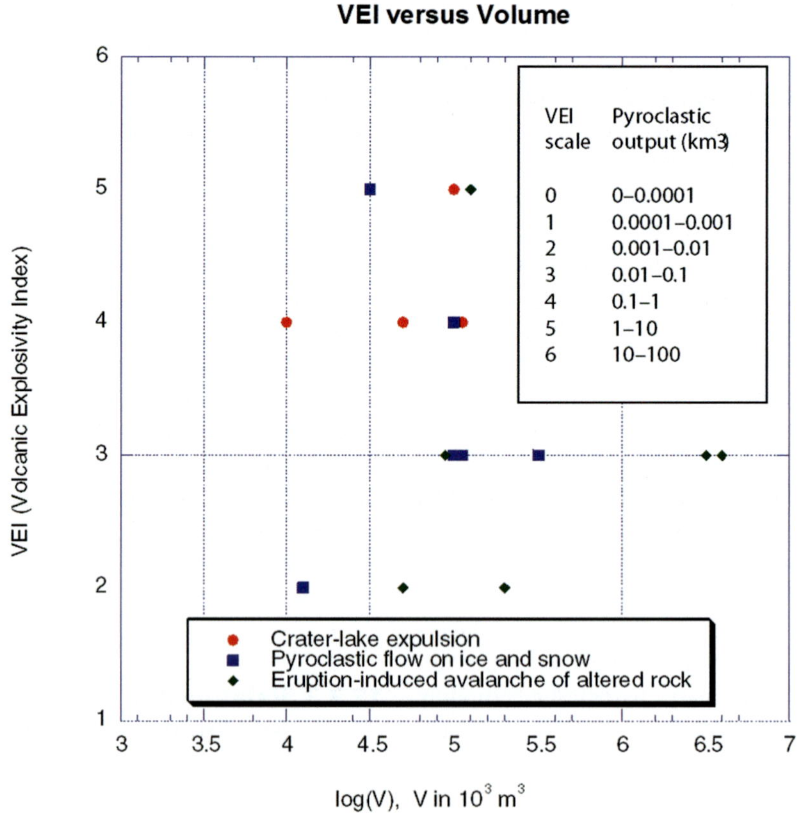

Fig. 12.2 Plot of VEI (volcano explosivity index) versus volume for primary lahars having three diverse origins

12.4 Downstream Behavior and Deposition

Lahars can change character as they move downstream because of their large sizes and long transport distances, thus some lahars show more pronounced changes with distance downstream than non-volcanic debris flows (Fig. 12.3). Small lahars (those less than about 10^6 m^3), like typical non-volcanic debris flows, commonly coalesce quickly, form bouldery fronts and levees through size-segregation processes (Johnson et al., 2012), and come to a halt once they reach low-gradient reaches of the drainage. Lahars larger than about 10^6 m^3 commonly do not fit the mold of such 'typical' debris flows. In the case of huge lahars that travel tens of kilometers, size-segregation processes can become relatively less important compared to downstream erosion and dilution

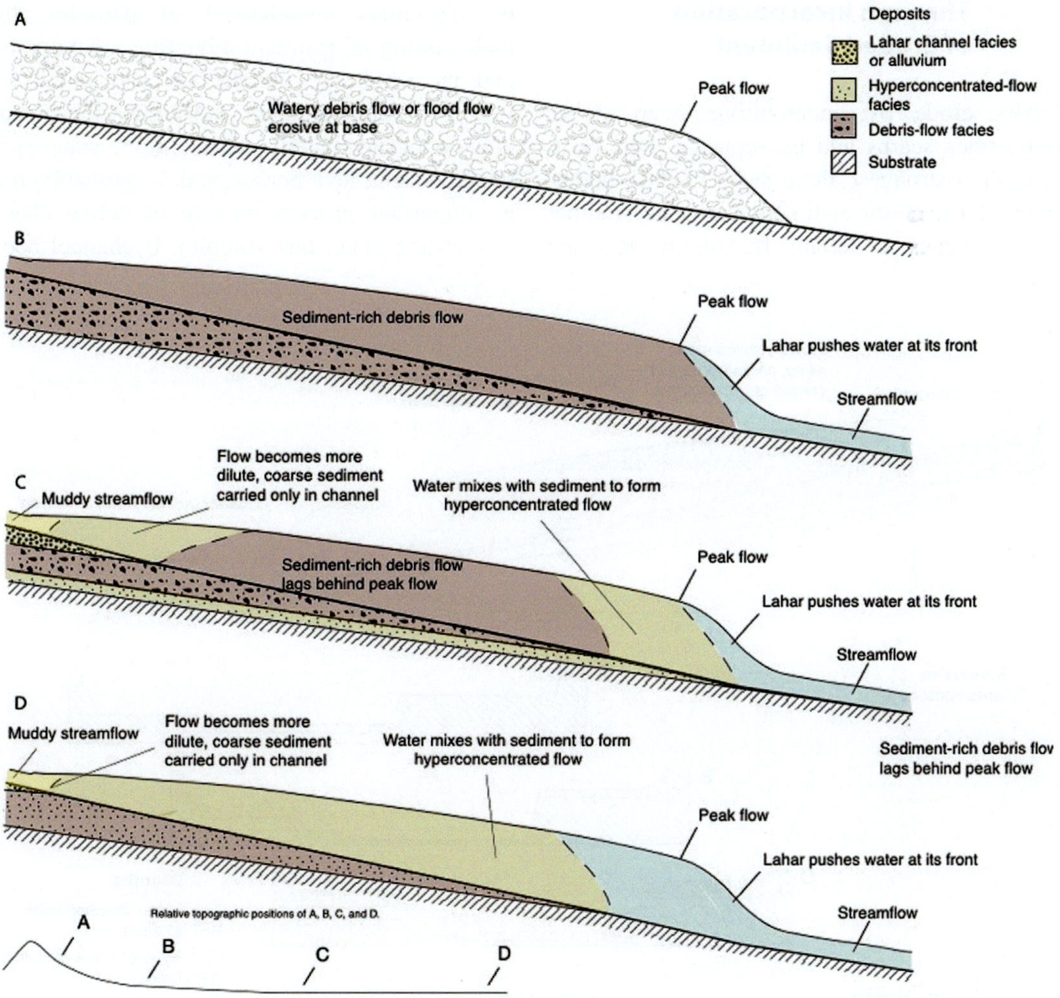

Fig. 12.3 Schematic models of a lahar moving down a river at four different places in the valley. **a** Watery flood-flow phase descending the steep slopes of the volcano is erosive. **b** Sediment rich debris-flow phase is beginning to push overrun river water at its front; flow is erosive at its front and begins depositing after peak-flow stage. Deposits are commonly inversely graded at the base. **c** Lahar moving down a river undergoing downstream dilution from debris-flow phase to hyperconcentrated-flow phase and depositing hyperconcentrated-flow facies beneath debris-flow facies. **d** Lahar moving down a river after it has completely transformed to hyperconcentrated flow (like type-2 debris flow to debris flood transition in non-volcanic terrain as per Church and Jakob (2020))

processes. In medial or distal reaches, such lahars can also transform from sediment-rich debris-flow phase to water-rich hyperconcentrated-flow or stream-flow phases. Another class of lahars that are rare in other environments is lahars that begin as huge (10^8 to 10^{10} m^3) debris avalanches caused by edifice collapse and transform to debris flows as they move downstream.

12.4.1 Erosion and Mass Growth Through Incorporation of Eroded Sediment

Lahars erode by undercutting steep slopes and terrace scarps and by scouring their beds. Erosion is strongest along steep reaches underlain by loose clastic sediment and weakest either along reaches underlain by highly resistant bedrock or reaches with very gentle gradients. Along a comparable reach, a debris flood or a hyperconcentrated-flow will be more erosive than a sediment-rich debris flow (Figs. 12.3a and 12.4a, b), but local erosion is always possible. The waxing stage of a lahar is likely to coincide with the most widespead and voluminous erosion and bulking. The final watery tail of a lahar is commonly erosive, causing incision of channels in fresh lahar deposits (Fig. 12.4).

Erosion at the base of lahars occurs chiefly by piecemeal dislodgment of particles, by undercutting at upstream migrating knickpoints, and by wholesale liquefaction and mobilization of water-saturated substrate. Piecemeal erosion occurs chiefly during more water-rich hyperconcentrated phases and is probably not an important process outside of active channels (Kang et al., this volume). If channel beds

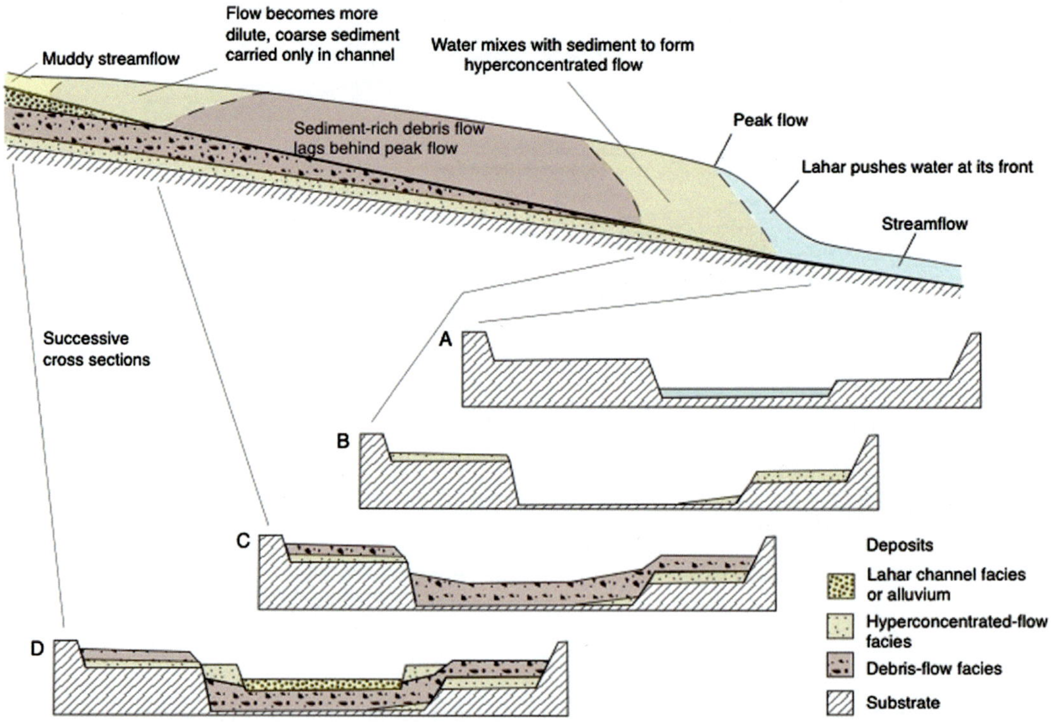

Fig. 12.4 Schematic model of a lahar moving down a river undergoing downstream dilution from debris-flow phase to hyperconcentrated-flow phase and deposit facies derived from it. The idealized model shows erosion and deposition of hyperconcentrated-flow facies from **A** to **B**; deposition of debris flow facies from **B** to **C**; and deposition of hyperconcentrated flow facies and subsequent erosion from **C** to **D**. Facies transitions can be abrupt or gradational. Adapted from Vallance (2005)

comprise erodible sediment, sequences of upstream migrating knickpoints are common during hyperconcentrated-flow phases (Dolan, 2004). Typically, knickpoint steps are a few tens of centimeters to a meter or so high and spaced on the order of tens to hundreds of meters apart. During more sediment-rich flows, knickpoints become higher and more widely spaced. Knickpoints may disappear during debris-flow phases, which, in fact, commonly aggrade rather than erode their beds. Apparently, this process is favored by a low gradient channel and water-rich lahar flow. Lahars voluminous enough to escape channels knock trees down and incorporate them. Root balls of falling trees drag sediment into the active flow and loosen even more sediment that is then available for erosion. Voluminous lahars that inundate large areas of forested terrain can incorporate significant amounts of sediment and huge amounts of wood in this way.

An important means by which lahars incorporate sediment is by compression of loose wet substrate debris or soil and coeval liquefaction of that material. A lahar that overruns wet or water-saturated sediment loads, compresses, and shears it. The stress drives up pore-fluid pressure and may cause "runaway liquefaction" (Iverson et al., 2011). Pervasive liquefaction of substrate permits easy erosion and entrainment of that sediment and lubricates the base of the flow, allowing it to accelerate downstream. The landslide-triggered Casita lahar occurred after more than a meter of rainfall at the end of the 1998 rainy season in Nicaragua. It probably owed its accelerated growth in volume (times 2–3) and mobility to loading and liquefaction of water saturated soil and debris along its path (Devoli et al., 2009; Scott et al., 2005).

Undercutting of steep slopes, fluvial terrace scarps and active stream banks is perhaps the most important way in which lahars erode and incorporate sediment. Undercutting is active during both debris-flow and hyperconcentrated-flow phases. Large lahars can incorporate megablocks (>10 m in dimension) of unconsolidated sediment, and sometimes even of bedrock, in this way (Fig. 12.5). Megaclasts formed of unconsolidated deposits may move tens of km downstream before they deform and ultimately break up into individual fragments.

Progressive downstream bulking imposes downstream changes of character on lahars. Bulking transforms flood flows and hyperconcentrated flows to more sediment-rich phases (Fig. 12.3a, b). If bulking continues, both waxing and waning stages of flow ultimately become debris-flow phases (Fig. 12.3b). With continued transportation, lahars become richer in exotic sediment, like alluvium, colluvium, and glacial drift. Commonly, the waxing flow-front and following peak stages of the flow are the most erosive and thus most readily incorporate exotic sediment. The sediment-rich falling flow that follows peak flow is less erosive and commonly actively deposits sediment rather than erodes it (Fig. 12.4). The final waning-stage flow is likely to be more water rich and more erosive, but less voluminous in discharge than the preceding stage. Final waning stages of lahars commonly incise previously deposited laharic sediment (Fig. 12.4).

12.4.2 Particle-Size and Particle-Density Segregation Processes

Particles in lahars can segregate by density or size. Particle segregation processes are mediated by solids fraction, proportion of coarse particles, and fluid density, as determined by the proportion of fine-grained particles in suspension (Vallance & Iverson, 2015). In dilute lahars, large dense grains readily settle gravitationally. By this process the largest grains collect in the lowest layer of the moving flow and progressively smaller ones above, developing a normally graded flow. Because the lowest parts of the flow where the largest grains are accumulating are moving slower than the average velocity of the flow, these coarse accumulations lag progressively farther behind the flow front.

If pumice less dense than the fluid is common in the flow, it may migrate toward the flow surface, sometimes forming a pumice raft. These surface rafts may move faster than the lahar's

Fig. 12.5 Photograph of cross section of 20-m thick lahar about 40 km downstream from Cotopaxi volcano (Ecuador). Megaclasts surrounded by matrix were derived locally by the lahar undercutting the incised river gorge with nearly vertical walls 20–50 m high. When the lahar undermines the vertical walls, megaclasts fall in as large clastic blocks. Megaclasts are mostly derived from earlier lahar deposits and are rounded by abrasion and rolling during flow. Bulldozer for scale. Photograph by Patty Mothes, Instituto Geofísico Nacional, 2013

average velocity and thus migrate toward flow fronts or margins (Dolan, 2004).

In flows with greater solids fraction, granular contacts inhibit settling and favor rise of large grains instead. In shearing granular flows with solids fraction >~0.4, even dense large clasts will rise rather than settle. Force imbalances or particle rotation can push large particles from one layer to a higher layer, and smaller grains are apt to fill spaces between the large ones and thus not rise. Once near the flow surface concentrations of large grains continually migrate forward then are pushed aside in levees (Fig. 12.6a). Accretion of large grains at flow perimeters causes bouldery flow fronts and margins that are resistant to flow (Fig. 12.6b). In contrast the mixture following the flow front remains fluid. This combination of liquid interior and resistant levee margins governs the behavior of granular high-solids-fraction lahars.

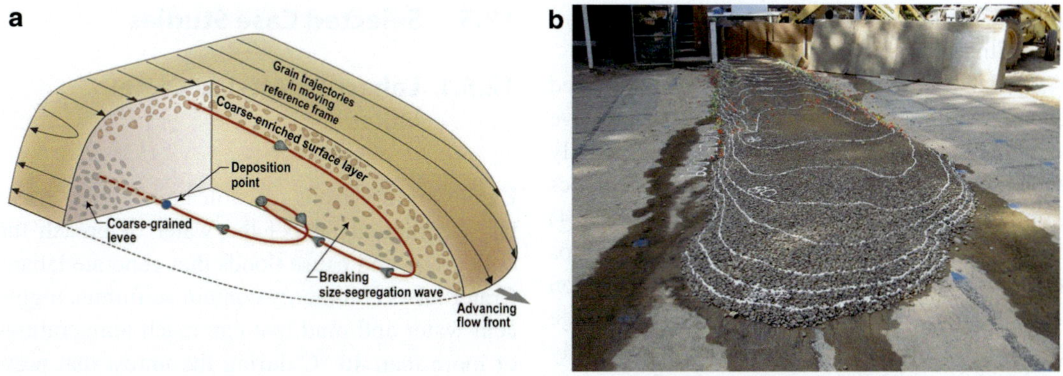

Fig. 12.6 **a** Schematic cutaway illustration of the process leading to formation of lateral levees behind a debris-flow front. Red path shows how coarse grain near surface migrates toward a levee. Reference frame moves at the speed of the advancing flow such that a less rapidly advancing grain appears to move backward (Johnson et al., 2012 give details). **b** Deposit of experimental debris flow illustrating coarse-grained margins and levees (Vallance & Iverson, 2015). Width of distal end of deposit is about 2.5 m. White lines are painted contours at 2.5 cm intervals

12.4.3 Downstream Processes: Dilution and Flow Transformation

Once off the flanks of the volcano and confined in river channels, lahars fill those channels to greater depths and thus move faster than normal stream flow. The lahars then push river water ahead of them and gradually, with distance downstream, begin to mix with that water (Fig. 12.3) (Cronin et al., 1997; Pierson & Scott, 1985; Scott, 1988). The front of the lahar becomes progressively more water rich and loses its capacity to carry larger gravel particles; then these gravels progressively lag behind the flow front (Fig. 12.3b, c). With time and distance downstream, a dilution front progresses from the front to the middle of the lahar, and eventually the entire lahar becomes more dilute (Fig. 12.3d). In lahars that occurred at Mount St. Helens in 1980 and 1982, downstream dilution occurred over tens of kilometers and completely transformed the lahars from debris-flow phase to hyperconcentrated-flow phase (Fig. 12.3d). In medial reaches, the hyperconcentrated-flow phase preceded the debris-flow phase of the lahar because the dilution process began at the flow front then gradually worked backward toward the tail as the flow migrated downstream (Pierson & Scott, 1985; Scott, 1988).

12.4.4 Depositional Processes

An increasing body of evidence suggests that deposits of both hyperconcentrated-flow and debris-flow phases aggrade progressively during significant periods of time (Vallance & Scott, 1997; Vallance, 2000). Major (1997) observed that levees in large-scale flume experiment accreted from bottom to top. Even in massive lahar deposits evidence of progressive aggradation is strong including (1) alignment of elongate clasts parallel to flow directions, (2) imbrication of elongate clasts in the upstream direction, (3) compositional variation with vertical position in outcrops, especially those which are graded, (4) marginal grain compositions similar to those at the base of thick valley-bottom deposits, and (5) stratification in deposits of transitional or hyperconcentrated flows. Lahars with massive valley-bottom deposits sometimes show bottom-to-top compositional trends that mirror trends in deposits from successively higher to lower positions along valley margins. A corollary observation is that remnants of peak flow on valley sides may indicate flow depths 5–10 times greater than the thickest valley bottom deposits. Compositional trends either within thick valley bottom deposits or with position along valley margins imply progressive aggradation of a flow that evolves compositionally with time as it passes.

In lahars, normally and inversely graded deposits commonly originate from progressive aggradation of a flow that is compositionally zoned in terms of its grain-size characteristics from front to tail of its hydrograph (Fig. 12.4) rather than a vertically graded flow that stops suddenly. Similarly, progressive aggradation from a lahar wave with a concentration of large particles at its front can generate a normally graded deposit (Vallance & Scott, 1997).

Aggradation from a dilute debris flow whose characteristic grain-size coarsens from its head toward its tail produces inversely graded deposits, which may be faintly stratified to massive (Fig. 12.7a, b and d). Farther downstream, where the entire lahar has transitioned to hyperconcentrated flow, progressive aggradation produces finer grained beds that may be inversely graded or both inversely and normally graded (Fig. 12.7c). In contrast, deposits near inundation limits commonly show little grading because the history of aggradation is very short.

Many small lahars, like bouldery non-volcanic debris flows, do not undergo downstream dilution but rather form cobble and boulder-rich perimeters owing to the size segregation process described above (Fig. 12.6). When a lahar of this type reaches sufficiently gentle slopes, the frictional boulder-cobble-rich perimeter slows to a stop, depositing a steep-fronted fines-poor margin and a partly liquefied fines-rich interior. Upstream, the fluid interior drains downslope, depositing coarse margins and forming levees (Johnson et al., 2012).

Avulsion can occur in lowland areas with gentle slopes where sediment-rich portions of lahars accrete and fill the active channel so that its cross-sectional capacity is diminished. When channel capacity is sufficiently diminished, the flow overtops this channel and cuts a new one (Fig. 12.8a–c). Once the new channel is established, the channel may incise to a new base level and tremendous volumes of sediment may be transported to river reaches far downstream (contrast Fig. 12.8b with c).

12.5 Selected Case Studies

12.5.1 Lahars Related to Crater Lakes

Explosive eruptions at volcanoes with crater lakes commonly expel the water or breach the outlet dam and cause floods that generate lahars. Crater lakes commonly contain sulfurous, highly acid water and mud that can reach temperatures of more than 40 °C during the unrest that presages eruptions. When explosive eruptions or outlet collapses release huge volumes of this nasty brew, it rushes down valleys and slopes, picks up abundant loose debris along the way, and becomes a particularly unpleasant variety of lahar comprising scalding hot acidic water, mud, and debris.

Kelut volcano, on the island of Java in Indonesia, is notorious for producing lahars in this fashion, and its lahars are notorious for killing people who live on its fertile slopes and for destroying villages within a 40-km radius. Of more than thirty eruptions in the past thousand years at Kelut, at least half have produced disastrous lahars (Bourdier et al., 1997; Simkin & Siebert, 1994). Since 1848, the volcano has erupted ten times, seven have blown out the crater lake, and five have produced devastating primary lahars. The most devastating of these lahars occurred when the largest volumes of water were expelled from the lake (Table 12.1). In 1919, an eruption blew 4×10^7 m^3 of water out of Kelut's lake and produced lahars that swept as far as 37 km downstream. These lahars covered more than 130 km^2 and killed more than 5,000 people.

Kelut volcano offers a spectacular example of lahar-hazard mitigation because of tunnels designed to lower its crater-lake level (Fig. 12.9) and check-dam installations designed to reduce downstream hazards. In 1905, an earthen sediment-retention dam was built about 10 km downstream in an initial effort to protect the town of Blitar, about 20 km south of Kelut. However, this effort proved futile when the 1919 lahars overwhelmed the dam and destroyed the

Fig. 12.7 Photographs showing depositional facies, common in lahars, that have progressively accreted with time. **a** Clay-poor lahar deposit at Mount Adams, USA that is inversely graded at its base and both massive and ungraded at its top (dashed line indicates base of deposit; shovel is 50 cm long). The schematic section in 3B illustrates the depositional scenario. **b** Transitional debris-flow to hyperconcentrated-flow deposits at Mount Rainier, USA. The basal unit (at the point of the ice axe) contains floating, rounded cobbles and boulders and probably reflects deposition from hyperconcentrated flow in a channel. **c** Hyperconcentrated-flow deposit that is inversely graded in the lowest two thirds of the outcrop and normally graded in the upper third of the outcrop at Mount St. Helens in 1982. The deposit comprises silt, sand, granules, and small pebbles with a single mode of coarse sand and granules. Deposit is about 1 m thick. Deposition scenario is represented in Fig. 12.3d. **d** Faintly stratified hyperconcentrated-flow deposit and overlying transitional debris-flow deposit at Mount St. Helens in 1982. Depositional scenario is represented in Figs. 12.3c and 12.4. **e** 1998 Casita lahar deposit, Nicaragua, illustrating lahar overbank and channel facies. The Casita lahar began as a watery flood like that shown in Fig. 12.3a, but because of its short travel distance, its evolution to debris-flow phase shown in Fig. 12.3b was incomplete. Photograph in panel C is by Kevin Scott, USGS; the others are by the author

town (Zen & Hadikusumo, 1965). The dam was rebuilt but was nearly full of sediment by 1925 (Blong, 1986). The 1919 tragedy prompted the Dutch colonial government to build a sequence of tunnels that lowered lake level by about 50 m and reduced the volume of the lake to about 2×10^6 m^3 (Fig. 12.9a) (Scrivenor, 1929; Zen & Hadikusumo, 1965) When the volcano erupted again in 1951, it produced only small lahars, and few lives were lost. Unfortunately, the 1951 eruption damaged the drainage tunnels, reamed out the crater, and lowered the lake bottom by 70 m (Fig. 12.9b) (Zen & Hadikusumo, 1965). The old drainage tunnel was repaired, but the

Fig. 12.8 Photographs downstream of Santiaguito volcano (Guatemala) **a**, **b** and **c**, showing the effects of lahar avulsion, followed by incision and finally downstream aggradation and **d** showing the sites (white arrows) of two towns swept away by 1998 Casita (Nicaragua) lahar. Photograph in **a** is El Palmar (Guatemala) after aggradation during a lahar event caused the river to change course, inundating the town. Several months later the river incised a channel 20 m deep. **b** and **c** are before and after the stream capture about 3 km downstream of El Palmar showing about 20 m of aggradation. All USGS photographs by the author

lake still contained at least 2×10^7 m^3 of water (Fig. 12.9b). In 1965, two Indonesian geologists recognized the threat posed by the inadequate lake drainage system and predicted another eruption within 5 years (Zen & Hadikusumo, 1965). Less than two years later, the volcano erupted again, and lahars caused by the ejection of lake water killed hundreds of people. This disaster prompted authorities to advance new drainage tunnels and build more diversion and

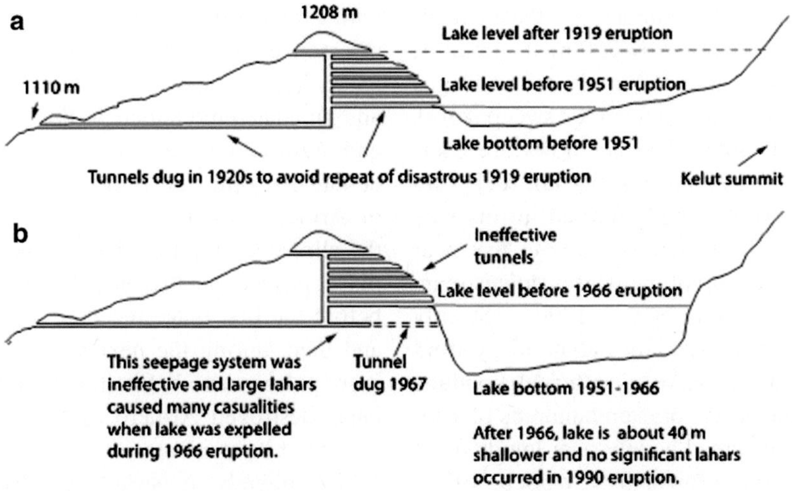

Fig. 12.9 Cross-sections of Kelut volcano crater area; **a** showing lake level after 1919 eruption and tunnel system constructed 1923–27 with resultant lowering of the lake; and **b** showing lake level and bottom before the 1966 eruption. Also shown are the ineffective seepage tunnel system and tunnel advanced after the 1966 eruption. Point 1208 m was the low point in the crater wall and Kelut's summit at 1731 m is to the right. The cross-sectional schematics are adapted from Zen and Hadikusumo (1965)

sediment retention structures. The new drainage tunnels and a fortuitous shallowing of the lake bottom reduced lake volume to less than 2×10^6 m^3, and no primary lahars occurred during the 1990 eruption of Kelut (Bourdier et al., 1997).

Sediment retention and diversion structures have been less effective for hazard mitigation than reducing lake volume. Sedimentation following the 1966 and 1990 eruptions filled most of these structures to or near capacity so that surrounding farms and towns 1–2 m lower than adjacent river channels were threatened with flooding (Blong, 1986; Rodolfo, 2000). Luckily, an eruption in 2007 formed a dome that entirely displaced the notorious crater lake without causing a catastrophic overflow; as a result, the subsequent explosive eruption of 2014 did not produce lake-ejection lahars but did leave the surrounding areas susceptible to secondary lahars owing to rainfall on pyroclastic deposits (Goode et al., 2019).

Crater-lake dams can fail during dormant periods owing to overflow and rapid incision, sapping and undermining, or collapse caused by weakening or earthquakes. At Kelut in 1875, a failure of the west crater wall during a dormant interval caused lahars that destroyed thirty villages and cost numerous lives (Escher, 1922). An infamous example is the Tangiwai disaster that occurred in 1953 when sapping and undercutting caused the crater-lake dam at Ruapehu, New Zealand to fail (Neall, 1976). Water gushed out of the lake, lowering it by 6 m, and caused a lahar downstream. Tragically, the resultant lahar destroyed a railroad bridge a few minutes before the arrival of a passenger train. The train could not be stopped in time, and it plunged into the valley below, killing all 151 passengers aboard. A lahar detection system was implemented after 1953. The system now automatically triggers warning signals for passing trains. In 2007 during a period of unrest, this crater lake broke out again causing a similar size lahar (Table 12.1). As the volcano and its drainages were well monitored, no loss of life and little property damage resulted.

12.5.2 Lahars Caused by Melting of Ice and Snow

Explosive eruptions of hot rock at ice- and snow-clad volcanoes have caused some of the largest known and most catastrophic primary lahars. The most voluminous of these occur when rapidly moving (10–200 m/s) mixtures of hot (typically >500 °C) rock, ash, and gas called pyroclastic flows and surges sweep across steep glaciated slopes. The hot flows are energetic, erosive, and efficiently mix hot rock with ice and snow. Water floods formed in this way bulk up rapidly by factors of three to five near source and form debris-flow phases downstream (Fig. 12.1c) (Pierson et al., 1990). Steeply incised valleys that typically radiate away from glaciated volcanoes serve as efficient conduits for lahars to travel tens or even hundreds of kilometers and then spread widely over hundreds of square kilometers in lowlands (Crandell, 1971). Moderate and large lahars can thus inundate unsuspecting populations with lethal effect in distal areas far from the source volcano.

Historically, Cotopaxi volcano, Ecuador, at 5897 m and clad by 20 km^2 of glacier ice has been the one of the world's most prolific producers of lahars that derive from pyroclastic flows sweeping across glaciers (Mothes, 1992). From 1532 until 1902, lahars had occurred on 27 occasions during explosive eruptions (Simkin & Siebert, 1994). The most widespread and destructive lahars occurred in 1744, 1768, and 1877 during explosive eruptions with VEI 4 (Table 12.1). In 1877 lahars extended more than 200 km down valleys, destroyed towns, killed hundreds of people and thousands of animals (Wolf, 1878). Based on past lahar inundations at Cotopaxi, Hall and von Hillebrant (1988) mapped a smaller, high-risk lahar hazard zone and a larger area, low-risk lahar hazard zone. Almost 50,000 people live in the high-risk hazard zone and almost 90,000 people live in the more extensive low-risk hazard zone.

The most tragic historical lahar event occurred in 1985 at Nevado del Ruiz (5321 m), Colombia, when pyroclastic flows swept across the ice cap of the volcano and produced lethal lahars. Shortly after 9 pm on 13 November a rather small (3.9×10^7 m^3 tephra; Naranjo et al., 1986) but explosive (VEI 3) pyroclastic eruption produced 2×10^7 m^3 of melt water (Pierson et al., 1990; Thouret, 1990). The resultant floods, augmented by eroded debris and overrun stream flow produced lahars in four valleys with a total volume of nearly 10^8 m^3 (Fig. 12.10) (Pierson et al., 1990). Within four hours these lahars had descended 5100 m, travelled 100 km, and inundated more than 50 km^2. The Lagunillas and Azufrado lahars flowed about 70 km and debouched onto the flood plain near the town of Armero. About two and a half hours after the climactic eruption, the first and largest lahar pulse arrived at Armero. About 15 min later, before the first pulse receded, a second arrived, and then, during the next hour and a half, 6–8 more pulses came before the disaster ended. The lahar destroyed Armero and killed more than 23,000 of its residents.

The tragedy at Nevado del Ruiz illustrated both scientific and sociopolitical lessons.

(1) Relatively small pyroclastic eruptions can generate large, destructive lahars, especially at ice-clad volcanoes (Fig. 12.10).
(2) Steep narrow valleys can focus lahars so that they flow tens of kilometers with little attenuation. In fact, such lahars are apt to grow within 10–20 km of the volcano.
(3) Land-use planning could have mitigated the disaster. Previous eruptions of Nevado del Ruiz in 1595 and 1845 produced lahars larger than those of 1985 and inundated the same valleys (Thouret et al., 1990). Armero was built on a lahar plain constructed of deposits from these past two eruptions. Furthermore, fatalities were recorded in the same places during the previous eruptions (Siebert et al., 2010). A land-use plan preventing construction in areas at risk could have saved many lives.
(4) Information about hazardous phenomena has little value if it is not shared in timely fashion with those in harm's way. A better understanding of the risk from lahars and an evacuation plan could have helped many to escape. The eruption sequence began at 15:06; by 17:00 ash began to fall in Armero; at 19:00 the Red Cross in Ibagué ordered evacuation of Armero, but this information was not communicated to the residents; at 21:08 the climactic eruption began; at 22:40 lahars arrived in Chinchiná, sweeping away hundreds of homes

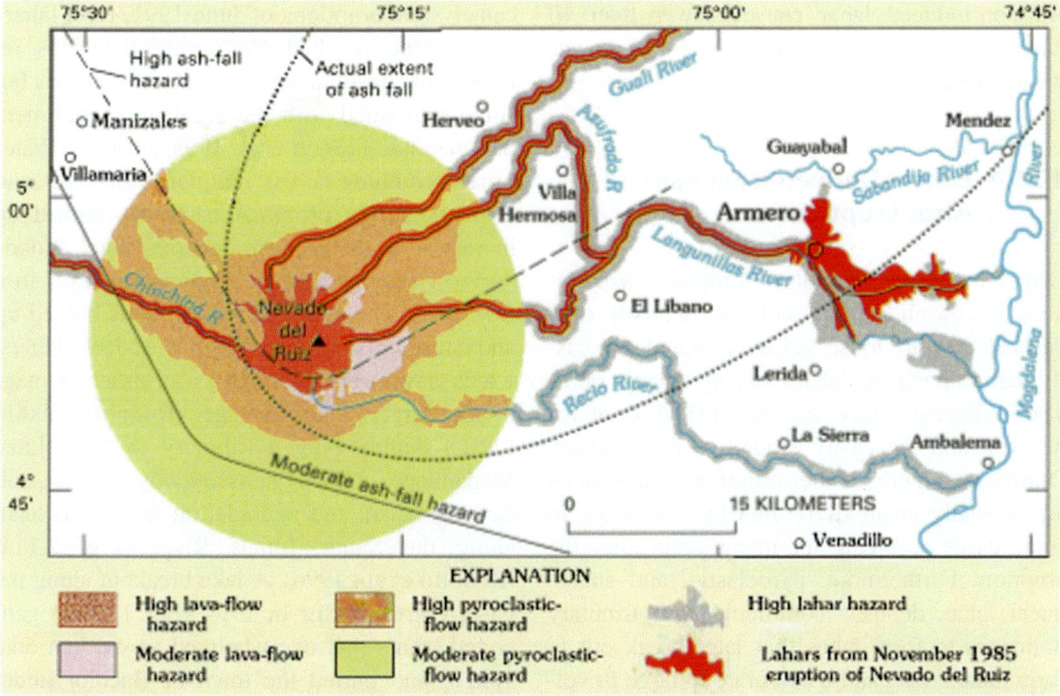

Fig. 12.10 Map showing hazards expected from an eruption of Nevado del Ruiz, Colombia, and the paths of lahars and distribution of ash from the volcano's November 1985 eruption adapted from Wright and Pierson (1992)

and about 1000 people; finally, at 22:35 lahars arrived in Armero. Residents of Armero were not told to evacuate, had little idea where to go, and did not understand the risk that confronted them (Hall, 1990).

(5) Even a rudimentary alarm system could have provided warning (Rodolfo, 2000). Lahars did not arrive until 4.5 h after ash began to fall in Armero, 2.5 h after the onset of the climactic eruption, and 55 min after the disaster at Chinchiná.

12.5.3 Lahars from Edifice Groundwater or Water Stored in Fissures

In 2007 and 2008 at Huila volcano, Colombia, three lahars ranging in size from 10^6 to 3×10^8 m^3, at least two of which are thought to have a novel origin, swept as far as 160 km downstream (Fig. 12.1e; Table 12.1). The first two of these originated high on the ice-clad slopes of the volcano, probably during phreatic events, and are inferred to have been caused by water floods that gushed out of fracture systems or perched groundwater reservoirs (Johnson et al., 2018; Worni et al., 2012). Exactly how so much water flooded out of the fractured edifice remains a mystery. In retrospect, it seems clear that magma, which began to intrude the edifice, melted glacier ice, that was then stored within the edifice. A trigger of unknown origin caused release of water floods and generated the lahars.

Monitoring of unrest and hazard response to these lahars had a much happier outcome than that at Nevado del Ruiz. The successful INGEOMINAS/VDAP responses to the 2007 Huila eruptions mark a complete reversal of the disastrous outcome of the 1985 Ruiz eruption ~20 years earlier, even though these eruptions were of comparable size and involved similar lahar hazards. Huila volcano was well monitored, and a communication plan was in place so that about 5000 people at risk could be evacuated with no lives lost. In 2008, a voluminous eruption-induced lahar caused fewer than 10 fatalities even though parts of several towns were inundated (Table 12.1).

12.5.4 Lahars Derived from Rain After Eruptions

Large explosive eruptions emplace voluminous unconsolidated pyroclastic deposits that, because they are loose and unvegetated, are easily remobilized as lahars during heavy rains. These deposits have no established drainage network and are easily eroded to form lahars. During rainy periods after eruptions, such lahars are typically small (~10^6 m^3) but common, in some cases occurring for many years after the eruption. Furthermore, pyroclastic, and subsequent lahar, deposits commonly dam tributary drainages to form lakes that later break out to form lahars that can be moderate to large in volume (~10^7–10^8 m^3).

The cataclysmic (VEI 5–6) 9–15 June 1991, eruptions of Mount Pinatubo, Philippines, generated about 5.5 km^3 of unconsolidated pyroclastic-flow deposits (Scott et al. 1996a, 1996b), but within 6 years, monsoon rains had remobilized 2.5 km^3 of that pyroclastic debris and generated hundreds of lahars (Umbal, 1997; Phivolcs staff, written communication, 2004) (Fig. 12.11). These lahars inundated villages, towns, and once-fertile farmland; they killed tens of people and displaced thousands of others; and they caused millions of dollars in damage to the local infrastructure (Bautista, 1996). Lahars began 12 June 1991, when rains mixed with tephra-fall deposits, and remobilized preexisting channel material. Large numbers of lahars occurred during the cataclysmic eruptions of 14–15 June when rainfall runoff from coeval typhoon Yunya incorporated tephra and eroded fresh pyroclastic deposits. These syneruptive lahars inundated all major drainages, destroyed bridges, and buried houses. Heavy rains during the 1991 monsoon season initiated all subsequent large lahars. By the end of 1991, lahars had inundated 360 km^2 of residential- and farmland (Rodolfo, 2000). In the years following the cataclysmic eruptions of June 1991, most lahars at Pinatubo resulted from torrential rains on loose unconsolidated pyroclastic debris but some originated with the breakout of sediment dammed lakes (Scott et al. 1996a, 1996b). Water torrents running across ignimbrite deposits and through narrow proximal barrancas bulked up in very short distances to become lahars. Floods during typhoons breached ephemeral lakes in a few cases. The resulting water floods bulked up and caused a small number of post-1991 lahars, which have were as much as an order of magnitude larger than average (Rodolfo, 2000). Lahars dammed up a tributary to form Lake Mapanuepe, which subsequently broke out, causing lahars and aggradation far downstream along the Santo Tomas River (Fig. 12.11a) (Rodolfo et al., 1996). A lake breakout along the Pasig-Potrero River in 1994 (Fig. 12.11b) generated lahars that overwhelmed an earthen dike system and buried the town of Bacolor under several meters of sediment.

Overall, the sedimentary response to the huge influx of pyroclastic debris was large in 1991 and diminished exponentially through 1997 (Fig. 12.11b), but the response in individual drainages was more complex owing to factors such as reestablishment of new drainage networks and stream capture. Most of this sedimentary transport was by lahars (Montgomery et al., 1999). In drainages with the largest proportions of pyroclastic deposits, the sediment transport increased in the first year or two as the drainage network developed (Fig. 12.11b). Because the eruption emplaced huge volumes of unconsolidated pyroclastic deposits in the Santo Tomas-Marella and the Balin-Baquero-Bucao River drainages systems, sedimentation remained high or increased during the first two years. Only after the drainage network developed fully, did sediment yield began to decline (Fig. 12.11b). The eruption emplaced less voluminous pyroclastic deposits in the Gumain and O'Donnell River drainages, the drainage pattern established itself within the first monsoon season, and sedimentary yield declined exponentially (Fig. 12.11b). In February 1992, secondary explosions blocked the Abacan

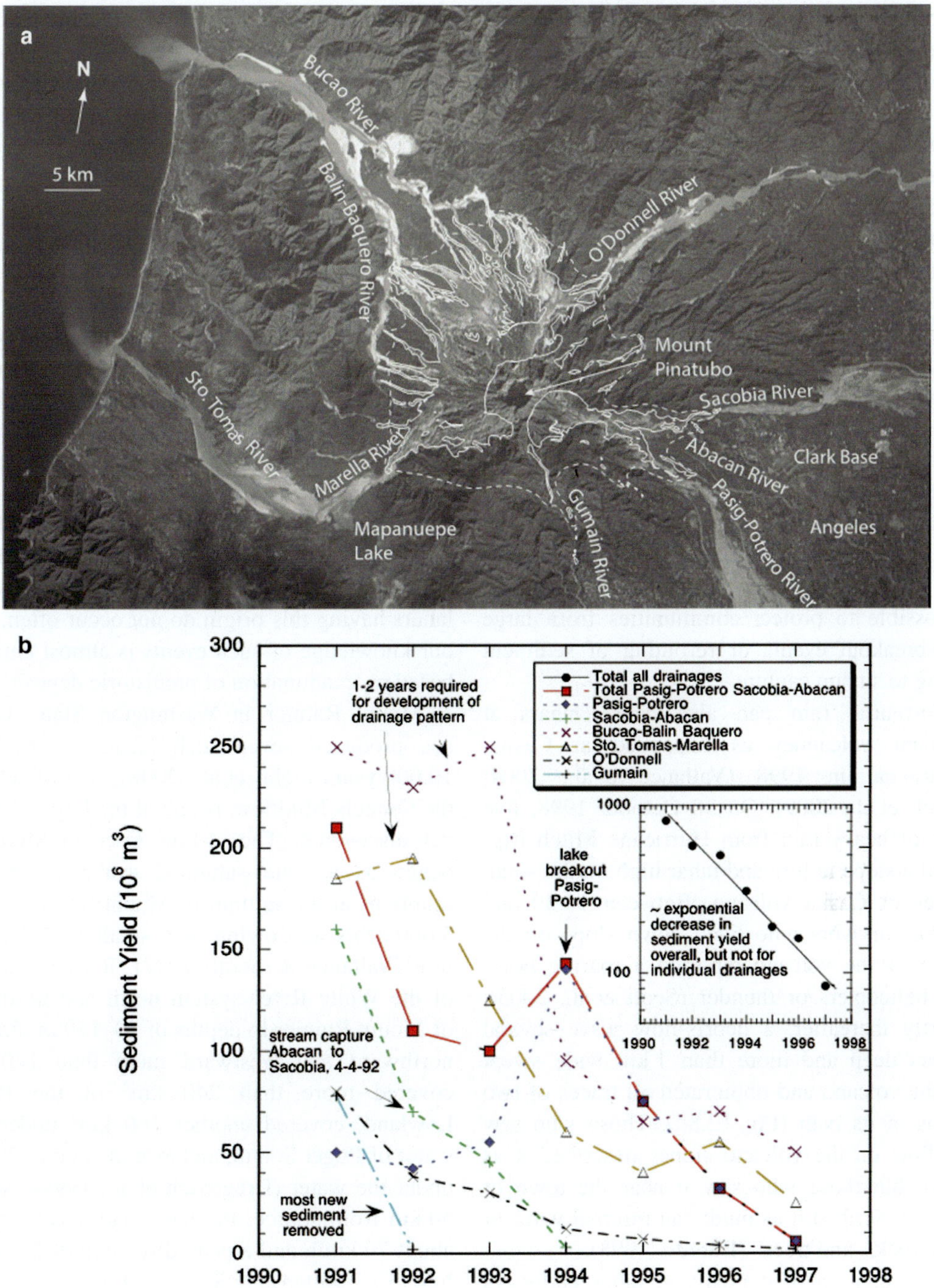

Fig. 12.11 **a** A 1996 NASA Space Shuttle image of Mount Pinatubo and vicinity, Philippines. The dashed line indicates the approximate extent of pyroclastic surges, and the solid lines outline the approximate extent of thick pyroclastic-flow fill deposits in 1991. The light gray color in the image along the drainages is mostly lahar deposits. **b** Plots of sediment yield versus time in individual drainages around Mount Pinatubo and for all its drainages overall (inset). During the time from 1991 to 1997 the bulk of sediment transport was by lahars. Adapted from Vallance (2005)

River and rerouted its lahars into the Sacobia River drainage. Because of sedimentation during a breached lake event in October 1993, the Pasig-Potrero River, which until then had been relatively unaffected by lahars, captured a large portion of the Sacobia River watershed as well as most of its lahar transport (Fig. 12.11b) (Umbal, 1997; Hayes et al., 2002).

In the first few years after 1991, the immense scale of the sedimentary response overwhelmed efforts to protect property and communities downstream. Dikes, typically constructed of laharic debris eroded easily (Rodolfo, 2000). Sediment retention structures were ineffective because they so rapidly filled up. Efforts to remove material from sediment-choked drainages simply could not keep pace. Diversion dams typically rerouted the threat of lahars from one populated area to another. It was nearly impossible to protect communities from large lake-breakout events or rerouting of sediment owing to stream capture.

Torrential rain can also cause lahars at dormant volcanoes as happened at Casita, Nicaragua, in 1998 (Vallance et al., 2004; Devoli et al., 2009). On 30 October 1998, five days of heavy rain from Hurricane Mitch triggered a slope failure and lahar high on the south slopes of Casita volcano. Between 10:30 and 11:00, survivors who lived down slope on the apron of the volcano reported a roaring noise like helicopters or thunder (Scott et al., 2005). Shortly thereafter, a debris-flow wave several meters deep and more than 1 km wide swept off the volcano and obliterated all traces of two towns in its path (Fig. 12.8d). Those who saw the flow on the volcano slopes described it as water, but those who saw it near the town or below described it as mud: "an infernal wave of mud, rocks, and trees" (Baretto, 1998).

Examination of deposits in lahar pathways at Casita and other Central American volcanoes show that such events occur periodically (Scott et al., 2005). In 1541, several days of heavy rain toward the end of the rainy season, when soils were saturated, caused a lahar that swept down the slopes of Agua volcano and destroyed the Capital of Guatemala, Cuidad Vieja (Schilling et al., 2001). Original accounts (Feldman, 1993) showed that this was rain-induced lahar not a lake breakout as is commonly believed.

12.5.5 Lahars Derived from Debris Avalanches

Some debris avalanches contain sufficient water and weakened rock that they transform to water-saturated flows as they deform and move down slope. Lahars of this type are most common at glaciated volcanoes whose rocks have been severely weakened by hydrothermal alteration and seem to be uncommon elsewhere. Debris-avalanche derived lahars are among the largest documented. Even at glaciated volcanoes, lahars having this origin do not occur often, and our knowledge of such events is almost entirely based on examination of prehistoric deposits.

Mount Rainier in Washington State, USA, has produced seven such lahars in the past 10,000 years (John et al., 2008), one of which, the Osceola Mudflow, is one of the largest lahars yet discovered. The ~ 4-km^3 Osceola Mudflow began as a water-saturated avalanche during eruptions at the summit of Mount Rainier about 5,600 years ago owing to a small VEI 3 eruption (Vallance & Scott, 1997). It filled valleys of the White River system north and northeast of Mount Rainier to depths of 80–150 m, flowed northward and westward more than 120 km, covered more than 240 km^2 of the Puget Lowland, covered another 160 km^2 under the water of Puget Sound, and extended up to 20 km under the water (Dragovich et al., 1994). About 50 km from source, the lahar had a velocity of about 70 km/h and a peak discharge of 2–3 million m^3/s (Vallance & Scott, 1997).

The Osceola Mudflow transformed completely from debris avalanche to clay-rich lahar within 2 km of its source because within the pre-avalanche mass there were large volumes of widely dispersed pore water, water residing in the hydrothermal system, and abundant weak hydrothermally altered rock (Vallance & Scott, 1997). Many lahars that transform from debris avalanches are rich in clay suggesting that the amount of hydrothermally altered rock in the pre-avalanche mass influences the probability that a debris avalanche will transform partly or wholly into a lahar or remain a largely unsaturated debris avalanche. The potential for transformation has important implications for hazards because lahars are more mobile than debris avalanches and therefore spread more widely and travel greater distances down valley. Furthermore, very large lahars of this type can be triggered by eruptions that are small in volume and modestly explosive (Fig. 12.2).

Some avalanches that are less fines-rich transform only partially to lahars. Examples include the enormous ~5000-year-old Chillos Valley Lahar from Cotopaxi (Ecuador) the 1980 North Fork Toutle River lahar (Washington, USA), and the 2010 avalanche and debris flow from Mount Meager (Canada) (Table 12.1). Eruptions triggered the first two, but the avalanche at Meager was unrelated to eruptions or unrest at the volcano. The Chillos Valley lahar resulted from the transformation of the water-saturated part of a debris avalanche into a lahar and flowed a remarkable 330 km downstream into the Pacific Ocean (Mothes et al., 1998). The North Fork Toutle River lahar occurred several hours after the 1980 Mount St. Helens eruption caused the voluminous debris avalanche. Several hours after emplacement, water-saturated parts of the Mount St. Helens avalanche deposit dewatered and coalesced and then flowed at least 60 km downstream (Scott, 1986; Major et al., 2005a, 2005b). Although neither of these avalanches contained the pervasive clay content of prehistoric Mount Rainier lahars, they did contain large portions of silt and some clay, which accounts for their mobility. Unlike the drier avalanche that followed, the leading part of the Mount Meager avalanche was watery enough to behave as a debris flow (Roberti et al., 2017). A lack of fines caused the Mount Meager debris flow to dewater quickly as it moved, and it flowed only 13 km.

12.6 Lahar Hazards: Assessment, Modeling, and Mitigation

12.6.1 Rationale for Lahar-Hazard Assessments

The paradigm most applied to assess lahar hazards at an individual volcano is that past events are the best indicators of what might potentially happen in the future. Lahar-hazard zones are readily delineated at volcanoes for which long historical records exist. Because many volcanoes lack such a detailed historical record, field mapping and stratigraphic reconstructions must serve to delineate temporal and spatial distribution of lahars. Correlation of individual lahar deposits to eruptive products like pyroclastic flows and tephra-fall deposits aids in the interpretation of lahar origins. Not all lahars at active volcanoes are related to volcanism, and many small to moderate sized lahars have causes similar to those of debris flows in non-volcanic terrain. Because of steep slopes and abundant loose debris on the slopes of volcanoes, lahars unrelated to eruptions are common. Triggers include glacial outburst floods, heavy precipitation events, and tectonic earthquakes (Table 12.1).

Inundation maps and understanding of origin are critical for delineating lahar-hazard zones, and accurate chronologies are essential to evaluate frequency-magnitude relationships and statistical patterns of lahar production1. At Mount Rainier, a high-profile volcano because of its large nearby population, about fifty Holocene lahars have been identified (Crandell, 1971; Scott et al., 1995). Lahar chronology and distribution allow delineation of lahar hazards (Scott & Vallance, 1995; Vallance & Sisson, 2022) (Fig. 12.12).

12.6.2 Lahar Modeling and Hazard Delineation

Because the time-consuming and expensive field investigations that allow lahar-hazard assessments are not done at many volcanoes, alternative methods of delineating hazard zones

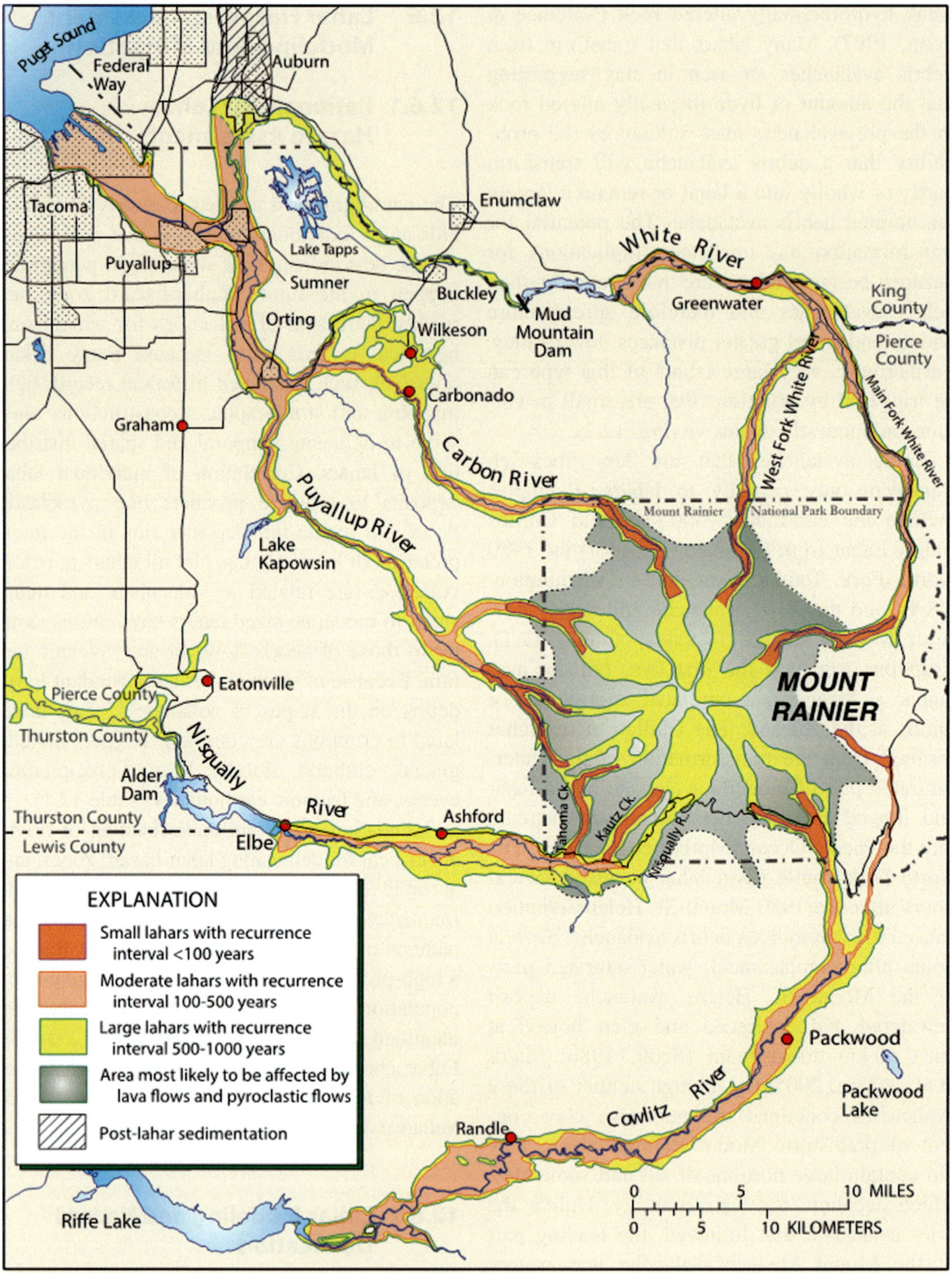

Fig. 12.12 Lahar-hazard zones for Mount Rainier, USA, after Hoblitt et al. (1998)

are needed. Iverson et al. (1998) developed a method of delineating lahar hazard zones based on inundation and size data for well-documented lahars worldwide (Fig. 12.13). The ratio *H*/*L*, where *H* is total vertical elevation dropped and *L* is runout distance, is commonly used to measure mobility. Although this measure is easy to understand and use, it disregards the important effects of lahar volume and three-dimensional path morphology. For example, *H*/*L* takes no account of lateral spreading, an important influence on downstream distance traveled. An alternative measure of mobility accounts for effects such as this by using scale-invariant relations for cross-sectional area inundated and total area inundated,

$$A = aV^{2/3} \text{ and } B = b\mathrm{V}^{2/3}, \qquad (12.1 \text{ and } 12.2)$$

where *A* is valley cross-sectional area inundated (wetted perimeter), *B* is valley planimetric area inundated, *V* is lahar volume, and *a* and *b* are coefficients that need to be empirically calibrated (Iverson et al., 1998). Values of *a* decrease and those of *b* increase as flows exhibit diminishing resistance to flow or enhanced mobility. By these measures, lahars are seen to be more mobile than volcanic debris avalanches and more mobile yet than rock-fall avalanches (Fig. 12.13). Mobility coefficients for lahars can vary considerably, but statistically calibrated values of $a = 0.05$ and $b = 200$ apply to a broad spectrum of lahars whose inundation parameters have been measured in the field. As a result, the empirically calibrated inundation-area relationships are useful for lahar-hazard assessment. Using these calibrated relationships and GIS-based topography, computer-generated lahar inundations can be calculated for given lahar volumes using a routine called LAHARZ (Iverson et al., 1998) The smallest-volume runs delineate areas where lahars are most likely and represent the largest risk, and largest-volume runs delineate the largest but lowest-risk areas. Combined with reconnaissance geology, the method allows expeditious generation of lahar-hazards maps. At Mount Rainier, for example, a reasonable range of potential lahar volumes would be from 10^7 to 10^9 m^3, and a reasonable geometric progression of volumes would be 1×10^7, 3×10^7, 1×10^8, 3×10^8, and 1×10^9 m^3 (Fig. 12.13).

$$A = 0.05V^{2/3} \text{ and } B = 200\mathrm{V}^{2/3} \qquad (12.3 \text{ and } 12.4)$$

Improvements in numerical procedures allow sophisticated physics-based simulations of lahar inundation (e.g., Iverson and George, this volume). The most practical tools currently available for understanding lahar dynamics are depth-averaged numerical models (Iverson & George, 2014). The equations used in these models are derived by applying the laws of mass and momentum conservation, but this application is simplified by assuming that momentum fluxes normal to the bed are negligible in comparison to downstream fluxes. If properly calibrated such models allow simulation of the primary features of lahar dynamics. Despite their many assumptions, modern depth-averaged lahar models are sophisticated enough to simulate the primary features of lahar dynamics. For example, the most sophisticated models would be able to account for excess pore-fluid pressure that may greatly influence mobility and account for entrainment of bed or bank material that may change total lahar mass and solid volume fraction.

There is no one physics-based model appropriate for simulating the wide variety lahar phenomena and the wide variety of their origins. Physics-based models are summarized by Thouret et al., (2020, their Table 4) and are the subject of another chapter (Iverson & George, this volume). Typically, physics-based models that have been applied to lahar inundation require knowledge of initial volume and a DEM of the source area and the area downstream. A key element in any lahar simulation is its flow resistance term. Such resistance terms range from viscous (essentially flood routing models) (Charbonnier et al., 2018; O'Brien & Julien, 2000) to dry Coulomb (Patra et al., 2005; Procter et al., 2012) and Voellmy models that combine viscous and Coulomb resistance

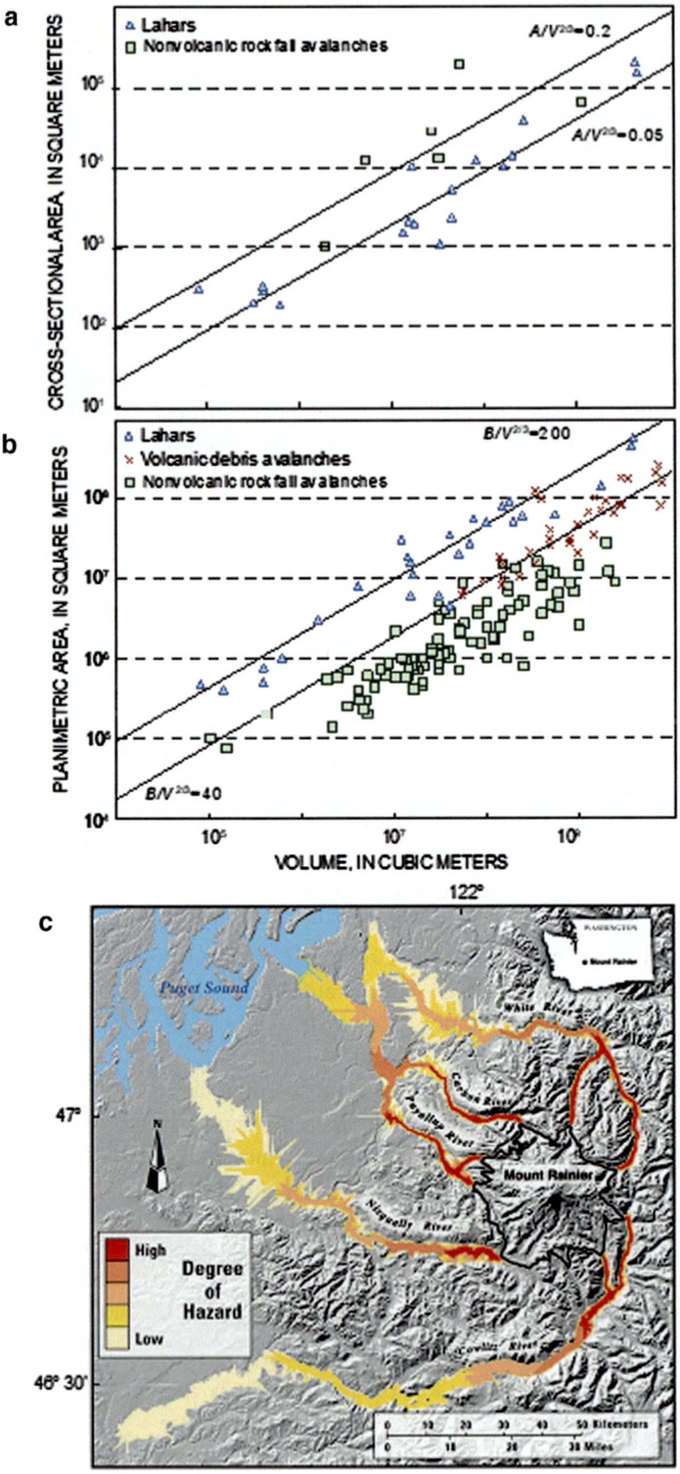

Fig. 12.13 Plots of **a** inundated flow-path, cross-sectional area, *A*, versus flow volume *V*; **b** planimetric area inundated, **b**, versus volume and **c** inundation areas calculated for valleys draining Mount Rainier, Washington, USA. The dark line encloses a proximal hazard zone which includes hazards from diverse processes including lahars. Further information on the LAHARZ method in Iverson et al. (1998); figure adapted from Vallance and Iverson (2015)

(Córdoba et al., 2015). A more elaborate simulation includes the mediating effects of pore-fluid pressure on flow resistance and allows for compressibility (George & Iverson, 2014; also see Iverson & George, this volume). For lahars originating as slope failures, the DCLAW model of Iverson and George (2014) can simulate a lahar from its onset to its deposition. Lahars with origins different from slope failure would require an empirical choice of initial conditions in their model. All models require at least some calibration to simulate a lahar flow well. At their best, these models yield realistic estimates of important flow parameters such as velocity and travel time between source and populated areas.

12.6.3 Mitigation of Lahar Hazards

Lahar hazards can be mitigated through land-use planning, engineering, or evacuation in time of crisis. An effective way to avoid risk from lahars is through land-use regulation (Blong, 1986; Crandell & Mullineaux, 1974). Once lahar-hazard maps are prepared, federal and local agencies can better manage residential and commercial building in areas potentially at risk. Unfortunately, dense populations that are already established in some threatened areas make land-use planning impractical. In many countries, people will choose to tolerate moderately high risk from lahars and other volcanic processes because the most fertile land and most healthy climates are located on or near volcano flanks (Blong, 1986).

Engineering strategies include sediment-retention dams, diversion dams and dikes, and draining of dangerous crater lakes as at Kelut. Draining crater lakes and other dangerous volcanically dammed lakes as at Mount St. Helens (Major et al., 2005a, 2005b) has proved to be effective in mitigating lahar hazards. Dams, diversions, and dikes work well for lahars of modest volume. Unfortunately, structures of these types cannot cope with the huge volume of volcanic debris that is commonly transported downstream during and after eruptions. In such cases, lahar-warning systems can give residents downstream time to escape; however, such systems do not protect property. Warning systems include those in which an observer is posted upstream with a radio, those in which a wire connection or series of connections are broken and those in which a seismic sensor sends a signal to emergency response personnel (LaHusen, 2005). More elaborate modern lahar warning systems process seismic or other geophysical indicators of dangerous flows and set off automated alarms.

In many cases, evacuation is the only mitigation option. At volcanoes, lahars are considerably more likely during or immediately following eruptions. Furthermore, the most voluminous lahars are associated with explosive eruptions. Fortunately, volcanoes most commonly give advanced warning of days to weeks prior to eruption and allow time for orderly evacuation of residents at risk. In the past forty years, tens of thousands of people have successfully been evacuated from areas at risk from lahars and other hazardous volcanic phenomena (Siebert et al., 2010).

12.7 Concluding Remarks

How climate change might affect lahar occurrence is a somewhat speculative proposition, but in two specific scenarios, some conclusions might be tenable. First at volcanoes that are experiencing glacier retreat and melting of permanent snow fields, climate warming progressively uncovers unconsolidated glacial drift on steep slopes. At Mount Rainier, for example, glacial outburst floods and atmospheric-river events interact with denuded and debuttressed glacial deposits to cause small lahars almost every year. Furthermore, with continued melting of snow and ice and glacier retreat, the locus of these debris flows has changed. In the early to mid-nineteenth century, the upper Nisqually River basin (Fig. 12.12) was the locus of these types of lahars (Richardson, 1968). By the 1980s as the Nisqually Glacier retreated far up valley, the locus of lahar activity shifted to Tahoma and Kautz Creeks (Fig. 12.12) (Walder

& Driedger, 1995). Since 2001, melting of permanent snow fields left upper Van Trump Creek, a tributary to the upper Nisqually River, vulnerable to lahars caused by intense rainfall events. Debris-flows initiated by intense rainstorms and outburst floods are common in any steep mountainous terrain that has been recently deglaciated. A second possible effect of climate warming and glacial retreat during the next 50–100 years could be complete removal of ice caps at volcanoes like Nevado del Ruiz in Colombia. Absence of glacial ice at the summit of a volcano like Nevado del Ruiz would preclude interaction of pyroclastic flows with glacier ice and snow during future explosive eruptions and thus make genesis of voluminous lahars originating by hot-rock-and-ice interaction unlikely.

The past two decades have seen a few key advances in better understanding lahar origins and processes, and a notable improvement in modeling lahar flows. Water stored within volcanic edifices influences lahar origins in two important ways. First, hydrothermal clay leads to widely dispersed water in the upper edifice of some volcanoes like Mount Rainier. When that weakened rock fails the resulting water-rich avalanches transform to voluminous lahars. Second, ice-clad volcanoes can store water in their upper edifices, and volcanic unrest can cause large volumes of that water to gush out and form lahars as it moves down the steep flanks as happened at Huila, Colombia in 2007.

Observations and detailed sedimentology show that far-traveled lahars transform downstream by eroding significant proportions of sediment, by incorporating overrun water, and by deposition. Understanding erosive processes has improved our understanding of how lahars bulk up. Iverson et al. (2011) showed how entrainment of water-saturated beds can lead to runaway growth of debris flows. The 1998 Casita example (Scott et al., 2005) showed that an analogous process is likely common in lahar genesis. Church and Jakob (2020) summarized how debris floods on steep slopes underlain by unconsolidated sediment can form debris flows, but a similar process can happen on a huge scale as water-rich debris floods descend steep slopes of volcanoes to form voluminous lahars (for example, Pierson et al., 1990; Scott, 1988). Proximal lahar bulking processes are not especially well understood, and this subject could benefit from future research in the field and laboratory.

Massive, graded debris-flow deposits have tempted geologists to interpret them as snap shots in time and interpret the deposits accordingly. However, experiments (Major, 1997) have shown that debris flows deposit incrementally rather than in mass, and an increasing body of field evidence shows that most lahars deposit incrementally too (Vallance & Iverson, 2015). A debris-flow-phase lahar overrunning water (a river, for example) evolves to a more dilute hyperconcentrated-flow phase at its front. The debris-flow phase then lags the more dilute flow. An evolving system of this sort will first deposit hyperconcentrated-flow then debris-flow sediment leading to inversely graded deposits. Normally graded deposits are also common when the leading part of lahars bulk up on coarse sediment, which then is overrun by debris less rich in coarse sediment during falling stages (Vallance & Scott, 1997). Interpretation of massive lahar deposits is possible but requires taking account of an accreting sedimentary system.

Development of more sophisticated lahar modeling has shown great improvement during the past two decades and is the most useful new tool for lahar hazard evaluation; it also has scope for future constructive development. Lahar flow models have evolved from empirical (LAHARZ) and single-phase physical models that require unrealistic empirical assumptions to sophisticated two-phase models that can simulate a lahar from initiation to deposition and theoretically include erosion of substrate (Iverson & George, this volume). Nonetheless, even the best models cannot simulate lahars that begin with floods of water from a volcano, for example, owing to crater lake expulsion or pyroclastic-flow interaction with snow and ice. Simulation of lahars that begin with voluminous water floods, which incorporate substrate debris and then bulk up to debris-flow-phase lahars, should be an important focus of future research.

Acknowledgements I thank Mary Benage, Maceij Obryk, and Larry Mastin for careful reviews of the manuscript. The manuscript also benefited from insightful comments of the editors, Matthias Jakob, Scott McDougall, and Paul Santi. This work would never have reached fruition without the enthusiasm and energy of Matthias Jakob; he will be greatly missed. Any use of trade, firm, or product names in this chapter is for descriptive purposes only and does not imply endorsement by the U.S. Government.

References

Baretto, P. E. (1998). Testimonios relacionados con los ruidos y imagines fatales segundos antes de la tragedia, llegada repentinamente desde el volcán Casita a las comunidades Rolando Rodriguez y El Porvenir. Unpublished Report, Managua, Nicaragua.

Bautista, C. B. (1996). The Mount Pinatubo disaster and the people of central Luzon. In C. G. Newhall & R. S. Punongbayan (Eds.), *Fire and Mud: Eruptions and lahars of Mount Pinatubo, Philippines* (pp. 151–164). University of Washington Press.

Blong, R. (1986). *Volcanic hazards* (p. 400). Academic Press.

Bourdier, J.-L., Pratomo, I., Thouret, J.-C., Boudon, G., & Vincent, P. M. (1997). Observations, stratigraphy and eruptive processes of the 1990 eruption of Kelut volcano, Indonesia. *Journal of Volcanology and Geothermal Research, 79*, 181–203.

Carrasco-Nuñez, G., Vallance, J. W., & Rose, W. I. (1993). A voluminous avalanche-induced lahar from Citlaltépetl volcano, Mexico: Implications for hazard assessment. *Journal of Volcanology and Geothermal Research, 59*, 35–46.

Charbonnier, S. J., Connor, C. B., Connor, L. J., et al. (2018). Modeling the October 2005 lahars at Panabaj (Guatemala). *Bulletin of Volcanology, 80*, 4. https://doi.org/10.1007/s00445-017-1169-x

Church, M., & Jakob, M. (2020). What is a debris flood. *Water Resources Research*, 1–17. https://doi.org/10.1029/2020WR027144

Córdoba, G., Villarosa, G., Sheridan, M. F., Viramonte, J. G., Beigt, D., & Salmuni, G. (2015). Secondary lahar hazard assessment for Villa la Angostura, Argentina, using Two-Phase-Titan modelling code during 2011 Cordón Caulle eruption. *Natural Hazards and Earth System Sciences, 15–4*, 757–766.

Crandell, D. R., & Mullineaux, D. R. (1974). Technique and rationale of volcanic-hazards appraisals in the Cascade Range, northwestern United States. *Environmental Geology, 1*, 23–32.

Crandell, D. R. (1971). *Postglacial lahars from Mount Rainier volcano, Washington* (75 pp.). U.S. Geological Survey Professional Paper 677

Cronin, S. J., Neall, V. E., Palmer, A. S., & Lecointre, J. A. (1997). Changes in Whangaehu River lahar characteristics during the 1995 eruption sequence, Ruapehu volcano, New Zealand. *Journal of Volcanology and Geothermal Research, 76*, 47–61.

Devoli, G., Cepeda, J., & Kerle, N. (2009). The 1998 Casita volcano flank failure revisited—new insights into geological setting and failure mechanisms. *Engineering Geology, 105*, 65–83. https://doi.org/10.1016/j.enggeo.2008.12.006

Dolan, M. T. (2004). Observations of lahars along the Sacobia-Bamban River systems, Mount Pinatubo, Philippines (48 pp.). Michigan Technological University MS thesis.

Dragovich, J. D., Pringle, P. T., & Walsh, T. J. (1994). Extent and geometry of the mid-Holocene Osceola mudflow in the Puget Lowland—implications for Holocene sedimentation and paleogeography. *Washington Geology, 22*(3), 3–26.

Driedger, C. L., & Fountain, A. G. (1989). Glacier outburst floods at Mount Rainier, Washington State, U.S.A. *Annals of Glaciology, 13*, 51–55. https://doi.org/10.3189/S0260305500007631

Escher, B. G. (1922). On the hot lahar of the Valley of Ten Thousand Smokes (Alaska). *Proceedings Koninklijke Akademie Van Wetenschappen, Amsterdam, 24*, 282–293.

Feldman, L. (1993). *Mountains of fire lands that shake* (295 pp.). Labyrinthos, Culver City, CA.

Fisher, R. V., & Schmincke, H.-U. (1984). *Pyroclastic rocks* (p. 472). Springer.

George, D. L., & Iverson, R. M. (2014). A depth-averaged debris-flow model that includes the effects of evolving dilatancy. II. Numerical predictions and experimental tests. *Proceedings of the Royal Society A, 470*. https://doi.org/10.1098/rspa.2013.0820

Goode, L. R., Handley, H. K., Cronin, S. J., & Abdurrachman, M. (2019). Insights into eruption dynamics from the 2014 pyroclastic deposits of Kelut volcano, Java, Indonesia, and implications for future hazards. *Journal of Volcanology and Geothermal Research, 382*, 6–23. https://doi.org/10.1016/j.jvolgeores.2018.02.005

Hall, M. L. (1990). Chronology of the principal scientific and governmental actions leading up to the November 13, 1985, eruption of Nevado del Ruiz, Colombia. *Journal of Volcanology and Geothermal Research, 42*, 101–115.

Hall, M. L., & von Hillebrant. (1988). Mapa de los peligros volcanicos potenciales asociads con el Volcan Cotopaxi. Instituto Geofisico, Escuela Politecnica Nacional, Quito.

Hayes, S. K., Montgomery, D. R., & Newhall, C. G. (2002). Fluvial sediment transport and deposition following the eruption of Mount Pinatubo. *Geomorphology, 45*, 211–224.

Hodgson, K. A., & Manville, V. R. (1999). Sedimentology and flow behavior of a rain triggered lahar, Mangatoetoenui Stream, Ruapehu volcano, New Zealand. *Geological Society of America Bulletin, 111*(5), 743–754.

Hoblitt, R. P., Walder, J. S., Driedger, C. L., Scott, K. M., Pringle, P. T., & Vallance, J. W. (1998). *Volcano hazards from Mount Rainier, Washington, Revised 1998* (11 pp.). U.S. Geological Survey Open-File Report 98-428, 2 plates.

Hungr, O., Leroueil, S., & Picarelli, L. (2014). The Varnes classification of landslide types, an update. *Landslides, 11*, 167–194. https://doi.org/10.1007/s10346-013-436-y

Iverson, R. M., & George, D. L. (2014). A depth-averaged debris-flow model that includes the effects of evolving dilatancy I. Physical basis. *Proceedings of the Royal Society A, 470*, 20130819. https://doi.org/10.1098/rspa.2013.0819

Iverson, R. M., & George, D. L. (this volume). Numerical modeling of debris flows: A conceptional assessment. In M. Jakob, S. McDougal, & P. Santi (Eds.), This volume.

Iverson, R. M., Reid, M. E., Logan, M., LaHusen, R. G., Godt, J. W., & Griswold, J. P. (2011). Positive feedback and momentum growth during debris-flow entrainment of wet bed sediment. *Nature Geoscience, 4*(2), 116–121.

Iverson, R. M., Schilling, S. P., & Vallance, J. W. (1998). Objective delineation of lahar-hazard zones. *Geological Society of America Bulletin, 110*, 972–984.

Kang, C., Imaizumi, F., & Theule, J. (this volume). Sediment entrainment and Deposition. In M. Jakob, S. McDougal, P. Santi (Eds.), This volume.

Jakob, M. (2005). Debris-flow hazard analysis. In M. Jakob & O. Hungr (Eds.), *Debris flow and debris avalanches—a practically-oriented overview of the state-of-the-art* (pp. 411–444).

John, D. A., Sisson, T. W., Breit, G. N., Rye, R. O., & Vallance, J. W. (2008). Characteristics, extent and origin of hydrothermal alteration at Mount Rainier Volcano, Cascades Arc, USA: Implications for debris-flow hazards and mineral deposits. *Journal of Volcanology and Geothermal Research, 175*, 289–314. https://doi.org/10.1016/j.jvolgeores.2008.04.004

Johnson, C. G., Kokelaar, B. P., Iverson, R. M., Logan, M., LaHusen, R. G., & Gray, J. M. N. T. (2012) Grain-size segregation and levee formation in geophysical mass flows. *Journal of Geophysical Research, 117*. https://doi.org/10.1029/2011JF002185

Johnson, P. J., Valentine, G. A., Stauffer, P. H., et al. (2018). Groundwater drainage from fissures as a source for lahars. *Bulletin of Volcanology, 80*, 39. https://doi.org/10.1007/s00445-018-1214-4

Lavigne, F., Thouret, J. C., Voight, B., Suwa, H., & Sumaryono, A. (2000). Lahars at Merapi volcano, Central Java: An overview. *Journal of Volcanology and Geothermal Research, 100*, 423–456.

Lavigne, F., & Thouret, J. C. (2003). Sediment transportation and deposition by rain-triggered lahars at Merapi Volcano, Central Java, Indonesia. *Geomorphology, 49*(1–2), 45–69.

LaHusen, R. (2005). Debris-flow instrumentation. In M. Jakob & O. Hungr (Eds.), *Debris flow and debris avalanches—a practically-oriented overview of the state-of-the-art* (pp. 291–304).

Major, J. J. (1997). Depositional processes in large-scale debris-flow experiments. *The Journal of Geology, 105*(3), 345–366.

Major, J. J., Pierson, T. C., & Scott, K. M. (2005a). Debris flows at Mount St. Helens, Washington, USA. In M. Jakob & O. Hungr (Eds.), *Debris flow and debris avalanches—a practically-oriented overview of the state-of-the-art* (pp. 685–731).

Major, J. J., Pierson, T. C., & Scott, K. M. (2005b). Volcanic debris flows at Mount St. Helens, Washington, U.S.A. In M. Jakob & O. Hungr (Eds.), *Debris flow and debris avalanches—a practically-oriented overview of the state-of-the-art* (pp. 685–731).

Major, J. J. (2022). *Subaerial volcaniclastic deposits-influences of initiation mechanisms and transport behavior on characteristics and distributions*. Geologic Society of London, Special Publication 520. https://doi.org/10.1144/SP520-2021-142

Major, J. J., & Newhall, C. G. (1989). Snow and ice perturbation during historical volcanic eruptions and the formation of lahars and floods–a global review. *Bulletin of Volcanology, 52*, 1–27.

Maizels, J. (1997). Jökulhlaup deposits in proglacial areas. *Quaternary Science Reviews, 16*(7), 793–819. https://doi.org/10.1016/S0277-3791(97)00023-1

Montgomery, D. R., Panfil, M. S., & Hayes, S. K. (1999). Channel-bed mobility response to extreme sediment loading at Mount Pinatubo. *Geology, 27*, 271–274.

Mothes, P. A. (1992). *Lahars of Cotopaxi volcano, Ecuador: Hazard and risk evaluation. Geohazards, Natural and Man-made* (pp. 53–63). Chapman and Hall.

Mothes, P. A., Hall, M. L., & Janda, R. J. (1998). The enormous Chillos Valley Lahar: An ash-flow-generated debris flow from Cotopaxi volcano, Ecuador. *Bulletin of Volcanology, 59*, 233–244.

Naranjo, J. L., Sigurdsson, H., Carey, S. N., & Fritz, W. (1986). Eruption of the Nevado del Ruiz Volcano, Colombia, on 13 November 1985: Tephra fall and lahars. *Science, 233*(4767), 961–963. https://www.istor.org/stable/1697896

Neall, V. E. (1976). Lahars as Major Geological Hazards. *Bulletin of Engineering Geology & the Environment, 14*, 233–240.

Newhall, C. G., & Self, S. (1982). The volcanic explosivity index (VEI): An estimate of explosive magnitude for historical volcanism. *Journal of Geophysical Research: Oceans, 87*, 1231–1238.

O'Brien, J. S., & Julien, P. Y. (2000). *Flo-2D user's manual, Version 2000.01; Flo-Engineering, Nutrioso, AZ, USA* (170 pp.).

Patra, A. K., Bauer, A. C., Nichita, C. C., Pitman, E. B., Sheridan, M. F., Bursik, M., Rupp, B., Webber, A., Stinton, A. J., Namikawa, L. M., & Renschler, C. S. (2005). Parallel adaptive numerical simulation of dry avalanches over natural terrain. *Journal of Volcanology and Geothermal Research, 139*, 1–21. https://doi.org/10.1016/j.jvolgeores.2004.06.014

Pierson, T. C. (1999). Transformation of a water flood to debris flood to debris flow following the eruption-triggered transient-lake breakout from the crater on March 19, 1982. In T. C. Pierson (Ed.), *Hydrologic consequences of hot-rock/snowpack interactions at Mount St. Helens volcano, 1982–84* (pp. 16–36). U.S. Geological Survey Professional Paper 1586.

Pierson, T. C. (2005). Hyperconcentrated flow—transitional process between water flow and debris flow. In M. Jakob & O. Hungr (Eds.), *Debris flow and debris avalanches—a practically-oriented overview of the state-of-the-art* (pp. 159–202).

Pierson, T. C., Janda, R. J., Thouret, J. C., & Borerro, C. A. (1990). Perturbation and melting of snow and ice by the 13 November 1985 eruption of Nevado del Ruiz, Colombia, and consequent mobilization, flow and deposition of lahars. *Journal of Volcanology and Geothermal Research, 41*, 17–66.

Pierson, T. C., & Scott, K. M. (1985). Downstream dilution of a lahar: Transition from debris flow to hyperconcentrated streamflow. *Water Resources Research, 21*, 1511–1524.

Prakash, N., Santi, P., Strouth, A., Sepulveda, S. A., & Dowling, C. (this volume). Fatalities from debris flow: Worldwide distribution and trends. In M. Jakob, S. McDougal, & P. Santi (Eds.), This volume.

Pringle, P. T., & Cameron, K. A. (1999). Eruption-triggered lahar on May 14, 1984. In T. C. Pierson (Ed.), *Hydrologic consequences of hot-rock/snowpack interactions at Mount St. Helens volcano, 1982–84, Reston, VA, U.S* (pp. 81–103). U.S. Geological Survey Professional Paper 1586.

Procter, J. N., Cronin, S. J., & Sheridan, M. F. (2012). Evaluation of Titan2D modelling forecasts for the 2007 Crater Lake break-out lahar, Mt. Ruapehu, New Zealand. *Geomorphology, 136–1*, 95–105. https://doi.org/10.1016/j.geomorph.2011.05.001

Richardson, D. (1968). *Glacial outbursts floods in the Pacific Northwest* (8 pp.). U.S. Geological Survey Professional Paper 600D.

Roberti, G., Friele, P., van Wyk de Vries, B., Ward, B., Clague, J. J., Perotti, L., & Giardino, M. (2017). Rheological evolution of the Mount Meager 2010 debris avalanche, southwestern British Columbia. *Geosphere, 13*(2), 369–390https://doi.org/10.1130/GES01389.1

Rodolfo, K. S. (2000). The hazards from lahars and jökulhlaups. In B. F. Houghton, S. R. McNutt, H. Rymer, & J. Stix (Eds.), *Sigurdsson, H* (pp. 973–995). Encyclopedia of volcanoes.

Rodolfo, K. S., & Arguden, A. T. (1991). Rain-lahar generation and sediment-delivery systems at Mayon volcano, Phillipines. In R. V. Fisher & G. A. Smith (Eds.), *Sedimentation in volcanic settings* (pp. 71–87). SEPM Special Publication 45.

Rodolfo, K. S., Umbal, J. V., Alonso, R. A., Remotigue, C. T., Paladio-Melosantos, M., Salvador, J. H. G., Evangelista, D., & Miller, Y. (1996). Two years of lahars on the western flank of Mount Pinatubo: Initiation, flow processes, deposits, and attendant geomorphic and hydraulic changes. In C. G. Newhall & R. S. Punongbayan (Eds.), *Fire and Mud: Eruptions and lahars of Mount Pinatubo, Philippines* (pp. 989–1014). University of Washington Press.

Schmidt, K. G. (1934). Die Schuttströme am Merapi auf Java nach dem Ausbruch von 1930. *De Ingenieur in Nederlandsch-Indië, 1*(7), 91–120.

Schilling, S. P., Vallance, J. W., Matías, O., & Howell, M. M. (2001). *Lahar hazards at Agua volcano, Guatemala* (9 pp.). U.S. Geological Survey Open-File Report 01-432, 1 plate.

Scott, K. M. (1988). *Origins, behavior, and sedimentology of lahars and lahar-runout flows in the Toutle-Cowlitz River System* (75 pp.). U.S. Geological Survey Professional Paper 1447-A.

Scott, K. M., Vallance, J. W., Kerle, N., Luis Macías, J., Strauch, W., & Devoli, G. (2005). Catastrophic precipitation-triggered lahar at Casita volcano, Nicaragua: Occurrence, bulking and transformation. *Earth Surface Processes and Landforms, 30*, 59–79. https://doi.org/10.1002/esp.1127

Scott, K. M., Macías, J. L., Naranjo, J. A., Rodríguez, S., McGeehin, J. P. (2001). *Catastrophic debris flows transformed from landslides in volcanic terrains: Mobility, hazard assessment, and mitigation strategies* (59 pp.). U.S. Geological Survey Professional Paper 1630.

Scott, K. M., Vallance, J. W., & Pringle, P. T. (1995). *Sedimentology, behavior, and hazards of debris flows at Mount Rainier, Washington* (106 p.). U.S. Geological Survey Professional Paper 1547, 1 plate.

Scott, W. E., Hoblitt, R. P., Torres, R. C., Self, L., Marinez, M., & Nillos, T., Jr. (1996a). Pyroclastic flows of the June 15, 1991, Climactic eruption of Mount Pinatubo. In C. G. Newhall & R. S. Punongbayan (Eds.), *Fire and Mud: Eruptions and lahars of Mount Pinatubo, Philippines* (pp. 151–164). University of Washington Press.

Scott, K. M., Janda, R. J., de la Cruz, E. G., Gabinete, E., Eto, I., Isada, M., Sexon, M., & Hadley, K. C. (1996b). Channel and sedimentation responses to large volumes of 1991 volcanic deposits on the east flank of Mount Pinatubo. In C. G. Newhall & R. S. Punongbayan (Eds.), *Fire and Mud: Eruptions and lahars of Mount Pinatubo, Philippines* (pp. 971–988). University of Washington Press.

Scott, K. M., & Vallance, J. W. (1995). *Debris flow, debris avalanche, and flood hazards at and downstream from Mount Rainier, Washington* (9 pp.). U.S. Geological Survey Hydrologic Investigations Atlas HA-729, 2 plates.

Siebert, L., Simkin, T., & Kimberly, P. (2010). *Volcanoes of the world* (p. 551). University of California Press.

Simkin, T., & Siebert, L. (1994). *Volcanoes of the world* (p. 349). Geoscience Press.

Smith, G. A. (1986). Coarse-grained nonmarine volcaniclastic sediment: Terminology and depositional process. *Geological Society of America Bulletin, 97*, 1–10.

Smith, G. A., & Fritz, W. J. (1989). Volcanic influences on terrestrial sedimentation. *Geology, 17*, 375–376.

Scrivenor, J. B. (1929). The mudstreams (lahars) of Gunoug Keleot in Java. *Geology Magazine, 66*, 433–434.

Thouret, J. C. (1990). Effect of the November 13, 1985, eruption in the snowpack and ice cap of Nevado del Ruiz, Colombia. *Journal of Volcanology and Geothermal Research, 41*, 177–202.

Thouret, J. C., Cantagrel, J. M., Salinas, R., & Murcia, A. (1990). Quaternary eruptive history of Nevado del Ruiz, Colombia. *Journal of Volcanology and Geothermal Research, 42*, 225–252.

Thouret, J. C., Magill, A., & Ollier, C. (2020). Lahars and debris flows: Characteristics and impacts. *Earth-Science Reviews, 201*, 103003. https://doi.org/10.1016/j.earscirev.2019.103003

Vallance, J. W. (2000). Lahars. In H. Sigurdsson, B. F. Houghton, S. R. McNutt, H. Rymer, & J. Stix (Eds.), *Encyclopedia of volcanoes* (pp. 601–616). Academic Press.

Vallance, J. W. (2005). Volcanic debris flows. In M Jakob & O. Hungr (Eds.), *Debris flow and debris avalanches—a practically-oriented overview of the state-of-the-art* (pp. 247–274).

Vallance, J. W., & Iverson, R. M. (2015). Lahars and their deposits. In H. Sigurdsson (Ed.), *The encyclopedia of volcanoes* (2nd ed., pp. 649–664). Academic Press. https://doi.org/10.1016/B978-0-12-385938-9.00037-7

Vallance, J. W., Schilling, S. P., Devoli, G., Reid, M. E., Howell, M. M., & Brien, D. L. (2004). *Lahar hazards at Casita and San Cristóbal volcanoes, Nicaragua* (12 pp.). U.S. Geological Survey Open-File Report 01-467, 3 plates.

Vallance, J. W., & Scott, K. M. (1997). *The Osceola mudflow from mount rainier: Sedimentology and hazard implications of a huge clay-rich debris flow. Geological Society of America Bulletin, 109*(2), 143–163.

Vallance, J. W., & Sisson, T. S. (2022). *Geologic field-trip guide to volcanism and its interaction with snow and ice at Mount Rainier, Washington*. U.S. Geological Survey Scientific Investigations Report 2017-5022A. https://doi.org/10.3133/sir20175022A

Van Bemmelen, R. W. (1949). *The geology of Indonesia 1A* (p. 732). Government Printing Office.

Umbal, J. V. (1997). Five years of lahars at Pinatubo volcano: Declining but still potentially lethal hazards. *Journal of the Geological Society of the Philippines, 52*(1), 1–19.

Walder, J. S., & Driedger, C. L. (1995). Frequent outburst floods from South Tahoma Glacier, Mount Rainier, USA—relation to debris flows, meteorological origin and implications for subglacial hydrology. *Journal of Glaciology, 41*, 1–10. https://doi.org/10.1086/627699

Wright, T. L., & Pierson, T. C. (1992). *Living with Volcanoes* (57 pp.). U.S. Geological Survey Circular 1073.

Wolf, T. (1878). Der Cotopaxi und seine letzte eruption am 26 Juni 1877. Neues Jahrbuch für Mineralogie. *Geologie und Paleontologie*, 113–167.

Waitt, R. B. (1989). Swift snowmelt and floods (lahars) caused by great pyroclastic surge at Mount St. Helens volcano, Washington, 18 May 1980. *Bulletin of Volcanology, 52*, 138–157.

Waythomas, C. F., Pierson, T. C., Major, J. J., & Scott, W. E. (2013). Voluminous ice-rich and water-rich lahars generated during the 2009 eruption of Redoubt Volcano, Alaska. *Journal of Volcanology and Geothermal Research, 259*, 389–413. https://doi.org/10.1016/j.jvolgeores.2012.05.012

Worni, R., Huggel, C., Stoffel, M., & Pulgarín, B. (2012). Challenges of modeling current very large lahars at Nevado del Huila Volcano, Colombia. *Bulletin of Volcanology, 74*, 309–324. https://doi.org/10.1007/s00445-011-0522-8

Zen, M. T, & Hadikusumo, D. (1965). The future danger of Mt. Kelut. *Bulletin Volcanologique, 74*, 275–282.

13 Regional Debris-Flow Hazard Assessments

Pascal Horton, Luigi Lombardo, Martin Mergili, Volker Wichmann, Ashok Dahal, Bastian van den Bout, Richard Guthrie, Manfred Scheikl, Zheng Han and Matthieu Sturzenegger

Abstract

Regional debris-flow hazard assessments provide consistent information on potential hazards over large areas, often with limited available data. Different approaches to regional debris-flow hazard assessment include heuristic, empirical, statistical, or physically-based techniques. The resulting product is often a debris-flow susceptibility map that identifies locations where such events are more likely to occur. These locations can also be related to frequent events that have occurred in the past and are therefore more likely to occur in the future under similar conditions. This chapter provides an overview of these different approaches and a description of the models available at the time of writing. It details the assessment of the potential source areas and the runout, as well as how these two steps can be coupled. Some data-related issues and other challenges are then discussed. The first common problem is data scarcity, which impacts methods in different ways. Physically-based methods are demanding in terms of input data, while data-driven models require event inventories that should be

P. Horton (✉)
Institute of Geography & Oeschger Centre for Climate Change Research, University of Bern, Bern, Switzerland
e-mail: pascal.horton@unibe.ch

P. Horton
Terranum Sàrl, Rue de l'industrie 35 bis, 1030 Bussigny, Switzerland

L. Lombardo · A. Dahal · B. van den Bout
Faculty of Geo-Information Science and Earth Observation (ITC), University of Twente, PO Box 217, 7500 Enschede, AE, Netherlands

M. Mergili
Institute of Geography and Regional Sciences, Faculty of Environmental, Regional and Educational Sciences, University of Graz, Heinrichstraße 36, 8010 Graz, Austria

V. Wichmann
Laserdata GmbH, Technikerstr. 21a, 6020 Innsbruck, Austria

R. Guthrie
Stantec, 200-325 25th Street SE, Calgary, AB, Canada

M. Scheikl
Alpinfra Engineering, Marktplatz 5, 5163 Mattsee, Austria

M. Scheikl
Canadian Hydrotech Corp., Vancouver, BC, Canada

Z. Han
School of Civil Engineering, Central South University, 68 Shaoshan Road, Changsha 410075, Hunan, China

M. Sturzenegger
BGC Engineering Inc., 980 Howe St, Vancouver, BC V6Z 0C8, Canada

M. Jakob et al. (eds.), *Advances in Debris-flow Science and Practice*, Geoenvironmental Disaster Reduction,
https://doi.org/10.1007/978-3-031-48691-3_13

as exhaustive as possible. Landslide inventories are important for every regional-scale approach, and compiling such an inventory is time-consuming and often biased due to mapping preferences. This raises questions about the impact of the completeness of these datasets and whether incomplete inventories still provide reliable predictive maps, mainly for data-driven models. As has been repeatedly demonstrated, the Digital Elevation Model (DEM) is an essential dataset for most methods, and its quality has an impact on the results of both the source areas assessment and the runout computation. In addition to the accuracy aspect, its resolution has a substantial impact on the calculated derivatives such as slope gradient or curvature, as well as on the modelled runout. Low-resolution datasets lack certain terrain features that may be relevant, but DEMs used at too high of a resolution are likely to add noise for regional-scale studies. These impacts may be overlooked by end-users. In light of these challenges, it is important to recognize that the data collected to run our models is a snapshot of the state of the environment at a given time. Yet we live in a changing environment, with, for example, changes in topography due to earthquakes, volcanic eruptions, avulsions, debris deposition, and human constructions. In addition, the climate is changing, affecting the frequency and intensity of extreme precipitation events, and consequently the frequency with which debris flows are triggered. As a result, new sites with little or no history of debris flows may be affected in the future. On the other hand, more frequent debris flows may reduce the volume of sediment transported in catchments with limited supply. A changing environment therefore brings uncertainties to the data used by our models, which may underestimate or overestimate potential susceptibility to debris flows in the future. With this in mind, we discuss the perspective of dynamic approaches. Finally, regional methods have several limitations. They rely on data available for large regions, but do not account for local watershed characteristics, such as sediment supply linked to specific geological processes. Site-specific conditions can also influence the runout computation, leading to atypical debris-flow trajectories. Additionally, regional models are not designed to provide reliable volume estimates or frequency-magnitude relationships, making them unsuitable for quantitative risk assessments, modelling specific events, or designing mitigation structures. Typically, the susceptibility map serves as the initial step in prioritizing areas requiring detailed debris-flow hazard mapping.

13.1 Introduction

Debris-flow hazard assessment at a regional scale aims to delineate areas potentially affected by an event over a large territory. While there is no formal definition of the regional scale in the context of natural hazards, the considered area ranges from 10 to 10^4 km^2 (Huggel et al., 2003; Pastorello et al., 2017). Working at such scales implies certain constraints on the work that can be done with a limited budget and on the suitability of methods or models. The first approach that should be systematically undertaken is a geomorphological analysis of the signs of debris flow in the landscape (e.g., fans and channels with lateral deposits) to understand the processes taking place in the region of interest. A good understanding of the type of events that occur and their frequency is of paramount importance. However, such analysis, or the production of comprehensive inventories, cannot be undertaken over large areas at a reasonable cost. One would instead characterize some subregions in detail, covering the different conditions (e.g., geology) present in the entire region, train (or calibrate) models to reproduce the observed past events over these subregions, and apply the models to the whole area of interest.

In terms of modeling, not all models are suitable for the regional scale. Physically-based modeling of debris flows requires substantial data and good characterization of the specific properties of a specific site (see Iverson and George, this volume). Applying such methods at a regional scale could require unreasonable costs to characterize specific local conditions (Carrara et al., 2008) or result in high uncertainty due to approximations of the modeling parameters. Such approximations are mainly due to the absence of parameters such as soil depth or hydraulic conductivity data, whose acquisition is inevitably discrete in space due to the surveying costs. To a lesser extent though, such approximations have also been expressed in the way physics-based solvers have been traditionally tasked with simulating debris flows, due to limited computational capacity. However, as computational power increased in recent years, together with the model algorithms, regional models are beginning to show considerable improvements from even a decade earlier.

Regional assessments generally rely on simpler, less data-intensive methods and provide less detailed output than mechanical models. Such approaches are often statistical or empirical, with parameters that can usually be transposed to similar environments. However, one should be aware that these approaches cannot fully account for site-specific characteristics (Blahut et al., 2010a, 2010b; Hürlimann et al., 2008; Kappes et al., 2011). The regional hazard assessment results in areas potentially affected by debris flows over the whole region and is often expressed in terms of susceptibility.

13.1.1 The Notion of Susceptibility

Landslide susceptibility was defined by Fell et al. (2008a) as "a quantitative or qualitative assessment of the classification, volume (or area), and spatial distribution of landslides which exist or potentially may occur in an area." It is usually expressed with a relative scale (Fell et al., 2008b) that does not include a temporal dimension or a notion of frequency. Although it can be expected that more events are likely to occur in areas with a higher susceptibility, there is no explicit consideration of the time frame (Fell et al., 2008a). Susceptibility zoning addresses both the sources and the travel of landslides within the study area and those outside that may travel into the study area (Fell et al., 2008a). The assessment of the potential source areas and the areas potentially affected by debris-flow travel consist most often of two independent analyses.

Susceptibility zoning primarily targets map scales between 1:25,000 (or even 1:5,000) to 1:250,000 (Fell et al., 2008a), and its interpretation should be in accordance with these scales. Modern, data-driven, statistical susceptibility models allow the user to work at much finer scales, driven mainly by the available input data (e.g., Chacón et al., 2006; Dahal et al., 2008; Kornejady et al., 2017; Regmi et al., 2010; Zhu et al., 2018; among others).

While classification as "susceptible" and "not susceptible" may be sufficient in some cases, most often, classes (e.g., high, medium, and low), probabilities, or degrees of susceptibility are provided. As the objective of the susceptibility map is to include all possible future events, quantifying how likely a specific area can be affected is useful in practice for the end-users.

Susceptibility mapping is less time-demanding than hazard mapping, which accounts for the intensity and frequency of the simulated event. It can serve different purposes: (a) provide a consistent assessment of a potential hazard over a large geographic domain with a single method, (b) map the areas exposed to natural hazards for remote locations where no detailed hazard mapping is necessary, and (c) as a first stage to prioritize the areas that require detailed debris-flow hazard mapping when infrastructure or people are exposed. Moreover, susceptibility zoning is recommended when data are unavailable for quantitative hazard- and risk zoning or are not of sufficient quality (Fell et al., 2008a). In such case, the analyses can be carried out at a local scale (1:5,000 to 1:25,000), but the end-users should be aware of the method's limitations.

Another approach that can be considered as hybrid between susceptibility mapping and medium-scale hazard assessment, with temporal and spatial probabilities, relies on the use of high-quality inventories (Blahut et al., 2010b; Guzzetti et al., 2006). Using multi-year inventories of debris-flow-related landslides and aerial photographs, temporal frequencies and spatial probabilities can be approximated for different classes of causative factors (Blahut et al., 2010b). Then, such a hazard initiation map can be propagated using empirical approaches to provide hazard classes (e.g., Blahut et al., 2010a, 2010b). However, this approach requires complete high-quality landslide inventories over the area where the model is developed.

The notion of susceptibility does not feature the expected temporal recurrence of landslides at a given location, nor the temporal recurrence of the trigger (earthquake, rainfall, snowmelt) that caused the landslides in the first place. It implies that the susceptibility (irrespective of the technique that generated it) is static; it is a modeled screenshot of reality that tells the user how spatially prone a landscape is to landslides under the assumption that the screenshot would be valid (or would not significantly change) for a certain amount of time. However, a static vision may be limited in a world that changes rapidly due to anthropic drivers (climate, land use) or natural factors (earthquakes, terrain deformations). We discuss an overview of static and dynamic susceptibility in Sect. 13.7.4.

Landslide susceptibility thus corresponds to an estimate of how likely a given portion of the landscape is to generate at least one landslide as a function of environmental factors, but without accounting for the triggering factors (van Westen et al., 2008). It can be assessed in various ways. For instance, one can estimate how prone a given location is to release a landslide by using either simple physically-based models, empirical ones, or data-driven approaches, such as statistical or machine-learning methods. In turn, these tools would estimate the probability of landslide occurrence. A major difference with physically-based models used for hazard zoning is the scale of the application. At a regional scale, insufficient data are available to accurately characterize each debris flow's conditions and behavior. Thus, assumptions must be made to regionalize the role of the environmental factors on the potential initiation of debris flows and travel behavior.

13.1.2 Mapping Units

An important element of the susceptibility definition is the representation of the geographic space, which can change depending on the model requirements or on the expectations of end-users. For instance, if the end-user is a national agency or a regional administration, one would partition the landscape according to fine-scale spatial objects (mapping units hereafter). These would typically correspond to grid cells (e.g., pixels with a squared lattice structure) or to slope units (e.g., Howes and Kenk, 1997) (see Fig. 13.1 for two mapping unit examples). The latter units are irregularly shaped and encompass the geographic space between ridges and streamlines of catchments that can be defined at different scales (although other discretizations exist; see Guzzetti et al., 1999).

Each of the two options has specific advantages. The former mapping unit is common because it supports both physically-based and data-driven methods. Grid cells are intuitive and straightforward to use. They directly represent the gridded structure of data collected from satellite sources, making this mapping unit the most convenient. Also, in the context of physically-based or empirical models, one grid cell represents a specific state at a given step as a function of the neighboring states. Thus, the grid cells are well suited to solve equations that explain the interactions among adjacent units (hexagons offer similar characteristics, see D'Ambrosio et al., 2003). However, these spatial objects can present limitations in representing discontinuous phenomena, such as slope failure. This becomes particularly evident in the data-driven context where one should train a model to recognize locations where the slope instability started. For fine grid cells, this is particularly challenging, and in most cases, the solution is to

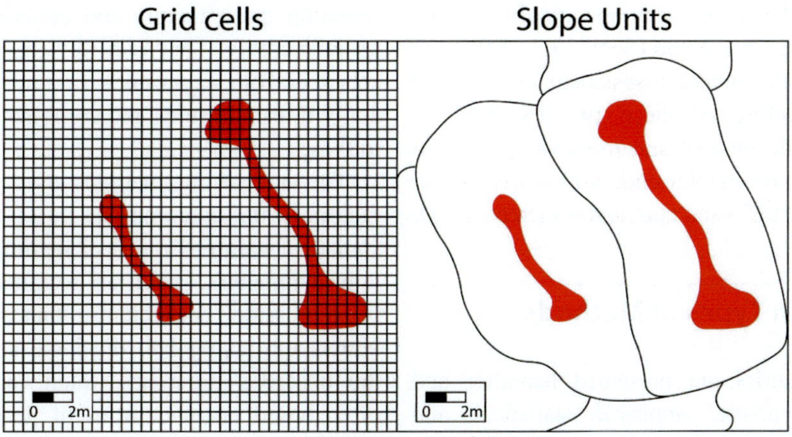

Fig. 13.1 The left panel shows a square lattice partitioning the geographic space into 40 × 40 grid cells. The right panel shows a slope unit partition, where the area is divided into eight polygons

approximate the failure at the highest grid cell along the polygonal landslide boundary (Cama et al., 2017; Tang et al., 2016).

As for the slope units, they are almost entirely used in the context of data-driven models (Tanyaş et al., 2022; Titti et al., 2021), with hardly any example available for physically-based models (Domènech et al., 2020). Their strength resides in the subdivision of a given landscape into portions that express the morpho-dynamic behavior of a landslide. However, they are technically more cumbersome to use since most environmental data are collected, managed, and shared in gridded formats. Therefore, choosing to work with slope units inevitably poses the question of how to represent the fine resolution of the available environmental data at the scale of much coarser spatial units.

In addition to the type of mapping unit used, their size is also important. Large units can hide a high variability of values for a given factor, imprecise morphologic information due to smoothing, and provide an inappropriate scale for runout modeling. On the other hand, small units are more difficult to manage, can be responsible for time-consuming modeling, and could provide a false impression of precision, not in accordance with the approximations made in the methods used. Practically, available data, most often the digital elevation model (DEM; note that we will use the term DEM also for digital terrain model), drives the choice of the size of these units.

13.1.3 Chapter Organization

This chapter describes the state of the art of several methods and models used for regional debris flow susceptibility mapping. It aims at presenting different approaches to the same problem—assessing the regional debris flow hazard—with tools specific to different fields, whether based on expert knowledge, modeling of physical processes, or data-driven. Lessons from practice and research are presented throughout the chapter.

An overview of the available tools for debris-flow susceptibility assessment is provided, covering methodological considerations and practical implications to set the stage for a holistic view of what landslide experts can achieve at the time of writing (2022). The chapter starts with a description of the main types of methods

(Sect. 13.2). Then, methods specifically for the source areas are presented (Sect. 13.3), followed by those for the runout assessment (Sect. 13.4) and the coupling of these two (Sect. 13.5). Then, available models and tools are presented (Sect. 13.6). The chapter ends with a discussion section (Sect. 13.7) and conclusions (Sect. 13.8).

13.2 Main Types of Methods

Method categories are presented hereafter and consist of heuristic, empirical, statistical, and physically-based methods. Overlap exists between some methods, and a model can often contain different approaches. For example, heuristic and empirical methods can use simple statistical tools or physically-based laws, and expert knowledge is often necessary to build a statistical or machine learning model, if only for selecting input features.

13.2.1 Heuristic and Empirical Methods

Heuristic approaches are arguably the most traditional category of methods and emerged before the widespread availability of numerical computing. They are mainly driven by expert knowledge and combine mapping and analysis of geomorphologic settings (Fell et al., 2008b; Fookes et al., 2007; Howes & Kenk, 1997). The expert relies on previous experiences to make decisions, which can vary with the local settings. In addition to direct geomorphic analyses, more quantitative approaches combine weighted input parameters to obtain susceptibility classes (e.g., Blais-Stevens & Behnia, 2016). The weighting is often subjective and based on experience.

As the name suggests, empirical methods rely on observational experience rather than physically-based modeling. They aim to predict a particular outcome based on observed process occurrences for a given set of circumstances or factors. Heuristic elements are often involved in selecting these factors and contain an element of subjectivity. They can rely on input features selected from experience or the literature and apply threshold rules and combination strategies to these.

The underlying methods used to relate the factors to the outcome can be of a heuristic or statistical nature. Empirical methods often contain parameters, such as weights, thresholds, or regression coefficients, that can be calibrated and adapted to the local conditions and the observed processes. Such calibrations are often based on inventories of past events, ideally located within the study area. The primary assumption is that "areas with similar topography, geology, and geomorphology as the areas which have experienced landsliding in the past are also likely to experience landsliding in the future" (Fell et al., 2008a). However, the climate is missing in the similarity factors from Fell et al. (2008a). The challenge with the latter is that, due to climate change, it is not constant over time for a given location, and thus what happened in the past might no longer be a key for the future (Sect. 13.7.3).

13.2.2 Statistical Methods and Machine Learning

A holistic overview of machine learning models should encompass statistical models, as even simple regression involves a machine learning process. Furthermore, even acclaimed machine learning models often are a combination of linear functions. For the sake of clarity, we will address two classes of models, statistical and machine learning (e.g., decision trees, support vector machines, neural networks) separately, as it is done in most of the available geoscientific literature. Both statistical and machine learning methods belong to a hierarchically higher family of data-driven models. They are not governed by physical laws but are built based on correlations between target and explanatory variables.

The models from both classes return probabilistic estimates of landslide occurrences by using presence/absence data as the respective target variables. They complete this task by deriving specific functional relations between landslide stable/unstable conditions and a set of causative factors. The main difference resides in their algorithmic architecture and its implications. Statistically-based models support interpretation throughout the analytical procedure, as they are built on parameters that carry physical meaning. However, they usually do not perform as well as their machine-learning counterparts. The latter tend to provide unmatched performance but at the expense of interpretability due to the nature and large number of their parameters. These characteristics have earned machine learning methods their "black box" label.

Statistical models are built so that the distribution of landslide presence and absence across space is assumed to behave according to a binomial probability distribution. This binary outcome is regressed against a set of explanatory variables (or covariates) in a multivariate space, where each model component interacts with the others. The most common operation is to limit the functional relations between landslides and covariates to be purely linear. In this case, the model corresponds to a Generalized Linear Model (GLM, hereafter), which works by linearly combining the estimated linear relations mentioned above.

If we were to imagine a spectrum of statistical and machine learning models sorted by complexity, GLMs would occupy one extreme because of their relative simplicity, whereas some of the available machine learning methods would be placed on the opposite side. The same ranking would also reflect the model interpretability and, in most cases, the resulting performance. In between these two extremes, Generalized Additive Models—another widespread statistical model in geoscience—would occupy the middle. They extend the GLM framework by providing greater flexibility, allowing one to estimate linear and nonlinear effects responsible for landslide occurrences. This combination of linear and nonlinear effects is then linearly combined. This framework still provides good levels of interpretability and rivals the performance obtained through machine learning. Some features of these models are part of machine learning approaches. For instance, GAMs estimate nonlinear functions analogously to Multivariate Adaptive Regression Splines. In both cases, one can opt to categorize a continuous explanatory variable in several classes. Then, one regression coefficient (or weight) is estimated for each class, and the combined effect across classes is retrieved through a spline function, which induces dependence between adjacent classes, respecting the ordinal structure of the original explanatory variable. As for the other extreme of the spectrum, the architecture of machine learning models varies enormously, and with it, their complexity. Machine learning tools can encompass linear combinations of linear effects, thus providing a relatively simple structure to rely on, up to nonlinear combinations of nonlinear functions, which implies highly complex structures. A constraint with complex machine learning methods is the large amount of data needed.

13.2.3 Physically-Based Methods

Physically-based methods aim at reproducing the physical processes involved in a phenomenon such as a debris flow. They often employ partial differential equations of mass and momentum conservation, which are solved in the time domain within a numerical scheme on the three-dimensional topography. Those differential equations consist of source terms and flux terms. The source terms consider the gain or loss of mass (by entrainment or deposition, respectively) and momentum (by gravitational forces on the one hand and frictional and viscous forces on the other). The flux terms govern the redistribution of mass and momentum between different entities in space (which might be bins in 1D approaches, raster cells in grid-based approaches, or points in Smoothed Particle Hydrodynamics (SPH) approaches). It is particularly the dissipation of momentum, which can be computed according to the implemented constitutive material model (rheologies). Detailed descriptions of physically-based models can be found in Iverson and George (this volume).

13.3 Potential Debris-Flow Source Areas

Takahashi (1981) defined two types of debris flows: one from the mobilization of the debris available in the channel (or a gully) and the other from landslides located in the vicinity of a channel that will slide into the stream and will thus incorporate more water and mobilize into a debris flow. Therefore, there are similarities in the methods used for landslide assessment and the assessment of potential debris flow source areas. For details on the causes and triggers of debris flows, please refer to Kaitna et al. (this volume).

Based on observations in Switzerland, Rickenmann and Zimmermann (1993) defined four types of debris-flow starting zones within two groups. The *slope-type starting zones* contain: (1) those in a steep and often weakly consolidated talus slope with gradients between 27 and 38° and (2) those located in a channel near a contact zone of a rock wall and a steep talus slope with similar gradients as the previous type. The *valley-type starting zones* contain: (3) those located in a rock gully filled with debris and with gradients between 24 and 35°, and (4) the mobilization of larger and temporary debris accumulations in the channel due to increased runoff conditions (bed slopes between 13 and 33°).

13.3.1 Common Causative Factors

Independently of the method, some typical environmental factors are often considered. We focus here on those that can be available at a regional scale, excluding, for example, site-specific characteristics retrieved from fieldwork campaigns, which are not practical over a large territory. Some key factors can be derived from the DEM. The factors and the corresponding criteria often considered for delineating source areas are listed hereafter, starting with DEM-derived ones.

Slope—The slope gradient is the most important parameter for the predisposition to debris flows (Guthrie et al., 2010; Rickenmann & Zimmermann, 1993), with a lower threshold from 15° (Takahashi, 1981) to 20–25° (Guinau et al., 2007; Ortigao & Kanji, 2004) or even 30–45° for some regions (Delmonaco et al., 2003). A decrease in the landslide and debris-flow activity for terrain slope angles higher than 40–45° has been observed by Dai and Lee (2001) and Guinau et al. (2007). This decrease is due to more stable slopes, rock scarps, or the difficulty for sediments to accumulate on steep terrain. Therefore, a higher (steeper) threshold is sometimes applied when selecting the slope gradient range.

Upslope contributing area—The amount of water that flows through a given point is related to the upslope contributing area, or flow accumulation. A minimum contributing area is needed for a debris-flow initiation, which can change depending on the climate and soil properties. A threshold of 1 ha was observed in the Swiss Alps (Rickenmann & Zimmermann, 1993). In addition, a relationship can be established between the upslope contributing area and the terrain slope. A lower water discharge (i.e., contributing area) is required to initiate a debris flow on steeper bed slopes (Takahashi, 1981). Based on the work of Rickenmann and Zimmermann (1993) and Heinimann et al. (1998), Horton et al. (2013) proposed two relationships that would correspond to rare and extreme events and a constant slope gradient threshold of 0.26 (15°) for catchments larger than 2.5 km^2:

$$\begin{cases} \tan\beta = 0.32A_{fa}^{-0.2} & \text{if } A_{fa} < 2.5\ \text{km}^2 \text{ and for rare events} \\ \tan\beta = 0.31A_{fa}^{-0.15} & \text{if } A_{fa} < 2.5\ \text{km}^2 \text{ and for extreme events} \\ \tan\beta = 0.26 & \text{if } A_{fa} \geq 2.5\ \text{km}^2 \end{cases} \quad (13.1)$$

where β is the slope gradient threshold and A_{fa} is the flow accumulation area (or contributing area).

Terrain curvature—Plan curvature is the degree of concavity/convexity perpendicular to the direction of the maximum slope. It can be used to identify areas where debris can accumulate (Blais-Stevens & Behnia, 2016) or the presence of gullies (Horton et al., 2013), thus concave areas (Delmonaco et al., 2003). Profile curvature, parallel to the direction of the maximum slope, is also sometimes considered (e.g., Blahut et al., 2010b). Care should be taken when using curvature, as the resolution and quality of the DEM strongly influence the calculations. Moreover, a given threshold cannot be transferred from one region to another and must be reassessed for every case study.

Aspect—The terrain aspect is sometimes considered to account for sunshine effects on snow, for example (Blahut et al., 2010b; Blais-Stevens et al., 2016), or potential differences in vegetation cover or rainfall intensity (Coe et al., 2014). If it was found to have minor to no influence on terrain instability in some regions (Guinau et al., 2007), strong signals were found in others (Coe et al., 2014; Jakob, 2000).

Geology—Surficial geological or lithological data can account for sediment availability. The difficulty is that the scale of such maps or the level of detail is often insufficient to provide useful information for a robust differentiation of the source areas.

Landuse—Landuse maps can differentiate areas more prone to debris-flow initiation than others. Forestry, for example, is a widespread landuse type with observed effects on debris-flow susceptibility (e.g., Guthrie, 2002). Such properties can change over time (e.g., change in landuse management practices, fire, hydrological recovery of logged areas; Sect. 13.7.3). Thus, susceptibility mapping should be conservative to avoid the need for frequent updating. On the other hand, debris-flow initiation can sometimes be excluded from some landuse types, such as built-up areas.

Proximity to drainage system—This information was considered important by Blais-Stevens and Behnia (2016), as debris flows are usually triggered in steep streams. It can be accounted for by using buffer zones around the drainage system.

Other datasets have been considered, such as internal relief and altitude (Blahut et al., 2010b) for DEM-derived information. Blahut et al. (2010b) also suggested considering the distance to faults, extracted from tectonics maps, due to their effect on landslide occurrence and, thus, the higher availability of loose material.

13.3.2 Heuristic and Empirical Methods

Heuristic and empirical methods for assessing potential debris-flow source areas often rely on the combination of environmental factors (Fell et al., 2008b; Fookes et al., 2007; Howes & Kenk, 1997). These factors are selected based on expert knowledge and well-known favorable settings (Sect. 13.3.1). The difference between the approaches is mainly how these factors are combined. Some are threshold-based, and others combine these factors with a weighting scheme (e.g., Blais-Stevens & Behnia, 2016). These methods can result in a binary selection of potential source areas (source/non-source), susceptibility values, or susceptibility classes.

Many approach variants exist that can differ from one expert to another. Thus, these will not be reviewed in detail here but illustrated with an example of how it is done in the Flow-R software (Sect. 13.6.1). Multiple factors can be combined to perform a binary selection. Based on thresholds, the grid cells of each input factor are classified as *favorable* when initiation is possible, *excluded* when initiation is unlikely, or *ignored* when no decision can be taken (Fig. 13.2). Then, datasets are combined according to the following rule: a cell is a source area if it is at least once selected as favorable but never excluded. The datasets typically considered are landuse, geology, terrain curvature, and a relationship between the slope and the flow accumulation (Sect. 13.3.1; see Horton et al., 2008, 2013 for more details). An example is shown in Fig. 13.3.

13.3.3 Statistical Methods and Machine Learning

In this section, we will introduce the two most common statistical models used to estimate landslide susceptibility. We will provide a more comprehensive description of Generalized Linear Models, introducing their structure and explaining best practices in this framework. Then, we will present Generalized Additive Models, which are an extension of the linear counterpart introduced before. Therefore, in the latter case, we will start from the linear case and simply focus on what a nonlinear extension implies.

13.3.3.1 Generalized Linear Models

Traditional regressions are built to describe the variability of a target variable as a function of a set of covariates. If we assume these covariates to be continuous in nature, then an ordinary Gaussian model can be denoted as follows:

$$Y = \beta 0 + \beta 1x1 + \beta 2x2 + \cdots + \beta nxn + \varepsilon, \tag{13.2}$$

where Y is the response variable, $x1$, …, xn are the n covariates, $\beta 0$ is the intercept, and βn are the regression coefficients. Then, the sum of $\beta 0 + \beta 1x1 + \beta 2x2 + \cdots + \beta nxn$ corresponds to the systematic component of this model, whereas ε is the random one.

This classical regression framework has then been extended to accommodate discrete covariates, nonlinear effects, and response variables that lie outside the Gaussian context. Concerning the latter, Generalized Linear Models offer a well-known and robust framework to describe the variability of target variables that fall in the class of linear exponential distributions, binomial being one of these. This distribution is commonly used when the target variable has only two expressions, typically addressing the presence or absence of a given phenomenon. The whole landslide susceptibility has been framed to distinguish stable from unstable slopes; thus, it is straightforward to use a GLM that assumes landslides behave across the geographic space according to a binomial probability distribution.

In this case, the GLM can be denoted as:

$$\eta(\pi) = \beta 0 + \beta 1x1 + \beta 2x2 + \cdots + \beta nxn, \tag{13.3}$$

where η is the logit link function, which links the mean of Y, or π, to the sum on the right-hand side in Eq. 13.3. From now on, we will refer to the sum mentioned above as the linear predictor, as it is a linear combination of several

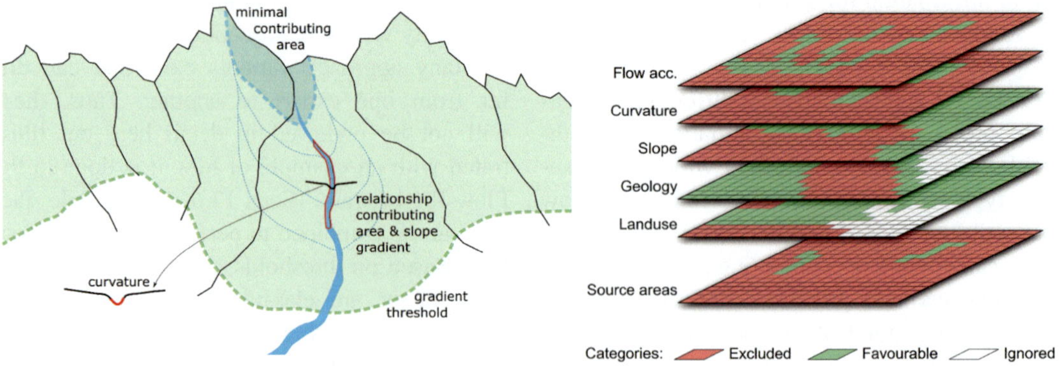

Fig. 13.2 Illustration of the combination of various factors for the source areas assessment (*Source* Horton et al., 2013)

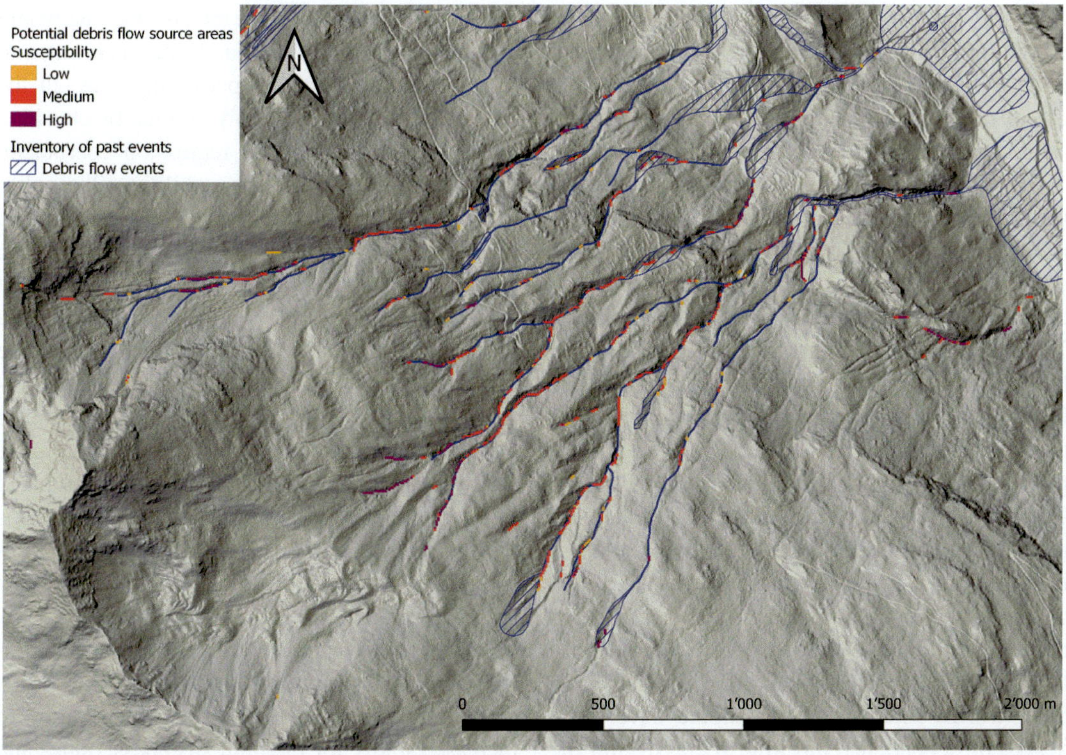

Fig. 13.3 Example of the source areas assessment for the Ticino Canton in Switzerland using a GIS-based combination of environmental factors. The main factors used to delineate the potential source areas were extracted from the 6 m DEM and consist of: (1) the slope gradient (between 15° and 40°), (2) a plan curvature of $-15/100\ \mathrm{m}^{-1}$, (3) the relationship between the slope and the upslope contributing area (Sect. 13.3.1), (4) a filtering based on a minimum threshold for the source area. The classes were established based on geology, the upslope contributing area, the terrain curvature, and the spatial density of source areas

model components. As regards to η, it is important to note that a number of link functions are generally possible. The simplest corresponds to the identity link function, which essentially leaves unchanged the input of a given GLM with respect to its output. In the context of a GLM built to return susceptibility estimates, the input consists of a series of 0/1 records, whereas the output consists of a series of probability values. Therefore, a link function other than the identity one is needed. The logit link is the traditional solution, and it can be further expressed as $\eta(\pi) = \log(\pi/(1-\pi))$. The latter is commonly referred to as the log-odds as the odds correspond to $\pi/(1-\pi)$ being equal to the probability of landslides being present at a given location, $P(Y=1)$, divided by the probability of a landslide to be absent, $P(Y=0)$. As a result of the logit link, Eq. 13.3 can be re-written, returning:

$$Susceptibility = \frac{\exp(\beta 0 + \beta 1x1 + \beta 2x2 + \cdots + \beta nxn)}{1 + \exp(\beta 0 + \beta 1x1 + \cdots + \beta nxn)} \tag{13.4}$$

13.3.3.2 Generalized Additive Models

The framework provided above features linear relations, where one single regression coefficient is estimated for each covariate. However, there are situations where a covariate may exhibit nonlinear behaviors. The simplest case, often easily addressed even in the GLM context, pertains to categorical covariates. It means that discrete classes of a given covariate do not share reciprocal dependence. This is, for instance, the case for the lithology or land use covariates, which are estimated with a regression coefficient per class, independently from the others. This use of covariates is commonly referred to as *iid* or independent and identically distributed.

Another way to model covariates—typical of the GAM framework and not part of the GLM—is achieved by modeling discrete covariates that retain some internal dependence. It is the case of discrete covariates with an ordinal structure. To provide a simple example, let us assume we have computed the slope steepness from the local DEM. To estimate the nonlinear effect of the slope steepness on landslides, we reclassify the slope map into 20 equidistant classes, from 0 to 90°, each class spanning 4.5°. Although we assume that the behavior of the slope steepness is nonlinear, we can also reasonably assume that the class 9° will most likely not have a much larger effect than the class 4.5° and will not have a much smaller effect than the class 13.5°. In other words, adjacent classes should have some degree of dependence left after reclassifying the original slope map. This adjacent class dependence can be captured using a spline function to estimate the regression coefficients of any discrete covariate with an internal ordinal structure. This, in turn, would modify Eq. 13.3 as follows:

$$\eta(\pi) = \beta 0 + f1(x1) + f2(x2) + \cdots + fn(xn), \quad (13.5)$$

where instead of a single regression coefficient as it appeared in Eq. 13.3, now the GAM framework leads to functions to be estimated, and where $x1$, $x2$, and xn have been transformed from continuous properties to discrete ordinal ones.

This class of statistical models thus offers the chance to retrieve the nonlinear influences of certain covariates over the target one (Brenning, 2009; Goetz et al., 2011; Lombardo et al., 2020). Because of this characteristic, at times, GAMs are referred to as interpretable machine learning tools. The difference from GLMs is that in the linear context, one can only interpret whether certain covariates behave overall as expected. For instance, expert knowledge tells us that one would expect the probability of landslide occurrence to increase as the slope steepness increases. In the GLM case, this can be easily verified by checking the sign of the estimated regression coefficient. The same level of interpretability still applies to GAMs, with the difference that the estimated regression coefficients associated with specific portions of a given covariate domain represent the relative contribution of each parameter. As a result, one can also interpret a specific covariate role and understand whether the associated function reasonably contributes to the model from a geomorphological standpoint. The example made above for slope steepness can be easily imagined in the nonlinear context. For instance, floodplains or most flat topographies cannot host slope failures. Thus, one should expect a nonlinear behavior of the slope steepness starting with a very negative contribution to the overall susceptibility. Then the assumption is that for debris flows, one should physically expect the contribution to increase rapidly for medium steepness values and transition back to negative values for very steep slopes as one should not expect thick soil covers in near-vertical slopes (a graphical sketch is provided in Fig. 13.4, right panel). This intuitive theoretical description of the role of slope steepness describes a behavior not too far from a sigmoidal function. It is precisely why a GLM would just approximate the role of the slope steepness, likely overestimating the contribution at low steepness values and overestimating the same for high steepness values (Fig. 13.4, left panel). Conversely, the strength of GAMs is that one can easily accommodate such behaviors and plot them to interpret whether they make sense, unlike the vast majority of other machine learning models.

13.3.3.3 Advanced Machine-Learning Techniques

With the advancement of machine learning, the past decade has seen several applications of different machine and deep learning techniques in many geoscientific fields, specifically for debris-flow modeling. Below, we will provide

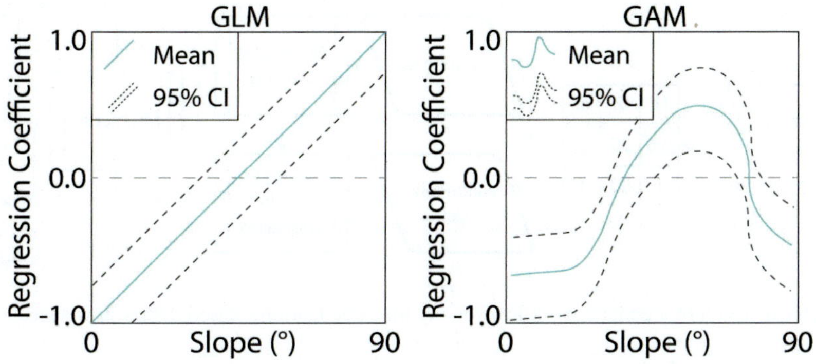

Fig. 13.4 Graphical representation of a theoretical linear and nonlinear effect estimated for the slope steepness. CI stands for the confidence interval

an overview of these methods, grouping them into classical machine learning and modern deep learning methods:

Classical Machine Learning

Machine learning gives the computer the "ability to learn and predict from data without being explicitly programmed to do so", which can be achieved in multiple ways, from simple mathematical regression models to complex neural network architectures. In debris-flow modeling, many machine learning models have been tested through the years: e.g., Random Forest, Support Vector Machin, Naive Bayes, Extreme Learning Machine, and Gradient Boosting (e.g., Dou et al., 2020; Qing et al., 2020; Xiong et al., 2020; Zhang et al., 2019; Zhao et al., 2021). These models are fitted using the training process, which finds the best coefficients (also referred to as weights and biases in deep learning) to classify or regress debris-flow presence and absence, conditionally on a series of input features. All these models are trained in a supervised manner, using an available inventory of debris flow and the different environmental factors that influence them (directly/indirectly). In geoscience, unsupervised training is quite a complex task and requires a good understanding of the underlying physics, which will be explained at the end of the following sub-section. As for the plethora of available machine learning tools, even though their architecture is different, their overall process of modeling debris-flow susceptibility is identical, as given in the example (1).

Example (1): let us summarize the key steps of a model meant to estimate debris-flow susceptibility for a given test site. The required components are past debris-flow inventories mapped from observations, environmental factors influencing the debris-flow occurrence, such as elevation, slope, topographic wetness index (TWI), geology, landuse, and the model itself. Depending on the study area, the correlation between observation and environmental factors may change, making the selection/removal of certain variables requisite. Assuming the dataset is available, a tabular structure must be created either based on the grid cells or slope units, where each row provides information on environmental factors as well as the presence or absence of debris flow. Once the table is ready, training and validation subset are used to fit the selected model and monitor its performance. At this point, the choice of parameters or hyperparameters (model parameters that are not observed physically but can influence the outcome) and predictors is adjusted to produce stable and reliable results.

Once the model shows acceptable performance with accuracy above a certain threshold (depending on the study area, past results, and data availability), the estimated weights are used to assign a susceptibility value to all the mapping units partitioning the test site. This general workflow is illustrated in Fig. 13.5.

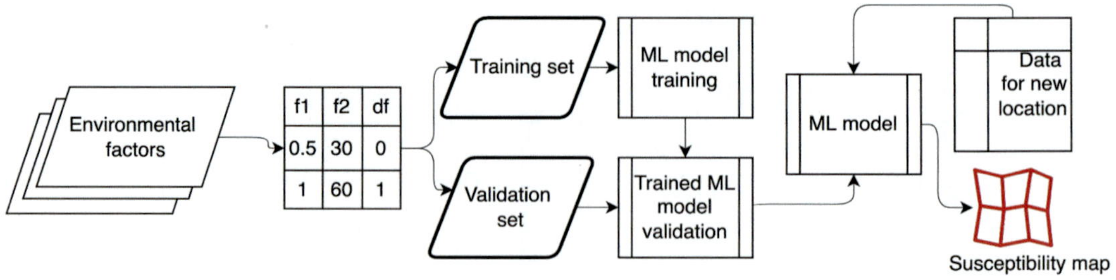

Fig. 13.5 Graphical sketch of a general workflow for any machine learning-based debris-flow susceptibility

The general representation provided above is valid for any machine-learning tool. To provide more detailed information, we selected two (Random Forest, RF; Support Vector Machines, SVM) out of the many available examples and summarized their analytical procedure below.

The RF (Hastie et al., 2009, p. 587) is an ensemble of many uncorrelated decision trees (hence 'forest'). Each tree is a trained classifier that can individually estimate the susceptibility. However, each of these basic decision trees is prone to overfit because of smaller and under-representative samples as well as model learning the noise in the data limiting its generalization capacity; thus, the results may be good when it comes to fitting, but they are usually unsuitable for prediction. For this reason, the RF decides the outcome (debris flow presence/absence) based on a majority rule.

The training procedure is carried out by creating a tree out of a random subset of the training data and a process called bagging (Bauer & Kohavi, 1999), then recursively increasing the tree nodes until the pre-defined minimum number of nodes is reached. Each node in the tree acts as a separation point where each factor domain is divided into two using a threshold value; finding that threshold value and the optimal number of trees is precisely the objective of the RF training procedure. The factors are selected at random for each tree as well as their position in node structure, keeping the stochastic nature of the model (hence 'random'). It is fine to select only a few factors for each tree, which is decided by the minimum number of nodes in the tree, usually for classification problems.

The SVM (Hastie et al., 2009, p. 417) model defines a hyperplane (imaginary plane with n-number of dimensions) based on observation to separate new data into two classes. In a case where only two variables are used, the separator is a line defined by the support vectors, which are the marginal points belonging to two classes (see Fig. 13.6 for clarity).

The margin line is then determined by the halfway distance between the support vectors, calculated using the maximal marginal classifier method (see Dash et al., 2007). Because SVM is sensitive to outliers, support vectors are determined using cross-validation to remove any adverse effect on the definition of the margin, which inevitably leads to misclassifications when used over new data. These imaginary lines parallel to the marginal line and pass-through support vectors are called soft margins, which give the sense of boundary for each class. This structure makes SVM relatively easy to train compared to other ML tools. However, defining a hyperplane becomes much more complex when it comes to multi-dimensional data with more than three factors. It is usually done through basis functions such as radial basis functions (Schölkopf et al., 1997). Notably, both SVM and RF models are very good at classifying the landscape susceptible to debris-flow. This classification can be either a binary classification with 0–1 values where 1 being susceptible to debris flow and 0 being not susceptible, or it can be defined as series of probability values ranging from 0 to 1 classified into multiple classes depending on the local classification systems adopted by governments. They

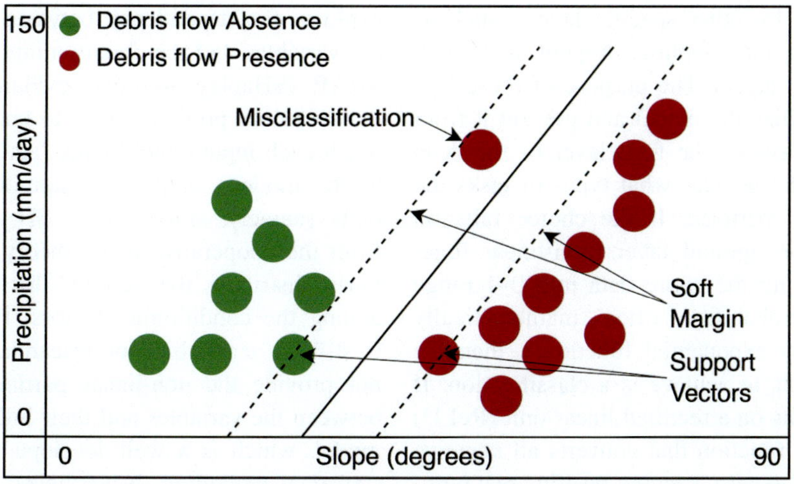

Fig. 13.6 Support vector classifier shows the separation of different classes using support vectors in a 2D data case

are extensively used due to their capacity and limited data demand, although they are prone to overfitting. The easiest way to observe such overfitting is by testing the model's generalization in the different test datasets. If the accuracy/performance drops significantly (such as 90% in the training data and 30% in the test) in the test dataset, the model is overfitted and not well generalized, but if the change is very low, it is well generalized (such as 90% in the training and 75% in the test dataset).

Deep Learning

Deep learning is an extension of traditional machine learning towards more advanced models capable of addressing various tasks such as computer vision, natural language processing, etc. The basic structure of the deep learning models is similar to an artificial neural network, thus relying on an input layer, several hidden layers, and an output layer. What makes them different is the types of hidden layers that can be used, how they are stacked, and the non-linear functions that connect them. In debris-flow susceptibility modeling (mainly in the case of identifying debris-flow-prone streams), deep learning has been attempted by several researchers, mostly testing Fully Connected Neural Networks and Convolutional Neural Networks (Bai et al., 2021; Bera et al., 2021; Chen et al., 2020; Lee et al., 2020; Li et al., 2021; Zhang et al., 2019). Both CNN and FCNN-based methods can then rely on several architectures, such as U-Net (Ronneberger et al., 2015), ResNet (He et al., 2016), and DeepLabv3 (Chen et al., 2017), etc.

In the context of debris-flow susceptibility assessment, FCNN-based methods are particularly suitable for slope-unit or any other irregular partition, whereas CNN is ideal for cases where a gridded partition is adopted.

The FCNN layers connect every neuron (or feature) in one layer to another layer without subsampling the data. It increases the number of parameters exponentially and limits the depth and width of the model because we cannot train a model with billions of parameters due to computational and data availability reasons. This type of layer is structure-agnostic, meaning that the spatial structure of the data does not influence the learning process.

The CNN-based method uses a convolution window (a moving window of a specific size, similar to a filter in remote-sensing image processing) to reduce the input size by linear function followed by a non-linear activation. CNN is not a spatial agnostic tool, and spatial representation of the input data influences the output. Each of these layers can be accompanied by other specific layers, such as the activation, max-pooling, dropout, and batch normalization layers. The purpose of those layers is to stabilize the model and prevent it from overfitting. Usually, the final layer of the deep learning model decides what types of tasks the model would perform. If the choice falls on a SoftMax or sigmoid layer (non-linear functions converting the entire data into 0–1 range of pseudo-probability, often mathematically formulated by exponential functions), then the task one wants to achieve is a classification. If the choice falls on a rectified linear unit (ReLU) (a non-linear function that converts all negative real numbers to zero, keeping positive real numbers unchanged), then the task corresponds to a regression.

The decision on which architecture to use depends mainly on the data. For debris-flow modeling without a temporal component, both FCNN and CNN can be considered; if the data are temporal or spatio-temporal, then Recurrent Neural Networks (RNN) are better suited.

Even though the architecture of deep learning models can differ based on the application cases, training of those models is done through backpropagation, where the weights and biases of the model are updated using some optimization algorithm such as stochastic gradient descent (Bottou, 2012) or Adam (Kingma & Ba, 2015) to minimize the prediction error. The error measurement function (loss function) is designed depending on the problem. For susceptibility mapping, binary cross-entropy is commonly used. This loss function evaluates whether the predicted susceptibility is closer to the observed data.

The significant difference between statistical and deep learning-based methods has always been the capacity of the former to explain the relationship between the variables and the black-box nature of the latter. However, in recent studies in landslide susceptibility modelling, Dahal and Lombardo (2023) have used Explainable Artificial Intelligence (XAI) methods to explain the relationship between the dependent variables and the conditioning factors. This approach explores the dependency between the dependent variables and the conditioning factors using SHAP (SHapley Additive exPlanations). The SHAP values provide a way to understand how much each input conditioning factor contributes to the model's output for individual mapping units (namely, slope units). SHAP originated from the cooperative game theory and aims to fairly distribute the "credit" for a prediction among the conditioning factors. However, this is still an active field of research, and it does not provide the non-linear partial dependence between the variables and their influence on the model, which is a well-developed approach in statistical modelling. Nevertheless, in scientific research, it is crucial to understand the effect of conditioning factors on the predictions generated by the model, because even though it cannot fully explain the non-linear relationships, it can at least provide an overview of their relations and be used as a sanity check. Therefore, we recommend using the XAI as a post-processing tool to understand the model's assumption to validate and check the underlying physical principles.

13.3.4 Physically-Based Methods

Debris flows often originate as shallow landslides. Engineers have used physically-based models to analyze shallow slope failures for several decades, most commonly applying limit equilibrium models based on a Mohr–Coulomb failure criterion: a factor of safety FoS is computed as the ratio between stabilizing and destabilizing forces (Carson & Kirkby, 1972; Crozier, 2010; Duncan & Wright, 2005). FoS > 1 would mean that a slope is stable, whereas FoS < 1 would characterize unstable slopes. Besides the slope angle, FoS is mainly determined by the sliding surface depth and geometry, the geotechnical parameters cohesion (soil + roots) and internal friction angle, and the groundwater level and flow direction. Specifically, for the analysis of shallow landslides, there is agreement that the infinite slope stability model, assuming an infinitely long slope and a plane, slope-parallel failure surface, represents an appropriate approach (Milledge et al., 2012). The infinite slope stability model can be used with GIS through computation of the factor of safety separately for each raster cell, and it can be coupled with models of slope hydraulics and thereby linked to rainfall or snowmelt patterns and the geo-hydraulic characteristics of the slope (Godt et al., 2008; Montgomery & Dietrich, 1994; Pack et al., 1998; Wilkinson et al., 2002).

GIS-based applications of the infinite slope stability model are mainly applied at local and catchment scales. Enhanced computational capacities make such approaches increasingly applicable at regional or even national scales from a technical point of view. van den Bout et al. (2018a), within the tool openLISEM (Sect. 13.6.4), have successfully used an extension to the infinite slope stability model to feed debris-flow simulations for the entire island of Dominica. In terms of data requirements, however, there are many situations where the acquisition of input data (i.e., the spatial distribution of geotechnical and geo-hydraulic parameters at a level of detail supporting and justifying the application of physically-based models) is challenging or even not feasible in practice. In such situations, physically-based approaches are not superior to statistical and machine learning methods, which can make use of better distinguishable spatial patterns such as lithology and therefore provide more transparent, though still less quantitative, results than physically-based model approaches. Inappropriate differentiation of geotechnical and geo-hydraulic parameters will lead to biased results, whereas assuming uniform parameters over a large area will provide nothing more than a reclassified slope map (de Lima Neves Seefelder et al., 2017).

In summary, we state that physically-based methods have some potential to be applied at a regional scale, but, in many cases, the lack of detailed, spatially distributed geotechnical and geo-hydraulic data (including information on soil depth) makes their appropriate application highly challenging in comparison to statistical and machine learning methods. Other physically-based methods exist but are not discussed here due to their current unsuitability for regional-scale applications.

13.4 Debris-Flow Runout Assessment

Modeling debris-flow propagation at a regional scale often consists of two components: algorithms defining the path (and the spreading) of the propagation and algorithms controlling the runout distance. Although these can interact differently according to the model, the basic principles are often the same, and the main approaches are presented below.

13.4.1 Routing

13.4.1.1 Flow Direction Algorithms

Flow direction algorithms originate from hydrological applications and determine how the flow distributes from a given cell of the DEM to its neighbors as it is routed down through the terrain from pixel to pixel. These algorithms are based on the slope gradient and follow different rules, as explained below for the most popular ones.

D8 (eight possible flow directions), ***maximum slope***, or ***single-flow-direction*** (O'Callaghan & Mark, 1984)—This is the simplest method and follows the steepest slope. The main problem with this approach is that there is no path divergence, so the flow is forced to follow directions with 45° steps (cardinal and diagonal), which results in unrealistic patterns on slopes with different orientations. In addition, it cannot deviate from the steepest descent direction, even in less steep areas, and therefore cannot accurately model debris-flow processes (Huggel et al., 2003). It has thus limited potential for realistic applications but can be used for rapid analyses (Horton et al., 2013; Wichmann, 2017). Huggel et al. (2003) introduced a certain flow diversion in the D8 approach using a function allowing the flow to divert from the steepest descent up to 45° on both sides, with a resistance factor.

D-infinity (Tarboron & Tarboton, 1997)—This algorithm distributes the flow to a single or two neighboring cells. It is based on the computation of downslope vectors of planar triangular facets. Although popular in hydrology, it is not often used for debris-flow propagation.

Multiple flow direction (Quinn et al., 1991)—This approach spreads the flow to all neighboring cells with a lower elevation, proportionally to the slope gradient. It includes a geometric weight factor to compute the fraction of flow draining to neighboring cells. Thus, it results in a divergence considered too high for debris flow processes (Huggel et al., 2003). Some variants of this algorithm slightly reduce the spreading, but not significantly (e.g., Freeman, 1991).

Adjustable multiple flow direction—Holmgren (1994) added an exponent in the multiple flow direction algorithm to control the convergence/divergence of the flow:

$$f_i = (\tan\beta_i)^x / \sum_{j=1}^{8} (\tan\beta_j)^x \text{ for all } \beta > 0 \quad (13.6)$$

where i,j are the flow directions (1..8), f_i is the flow proportion (0.. 1) in direction i, $\tan\beta_i$ is the slope gradient between the source cell and cell in direction i, and x is a variable exponent. The exponent x allows controlling the divergence so that higher values of x result in narrower and converging flow patterns. It thus provides great flexibility and the ability to reproduce most other flow direction algorithms: for $x = 1$, the spreading is similar to the multiple flow direction, and when $x \to \infty$, the behavior of the D8 is reproduced.

This algorithm has been extended by Horton et al. (2013), who introduced a raising of the source cell by a controllable factor *dh*, which changes the gradient values. It allows for ignoring small irregularities of the DEM and accounting for a certain thickness of the debris flow. Indeed, any cell just a few centimeters higher than the source cell cannot be reached by any other flow direction algorithm (if no deposition process modifies the DEM), which is unrealistic.

Multiple flow directions for debris flows (*mfdf*, Gamma, 2000)—This approach was specifically developed for the simulation of debris flows, usually implemented as a random walk model. Hydrological flow direction algorithms simulate lateral spreading over all slope gradients, while debris flows typically follow the dip line in steep slope sections. Thus, Gamma (2000) introduced a slope threshold to control when lateral spreading begins. In very steep sections of the process path, close to the slope threshold, only very steep neighbor cells are allowed. In sections with moderate slopes, additional (more gently dipping) neighbors are allowed. Finally, in flat sections, all lower neighbors are allowed.

For each neighbor cell i, a slope value γ_i, based on the slope threshold β_{thres}, is calculated:

$$\gamma_i = \tan\beta_i \tan\beta_{thres} \quad (13.7)$$

where β_i is the slope to the neighbor cell i ($\beta_i \geq 0$, i in range 1–8). The maximum value $\gamma_{\max} = \max(\gamma_i)$ is a measure of how close the slope of the steepest neighbor is to the slope threshold. If $\gamma_{\max} > 1$, the D8 single flow direction is used. Otherwise, the *mfdf* criterion is used to decide which other neighbor cells are additionally allowed:

$$\gamma_i >= \gamma_{\max}^{\alpha} \quad (13.8)$$

where α is an exponent to control the amount of divergent flow ($0 < \gamma_{\max} \leq 1$, $\alpha \geq 1$). If γ_i is greater than or equal to the *mfdf* criterion, the neighbor i is allowed as a potential flow path. The set of potential flow path cells N is thus given by:

$$N = \{i|\gamma_i \geq \gamma_{\max}^{\alpha}\} \text{ if } 0 < \gamma_{\max} \leq 1 \quad (13.9)$$

$$N = \{i|\gamma_i = \gamma_{\max}\} \text{ if } \gamma_{\max} > 1 \quad (13.10)$$

The *mfdf* algorithm is usually implemented in a random walk model combined with a Monte Carlo simulation. The *mfdf* criterion is used for each cell to determine a set N of potential flow directions. From this set N, a cell is picked at random. The probability for each cell to get selected is given by

$$prob_i = f_i * \tan \beta_i / \Sigma_j f_j * \tan \beta_j \quad (13.11)$$

where i is the currently processed neighbor cell (1–8), j depicts all neighbor cells in set N and f is a weighting factor. The latter can give a higher weight to the previous flow direction (see Sect. 13.4.1.2).

Because of the stochastic component, a high number of model runs (Monte Carlo simulation) from the same start cell will result in (slightly) different flow paths for each iteration. By aggregating all model runs, the full extent of the process area is reproduced. The final raster of the process area stores the transition frequency for each cell, i.e., the number of times each cell was selected as a process path.

13.4.1.2 Persistence Functions

Gamma (2000) introduced a notion of persistence to account for the inertia of the flow. It consists of a weighting based on a change in the direction of the flow:

$$p_i = w_{\alpha(i)} \quad (13.12)$$

where p_i is the persistence in direction i, and $\alpha(i)$ is the angle between the previous direction and the direction from the source cell to cell i. A key issue is knowing the weight $w_{\alpha(i)}$ for each direction. A debris flow will likely maintain its direction and cannot abruptly fold back. The weighting scheme proposed by Gamma (2000) simply puts 50% more weight in the same direction as the previous one. Horton et al. (2013) proposed different weighting schemes (see Table 13.1). This persistence factor is combined with the flow direction algorithm to provide a modified flow (or susceptibility) distribution (Fig. 13.7).

The variation of weighting schemes may be linked to the local topography that controls flow velocity. Han et al. (2017) proposed slope-angle-based weighting schemes using the data from a flume experiment. They demonstrated that weighting schemes approximate the cosines scheme by Horton et al. (2013) when the slope angle along the previous direction is greater than 10°; otherwise, it is close to Gamma's and Horton's proportional schemes.

13.4.1.3 Random Walk

Random walk approaches are commonly used in regional hazard assessments (e.g., Gamma, 2000; Mergili et al., 2015; Wichmann, 2017). They are a type of Monte-Carlo simulation that simulates lateral spreading through multiple propagations from the same source cell. At each step of the routing procedure, the flow is transferred to a single target cell (in contrast to flow direction algorithms). The selection of this target cell is based on probabilities as defined, for example, by the *mfdf* algorithm (Sect. 13.4.1.1). Strictly speaking, we talk about constrained random walk approaches, meaning that spreading is not equally likely in each direction. The routing must be repeated many times (e.g., 10^4, depending on the ratio between the event size and the pixel size) to achieve a realistic amount of lateral spreading.

Flow trajectories resulting from random walk procedures usually do not represent smooth lines but might show small-scale oscillations, which

Table 13.1 Weighting schemes for the persistence

Change in direction	Gamma (2000)	Horton et al. (2013)		Han et al. (2017)			
		Proportional	Cosines	<10°	10°–17°	17°–23°	>23°
w_0	0.22	0.28	0.42	0.30	0.40	0.50	0.60
w_{45}	0.13	0.24	0.29	0.24	0.20	0.18	0.15
w_{90}	0.13	0.12	0.00	0.11	0.10	0.07	0.05
w_{135}	0.13	0.00	0.00	0.00	0.00	0.00	0.00
w_{180}	0.00	0.00	0.00	0.00	0.00	0.00	0.00

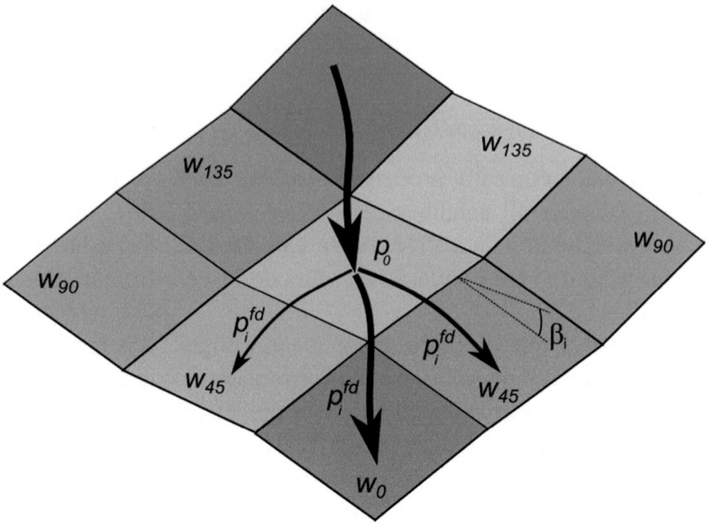

Fig. 13.7 Illustration of the combination of the flow direction algorithm and the persistence (from Horton et al., 2013)

are often invisible when visualizing the superposition of multiple random walk trajectories. However, they have an impact on the computation of the travel distance: for example, when using energy line approaches (angles of reach, see Sect. 13.4.2), an oscillating flow path would mean a longer travel distance and, therefore, a lower angle of reach, compared to a smooth flow path. Stopping criteria would therefore apply too early in the routing procedure, and the travel distance would be underestimated. Smoothing functions must be applied to counteract this issue (e.g., Mergili et al., 2015).

13.4.1.4 Agent-Based Approaches

Agent-based models were conceived several decades ago (Von Neumann, 1966; Wolfram, 1984) but were used in a limited capacity for regional landslide models (Guthrie & Befus, 2021; Guthrie et al., 2008; Turcotte et al., 2002). There is no universal definition of the term "agent" (Macal & North, 2009). Bonabeau (2002) provides the following description: "In agent-based modeling (ABM), a system is modeled as a collection of autonomous decision-making entities called agents. Each agent individually assesses its situation and makes decisions on the basis of a set of rules." They work in such a way that agents, or autonomous sub-routines, evolve through a series of timesteps and account for the values of neighboring agents (themselves evolving) and the matrix through which they travel (e.g., a DEM). Each agent evolution is based on a set of rules and the behavior of the local neighborhood of agents.

Agent-based models may, therefore, use any of the approaches herein to determine the path and spread: simple flow direction algorithms, persistence functions, random walk, and even energy line approaches described below. One approach to route selection (path direction and spread) has been to redistribute landslide mass held by individual agents to new cells, thus creating new agents, according to a probability density function, where the distributed mean is centered around the direction of travel of the agent in question, and its standard deviation (Guthrie & Befus, 2021).

Agents can therefore employ some aspects of the D8, persistence functions (related to the weighting of mass distribution), and multiple flow directions, all overseen by empirical or observed landslide behavior (manual input). Probabilistic results built into the agent spreading result in more realistic paths and limit path convergence. However, some path selection problems noted in the D8 model may still apply, particularly on steep slopes.

13.4.2 Energy Line Approaches

Energy line approaches are popular because of their simplicity and ability to be linked to field data. They rely on the vertical and horizontal distances between the source area and the most distant point reached by the debris flow. The most prevalent version for debris flows is the travel angle, also named Fahrböschung (Heim, 1932), angle of reach (Corominas, 1996), or α-line, where the horizontal distance is assessed along the path of the debris flow (i.e., not in a straight line):

$$\tan \alpha = d_v / d_h \tag{13.13}$$

where α is the angle to the horizontal and d_v and d_h are the vertical and horizontal offsets, respectively. Along a straight energy line from the start to the stopping position, the velocity can be calculated by (Körner, 1980):

$$v_i = \sqrt{2gh_v} \tag{13.14}$$

where v_i is the velocity [ms^{-1}] for the currently processed grid cell, g is the acceleration due to gravity [$ms{-}^{2}$] and h_v is the height difference [m] between the energy line and the current grid cell i.

The angle α has a characteristic value range for gravitational movements of a specific type. The calibration of the angle α is usually done by field observations and mapping as the angle can be measured easily. Based on an inventory of observed events, Rickenmann (2005) assessed that the lowest travel angle is about 4° ($\tan \alpha = 0.07$) for debris flows with a high proportion of fine sediments (less than sand size) and about 10–11° ($\tan \alpha = 0.19$) for coarse-grained debris flows.

The angle α can also be expressed as a function of the volume, when it can be assessed, or of the catchment area (Rickenmann, 2005). Based on field data, Zimmermann et al. (1997) proposed the following relationship to delineate a lower envelope of α as a function of the catchment area A [km^2]:

$$\tan \alpha_{\min} = 0.20A^{-0.26} \tag{13.15}$$

Another approach (ACS model—The average channel slope model—Prochaska et al., 2008) developed for modest-size debris lows is based on the angle between the midpoint elevation of the drainage channel and the end of the runout zone. This angle can be derived from a relationship with another angle: the one from the midpoint elevation and the fanhead. The benefits of this approach are that the initiation point is not needed (as it can be unknown or difficult to assess), and the reference point is objective and reproducible. The method showed good results but was developed and tested on a limited number of case studies and can thus be dependent on the site conditions of the sample. Nevertheless, it provided promising results when compared with traditional reach angle methods.

13.4.3 Physically-Based Methods

13.4.3.1 Voellmy Model

The first mass flow modeling approach particularly dedicated to gravitational processes in mountain areas which could be considered as physically-based (however, see Iverson and Geroge, this volume, for a further discussion on physically-based models), was the rheological approximation of Voellmy (1955). It was developed for snow avalanches in the aftermath of the catastrophic avalanche winters of 1951 and 1954 in the European Alps. As both flow-type snow avalanches and debris flows are often considered granular flows, similar in their principal dynamic behavior, the Voellmy (1955) flow formulation is equally relevant for debris flows. The rheological Voellmy approximation combines the Mohr–Coulomb material model for dry friction with the flow formulation of Chézy (1718–1798), where the Voellmy turbulence parameter ξ corresponds with the square of the Chézy-parameter C. The Voellmy flow approximation expresses the overall flow resistance as per Eq. 13.16, where μ is the tangent of the friction angle, ρ the unit weight of the mass flow, H the flow depth, g the gravitational acceleration, and φ the slope of the flow path. This part of the term represents the dry Mohr–Coulomb friction resistance of the flow. The second part represents the velocity (v) dependent flow resistance, based on the Chézy equation, where the turbulence parameter ξ is equal to C^2. The momentum source term S is computed as follows:

$$S = \mu\rho g H \cos\varphi + \left(\frac{\rho g v^2}{\xi}\right) \quad (13.16)$$

The Voellmy (1955) model and its direct derivatives and simplifications, such as Perla et al. (1980), are often used in regional-scale assessments.

13.4.3.2 PCM Model

The PCM model (Perla et al., 1980) is a two-parameter friction model based on the Voellmy (1955) model. Initially developed for snow avalanches, it has also been used successfully to simulate the runout of debris flows by various authors (Gamma, 2000; Horton et al., 2013; Rickenmann, 1990; Wichmann & Becht, 2004; Zimmermann et al., 1997). The center-of-mass model assumes that the motion is mainly governed by a sliding friction coefficient μ and a mass-to-drag ratio M/D. The value of μ dominates the velocity in the runout area. The velocity can be calculated along individual, more or less homogeneous, slope segments (e.g., Gamma, 2000). Other implementations (e.g., Horton et al., 2013; Wichmann, 2017) perform the calculations per grid cell of the process path. The velocity at a given location depends on the previous velocity, the slope, the slope length, and the two friction coefficients:

$$v_i = \sqrt{\alpha_i (M/D)_i (1 - \exp\beta_i) + (v_{(i-1)})^2 \exp\beta_i} \quad (13.17)$$

and

$$\alpha_i = g(\sin\theta_i - \mu_i \cos\theta_i) \quad (13.18)$$

$$\beta_i = -2L_i/(M/D)_i \quad (13.19)$$

where v_i is the velocity [ms$-^1$] on the currently processed grid cell, g is the acceleration due to gravity [ms^{-2}], θ is the local slope [°], L is the slope length [m], μ is the sliding friction coefficient [−], and M/D is the mass-to-drag ratio [m].

On concave transitions in slope direction, the velocity $v_{(i-1)}$ is corrected based on the conservation of linear momentum before v_i is calculated:

$$v_i = v_{(i-1)} \cos(\theta_{(i-1)} - \theta_i) \text{ if } \theta_{(i-1)} \geq \theta_i \quad (13.20)$$

The runout distance is calibrated by altering the two friction parameters. To overcome the problem of mathematical redundancy (various combinations of the two parameters can result in the same runout distance), the mass-to-drag ratio is usually constant along the process path. It is only calibrated once to obtain realistic maximum velocity ranges. The sliding friction coefficient can be set constant or vary along the process path, e.g., to model different rheological conditions due to different water content.

13.4.3.3 Recent Developments

Beginning in the late 1980s and accelerating from the early 2000s onwards, increasingly complex physically-based mass flow models emerged, most of which are in some way based on the Voellmy (1955) model. The models proposed by Savage and Hutter (1989), the DAN software (Hungr, 1995), Iverson (1997), and the RAMMS software (Christen et al., 2010) still considered the flows as homogeneous mixtures of solid and fluid material, without explicitly accounting for their two-phase nature. With TITAN-2D, Pitman and Long (2005) introduced a first two-phase model, followed by the Pudasaini (2012) two-phase mass flow model and the model of Iverson and George (2014), used in the software D-Claw (George & Iverson, 2014). The Pudasaini (2012) model was later extended to three phases (Pudasaini & Mergili, 2019), used by the software r.avaflow (Mergili & Pudasaini, 2022). All these models are depth-averaged, i.e., the vertical structure of the flow is largely disregarded. Recent approaches move towards full 3D simulations of mass flow processes using the material point method (Gaume et al., 2019). They are computationally very expensive and, therefore, currently off-limits for regional-scale applications. The same is true, though to a lesser extent, for the more complex depth-averaged models and tools, as they usually require detailed and complex computations for each point in a fine-spaced grid or point cloud, for very short time steps. However, the two-phase model of Pudasaini (2012) is employed in the software OpenLISEM (van den Bout et al., 2018a), which is one of the few software packages using complex physically-based models that can be efficiently applied at a regional scale (van den Bout et al., 2018b). Since many physically-based models are not well-suited for regional studies, this list of recent developments is not exhaustive. Interested readers should refer to literature specifically focused on this topic.

13.5 Coupling Source Areas and Runout

The most straightforward coupling of source areas and runout is by first assessing the potential source areas and using these in a propagation model. It can be done by considering either binary starting points (source/non-source) or by using the susceptibility or probability value of the starting zone, if available, and thus linking the reach susceptibility to the starting susceptibility (e.g., Michoud et al., 2012 for rockfall using the Flow-R software).

Few studies exist on coupling statistical landslide susceptibility analyses to corresponding runout models. Much has been written since the early 2000s on statistical (release) susceptibility analyses, neglecting the downslope consequences of the landslides. Only recently, more integrated approaches have been proposed (e.g., Goetz et al., 2021; Mergili et al., 2019).

Mergili et al. (2019) pointed out two possible ways of coupling, based on a simple statistical landslide release model and the r.randomwalk routing tool:

- Bottom-up approach: for each pixel in the study area (target pixel), the probability that there is at least one landslide in the catchment area is combined with the probability that such a landslide would reach the target pixel, based on the empirically-derived probability density function of the angle of reach for the study area.
- Top-down approach: routing is performed from each pixel in the study area (source pixel), and the likelihood of each target pixel to be affected by landsliding is a function of the release susceptibility and the probability that the target pixel is reached by the landslide, again based on an empirically-derived probability density function of the angle of reach. This is solved by scaling the number of random walks with the release susceptibility of each source pixel.

Mergili et al. (2019) claim that the bottom-up approach is statistically-mathematically cleaner and provides a real, quantitative, spatial probability. However, this approach is not very intuitive and involves a substantial level of smoothing of landslide susceptibility patterns, so that the result strongly correlates with the catchment size of each target pixel. The top-down approach, in contrast, is more intuitive and straightforward. However, the overlap of the trajectories starting from many source pixels complicates the derivation of a real probability and leads to a dominance of the runout component over the release component in the final results. Strategies to deal with such overlay issues will improve the statistical coupling of release and runout models.

Scenario-based methods are a practical alternative and are frequently coupled with susceptibility statistics. They involve running multiple debris-flow scenarios from susceptibility classes (and their respective likelihood metrics) using random, uniform, or manually assigned initiation locations within each class. Randomly- or uniformly generated starting zones require the user to apply a sufficiently dense network of potential landslide initiation locations, and number those locations from 1 to n such that any individual scenario will not unrealistically overlap. The user can run each scenario (scenario 1 … scenario n) multiple times to determine the difference between scenario runs.

Manually generated starting zones are similar but tend to be based on expert judgment (observable potential source locations) within a susceptibility class. The random or uniformly generated starting zones lend themselves better to statistical tests and are more practical across larger regions. Hazard data (rather than just susceptibility data) provides better limitations and discretization of scenarios (i.e., it is useful to know the number of expected debris flows/unit area/unit time for a given scenario in a given class).

13.6 Available Software for Runout Computation

This section presents some tools/software for assessing debris-flow susceptibility. They are software implementations of the model components described in the previous sections and cover mainly propagation calculations. The selection (Table 13.2) is restricted to relatively recent software (<15 years) that are publicly available (under different licenses).

Other pieces of software exist but are older or are not publicly available. Among these is a GIS-based software implemented by Huggel et al. (2003) to assess the hazard related to debris flows resulting from glacier lake outbursts at a regional scale. It was one of the first to combine the flow routing algorithms (D8 and multiple flow direction) and the energy-line approach (with a slope of 11°) and presents similarities with other software used nowadays. The work of Gamma (2000) also resulted in a software, *dfwalk*, which was not publicly released. Different concepts and algorithms from *dfwalk* are integrated into other software. The first regional cellular-automata tool for landslides (Guthrie et al., 2008) was never made publicly available; however, the concepts were also used for DebrisFlow Predictor (Guthrie & Befus, 2021). Another software that is not publicly available has been developed by Gregoretti et al. (2016) to simulate debris flow routing on a fan. It relies on a kinematic approach and works best with high-resolution DEMs (ideally 1 m), which can be a limitation at regional scales.

13.6.1 Flow-R

Flow-R (Flow path assessment of gravitational hazards at a Regional scale; Horton et al., 2013) was primarily developed for debris-flow susceptibility mapping at a regional scale. However, it has also been used for other gravitational hazards such as rock falls, rock avalanches, shallow landslides, and snow avalanches. It was

Table 13.2 List of publicly available software for debris flow susceptibility assessment, classified by year of publication

Software name	Release year	Author	Code	Availability	Source areas	Propagation
Flow-R v1	2009	P. Horton	Matlab	Freeware	Yes	Yes
r.randomwalk	2015	M. Mergili	C, Python	Open source	No	Yes
Gravitational Process Path	2017	V. Wichmann	C++	Open source	No	Yes
OpenLISEM	2018	B. van den Bout	C++	Open source		Yes
Flow-R v2	2020	P. Horton	C++	Commercial	No	Yes
DebrisFlow Predictor	2021	R. Guthrie	C#	Commercial	No	Yes
Flow-Py	2021	C. D'Amboise	Python	Open source	No	Yes

designed to provide flexibility in terms of data and algorithms so that its use can be adapted to the data availability and the process of interest.

The first version, written in Matlab, was released in 2009 and is freely available (Horton, 2022). A new version, Flow-R v2, written in C++, was released in 2020 and is commercial. This new version addressed several limitations of the first one. It provides a substantial increase in performance and efficient memory handling that facilitates working with unrestricted spatial extents. It is also significantly more user-friendly and data-friendly (supports many data formats).

13.6.1.1 Components

An identification of the potential source areas is provided in Flow-R v1, which consists of a combination of multiple data layers with classification rules (Sect. 13.3.2). For version 2, the source areas can be prepared in any GIS software. Flow-R then propagates the debris-flow susceptibility from each source area. While the initial value to propagate is most often the normalized value 1, specific initial susceptibility values can be assigned to each source area. This allows accounting for different initial failure susceptibility or frequency values in the resulting map (e.g., Michoud et al., 2012). Flow-R does not rely on a random walk approach but spreads the susceptibility from each source area through a single propagation, whose extent is meant to cover all possible events. The main result is a spatially-distributed susceptibility representing areas where debris flows are more likely.

The runout computed by Flow-R relies on spreading algorithms to control the path and the spreading of the susceptibility and simple friction laws determining the runout distance. Flow direction algorithms and persistence functions control routing and spreading. Several flow direction algorithms are implemented (see Sect. 13.4.1.1 for the details), such as the D8 (O'Callaghan & Mark, 1984), D-infinity (Tarboron & Tarboton, 1997), Rho8 (Fairfield & Leymarie, 1991), multiple flow direction (Quinn et al., 1991), Freeman (1991), Gamma (2000), and Holmgren (1994) algorithms, and its modified version (Horton et al., 2013). The modified version of Holmgren's algorithm is the most used as it allows reproducing most algorithms through its parameter controlling the spreading. Different weighting schemes (Table 13.1) are available for the persistence function (Sect. 13.4.1.2).

Two algorithms are available to account for the friction: the two parameters friction model by Perla et al., (1980; see Sect. 13.4.3) and a simplified friction-limited model (SFLM; Horton et al., 2013), which consists of an energy-line approach combined with a limitation of the maximum velocity (Fig. 13.8). This limitation aims at keeping the energy at reasonable values in steep catchments. With observed maximum velocities of debris flows in Switzerland being 13–14 ms^{-1} (Rickenmann & Zimmermann, 1993), a limit of 15 ms^{-1} is often used.

13.6.1.2 Data Needs

Flow-R is flexible in terms of data requirements. The minimum requirement is a DEM for version 1. For version 2, a file with the source areas (determined, for example, using GIS-based approaches as described in Sect. 13.3.2) is also required. As Flow-R v1 provides an identification of the potential source areas, any dataset considered relevant can be provided, such as land-use or geological maps or any DEM-derived product. As for the propagation, only the DEM is required. Optionally, a raster file providing spatially-variable travel angles to account for different land cover classes or mitigation measures, for example, can be provided.

13.6.1.3 Outputs

Most Flow-R results are exported as raster files in geotiff (Flow-R v2) or ascii (Flow-R v1) formats. The primary output is the susceptibility value for each cell of the study area, expressed as the maximum (Fig. 13.9, right) and/or the

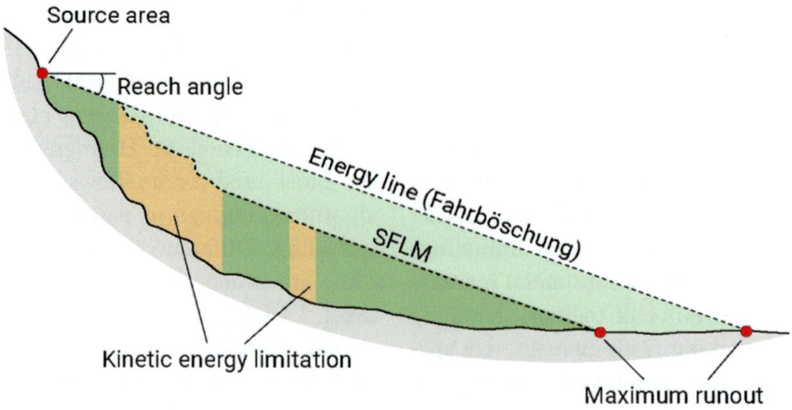

Fig. 13.8 Illustration of the reach angle and kinetic energy limitation of the SFLM friction model used (after Horton et al., 2013)

sum of the susceptibility values from all propagations originating from different source areas. For convenience, the overall extent of all propagations can also be exported, as well as the number of propagations passing through a given cell (v2 only). Additionally, a shapefile containing each propagation extent can be exported (v2 only). Rasters containing energy or velocity values (maximum values per grid cell) can also be exported for information purposes. However, they should be considered with care as they are based on calculations using a unit mass approach.

A multi-parameter approach is often applied and consists in computing propagations with different parameter sets representing varying scenarios—more or less conservative—to provide susceptibility classes. The results of these scenarios are combined by reclassification to assign a higher susceptibility class to the less conservative scenarios and a lower class to the more conservative scenarios (Fig. 13.9, left).

13.6.1.4 Calibration and Evaluation

Calibration of Flow-R primarily relies on a trial-and-error approach, using an inventory of observed events as a reference. The calibration can take place in a well-documented smaller area that is representative of the region. Then, the parameters can be applied to the entire region, if possible with additional observed events that can be used for validation. The objective of the calibration is to cover the extent of observed events while minimizing areas where debris flows are considered unlikely. Alternatively, an approach based on a confusion matrix proposed by Scheidl and Rickenmann (2010) has also been used to calibrate Flow-R (Pastorello et al., 2017). It consists of computing and maximizing accuracy indicators using the extent of past events.

The parameters controlling the spreading and those governing the runout distance can be calibrated independently as they do not interfere to a large extent. For debris flows, Perla's two-parameters model (PCM) and the SFLM provide similar results (see Horton et al., 2011, for examples). However, the SFLM is easier to calibrate and is thus more often used. Typical parameter ranges are provided in Table 13.3. When using the PCM model, classic parameters for debris flows can be used (e.g., Sect. 13.6.3.4).

13.6.2 r.randomwalk

r.randomwalk (Mergili, 2022; Mergili et al., 2015) represents a multi-functional tool for routing mass movement processes. It can be used as an extension to the open-source GRASS GIS software. The tool's architecture consists of a

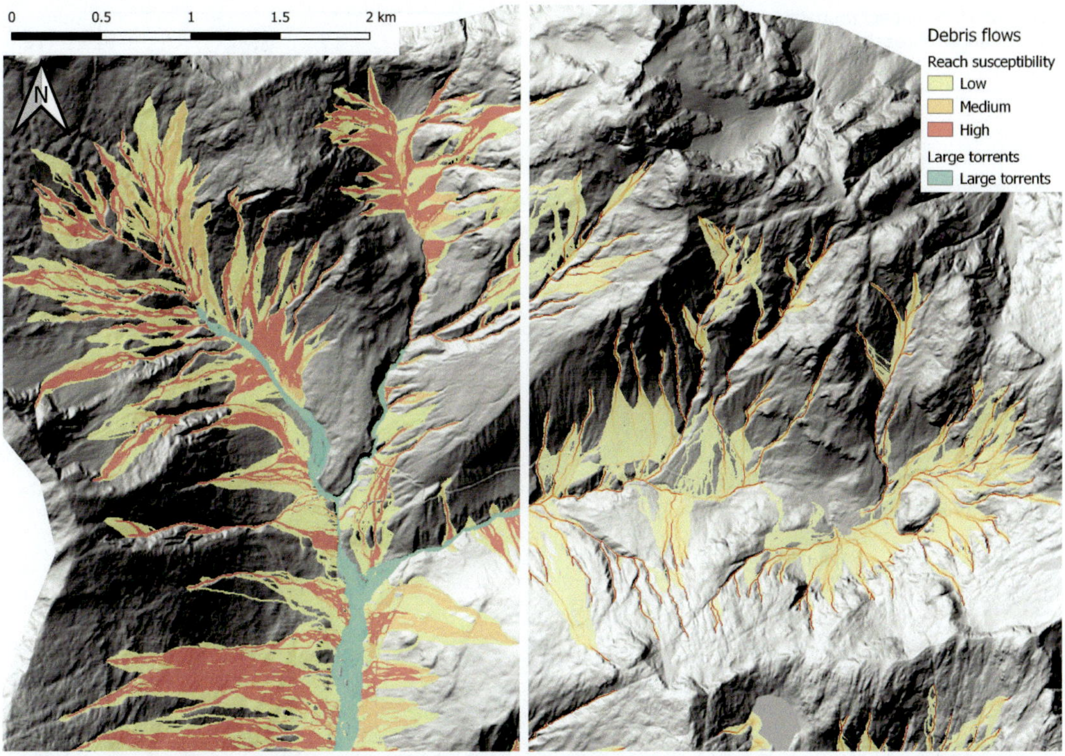

Fig. 13.9 Example of Flow-R outputs in the Mill Creek region (British Columbia, Canada) with (right) the spatial susceptibility originating from different source areas with a specific parameter set and (left) different reach susceptibility classes resulting from various parameter sets representing varying scenarios. The class "large torrents" is characterized by large contributing areas, hence potentially more water content, resulting in large debris flows or hyperconcentrated flows (with a lower angle of reach). The hillshade is based on the lidar DEM with a 1 m resolution, but the runout was computed at a 5 m resolution

Table 13.3 Flow-R configuration for debris-flow modeling at a regional scale and approximate parameter ranges. The reach angle can have much lower values for fine-grained debris flows or hyperconcentrated flows (Horton et al., 2011)

Model component	Algorithm	Parameter	Value range
Spreading	Holmgren modified	x	3–6
		dh	1–2 m
	Persistence	Weighing scheme	Proportional
Runout	SFLM	Reach angle	9–13°
		Maximum velocity	~15 m/s

core programmed in C, with a Python script wrapped around for pre- and post-processing of the data. Automated statistical procedures, visualization, and evaluation functionalities are included through R scripts which are directly called from the Python script.

13.6.2.1 Components

The core of r.randomwalk is a constrained random walk routing algorithm coupled to a selection of stopping criteria, including plain values and functions of angles of reach, as well as a simple physically-based model (2-parameter friction model). Random walks can be started

(a) Debris flow impact probability derived from PDF of angle of reach

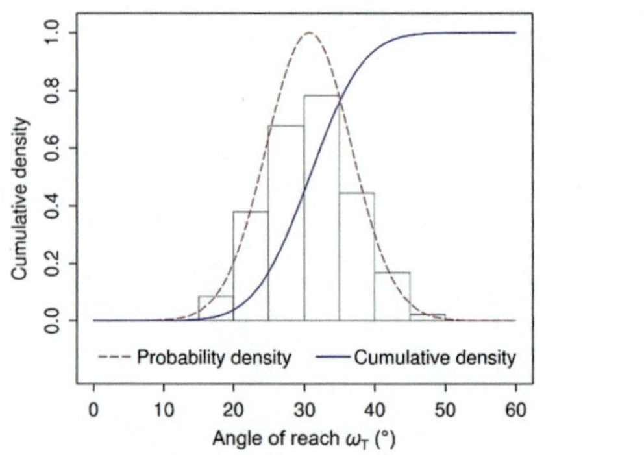

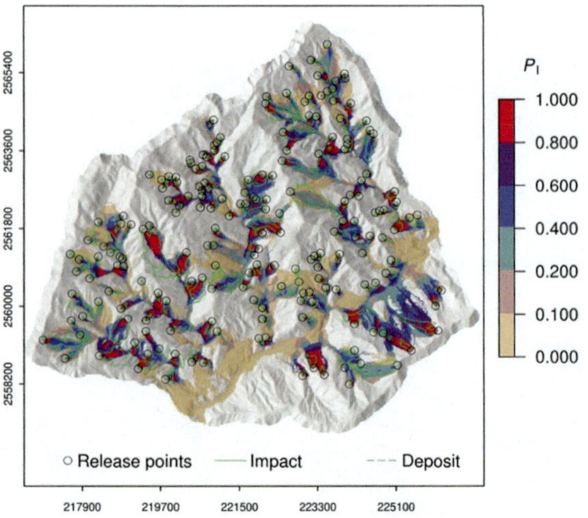

(b) GLOF impact indicator score derived from six empirical relationships

Equation	References	Process
1 $\omega_T = 11°$	Haeberli (1983); Zimmermann et al. (1997); Huggel et al. (2003, 2004)	Flood or debris flow
1 $\omega_T = 7°$	Zimmermann et al. (1997)	
2 $\omega_T = 18Q_p^{-0.07}$	Huggel (2004)	
3 $L_{max} = 1.9V^{0.16}Z^{0.83}$	Rickenmann (1999)	
4 $\omega_T = 6°$		Flood
5 $\omega_T = 4°$		
6 $\omega_T = 2°$	Haeberli et al. (1983); Huggel et al. (2004)	

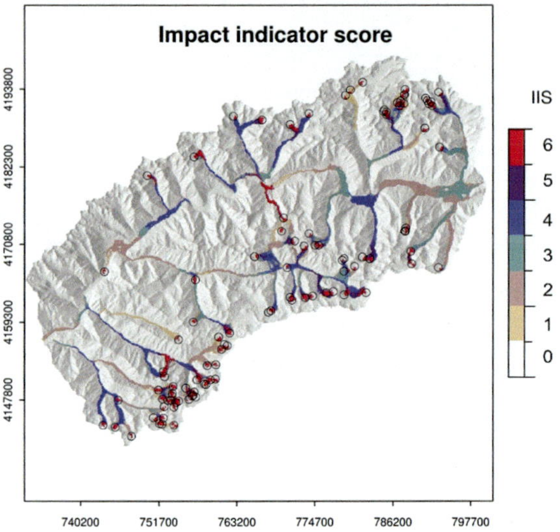

Fig. 13.10 Examples of impact probability and impact indicator score maps produced with r.randomwalk. **a** Impact probability curve derived from the back-calculation of a set of observed hillslope debris flows, and its application to a catchment-scale map of the spatial impact probability for the same type of process. **b** A set of six published empirical relationships for the angle of reach and travel distance potentially relevant for glacial lake outburst floods (GLOFs). The impact indicator score map shows, for a given study area, the number of relationships predicting an impact on each cell. *Source* Mergili et al. (2015)

from an arbitrary number of single points or landslide release areas, or from a landslide release susceptibility map, and are routed through the terrain until all applied stopping criteria are met. An arbitrary number of different stopping criteria of different types can be used to derive likelihood measures that a given target pixel is reached (impact indicator score). Simulations can also be run with large numbers of parameter combinations, and the fraction of parameter combinations predicting an impact on a target pixel is termed *impact indicator index* (Fig. 13.10b). Finally, an impact probability can be estimated based on back-calculated angles of reach of observed landslides (see Sect. 13.4.2).

13.6.2.2 Data Needs

The central piece of data needed for all types of use of r.randomwalk is a DEM. Apart from this, the required data depend on the type of use.

Source pixels can be defined through a text file providing coordinates and, optionally, volumes and other parameters characterizing the release of an arbitrary number of events (one event for each line). Optionally, the release file may be coupled to a release raster map, or the release raster map may be provided without a release file. Also, it is possible to define the release through a landslide release susceptibility map, which is important for coupling release and runout (see Sect. 13.5). In those cases where calibration or evaluation is involved, a raster map of the observed impact area is required as input. Further, the model needs sets of parameters controlling the random walks and the stopping criteria, as well as various other functionalities, depending on their use.

13.6.2.3 Outputs

Besides the raster maps outlined in Sect. 13.6.2.2 (as ascii rasters and cartographic visualizations through R-generated plots), r.randomwalk provides text files characterizing the routing parameters for each event under investigation, with parameters such as the size of the impact area, horizontal and vertical travel distances, and angles of reach. Additional files and graphs (such as ROC plots or heat maps, indicating the sensitivity of the output to the parameterization of the empirical equations and the random walk procedure) are produced, depending on which functionalities of the software are used (Fig. 13.10).

13.6.2.4 Calibration and Evaluation

r.randomwalk enables the back-calculation (angle of reach, travel distance) of the events depicted in the impact area map. This back-calculation can be used to automatically produce a probability density function of angles of reach or travel distances (Fig. 13.10a), which can then—again automatically—be used to produce an impact probability map based on the same source areas or any other arbitrary source areas. The impact probability map can then be evaluated against the impact area map, resulting in a ROC plot indicating the model performance. If applicable to the specific workflow, it is possible to divide the data into a training dataset (for the derivation of the probability density function) and an evaluation dataset (for the ROC plot).

13.6.3 Gravitational Process Path

The Gravitational Process Path (GPP) software has been designed to simulate the process path and runout area of gravitational processes based on a DEM. The first components of the software have been developed since 2000 (e.g., Wichmann & Becht, 2004), and a completely revised and open-source implementation was published in 2017 (Wichmann, 2017). The software is implemented in C++ as a tool for the System of Automated Geoscientific Analyses (SAGA, Conrad, et al., 2015). This facilitates input data preparation and the analysis of model outputs.

The conceptual model combines several components (process path, runout length, sink filling, and material deposition) to simulate the movement of a mass point from a source to the deposition area. Since version 1.1 (released in 2022), the software also supports detecting potentially endangered objects like infrastructure and reports from which process paths and source areas objects might be hit. Several modeling approaches are provided for each component, which makes the software configurable for different processes such as rock fall, debris flows, or snow avalanches. The software has been primarily developed for regional-scale studies such as natural hazard susceptibility mapping but also contains components for scenario-based modeling of single events.

13.6.3.1 Components

The design of the GPP software is modular, which makes it possible to combine different methods for process path, runout length, and material deposition (sink filling) computation. All components, besides the process path calculation, are optional. The *maximum slope* (O'Callaghan & Mark, 1984) and the *mfdf* (Gamma, 2000) flow direction algorithms are available for process path computation. The latter is implemented as a random walk as proposed by Gamma (2000). The lateral spreading of the process is simulated by calculating several iterations from the same start cell (Monte Carlo simulation). When combined with the sink-filling model component (which alters the DEM by the deposition of material), subsequent iterations can overcome (smaller) depressions in the DEM. The random walk provides three parameters to calibrate the extent of the simulated process area: a slope threshold above which no lateral spreading is modeled, an exponent which controls the amount of lateral spreading, and a persistence factor that adds a higher weight to the current flow direction (Sect. 13.4.1.2).

The software provides three kinds of energy line approaches to determine the runout length (velocity), as well as a one- and a two-parameter friction model. Available methods based on the energy line principle are the geometric gradient (Heim, 1932), the Fahrböschung (Heim, 1932), and the shadow angle (Hungr & Evans, 1988). The one-parameter model (Scheidegger, 1975) can be used to simulate the velocity based on sliding or rolling friction. The two-parameter model is the PCM model of Perla et al., (1980; Sect. 13.4.3). In the GPP model, it is possible to use spatially distributed friction coefficients (besides static ones), which can, for example, be used to simulate an increasing water content along the process path and thus reduced friction. For debris flow runout calculations, the Fahrböschung or the PCM model are usually used.

The GPP model implements various material deposition approaches, of which the sink filling is the most important. Once the process path arrives at a depression in the DEM, the sink is filled with the material available, using an algorithm that preserves a certain downward slope. This allows subsequent random walks to overcome the depression. Other deposition approaches simply deposit material on stop or deposit material along the process path based on slope and velocity thresholds. These deposits alter the DEM and thus influence subsequent random walks too. This can be used to simulate effects like channel plugging.

13.6.3.2 Data Needs

In its basic configuration, the GPP model only requires a DEM and a raster with potential source areas as input. The preparation of these datasets can be done directly within SAGA. For example, if no material deposition (sink filling) is modeled, sinks in the DEM should be filled beforehand, just like with hydrological analyses. Potential source areas can be identified by various approaches. If this is not done within SAGA, it is possible to import such a raster in various formats.

If a deposition approach is to be used, a raster with the material available at each start cell is required. For runout modeling, it is possible to use spatially distributed friction coefficients (or angles in case an approach based on the energy line principle is used). These rasters can be prepared, for example, from land cover maps or flow accumulation (catchment area). Gamma (2000) derived functional relationships between the catchment area and the sliding friction coefficient for scenarios of minimum, likely, and maximum runout distance.

13.6.3.3 Outputs

The GPP software outputs the results as raster layers. The process area is reported in a raster storing the transition frequency, i.e., the number of times (count) a cell has been part of the process path (Fig. 13.11). The maximum velocity for each cell of the process area is saved in another raster. If the material deposition is modeled, a raster with the height of the deposited material is produced. When added to the input DEM, this shows the final topography after the simulation run.

Optionally, a raster with the stopping positions can be output. It shows the cells in which the maximum runout length has been reached. If an input raster with potentially endangered objects has been provided as input (for exposure detection), an output raster will report the combination of classes that were hit from each raster cell. This makes it possible to analyze which object classes can be affected from which location.

13.6.3.4 Calibration and Evaluation

A typical configuration of the GPP software for regional-scale natural hazard susceptibility mapping uses the random walk model component for path finding and the PCM model for runout distance calculation. The parameters that require calibration and typical value ranges are shown in Table 13.4.

The parameters are usually calibrated by testing different values over a certain parameter range. The simulation result of each parameter combination is then compared to mapped debris-flow events, and the set of parameters showing the best agreement with the mapped process area extents is used. A procedure for automatic parameter optimization and validation of the GPP software for regional debris-flow runout modeling has been developed by Goetz et al. (2021).

13.6.4 OpenLISEM

13.6.4.1 Components

The Limburg Soil Erosion Model (LISEM; note that the original name is not used anymore nor representative of the current functionalities) software is a physically-based simulation tool for multi-hazard analysis. The software features extensive capabilities for simulation, calibration, and analysis of mass movements such as debris flows.

The debris-flow runout features in the LISEM software are an adaptation of the generalized two-phase debris flow equations of Pudasaini

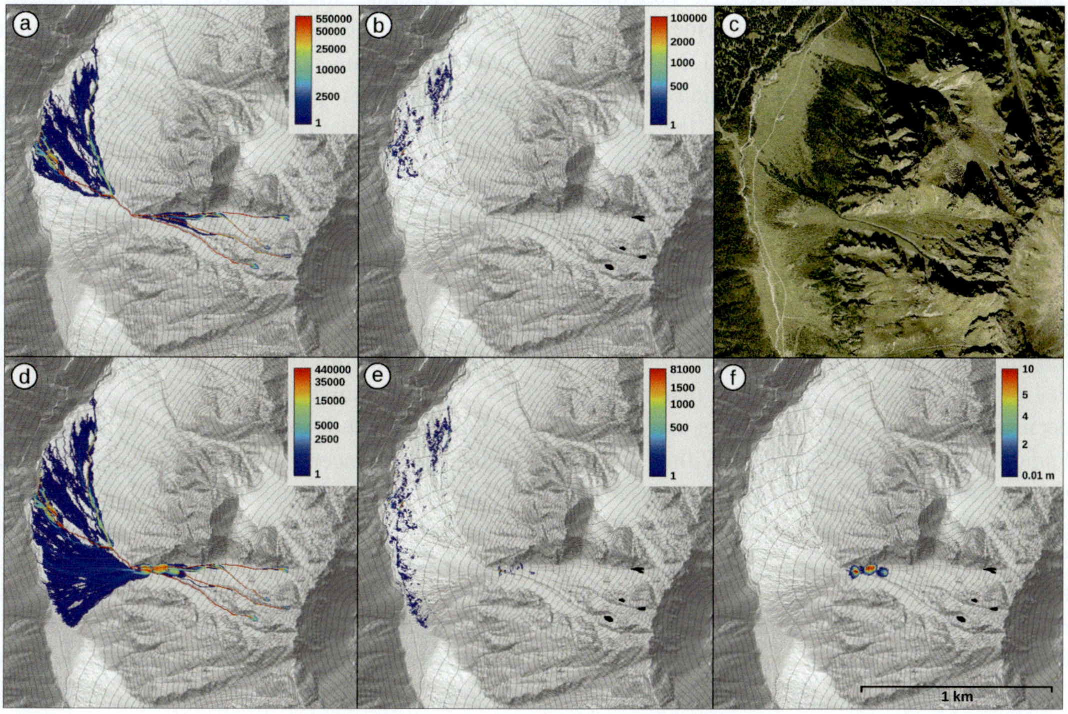

Fig. 13.11 Example of a GPP software output for a deeply incised debris flow channel without (top) and with (bottom) material deposition modeling. **a/d** transition frequencies [particle count], **b/e** stopping positions [particle count], **c** orthophoto, **f** material deposition by sink filling and slope/velocity thresholding [m]. Material deposition modifies the DTM and causes subsequent random walks to break out of the channel. This also classifies (older) parts of the fan as process areas that cannot be reached by the routing algorithm without terrain modification

Table 13.4 GPP software configuration for debris flow modeling on a regional scale and approximate parameter ranges (Wichmann, 2017)

Model component	Model approach	Parameter	Value range
Process path	Random walk	Slope threshold	20–40°
		Exponent of divergence	1.3–3.0
		Persistence factor	1.5–2.0
Runout	PCM model	μ	0.04–0.8, spatially distributed
		M/D ratio	20–150, spatially distributed

(2012). This set of equations distinctly considers mass and momentum conservation for both fluids and solids. Various forces act on these 'phases' during flow, resulting in a realistically-coupled flow. The originally implemented forces are gravity, pressure, buoyancy, Mohr–Coulomb friction, viscosity, non-Newtonian viscosity, and virtual mass. Within LISEM, the aim is to simulate a wide range of flow types, from clear water flow to debris flows, landslides, and rock avalanches. While the generalized two-phase equations effectively capture multiple types of mass movements, clear water flow lacked a frictional force at the interface between flow and terrain. This resulted in unrealistic flow velocities on sloped terrain, a problem that has since been addressed in software such as r.avaflow by Pudasaini and Mergili (2019). Within LISEM, the approach from the field of hydrological and hydraulics was utilized, where the Darcy-Weisbach friction law, an industry standard for flood models (Delestre et al., 2017), was implemented proportionally to the presence of fluids in the flow (van den Bout et al., 2018a).

As LISEM focuses heavily on multi-hazard research, the flow equations are integrated into a catchment-scale hydrological model that is capable of simulating infiltration, ground water, slope stability and failure, as well as river flows, erosion, tsunami propagation, and more (van den Bout et al., 2018a). This allows for more complex setups involving runoff-driven debris flows through erosion or debris flows evolving from slope failures, where landslide material is combined with surface runoff water. Because sub-pixel details are considered for channels, buildings, and roads, the hydrology, flow, and runout simulations can be carried out from plot-scale (grid spacing of several cm, with a total area of several square meters) to national-scale (grid spacing of 200 m, with a total area of tens of thousand square kilometers). Optimal results from the debris-flow simulations are often obtained with grid-spacing of 10 m or less, as relevant details in the elevation model then capture the common scale of debris-flow runout.

The numerical solutions to the equations are solved using a second-order accurate finite difference scheme on a square grid. To capture wave dynamics and maintain a higher timestep, flux-limiters are used in combination with adapted Riemann-solvers (Delestre et al., 2017). Riemann-solvers provide semi-analytical solutions to discontinuity problems at grid-cell interfaces. Effective mass and momentum flux can be calculated by solving for domains of influence using characteristic velocities. Where the Saint–Venant equations for shallow water flow allow for such a solution, the generalized debris-flow equations do not permit the derivation of characteristic velocities. Instead, the Harten-Lax-van Leer Riemann solver is employed for clear water, whereas a direct flux calculation is activated with higher solid concentrations. All calculations can be carried out either on CPU or GPU in parallel.

13.6.4.2 Data Needs

The central data requirement is the DEM, which strongly determines the simulations' quality. For flow modeling, the material properties must be available, namely density, grain size, basal friction angle of the material, and the amounts of fluids and solids. If the initial materials are estimated through slope stability modeling, the

Table 13.5 The most common parameters for the LISEM software

Model component	Algorithm	Parameter	Units
Flow	Generalized two-phase flow equations (Pudasaini, 2012)	Manning's surface roughness coefficient	m kg/s^2
		Drag coefficient	(–)
Boundary conditions	Initial release volume	Initial solid height	m
		Initial water height	m
		Initial material basal friction angle	radians
		Initial material density	kg/m^3
		Initial material grain size	m
Hydrology (optional)	Infiltration (Green & Ampt, 1912)	Saturated hydraulic conductivity	mm/h
		Porosity	m^3/m^3
		Moisture content	m^3/m^3
		Matric suction	kPa
Slope stability (optional)	Iterative failure method (van den Bout et al., 2021)	Soil depth	m
		Soil cohesion	kPa
		Soil internal friction angle	radians
		Soil density	kg/m^3
		Soil grain size	m
Erosion (optional)	Takahashi entrainment equation (Takahashi, 2009)	Soil cohesion	kPa
		Entrainment coefficient	(-)

density, grain size, and internal friction angle of the soil are required, as well as the soil depth (Table 13.5). The water content of the soil can be predicted through the hydrological components of the software. Finally, Manning's surface roughness coefficient is used to estimate the frictional force of the fluid phase. Depending on the additional processes that are simulated, additional data about surface properties (vegetation cover, root cohesion, and building presence), boundary conditions (river discharge, boundary condition for flow height or velocity, precipitation or snowmelt), or channel information (channel network, channel dimensions) are required.

13.6.4.3 Outputs

The software can output a variety of model variables in Geotiff or any format supported by GDAL. The possible outputs for each timestep (dynamic) include flow heights, velocities, phase heights, phase velocities, slope failures, water contents, infiltration amounts, factor of safety, and many more, depending on the processes that are activated. For the entire simulation, maximum flow height, velocities, and minimum factors of safety can further be provided. As the entire functionality of the LISEM software is available from the related scripting environment, the user has full access to all internal model parameters and can specify custom outputs. An example of a typical output is shown in Fig. 13.12.

13.6.4.4 Calibration and Evaluation

The LISEM software might be calibrated similarly to many other physically-based modeling methods. As long as observation data on some parameters are available, the user can specify a custom simulation error metric and run an automated calibration sequence. LISEM implements both brute-force and gradient-descent methods. The brute-force approach varies input parameters in a regular manner to attempt to cover all possible input values and find optimal model performance. While this method is reliable, it can increase computation costs exponentially with the number of parameters. The gradient-descent method attempts to move toward lower model error by repeatedly determining the gradient of the model error function (depending on parameter values) and altering the parameters to obtain lower model errors. LISEM uses the Armijo-backtracking method to determine step sizes, ensuring a stable search for minimum model errors. The most common approach is a calibration using an inventory of debris-flow extents. Here, debris-flow outlines are mapped from satellite or areal imagery. The model outcomes are compared as two binary classifiers using a metric such as Cohens Kappa or Precision-Recall. The second common method of calibration uses flow height or discharge time series. Observations of such variables can be compared to model output using, for example, the Root Mean Square Error (RMSE) or Nash–Sutcliffe error metrics.

13.6.5 DebrisFlow Predictor

DebrisFlow Predictor is an agent-based landslide simulation package that models the complex behavior of debris flows and debris avalanches (Guthrie & Befus, 2021). It was developed to allow geohazards specialists to model debris-flow runout at a regional scale and collect important information about the characteristics of individual landslides within the modeled domain. That information includes runout length, probability of inundation, and scour and deposition depth along the debris-flow paths.

DebrisFlow Predictor deploys agents, 5 m by 5 m sub-autonomous programs, over a digital elevation model of the same resolution. Each agent enacts a set of physically and empirically derived rules that describe its behavior at each timestep based on input data provided by the DEM and the neighborhood of agents for each model run.

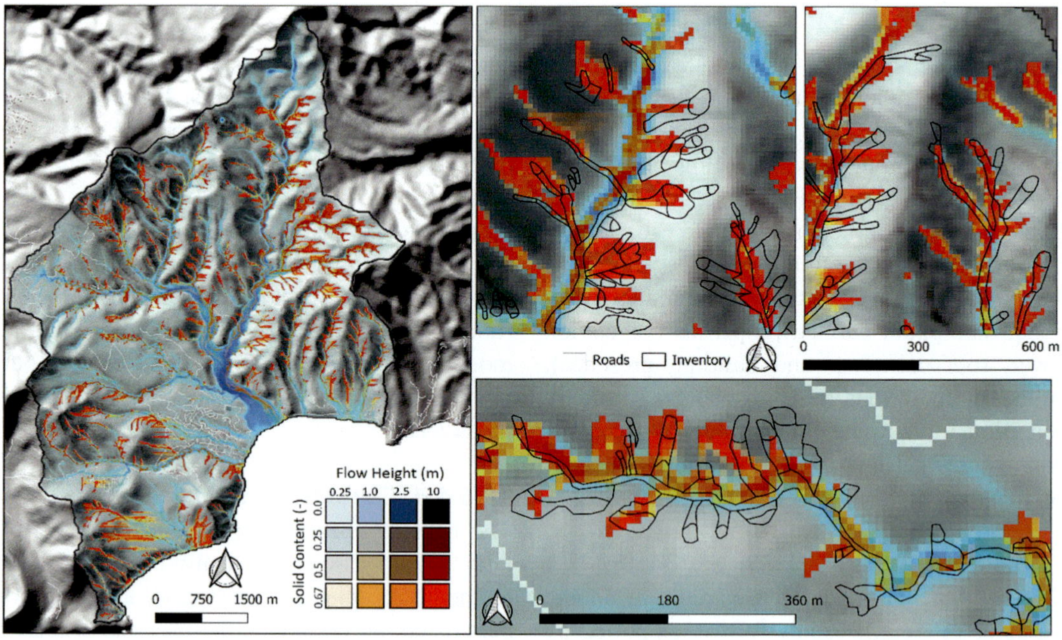

Fig. 13.12 An application of the LISEM software for the impact of Hurricane Maria on Dominica. This category-5 hurricane caused an abundance of slope failures, landslides, and debris flows, that merged with flood waters. The color legend indicates both the flow height of the water and the solids concentration of the peak flow during the event

It was written in C# and Extensible Application Markup Language (XAML) and is a standalone program with a fully developed user interface that includes the ability to import shapefiles and a DEM (ascii files) and to export data to Excel or as Geotiffs and shapefiles. It is expected to be available for commercial use through the creator's website (Stantec, 2022) in early 2024.

13.6.5.1 Components

The source areas can be manually selected in the software and often consist of nine agents (15 m × 15 m) or larger. Alternatively, random, uniformly, or manually selected initiation locations (by scenario) can be imported as point files into the software. Each point file can be individually assigned a 15 m × 15 m initiation location and can be turned on or off to run different scenarios.

Once initiated, agents follow a set of rules at each timestep:

- Scour and deposit: agents refer to reference curves representing the probability that they will scour and deposit (independently) at the current slope. These curves are based on an empirical dataset (Guthrie & Befus, 2021; Guthrie et al., 2008).
- Mass loss with direction changes: agents lose mass with direction changes as a percentage of the net scour (weighted so that larger angular changes lose more mass)
- Merge: when an agent runs into another agent during a timestep, they merge into a single agent and combine their mass.
- Turn: agents turn to the lowest three elevations and move into the lowest unoccupied space with some additional rules to account for ambiguous scenarios (no spaces unoccupied, equal elevations, etc.).
- Spread: agents spread according to a PDF function:

$$\sigma = ((\frac{m_{MAX} - m}{m_{MAX}})^n \times ((\sigma_L - \sigma_S) + \sigma_S)) \tag{13.21}$$

where: m_{MAX} is the maximum fan slope, m is the DEM slope, n,σ_L, and σ_S are the skew, low slope, and steep slope coefficients, respectively. Spreading behavior is controlled by sliders within the program (see Guthrie & Befus, 2021 for more discussion) and calibrated for local conditions.

Runout continues while any agent has mass. New agents are created during the spreading function, and runout stops when all agents are terminated (mass = 0 for each agent). Because the results are stochastic, the user is generally advised to conduct multiple runs per scenario.

13.6.5.2 Data Needs

DebrisFlow Predictor needs, at a minimum, a 5 m DEM imported as an ASCII file and a landslide initiation location (the latter can be chosen using the in-program tool). Other authors have discussed model limitations with respect to DEM cell size (e.g., Horton et al., 2013). DebrisFlow Predictor is not designed for DEMs finer than 5 m. For regional scales, and even at the scale of a single watershed, the 5 m resolution provides considerable detail.

13.6.5.3 Outputs

DebrisFlow Predictor produces results from any combination of single/multiple run(s) and single/multiple landslide(s) (Fig. 13.13). Each pixel can be queried in the software to obtain information about the debris depth and the landslide number. Basic statistics are provided, including the number of times a pixel was occupied by an agent and the average, minimum, and maximum debris depths over all runs. Landslide-specific information, including the area and volumes of individual landslides (for single runs), can be exported to Excel. Landslide scenarios, the results of single or multiple runs, can be exported as shapefiles. Finally, the results can also be exported as geotiff to enable viewing in other software. The software does not output velocities.

13.6.5.4 Calibration and Evaluation

Like the other models described in this section, the calibration of DebrisFlow Predictor relies on trial and error. It is completed iteratively using the controls within the program.

The user (a landslide professional) compares the results of initial model runs to observed or mapped landslides or to morphological data provided by the DEM. Field evidence, geology, and geomorphology, combined with expert knowledge of debris flow processes, are all useful in distinguishing a credible model result from an unrealistic one. Calibration is also possible by comparing magnitude-frequency and volume-area relationships to existing landslide inventories. Once calibrated, the model can be deployed across a larger region. Recalibration may be necessary if local conditions result in different flow behavior from one region to the next.

13.6.6 Flow-Py

The Flow-Py modeling tool provides numerical run-out estimates of gravitational mass flows on the topography. It was developed at the Austrian Research Center for Forests and the University of Innsbruck and was published in 2022. The software is classified as a topography-based, empirical model, using the α-line approach (Lied & Bakkehøi, 1980) and the minimum flux criterion for maximum flow propagation, the algorithm of Holmgren (1994) for flow-path routing, and the persistence approach of Gamma (2000). The overall modeling philosophy is based on the energy line concept, where the energy line connects the starting zone with the toe of the mass flow, derived from the principle of conservation of energy. Derivation of the local energy content of the mass flow, however, would only yield plausible results for perfectly even run-off surfaces with uniformly distributed friction parameters.

The software was coded in Python, and the source code is freely available (D'Amboise et al., 2022). It has been applied for avalanches, rock falls, and shallow landslides for hazard identification at a regional and indicative scale. The authors suggest that debris flows and wet avalanches could also be assessed based on experience with similar modeling tools.

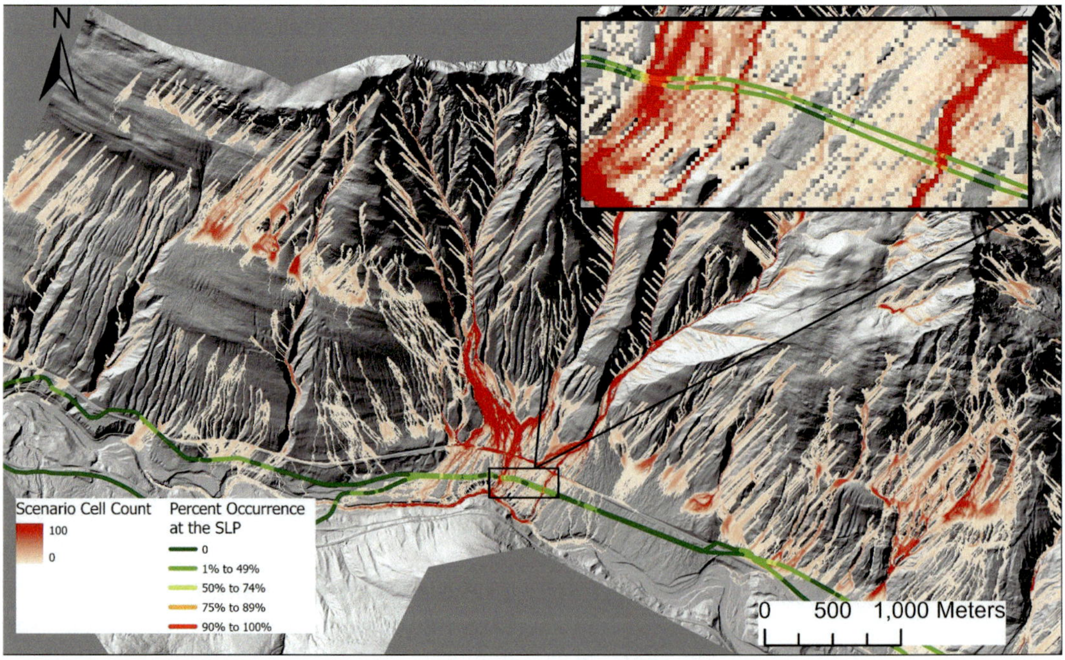

Fig. 13.13 An example output combining multiple scenarios and multiple runs to show the zonal susceptibility of a road to upslope debris-flow hazards

13.6.6.1 Components

Three approaches are used to compute flow propagation, accounting for the principle of the conservation of mass, topography-associated flow path identification (according to Holmgren, 1994), and consideration of the momentum of mass elements (persistence approach from Gamma, 2000). Conservation of mass is satisfied by balancing volumes stopping at cells as per stopping criteria and volume flux at each cell experiencing contact with the mass flow. Preliminary tests confirm that the selected persistence approach (Gamma, 2000) provides a workaround solution to represent the principle of conservation of momentum.

Two stopping criteria are used. First, to limit maximum flow propagation, the software uses an approximation of the bulk energy dissipation of the mass flow, represented by the α-line (Lied & Bakkehøi, 1980; Heim, 1932; Bakkehoi et al., 1983) as discussed in Sect. 13.4.2. Second, a minimum flux criterion is used to account for a required minimum energy content of flux allowing for further propagation.

13.6.6.2 Data Needs

Only two datasets are required to conduct runout computations: a DEM and a starting zone raster data set, as ascii file (.asc) or geotiff (.tiff).

13.6.6.3 Outputs

Flow-Py produces a set of output raster files with the following content:

- The local maximum flow energy.
- Local maximum routing flux for a cell considering all calculated releases and paths, respectively.
- Sum of locally detected residual energy heights of all mass flows passing a cell.
- Cell counts representing the number of paths that route flux through a location, where one path is produced by one release (one release raster cell) at the starting zones.
- Local maximum flow path travel angle detected at a cell, produced by all flow paths passing this cell. This value provides the spatial distribution of the maximum residual energy heights.

13.7 Discussion

13.7.1 Impact of the DEM Resolution and Accuracy

The DEM is often the primary resource for the different methods, and various terrain characteristics are derived from it, such as slope, curvature, or upslope contributing area. These characteristics can be heavily influenced by the DEM accuracy and resolution. This impacts the outputs of the source areas assessment and the runout computation. It was shown on multiple occasions that the DEM resolution, quality, and accuracy are critical factors in the modeling process (Huggel et al., 2003; Kappes et al., 2011). A poor DEM resolution has, for example, a substantial impact on the runout computation as some landscape features, such as a narrow ridge, might not be accurately represented, resulting in propagations in locations where they would be physically impossible. End-users might not be aware of such limitations and can put too much trust in the extent delineated by the models.

Although LiDAR DEMs nowadays provide high-quality data with high accuracy, such products are not available everywhere or can be expensive. Global DEM products available for free have a 30 m resolution at the time of writing and contain non-negligible errors, particularly in mountainous regions, which are the regions of interest. The DEM resolution is less of an issue for data-driven methods, though, where a 30 m resolution can be sufficient, particularly when working with slope units.

Regardless of the DEM quality, the resolution selected for a study impacts the results. DEMs are discrete representations of continuous surfaces, and their derivatives remain approximations of natural characteristics (Claessens et al., 2005). Their derived characteristics depict different spatial patterns and statistical distributions with different resolutions (Wilson, 1996). For example, the distribution of the slope gradients varies with the grid resolution as a coarser resolution smooths the local topographic variability (Wichmann et al., 2008). On the other hand, small-scale variability might be undesirable, such as high slope gradients on a single 1 m^2 grid cell for the identification of source areas. We aim at topographic characteristics representative of the phenomenon's scale. The topographic form variability for hillslopes was well captured by DEMs with a resolution of 10 m or smaller (Quinn et al., 1995; Zhang & Montgomery, 1994).

The DEM resolution also impacts the runout results, as shown by Wichmann et al. (2008) and Horton et al. (2013). Wichmann et al. (2008) showed that, for a few case study sites, small slope-type debris flows were best reproduced with DEM resolutions up to 5 m, but large debris fans were better modelled using lower grid resolutions, due to small-scale topographical features preventing the model from reaching certain parts of the debris fan. Horton et al. (2013) assessed the impact of the DEM resolution using Holmgren's enhanced flow direction algorithm with five different resolutions. While coarser resolutions (25 and 50 m) deteriorated the quality of the results, resolutions that were too high (in this case, 2 m) resulted in overly channelized propagations. The use of Holmgren's enhanced flow direction algorithm helps mitigate issues related to high-resolution DEMs by reducing the effect of small-scale roughness (Horton et al., 2013). However, the use of such high resolutions leads to time-consuming computations and may provide a false sense of precision. A resolution of 5 or 10 m was shown to be appropriate for regional modeling (Fischer et al., 2012; Horton et al., 2013).

Another aspect related to the impact of DEMs is their content, i.e., which type of surface is represented. In high-resolution DEMs, objects at the surface, such as buildings or forests, are removed, but some can remain in certain products, such as bridges. These objects can generate artifacts in the computation of debris-flow runouts.

13.7.2 Data Scarcity Issues

Data requirements differ from one method to another. Physically-based methods are the most demanding in terms of data quantity and accuracy at high resolution. Heuristic and empirical methods usually need less data but still with sufficient resolution and accuracy. In the case of data-driven models, both the explanatory and the target variables' information are required. The target variable relates to landslide inventories in this context. As primarily data-driven methods are sensitive to inventory completeness and quality, special attention will be given to them.

Nowadays, the availability of data representing landscape characteristics has drastically increased. However, it is still difficult to access quality data related to geological and lithological characteristics or anthropic effects, such as road cuts or land-use changes. For the latter, cloud computing resources can help explore past data and monitor changes in space and time with relative ease. Collecting predictors for data-driven models nowadays is a problem that can be largely addressed unless specific project requirements demand very high-resolution data that cannot be derived even from commercial satellites.

The main issue in landslide susceptibility modeling is the availability of reliable inventories. Although one can use semi-automated landslide mapping techniques, these are prone to error type I (false positives—they tend to map landslides where there are none). Thus, these techniques have not yet reached a level of accuracy for which one can entirely rely on remote sensing data (orthophotos or satellite scenes) to generate inventories of any site on earth. In data-scarce environments, or in places where reliable landslide inventories do not exist, establishing such an inventory requires considerable effort.

In addition, the qualification of an inventory as a good one is not obvious. Extensive literature on that topic only exists in the context of co-seismic landslides, where an inventory is defined as good or bad as a function of two semi-quantitative metrics: quality and completeness. However, due to different triggering factors, their definitions are not directly transferable to the context of debris flows. Irrespective of the current literature status, quality and completeness indicators pertain to the mapping effort and to the data used to support the mapping itself.

Hand-mapping over large areas is extremely time-consuming. Therefore, the question arises whether it would be possible to map fewer landslides than the theoretical total and still provide useful predictive maps for data-driven models. And if so, would the generated susceptibility map be reliable? These questions have been partially addressed in the geoscientific literature under the umbrella of inventory incompleteness and bias propagation issues (Guzzetti et al., 2012; Loche et al., 2022).

Some of the most evident sources of incomplete inventories are due to mapping preferences. Some inventories are established preferably or exclusively along road networks or in the vicinity of urban areas. Perhaps one could train a data-driven model using an inventory with the above characteristics. One can assume that if an explanatory variable exists in the model that is sensitive to the uneven or misrepresented distribution of landslides in the geographic space, then this variable would carry this signal onto the final susceptibility map, biasing the results. It can be, for instance, the case of land use, a common categorical property. This is one of the situations where the inappropriate use of any data-driven model would lead to analogous conclusions. Conversely, physically-based models would not induce the same problems.

As for when one should consider a landslide inventory complete enough for a data-driven model, the answer lies in the capacity of the inventory subset to mimic the covariates' variability. In general, it is simply the safest and best practice to systematically cover the whole study site. Generally, the coarser the mapping unit, the less the inventory needs to be complete.

13.7.3 Working in a Changing Environment

As shown in Sect. 13.6, several models are currently available for regional hazard assessment. These models commonly require DEM data to predict debris-flow paths. Regardless of the influence of the DEM spatial resolution, changes in the topography due to earthquakes, volcanic eruptions, avulsions, deposition of debris mass, and human constructions, lead to a discrepancy in debris-flow paths. Besides the DEM data, the potential source areas depend to some extent on the climate and whether rainfall-based parameters are considered in the chosen method. Thus, changing environments bring uncertainties to the key data used in regional hazard assessment models.

Topographic change due to earthquakes—Earthquakes reshape the landscape in two ways: by building the topography through surface uplift or by destroying it through the landslides they cause. Either way, the differences between pre- and post-seismic DEM cannot be ignored (Cucchiaro et al., 2019; Pineux et al., 2017). Li et al. (2018) compared the DEM data between 2002 and 2010 in the Wenchuan earthquake area, showing that 45% of the area experienced changes in topography, with fewer steep slopes. Topographic changes can influence both the assessment of the potential source areas and the debris-flow routing.

Topographic change due to avulsions—Avulsions are episodic shifts in the position of debris flow channels on fans. Some avulsions occur due to the deposition of sediment plugs and associated backstepping processes, which locally block channels and reduce the gradient, forcing later parts of the flow or subsequent flows into alternative flow paths (de Haas et al., 2018 and this volume). Temporal topographic changes associated with such mechanisms cannot be captured on a single DEM used for runout susceptibility, as runout susceptibility would need to be re-analyzed on DEMs featuring sediment plugs and backstepping sediment accumulations. This limitation may be addressed to a certain extent by adjusting flow direction parameters and allowing for more lateral spreading, or by introducing topographic plugs at specific locations, both of which could provide conservative results.

Volcanic ash deposition—Potential debris flow source areas are often related to assessing potential landslide release areas. In volcanic regions, a volcanic eruption can drastically change the sediment availability over a large territory. For example, volcanic ash accumulated in a catchment of the Sakurajima Island in Japan over a short period resulted in an increase in debris flows. According to Kisa et al. (2014), the volcanic ash accumulated to 60 kg/m^2 in 2012, which resulted in 55 reported debris flows in the Sakurajima region, many of which were unexpected. It shows that the present delineation of potential source areas may limit predicting future debris-flow paths, particularly in environments subject to such drastic changes.

Impact of climate change—Climate change impacts the environment in various ways. First, extreme precipitation events are expected to increase in frequency and intensity (Witze, 2018). Whether or not climatic factors, such as extreme precipitation values for certain return periods, are explicitly accounted for when assessing potential source areas, this increase may result in debris-flow triggering from locations where they had not been observed. An increase in hydroclimatic threshold exceedances for watersheds with supply limitations will produce more debris flows but with lower respective volumes resulting in lesser runout and lower overall susceptibility (see Stoffel et al., this volume). Historical data used to train the models, both in terms of climatic factors and debris-flow inventories, might underestimate or overestimate potential future debris flow susceptibility. Climate change is also affecting the cryosphere. For example, glacier retreat leaves behind potential instabilities and erodible material, and the melting permafrost could potentially release new material in locations where no debris flow has been observed yet.

Human activities—In some cases, human activities can contribute to causing debris flows (Sun et al., 2021). A classic example are cutslopes for road construction that reduce slope stability and alter surface hydrology. Soil denudation also increases erosion and, thus, sediment availability for debris flows. Such anthropic changes in the catchment can locally invalidate previously established susceptibility maps. Therefore, continuous efforts to update the susceptibility map might be necessary for some rapidly-changing environments.

13.7.4 Static Versus Dynamic and Susceptibility Versus Intensity Approaches

A general distinction between physically-based and data-driven models in the context of debris-flow prediction is that the former are capable of dynamically returning locations where the failure initiated as well as the evolution of the mass movement (Formetta et al., 2014; Tufano et al., 2021). Thus, one may obtain essential information about the threat a given debris flow may pose by estimating quantities such as extent, velocity, kinetic energy, force, arrival time, and runout distance (van den Bout et al., 2021). In the data-driven context, most of these quantities cannot be estimated simply because no observational big data exists on landslide kinematics. It is exactly for this reason that the part of the community working on physically-based models uses the most recent definition of landslide hazard proposed by Corominas et al. (2014), where the hazard is estimated in light of the dynamics of the phenomenon itself. Conversely, the part of the geoscientific community working on data-driven models leans towards the earlier definition proposed by Guzzetti et al. (1999). There, landslide hazard is the result of probabilistic analyses from which the expectation of where landslides may initiate and how frequently they may occur is combined with landslide size characteristics (mainly landslide area, although other metrics can also be found in the literature; e.g., Taylor et al., 2018), which can be easily derived from any landslide inventory.

This distinction between two scientific sides of the same coin has dominated the relevant scientific literature in the last five decades, with clear distinctions also on the applicability of the corresponding models. Physically-based ones were mostly employed for site- to basin-scale analyses, where the required geotechnical setup could be achieved. As for the data-driven counterpart, these methods were traditionally employed for large-scale analyses, from catchment to regional or even national and global scales. However, technological advancements are blurring the line drawn between the two. Physically-based models can nowadays rely on global estimates of soil parameters such as bulk density, thickness (de Sousa et al., 2020), or other parameters acquired through remote sensing. As a result, physically-based simulations of debris flows have become commonplace even over wide regions (Lehmann et al., 2019; Leonarduzzi et al., 2021).

The data-driven context is moving forward, touching upon what just a decade ago was uncharted territory and is currently being shaped in two different directions. The first attempts to reproduce the dynamic aspect typical of the physically-based context. This is achieved by data-driven modeling architectures and by training models that estimate the effect of landscape and rainfall characteristics on the landslide susceptibility pattern. Then, a subsequent simulation step recomputes the susceptibility by substituting previous rainfall measurements with future ones (Wang et al., 2022). This can be done at different temporal units depending on how the model has been trained. In turn, these new space–time or dynamic probabilistic estimates gain characteristics typical of susceptibility (featuring geomorphology and geological signals in the prediction) and of rainfall-triggering thresholds (Monsieurs et al., 2019; Segoni et al., 2018) models. It is worth noting that currently, these dynamic susceptibility models have been mostly tested over large temporal units (yearly or above), although their applicability is theoretically suitable also for near-real-time applications, as shown through the work of Khan et al. (2021).

The second research direction relates to the possibility of building artificially intelligent models capable of moving past the pure landslide occurrence probability framework behind the susceptibility concept and focusing instead on modeling other landslide characteristics. For instance, instead of modeling presence/absence data, one can extend this paradigm toward modeling counts. Lombardo et al. (2018) proposed a method to estimate how many debris flows may be triggered within a given slope unit, thus deriving estimates much closer to landslide intensity than to susceptibility (Hungr, 2018). Taking this to the spatio-temporal context returns dynamic estimates of landslide intensities (Lombardo et al., 2020). The same can be done to model landslide size instead. In this case, Lombardo et al. (2021) proposed the first model able to predict landslide areas per mapping units, and Dahal et al. (2022) have extended it towards doing the same in space and time.

Such extensions are important because landslide susceptibility has been largely assumed to be time-invariant. However, the recent space–time extensions have shown that we can now estimate susceptibility both in space and time. This brings the question of how much data-driven models (i.e., trained on past events) that have a spatio-temporal connotation can be relied on when estimating future occurrences. In fact, in a changing climate, the climatic conditions responsible for past debris flows may not be the same in a few years. In other words, even if the physics behind debris flow occurrences will stay the same, data-driven models may learn to replicate the landscape response from average rainfall events, whereas the future can see much more intense precipitation regimes. Overall, the capacity to extrapolate toward different climate scenarios will need to be tested as space–time data-driven models will consolidate their use within the geoscientific community. This will have to happen in accordance with an understanding of debris supply limitations and recharge rates (see Stoffel et al., this volume).

13.7.5 Limitations of Regional Methods

The regional hazard assessment methods introduced in this chapter are based on limited data that are available for large areas. Local watershed characteristics are typically not included in such assessments. For example, information about sediment supply in specific watersheds may not be available without more detailed site investigations or at least based on assumptions about the local geological conditions. If the sediment supply is small in a given watershed, the predicted debris flow source areas may overestimate the actual hazard for this watershed.

Runout susceptibility may not be able to account for certain site-specific conditions either. Sturzenegger et al. (2021) highlighted limitations of regional susceptibility results for: (1) atypical debris-flow paths, e.g., with sharp bends where debris-flow superelevation may cause overtopping of the channel (Nicol et al., 2013); (2) atypical watersheds with flatter sections/terraces alternating with steeper reaches; (3) atypical processes, such as debris flows initiating in volcanic complexes (Jakob et al., 2020); and (4) fans subject to different processes (e.g., debris flow and debris flood), and whose runout would be better calibrated and assessed separately.

In regional debris-flow runout assessments, the volume/mass is either ignored or roughly estimated using regional data. This is because the availability and characteristics of sediments at the torrent scale are difficult to assess over a large area. Therefore, regional hazard assessments should not be used to model specific events or design mitigation structures. Neither can regional models provide frequency-magnitude relationships (see Jakob and Nolde, this volume), which are required to generate composite hazard maps (i.e., Jakob et al., 2022) or to conduct quantitative risk assessments (see Strouth et al., this volume).

As defined in Sect. 13.1.1, the objective of the susceptibility map is to encompass all possible future events. The susceptibility map can be a first step in prioritizing areas requiring detailed debris flow risk mapping. As such, to avoid missing potential hazardous zones, conservatism may be applied. For example, one may calibrate runout susceptibility models based on the largest credible events, which correspond to the ones with long return periods (low frequency). Due to their low frequency, geomorphological evidence for such events may not be available within the study area without detailed field mapping. Consequently, it may be difficult to compare calibration scenarios with the potential runout and avulsion paths of large events. Either way, the practitioner must make it very clear where the limitations of the chosen regional models lie so that the end user does not misinterpret the maps.

13.8 Conclusions

Regional hazard analyses have to deal with broad scales, which come with data limitations and computational restrictions. Complex physically-based models are challenging to use in such contexts due to unknown local characteristics and long modeling times. Therefore, regional approaches are usually based on fewer available data over an entire region and simpler models. However, due to technological advances, more detailed and better data become available for large territories, and the growing computing power allows for increasing model complexity even at a regional scale. Therefore, a trend can be discerned toward increasing complexity both in physically-based models and data-driven ones.

Despite these recent evolutions, regional debris-flow hazard assessment is not likely to match detailed case study analyses even in the future due to a lack of information on the local environmental factors, which are mostly assessed doing fieldwork. However, although less precise, regional models provide valuable predictions that can be reasonably accurate at the appropriate scale and offer practical solutions to work over large areas. Such approaches are usable in data-scarce regions and offer spatially homogeneous susceptibility assessments. They are also valuable as a first assessment of potential hazards for subsequent refinement in areas of interest, such as inhabited areas. In this sense, they help prioritize areas where more detailed studies are needed. Moreover, they provide hazard assessments for remote locations where there is not enough at stake to invest money in an extensive study.

Several types of models and methods exist, such as empirical, statistical, or physically-based ones. These can target different applications: assessing potential debris-flow source areas, modeling the runout, or quantifying the overall slope unit susceptibility. All methods need a DEM, whose quality and precision can influence the results. Additional data often used in this context are land-use and geological (or lithological) maps. The models then result in the zoning of areas susceptible to debris flows, often with additional information. This zoning is most often expressed in terms of susceptibility, which quantifies less the temporal frequency associated with potential events than the probabilities. However, in a changing environment—because of land cover changes or climate change, for example—static susceptibility maps can show some limitations. Dynamic approaches are being considered to account for these changes.

Acknowledgements Funds from the Geological Survey of Canada (A. Blais-Stevens' Pipeline Geohazards project GSC-19-103) contributed to Flow-R analyses for Mill Creek.

References

Bai, T., Jiang, Z., & Tahmasebi, P. (2021). Debris flow prediction with machine learning: Smart management of urban systems and infrastructures. *Neural Computing and Applications, 33*(22), 15769–15779. https://doi.org/10.1007/s00521-021-06197-y

Bakkehoi, S., Domaas, U., & Lied, K. (1983). Calculation of snow avalanche runout distance. *Annals of Glaciology, 4*, 24–29.

Bauer, E., & Kohavi, R. (1999). Empirical comparison of voting classification algorithms: Bagging, boosting, and variants. *Machine Learning, 36*(1), 105–139. https://doi.org/10.1023/a:1007515423169

Bera, S., Upadhyay, V. K., Guru, B., & Oommen, T. (2021). Landslide inventory and susceptibility models considering the landslide typology using deep learning: Himalayas, India. In *Natural Hazards* (Vol. 108, Issue 1). Springer Netherlands. https://doi.org/10.1007/s11069-021-04731-8

Blahut, J., Horton, P., Sterlacchini, S., & Jaboyedoff, M. (2010a). Debris flow hazard modelling on medium scale: Valtellina di Tirano, Italy. *Natural Hazards and Earth System Sciences, 10*(11), 2379–2390. https://doi.org/10.5194/nhess-10-2379-2010

Blahut, J., van Westen, C. J., & Sterlacchini, S. (2010b). Analysis of landslide inventories for accurate prediction of debris-flow source areas. *Geomorphology, 119*(1–2), 36–51. https://doi.org/10.1016/j.geomorph.2010.02.017

Blais-Stevens, A., & Behnia, P. (2016). Debris flow susceptibility mapping using a qualitative heuristic method and Flow-R along the Yukon Alaska Highway Corridor, Canada. *Natural Hazards and Earth System Sciences, 16*(2), 449–462. https://doi.org/10.5194/nhess-16-449-2016

Bonabeau, E. (2002). Agent-based modeling: Methods and techniques for simulating human systems. *Proceedings of the National Academy of Sciences of the United States of America, 99*(SUPPL. 3), 7280–7287. https://doi.org/10.1073/pnas.082080899

Bottou, L. (2012). Stochastic gradient descent tricks. In G. Montavon, G. B. Orr, & K. Müller (Eds.), *Neural networks: Tricks of the trade*. Lecture Notes in Computer Science (Vol. 7700, pp. 421–436). Springer. https://doi.org/10.1007/978-3-642-35289-8_25

Brenning, A. (2009). Benchmarking classifiers to optimally integrate terrain analysis and multispectral remote sensing in automatic rock glacier detection. *Remote Sensing of Environment, 113*(1), 239–247. https://doi.org/10.1016/j.rse.2008.09.005

Cama, M., Lombardo, L., Conoscenti, C., & Rotigliano, E. (2017). Improving transferability strategies for debris flow susceptibility assessment Application to the Saponara and Itala catchments (Messina, Italy). *Geomorphology, 288*, 52–65. https://doi.org/10.1016/j.geomorph.2017.03.025

Carrara, A., Crosta, G., & Frattini, P. (2008). Comparing models of debris-flow susceptibility in the alpine environment. *Geomorphology, 94*(3–4), 353–378. https://doi.org/10.1016/j.geomorph.2006.10.033

Carson, M. A., & Kirkby, M. J. (1972). *Hillslope form and process*. Cambridge University Press.

Chacón, J., Irigaray, C., Fernández, T., & El Hamdouni, R. (2006). Engineering geology maps: Landslides and geographical information systems. *Bulletin of Engineering Geology and the Environment, 65*(4), 341–411. https://doi.org/10.1007/s10064-006-0064-z

Chen, L.-C., Papandreou, G., Schroff, F., & Adam, H. (2017). *Rethinking Atrous convolution for semantic image segmentation*. http://arxiv.org/abs/1706.05587

Chen, W., Chen, Y., Tsangaratos, P., Ilia, I., & Wang, X. (2020). Combining evolutionary algorithms and machine learning models in landslide susceptibility assessments. *Remote Sensing, 12*(23), 1–26. https://doi.org/10.3390/rs12233854

Christen, M., Kowalski, J., & Bartelt, P. (2010). RAMMS: Numerical simulation of dense snow avalanches in three-dimensional terrain. *Cold Regions Science and Technology, 63*(1–2), 1–14. https://doi.org/10.1016/j.coldregions.2010.04.005

Claessens, L., Heuvelink, G. B. M., Schoorl, J. M., & Veldkamp, A. (2005). DEM resolution effects on shallow landslide hazard and soil redistribution modelling. *Earth Surface Processes and Landforms, 30*(4), 461–477. https://doi.org/10.1002/esp.1155

Coe, J. A., Kean, J. W., Godt, J. W., Baum, R. L., Jones, E. S., Gochis, D. J., & Anderson, G. S. (2014). New insights into debris-flow hazards from an extraordinary event in the Colorado front range. *GSA Today, 24*(10), 4–10. https://doi.org/10.1130/GSATG214A.1

Conrad, O., Bechtel, B., Bock, M., Dietrich, H., Fischer, E., Gerlitz, L., Wehberg, J., Wichmann, V., & Böhner, J. (2015). System for automated geoscientific analyses (SAGA) v. 2.1.4. *Geoscientific Model Development, 8*, 1991–2007. https://doi.org/10.5194/gmd-8-1991-2015

Corominas, J. (1996). The angle of reach as a mobility index for small and large landslides. *Canadian Geotechnical Journal, 33*(2), 260–271. https://doi.org/10.1139/t96-005

Corominas, J., van Westen, C. J., Frattini, P., Cascini, L., Malet, J.-P., Fotopoulou, S., Catani, F., Van Den Eeckhaut, M., Mavrouli, O., Agliardi, F., Pitilakis, K., Winter, M. G., Pastor, M., Ferlisi, S., Tofani, V., Hervás, J., & Smith, J. T. (2014). Recommendations for the quantitative analysis of landslide risk. *Bulletin of Engineering Geology and the Environment, 73*(2), 209–263. https://doi.org/10.1007/s10064-013-0538-8

Crozier, M. J. (2010). Deciphering the effect of climate change on landslide activity: A review. *Geomorphology, 124*(3–4), 260–267. https://doi.org/10.1016/j.geomorph.2010.04.009

Cucchiaro, S., Cavalli, M., Vericat, D., Crema, S., Llena, M., Beinat, A., Marchi, L., & Cazorzi, F. (2019). Geomorphic effectiveness of check dams in a debris-flow catchment using multi-temporal topographic surveys. *Catena, 174*(February 2018), 73–83. https://doi.org/10.1016/j.catena.2018.11.004

D'Amboise, C. J. L., Neuhauser, M., Teich, M., Huber, A., Kofler, A., Perzl, F., Fromm, R., Kleemayr, K., & Fischer, J. (2022). Flow-Py v1.0: a customizable, open-source simulation tool to estimate runout and intensity of gravitational mass flows. *Geoscientific Model Development, 15*(6), 2423–2439. https://doi.org/10.5194/gmd-15-2423-2022

D'Ambrosio, D., Di Gregorio, S., & Iovine, G. (2003). Simulating debris flows through a hexagonal cellular automata model: SCIDDICA S3–hex. *Natural Hazards and Earth System Sciences, 3*(6), 545–559. https://doi.org/10.5194/nhess-3-545-2003

Dahal, R. K., Hasegawa, S., Nonomura, A., Yamanaka, M., Dhakal, S., & Paudyal, P. (2008). Predictive modelling of rainfall-induced landslide hazard in the Lesser Himalaya of Nepal based on weights-of-evidence. *Geomorphology, 102*(3–4), 496–510. https://doi.org/10.1016/j.geomorph.2008.05.041

Dahal, A., Tanyas, H., van Westen, C. J., van der Meijde, M., Mai, P. M., Huser, R., & Lombardo, L. (2022). Space-time landslide hazard modeling via Ensemble Neural Networks. Earth-arxiv. https://doi.org/10.31223/X5B075

Dahal, A., & Lombardo, L. (2023). Explainable artificial intelligence in geoscience: A glimpse into the future of landslide susceptibility modeling. *Computers & Geosciences, 176*, 105364. https://doi.org/10.1016/j.cageo.2023.105364

Dai, F. C., & Lee, C. F. (2001). Terrain-based mapping of landslide susceptibility using a geographical information system: A case study. *Canadian Geotechnical Journal, 38*(5), 911–923. https://doi.org/10.1139/cgj-38-5-911

Dash, P. K., Samantaray, S. R., & Panda, G. (2007). Fault classification and section identification of an advanced series-compensated transmission line using support vector machine. *IEEE Transactions on Power Delivery, 22*(1), 67–73. https://doi.org/10.1109/TPWRD.2006.876695

de Haas, T., Densmore, A. L., Stoffel, M., Suwa, H., Imaizumi, F., Ballesteros-Cánovas, J. A., & Wasklewicz, T. (2018). Avulsions and the spatio-temporal evolution of debris-flow fans. *Earth-Science Reviews, 177*(November 2017), 53–75. https://doi.org/10.1016/j.earscirev.2017.11.007

de Sousa, L., Poggio, L., Batjes, N., Heuvelink, G., Kempen, B., Riberio, E., & Rossiter, D. (2020). *SoilGrids 2.0: Producing quality-assessed soil information for the globe*. SOIL Discussions, November, 1–37.

Delestre, O., Darboux, F., James, F., Lucas, C., Laguerre, C., & Cordier, S. (2017). FullSWOF: Full shallow-water equations for overland flow. *The Journal of Open Source Software, 2*(20), 448. https://doi.org/10.21105/joss.00448

Delmonaco, G., Leoni, G., Margottini, C., Puglisi, C., & Spizzichino, D. (2003). Large scale debris-flow hazard assessment: A geotechnical approach and GIS modelling. *Natural Hazards and Earth System Sciences, 3*, 443–455.

Domènech, G., Alvioli, M., & Corominas, J. (2020). Preparing first-time slope failures hazard maps: From pixel-based to slope unit-based. *Landslides, 17*(2), 249–265. https://doi.org/10.1007/s10346-019-01279-4

Dou, J., Yunus, A. P., Bui, D. T., Merghadi, A., Sahana, M., Zhu, Z., Chen, C. W., Han, Z., & Pham, B. T. (2020). Improved landslide assessment using support vector machine with bagging, boosting, and stacking ensemble machine learning framework in a mountainous watershed, Japan. *Landslides, 17*(3), 641–658. https://doi.org/10.1007/s10346-019-01286-5

Duncan, J. M., & Wright, S. G. (2005). *Soil strength and slope stability*. Wiley.

Fairfield, J., & Leymarie, P. (1991). Drainage networks from grid digital elevation models. *Water Resources Research, 27*(5), 709–717. https://doi.org/10.1029/90WR02658/full

Fell, R., Corominas, J., Bonnard, C., Cascini, L., Leroi, E., & Savage, W. Z. (2008a). Guidelines for landslide susceptibility, hazard and risk zoning for land use planning. *Engineering Geology, 102*(3–4), 85–98. https://doi.org/10.1016/j.enggeo.2008.03.022

Fell, R., Corominas, J., Bonnard, C., Cascini, L., Leroi, E., & Savage, W. Z. (2008b). Guidelines for landslide susceptibility, hazard and risk zoning for land-use planning. *Engineering Geology, 102*(3–4), 99–111. https://doi.org/10.1016/j.enggeo.2008.03.014

Fischer, L., Rubensdotter, L., Sletten, K., Stalsberg, K., Horton, P., & Jaboyedoff, M. (2012). Debris flow modeling for susceptibility mapping at regional to national scale in Norway. In *Proceedings of the 11th International and 2nd North American Symposium on Landslides* (pp. 723–729).

Fookes, P. G., Lee, E. M., & Griffiths, J. S. (2007). *Engineering geomorphology: Theory and practice* (p. 288). Whittles Publishing.

Formetta, G., Rago, V., Capparelli, G., Rigon, R., Muto, F., & Versace, P. (2014). Integrated physically based system for modeling landslide susceptibility. *Procedia Earth and Planetary Science, 9*, 74–82. https://doi.org/10.1016/j.proeps.2014.06.006

Freeman, T. G. (1991). Calculating catchment area with divergent flow based on a regular grid. *Computers & Geosciences, 17*(3), 413–422.

Gamma, P. (2000). dfwalk-Ein Murgang-Simulationsprogramm zur Gefahrenzonierung. Geographisches Institut der Universität Bern.

Gaume, J., van Herwijnen, A., Gast, T., Teran, J., & Jiang, C. (2019). Investigating the release and flow of snow avalanches at the slope-scale using a unified model based on the material point method. *Cold Regions Science and Technology, 168*(April), 102847. https://doi.org/10.1016/j.coldregions.2019.102847

George, D. L., & Iverson, R. M. (2014). A depth-averaged debris-flow model that includes the effects of evolving dilatancy. II. Numerical predictions and experimental tests. *Proceedings of the Royal Society A: Mathematical, Physical and Engineering Sciences, 470*(2170). https://doi.org/10.1098/rspa.2013.0820

Godt, J. W., Baum, R. L., Savage, W. Z., Salciarini, D., Schulz, W. H., & Harp, E. L. (2008). Transient deterministic shallow landslide modeling: Requirements for susceptibility and hazard assessments in a GIS framework. *Engineering Geology, 102*(3–4), 214–226. https://doi.org/10.1016/j.enggeo.2008.03.019

Goetz, J. N., Guthrie, R. H., & Brenning, A. (2011). Integrating physical and empirical landslide susceptibility models using generalized additive models. *Geomorphology, 129*(3–4), 376–386. https://doi.org/10.1016/j.geomorph.2011.03.001

Goetz, J., Kohrs, R., Parra Hormazábal, E., Bustos Morales, M., Araneda Riquelme, M. B., Henríquez, C., & Brenning, A. (2021). Optimizing and validating the gravitational process path model for regional debris-flow runout modelling. *Natural Hazards and Earth Systems Sciences, 21*(8), 2543–2562.

Green, H., & Ampt, G. A. (1912). Studies on soil physics: Part II—the permeability of an ideal soil to air and water. *The Journal of Agricultural Science, 5*(1), 1–26. https://doi.org/10.1017/S0021859600001751

Gregoretti, C., Degetto, M., & Boreggio, M. (2016). GIS-based cell model for simulating debris flow runout on a fan. *Journal of Hydrology, 534*, 326–340. https://doi.org/10.1016/j.jhydrol.2015.12.054

Guinau, M., Vilajosana, I., & Vilaplana, J. M. (2007). GIS-based debris flow source and runout susceptibility assessment from DEM data—A case study in NW Nicaragua. *Natural Hazards and Earth System Sciences, 7*(6), 703–716. https://doi.org/10.5194/nhess-7-703-2007

Guthrie, R. H. (2002). The effects of logging on frequency and distribution of landslides in three watersheds on Vancouver Island, British Columbia. *Geomorphology, 43*(3–4), 273–292. https://doi.org/10.1016/S0169-555X(01)00138-6

Guthrie, R. H., & Befus, A. (2021). DebrisFlow Predictor: An agent-based runout program for shallow landslides. *Natural Hazards and Earth System Sciences, 21*(3), 1029–1049. https://doi.org/10.5194/nhess-21-1029-2021

Guthrie, R. H., Deadman, P. J., Cabrera, A. R., & Evans, S. G. (2008). Exploring the magnitude-frequency distribution: A cellular automata model for landslides. *Landslides, 5*(1), 151–159. https://doi.org/10.1007/s10346-007-0104-1

Guthrie, R. H., Hockin, A., Colquhoun, L., Nagy, T., Evans, S. G., & Ayles, C. (2010). An examination of controls on debris flow mobility: Evidence from coastal British Columbia. *Geomorphology, 114*(4), 601–613. https://doi.org/10.1016/j.geomorph.2009.09.021

Guzzetti, F., Carrara, A., Cardinali, M., & Reichenbach, P. (1999). Landslide hazard evaluation: A review of current techniques and their application in a multi-scale study, Central Italy. *Geomorphology, 31*(1–4), 181–216. https://doi.org/10.1016/S0169-555X(99)00078-1

Guzzetti, F., Reichenbach, P., Ardizzone, F., Cardinali, M., & Galli, M. (2006). Estimating the quality of landslide susceptibility models. *Geomorphology, 81*(1–2), 166–184. https://doi.org/10.1016/j.geomorph.2006.04.007

Guzzetti, F., Mondini, A. C., Cardinali, M., Fiorucci, F., Santangelo, M., & Chang, K. T. (2012). Landslide inventory maps: New tools for an old problem. *Earth-Science Reviews, 112*(1–2), 42–66. https://doi.org/10.1016/j.earscirev.2012.02.001

Han, Z., Li, Y., Huang, J., Chen, G., Xu, L., Tang, C., Zhang, H., & Shang, Y. (2017). Numerical simulation for run-out extent of debris flows using an improved cellular automaton model. *Bulletin of Engineering Geology and the Environment, 76*(3), 961–974. https://doi.org/10.1007/s10064-016-0902-6

Hastie, T., Tibshirani, R., & Friedman, J. (2009). The elements of statistical. *Learning*. https://doi.org/10.1007/978-0-387-84858-7

He, K., Zhang, X., Ren, S., & Sun, J. (2016). Identity mappings in deep residual networks. In B. Leibe,

J. Matas, N. Sebe, & M. Welling (Eds.), *Computer Vision – ECCV 2016*. ECCV 2016. Lecture Notes in Computer Science (Vol. 9908, pp. 630–645). https://doi.org/10.1007/978-3-319-46493-0_38

Heim, A. (1932). Bergsturz und Menschenleben. Beiblatt zur Vierteljahrsschrift der Naturforschenden Gesellschaft in Zurich, 1932 (in German).

Heinimann, H. R., Hollenstein, K., Kienholz, H., Krummenacher, B., & Mani, P. (1998). Methoden zur Analyse und Bewertung von Naturgefahren. *Eine Risikoorientierte Betrachtungsweise. Umwelt-Materialien, 85*, 248. (in German).

Holmgren, P. (1994). Multiple flow direction algorithms for runoff modelling in grid based elevation models: An empirical evaluation. *Hydrological Processes, 8*, 327–334. https://doi.org/10.1002/hyp.3360080405

Horton (2022). Flow-R web page. Retrieved December 15, 2022 from https://www.flow-r.org/

Horton, P., Jaboyedoff, M., & Bardou, E. (2008). Debris flow susceptibility mapping at a regional scale. In *4th Canadian Conference on Geohazards* (pp. 399–406).

Horton, P., Jaboyedoff, M., Zimmermann, M., Mazotti, B., & Longchamp, C. (2011). Flow-R, a model for debris flow susceptibility mapping at a regional scale - some case studies. In *Proceedings of the 5th International Conference on Debris-Flow Hazards Mitigation: Mechanics, Prediction and Assessment - Italian Journal of Engineering Geology and Environment*, 875–884. https://doi.org/10.4408/IJEGE.2011-03.B-095

Horton, P., Jaboyedoff, M., Rudaz, B., & Zimmermann, M. (2013). Flow-R, a model for susceptibility mapping of debris flows and other gravitational hazards at a regional scale. *Natural Hazards and Earth System Sciences, 13*(4), 869–885. https://doi.org/10.5194/nhess-13-869-2013

Howes, D. E., & Kenk, E. (1997). Terrain classification system for British Columbia, Version 2. Resource Inventory Branch, Ministry of Environment Lands and Parks, Victoria, BC, 114 pp. https://www2.gov.bc.ca/assets/gov/environment/natural-resource-stewardship/nr-laws-policy/risc/terclass_system_1997.pdf

Huggel, C., Kääb, A., Haeberli, W., & Krummenacher, B. (2003). Regional-scale GIS-models for assessment of hazards from glacier lake outbursts: Evaluation and application in the Swiss Alps. *Natural Hazards and Earth System Sciences, 3*(6), 647–662. https://doi.org/10.5194/nhess-3-647-2003

Hungr, O. (1995). A model for the runout analysis of rapid flow slides, debris flows, and avalanches. *Canadian Geotechnical Journal, 32*(4), 610–623. https://doi.org/10.1139/t95-063

Hungr, O. (2018). Some methods of landslide hazard intensity mapping. In *Landslide risk assessment* (pp. 215–226). Routledge.

Hungr, O., & Evans (1988): Engineering evaluation of fragmental rockfall hazards. In *Proceedings of the 5th International Symposium on Landslides*, Lausanne, Switzerland, 10–15 July 1988 (pp. 685–690).

Hürlimann, M., Rickenmann, D., Medina, V., & Bateman, A. (2008). Evaluation of approaches to calculate debris-flow parameters for hazard assessment. *Engineering Geology, 102*(3–4), 152–163. https://doi.org/10.1016/j.enggeo.2008.03.012

Iverson, R. M. (1997). The physics of debris flows. *Reviews of Geophysics, 35*(3), 245–296. https://doi.org/10.1029/97RG00426

Iverson, R. M., & George, D. L. (2014). A depth-averaged debris-flow model that includes the effects of evolving dilatancy. I. Physical basis. *Proceedings of the Royal Society A: Mathematical, Physical and Engineering Sciences, 470*(2170). https://doi.org/10.1098/rspa.2013.0819

Jakob, M. (2000). The impacts of logging on landslide activity at Clayoquot Sound, British Columbia. *CATENA, 38*(4), 279–300. https://doi.org/10.1016/S0341-8162(99)00078-8

Jakob, M., Davidson, S., Bullard, G., Busslinger, M., Collier-Pandya, B., Grover, P., & Lau, C. A. (2022). Debris-flood hazard assessments in steep streams. *Water Resources Research, 58*(4), 1–31. https://doi.org/10.1029/2021WR030907

Jakob, M., Mark, E., McDougall, S., Friele, P., Lau, C. A., & Bale, S. (2020). Regional debris-flow and debris-flood frequency–magnitude relationships. *Earth Surface Processes and Landforms, 45*(12), 2954–2964. https://doi.org/10.1002/esp.4942

Kappes, M. S., Malet, J.-P., Remaître, A., Horton, P., Jaboyedoff, M., & Bell, R. (2011). Assessment of debris-flow susceptibility at medium-scale in the Barcelonnette Basin, France. *Natural Hazards and Earth System Sciences, 11*(2), 627–641. https://doi.org/10.5194/nhess-11-627-2011

Khan, S., Kirschbaum, D. B., & Stanley, T. (2021). Investigating the potential of a global precipitation forecast to inform landslide prediction. *Weather and Climate Extremes, 33*(September 2020), 100364. https://doi.org/10.1016/j.wace.2021.100364

Kingma, D. P., & Ba, J. L. (2015). Adam: A method for stochastic optimization. In *3rd International Conference on Learning Representations, ICLR 2015 - Conference Track Proceedings* (pp. 1–15). https://doi.org/10.48550/arxiv.1412.6980

Kisa, H., Yamakoshi, T., & Ishizuka, T. (2014). Impact of short-term temporal changes in volcanic ash fall on rainfall threshold for debris flow occurrence in Sakurajima, Japan. *International Journal of Erosion Control Engineering, 7*(3), 75–84. https://doi.org/10.13101/ijece.7.75

Kornejady, A., Ownegh, M., & Bahremand, A. (2017). Landslide susceptibility assessment using maximum entropy model with two different data sampling methods. *CATENA, 152*, 144–162. https://doi.org/10.1016/j.catena.2017.01.010

Körner, H. J. (1980). The energy-line method in the mechanics of avalanches. *Journal of Glaciology, 26*, 501–505.

Lee, S., Baek, W. K., Jung, H. S., & Lee, S. (2020). Susceptibility mapping on urban landslides using deep learning approaches in mt. Umyeon. *Applied Sciences (switzerland), 10*(22), 1–18. https://doi.org/10.3390/app10228189

Lehmann, P., von Ruette, J., & Or, D. (2019). Deforestation effects on rainfall-induced shallow landslides: Remote sensing and physically-based modelling. *Water Resources Research, 55*(11), 9962–9976. https://doi.org/10.1029/2019WR025233

Leonarduzzi, E., McArdell, B. W., & Molnar, P. (2021). Rainfall-induced shallow landslides and soil wetness: Comparison of physically based and probabilistic predictions. *Hydrology and Earth System Sciences, 25*(11), 5937–5950. https://doi.org/10.5194/hess-25-5937-2021

Li, C., Wang, M., Liu, K., & Xie, J. (2018). Topographic changes and their driving factors after 2008 Wenchuan earthquake. *Geomorphology, 311*, 27–36. https://doi.org/10.1016/j.geomorph.2018.03.019

Li, Y., Chen, W., Rezaie, F., Rahmati, O., Davoudi Moghaddam, D., Tiefenbacher, J., Panahi, M., Lee, M. J., Kulakowski, D., Tien Bui, D., & Lee, S. (2021). Debris flows modeling using geo-environmental factors: Developing hybridized deep-learning algorithms. *Geocarto International*, 1–25. https://doi.org/10.1080/10106049.2021.1912194

Lied, K., & Bakkehøi, S. (1980). Empirical calculations of snow-avalanche run-out distance based on topographic parameters. *Journal of Glaciology, 26*(94), 165–177.

Loche, M., Alvioli, M., Marchesini, I., Bakka, H., & Lombardo, L. (2022). Landslide susceptibility maps of Italy: Lesson learnt from dealing with multiple landslide types and the uneven spatial distribution of the national inventory. *Earth-Science Reviews, 232*(February), 104125. https://doi.org/10.1016/j.earscirev.2022.104125

Lombardo, L., Opitz, T., Ardizzone, F., Guzzetti, F., & Huser, R. (2020). Space-time landslide predictive modelling. *Earth-Science Reviews, 209*(6), 103318. https://doi.org/10.1016/j.earscirev.2020.103318

Lombardo, L., Opitz, T., & Huser, R. (2018). Point process-based modeling of multiple debris flow landslides using INLA: An application to the 2009 Messina disaster. *Stochastic Environmental Research and Risk Assessment, 32*(7), 2179–2198. https://doi.org/10.1007/s00477-018-1518-0

Lombardo, L., Tanyas, H., Huser, R., Guzzetti, F., & Castro-Camilo, D. (2021). Landslide size matters: A new data-driven, spatial prototype. *Engineering Geology, 293*(May). https://doi.org/10.1016/j.enggeo.2021.106288

Macal, C. M., & North, M. J. (2009). Agent-based modeling and simulation. In *Proceedings of the 2009 Winter Simulation Conference (WSC)* (pp. 86–98). https://doi.org/10.1109/WSC.2009.5429318

Mergili. (2022). r.randomwalk - The landslide routing tool. Retrieved December 15, 2022, from https://www.landslidemodels.org/r.randomwalk/

Mergili, M., Krenn, J., & Chu, H. J. (2015). R.randomwalk v1, a multi-functional conceptual tool for mass movement routing. *Geoscientific Model Development, 8*(12), 4027–4043. https://doi.org/10.5194/gmd-8-4027-2015

Mergili, M., Schwarz, L., & Kociu, A. (2019). Combining release and runout in statistical landslide susceptibility modeling. *Landslides, 16*(11), 2151–2165. https://doi.org/10.1007/s10346-019-01222-7

Mergili, & Pudasaini. (2022). r.avaflow - The mass flow simulation tool. Retrieved December 15, 2022, from https://www.landslidemodels.org/r.avaflow/

Michoud, C., Derron, M.-H., Horton, P., Jaboyedoff, M., Baillifard, F.-J., Loye, A., et al. (2012). Rockfall hazard and risk assessments along roads at a regional scale: Example in Swiss Alps. *Natural Hazards and Earth System Sciences, 12*(3), 615–629. https://doi.org/10.5194/nhess-12-615-2012

Milledge, D. G., Griffiths, D. V., Lane, S. N., & Warburton, J. (2012). Limits on the validity of infinite length assumptions for modelling shallow landslides. *Earth Surface Processes and Landforms, 37*(11), 1158–1166. https://doi.org/10.1002/esp.3235

Monsieurs, E., Dewitte, O., & Demoulin, A. (2019). A susceptibility-based rainfall threshold approach for landslide occurrence. *Natural Hazards and Earth System Sciences, 19*(4), 775–789. https://doi.org/10.5194/nhess-19-775-2019

Montgomery, D. R., & Dietrich, W. E. (1994). A physically based model for the topographic control on shallow landsliding. *Water Resources Research, 30*(4), 1153–1171. https://doi.org/10.1029/93WR02979

Nicol, D., Jordan, P., Boyer, D., & Yonin, D. (2013). Johnsons landing landslide hazard and risk assessment. Report prepared for the Regional District of Central Kootenay.

O'Callaghan, J. F., & Mark, D. M. (1984). The extraction of drainage networks from digital elevation data. *Computer Vision Graphics and Image Processing, 28*(3), 323–344. https://doi.org/10.1016/S0734-189X(84)80011-0

Ortigao, J. A. R., & Kanji, M. A. (2004). Landslide classification and risk management. In *Handbook of slope stabilisation* (pp. 27–66). https://doi.org/10.1007/978-3-662-07680-4_3

Pack, R. T., Tarboton, D. G., & Goodwin, C. N. (1998). The SINMAP approach to terrain stability mapping. In *8th Congress of the International Association of Engineering Geology*, Vancouver, BC (8 pp.).

Pastorello, R., Michelini, T., & D'Agostino, V. (2017). On the criteria to create a susceptibility map to debris flow at a regional scale using Flow-R. *Journal of Mountain Science, 14*(4), 621–635. https://doi.org/10.1007/s11629-016-4077-1

Perla, R., Cheng, T. T., & McClung, D. M. (1980). A two-parameter model of snow-avalanche motion. *Journal of Glaciology, 26*, 197–207.

Pineux, N., Lisein, J., Swerts, G., Bielders, C. L., Lejeune, P., Colinet, G., & Degré, A. (2017). Can DEM time series produced by UAV be used to quantify diffuse erosion in an agricultural watershed? *Geomorphology, 280*, 122–136. https://doi.org/10.1016/j.geomorph.2016.12.003

Pitman, E. B., & Long, L. E. (2005). A two-fluid model for avalanche and debris flows. *Philosophical Transactions of the Royal Society a: Mathematical, Physical and Engineering Sciences, 363*(1832), 1573–1601. https://doi.org/10.1098/rsta.2005.1596

Prochaska, A. B., Santi, P. M., Higgins, J. D., & Cannon, S. H. (2008). Debris-flow runout predictions based on the average channel slope (ACS). *Engineering Geology, 98*(1–2), 29–40. https://doi.org/10.1016/j.enggeo.2008.01.011

Pudasaini, S. P. (2012). A general two-phase debris flow model. *Journal of Geophysical Research: Earth Surface, 117*(3), 1–28. https://doi.org/10.1029/2011JF002186

Pudasaini, S. P., & Mergili, M. (2019). A multi-phase mass flow model. *Journal of Geophysical Research: Earth Surface, 124*(12), 2920–2942. https://doi.org/10.1029/2019JF005204

Qing, F., Zhao, Y., Meng, X., Su, X., Qi, T., & Yue, D. (2020). Application of machine learning to debris flow susceptibility mapping along the China-Pakistan Karakoram Highway. *Remote Sensing, 12*(18). https://doi.org/10.3390/RS12182933

Quinn, P., Beven, K., Chevallier, P., & Planchon, O. (1991). The prediction of hillslope flow paths for distributed hydrological modelling using digital terrain models. *Hydrological Processes, 5*(1), 59–79. https://doi.org/10.1002/hyp.3360050106

Quinn, P. F., Beven, K. J., & Lamb, R. (1995). The in(a/tan/β) index: How to calculate it and how to use it within the topmodel framework. *Hydrological Processes, 9*(2), 161–182. https://doi.org/10.1002/hyp.3360090204

Regmi, N. R., Giardino, J. R., & Vitek, J. D. (2010). Modeling susceptibility to landslides using the weight of evidence approach: Western Colorado, USA. *Geomorphology, 115*(1–2), 172–187. https://doi.org/10.1016/j.geomorph.2009.10.002

Rickenmann, D. (1990). Debris flows 1987 in Switzerland: Modelling and fluvial sediment transport. IAHS-AISH P., (pp. 194, 371–378).

Rickenmann, D. (2005). Runout prediction methods. In *Debris-flow hazards and related phenomena* (pp. 305–324).

Rickenmann, D., & Zimmermann, M. (1993). The 1987 debris flows in Switzerland: Documentation and analysis. *Geomorphology, 8*(2–3), 175–189.

Ronneberger, O., Fischer, P., & Brox, T. (2015). U-Net: Convolutional networks for biomedical image segmentation. In N. Navab, J. Hornegger, W. M. Wells, & A. F. Frangi (Eds.), *Medical Image Computing and Computer-Assisted Intervention -- MICCAI 2015* (pp. 234–241). Springer International Publishing. https://doi.org/10.1007/978-3-319-24574-4_28

Santec. (2022). DebrisFlow Predictor. Retrieved June 27, 2022, from https://www.stantec.com/en/services/debrisflow-predictor

Savage, S. B., & Hutter, K. (1989). The motion of a finite mass of granular material down a rough incline. *Journal of Fluid Mechanics, 199*(2697), 177–215. https://doi.org/10.1017/S0022112089000340

Scheidegger, A. E. (1975). *Physical aspects of natural catastrophes*. Elsevier Publishing Company.

Scheidl, C., & Rickenmann, D. (2010). Empirical prediction of debris-flow mobility and deposition on fans. *Earth Surface Processes and Landforms, 35*(2), 157–173. https://doi.org/10.1002/esp.1897

Schölkopf, B., Sung, K. K., Burges, C. J. C., Girosi, F., Niyogi, P., Poggio, T., & Vapnik, V. (1997). Comparing support vector machines with Gaussian kernels to radial basis function classifiers. *IEEE Transactions on Signal Processing, 45*(11), 2758–2765. https://doi.org/10.1109/78.650102

de Lima Neves Seefelder, C., Koide, S., & Mergili, M. (2017). Does parameterization influence the performance of slope stability model results? A case study in Rio de Janeiro, Brazil. *Landslides, 14*(4), 1389–1401. https://doi.org/10.1007/s10346-016-0783-6

Segoni, S., Piciullo, L., & Gariano, S. L. (2018). A review of the recent literature on rainfall thresholds for landslide occurrence. *Landslides, 15*(8), 1483–1501. https://doi.org/10.1007/s10346-018-0966-4

Sturzenegger, M., Holm, K., Lau, C.-A., & Jakob, M. (2021). Debris-flow and debris-flood susceptibility mapping for geohazard risk prioritization. *Environmental Engineering Geoscience, 27*(2), 179–194. https://doi.org/10.2113/EEG-D-20-00006

Sun, L., Ma, B., Pei, L., Zhang, X., & Zhou, J. L. (2021). The relationship of human activities and rainfall-induced landslide and debris flow hazards in Central China. *Natural Hazards, 107*(1), 147–169. https://doi.org/10.1007/s11069-021-04577-0

Takahashi, T. (1981). Estimation of potential debris flows and their hazardous zones: Soft countermeasures for a disaster. *Journal of Natural Disaster Science, 3*(1), 57–89.

Takahashi, T. (2009). A review of Japanese debris flow research. *International Journal of Erosion Control Engineering, 2*(1), 1–14. https://doi.org/10.13101/ijece.2.1

Tang, C., Van Westen, C. J., Tanyas, H., & Jetten, V. G. (2016). Analysing post-earthquake landslide activity using multi-temporal landslide inventories near the epicentral area of the 2008 Wenchuan earthquake. *Natural Hazards and Earth System Sciences, 16*(12), 2641–2655. https://doi.org/10.5194/nhess-16-2641-2016

Tanyaş, H., Hill, K., Mahoney, L., Fadel, I., & Lombardo, L. (2022). The world's second-largest, recorded landslide event: Lessons learnt from the landslides triggered during and after the 2018 Mw 7.5 Papua New Guinea earthquake. *Engineering Geology, 297*(December 2021). https://doi.org/10.1016/j.enggeo.2021.106504

Tarboron, G., & Tarboton, D. G. (1997). A new method for the determination of flow directions and upslope areas in grid digital elevation models. *Water Resources Research, 33*(2), 309–319. https://doi.org/10.1029/96WR03137

Taylor, F. E., Malamud, B. D., Witt, A., & Guzzetti, F. (2018). Landslide shape, ellipticity and length-to-width ratios. *Earth Surface Processes and Landforms, 43*(15), 3164–3189. https://doi.org/10.1002/esp.4479

Titti, G., van Westen, C., Borgatti, L., Pasuto, A., & Lombardo, L. (2021). When enough is really enough? On the minimum number of landslides to build reliable susceptibility models. *Geosciences, 11*(11), 1–26. https://doi.org/10.3390/geosciences11110469

Tufano, R., Formetta, G., Calcaterra, D., & De Vita, P. (2021). Hydrological control of soil thickness spatial variability on the initiation of rainfall-induced shallow landslides using a three-dimensional model. *Landslides, 18*(10), 3367–3380. https://doi.org/10.1007/s10346-021-01681-x

Turcotte, D. L., Malamud, B. D., Guzzetti, F., & Reichenbach, P. (2002). Self-organization, the cascade model, and natural hazards. *Proceedings of the National Academy of Sciences of the United States of America, 19*, 2530–2537. https://doi.org/10.1073/pnas.012582199

van den Bout, B., Lombardo, L., van Westen, C. J., & Jettena, V. G. (2018a). Integration of two-phase solid fluid equations in a catchment model for flashfloods, debris flows and shallow slope failures. *Environmental Modelling & Software, 105*, 1–16. https://doi.org/10.1016/j.envsoft.2018.03.017

van den Bout, B., Emtehani, S., Brenner, F., Dibaba, M., Orr, A., Westen, C. J. V., & Jetten, V. (2018b). Physically-based multi-hazard modelling and its influence on predicted hazard and risk for Dominica. American Geophysical Union, Fall Meeting 2018, abstract #NH33D-1031

van den Bout, B., Lombardo, L., Chiyang, M., van Westen, C., & Jetten, V. (2021). Physically-based catchment-scale prediction of slope failure volume and geometry. *Engineering Geology, 284*(July 2020), 105942. https://doi.org/10.1016/j.enggeo.2020.105942

van Westen, C. J., Castellanos, E., & Kuriakose, S. L. (2008). Spatial data for landslide susceptibility, hazard, and vulnerability assessment: An overview. *Engineering Geology, 102*(3–4), 112–131. https://doi.org/10.1016/j.enggeo.2008.03.010

Voellmy, A. (1955): Über die Zerstörungskraft von Lawinen, Schweizerische Bauzeitung (pp. 73, 159–165, 212–217, 246–249, 280–285) (in German)

Von Neumann, J. (1966). *Theory of self-reproducing autamata*. University of Illinois Press.

Wang, N., Cheng, W., Marconcini, M., Bachofer, F., Liu, C., Xiong, J., & Lombardo, L. (2022). Space-time susceptibility modeling of hydro-morphological processes at the Chinese national scale. *Engineering Geology, 301*(September 2021), 106586. https://doi.org/10.1016/j.enggeo.2022.106586

Wichmann, V. (2017). The gravitational process path (GPP) model (v1.0) – a GIS-based simulation framework for gravitational processes. *Geoscientific Model Development, 10*(9), 3309–3327. https://doi.org/10.5194/gmd-10-3309-2017

Wichmann, V., & Becht, M. (2004): Spatial modelling of debris flows in an alpine drainage basin. In IAHS-AISH P. (pp. 288, 370–376).

Wichmann, V., Rutzinger, M., & Vetter, M. (2008). Digital terrain model generation from airborne laser scanning point data and the effect of grid-cell size on the simulation results of a debris flow model. In J. Böhner, T. Blaschke, & L. Montanarella (eds.), *SAGA - Seconds Out*. Hamburger Beiträge zur Physischen Geographie und Landschaftsökologie (Vol. 19, pp. 103–113).

Wilkinson, P. L., Anderson, M. G., & Lloyd, D. M. (2002). An integrated hydrological model for rain-induced landslide prediction. *Earth Surface Processes and Landforms, 27*(12), 1285–1297. https://doi.org/10.1002/esp.409

Wilson, J. (1996). GIS-based land surface/subsurface modeling: new potential for new models. In *Third International Conference/Workshop on Integrating GIS and Environmental Modeling* (pp. 80–106).

Witze, A. (2018). Why extreme rains are gaining strength as the climate warms. *Nature, 563*(7732), 458–460. https://doi.org/10.1038/d41586-018-07447-1

Wolfram, S. (1984). Cellular automata as models of complexity. *Nature, 311*(5985), 419–424. https://doi.org/10.1038/311419a0

Xiong, K., Adhikari, B. R., Stamatopoulos, C. A., Zhan, Y., Wu, S., Dong, Z., & Di, B. (2020). Comparison of different machine learning methods for debris flow susceptibility mapping: A case study in the Sichuan Province, China. *Remote Sensing, 12*(2), 1–20. https://doi.org/10.3390/rs12020295

Zhang, W., & Montgomery, D. R. (1994). Digital elevation model grid size, landscape representation, and hydrologic simulations. *Water Resources Research, 30*(4), 1019–1028. https://doi.org/10.1029/93WR03553

Zhang, Y., Ge, T., Tian, W., & Liou, Y.-A. (2019). Debris flow susceptibility mapping using machine-learning techniques in Shigatse area, China. *Remote Sensing, 11*(23), 2801. https://doi.org/10.3390/rs11232801

Zhao, Y., Meng, X., Qi, T., Chen, G., Li, Y., Yue, D., & Qing, F. (2021). Modeling the spatial distribution of debris flows and analysis of the controlling factors: A machine learning approach. *Remote Sensing, 13*(23). https://doi.org/10.3390/rs13234813

Zhu, A. X., Miao, Y., Yang, L., Bai, S., Liu, J., & Hong, H. (2018). Comparison of the presence-only method and presence-absence method in landslide susceptibility mapping. *CATENA, 171*(July), 222–233. https://doi.org/10.1016/j.catena.2018.07.012

Zimmermann, M., Mani, P., Gamma, P., Gsteiger, P., Heiniger, O., & Hunziker, G. (1997): Murganggefahr und Klimaänderung – ein GIS-basierter Ansatz. Hochschulverlag an der ETH, Zürich, Switzerland, Schlussbericht NFP 31. (in German).

14 Debris-Flow Risk Assessment

Alex Strouth, Philip LeSueur, Sophia Zubrycky, Saskia de Vilder, Frankie Lo, Ken Ho and Scott McDougall

Abstract

Risk assessments can inform a wide variety of decisions related to debris flows, such as: Is the community safe enough? Where is mitigation needed? How much should be invested? However, risk assessments tend to be complicated. They can be structured in innumerable ways, requiring practitioners to navigate through many options, nuances, and pitfalls. This chapter is a guide for that journey, intended for both new and experienced debris-flow risk management practitioners. In it, we describe the three components (hazard, exposure, and vulnerability) that are combined to create debris-flow risk, highlighting key points and case examples. We then detail the three phases of a risk assessment: scope definition, risk analysis, and risk evaluation. In the scope definition section, we describe how to identify the most effective way to inform and communicate a decision, including identifying where risk assessments are likely to add unnecessary complexity and uncertainty rather than clarity. In the risk analysis section, we describe various methods for estimating and describing debris-flow risk, including the typical uses, advantages, and disadvantages of each method. In the risk evaluation section, we describe how risk estimates can be compared with uncertainty, available resources, and perceptions of tolerable risk to decide what actions, if any, are needed. The chapter concludes with a summary of key lessons and a list of research that is needed to improve future risk assessments.

A. Strouth (✉)
BGC Engineering Inc., Golden, CO, USA
e-mail: AStrouth@bgcengineering.ca

A. Strouth · S. McDougall
University of British Columbia, Vancouver, Canada

P. LeSueur
Geological Survey of Canada, Vancouver, Canada

S. Zubrycky
BGC Engineering Inc., Vancouver, Canada

S. de Vilder
GNS Science, Lower Hutt, New Zealand

F. Lo · K. Ho
Geotechnical Engineering Office, Civil Engineering and Development Department, Kowloon, Hong Kong

14.1 Why Risk Assessment?

Risk assessment is a powerful tool for guiding and communicating decisions. For the risk analyst, it can provide a new or more complete understanding of the problem. For the decision maker, it can be a repeatable and defensible method for comparing options and guiding a decision. To the public, risk assessment can be a transparent framework for communicating the problem, the factors that influenced the decision, the options that were considered, and ultimately 'why' the final decision was made.

M. Jakob et al. (eds.), *Advances in Debris-flow Science and Practice*, Geoenvironmental Disaster Reduction,
https://doi.org/10.1007/978-3-031-48691-3_14

Assessment of risks related to debris-flow hazards has become increasingly common in the last two decades in the places where we work. These debris-flow risk assessments allowed a wide variety of decisions (Fig. 14.1) to be well supported by logic and clearly communicated, including:

- Is this community, highway, railway, or pipeline *safe enough* from debris flows?

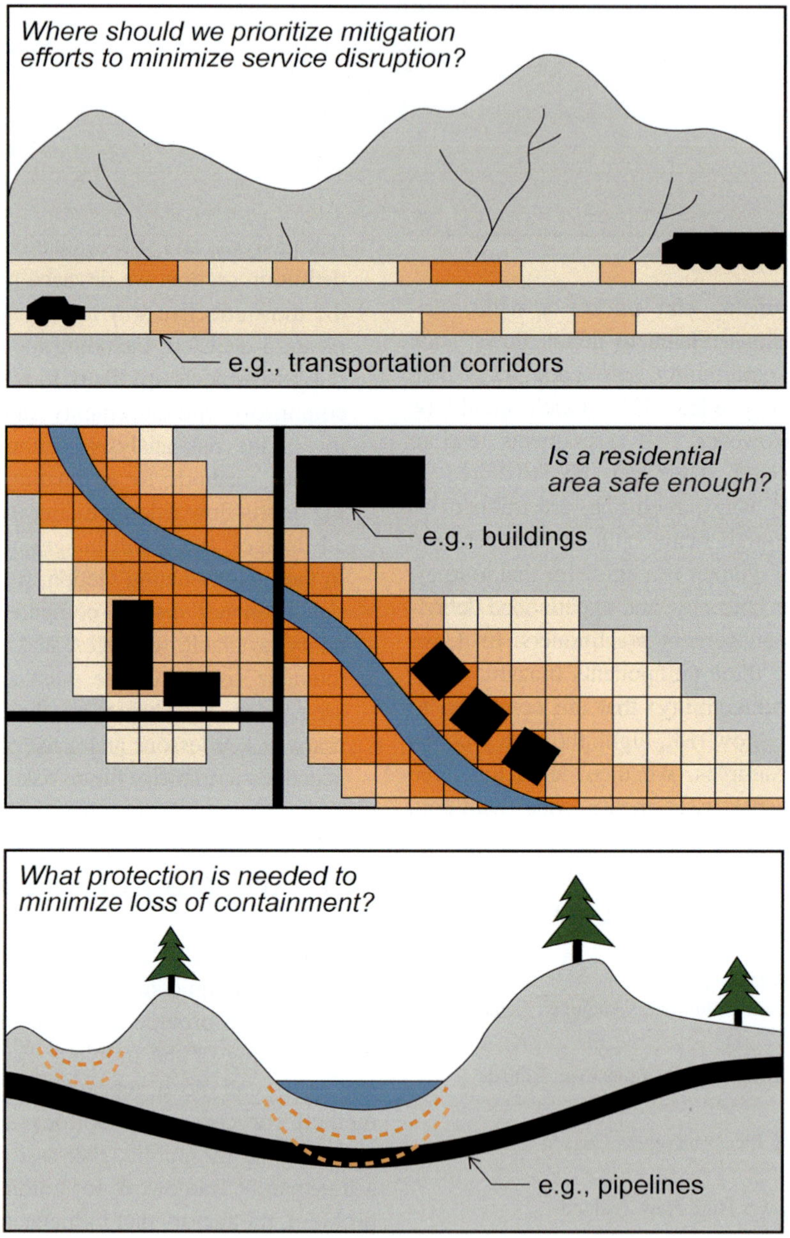

Fig. 14.1 Example debris-flow risk settings and decisions. The examples illustrate the breadth of decisions that a risk assessment can inform; it is not exhaustive. Deepening shades of orange represent increasing debris-flow hazard

- How should land be used in debris-flow hazard zones?
- When and from where should people be evacuated?
- When can the evacuees return home?
- How important is the debris-flow hazard compared to other hazards that affect the community?
- Which part of the highway, railway, or community should be prioritized for debris-flow protection?
- How should resources be allocated across many hazard sites?
- Which debris-flow mitigation options should be implemented?
- What should the scale of the debris-flow protection be?
- Was the investment in debris-flow mitigation justified?
- What needs to be done to prepare for climate change?

However, risk assessment is not a panacea. It tends to be complex, expensive, and uncertain, due to the many subjective judgements that are required throughout the process. Simpler methods of supporting decisions, like referencing guidelines or following precedent, may be more familiar, more cost effective, and more than sufficient. So, when should a risk assessment be applied? What method is best? What are the opportunities? What are the pitfalls?

In this chapter, we explore these questions by compiling lessons-learned from the literature and the authors' practical experience, and by adapting more generic risk assessment guidance (e.g., AGS, 2007; ISO, 2019; Lee & Jones, 2014; Saunders & Glassey, 2007) to debris-flow hazards. The chapter is organized as follows:

- Section 14.2 describes how risk is derived from the combination of debris-flow hazard, exposure of elements at risk, and vulnerability of those elements to debris flows.
- Section 14.3 describes the risk assessment framework and defines key terms.
- Section 14.4 describes how to define the scope of a risk assessment, and highlights situations where risk assessment may not be appropriate.
- Section 14.5 describes qualitative and quantitative methods for risk analysis with uses, advantages, and disadvantages of each.
- Section 14.6 describes how risk estimates can be used to support decisions, including techniques for evaluating the significance of debris-flow risks and selecting between risk reduction options.
- Section 14.7 summarizes key messages.
- Section 14.8 summarizes research needs.

14.2 What is Risk?

The word *risk* has several meanings that vary with context (Lee & Jones, 2014). For this article, *risk* is the potential for loss caused by future debris flows. Quantitatively, *risk* is the product of some adverse consequence (e.g., number of lives lost, economic loss, days of service disruption) and the probability of that loss occurring (Eq. 14.1). The adverse consequence and the probability of it occurring are both a function of the interaction between the debris-flow hazard, the exposed elements at risk, and the vulnerability of those elements to the hazard (Fig. 14.2, Fell et al., 2005).

$$\text{Risk} = \sum \text{Consequence} \times \text{Probability} \tag{14.1}$$

Hazard refers both to the likelihood of the debris flow impacting the exposed elements and the intensity of flow when the impact occurs. *Exposure* describes the quantity, value, and characteristics of elements that are at risk, like people, buildings, infrastructure, vehicles, economic activities, ecosystems, and other things that could be adversely affected by a debris flow. *Vulnerability* describes how much is damaged or lost when an exposed element is impacted by a debris flow of a given intensity.

Consequence is the magnitude of loss caused by the debris flow, and it arises due to a chain-of-events where a debris flow occurs, reaches an element at risk, and causes a loss depending on the vulnerability of that element and intensity

Fig. 14.2 Risk occurs at the intersection of hazard, exposure, and vulnerability

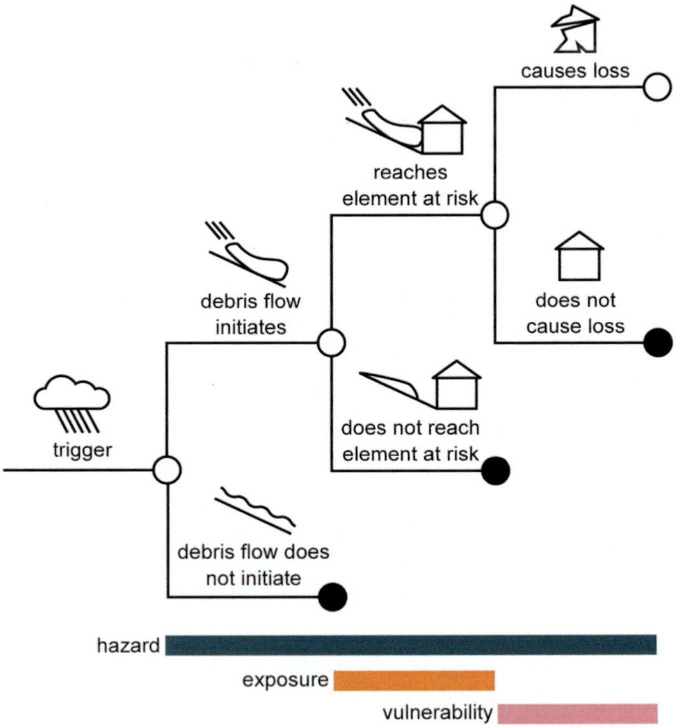

Fig. 14.3 An example debris-flow risk assessment event tree

of the hazard (Fig. 14.3). Consequence is estimated by combining hazard, exposure, and vulnerability. A single debris-flow event can affect many elements and, therefore, the consequence is the sum of losses experienced by the various elements.

Debris flows cause physical damage through burial, erosion, dynamic impact, abrasion, and flooding. Consequences of this damage can be direct, indirect, or intangible.

- *Direct consequences* are related to physical damage to elements in the debris-flow hazard zone like death, injury, property destruction, and economic loss. An element's vulnerability to direct consequences is a function of the debris-flow impact intensity, commonly described by the flow depth, flow velocity, and/or impact pressure.
- *Indirect consequences* are quantifiable losses that occur within and outside of the debris-flow hazard zone as a secondary effect of the direct consequences. Indirect consequences tend to be economic losses related to interruption of services provided by infrastructure like roads, railroads, powerlines, and pipelines and related business interruption. Indirect losses can often exceed direct losses, but the indirect losses are considerably more difficult to quantify (MacLeod et al., 2005).
- *Intangible consequences* are the vague and diffuse losses that cannot be quantified or described with a simple metric, like loss of cultural or historical sites, archeological sites, or ecosystems, and a lost sense of well-being among residents (Fell et al., 2005; Lee & Jones, 2014). These losses tend to be overlooked and, in some settings, deserve more attention.

The *probability* component of risk is in relation to a specific consequence. In a debris-flow risk assessment, it is the probability of a debris flow occurring in a given time period (such as within one year) and impacting the elements at risk with an intensity that causes the estimated loss. For debris-flow hazards, there are rarely comprehensive databases that allow for statistical probability of certain consequences to be estimated. Therefore, it is ultimately a *subjective* or *degree of belief* probability, which is a measure of belief, judgement, or confidence in the likelihood of an outcome (Fell et al., 2005; Baecher, 2023). The assigned probability is largely based on expert judgement informed by the hazard assessment methods described throughout this book.

A risk estimate is specific to a given set of conditions. It is a 'snapshot' of one set of hazard probability, hazard intensity, exposure, and vulnerability assumptions. Risk changes as conditions like topography, vegetation cover, rainfall intensity, debris-flow hazard, population size, and development layout change. Risk assessment is therefore a useful tool for evaluating the importance of changes and comparing options.

The risk concept is also versatile. It can describe the interaction of debris flows with almost anything that people value and inform decisions in a wide range of settings (Fig. 14.1). Given this broad applicability, this section defines the three key components of risk that remain consistent across the range of settings and methods:

1. Debris-flow hazard characteristics
2. Exposure of elements at risk
3. Vulnerability of elements at risk

14.2.1 Hazard

What is the likelihood and intensity of debris flows?

Hazard refers to the characteristics of debris-flows that could occur at a given location. In a risk assessment, the hazard is represented by a probability (or likelihood) of a given impact intensity. As debris-flow depths and velocities at a given location vary depending on the size and character of the specific event, the full debris-flow hazard is sometimes discretized into a set of possible scenarios, with each scenario having its own probability of occurrence and impact intensity. Geomorphic mapping, empirical runout assessment, and numerical modelling are used to estimate the spatial distribution of debris-flow impacts.

14.2.1.1 A Matter of Time

Hazard probability is always with respect to a given period of time, and this period must be stated along with the risk estimate. Many risk estimates are given *per year*, likely because they begin with a frequency-magnitude relationship that is organized around return periods. Other time periods can also be used and can often improve risk communication. For example:

- Risk *per visitor day* was used to assess geohazard risks to visitors of National Parks in New Zealand because visitors tend to spend only a few days per year in the parks (de Vilder et al., 2022; Massey et al., 2020, 2022; Taig, 2022a, 2022b).
- Risk *per 50 years* could be used to assess risk at a permanent settlement or community, similar to earthquake hazards, to help people understand the potential for disaster in their lifetime.
- Risk *per service life* could be used for both long-term (e.g., 25 years) or short-term installations (e.g., 3 months), to help people understand the potential for disaster during a structure's service life.

14.2.1.2 Probability of Debris-Flow Impact

In a risk assessment, *hazard probability* refers to more than an estimate of debris-flow initiation. Many debris flows remain within the channel (Zubrycky et al., 2021) and do not lead to loss on the debris-flow fan. For example, a hazard assessment might conclude that a debris flow is initiated and travels down the creek channel for some distance every 5 years. But if the assessment stops there, important questions remain unanswered: *What is the probability that the exposed element is impacted? What is the intensity of debris-flow impact?* Answers to these questions allow the risk analyst to estimate the *probability of debris-flow impact* needed for risk assessment. Probability of debris-flow impact is a combination of (Fig. 14.4):

1. *Initiation behaviour*: Probability of debris-flow initiation; and
2. *Runout behaviour*: Conditional probability of impacting an element at risk, given debris-flow initiation.

Initiation behaviour: The probability of debris-flow initiation is an outcome of debris-flow hazard assessments and is a function of source zone geology, topography, hydrology, and other factors (see Kaitna et al., Horton et al., and Jakob and Nolde, this volume). The probability of initiation changes with time in response to changes in these conditions, like vegetation loss, sediment accumulation, and rainfall frequency or intensity. A given risk estimate is therefore specific to the one set of conditions assumed in the analysis.

Runout behaviour: The conditional probability of impact is a function of the debris-flow behaviour after it initiates, called debris-flow runout. It is the probability that the debris flow reaches the exposed element with a flow intensity (e.g., flow depth, velocity, pressure) that causes the estimated loss, given a debris flow occurs.

In many settings, the two behaviours need to be assessed independently and then combined. For example, the probability of a debris flow of a given size is estimated based on triggering factors (e.g. rainfall intensity, USGS, 2022) or from frequency-magnitude relationships (See Jakob and Nolde, this volume). Then the probability of the debris flow reaching the element(s) at risk is estimated using runout analysis. The probability of a debris flow occurring and reaching the element at risk is sometimes referred to as "encounter probability" or "partial risk", although the authors prefer the former.

In other settings, the two parameters may be inseparable. For example, a highway owner may have a historical record of debris-flow impacts to the highway that does not recognize debris flows that stop before hitting the highway. The probability of highway impact derived from this record would be the combined probability parameter needed for risk assessment.

The debris-flow intensity at the exposed element, which is a function of both the initiating conditions and the runout behaviour, is used in later stages of the assessment to estimate consequences using the concept of *vulnerability*.

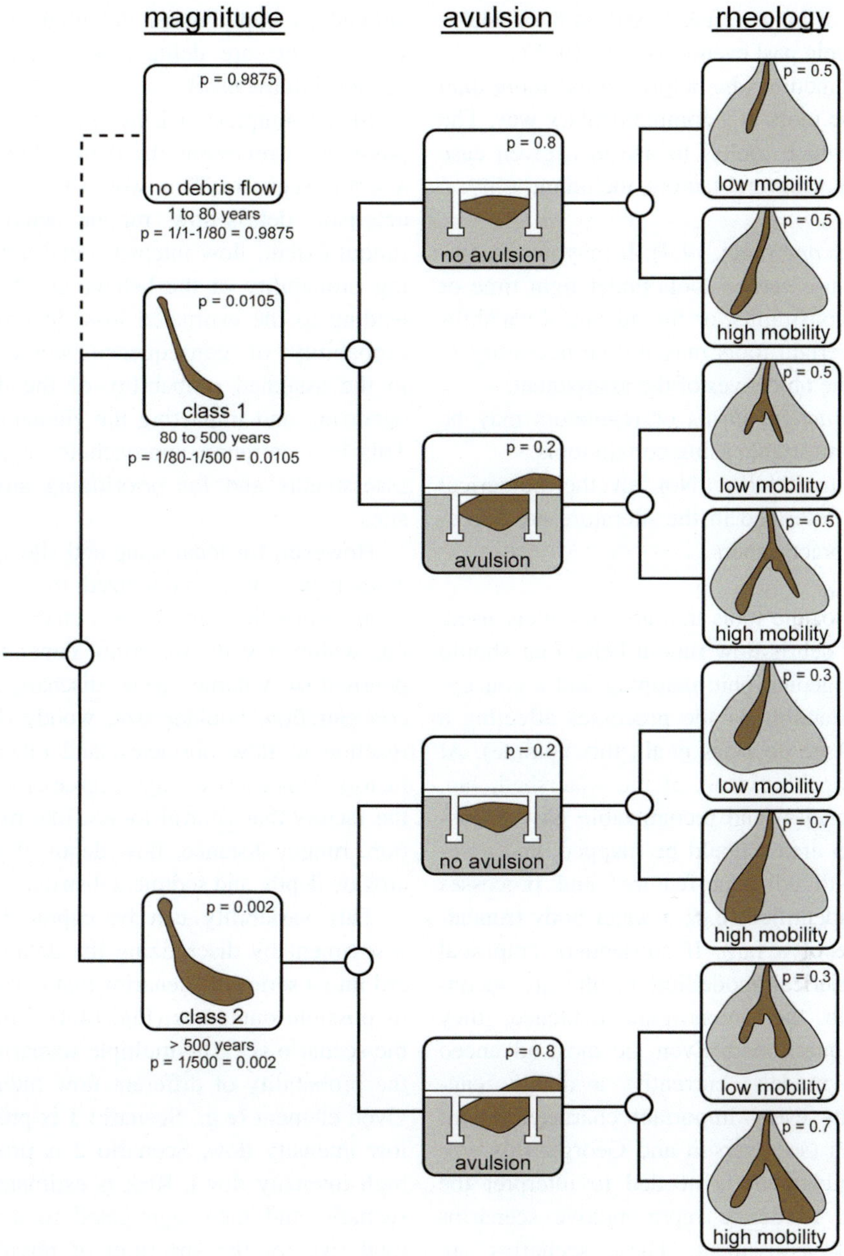

Fig. 14.4 Example of a logic tree considering two debris-flow magnitudes, two avulsion scenarios, and two rheologies. Note that the probabilities (p) sum to 1 at each node

14.2.1.3 Debris-Flow Runout Analysis

Many different analysis tools are available to inform predictions of debris-flow runout behaviour. They range from simple empirical relationships based on observations of past events (e.g., Iverson et al., 1998; Rickenmann, 1999) to advanced two-phase (solid + fluid) numerical models that require the input of measurable physical parameters (see Iverson and George, Horton et al. this volume). Along the spectrum between these two end members lie "semi-empirical" numerical models that require the calibration of input values based on back-analyses of multiple past events (Hungr, 1995).

It can sometimes be helpful to use more than one of these tools in a complementary way. The choice of which tool(s) to use in a given case can depend on several factors, including:

1. *Project scope* (Sect. 14.4): It may not be feasible to use certain tools under tight time or budget constraints, or the advanced capabilities of certain tools may not be necessary to satisfy the objectives of the assessment.
2. *Local practice*: Firms or regulators may be more comfortable using certain tools.
3. *Model availability*: Not all the numerical models described in the literature are accessible to practitioners.

Regardless of the tools that are ultimately used, analysis of debris-flow runout behaviour should begin with geomorphic mapping and a conceptual understanding of the processes affecting a given site (see de Haas et al., this volume). At a minimum, the extents of the watershed, fan, active channel(s) and recognizable past debris-flow impact areas should be mapped, and associations with adjacent features and processes should be identified (e.g., a water body truncating the toe of a fan). If subsequent empirical and/or numerical modelling results are inconsistent with the geomorphic evidence, they should be questioned. Even the most advanced numerical models currently available cannot simulate every important characteristic of debris-flows (see Iverson and George, this volume), so judgment is needed to interpret the results and to define representative scenarios in the risk assessment. These scenarios are explained in the subsequent section.

14.2.1.4 Debris-Flow Hazard Scenarios

The next debris flow to occur at a given site has a spectrum of possible sizes, flow mobilities, and flow paths. For the risk assessment, this spectrum of possibilities can be divided into scenarios, each with a specific probability and corresponding flow intensity and flow distribution across the fan (for this chapter, *scenarios* are defined as hypothetical potential debris flows, while *events* are defined as debris flows that occurred in the past).

In the simplest risk assessments, it may be possible to represent the debris-flow hazard as a single scenario. This would be a single characteristic debris-flow runout behaviour (e.g., runout extent, flow intensity) and a corresponding probability of the behaviour occurring and leading to the estimated loss. In this case, the probability of consequence would be equal to the assigned probability of the debris flow occurring and impacting the element(s) at risk. This is a common approach for regional-scale assessments and for prioritizing among many sites.

However, for local-scale and site-specific risk assessments, it is recognized that each subsequent debris flow has its own characteristics that fall within a wide spectrum, depending on the debris-flow volume, peak discharge, sediment concentration, boulder size, woody debris load, reaction to flow obstacles, and other unknown factors. This causes a high degree of variability in the factors that control losses, like runout direction, runout distance, flow depth, flow velocity, erosion depth, and sediment distribution.

This variability can be captured in a risk assessment by discretizing the debris-flow hazard into a series of scenarios that cover the range of possible outcomes (Fig. 14.4). This is called the scenario set. The multiple scenarios describe the probability of different flow intensities at a given element (e.g., Scenario 1 is probability of low intensity flow, Scenario 2 is probability of high intensity flow). Risk is estimated for each scenario and then aggregated to arrive at the total risk for the spectrum of possible debris-flow behaviours.

Dividing the debris-flow hazard into a spectrum of scenarios is an important way of exploring uncertainty caused by random variability of the debris flow and lack of knowledge about the debris-flow hazard, exposure, and vulnerabilities. In some settings, the total risk is dominated by high probability, low magnitude events. In others, low probability, high magnitude events control total risk (Strouth & McDougall, 2022).

The scenario set should cover a representative set of credible and relevant cases that could occur across a range of debris-flow frequencies and magnitudes, but fewer scenarios are generally preferrable to more. Working with fewer scenarios simplifies the analysis and reduces effort and the potential for errors. Trial-and-error is typically needed to optimize the number of scenarios. The following steps can be followed to define scenarios (Strouth & McDougall, 2022):

1. *Lower-event*: Identify the smallest debris-flow depth and velocity that could result in a loss to elements at risk. Then estimate the debris-flow magnitude that results in that flow behaviour, using numerical modelling, empirical relationships, and/or historical data. This is the smallest event that needs to be considered in the risk estimate and has a strong control on the estimated risk value.
2. *Lower-event exceedance probability*: Estimate the probability of initiation of the smallest debris flow that could result in loss. The frequency-magnitude relationship provides the return period of this debris-flow magnitude. The return period is the inverse of the *annual exceedance probability* (Lee & Jones, 2014). As such, it is describing the probability of the smallest (most frequent) debris flow or *any larger (less frequent) debris flow* occurring. This value has a strong control on the estimated risk value.
3. *Upper-event*: Identify the maximum credible debris-flow, in terms of extent and flow intensity at the exposed element(s). Estimate the smallest debris-flow magnitude that could result in that flow behaviour, using numerical modelling, empirical relationships, and/or historical data. Avoid using an arbitrary value, like the 1 in 200-year or 1 in 10,000-year debris flow set by local custom, to define the upper-event.
4. *Magnitude classes*: Divide the lower-event to upper-event range into magnitude classes. Use the fewest number of classes possible. If the lower-event causes approximately the same flow intensity as the upper-event, then only a single class is needed. Step changes or inflection points in the runout area and impact intensity value are useful markers for differentiating classes as are changes in triggering mechanism (e.g., in-channel initiation versus rock avalanche initiation).
5. *Incremental probability*: Calculate the incremental probability of each magnitude class based on the frequency-magnitude relationship. For example (Fig. 14.4), if the lower-event is 80-year return period and the upper-event is 500-year return period, magnitude classes could be defined as Class 1, 80 to 500 years; Class 2, >500 years. The incremental probability of Class 2 is (1/500), 0.002. The incremental probability of Class 1 is (1/80–1/500), 0.0105. Addition or subtraction of individual classes must not change the total probability of a debris flow occurring, which is represented by the lower-event exceedance probability. Therefore, the sum of the incremental probabilities of all scenarios must equal the lower-event exceedance probability.
6. *Sub-classes*: Additional scenarios can explore other factors that affect losses, like avulsion paths or flow mobility (see Iverson and George, de Haas et al., this volume, Zubrycky et al., 2021). These are added by discretizing a magnitude class into two or more sub-classes. The incremental probability for the magnitude class is multiplied by a conditional probability that describes how likely the mobility or avulsion scenario is to occur, given the magnitude class occurs. The conditional probabilities of the sub-classes must sum to 1.0 (Fig. 14.4).

14.2.1.5 Scenario Example: Cheekeye Debris-Flow Risk Assessment

The Cheekeye Fan, located in southwestern British Columbia, Canada, is prone to debris flows that pose a risk to existing and future development on the fan. LeSueur et al. (in review) carried out an assessment of the risk to life from these debris flows to determine if the associated risks to life were tolerable, and if not, to inform design requirements for debris-flow protection measures.

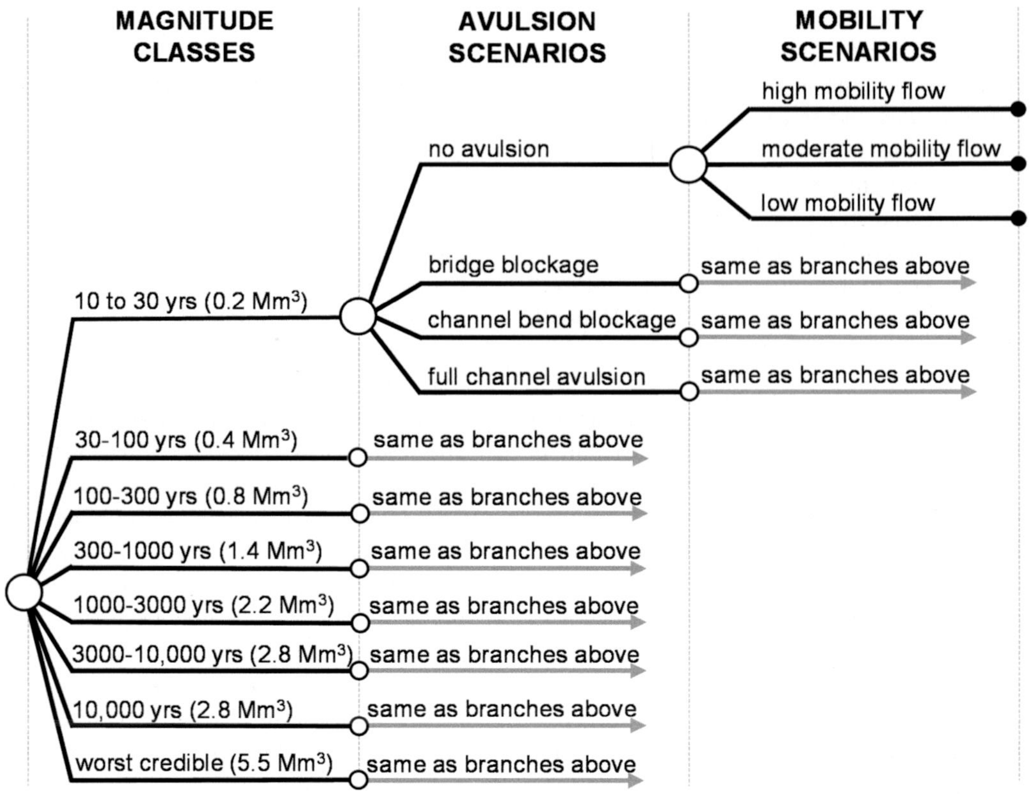

Fig. 14.5 Logic tree "branches" used to develop hazard scenarios modelled for the Cheekeye Fan debris flow risk assessment. 8 magnitude classes × 4 avulsion scenarios × 3 mobility scenarios results in 96 possible scenarios. 84 scenarios were carried forward into the risk assessment by "pruning" the logic tree to remove unlikely combinations

In this study, the debris-flow hazard was discretized into 84 scenarios (Fig. 14.5). The purpose was to account for the variability in the potential behaviour of Cheekeye debris flows, and to develop a range of scenarios that could be used to inform debris-flow mitigation design. These scenarios were generated by identifying potential Cheekeye debris-flow volume, avulsion, and flow mobility cases.

The probabilities of different debris-flow volumes were estimated using a previously derived debris-flow frequency-magnitude curve (Jakob & Friele, 2010). This curve was divided into eight event classes with return periods ranging from 1:10 to 1:10,000. This broad range of events were accounted for because of uncertainties about the impact area in high frequency events, and because the Cheekeye fan is prone to low-frequency, high-magnitude rock-avalanche generated debris flows.

Four avulsion cases were considered (Fig. 14.5) to account for uncertainty in the travel path of debris-flow scenarios. These cases were identified primarily based on the evidence of past debris flows and geometry of the channel, fan, and infrastructure on the fan. For example, a highway bridge crosses over the Cheekeye channel near the fan apex and was assumed to be a potential location for channel blockages leading to an avulsion. Probabilities of each avulsion case were assigned using expert judgement, informed by an understanding of the channel capacity, geometry, preliminary debris-flow modelling results, and sensitivity analyses.

Field evidence of past debris flows reaching Cheekeye Fan indicated a wide range of flow behaviours and compositions. Cases for low mobility (i.e., coarse-grained and/or low water content debris flows), moderate mobility, and high debris-flow mobility (i.e., fine-grained and/or high water content debris flows) were defined to represent the full spectrum of potential flow behaviour (Fig. 14.5). The relative probability of each occurring was assigned based on evidence of the flow behaviour of historical debris flows.

The combined scenario occurrence probability was estimated as the product of event probability, avulsion probability, and the flow mobility probability. The resultant set of 84 scenarios had occurrence probabilities that ranged by six orders of magnitude. This set of scenarios then provided the basis for runout modelling and associated consequence estimation in the risk assessment. Total debris flow risk on Cheekeye Fan was quantified by summing the product of consequence and occurrence probability for each scenario.

The assigned scenario occurrence probability had a strong control over the final risk value. The magnitude class probability had the strongest control because it varied over several orders of magnitude, for example from 0.067 (1/10–1/30) for the smallest class to 0.0001 (1/10,000) for the 1 in 10,000 year magnitude class. This highlights the importance of the debris-flow frequency-magnitude relationship for risk estimation. The assigned avulsion and mobility probabilities also varied with magnitude class. For example, the smallest magnitude class was assumed to have 100% probability of "no avulsion" because the peak discharge is much smaller than the channel capacity, while the largest magnitude class was assumed to have only 10% probability of "no avulsion" because the peak discharge is likely to be much larger than the channel capacity and will carry more large boulders and woody debris. Assigning probabilities to the avulsion and mobility scenarios was challenging and uncertain, and it highlighted the need for more research along the lines of Zubrycky et al. (2021) and Aaron et al. (2022).

14.2.2 Exposure

What could be impacted?

Exposure refers to the elements at risk, such as people, buildings, or critical infrastructure that are located within the debris-flow hazard zone and could experience loss. Assessing exposure is a crucial component of the risk assessment because it identifies what could be impacted by the debris flow and provides a basis for estimating debris-flow consequences.

14.2.2.1 Exposure Models

Exposure is represented in a risk assessment with an *exposure model*, which is a spatial–temporal inventory of the elements at risk attributed with information about their value (e.g., number of building occupants, economic value of the building and contents, quantity and value of oil in a pipeline), temporal characteristics (e.g., permanent, transitory, proportion of time in hazard zone), and characteristics that define their vulnerability to debris flows (e.g., building type, building age, presence of basement; pipeline burial depth, diameter, and wall thickness). An exposure model is a subset of an *asset model*, which identifies and characterizes all elements within a given planning or consultation region (Porter et al., 2009).

Exposure models can be represented spatially as continuous grids (e.g., a raster summarizing the distribution of building density), defined areas (e.g., polygons defining collections of buildings), or according to each element at risk (e.g., building footprints) (Corbane et al., 2017). The choice of spatial representation depends on the available data and risk assessment objectives.

There are many types of elements at risk and associated attributes that can be included in an exposure model. For debris-flow risk assessments, the most common include (Van Western, 2004):

- *Buildings*, including information about occupancy type (e.g., residential, commercial, industrial), construction type (e.g., masonry, wood, reinforced concrete), number of floors, design level (e.g. low-code, moderate-code, high-code), presence of basement, height of first floor, and location and height of openings (see FEMA, 2021a; Papathoma-Köhle et al., 2017).

- *Population at Risk*, including the numbers and location of people at different times of day.
- *Roads and Railways*, including information about the type (e.g., highway, local road, rail) and use rate at different times of day (see Winter et al., 2014).
- *Utilities*, including pipelines, water mains, and power lines, including information about depth of burial, diameter, wall thickness, and other factors that affect vulnerability.
- *Critical Facilities*, such as hospitals or emergency facilities.
- *Industrial Facilities,* such as mining facilities or factories.

Elements at risk with common vulnerability characteristics are often grouped into categories in the exposure model to facilitate the risk assessment (FEMA, 2021a). For example, all reinforced concrete buildings within an exposure model could be assigned one vulnerability value, while all wood frame buildings are assigned a different vulnerability value (e.g., Lan et al., 2013). The purpose is to provide a basis for assigning vulnerability relationships to groups of similar elements and reduce the burden of assessing the consequences to each element individually based on its specific characteristics.

14.2.2.2 Key Exposures

Key exposures are the subset of elements at risk and associated exposure attributes (e.g., building value, building use, building construction type) that are included in a risk assessment to meet an objective or reach a decision. Some risk assessments may attempt to consider many key exposures, such as for a cost–benefit assessment of debris-flow protection on a highly developed fan. Other risk assessments have a narrower scope and only consider a few key exposures, such as a safety risk assessment for people in a hazard zone. Input from stakeholders and decision makers is needed at the start of the risk assessment process to select key exposures, and the choice depends on:

- The element at risk of interest
- The consequence types of interest
- The availability of data

Most decisions can be made without considering all elements at risk. For example, the owner of a road may only be interested in service disruption risk along their road, even if commuters along the road are also exposed to safety risks from debris flows. Alternatively, a local government could focus a prioritization study on life loss risk without also assessing economic risks.

Certain key exposures must be included in the exposure model to assess specific consequences of interest. For example, to assess if debris-flow risks to people are tolerable, the number and location of persons must be included in the exposure model (e.g., Jakob & Holm, 2012). Similarly, to estimate the economic risk of debris flows to buildings, the value of buildings must be included (Tang et al., 2022). These key exposures are commonly identified by working backwards from the key risks identified in risk assessment scoping.

Data is needed to estimate risk. Sometimes the key exposures are those which have the most robust and reliable data set. For example, assessment value of buildings is commonly available and is often used as a key exposure. Sometimes a mitigation project can be justified solely based on the risk of direct building losses, so that other elements do not need to be considered. In other settings, additional exposures, such as building contents or business interruption, are added to the assessment to enhance the justification. In the absence of data, these additional exposures can be approximated as a function of the building value (FEMA, 2018, 2021b).

In cases where an owner is responsible for only a few element types, such as a road owner, identifying key exposures can be straightforward (Winter & Wong, 2020). However, in cases where an owner is responsible for a variety of element types, identifying key exposures can be more complex, and some may be overlooked. For example, in western Canada, many municipalities have historically focused on assessing life-loss risk, even though economic risks may have provided a clearer basis for decision making (Strouth & McDougall, 2022). Creating an inventory of potential exposures and reviewing the inventory with stakeholders prior to a risk assessment can help identify the key exposures.

14.2.2.3 Exposure Analysis

The exposure model is produced during an exposure analysis, which can be very time-intensive depending on the complexity of the model. The selected key exposures determine the scope, level of effort, and complexity of the exposure assessment.

Although less advanced, approaches for assessing exposure to debris flows have mainly evolved from earthquake, flood, tsunami, and hurricane risk assessment practices (FEMA, 2021b; Schneider & Schauer, 2006).

The exposure analysis has two main objectives:

1. Identify and map the elements at risk
2. Characterize the elements at risk with information pertinent to the risk assessment

Identification and mapping of elements at risk can usually be achieved through compilation of available data. Common outputs of this step include delineated built-up areas, land use maps, land parcel boundaries, building footprints, and transport and utility infrastructure centrelines. This spatial data is often available through government or infrastructure owner databases, open-source databases (e.g. OpenStreetMap), or through commercially available databases (e.g. business registries). However, when this information is not available, or when the quality is insufficient, site surveys or remote sensing techniques can be used to identify and map elements at risk (Corbane et al., 2017). Site surveys are carried out to manually map the location and footprint of elements at risk. This step can greatly reduce uncertainties in the exposure model (Huyck et al., 2022), but is often impractical when there are many elements at risk to be mapped (Van Western, 2004). Remote sensing approaches include manual or automated interpretation of remotely sensed data like satellite or aerial images, lidar or insar to identify and map elements.

The identified elements at risk can be characterised using a combination of data compilation, site surveys, and modelling. There are rarely reliable and complete datasets that summarize the characteristics that are most relevant to a debris-flow risk assessment. In most cases, surveys based on site visits or site photographs, coupled with modelling approaches is required. For example, a sample of buildings can be surveyed in the field or remotely to identify their occupancy type, construction type, elevation of living spaces, and other characteristics pertinent to the risk assessment. This sample can then be used as a basis for statistical modelling to infer the characteristics of other buildings in the model. FEMA (2018, 2021a) provide useful resources that summarize approaches for characterizing elements at risk for natural hazard risk assessment.

The following are suggestions for exposure analysis:

- *The exposure model and hazard model precision should be similar*: Risk occurs at the intersection of hazard and exposure, so the risk assessment is limited by the lowest resolution input model.
- *Focus effort on high-risk elements*: Often, a small proportion of the elements at risk are responsible for most of the risk, because those elements are disproportionately valuable, or they are disproportionately impacted by debris flows. Exposure models that are focused on high-risk elements reduce uncertainty in the risk value and modelling effort.
- *Beware of differences between replacement cost & economic cost*: Replacement costs are the cost to replace, repair, maintain, or 'clean-up' lost or damaged property to an equal quality or utility. Economic costs are the value of the asset accounting

for depreciation. Building assessment values assigned by governments for taxation tend to be economic costs and tend to be less than replacement costs in most markets. Replacement cost and economic cost can be very different from one another, and the choice of which to apply must be made intentionally.

- *Measure the first-floor height above ground*: The results of debris-flow and debris-flood risk assessments are often most sensitive to the height of the first floor (which is often assumed to be a primary living space) relative to the flow depth. Measure the first-floor height rather than applying an uninformed assumption. This parameter can be obtained through site surveys, or through interpretation of site photographs (e.g., Google Streetview).

14.2.2.4 Temporal Changes to Exposure

The risk from debris flows depends on the length of time an exposed element is within the debris-flow hazard zone. For example, a person walking across a debris-flow fan once in a year would be at several orders of magnitude lower risk than someone who lives year-round on the same fan. In a risk assessment, this is expressed as the temporal probability of impact in the risk equation (Eq. 14.2).

The temporal probability of impact is the fraction of time that an element at risk is within the hazard zone relative to the risk period (Sect. 14.2.1.1). It varies from 0 to 1, representing the conditional probability an exposed element is within the hazard zone when the debris-flow scenario occurs.

Temporal probability of impact for permanent infrastructure, such as buildings, roads, or bridges, is one (1.0). Temporal probability of impact for transient elements, such as people or vehicles that travel in and out of a hazard zone, is less than one.

Temporal probability of impact can vary for the same element depending on how the risk scenarios are defined. For example, consider a person who is home every night (12 h) but away from home every day (12 h). A risk scenario that estimates risk over the year would assign a temporal probability of impact of 0.5 to the person. However, a risk scenario that estimates risk during the night would assign a temporal probability of impact of 1 to the person.

Temporal probability of impact is commonly estimated using judgement, but it can also be measured, for example using surveys of traffic frequency or building occupancy rates. Jakob and Holm (2012) provide a useful example of estimating temporal probability of impact for person in buildings and roads, considering risk to individuals and groups.

14.2.2.5 Exposure Model Example: Cougar Creek Debris-Food Risk Assessment

Cougar Creek, located in Canmore, Alberta, crosses one of Canada's most developed alluvial fans and is prone to debris floods. In 2013, a debris flood caused by a high intensity, long duration rainstorm damaged many elements on this fan, including buildings, roads, railways, and other infrastructure.

An engineering consulting firm carried out an assessment of debris-flood hazards and risks to residential and commercial development to support cost–benefit assessments of debris-flood mitigation options (BGC, 2014; Jakob et al., 2017). The exposure model was complex, as it identified and characterized many key exposures. The model was developed at the scale of individual buildings and facilities because the hazard assessment was also at a very detailed resolution and because detailed exposure data were available.

The assessment quantified economic loss to buildings and business interruption for cost–benefit assessment using depth-damage functions developed by FEMA (2013). Other key infrastructures that would be within the path of debris-flood scenarios were identified, which qualitatively supported justification for debris-flood mitigation. Table 14.1 summarizes how the exposure model was organized.

The risk assessment concluded that annual risk to individuals (annual probability of death to an individual) exceeded 1 in 10,000 at 190

Table 14.1 Key exposures in the Cougar Creek risk assessment exposure model

Element	Characteristics	Assessed consequence
Buildings	– Building occupancy type (single family residential, multifamily residential, commercial, or industrial) – Building value (in dollars) – Number of building occupants – Annual probability the individual most at risk is within the building – Annual probability any individual is within the building	Direct consequences – Economic loss – Life loss to individuals, life loss to groups
Businesses	– Business type (commercial, industrial, government, education) – Annual revenue (in dollars)	Indirect consequences – Economic risk from lost business revenue
Roads	– Type (local roads, and highways)	Exposure potential to debris-flood scenarios
Utilities	– Type (sewerage, storm water management, gas distribution, transmission lines)	Exposure potential to debris-flood scenarios
Critical facilities	– Type (schools, heliport, police building, industrial storage yard)	Exposure potential to debris-flood scenarios

parcels on Cougar Creek Fan, and that approximately 60 people could be killed by a 1 in 1,000-year to 1 in 3,000-year debris flood. The annualized building damage risk (direct consequences only) was on the order of $400,000 to $700,000 (CAD).

14.2.3 Vulnerability

How much is damaged or lost?
Vulnerability depends on the characteristics of the exposed element. It is the component of risk that allows an estimation of consequences given the hazard and exposure models. Two types of vulnerability can be considered: *physical vulnerability* and *social vulnerability*.

14.2.3.1 Physical Vulnerability

Physical vulnerability governs the amount of loss or damage an element at risk may experience when it is impacted by a debris flow of given intensity. It can be the proportion that is lost, or the probability of a given loss (e.g., probability of death). Physical vulnerability depends on the debris-flow intensity and physical characteristics of the exposed element (Safeland, 2011). Many of these characteristics are highly specific to a particular exposed element, so physical vulnerability can vary considerably between different elements. For example, the physical vulnerability of a building depends on building material, height, use, age/condition, foundation type, orientation, presence and location of doors and windows, presence of basement, height of living space above the ground, and roof type (Papathoma-Köhle et al., 2017). These factors can vary regionally based on local construction practices, but can also vary locally based on design and use differences between buildings in the same neighbourhood.

Physical vulnerability is represented in a risk assessment with two conceptually distinct groups of models: *physical vulnerability models* and *fragility models*. These models can be expressed as a matrix, curve, or index that relate a given hazard intensity to a level of damage or loss (Papathoma-Köhle et al., 2017). When represented as a matrix, fragility or vulnerability values increase in steps, which can lead to step increases in the estimated losses and uncertainties when hazard intensity values are near the boundary of the matrix bins. Vulnerability or fragility curves or indexes are not prone to this limitation.

Physical vulnerability models (Fig. 14.6) are used to estimate expected losses to a given element as a function of debris-flow intensity,

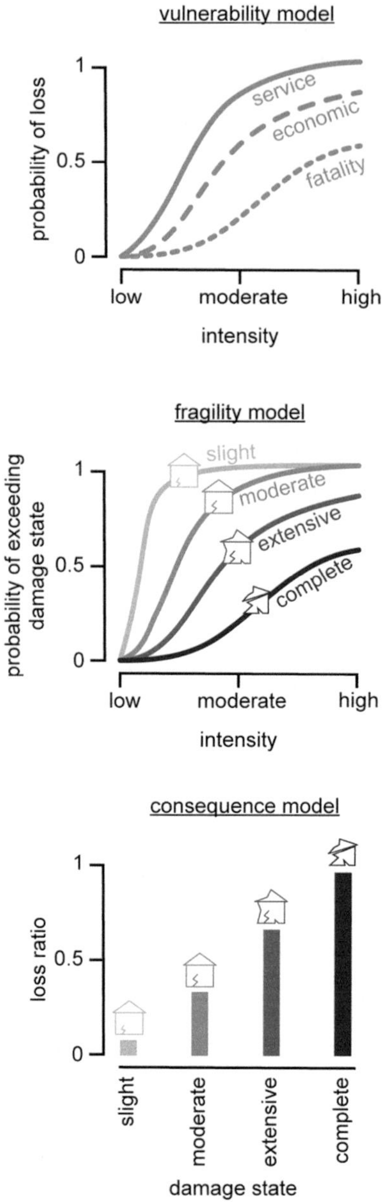

Fig. 14.6 Example vulnerability (top), fragility (middle), and consequence (bottom) models. The vulnerability model relates the debris-flow intensity directly to the proportion or probability of loss. Alternatively, the proportion or probability of loss can be obtained using a two-step process. First, use a fragility model to relate the debris-flow intensity to a damage state. Then, use a consequence model to relate the damage state to the proportion or probability of loss

typically on a scale of 0 (no loss) to 1 (total loss) (Fell et al., 2005). The losses can refer to many consequence types, but most commonly represent the proportion of economic loss (e.g. percent of building value) or conditional probability of death. Because these models relate a hazard intensity directly to an expected loss, the physical processes that govern the loss are not explicitly accounted for. For example, the processes that lead to someone being killed by a debris flow in a building do not only depend on hazard intensity, but also on other factors, such as the person's age and response behaviour (Pollock & Wartman, 2020). For this reason, physical vulnerability models are often associated with very wide uncertainty bounds and can be a large source of uncertainty in debris-flow risk assessments. Despite this drawback, these models are widely used due to their simplicity to develop, describe, and apply.

Fragility models (Fig. 14.6) are used to estimate the probability of a given element reaching or exceeding a certain damage state as a function of debris-flow intensity. The damage state classifications (e.g. minor damage, moderate damage) and accompanying definitions are specific to each fragility model. This class of models allows for more explicit consideration of the physical processes that lead to damage, and therefore can be used to reduce uncertainties. However, because fragility models only provide a relationship between hazard intensity and damage state, they do not directly provide an estimate of losses.

Consequence functions provide a means to relate damage state to a level of loss (e.g. business interruption, economic loss), and are well established in earthquake engineering (Fig. 14.6). For example, So (2016) describes a set of consequence functions that relate the probability of fatality for persons within buildings of different damage states, given they were impacted by earthquake ground shaking. The authors are not aware of published consequence functions for debris-flow hazards, likely due to a lack of available data. In absence of published functions, some authors have derived these for use in debris-flow risk assessment based on judgement (Jakob & Holm, 2012; Winter & Wong, 2020). The consequence function can have a large influence on the assessment results, and sometimes minor adjustments to the function can make the difference between an acceptable and unacceptable risk estimate. Therefore, consequence functions that are derived from judgement should be used with caution and the uncertainties should be transparently communicated.

There are many published physical vulnerability and fragility models that can be applied in a debris-flow risk assessment (Table 14.2). Papathoma-Köhle et al. (2017) and Prieto et al (2018) provide reviews of many of these models. Consider the following when selecting a model:

- The method used to develop the model
- The types of elements at risk to which the model applies
- The measure(s) used to describe hazard intensity
- The damage state classification system used for fragility models
- The loss measures reflected in physical vulnerability models

Methods that can be used to derive physical vulnerability or fragility models include: empirical (e.g. based on observational damage data), mechanics-based (e.g. using numerical modelling or statics/dynamics analysis), expert judgement, or using a combination of methods (SafeLand, 2011). There are benefits and drawbacks to each method that influence the quality and usability of the model. For example, most physical vulnerability or fragility models are derived empirically. However, there are limitations in their transferability between studies. Uncertainty is also large due to the limited availability of reliable data and the subjective judgements used to estimate the amount of loss and state of damage (Eidsvig et al., 2014; Papathoma-Köhle et al., 2017). Mechanics-based approaches allow for more systematic accounting for different debris-flow intensities or characteristics of elements at risk, however, may not capture the full range of physical processes that could occur. Expert judgement approaches can be useful in cases where empirical data is lacking but are prone to uncertainties that arise from expert biases and incomplete knowledge.

Each vulnerability or fragility model applies to a specific type of element. For example, models can apply to specific types of infrastructure (e.g., buildings, roads) and subtypes of infrastructure (e.g., reinforced concrete or wood-frame buildings; residential or commercial buildings). The vulnerability or fragility model used in risk assessment should apply to the types of elements at risk for which risk is being assessed. In the case of empirically derived curves, it is important to review the types and locations of elements used to derive the model, as these can be highly specific. For example, Tsao et al. (2019) generated debris-flow fragility curves based on observed damages to reinforced concrete, reinforced brick, and brick buildings that were impacted by a debris flow in northern Taiwan. This model would not be applicable for wood-frame buildings that are common in Canada.

There are many intensity measures that are used in physical vulnerability or fragility models, including: inundation depth (d), velocity (v), momentum flux (dv^2), overturning moment (dv or d^2v^2), impact pressure (kPa), and relative intensity (Li et al., 2010; Jakob et al., 2012a, 2012b; Zhang et al., 2018). Most early models of physical vulnerability used inundation depth to represent hazard intensity, largely because it could be readily measured in the field following a debris flow. However, the intensity of debris flows also varies with velocity, and work by Luo et al (2021) has indicated that considering inundation depth alone leads to an inaccurate estimation of building damages. Composite intensity measures allow for both velocity and depth to be considered simultaneously, and they appear to predict damage or loss more accurately for most

Table 14.2 Example models of physical vulnerability and fragility to debris flow

References	Model type	Approach	Intensity parameter	Element type	Consequence/damage classification
Ciurean et al. (2017)	Vulnerability Buildings	Empirical	Flow depth (cm)	Based on buildings in the Italian Alps	Maximum, average, and minimum economic loss ratio
Jakob et al., (2012a, 2012b)	Fragility Buildings	Empirical	Momentum flux (m^3/s^2)	Any building type	Probability of exceeding some sedimentation; some structural damage; major structural damage; and complete destruction
Prieto et al. (2018)	Fragility Buildings	Mechanics-based	Momentum flux (m^3/s^2)	Wood, Concrete, and Masonry	Probability of exceeding moderate, extensive, and complete damage
Winter et al. (2014)	Fragility Road	Expert Judgement	Volume (m^3)	High speed road, local road	Probability of exceeding limited, serious, or destroyed damage state
Daga (2018)	Fragility Bridge	Mechanics-based	Flow depth (m)	Single span bridge, multi span bridge	Probability of pier overturning, abutment overturning, or deck sliding
Pollock and Wartman, (2020)	Vulnerability to People in buildings	Empirical	Flow depth (m)	People in buildings	Probability of death

damage types (Luo et al., 2021; Zhang et al., 2018). Inundation depth and velocity can also be considered independently to overcome the difficulty of selecting a composite measure.

Each vulnerability model applies to a particular type of loss, and each fragility model applies to a particular damage state. Awareness of how loss metrics were derived, and the specific definition of each damage state in fragility models, is important to interpreting the output of a risk assessment. For example, Fuchs et al. (2007) and Ciurean et al. (2017) derived vulnerability models to estimate the economic loss of buildings; however, the former is based on the ratio of damage costs to replacement costs, and the latter is based on the ratio of damage costs to building market value. Similarly, many distinct classification systems exist that define damage states after debris-flow impact, and the definition of minor, moderate, extensive, or complete damage is inconsistent between many of them (Luo et al., 2021).

14.2.3.2 Social Vulnerability

Social vulnerability refers to the susceptibility of certain populations to the effects of natural hazards (Cutter & Finch, 2008). This factor depends on social characteristics of the population that influence their capacity to identify, respond to, and recover from events, such as demographics, land tenure, living conditions, socioeconomic status, health, and risk perceptions (Drakes & Tate, 2022; Journeay et al., 2022). Ample evidence shows that socially vulnerable populations often bare disproportionate impacts from natural hazards (Smith, 2013; Wisner et al., 2004), so social vulnerability is being increasingly used in natural hazard risk assessments to identify communities or areas that could be prone to increased risk.

There are many ways to represent and assess social vulnerability (Burton et al., 2019). In the context of debris-flow risk assessment, it is most commonly expressed as an index value (Cutter et al., 2003; Diana et al., 2021; Park et al., 2016; Sung & Liaw, 2021), where higher values represent relatively higher social vulnerability. These indices are often derived based on a weighted linear combination of variables, where each variable represents a characteristic of the population that may influence social vulnerability. Example characteristics that contribute to social vulnerability to debris flows include: age, gender, ethnicity, income, family structure, education, employment status, and urban/rural setting (Diana et al., 2021).

Social vulnerability can be applied in a qualitative risk assessment. For example, this could include deriving a qualitative relative risk index that combines estimates of the physical consequences of debris flows (e.g. economic damage, life loss) with a social vulnerability index. FEMA (2021a, 2021b) provides a good example of this, where they define relative risk, in part, as the product of the annual economic loss from natural hazards and a social vulnerability index ranging from 0 to 100.

In general, social vulnerability concepts are less well developed than physical vulnerability concepts for debris flow risk assessments, and more work is needed to develop methods that better incorporate social vulnerability, particularly for quantitative risk assessments.

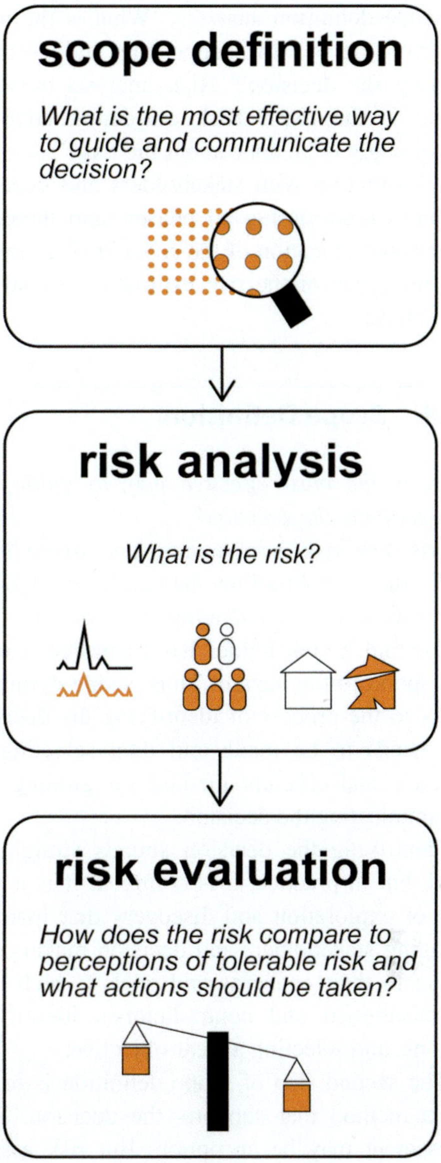

Fig. 14.7 Debris-flow risk management framework (adapted from Fell et al., 2005; ISO, 2019; Saunders & Glassey, 2007; VanDine, 2012). Each step requires open communication, collaboration, and review

14.3 What is Risk Assessment?

Although many risk management frameworks exist with competing notations (e.g., AGS, 2007; Fell et al., 2005; ISO, 2019; Lee & Jones, 2014; Saunders & Glassey, 2007; VanDine, 2012), all describe the same general sequence. For this article, the risk assessment steps are grouped into three broad phases (Fig. 14.7). The first phase, *Scope Definition*, involves selecting a method and boundaries that can effectively and efficiently inform the decision that needs to be made. If a risk assessment method is employed, the second phase, *Risk Analysis*, involves making an estimate of the risk. The third phase, *Risk Evaluation*, involves comparing the risk estimate with available resources and perceptions of tolerable risk to decide what actions, if any, are needed.

Scope definition answers, 'What is the most effective method for supporting and communicating the decision?' Risk analysis answers, 'What is the risk?' Risk evaluation answers, 'What ought to be done about the risk?'.

Consultation with stakeholders and decision makers occurs during all phases, and therefore clear communication of the scope, risk estimate, and relevance of the risk estimate are goals of each phase.

14.4 Scope Definition

What is the most effective way to guide and communicate the decision?

Debris-flow risk assessments tend to be initiated after a debris-flow hazard is recognized (either following a damaging event or screening study) and a stakeholder has identified a need to respond to the hazard. Thus, scope definition refers to the process of identifying the decision that needs to be made and then selecting an effective and efficient method for guiding and communicating the decision.

Identifying the decision sounds straightforward, but in practice, it is complex. It is a process of exploration and discovery that involves engaging stakeholders and decision makers, listening to their concerns and goals (which may be misaligned and contradictory), identifying options, and selecting a clear objective.

The second step of scope definition is to lay out a method that supports the decision. Risk assessment may be an option. But risk assessment is not always, or even usually, the best method. Risk assessments tend to be more onerous and expensive than other methods and can add a level of complication that is not needed. Decisions are easiest to communicate when they have a clear and simple basis with just enough detail to support the conclusion, without added complexity.

For example, many decisions can be fully supported by a hazard assessment. This is particularly true for zoning and prioritization studies where elements at risk and potential losses are similar from one zone to the next. For example, the Fraser Valley Regional District (FVRD) in southwestern British Columbia, Canada, has been using a hazard-based approach to fluvial and landslide threats since 1993 with reasonable success. Notably, decisions on the FVRD's development permit applications are a function of frequency of the hazard and the development type, which could range from a house renovation to a new subdivision (FVRD, 2020). While this avoids adding exposure and vulnerability information and the associated complexity and uncertainty, it does recognize that the scale of proposed or existing development matters. In other cases, more common engineering approaches, like applying factors of safety against a standard design event, can be sufficient. This is particularly true where these methods have been shown to generally meet performance objectives and there are well established best practices.

Risk assessments have a place in more complex or novel situations. Risk assessments tend to be most useful where (building from Wong, 2005):

- *Hazard comparison*: Debris-flow hazard needs to be compared with other hazard types, such as for prioritization or resource allocation.
- *Exposure or vulnerability varies* significantly within or between areas of regional prioritization or zoning.
- *Consequence mitigation*: It is not feasible to prevent debris-flow initiation, and risk is managed by mitigating the debris-flow consequences (which is common in debris-flow risk management settings).
- *Mitigation option comparison*: The risk concept can aide comparison of mitigation options, for example to quantify or highlight benefits of risk reduction in a cost-benefit assessment.
- *Level of safety*: There is a need to identify land that is *safe enough for the intended use* in areas that are subject to low probability hazards.

- *Factor of safety selection*: Debris-flow initiation can be controlled by a factor of safety (e.g., at a road embankment in a watershed), but assessment of risk and uncertainty is needed to support selection of a design factor of safety.
- *Residual risk*: A design factor of safety is met, but consequences of debris-flow initiation are so severe that assessment of residual risk of failure is needed.
- *Familiar terms*: The potential risks of a debris flow need to be communicated in a context that is more familiar to stakeholders, such as in terms of monetary loss or probability of death.
- *Alternative perspective*: A hazard-based decision is contested in court and risk assessment allows the decision to be viewed from a different angle.

If a risk assessment is selected as a decision support tool, the third step is to define the scope of the assessment, including the study area, key exposures for which risk will be estimated, analysis methods, and the risk metric that will be used for risk analysis, risk evaluation, and risk communication. There is no single best risk assessment method. Each method is fit for purpose to support a decision. The following factors and questions help to define the scope of a risk assessment:

- *Decision that needs to be made*: What are the objectives of stakeholders (e.g., life safety protection, reduced economic loss, reduced service disruption)? What is the purpose of the risk assessment?
- *Who the stakeholders are*: Who is affected by the potential debris flow? Who is affected by the potential risk management plans? Who is taking the risk? What are the benefits of taking the risk? Who bears the losses? Who bears the cost of risk reduction? Who needs to provide input to the risk assessment? Who is making the ultimate decision?
- *Prior experience and preferences of stakeholders*: How have decisions been made in the past? Will an existing risk evaluation framework be used or does one need to be introduced? How much technical knowledge and risk experience do decision makers and stakeholders have? What format will be most transparent and compelling to decision makers and stakeholders?
- *Key exposures at risk and the risk metric*: What are the key exposures at risk and key consequences? Buildings, persons, critical facilities? Life loss, economic loss, service loss, business interruption, intangible losses? How will the risk be measured and described?
- *Geographic extent of the decision*: Is it a site specific or regional scale? What is the maximum credible extent of the debris flow? What is the extent of elements at risk?
- *Information availability*: What information about the debris-flow hazard, elements at risk, and vulnerability is available for risk estimation? How will information be collected? What information is available to support risk evaluation, such as costs and benefits of mitigation options or information that can put risk estimates into context?
- *Time available for decision making*: Is a decision needed immediately? Is the hazard imminent? Does the decision support long-term investment planning?
- *Budget available for risk assessment and risk communication*: Is sufficient budget available for a detailed risk estimate and to reduce uncertainties? Is there a budget limitation that requires a simplified method with broad assumptions?
- *Options available for managing risk*: Are there options for managing risk or is the risk management plan predetermined? Are there resources available for risk reduction? Are stakeholders willing to bear the cost of risk reduction measures?
- *Experience of the risk analyst*: Do the risk analysts have a preferred methodology developed for risk analysis and risk evaluation that they have used for previous decisions? Are those analysis and evaluation methods the most effective and efficient option for guiding the decision and communicating the risk and risk management options?

- *Hazard context*: What other hazards (e.g., forest fire, flooding, earthquake), in addition to debris flows, exist? How important are the other hazards relative to debris flows? How should the risks be compared to each other? How should resources be allocated across hazard types?

14.5 Risk Analysis

What is the risk?
After the risk assessment scope is set, the next phase is to estimate risk by combining estimates or models of hazard, exposure, and vulnerability. These parameters can be combined in many ways, and no one method is universally better or more accurate than others. All are uncertain forecasts of future events.

This section describes the most common methods for estimating and describing debris-flow risk, including typical uses, advantages, disadvantages, and how to choose between methods.

Methods are grouped into two broad categories: qualitative and quantitative. Qualitative methods describe risks in a relative way using adjectives (i.e., words) or numerical or letter indices. Quantitative methods describe risk as the expected loss over a period of time. There are many possible metrics for loss, related to the elements at risk and objectives of the risk assessment, but the most common are life loss and monetary units.

14.5.1 Qualitative

Qualitative risk analyses measure risks in relative terms, using a metric that is:

- *Descriptive*, using words like Low, Medium, High;
- *Indexed*, using numeric values (e.g., 0 for no risk, and 100 for maximum risk); or
- *Semi-quantitative*, connecting descriptors to real quantities (e.g., 'High' likelihood meaning greater than 90% chance of occurrence in a year).

Qualitative risk metrics allow relative comparisons of one hazard or hazard site to others that are assessed within the same scheme. They are used for:

- *Screening assessments* to identify where significant risks exists, for example, to define the scope of more detailed studies.
- *Ranking* many different hazard sites, for example along a road or within a region.
- *Tracking* how risks and risk controls change with time.
- *Identifying* tolerable and intolerable risks.
- *Demonstrating* the risk reduction achieved by risk controls and comparing risk control options.

Common formats for qualitative risk analyses are the risk index and risk matrix.

A *risk index* is a value assigned to a given hazard site that describes the relative level of risk compared to other hazard sites. The index value is with respect to a pre-determined scale, such as 0 to 1, 0 to 100, or a letter grade (e.g., A to F). The index value has no meaning except in relation to other index values within the same scheme. Risk indexes are most useful where it is advantageous to have the risk level summarized in a single, precise value, such as for ranking of many hazard sites, especially when input data is limited. They also provide a simple framework to account for many factors that may contribute to risk simultaneously. Risk indexes can also be used to explore effectiveness of different risk reduction options, for example by demonstrating the change in index value or ranking caused by implementation of the option. Risk tolerance thresholds or risk reduction objectives can also be set within a given scheme. Key disadvantages of a risk index are that the risk components are hidden, and each risk index scheme is unique. These disadvantages may not be felt by the risk analyst and stakeholders who are familiar with the scheme and hazard sites, but they inhibit

Table 14.3 Risk index examples in literature, all at a regional scale

Location	Index scale	Description	References
Sichuan, China	0 to 1	Risk index value assigned to areas based on detailed analysis of hazard likelihood, flow depth, and vulnerability	Cui et al. (2013a, 2013b)
Western Canada	0 to 1	Risk index values used to rank and prioritize hundreds of fans that are crossed by a highway system	Holm et al. (2016)
Sichuan-Tibet, China	0 to 1	A vulnerability index is calculated to prioritize segments of a highway affected by debris flows	Zou et al. (2018)
Sichuan, China	0 to 1	Several methods for ranking many hazard sites with risk indexes are compared, and a method is used to rank sites along a highway	Li et al. (2021a, 2021b)
Pakistan	33 to 297–	Debris-flow risk rating is assigned to sites along a highway using a system that was modelled after the Rockfall Hazard Rating System	Ali et al. (2021)

external comparisons and risk communication to less familiar stakeholders and the public.

Table 14.3 summarizes several risk index examples from literature.

A *risk matrix* is a common tool used to organize hazard and consequence information and to structure selection of a risk value. The risk value is typically a relative risk descriptor, on a Low, Medium, High scale. The descriptor is derived from the intersection of a hazard likelihood factor and a consequence factor, which make up the two axes of the matrix. Sometimes a series of matrices is used. For example, a hazard matrix may arrive at a single 'hazard factor' by combining the likelihood of debris-flow initiation with likelihood of impact. A second vulnerability matrix may arrive at a 'vulnerability factor' by combining physical and social vulnerability. A third consequence matrix may combine the elements at risk with vulnerability to arrive at a single 'consequence factor'. Risk is then estimated by combining the hazard and consequence factors in a fourth risk matrix.

The key advantage of the risk matrix approach is risk communication. Stakeholders and decision makers can clearly see how hazard and consequence were combined to arrive at risk. This transparency is particularly true for semi-quantitative risk matrices, which discretize the matrix axes based on real values (see Fig. 14.8). Many risk matrices describe hazard (e.g., likelihood of impact) and consequence (e.g., economic loss) in a way that allows many different types of hazards and hazard sites to be compared within the same framework. Risk matrices are simple and efficient to apply, so they are useful in regional risk-based prioritization assessments when many sites need to be compared.

A key disadvantage of the risk matrix approach is that many of the hazard sites and hazard types can end up with the same risk value, making it difficult to rank many hazard sites. This disadvantage can be overcome by using a risk matrix to demonstrate how the broad risk categories are defined, and a risk index to rank hazard sites that fall within the same broad category.

Porter et al. (2019) describes other challenges of using matrices, including:

- *When hazard and consequence descriptors are subjective*, users interpret hazard likelihood, consequence, and risk ratings differently from each other, leading to inconsistency in estimated risks.

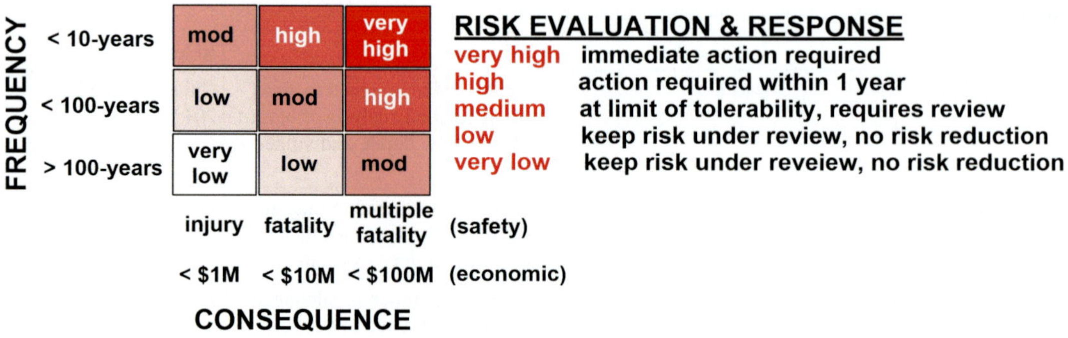

Fig. 14.8 Example semi-quantitative risk matrix

- *Risk descriptors are subjective* and often unique to the organization that developed and uses the risk matrix. Therefore, risk tolerance thresholds set by one organization are not easily transferrable to other organizations, which leads to inconsistency in the level of safety between organizations.
- *It is difficult to account for multiple scenarios*: A debris flow at a specific site can have many different magnitudes and possible impacts (See Sect. 14.2.1). Risk at a single site may need to be represented by multiple scenarios, but it is not possible to aggregate the risk from these scenarios in a single risk matrix.
- *Low-probability/high-consequence events tend to be omitted.* Many commonly used industrial risk matrices that compare risk from many different hazard types consider a 100-year return period to be 'very low' likelihood, but the highest consequence debris flows often have a return period of thousands of years.
- *Fractional fatalities are omitted*: In some risk scenarios, expected life loss is a fraction of one fatality (e.g., one person is exposed and has vulnerability < 1), but the consequence axis of a risk matrix often skips from injury to one fatality to multiple fatalities. The explanatory text accompanying most matrices lack guidance for how to treat a fractional fatality.
- *Overlapping hazards*: A single triggering event, such as a heavy rainfall, can cause many concurrent debris flows at most or all hazard sites within a system. The loss caused by these concurrent events, and the options for managing such events, can be much different than what is assumed when each site is assessed independently using a risk matrix.

Table 14.4 summarizes several risk matrix examples from the literature.

14.5.2 Quantitative

Quantitative risk analyses measure risks in absolute terms as the expected loss over a period of time. The metric of loss depends on the exposed elements at risk and the objectives of the risk assessment. Common metrics include:

- Economic loss in monetary units. Most losses (e.g., life loss, business interruptions, environmental impacts, cultural impacts) can be converted into a monetary value to allow comparison of the many different consequence types.
- *Individual life loss* that describes the probability of death to a specific individual.
- *Group life loss* that describes the probability and number of people killed by a hazard.
- *Duration of service loss* that describes the probability and number of days that a critical service, like road use, fuel, or electrical power is unavailable.

Table 14.4 Debris-flow risk matrix examples in literature

Location	Scale	Description	References
Switzerland	Site-specific or regional	Qualitative risk level is used to guide land use planning and development	Zimmerman (2005)
Western Canada	Site-specific	Semi-quantitative risk matrices, considering occurrence, spatial, and temporal probability and vulnerability	Arksey and VanDine (2008)
Western Canada	Site-specific or regional	Qualitative risk rating system and examples for debris-flow risk after wildfires	Hope et al. (2015)
Korea	Regional	Qualitative risk value assigned to rank a large number of individual hazard sites on a highway	Kim et al. (2018)
Eastern Canada	Regional	Multiple matrices are used to assign a hazard level, expected consequence and qualitative risk value to many sites along a railway line	Lato et al. (2016)
Canada	Site-specific or regional	Discusses challenges of using risk matrices for risk management at resource development projects	Porter et al. (2019)
Gansu, China	Site-specific	Detailed analysis of intensity, probability of occurrence, and vulnerability are combined in a matrix to define risk zones on a fan	Ouyang et al. (2019)

Quantitative risk metrics allow absolute comparison of a hazard or hazard site to any number of external measures. For example, quantitative risk estimates can be used to compare:

- Debris-flow risk with other unrelated risks (e.g. risk of being killed in a car or drowning in a tsunami) and background risk levels (e.g., risk of dying from all causes in a year)
- Debris-flow risk with risk tolerance thresholds (including thresholds established in other risk settings)
- One type of consequence (e.g., environmental loss) with other unrelated types (e.g., economic loss)
- Benefits of tolerating the debris-flow risk with the risk itself
- Benefits of different risk reduction options
- Cost effectiveness of different risk reduction options

These comparisons can be used to assess tradeoffs that inform decisions and facilitate communication, including:

- Level of safety assessments to determine if a site is 'safe enough' from debris-flow hazards
- Assessments to justify the level of investment in debris-flow risk control
- Resource allocation planning to determine which hazards and hazard sites receive risk control funding
- Mitigation options assessment and sizing of final structural mitigation designs

Risk is estimated quantitatively as the product of the potential loss and probability of the loss occurring (Fig. 14.9). Equation 14.2 is a general form of the risk equation that can be used for various metrics and to estimate total risk from multiple debris-flow scenarios (adapted from Jakob et al., 2016).

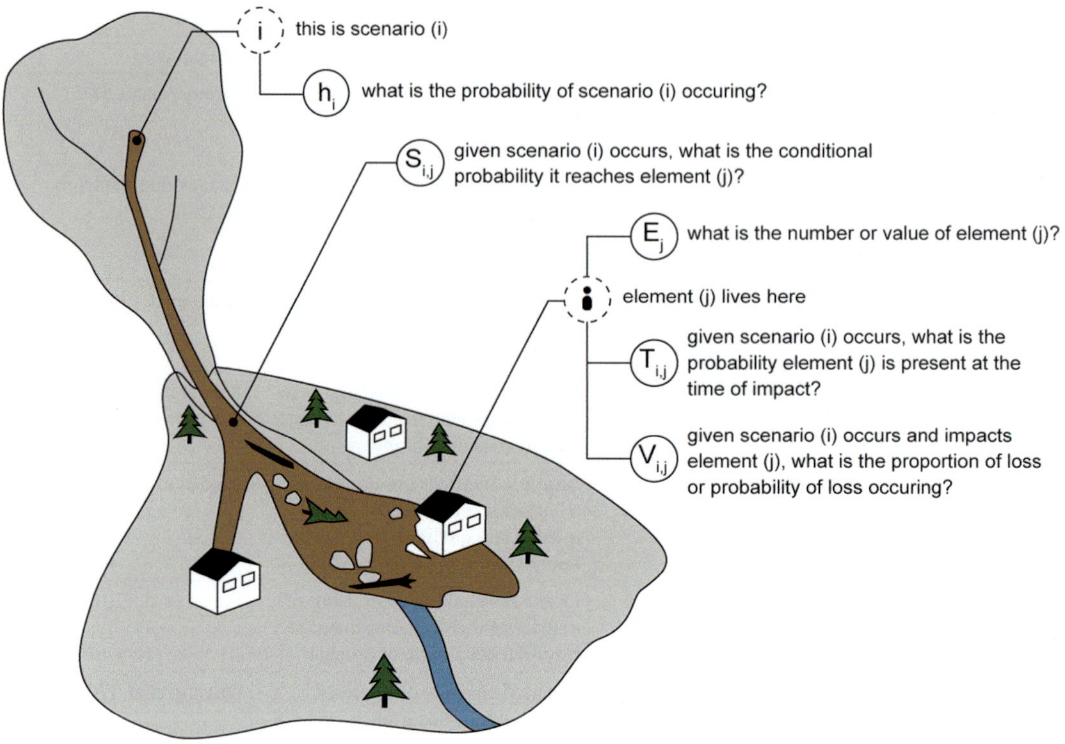

Fig. 14.9 Example debris-flow risk calculation for one element (j) in terms of life-loss or economic loss

$$R_j = \sum_{i=1}^{n} h_i * S_{i,j} * T_{i,j} * E_j * V_{i,j} \tag{14.2}$$

where:

- $\boldsymbol{R_j}$ *is the risk to element (j)* from (n) debris-flow scenarios. Risk at multiple elements can be summed to quantify total risk within a hazard zone, across a study region, or within an organization.
- $\boldsymbol{h_i}$ *is the probability of debris-flow scenario* (*i*), out of (*n*) total scenarios, in a given time period. When summed, the total probability of all scenarios, including the 'no debris flow occurs' scenario, should equal 1 (certainty). This parameter is related primarily to the return period or probability of debris flows that are large enough to cause loss.
- $\boldsymbol{S_{i,j}}$ *is the spatial probability of impact*, which given that scenario (*i*) occurs, is the conditional probability that the debris flow reaches element (j). This parameter is related primarily to the runout behavior of the debris flow. For a building, it could be the probability that the debris flow travels far enough down the fan to impact the building. For a buried pipeline, it could be the probability that the debris flow scours deep enough to impact the pipe.
- $\boldsymbol{T_{i,j}}$ *is the temporal probability of impact*, which given that scenario (*i*) occurs, is the conditional probability that the element (*j*) is present at the time of impact. This parameter is related to the behavior of the element at risk. It could quantify, for example, the proportion of time a person occupies a building, the probability that oil is present in a pipeline, or the proportion of time that a moving vehicle

occupies a hazard zone. The temporal probability of impact is 1 for stationary elements.

- E_j *is the exposure*, which is a quantity that represents the element at risk. For example, it can be the number of exposed people (or 1 for individuals) or the total economic value of infrastructure in a hazard zone.
- $V_{i,j}$ *is the physical vulnerability*, which given that scenario (*i*) occurs and impacts element (j), is the conditional probability of loss occurring or the proportion of loss. This parameter is related to both the characteristics of the element at risk and the flow intensity of the debris-flow scenario.

14.5.2.1 Economic Risk

For economic risk, it is common to assess risk for groups of similar elements, like roads, buildings, or businesses. The risk at each of these elements is then summed to estimate the total economic risk from debris flows. In this case, exposure is the economic value of the elements, and physical vulnerability is the proportion of loss caused by the debris-flow scenario at each element. Exposure is often the value of infrastructure (e.g. roads or buildings), but other elements, like life or environmental values, can be described in monetary terms and be included in the risk assessment.

The most common outputs are the annual average loss, or the probability of exceeding a given amount of economic loss, often plotted as an exceedance probability curve. Annual average loss is useful for cost benefit comparisons, while exceedance probability curves for economic loss are useful for identifying low frequency but high magnitude losses that may exceed what a community or owner can absorb.

Most economic risk analyses underestimate the risk. There are often several indirect and intangible losses that arise as a result of complex interdependencies between exposed elements (e.g. litigation costs, impacts on tourism, loss of employment) that are difficult to estimate and often not accounted for. These additional diffuse impacts can be represented or assessed qualitatively to further inform the risk assessment. Alternatively, economic values can be assigned to the indirect and intangible losses, usually following significant assessment and debate, so that they can be directly incorporated into the quantitative economic risk assessment.

14.5.2.2 Individual Life Loss Risk

Individual life loss risk is often presented as the Probability of Death to an Individual (PDI) or Annual Individual Fatality Risk (AIFR), typically for the individual who is most at risk. This is usually the person who spends the most time in the hazard zone or could be the person who is most vulnerable to the hazard, such as an infant, elderly person, or housekeeper. Spatial probability describes the likelihood that the facility (e.g., building, trail, road) occupied by the person is impacted during the debris-flow scenario. Temporal probability describes the proportion of time that the person spends in that facility.

Annual PDI values tend to be small numbers, written in various forms like 0.0006, 3×10^{-5}, or 2E−4. Most stakeholders, decision makers, and the public, including those who are highly educated, have a difficult time interpreting scientific notation. Even for an experienced risk analyst, the absolute size of small risks is difficult to grasp. A risk unit called "micromort", helps to overcome some of these challenges. A micromort is a one-in-a-million risk of sudden death (Ahmad et al., 2015; Howard, 1980). Multiply the PDI value by 1 million to convert it to micromorts (Fig. 14.10). For example, a risk value of 0.0006 becomes 600 micromorts, 3×10^{-5} becomes 30 micromorts, and 2E−4 is 200 micromorts. Risk values in micromorts may help risk analysts, stakeholders, and decision makers have a more intuitive feel for the magnitude of risks and uncertainties (Strouth & McDougall, 2022).

14.5.2.3 Group Life Loss Risk

Group life loss risk can be expressed as probable life loss (PLL) or graphically using an fN or FN diagram (Strouth & McDougall, 2021).

Probable life loss is calculated directly using the risk equation (Eq. 14.2). It describes the expected number of deaths over a period of time, such as annual probable life loss, or

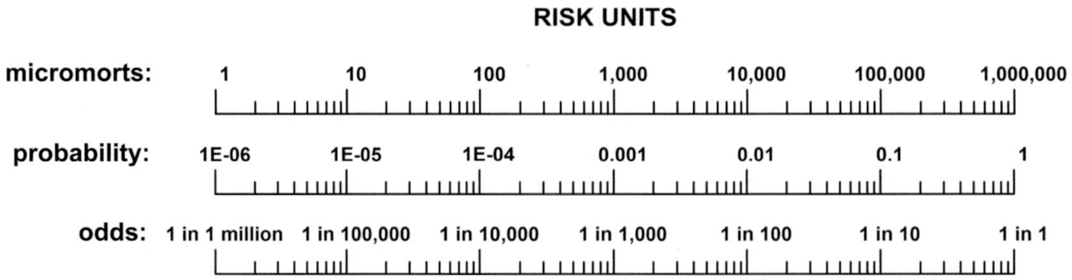

Fig. 14.10 A comparison of micromorts with other units used to express risk

probable life loss over the design life of a structure (e.g., 100-year period). Probable life loss expresses group risk as a single value, such as '3 lives lost per 100 years'. The single value can be useful for concisely comparing many different hazards or hazard sites, for cost–benefit analysis, and for comparing debris-flow mitigation options. For example, PLL from a single risk scenario can be compared against other scenarios, or PLL from various independent risk scenarios can be summed to yield the total PLL for a debris-flow hazard area. However, PLL also hides information, as it can include many combinations of possible scenarios and does not identify if risk is controlled by high or low probability scenarios.

An fN diagram is a graphical representation of group risk. It displays risk like a risk matrix. The x-axis of an fN diagram represents the number of fatalities (N) in a particular debris-flow risk scenario. The y-axis represents the probability (f) of the scenario. Each risk scenario is plotted as a point on the fN diagram to display its relative importance. fN diagrams are important, practical tools for prioritizing landslide scenarios and hazard sites and designing mitigation measures. fN diagrams inform many decisions, such as:

- Which scenario contributes most to total risk?
- Which scenario should be the focus of risk management actions?
- How much risk reduction is achieved by the proposed mitigation measures?
- What is a justifiable risk reduction expense?

The risk equation is separated into two parts to derive the f and N values. There are multiple ways that the equation can be divided depending on the objective of the risk analysis and the way in which debris-flow scenarios are defined. The key is that N represents the number of people killed by a single debris-flow scenario. Two examples include:

1. The number of people killed (N) is the product of the number of people exposed (E), vulnerability (V), the proportion of people present when the debris flow occurs (T), and the proportion of area impacted by the flow (S). The scenario probability (f) is represented only by the probability of debris flow scenario parameter (h_i). This formulation might apply for estimating group risk for people distributed across an entire fan (Fig. 14.11).
2. The number of people killed (N) is the product of number of people exposed (E) and vulnerability (V). The scenario probability (f) is the product of the debris flow occurring (h_i) and reaching the element at risk (S) at a time when people are present (T). This might apply for estimating group risk at a single building that contains many people, like a school.

An FN diagram combines many debris-flow risk scenarios into a single curve (Ball & Floyd, 1998; Baecher et al., 2015). It displays the probability (F) of *N or more* fatalities. F is calculated by summing fN pairs. For example, F associated with N = 1 is the sum of the probabilities (f) of all risk scenarios that resulted in 1 or more

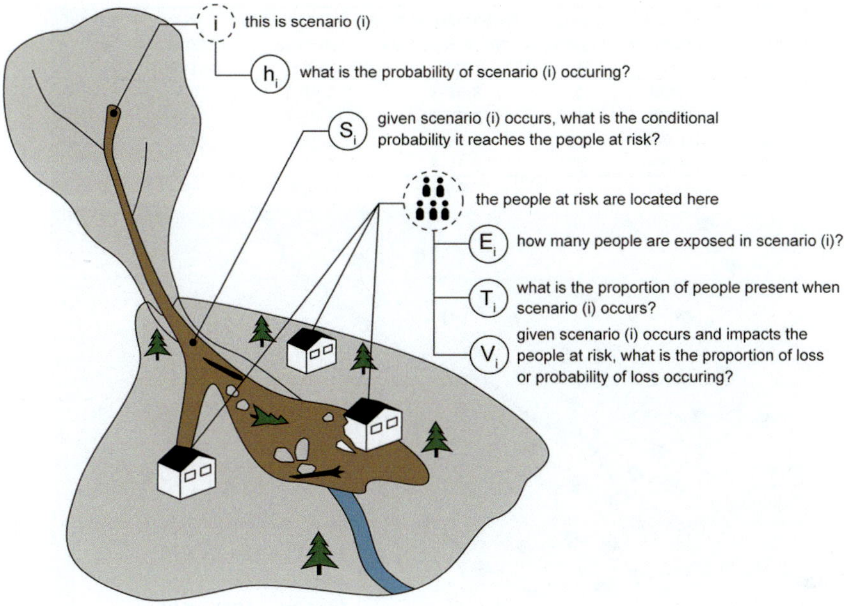

Fig. 14.11 An example group risk calculation

fatalities. The curve is plotted on a diagram with the x-axis being the number of fatalities (N) and the y-axis being the probability of N or more fatalities (Fig. 14.12). The curves illustrate the differences between high probability and low probability risks and are commonly used for comparing group risk at a given site to tolerable risk thresholds.

14.5.2.4 Key Points

Regardless of the loss metric that is used, the following key points apply to quantitative risk analysis (Strouth & McDougall, 2021, 2022):

- *Avoid risk estimate conservatism*: Engineers are trained to incorporate conservatism into their design assumptions. However, risk is overestimated when this conservative attitude is applied to the inputs in a risk analysis. Inflated risk estimates are not appropriate for risk evaluation.
- *Present uncertainty transparently*: Instead of 'being conservative' with parameter values, use best estimates and provide uncertainty bounds. USBoR-USACE (2019) provide several methods for estimating and portraying risk and uncertainty.
- *Present assumptions transparently*: Judgment is used to derive each parameter that contributes to the quantitative risk estimate. Each risk analyst will apply different assumptions. It is important that these assumptions are clearly documented so that they can be reviewed and debated.
- *Avoid a false degree of precision*: Quantitative risk estimates often imply a level of precision that is not warranted by the uncertainties that are involved. Use rounding of risk values and provide uncertainty ranges to avoid false precision. Conversely, if not enough information is known to quantify the uncertainty ranges, then clearly communicate that the singular risk estimate represents only one point in an unknown range of uncertainty.
- *Carefully define and combine scenarios*: The final risk estimate value is largely controlled by how the scenarios are defined and combined. Each risk setting is unique, and scenarios used to estimate risk should be fit

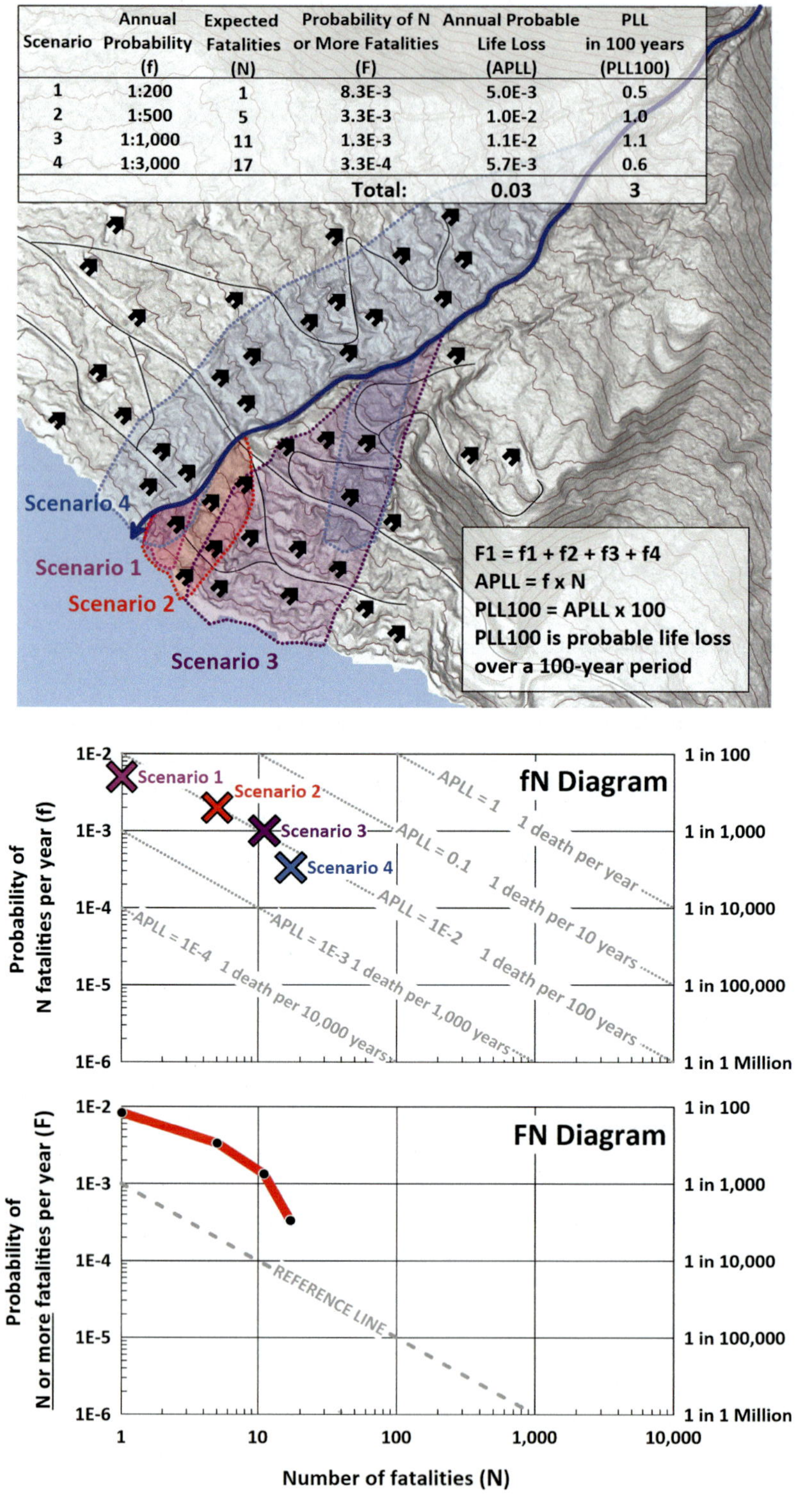

Scenario	Annual Probability (f)	Expected Fatalities (N)	Probability of N or More Fatalities (F)	Annual Probable Life Loss (APLL)	PLL in 100 years (PLL100)
1	1:200	1	8.3E-3	5.0E-3	0.5
2	1:500	5	3.3E-3	1.0E-2	1.0
3	1:1,000	11	1.3E-3	1.1E-2	1.1
4	1:3,000	17	3.3E-4	5.7E-3	0.6
			Total:	0.03	3

Fig. 14.12. A hypothetical example of societal risk at a debris-flow fan (Strouth & McDougall, 2021). APLL is annual probable life loss

for purpose. In some settings the total risk is dominated by high probability events (e.g., where vulnerability is high, or at short duration installations). In others, low probability/high consequence events control total risk (e.g., where structures are resistant to most impacts, or where low probability events have much higher intensity than more frequent events). Develop an understanding of the factors that control the total risk value before defining scenarios at a particular site.

- *Recognize low probability/high consequence scenarios*: Do not ignore low probability/high consequence scenarios. Where possible, quantify risk from the full range of possible debris-flow magnitudes. If it is not possible to quantify these risks or if the risk estimate uncertainty is too high, then assess low probability risks and uncertainties qualitatively and consider them in the risk evaluation decision. Alternatively, use a different method (e.g., stress testing) to supplement the risk estimate.

14.5.2.5 Quantitative Risk Examples

Table 14.5 summarizes quantitative debris-flow risk analysis examples from literature. These examples illustrate how economic, individual, and group life loss risks can be estimated for a variety of scales and settings.

14.5.3 Choosing a Method

No one method for estimating and presenting risk is inherently better than the others. For example, quantifying a risk does not make a risk estimate better or more accurate. On the contrary, a precise-looking risk value may be wildly inaccurate if input parameters are false. Each method has an application that depends on the nature of the problem to be solved and the available time and resources (Van Westen et al., 2006; Wong, 2005, 2013). This section compares the risk analysis methods and provides guidance for selecting a method.

14.5.3.1 Differences Between Qualitative and Quantitative Methods

Although exceptions can be found, information in Table 14.6 generalizes the differences between qualitative and quantitative methods. These methods can be used together in a complementary way to overcome the limitations of each. For example, a qualitative method could be used for screening and initial ranking of hazard sites, followed by quantitative analysis at the most critical sites.

14.5.3.2 Uses, Advantages, Limitations of Each Method

Similarly, each qualitative and quantitative method has specific uses, advantages, and limitations, as summarized in Table 14.7.

14.5.3.3 Examples

The following examples illustrate a few of the many factors that are involved in selecting a risk analysis method.

- *Screening assessment to identify sites for further study*: A transportation agency might want to identify debris-flow risks sites among thousands of kilometres of highway. Quantitative risk estimation would not generally be feasible because too little is known about the hazard characteristics at each debris-flow fan, and the effort and expense required to attempt a quantitative assessment would not be justified. A simple qualitative risk matrix could be used to assign risk ratings to each site based on hazard likelihood (e.g., activity of the fan and position of the road on the fan) and consequence (e.g., closure duration, number of people affected, relative importance of the highway).

Table 14.5 Quantitative debris-flow risk analysis examples in literature

Location	Scale	Description	References
Washington, USA	Site-specific	Group life loss risk assessment for up to 5,000-year return period, presented on FN diagram	Jakob and Weatherly (2005)
Western Canada	Site-specific	Detailed explanation and examples of estimating and presenting individual and group life loss risks and economic risks	Jakob and Holm, (2012); Jakob et al. (2012a, 2012b, 2016)
Qinghai, China	Regional	Economic and life loss risk estimates for people in buildings, with vulnerability varying by building construction type	Lan et al. (2013)
New Zealand	Site-specific	Individual and group life loss risk assessment for residential development, presented on maps and an FN diagram	Tonkin &Taylor, (2015)
Sichuan, China	Regional	Individual and societal life loss risk estimates for road users, and assessment of how redistribution of people affects risk	Chen et al. (2016, 2017)
Western Canada	Site-specific	Examples of individual and group life loss risk estimates, presented on FN diagrams	BGC (2015); Porter et al. (2017)
Sichuan, China	Site-specific	Virtual geographic environment system to support rapid life loss risk analysis and communication during a disaster	Yin et al. (2017)
Western Canada	Site-specific	Loss of containment risk assessment to support debris-flow protection designs at an oil and gas pipeline	Gartner and Jakob (2019)
Chuncheon, Korea	Regional	Economic risk to buildings for prioritization of risk areas, with vulnerability varying by building construction type	Choi et al. (2021)
New Zealand	Site-specific	Individual and group life loss risk assessment for visitors and workers	De Vilder et al. (2022); Massey et al. (2022)

- *Regional prioritization of many hazard sites*: The screening assessment described in the previous example could identify hundreds of 'high' risk sites. The transportation agency might want to know which of the high-risk sites are most important. The simple qualitative risk matrix does not provide enough information for ranking the sites, and a quantitative approach is likely to still be impractical due to effort, expense, and data limitations. A risk index approach could be used to rank the sites based on qualitative or semi-quantitative factors that describe the hazard likelihood and consequence at each site.
- *Level of safety for an individual*: Hazards on a debris-flow fan are often of the low probability, high consequence variety. They can be like a commercial airplane crash; they do not happen often but are highly destructive when they do occur. So, which risk analysis method best helps us determine if a person is *safe enough*? A risk matrix is not typically helpful. For example, imagine how commercial aviation risk would plot on the risk matrix in Fig. 14.8. No matter how many safety improvements are made, risk will always be "very high" because it is impracticable to eliminate all accidents. A risk index would be able to demonstrate the risk reduction associated with the safety

Table 14.6 Comparison of qualitative and quantitative risk analysis methods

Item	Method	
	Qualitative	Quantitative
Use	Commonly used for risk-based screening, ranking and tracking of hazard sites, or as a screening tool prior to a more detailed quantitative assessment	Commonly used for resource allocation, risk tolerance decisions and mitigation decisions (e.g., options assessments, cost/benefit assessment, and design sizing)
Inputs	Ordinal qualitative values, or quantitative values typically expressed as a range	Precise quantitative values
Outputs	Relative risk estimates, represented with descriptors (e.g. high, medium, low, or index values)	Quantitative risk estimates (e.g., lives lost, dollars lost, time lost)
Effort	Requires less input data, so less effort and less time to complete	Requires more input data and more precise input data, so are more effort to complete. Are not always feasible due to limitations in data, time, or budget
Scenarios Considered	Outputs commonly represent risk from one scenario, because multiple scenarios are not easily cumulated	More readily allows for cumulative risk estimates, such as the total risk from many debris-flow scenarios
Risk comparison	Are efficient for comparing hazard sites or hazard types within the same risk analysis scheme. Outputs do not have universal meaning so cannot be compared to risks estimated with another method	Outputs have an objective and universal meaning, so can be compared with many different hazard sites and hazard types
Communication	Outputs are subjective, and can contribute to misunderstanding between the risk analyst, decision makers, and the public	Outputs can be compared to other risk measures to support understanding but can sometimes imply a level of certainty in the risk estimate that is not warranted
Application	Are sufficient for informing and communicating many common decisions. More common than quantitative methods for large-scale studies that involve many different hazard sites	Are helpful for communicating and supporting more complex decisions, such as mitigation of a highly developed debris-flow fan. Not always needed if a simpler method can adequately support and communicates the decision

improvements but would not give the context needed to decide what risk level can be considered '*safe*'. In contrast, a quantitative individual life loss risk method could be used to estimate risk in terms of probability of life loss, and this risk value could then be compared to other types of hazards. This type of risk analysis would show, for example, that commercial airplanes are the safest mode of transportation (Allianz, 2020) despite rating as "very high" on the risk matrix in Fig. 14.8.

- *Level of safety at a tourist attraction*: The operator of a tourist attraction on a debris-flow fan might want to evaluate if visitors are *safe enough*. Quantitative individual life loss risk might not be helpful because each tourist spends such a small proportion of time in the hazard zone that individual risk is insignificant. A group life loss risk analysis might offer a better alternative, as it would provide an estimate of the probability and number of people killed, which could be compared to group risk from other types of hazards or risk tolerance thresholds.
- *Mitigation investment decision*: A decision maker in a local government might need to decide how much money to allocate towards debris-flow mitigation from a

Table 14.7 Uses, advantages, and limitations of the risk analysis methods

Method	Uses and advantages	Limitations
Qualitative methods		
Risk index	• Provides a single value that is easy to work with • Used where data is limited • Used where a high degree of precision is needed, such as: – Ranking & prioritizing many sites – Quantifying changes (e.g., from mitigation)	• Methods and schemes are fit for purpose and difficult to transfer to new settings • Index value has no meaning outside of scheme, and is difficult to compare to other schemes • Relative & subjective weighting factors may not represent risk for all stakeholders
Risk matrix	• Efficient method for assigning a risk level (e.g., high, moderate, low) • Good risk communication tool that shows how the risk value is derived from consequence and probability • Used where data is limited • Used for comparing risk estimates to risk tolerance thresholds (e.g., 'high' risks require mitigation)	• Multiple sites tend to have the same risk value, making it difficult to rank many sites • Hazard probability, consequence, and risk descriptors tend to be subjective, which leads to inconsistency in how they are assigned and interpreted • Risks from multiple scenarios or overlapping hazards cannot be easily aggregated
Quantitative methods		
Economic	• Used for estimating direct, indirect, and intangible consequences • Most consequence types (e.g., life loss, service disruption, business interruption, environmental harm) can be converted to monetary units • Used for cost–benefit assessment, resource allocation, mitigation sizing	• Loss estimates tend to be highly uncertain due to limited historical data and the many types of indirect and intangible consequences • It is challenging to convert and defend conversion of other consequence types to monetary units • Decisions made solely based on economic risk can promote societal inequalities by preferentially allocating resources to wealthy communities
Individual life loss	• Used for level of safety assessments to establish if a hazard zone is 'safe enough' • Most applicable where a person spends a large amount of time in a hazard zone (e.g., in a building) • Individual risk thresholds have a clearer basis than group risk thresholds and are easier to apply	• Input parameters and risk values can be highly uncertain due to limited historical data
Group life loss	• Used for level of safety assessments to establish if a hazard zone is 'safe enough' • Used where no individual occupies the hazard zone for a large proportion of time (e.g., public areas) • Most applicable where 10 or more people could be killed in a single scenario	• Input parameters and risk values can be highly uncertain due to limited historical data • Group risk tolerance thresholds are difficult to understand and typically lack a strong basis for application

community's limited annual budget. A quantitative economic risk assessment could provide the amount of debris-flow risk reduction achieved by the proposed mitigation in monetary terms, which could set the justifiable budget for the mitigation.

14.6 Risk Evaluation

What ought to be done about the risk?

The third phase of risk assessment is risk evaluation. Risk evaluation is the comparison of estimated debris-flow risks with uncertainty, available resources, and perceptions of tolerable risks to decide what actions, if any, are needed. This is the phase in which decisions are made, with the goal of achieving an acceptable balance between the risk and the benefits that come with taking the risk.

This section describes key components of risk evaluation and methods for communicating risks and risk reduction options to facilitate stakeholder engagement and guide decisions, including: (1) identifying and using *uncertainty* to guide next steps; (2) the concept of *risk tolerance* and conditions under which people are willing to take a risk; and (3) methods for comparing and describing the benefits of *risk reduction measures*.

14.6.1 Uncertainty

Risk estimates are forecasts of future events and, by their very nature, are uncertain. Therefore, a first step of risk evaluation is to determine if enough is known about the risk to support a decision. If uncertainty is too high, the uncertainty may need to be reduced before decisions are made (Fig. 14.13).

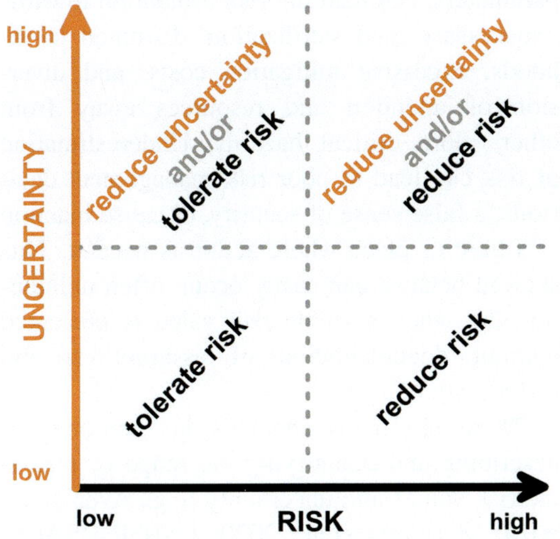

Fig. 14.13 Risk evaluation actions as a function of risk level and uncertainty level. Adapted from USBoR-USACE (2019)

The amount of uncertainty that can be tolerated is specific to the setting, decision, and risk level. Some decisions can tolerate a high level of uncertainty (e.g., screening assessment that identifies sites for more detailed study), while others need low uncertainty (e.g., a site-specific risk assessment that informs mitigation design). Similarly, a decision may be clear in the face of high uncertainty if all plausible risk estimates support the same decision (e.g., the full range of risk estimates is unacceptably high). In other situations, a risk estimate with relatively low uncertainty may fall on a boundary between two decisions, requiring uncertainty to be further reduced.

Uncertainty in debris-flow risk estimates tends to be a function of both the random variability of debris flows and a lack of knowledge about the debris-flow hazard, exposure, and vulnerabilities (Vose, 2009). It can be helpful to identify which type of uncertainty is coming from the various uncertainty sources. For example, are input data showing natural variability or measurement errors? Are modelled scenarios representing the natural variability or are they testing a lack of knowledge about behaviour?

The random variability of debris flows cannot be eliminated. Within the same creek channel, each debris flow can have different compositions, flow phases, flow velocities, depths, discharges, and intensities. For example, eight debris flows were recorded at Illgraben, Switzerland in 2005 and 2006, with a front velocity ranging from 0.7 to 7.9 m/s, flow depth ranging from 0.8 to 2.7 m, and volume ranging from 4,000 to 55,000 m^3 (Wendeler, 2016). Perhaps most interestingly, the maximums in each of these ranges did not occur during the same event. Similarly, the random variability of exposure (e.g., number of people present) and vulnerability (e.g., people located outdoors or indoors) cannot be eliminated.

However, uncertainty in debris-flow risk estimates caused by lack of knowledge can be reduced through improved data collection, model validation, and expert input. Sometimes the reduction in uncertainty can alter the risk estimate in a way that controls a decision.

Therefore, uncertainty is explored in the risk evaluation by asking:

1. Is knowledge supporting the assessment strong enough to guide the decision?
2. Do all plausible risk estimates support the same decision?
3. If not, is it feasible to reduce the uncertainty?
4. What is known with confidence that can support clear communication?

14.6.1.1 Strength of Knowledge

Is knowledge supporting the assessment strong enough to guide the decision?

Risk assessment knowledge is strong if the assessment is based on reasonable assumptions for which there is reliable data, well understood phenomena, reliable models, and broad agreement among experts. Risk assessment knowledge is weak if the assessment is based on strong simplifications, unreliable data, and unreliable models, and has a lack of consensus among experts (Aven et al., 2014).

For example, knowledge of a debris-flow risk estimate might be considered 'strong' if it is based on a detailed hazard assessment to establish the frequency-magnitude relationship and a variety of hazard scenarios, detailed exposure model developed by field surveys, well-constrained numerical modelling, and technical review by experts. Knowledge of a debris-flow risk estimate would be weak if it is based on a regional debris-flow hazard model, regional exposure model, and no field validation.

Risk assessments based on strong knowledge tend to have lower uncertainty, while those based on weak knowledge tend to have higher uncertainty and higher unrecognized uncertainty (i.e., unknown unknowns).

While many decisions require a strong basis of knowledge, some do not. Regardless, it is important to review the state of knowledge, present it transparently, and determine if it sufficient for the decision that needs to be made.

If strong knowledge is needed, and guidance is not readily available in published literature, it can be helpful to engage a panel of experts. It is rare that each expert would estimate the same risk value. Bringing experts together allows areas of consensus to be identified and areas of disagreement to be explored, with an ultimate goal of defining the range of reasonable input parameters to be used in risk calculations. The US Bureau of Reclamation and Army Corps of Engineers (2019) have developed best practices for expert elicitation in risk assessments that can be adapted to debris-flow risk assessments.

14.6.1.2 Risk Estimate Range

Do all plausible risk estimates support the same decision?

As risk estimates are inherently uncertain, it is best to present them as a range that illustrates the plausible risk values. If a single risk value is needed for a risk evaluation comparison, use the best estimate value while also presenting the range of uncertainty separately and transparently. Presenting risk as a range satisfies the decision maker's need to understand the possible range of outcomes as well as the risk analyst's need to convey uncertainty in the risk estimate.

As the goal of the risk evaluation is to achieve a balance between risk and benefits of taking the risk, it follows that a risk estimate that is over- or underestimated will skew this balance. Overestimation of risk, for example by applying 'conservative' assumptions to input parameters, can lead the risk evaluation towards unnecessary land sterilization, disrupted livelihoods, excessive mitigation costs, and diversion of attention and resources away from other, more critical hazards. Underestimation of risk can lead to poor risk management decisions, a false sense of security, or cause inaction at a time or place where action is needed. This skewed balance can easily occur, often unintentionally, when a single risk value is presented without documentation of assumptions and uncertainty bounds.

There are many methods in literature for describing and quantifying the range of plausible risk values and uncertainty (e.g., Aven et al., 2014; Macciotta et al., 2020; USBoR-USACE, 2019; You & Tonon, 2012;).

The most simplistic approach is to repeat the risk analysis with 'best-case', 'worst-case', and 'most-likely' parameter values to arrive at 'lower-bound', 'upper-bound', and 'best-estimate' of risk. Some decisions may be based on the upper-bound (e.g., prioritizing sites based on the 'worst-case'), others on the best-estimate (e.g., cost–benefit assessment for a mitigation investment), and others on the lower-bound (e.g., if the lower-bound risk is intolerable, this can be clear justification for risk reduction). Vose (2009) describes how this simplistic approach can dramatically over-estimate (or under-estimate) risk if the 'worst-case' (or 'best-case') of each parameter is simultaneously applied.

Alternatively, a logic-tree approach can be used to define a wide range of plausible scenarios and to assign a subjective probability to each scenario. An example of this approach was provided in Sect. 14.2.1, in which a debris-flow hazard was divided into 84 different scenarios to explore how consequences changed with plausible variations in debris-flow magnitude, avulsion location, and rheology. Where possible, to reduce the opportunity for calculation errors and minimize effort, the logic tree should be "pruned" to exclude analysing scenarios with low conditional probabilities that have an inconsequential bearing on the risk estimate.

Another approach is sensitivity analysis (e.g., Monte Carlo analysis), where different values for input parameters are sampled and used to derive the risk estimate and constrain uncertainty (Vose, 2009). While this approach can provide a detailed understanding of the uncertainty, estimating distributions for input parameters is challenging.

Regardless of the method used, risk analysts and decision makers need to decide if enough is known about the risk to support a decision. The decision is clear if the full range of plausible risk values supports the same decision. The decision may also be clear if factors like availability (or lack) of resources, perceived benefits of taking the risk, and application of the precautionary principle, dictate the decision as opposed to the risk estimate. However, sometimes the uncertainty range spans multiple decisions, in which case a next step may be to reduce uncertainty prior to acting on the risk.

14.6.1.3 Uncertainty Reduction

Is it feasible to reduce the uncertainty?

The process for estimating risk and reducing uncertainty tends to be iterative, as follows:

- *Step 1*: Estimate risk with uncertainty bounds.
- *Step 2*: Evaluate if a decision can be made despite the uncertainty. End at this step if the decision can be well supported. Continue to Step 3 if uncertainty is clouding the decision.
- *Step 3*: Identify which risk estimate input parameters cause the largest uncertainty, and methods for reducing the uncertainty in those input. If no feasible methods for reducing uncertainty are found, make a decision and use other methods or risk control measures to manage the uncertainty. Continue to Step 4 if uncertainty reduction is feasible.
- *Step 4*: Carry out data collection, analyses, expert elicitation, or other actions to reduce the uncertainty in risk estimate input parameters. Return to Step 1.

Reducing the uncertainty in a risk estimate usually requires more data, more time, more extensive analysis, and input from experts. Sometimes it can cost more to reduce uncertainty than to implement a risk control. *Analysis paralysis* (i.e., over-analysing a problem without making a decision) is similarly a legitimate concern, particularly because uncertainty can never be eliminated for debris-flow risk. Therefore, when uncertainty is clouding a decision, the next step is to explore the feasibility of reducing the uncertainty, prior to beginning a blind pursuit of more data and better analyses. Feasibility may be related to:

- *The strength of knowledge*: There may be few opportunities to effectively reduce uncertainty if the risk estimate is based on strong knowledge.
- *Availability of funding*: Some decisions have larger impacts and affect larger, more expensive infrastructure. Larger scale decisions tend to have more budget available for reducing uncertainty. In other cases, the risk

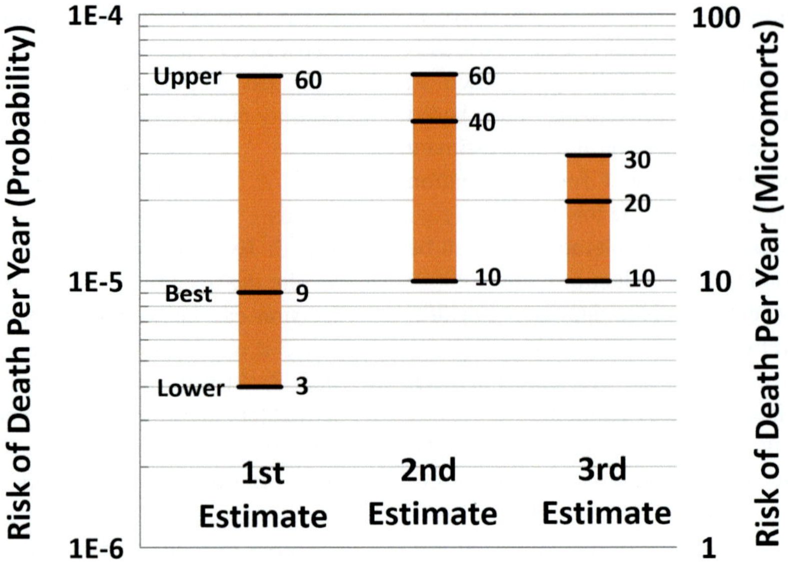

Fig. 14.14 An example of three iterations of a risk estimate. Uncertainty reduces with each iteration of the estimate, but is never eliminated, and the best estimate falls within the uncertainty boundaries of previous iterations

assessment budget is fixed and insufficient for reducing uncertainty.

- *Availability of data*: Data that support hazard scenario probability, exposure, and vulnerability estimates tend to be scarce and subject to the law of diminishing returns. Initial data collection can be highly cost-effective and important for reducing uncertainty. At some point, additional data can become more expensive to collect and have little bearing on the final decision. Sensitivity analysis is a useful tool to identify parameters that drive the risk estimate, and therefore which data can be collected for the largest reduction in uncertainty.
- *Availability of time*: Some decisions (e.g., during an emergency) need to be made quickly when there is little time to reasonably reduce the uncertainty.
- *Availability of options*: At some point, it becomes impossible to further reduce uncertainty. Even if knowledge could be complete, uncertainty related to random variability would remain.

When it is not feasible to reduce uncertainty, a decision needs to be made despite the uncertainty. Sometimes the decisions rely on other methods (e.g., stress testing, factor of safety calculations) or incorporate risk control measures (e.g., slope monitoring, early warning, and evacuation) to manage the uncertainty. When risk and uncertainty are estimated and presented properly, the uncertainty bounds should shrink with each iteration of risk analysis, and each subsequent risk estimate should fall within the uncertainty bounds of previous estimates (Fig. 14.14).

14.6.1.4 Uncertainty Communication

What is known with confidence that can support clear communication?

Unresolved questions tend to be what scientists and risk analysts find most interesting, and as presented above, uncertainty is an important component of the risk analysis. However, in risk communication with decision makers and the public, a focus on uncertainty conveys that no one understands what is happening and can undermine important messages about hazards and risk. This leads to confusion, contributes to fear, and undermines a decision maker's ability to make informed, logical decisions about risks (Jones, 2020; Lerner et al., 2015). Saunders et al. (this volume) presents an interesting case study pertaining to this point.

The range of risk estimates and uncertainty bounds that we advocate for in previous sections are not intended to undermine confidence in the risk assessment. Instead, they can be used to build confidence in an assessment. For example, they overcome the false precision that a single risk estimate suggests and that will likely be seen as inaccurate when the next debris flow occurs. False precision undermines faith in the risk assessment. Focus risk communication on what is known about the risk, rather than the uncertainties. Risk analysts, engineers, and earth scientists tend to know a lot about debris-flow risk even at sites with large uncertainty bounds. Clear communication of what is known guides decisions and leads to action. For example:

- Although the timing is uncertain, debris flows will occur and will impact the debris-flow fan.
- Sometimes, even the lower-bound of the risk estimate range is too high and needs to be reduced.
- Other times, a risk control is clearly a cost-effective and prudent measure to protect against the full range of possible debris flows.

14.6.2 Risk Tolerance

14.6.2.1 Tolerability of Risk Framework

The tolerability of risk framework developed by the UK Health and Safety Executive (HSE) (2001) is a model for conceptualizing risk levels that has been widely adopted for debris-flow risk (Macciotta & Lefsrud, 2018; Porter & Morgenstern, 2013). It categorizes risks as broadly acceptable, unacceptable, or tolerable.

- *Broadly Acceptable* risks are low enough that additional risk reduction is not necessary. Risks in this region are regarded as insignificant or trivial and similar to risks taken in daily life.
- *Unacceptable* risks are the opposite extreme and are so high that they are not taken regardless of the associated benefits. Taking the risk is ruled out unless the risk can be modified to reduce it to a tolerable level.
- *Tolerable* risks fall between these two extremes and are worth pursuing to secure certain benefits.

The tolerability of risk framework is a structure for dialogue about challenging risk, cost, and benefit trade-offs in each decision (Bandle, 2007; Le Guan, 2007). HSE developed the tolerability of risk framework as a basis for flexible and adaptable risk-informed decision making, recognizing that the factors that govern risk decisions vary by time, place, and circumstance.

The boundary between unacceptable and tolerable risks is called the tolerable risk threshold. The following sections describe options for this threshold, as well as options for tolerating a risk.

14.6.2.2 Conditions for Tolerating a Risk

Most risk assessments deal with risks that fall in the transitional 'grey-zone' between clearly safe enough and clearly unacceptable. These risks can be lived with to acquire benefits that come with taking the risk, if certain conditions are met. Risk tolerance decisions hinge on the balance between the level of risk, benefits of taking the risk, and cost of reducing the risk. Therefore, in each risk assessment, it is important to explicitly identify the benefits, costs, and risks along with the stakeholder who bears the costs and reaps the benefits.

Debris-flow hazards occur in mountainous areas with limited land available for development. Therefore, example benefits of taking a debris-flow risk include:

- To avoid other more critical hazards like flooding or rock fall that affect adjacent areas
- To take advantage of elevated, well-drained, gently sloping ground that is easier to develop than river flood plains or steeply sloping terrain
- To protect more critical environmental or ecological resources in adjacent areas
- To remain in a familiar and cherished home
- To be near population centres that provide schools, resources, and economic opportunities
- To provide infrastructure and services to communities in mountainous settings

Risk reduction is sometimes, but not always, desirable. The costs that come with risk reduction measures are immediate, tangible, and easily overlooked. Example costs include:

- Diversion of energy and resources away from other hazards, which are sometimes more critical than debris flows
- Capital cost of mitigation structures
- Operation and maintenance costs of mitigation structures
- Where evacuation is used, costs can include families being forced from their homes and lost business revenues
- Environmental impacts of mitigation structures

Many conditions exist under which risks are tolerated. In literature, the most commonly cited condition is that risk is reduced until it is as low as reasonably practicable (ALARP). However, the definition of ALARP is inconsistent in literature, with at least 3 meanings (Strouth et al., 2022):

1. Gross disproportion
2. Cost-effective
3. Synonym for tolerable

The original *gross disproportion* definition of ALARP originates in the UK and is related to the obligations of employers to protect the safety of workers. It requires risk to be lowered until the point of gross disproportion between the cost and benefits of risk reduction. In effect, a risk is ALARP if it is insignificant in relation to the cost in money, time, or trouble required to reduce it further. This definition is not one of balancing cost and benefits of measures, nor of finding cost-effective, pragmatic measures. Rather, it requires adopting measures except where they are ruled out because they involve grossly disproportionate sacrifices (Baecher et al., 2015; HSE, 2001;). This definition applies where those who benefit from taking the risk (e.g., a factory owner) are different from those who bear the potential for loss (e.g., factory workers), and in settings where there are financial resources to pay for grossly disproportionate risk control measures, and a regulatory framework to force these investments. In the context of debris-flow hazards, this definition could apply at new developments in recognized hazard zones or where the debris flow hazard is caused by land use changes (e.g., mining infrastructure, forestry roads). This definition is illogical for economic risks faced by businesses and for many settings with constrained resources.

The second, *cost-effective* definition of ALARP is commonly used or implied in geohazards literature. In this usage, a risk is ALARP if reasonable and cost-effective measures are implemented (FERC, 2016; Macciotta & Lesfrud, 2018). This focus on pragmatism and cost-effectiveness is often applicable where the resources needed for the desired risk reduction are not available and where a policy of disproportional investment in risk reduction is illogical. For example, this would apply in cases where a business bears both the benefits of taking the risk and the potential for loss. This definition would typically be applied where debris-flow hazards are recognized at existing developments and where public funding sources are used to manage a wide variety of hazards (e.g., local governments, transportation agencies).

The third usage of ALARP is as an implied *synonym for tolerable* (Lee & Jones, 2014). For example, by saying that "risk is currently unacceptable, but could be ALARP if an early warning system is implemented." This is a symptom of the confusion that surrounds the ALARP principle, and it contributes to further confusion and miscommunication.

In reality, there are many conditions under which people tolerate debris-flow risks, and sometimes several conditions apply. Specific, clear, and achievable conditions for tolerating a risk should be identified, debated, and applied for each new risk scenario. Depending on the setting, risk can be tolerated if:

1. *Risk is kept under review*: A review might include identifying changes to: benefits of tolerating the risk, costs of risk control, effectiveness of risk control, opportunities for implementing new risk controls, relative importance of other hazards, uncertainty in the risk estimate, hazard characteristics, elements at risk, and perceptions of the risk.
2. *Good practice* is followed in managing risk. Good practice is specific to the context, and varies by region, industry, and hazard. Where good practice is well defined (e.g., professional standards, regulations), this can be the simplest and most effective condition for tolerating a risk, and the only condition that is needed.
3. *Cost of risk reduction is unaffordable*: Many risks are tolerated simply because the resources needed to reduce risk are not available, the intangible costs of risk reduction are unacceptable, and the benefits of taking the risk outweigh the costs of avoiding the risk. Willingness-to-pay (WTP) describes the maximum price a person or community is willing and able to pay for risk reduction and is highly variable (e.g., Viscusi & Masterman, 2017).
4. *Risk is reduced as far as possible with available resources*: Reducing risk as far as possible using whatever resources are at hand is a pragmatic approach to tolerating a risk. It is a condition that applies when the desired risk reduction is unacceptable, unaffordable, or takes a long period of time to be implemented.
5. *Cost-effective measures* are implemented to manage risk. Requiring that cost-effective mitigation measures are implemented is a common condition for tolerating geohazard risks by governments and by large businesses, like mining or hydropower, particularly when the focus is on economic risks. These entities tend to bear both the benefits of taking the risk and the potential for loss, and they use a policy of balancing risk and reward to manage risks. It is particularly important within organizations (such as governments) that have more hazards and hazard sites than can be managed with available resources.
6. *Cost of further risk reduction is grossly disproportionate to the improvement gained*: It is relatively uncommon to tolerate a risk only when the risk is lowered to the point where the costs of further risk reduction are grossly disproportionate to the benefits of risk reduction. This condition is applied most often in relation to life loss and safety risks and could also be applicable to environmental risks.
7. *Tolerable risk threshold*: Risk is tolerated if it is less than a tolerable risk threshold.

14.6.2.3 Tolerable Risk Thresholds: Qualitative

In qualitative risk assessments, each risk matrix, index, and project is unique, but the risk descriptors are often defined around the concept of tolerable risk. For example, each degree of risk may be tied to a risk evaluation and have a prescribed risk control, such as (see also Cui et al., 2013a, 2013b; Zimmerman, 2005, Fig. 14.8):

- 'High' risks are unacceptable and must be reduced to moderate or lower
- 'Moderate' risks are tolerable and must be kept under review and employ cost-effective risk control measures
- 'Low' risks are acceptable, but must be kept under review

An advantage of the project-specific tolerable risk thresholds is that they reflect the risk perception, willingness-to-pay to avert the risk, and resources available to the stakeholders who create the thresholds. However, it can be challenging to directly transfer tolerable risk thresholds from one setting to another or across hazard types due to the relative, subjective, and inconsistent nature of the risk descriptors.

14.6.2.4 Tolerable Risk Thresholds: Quantitative

This section focuses on quantitative risk tolerance thresholds applied to life-loss risks, as this is the most common type of quantitative threshold applied in debris-flow risk evaluation. Thresholds for other types of losses (e.g., maximum closure period for a highway or utility, maximum economic loss) could also be set. For example, Gartner and Jakob (2019) describe a threshold for risk of product loss for buried oil and gas pipelines on debris-flow fans.

Quantitative risk tolerance thresholds can be defined for both individual and group (societal) life-loss risk. In both cases, the thresholds are absolute values that describe the probability of life loss. This leads to strengths and limitations that are nearly opposite to those of qualitative thresholds.

The key advantage of quantitative risk tolerance thresholds is that they are transferable and can be referenced for a wide range of settings and hazard types. This advantage is critical for comparing and prioritizing across hazard types within a single region. The consistency and transparency of quantitative risk tolerance thresholds also facilitates communication of risks and decisions, as stakeholders can more easily rationalize about debris-flow risks when they are compared to risks that people take in daily life (Fig. 14.15).

On the other hand, this ease of use can lead to misapplication (Strouth & McDougall, 2022). No risk tolerance threshold is universally applicable. Each author who originally proposed quantitative thresholds for landslides (e.g., ERM, 1998; Fell, 1994; Finlay & Fell, 1997; Leroi et al., 2005) describe the importance of engaging stakeholders, decision makers, and the public in defining tolerable risk thresholds for a given region, hazard type, and set of conditions. Ultimately, it should be the stakeholders, decision makers and public who determine the tolerable risk thresholds, with the practitioners providing the risk estimate, uncertainty estimate, and advice for risk reduction measures. The thresholds should reflect the perception of debris-flow risk relative to other hazard types, the willingness-to-pay to avert risks, and the resources and options available for managing the risks. As risk perception, willingness-to-pay, and resources vary between regions and circumstances (e.g., cultural, economic, educational), it follows that tolerable risk thresholds should also vary. Therefore, beware of risk assessments that refer to 'international guidance' on risk tolerance or blindly adopt risk tolerance thresholds from other regions. Instead, use these thresholds as benchmarks to facilitate discussions of risk tolerance with stakeholders.

Depending on the setting and circumstances, risk tolerance thresholds can promote equality or exacerbate inequalities. For example, if a government commits to funding risk reduction measures, a risk tolerance threshold can be used to set an equal level of safety for all residents. However, inequality can increase if a risk tolerance threshold is set without resources committed for risk reduction. Strouth and McDougall (2021) show how wealthy communities in Western Canada were able to acquire mitigation, whereas poorer communities were left to tolerate debris-flow risks that exceeded the risk tolerance thresholds because mitigation funding was not available. Inequality can be further exacerbated by using economic risk assessments for decision making because economic risks and benefits of risk reduction are greatest for wealthy communities. The concept of social vulnerability (See Sect. 14.2.3) can be used to overcome these challenges and to promote equality.

Individual life-loss risk tolerance thresholds (e.g., Bründl et al., 2009; ERM, 1998; Fell, 1994; Porter & Morgenstern, 2013) tend to be in the range of 10 to 1,000 micromorts per year (1 in 100,000 to 1 in 1,000 risk of death per year; Fig. 14.15). The thresholds are typically derived from the background risk of death from other causes for the exposed people. Background risk of death varies substantially by age, world region, occupation, and life-style choices (Fig. 14.15).

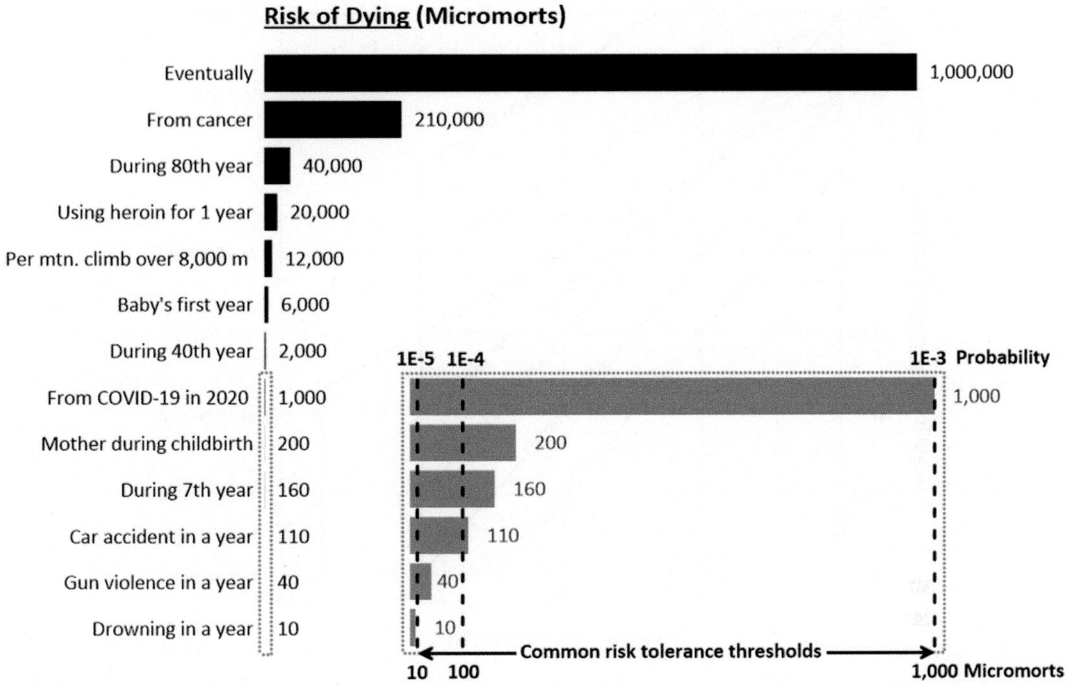

Fig. 14.15 Risk of death in the USA for several populations, causes, and activities (CDC, 2020; Heron, 2021), except heroin use and Mt. Everest which are from Blastland and Spiegelhalter (2014)

Generally, higher risks are tolerated if risk is taken voluntarily, risk is related to an existing (rather than new) development or installation, and if (by necessity) there are no acceptable or affordable methods to reduce risk. Lower risk tolerance thresholds are set if risk is taken involuntarily or affects children, if there are resources and regulatory structures that promote risk reduction, and if one entity is liable for risks faced by others. For example, Fell (1994) reported that acceptable levels of voluntary risks were two orders of magnitude higher than acceptable levels of involuntary risk.

Multiple thresholds can be defined within a given region to account for this variability. For example, several jurisdictions tolerate higher risk for existing development compared to new development (e.g., ERM, 1998; Tappenden, 2014). Similarly, different thresholds have been defined for the different groups of National Park users in New Zealand, as illustrated by the case study in Sect. 14.6.2.5.

Group risk tolerance thresholds are usually defined using an fN plot or FN plot (Fig. 14.16) and are called *FN criteria*. They are based on the idea that societies have a lower tolerance for incidents that cause many simultaneous casualties (Fell et al., 2005). The threshold line is defined by an anchor point and a slope. The anchor point is the tolerable probability of a disaster that kills several people. For example, if the death of 10 or more people in a single debris flow is thought to be a disaster that would provoke a societal response, and the tolerable annual probability of a disaster is 1 in 10,000, then the anchor point would be at $N = 10$, probability $= 10^{-4}$. The slope of the line is set by a society's relative aversion to multiple fatality events. In most examples, the slope is -1 or -2 in log–log space (Ball & Floyd, 1998; Sim et al., 2022).

Group risk tolerance thresholds vary widely based on the scale of application, purpose, and region, and are not directly transferable from

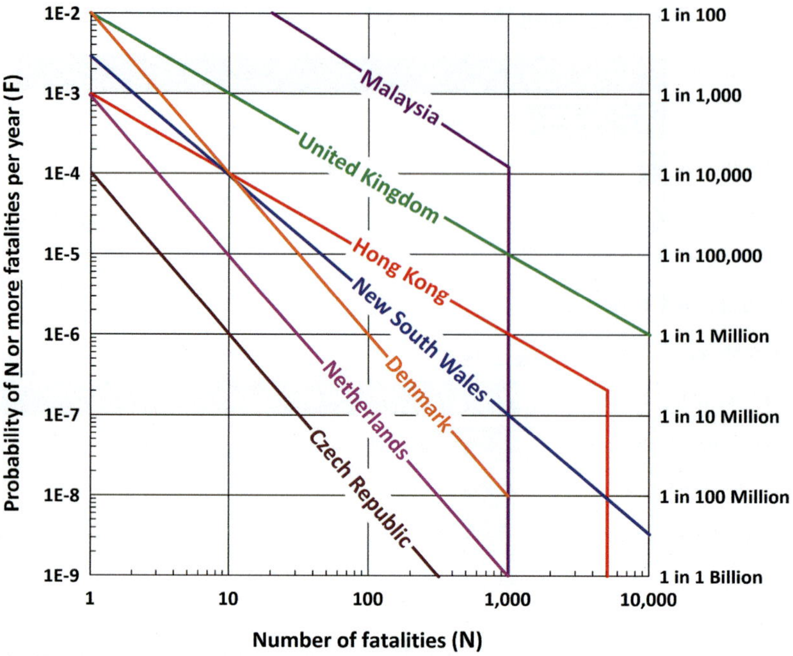

Fig. 14.16 Comparison of FN criteria developed in different places for different hazards and different scales, adapted from Sim et al. (2022)

one region to another (e.g., Fig. 14.16, Ball & Floyd, 1998; Hungr & Wong, 2007; Strouth & McDougall, 2021; Sim et al., 2022). For example, the Hong Kong criteria are intended for evaluating risk at individual landslide sites that are densely populated and only 7 ha (500 m by 150 m) in size (ERM, 1998). The Hong Kong criteria are not applicable for evaluating debris-flow risk for an entire city, region, or country.

Due to the inherent complexity and uncertainty of both quantitative risk assessment and public perception, quantitative risk tolerance thresholds should be seen as guidelines to inform a decision rather than a precise dividing line (Ball & Floyd, 1998). For example, it can be helpful to categorize risk broadly as clearly above the threshold, near the threshold, or clearly below the threshold. A US agency that manages dam safety goes so far as to intentionally omit terms like "risk tolerance criteria" and "tolerable risk". Instead, they replace the tolerable risk threshold with a "reference line". Justification to reduce risk increases for risks that plot farther above the reference line and decreases for risks that plot below the reference line (Fig. 14.17, USBR, 2011).

Similarly, recognize that in many settings, the individual or group risk tolerance "threshold" that is set by a government or managing agency is not really a dividing line between risks that are so high that they are not taken (i.e., unacceptable) and risks that can be tolerated. Instead, the "threshold" represents an aspirational goal for risk reduction. In these cases, many people or groups may be tolerating risks that are above the "threshold" for many years or decades, while authorities prioritize hazard sites and procure resources for reducing risk.

If the risk reduction goal is set so low that resources are not available to reach it or people are not willing to bear the cost of risk reduction, then the goal has limited use for decision making, and can lead to challenging political situations. For example, after a devasting debris flow in 2005, a lengthy risk reduction and risk assessment process was undertaken to quantify the debris-flow risk to the town of Matata, New Zealand (see Saunders et al., this volume,

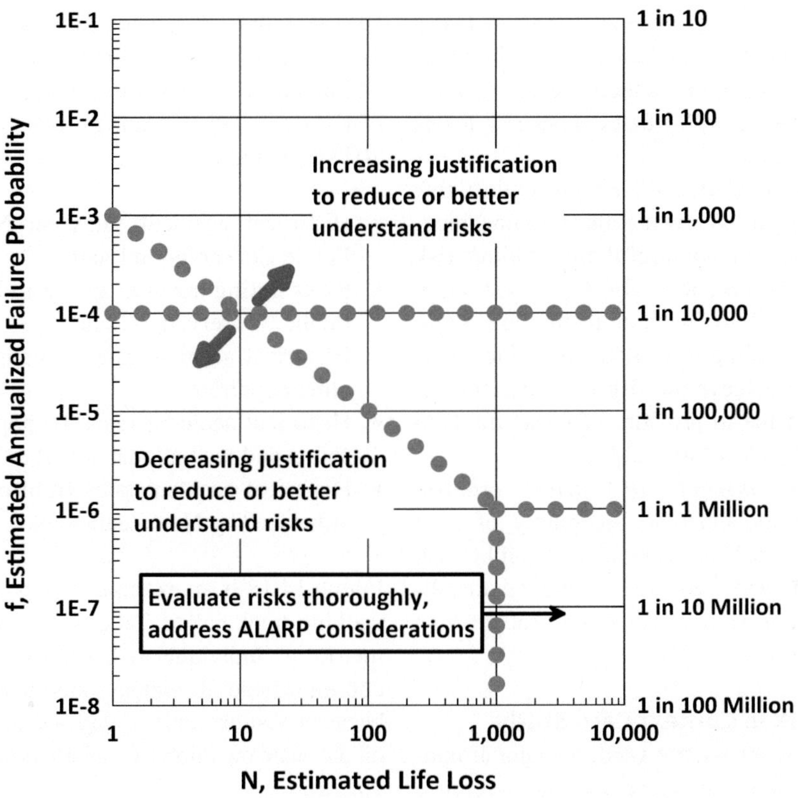

Fig. 14.17. A fN criteria adapted from US Bureau of Reclamation for dam safety (USBR, 2011)

Farrell & Davies, 2019). This process revealed life-loss risk estimates greater than 100 micromorts per year (1×10^{-4}) and that there were no suitable risk reduction measures that could be adopted apart from managed retreat. Some property owners did not accept either the risk assessment or risk reduction conclusions, resulting in a legal process between the residents and the local authority.

The following are summary recommendations related to quantitative risk tolerance thresholds:

- *Individual risk tolerance thresholds* have a stronger basis than group risk thresholds and are easier to explain and apply. FN criteria are challenging to grasp and can contribute more to confusion than a transparent decision. Therefore, evaluate individual risk as a first step, and evaluate group risk only if the individual risk is not sufficient to support a decision.
- *There are no universal risk tolerance thresholds*: Risk perceptions, scale of hazard and number of people at risk, background risk levels, and available resources influence thresholds that should be applied, and are highly variable between regions. It is more rational to apply the same threshold to multiple similar hazard types within the same region than to apply the same debris-flow risk tolerance threshold across different regions.
- *Stakeholder participation*: Involve stakeholders, decision makers, and the public in setting risk tolerance thresholds and risk reduction goals. Thresholds represent stakeholder and public attitudes and must be informed by availability of options for risk reduction, funding availability, and the relative importance of debris-flow hazards to other risks.

- *Policy versus aspirational goal*: Clearly define if the risk tolerance threshold is an aspirational goal for reducing existing risks or if it is a rigid policy that defines risk levels that must be met.
- *Set an achievable goal*: Avoid setting risk reduction goals so low that they are unachievable. These are not useful for guiding risk management decisions. In the worst case, they can lead to inaction if funding agencies are unwilling to provide available funds that do not achieve the risk tolerance threshold and unable to provide sufficient funds to reduce risk below threshold.
- *Guidelines instead of strict criteria*: Recognize the inherent uncertainty of both quantitative risk assessment and public risk perception, and view risk tolerance thresholds as approximate guidelines rather than strict criteria.

14.6.2.5 Risk in Context Case Study

Although they are widely used, a major limitation of risk tolerance thresholds and reference lines is that decision makers and the public rarely have an intuitive feel for the risk level that is represented by the threshold, which impedes risk communication. This limitation can be overcome by presenting the risk tolerance thresholds or estimated risks in context with other common risks that are familiar to most people. The following case study from New Zealand provides an example of how risk comparison data was used to develop risk tolerance thresholds and communicate risk to end-users and risk managers (Taig, 2022a, 2022b).

The Department of Conservation (DOC) manages New Zealand's public conservation lands and waters, including visitor safety. DOC assesses risk at visitor destinations, controls risk to levels deemed appropriate for the predominant visitor group, and informs visitors of the risks. Visitors are responsible for decisions about the risks they take (DOC, 2017).

An individual's tolerance to risk is highly variable based on their perception of different hazard types, life choices, and experiences. Therefore, estimated risks at parks and conservation lands were compared against a variety of risks that are relevant to most visitors (Taig, 2022a), including:

- Historical accident rates on New Zealand Public Conservation Land
- Risks during travel to and from New Zealand Public Conservation Land
- Historical accident rates at national parks in other countries
- Historical accident rates for popular outdoor sports and recreational activities
- Historical accident rates from common hazards faced by New Zealanders

Figure 14.18 is an example of landslide risk at a walking track called Cape Kidnappers. The risk metric of 'individual risk per trip per day' was chosen instead of 'annual individual fatality risk' because visitors only spend a small proportion of the year on Public Conservation Land (Taig, 2022a).

Graphical representation of estimated risk and comparisons helped communicate the small numbers and scientific notation associated with risk results (e.g., 10^{-4}), the inherent uncertainty in the data, and to overcome language barriers required to understand the risk results (Taig, 2022a). A risk range that illustrates uncertainty is shown instead of precise values. Uncertainty results from variability of the data, the statistical uncertainty associated with small datasets, and imprecision of the source data (e.g., an individual's participation in an activity is recorded as "3–4 times a week').

This type of information was used by DOC to develop risk tolerance thresholds (Taig, 2022a, 2022b). Destinations were grouped into three categories based on the predominant visitor group, and a threshold was assigned to each category (Fig. 14.19). These thresholds were applied only to destinations where DOC maintains facilities for visitors, such as tracks and

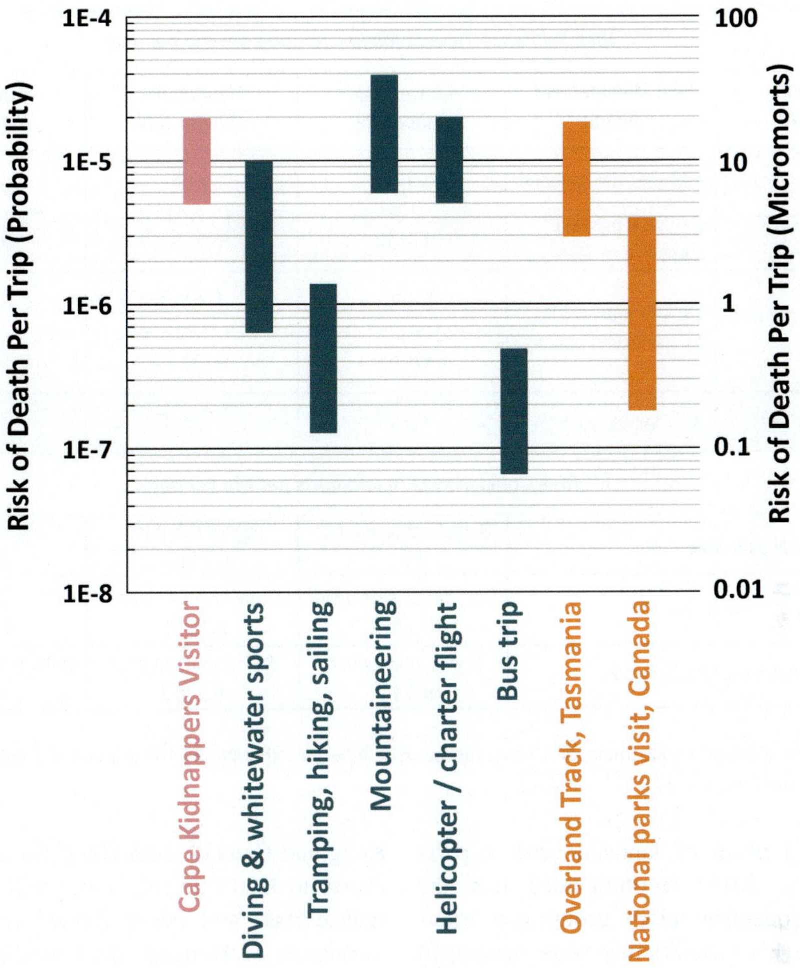

Fig. 14.18 Quantitative estimates of individual risk per trip per day for Cape Kidnappers visitors compared against popular tourist activities, modes of transportation used to access parks, and individual risk per trip to national parks overseas (adapted from de Vilder et al., 2022; Massey et al., 2020; Taig, 2022a, 2022b)

huts, and were not applicable to higher risk activities like mountaineering.

1. *Lower risk visitor sites*: Tolerable risk is similar to low-risk activities like playing golf, as visitors to these sites do not expect harm.
2. *Medium risk visitor sites*: Tolerable risk is similar to activities like swimming, biking, or jet boating.
3. *Higher risk visitor sites*: Tolerable risk is similar to relatively high-risk activities like rafting, caving, and climbing, which participants recognize as potentially dangerous.

14.6.2.6 Risk Tolerance Case Study

The following case study from British Columbia, Canada illustrates recent lessons related to applying quantitative risk tolerance thresholds, including the challenges of importing a foreign risk tolerance framework and the benefits of applying different risk tolerance frameworks to different decisions within a single region.

Risk Tolerance Thresholds (micromorts per trip per day)				
Evaulation Category	***Risk Management Action***	***Lower Risk Visitor Site***	***Medium Risk Visitor Site***	***Higher Risk Visitor Site***
Intolerable	Halt until risk reduced	>10	>30	>100
	Continue only after corporate review	1 to 10	3 to 30	30 to 100
Tolerable if reduced to ALARP	Explore practicable risk reduction options	0.1 to 1	0.3 to 3	3 to 30
		0.03 to 0.1	0.1 to 0.3	0.3 to 3
Tolerable	Monitor situation	< 0.03	< 0.1	< 0.3
Risk Comparisons (micromorts per trip per day)				
Typical national park walk		Safest national park 0.1	Higher risk park 0.3 to 3	Remote hike 3 to 30
Typical mode of travel to location		Bus, car w/ safe driver 0.1 to 1	Car w/ typical driver 0.3 to 3	Car w/ young driver 3 to 30
Typical activities visitor might enjoy		Bowls, golf, tennis < 0.1	Swim, bike, jet boat 0.3 to 3	Rafting, climbing,caving 1 to 30

Fig. 14.19 New Zealand Department of Conservations risk tolerance thresholds for individual fatality risk for the three visitor groups (from Taig, 2022a, 2022b)

In 2015, a panel of landslide risk experts (Clague et al., 2015) recommended that the District of Squamish adopt group risk tolerance criteria (FN criteria) that were developed in Hong Kong (Fig. 14.20a) for managing risk to new development on the Cheekeye debris-flow fan. These criteria are prescriptive and risk averse. Risks that plot above the tolerable risk threshold are not tolerated and must be reduced to a tolerable level without regard for cost. Risks in the tolerable zone must be further reduced until the cost of further risk reduction is grossly disproportional to the benefits gained.

Prior to adoption of these criteria, risk in the debris-flow hazard area was managed by restricting new development. But adoption of the risk tolerance criteria allowed new development to occur if the risk is managed by debris-flow mitigation structures, like a barrier or diversion berm. It was recognized that perceptions of landslide risk, attitudes about the role of government, funding availability, and risk management mechanisms are very different in Hong Kong and Canada, with Hong Kong being more averse to landslide risk, more willing to pay to reduce risk, and better funded (e.g., compare Strouth & McDougall, 2021 to Cheung, 2021). Regardless, the risk averse criteria were deemed to be appropriate for this application because the development was new, the local government had the authority to enforce the criteria, and the developer had resources to construct the required debris-flow protection. In 2022, at the time of writing, adoption of these criteria appears to be successful. A large debris-flow barrier designed to protect existing and new development on Cheekeye Fan is in the final stages of design and permitting and is widely supported by all levels of government and the public.

However, large new developments on debris-flow fans are rare, and most debris-flow risk scenarios in British Columbia involve existing development (including infilling, renovations, home additions) located on debris-flow fans that were unrecognized at the time of construction. In these more common scenarios, the risk

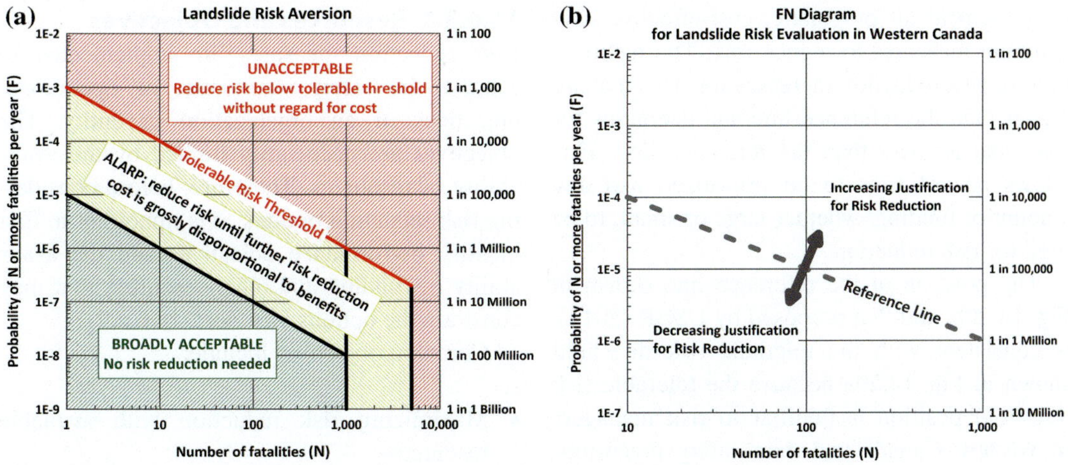

Fig. 14.20 Two group risk evaluation criteria that have been referenced for projects in British Columbia, Canada (Strouth & McDougall, 2021). **a** Rigid and prescriptive criteria used for new development. **b** Flexible criteria used for implementing cost-effective measures at existing development. Note the differing limits on the axes and that the reference line in part B is at the same position as the tolerable risk threshold in part A

management process is usually triggered following a debris flow that alerts residents and local governments of the hazard. While the District of Squamish contemplated using the Hong Kong risk tolerance criteria for new development, other nearby communities and consultants referenced them for debris-flow risk scenarios at existing development.

In hindsight, application of these risk tolerance criteria at existing development appears to be mostly unsuccessful. Rather than facilitating risk communication, decision making, and risk reduction, the criteria led to designs that were unaffordable and failed to persuade funding authorities that proposed mitigations were a justifiable allocation of government's limited resources. In several cases, no risk reduction was implemented because the proposed risk reduction measures were inadequate when compared to the tolerable risk threshold and requirements for 'gross disproportion' between costs and benefits (Strouth & McDougall, 2021).

In one example, group risk from debris-flow hazards at an existing community plotted well into the Unacceptable region (Fig. 14.20a). A debris-flow mitigation structure was proposed that reduced risk to near (but above) the tolerable risk threshold and fell within the budget of available disaster mitigation grants. The proposed structure was a pragmatic and cost-effective method for reducing risk. However, without mentioning funding limitations, the government regulator who reviewed the mitigation funding application responded:

> Overall, the regulatory authority is required to hold paramount the health and safety of the public ("Professional Governance Act"). If this cannot be done to a tolerable level, then there is one view that suggests that we should not be doing the risk mitigation work to a lesser threshold. Would a decision maker be viewed to be holding public safety paramount if the project knowingly did not meet the tolerable threshold and there were no other plans to reduce the residual risk in another manner?

Since that time, a more flexible risk evaluation framework has been found to be more applicable at existing developments in debris-flow hazard zones in that region of Western Canada. This new approach, represented by FN criteria in Fig. 14.20b, focuses on cost-effectiveness and pragmatism rather than risk tolerance thresholds. It is adapted from USBR (2011) and uses a reference line instead of a tolerable risk threshold and omits terms like 'tolerable' and 'unacceptable'. Instead, the same condition is applied to tolerate any level of risk. This condition is to implement all pragmatic, cost-effective, and common measures to reduce risk. The justification for risk reduction increases for risks that are higher than the reference line and decreases for risks that are less than the reference line. This allows hazard sites to be prioritized and any amount of funding, whether large or small, to be used for risk reduction.

The position of the reference line shown in Fig. 14.20b, and that proposed by USBR (2011), is consistent with the tolerable risk threshold shown in Fig. 14.20a because the tolerable risk threshold position is familiar to risk managers in Western Canada and dam safety practitioners. However, the position of the reference line is not a critical factor because it is not a rigid boundary. The reference line could be shifted to any other location that would be meaningful to the users, such as to represent an average, maximum, or minimum risk level for a portfolio of dams or debris-flow hazard sites.

14.6.3 Risk Reduction Measures

Many risk evaluation decisions are related to choosing between risk reduction measures. This section describes *how* to include risk reduction measures in risk assessments. It is organized around important questions that often need to be addressed:

1. What are the risk reduction objectives and resources available?
2. What are the options for reducing risk?
3. How much risk reduction is achieved?
4. What are the costs associated with each option?
5. Is the cost an effective use of resources?
6. How are risks changing with time?

Guidance on selection of a risk reduction measure for different settings is provided elsewhere, in Piton et al. and Kwan et al. (this volume), and in literature (e.g., Holub & Hubl, 2008, Santi et al., 2010).

14.6.3.1 Resources and Objectives

Risk reduction measures are implemented to achieve the safety, economic, or operational objectives of an organization or entity. The objectives are necessarily limited by the availability of resources. Therefore, before considering risk reduction options, it is important to first identify performance objectives and the availability of funding and other resources that may constrain the options (see Sect. 14.6.2.2).

Common objectives include:

- Maximizing risk reduction with available resources
- Setting a budget for risk management
- Reducing risk to below a risk tolerance threshold

Other factors influence and add detail to the objectives, such as:

- How much loss has historically been tolerated? Are these losses tolerable or do they need to be reduced?
- What are achievable performance goals?
- Is there a risk tolerance threshold that must be met?
- Can some areas tolerate higher risk levels than others?
- Are some areas a higher priority than others for risk reduction?
- Is risk evenly distributed or concentrated?
- Who are the stakeholders and what are their goals?

Limitations and constraints on risk reduction options are related to:

- What funding is available?
- Where does funding come from?
- What land is available for physical structures and who owns it?
- Who is available to carry out assessments and design risk reduction measures?
- Is there a funding limitation that sets a maximum cost of risk reduction?

- Who will own the structure(s) in perpetuity and fund their maintenance, including clean-out of large volumes of debris?
- Is there a schedule limitation that limits feasibility of mitigation options?
- Do measures need to be implemented immediately?

14.6.3.2 Risk Reduction Options

One useful method for identifying risk reduction options is to review the event tree, risk analysis method, or risk equation that was used to estimate risk. Each branch of the risk tree or each parameter of the risk equation represents an opportunity to sever the chain of events that leads to loss and thereby reduce risk (Fig. 14.21). This method is not only a useful framework for brainstorming options, but also provides a clear basis for estimating the risk reduction and communicating the benefits of the risk reduction measure.

After identifying general options, it is important to fill out details of each option that relate to the performance objectives, constraints, and behaviours of people during a debris flow (e.g., Pollock & Wartman, 2020). Options need to be specific before the risk reduction or implementation cost can be estimated. For example:

- What type of protective structure will be implemented?
- What is the size of the protection structure? Where will it be located?
- Will there be auxiliary structures?
- How will structures work together to create a functional chain of elements?
- How will rainfall and watershed conditions be monitored? How will monitoring data be transmitted to analysts? Who will review the data? Who owns the instrumentation array and is responsible for its maintenance or replacement?
- Which areas will be evacuated? How will the evacuation notice be given? How likely are people to follow the evacuation notice? (e.g., Goto et al., 2020)

14.6.3.3 Risk Reduction and Residual Risk

Although a few isolated cases may exist where debris-flow risk can be eliminated by avoiding the hazard areas, for most settings it is not feasible to eliminate debris-flow risk. The risk is taken because the benefits outweigh the risks, and because there are not enough resources available to eliminate all the risks that are faced. Therefore, it becomes important to estimate how far a risk can be reduced by the various risk reduction measures.

Risk reduction is the risk removed by a mitigation measure. It is the *difference between* the *baseline risk*, which is the risk without mitigation in place, and the *residual risk*, which is the risk that remains following mitigation. Risk reduction and residual risk are measured in the same units as those of the baseline risk estimates. For example, in a qualitative framework, the risk reduction could be a change in risk from a 'high' to a 'low' category. In a quantitative framework, the risk reduction could be lives saved or dollars saved over a given time.

Usually, the same risk analysis framework will be used to estimate both the baseline and residual risk. The residual risk is estimated using parameters that are modified by the mitigation measures. Some examples include:

- A qualitative risk matrix analysis might assume the likelihood of debris-flow initiation reduces from high to moderate when a reforestation program is implemented.
- Where a debris-flow barrier is constructed and sized for a 100-year return period debris flow, an individual risk calculation might assume that risk from events smaller than the 100-year return period are eliminated, but residual risks from larger events remain.
- A group risk calculation where a warning and evacuation system is implemented might assume that the evacuation is 50% successful. In other words, if 100 people were present in the baseline assessment, only 50 people are present in the residual assessment.

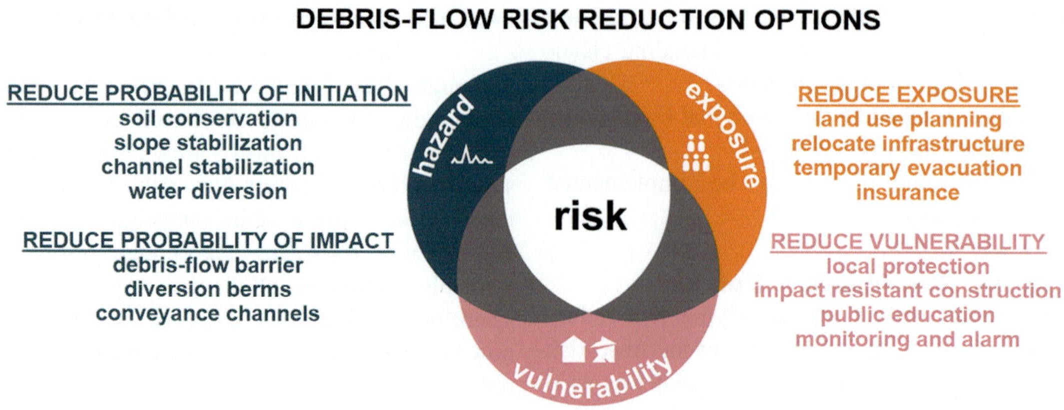

Fig. 14.21 Risk reduction options can be derived from each component that contributes to the risk

Typically, there are performance objectives related to the risk reduction or residual risk. For example:

- Where a rigid risk tolerance threshold must be met, risk reduction measures are selected and sized to achieve the required risk reduction, and the risk assessment is used to demonstrate that residual risk is less than the threshold. Using quantitative analysis methods, this can be done by comparing the risk associated with each scenario to the risk tolerance threshold. Scenarios that exceed the threshold need to be mitigated.
- For regional prioritization studies, there is often a need to maximize the risk reduction achieved with available resources. Maximum risk reduction is not necessarily achieved by addressing the highest risk site first, in a "worst-first" approach. It may be more cost effective and efficient to reduce risk at many lower risk sites (e.g., NASEM, 2019).

14.6.3.4 Costs

For risk evaluation, the *costs* of each risk reduction option need to be considered in the broadest sense. A comparison only of the immediate construction costs is rarely sufficient. For example, some measures (e.g., forced evacuation) may have no capital cost, but very high opportunity costs and social costs. All the costs need to be considered together for the measures to be compared.

The monetary value of construction or initial implementation can be combined with operations, maintenance, and any future recurring costs using net present value. This is called a life-cycle cost and can be used to compare the financial aspects of multiple options. However, other risk reduction costs may be less tangible but have a stronger control on decisions, such as:

- Opportunity costs
- Where are funds coming from?
- What is not being funded due to this risk reduction program?
- Business interruption costs (e.g., related to pre-emptive evacuation or road closure)
- Lost economic opportunities due to permanent evacuation
- Social costs
- What are residents losing?
- Lost homes, livelihoods, businesses, educational opportunities due to evacuation or property acquisition
- Environmental costs
- Debris-flow barriers may disrupt aquatic habitats and natural sediment movement
- All physical structures have a physical footprint with associated impacts
- Reputational costs

- How will repeated evacuation without a debris flow affect public trust in experts and in government?

14.6.3.5 Cost Effectiveness

Cost effectiveness is the comparison of benefits of a risk reduction measure with the cost(s). Cost effectiveness can be assessed prior to implementation of measures to prioritize and guide selection of those measures, or after implementation to evaluate if mitigation funding was well spent. Evaluation of cost effectiveness is important for any risk evaluation that compares risk reduction options. It is a key element for supporting and communicating decisions because it summarizes the risk reduction options in well understood terms.

Common methods for presenting cost effectiveness include:

- In a qualitative framework, cost effectiveness would usually be considered separately for each risk scenario, for example, by comparing the cost of several measures that reduce one risk scenario from 'high' to 'moderate'
- In a quantitative life loss risk framework, it is common to present the cost to save a statistical life (CSSL) (e.g., Leroi et al., 2005; Strouth & McDougall, 2021). CSSL is the mitigation cost over a given period divided by the statistical lives saved over the same period. In a group risk analysis, the statistical lives saved is the difference between the probable life loss (PLL) with and without mitigation.
- In a quantitative economic risk assessment, the costs and benefits of each option can be compared using a benefit to cost ratio. Life loss risk reduction, which is a common benefit of risk reduction options, can be monetized using the value of a statistical life concept, which quantifies the amount people are willing to pay for increased safety (Viscusi & Aldy, 2003; Viscusi & Masterman, 2017).

14.6.3.6 Risk Tracking

Risks change with time due to changing hazard conditions, changes to exposure, and in response to risk reduction measures. A method for tracking how risks change with time is often needed, and a *risk register* is a common tool for this purpose.

A risk register is an active, continuously updatable document that lists all known risks in an area, together with the decisions taken about how to monitor and manage them. Risk registers are useful for identifying hazard sites, screening, prioritizing, and tracking changes. They typically include written descriptions of the hazard type, frequency of occurrence, likelihood of impact, elements at risk, possible consequences, risk controls, and a risk-based priority ranking. The risk register is reviewed and updated periodically, as new risks are identified and as risks and risk controls change. Risk registers provide a means of tracking low priority, and low-probability/ high-consequence risks that could otherwise be ignored and forgotten. Risk registers are mostly used by entities who manage infrastructure that is exposed to many different hazard types at many different locations, such as mine operators and utility owners.

A risk estimate is a "snapshot" in time for a single set of assumptions. But risk estimates change as conditions change (e.g., Fig. 14.22). In most settings, risk estimates need to be updated if there are substantial or sudden changes in the hazard, exposure, or vulnerability conditions, like an earthquake, wildfire, or new development (e.g., Zhang & Zhang, 2017). Periodic updates may also be needed in response to gradual changes, like population growth or climate change (e.g., Cannon & DeGraff, 2009; Stoffel et al., 2014; Stoffel et al., this volume). A risk management plan should outline when a re-analysis of risk is required or triggered, particularly for long-term applications like land use planning, and long-term applications should allow for changes of risk with time.

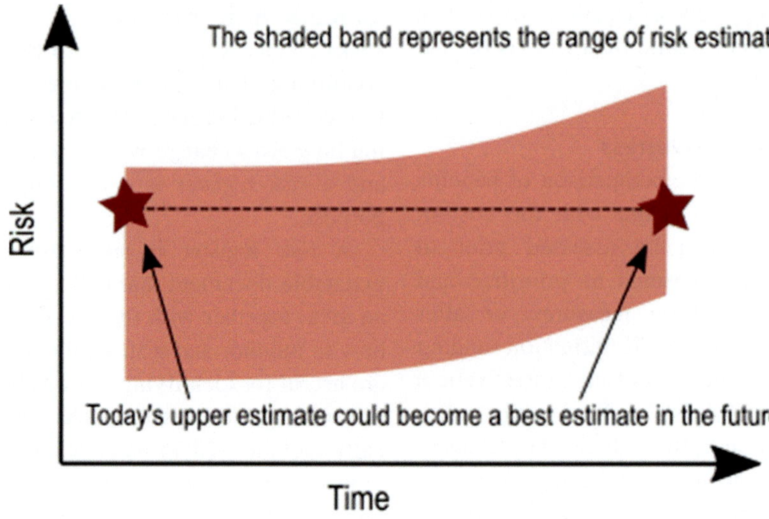

Fig. 14.22 Gradual changes in a group risk estimate that occur due to population growth in a hazard zone (adapted from Taig et al., 2012)

14.6.3.7 Hong Kong Risk Management Case Study

Quantitative risk assessments have informed landslide risk management in Hong Kong since the 1990's across a large portfolio of hazard sites (Malone, 2005). These assessments inform decisions related to risk management strategy, site prioritization, and resource allocation by:

1. Identifying the overall scale of the problem
2. Examining risk distribution and relative contributions from different components
3. Setting quantified risk reduction targets
4. Facilitating risk communication with stakeholders
5. Demonstrating cost-benefit of risk management actions
6. Identifying changes in risk level with time
7. Demonstrating system performance

In the 1970's the Geotechnical Engineering Office of the Government of Hong Kong began implementation of the Landslip Preventive Measures (LPM) program. In 2000, a quantitative risk assessment that considered 40,000 man-made slope hazard sites suggested that LPM measures implemented between 1977 and 2000 had reduced risk by 50%. This led to adopting a target for reducing risk an additional 50% by the year 2010.

The quantitative risk assessment also provided invaluable insight on the risk distribution among various slope types and facilities at risk and on cost-effectiveness of the LPM program. For example, it indicated that overall risk could be most effectively reduced by focusing mitigation resources at approximately 10% of the slopes that pose the highest risk. This allowed the risk reduction goal to be met cost effectively. In 2010, the assessment showed that that LPM program operated at a cost of HK$ 15 million (~US$ 2 million) per statistical life saved between the year 2000 and 2010, which was considered a justifiable expenditure (Wong, 2005).

Portfolio-level quantitative risk assessments have been repeated about every 5 years, and they continue to guide landslide risk management plans. For example, in 2010, the assessment suggested that the risk from natural hillsides was similar to risk from man-made slopes (Ho & Ko, 2009). This led to more efforts to systematically study and mitigate landslides from natural hillsides and allocation of approximately half of resources to manage risk from natural hillsides under the post-2010 rolling Landslip Prevention and Mitigation Programme.

14.7 Concluding Remarks

When used appropriately, a risk assessment can be a valuable tool to inform decisions about debris-flow hazards. A well-formed risk assessment benefits decision makers by providing transparent justification for decisions and a framework for explaining the decisions to the public. But the greatest benefit may be the deep understanding of the problem and potential solutions that all stakeholders develop as they work through the risk assessment process together.

The goal of this chapter was to lay out the components of debris-flow risk and risk assessments to support others who choose to employ risk assessment techniques. We hope this chapter supports more rational and consistent decisions that save lives and resources in debris flow-prone regions of the world. Keep the following points in mind as you apply risk assessment techniques:

- *Collaborate*: The risk assessment process requires active stakeholder input, particularly during the scope definition and risk evaluation phases. Collaboration and regular communication between risk analysts, decision makers and the public are essential.
- *Less is more*: Risk assessments tend to be complicated. Avoid doing a risk assessment if the problem can be addressed with a simpler method. If risk assessment is chosen, make the assessment as simple as possible.
- *Adapt*: Risk assessments need to be fit to the decision that needs to be made. There is no 'one-size-fits-all' solution.
- *Balance*: All components of risk have a strong influence on the final risk values. Similarly, all components of the risk assessment have a strong influence on the decision. Therefore, balance risk estimation efforts evenly between hazard, elements at risk, and vulnerability characterization. Balance risk assessment efforts evenly between scope definition, risk analysis, and risk evaluation.
- *Be transparent*: Clearly document assumptions and uncertainties.
- *Communicate clearly*: Consider risk communication at the start of the study. Design the risk assessment to optimize clear communication.
- *Focus on objectives, decisions, resources*: A risk value is not usually an end in itself. At each step of the risk assessment, take a step back from 'risk' and review the objectives that need to be achieved, the decisions that need to be made, and resources that are available.
- *Avoid false precision*: Be careful that the presentation of the results does not convey an unwarranted degree of precision. False precision can mislead decisions.

14.8 Research Needs

The following research is needed to reduce uncertainties and improve debris-flow risk assessments:

- *Spatial empirical relationships*: The spatial distributions relative to the debris-flow fan boundaries of debris-flow depth, velocity, intensity, impact frequency, and estimated risk are needed for developing spatial-empirical relationships that can be used for quantifying likelihood of avulsion and flow path length for risk estimation, and for runout model calibration (e.g., see Zubrycky et al., 2021; Aaron et al., 2022).
- *Building classification for debris flows*. No standard classification exists for building types in the context of debris-flow vulnerability. Building classifications used in practice are inconsistent, and some are adapted from classification developed for other hazards (e.g., earthquakes or floods). More research is needed to understand building characteristics that control physical vulnerability to

debris flows, and a building classification that accounts for these factors should be developed.

- *Life loss vulnerability*: Vulnerability estimates tend to be a large source of uncertainty in debris-flow risk estimates. Although the body of literature that documents correlations between debris-flow intensity and building damage (i.e., fragility) is growing, little information is available that correlates debris-flow intensity or building damage to life loss probability (i.e., consequence functions). Similarly, there is a paucity of data on the physical vulnerability of people outside of buildings, in vehicles or on foot. More documentation of the number and location of people present at the time of impact and the number of people killed is needed.
- *Fragility and vulnerability curve consistency*: Many fragility and vulnerability models for debris flows have been published, however they are inconsistent. They vary in terms of the hazard intensity parameter used, the classification and definition of damage states, and the definition of loss. This makes it difficult to compare curves, transfer values between studies, and they are prone to misuse. Consistent definitions of hazard intensity, and debris-flow damage and loss are needed for different infrastructure types.
- *Social vulnerability*: More work is needed to develop methods for incorporating social vulnerability concepts into debris flow risk assessments, particularly into quantitative risk assessments.
- *Consequence database*: More robust data collection after debris-flow disasters is needed to fully quantify the consequences. Important factors to document include life loss, number of people in the impact zone, building damage states and economic losses, utility and road service disruptions and durations, indirect economic losses related to business interruptions, and intangible losses. These data are needed to develop consequence functions, improve risk estimates, and build a stronger case for risk management measures in cost-benefit analyses.
- *Risk in context*: More analysis of economic and life-loss risks caused by other hazards is needed so that debris-flow risks can more readily be put into clear context. This should include more granular analysis of familiar hazards (e.g., driving in clear conditions versus in the snow) and natural hazards that people perceive similarly to debris flows (e.g., forest fires, floods, tornadoes, earthquakes).
- *Risk creep*: More rigorous methods are needed for quantifying changes in debris-flow risk due to climate change and land use changes. While Stoffel et al. (this volume) has attempted to do so for a fictional case, integrating such changes in debris-flow risk assessments is still the exception rather than the rule.

References

Aaron, J., McDougall, S., Kowalski, J., Mitchell, A., & Nolde, N. (2022). Probabilistic prediction of rock avalanche runout using a numerical model. *Landslides, 19*(12), 2853–2869.

Ahmad, N., Peterson, N., & Torella, F. (2015). The micromort: A unit for comparing and communicating risk to patients. *International Journal of Clinical Practice, 69*(5), 515–517. https://doi.org/10.1111/ijcp.12643

Ali, S., Haider, R., Abbas, W., Basharat, M., & Reicherter, K. (2021). Empirical assessment of rockfall and debris flow risk along the Karakoram Highway, Pakistan. *Natural Hazards, 106*(3), 2437–2460.

Allianz. (2020). Aviation risk 2020—safety and the state of the nation. https://www.agcs.allianz.com/content/dam/onemarketing/agcs/agcs/reports/AGCS-Aviation-Risk-Report-2020.pdf

Arksey, R., & VanDine, D. (2008). Example of a debris-flow risk analysis from Vancouver Island, British Columbia, Canada. *Landslides, 5*(1), 121–126.

Australian Geomechanics Society (AGS). (2007). Practice note guidelines for landslide risk management. ISSN 0818-9110.

Aven, T., Baraldi, P., Flage, R., & Zio, E. (2014). *Uncertainty in risk assessment: The representation and treatment of uncertainties by probabilistic and non-probabilistic methods* (p. 200). Wiley. ISBN: 978-1-118-48958-1.

Ball, D. J., & Floyd, P. J. (1998). *Societal risks: A report prepared for the Health and Safety Executive*. Health and Safety Executive.

Baecher, G., Abedinisohi, F., & Patev, R. (2015). *Societal risk criteria for loss of life concepts, history, and mathematics*. University of Maryland.

Baecher, G. B. (2023). 2021 Terzaghi lecture: Geotechnical systems, uncertainty, and risk. *Journal of Geotechnical and Geoenvironmental Engineering, 149*(3), 03023001.

Bandle, T. (2007). Tolerability of risk: The regulator's story. In *The tolerability of risk, a new framework for risk management*. Earthscan.

BGC Engineering Inc. (BGC) (2014). Cougar creek debris flood risk assessment. Report prepared for the Town of Canmore.

BGC Engineering Inc. (BGC). (2015). Catline Creek debris-flow hazard and risk assessment. Final report issued to Squamish-Lillooet Regional District, dated Jan 22, 2015.

Blastland, M., & Spiegelhalter, D. (2014). *The norm chronicles: Stories and numbers about danger and death*. Basic Books.

Bründl, M., Romang, H. E., Bischof, N., & Rheinberger, C. M. (2009). The risk concept and its application in natural hazard risk management in Switzerland. *Natural Hazards and Earth System Sciences, 9*(3), 801–813.

Burton, C., Rufat, S., & Tate, E. (2019). *Social vulnerability: Conceptual foundations and geospatial modeling. Vulnerability and resilience to natural hazards*. Cambridge University Press. https://doi.org/10.1017/9781316651148

Cannon, S., & DeGraff, J. (2009). The increasing wildfire and post-fire debris-flow threat in western USA, and implications for consequences of climate change. In K. Sassa & P. Canuti (eds.), *Landslides—disaster risk reduction*. Springer. https://doi.org/10.1007/978-3-540-69970-5_9

Centers for Disease Control and Prevention (CDC). (2020). CDC COVID data tracker, coronavirus disease 2019 (COVID-19). https://covid.cdc.gov/covid-data-tracker/#cases_casesper100klast7days. Accessed 16 Dec 2020.

Chen, H. X., Zhang, S., Peng, M., & Zhang, L. M. (2016). A physically-based multi-hazard risk assessment platform for regional rainfall-induced slope failures and debris flows. *Engineering Geology, 203*, 15–29.

Chen, H. X., Zhang, L. M., & Feng, S. J. (2017). Risk assessment of debris flows along a road considering redistribution of elements at risk. In *Geo-risk 2017* (pp. 206–214).

Cheng, P. F. K., & Ko, F. W. Y. (2010). An updated assessment of landslide risk posed by man-made slopes and natural hillsides in Hong Kong. GEO Report No. 252, Geotechnical Engineering Office, Hong Kong, 46 p.

Cheung, R. W. (2021). Landslide risk management in Hong Kong. *Landslides, 18*(10), 3457–3473.

Choi, G. M., Lee, I. W., & Yune, C. Y. (2021). Risk assessment of debris flow at regional scale considering catchment area in Chuncheon, Korea. *KSCE Journal of Civil Engineering, 25*(4), 1176–1190.

Ciurean, R. L., Hussin, H., Van Westen, C. J., Jaboyedoff, M., Nicolet, P., Chen, L., ... & Glade, T. (2017). Multi-scale debris flow vulnerability assessment and direct loss estimation of buildings in the Eastern Italian Alps. *Natural hazards, 85*, 929–957.

Clague, J. J., Hungr, O., Morgenstern, N. R., & VanDine, D. (2015). Cheekye river (Ch'kay Stakw) and fan landslide risk tolerance criteria. Report of the Cheekeye Expert Review Panel #2. Prepared for Province of British Columbia, Squamish Nation, and District of Squamish, Canada.

Corbane, C., Gamba, P., Pesaresi, M., Pittore, P., & Wieland, M. (2017). Current and innovative methods to define exposure. In K. Poljansek, M. M. Ferrer, T. De Groeve, & I. Clark (eds.), *Science for disaster risk management 2017. Knowing better and losing less*. European Commission, DG-JRC.

Cui, P., Xiang, L. Z., & Zou, Q. (2013a). Risk assessment of highways affected by debris flows in Wenchuan earthquake area. *Journal of Mountain Science, 10*(2), 173–189.

Cui, P., Zou, Q., Xiang, L. Z., & Zeng, C. (2013b). Risk assessment of simultaneous debris flows in mountain townships. *Progress in Physical Geography, 37*(4), 516–542.

Cutter, S., Boruff, B., & Shirley, L. (2003). Social vulnerability to environmental hazards. *Social Science Quarterly., 84*(2), 242–261.

Cutter, S., & Finch, C. (2008). Temporal and spatial changes in social vulnerability to natural hazards. *PNAS, 105*(7), 2301–2306.

Dagá, J., Chamorro, A., de Solminihac, H., & Echaveguren, T. (2018). Development of fragility curves for road bridges exposed to volcanic lahars. *Natural Hazards and Earth System Sciences, 18*(8), 2111–2125.

de Vilder, S., Massey, C., Lukovic, B., Taig, T., & R. Morgenstern. (2022). What drives landslide risk: Disaggregating risk analyses, an example from the Franz Josef and Fox Glacier Valleys, New Zealand. *Natural Hazards and Earth System Sciences Discussions*.

Department of Conservation (DOC). (2017). Visitor risk management policy, Wellington, New Zealand.

Diana, M., Nurfashareena, M., Taha, M., et al. (2021). Social vulnerability assessment for landslide hazards in Malaysia: A systematic review study. *Land, 2021*(10), 315.

Ding, M., & Huang, T. (2019). Vulnerability assessment of population in mountain settlements exposed to debris flow: A case study on Qipan gully, Wenchuan County, China. *Natural Hazards, 99*(1), 553–569.

Drakes, O., & Tate, E. (2022). Social vulnerability in a multi-hazard context: a systematic review. *Environmental Research Letters, 17*(3).

Ebbwater Consulting (Ebbwater). (2021). Boundary region flood and geohazard risk assessment. Appendix G—Risk Assessment Map Book. For Regional District of Kootenay Boundary, dated 14 May 2021.

Eidsvig, U., Papathoma-Kohle, M., Du, J., et al. (2014). Quantification of model uncertainty in debris flow vulnerability assessment. *Engineering Geology, 181*(2014), 15–26.

ERM-Hong Kong Ltd. (ERM). (1998). Landslides and boulder falls from natural terrain: Interim Risk Guidelines. GEO Report No. 75. Geotechnical Engineering Office, Hong Kong.

Farrell, J., & Davies, T. (2019). Debris-flow risk management in practice: A New Zealand case study. In *7th international conference on debris-flow hazards mitigation*, Golden, Colorado, USA.

Fell, R. (1994). Landslide risk assessment and acceptable risk. *Canadian Geotechnical Journal, 31*, 261–272.

Fell, R., Ho, K. K., Lacasse, S., & Leroi, E. (2005). A framework for landslide risk assessment and management. In *Landslide risk management* (pp. 13–36). CRC Press.

Federal Emergency Management Agency (FEMA). (2018). *Seismic performance of assessment of buildings, volume 1—methodology* (2nd edn). FEMA P-58-1, December 2018.

Federal Emergency Management Agency. (2013). *Multi-hazard loss estimation methodology, flood model, HAZUS MR3, Technical Manual*. Department of Homeland Security.

Federal Emergency Management Agency. (2021a). *Hazus inventory technical manual*.

Federal Emergency Management Agency (FEMA). (2021b). *National risk index, technical documentation*.

Federal Energy Regulatory Commission (FERC). (2016). *Risk-informed decision making (RIDM)—risk guidelines for dam safety (Interim Guidance)*, Version 4.1, USA.

Finlay, P. J., & Fell, R. (1997). Landslides: Risk perception and acceptance. *Canadian Geotechnical Journal, 34*, 169–188.

Fraser Valley Regional District (FVRD). (2020). Hazard acceptability thresholds for development approvals. https://www.fvrd.ca/assets/Services/Documents/Planning~and~Development/Application~Forms~and~Resources/2020%20Hazard%20Acceptability%20Thresholds.pdf

Fuchs, S., Keiler, M., Ortlepp, R., Schinke, R., & Papathoma-Köhle, M. (2019). Recent advances in vulnerability assessment for the built environment exposed to torrential hazards: Challenges and the way forward. *Journal of Hydrology, 575*, 587–595. https://doi.org/10.1016/j.jhydrol.2019.05.067

Gartner, J. E., & Jakob, M. (2019). Debris-flow risk assessment and mitigation design for pipelines in British Columbia, Canada. In *7th international conference on debris-flow hazards mitigation*, Golden, Colorado, USA.

Goto, E. A., Gray, S., Keller, E., & Clarke, K. C. (2020). Using mixed-methods to understand community vulnerability to debris flows in Montecito, CA. In *Workshop on world landslide forum* (pp. 435–440). Springer.

Health and Safety Executive (HSE). (2001). *Reducing risks, protection people—HSE decision-making process*.

Heron, M. (2021). Deaths: Leading causes for 2019. National vital statistics report volume 70, number 9. U.S. Department of Health and Human Services.

Ho, K. K. S., & Ko, F. W. Y. (2009). Application of quantified risk analysis in landslide risk management practice: Hong Kong experience. *Georisk: Assessment and Management of Risk for Engineered Systems and Geohazards*, 3(3), 134–146.

Hope, G., Jordan, P., Winkler, R., Giles, T., Curran, M., Soneff, K., & Chapman, B. (2015). Post-wildfire natural hazards risk analysis in British Columbia. *Land Management Handbook-Ministry of Forests and Range, British Columbia, 69*.

Holm, K., Jakob, M., Scordo, E., Strouth, A., Wang, R., & Adhikari, R. (2016). Identification, prioritization, and risk reduction: Steep creek fans crossed by highways in Alberta. In *Proceedings, GeoVancouver 2016 conference*.

Holub, M., & Hübl, J. (2008). Local protection against mountain hazards? state of the art and future needs. *Natural Hazards and Earth System Sciences, 8*(1), 81–99.

Howard, R. (1980). On making life and death decisions. In Schwing et al. (eds.), *Societal risk assessment* (pp. 89–113). Springer Science + Business Media.

Hungr, O. (1995). A model for the runout analysis of rapid flow slides, debris flows, and avalanches. *Canadian Geotechnical Journal, 32*(4), 610–623.

Hungr, O., & Wong, H. N. (2007). Landslide risk acceptability criteria: Are F-N plots objective? *Geotechnical News*.

Huyck, C., Hu, Z., & Eguchi, M. et al. (2022). Characterizing uncertainty of general building stock exposure data. *Earthquake Spectra*. Online. https://doi.org/10.1177/87552930221079852

International Organization for Standardization (ISO). (2019). *Risk management—Risk assessment techniques*. ISO.IEC 31010 Edition 2.0 2019-06.

Iverson, R. M., Schilling, S. P., & Vallance, J. W. (1998). Objective delineation of lahar-inundation hazard zones. *Geological Society of America Bulletin, 110*(8), 972–984.

Jakob, M., & Friele, P. (2010). Frequency and magnitude of debris flows on Cheekye River, British Columbia. *Geomorphology, 114*(3), 382–395.

Jakob, M., & Holm, K. (2012). Risk assessments for debris flows. In J. Clague & D. Stead (eds.), *Landslides: Types, mechanisms and modeling*. Cambridge University Press.

Jakob, M., Holm, K., Weatherly, H., Liu, S., & Ripley, N. (2012a). Debris flood risk assessment for Mosquito

Creek, British Columbia, Canada. *Natural Hazards, 2013*(65), 1653–1681. https://doi.org/10.1007/s11069-012-0436-6

Jakob, M., Holm, K., & McDougall, S. (2016). Debris-flow risk assessment. In *Oxford research encyclopedia of natural hazard science*.

Jakob, M., Stein, D., & Ulmi, M. (2012b). Vulnerability of buildings to debris flow impact. *Natural Hazards, 60*(2), 241–261.

Jakob, M., & Weatherly, H. (2005). Debris flow hazard and risk assessment, Jones Creek, Washington. In *Landslide risk management* (pp. 543–552). CRC Press.

Jakob, M., Weatherly, H., Bale, S., Perkins, A., & MacDonald, B. (2017). A multi-faceted debris-flood hazard assessment for Cougar Creek, Alberta, Canada. *Hydrology, 4*(1), 7. https://www.mdpi.com/2306-5338/4/1/7

Jones, L. M. (2020). Empowering the public with earthquake science. *Nature Reviews Earth & Environment, 1*(1), 2–3.

Journeay, M., Yip, J. Z. K., Wagner, C. L., LeSueur, P. L., & Hobbs, T. (2022). Social vulnerability to natural hazards in Canada. *Geological Survey of Canada, 8902*, 1. https://doi.org/10.4095/330295

Kean, J. W., Staley, D. M., Lancaster, J. T., Rengers, F. K., Swanson, B. J., Coe, J. A., & Lindsay, D. N. (2019). Inundation, flow dynamics, and damage in the 9 January 2018 Montecito debris-flow event, California, USA: Opportunities and challenges for post-wildfire risk assessment. *Geosphere, 15*(4), 1140–1163.

Kim, B. J., Yune, C. Y., Kim, K. S., & Lee, S. D. (2018). Risk assessment of debris flow on expressway in Korea using GIS. In *Landslides and engineered slopes. Experience, theory and practice* (pp. 1183–1187). CRC Press.

Lan, H. X., Li, L. P., Zhang, Y. S., Gao, X., & Liu, H. J. (2013). Risk assessment of debris flow in Yushu seismic area in China: A perspective for the reconstruction. *Natural Hazards and Earth System Sciences, 13*(11), 2957–2968.

Lato, M., Quinn, P., Pritchard, M., & Sirois, D. (2016). Multi-geohazard risk assessment along railway corridors. In *Proceeding of Geo-Vancouver 2016 conference*.

Lee, E. M., & Jones, D. K. C. (2014). *Landslide risk assessment* (2nd ed.). ICE Publishing.

Lerner, J. S., Li, Y., Valdesolo, P., & Kassam, K. S. (2015). Emotion and decision making. *Annual Review of Psychology, 66*, 799–823.

Leroi, E., Bonnard, C., Fell, R., & McInnes, R. (2005) *Risk assessment and management. In: Landslide risk management* (pp. 169–208). CRC Press.

Le Guan. (2007). Applying the HSE's risk decision model: Reducing risks, protecting people. In *The tolerability of risk, a new framework for risk management*. Earthscan.

LeSueur, P., Strouth, A., Mark, E., Collier-Pandya, B., & Zubrycky, S. (in review). *Risk-informed design of debris-flow mitigation at Cheekeye Fan*. Canadian Geotechnical Journal

Li, Z., Nadim, F., Huang, H., Uzielli, M., & Lacasse, S. (2010). Quantitative vulnerability estimation for scenario-based landslide hazards. *Landslides, 7*, 125–134.

Li, Y. M., Su, L. J., Zou, Q., & Wei, X. L. (2021a). Risk assessment of glacial debris flow on alpine highway under climate change: A case study of Aierkuran Gully along Karakoram Highway. *Journal of Mountain Science, 18*(6), 1458–1475.

Li, M., Tian, S., Huang, C., Wu, W., & Xin, S. (2021b). Risk assessment of highway in the upper reaches of Minjiang River under the stress of debris flow. *Journal of Geoscience and Environment Protection, 9*(7), 21–34.

Luo, H. Y., Fan, R. L., Wang, H. J., & Zhang, L. M. (2020). Physics of building vulnerability to debris flows, floods and earth flows. *Engineering Geology, 271*, 105611.

Luo, H. Y., Zhang, L. M., He, J., &Yin, K. S. (2021). Reliability-based formulation of building vulnerability to debris flow impacts. *Canadian Geotechnical Journal*. https://doi.org/10.1139/cgj-2020-0120

Macciotta, R., Grapel, C., Keegan, T., Duxbury, J., & Skirrow, R. (2020). Quantitative risk assessment of rock slope instabilities that threaten a highway near Canmore, Alberta, Canada: Managing risk calculation uncertainty in practice. *Can Geotech Journal, 57*(3), 337–353.

Macciotta, R., & Lefsrud, L. (2018). Framework for developing risk to life evaluation criteria associated with landslides in Canada. *Geoenvironmental Disasters, 5*(1), 1–14.

Macleod, A., Hofmeister, J., Wang, Y., & Burns, S. (2005). *Landslide indirect losses: methods and case studies from Oregon*. State of Oregon, Department of Geology and Mineral Industries.

Malone, A. W. (2005). The story of quantified risk and its place in slope safety policy in Hong Kong. In *Landslide hazard and risk* (pp. 643–674).

Massey, C. I., de Vilder, S., Taig, T., Lukovic, B., Archibald, G., & Morgenstern, R. (2022). Landslide hazard and risk assessment for the Fox and Franz Josef Glacier Valleys. GNS Sci. Science. Rep., 2022/29, 101 p.

Massey, C. I., Lukovic, B., de Vilder, S., & Archibald, G. C. (2020). Landslide risk analysis for Clifton Beach, Cape Kidnappers, Hawke's Bay. Lower Hutt (NZ): GNS Science. 101 p. Consultancy Report 2020/28.

National Academies of Sciences, Engineering, and Medicine (NASEM). (2019). Geotechnical asset management for transportation agencies. The National Academies Press. https://doi.org/10.17226/25363

Nieto, N., Chamorro, A., Echaveguren, T., Sáez, E., & González, A. (2021). Development of fragility curves for road embankments exposed to perpendicular debris flows. *Geomatics, Natural Hazards and Risk, 12*(1), 1560–1583.

Ouyang, C., Wang, Z., An, H., Liu, X., & Wang, D. (2019). An example of a hazard and risk assessment for debris flows—a case study of Niwan Gully, Wudu, China. *Engineering Geology, 263*, 105351.

Papathoma-Köhle et al. (2017). Matrices, curves and indicators: A review of approaches to assess physical vulnerability to debris flows. *Earth-Science Reviews, 171*, 272–288.

Park, Y., Pradhan, A, & Kim, U. et al. (2016). Development and application of urban landslide vulnerability assessment methodology reflecting social and economic variables. *Advances in Meteorology, 2016*, 4572498.

Pollock, W., & Wartman, J. (2020). Human vulnerability to landslides. *GeoHealth, 4*(10), e2020GH000287.

Porter, M., Jakob, M., & Holm, K. *Proposed landslide risk tolerance criteria* (9 pp). GeoHalifax.

Porter, M., Jakob, M., Holm, K., & McDougall, S. (2017). Risk-based landslide safety assessments in Canada. In: *Proceedings of the North American symposium on landslides, Roanoke, Virginia*, (pp. 4–8).

Porter, M., Lato, M., Quinn, P., & Whittall, J. (2019). Challenges with use of risk matrices for geohazard risk management for resource development projects. In J. Wesseloo (ed.), *Proceedings of the first international conference on mining geomechanical risk, Australian centre for geomechanics, Perth*, (pp. 71–84).

Porter, M., & Morgenstern, N. (2013). Landslide risk evaluation: Canadian technical guidelines and best practices related to landslides: A national initiative for loss reduction. *Natural Resources Canada*.

Prieto, J. A., Journeay, M., Acevedo, A. B., Arbelaez, J. D., & Ulmi, M. (2018). Development of structural debris flow fragility curves (debris flow buildings resistance) using momentum flux rate as a hazard parameter. *Engineering Geology, 239*, 144–157.

Rickenmann, D. (1999). Empirical relationships for debris flows. *Natural Hazards, 19*(1), 47–77.

SafeLand. (2011). D2.5: Physical vulnerability of elements at risk to landslides: Methodology for evaluation, fragility curves and damage states for buildings and lifelines. Work Package 2.2—Vulnerability to landslides. In: *SafeLand, living with landslide risk in Europe: Assessment effects of global change, and risk management strategies*. Seventh Framework Programme.

Santi, P. M., Hewitt, K., Van Dine, D., & Barillas, E. M. (2010). Debris-flow impact, vulnerability, and response. *Natural Hazards*. https://doi.org/10.1007/s11069-010-9576-8

Saunders, W., & Glassey, P. (2007). Guidelines for assessing planning, policy and consent requirements for landslide-prone land. In *GNS Science Miscellaneous Series* (vol. 7).

Schneider, P., & Schauer, B. (2006). HAZUS—its development and its future. *Natural Hazard Review 7*(2), 40.

Shu, H., Ma, J., Zhang, P., Yu, H., Ren, S., Qi, S., & Yang, H. (2019). Debris-flow risk assessment: From catchment to regional scale: A case study from southern Gansu Province, China. *Ekoloji, 28*(107), 2319–2333.

Sim, K. B., Lee, M. L., & Wong, S. Y. (2022). A review of landslide acceptable risk and tolerable risk. *Geoenvironmental Disasters, 9*, 3. https://doi.org/10.1186/s40677-022-00205-6

Smith, K. (2013). *Environmental hazards*. Routledge.

So, E. (2016). *Estimating fatality rates for earthquake loss models*. Spinger. https://doi.org/10.1007/987-3-319-26838-5

Stoffel, M., Mendlik, T., Schneuwly-Bollschweiler, M., et al. (2014). Possible impacts of climate change on debris-flow activity in the Swiss Alps. *Climatic Change, 122*, 141–155. https://doi.org/10.1007/s10584-013-0993-z

Strouth, A., LeSueur, P., & McDougall, S. (2022). ALARP and other conditions for tolerating geohazard risks. In Cloutier et al. (eds.), *Proceedings of the 8th Canadian conference on geotechnique and natural hazards*.

Strouth, A., & McDougall, S. (2021). Societal risk evaluation for landslides: Historical synthesis and proposed tools. *Landslides, 18*, 1071–1085. https://doi.org/10.1007/s10346-020-01547-8

Strouth, A., & McDougall, S. (2022). Individual risk evaluation for landslides: Key details. *Landslides, 19*, 977–991. https://doi.org/10.1007/s10346-021-01838-8

Sturzenegger, M., Holm, K., Lau, C. A., & Jakob, M. (2021). Debris-flow and debris-flood susceptibility mapping for geohazard risk prioritization. *Environmental & Engineering Geoscience, 27*(2), 179–194.

Sung, C., & Liaw, S. (2021). A GIS-based approach for assessing social vulnerability to flood and debris hazards. *International Journal of Disaster Risk Reduction, 46*.

Taig, T., Massey, C., &Webb, T. (2012). *Canterbury earthquakes port hills slope stability: Principles and criteria for the assessment of risk from slope instability in the port hills, Christchurch*. GNS Science Consultancy Report 2011/319.

Taig, T., Massey, C. I., & Webb, T. (2011). Canterbury earthquakes 2010/11 Port Hills slope stability: Principles and criteria for the assessment of risk from slope instability in the Port Hills, Christchurch. In *GNS Science Consultancy Report 2011/399, GNS* (47 pp).

Taig, T. (2022a). *Risk comparisons for DOC visitors and staff* (131 p.). Cheshire (GB): TTAC Ltd.

Taig, T. (2022b). *Guidelines for DOC on dealing with natural hazard risk* (91 p.). Cheshire (GB): TTAC Ltd.

Tappenden, K. M. (2014). The district of North Vancouver's landslide management strategy: Role of public involvement for determining tolerable risk and increasing community resilience. *Natural Hazards, 72*, 481–501.

Tang, Y., Guo, Z., & Wi, L. et al. (2022). Assessing debris flow risk at a catchment scale for an economic decision based on the LiDAR DEM and numerical simulation. *Frontiers in Earth Science*. https://doi.org/10.3389/feart.2022.821735

Tonkin & Taylor Ltd. (2015). Supplementary risk assessment, debris flow hazard, Matatā, Bay of Plenty. Final report to Whakatāne District Council dated July 2015.

Tsao, T. C., Hsu, C. H., Yin, H. Y., & Cheng, K. P. (2019). Debris flow building damage level and vulnerability curve: A case study of a 2015 typhoon event in northern Taiwan. *Association of Environmental and Engineering Geologists*. Special publication 28.

United States Bureau of Reclamation (USBR). (2011). Rationale used to develop reclamation's dam safety public protection guidelines.

U.S. Department of the Interior, Bureau of Reclamation, and U.S. Army Corps of Engineers (USBoR-USACE). (2019). Best practices in dam and levee safety risk analysis. Version 4.1, July 2019. https://www.usbr.gov/ssle/damsafety/risk/methodology.html

U.S. Geological Survey (USGS). (2022). Post-fire debris-flow hazards, scientific background. Accessed 29 Aug 2022. https://landslides.usgs.gov/hazards/postfire_debrisflow/background2016.php

VanDine, D. (2012). Risk management—Canadian technical guidelines and best practices related to landslides: A national initiative for loss reduction. *Geological survey of Canada*, 6996.

Van Western, C. J. (2004). Geo-Information tools for landslide risk assessment: An overview of recent developments. In *Proceedings of the 9th International Symposium on Landslides*.

Van Westen, C. J., Van Asch, T. W., & Soeters, R. (2006). Landslide hazard and risk zonation—why is it still so difficult? *Bulletin of Engineering Geology and the Environment, 65*(2), 167–184.

Viscusi, W. K., & Aldy, J. E. (2003). The value of a statistical life: A critical review of market estimates throughout the world. *Journal of Risk and Uncertainty, 27*, 5–76.

Viscusi, W. K., & Masterman, C. J. (2017). Income elasticities and global values of a statistical life. *Journal of Benefit-Cost Analysis, 8*, 226–250.

Vose, D. (2009). *Risk analysis—a quantitative guide* (3rd ed.). John Wiley & Sons.

Vrijling, J. K., Van Hengel, W., & Houben, R. J. (1995). A framework for risk evaluation. *Journal of Hazardous Materials, 43*, 245–261.

Wendeler, C. (2016). Debris-flow protection systems for mountain torrents. Basic principles for planning and calculation of flexible barriers. *WSL Berichte, 44*, 267. ISSN 2296-3456.

Wisner, B., Blaikie, P., Cannon, T., & Davis, I. (2004). *At risk: Natural hazards*. Routledge.

Winter, M. G., Smith, J. T., Fotopoulou, S., Pitilakis, K., Mavrouli, O., Corominas, J., & Argyroudis, S. (2014). An expert judgement approach to determining the physical vulnerability of roads to debris flow. *Bulletin of Engineering Geology and the Environment, 73*, 291–305.

Winter & Wong. (2020). The assessment of quantitative risk to road users from debris flow. *Geoenvironmental Disasters, 7*(4).

Wong. (2005). Landslide risk assessment for individual. In *Proceedings of the international conference on landslide risk management, Vancouver, Canada* (pp. 237–296).

Wong (2013). Is landslide risk quantifiable and manageable? In L. Zhang, Y. Wang, G. Wang & D.Q. Li (eds.), *Geotechnical safety and risk IV* (pp. 101–111). CRC Press.

Yin, L., Zhu, J., Li, Y., Zeng, C., Zhu, Q., Qi, H., & Zhang, P. (2017). A virtual geographic environment for debris flow risk analysis in residential areas. *ISPRS International Journal of Geo-Information, 6*(11), 377.

You, X., & Tonon, F. (2012). Event-tree analysis with imprecise probabilities. *Risk Analysis, 32*(2), 330–344.

Zhang, S., & Zhang, L. M. (2017). Impact of the 2008 Wenchuan earthquake in China on subsequent long-term debris flow activities in the epicentral area. *Geomorphology, 276*, 86–103.

Zhang, S., Zhang, L., Li, X., & Xu, Q. (2018). Physical vulnerability models for assessing building damage by debris flows. *Engineering Geology, 247*, 145–158.

Zimmermann, M. N. (2005). Analysis and management of debris-flow risks at Sörenberg (Switzerland). In *Debris-flow hazards and related phenomena* (pp. 615–634). Springer.

Zou, Q., Cui, P., Zhou, G. G., Li, S., Tang, J., & Li, S. (2018). A new approach to assessing vulnerability of mountain highways subject to debris flows in China. *Progress in Physical Geography: Earth and Environment, 42*(3), 305–329.

Zou, Q., Cui, P., He, J., Lei, Y., & Li, S. (2019). Regional risk assessment of debris flows in China—An HRU-based approach. *Geomorphology, 340*, 84–102.

Zubrycky, S., Mitchell, A., McDougall, S., Strouth, A., Clague, J. J., & Menounos, B. (2021). Exploring new methods to analyse spatial impact distributions on debris-flow fans using data from south-western British Columbia. *Earth Surface Processes and Landforms, 46*(12), 2395–2413.

15 Functional Design of Mitigation Measures: From Design Event Definition to Targeted Process Modifications

Guillaume Piton, Vincenzo D'Agostino, Toshiyuki Horiguchi, Akihiko Ikeda and Johannes Hübl

Abstract

Structural and non-structural measures might be implemented to protect elements at risk against debris-flow hazards. However, despite centuries of forestry and empirical soil conservation works, as well as decades of research on debris-flow processes, defining protection strategies against debris flows remains complicated. How to select and tailor protection measures is still a very active research topic. This chapter covers recent advances regarding the design and maintenance of structural mitigation measures. In essence, we provide a framework and elements to help define mitigation strategies. We briefly describe how design events can be selected in view of the mitigation of adverse consequences and risk (see Strouth et al., this volume). We also discuss the importance of accounting for routine events and rare events stronger than the design events to increase the robustness of the system against operational failure (e.g., excessive maintenance costs and environmental side effects), sudden failure or unexpected behaviour during overloading. The second part explains how functional analysis of the current debris-flow channel must be conducted to understand the initiation of channel malfunctioning and the associated cascading processes leading to widespread debris-flow hazards. This step enables one to identify the adaptations required to mitigate them with minimal actions. The main part of the chapter is then a review of the various types of structural measures than can be implemented, explaining in detail their main function, and how they can be used to cope with specific, targeted malfunctions. This framework and catalogue will help users select the type, location and main features of the measures to implement.

G. Piton (✉)
Université Grenoble Alpes, INRAE, CNRS, IRD, Grenoble INP, IGE, 38000 Grenoble, France
e-mail: guillaume.piton@inrae.fr

V. D'Agostino
Department of Land, Environment, Agriculture and Forestry, University of Padova, Agripolis, Viale Dell'Università 16, 35020 Padua, Legnaro, Italy

T. Horiguchi
National Defense Academy, Yokosuka, Kanagawa 239-8686, Japan

A. Ikeda
Sabo and Landslide Technical Center, 2-7-5 Hirakawacho, Chiyoda-Ku, Tokyo 102-0093, Japan

J. Hübl
Institute of Mountain Risk Engineering, University of Natural Resources and Life Sciences Vienna, Peter Jordan Strasse 82, 1190 Vienna, Austria

M. Jakob et al. (eds.), *Advances in Debris-flow Science and Practice*, Geoenvironmental Disaster Reduction,
https://doi.org/10.1007/978-3-031-48691-3_15

15.1 Introduction

Debris flows and debris floods pose great risks to people and infrastructure. When relocation of assets and avoidance of areas at risk are not possible (Saunders et al., this volume) or insufficient to reach an acceptable risk level (Strouth et al., this volume), implementing mitigation measures can help.

Planning and designing mitigation measures against debris flows and debris floods is more complex than against river floods, because (i) the processes are more variable and include large sediment volumes in a short time period, (ii) the processes are less well characterized (e.g. in term of their fluid mechanics and frequency-magnitude relationship—see Jakob & Nolde, *this volume*), but also because (iii) mitigation measures may act remotely (e.g., a landslide drainage system reduces the volume of debris flows downstream) or reforestation may change the frequency and magnitude of events by reducing soil erosion. In addition, many types of measures can be used to mitigate, in various ways, the natural processes involved in the sediment cascade leading to debris-flow hazards. General presentations of these various measures can be found in the literature (e.g. Carladous et al., 2016; Chatwin et al., 1994; Eisbacher, 1982; Hübl & Fiebiger, 2005; Ikeya, 1989; Kronfellner-Kraus, 1983; Marchi et al., 2019; Mark, 2017; Patel, 2012; VanDine, 1996) with interesting examples and considerations on effects of measures on flows and sediment connectivity.

A large body of literature focuses on all aspects of the interactions between mitigation measures and debris-flow processes has been published since the previous version of this book (Jakob & Hungr, 2005). The literature on the topic was so large that it was decided to include chapters on structural mitigation measures in this volume. The present chapter is an overview of functional designs. It should help to select the types of structures, and to define their shapes. The chapter by Kwan et al. (*this volume*) covers structural design, i.e. the computation of impact forces and loading diagrams helping to define the resistance of the structures. Many lessons can also be learnt through forensic analyses of examples of failures of mitigation structures, which is the topic of another chapter (Hübl et al., *this volume*).

Once a comprehensive debris-flow risk assessment has been completed and some protection objectives are defined (Strouth et al., this volume), it is, however, often difficult to plan the relevant measures to be used in a given catchment, and to design the selected structures in detail (Piton, 2023). The present chapter tries to bridge this gap by providing a framework of analysis to identify cascading processes. This step seeks to highlight which features of the events that can be transferred without adverse consequences, and which features lead to damages (e.g., peak discharge, size of boulders and logs, cumulated solid volume). Meanwhile, the main types of mitigation measures, their function and the associated effects on debris-flow processes are reviewed. The planning and design of mitigation measures is then tailored to prevent the cascading processes leading to hazards by relevant transformation, conveyance and addressing the hazard sources.

The first section of the chapter briefly recalls that the planning of an integrated mitigation strategy should account for the whole range of events: from the most frequent events (to limit maintenance effort) to the rarest events (to be aware of the protection strategy limits). The second section of the chapter focuses on channel malfunctions, i.e. it defines the main reasons why debris flows may lead to adverse consequences. Understanding the underlying processes helps to adjust and custom tailor a relevant suite of mitigation measures. The third section provides a long catalogue of the various measures typically used in debris-flow hazard mitigation. They are gathered within three main objectives and several functions: (i) preventing sediment supply to the channel (function: soil retention on hillslopes, hillslope buttressing, channel stabilization and flow diversion), (ii) changing flow features using open check dams (flow transformation, large wood filtration, debris buffering, debris deposition, water retention), and (iii) guiding structures (conveyance and deflection structures). The type of structures that can achieve these functions are highlighted and their main design criteria are explained.

In a given catchment with its peculiarities as it pertains to the sediment supply from upstream and the channel capacity near the elements at risk, we hope that this review will help readers to tailor relevant measures to address the hazards and detrimental processes acting in debris-flow prone channels.

15.2 Routine, Project, Safety-Check and Danger Events

Focusing almost exclusively on one single event is insufficient to plan an integrated debris-flow hazard mitigation strategy. A comprehensive design study should address the whole range of events that the system might experience, from the most frequent to the rarest and including expert-informed avulsion scenarios. The features of the *project design event* (e.g., its peak discharge and volume) are used to define the mitigation measure geometry and characteristics (e.g., a conveyance capacity with a certain safety factor or freeboard, or a barrier crest level enabling it to trap the volume).

Safety-check events correspond to the range of events more intense than project design events, for which the structure should nonetheless not fail (Royet et al., 2009; CFBR, 2013, p. 174). A lower mitigation efficacy is accepted for these events since the system is functionally overloaded, but it should be implausible that the structures fail, both functionally (e.g., bank protections should not be outflanked or scoured) and structurally (e.g., check dams should not overturn or collapse under the impact force). These events are used, for instance, to choose the structures' freeboards and to design safety components of the mitigation structures (e.g., side dikes of debris basins preventing lateral overflowing, or spillways of barriers preventing barrier outflanking).

In case of an extreme event, structures might fail at a probability that ought to be estimated if possible. This raises the question of *danger events*, i.e., event/flood/level conditions over which the structure's safety is no longer guaranteed (the terminology "danger events" originates from dam reservoirs:– Royet et al., 2009; CFBR, 2013, p.174). In case of an event stronger than a danger event, structural failure is plausible. Qualifying danger events raises the awareness of structure managers and citizens that some residual risk always remains. It paves the way towards evacuation plans. It also provides an opportunity to study the failure modes of the structures and, eventually, to define slight adjustments or adaptations to improve the robustness of the system.

The simplest mitigation systems merely trap, convey or divert flows. Complex mitigation systems might include structures seeking to modify the processes by interacting with solid transport (e.g. open check dams). Appropriate maintenance of structures, including monitoring and eventual excavation, is mandatory to avoid operational failure (i.e., structures not fulfilling their function because they are not properly maintained). Carladous et al. (2022) suggest calling *routine events* those events that would not endanger the elements at risk because of sufficiently low magnitude. In many debris flow catchments, routine events might produce normal flood-related bedload transport or debris floods sensu stricto (Church & Jakob, 2020) or a combination of these. Limiting maintenance cost, in particular excavation operations associated with routine events, deserves heightened attention by mitigation works designers (Cánovas et al., 2016). Structures excessively trapping solid material can become unaffordable and might decrease in protection efficacy, or even become abandoned (Carladous et al., 2022; SedAlp, 2015). Full transfer of water and solid discharge must be sought for these events to minimize maintenance. Guiding channels transferring low flows downstream or bottom openings in barriers are, for instance, designed using typical event parameters such as peak discharge, solid concentration and potential presence of large wood pieces (Moldenhauer-Roth et al., 2021; Simoni et al., 2017).

15.3 Processes of Channel Malfunction Leading to Debris-Flow Hazards

15.3.1 Why Perform a Functional Analysis of the Channel?

Overall, mitigation measures seek to prevent or to decrease the frequency of debris-flow deposition and spreading in vulnerable areas. Since people expect that channel beds convey water and solid transport without adverse consequences, we call "malfunction" any negative consequences associated with debris flows along their channel (e.g., obstruction, deposition, bank overtopping, bank collapse). In areas with poor sediment connectivity between the debris-flow catchment, its debris fan and the downstream river system, chronic deposition of the sediment supplied is the rule (Harvey, 2012, de Haas et al. *this volume*). In better connected catchment—fan—river systems, some solid transport is transferred to the river and the initiation and occurrence of malfunction can be more complicated. Widespread deposition of debris flows might be initiated by different types of malfunctions (see next section). Trying to understand which features of the systems and processes tend to trigger deposition and cascading effects (e.g. channel filling, bank overtopping, spreading and avulsion) is the first step to tailor a relevant mitigation system (Hübl, 2018a; Simoni et al., 2017). In essence, it is sometimes possible to design more affordable mitigation systems in cases where slight or moderate changes of the flood process might suffice to prevent cascading consequences (e.g., removing only large boulders and large wood pieces or reducing the peak flow). Conversely, if any supplied solid matter tends to deposit and threaten elements at risk, more drastic strategies are required.

The next sections synthesize the main types of basic malfunctions that might trigger deposition. These malfunctions mirror natural processes occurring in wild areas, some artificial structures simply increase the probability of the process to occur, e.g. bridges are jammed by boulders as can be canyons and rock arches (Piton et al., 2022b). Readers interested in more details regarding the processes leading to avulsions are referred to de Haas et al. (*this volume*). Compound cases can often be found. Although we did our best, this list might not be exhaustive. These simplified categories seek to help better understand the processes leading to debris flow hazards to later tailor a relevant mitigation strategy. Figure 15.1 illustrates these types of malfunctions. Later, we discuss which types of mitigation measures are suitable to prevent these different types of malfunctions.

15.3.2 Jamming of Bottleneck Sections

15.3.2.1 Boulder Jams (Malfunction Type-B)

Apart from the impressively large volumes of solid material that debris flows can transport, the large size of the boulders that are transported by some events can be astonishing (e.g., Baggio & D'Agostino, 2021; Jakob, 2005; Perov et al., 2017). In areas where geology makes them available, boulders with diameters above one meter are quite common in debris-flow deposits. Even bigger boulders, sometimes up to five meters and more can be transported and play a key role in the destructive power of this geophysical process (Horiguchi & Richefeu, 2020; Horiguchi et al., 2021a; Kaitna & Hübl, 2012; Kean et al., 2019; Shima et al., 2016).

Jamming of bottleneck sections is mainly a granular process driven by the ratio between the channel or opening size (width w_0 or height

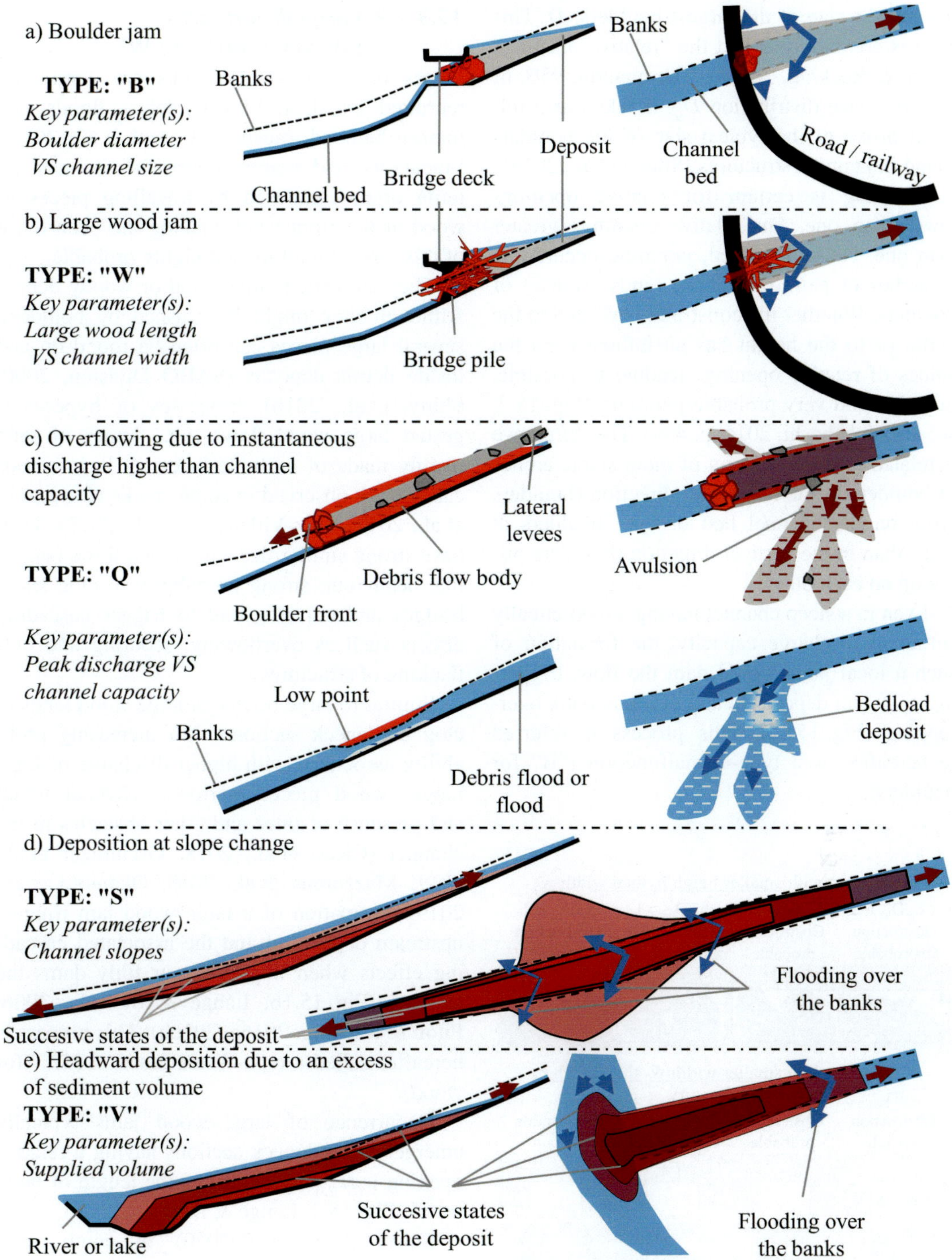

Fig. 15.1 Main types of channel functional failures or "malfunctions", the key driver being drawn in red, water in blue, and sediment in grey: jamming of bottleneck section by **a** boulders (type-B failure) or **b** large wood (type-W failure); **c** insufficient channel discharge capacity for debris flows or debris floods (type-Q failure), **d** deposition at slope change due to excess in sediment concentration (type-S failure); and deposition due to excess of sediment volume as compared to the system buffering capacity (type-V failure)

h_0) and the size of the largest boulders, D. This ratio is commonly called the "relative opening" (Piton & Recking, 2016a). The quantile 95% of the grain size distribution D_{95} is taken as a relevant proxy of the typical size of the boulders found in jammed structures (Shima et al., 2016).

Jamming is certain for relative openings lower than one. For relative openings greater than one, but still limited, jamming occurs due to arches of pairs (and more rarely triplets) of boulders. Whether the constraint is related to the width or to the height has an influence on the values of relative openings leading to possible, probable and very probable jamming (Fig. 15.2, see also Takahashi, 2014, p. 454). The difference is related to the formation of more stable arches of boulders in the horizontal direction (boulders lying on the channel bed or over boulders at rest), than in the vertical direction (boulders piling up on each other).

Even in a steep channel having a theoretically sufficient discharge capacity, the formation of such a local jams would dam the flow, triggering upstream deposition and eventual bank overtopping (Fig. 15.1a). This process is referred to hereafter as a type-B malfunction ("B" for boulders).

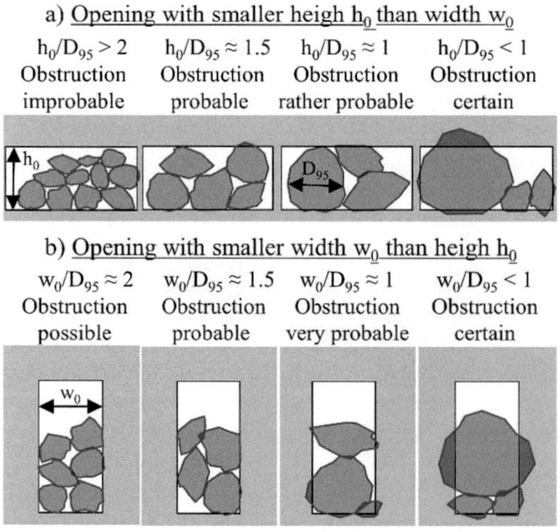

Fig. 15.2 Effect of the constraint orientation on the jamming probability and emergence of type-B malfunction

15.3.2.2 Large Wood Jams (Malfunction Type-W)

Pieces of large wood chunks are commonly recruited by debris flows, debris floods and intense bedload transport (Comiti et al., 2016). Precursory and mature debris-flow surges are quite often composed by travelling pieces of wood at the front head, making the occurrence of this type of malfunction highly probable.

They are called large wood or woody debris. Although they might be crushed by boulders, several large pieces can often be found on and inside debris deposits (SABO Division, 2000; Shima et al., 2016). Processes of hyper-congested large wood flows (i.e., surges of rafts mainly made of multiple layers of large wood) can also be observed in some creeks (Ravazzolo et al., 2017; Ruiz-Villanueva et al., 2019). They have strong similarities with debris flows (surge-like behaviour, strong probability to clog small bridges and culverts, and to trigger cascading effects such as overflowing, scouring and outflanking of structures).

Similar to large boulders, large wood tends to clog bottleneck sections with increasing probability associated with higher discharge of logs, longer wood pieces, narrower channel width and presence of piles and other obstacles in the channel (Cicco et al., 2018; Gschnitzer et al., 2017; Mazzorana et al., 2018; Okamoto et al., 2019). Formation of a large wood jam triggers upstream deposition and the associated cascading effects when it partially or fully dams the channel (Fig. 15.1b; Lange & Bezzola, 2006; Piton & Recking, 2016a, 2016b). This process is hereafter called a type-W malfunction ("W" for wood).

Occurrence of large wood jams typically emerges at bottleneck sections having a relative opening $w_0/L_{LW}<1$, with L_{LW} the length of large wood (Fig. 15.3, Lange & Bezzola, 2006).

Storm Alex (10/2/2020) in south eastern France impacted over 140 bridges, 1/3 of which were destroyed or heavily damaged. Unpublished observations showed that the 95th quantile of the large wood length distribution L_{95} is considered a relevant proxy of the

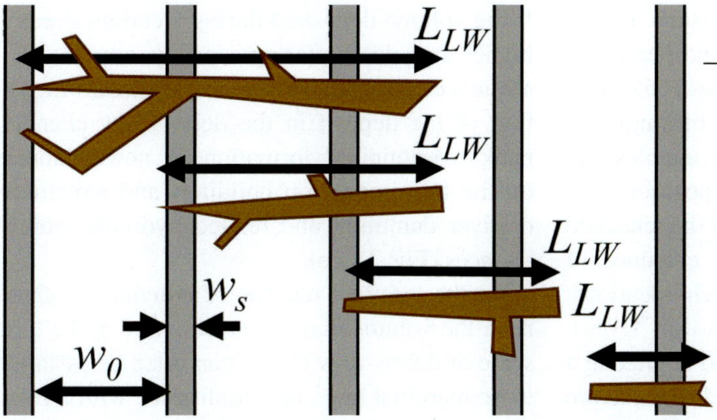

w_0/L_{LW}	Trapping
<1/3	Very probable
1/3 – 1/2	Probable
1/2 – 1	Uncertain
>1	Very improbable

Fig. 15.3 Conditions of blockage of large wood pieces and type-W malfunction

parameter L_{LW} when assessing the relative opening of structures.

It is worth stressing that when falling and being transported, trees break into pieces smaller than their standing tree height (Rickli et al., 2018). Field surveys and analyses accounting for this breaking are advised to predict the size and quantity of large wood to be supplied by rare events. An approach using sub-scenarios of events with variable amounts and eventual sizes of wood is often necessary because strong random variation of the supply is observed in the field (Mazzorana et al., 2011, 2012; Steeb et al., 2017).

15.3.3 Insufficient Channel Discharge Capacity (Malfunction Type-Q)

Insufficient channel capacity to convey the peak discharge is the most intuitive channel malfunction (type-Q malfunction). It may be related to a local lower bank level, as in the case of river floods (Fig. 15.1c), or to a transition of cross-section shape (e.g. from a U shape to a less efficient shape). It may also be caused by a relatively long channel reach with a thalweg slope angle prone to debris flow deposition (e.g. <8°), which is an intermediate case with the malfunction type described in the following section. Deposition on banks especially due to super-elevation in curves, building of levees, spreading on debris fans and eventual avulsion can also occur in case of debris flow surges having a peak discharge higher than the bankfull channel capacity (see De Haas et al., *this volume*).

The occurrence of this kind of malfunction is merely related to the peak discharge value as compared to the channel capacity. The hydrograph shape, but also the viscosity and more generally the rheology of the process (see Iverson & George, *this volume*), drive the peak discharge. The channel slope, width and depth, and additional energy-dissipating factors are the key factors driving the channel capacity.

15.3.4 Deposition at Slope Break (Malfunction Type-S)

The channel slope is a strong driver of solid transport and debris flow transfer (Hungr et al., 2005; Rickenmann & Koschni, 2010). Channel gradient naturally decreases from headwaters to lowlands. If the channel slope has inflection points, deposition is generally to be expected at these locations, especially in transport-limited reaches (Bovis & Jakob, 1999). Whether deposition, transfer or erosion occurs over various slopes is strongly driven by the solid concentration of the flows.

Strong deposition occurring at such inflection point, along with downward and headward deposition, is hereafter called a type-S malfunction (Fig. 15.1d). Detailed analyses of longitudinal profiles of debris-flow channels enables one to pre-identify such locations of deposition (e.g. Theule et al., 2015). They should be checked with field visits aimed at finding evidence of past deposition and erosion at such locations. Clusters of boulders, former debris-flow levees, gravel wedges perched on banks and aligned at a certain deposition slope or even old terraces are examples of the "silent witnesses" testifying that depositions did occur in such locations (Kaitna & Hübl, 2012).

It is worth stressing that if deposition often occurs at various sites during high magnitudes and/or under high solid concentration, subsequent flows may erode and wash away the deposited volumes, especially in supply-limited reaches (de Haas et al., 2020; Nishimoto, 2011; Piton et al., 2017). That is why inflection points can remain in the landscape despite many past debris flow events smoothing them transiently.

15.3.5 Insufficient Volume Buffering Capacity (Malfunction Type-V)

Debris flow materials reaching downstream river systems at confluences are rarely instantaneously transported by the rivers (and even often dam them, e.g. Chen et al., 2015). Transient deposition often occurs and will last some time, depending on the hydrology and transport capacity of the main river channel, and on the overall topographical features of the junction river-debris flow channel. Relatively high transient storage of solid material (i.e., cycles of in-channel erosion and deposition) is a key feature of steep active streams (Cucchiaro et al., 2019; de Haas et al., 2020; Theule et al., 2012, 2015). These "channel plugs" are key drivers of avulsion dynamics (de Haas et al., *this volume*). Each channel reach has consequently a certain capacity to absorb, and later release, some sediment. This can be called a "buffering capacity". If the volume deposited during a certain event is higher than this buffering capacity, adverse consequences occur as progressive upstream migration of the deposit in the debris flow channel, bank overtopping, formation of new channels on the fan, impact on buildings and structures, or river damming and related hydraulic consequences (Fig. 15.1e).

In the terminal reaches of alluvial fan channels, the volume that can be exported at the time scale of debris-flow events can often be assumed to be marginal (e.g., at a confluence with a river, at an estuary in a lake or the sea, or where the debris flow channel disappears in fully disconnected systems). In such reaches, adverse consequences occur each time the supplied volume is higher than the buffering volume. The occurrence of this type-V malfunction is thus directly driven by the supplied solid volume as compared to the system buffering capacity.

15.3.6 Functional Analysis of the Channel System

Several types of elementary triggering processes of sediment deposition are presented in the previous paragraphs. Debris flow channels, particularly the reaches where there are some elements at risk, can be analysed to map the homogenous reaches and local bottleneck sections where one or several of these malfunctions may occur for some or all debris flow events (Hübl, 2018a). Highlighting where and how solid transport might be disturbed for each class of events (e.g., routine, project design and danger events) is a prerequisite to the definition of well-adjusted mitigation strategies (i.e., tailored to the catchment peculiarities) (Simoni et al., 2017). Not only is the best possible estimation of the debris-flow processes coming from upstream necessary, but also a clear vision of the type and quantity of material that can be transferred downstream.

This preliminary analysis should clarify the allowable maximum discharge and size of boulders and large wood that can be conveyed. For given hypotheses on the supplied hydrographs and process types, the analysis should also try to predict the typical volume transferred and buffered. For example, detailed field and archive analyses of evidence of channel plugs or cycles of erosion and deposition will help to quantify potential buffering volumes. Numerical modelling of events with various magnitudes, varying secondary parameters across their natural ranges (e.g. peak discharge, rheology, and solid concentration) can help to quantify potential transfer capacity, obviously after a suitable calibration phase (Jakob, 2021). If the channel section and slope and/or the drained area vary significantly, this should be repeated for each homogenous reach of the stream.

This step deserves sufficient attention and rigor to more clearly define mitigation strategies, and it should be merged with the analysis of sediment source areas and their degree of connectivity to a given outlet (D'Agostino & Bertoldi, 2014). Such strategies will try to transform solid transport processes to precisely mitigate the adverse consequences of debris flows, while limiting costly side effects both in terms of money and environmental damage (Marchi et al., 2019; Rinaldi et al., 2009; Simoni et al., 2017).

15.4 Knowing Mitigation Measure Functions to Tailor a Strategy

This section is the core of this chapter. It is an attempt to provide a view of the effects and functions of the various measures used in mitigation strategies. It will hopefully support readers to select suitable measures to cope with the malfunctions they assess in their catchment. Figure 15.4 lists the main functions defined in this chapter and the associated effects on the processes (Hübl, 2018b). The subsequent subsections provide general comments, the main relevant design criteria and advice on these measures reported in the literature or gathered by the authors during their exchanges with practitioners.

15.4.1 Preventing Sediment Supply to the Channel

Whenever possible and affordable, it is useful to address the causes of debris flow hazards by preventing erosion.

15.4.1.1 Soil Conservation

Vegetation cover is the primary barrier to soil erosion and substantially affects geomorphology (Gibling & Davies, 2012; Zeng et al., 2009). Techniques to restore or enhance the protective effect of vegetation have been tried and tested for centuries. Forestry and soil-bioengineering were much used techniques long before heavy civil engineering (Demontzey, 1882; Wang, 1903; Kronfellner-Kraus, 1983; Schiechtl & Stern, 1994; Sheng & Liao, 1997; Sabo Department, 2000; Evette et al., 2009), and geotechnical slope stability analyses have also been developed to assess their effectiveness in shallow landslide stabilization (Bischetti et al., 2010).

Forestry and, more generally, soil conservation measures used to be very popular. They need a substantial labour workforce and have become unaffordable in some regions. They also need time to be fully effective and are thus long-term measures. Their landscape regeneration effects are very impressive (Fig. 15.5). In the context of climate change, soil conservation masterplans should consider this point in the selection of the species and techniques used (Kostadinov, 2010).

These techniques typically use vegetation and small structures (e.g. terraces, small log dams or fascines) often coupled with drainage systems to help the vegetation establish and grow. In the most active erosion areas, heavier civil engineering proved to be necessary to slow down erosion processes and decrease debris-flow activity (Kostadinov, 2007). Civil engineering techniques also help vegetation recover in reforested areas, but also spontaneously (Zeng et al., 2009), as discussed in the next section. Providing design criteria for soil conservation is beyond

General objective	Logo	Function	Effect on the processes
Prevent sediment supply		Soil conservation	Prevent erosion on hillslopes using vegetation protective capacity
		Channel stabilization	Prevent long term incision (temporal) and extented channel erosion events (spatial)
		Hillslope buttressing	Decrease sediment supply by decreasing landslide or rock avalanche movements
		Diversion	Divert runoff and channel flows from erosion-prone areas toward stabler areas
Change certain features of the processes (type, size, discharge, volume)		Transformation	Break debris flow fronts by trapping boulders
		Wood filtration	Trap large wood pieces
		Debris buffering	Transiently store solid transport to reduce solid peak discharge
		Debris deposition	Trap a certain volume of sediment
		Water retention	Transiently store water to reduce water peak discharge
Guide flows		Conveyance	Convey flows along a chosen path
		Deflection	Deflect flows toward less vulnerable areas

Fig. 15.4 Main objectives, functions and effects of debris-flow hazard mitigation measures (adapted from Hübl, 2018b)

the scope of this chapter and has been covered by manuals for more than a century (e.g. Brown, 1876; Chatwin et al., 1994; Demontzey, 1882 and many national guidelines).

15.4.1.2 Channel Stabilization

Channel bed erosion, especially incision, is a major source of sediment. We call "channel stabilization" the prevention or reduction of channel bed erosion (mostly vertically, but also laterally, see Piton et al., 2017). The most common type of works to achieve this function are a series of check dams or artificial step-pool sequences, but aprons and lined channels can also be used in especially sensitive erosion areas (Fig. 15.6). This chapter cannot provide comprehensive guidelines to design channel stabilization (but see Lòpez Cadenas de Llano, 1988; USACE, 1994a; NILIM, 2007; Conesa-Garcia & Lenzi, 2013 and some comments in Fig. 15.6).

Channel stabilization of erosion-prone reaches is normally effective in reducing debris-flow hazards. It is, however, complicated to accurately model its efficacy because it involves many complex feedback loops (Lucas-Borja et al., 2021), including: (i) between channel stability, vegetation establishment and bank erosion (less active channels allow vegetation to recover, reduce sediment production and bank erosion and thus debris flow activity), (ii) between channel stability and hillslope stability (more stable channels will less frequently destabilize their side hillslopes), and (iii) between channel stability, channel morphology and sediment transport efficacy. More stable channels allow stronger

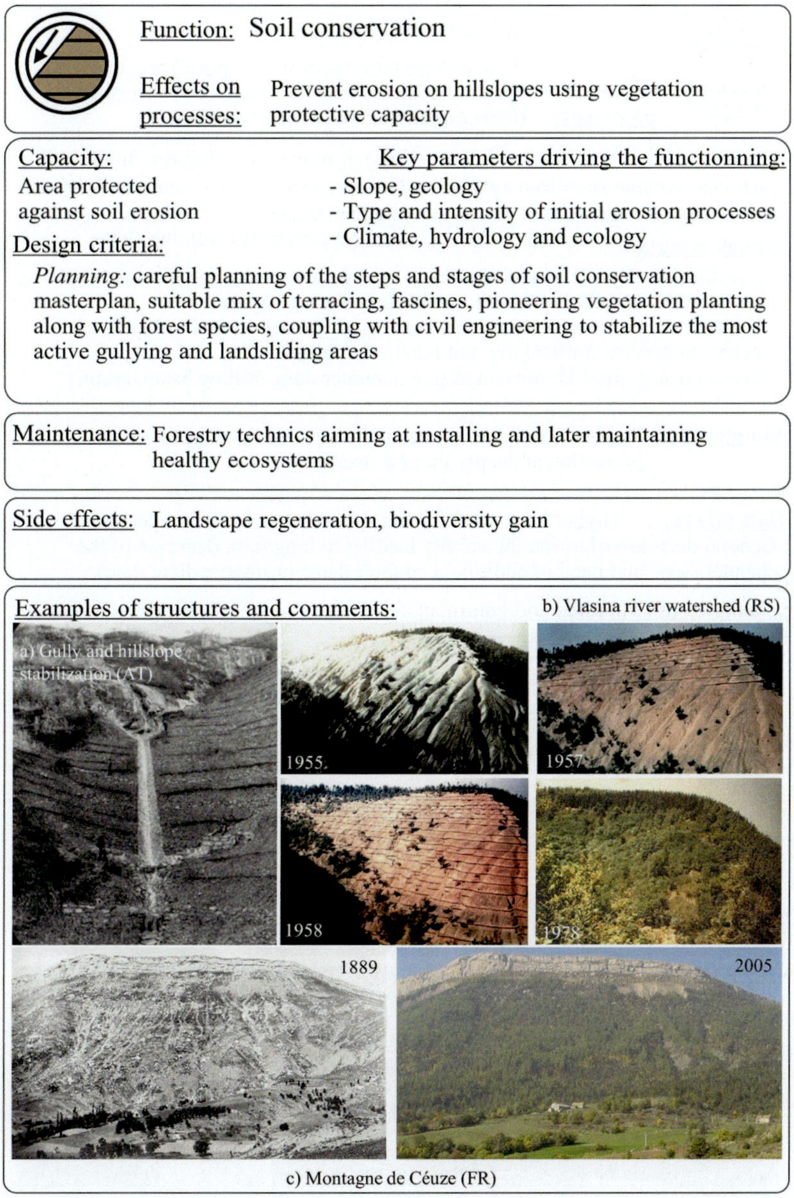

Fig. 15.5 Main features of measures targeted for soil conservation (images courtesy of **a** BOKU, **b** V. Milojević in Kostadinov (2010), **c** ONF-RTM)

channel armouring to develop. The extension of erosion events is also reduced by check dams (Piton & Recking, 2014), which promote less intense erosion and sediment transport events, particularly if corroborated by energy dissipation of a staircase check dam profile.

Stabilized channels should no longer experience long-term channel incision. When designing channel stabilization, one can choose to shift the new channel axis on a given bank to provide some distance from a lateral sediment source (landslide, erosion-sensitive bank).

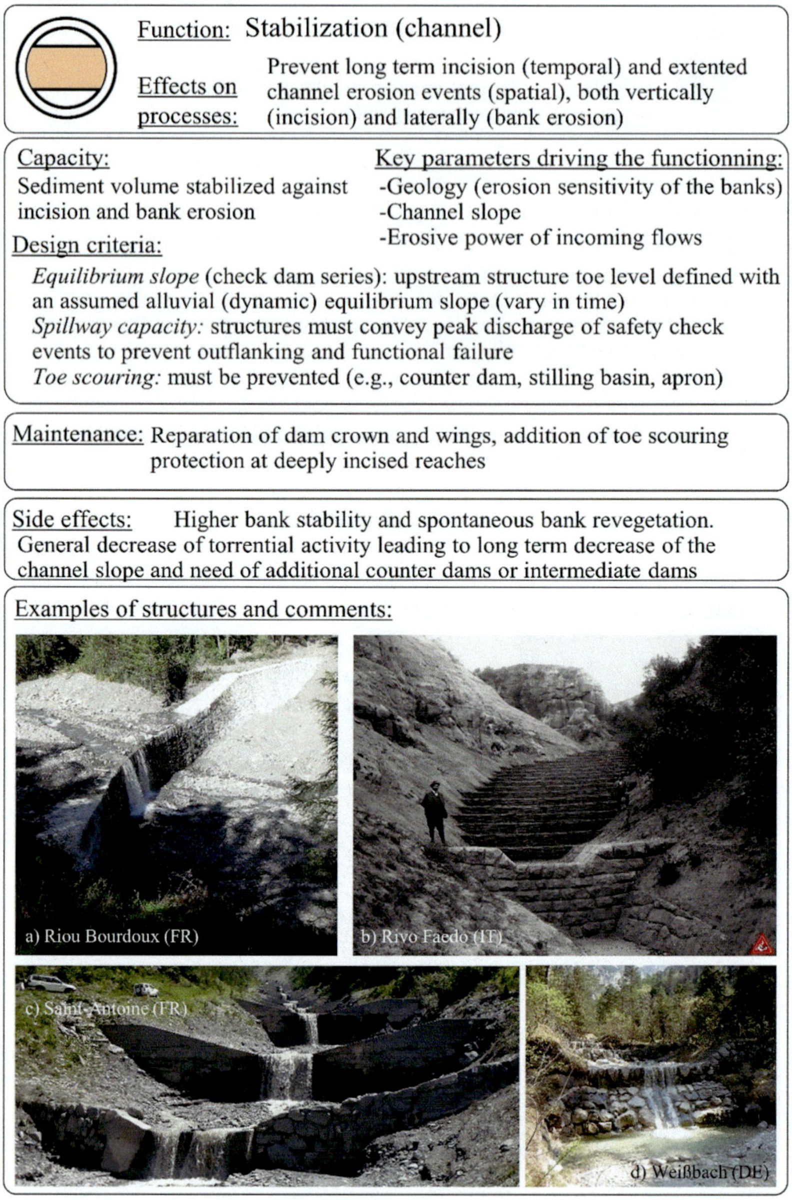

Fig. 15.6 Main features of measures targeted for channel stabilization (images courtesy of **a** ONF-RTM, **b** BOKU, **c** and **d** the authors)

It is also worth stressing another effect emerging in the long term and related to the natural buffering effect of channels, effect that is generally neglected by practioners. During debris-flow events, both deposition and erosion of sediment in the channel can be reduced in reaches segmented by check dams (Piton et al., 2017). The prevention of extended bed erosion events likely reduces the magnitude of large events, though it is difficult to accurately predict or appraise without long-term field observations or finely tuned models (but see Remaître et al., 2008). Nonetheless, after decades of field works, Poncet (1995, p. 713) stated that *"Check dams are built to deliver in small amounts, what the torrent would have release abruptly en masse"*.

15.4.1.3 Hillslope Buttressing

Large instabilities on hillslope, such as deep-seated landslides or rock avalanches, are major sediment sources. The incision of their toe by channel erosion is generally a key driver of their activity. Even better than stopping this incision, refilling the valley bottom at a higher level usually has a strong positive buttressing effect against hillslope movement and thus reduces sediment supply to the torrent (Zeng et al., 2009). Series of check dams are sometimes used specifically to fulfil this function (called here "hillslope buttressing"—Fig. 15.7a, b—and also called elsewhere "hillslope consolidation", e.g.

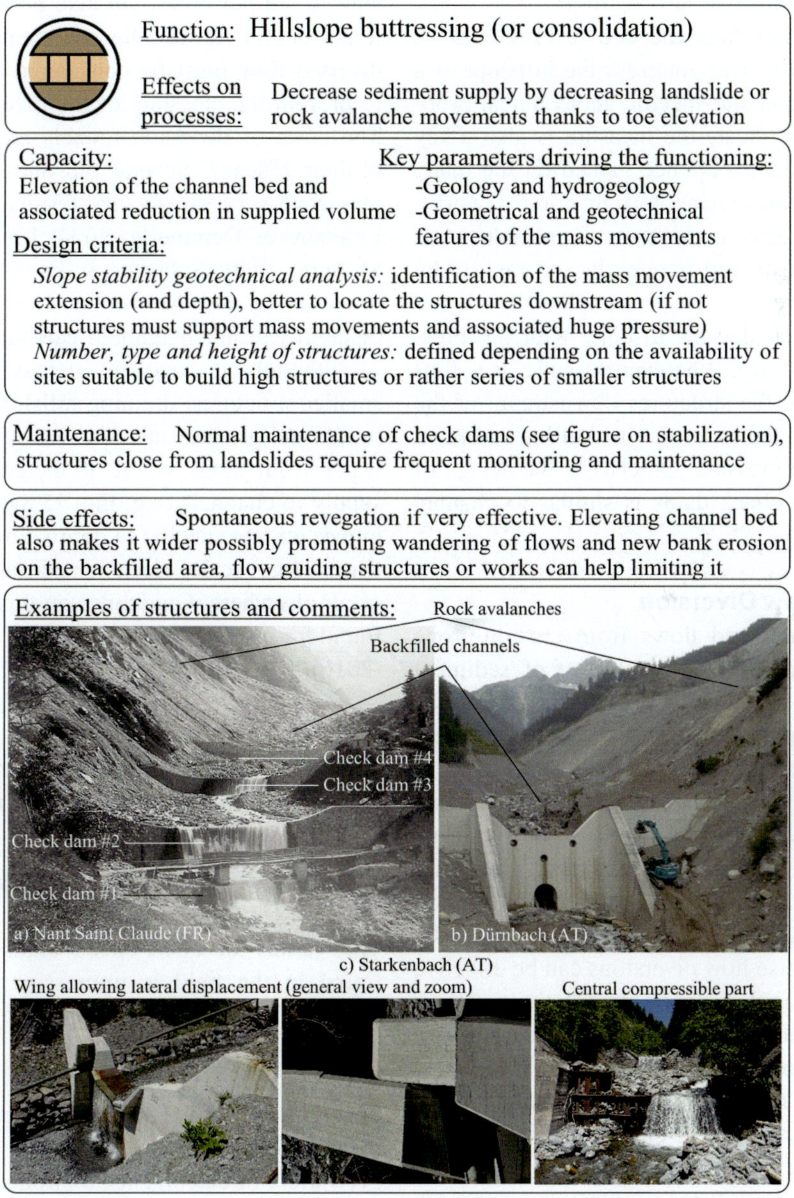

Fig. 15.7 Main features of measures targeted for hillslope consolidation (images courtesy of **a** ONF-RTM, **b-d** the authors)

in Hübl, 2018a; Piton et al., 2017). There are examples of bed level rising by several dozens of meters using check dams (e.g., Eisbacher, 1982; Piton et al., 2017; Yang et al., 2021). Retaining walls also serve the function to buttress a landslide toe, but are less suitable for deep-seated and/or large instabilities. Hillslope buttressing check dams are often used in conjunction with drainage systems to stabilize side slope instabilities (see next section).

Defining how high the bed level should be elevated to effectively buttress the hillslope is a complicated and site-specific geotechnical problem. Where to locate the measure is also a key point. From the experience gained on the many landslides treated using torrent control works, structures located at the foot of unstable areas tend to get destroyed frequently and rapidly by the lateral pressure (see examples of adaptations of typical designs to cope with the pressure in Fig. 15.7c). Wherever possible, it is thus better to build the structures downstream of the unstable area and to let the backfilling sediment take the earth pressure. Otherwise, the design of consolidation check dams is similar to channel stabilization.

15.4.1.4 Flow Diversion

Diverting water and flows from erosion-prone area can help prevent recruitment of sediment and slope instabilities. Flow diversion can be diffuse and/or punctual and involves relatively small discharges (e.g., drainage networks in landslide prone areas). It has been proven that threshold discharges for debris flow triggering are very low (e.g. from some tenths to few hundreds of L/s, Pastorello et al., 2020; Kaitna et al. *this volume*). The management of runoff by means of precise flow diversions can be effective in the reduction of debris-flow activity (lower sediment volumes occurring less frequently). Localised flow diversion involves generally much higher discharges (e.g., diversion tunnels by-passing landslides or channels sending flows into other catchments) (Fig. 15.8).

Although the hillslope buttressing approach presented in the previous section can be effective, flows are maintained at the foot of the instabilities and may still erode material. When the geology is suitable to construct tunnels or diversion channels, the most effective way to curtail the sediment supply of a torrent by an unstable area is to by-pass it. The unstable area is then no longer regularly incised and may fill up the valley until it reaches a stable level. The diversion structures must then be designed to convey normal flows, but also floods and sometimes debris flows. The sediment load of the diverted flow must be considered in the design to prevent the clogging of the diversion system. The costs of diversion tunnels are high but so is their efficacy. Several old and recent examples gave very satisfactory results (Kuss, 1900; Lefebvre & Demmerle, 2004; Liu et al., 2017; Mougin, 1900; Schuster & Evans, 2011). The flows can be managed further downstream of the instability or in a neighbour catchment.

Although it involves networks of much smaller structures, draining hillslopes of surface runoff and groundwater also assists in removing water from areas where it increases sediment supply to channels (e.g. Fig. 15.8g). These techniques are not specific to debris flow catchments since large scale, deep-seated landslides also occur elsewhere. Guidelines on this topic can be found for instance in PWRI (2007) and in FOEN (2016).

15.4.2 Changing Flow Features Using Open Check Dams

15.4.2.1 Introduction to Open Check Dams

The variety of open check dams that can be found in the field is impressive (Bergmeister et al., 2009; Hübl & Fiebiger, 2005; Leys, 1969; López et al., 2010; Wehrmann et al., 2006; Zollinger, 1983). It can be confusing to find so many shapes and types. The great variety of open structures is likely related to the even bigger variety of catchments, as well as to the creativity of designers who tried and tested strategies to change debris-flow processes in a certain way.

Function: Diversion

Effects on processes: Divert runoff and channel flows from erosion-prone areas toward stabler areas

Capacity:
Maximum discharge possibly diverted

Key parameters driving the functionning:
-Hydrology
-Sediment load of the discharge to divert
-Mass movements at diversion structures

Design criteria:
Extension of instable areas: Diversion structure must capture flows approaching the instable area, but not be threatened by its movements (i.e. located upstream)
Hydrogeology drives the implantation of landslide drainage networks
Maximum discharges: that should be diverted drives the design of the intake, diversion tunnel or channel and the eventual downstream stilling structure

Maintenance: Keep intake structure capacity to divert flows, keep clean drains, repair sections broken by mass movement and flow erosion and abrasion

Side effects: Very strong decrease of the erosion activity and eventual mass movement in the by-passed instable area.

Examples of structures and comments:

Fig. 15.8 Main features of measures targeted for flow diversion (all images courtesy of the authors)

Open check dams can be complicated to design. At least four effects leading to sediment deposition may emerge within these structures (Fig. 15.9). These effects are analogous to the malfunctions listed previously. Open check dams are placed at dedicated sites where stream managers choose to manage and transform those torrential processes that would otherwise occur at undesirable areas in a less controlled manner. The size and shape of the deposition basin upstream of the dam are chosen to store a certain volume but also to transfer routine events. Its bottom slope can create an inflexion point in the deposition area (type-S 'malfunction' according to Fig. 15.1), thus preventing its emergence further downstream (Fig. 15.9a).

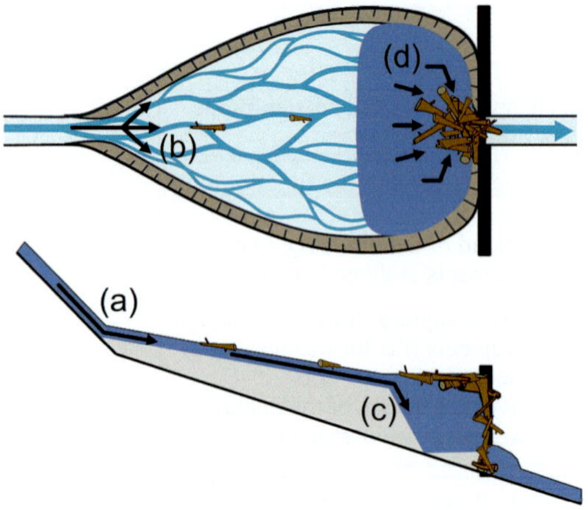

Fig. 15.9 Top and side views of an open check dam with sketches of the four main processes leading to deposition: **a** slope breaking analogous to type-S malfunction, **b** flow spreading, analogous to type-Q malfunction, **c** delta-like hydraulic control, analogous to type-V malfunction and **d** mechanical control at outlet section, analogous to type-B and/or type-W malfunctions

The basin width and the guiding channel passing through control flow spreading and the flow discharge at which a type-Q malfunction triggers deposition in the basin, and reducing sediment deposition in the downstream reaches (Fig. 15.9b). Maintenance efforts of open check dams, especially regarding the excavation of routine events, are primarily driven by these type-S and type-Q malfunctions or their prevention using a suitable guiding channel (sufficiently steep and with the relevant discharge capacity).

Meanwhile, the barrier is chosen to drive a more or less strong "hydraulic-control" with its discharge capacity (Piton & Recking, 2016a), a process analogous to type-V malfunction (Fig. 15.9c). Also, possible jamming of the opening, driven by shape, size and number of openings, are analogous to type-B and/or type-W malfunctions (Fig. 15.9d).

The following paragraphs seek to link the features of these structures with the driving processes, and the function a designer may achieve. Managers and designers can then select and adjust the features of the basin and of the barrier openings to promote or restrain some processes to change debris flow behaviour in a certain way.

Debris flows are subject to random variations (e.g. changes in the size of boulders and grains transported, varying rheology, presence/absence of large wood), eventually leading to surprising effects (e.g. clogging of structures that varies between events, Schwindt et al., 2017 and Carladous et al., 2022, or even within the event, Bezzola, 2008). Comprehensive hazard assessment, suitable structure monitoring with rigorous recording of observations of the structure life, and an adaptive management approach are necessary to best adjust the structures to each catchment's peculiarities (Carladous et al., 2022). Despite the complexity and variability of the challenge, we hope that the next paragraphs will help define relevant mitigation strategies.

15.4.2.2 Flow Transformation: Trapping Boulders and Debris Flow Fronts

General Remarks on Debris Flow Breakers

We call "transformation" the function of trapping boulders, especially those clustering at debris flow fronts. The purpose is to break up the debris-flow front and to release most of the material partially unloaded from the very coarse components of the debris flow. It is usually achieved by structures called "debris-flow breakers" (Proske et al., 2011). These structures only seek to interfere with boulders to reduce the destructive power of debris flows, and do not intend to trap other material (cobbles, gravel and mud), although marginal (and sometimes even significant) trapping occurs.

Breaking up debris-flow fronts and removing large boulders from debris flows is relevant in channels experiencing type-B malfunctions, but also to protect downstream mitigation measures from extreme impact forces (e.g., a large wood trapping device with a steel grill or series of check dams whose crowns could be damaged by the passage of giant boulders).

In Japan, debris-flow breakers are mostly frame-structures made of large steel pipes located at the centre of a gravity dam with a very large slit (e.g. Fig. 15.10a). In Europe, they are mostly sectional barriers with several fins (also called teeth in Europe) made of reinforced concrete (e.g. Fig. 15.10b–d). If not properly designed, classical planar steel grills or single bars tend to get destroyed by the extremely high impact forces of debris flows (e.g. Piton et al., 2020a; Weinmeister, 2007), which can have the same order of magnitude of the weight of the largest impacting boulders (D'Agostino et al., 2016). Several guidelines exist regarding load cases and computation of impact forces, but this topic is outside the scope of this chapter (see Kwan et al., *this volume,* as well as Osanai et al., 2010; Rudolf-Miklau & Suda, 2011, 2013; Hübl et al., 2017).

Very large boulders tend to deposit and debris-flow fronts disaggregate when the surges run into wide, unconfined areas. With flow being less confined (eventually helped by a slope

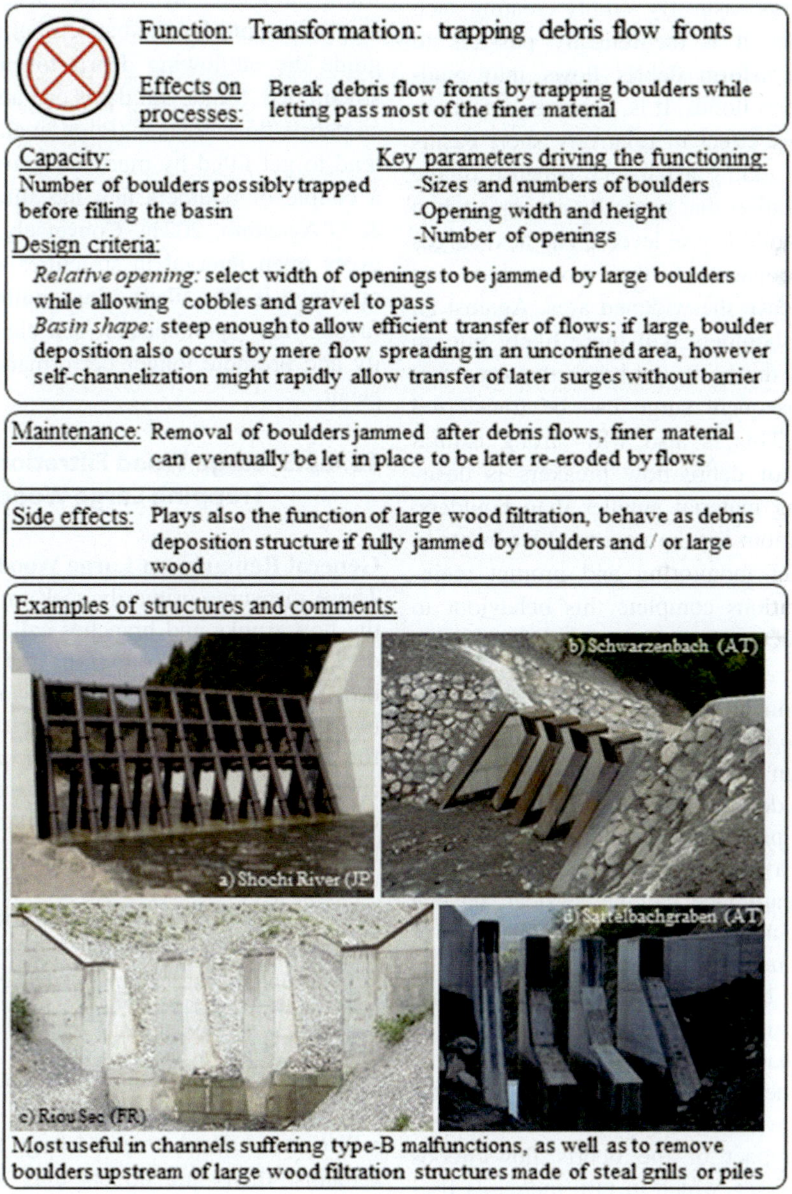

Fig. 15.10 Main features of measures targeted for debris flow transformation (images courtesy of the authors)

inflexion), a size-selective deposition can also emerge with large boulders and cobbles becoming less mobile (e.g. Kean et al., 2019). This is a natural driving process of the transition experienced by transport processes from debris flows to debris floods (SedAlp, 2015). Clusters of large boulders are thus typically found near the inlet of the basin of open check dams (e.g., Piton et al., 2018). This is evidence of the flow transformation occurring when a first surge reaches an empty debris basin. By simply creating such widened areas, it is theoretically possible to break and transform debris flows into mudflows and debris floods. It is, however, uncertain to rely on this effect in relatively short basins (i.e., without adding a barrier), because of the self-channelization that naturally emerges when debris flows build lateral levees. The next surges will likely propagate further downstream and may finally leave the widened area. Against an open barrier, boulders will more likely pile-up until they fill the basin (or break the structure) before a subsequent surge can be transferred downstream. This is also why strong natural self-cleaning of debris-flow breakers is desirable regarding material smaller than boulders: it maintains room for the next boulder fronts to come. Regular monitoring and prompt maintenance operations complete this behaviour to empty the structure from the trapped boulders.

Design Approach

The *relative opening* concept of Fig. 15.2 is the primary design criterion of debris-flow breakers. It is used to define the interval space between fins or steel pipes. The barrier height is chosen to store a given volume of boulders. The barrier upstream face can be vertical, have a homogenous inclination (e.g. Fig. 15.10a, b) or even several inflexion points (e.g. Fig. 15.10c-d). Adding a bottom part with a low inclination (e.g. Fig. 15.10d) gives a better capacity to transfer bedload transport and the debris flow body below the trapped coarse grains. This part also facilitates water drainage (from below) during the impact of the debris flows/floods against the barrier, which in turn increases friction between grains and empowers the retention capacity of the breaker. Boulders (if not too numerous) will pile-up and get perched on this part, with the remaining room below allowing better self-cleaning (Ishikawa & Mizuyama, 1988; Piton & Recking, 2016b). Inclined upstream faces also suffer slightly lower impact forces by promoting partial run-up (Horiguchi & Komatsu, 2018) and putting in play the stabilising weight force of the sediments.

The number of fins or total width of the frame structure are defined according to the available room for the barrier, but also to safely guide the outflowing debris toward the downstream bed. Single slit dams can also play a role as debris-flow breaker (Piton et al., 2020a), but tend to get filled by mud and gravel as soon as a couple of boulders jam the slit (e.g. Baggio & D'Agostino, 2021). Conversely, jamming of every open interval in structures with multiple openings is less likely, thus multiple openings favour the spontaneous self-cleaning capacity and promote longer performance during the event.

15.4.2.3 Large Wood Filtration: Trapping Large Woody Debris

General Remarks on Large Wood Traps

These measures primarily seek to remove from the flow trunks and branches called large wood or large woody debris (Fig. 15.11). They are best suited for channels experiencing type-W malfunctions and are particularly necessary for those channel reaches upstream of bridges and culverts. Such structures are also useful upstream of sediment buffering structures (e.g., barriers with large slits or slots aiming at transiently storing sediment; see later). Where large woody debris is present, buffering structures tend to get clogged and then behave as sediment deposition structures until emptied (a type-W malfunction of the buffering barriers—Bezzola et al., 2004).

A large wood filtration measure is just a way to choose where the large wood accumulation will occur. Unless removed large woody debris remains; their side effects thus leading to an increase in flow depth where they accumulate

Function: Wood filtration

Effects on processes: Trap large wood pieces to prevent clogging of bottleneck sections (bridges, dams, narrow sections)

Capacity:
Volume large wood possibly trapped before release by overtopping

Key parameters driving the functioning:
-Length and number of logs
-Opening width
-Basin shape (slope & width)

Design criteria:
Relative opening: select width of openings to ensure the trapping of (possibly) dangerous large wood chunks, typically as long as the channel is wide
Basin shape: if steep, large wood pile up against the barrier; in gentler basin where a ponding effect emerge, large wood is stored as a floating single layer carpet

Maintenance: Removal of large wood or pruning of large elements in pieces small enough not to be dangerous

Side effects: Sediment deposition by hydraulic and mechanical control. Bottom outlet unclogged allows self-cleaning, strong trapping to be expected otherwise

Examples of structures and comments:

Useful to cope with type-W malfunctions and to remove large wood upstream of debris buffering

Fig. 15.11 Main features of measures targeted for large wood filtration (images courtesy of (**a**, **c-e** the authors **b, f** E. Mark, **g** Kane Geotech Inc.)

(especially if they dam the whole barrier section), and associated deposition of sediment in the upstream ponding area (Schalko et al., 2019a, 2021) or against the large woody debris jam (Shima et al., 2016). However, the accumulation occurs in accordance with the stream manager's decision regarding the location, i.e. in the trap, and eventually the intensity (either high or low sediment trapping are possible; see later).

Several types of structures can achieve a large woody debris filtration function (Piton & Recking, 2016b): (i) heavy structures suitable for steep creeks (e.g., Austrian-type inclined grills or type-A SABO dams, see

Fig. 15.11a–c), (ii) light structures more suitable for gentle slopes and less intense bedload transport conditions (racks made of vertical piles, see Fig. 15.11d–f), or (iii) flexible barriers for small creeks (e.g. Fig. 15.11g; though quite large structures can be built with high barriers or intermediate posts, e.g., Wendeler et al., 2018).

Scientific works on large wood filtration structures were first published in Japan (Ishikawa & Mizuyama, 1988; Kasai et al., 1996, among many others) where some SABO dams made of steel pipes are specifically designed to trap large wood for decades (e.g., Fig. 15.11a). Meanwhile, Austrian and Italian steep creek engineers developed inclined grill structures for debris flood creeks (Cerato & Coali, 1997; D'Agostino et al., 2000; Hübl & Fiebiger, 2005; ÖWAV, 2021). These structures are quite similar to the debris-flow breakers presented in the previous section (e.g., Fig. 15.11b). They differ mostly by: (i) using larger basins accommodating bigger volumes of bedload and large wood; and (ii) being made of steel grills with particular shapes. However, they are unsuitable for direct debris-flow impacts (see e.g. Weinmeister, 2007). Rack structures used in isolation (e.g., Fig. 15.11d-e) or added upstream of slot dams were also developed (Bezzola et al., 2004; Koulinski & Richard, 2008; Lange & Bezzola, 2006, p.49).

In active catchments experiencing regular soil erosion and sediment transport, the maintenance frequency of these structures is lower than that of debris deposition structures. Indeed, massive transport of large woody debris is usually even less frequent than transport of sediment. The critical discharge for the incipient motion of sediment is lower than the range of discharge triggering massive erosion on banks and landslide toes, which leads to large woody debris recruitment (Comiti et al., 2016; Wohl et al., 2019). In forested catchments, where sediment transport is very limited except during extreme rainfall events (e.g. typhoons) triggering abundant landslides, sediment and large woody debris transport occur simultaneously. Therefore, the cleaning of large woody debris occurs infrequently. In addition, excavation and dredging can be directed towards accumulations of woody debris only; especially in catchments with sufficient downstream sediment connectivity (sediment is left in place and is expected to be re-eroded later by natural flows). Conversely, in catchments where minimal sediment transport is desirable, full dredging of every trapped material should be done. Intermediate cases exist.

Design Approach

Rimböck (2004) performed pioneering works on the use of flexible barriers aiming at trapping large woody debris (see also recommendations by Lange & Bezzola, 2006, p. 76–81). Rimböck (2004) showed that the typical mesh size of flexible structures being in the range 0.3–0.5 m, it is smaller than large wood length L_{LW}, ensuring their trapping. He also stated that the elevation of the barrier lowest cable should be sufficiently above the channel bed to prevent fast clogging even during low flows that would greatly increase maintenance effort. For instance, a 40 m-wide, 6 m-high flexible barrier was installed on Cougar Creek (AB, Canada) after the 2013 debris flood to trap large woody debris and sediment. It had initially a bottom cable very close from the channel bed. It was trapping small debris and bedload transport associated to minor routine events and had to be elevated to prevent expensive and heavy maintenance (personal communication with Félix Camiré, 2021). Piton et al. (2022a) provided a criterion to define this bottom clearance depending of the targeted trapping efficacy (partial, probable or certain) and of unit discharge and large wood diameter.

Rimböck (2004) recommended to install flexible barriers in reasonably wide streams (channel width<15 m), and in relatively straight reaches to prevent uneven accumulation of large woody debris and barrier loading (radius of curve>10 channel width, barrier located at least 5 channel width downstream of upstream curves). With the barrier size he used in his full scale and small scale experiments, Rimböck (2004) recommended not to use flexible barriers in channels with unit discharge>5 m^3/s.m, and with large wood unit volume>20 m^3/m and/or with sediment unit volume > 100 m^3/m. In channel steeper than 5%, Rimböck (2004) hypothesized that flexible barriers intending to trap large wood would lack sufficient upstream deposition area to buffer large wood and debris deposition, especially because the sediment trapped with the large wood may interfere with the flow and large wood accumulation, potentially easing the release of the large wood over the barrier. This would trigger a functional failure: the barrier no longer achieving its function even though still in a satisfying structural state (see later regarding criteria to assess such large wood releases by overtopping).

For rigid structures, the *relative opening* concept of Fig. 15.3 is the primary design criterion of large wood traps to define the interval space between bars. It should be applied accounting also for the fact that the L_{LW} choice is preferable to be slightly overestimated because the branches tend to magnify the wood blockage. The structure height governs the basin capacity. According to Japanese guidelines (SABO Division, 2000), the large woody debris storage capacity of basins subject to debris flows can reach 10–30% of the total trapped volume of debris. This is because large wood is mixed with sediment. Meanwhile, during floods transporting bedload, a ponding effect occurs in the basin and large wood is mostly stored as a single layer forming a floating carpet. The trapping capacity is then approximately computed as the surface area of the basin multiplied by the average diameter of large wood.

Head losses at racks jammed by large wood were thoroughly studied at the ETHZ (Schalko, 2020; Schalko et al., 2019a, 2019b; Schmocker & Hager, 2013, among others), at slit, slot and racks by Piton et al., (2020b) and at flexible barriers by Piton et al. (2022a). The authors mainly focused on the large woody debris energy head losses by measuring the energy head $H = h + v^2/2g$, with and without large woody debris, upstream of the structures. h the flow depth, v the mean flow velocity and g the acceleration due to gravity. Many repetitions and sensitivity analyses were performed. The authors observed, in presence of large wood jamming the structure, increases in upstream energy head compared to energy head that can be computed using pure-water hydraulics at orifices, slits or racks ranging from 5 to 60% at slit and slot dams, from 20 to 140% at racks and from 25 to 175% at flexible barriers (see Piton et al., 2020b and 2022a for more details). Accounting for the non-linearity between stage and discharge at these structures ($Q \propto h^{1.5}$), this is equivalent to decreases of discharge capacities of 7% to 50% for slit and slot barriers, of 24% to 73% for racks and of 28% to 78% for flexible barriers, for a given flow stage. Such variability is both due to the randomness of the log arrangement upstream of the structures, but also to variation of parameters that are hardly predictable (e.g. the amount of large wood supplied, the size of large wood pieces or their density, or the amount of leaves and branches).

It is worth stressing that these studies did not include massive sediment transport. They are thus analogous to debris-flood processes in structures sufficiently large to trigger bedload deposition in the upstream part of the basin at some distance from the logs floating in front of the barrier (as sketched in Fig. 15.9, see also Bezzola et al., 2004; Piton & Recking, 2016b). Structures tested on mobile beds (no apron below the rack) experienced deep scouring and energy head losses reduced by half as compared to structures with an apron (Schalko et al., 2019b). These two peculiarities (scour hole below the large wood jam and less ponding effect) are interesting features to minimize the bedload trapping at the structure. Inclined racks are interesting because scouring seems in addition shallower while head losses are even more reduced (up to 60% as compared to vertical racks on an erodible bed for racks inclined at 20° with horizontal—Schalko, 2020). Conversely, racks intending to maximize sediment trapping must be built on a weir to prevent the scouring and passage of bedload below the jam and use vertical piles (D'Agostino et al., 2000; Koulinski & Richard, 2008).

In case of high overflow over the structure, uncontrolled releases of the large wood carpet may occur (Horiguchi et al., 2021b). Piton et al.,

(2020b) demonstrated that this happens for depths above the crest reaching two times the mean large wood diameter on closed-type check dams (similar to reservoir dam spillways—Bénet et al., 2021). On open structures, releases occur for higher overflowing depths because of the logs entangling in the opening and stuck against the structure by the flow passing in the openings. Releases were nonetheless observed for overflowing depths of three to eight times the large wood mean diameter (see Piton et al., 2020b, 2022a for a simple criterion to refine the range). It means that large wood may accumulate above the crest of the structure as shown in Fig. 15.11g. Computing the range of discharge that may trigger uncontrolled releases helps to define the safety-check events over which structures no longer achieve their mitigation function.

A 2 m-long horizontal grill is recommended to be installed at the top of the Austrian-type large wood traps to allow the logs to be deposited on this horizontal grill and thus prevent such releases (see for instance Fig. 15.11b). Hooks or racks can be installed on the crest of other structures for the same reason (e.g., Fig. 15.10b, Fig. 15.11f, or Horiguchi et al., 2021b). Designers need to reconsider the hydraulic design of the spillway of rigid structures trapping large woody debris due to its possible partial clogging.

Only transverse structures were mentioned in this section. Side structures storing large wood on side terraces located on the outer bank of bends and meanders have been implemented in rivers (e.g., Schmocker & Weitbrecht, 2013 in a 0.6%-steep river). Such structures use secondary currents in bends to separate large wood (sent to the floodplain and then retained by racks) from bedload (kept in the channel so as not to interfere with large wood and the racks). In debris-flood channels, the strong mobility of the bed and the limited water depth are not suitable for such separation. That is why transverse structures are preferably used, while fully or partially damming the channel (e.g., Hasegawa et al., 2010; Lange & Bezzola, 2006, p.85).

15.4.2.4 Debris Buffering: Dosing Solid Peak Discharge

Generalities on Debris Buffering Measures

Debris buffering structures, also called dosing structures (Hübl, 2018a), aim at storing sediment to reduce solid peak discharge, but are also expected to self-clean during the flow recession and during later smaller events (Armanini et al., 1991). Such measures are well adapted to channels having a type-Q malfunction. These measures thus require a detailed knowledge of the downstream discharge capacity.

These measures are often barriers with large slits or slots (Fig. 15.12). They seek to mimic and concentrate in a single location the process of natural buffering of intense sediment pulses (i.e. becoming more regular sediment transport—though less intense) by transient storage in the channel bed where the slope decreases and the bed widens, i.e., type-S and type-V good functioning (Marchi et al., 2019). Widening channels and/or changing channel slopes using earth-moving machinery will also achieve this function. These artificial channel geometries must then be maintained after each major event. To prevent excessive maintenance requirements, adding a guiding channel for routine events is strongly recommended. Then, widening of the basin above this level will simply allow debris-flow spreading and eventual avulsion in a dedicated area that, without maintenance, will progressively lose its buffering effect. To the best of our knowledge, no general guidelines on the design of artificial channel geometries have been published. We suggest adopting a functional analysis as introduced previously and to adjust the geometry of the guiding channel and of the basin to: (i) transfer the amount of sediment that the most constrained part of the downstream torrent system is able to export, while (ii) promoting a type-Q malfunctioning in a selected area (Piton et al., 2019). Debris buffering is also a side effect of check dam series: deposits from intense sediment transport are later re-eroded during less concentrated flows (Bramberger & Hübl, 2018; de Haas

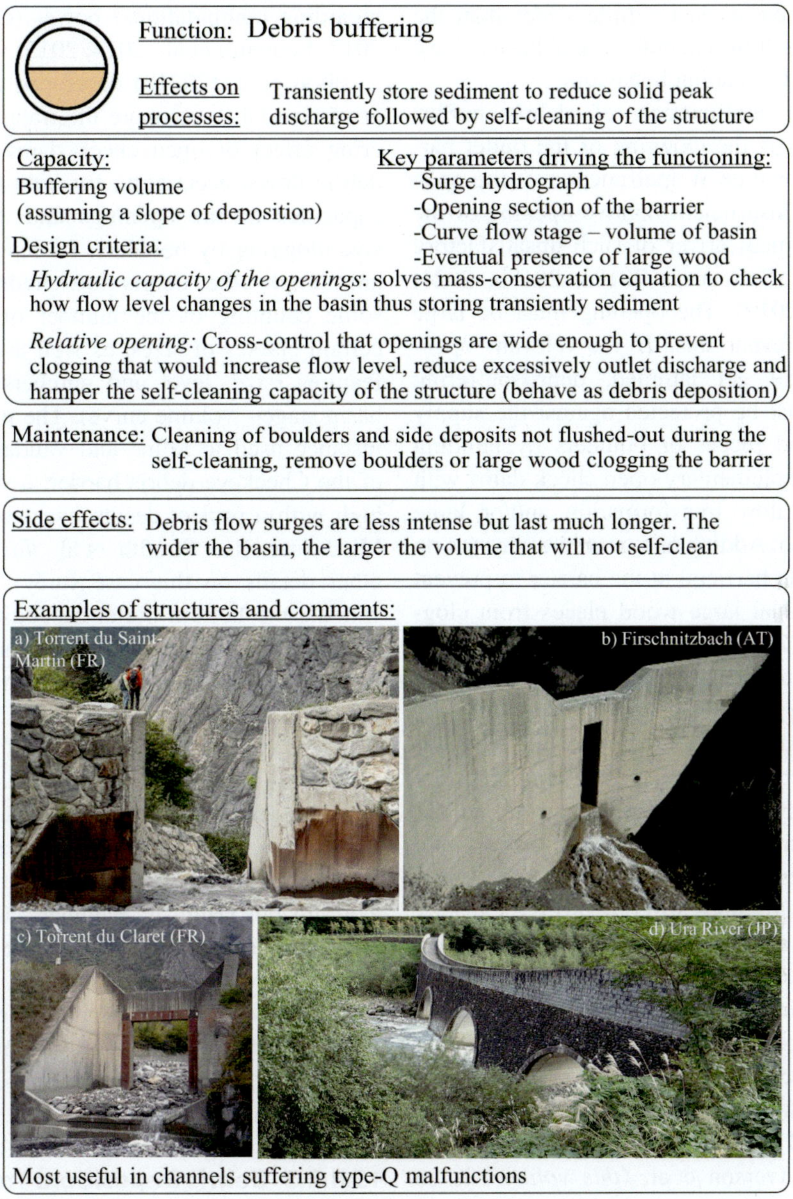

Fig. 15.12 Main features of measures targeted for debris buffering (images courtesy of (**a-c**) the authors, (**d**) E. Mark)

et al., 2020; Nishimoto, 2011; Piton & Recking, 2016c; Piton et al., 2017).

Design Approach

Being equipped either with or without an outlet barrier, the basin of debris buffering structures must be able to convey routine events with marginal (or better, no) deposition. For events of higher magnitude than routine events, flow spreading out of the guiding channel will naturally store debris by building paired terraces. Enhancing this process tends to increase the amount of debris that will not self-clean (Piton & Recking, 2016a). Conversely, flow spreading is limited in basins just a little wider than the natural debris-flow channel. Such basins have the highest self-cleaning behaviour.

The main malfunction of debris buffering structures is the clogging of the outlet barrier (type-B or type-W malfunction of the open check dam). Insufficient relative opening of the outlet is the main driver of such unsatisfactory self-cleaning (Sun et al., 2018a, 2018b, 2021; Yuan et al., 2019). The opening must be large enough to prevent this. If the relevant opening sizes will get jammed, debris-buffering structures must be protected against the supply of large wood pieces or boulders by building upstream complementary open check dams with functions of flow transformation and/or large wood filtration. Adding one or several fins (as on wood filtration barriers) at the barrier to prevent eventual residual large wood pieces from clogging the whole barrier is a common precautionary measure used in debris buffering structures (Hübl & Fiebiger, 2005).

Debris buffering structures are mainly designed according to the hydraulic capacity of the barrier and the storage capacity of the upstream basin, such that the outflowing solid discharge does not exceed the downstream channel capacity. Methods to compute the capacity of slits, slots and fins can be found in Piton and Recking (2016a). Armanini and Larcher (2001) and D'Agostino (2013) provide simple methods to compute the associated deposit height in case of debris floods (see Larcher & Armanini, 2000, for debris flows).

Using 1D or 2D numerical models, as described in Iverson et al. (*this volume*) is an alternative approach to compute how such structures store sediment and dose the downstream peak discharge (Busnelli et al., 2001; Chen & Tfwala, 2018; Nakatani et al., 2008). For this purpose, the main limitation of classical 1D and 2D numerical models based on computational fluid dynamics is that they do not explicitly compute the mechanical blockage of the opening by boulders and logs. More sophisticated models involving coupled granular physics with complex fluids are emerging to overcome this challenge, they are however very costly regarding computational power (Canelas et al., 2017; Leonardi et al., 2014, 2015).

Piton et al. (2022b) developed a simple reservoir model to compute the trapping and buffering effect of open check dams subjected to debris flows, accounting for both the hydraulic capacities of the openings and their progressive clogging by boulders. Its input parameters are simply the debris-flow hydrographs and some counting of the number of boulders of certain classes of sizes, as well as the structure features (size, level and numbers of opening, basin stage—volume curve). The model was for instance used to refine and validate the design of the Cheekeye debris barrier: a slit dam 25-m high with a project design event volume of 2.8 Mm3 (see also Strouth et al. *this volume,* for other details on this case study). It helped to validate the bottom outlet capability to transfer routine events and to study the debris deposition or buffering function depending on the presence of bars in the 6-m wide slit.

Even though debris buffering barriers are theoretically expected to self-clean, minor to moderate excavation of such structures is usually necessary if they are correctly designed; strong to almost full excavation is required if they are not (Hübl & Fiebiger, 2005). In such cases, adaptations of the structure can be performed by cutting large opening in the barrier (Carladous et al., 2022; Chahrour et al., 2022a). Full excavation is also necessary in case of barrier clogging (i.e. in case of Type-B and/or –W malfunction of the barrier).

15.4.2.5 Debris Deposition: Trapping Solid Volume

Debris deposition measures seek to trap a given volume of sediment. This function was formerly performed with large closed check dams (e.g., Fig. 15.13b, c). Such "retention check dams" have generally similar aspects to check dams with a function of hillslope buttressing or channel stabilization, but their main function is neither to stabilize hillslopes, nor to fix the longitudinal profile at a given level. They seek to trap the supplied sediment in their deposition area. In some locations, the deposition

Function: Debris deposition

Effects on processes: Trap a certain volume of sediment but allow water and suspended sediment to pass

Capacity:
Trapping volume
(assuming a slope of deposition)

Key parameters driving the functioning:
-Basin shape
-Curve flow stage – volume of basin
-Opening size

Design criteria:

Relative opening: Opening interval small enough to ensure clogging
Hydraulic capacity of the openings: can be small to force a ponding effect
Bottom outlet: allow passage of routine events with marginal interference
Basin: possibly wide, with a gentle slope just steep enough to convey routine events in a guiding channel (almost full trapping of all events if slope very low)

Maintenance: Usually heavy, removal of trapped material to recover trapping capacity (more limited with guiding channel and bottom outlet)

Side effects: "Hungry water effect", sediment starving of the downstream reaches triggering incision and bank erosion, often require a stabilization masterplan of the dowsntream channel

Examples of structures and comments:

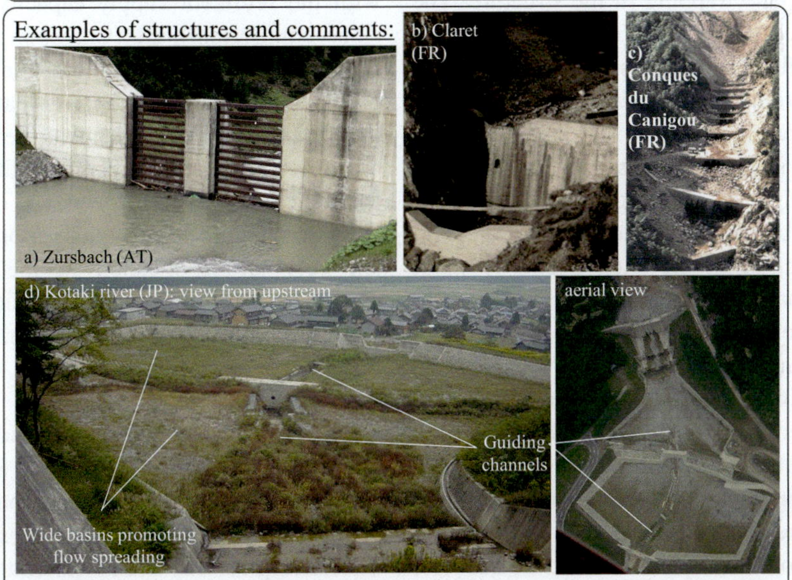

Most useful in channels suffering type-V (and type-S too eventually) malfunctions

Fig. 15.13 Main features of measures targeted for debris deposition (images courtesy of (**a**) E. Mark, (**b** and **c**) ONF-RTM, (**d**) the authors)

area of retention check dams nonetheless affects hillslope buttressing (Yang et al., 2021). Most check dams, especially in very active debris-flow catchments, do not have debris deposition as the main function; a quick computation shows that their trapping capacity is negligible when compared to the potential catchment sediment yield. Even in the rare cases of general collapse of a whole check dam series, the great majority of the debris flow was recruited in the torrent natural channel and banks, and not in the sediment wedge formerly trapped by the check dams

(Baggio & D'Agostino, 2021, see also Hübl et al., *this volume*).

Check dams seeking to achieve a retention function are generally characterized by elongated and deep deposition areas. Such high closed check dams (Fig. 15.13b) or series of check dams (Fig. 15.13c) seeking to maximize their trapping capacity are less frequently constructed since the advent of earth moving machinery. This function can now be performed by smaller structures that are regularly excavated. Flexible barriers can play this role in small creeks.

Closed check dams can encourage debris deposition (López et al., 2010), but open structures are now preferred. They transfer suspended sediment and fine bedload material and prevent their siltation and trapping in the structure. The design of a debris deposition structure fully trapping debris flow and debris flood sediment is focused on promoting every possible deposition process by (recall Fig. 15.9): (i) using the gentlest possible slope in the basin; (ii) maximizing flow spreading; (iii) causing a ponding effect due to a low barrier discharge capacity; and (iv) inducing clogging of the latter thanks to small openings. Small space between bars is the main feature of open check dams with a debris deposition function (e.g., Fig. 15.13a).

The trapping capacity of a sediment deposition barrier is driven by both the height of the structure and by the slope of deposition (Piton, 2016, Piton & Recking, 2016a, 2016c, 2016d). Several theoretical methods were proposed to predict the slope of deposition (e.g. Kostadinov, 1993). In Japan, a simple empirical rule considers that the slope of deposition is one-half of the initial channel slope during low flows or after several small floods, this value is used in the design to represent initial conditions prior to project design event. A steeper slope of deposition of two thirds of the initial channel slope is chosen for deposition occurring during events of strong magnitude with higher solid concentrations (SABO Division, 2000; Osanai et al., 2010).

However, field data show that the slope of deposition varies strongly and can exceed this range (in both Japan—Mizuyama, 1990, and elsewhere—Piton and Recking, 2016d). The range of variation of this slope is larger than many practitioners think. It is affected by sediment mixture composition (lower for muddy debris flows, higher for stony debris flows), by the flow confinement (steeper slope of deposition in unconfined deposition area—Hungr et al., 1984), and by the interactions of the front with the openings (larger opening, larger water drainage, steeper deposition slope, possibly due to steepening occurring during the flood recession and the associated self-cleaning—Chiu et al., 2021). It is also affected by time: the longer the depositional slope exists behind a barrier and the sediment is not removed, the more the slope will approach a slope consistent with the sediment supply capacity (an alluvial slope equilibrium), e.g. Chambon et al. (2010). In the opinion of the authors, the best way to provide a prediction in a given catchment is to measure it on sparse deposits near and along the torrent and ideally estimate the time since deposition.

Downstream sediment starvation and accordant very strong incision in the downstream reaches are the most common side effects of these structures. In the long term, the required stabilization of the downstream reaches and the environmental damage can, however, be more costly than the damage avoided by debris deposition (Brochot et al., 2003; Simoni et al., 2017). It is more sustainable, but also more complicated, to design structures capable of transferring routine events (to restrain the sediment starving side effect), while behaving as debris deposition structures for project design events. Using a guiding channel in the basin is, as always, a prerequisite (e.g., Fig. 15.13d). A relevant solution can be to excavate the basin on the side of the channel (Carladous et al., 2022; Ghilardi et al., 2012; Piton & Recking, 2016a). The channel is then equipped with a side weir where non-routine events are supposed to avulse. A widely opened interception structure is generally built in the channel to ensure this functioning (see also section on deflection structures).

Whether the basin is aligned with the channel or located on the side, the basin outlet barrier should be equipped with a finely tuned bottom outlet. In debris-flow catchments, it is possible to rely on the increased number and size of boulders (and eventually large wood) to clog orifices or SABO dams (relative opening criteria with assumptions on the different size of material supplied by routine- and project design-events—D'Agostino, 2013). In debris-flood catchments, grain sizes are usually less variable between routine- and project design-events and thus cannot be the only parameter driving the change in the structure behaviour. In addition, relying only on the hydraulic control of the trapping is unsafe (Fig. 15.9c) because it enables self-cleaning during the flood recession (if large wood does not clog the opening, an effect that is partially random). Hybrid structures having an upstream grill with a bottom clearance, installed upstream of an orifice having the precise hydraulic capacity of the guiding channel, proved to be a promising solution to design debris deposition basins allowing the transfer of routine events (Horiguchi et al., 2021b; Moldenhauer-Roth et al., 2021). Regular monitoring visits of such structures remain necessary; especially to remove the large wood that may interfere with its functioning (Schalko et al., 2021). Insufficient maintenance and dredging of such structures, related either to insufficient monitoring, lack of funding, environmental issues or any other barrier, lead to the progressive filling of the basin and to an equivalent decrease of the trapping efficacy for large events (Chahrour et al., 2022b).

Functional design guidelines for flexible barriers intending to achieve a debris deposition function were proposed by Berger et al. (2021) (see also previous versions by Wendeler, 2008; Volkwein et al., 2011; Volkwein, 2014; and Wendeler & Volkwein, 2015): (i) mesh opening should be approximately the size of the biggest boulders (quantile 90% of the grain size distribution), and (ii) the barrier bottom clearance, i.e., the height from the channel bed to the bottom cable, should be about 2/3 of the expected initial maximum flow depth. As for rigid barriers, the capacity of flexible barriers is controlled by their height, the geometry of the upstream channel and the slope of deposition.

Flexible barriers were used in many small creeks to trap debris flows with strong trapping efficacy but associated volume that are rather small, typically a few hundreds of m^3 to a couple of thousands m^3, notably because of the steepness of the channels where they are implemented (e.g. Victoriano et al., 2018). Larger barriers can certainly store bigger volume (see e.g. Wendeler et al., 2018 or Berger et al.,; 2021, p. 69–73). These flexible structures are usually faster and less expensive to install than rigid barriers. However, their steal components need higher monitoring and repairing despite being protected by some surface treatments, they must cope with the harsh climate of mountain and most of all the destructive passage of debris flows and debris floods against the barrier and over their crest once filled. As recalled by Berger et al. (2021), flexible barriers must be emptied and repaired after each event filling them. Without event, a service life of 25 years can typically be assumed. As for old, insufficiently maintained check dams that can also become sediment sources contributing to the sudden bulking of debris flows in case of failure, a strategy must be defined during their initial planning regarding how to deal with such structures once they are filled and how to access these sites with the relevant machinery allowing cleaning these structures.

Numerous rigid structures were equipped with bars fixed on devices allowing the spacing to be changed to increase or decrease the trapping efficacy. However, in our experience, in most cases this possibility was infrequently used because of: (i) the rarity of the structure functioning (lack of observations justifying changing the bar interval), (ii) the variability of the debris flows, notably regarding grain sizes and amount of large wood, and (iii) the poor understanding of the processes at work. However, progress in monitoring techniques enables nowadays to use structures with shutters on sites with frequent sediment transport and sufficient monitoring (e.g., on the Jo-Gan-Ji River in

Japan—Nagayama et al., 2019). Such structures can behave as debris deposition or debris buffering structures according to the project manager's decision.

Transformation of structures initially planned to be debris deposition measures into more open debris buffering structures is an adaptation often performed in catchments where excessive deposition was observed, along with strong side effects (downstream incision and/or unaffordable maintenance), or in catchments where other consolidation measures were under construction upstream (at the time the debris deposition basin was constructed) and then, some years later, the sediment supply regime changed. Chen and Tfwala (2018), for instance, numerically modelled the effect of changing a debris deposition structure into a wide slit dam. Chiu et al. (2021) studied the same operation with experiments and monitored its actual implementation in the field (see also Carladous et al., 2022 for other examples and Chahrour et al., 2022b for the modelling of the associated shift of function from debris deposition to debris buffering).

15.4.2.6 Water Retention: Flood Control Dams

Water retention dams are common structures in river and urban engineering. They are used to prevent type-Q malfunctions in debris flood catchments. They are similar to the debris buffering type of structure, but seek to moderate water peak discharge rather than solid peak discharge. Their basin volumes are usually very large since water runoff volumes are typically about one order of magnitude larger than solid volumes (even more in large catchments). Such structures are thus rarely used in debris flood catchments because the steep gradients make it difficult to find a reservoir location for giant dams to store the required volumes of water (e.g., Fig. 15.14). Since the objective of such a structure is to store significant volume of water, it also has the side effect of trapping almost all bedload transport in the ponding area.

The same design criteria as in river engineering must be used for these structures (e.g., ICOLD, 2014, 2019 and national guidelines). Special care must be taken for their use in debris flood contexts where relatively high volumes of large wood can be expected and require specific protection measures at the bottom outlet and spillway (Bénet et al., 2021; SCD, 2017). Protecting the outlet against large woody debris is especially necessary because large wood jamming can trigger an early filling of the basin, making it useless if it is already full at peak flows.

15.4.3 Guiding Structures

15.4.3.1 Conveyance Structures

Conveyance structures aim at guiding and confining flows along a chosen path. This includes keeping the path relatively stable in location. We thus include bank erosion mitigation in this broad function. It was the first type of flood mitigation method to be implemented historically in many regions and it was often placed on alluvial fans to protect villages and infrastructure. Complementary methods aimed at curtailing or controlling sediment transport emerged when conveyance structures proved to be insufficient and unsuitable in geomorphologically active catchments (e.g., Okamoto, 2007; Piton et al., 2017; Vischer, 2003). Many structures having a conveyance function (e.g. dykes, groynes and bank protections) are also widely used in lowland river management (CIRIA et al., 2013; USACE, 1994b). These structures play the role of *lateral/sideways* walls against flooding (dykes) or bank erosion, either locally (groynes) or continuously (bank protections). In debris-flow- and debris-flood channels, the intense sediment transport regime often requires the stabilization of the chosen path to prevent *vertical* erosion (re-loading the flows with sediment), and scouring of the lateral guiding structures. That is why channels, especially on alluvial fans, are also equipped with bed sills (locally guiding the flow on their spillway and also fixing the longitudinal profile—Fig. 15.15a) or fully lined channels with bank

Function: Water retention

Effects on processes: Transiently store water to reduce water peak discharge

Capacity:
Basin storage capacity

Key parameters driving the functionning:
-Flood hydrograph
-Outlet type and capacity
-Presence of large wood

Design criteria:
Hydraulic capacity of the openings: should be large enough to allow routine events passing without interference and to prevent early filling of the basin in order to keep sufficient room to store water volume at peak flows
Outlets shape and type: selected to prevent clogging and disruption by large wood, designed to transfer sediment transport of routine events

Maintenance: Removal of large wood and eventual sediment trapped

Side effects: Almost full trapping of bedload transport in the ponding area at peak flows, deposit likely strongly reworked during the flood recession

Examples of structures and comments:

Most useful in channels suffering type-Q malfunctions

Fig. 15.14 Main features of measures targeted for water retention (images courtesy of (a) ONF-RTM, (b-d) the authors)

protection and aprons in the most active reaches (Fig. 15.15b–e).

Check dam crests are usually set above the bed level when built in headwaters for stabilization and buttressing of sediment sources, or in gorges for debris deposition. Conversely, for flow guidance and stabilization of the channel in downstream reaches, check dam crests must be set at the initial bed level to prevent dangerous, uncontrolled aggradation and an increase in avulsion probability (Piton et al., 2017). This solution is also known as a bed sill sequence, which is preferable in all those cases for which the drop-associated energy dissipation (standard check dams) during the transit of dense sediment flows must be avoided and maximisation

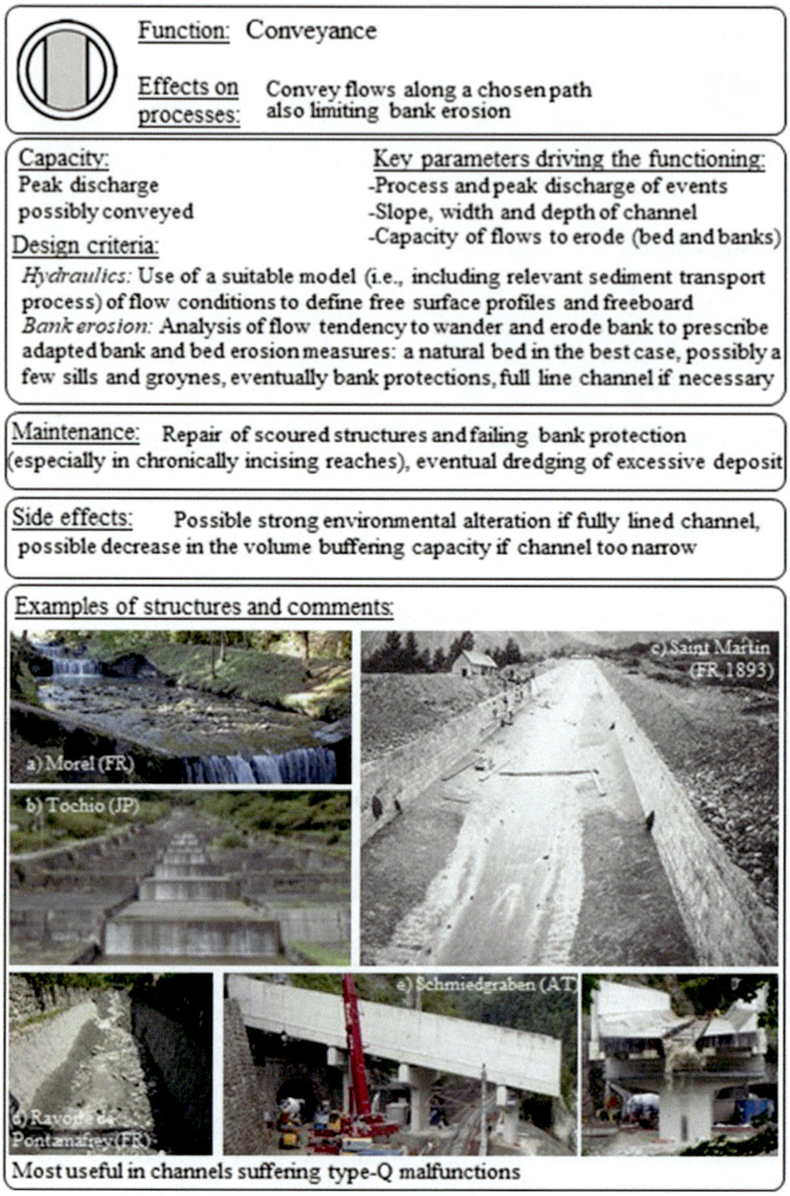

Fig. 15.15 Main features of measures targeted for flow conveyance (images courtesy of **a** and **d**-**e** the authors, **b** E. Mark, **c** C. Kuss, public domain)

of the channel conveyance capacity is needed (Marchi et al., 2019).

Bridges and crossing structures often create bottleneck sections along channels, inducing possible local type-Q malfunctions, but are also often jammed by boulders and/or large wood (type-B and type-W malfunctions). Increase of bridge capacity and the removal of intermediate piles decreases the likelihood of all these malfunctions. Figure 15.16 shows examples of adaptations made to decrease the probability that an obstruction or an overflow occurs at a bridge. To reduce flow levels at the bridge deck, an upstream steeper ramp can be installed (e.g. Fig. 15.16a). This accelerates the flows below the bridge deck but requires deepening the

Fig. 15.16 Examples of bridge adaptation decreasing the probability of obstruction: **a** upstream ramp to speed up the flow below the deck, **b** baffle on the bridge deck to ease pressure flows and guide boulders and woody debris below the deck, **c** sacrificial bridge whose deck can be transported by debris flows, **d** sliding bridge deck that move downstream opening a larger section, and **e** moveable bridge that can open up when a debris flow is detected. Black dotted arrows show possible movement of the bridge elements (images courtesy of (**a**, **b** and **e**) the authors, (**c** and **d**) E. Mark).

channel, which is possible without decreasing the slope (that would trigger a type-S malfunction) only if the channel shows a drop further downstream, e.g. a check dam or a steeper section. Alternatively, reinforced baffle can be installed on the upstream face of a bridge deck to guide debris flows below the deck and to decrease the likelihood that boulders or woody debris get blocked against the deck upstream face (e.g. Fig. 15.16b).

Alternatively, it is sometimes easier to use small sacrificial bridges that are designed to be carried away by the debris flows rather than to hold the impact and block debris (e.g. Fig. 15.16c). For bridge decks that must be strong enough to support high traffic and cannot be designed as sacrificial-bridges, another option is to mount the bridge deck on sliding rails. In the case of the Trachtbach in Switzerland shown in Fig. 15.16d, fire fighters can move manually the bridge in one hour, but it is also designed so that a debris flow hitting the bridge deck will move it forward on the sliding rails. By shifting its position over this very steep channel, the bridge—that in normal time has a 1.8 m height and 10 m^2 section—will open a larger flow section—up to 24 m^2 and a flow depth of 4.4 m (see Mark, 2017, Appendix C for other details). Some sites were equipped with movable bridges that are even operated by hydraulic jacks withdrawing the bridge deck from the channel (e.g. Fig. 15.16e, see also Hübl & Fiebiger, 2005). This option requires detecting the debris flow occurrence sufficiently early to operate the system.

Flow conveyance structures are first designed to transfer peak discharges with a certain freeboard. Since sediment transport is intense in debris-flow- and debris-flood channels and flow rheology is sometimes very different from the Newtonian flows of rivers, specific numerical or physical models must be used to assess or predict the channel capacity. It is out of the scope of this chapter to review these methods (see Iverson et al., *this volume*). Free surface profiles are, in addition, quite uncertain because: (i) channel bed levels can be variable due to partially random deposition and erosion, (ii) flows have much higher inertia terms than in rivers, and (iii) large woody debris is often transported during powerful events. Consequently, freeboard of conveyance structures in debris flow creeks must be high enough, typically of 1–1.5 m, down to 0.5 m in the smallest and quietest creeks and more than 1.5 m in the most active debris-flow torrents (see Hunzinger, 2014 for a method to calculate freeboard). For boulder-rich debris flows, we suggest taking values of at least 2 m when there are no constraints, depending on pre-existing structures. For muddy debris flows, this value could be reduced to a minimum of 1 m.

The second key criterion of conveyance measures is their capacity to resist the erosive power of the flows. Channel erosion and mobility vary hugely between catchments depending on the type and intensity of processes and features of the channel (slope, width, depth, connectivity with the river). It is worth stressing that a sufficiently wide free space in which debris flows and debris floods are transferred naturally is the most viable, robust and affordable solution in the long term. In many cases, space is lacking and bank erosion must be limited. Debris floods are known to have strong bank and bed erosion power (Church & Jakob, 2020; Eaton et al., 2017). Lateral protection and bed sills must thus have very strong scouring protection. Debris-flow channels used to be protected with cut-stones (e.g., Fig. 15.15c) or concrete, but suffered fast abrasion and failed (see e.g., Chen et al., 2014). Rip-rap of sufficiently large size, buried at the toe of dykes and banks, is a widely used solution in debris flood channels (see the guidelines of CIRIA et al., 2007, but note that debris floods and debris flows have much higher flow density than river floods, which must be accounted for in design criteria—Church & Jakob, 2020). In regions where large boulders of strong rock are not readily available, structures made of concrete are used to protect against scouring. Bank protection made of grouted riprap are now preferred in Europe. In any case, the base level of side walls must be buried sufficiently deep to prevent scouring. Full channel lining with grouted riprap or a continuous series of check dams with bank protection in between is used in the most active creeks (Fig. 15.15b).

Although it can be tempting to maximize the channel capacity using a near-rectangular section, deposits often remain at the toe of such vertical side-walls (e.g. Fig. 15.15d) due to the friction with both the channel bed and side walls, and to the Non-Newtonian nature of the flows. Transversal sections with "U"-shape or parabolic shape tend to better self-clean and probably to discharge debris flows more efficiently. Also, to convey debris flows, it is better to keep them well confined. It is advisable (as a first approximation) to adopt a channel aspect ratio (width / depth) lower than 5 and, at the same time, larger than 3 times the diameter of the largest transported boulders (to prevent type-B malfunctions by boulders clogging the channel).

15.4.3.2 Deflection

Where sufficient room is available in valleys and on alluvial fans, simple deflection of debris flows and debris floods toward areas with low vulnerability is a relevant strategy (Marchi et al., 2019). Deflection berms are dykes intended to achieve this purpose (e.g., Fig. 15.17c). Deflection berms are also useful in creeks experiencing snow avalanches or rock falls. These structures are sometimes primarily implemented for this purpose and must also withstand debris flows. Smaller deflection structures (in groups or isolated) as baffles or barriers are also used to enhance flow spreading in open check dams, especially those having a lateral basin (e.g., Fig. 15.17a).

Design of defection berms is similar to the approach for guidance structures (see previous section). The possible run-up height associated with the deflection must be carefully accounted for (Prochaska et al., 2008; Scheidl et al., 2019). 2D numerical modelling is generally used to assess or predict the flow spreading and deflection berm efficacy (Egashira et al., 2016; Marchi et al., 2010).

15.5 Key Challenges and Opportunities for Further Research

There are still many scientific questions that remain open regarding the effect and efficacy of protection structures against debris flows and debris floods. In this short section, we outline a few topics that will probably strongly progress in the next decades thanks to emerging approaches:

- Debris flows and debris floods are heavily influenced by feedback-loops processes coupling hillslopes (where mass wasting occur) and channels (where erosion, transport and deposition occur). Estimating the protection efficacy of a torrent control masterplan involving works on the sources (soil conservation measures, channel stabilisation and hillslope buttressing and landslide drainage) is consequently a very complicated task: we should ultimately model every process and the feedback loops between them (Lucas-Borja et al., 2021). Landscape evolution models are numerical models including both mass wasting and channel processes. Such models are beginning to be used in pioneer works, for instance to show how check dams and their maintenance policy affect the catchment sediment yield and the channel morphology (Mertin, 2018; Ramirez et al., 2022). Increasingly realistic simulations of long term trends and / or of extreme events can be performed by including in models modules refining the mechanics of landslides and of channel erosion and sediment transport, as well as computing (i) the dynamics of vegetation establishment and growing depending on hillslope erosion processes (e.g. Latella et al., 2020; Carriere et al. 2020), and (ii) how structures fail by, for instance, toe scouring or structure outflanking depending on the channel morphodynamics (e.g. Baggio & D'Agostino, 2021; Chahrour et al., 2021). These sophisticated simulations, however, will always model simplification of the true complexity of nature. It is probable that their use for practical case studies will require extremely comprehensive field surveys and catchment monitoring to perform relevant calibrations of such models. These field observations will also be good opportunities to learn new things on debris flow dynamics.
- The internal dynamics of debris flows cannot be directly measured because of their destructive power. Multiphase numerical models simulating complex flow rheology in a coupled way with boulder dynamics are emerging (Canelas et al., 2017; Leonardi

Function: Deflection

Effects on processes: Deflect flows toward less vulnerable areas

Capacity:
Maximum discharge possibly deflected

Key parameters driving the functionning:
-Incoming flow peak discharge and momentum (run-up potential)
-Angle of deflection

Design criteria:

Hydraulics: Analysis of flow conditions at the deflection structure accounting for run-up height associated with the flow deflection as well as for the varying bed level (former deposition or erosion associated with previous events)
Bed and bank erosion: Assessment of flow capacity to erode the structure or its toe and use of relevant erosion protection

Maintenance: Eventual dredging if deposition at the structure becomes sufficiently high to disrupt its deflection capacity and make overtopping likely

Side effects: Deflect also avalanches or rock falls that may follow creek channels (sometime the primary function), eventual heavy landscape footprint

Examples of structures and comments:

Fig. 15.17 Main features of measures targeted for flow deflection (images courtesy of (**a**) E. Mark, (**b-d**) the authors)

et al., 2014, 2015). Such models open new perspectives to explore the internal dynamics, such as the effects of the interstitial fluid and of the type and size of coarse elements that are transported. We will thus learn about effects on the flow mobility, its interaction with the channel bed material to better understand debris-flow bulking and also interactions with structures such as open check dams. Such sophisticated models are costly regarding computational power and cannot be used on large scale with sufficiently fine resolution. Coupling depth-averaged models on some reaches with 3D multiphase models on singular sections where flow—structures interactions must be refined is also an emerging and promising approach (e.g. Pasqua et al., 2022).

- These multiphase models are also particularly needed because the small scale modelling of debris flows is subject to strong scale effects (Iverson, 2015). Numerous research results coming from small scale modelling of debris flows still to be refined or validated with numerical studies and field observations. Hybrid approaches using in parallel physical and numerical twin cases with relevant consideration for scale effects and innovative ways to address similitudes will remain very useful, especially to study complicated structures like flexible barriers (e.g. Lambert et al., 2022).

We explicitly focus on functions of structures in this chapter. However, aesthetics of structures and their landscape integration have become topics of interest for structure managers, especially in touristic areas. The perception of the structure aspect is culture-dependent. Addressing this question requires using surveys and approaches that are out of the scope of this chapter (see Peng and Tang, 2015, for an example of application on torrent control works). More generally, understanding natural processes to moderate them in selected areas, while mimicking them in others, is the backbone of the approach we suggest to follow in this chapter. Lenzi (2002) describes how check dams made of pitched riprap mimicking a step-pool morphology can be used to replace classical check dams: better landscape-integrated structures were then also better adapted to cope with the processes. At a longer time scale, the pioneers of torrent control works in the late nineteenth century and early twentieth century were mostly foresters who understood by empirical observation how to promote and accelerate reforestation and gully stabilisation using bioengineering helped by civil- and drainage-engineering where necessary (Evette et al., 2009; Piton et al., 2017). Studying the physics of debris flow processes in their natural environment, along with the form of channels, i. e. their morphology, has certainly a great potential to help designing in the future more effective, efficient and integrated debris-flow protection measures.

15.6 Concluding Remarks

The present chapter presents a framework that will hopefully help to link more tightly debris-flow hazard studies and the selection of appropriate measures to be used in debris-flow mitigation masterplans. The main functions that can be achieved with a wide range of structural measures are reviewed considering their key design criteria and accounting for the main differences in addressing debris flows and debris floods. There cannot be a one-type-fits-all strategy to mitigate debris-flow hazards; each catchment is unique and a strategy making sense in one watershed could sometimes make much less sense in the neighbouring watershed. Being a designer for debris-flow mitigation measures will thus always remain a highly technical and stimulating job that requires much hydraulic, hydrological and geomorphological and geological knowledge often paired with an understanding of forestry, plant species and landuse planning. We hope this contribution will help to define tailored strategies.

It is finally worth stressing that an enlightened risk mitigation should not be based only on structural mitigation measures. It should find the right balance between relocation (in essence the safest solution), hazard mapping and land use planning, and alert and crisis management and preparation (Saunders et al., *this volume;* Strouth et al., *this volume*).

Acknowledgements The work of G.P. was funded by the European Union through the FEDER-POIA program and by state funds through the FNADT-CIMA program both within the research project HYDRODEMO. We thank Stanimir Kostadinov, Emily Erin Mark and the French torrent control service—ONF-RTM—for sharing and allowing us to reuse pictures of structures, Matthias Busslinger and Damien Kuss for comments on previous versions of this chapter as well as Matthias Jakob, Scott McDougall and Paul Santi for a thorough review and numerous stimulating comments on the first version of this chapter. We dedicate this chapter to the memory of Matthias Jakob who invited us to prepare it in 2019: we made our best to honour his trust.

References

Armanini, A., Della Giacoma, F., & Ferrari, L. (1991). From the check dam to the development of functional check dams. *Fluvial Hydraulics of Mountain Regions, 37*, 331–344. https://doi.org/10.1007/BFb0011200

Armanini, A., & Larcher, M. (2001). Rational criterion for designing opening of slit-check dam. *Journal of Hydraulic Engineering, 127*, 94–104. https://doi.org/10.1061/(ASCE)0733-9429(2001)127:2(94)

Baggio, T., & D'Agostino, V. (2021). Simulating the effect of check dam collapse in a debris-flow channel. *Science of The Total Environnent*, 151660. https://doi.org/10.1016/j.scitotenv.2021.151660.

Bénet, L., De Cesare, G., & Pfister, M. (2021). Reservoir level rise under extreme driftwood blockage at ogee crest. *Journal of Hydraulic Engineering, 147*, 04020086. https://doi.org/10.1061/(ASCE)HY.1943-7900.0001818

Berger, C., Denk, M., Graf, C., Stieglitz, L., & Wendeler, C. (2021). Practical guide for debris flow and hillslope debris flow protection nets. Swiss Federal Institute for Forest, Snow and Landscape Research WSL: Birmensdorf, Switzerland. Retrieved October 20, 2022, from https://www.wsl.ch/de/publikationen/default-d17f18e4e2.html.

Bergmeister, K., Suda, J., Hübl, J., & Rudolf-Miklau, F. (2009). *Schutzbauwerke gegen Wildbachgefahren: Grundlagen, Entwurf und Bemessung, Beispiele*. Ernst & Sohn. https://doi.org/10.1002/9783433600283.

Bezzola, G. R. (2008). Unexpected Processes In A Sediment Retention Basin—The "Stiglisbrücke" Basin On The Schächen Torrent During The Flood Of August 2005. INTERPRAEVENT Conference Proceedings (pp. 271–282). Retrieved October 30, 2022, from http://www.interpraevent.at/palm-cms/upload_files/Publikationen/Tagungsbeitraege/2008_1_271.pdf.

Bezzola, G. R., Sigg, H., & Lange, D. (2004). Driftwood retention works in Switzerland [Schwemmholzrückhalt in der Schweiz]. In *INTERPRAEVENT Conference Proceedings* (pp. 29–40). Retrieved December 30, 2022, from http://www.interpraevent.at/palm-cms/upload_files/Publikationen/Tagungsbeitraege/2004_3_VII-29.pdf.

Bischetti, G. B., Chiaradia, E. A., D'Agostino, V., & Simonato, T. (2010). Quantifying the effect of brush layering on slope stability. *Ecological Engineering, 36*, 258–264. https://doi.org/10.1016/j.ecoleng.2009.03.019

Bovis, M. J., & Jakob, M. (1999). The role of debris supply conditions in predicting debris flow activity. *Earth Surface Processes and Landforms, 24*, 1039–1054. https://doi.org/10.1002/(SICI)1096-9837(199910)24:11<1039::AID-ESP29>3.0.CO;2-U

Bramberger, J; Hübl,J (2018): Geschiebedosierung in Sperrenstaffelungen. Wildbach- und Lawinenverbau, 181, 260–272; ISSN 978–3–9504159–5–7

Brochot, S., Duclos, P., & Bouzit, M. (2003). L'évaluation économique des risques torrentiels: Intérêts et limites pour les choix collectifs de prévention. *Ingénieries* **Numéro Spécial**, 53–68. Retrieved December 30, 2022, from http://www.set-revue.fr/sites/default/files/articles-eat/pdf/GR2003-PUB00011910.pdf.

Brown, J. C. (1876). Reboisement in France: or, records of the replanting of the Alps, the Cevennes, and the Pyrenees with trees, herbage, and bush, with a view to arresting and preventing the destructive consequences and effects of torrents. . Henry S. King and Co. https://archive.org/details/reboisementinfr00brow/page/n9.

Busnelli, M., Stelling, G., & Larcher, M. (2001). Numerical morphological modeling of open-check dams. *Journal of Hydraulic Engineering, 127*, 105–114. https://doi.org/10.1061/(ASCE)0733-9429(2001)127:2(105)

Canelas, R. B., Domínguez, J. M., Crespo, A. J. C., Gómez-Gesteira, M., & Ferreira, R. M. L. (2017). Resolved simulation of a granular-fluid flow with a coupled SPH-DCDEM model. *Journal of Hydraulic Engineering, 143*, 06017012. https://doi.org/10.1061/(asce)hy.1943-7900.0001331

Cánovas, J. A. B., Stoffel, M., Corona, C., Schraml, K., Gobiet, A., Tani, S., Sinabell, F., Fuchs, S., & Kaitna, R. (2016). Debris-flow risk analysis in a managed torrent based on a stochastic life-cycle performance. *Science of the Total Environment, 557–558*, 142–153. https://doi.org/10.1016/j.scitotenv.2016.03.036

Carladous, S., Piton, G., Kuss, D., Charvet, G., Paulhe R., Morel, M., & Quefféléan, Y. (2022). Chap. 13: French experience with open check dams: Inventory and lessons learnt through adaptive management. In *Check dam construction for sustainable watershed management and planning* (pp. 247–266). Wiley Online Library. https://doi.org/10.1002/9781119742449.ch13

Carladous, S., Piton, G., Recking, A., Liebault, F., Richard, D., Tacnet, J. M., Kuss, D., Philippe, F., Quefféléan, Y., Marco, O. (2016). Towards a better understanding of the today French torrents management policy through a historical perspective (pp.1–11). https://doi.org/10.1051/e3sconf/20160712011.

Carriere, A., Le Bouteiller, C., Tucker, G. E., Klotz, S., & Naaim M. (2020). Impact of vegetation on erosion: Insights from the calibration and test of a landscape evolution model in alpine badland catchments. *Earth*

Surface Processes and Landforms, 45(5), 1085-1099. https://doi.org/10.1002/esp.v45.5. https://doi.org/10.1002/esp.4741

Cerato, M., & Coali, R. (1997). Open Check Dams in the Italian Alps. *Structural Engineering International, 7*, 89–91. https://doi.org/10.2749/101686697780495157

CFBR. (2013). *Dam spillway design guidelines*. Working group for the design of dam spillways (Ed.), French Committee for Dams and Reservoirs. Retrieved October 22, 2019, from http://www.barrages-cfbr.eu/IMG/pdf/recommandations_cfbr_2013_evc.pdf.

Chahrour, N., Piton, G., Tacnet, J.-M., & Bérenguer, C. (2022a). Designing protection systems in mountains for reduced maintenance costs: claret's retention dam case study. In *32nd European Safety and Reliability Conference (ESREL 2022), TU Dublin; European Safety and Reliability Association, Trinity College Dublin, Aug 2022, Dublin, Ireland* (pp. 2797–2804). https://doi.org/10.3850/978-981-18-5183-4_S21-05-576-cd.

Chahrour, N., Piton, G., Bérenguer, C., & Tacnet, J. M. (2022b). Designing protection systems in mountains for reduced maintenance costs: claret's retention dam case study. Presented at the ESREL 2022 - 32nd European Safety and Reliability Conference (ESREL 2022). Dublin, Ireland. 2797–2804 pp. Retrieved Stemper 1, 2023, from https://hal.inrae.fr/hal-03767911v2.

Chahrour, N., Nasr, M., Tacnet, J.-M., & Bérenguer, C. (2021). Deterioration modeling and maintenance assessment using physics-informed stochastic Petri nets: Application to torrent protection structures. *Reliability Engineering & System Safety, 210*, 107524. https://doi.org/10.1016/j.ress.2021.107524

Chambon, G., Richard, D., & Segel, V. (2010). Diagnostic du fonctionnement d'un bassin versant générateur de laves torrentielles et estimation de l'aléa: Le cas du Réal (Alpes-Maritimes, France). *Sciences Eaux Et Territoires, 2*, 140–151. https://doi.org/10.14758/SET-REVUE.2010.2.18

Chatwin, S. C., Howes, D. E., Schwab, J. W., & Swanston, D. N. (1994). A guide for management of landslide-prone Terrain in the Pacific Northwest. Retrieved July 15, 2021, from https://www.for.gov.bc.ca/hfd/pubs/docs/lmh/Lmh18.htm.

Chen, J., Chen, X., Wang, T., Zou, Y., & Zhong, W. (2014). Types and causes of debris flow damage to drainage channels in the Wenchuan earthquake area. *Journal of Mountain Science, 11*, 1406–1419. https://doi.org/10.1007/s11629-014-3045-x

Chen, S.-C., & Tfwala, S. (2018). Evaluating an optimum slit check dam design by using a 2D unsteady numerical model. In A. Paquier, & N. Rivière (Eds.), *E3S Web of Conferences* (vol. 40, p. 03027). https://doi.org/10.1051/e3sconf/20184003027. Retrieved November 29, 2021, from https://www.e3s-conferences.org.

Chen, X., Cui, P., You, Y., Chen, J., & Li, D. (2015). Engineering measures for debris flow hazard mitigation in the Wenchuan earthquake area. *Engineering Geology, 194*, 73–85. https://doi.org/10.1016/j.enggeo.2014.10.002

Chiu, Y., Tfwala, S. S., Hsu, Y., Chiu, Y., Lee, C., & Chen, S. (2021). Upstream morphological effects of a sequential check dam adjustment process. *Earth Surface Processes and Landforms, 46*(13), 2527–2539. https://doi.org/10.1002/esp.5178

Church, M., & Jakob, M. (2020). what is a debris flood? *Water Resources Research, 56*. https://doi.org/10.1029/2020WR027144.

Cicco, P. N. D., Paris, E., Ruiz-Villanueva, V., Solari, L., & Stoffel, M. (2018). In-channel wood-related hazards at bridges: A review. *River Research and Applications, 34*, 617–628. https://doi.org/10.1002/rra.3300

CIRIA, CUR, CETMEF. (2007). The rock manual—The use of rock in hydraulic engineering (2nd edition). CIRIA, https://www.ciria.org/ItemDetail?iProductCode=C683.

CIRIA, Ministry of Ecology, USACE. (2013). *International Levee handbook*. CIRIA. Retrieved January 5, 2022, from https://www.ciria.org/ItemDetail?iProductcode=C731.

Comiti, F., Lucía, A., & Rickenmann, D. (2016). Large wood recruitment and transport during large floods: A review. *Geomorphology, 269*, 23–39. https://doi.org/10.1016/j.geomorph.2016.06.016

Conesa-Garcia, C., & Lenzi, M. A. (Eds.). (2013). *Check dams, morphological adjustments and erosion control in torrential streams*. Nova Science Publishers, Inc.

Cucchiaro, S., Cavalli, M., Vericat, D., Crema, S., Llena, M., Beinat, A., Marchi, L., & Cazorzi, F. (2019). Geomorphic effectiveness of check dams in a debris-flow catchment using multi-temporal topographic surveys. *CATENA, 174*, 73–83. https://doi.org/10.1016/j.catena.2018.11.004

D'Agostino, V. (2013). Filtering-retention check dam design in mountain torrents. In C. Conesa-Garcia, & M. A. Lenzi (Eds.), *Check dams, morphological adjustments and erosion control in torrential streams* (pp. 185–210). Nova Science Publishers, Inc.

D'Agostino, V., Degetto, M., & Righetti, M. (2000). Experimental investigation on open check dam for coarse woody debris control. In *Dynamics of water and sediments in mountain basins*. Bios, 201–212. Retrieved December 30, 2019, from http://intra.tesaf.unipd.it/people/dagostino/Pubblicazioni/P43_2000.pdf.

D'Agostino, V., & Bertoldi, G. (2014). On the assessment of the management priority of sediment source areas in a debris-flow catchment: Management priority of sediment source areas in debris-flow catchment. *Earth Surface Processes and Landforms, 39*, 656–668. https://doi.org/10.1002/esp.3518

D'Agostino, V., Bertoldi, G., & Bettella, F. (2016). 'Back Analysis' Dell'efficienza Di Un Sistema Di Difesa Dalle Colate Detritiche. *Quaderni Di Idronomia Montana, 34*, 283–294.

Demontzey, P. (1882). Traité pratique du reboisement et du gazonnement des montagnes. Rothschild, J. (Ed.), *Ministères de l'agriculture et du commerce et des travaux publiques*. Retrieved December 28, 2021, from https://gallica.bnf.fr/ark:/12148/bpt6k9609991O.

Eaton, B., MacKenzie, L., Jakob, M., & Weatherly, H. (2017). Assessing erosion hazards due to floods on fans: Physical modeling and application to engineering challenges. *Journal of Hydraulic Engineering, 143*, 04017021. https://doi.org/10.1061/(asce)hy.1943-7900.0001318

Egashira, S., Takebayashi, H., Sekine, M., & Osanai, N. (2016). Sediment run-out processes and possibility of sediment control structures in the 2013 Izu-Ohshima event. *International Journal of Erosion Control Engineering, 9*, 155–164. https://doi.org/10.13101/ijece.9.155

Eisbacher, R. H. (1982). Slope stability and land use in mountain valleys. *Geoscience Canada, 9*, 14–27. https://journals.lib.unb.ca/index.php/GC/article/view/3280.

Evette, A., Labonne, S., Rey, F., Liebault, F., Jancke, O., & Girel, J. (2009). History of bioengineering techniques for erosion control in rivers in western Europe. *Environmental Management, 43*, 972–984. https://doi.org/10.1007/s00267-009-9275-y

FOEN. (2016). Protection against Mass Movement Hazards. Guideline for the integrated hazard management of landslides, rockfall and hillslope debris flows. Federal Office for the Environment: Bern, CH. Retrieved January 5, 2022, from https://www.bafu.admin.ch/bafu/en/home/topics/natural-hazards/publications-studies/publications/protection-against-mass-movement-hazards.html.

Ghilardi, T., Boillat, J. L., Schleiss, A. J., De Montmollin, G., & Bovier S. (2012). Gestion du risque d'inondation sur l'Avançon par rétention de sediments - Optimisation sur modèle physique [Flood risk management of the Avançon river by means of sediment retention - physical model based study]. In *INTERPRAEVENT Conference Proceedings* (pp. 687–698). Retrieved December 30, 2020, from http://www.interpraevent.at/palm-cms/upload_files/Publikationen/Tagungsbeitraege/2012_2_687.pdf.

Gibling, M. R., & Davies, N. S. (2012). Palaeozoic landscapes shaped by plant evolution. *Nature Geoscience, 5*, 99–105. https://doi.org/10.1038/ngeo1376

Gschnitzer, T., Gems, B., Mazzorana, B., & Aufleger, M. (2017). Towards a robust assessment of bridge clogging processes in flood risk management. *Geomorphology, 279*, 128–140. https://doi.org/10.1016/j.geomorph.2016.11.002

de Haas, T., Nijland, W., de Jong, S. M., & McArdell, B. W. (2020). How memory effects, check dams, and channel geometry control erosion and deposition by debris flows. *Scientific Reports, 10*, 14024. https://doi.org/10.1038/s41598-020-71016-8

Harvey, A. M. (2012). The coupling status of alluvial fans and debris cones: A review and synthesis. *Earth Surface Processes and Landforms, 37*, 64–76. https://doi.org/10.1002/esp.2213

Hasegawa, Y., Sugiura, N., Shouzawa, M., & Mizuyama, T. (2010). An investigation of measures against woody debris through hydraulic model experiments. Presented at the INTERPRAEVENT Conference Proceedings (pp. 135–143). Retrieved December 28, 2021 from http://www.interpraevent.at/palm-cms/upload_files/Publikationen/Tagungsbeitraege/2010__135.pdf.

Horiguchi, T., Kokuryo, H., & Ishikawa, N. (2021a). Failure analysis of steel open dam against an extreme boulder debris flow. *International Journal of Protective Structures, 12*, 263–283. https://doi.org/10.1177/2041419620970566

Horiguchi, T., & Komatsu, Y. (2018). Method to evaluate the effect of inclination angle of steel open-type check dam on debris flow impact load. *International Journal of Protective Structures, 10*, 95–115. https://doi.org/10.1177/2041419618789702

Horiguchi, T., Piton, G., Munir, M. B., & Mano, V. (2021b). Driftwood and hybrid debris barrier interactions: process of trapping and prevention of releases during overtopping of releases during overtopping. In *Proc. 14th INTERPRAEVENT Congress* (pp. 206–215). https://hal.archives-ouvertes.fr/hal-03363332/.

Horiguchi, T., & Richefeu, V. (2020). Post-analysis simulation of the collapse of an open sabo dam of steel pipes subjected to boulder laden debris flow. *International Journal of Sediment Research, 35*(6), 621–635. https://doi.org/10.1016/j.ijsrc.2020.05.002

Hübl, J. (2018a). Conceptual framework for sediment management in torrents. *Water, 10*, 1718. https://doi.org/10.3390/w10121718

Hübl, J. (2018b). Schutzstrategien und Funktionen von Schutzbauwerken. *Wildbach- und Lawinenverbau, 181*, 34–44, ISSN 978-3-9504159-5-7.

Hübl, J., Fiebiger, G. (2005). Chap. 18 - Debris-flow mitigation measures. In Debris-flow hazards and related phenomena. In M. Jakob, & O. Hungr (Eds.) (pp. 445–487). Berlin, Heidelberg: Springer. https://doi.org/10.1007/3-540-27129-5_18.

Hubl, J., Nagl, G., Suda, J., & Rudolf-Miklau, F. (2017). Standardized stress model for design of torrential barriers under impact by debris flow (according to Austrian Standard Regulation 24801). *International Journal of Erosion Control Engineering, 10*, 47–55. https://doi.org/10.13101/ijece.10.47

Hungr, O., Morgan, G. C., & Kellerhals, R. (1984). Quantitative analysis of debris torrent hazards for design of remedial measures. *Canadian Geotechnical Journal, 21*, 663–677. https://doi.org/10.1139/t84-073.

Hungr, O., McDougall, S., & Bovis, M. (2005). Chap. 7: Entrainment of material by debris flows. In *Debris-flow hazards and related phenomena* (pp.135–158). Berlin, Heidelberg: Springer Berlin Heidelberg. https://doi.org/10.1007/3-540-27129-5_7.

Hunzinger, L. (2014). Freeboard analysis in river engineering and flood mapping-new recommendations. Presented at the Special Session on Swiss Competences in River Engineering and Restoration of the 7th International Conf. on Fluvial Hydraulics, RIVER FLOW 2014. Lausanne, Switzerland, 31–37 pp.

ICOLD. (2014). Integrated flood risk management. *ICOLD Bulletin, 156*, 288.

ICOLD. (2019). Blockage of reservoir spillways, intakes and bottom outlets by floating debris. International Commission on Large Dams (vol. 176, p. 50). Bulletin.

Ikeya, H. (1989). Debris flow and its countermeasures in Japan. *Bulletin - International Association of Engineering Geology, 40*, 15–33.

Ishikawa, Y., & Mizuyama, T. (1988). An experimental study of permeable sediment control dams as a countermeasure against floating logs. In *Proceedings of the 6th congress Asian and Pacific regional division—IAHR*, Kyoto, Japan, 20 July (pp. 723–730).

Iverson, R. M. (2015). Scaling and design of landslide and debris-flow experiments. *Geomorphology, 244*, 9–20. https://doi.org/10.1016/j.geomorph.2015.02.033

Jakob, M. (2005). Chap. 17: Debris-flow hazard analysis. In *Debris-flow hazards and related phenomena* (pp. 411–443). Berlin, Heidelberg: Springer Berlin Heidelberg. https://doi.org/10.1007/3-540-27129-5_17.

Jakob, M. (2021). Debris-flow hazard assessments: A practitioner's view. *Environmental and Engineering Geoscience, 27*(2), 153–166. https://doi.org/10.2113/EEG-D-20-00110

Jakob, M., & Hungr, O. (2005). *Debris-flow hazards and related phenomena*. Berlin Heidelberg: Praxis Springer.

Kaitna, R., Hübl, J. (2012). Chap. 7 - silent witnesses for torrential processes. In M. Schneuwly-Bollschweiler, M. Stoffel, & F. Rudolf-Miklau (Eds.), *Dating torrential processes on fans and cones* (pp. 111–130). Springer Netherlands.

Kasai, S., Ohgi, Y., Mizoguchi, I., Matsuda, A., Aramaki, H., & Tanami, M. (1996). Structural characteristics of wood-debris entrapment facilities. In *INTERPRAEVENT conference proceedings*, Garmisch-Partenkirchen, Germany. Retrieved December 3, 2020, from http://www.interpraevent.at/alm-cms/upload_files/Publikationen/Tagungsbeitraege/1996_5_271.pdf.

Kean, J. W., Staley, D. M., Lancaster, J. T., Rengers, F. K., Swanson, B. J., Coe, J. A., Hernandez, J. L., Sigman, A. J., Allstadt, K. E., & Lindsay, D. N. (2019). Inundation, flow dynamics, and damage in the 9 January 2018 Montecito debris-flow event, California, USA: Opportunities and challenges for post-wildfire risk assessment. *Geosphere, 15*, 1140–1163. https://doi.org/10.1130/GES02048.1

Kostadinov, S. (1993). Possibility of assessment of the slope of siltation based on the some hydraulic characteristics of the torrential flows. *Journal of the Japan Society of Erosion Control Engineering, 45*, 28–33. https://doi.org/10.11475/sabo1973.45.5_28

Kostadinov, S. (2007). Erosion and torrent control in Serbia: hundred years of experiences. presented at the International conference "Erosion and torrent control as a factor in sustainable river basin management." Belgrade. Serbia. 1–17 pp.

Kostadinov, S. (2010). Forests in Serbia as the factor of soil and water protection against degradation in the conditions of global climate change, monograph "global environmental change: challenges to science and society in Southeastern Europe": Editors: Vesselin Alexandrov, Martin Felix Gajdusek, C. Gregory Knight, Antoaneta Yotova. Springer Science+Business Media B.V., pp.177–190.

Koulinski, V., & Richard, P. (2008). Apports des modéles réduits pour la gestion des sédiments et des flottants en torrents et rivières torrentielles [Scale models contribution to the sedimentation processes and floating debris transit]. *Houille Blanche, 4*, 90–97. https://doi.org/10.1051/lhb:2008044

Kronfellner-Kraus, G. (1983). Torrent Erosion And Its Control In Europe And Some Research Activities In This Field In Austria. *Sabo Gakkaishi - Journal of the Japan Society of Erosion Control Engineering, 35*, 33–44. https://doi.org/10.11475/sabo1973.35.3_33

Kuss, C. (1900). Restauration et conservation des terrains en montagne. Éboulements, glissements et barrages. Imprimerie Nationale. http://gallica.bnf.fr/ark:/12148/bpt6k64593256.

Lambert, S., Bourrier, F., Ceron Mayo, A.R., Dugelas, L., Dubois, F., & Piton, G. (2022). Small scale modelling of flexible barriers. I: Achieving mechanical similitude. *Journal of Hydraulic Engineering, 149*, 04022043. https://doi.org/10.1061/JHEND8.HYENG-13070.

Lange, D., & Bezzola, G. (2006). Schwemmholz - Probleme und Lösungsansätze [Driftwood - Problems and solutions]. Versuchsanstalt für Wasserbau Hydrologie und Glaziologie der Eidgenössischen Technischen Hochschule (VAW) Zürich. Retrieved December 30, 2020, from https://ethz.ch/content/dam/ethz/special-interest/baug/vaw/vaw-dam/documents/das-institut/mitteilungen/2000-2009/188.pdf.

Larcher, M., & Armanini, A. (2000). Design criteria of slit check dams and downstream channels for debris flows. Presented at the International Workshop on the debris flow disaster of December 1999 in Venezuela. Caracas: Venezuela.

Latella, M., Bertagni, M. B., Vezza, P., & Camporeale, C. (2020). An Integrated methodology to study riparian vegetation dynamics: from field data to impact modeling. *Journal of Advances in Modeling Earth Systems, 12*(8), 1–23. https://doi.org/10.1029/2020MS002094

Lefebvre, B., & Demmerle, D. (2004). Protection du village du Tour contre le glissement des Posettes à l'amont de la vallée de Chamonix Mont-Blanc [Protection of the Le Tour village against the Posettes landslide in the upstream part of Chamonix Valley]. *Houille Blanche, 3i*, 31–36. https://doi.org/10.1051/lhb:200403003.

Leonardi, A., Wittel, F. K., Mendoza, M., & Herrmann, H. J. (2014). Coupled DEM-LBM method for the free-surface simulation of heterogeneous suspensions. *Computational Particle Mechanics, 1*, 3–13. https://doi.org/10.1007/s40571-014-0001-z

Leonardi, A., Wittel, F. K., Mendoza, M., Vetter, R., & Herrmann, H. J. (2015). Particle-fluid-structure interaction for debris flow impact on flexible barriers. *Computer-Aided Civil and Infrastructure Engineering, 31*, 323–333. https://doi.org/10.1111/mice.12165

Lenzi, M. A. (2002). Stream bed stabilization using boulder check dams that mimic step-pool morphology features in Northern Italy. *Geomorphology, 45*, 243–260. https://doi.org/10.1016/S0169-555X(01)00157-X

Leys, E. (1969). Die technischen und die wirtschaftlichen Grundlagen der großdoligen und kronenoffenen Bauweise in der Wildbachverbauung, Dissertation, Universität für Bodenkultur Wien (unpublished).

Liu, F. Z., Xu, Q., Dong, X. J., Yu, B., Frost, J. D., & Li, H. J. (2017). Design and performance of a novel multi-function debris flow mitigation system in Wenjia Gully, Sichuan. *Landslides, 14*, 2089–2104. https://doi.org/10.1007/s10346-017-0849-0

Lòpez Cadenas de Llano, F. (1988). *Torrent control and streambed stabilization*. FAO (Food and Agriculture Organization of the United Nations). 177 pp.

López, J. L., Hernandez-Perez, D., & Courtel, F. (2010). Monitoreo y evaluación del comportamiento de las presas de retención de sedimentos en el estado Vargas. In J. L. López (Ed.), Lecciones aprendidas del desastre de Vargas. Aportes científico-tecnológico y experiencias nacionales en el campo de la prevención y mitigación de riesgos. Fundación Polar/Universidad Central De Venezuela, Caracas: Venezuela (pp. 479–459).

Lucas-Borja, M. E., Piton, G., Yu, Y., Castillo, C., & Antonio, Z. D. (2021). Check dams worldwide: Objectives, functions, effectiveness and undesired effects. *CATENA, 204*, 105390. https://doi.org/10.1016/j.catena.2021.105390

Marchi, L., Cavalli, M., & D'Agostino, V. (2010). Hydrogeomorphic processes and torrent control works on a large alluvial fan in the eastern Italian Alps. *Natural Hazards and Earth System Sciences, 10*, 547–558. https://doi.org/10.5194/nhess-10-547-2010

Marchi, L., Comiti, F., Crema, S., & Cavalli, M. (2019). Channel control works and sediment connectivity in the European Alps. *Science of the Total Environment, 668*, 389–399. https://doi.org/10.1016/j.scitotenv.2019.02.416

Mazzorana, B., Comiti, F., Scherer, C., & Fuchs, S. (2012). Developing consistent scenarios to assess flood hazards in mountain streams. *Journal of Environmental Management, 94*, 112–124. https://doi.org/10.1016/j.jenvman.2011.06.030

Mazzorana, B., Comiti, F., Volcan, C., & Scherer, C. (2011). Determining flood hazard patterns through a combined stochastic–deterministic approach. *Natural Hazards, 59*, 301–316. https://doi.org/10.1007/s11069-011-9755-2

Mazzorana, B., Ruiz-Villanueva, V., Marchi, L., Cavalli, M., Gems, B., Gschnitzer, T., Mao, L., Iroumé, A., & Valdebenito, G. (2018). Assessing and mitigating large wood-related hazards in mountain streams: Recent approaches. *Journal of Flood Risk Management, 11*, 207–222. https://doi.org/10.1111/jfr3.12316

Mizuyama, T., Irasawa, M., Amada, T., & Kobayashi, M. (1990). Analyses of Sedimentation Data on Sabo Dams. *Sabo Gakkaishi, 43*, 33–35. https://doi.org/10.11475/sabo1973.43.4_33

Mark, E. E. (2017). Guidance for debris-flow and debris-flood mitigation design in Canada, Master's Thesis, Simon Fraser University. http://summit.sfu.ca/item/17457.

Mertin, M. (2018). What is the effect of check dams? Simulating the impact of check dams on landscape evolution at centennial time scale, Master's Thesis, Univ. Bern.

Moldenhauer-Roth, A., Piton, G., Schwindt, S., Jafarnejad, M., & Schleiss, A. J. (2021). Design of sediment detention basins: Scaled model experiments and application. *International Journal of Sediment Research, 36*, 136–150. https://doi.org/10.1016/j.ijsrc.2020.07.007

Mougin, P. (1900). Restauration et conservation des terrains en montagne - consolidation des berges par dérivation d'un torrent (torrent de Saint Julien). *Imprimerie Nationale*. http://gallica.bnf.fr/ark:/12148/bpt6k6462342t.

Nagayama, T., Furuya, T., Matsuda, S., Itoh, T., Fujita, M., & Mizuyama, T. (2019). Monitoring of sediment runoff and observation basin for sediment movements focused on active sediment control in Jo-Gan-Ji River. Presented at the International Conference on Debris-Flow Hazards Mitigation: Mechanics, Prediction, and Assessment. Boulder. CO. Retrieved January 5, 2022, from https://doi.org/10.25676/11124/173222.

Nakatani, K., Wada, T., Satofuka, Y., & Mizuyama, T. (2008). Development of "Kanako 2D (Ver. 2.00)", a user-friendly one-and two-dimensional debris flow simulator equipped with a graphical user interface. *International Journal of Erosion Control Engineering, 1*, 62–72. https://doi.org/10.13101/ijece.1.62

NILIM. (2007). Manual of technical standards for establishing sabo master plans for debris flows and driftwood. National Institute for Land and Infrastructure Management. Tsukuba, Japan. http://www.nilim.go.jp/lab/bcg/siryou/tnn/tnn0364pdf/ks0364.pdf.

Nishimoto, H. (2011). Discussion on the transition of idea about sediment control effect functioned by check dam. *Journal of the Japan Society of Erosion Control Engineering, 64*, 46–51.

Okamoto, M. (2007). The structure of Sabo Administration. Japan Sabo Association. Retrieved January 5, 2022, from http://www.sabo-int.org/guideline/pdf/structure_administration.pdf.

Okamoto, T., Takebayashi, H., Sanjou, M., Suzuki, R., & Toda, K. (2019). Log jam formation at bridges and the effect on floodplain flow: A flume experiment. *Journal of Flood Risk Management, 13*, e12562. https://doi.org/10.1111/jfr3.12562

Osanai, N., Mizuno, H., & Mizuyama, T. (2010). Design standard of control structures against debris flow in Japan. *Journal of Disaster Research, 5*, 307–314.

ÖWAV. (2021). Holz in und an Fließgewässern—Wildholzmanagement. Österreichischen Wasser- und Abfallwirtschaftsverbandes (ÖWAV): Wien, AUT. Retrieved February 25, 2022, from https://www.oewav.at/Publikationen?current=410953&mode=form.

Patel, A. (2012). Mountain erosion and mitigation: Global state of art. *Environmental Earth Sciences, 66*, 1631–1639. https://doi.org/10.1007/s12665-012-1524-3

Pasqua, A., Leonardi, A., & Pirulli, M. (2022). Coupling depth-averaged and 3D numerical models for the simulation of granular flows. *Computers and Geotechnics, 149*, 104879. https://doi.org/10.1016/j.compgeo.2022.104879

Pastorello, R., D'Agostino, V., & Hürlimann, M. (2020). Debris flow triggering characterization through a comparative analysis among different mountain catchments. *CATENA, 186*, 104348. https://doi.org/10.1016/j.catena.2019.104348

Peng, S-H., & Tang, C. (2015). Blending the analytic hierarchy process and fuzzy logical systems in scenic beauty assessment of check dams in streams. *Water, 7*(12), 6983–6998. https://doi.org/10.3390/w7126670

Perov, V., Chernomorets, S., Budarina, O., Savernyuk, E., & Leontyeva, T. (2017). Debris flow hazards for mountain regions of Russia: Regional features and key events. *Natural Hazards, 88*, 199–235. https://doi.org/10.1007/s11069-017-2841-3

Piton, G. (2016). Sediment transport control by check dams and open check dams in Alpine torrents, PhD Thesis, Univ. Grenoble Alpes. https://tel.archives-ouvertes.fr/tel-01420209.

Piton, G. (2023). Keynote lecture. Defining Protection Works Against Debris-Flow Hazards: Industrial Standard, Tailor-made or Haute-couture? Presented at the 8th International Conference on Debris Flow Hazard Mitigation (DFHM8) (vol. 415, p. 07011). https://doi.org/10.1051/e3sconf/202341507011.

Piton, G., Carladous, S., Recking, A., Liebault, F., Tacnet, J., Kuss, D., Quefféléan, Y., & Marco, O. (2017). Why do we build check dams in Alpine streams? An historical perspective from the French experience. *Earth Surface Processes and Landforms, 42*, 91–108. https://doi.org/10.1002/esp.3967

Piton, G., Ceron Mayo, A. R., & Lambert, S. (2022a). Small scale modelling of flexible barriers. II: Interaction with Large Wood. *Journal of Hydraulic Engineering, 149*, 04022044. https://doi.org/10.1061/JHEND8.HYENG-13071.

Piton, G., Charvet, G., Kuss, D., & Carladous S. (2020a). Putting a grill (or not) in slit dams aiming at trapping debris flows? Lessons learnt from France. AGHP Technical Notes, 1–5. https://hal.archives-ouvertes.fr/hal-02701076.

Piton, G., Fontaine, F., Bellot, H., Liébault, F., Bel, C., Recking, A., & Hugerot, T. (2018). Direct field observations of massive bedload and debris flow depositions in open check dams. Presented at the RiverFlow 2018 (pp. 1–8).

Piton, G., Goodwin, S. R., Mark, E., & Strouth, A. (2022b). Debris flows, boulders and constrictions: A simple framework for modeling jamming, and its consequences on outflow. *Journal of Geophysical Research: Earth Surface, 127*(5), 1–27. https://doi.org/10.1029/2021JF006447

Piton, G., Horiguchi, T., Marchal, L., & Lambert, S. (2020b). Open check dams and large wood: Head losses and release conditions. *Natural Hazards and Earth System Sciences, 20*, 3293–3314. https://doi.org/10.5194/nhess-20-3293-2020

Piton, G., Mano, V., Richard, D., Evin, G., Laigle, D., Tacnet, J. M., & Rielland, P. A. (2019). Design of a debris retention basin enabling sediment continuity for small events: the Combe de Lancey case study (France). Presented at the International Conference on Debris-Flow Hazards Mitigation: Mechanics, Prediction, and Assessment. Golden, Colorado, USA (pp. 1019–1026).

Piton, G., & Recking, A. (2014). The dynamic of streams equipped with check dams. In *Proceedings of the International Conference on Fluvial Hydraulics, RIVER FLOW 2014, Lausanne, Switzerland*, (pp. 1437–1445).

Piton, G., & Recking, A. (2016a). Design of sediment traps with open check dams. I: Hydraulic and deposition processes. *Journal of Hydraulic Engineering, 142*, 1–23. https://doi.org/10.1061/(ASCE)HY.1943-7900.0001048

Piton, G., & Recking, A. (2016b). Design of sediment traps with open check dams. II: Woody debris. *Journal of Hydraulic Engineering, 142*, 1–17. https://doi.org/10.1061/(ASCE)HY.1943-7900.0001049

Piton, G., & Recking, A. (2016c). Effects of check dams on bed-load transport and steep slope stream morphodynamics. *Geomorphology, 291*, 94–105. https://doi.org/10.1016/j.geomorph.2016.03.001

Piton, G., & Recking, A. (2016d). Closure to "design of sediment traps with open check dams. i: hydraulic and deposition processes" by guillaume piton and alain recking. *Journal of Hydraulic Engineering, 142*, 07016009. https://doi.org/10.1061/(ASCE)HY.1943-7900.0001207

Poncet, A. (1995). Restauration et conservation des terrains en montagne. Service RTM (Restauration des Terrains de Montagne). Office National des Forêts. 1000 pp.

Prochaska, A. B., Santi, P. M., & Higgins, J. D. (2008). Debris basin and deflection berm design for fire-related debris-flow mitigation. *Environmental and Engineering Geoscience, 14*, 297–313. https://doi.org/10.2113/gseegeosci.14.4.297

Proske, D., Suda, J., & Hübl, J. (2011). Debris flow impact estimation for breakers. *Georisk, 5*, 143–155. https://doi.org/10.1080/17499518.2010.516227

PWRI. (2007). Guidelines for landslide prevention technologies. Landslide Research Team. Erosion and Sediment Control Research Group. Public Works Research Institute (PWRI). Retrieved January 5, 2022, from http://www.sabo-int.org/guideline/pdf/Landslide_20140110_01.pdf.

Ramirez, J. A., Mertin, M., Peleg, N., Horton, P., Skinner, C., Zimmermann, M., & Keiler, M. (2022). Modelling the long-term geomorphic response to check dam failures in an alpine channel with CAESAR-Lisflood. *International Journal of Sediment Research, 37*, 687–700. https://doi.org/10.1016/j.ijsrc.2022.04.005

Ravazzolo, D., Mao, L., Mazzorana, B., & Ruiz-Villanueva, V. (2017). Brief communication: The curious case of the large wood-laden flow event in the Pocuro stream (Chile). *Natural Hazards and Earth System Sciences, 17*, 2053–2058. https://doi.org/10.5194/nhess-17-2053-2017

Remaître, A., Van Asch, W. J. T., Malet, J.-P., & Maquaire, O. (2008). Influence of check dams on debris-flow run-out intensity. *Natural Hazards and Earth System Science, 8*, 1403–1416.https://doi.org/10.5194/nhess-8-1403-2008.

Rickenmann, D., & Koschni, A. (2010). Sediment loads due to fluvial transport and debris flows during the 2005 flood events in Switzerland. *Hydrological Processes, 24*, 993–1007. https://doi.org/10.1002/hyp.7536

Rickli, C., Badoux, A., Rickenmann, D., Steeb, N., & Waldner, P. (2018). Large wood potential, piece characteristics, and flood effects in Swiss mountain streams. *Physical Geography, 39*, 542–564. https://doi.org/10.1080/02723646.2018.1456310

Rimböck, A. (2004). Design of rope net barriers for woody debris entrapment: introduction of a design concept. In *INTERPRAEVENT Conference Proceedings* (pp. 265–276). Retrieved October 10, 2022 from http://www.interpraevent.at/palm-cms/upload_files/Publikationen/Tagungsbeitraege/2004_3_VII-265.pdf.

Rinaldi, M., Simoncini, C., & Piégay, H. (2009). Scientific design strategy for promoting sustainable sediment management: The case of the Magra River (Central-Northern Italy). *River Research and Applications, 25*, 607–625. https://doi.org/10.1002/rra.1243

Royet, P., Degoutte, G., Peyras, L., Lavabre, J., & Lemperrière, F. (2009). Cotes et crues de protection, de sûreté et de danger de rupture. In F. Lyon, Royet, P., & G. Degoutte (Eds.), *Proceedings of the Colloque CFBR-SHF: Dimensionnement et fonctionnement des évacuateurs de crues, 20–21 janvier 2009* (8 p.). Retrieved October 10, 2022, from: https://hal.archives-ouvertes.fr/hal-00468565.

Rudolf-Miklau, F., & Suda, J. (2011). Technical standards for debris flow barriers and breakers. *Italian Journal of Engineering Geology and Environment, 201103*, 1083–1091. https://doi.org/10.4408/IJEGE.2011-03.B-117

Rudolf-Miklau, F., & Suda, J. (2013). Chap. 26—Design criteria for torrential barriers. In Schneuwly-Bollschweiler M, Stoffel M, and Rudolf-Miklau F (Eds.), *Dating Torrential Processes on Fans and Cones* (pp. 375–389). Springer Netherlands. https://doi.org/10.1007/978-94-007-4336-6_26.

Ruiz-Villanueva, V., et al. (2019). Characterization of wood-laden flows in rivers. *Earth Surface Processes and Landforms, 44*, 1694–1709. https://doi.org/10.1002/esp.4603

Sabo Department. (2000). *Sabo methods and facilities made of natural materials*. Ministry of Construction. Japan. Retrieved January 5, 2022, from http://www.sabo-int.org/guideline/japan.html.

SABO Division (2000). *Guideline for driftwood countermeasures (proposal and design)*. In Sediment Control (Sabo) Department (Ed). Ministry of construction. Japan. Retrieved January 5, 2022, from http://www.sabo-int.org/guideline/pdf/driftwoodCountermeasure-Guideline.pdf.

SCD. (2017). floating debris at reservoir dam spillways. Techreport. Swiss Committee on Dams on the state of floating debris issues at dam spillways. Retrieved October 22, 2019, from http://www.swissdams.ch/fr/publications/publications-csb/2017_Floating%20debris.pdf.

Schalko, I. (2020). Wood retention at inclined racks: Effects on flow and local bedload processes. *Earth Surface Processes and Landforms, 45*, 2036–2047. https://doi.org/10.1002/esp.4864

Schalko, I., Lageder, C., Schmocker, L., Weitbrecht, V., & Boes, R. M. (2019a). Laboratory flume experiments on the formation of spanwise large wood accumulations Part I: Effect on backwater rise. *Water Resources Research, 55*, 4854–4870. https://doi.org/10.1029/2018wr024649

Schalko, I., Lageder, C., Schmocker, L., Weitbrecht, V., & Boes, R. M. (2019b). Laboratory flume experiments on the formation of spanwise large wood accumulations Part II: Effect on local scour. *Water*

Resources Research, 55, 4871–4885. https://doi.org/10.1029/2019WR024789

Schalko, I., Ruiz-Villanueva, V., Maager, F., & Weitbrecht, V. (2021). Wood retention at inclined bar screens: Effect of wood characteristics on backwater rise and bedload transport. *Water, 13*, 2231. https://doi.org/10.3390/w13162231

Scheidl, C., McArdell, B., Nagl, G., Rickenmann, D. (2019). Debris flow behavior in super- and subcritical conditions. In Arthur Lakes Library (Ed.), *Seventh International Conference on Debris-Flow Hazards Mitigation—Proceedings*. Colorado School Of Mines.https://doi.org/10.25676/11124/173187

Schiechtl, H. M., & Stern, R. (1994). *Water bioengineering techniques for watercourse bank and shoreline protection* (p. 186). Blackwell Publishing.

Schmocker, L., & Hager, W. H. (2013). Scale modeling of wooden debris accumulation at a debris rack. *Journal of Hydraulic Engineering, 139*, 827–836. https://doi.org/10.1061/(ASCE)HY.1943-7900.0000714

Schmocker, L., & Weitbrecht, V. (2013). Driftwood: Risk analysis and engineering measures. *Journal of Hydraulic Engineering, 139*, 683–695. https://doi.org/10.1061/(ASCE)HY.1943-7900.0000728

Schuster, R. L., & Evans, S. G. (2011). engineering measures for the hazard reduction of landslide dams. In S. G. Evans, R. L. Hermanns, A. Strom, & G. Scarascia-Mugnozza (Eds.), Natural and artificial rockslide dams (pp. 77–100). Springer Berlin Heidelberg: Berlin, Heidelberg. Retrieved November 25, 2021, from http://link.springer.com, https://doi.org/10.1007/978-3-642-04764-0_2.

Schwindt, S., Franca, M., De Cesare, G., & Schleiss, A. (2017). Analysis of mechanical-hydraulic bedload deposition control measures. *Geomorphology, 295*, 467–479. https://doi.org/10.1016/j.geomorph.2017.07.020

SedAlp. (2015). Work Package 6—Interactions with structures. Alpine Space European project. http://www.sedalp.eu/download/reports.shtml.

Sheng, J., & Liao, A. (1997). Erosion control in South China. *CATENA, 29*, 211–221. https://doi.org/10.1016/s0341-8162(96)00057-4

Shima, J., Moriyama, H., Kokuryo, H., Ishikawa, N., & Mizuyama, T. (2016). Prevention and mitigation of debris flow hazards by using steel open-type sabo dams. *International Journal of Erosion Control Engineering, 9*, 135–144. https://doi.org/10.13101/ijece.9.135

Simoni, S., Vignoli, G., & Mazzorana, B. (2017). Enhancing sediment flux control and natural hazard risk mitigation through a structured conceptual planning approach. *Geomorphology, 291*, 159–173. https://doi.org/10.1016/j.geomorph.2017.01.026

Steeb, N., Rickenmann, D., Badoux, A., Rickli, C., & Waldner, P. (2017). Large wood recruitment processes and transported volumes in Swiss mountain streams during the extreme flood of August 2005. *Geomorphology, 279*, 112–127. https://doi.org/10.1016/j.geomorph.2016.10.011

Sun, H., You, Y., & Liu, J. (2018a). Experimental study on blocking and self-cleaning behaviors of beam dam in debris flow hazard mitigation. *International Journal of Sediment Research, 33*, 395–405. https://doi.org/10.1016/j.ijsrc.2018.04.004

Sun, H., You, Y., & Liu, J. (2018b). Experimental study on characteristics of trapping and regulating sediment with an open-type check dam in debris flow hazard mitigation. *Journal of Mountain Science, 15*, 2001–2012. https://doi.org/10.1007/s11629-017-4619-1

Sun, H., You, Y., Liu, J., Zhang, G., Feng, T., & Wang, D. (2021). Experimental study on discharge process regulation to debris flow with open-type check dams. *Landslides, 18*, 967–978. https://doi.org/10.1007/s10346-020-01535-y

Takahashi, T. (2014). *Debris flow: Mechanics, prediction and countermeasures*. CRC Press.

Theule, J., Liébault, F., Laigle, D., Loye, A., & Jaboyedoff, M. (2015). Channel scour and fill by debris flows and bedload transport. *Geomorphology, 243*, 92–105. https://doi.org/10.1016/j.geomorph.2015.05.003

Theule, J. I., Liébault, F., Loye, A., Laigle, D., & Jaboyedoff, M. (2012). Sediment budget monitoring of debris-flow and bedload transport in the Manival Torrent, SE France. *Natural Hazards and Earth System Science, 12*, 731–749. https://doi.org/10.5194/nhess-12-731-2012

USACE. (1994a). *Channel stability assessment for flood control projects*. U.S. Army Corps of Engineers. Retrieved January 5, 2022, from http://140.194.76.129/publications/eng-manuals/EM_1110-2-1418_sec/EM_1110-2-1418.pdf.

USACE. (1994b). *Hydraulic design of flood control channels*. U.S. Army Corps of Engineers. Retrieved January 5, 2022, from http://www.publications.usace.army.mil/Portals/76/Publications/EngineerManuals/EM_1110-2-1601.pdf.

VanDine, D. F. (1996). *Debris flow control structures for forest engineering*. Res. Br., B.C. Min. For., Victoria, British Colombia, Canada. Retrieved January 5, 2022, from https://www.for.gov.bc.ca/hfd/pubs/docs/wp/wp22.pdf.

Victoriano, A., Brasington, J., Guinau, M., Furdada, G., Cabré, M., & Moysset, M. (2018). Geomorphic impact and assessment of flexible barriers using multi-temporal LiDAR data: The Portainé mountain catchment (Pyrenees). *Engineering Geology, 237*, 168–180. https://doi.org/10.1016/j.enggeo.2018.02.016

Vischer, D. L. (2003). *Histoire de la protection contre les crues en Suisse, Des origines jusqu au 19e siècle. OFEG, Série Eaux*. Retrieved December 30, 2020, from https://www.bafu.admin.ch/dam/bafu/fr/dokumente/wasser/uw-umwelt-wissen/die_geschichte_deshochwasserschutzesinderschweizvondenanfaengenb.pdf.download.pdf/

histoire_de_la_protectioncontrelescruesensuisse-des-originesjusqu.pdf.
Volkwein, A. (2014). *Flexible debris flow barriers: Design and application: WSL Berichte, 18.* Birmensdorf, Switzerland: Swiss Federal Institute for Forest, Snow and Landscape Research WSL. Retrieved October 10, 2022, from https://www.wsl.ch/de/publikationen/flexible-debris-flow-barriers-design-and-application.html.
Volkwein, A., Wendeler, C., & Guasti, G. (2011). Design of flexible debris flow barriers. In *Proceedings of the International Conference on Debris-Flow Hazards Mitigation: Mechanics, Prediction, and Assessment* (pp. 1093–1100). Università La Sapienza.
Wang, F. (1903). *Grundriss der Wildbachverbauung, Zweiter Theil, Verlag S. Hirzel, Leipzig Schiechtl, H.M. 1973: Sicherungsarbeiten im Landschaftsbau.* Verlag Georg D.W. Callwey, München.
Wehrmann, H., Hübl, J., & Holzinger, G. (2006). Classification of dams in torrential watersheds. In *Interpraevent Conference Proceedings* (pp. 829–838). Retrieved December 28, 2021, from www.interpraevent.at/palm-cms/upload_files/Publikationen/Tagungsbeitraege/2006_2_829.pdf.
Weinmeister, H. W. (2007). Integrated debris flow disaster mitigation. *Journal of Mountain Science, 4*, 293–308. https://doi.org/10.1007/s11629-007-0293-z
Wendeler, C., & Volkwein, A. (2015). Laboratory tests for the optimization of mesh size for flexible debris-flow barriers. *Natural Hazards and Earth System Sciences, 15*, 2597–2604. https://doi.org/10.5194/nhess-15-2597-2015
Wendeler, C. S. I. (2008). *Murgangrückhalt in Wildbächen: Grundlagen zu Planung und Berechnung von flexiblen Barrieren*, 1 Band pp., ETH Zurich. Retrieved February 21, 2022 from http://hdl.handle.net/20.500.11850/150736.
Wendeler, C., Volkwein, A., McArdell, B. W., & Bartelt, P. (2018). Load model for designing flexible steel barriers for debris flow mitigation. *Canadian Geotechnical Journal, 56*, 893–910. https://doi.org/10.1139/cgj-2016-0157
Wohl, E., et al. (2019). The natural wood regime in rivers. *BioScience, 69*, 259–273. https://doi.org/10.1093/biosci/biz013
Yang, Z., Duan, X., Huang, J., Dong, Y., Zhang, X., Liu, J., & Yang, C. (2021). Tracking long-term cascade check dam siltation: Implications for debris flow control and landslide stability. *Landslides*. https://doi.org/10.1007/s10346-021-01755-w
Yuan, D., Liu, J., You, Y., Zhang, G., Wang, D., & Lin, Z. (2019). Experimental study on the performance characteristics of viscous debris flows with a grid-type dam for debris flow hazards mitigation. *Bulletin of Engineering Geology and the Environment, 78*, 5763–5774. https://doi.org/10.1007/s10064-019-01524-z
Zeng, Q. L., Yue, Z. Q., Yang, Z. F., & Zhang, X. J. (2009). A case study of long-term field performance of check-dams in mitigation of soil erosion in Jiangjia stream, China. *Environmental Geology, 58*, 897–911. https://doi.org/10.1007/s00254-008-1570-z
Zollinger, F. (1983). Die Vorgänge in einem Geschiebeablagerungsplatz (ihre Morphologie und die Möglichkeiten einer Steuerung), PhD Thesis, ETH Zürich. https://doi.org/10.3929/ethz-a-000318964.

Advances in Design of Barriers for Debris Flow Impact

16

Julian S. H. Kwan, Carlos Lam
and Clarence Edward Choi

Abstract

Debris flows, which often contain large boulders, may impact and destroy downstream facilities such as buildings and bridges. To protect human lives and infrastructures in mountainous regions from debris flows, engineering countermeasures such as barriers and dams are often installed along the predicted flow paths. The design of safe countermeasures relies on robust predictions of the impact force and provision of adequate retention capacity of the measures. However, impact forces are difficult to estimate because of limited field evidence associated with the prevailing impact mechanisms, diversity of flow types that may exist (e.g., stony and muddy debris flows), and the scale-dependent nature of debris flows. Consequently, the recommendations provided by design guidelines in different countries differ. In this chapter, we provide an overview of the existing design practices, including the commonly used impact models and assumed impact scenarios, and state-of-the-art research findings that enable the optimization of the design of engineering countermeasures against debris-flow impact. The practice of designing barriers against debris flows is still evolving as new research findings become available. Future research may focus on topics such as the dynamic interaction between debris flows and mitigation structures.

J. S. H. Kwan (✉)
Civil Engineering and Development Department, Government of the Hong Kong Special Administrative Region, Hong Kong SAR, China
e-mail: juliankwan@cedd.gov.hk

C. Lam
Geotechnical Engineering Office, Civil Engineering and Development Department, Government of the Hong Kong Special Administrative Region, Hong Kong SAR, China

C. E. Choi
Department of Civil Engineering, The University of Hong Kong, Hong Kong SAR, China

Notation

A_d	Equivalent static boulder impact force
A	Contact area between barrier base and ground
C	Coefficient accounting for failure type of structural members
COR	Coefficient of restitution
D_{95}	Diameter at which 95% of the boulders are smaller
EI	Flexural rigidity
E	Impact energy
e	Thickness of cushioning layer
F_B	Boulder impact force
F_d	Dynamic impact force of debris

M. Jakob et al. (eds.), *Advances in Debris-flow Science and Practice*, Geoenvironmental Disaster Reduction,
https://doi.org/10.1007/978-3-031-48691-3_16

F_{drag} Drag force of debris
F_k Characteristic boulder impact force on cushioning layer
F_s Static lateral force of debris
Fr Froude number
g Gravitational acceleration
I_θ Mass moment of inertia of barrier
h Flow depth or thickness of debris
h Height of stem wall
h_i Vertical distance between point of rotation and impact location
K_c Force-reduction factor
k Static lateral pressure coefficient
k Flexural stiffness of structural member
k_{eff} Effective flexural stiffness of structural member
l Length of base slab
L Cantilever length
$M_{E,k}$ Characteristic modulus of cushioning layer
M Mass of barrier
M_{stem} Mass of stem wall
M_{base} Mass of base slab
m Mass of boulder
m_k Characteristic mass of boulder
n Number of debris surge
p_d Dynamic impact pressure of debris
p_s Static lateral pressure of debris
R Distance between point of rotation and impact location
R_e Relative density of cellular glass
r_B Radius of boulder
t Ending time of debris surge
u Water uplift pressure
v Velocity of debris surge
v_0 Initial velocity of boulder
v_1 Rebound velocity of boulder after impact
v_2 Velocity of barrier after impact
v_p Peak velocity of debris surge
V Volume of debris surge
v_B Velocity of boulder
W_d Self-weight of debris flow
W_s Self-weight of deposited debris
w Width of debris flow
w_{stem} Thickness of stem wall
w_{base} Thickness of base slab
α Dynamic pressure coefficient
δ Interface friction angle between barrier base and ground
δ Penetration depth into cushioning layer
δ_e Required thickness of cushioning layer
Δ Translational movement of barrier
$\Delta_{C.G.}$ Rise in center of gravity of barrier
ε_D Limiting strain of cellular glass
ϕ_k Characteristic internal friction angle of cushioning layer
κ Ratio accounting for the inertial effect of barrier undergoing rotational movement
λ Mass ratio of barrier to boulder
θ Angle between the impact face of barrier and the direction of debris motion
θ Angle of rotation of barrier
θ_{max} Maximum grain diameter of cushioning material
ρ Bulk density of debris

16.1 Assessment of Impact Forces

Boulders entrained by a debris flow tend to migrate to the free surface of the flow and are transported to the front through particle-size segregation (Johnson et al., 2012). In the assessment of debris-flow impact forces against mitigation structures (e.g., barriers, dams, breakers and baffles), the force exerted by the fluidized assembly of soil particles and water (i.e., debris) is considered separately from those exerted by boulders. The debris impact force is treated as a distributed force, whereas that induced by a boulder is treated as a concentrated force. Field measurement data have demonstrated that boulders induce concentrated transient forces that are many times larger than the distributed force exerted by the debris (Fig. 16.1). The assessments of debris and boulder impact forces are discussed in the following sections.

16.1.1 Debris Impact

16.1.1.1 Hydrostatic Model

In the hydrostatic model, the dynamic debris impact pressure on a barrier is calculated as follows:

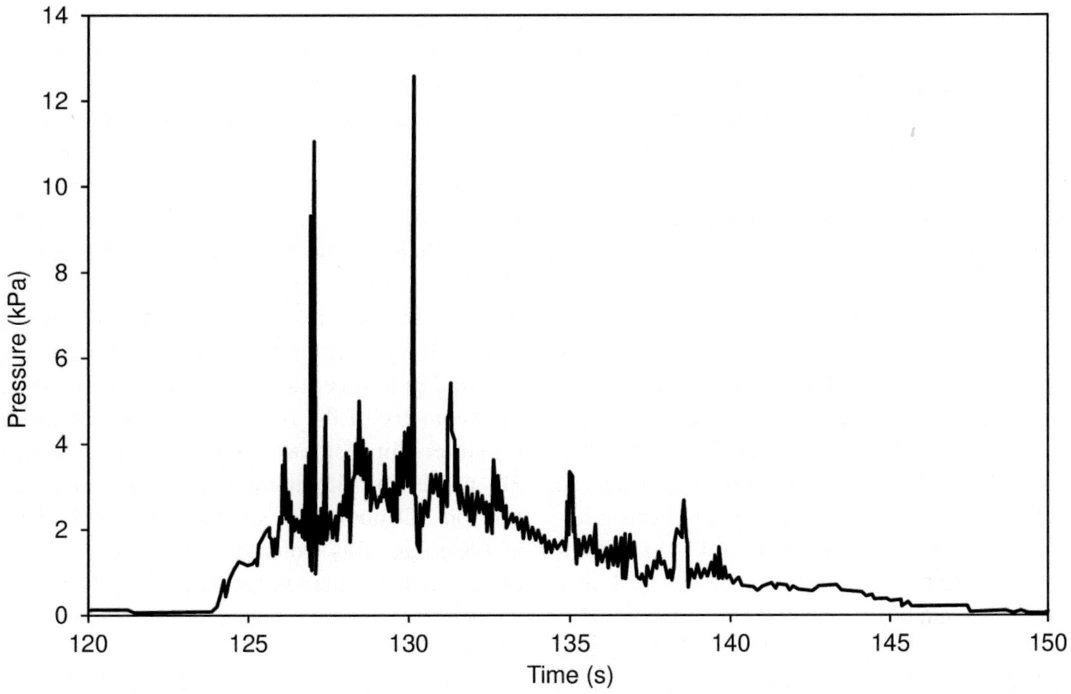

Fig. 16.1 Time histories of debris-flow impact pressure on a rigid obstacle in Jiangjia Ravine, China (modified after Hu et al., 2006)

$$p_d = k\rho g h \tag{16.1}$$

where p_d is the dynamic impact pressure (N/m^2), k is the static pressure coefficient (dimensionless), ρ is the bulk density of the debris (kg/m^3), g is the gravitational acceleration (m/s^2), and h is the flow depth of the impacting debris (m). Equation 16.1 is based on a triangular pressure distribution, in which the pressure increases with the depth. Moreover, Eq. 16.1 is commonly used for practical design because only the bulk density and impacting flow depth of the debris are required (Hübl et al., 2009; Proske et al., 2011; Vagnon & Segalini, 2016). It is often misconstrued that Eq. 16.1 is only used to predict a static pressure. However, the hydrostatic model is just another semi-empirical approach to predict the dynamic impact force by only explicitly considering the static component of a debris flow via ρgh and implicitly considering the dynamic effects via k.

Equation 16.1 represents a simple framework; however, as reported by Vagnon (2020), the hydrostatic model may only be applicable to low-velocity debris flows on shallow slopes. The model involves uncertainties owing to the empirical nature of the coefficient k that is used to reflect the dynamic effect of a debris flow during impact. Several researchers have correlated the value of k to the properties of a debris flow. For example, Lichtenhahn (1973) back-analyzed field events and reported that the debris impact pressure could be approximately 7 to 10 times the hydrostatic pressure. If the density of the debris was assumed to be two times that of water, the corresponding k would be 3.5–5. Armanini (1997) performed a theoretical analysis and reported the value of k to be 9. Furthermore, Scotton and Deganutti (1997) conducted laboratory-scale experiments and reported that the value of k ranged from 2.5 to 7.5 depending on the viscosity of the interstitial fluid and hydraulic conductivity of the debris material. Proske et al. (2011) and Scheidl et al. (2013) correlated the value of k to the debris velocity (v) by using the Froude number

($Fr = v/\sqrt{gh}$ Fr) and their results showed that the value of k typically increased linearly with Fr. Nagl et al. (2022) back-analyzed field events and reported the value of k ranging from 0.1 to 3.5, with the higher values (>2) associated with boulder impacts. Overall, the value of k depends on many factors, and thus, caution must be exercised when determining the appropriate value for design.

In addition to being used to calculate debris impact pressures, the hydrostatic model can be used to calculate the static earth pressure of the debris deposited behind a load-bearing surface (p_s). Under such a circumstance, the depth h may be considered as the deposit depth instead of the flow depth. If the debris has been intercepted only a short while ago and is in a fluid-like state, the value of k may be set as 1 in a total stress analysis, which is the approach currently used in Hong Kong (Kwan, 2012; Kwan & Cheung, 2012) and Austria (ASI, 2013). If the debris has been deposited for a considerable period and is in a consolidated state, the value of k may be determined using Rankine's or Coulomb's earth pressure theories in an effective stress analysis (Atkinson, 2017). To examine the stability of a barrier, the deposited debris acting on a load-bearing face may be assumed to be in an active stress state (CACHP 2018; SWCB, 2017). Because the deposited debris is typically assumed to be in a drained condition, the effective unit weight must be used in the calculations and the static water pressure must be considered, except in cases involving open-type barriers (e.g., slit or grid dams) that allow water to drain through them (NILIM, 2016a).

16.1.1.2 Hydrodynamic Model

As an alternative to the hydrostatic model, the hydrodynamic equation can be used to estimate the dynamic debris impact pressure:

$$p_d = \alpha \rho v^2 \quad (16.2)$$

where α is the dynamic pressure coefficient (dimensionless), and v is the velocity of the impacting debris (m/s). An additional $\sin\theta$ term is introduced in Eq. 16.2 in the design guidelines pertaining to Hong Kong (Kwan, 2012) and mainland China (MLR, 2006; CAGHP 2018a) to consider the different impact angles θ with respect to the flow direction and force-bearing surface. Equation 16.2 is favoured in practice because it only requires the bulk density and velocity of the impacting debris and an empirical coefficient. The debris reaching the barrier at different times may have different combinations of velocity (v), width (w) and depth (h) values. To convert the pressure determined using Eq. 16.2 to a force, w can be estimated based on the geometry of the runout path or by performing three-dimensional (3D) or depth-averaged 3D numerical simulations (for more information on 3D numerical simulations, see Iverson and George, this volume). Depending on the local practice, h may be predicted based on prior knowledge, empirical calculations, or numerical simulations.

Despite the simplicity of Eq. 16.2, the dynamic pressure coefficient involves some uncertainties. This coefficient has been extensively studied by researchers using experimental (Ng et al., 2021a; Song et al., 2017), analytical (Li et al., 2021), and numerical (Di Perna et al., 2022) approaches, and a range of values has been reported. Watanabe and Ike (1981) proposed a dynamic pressure coefficient of 2 for laminar fine-grained flows. Zhang (1993) reported dynamic pressure coefficients of between 3.0 and 5.0 based on the field measurements of over 70 heterogeneous debris flows with large and hard inclusions. Egli (2005) proposed a dynamic pressure coefficient of up to 4 for coarse-grained flows. Furthermore, Proske et al. (2011) back-analyzed eight field cases and reported that the dynamic pressure coefficient ranged from 0.38 to 18.76. Moreover, the dynamic pressure coefficient has been reported to be dependent on the flow type.

The dynamic pressure coefficient values differ across design guidelines for different regions. In Taiwan (SWCB, 2017), the coefficient is recommended to be from 0.5 to 2.0 for a rigid barrier. In the early versions of the design guidelines in Hong Kong (Kwan, 2012), this coefficient was specified as 2.0 and 2.5 for flexible and rigid debris-resisting barriers, respectively. These recommended dynamic pressure coefficients implicitly allow for the presence of boulders smaller than 0.5 m in diameter (Kwan et al., 2018). Flexible barriers typically correspond to a lower dynamic pressure coefficient than those of rigid barriers because of their deformable and pervious nature that allows certain debris to pass through upon impact. Wendeler (2008) conducted laboratory-scale experiments and recommended dynamic pressure coefficients ranging from 0.7 to 2.0 for the design of flexible barriers. Recently, the dynamic pressure coefficient for the design of rigid barriers in Hong Kong was decreased from 2.5 to 1.5 (GEO, 2020) and Lam et al. (2022) also recommended a value of 1.0 for the design of flexible barriers. In Austria (ASI, 2013) and Japan (NILIM, 2016b), the dynamic pressure coefficient is set as unity for designing rigid barriers. In mainland China (MLR, 2006; CAGHP 2018a), the dynamic pressure coefficient is recommended for obstacles instead of barriers. For example, values of 1, 1.33, and 1.47 are recommended for circular, rectangular, and square obstacles, respectively. Notably, for obstacles, the dynamic pressure coefficient can be considered a drag coefficient, and these coefficients may not be directly comparable to those recommended for barriers in channelized conditions. CAGHP (2018a) also allowed the designer to use different coefficient values to account for entrained boulders. For example, a coefficient value of 1.47 is suggested for debris flows entraining boulders with a maximum diameter of no more than 0.5 m, whereas a value of 8.0 is proposed if the expected boulder diameter is greater than 3.0 m. Bugnion et al. (2012) conducted a series of large-scale experiments of debris flows impacting rigid obstacles and reported dynamic pressure coefficients ranging from 0.39 to 2. In that study, the debris was allowed to overtop and also flow around the obstacles.

Evidently, the recommended dynamic pressure coefficients in design guidelines are inconsistent. Some guidelines (Kwan, 2012) suggest that the dynamic pressure coefficients depend on the type of structure being impacted (i.e., rigid or flexible), whereas other guidelines (MLR, 2006; CAGHP 2018a) suggest that the dynamic pressure coefficients depend on the shape of the obstacles being impacted. Furthermore, the measured pressure coefficients for flexible barriers may not be comparable to those for rigid barriers.

16.1.1.3 Other Debris Impact Models

Several researchers have proposed empirical impact models by fitting experimental data (Hübl & Holzinger, 2003; Hübl et al., 2009), in which experimental and field measurements are scaled using the Froude number ($Fr = v/\sqrt{gh}$) and the impact force is normalized by either the theoretical static force (i.e., ρgh) or theoretical dynamic force (i.e., ρv^2) to allow the data to be comparable across different scales. Hübl and Holzinger (2003) and Proske et al. (2011) proposed the following model:

$$p_{\mathrm{d}} = 5\rho v^{0.8}(gh)^{0.6} \tag{16.3}$$

The limitation of models obtained using regression analyses is the assumption that complex field debris flows can be characterized using Fr (Ng et al., 2019a). This assumption may be an idealization, noting that debris flows are highly heterogeneous and vary spatially and temporally. Specifically, in most reported data, it is unclear how Fr was calculated or defined. For example, the velocity and flow depth can be the frontal, maximum, or depth-averaged values. Furthermore, in the fitted models, the impact force is difficult to define because it varies with time and may exhibit strong fluctuations in the force–time history, especially if the debris involves hard inclusions such as large rocks and wood, and the load measurement system requires special considerations (Goodwin & Choi, 2021). These considerations include estimating potential inertial effects between the impactor and load-measuring system. Additionally, it is unclear whether the impact force should be normalized by the theoretical static or dynamic load. The normalization of the measured force by the theoretical dynamic force

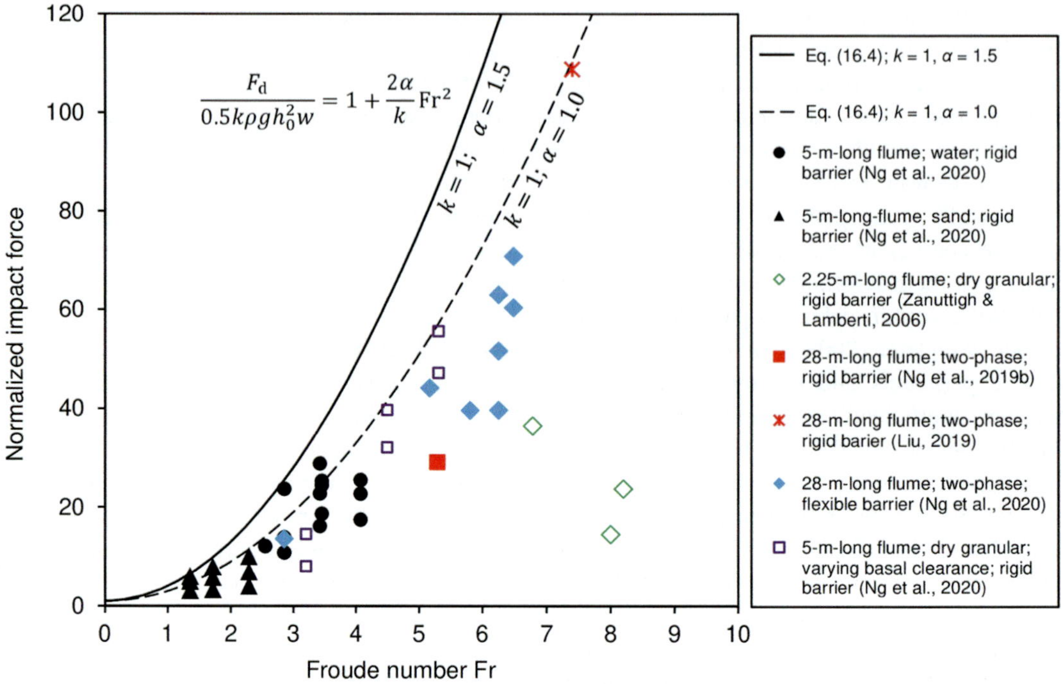

Fig. 16.2 Comparison of impact forces for dry granular flow, water, and debris flows at different scales using Eq. 16.4 (modified after Ng et al., 2020)

leads to a dynamic pressure coefficient that tends toward infinity when Fr approaches zero, which lacks physical meaning. Certain models simultaneously consider the static and dynamic forces (Arattano & Franzi, 2003). Ng et al. (2020) proposed an impact model based on the normalization of the dynamic impact force (F_d) by the theoretical static force ($F_s = 0.5k\rho g h^2 w$) while incorporating dynamic and static coefficients:

$$\frac{F_d}{0.5k\rho g h^2 w} = 1 + \frac{2\alpha}{k}\mathrm{Fr}^2 \quad (16.4)$$

Equation 16.4 enables the comparison of design recommendations by combining static and dynamic forces. Based on Eq. 16.4, Ng et al. (2020) and Ng et al. (2021b) compared the experimental data of various scales and different debris mixtures, and recommended a dynamic pressure coefficient of 1.5 and 1.0 for rigid and flexible barriers, respectively. Figure 16.2 summarizes the impact coefficients considering the small- and large-scale experimental data of water, dry sand, and debris flows. Evidently, $k = 1$ and $\alpha = 1$ are the upper-bound values for the design of flexible barriers, and $k = 1$ and $\alpha = 1.5$ are the corresponding values for rigid barriers. Considering these results, Ng et al. (2020) recommended the use of $\alpha = 1.5$ and$=\alpha 1.0$ for the design of rigid and flexible barriers, respectively.

16.1.2 Boulder Impact

16.1.2.1 Hertz Model

Elastic Hertzian mechanics (Hertz, 1882) is used to estimate the impact force of a boulder on a load-bearing surface in mountainous regions globally (ASI, 2013; NILIM, 2016b; SWCB, 2017; VanDine, 1996). Kwan (2012) reported a simplified Hertz equation based on typical values of Poisson's ratios and the elastic moduli for a spherical granite boulder impacting a reinforced concrete surface:

$$F_B = K_c 4000 v_B^{1.2} r_B \quad (16.5)$$

where F_B (kN) is the boulder impact force, K_c is a force-reduction factor, v_B (m/s) is the boulder impact velocity vector component normal to the barrier, and r_B (m) is the boulder radius. Notably, Eq. 16.5 does not consider any plastic deformation during impact, which may lead to overpredicted boulder impact forces. For this reason, Kwan (2012) applied a simplified Hertz equation first reported by Zhang et al. (1996) to compare the boulder impact force estimated using Eq. 16.5 with field data and reported that the expression overpredicts the boulder impact force by approximately two orders of magnitude. Hungr et al. (1984) recommended a force-reduction factor (K_c) of 0.1 for the Hertz equation to account for the plastic deformation that occurs during impact to enable more realistic predictions of the boulder impact force. Kwan (2012) reviewed the value of K_c based on case histories presented by Lo (2000). In the design guidelines of Hong Kong, the boulder is assumed to travel at the same velocity as that of the debris flow. This assumption is conservative as Prochaska et al. (2008) reported that the largest boulders travel slower than the debris in their analysis of the relationship between debris and boulder velocity using video analysis of field flows. To achieve a cost-effective design of a mitigation structure, the design velocity of the debris can be estimated using numerical simulations considering the actual site settings. Moreover, the frontal velocity should be used because boulders are often transported to the flow front via particle-size segregation.

VanDine (1996) provided a design chart for estimating the boulder impact forces based on the size of the boulder (i.e., weight or diameter) and normal component of the impact velocity. The Japanese design guidelines (NILIM, 2016a) present a modified Hertz equation with a force-reduction factor calculated based on the boulder impact velocity and ratio of the boulder mass to the barrier mass. Moreover, the elastic modulus of concrete is recommended to be decreased by 90% to mimic the effects of local plastic deformation on the concrete surface (NILIM, 2016a). According to the Taiwanese guidelines (SWCB, 2017), the force-reduction factor in the Hertz equation ranges from 0.2 to 0.5. Notably, although the upper bound of the range is larger than that considered in other countries, the Taiwanese guidelines involve stability calculations in which the impact forces of the boulders and debris flows are considered separately, i.e., the forces are not additive.

16.1.2.2 Cushioning Materials

The boulder impact force determined using the Hertz equation is generally applicable for reinforced concrete surfaces. To protect these surfaces from structural damage, cushioning materials are commonly used to shield protection structures (e.g. rockfall protection galleries and embankments) in mountainous regions globally (ASTRA 2008; Liu and Jan 2013; Heymann et al., 2011; Lambert et al., 2014; Ng et al., 2016) (Fig. 16.3). The installation of cushioning layers in front of rigid barriers is typically prescriptive; in other words, the added benefits of the cushioning layer in attenuating the impact force are not considered when assessing the geotechnical and structural stabilities of a barrier.

Rock-filled gabions are used as the cushioning material in mountainous regions worldwide (Heymann et al., 2011; Lambert et al., 2014; Ng et al., 2016). These structures consist of rock fragments placed inside steel wire baskets, which are stacked on top of each other. The impact force exerted on rock-filled gabions is difficult to predict because it depends on the particle shape (Su & Choi, 2021) and size (Su et al., 2019), extent of particle crushing (Su et al., 2021), and bulk density of the rock fragments in a gabion cell. Despite substantial research, design guidelines for rock-filled gabion cushioning layers are limited. Detailed design recommendations for cushioning layers for protection galleries are provided in the Swiss design guidelines (ASTRA 2008), where a static equivalent force A_d is used to consider the characteristic dynamic boulder impact force on a cushioning layer F_k:

Fig. 16.3 Rock-filled gabions used as a cushioning layer to protect the surface of a rigid barrier in Hong Kong SAR, China

$$A_{\mathrm{d}} = CF_k \tag{16.6a}$$

$$F_{\mathrm{k}} = 2.8e^{-0.5}r^{0.7}M_{\mathrm{E,k}}^{0.4}\tan\phi_{\mathrm{k}}\left(\frac{m_{\mathrm{k}}v_{\mathrm{k}}^2}{2}\right)^{0.6} \tag{16.6b}$$

where A_{d} is the static equivalent force for design (kN), C is the coefficient to account for ductile (0.4) or brittle (1.2) failure of the structural member, e is the thickness of the cushioning layer, $M_{\mathrm{E,k}}$ is the characteristic modulus of the cushioning layer (kN/m^2), ϕ_{k} is the characteristic internal friction angle of the cushioning layer (°), m_{k} is the characteristic boulder mass (t), and v_{k} is the characteristic boulder impact velocity (m/s). Notably, Eq. 16.6 was developed based on falling weight experiments (Montani-Stoffel, 1998), and the effects of local damage, especially punching, are not considered in the formulation. Furthermore, Eq. 16.6 is limited to a penetration depth in the cushioning layer that is smaller than half its thickness.

Substantial research on cushioning layers has been performed using the existing design equations, such as the simplified Hertz equation (Eq. 16.5). For example, when using Eq. 16.5 to estimate the boulder impact force on a 1-m-thick rock-filled gabion cushioning layer, a force-reduction factor of 0.05 can be used for up to six successive impacts on the same point (Ng et al., 2016). After six successive impacts, some rock fragments spilled out of the gabion wire baskets. However, the gabion cells remained largely intact. Table 16.1 summarizes the recommendations for various types of cushioning materials tested using the same impact setup and conditions (i.e., the same boulder diameter and energy levels).

Table 16.1 Design recommendations for different types of cushioning materials

Cushioning material (reference)	Energy level (kJ)	Number of successive impacts	Design thickness, δ_e (m)	Force-reduction factor in Eq. 16.5, K_c	Remarks
Rock-filled gabion (Ng et al., 2016)	20 40 70	6 6 6	$\delta_e = \min\{0.5, \delta + 3\theta_{max}, 2\delta\}$ δ: penetration depth (in m) θ_{max}: maximum grain diameter of the cushioning material (in m)	0.014–0.05 (for 20 kJ) 0.012–0.04 (for 40 kJ) 0.016–0.038 (for 70 kJ) no more than **0.05** for repeated impacts	The peak boulder impact force determined using the Hertz equation is at least four times the measured values The recommended value of K_c (0.1) used in practice can be decreased by a factor of two
Cellular glass (Ng et al., 2018)	20 70	5 2	$\delta_e = e \times \varepsilon_D = e \times (1 - 1.4R_e)$ e = cushion layer thickness (1 m) ε_D = limiting strain R_e = relative density (0.06) Therefore, $\delta_e = 0.92$ m	0.01–0.037 (for 20 kJ) 0.012–0.022 (for 70 kJ) no more than **0.037** for repeated impacts	The penetration depths are larger than those into rock-filled gabions owing to crushing
Foam concrete (Su et al., 2020)	20 70	6	The cushioning thickness of foam concrete does not considerably influence the maximum penetration depth at impact energy of 20 kJ $\delta_e = 0.58$ m (for impact energy of 70 kJ)	0.09–0.19 (20 kJ, 0.4 m foam) 0.06–0.15 (20 kJ, 0.6 m foam) 0.07–0.16 (70 kJ, 0.4 m foam) 0.08–0.15 (70 kJ, 0.6 m foam) no more than **0.19** for repeated impacts	Consistent cushioning performances can be achieved as long as the normalized penetration depth is less than the theoretical limiting strain of concrete foam
Ethylene–vinyl acetate (EVA) foam (Lam et al., 2018a)	20 70	6	$\delta_e = 0.75$ m (for impact energy of 70 kJ)	-	EVA is susceptible to degradation under ultraviolet radiation; little to no plastic deformation after each impact

(continued)

Table 16.1 (continued)

Cushioning material (reference)	Energy level (kJ)	Number of successive impacts	Design thickness, δ_e (m)	Force-reduction factor in Eq. 16.5, K_c	Remarks
Recycled crushed glass (RCG) (Su et al., 2019)	20 70	6	A thinner cushioning layer is required than in rock-filled gabions (Ng et al., 2016)	0.03	The load-diffusion angle of RCG is three times larger and the transmitted load is distributed more uniformly on the barrier wall than in the case of cellular glass aggregates More effective in decreasing the boulder impacts at 70 kJ than rock fragments and cellular glass aggregates

16.1.3 Debris-Flow Impact Scenarios

Figure 16.4 shows the side view of a typical loading scenario on a rigid barrier, along with the dynamic impact pressure from the debris (p_d) and static pressure from the retained debris soon after filling (p_s). The overburden pressure associated with the debris overflowing the barrier or the boulder impact force is not shown. Depending on the design guideline and whether a barrier is open (i.e., slits, slots, grids, or apertures) or closed, various combinations of loading have been reported for assessing the geotechnical and structural limit states.

The loading scenario considered in the Austrian design guidelines (ASI, 2013) is similar to that in the Hong Kong guidelines shown in Fig. 16.4. The largest forces are assumed to act orthogonally on the load-bearing face of a barrier during initial impact. Any deposits present behind the barrier provide a cushioning effect for subsequent debris impact on the barrier. For the debris that impacts the barrier and is not consolidated, a total stress approach (in which the lateral pressure coefficient is unity) is used to calculate the static pressure. For sediments or debris that has been retained prior to impact and has consolidated, the static earth pressure is calculated using an effective stress approach. In the case of overflow, its surcharge must be considered. Furthermore, the impact forces owing to boulders and trees are only considered for the structural limit states. Typical boulder impact forces are provided (e.g., 1000 kN for a 0.7 m × 0.7 m block). The Hong Kong guidelines typically assume that debris-flow impact occurs as pulses (GEO, 2015, 2016, 2020; Kwan, 2012). The debris

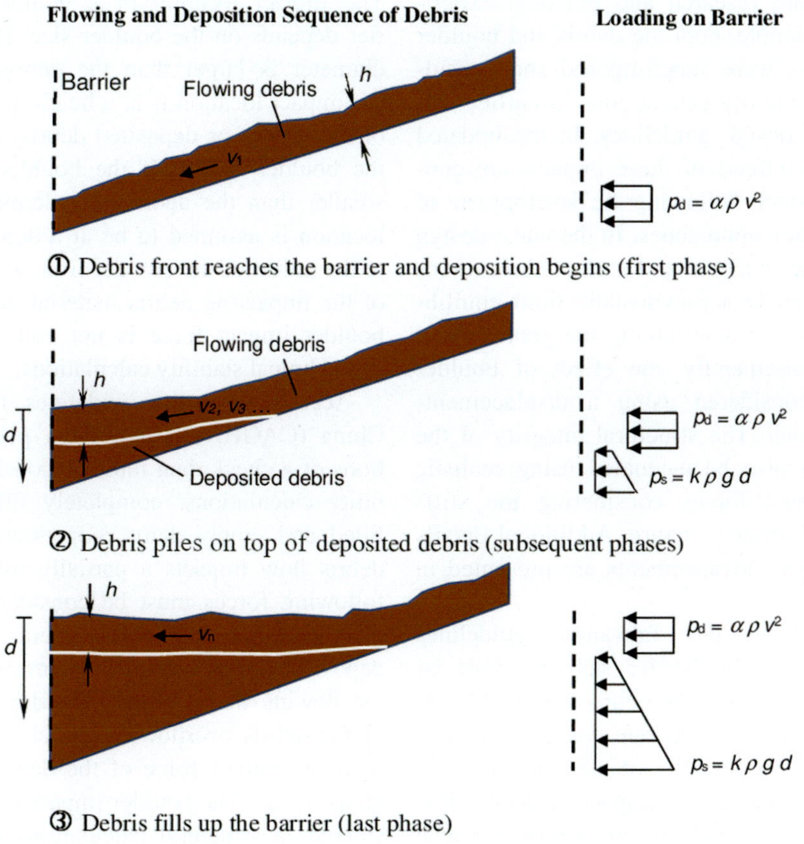

Fig. 16.4 Side view of a typical loading scenario for a rigid barrier

impact pressure is assumed to be orthogonal to the load-bearing surface. Experience shows that the maximum impact forces are induced from the initial impact as the velocity is typically the highest near the debris front. After an impact event, the debris is assumed to be at rest but not yet consolidated. In this scenario, the total stress approach (i.e., $k = 1$) is used to calculate the static debris pressure (p_s). Unlike situations in which the barriers may be allowed to remain filled with debris or sediment, barriers in Hong Kong are regularly maintained, and thus, the effective stress approach is not used for calculating p_s. Moreover, unlike several European guidelines (e.g., ASI, 2013) that consider the static pressure component to extend to the top of the flowing (impacting) debris, in the Hong Kong practice the static pressure diagram is assumed to be trapezoidal (Fig. 16.4). The design guidelines in Hong Kong have continued to evolve with research and practical experience. For example, both the debris and boulder impact forces were superimposed and considered in the stability calculations in earlier versions of the design guidelines. In the updated versions, the effects of these impacts are considered separately following the development of new calculation approaches. In the latest design approach, the dynamic and static debris forces are considered in a pseudostatic limit equilibrium analysis for determining the geotechnical stability. Consequently, the effect of boulder impact is considered using a displacement-based approach. The structural integrity of the barrier must also be evaluated using realistic boulder impact forces considering the stiffness of the barrier structure. Additional details regarding these advancements are presented in Sect. 16.2.

According to the Taiwanese guidelines (SWCB, 2017), the barrier stability must be examined considering the debris-flow and boulder impact forces in separate design scenarios. For example, the stability of a barrier must be examined considering (*i*) impact of debris flow without boulders, and (*ii*) impact of a single boulder at the top of the barrier. In particular, the impact forces of both the debris and boulders are considered to examine the barrier stability, whereas only the boulder impact is considered to evaluate the structural capacity of the barrier. The guidelines state that the boulder size cannot be known in advance, and thus, the boulder impact force is calculated assuming the diameter to be the D_{95} value for all boulders larger than 0.2 m in a representative survey area that is at least 10 to 20 times the width of the channel. Moreover, the drag force from overflowing debris is not considered (SWCB, 2017).

According to the Japanese guidelines (NILIM, 2016b), the design force must be determined by superposing the boulder and debris impact forces to design the upper portion of a debris barrier. For open structures (e.g., slit or grid dams), the hydrostatic pressure from impounded water is not considered, although this pressure must be considered for closed structures (i.e., rigid barriers and check dams). The impact location of a boulder on a barrier depends on the boulder size. If the boulder diameter is larger than the debris-flow depth, the impact location is at a height (from the base of the barrier or deposited debris) equivalent to the boulder radius. If the boulder diameter is smaller than the debris-flow depth, the impact location is assumed to be at a depth equivalent to the radius of the boulder from the free surface of the impacting debris material. Moreover, the boulder impact force is not considered in the geotechnical stability calculations.

According to the guidelines for mainland China (CAGHP 2018a), three possible conditions of a check dam must be considered in stability calculations: completely filled, partially filled and empty dams. For example, when a debris flow impacts a partially filled dam, the following forces must be considered: the self-weight of the dam, static lateral force of the deposited debris behind the dam, self-weight of the flowing debris behind the dam, self-weight of the debris overflowing the crest of the dam, dynamic impact force of the debris, and earthquake loads. The boulder impact forces are considered in structural calculations only when a debris flow impacts bridge piers or the structural elements of open check dams.

16.2 Design Optimization of Rigid Barriers

16.2.1 Background

Owing to the heterogeneous nature of debris flows, barrier designers typically incorporate assumptions regarding the flow properties and debris–barrier interaction mechanisms. The required parameters, such as the debris volume, physical properties of the debris, and magnitude of the dynamic impact force, are often determined semi-empirically. Although these empirical approaches represent a practical solution, the dynamic characteristics of debris flows are often ignored, so that their interaction with the barriers may be oversimplified, or conservative values may be obtained. In view of the intrinsic uncertainties associated with debris flows and the design simplifications, the design may be conservative and the resulting barriers may have large dimensions and require heavy steel reinforcement, leading to construction difficulties in remote areas or on steep hillsides with limited access (Wong, 2021). For example, Fig. 16.5 shows a reinforced concrete rigid barrier being constructed on a natural hillside in Hong Kong for intercepting debris flows. Although this barrier was designed using the best methods available at the time, its construction still required extensive site formation works, introduction of heavy steel reinforcement, and use of long ground anchors as tie-backs. There is room for enhancing the guidelines to optimize the barrier design. In the past two decades, considerable research has clarified the characteristics of debris flows and their interaction with barriers. Significant effort has been expended to rationalize the design of rigid barriers for arresting

Fig. 16.5 Reinforced concrete barrier being constructed on a natural hillside in Hong Kong

debris flows. The domains in which significant advancements have been made are discussed in the following sections.

16.2.2 Attenuation of Debris-Flow Velocity

Debris-flow velocity is a key parameter for calculating the debris impact force using the hydrodynamic model or other impact models that require the velocity as an input. The flow velocity is commonly estimated using semi-empirical equations such as Manning's formula for open channel flows (Mizuyama and Ishikawa 1998; NILIM, 2016b; SWCB, 2017; CAGHP 2018b). The use of these flow equations requires the flow depth, channel bed gradient, and empirical coefficients representing the flow characteristics. When these equations are used to design barriers, the flow velocity is typically assumed to remain constant during the impact process. If necessary, a designer may vary the input parameters (e.g., Manning's roughness coefficient) to model different velocities between different flows (Mizuyama and Ishikawa 1998). Over the past decade, numerical tools are increasingly used to predict the mobility of a debris flow in the design of barriers (Kwan et al., 2015, 2021). These numerical tools can simulate the dynamic motion of the debris to provide useful information regarding the debris velocity, thickness, and volume at any point along the predicted debris flow path. The simulation outputs are typically presented in hydrographs to show the temporal variations in the debris properties.

According to VanDine (1996), the debris-flow velocity may decrease before deposition owing to a decrease in the channel gradient, loss of channel confinement of the debris, or changes in the bed friction, among other factors. Field observations of landslides and physical flume test results suggest that the velocity of the debris front may be higher than those of the surges behind it in free-field conditions (Koo et al., 2017). In these scenarios, the debris pressure from subsequent impacts is expected to decrease. Wendeler et al. (2018) discussed the variation in the dynamic impact force owing to changes in the debris velocity and thickness.

Given the improved understanding of debris-flow behaviour, the Geotechnical Engineering Office (GEO) in Hong Kong developed a rationalized design approach to calculate the debris impact velocity and thickness for the design of barriers (GEO, 2015). The approach is based on the multi-surge impact scenario, in which the landslide debris is regarded as an elongated mass during its travel until it impacts the barrier in surges. Debris that arrives from subsequent surges overrides the debris deposited behind the barrier and impacts the barrier at a higher elevation in a discretized impact-and-filling process (Kwan, 2012; Wendeler, 2008). In general, the debris reaching the barrier at different time points may have different combinations of velocity and thickness values. However, in the original design recommendations proposed by Kwan (2012), the change in the debris velocity was ignored, and the impact load calculated for the first surge was applied to all subsequent surges. Later, to optimize barrier designs, GEO (2015) proposed a calculation procedure to consider the temporal changes in the debris velocity.

The calculation procedure proposed by GEO (2015) is briefly described as follows. The design debris impact velocity of each surge is calculated using (*i*) the volume of debris deposition after each phase of impact, (*ii*) the cumulative debris volume hydrograph, and (*iii*) debris velocity hydrograph as illustrated in Fig. 16.6. The two hydrographs show the conditions of the debris at the barrier location, established from a debris mobility analysis considering the free-field situation (i.e., without the barrier). The cumulative volume of debris deposited behind the barrier after each surge ($V_1, V_2 \cdots V_n$) is first calculated using the design debris thickness ($h_1, h_2 \cdots h_n$). For this purpose, the debris deposition wedge is assumed to be triangular in shape for the purpose of calculating the debris volume of each surge stopped by the barrier, and a suitable deposition angle is assumed (Fig. 16.6a). With the cumulative volume of the deposited debris calculated after each surge, the time interval of each surge ($t_1, t_2 \cdots t_n$) can be read

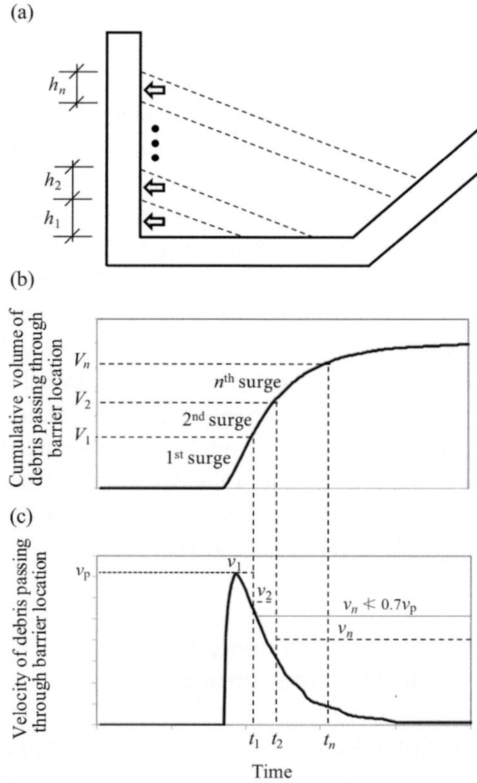

Fig. 16.6 A calculation approach for establishing design debris velocities for barrier design: **a** idealized shape of debris deposited behind a barrier; **b** cumulative debris volume hydrograph; **c** velocity hydrograph and design velocities (GEO, 2015; Koo et al. 2016)

off from the cumulative debris volume hydrograph (Fig. 16.6b), which can be obtained from a debris mobility analysis. Finally, the design debris velocity of each surge ($v_1, v_2 \cdots v_n$) is then determined from the velocity hydrograph (Fig. 16.6c), which can also be obtained from a debris mobility analysis.

To ensure the robustness of the barrier, the highest debris velocity in each surge is used to define the corresponding design value, and a reduction limit (a lower bound) is imposed to ensure that the design velocity of any surge is not less than 70% of the peak velocity, i.e., $v_n \nless 0.7v_p$. Moreover, most of the existing debris mobility models do not consider the magnification of the flow depth at the debris front owing to the accumulation of boulders (Hungr, 2000). To address this issue, the design debris thickness ($h = h_1, h_2 \cdots h_n$) is set as the largest value of (*i*) the estimated thickness of the debris front, (*ii*) the maximum thickness of the landslide debris that would pass through the barrier location in the free-field condition, or (*iii*) a nominal thickness of 0.5 m. The abundance of boulders should be considered when determining the debris thickness. Specifically, the design impact thickness in the first surge or the first several surges may be controlled by the boulder-enriched debris front, although this effect may be absent in the following surges.

To illustrate the effect of the improved approach on the design of rigid barriers, Kwan and Koo (2018) compared this approach with that developed by Kwan (2012) using several notable landslide cases in Hong Kong. They noted that the new approach can decrease the design impact forces and resulting destabilizing moments by 15%–20% and 25%–30%, respectively. Thus, a more cost-effective, yet robust design can be achieved using the rationalized approach.

Notably, in this design approach, only the debris velocity hydrograph simulating the free-field condition is used to establish the design velocity of each surge. The design can be further optimized by considering the debris–barrier interaction. In the past decade, many researchers have investigated the interaction of debris with various types of structural mitigation measures, and different load models have been proposed. Research on this aspect is still evolving, and thus, definitive guidance is not yet available. Nevertheless, Koo et al. (2017) and Faug (2021) have developed analytical models that consider the change in the flow velocity owing to the conversion between kinetic and potential energies when the debris runs up against a barrier wall.

16.2.3 Self-Weight of Debris

The self-weight of debris (W_d) can increase or reduce barrier stability. When the debris arrives at a barrier, it may induce several forces on the structure, including: the horizontal impact force F_d, a horizontal drag force F_{drag} if the debris is overtopping the barrier, and the self-weight of the debris surge. The horizontal impact force and drag force always reduce the stability of the barrier, whereas the self-weight of the debris surge may improve the stability depending on the condition and geometry of the barrier. If the retention basin is partially or completely filled such that sediment or debris is deposited on the upstream side, the self-weight of the newly arriving debris surge will induce a surcharge on the deposited materials with self-weight W_s. This surcharge can increase the static lateral force (F_s) acting on the barrier wall, which reduces the stability of the barrier. However, if the barrier is designed such that the self-weight of a debris surge is transferred onto the barrier, the sliding resistance of the barrier will also increase owing to the additional weight, thereby enhancing the stability. Barriers with an inclined load-bearing face or a base slab that is structurally connected to the barrier wall exhibit a higher geotechnical stability than those without such surfaces or slabs. Moreover, debris overflowing the barrier's crest can increase the barrier weight and thus its sliding resistance, though at the same time the debris will also induce a drag force on the crest. Figure 16.7 illustrates the scenario in which a debris surge overtops a partially filled barrier with an inclined load-bearing face.

The design guidelines for self-weight presented by different authorities differ in several aspects. For example, ASI (2013), NILIM (2016a) and CAGHP (2018a) consider the self-weight of the debris flow as a surcharge acting on both the deposited materials and barrier body; however, SWCB (2017) does not explicitly consider the potential beneficial effects. In the original Hong Kong design guidelines, Kwan (2012) recommended that the beneficial effect of the weight of the debris surge should not be considered when examining the geotechnical stability. However, this recommendation was later modified to account for the beneficial effect of the self-weight of the debris in enhancing the geotechnical stability of the barrier (GEO, 2016; Kwan et al., 2018).

16.2.4 Superimposing Debris and Boulder Impact Loads

Debris flows may have a boulder-rich front owing to particle-size segregation. Therefore, boulder impact forces must be considered in the design of mitigation structures. According to guidelines such as ASI (2013), NILIM (2016a), and CAGHP (2018a), the boulder impact forces should be considered only in structural assessments because these forces, unlike the fluid

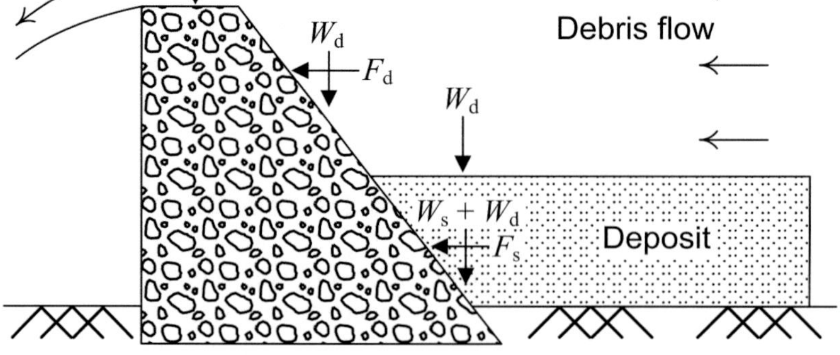

Fig. 16.7 A partially filled barrier being overtopped by a debris flow surge showing the effect of the self-weight of debris (modified after CAGHP 2018a)

thrust, are highly transient point loads that are more likely to locally damage a structure than to cause global failure. In contrast, SWCB (2017) suggests that barrier stability should be evaluated considering both the debris and boulder impact forces in separate design scenarios. In the design example presented in SWCB (2017), barrier stability is evaluated when the barrier is impacted by either flowing debris without any boulders or a single boulder without any debris. In other words, the two impact forces are not superimposed. In Hong Kong, recommendations for the treatment of boulder impact forces have evolved with new research findings. In the original design recommendations presented by Kwan (2012), the debris and boulder impact forces were superimposed in stability calculations. However, a review of the published data, such as those reported by Cui et al. (2015), indicated that only boulders larger than 0.5 m need to be considered separately, as smaller boulders can be considered as part of the fluidized debris (GEO, 2016; Kwan et al., 2018). Considering the highly transient nature of boulder impact forces, researchers have examined the effect of these dynamic loads and noted that boulder impact loads potentially do not need to be considered in geotechnical stability assessments considering force and moment equilibrium, provided that the ratio of the barrier mass to boulder mass is sufficiently high. However, boulder impact forces should still be considered in structural integrity assessments. This brief review of design practices worldwide indicates general agreement among published guidelines that the debris and boulder impact loads do not need to be superimposed in stability evaluation, although they may need to be considered separately.

16.2.5 Displacement-Based Approach for Geotechnical Stability Assessment

The boulders entrained in a debris flow can induce dynamic impact forces that are considerably larger than those exerted by the fluidized debris (Fig. 16.1). Boulder impact forces are short-lived, typically lasting for a fraction of a second. Therefore, the use of a pseudostatic approach to represent these forces in stability assessments may yield overly conservative designs as the transient nature of the forces is neglected, and the degree of conservatism would only manifest itself if the barriers are impacted by many boulders at the same time. In contrast, if the boulder impact forces are ignored in stability evaluations, the resulting designs may be unsafe as these loads, although transient, typically have high magnitudes. In this context, determination of the barrier's movement after an impact provides a more rational criterion than factors of safety calculated based on force and moment equilibrium analyses. Therefore, a novel displacement-based approach has been developed for assessing the stability of barriers in Hong Kong (GEO, 2020). The analytical solutions behind this approach have been derived based on the conservation of momentum and energy to consider the effect of energy loss and contribution of the inertial resistance of the barrier in resisting an impact. Details regarding the mathematical formulation, experimental verification and limitations have been presented by Lam et al. (2018b) and Yong et al. (2019), and thus, only the key features are described here.

Two modes of barrier movement are considered in this approach: translational (sliding) and rotational modes. If a barrier is assumed to undergo sliding displacement as a rigid body after a boulder impact (Fig. 16.8a), the amount of sliding displacement can be estimated as

$$\Delta = \frac{mv_0^2}{2(Mg - uA)\tan\delta} \times \lambda \left(\frac{1 + \mathrm{COR}}{1 + \lambda} \right)^2 \tag{16.7}$$

where Δ is the translational movement of the barrier (m), m is the boulder mass (kg), M is the barrier mass ($=\lambda m$) (kg), v_0 is the impact velocity of a boulder (m/s), λ is the mass ratio of the barrier to boulder ($M/m \geq 1$), g is the gravitational acceleration (m/s^2), u is the water uplift pressure acting on the barrier base (N/m^2), A is the contact area between the barrier base and ground (m^2), δ

is the interface friction angle between the barrier base and ground (°), and COR is the coefficient of restitution (dimensionless).

The COR is the only parameter in Eq. 16.7 that cannot be easily obtained in a preliminary design analysis. This parameter is used to consider the effect of energy loss and is defined as the ratio of relative speeds before and immediately after an impact, i.e., $(v_1 + v_2)/v_0$, where v_0 is the initial impact velocity of the boulder, v_1 is the rebound velocity of the boulder, and v_2 is the velocity of the barrier immediately after the impact. The COR typically ranges between 0.1 and 0.5 for impacts between rock and concrete. Equation 16.7 assumes that the energy gained by a barrier is dissipated solely through sliding, and no energy is lost because of other types of response, i.e., localized damage, bending, and overturning. In reality, a barrier is expected to experience all four types of response, and therefore, a smaller sliding displacement can be expected. Hence, Eq. 16.7 can be used to provide a conservative estimate of the sliding displacement of a barrier after a boulder impact. Additionally, when calculating the mass ratio λ, the mass of the whole barrier must be considered because the entire structure is assumed to be a rigid body.

If the barrier is impacted by a boulder near the wall crest and assumed to undergo pure rotational movement after the impact (Fig. 16.8b), the rise in the barrier's center of gravity can be estimated using Eq. 16.8:

$$\Delta_{\text{C.G.}} = \frac{mv_0^2}{2Mg} \times \frac{\kappa h_{\text{i}}}{R}\left(\frac{1+\text{COR}}{1+\kappa}\right)^2 \tag{16.8}$$

where $\Delta_{\text{C.G.}}$ is the rise in the barrier's center of gravity (m), κ equals $\frac{I_\theta}{mh_{\text{i}}R}$ (dimensionless, ≥ 1), h_{i} is the vertical distance between the point of rotation and impact location (m), and R is the distance between the point of rotation and impact location (m) (not shown for clarity). For L-shaped barriers similar to that shown in Fig. 16.8b, $R = \sqrt{h_{\text{i}}^2 + w_{\text{stem}}^2}$.

The barrier's mass moment of inertia, I_θ, must be determined to calculate the dimensionless number κ that considers the effect of a barrier's inertial resistance. For L-shaped barriers with cross-sections similar to that shown in Fig. 16.8b, I_θ can be calculated using the following approximate relationship:

$$I_\theta \approx \frac{M_{\text{stem}}}{3}\left(h^2 + hw_{\text{base}}\right) + \frac{M_{\text{base}}}{3}l^2 \tag{16.9}$$

where I_θ is the mass moment of inertia (kg·m^2), M_{stem} is the mass of the stem wall (kg), M_{base} is the mass of the base slab (kg), h is the height of the barrier wall (m), w_{base} is the thickness of the base slab (m), and l is the length of base slab (m). Equation 16.9 can be extended to include the beneficial effect of any side walls. After determining $\Delta_{\text{C.G.}}$ using Eqs. 16.8 and 16.9, the corresponding angle of rotation (θ) can be

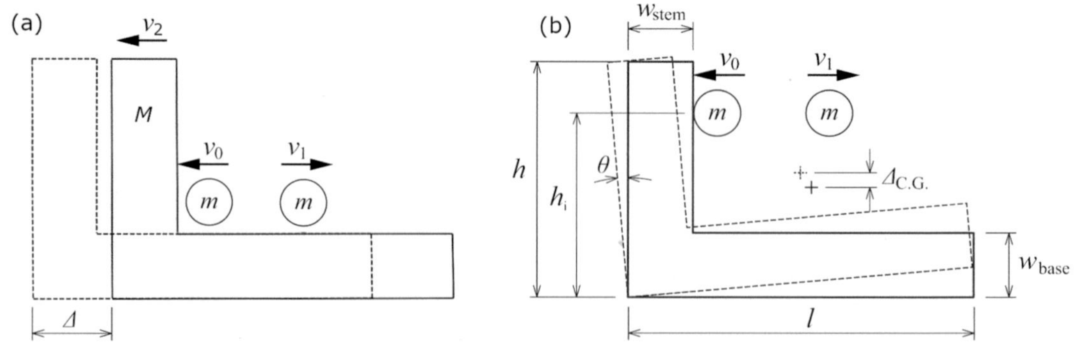

Fig. 16.8 Types of movements of an L-shaped barrier after a boulder impact: (a) sliding, (b) rotation

Table 16.2 Cases in which the barrier movement does not need to be evaluated in Hong Kong (GEO 2020)

Boulder impact velocity (v, m/s)	$v \leq 8$	$8 < v \leq 10$	$10 < v \leq 12$
Mass ratio of barrier to boulder (M/m)	$M/m > 20$	$M/m > 25$	$M/m > 30$

calculated using the geometrical relationships specified by Lam et al., (2018a, 2018b). Because Eq. 16.9 assumes that the kinetic energy gained by a barrier after an impact is dissipated solely through rotational movements and other modes of energy loss are ignored, the calculated θ is expected to err on the conservative side.

A parametric study was performed using Eqs. 16.7 to 16.9 to estimate the sliding and rotational displacements of rigid barriers in different design and impact conditions. The results show that the barrier movement will be insignificant if its mass (M) is sufficiently larger than that of the boulder (m). Therefore, for routine designs in Hong Kong, barrier movement does not need to be evaluated in cases satisfying the conditions listed in Table 16.2 (GEO, 2020). If these conditions are not met, the acceptability of the barrier movement must be evaluated. It may be noted that, for the purpose of developing this design guideline, a sliding displacement of < 20 mm and an angle of rotation of < 0.5° were chosen as the maximum acceptable movements in the parametric study. Other values may be used for site-specific designs. Generally speaking, the amount of tolerable movement will depend on a range of factors, such as the site conditions (e.g., ground profile and the distance between the barrier and the facilities immediately downstream of the barrier), the structural integrity of the barrier, maintenance considerations and the local design tradition including risk perception. So far, the displacement-based approach to dealing with boulder impacts on rigid barriers has only been adopted in routine designs in Hong Kong.

16.2.6 Enhanced Flexural Stiffness Method for Structural Assessment

The Hertz equation and its simplified form are commonly cited in the literature as a potential method for estimating the boulder impact forces acting on structures (Hungr et al., 1984; Kwan, 2012; NILIM, 2016a; SWCB, 2017) (Sect. 16.1.2). However, for structural elements such as beams and piers, the deflection more notably influences the magnitude of the impact force than the contact deformation which is the basis of the Hertz equation. The impact force acting on such structural elements can be obtained by equating the kinetic energy of the boulder with the work expended by bending (Hungr et al., 1984; Zhang et al., 1996):

$$F_B = v_0\sqrt{mk} \quad (16.10)$$

where F_B is the boulder impact force (N), v_0 is the boulder impact velocity (m/s), m is the boulder mass (kg), and k is the flexural stiffness of the structural member (N/m). For a cantilevered beam, k is $3EI/L^3$, where EI is the flexural rigidity (N·m^2) and L is the cantilever length (m). This expression indicates that flexible structural members can more efficiently resist impacts than stiff members. Flexural stiffness equations developed based on the flexural stiffness method (FSM) (i.e., Eq. 16.10) have been introduced in design guidelines such as MLR (2006) and CAGHP (2018a) for calculating boulder impact forces acting on beams and bridge piers. A key assumption of Eq. 16.10 is that the kinetic energy of the boulder is completely transferred to the structural element. Therefore, the effect of energy loss is ignored, and the predicted forces are typically conservative.

Recognizing the limitations of the Hertz equation and FSM, the GEO in Hong Kong adopted the enhanced flexural stiffness method (EFSM) for structural designs of rigid barriers subjected to boulder impacts (GEO, 2020). The EFSM was proposed by Lam and Gad

(2016) and has since been verified by Yong et al. (2020) using large-scale impact experiments. The advantage of the EFSM is that it considers energy loss during an impact and the inertial resistance of the structure. By considering energy and momentum conservation, this method incorporates the participating mass ratio between the barrier and boulder and the COR of the two objects during an impact. The boulder impact force is expressed as follows:

$$F_B = \sqrt{\lambda\left(\frac{1+\mathrm{COR}}{1+\lambda}\right)^2} v_0\sqrt{mk} \tag{16.11}$$

where λ is the ratio between the participating (generalized) mass ratio between the barrier and boulder (dimensionless). Notably, the participating mass for a cantilever element is only a quarter of the total mass (Ali et al. 2014; Sun et al. 2016). For a concrete stem wall, the total mass can be determined by considering the width of a single bay between the movement joints (GEO, 2020). When using Eq. 16.11, the effective flexural stiffness must be used such that $k = k_{\mathrm{eff}}$, and this value can be obtained from a moment–curvature analysis or by using simplified equations (Yong et al., 2020). According to the results of large-scale impact tests on cantilevered reinforced concrete walls, the COR between concrete and rock may be set as 0.3 in routine designs (GEO, 2020).

A comparison of Eqs. 16.10 and 16.11 indicates that the effect of energy loss and the contribution of the inertial resistance are taken into account by the additional term $\sqrt{\lambda\left(\frac{1+\mathrm{COR}}{1+\lambda}\right)^2}$. To verify the accuracy of the EFSM, Wong (2021) conducted a series of dynamic finite-element analyses and found that Eq. 16.11 can provide a more realistic prediction of the boulder impact force and the associated wall deflection than the traditional calculation methods (i.e. modified Hertz equation and FSM). It must be emphasized that the impact force calculated using EFSM is a quasi-static reaction force, which is defined as the force to emulate the maximum deflection demand of the impact. This force should not be confused with the contact force developed at the point of contact between the boulder and the surface of the structure, which is often an order of magnitude larger than the reaction force experienced by the wall at its base. The difference between the two types of force is the result of the inertia force developed from within the structure (Yong et al., 2020). Therefore, although forces obtained from the EFSM can be used in structural designs, they are not suitable for assessing localized damage to a structure at the point of impact.

16.2.7 Localized Damage of Barrier Wall

Boulder impacts on rigid barriers may cause localized damage to the barrier walls if they are unprotected by cushioning materials. However, current design guidelines such as VanDine (1996), SWCB (2005), NILIM (2016a), ASI (2011, 2013) and CAGHP (2018a) provide little guidance on the assessment or prevention of localized damage. To examine the structural response of rigid barriers subjected to hard impacts, Yong et al. (2020) and Majeed et al. (2021) conducted a series of impact tests on a cantilevered reinforced concrete wall that was 1.5 m high, 3 m wide and 0.23 m thick. A granite boulder and a torpedo-shaped impactor made of solid steel was used. The test results showed that for an impact energy (E) of up to about 4 kJ, the response of the wall was dominated by flexural deformation and localized damage was minor. The indentation depths were less than 4 mm and there were only minimal surface cracks in the vicinity of the point of impact. No spalling of concrete, perforation or penetration were observed. Figure 16.9 shows the condition of the wall after four of the tests.

To further investigate the possible effect of boulder impact on the structural integrity of barrier walls typically constructed for hillside risk mitigation projects in Hong Kong, the GEO conducted a series of dynamic finite-element analyses for a range of impact scenarios and wall configurations. The accuracy of the finite-element model was first validated using the

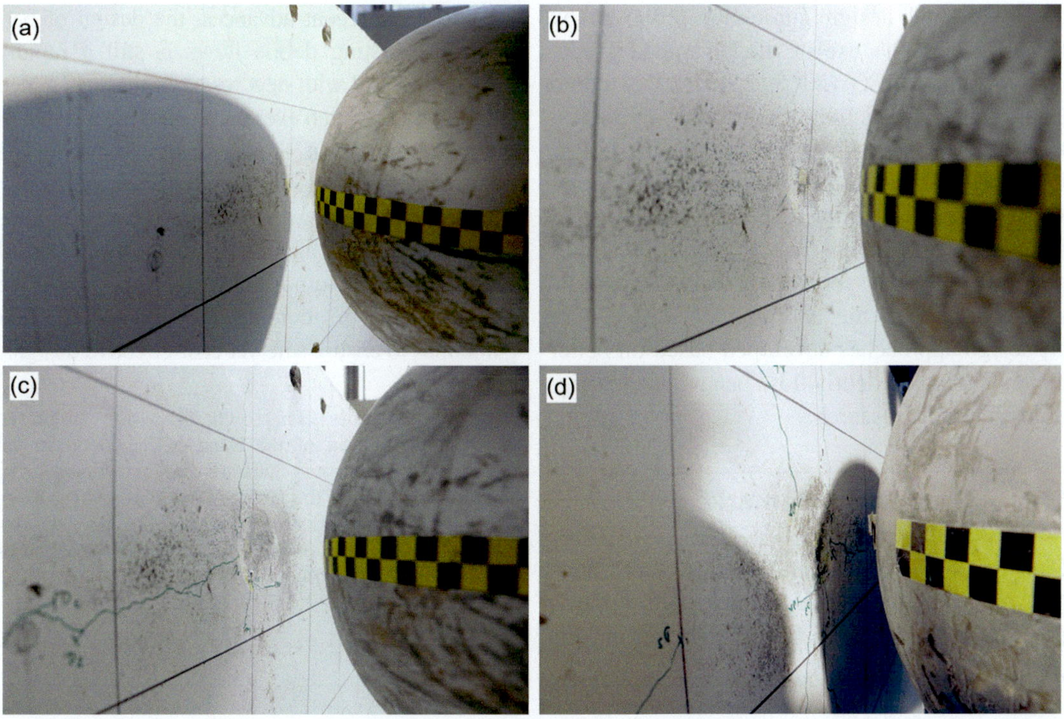

Fig. 16.9 Localized damage to a 230-mm thick reinforced concrete wall after hard impacts: (a) $E=0.6$ kJ, (b) $E=1.4$ kJ, (c) $E=2.5$ kJ, and (d) $E=3.9$ kJ (Photographs courtesy of ACY Yong)

physical test results given in Yong et al. (2020) and Majeed et al. (2021). The barrier wall considered in the analyses was 600 mm thick and was subjected to boulder impacts with an energy of 70 kJ, which can be brought about by a 1-m diameter boulder travelling at a velocity of 10 m/s. The results showed that at this impact energy level, the wall remained intact with only some local cracking at the back side of the wall in all the impact scenarios considered. The steel reinforcement in the concrete also did not yield.

With an improved understanding of the dynamic response of reinforced concrete structures subjected to boulder impacts, assessment of boulder impact forces is no longer required for boulders smaller than 1 m in diameter for the structural design of rigid barriers in Hong Kong (GEO, 2020). The current design philosophy is that it is more cost-effective to adopt a performance-based approach in the design of rigid barriers. Localized or minor damage that can be repaired after a debris flow is generally tolerable as long as the barrier would not collapse or fail to retain the design volume of debris. It is understood that this design philosophy is also increasingly being adopted for the design of barriers in other places, where some structural damage may be tolerable in certain scenarios and capital versus maintenance costs are weighed during the design stage.

16.2.8 Effect of Using Improved Guidelines on the Design of Rigid Barriers

The benefits of using the improved guidelines for rigid barrier design in Hong Kong have been demonstrated by Wong (2021) in an illustrated design example. The concerned rigid barrier was subjected to a design debris-flow event, which comprised 1-m diameter boulders travelling in surges at an impact velocity ranging from 4.7 to 7.0 m/s. Wong (2021) found that, compared to the original design guidelines proposed by Kwan (2012), the use of the revamped guidelines given in GEO (2015, 2016, 2020) could lead to a 30% reduction in concrete volume and steel reinforcement (by weight), and the use of long soil nails as tie-backs is no longer necessary as a small amount of barrier movement is allowed in the design. This example illustrates that cost-effective yet robust designs can be achieved through the adoption of state-of-the-art design approaches developed based on better understanding of debris flow behaviour and rigorous scientific research.

16.3 Concluding Remarks

Over the last decade, significant advances have been made in the design of debris-flow barriers. The main accomplishments covered in this chapter include:

(i) More accurate assessment of debris-flow impact forces.
(ii) Use of cushioning materials for shielding protection structures.
(iii) Development of an idealized multi-surge impact scenario.
(iv) Consideration of the attenuation of debris-flow velocity during the impacting process.
(v) Consideration of the self-weight of debris in stability assessment.
(vi) Cancellation of the requirement to superimpose debris and boulder impact loads in stability assessment.
(vii) Development of a displacement-based approach for assessing the effect of boulder impact in stability assessment.
(viii) Development of an enhanced flexural stiffness method for structural assessment.
(ix) Better understanding of localized damage of barrier wall due to boulder impacts and the adoption of a performance-based design approach where minor damage is tolerable.

Despite these recent advances, the design of barriers for resisting debris flows is still a rapidly developing field with new research findings added to the literature at an ever-increasing rate. To allow further optimization of the design of barriers, a fuller understanding of the dynamic interaction between debris flows and mitigation structures will be necessary and this will be the focus of future research. Authorities around the world are also expected to continue to rationalize their guidelines to ensure safe yet cost-effective designs.

Acknowledgements This chapter is published with the permission of the Head of the Geotechnical Engineering Office and Director of Civil Engineering and Development of the Government of the Hong Kong SAR, China. The third author is grateful for the generous financial sponsorship of the Research Grants Council of Hong Kong (27205320; AoE/E-603/18).

References

Armanini, A. (1997). On the dynamic impact of debris flows. In: A. Armanini, & M. Michiue (Eds.), *Recent developments on Debris flows* (pp. 208–226). Berlin, Heidelberg: Springer. https://doi.org/10.1007/BFb0117770

Arattano, M., & Franzi, L. (2003). On the evaluation of debris flows dynamics by means of mathematical models. *Natural Hazards and Earth System Sciences, 3*(6), 539–544. https://doi.org/10.5194/nhess-3-539-2003

ASI. (2013). ONR 24801: Schutzbauwerke der Wildbachverbauung – Statische und dynamische Einwirkungen. Austrian Standards Institute (ASI), Vienna, Austria (in German).

ASI. (2011). ONR 24802: Schutzbauwerke der Wildbachverbauung – Entwurf, Bemessung und konstruktive Durchbildung. Austrian Standards Institute (ASI), Vienna, Austria (in German).

ASTRA. (1998). Einwirkungen infolge Steinschlags auf Schutzgalerien, Bundesamt für Strassen ASTRA, Bern, Switzerland (in German).

Atkinson, J. (2017). *The Mechanics of Soils and Foundations* (2nd ed., p. 480). CRC Press.

Bugnion, L., McArdell, B. W., Bartelt, P., & Wendeler, C. (2012). Measurements of hillslope debris flow impact pressure on obstacles. *Landslides, 9*(2), 179–187. https://doi.org/10.1007/s10346-011-0294-4

CAGHP. (2018a). T/CAGHP 021-2018: Specification of design of debris flow prevention. China Association of Geological Hazard Prevention (CACHP), China (in Chinese).

CAGHP. (2018b). T/CAGHP 006-2018: Specification of geological investigation for debris flow stabilization. China Association of Geological Hazard Prevention (CAGHP), China (in Chinese).

Cui, P., Zeng, C., & Lei, Y. (2015). Experimental analysis on the impact force of viscous debris flow. *Earth Surface Processes and Landforms, 40*(12), 1644–1655. https://doi.org/10.1002/esp.3744

Di Perna, A., Cuomo, S., & Martinelli, M. (2022). Empirical formulation for debris flow impact and energy release. *Geoenvironmental Disasters, 9*, 8. https://doi.org/10.1186/s40677-022-00210-9

Egli, T. (2005). Wegleitung Objektschutz gegen gravitative Naturgefahren. Kapitel Murgänge. Kantonale Gebäudeversicherungen, Bern, Switzerland, 11 p. (in German).

Faug, T. (2021). Impact force of granular flow on walls normal to the bottom: Slow versus fast impact dynamics. *Canadian Geotechnical Journal, 58*(1), 114–124. https://doi.org/10.1139/cgj-2019-0399

GEO. (2015). Assessment of landslide debris impact velocity for design of debris-resisting barriers. GEO Technical Guidance Note No. 44, Geotechnical Engineering Office (GEO), Hong Kong.

GEO. (2016). Updates of design guidance of rigid debris-resisting barriers. GEO Technical Guidance Note No. 47, Geotechnical Engineering Office (GEO), Hong Kong.

GEO. (2020). Enhanced technical guidance on design of rigid debris-resisting barriers. GEO Technical Guidance Note No. 52, Geotechnical Engineering Office (GEO), Hong Kong.

Goodwin, G. R., & Choi, C. E. (2021). Translational inertial effects and scaling considerations for coarse granular flows impacting landslide-resisting barriers. *Journal of Geotechnical and Geoenvironmental Engineering, 147*(12), 04021153. https://doi.org/10.1061/(ASCE)GT.1943-5606.0002661

Hertz, H. (1882). On the contact of rigid elastic solids and on hardness. In P. Lenard (Ed.), *Miscellaneous Papers, Chapter 6* (pp. 163–183). Macmillan and Co.

Heymann, A., Collombet, M., Lambert, S., & Gotteland, P. (2011). Use of external testing methods to assess damage on rockfall protection structures. In: *Applied mechanics and materials* (Vol. 82, pp. 704–709). Trans Tech Publications Ltd, Switzerland. https://doi.org/10.4028/www.scientific.net/AMM.82.704.

Hu, K., Wei, F., Hong, Y., Li, X. (2006). Field measurement of impact force of debris flow. *Chinese Journal of Rock Mechanics and Engineering, 25*(suppl 1), 2813–2819 (in Chinese).

Hu, K., Wei, F., & Li, Y. (2011). Real-time measurement and preliminary analysis of debris-flow impact force at Jiangjia Ravine. *China. Earth Surface Processes and Landforms, 36*(9), 1268–1278. https://doi.org/10.1002/esp.2155

Hübl, J., & Holzinger, G. (2003). Kleinmaßstäbliche Modellversuche zur Wirkung von Murbrechern. WLS Report 50, Universität für Bodenkultur Wien (in German).

Hübl, J., Suda, J., Proske, D., Kaitna, R., & Scheidl, C. (2009). Debris flow impact estimation. In: *Proceedings of the 11th International Symposium on Water Management and Hydraulic Engineering, Ohrid, North Macedonia*, 1–5 September 2009 (pp. 137–148). University of Ss. Cyril and Methodius, Skopje, Macedonia.

Hungr, O., Morgan, G. C., & Kellerhals, R. (1984). Quantitative analysis of debris torrent hazards for design of remedial measures. *Canadian Geotechnical Journal, 21*(4), 663–677. https://doi.org/10.1139/t84-073

Hungr, O. (2000). Analysis of debris flow surges using the theory of uniformly progressive flow. *Earth Surface Processes and Landforms, 25*(5), 483–495. https://doi.org/10.1002/(SICI)1096-9837(200005)25:5<483::AID-ESP76>3.0.CO;2-Z

Johnson, C. G., Kokelaar, B. P., Iverson, R. M., Logan, M., LaHusen, R. G., & Gray, J. M. N. T. (2012). Grain-size segregation and levee formation in geophysical mass flows. *Journal of Geophysical Research: Earth Surface, 117*(F1), F01032. https://doi.org/10.1029/2011JF002185

Koo, R. C. H., Kwan, J. S. H., Ng, C. W. W., Lam, C., Choi, C. E., Song, D., & Pun, W. K. (2017). Velocity attenuation of debris flows and a new momentum-based model for rigid barriers. *Landslides, 14*(2), 617–629. https://doi.org/10.1007/s10346-016-0715-5

Kwan, J. S. H. (2012). Supplementary technical guidance on design of rigid debris-resisting barriers. GEO Report No. 270. Geotechnical Engineering Office, Hong Kong.

Kwan, J. S. H., & Cheung, R. W. M. (2012). Suggestion on design approaches for flexible debris-resisting barriers, discussion note DN 1/2012. Geotechnical Engineering Office, Hong Kong.

Kwan, J. S. H., Koo, R. C. H., & Ng, C. W. W. (2015). Landslide mobility analysis for design of multiple debris-resisting barriers. *Canadian Geotechnical Journal, 52*(9), 1345–1359. https://doi.org/10.1139/cgj-2014-0152

Kwan, J. S. H., & Koo, R. C. H. (2018). Enhanced technical guidelines for design of debris-resisting barriers. GEO Report No. 333, Geotechnical Engineering Office, Hong Kong.

Kwan, J. S. H., Koo, R. C. H., & Lam, C. (2018) A review on the design of rigid debris-resisting barriers. GEO Report No. 339, Geotechnical Engineering Office, Hong Kong.

Kwan, J. S. H., Sze, E. H. Y., Lam, C., Law, R. P. H., & Koo, R. C. H. (2021). Development and applications of debris mobility models in Hong Kong. *Proceedings of the Institution of Civil Engineers – Geotechnical Engineering, 174*(5), 593–610. https://doi.org/10.1680/jgeen.21.00008.

Lam, C., Kwan, J. S. H., Su, Y., Choi, C. E., & Ng, C. W. W. (2018a). Performance of ethylene-vinyl acetate foam as cushioning material for rigid debris-resisting barriers. *Landslides, 15*(9), 1779–1786. https://doi.org/10.1007/s10346-018-0987-z

Lam, C., Yong, A. C. Y., Kwan, J. S. H., & Lam, N. T. K. (2018b). Overturning stability of L-shaped rigid barriers subjected to rockfall impacts. *Landslides, 15*(7), 1347–1357. https://doi.org/10.1007/s10346-018-0957-5

Lam, H. W. K., Sze, E. H. Y., Wong, E. K. L., Poudyal, S., Ng, C. W. W., Chan, S. L., & Choi, C. E. (2022). Study of dynamic debris impact load on flexible debris-resisting barriers and the dynamic pressure coefficient. *Canadian Geotechnical Journal, Published Online.* https://doi.org/10.1139/cgj-2021-0325

Lam, N. T. K., & Gad, E. (2016). The estimation of impact forces based on first principles. In: *Proceedings of the Australasian Structural Engineering Conference: ASEC 2016*, 23–25 November 2016, Engineers Australia.

Lambert, S., Heymann, A., Gotteland, P., & Nicot, F. (2014). Real-scale investigation of the kinematic response of a rockfall protection embankment. *Natural Hazards and Earth System Sciences, 14*(5), 1269–1281. https://doi.org/10.5194/nhess-14-1269-2014

Lichtenhahn, C. (1973). Die Berechnung von Sperren in Beton und Eisenbeton. Kolloquium über Wildbachsperren, Mitteilungen der Forstlichen Bundesanstalt. *Wien. Heft, 102*, 91–127. (in German).

Li, X., Zhao, J., & Soga, K. (2021). A new physically based impact model for debris flow. *Géotechnique, 71*(8), 674–685. https://doi.org/10.1680/jgeot.18.P.365

Liu, H. M. (2019). Impact mechanisms of debris flow against multiple rigid barriers with basal clearance. PhD thesis, The Hong Kong University of Science and Technology, Hong Kong.

Liu, K. F., Jan, C. D., Lin, P. S., & Li, H. C. (2013). Advances of geo-disaster mitigation technologies in Taiwan. *Progress of Geo-Disaster Mitigation Technology in Asia* (pp. 77–103). Springer.

Lo, D. O. K. (2000). Review of natural terrain landslide debris-resisting barrier design. GEO report No. 104, Geotechnical Engineering Office, Hong Kong.

Majeed, Z. Z. A., Lam, N. T. K., & Gad, E. F. (2021). Predictions of localised damage to concrete caused by a low-velocity impact. *International Journal of Impact Engineering, 149*, 103799. https://doi.org/10.1016/j.ijimpeng.2020.103799

Mizuyama, Y., & Ishikawa, Y. (1988). Technical standard for the measures against debris flow (draft). Technical Memorandum of PWRI No. 2632, Sabo (Erosion Control) Division, Sabo Department, Public Works Research Institute (PWRI), Ministry of Construction, Japan.

Montani-Stoffel, S. (1998). Sollicitation dynamique de la couverture des galeries de protection lors de chutes de blocs, Dissertation, Nr. 1899, École Polytechnique Fédérale de Lausanne, p. 190 (in French).

MLR. (2006). DZ/T 0220-2006: Specification of geological investigation for debris flow stabilisation. Ministry of Land Resources (MLR), China (in Chinese).

Nagl, G., Hübl, J., & Kaitna, R. (2022). Stress anisotropy in natural debris flows during impacting a monitoring structure. *Landslides, 19*(1), 211–220. https://doi.org/10.1007/s10346-021-01779-2

Ng, C. W. W., Choi, C. E., Su, A. Y., Kwan, J. S. H., & Lam, C. (2016). Large-scale successive impacts on a rigid barrier shielded by gabions. *Canadian Geotechnical Journal, 53*(10), 1688–1999. https://doi.org/10.1139/cgj-2016-0073

Ng, C. W. W., Su, A. Y., Choi, C. E., Lam, C., Kwan, J. S. H., Chen, R., & Liu, H. (2018). Comparison of cushion mechanisms between cellular glass and gabions subjected to successive boulder impacts. *Journal of Geotechnical and Geoenvironmental Engineering, 144*(9), 04018058. https://doi.org/10.1061/(ASCE)GT.1943-5606.0001922

Ng, C. W. W., Choi, C. E., & Goodwin, G. R. (2019a). Froude characterization for unsteady single-surge dry granular flows: Impact pressure and runup height. *Canadian Geotechnical Journal, 56*(12), 1968–1978. https://doi.org/10.1139/cgj-2018-0529

Ng, C. W. W., Choi, C. E., Majeed, U., Poudyal, S., De Silva, W. A. R. K. (2019b). Fundamental framework to design multiple rigid barriers for resisting debris flows. In: *Proceedings of the 16th Asian Regional Conference on Soil Mechanics and Geotechnical Engineering, Taipei, Taiwan.*

Ng, C. W. W., Choi, C. E., Liu, H., Poudyal, S., Kwan, J. S. H. (2020). Design recommendations for single and dual debris flow barriers with and without basal clearance. In: *Proceedings of Workshop on World Landslide Forum, Springer, Cham* (pp. 33–53). https://doi.org/10.1007/978-3-030-60196-6_2.

Ng, C. W. W., Liu, H., Choi, C. E., Kwan, J. S. H., & Pun, W. K. (2021a). Impact dynamics of boulder-enriched debris flow on a rigid barrier. *Journal of Geotechnical and Geoenvironmental Engineering, 147*(3), 04021004. https://doi.org/10.1061/(ASCE)GT.1943-5606.0002485

Ng, C. W. W., Choi, C. E., Liu, H., Wang, C., & Kwan, J. S. H. (2021b). Impact mechanisms of debris flow on barriers: modelling, analysis and design. In: *Proceedings of the 13th International Symposium of Landslides, Cartegena, Colombia.*

NILIM. (2016a). *Manual of technical standard for designing Sabo facilities against debris flow and driftwood.* Technical Note of NILIM No. 905. Natural Institute for Land and Infrastructure Management (NILIM), Japan (in Japanese).

NILIM. (2016b). *Manual of technical standard for establishing Sabo master plan for debris flow and driftwood.* Technical Note of NILIM No. 904. Natural Institute for Land and Infrastructure Management (NILIM), Japan (in Japanese).

Proske, D., Suda, J., & Hübl, J. (2011). Debris flow impact estimation for breakers. *Georisk, 5*(2), 143–155. https://doi.org/10.1080/17499518.2010.516227

Prochaska, A. B., Santi, P. M., & Higgins, J. D. (2018). Relationships between size and velocity for particles within debris flows. *Canadian Geotechnical Journal, 45*(12), 1778–1783. https://doi.org/10.1139/T08-088

Scheidl, C., Chiari, M., Kaitna, R., Müllegger, M., Krawtschuk, A., Zimmermann, T., & Proske, D. (2013). Analysing debris-flow impact models, based on a small scale modelling approach. *Survey in Geophysics, 34*(1), 121–140. https://doi.org/10.1007/s10712-012-9199-6

Scotton, P., & Deganutti, A. M. (1997) Phreatic line and dynamic impact in laboratory debris flow experiments. In: *Proceedings of the 1st ASCE International Conference on Debris-flow Hazard Mitigation: Mechanics, Prediction and Assessment, San Francisco, USA*, 7–9 August 1997 (pp. 777–786).

Song, D., Ng, C. W. W., Choi, C. E., Zhou, G. G. D., Kwan, J. S. H., & Koo, R. C. H. (2017). Influence of debris flow solid fraction on rigid barrier impact. *Canadian Geotechnical Journal, 54*(10), 1421–1434. https://doi.org/10.1139/cgj-2016-0502

Su, Y., & Choi, C. E. (2021). Effects of rock shape on the cushioning mechanics of rock-filled gabions. *Acta Geotechnica, 16*(4), 1043–1052. https://doi.org/10.1007/s11440-020-01080-x

Su, Y., Choi, C. E., Ng, C. W. W., Lam, H. W. K., Wong, L. A., & Lee, C. (2020). New light-weight concrete foam to absorb debris-flow-entrained boulder impact: Large-scale pendulum modelling. *Engineering Geology, 275*(20), 105724. https://doi.org/10.1016/j.enggeo.2020.105724

Su, Y., Choi, C. E., Lv, Y. R., Wang, Y., & Li, X. (2021). Towards realistic simulations of the impact dynamics of boulders on rock-filled gabion: Combined effects of rock shapes and their crushing strength. *Engineering Geology, 283*, 106026. https://doi.org/10.1016/j.enggeo.2021.106026

Su, Y., Choi, C. E., Ng, C. W. W., Lam, C., Kwan, J. S. H., Wu, G., Huang, J., & Zhang, Z. (2019). Eco-friendly recycled crushed glass for cushioning boulder impacts. *Canadian Geotechnical Journal, 56*(9), 1251–1260. https://doi.org/10.1139/cgj-2018-0200

SWCB. (2017). Soil and Water Conservation Handbook. Soil and Water Conservation Bureau (SWCB), Council of Agriculture, Executive Yuan. Taiwan, R.O.C (in Chinese).

Vagnon, F., & Segalini, A. (2016). Debris flow impact estimation on a rigid barrier. *Natural Hazards and Earth System Sciences, 16*(7), 1691–1697. https://doi.org/10.5194/nhess-16-1691-2016

Vagnon, F. (2020). Design of active debris flow mitigation measures: A comprehensive analysis of existing impact models. *Landslides, 17*(2), 313–333. https://doi.org/10.1007/s10346-019-01278-5

VanDine, D. F. (1996). *Debris flow control structures for forest engineering, working paper 22/1996*. B.C. Ministry of Forests, Research Program, Victoria, British Columbia, Canada (68 p.

Watanabe, M., & Ikeya, H. (1981). Investigation and analysis of volcanic mud flows on Mt Sakurajima, Japan. In: *Proceedings of the Symposium on Erosion and Sediment Transport Measurement, International Association of Hydrological Sciences*, IAHS Publ. No. 133 (pp. 245–256).

Wendeler, C. (2008). Murgangrückhalt in Wildbächen—Grundlagen zu Planung und Berechnung von flexiblen Barrieren. PhD thesis, ETH Zurich, Switzerland (in German).

Wendeler, C., Volkwein, A., Roth, A., Herzog, B., & Hahlen, N., Wenger M (2008) Hazard prevention using flexible multi-level debris flow barrier—Protection against debris flows by installation of 13 flexible barriers in the Milibach River (Canton Berne, Switzerland). *Proceedings of Interpraevent 2008, 1*, 547–554.

Wendeler, C., Volkwein, A., McArdell, B. W., & Bartelt, P. (2018). Load model for designing flexible steel barriers for debris flow mitigation. *Canadian Geotechnical Journal, 56*(6), 893–910. https://doi.org/10.1139/cgj-2016-0157

Wong, A. L. (2021). Development of a novel design approach for rigid landslide debris-resisting barriers. *HKIE Transactions, 28*(4), 199–212. https://doi.org/10.33430/V28N4THIE-2021-0010

Yong, A. C. Y., Lam, C., Lam, N. T. K., Perera, J. S., Disfani, M. M., & Kwan, J. S. H. (2019). Analytical solution for estimating sliding displacement of rigid barriers subjected to boulder impact. *Journal of Engineering Mechanics, 145*(3), 04019006. https://doi.org/10.1061/(ASCE)EM.1943-7889.0001576

Yong, A. C. Y., Lam, N. T. K., Menegon, S. J., & Gad, E. F. (2020). Experimental and analytical assessment of flexural behavior of cantilevered RC walls subjected to impact actions. *Journal of Structural Engineering, 146*(4), 04020034. https://doi.org/10.1061/(ASCE)ST.1943-541X.0002578

Zanuttigh, B., & Lamberti, A. (2006). Experimental analysis of the impact of dry avalanches on structures and implication for debris flows. *Journal of Hydraulic Research, 44*(4), 522–534. https://doi.org/10.1080/00221686.2006.9521703

Zhang, S. (1993). A comprehensive approach to the observation and prevention of debris flow in China. *Natural Hazards, 7*, 1–23. https://doi.org/10.1007/BF00595676

Zhang, S., Hungr, O., Slaymaker, O. (1996). The calculation of impact force of boulders in debris flow. In: *Debris flow observation and research* (pp. 67–72) (in Chinese).

17 Check Dam Failures

Johannes Hübl, Jürgen Suda, Taro Uchida
and Georg Nagl

Abstract

Check dams are used in harsh environments to withstand enormous forces and to operate for long periods of time. Failure of check dams is a rare but dangerous event that may result in a higher degree of damage than in the condition without structures. A few failures are documented in literature that are related to a variety of failure mechanisms such as overload of the structure or poor maintenance. Nevertheless, even small defects, such as cracks, can cause damage and lead to structural failure or increased repair costs. In order to maintain structures for their designed lifetimes, it is important to consider the full range of possible failure mechanisms. This additionally highlights the need to investigate and assess the origin of damages and deterioration of check dams. Some examples from different countries are discussed in this chapter to provide insight into possible failure mechanisms as well as basis for maintenance management. Therefore, life cycle management comprises periodic inspection and documentation, which may reveal future failure mechanisms that are often caused by progressive construction material deterioration. To increase the lifetime of structural measures and to restore the functional condition, maintenance measures must be completed. This, in turn, can lead to improved designs and improved material use.

J. Hübl (✉) · G. Nagl
Institute of Mountain Risk Engineering, University of Natural Resources and Life Sciences Vienna, Peter-Jordan-Str. 82, 1190 Vienna, Austria
e-mail: johannes.huebl@boku.ac.at

G. Nagl
e-mail: georg.nagl@boku.ac.at

J. Suda
Alpinfra Engineering, Lilienberggasse 13/2, 1130 Vienna, Austria
e-mail: juergen.suda@gmail.com

T. Uchida
Faculty of Life and Environmental Sciences, University of Tsukuba, Tennodai 1-1-1, Tsukuba, Japan
e-mail: uchida.taro.fw@u.tsukuba.ac.jp

17.1 Introduction

The failure or collapse of mitigation structures often results in serious consequences for people and infrastructure. Failure not only removes the protection offered by the dam, but it can also lead to an increased supply of sediment downstream (Baggio & D'Agostino, 2022; Benito et al., 1998; Sodnik et al., 2015; Wang, 2013). Since hazard and risk maps account for mitigation structures (Fuchs et al., 2007), a failure of check dams can increase life loss and economic risks. As a result, areas may be affected

M. Jakob et al. (eds.), *Advances in Debris-flow Science and Practice*, Geoenvironmental Disaster Reduction,
https://doi.org/10.1007/978-3-031-48691-3_17

that would otherwise not have been subject to risk. Some countries include extreme, very rare events in the generation of hazard and risk maps (see Horton et al., this volume), but a specific consideration of mitigation failure is typically missing in the risk assessment as it is assumed that the mitigation works will perform as intended. From a risk perspective, it is implicitly assumed that the chance of mitigation failure is zero.

In this document, the term "check dam" refers to transverse structures across channels in torrential catchments. They may be made of logs, gabions, dry stone, masonry or/and reinforced concrete (Piton & Recking, 2016). A close examination of failure modes can help to prevent future disasters. In the literature, only a few reports on check dam failures can be found (Table 17.1). This is understandable, as failures are tantamount to an admission of guilt at worst, or poor design work at best. Failures are usually buried within the responsible organization. It is a sign of an organization's level of development whether a culture of error admission, analysis and improvement is encouraged or not. In the last 150 years, several thousand different debris-flow mitigation structures have been constructed, and the design of today's structural

Table 17.1 Examples of check dam failures and damages

Country	Region	Year	Damage and fatalities	Literature	Failure type
Austria	Village See (Schallerbach)	2015	70 houses were damaged, 4 houses destroyed	(Hübl et al., 2016b)	Erosion of lateral foundation of check-dam
Austria	Pinnisbach	2015	no damage	(Hübl et al., 2016a)	Check-dam failure
China	Wenjia Gully	2010	7 dead; 5 missing; 39 injured; 479 houses buried	(Yu et al., 2013)	Check-dam collapse due to extreme event
China	Zechawa Gully	2017	damaged pedestrian walkways and scenic roads	(Gong et al., 2020)	Check-dam failure due to overload
China	Wenchuan	2010	13 deaths; 59 missing; 8000 local residents were displaced	(Chen et al., 2015; Yu et al., 2013)	Check dam failure due overload; sediment supply through earthquake
China	Zhouqu	2010	1756 dead; 9 destroyed check dams	(Wang, 2013)	poor construction and inappropriate positioning of the check dams
France	Faucon torrent	2003	22 structures out of 82 were destroyed and 21 damaged	(Remaître et al., 2008)	partial structure damaged and structures were destroyed
Italy	Rotian	2018	1 dead	(Baggio & D'Agostino, 2022; Borga & Zaramella, 2020)	Check dam failure due overload
Japan	Nagano	2014	1 dead	(Hiramatsu et al., 2014)	partial structure collapse
Spain	Biescas	1996	87 dead	(Benito et al., 1998)	Overload and bad maintenance

measures is based on this longtime experience. After each damaging torrential event, the design of the structures has likely been scrutinized and improved. Unfortunately, this incremental knowledge increase was kept within the responsible organizations and were used for the further development of structural mitigation measures. Hence the design rules, which are published nowadays, are the result of a long time of trial and error.

Based on the fundamental ongoing geomorphological processes and environmental conditions, failure modes can be associated with various damaging mechanisms (Suda, 2013). These boundary conditions (Fig. 17.1) to which a barrier structure is subjected to can be divided into 7 groups (Rudolf-Miklau et al., 2007). They have a direct or indirect influence on the temporal development of the barrier state. These boundary conditions are:

1. Engineered boundary conditions (e.g. design, working process, maintenance)
2. Climatic, physical and chemical boundary conditions (e.g. weather, chloride environment, climate environment)
3. Biotic boundary conditions (e.g. plants)
4. Geotechnical, geological and tectonically boundary conditions (e.g. earth pressure, rock resistant, groundwater, landslides, earthquakes, rock fall)
5. Process oriented boundary conditions (e.g. discharge, sediment concentration, debris flows)
6. Structural boundary conditions (e.g. dimensions, concrete, durability, protective measures)
7. Functional boundary conditions (e.g. function type, location, design)

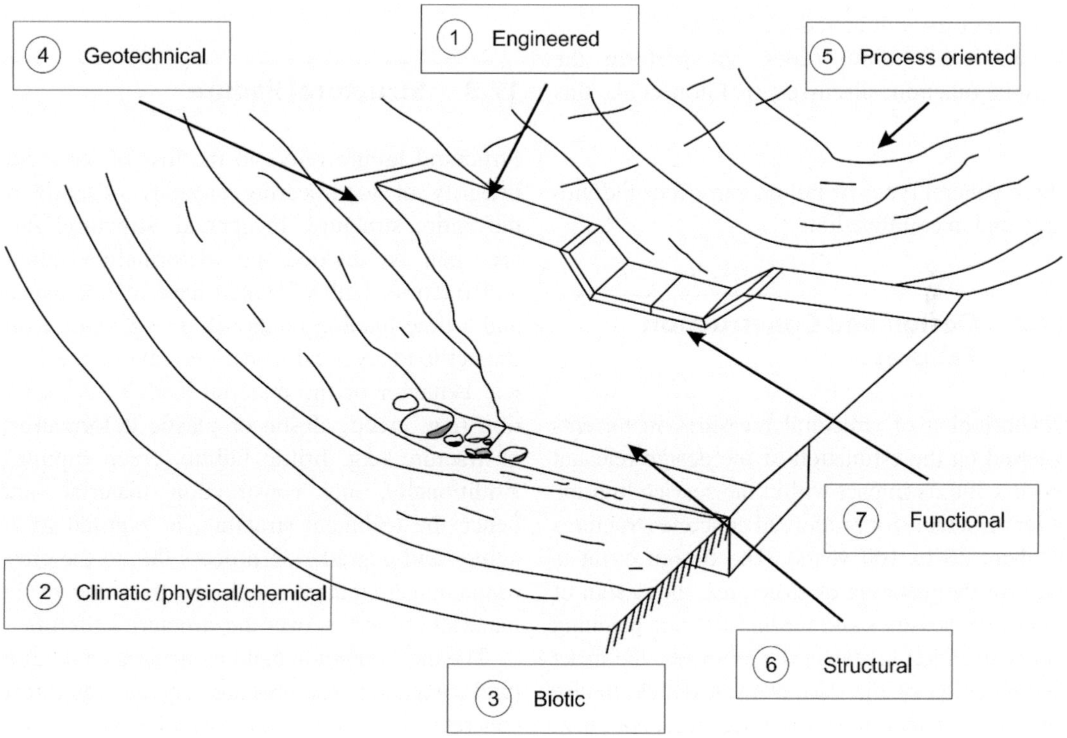

Fig. 17.1 Boundary conditions acting on structural barriers, adapted after (Rudolf-Miklau et al., 2007)

In addition to the consequences of overloading and destruction of mitigation works, minor damages occurring over a long time period (decades) can result in significant damage and increased total life cycle costs (Lee et al., 2022; Suda, 2013). For example, small cracks favor the infiltration of water and accelerate the corrosion of reinforcement and the deterioration of masonry. Instead, overloading of structures, i.e., loading in greater magnitude than the load-bearing capacity of the structure, leads to deformation of structural members, cracking, and collapse of the structure.

In general, failure types can be separated into four main categories:

1. Design and construction failure (planning error, poor construction practice) (Sect. 17.2).
2. Structural failure (failure due to degradation and/or unexpected loading) (Sect. 17.3)
3. Maintenance failure (without maintenance the lifetime will be reduced) (Sect. 17.4)
4. Functional failure (does not perform the right function, discussed in Piton et al., this volume)

These general types of failure can occur individually and in combination.

17.2 Design and Construction Failures

Dimensioning of structural measures in torrents is based on the estimation of the design-relevant process and its impact within the service lifetime of the structure (for reinforced concrete structures of about 50 to 100 years). The design event is used for the planning of structures, the design of protective measures and for hazard zone planning and is specified in standards. However, the exact determination of the design event defies precise calculation; rather, it is also a decision about hazard scenarios based on numerous factors.

During the design stage, the function, structure, type of construction, static system, building material and functional parts must be defined (Table 17.2).

Incorrect planning of these issues can often result in failures because the design and the governing design parameters are often fraught with significant uncertainties. Those are partly connected to the unpredictability of the future behavior of torrential catchments under changing climatic conditions or prospective land-use changes. In addition, unpredictable earthquakes can weaken material and condition it for catastrophic events (Yu et al., 2013). Based on this, the maintenance and life cycle assessment becomes even more important (Mazzorana et al., 2014; Paratscha et al., 2019). Life cycle management comprises periodic inspection and documentation, which may reveal future failure mechanisms that are often caused by progressive construction material deterioration. To increase the lifetime of structural measures and to restore the functional condition, maintenance measures must be completed. This, in turn, can lead to improved designs and improved material use.

17.3 Structural Failure

Structural failure refers to the loss of structural integrity or load-bearing capacity of parts or the entire structure. In general, structural failures can be divided into deformation failure and fracture, both of which have a time aspect and a time-loading relationship (e.g. static load, impact load, cyclical load). The related mechanical behavior of the material leads to deformation (e.g. creep, elastic or plastic deformation) or fracture (e.g. brittle failure, creep rupture). Additionally, any construction material, and hence the technical structure, is exposed to an aging- and degradation process due to the environmental boundary conditions that change the material properties over the structure's lifetime.

The most common building materials for structural mitigation measures are concrete, reinforced concrete, steel, and wood. Nevertheless, due to a long tradition in the Alps, many structures have been built with different working methods and materials such as dry-masonry, cement-mortar masonry and wood-and-stone dams (Piton & Recking, 2016). It is beneficial to include the

Table 17.2 Classification of structural measures in torrents (in accordance with Austrian Standards ÖNORM B4800)

<table>
<tr><td>Function</td><td colspan="2">Stabilization consolidation</td><td colspan="2">Deposition</td><td colspan="3">Buffering</td><td>Filtration</td><td>Transformation</td></tr>
<tr><td>Structure</td><td colspan="2">Check dam sill bed
sill ramp</td><td colspan="2">Deposition structure</td><td colspan="3">(Debris) buffering structure</td><td>(Debris or wood) filter structure</td><td>Debris flow breaker drop structure</td></tr>
<tr><td rowspan="3">Type of construction</td><td colspan="2">Solid body check dam</td><td colspan="5">Single solid body check dam
Compound dam</td><td colspan="2">Multiple solid body check dam</td></tr>
<tr><td colspan="2" rowspan="2">Open check dam</td><td colspan="5">Crest closed open check dam</td><td colspan="2">Crest slotted open check dam</td></tr>
<tr><td colspan="3">Small slot dam
Big slot dam</td><td colspan="2">Lattice dam
Frame dam
Net barrier
Rope barrier</td><td>Slit dam</td><td>Sectional dam</td></tr>
<tr><td rowspan="3">Static system</td><td rowspan="3">Gravity dam</td><td rowspan="3">Arch dam</td><td colspan="4">Reinforced Flat slab dam</td><td colspan="3">Girder structure</td></tr>
<tr><td rowspan="2">Flat slab dam</td><td rowspan="2">Buttress dam</td><td colspan="2" rowspan="2">Cantilever retaining dam</td><td>Mass oriented structures</td><td colspan="2">Force path oriented structures</td></tr>
<tr><td>Sectional dam</td><td>Frame barrier (rigid)</td><td>Net barrier
Rope barrier (flexible)</td></tr>
<tr><td>Material</td><td colspan="9">Wood
Stone
Cement-mortar masonry
Concrete fixed stone pitching
Concrete, reinforced concrete
Steel</td></tr>
<tr><td>Functional parts</td><td colspan="9">Beams
Rakes
Grids (framework)</td></tr>
</table>

historical aspect in the analysis of the previous working methods and design codes (Armanini et al., 1991) to assess the condition of mitigation structures. The analysis of failures is then part of forensic engineering and comprises data of both the relevant process and the structure. But before numerical modelling, a clear idea of the governing failure process is necessary (Fig. 17.2).

17.3.1 Primary Failure Modes of the Structure

According to technical standards (e.g. EN, 1990), the primary failure mode of structures can be separated into: (1) the strength of the structure or structural members corresponding to the maximum load carrying resistance (ultimate limit state), and (2) the state of design beyond which a structural system loses operationally its serviceability for the actual service load that the structure is subjected to (serviceability limit state). Furthermore, the failure of a structure can be considered as internal failure or excessive deformation of the structure (STR), or failure or excessive deformation of its foundation (GEO), see Figs. 17.3 and 17.4, and (Table 17.3).

A failure of process-related serviceability exists if the individual structure or the overall system was not able to sufficiently control the event to the projected extent of the design event (see Piton et al., this volume).

Fig. 17.2 Different failures modes of structures of a height of about 10 m: inadequately installed reinforcement (Pinnisbach, Tyrol, Austria) (left); scouring and too little reinforcement (Fischbach, Salzburg, Austria) (right)

Fig. 17.3 Bearing resistance failure of the soil below the base and toppling GEO (Lorenzerbach, Styria, Austria) (left), Toppling failure GEO (Tschenglserbach, Southern Tyrol, Italy) (right)

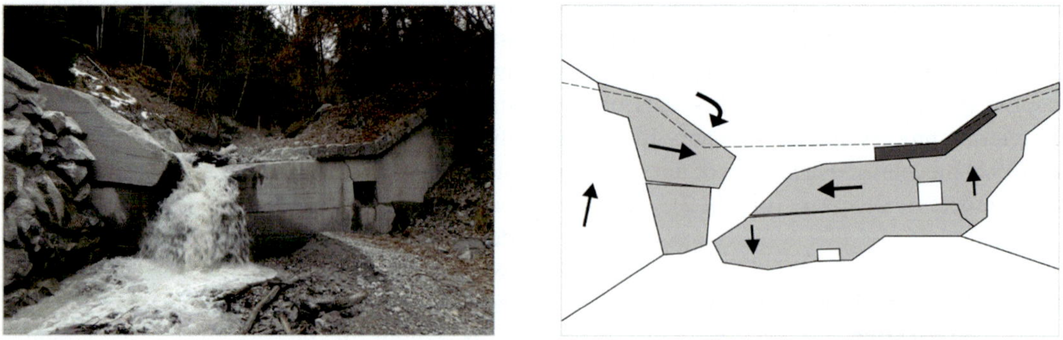

Fig. 17.4 Shear resistance failure of structure STR (Lattenbach, Tyrol, Austria) with schematic illustration

Table 17.3 Limit state design used in structural engineering according to Eurocode 2 (design of concrete structures) and Eurocode 7 (geotechnical design)

Ultimate limit state		
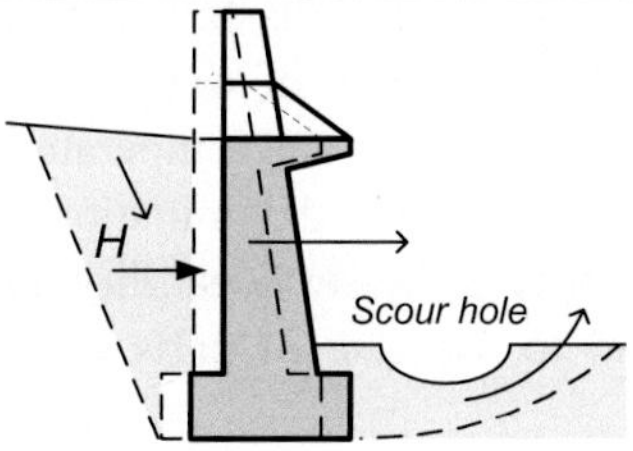	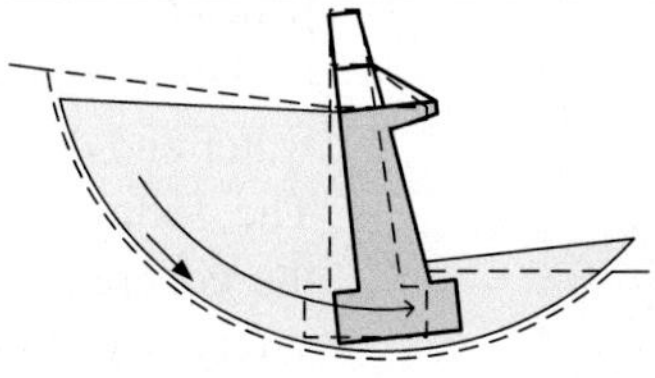	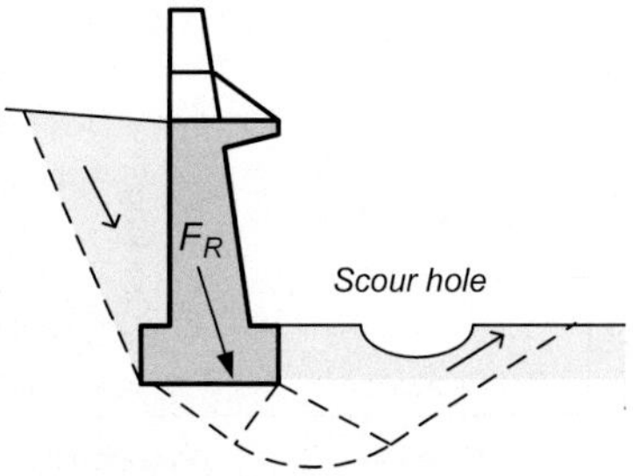
Sliding at the base GEO (EN 1997-1:2004)	**Loss of overall stability GEO** (EN 1997-1:2004)	**Bearing resistance failure of the soil below the base GEO** (EN 1997-1:2004)
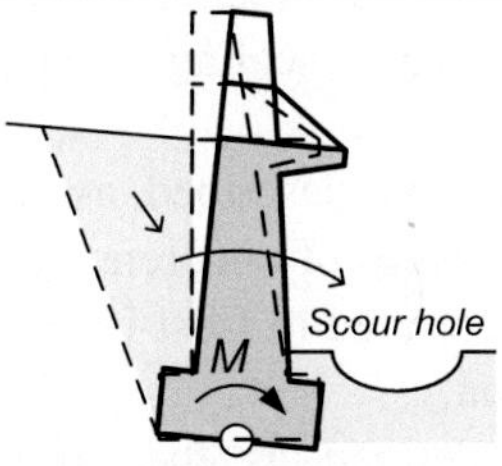	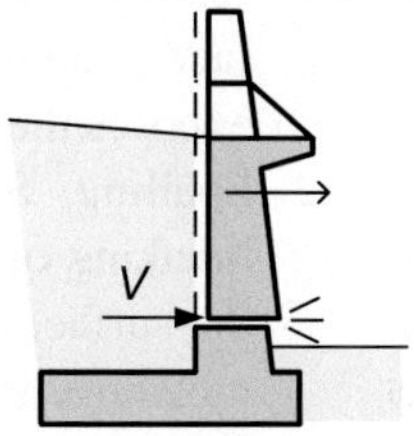	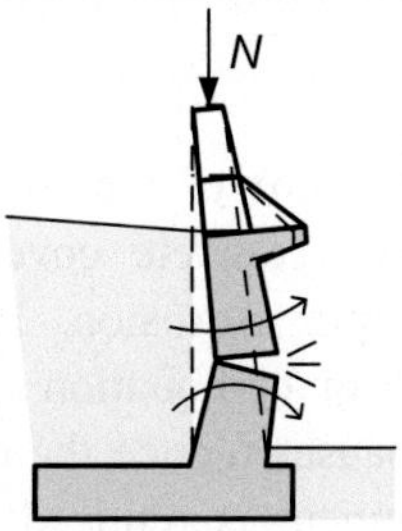
Toppling GEO (EN 1997-1-1:2004)	**Shear resistance failure STR** (EN 1992-1-1: 2004)	**Bearing resistance failure STR** (EN 1992-1-1:2004)
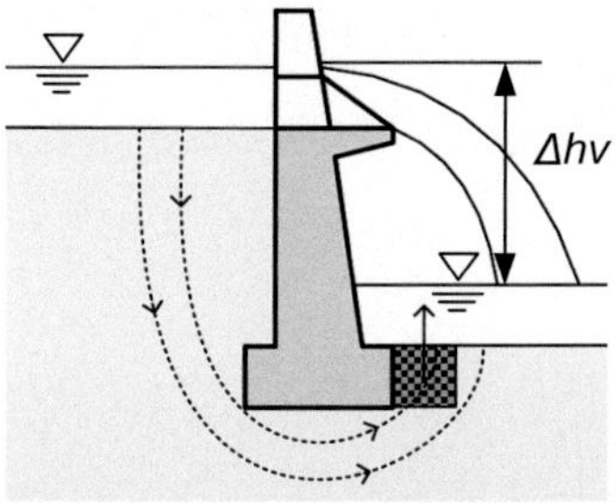	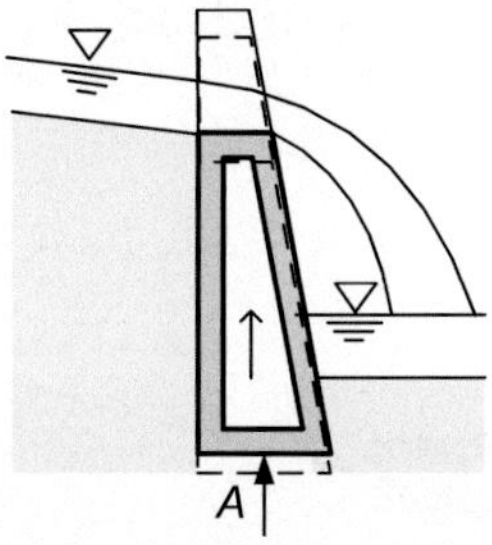	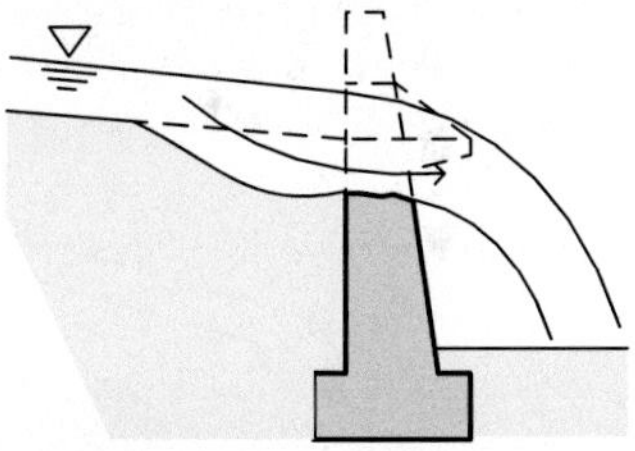
Hydraulic heave and piping HYD (EN 1997-1:2004)	**Uplift by water pressure UPL** (EN 1997-1:2004)	**Abrasion STR** (EN 1992-1-1:2004)

17.3.2 Deterioration of Concrete

In general, cracks impair the durability of concrete by increasing the ingress of aggressive agents and affecting the water tightness. Starting in the first hours after the structure has been completed, cracks induced by plastic settlement, drying shrinkage or creep can lead to deterioration. A close interconnection exists between concrete deterioration and the corrosion of the steel reinforcement. The increased water infiltration in cracks provokes corrosion of reinforcement, but vice versa, corrosion additionally provokes cracking and destruction of the concrete cover.

The deterioration can work in different ways and lifetimes and produces different damage patterns (Figs. 17.5 and 17.6). In the first phase

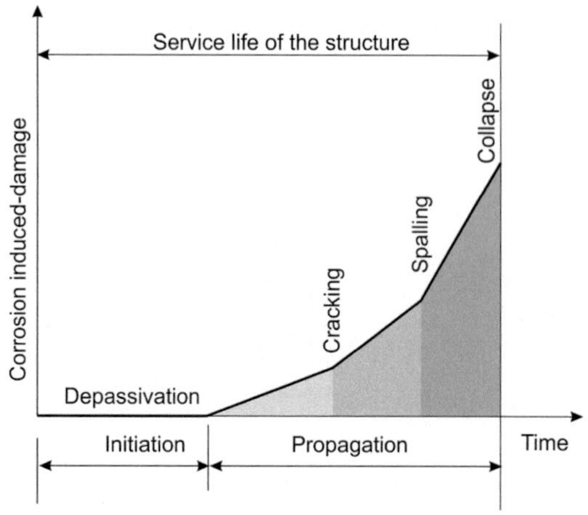

Fig. 17.5 Deterioration of concrete during service life of structure, adapted after (Bertolini et al., 2013)

of the service lifetime of the structure, the initiation phase, the concrete cover protects the reinforcement from corrosion. If the penetration of chloride or carbonation is sufficient or other physical causes reduce the cover, the passivation environment is reduced and corrosion initiates. This breakdown of the protective layer is called depassivation. The duration of the initiation phase depends on the concrete cover and penetration rate. In the second phase, corrosion initiation starts in the presence of oxygen and water and depends on humidity and temperature. The following spalling and/or delamination of the concrete cover affects the serviceability and safety and can lead to collapse of parts or the entire structure.

Some examples are listed below:

1. Scaling: The finished surface flakes or peels off. It starts mostly as small patches and does not expose the coarse aggregate. Moderate scaling leads to loss of up to 10 mm of surface material and severe scaling clearly exposes the aggregate.
2. Spalling: Spalling can be described as the breaking of layers or pieces of concrete from the surface. It is often related to frost-impact, moisture, and corrosion.
3. Curling: Curling is defined as the up or downward bending of edges of concrete

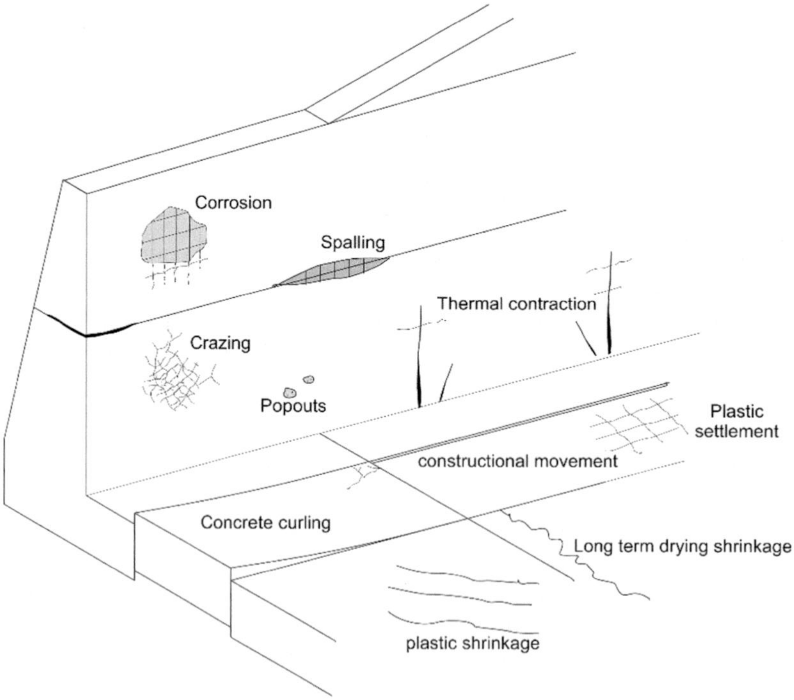

Fig. 17.6 Schematic illustration of deterioration and crack patterns

elements. If the edges are unsupported and subjected to loads, they are prone to cracking.
4. Cracking: Cracks can have several sources, but can also result from the simultaneous action of different mechanisms (see Fig. 17.7).

Therefore, it is necessary to identify the origin and to distinguish between the following two categories (Fig. 17.7):

1. Cracks before hardening.
2. Cracks after hardening.

17.3.2.1 Cracks Before Hardening

After mixing the concrete, the water reacts exothermically with the clinker minerals (Stage 1). In the dormancy period (Stage 2), the temperature is relatively low and the concrete can be placed and is still workable. At the end of the dormancy period, the concrete strength increases and the main period of hydration begins (Stage 3). The period of maximum heat evolution occurs after 10 to 20 h. In this stage, the risk for crack debonding is significant. In the deceleration period (Stage 4), temperature decreases. In the diffusion period (Stage 5), the hydration process slows and concrete reaches the final strength after 28 days (Fig. 17.8).

Different types of crack patterns are related to the different stages of fresh concrete:

- **Craze cracks or crazing** is the development of fine irregular cracks on the surface**.** Early surface drying causes the immediate surface to shrink differently. Small cracks form typically in grids of about 6 cm.
- **Plastic settlement cracks** form before hardening in still plastic concrete and result from poor compaction or a high water/cement ratio

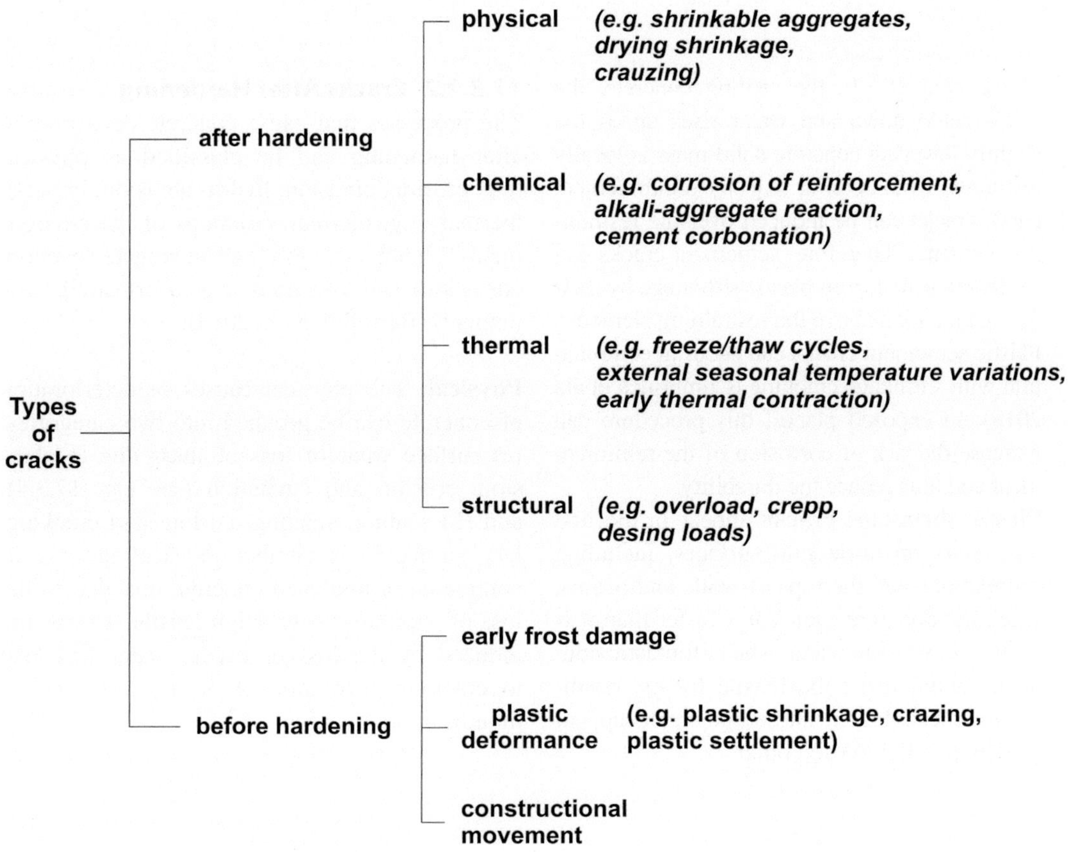

Fig. 17.7 Classification of cracks adapted after Concrete Society (2010)

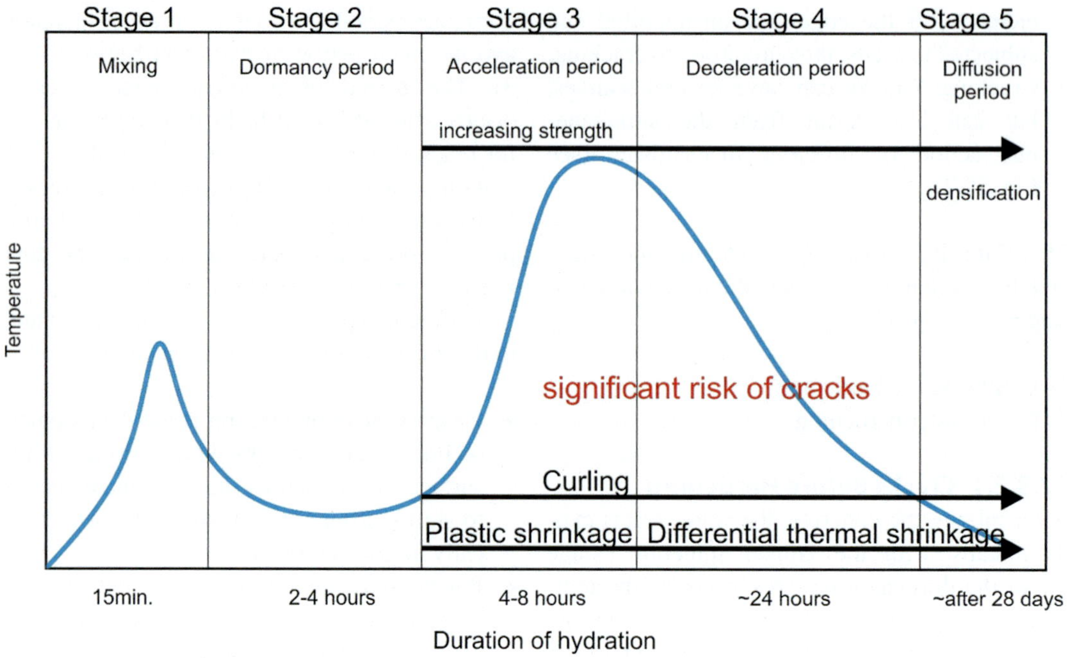

Fig. 17.8 Hydration of concrete and related risk of cracks, according to (Neville, 2011)

of the concrete. In the settling concrete, the solids settle down and water rises up. If the slightly lowering concrete solid mass is locally restrained from settling (e.g. by reinforcement bars), cracks can be induced over the restraining elements. Therefore, settlement cracks can be distinguished from plastic shrinkage by their distinct location above the restraining elements. Plastic settlement cracks can occur in combination with shrinkage cracking (Combrinck et al., 2018). In exposed places, this procedure can increase the risk of corrosion of the reinforcement and thus reduce the durability.

- **Plastic shrinkage cracks** appear in the first few hours on horizontal surfaces, including foundations and the tops of walls and beams, when the concrete evaporates faster than it is replaced by rising water. The still plastic concrete cannot resist the tensile forces, resulting in cracks. Plastic shrinkage cracks appear usually parallel to each other.

17.3.2.2 Cracks After Hardening

The processes that cause concrete deterioration after hardening can be classified as physical (e.g. erosion, abrasion, hydro- abrasion, impact), thermal (e.g. thermal variations of the environment), chemical (e.g. alkali aggregate reaction, corrosion) and structural (e.g. overloading, settlement) (Bertolini et al., 2013).

Physical: The physical causes of deterioration of concrete can be grouped into two categories: (a) surface wear or loss of mass due to abrasion, erosion and cavitation (see e.g. 17.3.4), and (b) scaling, spalling, curling and cracking. Dry shrinkage is another physical process of contracting a hardened concrete mix due to the loss of capillary water. High tensile stresses are induced by the loss of excess water and lead to cracking. The time of drying lasts several months up to 4 years.

Chemical: Steel is used in protective structures as both structural and reinforcing components. In civil engineering, especially in reinforced concrete, electrochemical corrosion is predominant (Neroth et al., 2011). Once the passive layer is lost, the corrosion starts by anodic and cathodic reactions (Fig. 17.9).

The consequences of corrosion of steel reinforcement (Fig. 17.10) are related to several aspects of the structure itself, like the condition of the structure, serviceability and safety aspect. Cracking and the subsequent spalling of concrete cover is the main consequence accompanied with rust spots on the surface.

As a result, the load-bearing capacity and fatigue strength of the steel reinforcement may be reduced. Under special conditions, brittle failure can take place.

There are different chemical causes that lead to the corrosion of reinforcement in concrete:

- *Carbonation-induced corrosion* occur when carbon dioxide reacts with the alkaline constituents of concrete, the normal pH of concrete of about 13 to 14 drops down to 7 and provides a suitable environment for corrosion. Carbonation damage occurs rapidly in cases of low concrete cover, cracks, open pore structure and low alkaline reserves (by low cement content, high water cement ratio and poor curing). This allows the CO_2 to quickly penetrate and transform in the environment. The corrosion product's volume is 2 to 6 times higher than the iron volume. Consequently, spalling or even delamination of the concrete occurs due to the tensile strength generated.
- *Chloride-induced corrosion* based in some cases of structural mitigation structures (exposed to salt-spray or maritime environment), chloride-induced corrosion may be a cause of damage to structural steel or reinforcement. Chlorides penetrate the pore system either by diffusion or by capillary suction, although concrete constituents can also bind chlorides and thus an anode and cathode reaction can occur.
- *Alkali-aggregate reaction* is the chemical reaction of mortar or concrete and alkalis under environmental moisture may lead to cracks. Alkali-aggregate reactions can be either alkali-carbonate reactions (ACRs) or alkali-silica reactions (ASRs).
- *Sulfate Impact* occur if the cement that is used does not have sufficient sulphate resistance, sulphate—containing water

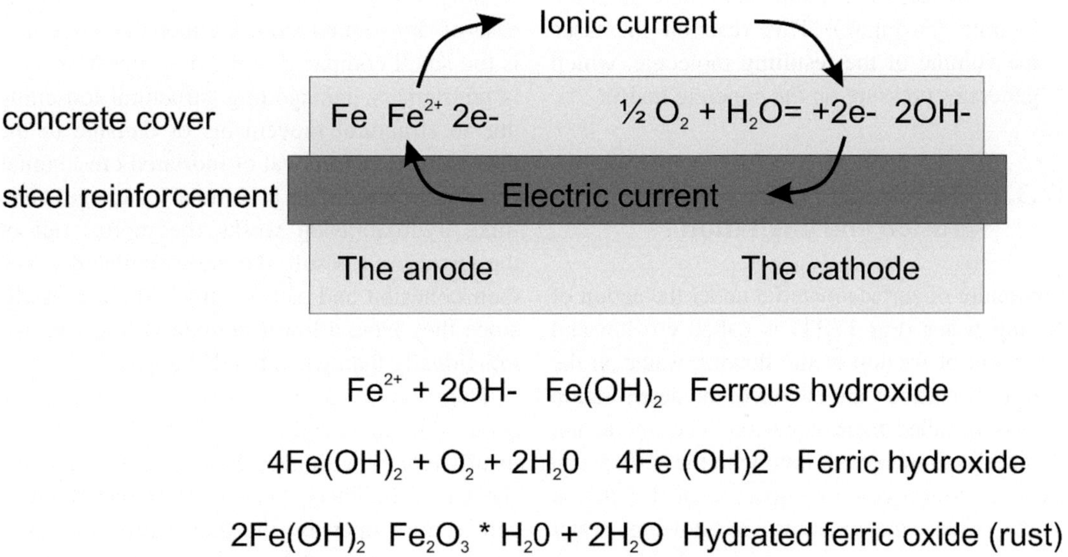

Fig. 17.9 Corrosion of steel reinforcement

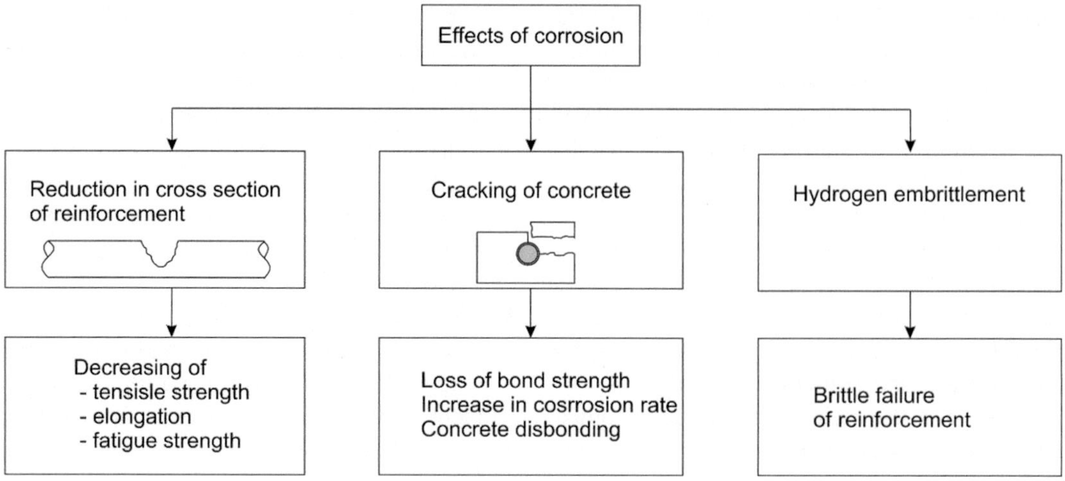

Fig. 17.10 Effects of corrosion, adapted after (Bertolini et al., 2013)

(groundwater or stream water) can cause an attack on concrete parts. This attack becomes visible through selective spalling (pop-outs) and/or a concrete surface that appears "soapy". The surface is then more vulnerable to other damage mechanisms (e.g. hydroabrasion). The dissolved sulphates have a high wetting capacity and penetrate quickly deep into the concrete. They react with the calcium aluminates or calcium aluminum hydrates in the cement, with the inclusion of crystal water to form tricalcium aluminate sulphate hydrate (ettringite). This reaction increases the volume of the resulting molecule, which generates pressure on the concrete matrix.

17.3.3 Mechanical Hydro-abrasion, Erosion and Cavitation

The failure of surface material under the action of flowing water (Fig. 17.11) is called erosion and is a result of friction of the flowing water on the surface. If particles are included, the deterioration process is called hydro-abrasion. The interaction of particles immerged in fluid with the surface can act as grinding wear or impact wear. Cavitation occurs when the pressure in flowing water changes abruptly, for example, air bubbles collapse in the flowing water and generate impulses.

In hydro-abrasion, the structure is stressed by flowing water containing solid particles (typically sand and gravel). Depending on the type of substances involved (liquid, solids) and the type of stress (flowing, impacting, oscillating), a distinction is made between different wear processes of erosion (fluid) and abrasion (fluid with particles). An example of strong abrasion in combination with non frost-resistant concrete is shown in Fig. 17.12.

The erosion of crest stones occurs as a result of drag stress of the water (Fig. 17.13). In the case of dry-jointed stones, either the stone size is too small compared to the drag stress or there is preliminary damage (e.g. structural loosening due to structural movements or damage to the fastening). The removal of mortared crest stones usually occurs due to preliminary damage. Over time, hydroabrasion works the mortar out of the joints. As a result, the individual stones lose their cohesion and can be removed more easily, since they have a lower boundary drag strength individually than when bonded together.

Degradation of protective steel plates can occur due to decades of hydro-abrasion, and insufficient steel plate thickness. If the steel sheet is completely shaved off at one point, it can be torn open by the drag stress and worn away more quickly. Protruding parts provide a point of attack for runoff, bedload and woody

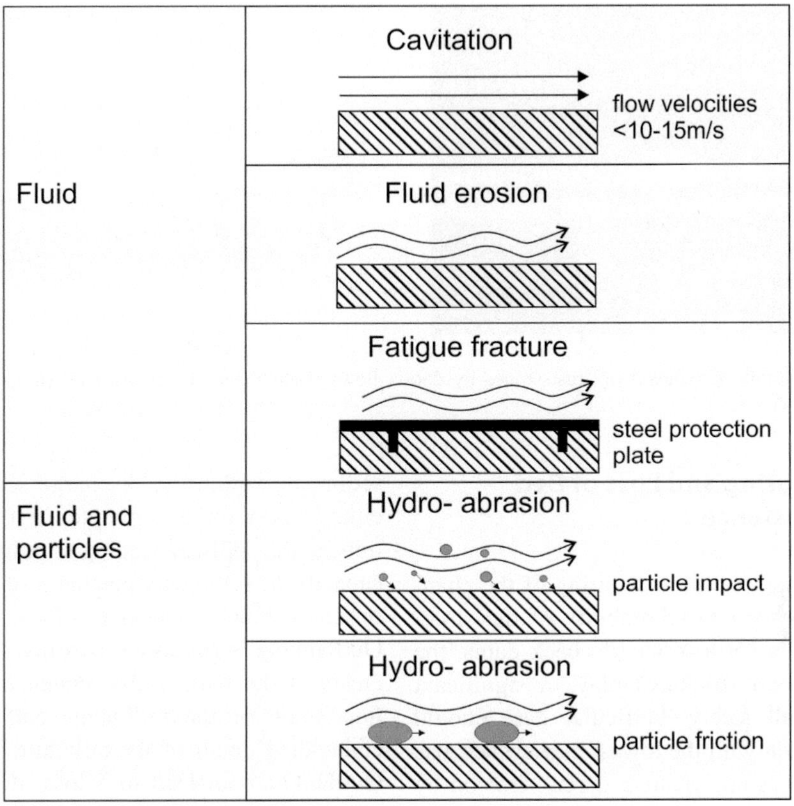

Fig. 17.11 Mechanical hydro-abrasion, erosion and cavitation (Jacobs et al., 2001)

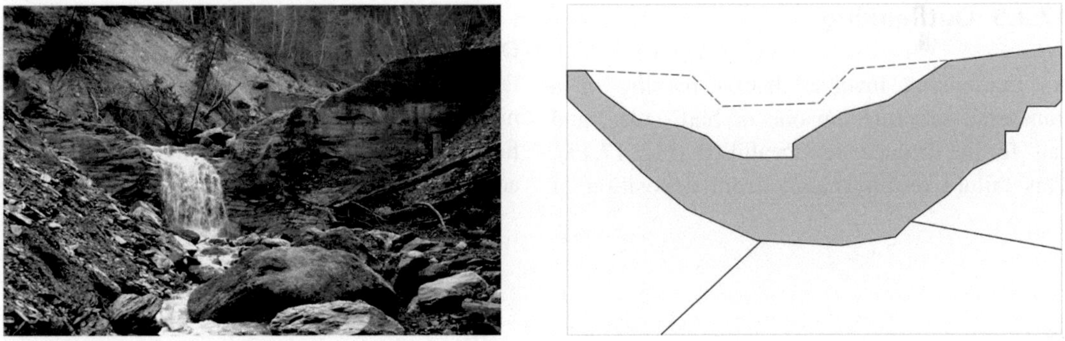

Fig. 17.12 Example of strong abrasion and frost thaw impact STR (Rettenbach, Tyrol, Austria) (left) with schematic illustration (right)

debris. These parts are rapidly torn away. If sheets are insufficiently anchored, they can be torn from their anchorage by the drag force of the water and swept downstream.

The abrasion of structural elements is caused by the interaction of the mechanisms of abrasion, drag stress and the effect of dynamic water pressure or mud pressure.

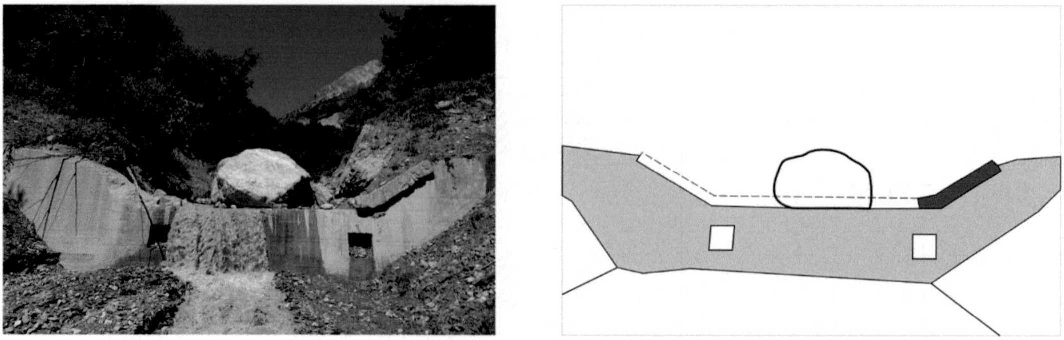

Fig. 17.13 Example of removal of crest stones by debris flows (Lattenbach, Tyrol, Austria) (left) with schematic illustration (right)

17.3.4 Scouring and Loss of Bed Resistance

In general, nappe-induced scouring of the channel bed leads to reduced stability of the structure. Especially for a series of check dams, the spacing between structures plays a significant role in overall safety. Particular care should always be taken with the lowest barrier, because its foundation is not secured against erosion by the deposits of an upstream barrier (Figs. 17.14, 17.15 and 17.16).

17.3.5 Outflanking

An outflanking involves lateral passing of a transverse structure, on one or both sides, and can further reduce serviceability (Fig. 17.17). This failure results mainly from deposition of sediments within the discharge section. This is either based on the complex hydraulic conditions at the spillway leading to reduced transport capacity of sediments upstream of the structure or on sediment deposition from downstream. Outflanking is supported by only small inclined crests of the wing walls related to the deposition angle of the sediments and a too small embedding depth of the structure in the flanks. Outflanking can lead to a loss of resistance of the structure.

17.3.6 Impacts and Overloading

Dynamic pressure builds up on impinged surfaces during mass wasting processes. Its magnitude depends on the discharge height, the flow velocity and the density of the process. In addition, a single impact of a boulder can cause

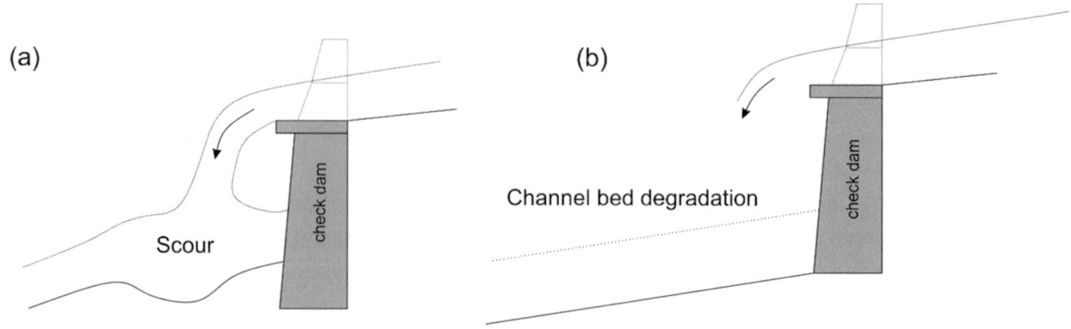

Fig. 17.14 Illustration of scouring (**a**) and loss of bed resistance (**b**)

Fig. 17.15 Example of loss of bed resistance: Gimbach, Upper Austria, Austria (left), Gadria, Southern Tyrol, Italy (right)

Fig. 17.16 Eroded basal and lateral embankment and bending of the right wingwall (crack line) because of lateral slope movement (Rettenbach, Tyrol, Austria) (left), and schematic illustration of failure type (right)

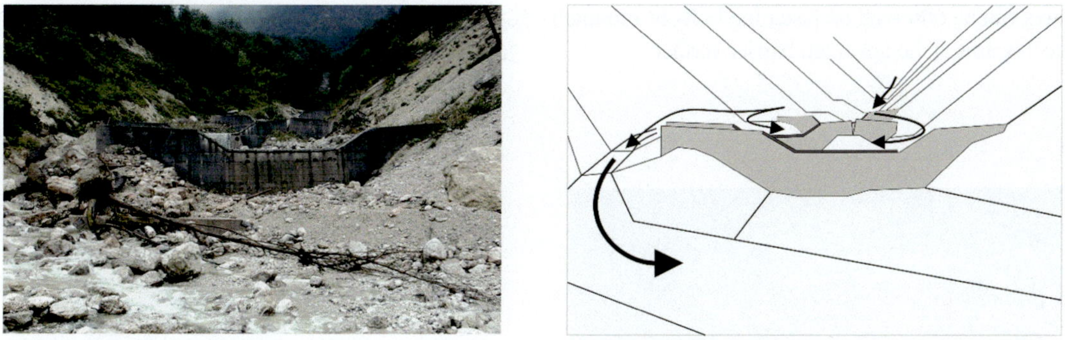

Fig. 17.17 Example of outflanking and loss of lateral resistance (Gimbach, Upper Austria, Austria) (left) with schematic illustration of failure type (right)

a point load of up to several megatons. These loads can cause loss of internal stability (bending, shear failure) and/or tilting of structural components. Tilted components can be further displaced by the drag stress of the discharge (Figs. 17.4 and 17.18).

Cracks occur when barriers made from cement mortar masonry, concrete or reinforced concrete are overloaded or improperly built. They occur in component areas with tensile stresses due to loading, constraint, or residual stresses as soon as the concrete tensile strength is exceeded.

Compared to microcracks, macrocracks penetrate larger areas of the cross-section. Their widths are larger and their orientation is directional because it is determined by the main tensile stresses.

The parts of the wing walls above the bottom edge of the spillway are particularly exposed. In the case of structures filled with sediment, like consolidation (slope buttressing) barriers, these parts are usually the only ones that are directly subjected to debris flows and runoff.

Wingwalls of consolidation barriers are particularly exposed or overstressed when:

- the discharge section is too small or unfavorably dimensioned
- the discharge exceeds the designed discharge
- debris deposits in the upstream apron and the stream axis shifts
- the wings are overloaded by a debris flow.

Plastic deformations of functional parts of check dams, like steel rakes or grades, occur quite frequently. These usually local deformations are caused by the impact of large boulders. Shear fractures can be observed on fastening bolts of mounted steel beams at the upstream barrier surface.

17.4 Inspections, Maintenance and Life Cycle Assessment

Damage to protective structures occurs mainly due to the interaction of the structure with natural processes, depending on the boundary conditions, and leads to a reduction of the load-bearing capacity, serviceability or durability of the structure. Severe damage occurs when a certain limit value of the wear reserve is exceeded and a critical condition in the functional capability is achieved (Rudolf-Miklau et al., 2007). After construction, technical protection measures usually show their highest efficiency. With the decrease of the structural condition, the safety level also decreases.

To achieve the planned optimal service life of the structure, it must be maintained on a regular basis. Therefore, life cycle assessment with inspection/condition assessment and maintenance measures is necessary. In the first stage, regular inspections and condition assessments serve to assess the current condition of the structure by identifying failures and damage as well as impairments of the protective function, especially after events. The second stage involves direct structural or organizational measures. Maintenance measures prevent the critical condition of protective structures and extend the service life of the structures (ÖNORM B4800, 2022; Suda & Strauss, 2007).

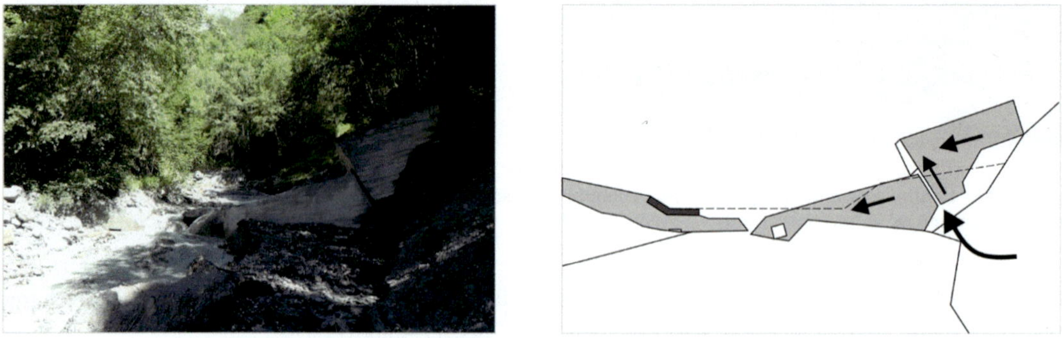

Fig. 17.18 Multiple failure mechanism (lateral earth pressure, uplift caused by shear zones below the structure) (left) and schematic illustration of failure type (right)

17.5 Examples of Structural Failures

17.5.1 Pinnisbach, Tyrol, Austria

On August 7, 2015, after heavy rainfall in the catchment area of the Pinnisbach Tyrol Austria, an event with flooding and heavy bedload transport occurred. This event largely destroyed the filter barrier. The rupture of the barrier triggered a runoff surge, which, however, was caught in the downstream retention basin. The remobilized bedload from the retention area of the destroyed dam was partly deposited in the downstream basin and partly transported to the fan. The failure of the large slot barrier was due to an internal stability failure caused by bond fractures between the reinforcement and the concrete, which led to the pull-out of the reinforcement (Fig. 17.19).

17.5.2 Lattenbach, Tyrol, Austria

Starting in 1965 a series of check dams were built in the middle reach of Lattenbach, Tyrol, Austria. Nearly every year, these structures are impacted by at least one debris flow or debris flood. In addition to the impact forces of these flows, deep-seated landslides on one or both sides affect the stability of these check dams. According to the technology of the 1960's, the steel reinforcement is quite small, leading to different types of failures (Fig. 17.20, 17.21, 17.22, 17.23, and 17.24).

17.5.3 Rumpelgraben, Salzburg, Austria

In the summer 1991, a debris flow occurred in the catchment of Rumpelgraben, Salzburg after a heavy cloudburst. The total volume of the debris flow was estimated to be 12,000 m^3. The slit dam, built at the apex of the fan below a 200 m tall waterfall was impacted by the debris flow and it was reported that a boulder the size of a mini bus (about 25 m^3) destroyed the structure due to an internal stability failure (Fig. 17.25). This slit dam was the only transversal structure in this catchment, but because of a large fan the sediments were widely spread and only a few houses were slightly damaged.

17.5.4 Eisbachgraben, Styria, Austria

On August 3rd, 1981, heavy rainfall in the catchment area of the Eisbachgraben, Styria, triggered a debris flow. This event destroyed the slit dam at the fan apex. The failure of the slit dam, that was designed as a gravity dam, was due to an internal stability failure in the horizontal and vertical construction joints (Fig. 17.26).

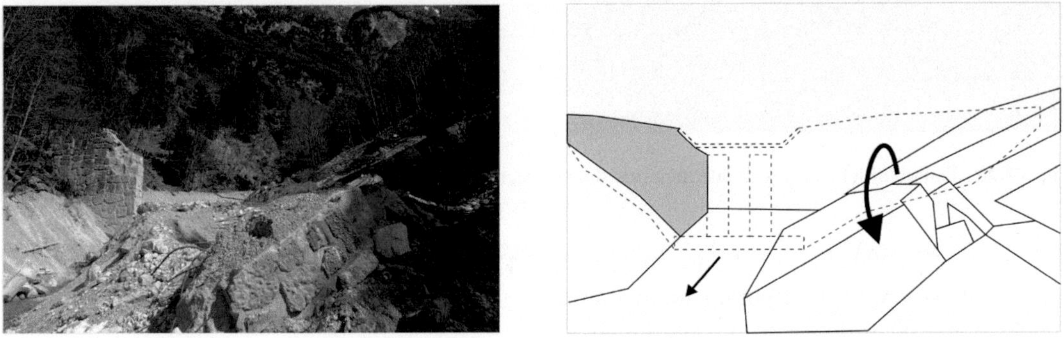

Fig. 17.19 Check dam broken in 6 parts (Hübl et al., 2016a) (left) and schematic illustration of failure (right)

Fig. 17.20 Toe of a landslide blocking the channel, forcing subsequent flows to bypass the structure and directly impact the right wing wall. Schematic illustration of failure type (right)

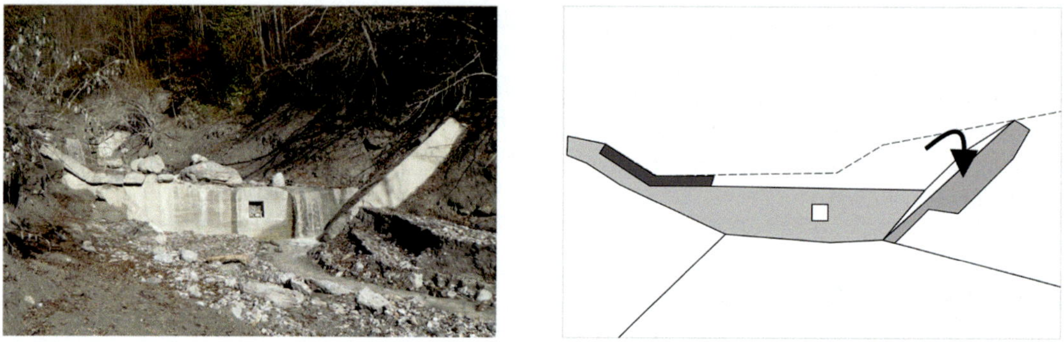

Fig. 17.21 Failure of left wing wall and removal of crest stones due to debris-flow impact load. Schematic illustration of failure type (right)

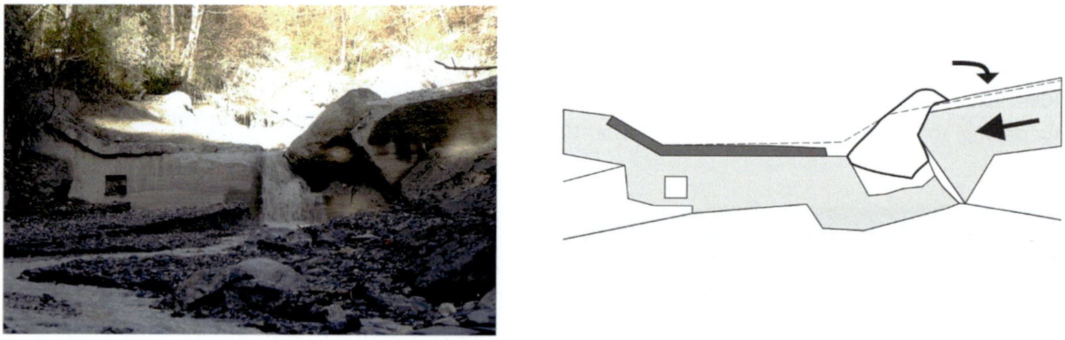

Fig. 17.22 Failure of left wing wall due to debris-flow impact load, (left). Scematic illustration of failure type (right)

Approximately 400 tons of heavy fragments of the slit dam were transported to two areas about 30 and 60 m downstream. The failure of the barrier initiated a subsequent surge, eroding the channel on the fan and leading to sediment deposition in the settlement of Gleiming.

17.5.5 Asane River, Kyoto Prefecture, Japan

A debris flow occurred in the Asane River, located in Kyoto prefecture, Japan, due to heavy rainfall on October 21 to 23, 2017.

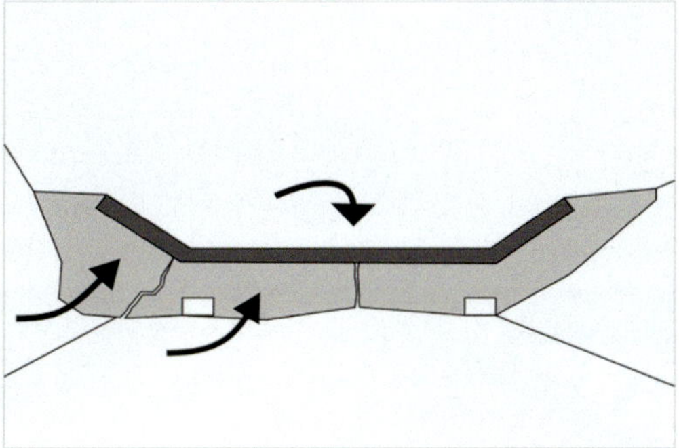

Fig. 17.23 Multiple failure mechanisms (lateral earth pressure, direct impact), (left). Schematic illustration of failure type (right)

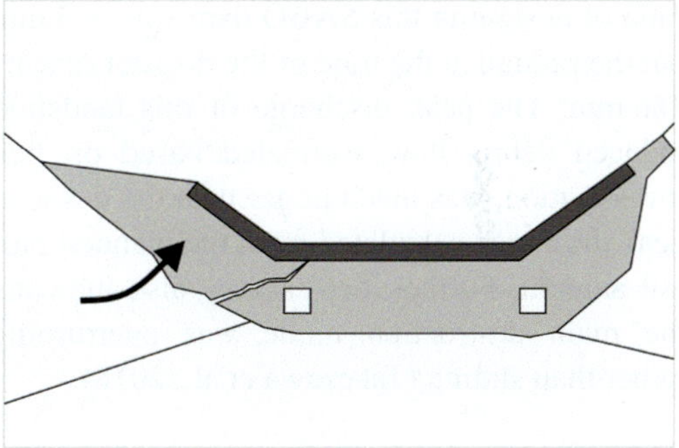

Fig. 17.24 Multiple failure mechanisms (lateral pressure and uplift caused by shear zones below the structure) (left). Schematic illustration of failure type

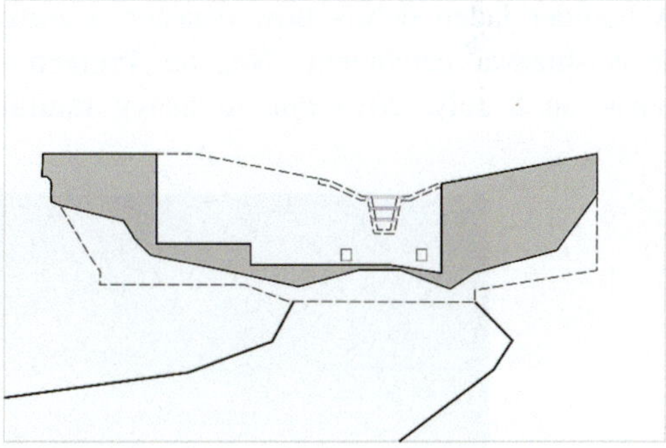

Fig. 17.25 Destroyed slit dam (left), and schematic illustration (right)

The debris flow was initiated by a landslide of about 30 m in width, 60 m in length, and 6 m in depth. Although there were no casualties due to the debris flow, the SABO dam (closed gravity concrete type) in the mountain stream was destroyed. The height of the dam was 9 m and the length was 70 m. The SABO dam was destroyed in the joint-to-joint section (total width of 15 m) in the wing section on the right bank side of the dam due to the debris flow and disintegrated into four concrete blocks (Fig. 17.27). The estimated 24-h rainfall at the

Fig. 17.26 Destroyed slit dam in the catchment area of the Eisbachgraben, before (left) and after event (right)

time of designing this SABO dam was 313 mm, but the rainfall at the time of the disaster reached 486 mm. The peak discharge of this landslide-induced debris flow, estimated based on field investigation, was much larger than the designed peak discharge calculated from the planned rainfall amount. Further, field survey also indicated the main destruction mode was overturning, rather than sliding (Takezawa et al., 2018).

17.5.6 Nashizawa Catchment, Nagano Prefecture, Japan

A boulder laden debris flow disaster occurred at Nashizawa catchment, Nagano Prefecture, Japan on 8 July, 2014 due to heavy rainfall. The drainage area of the Nashizawa catchment is 3.35 km^2, and the average bed slope is approximately 16 degree. One steel pipe open SABO dam was damaged owing to this boulder laden debris flows. It has been considered that the front part of debris flow, which consisted of large boulders, hit the upstream vertical steel pipes, causing the joints of these steel pipes to break in several places. Then the debris flow overtopped the steel pipe SABO dam, damaging top and downstream steel pipes, the upper parts of the steel pipe dam (Fig. 17.28) was shared off along the coupling joint and upper parts of the dam flowed to downstream. The remaining steel pipe components captured the boulders and sediment (Fig. 17.28). The capturing volume was evaluated to be approximately 5,000 m^3.

Fig. 17.27 Destroyed barrier due to debris-flow impact (Takezawa et al., 2018)

Field surveys and numerical simulations were conducted to decipher the failure process. Field surveys showed that huge boulders (5 m diameter) were part of the debris flow, although the maximum boulder diameter was predicted to only be around 1 m at the design stage. In order to design a SABO dam, a field survey was conducted at a range of about 200 m upstream and downstream of the dam site to evaluate the size of boulders that were delivered by past debris flows and set planned maximum boulder diameter. However, this experience indicated that a more detailed field survey was necessary in creeks with large boulders in the debris flow.

17.5.7 Rotian Creek, Trento, Italy

Between 27 and 29 October 2018, north-eastern Italy was affected by a severe cyclonic storm named "Vaia", characterized by exceptionally high cumulative rainfall depths, and locally by high rainfall intensities and record wind velocities. Such conditions triggered a debris flow event, consisting of a series of three surges, in the Rotian Creek (Autonomous Province of Trento) with a watershed of about 2.4 km^2. Here, the initial debris-flow volume was greatly amplified by the collapse and/or undermining of most of the control works built in the

Fig. 17.28 Barrier in the Nashizawa catchment, Japan. (Hiramatsu et al., 2014; Horiguchi & Richefeu, 2014), before (left) and after event (right)

Fig. 17.29 Scoured and subsequently emptied check dam (left); Heavily damaged check dams in the upper part of the channel with loss of bed resistance (right)

1970s, which released an enormous amount of sediment formerly stored in the channel. About 15 consolidation check-dams present in the medium-upper channel were heavily damaged or destroyed (Fig. 17.29). The final debris-flow volume was estimated to be around 150,000 m^3, which included very large natural rocks blocks and pieces of consolidation check-dams transported from the upstream collapses. Downstream, a more recent open check-dam was severely damaged (Fig. 17.30), but it still retained about 25,000 m^3 of sediment. The statical reasons for the extensive damages and the collapses of the consolidation check-dams in the Rotian are still being investigated in detail. However, they were most likely caused by the very high-magnitude of the 2018 debris-flow coupled to the poor conditions of the structures built in the 1970s.

17.5.7.1 Schallerbach, Tyrol Austria

After heavy rainfalls on 7 June 2015 in the watershed of Schallerbach, a catchment of 8.3 km2, the deposition basin at the fan apex was filled with sediments of a debris flood. A subsequent debris flow early in the morning of June 8 with an estimated discharge of about 90–120 m^3/s overtopped the dam crest and redirected the flow to the lateral embedding (Fig. 17.31). The earth dam was eroded on one side and the hence the already deposited sediments were remobilized and transported downstream to the fan. The open check dam with a buffering function remained undamaged. Of the total mobilized 120,000 m^3 of sediments, about 25,000 m^3 were nevertheless retained in the basin, 35,000 m^3 were deposited at the fan and about 60,000 m^3 were discharged to the receiving river Trisanna. (Hübl et al., 2016b).

17.6 Conclusion

Structural failures often occur in combination with disastrous events, since torrents usually include structures from different generations that resist the processes to varying degrees. A total failure of a check dam is in many cases related to factors such as engineering design, geological and geotechnical conditions, processes, and structural and functional boundary conditions (Table 17.1).

It is therefore necessary to determine the correct processes, magnitude of the events, (e.g. F–M Curve development in Jakob and Nolde, this volume), and resulting forces, to develop the appropriate counter measure system (building material, structural design) for the specific catchment (Piton et al., this volume).

Additionally, continuous monitoring of structures should help to prevent total structural failure, since maintenance measures can be implemented as needed to reflect the state of the art in construction and can be designed

Fig. 17.30 Structural failure of the left wing wall (left), and deposited boulder (estimated volume of 18 m^3) in the slit of the open check dam (right)

Fig. 17.31 Overtopping and erosion of an earth dam of a sediment deposition basin (Schallerbach, Tyrol, Austria)

specifically for the age of the mitigation system. To maintain the function of the structures, the hydrology and topographic conditions of the catchment area must be continuously monitored, accounting for the changes caused by climate change.

Acknowledgements We would like to give thanks to the Austrian Forest Engineering Service in Torrent and Avalanche control and the Autonomous Province of Trento for providing examples of failures.

References

Armanini, A., Dellagiacoma, F., & Ferrari, L. (1991). From the check dam to the development of functional check dams. In A. Armanini & G. Di Silvio (Eds.), *Fluvial Hydraulics of Mountain Regions* (pp. 331–344). Springer.

Baggio, T., & D'Agostino, V. (2022). Simulating the effect of check dam collapse in a debris-flow channel. *Science of the Total Environment, 816*, 151660. https://doi.org/10.1016/j.scitotenv.2021.151660

Benito, G., Grodek, T., & Enzel, Y. (1998). The geomorphic and hydrologic impacts of the catastrophic failure of flood-control-dams during the 1996-Biescas flood (Central Pyrenees, Spain). *Z Für Geomorphol, 42*, 417–437. https://doi.org/10.1127/zfg/42/1998/417

Bertolini, L., Elsener, B., Pedeferri, P., et al. (2013). *Corrosion of steel in concrete prevention, diagnosis, repair*. Wiley-VCH.

Borga, M., & Zaramella, M. (2020). Evento di piena del 27–29 ottobre 2018 sul bacino del Rio Rotian: stima della precipitazione e valutazione della sua severità.

Chen, X., Cui, P., You, Y., et al. (2015). Engineering measures for debris flow hazard mitigation in the Wenchuan earthquake area. *Engineering Geology, 194*, 73–85. https://doi.org/10.1016/j.enggeo.2014.10.002

Combrinck, R., Steyl, L., & Boshoff, W. P. (2018). Interaction between settlement and shrinkage cracking in plastic concrete. *Construction and Building Materials, 185*, 1–11. https://doi.org/10.1016/j.conbuildmat.2018.07.028

Concrete Society. (2010). Non-structural cracks in concrete. A concrete Society Report.

EN 1990. (2002). Eurocode - Basis of structural design [Authority: The European Union Per Regulation 305/2011, Directive 98/34/EC, Directive 2004/18/EC].

EN 1992-1-1. (2004). Eurocode 2: Design of concrete structures - Part 1–1: General rules and rules for buildings [Authority: The European Union Per Regulation 305/2011, Directive 98/34/EC, Directive 2004/18/EC].

EN 1997-1. (2004). Eurocode 7: Geotechnical design Part 1: General rules [Authority: The European Union Per Regulation 305/2011, Directive 98/34/EC, Directive 2004/18/EC].

Fuchs, S., Heiss, K., & Hübl, J. (2007). Towards an empirical vulnerability function for use in debris flow risk assessment. *Natural Hazards and Earth Systems Sciences, 7*, 495–506.

Gong, X.-L., Chen, K.-T., Chen, X.-Q., et al. (2020). Characteristics of a debris flow disaster and its mitigation countermeasures in Zechawa Gully, Jiuzhaigou Valley China. *Water, 12*, 1256. https://doi.org/10.3390/w12051256

Hiramatsu, S., Fukuyama, T., Yamada, T., et al. (2014). Disaster report on the debris flow occurred on 9 July 2014 in Nagiso, Nagano prefecture.

Horiguchi, T., & Richefeu, V. (2014). Post-analysis simulation of the collapse of an open sabo dam of steel pipes subjected to boulder laden debris flow. *International Journal of Sediment Research,* 621–635.

Hübl, J., Beck, M., Dörler, S., et al. (2016a). IAN Report 175: Band 2: Analyse Sperrenrbuch Pinnisbach 2015.

Hübl, J., Beck, M., Zöchling, M., et al. (2016b). IAN Report 175: Band 1 Ereignisdokumentation 2015.

Jacobs, F., Winkler, K., Hunkeler, F., Volkart, P. (2001). Betonabrasion im Wasserbau. VAW Versuchsanstaöt für Wasserbau, Zürich, Schweiz.

Lee, K.-H., Ewane, E. B., Uchida, T., & Woo, C.-S. (2022). Damage types and deterioration characteristics of check dams built on mountain streams in Southeast Korea. *Frontiers in Earth Science, 9*, 788029. https://doi.org/10.3389/feart.2021.788029

Mazzorana, B., Trenkwalder-Platzer, H. J., Fuchs, S., & Hübl, J. (2014). The susceptibility of consolidation check dams as a key factor for maintenance planning. *Österr Wasser-Abfallwirtsch, 66*, 214–216. https://doi.org/10.1007/s00506-014-0160-4

Neroth, G., Vollenschaar, D., Wendehorst, R. (Eds.). (2011). Wendehorst Baustoffkunde: Grundlagen - Baustoffe - Oberflächenschutz, 27., vollständig überarbeitete Auflage. Vieweg + Teubner, Wiesbaden.

Neville, A. M. (2011). *Properties of concrete* (5th ed.). Pearson, Harlow, England.

ÖNORM B4800. (2022). Schutzbauwerke der Wildbachverbauung Entwurf, Einwirkungen, Bemessung und Instandhaltung. ICS 13.200.

Paratscha, R., von der Thannen, M., Smutny, R., et al. (2019). Screening LCA of torrent control structures in Austria. *International Journal of Life Cycle Assessment, 24*, 129–141. https://doi.org/10.1007/s11367-018-1501-5

Piton G, Recking A (2016) Design of Sediment Traps with Open Check Dams. I: Hydraulic and Deposition Processes. J Hydraul Eng 142:04015045. https://doi.org/10.1061/(ASCE)HY.1943-7900.0001048

Remaître, A., van Asch, Th. W. J., Malet, J.-P., & Maquaire, O. (2008). Influence of check dams on debris-flow run-out intensity. *Natural Hazards and Earth Systems Sciences, 8*, 1403–1416. https://doi.org/10.5194/nhess-8-1403-2008

Rudolf-Miklau, F., Suda, J., & Sicher, P. (2007). Erhaltungsmanagement der Wildbachverbauung. *Bautechnik, 84*, 838–848. https://doi.org/10.1002/bate.200710070

Sodnik, J., Martinčič, M., Mikoš, M., Kryžanowski, A. (2015). Are torrent check-dams potential debris-flow sources? In: G. Lollino, D. Giordan, G. B. Crosta, et al. (Eds.), *Engineering geology for society and territory* (vol. 2, pp 485–488). Cham: Springer International Publishing.

Suda, J. (2013). *Instandhaltung von Schutzbauwerken gegen alpine Naturgefahren =: Maintenance strategies for protection works*. Guthmann-Peterson.

Suda, J., & Strauss, A. (2007). Schadensklassifizierung für Schäden an Schutzbauwerken der Wildbachverbauung.

Takezawa, N., Akazawa, F., Ishii, Y., & Nishii, H. (2018). Survey of sabo dam destruction caused by debris flow in Ayabe City, Kyoto Prefecture (prompt report). In: *Proceedings of Annual Meeting of The Japan Society of Erosion Control Engineering*.

Wang, G. L. (2013). Lessons learned from protective measures associated with the 2010 Zhouqu debris flow disaster in China. *Natural Hazards, 69*, 1835–1847. https://doi.org/10.1007/s11069-013-0772-1

Yu, B., Ma, Y., & Wu, Y. (2013). Case study of a giant debris flow in the Wenjia Gully, Sichuan Province, China. *Natural Hazards, 65*, 835–849. https://doi.org/10.1007/s11069-012-0395-y

Warning Systems and Instrumentation

18

Lorenzo Marchi, Velio Coviello and Marcel Hürlimann

Abstract

In the last two decades, sensors and techniques for debris-flow monitoring and their application for warning purposes have progressed significantly and many operational warning systems are working worldwide. Such advances motivate presenting an updated overview of the state of the art of early warning systems tailored towards debris-flow risk reduction. In this chapter, we classify debris-flow warning systems into advance warning systems, which are based on the monitoring of meteorological factors and issue a warning before a debris flow is triggered, and event warning systems, which detect debris flows after they have initiated. Advance warning systems typically rely on rainfall thresholds that are based on rain gauge data related to past debris flows. We discuss the main issues related to the rainfall thresholds and the prospects for the use of radar-based rainfall estimates. Several sensors and devices can be used in event warning systems. In this work, we focus on the most widespread ones: (i) radar and ultrasonic sensors, which detect debris flows by measuring changes in the flow level, (ii) impact sensors such as trip-wires and pendulums, (iii) seismic and infrasound sensors, whose operation is based on the detection of ground vibrations and air pressure waves induced by debris flows, and (iv) video cameras. We outline some basic issues related to the implementation of protocols for debris-flow warning, emphasizing the importance of focusing the warning on end-users with a different technical background and the need to inform the population about debris-flow risks, the expected performance of the warning system, and the action following a warning. In the concluding remarks, we stress that debris flows are often concurrent with other meteorological-induced hazards and risks, such as flash floods. This implies that early warning systems for debris flows need to be developed in a multi-risk framework.

L. Marchi (✉) · V. Coviello
Research Institute for Geo-Hydrological Protection, National Research Council (CNR IRPI), Padova, Italy
e-mail: lorenzo.marchi@cnr.it

M. Hürlimann
Department of Civil and Environmental Engineering, UPC BarcelonaTech, Barcelona, Spain

18.1 Introduction

Hazard and risk caused by debris flow have urged the development of an array of control measures, both structural and non-structural (Davies, 1997; Hübl & Fiebiger, 2005; Hungr

M. Jakob et al. (eds.), *Advances in Debris-flow Science and Practice*, Geoenvironmental Disaster Reduction,
https://doi.org/10.1007/978-3-031-48691-3_18

et al., 1987; Ikeya, 1989); among the latter, early warning systems play an important role (cf. Baum & Godt, 2010, and Jakob et al., 2012 for a list of select debris-flow warning systems in the USA and worldwide, respectively, and Guzzetti et al., 2020 for a review of landslide warning systems, including debris flows). Some features of debris flows, however, challenge the implementation of warning systems: the high velocity of debris flows results in very short lead times, and the low frequency of the events in a given channel limits the opportunity of testing the detection devices and algorithms. These issues need to be accounted for in developing effective warning systems, which differ from those developed for other types of landslides (Chae et al., 2017).

The NOAA-USGS Debris Flow Task Force (2005) extended to debris flows the terminology used by the NWS (National Weather Service) for hazardous weather messages. Debris-flow warning systems are thus based on possible development of hazardous meteorological or hydrological events (Outlook), the increased probability of such events (Watch), and the occurrence, imminence, or very high probability of a hazardous event (Warning). The focus of this contribution is on the Warning level.

Hungr et al. (1987) proposed subdividing warning systems for debris flows into Advance Warning Systems and Event Warning Systems. The aim of Advance Warning Systems, which are based on monitoring causative hydrometeorological processes (usually rainfall), is to issue a warning before a debris flow is triggered. Event Warning Systems detect debris flows after they have started and aim at issuing warnings before they affect vulnerable areas. Advance Warning Systems enable longer lead times than event warning systems but at the cost of uncertainty on the actual occurrence of a debris-flow (Fig. 18.1). In that, false warnings (false positives) are much more likely than for Event Warning Systems. Where such warnings are linked to evacuations, false warnings may eventually lead to "evacuation fatigue" meaning that such warnings will be ignored. In this work, we will follow the classification by Hungr et al.

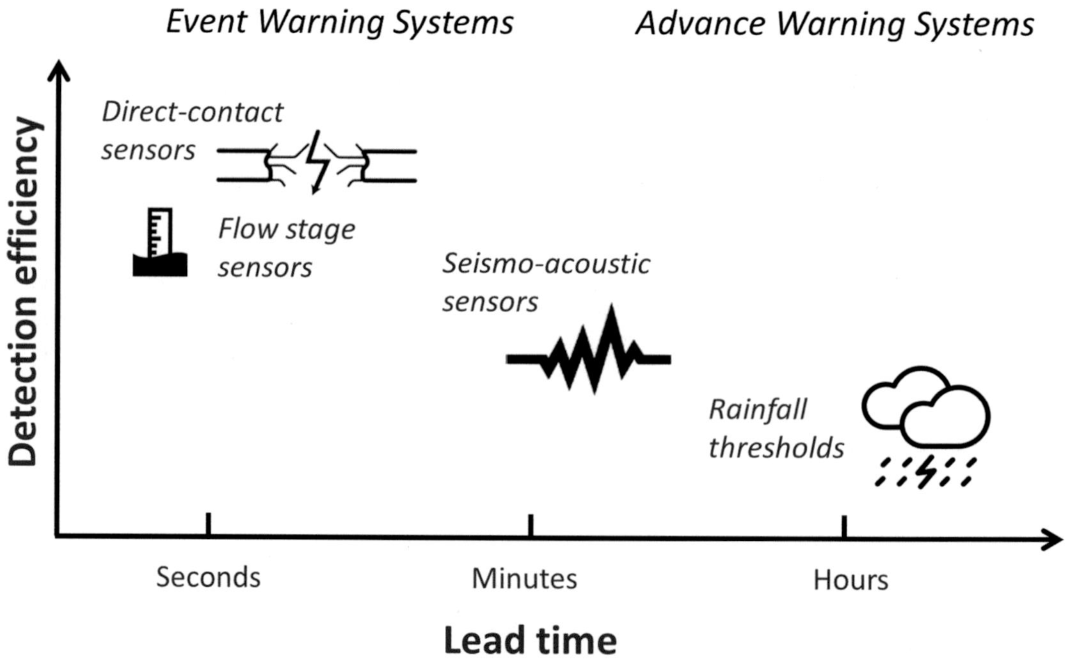

Fig. 18.1 Detection efficiency versus lead time of the most common sensors for debris-flow warning (modified from Coviello, 2023)

(1987). The sketch of Fig. 18.1 refers to the detection of a single event, e.g. the debris flow main front. In the case of secondary surges, the performance of direct-contact sensors dramatically decrease (cf. Sect. 18.3.2).

The systems for debris-flow monitoring for research and for applied warning systems share several common features, including the use of the same sensors for flow stage and ground vibration measurement (Arattano & Marchi, 2008; Itakura et al., 2005), but differ by the objectives of the systems, which also affect the deployment of the sensors and the collection and transmission of the data. A recent review (Hürlimann et al., 2019) encompasses both research and warning tasks. In this chapter, we focus on the warning systems, with attention to both the sensors for debris-flow detection and the implementation of warning procedures. According to the objectives of the systems, the choice of the sites to be instrumented follows different criteria. The debris-flow catchments and channels to be monitored for research purposes can be selected accounting for the high frequency of the events, enabling effective data collection, as well as practical conditions, such as easy access to the monitored sites or the availability of electrical power. The choice of the sites to be instrumented for debris-flow warning is dictated by the presence of elements at risk, including the need to deploy detection sensors also under less favorable logistic conditions. The debris-flow warning systems usually consist of permanent installations that must be based on a solid design and are aimed at operating for several years. Temporary installations can be devised to protect temporary elements at risk, such as workplaces within or near a debris-flow channel. Mobile stations for debris-flow monitoring and warning, consisting of vehicles platforms equipped with sensors for debris-flow detection, signal processing, and transmitting units, have also been developed (Yin et al., 2011). Mobile stations can be deployed to enhance debris-flow detection and warning at sites where no permanent warning systems are located. The use of mobile stations is useful in debris-flow-prone regions, such as Taiwan and other regions hit by tropical typhoons, where debris-flow triggering storms follow variable but at least partly predictable paths (Yin et al., 2011).

In the last decades, monitoring and warning techniques have significantly progressed (Hürlimann et al., 2019), and many operational warning systems are working worldwide. In a previous book on debris-flow hazards (Jakob & Hungr, 2005), the topic of warning was only dealt with indirectly in the chapter "Debris-flow instrumentation" (LaHusen, 2005) and some case studies (Hübl & Fiebiger, 2005; Zimmermann, 2005). This chapter presents an overview of the state of the art and the recent advances of early warning systems aimed at debris-flow risk reduction.

This chapter is organized as follows: Sect. 18.2 describes Advance Warning Systems, which are based on precipitation monitoring and rainfall thresholds, Sect. 18.3 presents the main types of sensors for the detection of debris flows in event warning systems, Sect. 18.4 briefly outlines the protocols for warning, while Sect. 18.5 discusses open issues and prospects, and draws some concluding remarks.

18.2 Advance Warning Based on Rainfall Monitoring

18.2.1 Spatial Scale of Debris-Flow Triggering

Debris flows typically occur in small catchments, usually up to a few square kilometers in most geographical regions, and less frequently reach or exceed a few tens of square kilometers (e.g. García-Martínez & López, 2005; Guo et al., 2016a). Debris-flow triggering precipitation, however, may occur at different space and time scales (from convective cells to mesoscale convective systems and fronts), which imply different monitoring devices and strategies (refer to Kaitna, et al., this volume, for detailed analysis of triggering mechanisms). The rain gauge networks operated by hydrological and meteorological offices or environmental agencies are suitable for assessing debris-flow triggering rainfall at the regional scale for frontal precipitation and mesoscale convective systems, although in this second case the use of radar rainfall estimates would help better detecting the strong spatial variability of rainfall. In the case of debris flows caused by convective cells, their limited extent often causes regional rain gauge networks to miss-high-intensity cells, thus underestimating debris-flow triggering rainfall, while radar rainfall estimates based on a complete correction algorithm lead to rainfall at the debris-flow initiation sites higher than rain gauge-recorded values (Marra et al., 2014). The main limitation in the use of quantitative radar rainfall estimates in early warning systems is that the implementation of a full set of corrections to the raw radar signal is a complex task. In particular, the integration with rain gauge data commonly requires cumulated data at a multi-hour scale or for the whole event (Marra et al., 2016). However, recent advances permitted the incorporation of a combined product in the landslide early warning system over Catalonia, which can be updated every 30 min (Palau et al., 2022). In addition, weather radar observations permit tracking the path of convective cells, thus contributing to identify the debris-flow-prone catchments at risk to be impacted by intense rainfall and extending the alert lead time (Badoux et al., 2009; LaHusen, 2005; Tiranti et al., 2014).

When trying to predict debris flows caused by convective cells (the debris-flow history of a catchment permits recognizing the atmospheric processes responsible for debris-flow occurrence—e.g. Nikolopoulos et al., 2015b; Toreti et al., 2013), installation of rain gauges close to the debris-flow initiation sites in the headwaters of the concerned catchments is ideal (Bernard & Gregoretti, 2021). This is typically done in catchments instrumented for research purposes (Hürlimann et al., 2019). However, it should be considered to improve warning effectiveness in catchments with frequent debris flows and substantial downstream risk. This coupling of Event Warning Systems with an Advance Warning System will extend the warning time. However, even when rain gauges are installed close to debris-flow initiation areas, the limited extent of the convective cells and the associated strong spatial gradients of rainfall challenge the recognition of triggering rainfall.

The Gadria catchment (eastern Italian Alps), where two rain gauges have been installed close to the debris-flow initiation area (Coviello et al., 2021b), provides an example of the variability in space of debris-flow triggering rainfall. Figure 18.2 plots rainstorm duration (from the onset of the rainfall to the occurrence of the debris flow) for 12 debris-flow triggering rainstorms between 2014 and 2020. Some events are similar in rainfall duration and mean intensity, whereas others differ substantially. The latter is especially common for short-duration rainstorms, related to small convective cells. It is possible to argue that one of the two rain gauges (or perhaps both of them) were not able to measure the highest rainfall amounts and intensities that caused debris-flow initiation. It is especially remarkable that such profound differences were measured between two neighboring rain gauges (horizontal distance: 2.5 km, vertical distance: 160 m) in a catchment that is much more densely instrumented than most debris-flow catchments. The uncertainties in recording the triggering rainfall affect the assessment of rainfall thresholds for debris-flows: we refer to Nikolopoulos et al. (2014) for a study at a regional scale that analyzes the mismatch between the location of the rain gauges and the initiation areas of debris flows. Data gathered in monitored basins confirm that the actual area contributing to debris-flow initiation is often considerably smaller than the whole basin (e.g., Berti et al., 2020; Coviello et al., 2021b).

18.2.1.1 Rainfall Thresholds for Debris-Flow Warning

Some of the first and most known studies aimed at identifying rainfall thresholds for the occurrence of debris flows date back to the early 1980s with the pioneering works by Caine (1980) and Innes (1983). Caine (1980) pooled debris flows with other landslide types using a global literature review, while Innes (1983) only included debris flows.

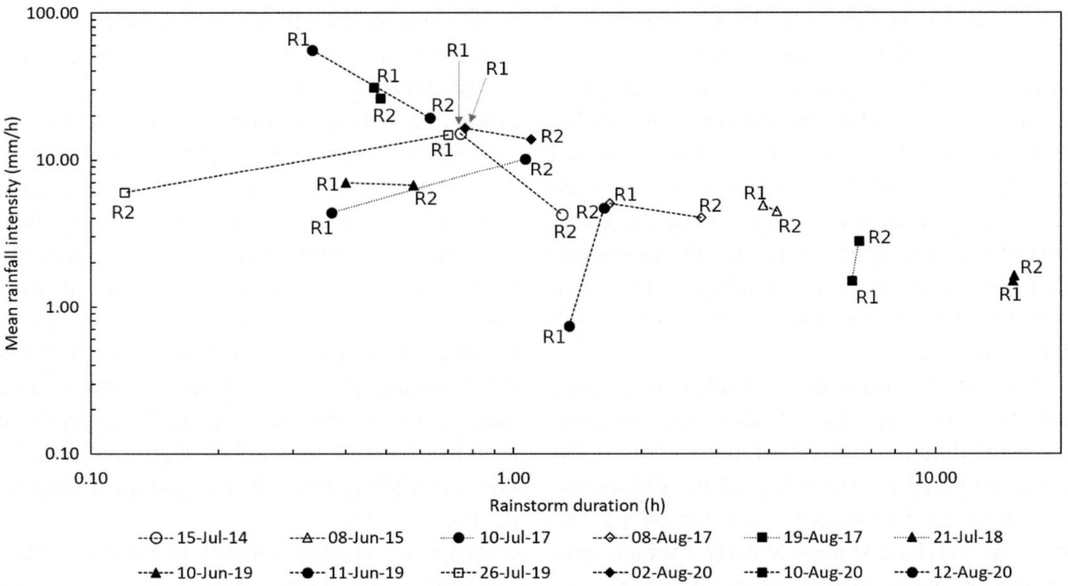

Fig. 18.2 Duration and mean intensity of debris-flow triggering rainstorms recorded at two rain gauges (R1 and R2) in the upper part of the Gadria catchment

The equations proposed by Caine (1980), Eq. 18.1, and Innes (1983), Eq. 18.2 are reported as archetypes of many equations for limiting rainfall curves developed afterward by many authors.

$$I = 14.82 \cdot D^{-0.39} \quad (18.1)$$

$$E = 4.9355 \cdot D^{0.5041} \quad (18.2)$$

where D is the rainfall duration (h), and I and E are the mean rainfall intensity of the rainstorm (mm/h) and total rainfall amount (mm), respectively.

Equation 18.1 is the prototype of critical rainfall thresholds on plots of mean intensity versus rainfall duration. These equations are very common, but critical rainfall equations based on cumulated rainfall versus rainfall duration are becoming more and more popular as they avoid having the same variable (i.e., rainstorm duration), on the x-axis and as a component of the variable on the y-axis.

Two review papers (Guzzetti et al., 2008; Segoni et al., 2018) and several other scholars (Marra et al., 2014; Nikolopoulos et al., 2015a; Pavlova et al., 2014; Winter et al., 2010) identified rainfall thresholds at the regional scale and considered debris flows together with other types of landslides. One reason for this choice is that in many sources debris flows are not distinguished from other landslides, moreover pooling debris flows and other landslides together permits building larger datasets that enable more robust statistical analysis. Such choice is surely acceptable because the statistical approaches to rainfall thresholds assessment do not imply the assessment of the physical process leading to landslide or debris-flow initiation. When possible, however, a separate analysis for debris flows (or, at least, for flow-type landslides) is recommended to better identify the hydrometeorological controls on the triggering of these phenomena.

Many thresholds (41% of those reviewed by Segoni et al., 2018) are merely empirical and fitted manually, but there is a strong need for objective and reproducible criteria for defining rainfall threshold (regression, probabilistic or physically-based models). Such methods require large datasets of debris-flow triggering and not-triggering rainfall but may provide more robust and objective results (e.g. Berti et al.,

2012; Melillo et al., 2018). In addition, a validation of the threshold is recommended using confusion matrices and receiver operating characteristic curves (e.g. Hirschberg et al., 2021). Additional information on many aspects related to rainfall thresholds can be found in the recent review by Segoni et al. (2018), while the selection of the most appropriate rainfall variables is still being debated. Refer also to Gartner, et al. (this volume) for post-wildfire effects on rainfall thresholds.

Most of the available thresholds focus principally on rainfall characteristics, such as mean intensity, duration, antecedent precipitation. The concept of hydro-meteorological thresholds was re-introduced by Bogaard and Greco (2018), as some previous studies already applied similar approaches (Baum & Godt, 2010). Several studies have incorporated hydro-meteorological thresholds using the information of soil moisture gathered by in-situ monitoring, hydrological models, or satellite observations (e.g. Jakob et al., 2012; Krøgli et al., 2018; Mirus et al., 2018; Oorthuis et al., 2023; Thomas et al., 2019).

Figure 18.3 presents an example of critical rainfall thresholds for landslides (including debris flows) in the Italian Alps. The thresholds were represented by power-law equations and

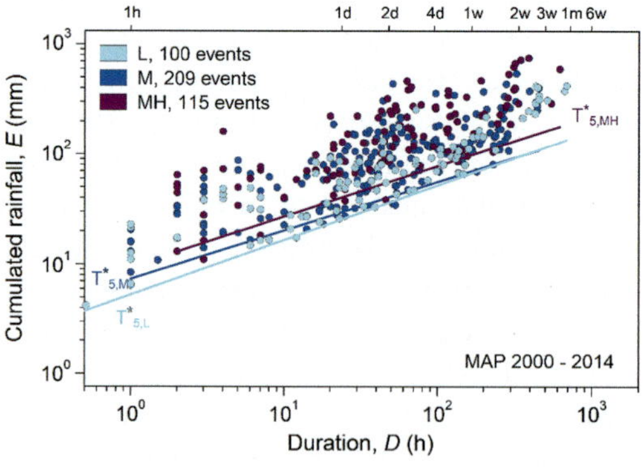

Fig. 18.3 Log–log plot of cumulated rainfall versus rainstorm duration for landslide-triggering rainstorms in the Italian Alps (modified from Palladino et al., 2018). The threshold lines at 5% exceedance probability levels ($T^*_{5,MH}$, $T^*_{5,M}$, and$T^*_{5,L}$) have been computed for three classes of Mean Annual Precipitation (MAP): low (L), medium (M), and medium–high (MH)

were calculated at the 5% probability exceedance level. To consider climatically homogeneous domains, the sample was stratified into three classes according to Mean Annual Precipitation (MAP) in the period 2000–2014: L (low): MAP 501–1000 mm, M (medium) MAP 1001–1500 mm, MH (medium–high): MAP 1501–2000 mm. Higher rainfall amounts are required for triggering landslides and debris flows in areas where mean annual precipitation is higher. A possible explanation (Palladino et al., 2018) is the existence of an equilibrium between landscape settings (for example the occurrence of rainfall-triggered processes that can modify them, including debris flows) and local climatic regime.

More in general, rainfall thresholds reflect catchment conditions, and they need to be modified if such conditions undergo significant changes. A widely studied example (Zhou & Tang, 2014, and references therein) is the Wenchuan earthquake in southwestern China. In this example, lowered debris-flow rainfall thresholds observed after the earthquake were ascribed to increased availability of loose material induced by the seismic shocks. In later years, several processes, including depletion of erodible debris and revegetation of bare slopes, resulted in the recovery of rainfall thresholds close to pre-earthquake values (Guo et al., 2016b; Jiang et al., 2021).

18.3 Event Warning Based on Debris-Flow Detection

An event debris-flow warning system consists of three main components: detection devices, a processing unit, and devices for warning dissemination (Fig. 18.4). The detection devices should be placed as much upstream as possible in order to maximize warning time. However, since it is possible that a debris flow that originated in the upper part of the channel does not reach the lower part of the catchment or the alluvial fan where the elements at risk are commonly located (Bel et al., 2017; Marchi et al., 2021), debris-flow detection in the headwaters should

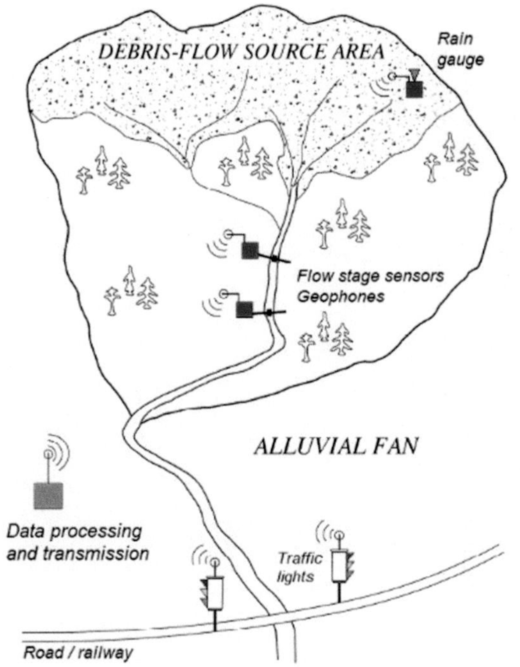

Fig. 18.4 Sketch of a warning system for debris flows (modified from Arattano & Marchi, 2008)

be integrated with sensors closer to the elements at risk. Depending on the size of the catchment and the pattern of the debris-flow channels, two to four debris-flow detection sites are recommended along the debris-flow path; more sites may add unnecessary system complexity. Such a detection system helps minimize false alarms thanks to the sensors in the lower part of the channel, but also provides an earlier warning—although it is more prone to be a false positive—attributable to the sensors in the headwaters. The processing unit should be installed in a safe location with a reliable connection to the detection sensors and the dissemination devices. The devices and tools for alarm dissemination depend on the infrastructure(s) to be protected, and encompass traffic lights, sirens, and mobile telephone messages. The type and location of the elements at risk influence the addressees of the warnings. In addition to the technical office in charge of the management of the warning system, the addressees may include local authorities, if different from the technical managers, and the general public.

According to the scope of this contribution, we will focus on the detection sensors. Table 18.1 lists the most common sensors and devices that can be used to detect debris flows for warning purposes. The sensors listed in Table 18.1 have different stages of maturity. While there were not major advances in the technology of seismic sensors, the algorithms for debris-flow detection based on ground vibrations sensors have evolved significantly in the last two decades (cf. Sect. 18.3.3). Infrasound sensors and microphones are a promising technology that will benefit from further research in the assessment of warning thresholds. Trip-wires and stage sensors are mature devices, still effective in the context of debris-flow warning.

Fiber optic sensors, namely distributed acoustic sensors (DAS), have been proposed recently for debris flow monitoring. In DAS, the optical fiber cable acts as a coherent array of equivalent geophones (Schenato & Pasuto, 2022). A large number of equivalent sensing points, in the order of many hundreds per kilometer of fiber, along with the high sensitivity of this technology, promises to constitute an effective platform capable of precisely following the flow, providing the distributed characterization of the spectral response. This approach could become part of a new generation of systems for debris-flow warning.

18.3.1 Flow Level Sensors

The operation of the sensors for the measurement of flow level is the same as the sensors for the water level in rivers and lakes or the level of liquids (or granular material) in silos and tanks. However, because of the fast flow level variations occurring in debris flows, ultrasonic and radar sensors must record data at time intervals of approximately one value per second.

Regarding the type of sensor to be used in debris-flow warning systems, Badoux et al. (2009) stress the advantages of radar sensors over ultrasonic sensors, which sometimes miss returning a signal under rapidly changing flow depth associated with debris-flow surges.

Table 18.1 Sensors for debris-flow detection for warning purposes

Sensor	Measured parameter/observed phenomenon	Advantages	Limitations and possible drawbacks
Ultrasonic, radar, and laser sensors	Flow stage	Easy to set and change warning thresholds, non-contact sensor	Need for stable banks (the sensors have to be suspended over the channel)
Trip-wire sensors	Debris-flow occurrence	Simple and robust device	Need for restoration after activation. Channel erosion or aggradation may undermine the sensor's operation
Pendulums	Debris-flow occurrence	Simple device	Channel erosion or aggradation may undermine the device's operation
Geophones and seismometers	Ground vibration velocity	Low cost (geophones), easy and safe installation, non-contact sensor	Setting warning thresholds can be complicated
Infrasound sensors and microphones	Air pressure waves	Low cost (microphones), safe installation, non-contact sensor	Data analysis and interpretation is complex. Threshold definition is difficult
Videocameras	Debris-flow occurrence/debris-flow features recognition	Low cost, helpful in event recognition and interpretation	Not useful for detection, unless a sophisticated algorithm of image processing is available

An alarm can be issued when the increase of flow stage exceeds a pre-defined threshold in a given time interval. The maximum level of past debris flows visible on the channel banks is fundamental for setting the warning thresholds, which must obviously be lower, and for installing the sensors at a safe elevation. Many debris-flow channels are affected also by water flows: when setting the warning threshold, a short time interval for the rise in flow level permits discriminating debris flows from water floods that may reach flow depth similar to small debris-flow surges but show a much slower rise. For a system recording flow levels every second, the duration for assessing the critical increase of flow depth can be set tentatively to 5 or 10 s. The level threshold for warning must be set relative to the elevation of the channel bed before the event, which accounts for prior channel bed incision or aggradation. A time threshold should also be set to discriminate small debris-flow surges from water floods, which have a much slower rise. Spikes in the records of flow depth may occur because of splashes in turbulent water flow: to avoid false alarms, the detection algorithms should filter out short-duration (<5 s) spikes occurring only at one sensor measuring the flow depth. Real alarms can be issued if an increase of the flow depth occurs at the same times at two sensors installed on the same cross-sections, or progressively from upstream to downstream if two or more sensors are installed at different cross-sections. When available, recorded debris-flow hydrographs should be used to set the level thresholds for debris-flow warnings. As an example, in the Moscardo Torrent (eastern Italian Alps), 62 surges were observed during 26 debris-flow events recorded between 1990 and 2019 (Marchi et al., 2021). The flow depth, measured as the rise from the previous level, varies from 0.51 m to 3.96 m, with a mean and median of 1.48 and 1.46 m, respectively, while the lower quartile is 0.78 m. While the detection of debris-flow surges using flow stage sensors would be relatively easy for major events, a careful selection of the threshold level and the time of rise is necessary for small surges, whose depth is similar to that of water floods with bedload and suspended load occurring in the same channel. In the case of Moscardo Torrent, the detection threshold could be set close to 0.5 m, whereas a threshold of 0.8 m would leave 25% of the debris-flow surges undetected.

An alarm ceases after a pre-defined time interval, which can be chosen based on the duration of previous debris flows in the same channel, or when flow-level measurements become stable; that is, they show fluctuations smaller than a given level for a selected time interval. Figure 18.5 shows a multiple-surge debris-flow hydrograph recoded in the Gadria catchment: the red and green arrows indicate the ideal times of debris-flow detection and end of the event. Possible early interruption of the alarm—occurring when the flow depth shows small

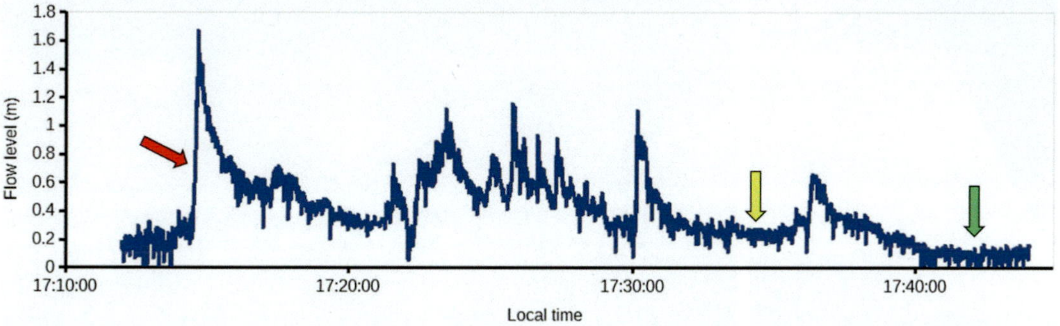

Fig. 18.5 Example of debris-flow hydrograph with optimal times for the detection of the onset (red arrow) and the end (green arrow) of the debris flow event; the yellow arrow indicates a temporary phase of almost steady flow that could result in an early cessation of the alarm

variations between two surges (yellow arrow in Fig. 18.5)—are not detrimental for the management of warning, provided that the system can detect new, even minor surges: a rise of 0.4 m occurring in no more than 5 s could be an adequate threshold in this case.

As pointed out in Table 18.1, sensors for flow stage measurement must be installed above the channel, and their installation requires stable side slopes. If present, a bridge (Fig. 18.6a) can be a satisfactory site for the installation of flow stage sensors, provided that the span below the bridge is large enough to permit the passage of the debris flow without damaging the sensors (Fig. 18.6b). Since the transversal cross section of the channel can change over time and the deepest point may shift laterally, the flow stage sensor should be checked after each debris-flow event and maybe moved to this deepest point of the cross section.

18.3.2 Trip-Wire Sensors and Pendulums

Trip-wire sensors (Fig. 18.7) are simple devices that consist of a steel wire installed across a channel cross-section and connected to an on/off switch: the passage of the debris flow causes the wire to be ripped-off and results in the activation of the switch.

The need for sensor restoration after a debris flow (Table 18.1) is a severe limitation of trip-wire sensors in case of multiple-surges events (only the first one gets detected) or if a second debris flow occurs after the one that severed the wire. Moreover, channel erosion from minor intermittent debris flows can compromise the functioning of the system if the span between the channel bed and the wire installed across the cross-section is too large. Channel aggradation can also be detrimental for the operation of trip-wire sensors because it can cause the wire to be buried under the sediment, and thus be no longer of use for debris-flow detection.

The problems related to channel erosion can be avoided (and those of aggradation at least reduced) by installing trip-wire sensors on a stable cross-section, such as a lined channel or the spillway of a check dam. Trip-wire sensors are also moderately sensitive to accidental activation caused by the fall of trees from the side slopes or tripping by large animals. To cope with these problems, two or more wires can be installed in

Fig. 18.6 Examples of radar sensors installed for debris-flow detection in northern Italy. **a** installed under a bridge (Cancia channel, eastern Italian Alps); **b** wire-suspended (Gadria catchment)

Fig. 18.7 Trip-wire installed at Erh-Pu-Keng Gully (Taiwan) (**a**) and detail of the cable connection to the check dam (**b**)

sequence so that their snapping in downstream sequence is required to announce a debris flow. Despite these drawbacks, wire sensors are rather common in warning systems for roads and railways because of the simplicity of the system and easy installation. This simple detection device is also suitable for temporary installations, such as those intended to protect worksites for the construction of mitigation works in debris-flow-prone channels. The presence of skilled personnel at permits fast reinstallation of the trip-wire after activation.

Pendulums (Fig. 18.8), similar to those used for snow avalanches warning systems (Bossi et al., 2015), have similar advantages and limitations as trip-wire sensors, although ideally they do not need restoration after activation. However, experience with these devices has shown that they can be entrained and destroyed by debris flows or damaged by their impact, thus requiring reinstallation after an event.

Fig. 18.8 Example of cable-suspended pendulum for debris-flow detection (Rotwandbach, eastern Italian Alps)

18.3.3 Seismo-Acoustic Sensors

Debris flows generate seismic, sonic, and infrasound waves that propagate in the ground and in the air. Seismic waves are mainly produced by solid particles impacting the channel bed and banks as well as by the collisions between particles within the moving debris flow (e.g., Kean et al., 2015). Sonic and infrasound waves are formed by the splashing that occurs at the surface of the flow (e.g., Coco et al., 2021). In recent years, the use of seismo-acoustic sensors for debris-flow warning is experiencing a growing interest among scientists and local authorities. These sensors are attractive because they can provide continuous streams of data at a safe distance from the active channel. Low-cost sensors like geophones and microphones represent a unique opportunity to monitor remote areas with a dense and easy-to-install sensor network. On the other hand, instrument installation, maintenance, and data analysis require efforts and competencies in signal processing. Current research on seismo-acoustic sensors is devoted to detect runoff processes occurring in the source areas that may evolve into debris flows. This detection can be used as an advance early warning.

Three main approaches have been developed so far for the seismo-acoustic detection of debris flows:

1. linear array of geophones along the debris-flow channel;
2. network of seismometers in an active basin;
3. array of infrasound sensors.

A linear array of geophones installed along a debris-flow channel was tested by Arattano (1999) in the Moscardo basin, the observation basin providing the longest debris-flow monitoring dataset of the Italian Alps (Marchi et al., 2021). Data collected at Moscardo more than twenty years ago demonstrated that the passage of a debris flow can be identified using seismic devices placed on the channel banks and that the estimation of the flow velocity is also possible. This finding also suggested that an amplitude threshold could be identified to detect the passage of a debris flow at a monitored cross-section. A similar approach was developed at Illgraben (Switzerland), where the detection is based on a different signal metric, i.e. the number of impulses (Abancò et al., 2012) An impulse is computed when an amplitude threshold is exceeded and it lasts until the signal crosses the line of zero amplitude. This computation requires the selection of a proper amplitude threshold that controls the possibility to detect a precursory surge and the first signal glimpse produced by the arrival of the debris-flow front (Arattano et al., 2014). Coviello et al. (2019) developed an automatic algorithm based on the Short Time Average over Long Time Average (STA/LTA) ratio of the seismic signal for early detection of debris-flow occurrence in a continuous stream of seismic data and to filter out different seismic sources (anthropic noise, earthquakes, small rockfalls). In Fig. 18.9 we show an example of application of the STA/LTA detection algorithm to the debris flow that occurred on 18 July 2013 at Gadria. The different surges of the debris flow are well visible in the amplitude curves (Fig. 18.9a), and the main front (17:22:50) is detected about one minute in advance using the STA/LTA (Fig. 18.9b and c). This approach of an array of sensors parallel to the channel is very effective in protecting vulnerable sites located on a specific alluvial fan. The greater the distance between the geophone array and the vulnerable site, the greater the lead time.

Broadband seismometers commonly employed for tectonic and volcano monitoring also detect signals produced by large lahars (Capra et al., 2018; Kumagai et al., 2009; Walsh et al., 2016). Arrival-time methods and Amplitude Source Location (ASL) are established techniques that can locate and track seismic sources produced by debris flows. Uncertainties in source location due to path and site effects can be reduced by field surveys but this is an onerous task that also limits the transferability to other locations (Allstadt et al., 2019). The ASL method was tested at the scale of an Alpine basin (Illgraben, Switzerland) to estimate the location of the debris-flow front using data gathered by a temporary network of seismometers (Walter et al., 2017). More recently, machine learning techniques based on the random forest algorithm have been proposed to identify debris flows and extend warning times at Illgraben (Chmiel et al., 2021).

Low-frequency sound monitoring, typically in the infrasound (<20 Hz) band, is an emerging technique for debris-flow detection. An array of infrasound sensors and microphones can be installed at hundreds of meters from the trajectories of debris flows (Johnson & Palma, 2015; Marchetti et al., 2019). Also, the combination of a low-cost microphone with a geophone installed at a few meters from the active channel has been proposed as a compact and single-station solution for debris-flow detection (Schimmel et al., 2018).

18.3.4 Video Cameras

Video cameras are an important component of a debris-flow warning system as they permit

Fig. 18.9 Example of application of the STA/LTA algorithm in the Gadria catchment, debris flow of 18 July 2013: **a** amplitude curves recorded by two vertical geophones located along the channel; **b** STA/LTA curves calibrated for the Gadria site; **c** STA/LTA* curves using the last computed LTA (*) after the first detection (dashed line at 17:21:41), the second dashed line indicates the end of the alarm status (17:58:44)

visual interpretation of the flow processes, thus integrating the information provided by detection sensors. Figure 18.10 presents an example of video cameras set-up aimed at monitoring a sedimentation basin for debris flows. Although these video cameras have been installed for research purposes (Coviello et al., 2021b), their installation can serve as an example for warning systems too. CAM1 shoots the debris flow spreading in the debris basin, CAM2 enables observing the interaction of the debris flow with the filter of the check dam, while CAM3 records the debris flow in the inlet channel equipped with grade-control check dams.

Security outdoor surveillance video cameras are suitable for debris-flow warnings, provided that they can face the environmental conditions of the monitoring site. In general, a short focal length is preferred because it permits monitoring a wider area.

Different from monitoring for research purposes, which requires high-resolution and high frame rate videos, warning systems should favor the effective remote control of the scene shot by the camera, even at the expense of video quality. A 360p resolution can be deemed acceptable, 720p is fully satisfactory. Five frames per second is a rate acceptable in debris-flow warning systems.

Fig. 18.10 Example of video cameras set-up at a debris basin (Gadria catchment)

To overcome the limitations of the night vision of surveillance video cameras, the installation of an infrared illuminator is recommended. If the area to be monitored is large or it is located away from the video camera, more complex lighting systems can be devised. Since permanent illumination of the scene is not recommended, even if only to avoid light pollution, the activation of the lighting system can be linked to the monitoring of a variable related to debris-flow occurrences, such as the rainfall rate, or debris-flow detection at an upstream site.

While video processing aimed at quantitative analysis (e.g., large-scale particle image velocimetry) greatly benefits from the nadir view of the channel, the lateral view is fully acceptable for debris-flow warning: the video cameras can be set up on a pole or a cantilever installed on a safe position on the top of a channel bank. Video recordings can prove particularly valuable for warning if the cameras are installed to shoot protection structures, such as debris-flow breakers or debris basins, thus permitting to monitor their performance, or elements at risk, such as bridges or roads. Other features linked to debris-flow hazards that can be observed by video cameras are the presence of large wood or very large boulders in the flowing mass.

The reliability of the remote access to the video camera, which is of great importance in warning systems, can be accomplished through a wired network connection; if this is not available, modems ($\geq$3G) or radio links can be used. A Virtual Private Network increases system security and enables access to the video camera regardless of possible limitations of the internet service provider.

18.4 Protocols for Warnings

Protocols for defining and issuing warnings are fundamental to the implementation of every warning system. The protocols must achieve a trade-off between the need for simple and robust warning procedures and the importance of accounting for all the signals that detect debris-flow occurrence. The protocols must be adapted to the local conditions (length of the debris-flow path, type and location of the elements at risk). Here we provide some general guidelines deemed relevant for every debris-flow warning system.

Because of the high debris-flow velocity, a basic requirement is fast data acquisition and processing: the sensors must be queried at very short time intervals. Evidence from past events, including accounts of eyewitnesses, help determine the time interval for sensors queries.

The warnings must be focused on the target; it is possible to recognize three classes of end-users with different skills: (I) staff of civil protection offices and companies involved in the management of roads and railways (high technical skills); (II) local administrators (usually low to moderate technical skills, but deep knowledge of the terrain and its hazards); (III) the general public (assuming that the general public has no or little knowledge of the processes and limited knowledge of the terrain).

A strategy for informing the population about debris-flow risks, the expected performance of the warning system, and the action following a warning is a fundamental part of the design of a debris-flow warning system (see Coviello, 2023 for opportunities and issues related to the management of information gathered by debris-flow detection systems). Ideally, the team in charge of developing a warning system should include personnel with public communication skills, which are uncommon amongst debris-flow and instrumentation specialists.

Event warnings issued in different parts of the channel can be associated with different warning levels, the highest level pertaining to the warning activated close to the elements at risk. For instance, if a large debris basin is located upstream of the elements at risk, a debris flow detected before it reaches the debris basin can have a lower alert severity than the detection of the passage of the wave downstream of the debris basin, which implies the filling of the debris basin and its failure in retaining the entire material. Such a differentiation of the event warning levels, however, must be supported by the presence of defense works or morphological features (for instance, a low-slope channel reach that favors deposition) that indicate the likelihood of debris-flow to stop or transform into a more diluted flow. Moreover, the need to keep the management of the system simple suggests avoiding more than two warning levels. Advance warnings based on rainfall threshold can admit more levels: as an example, Huang et al. (2015) propose four warning levels, but only the two highest, referred to as "Attention" and "Warning", respectively, imply alert for authorities and population. Jakob et al. (2012) consider four warning levels (in addition to the "no watch" state) for the District of North Vancouver (Canada). The two highest (Warning I and Warning II) indicate that it is likely or very likely that debris flows will occur in the area considered, while the two lowest (Watch I and Watch II) correspond to weather conditions that require attention, but debris flows are unlikely to occur.

After a warning has been issued, it is important to determine when it is safe to cancel it. We distinguish between an alert based on the exceedance of a rainfall threshold (Advance Warning System) and an alert from an Event Warning System. Advance warnings should remain active until the variable used for issuing the warning returns below the threshold. Event warnings must be treated differently for false positives and real warnings. Alarms from false positives should be cancelled as soon it is possible to ascertain that no debris flow occurred, and possibly to determine the cause of the false alarm. Warnings caused by a real debris flow should not be cancelled until field inspections aimed at evaluating the conditions of the catchment are carried out and the warning system is restored as necessary.

Early warning systems must include diagnostics for detecting malfunctioning and pre-defined procedures to manage them. The protocols for warning should be robust against the failure of one or more sensors, enabling warning based on the other sensors, and an alert must be issued to permit immediate repair. A design based on redundancy criteria enables meeting the basic requirement of robustness against the malfunctioning of sensors or every other part of the system.

In the following, we report two applications of warning systems in the Italian Alps, which are useful to present the state of the art and the future challenges of operational, event-based warning systems for debris flows. On 5 August 2022, two debris flows occurred in two different basins of the Italian Alps: the Rotwandbach, in South Tyrol (eastern Alps) and the Rochefort basin, in Valle d'Aosta (western Alps). In both cases, the debris flows invaded the roads crossing the channels and damaged the bridges (Fig. 18.11). In the Rotwandbach, the road is protected by a warning system controlled by pendulums installed at two cross-sections (one of them is shown in Fig. 18.8) while geophones are adopted in the Rochefort basin. In both cases, video cameras provide images to local authorities of the channel intercepting the roads. The warning system of Rotwandbach detected the debris flow a few tens of seconds before the debris-flow front reached the bridge. However, this first surge damaged the cross sections where the two pendulums were installed and, consequently, they were not able to detect the subsequent surges of the same event and would have failed in detecting debris flows until the system was repaired. This situation pushed the manager of the warning system (Road Office of the Autonomous Province of Bozen-Bolzano) to rapidly add a geophone to the warning system to keep it operational for the rest of the summer season, as already planned as possible future development (Coviello et al., 2021a). The warning system of the Rochefort basin adopts an algorithm based on the variation of the STA/LTA parameter that is mentioned in Sect. 18.3.3 of this chapter. On 5 August 2022, the warning system activated the traffic lights and stopped the traffic about three minutes before the debris flow invaded the road (Arattano et al., 2023). The geophone system kept working after the debris flow and the local authorities were informed about the road damage thanks to the images provided by the video camera (Fig. 18.11b).

Fig. 18.11 Debris flows in channels equipped with warning systems for the protection of roads: **a** Rotwandbach, **b** Rochefort (images courtesy of Civil Protection Agency of the Autonomous Province of Bozen-Bolzano and municipality of Courmayeur)

18.5 Discussion and Concluding Remarks

Reliability and effectiveness of a warning system are commonly evaluated empirically, by observing their performance in terms of probability of detection and probability of false alarms. For debris flows, such an assessment is often hampered by the low frequency of events at the local scale (i.e. in the same catchment), while it is feasible in larger areas encompassing multiple debris-flow sites (e.g., Restrepo et al., 2008). On one hand, this problem stresses the importance of the catchments equipped for debris-flow monitoring for research purposes, which are chosen because of the high frequency of the events, as test areas for warning technologies. On the other hand, warning systems may greatly benefit from probabilistic models that permit assessing their technical and inherent reliability and effectiveness. Sättele et al. (2015) proposed a framework to assess the reliability of alarm systems and tested it in the Illgraben. The framework is based on a Bayesian network, that is a graphical probabilistic model consisting of a set of random variables (nodes) and their stochastic dependence (links). The Bayesian network calculates the probability of detection and the probability of false alarms as a function of the thresholds of all sensors, thus enabling assessment of system reliability.

Warning systems should be evaluated in the wider context of debris-flow control measures, which encompass other non-structural measures (e.g. land-use limitation in debris-flow prone areas), as well as structural measures, typically consisting of hydraulic works. Moreover, data recorded by warning systems may effectively contribute to the design of debris-flow control works: video-derived images and data from level sensors or geophones provide data to estimate flow velocity and discharge rate for future mitigation design.

Structural measures attenuate the severity of debris flows, for instance by retaining part of the volume before it reaches the areas at risk, or reducing the flow velocity. These positive effects, however, could be not sufficient if unexpected processes cause debris flows to exceed the magnitude of the design event. Therefore, warning systems would help manage the residual risk that results from the decrease of effectiveness of structural measures. For instance, monitoring a debris basin through level sensors and/or videocameras permits early detection of the filling of the basin or the overtopping of its dykes, or enable checking the operation of critical structures like debris-flow breakers (Fig. 18.12). Even when the performance of structural measures is satisfactory, warning systems greatly contribute to risk management by providing real-time data on the evolution of the processes. Finally, when structural measures are not feasible (for instance, because of topographic constraints that do not permit the construction of debris basins, or because of economic restrictions), warning systems can prove the only solution to cope with debris-flow risk. This is not an optimal scenario because warning systems do not prevent damage to transport routes and buildings, but, if linked to effective procedures for alert and warning, they permit at least to avoid the loss of lives.

The co-occurrence of debris flows and flash floods, which are caused by the same meteorological forcing and share—at least partly—the same space and time scales (Fig. 2 in Alfieri et al., 2012), generate multiple risks that are especially difficult to manage because of the different processes to be taken into account. Debris-flow warning systems must then be developed within the framework of multi-risk management. Common input data (e.g. rain gauge and radar rainfall estimates) can serve for rainfall-runoff models aimed at flash-flood forecasting and for thresholds aimed at the debris-flow warning. Closer integration may consist of the development of critical runoff thresholds for debris flows that share common model calibration with real-time flash flood forecasting. Destro et al. (2018), in a study on a rainstorm that triggered both a flash flood in the main river and debris flows in small tributaries in an Alpine valley, have found that thresholds based on critical runoff perform better than the traditional

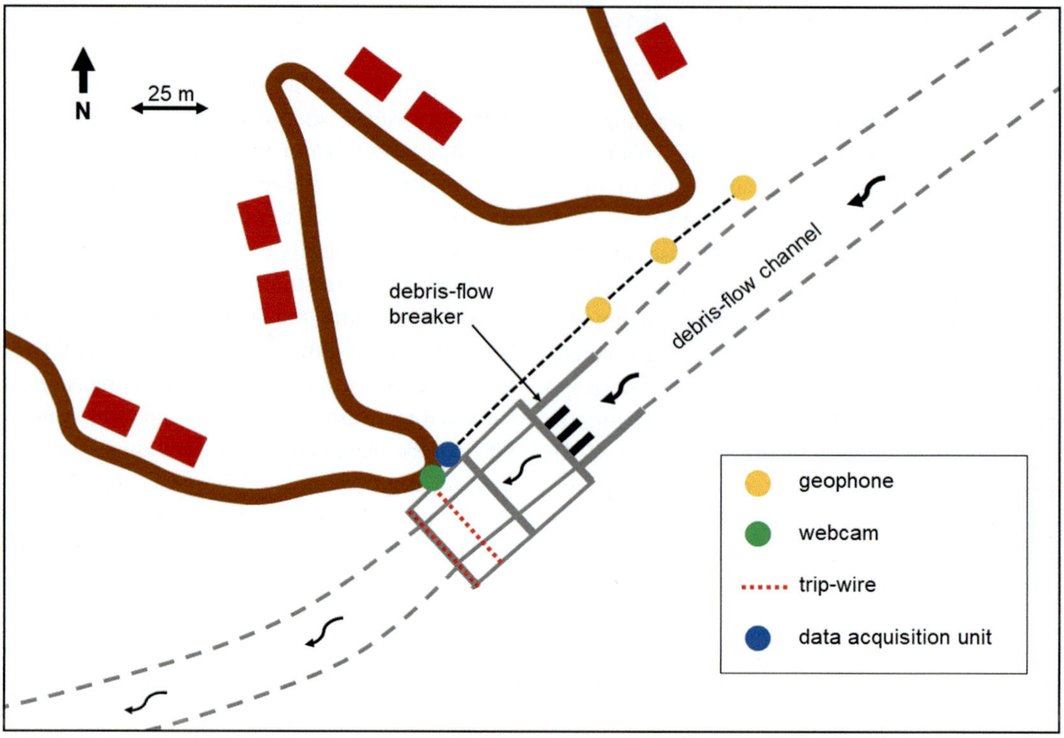

Fig. 18.12 Sketch of a warning station at a debris-flow breaker located some hundreds of meters upstream of the areas at risk. The geophones array is intended to detect debris flows, while the webcam permits observing the performance of the debris-flow breaker

rainfall thresholds in identifying the catchments where debris-flows were triggered.

In this work, because of the different sensors and technologies adopted, we have treated separately Advance Warning and Event Warning Systems, but in the practice the two approaches require integration. If both rain gauges and debris-flow detection devices are installed in the same catchment (typically on a scale of a few km^2), data must be transmitted to the same processing unit, and the warnings must be coordinated. However, a rain gauge network—ideally integrated with a weather radar—can provide debris-flow advance warnings at river basin or regional scale: in this case, advance warnings can be managed differently from event warnings based on debris-flow detection at local (i.e. single catchment) scale.

As we mentioned in the introduction, the high velocity of debris flows results in short warning times. These factors condition the types of elements at risk that can be protected by means of warning systems. Warning systems for debris flows are usually adequate for low-traffic roads and low to medium speed railways, whereas they are unfit for motorways and high-speed railways. Regarding urban areas, the lead times are usually insufficient for evacuation, but warning systems may enable an attenuation of the risk if the alert is followed by good practices (e.g. moving away from bridges and the areas adjacent to the channel) that should be prescribed in civil protection plans.

Acknowledgements We would like to dedicate this work to the memory of our co-author Velio Coviello, who died in a snow avalanche in the Alps at the beginning of April 2023. The authors wish to thank the colleagues Massimo Arattano, Maria Teresa Brunetti, Marco Cavalli and Stefano Crema, (CNR IRPI), Francesco Marra and Luca Schenato (University of Padova), Sandro Gius and Götz Florian Rufinatscha (Autonomous Province of Bozen-Bolzano), and Matteo

Cesca (ARPAV—Regional Agency for Environmental Protection and Prevention of Veneto Region) for stimulating discussions and useful suggestions. We acknowledge the Civil Protection Agency of the Autonomous Province of Bozen-Bolzano for data and video frames from the Gadria debris-flow monitoring station.

References

Abancó, C., Hürlimann, M., Fritschi, B., Graf, C., & Moya, J. (2012). Transformation of ground vibration signal for debris-flow monitoring and detection in alarm systems. *Sensors, 12*(4), 4870–4891.

Alfieri, L., Salamon, P., Pappenberger., F., Wetterhall, F., & Thielen, J. (2012). Operational early warning systems for water-related hazards in Europe. *Environmental Science and Policy, 21*, 35–49.

Allstadt, K. E., Farin, M., Lockhart, A. B., McBride, S. K., Kean, J. W., Iverson, R. M., Logan, M., Smith, J. B., Tsai, V. C., & George, D. (2019). Overcoming barriers to progress in seismic monitoring and characterization of debris flows and lahars. In: J. W. Kean, J. A. Coe, P. M. Santi, & B. K. Guillen (Eds.), *7th International Conference on Debris-Flow Hazards Mitigation*. Association of Environmental and Engineering Geologists; special publication (Vol. 28, pp. 77–84). https://repository.mines.edu/handle/11124/173234.

Arattano, M., & Marchi, L. (2008). Systems and sensors for debris-flow monitoring and warning. *Sensors, 8*, 2436–2452.

Arattano, M. (1999). On the use of seismic detectors as monitoring and warning systems for debris flows. *Natural Hazards, 20*, 197–213.

Arattano, M., Abancó, C., Coviello, V., & Hürlimann, M. (2014). Processing the ground vibration signal produced by debris flows: The methods of amplitude and impulses compared. *Computers & Geosciences, 73*, 17–27.

Arattano, M., Chiarle, M., Coviello, V., and Nigrelli, G. (2023). Performance of the debris flow alarm system ALMOND-F on the Rochefort Torrent (Val d'Aosta) on August 5, 2022. In: M. Pirulli, A. Leonardi, & F. Vagnon (Eds.), *8th International Conference on Debris Flow Hazard Mitigation (DFHM8), E3S Web of Conferences, EDP Sciences Les Ulis* (vol. 415). https://doi.org/10.1051/e3sconf/202341503002.

Badoux, A., Graf, C., Rhyner, J., Kuntner, R., & McArdell, B. W. (2009). A debris-flow alarm system for the Alpine Illgraben catchment: Design and performance. *Natural Hazards, 49*, 517–539. https://doi.org/10.1007/s11069-008-9303-x

Baum, R. L., & Godt, J. W. (2010). Early warning of rainfall-induced shallow landslides and debris flows in the USA. *Landslides, 7*(3), 259–272. https://doi.org/10.1007/s10346-009-0177-0

Bel, C., Liébault, F., Navratil, O., Eckert, N., Bellot, H., Fontaine, F., & Laigle, D. (2017). Rainfall control of debris-flow triggering in the Réal Torrent, Southern French Prealps. *Geomorphology, 291*, 17–32. https://doi.org/10.1016/j.geomorph.2016.04.004

Bernard, M., & Gregoretti, C. (2021) The use of rain gauge measurements and radar data for the model-based prediction of runoff-generated debris-flow occurrence in early warning systems. *Water Resources Research, 57*, e2020WR027893, https://doi.org/10.1029/2020WR027893.

Berti, M., Martina, M. L. V., Franceschini, S., Pignone, S., Simoni, A., & Pizziolo, M. (2012). Probabilistic rainfall thresholds for landslide occurrence using a Bayesian approach. *Journal of Geophysical Research: Earth Surface, 117*(4), F04006. https://doi.org/10.1029/2012JF002367

Berti, M., Bernard, M., Gregoretti, C., Simoni, A. (2020) Physical interpretation of rainfall thresholds for runoff-generated debris flows. Journal of Geophysical Research: Earth Surface, 125, e2019JF005513. https://doi.org/10.1029/2019JF005513

Bogaard, T. A., & Greco, R. (2018). Invited perspectives: Hydrological perspectives on precipitation intensity-duration thresholds for landslide initiation: Proposing hydro-meteorological thresholds. *Natural Hazards and Earth System Sciences, 18*(1), 31–39. https://doi.org/10.5194/nhess-18-31-2018

Bossi, G., Crema, S., Frigerio, S., Mantovani, M., Marcato, G., Pasuto, A., Schenato, L., & Cavalli, M. (2015). In: G. Lollino, M. Arattano, M. Rinaldi, O. Giustolisi, J. C. Marechal, & G. Grant (Eds.), *The rotolon catchment early-warning system* (vol. 3, pp. 91–95), Engineering Geology for Society and Territory. Cham: Springer. https://doi.org/10.1007/978-3-319-09054-2_18.

Caine, N. (1980). The rainfall intensity-duration control of shallow landslides and debris flows. *Geografiska Annaler: Series a, Physical Geography, 62*, 23–27. https://doi.org/10.1080/04353676.1980.11879996

Capra, L., Coviello, V., Borselli, L., Márquez-Ramírez, V.-H., & Arámbula-Mendoza, R. (2018). Hydrological control of large hurricane-induced lahars: Evidences from rainfall, seismic and video monitoring. *Natural Hazards and Earth Systems Sciences, 18*(October), 781–794. https://doi.org/10.5194/nhess-2017-354

Chae, B. G., Park, H. J., Catani, F., Simoni, A., & Berti, M. (2017). (2017) Landslide prediction, monitoring and early warning: A concise review of state-of-the-art. *Geosciences Journal, 21*, 1033–1070.

Chmiel, M., Walter, F., Wenner, M., Zhang, Z., McArdell, B. W., & Hibert, C. (2021). Machine learning improves debris flow warning. *Geophysical Research Letters, 48*(3), 1–11. https://doi.org/10.1029/2020GL090874

Coco, M., Marchetti, E., & Morandi, O. (2021). Numerical modeling of infrasound energy radiation by debris flow events. *Pure and Applied Geophysics,*

178(6), 2301–2313. https://doi.org/10.1007/s00024-021-02759-2

Coviello, V. (2023) Debris-flow detection for early warning purposes: Recent advances, open problems, and future challenges. In: M. Pirulli, A. Leonardi, & F. Vagnon (Eds.), *8th International Conference on Debris Flow Hazard Mitigation (DFHM8), E3S Web of Conferences, EDP Sciences Les Ulis* (Vol. 415). https://doi.org/10.1051/e3sconf/202341503007.

Coviello, V., Arattano, M., Comiti, F., Macconi, P., & Marchi, L. (2019). Seismic characterization of debris flows: Insights into energy radiation and implications for warning. *Journal of Geophysical Research: Earth Surface, 124*(6), 1440–1463. https://doi.org/10.1029/2018jf004683

Coviello, V., Gius, S., Rufinatscha, G.-F., Macconi, P., & Comiti, F. (2021a). Recent advances on debris flow early warning and risk management in Alto Adige/South Tyrol, Italy. In *14th Congress INTERPRAEVENT* (pp. 302–304).

Coviello, V., Theule, J. I., Crema, S., Arattano, M., Comiti, F., Cavalli, M., Lucía, A., Macconi, P., & Marchi, L. (2021b). Combining instrumental monitoring and high-resolution topography for estimating sediment yield in a debris-flow catchment. *Environmental and Engineering Geoscience, 27*(1), 95–111. https://doi.org/10.2113/EEG-D-20-00025

Davies, T. R. H. (1997). Using hydroscience and hydrotechnical engineering to reduce debris flow hazards. Debris-flow hazard mitigation: Mechanics, prediction, and assessment. In: *Proceedings of First International Conference, San Francisco, California* (pp. 787–810), Aug. 7–9, 1997.

Destro, E., Amponsah, W., Nikolopoulos, E. I., Marchi, L., Marra, F., Zoccatelli, D., & Borga, M. (2018). Coupled prediction of flash flood response and debris flow occurrence: Application on an alpine extreme flood event. *Journal of Hydrology, 558*, 225–237. https://doi.org/10.1016/j.jhydrol.2018.01.021

García-Martínez, R, & López, J. L. (2005). Debris flows of December 1999 in Venezuela. In: M. Jakob, O. Hungr (Eds.), *Debris-flow hazards and related phenomena* (pp. 519–538). Berlin/Heidelberg: Praxis, Springer.

Guo, X., Cui, P., Li, Y., Zou, Q., & Kong, Y. (2016a). The formation and development of debris flows in large watersheds after the 2008 Wenchuan Earthquake. *Landslides, 13*, 25–37. https://doi.org/10.1007/s10346-014-0541-6

Guo, X., Cui, P., Li, Y., Ma, L., Ge, Y., & Mahoney, W. B. (2016b). Intensity-duration threshold of rainfall-triggered debris flows in the Wenchuan Earthquake affected area, China. *Geomorphology, 253*, 208–216. https://doi.org/10.1016/j.geomorph.2015.10.009

Guzzetti, F., Peruccacci, S., Rossi, M., & Stark, C. P. (2008). The rainfall intensity–duration control of shallow landslides and debris flows: An update. *Landslides, 5*, 3–17. https://doi.org/10.1007/s10346-007-0112-1

Guzzetti, F., Gariano, S. L., Peruccacci, S., Brunetti, M. T., Marchesini, I., Rossi, M., & Melillo, M. (2020). Geographical landslide early warning systems. *Earth-Science Reviews, 200*, 102973. https://doi.org/10.1016/j.earscirev.2019.102973

Hirschberg, J., Badoux, A., McArdell, B. W., Leonarduzzi, E., & Molnar, P. (2021). Evaluating methods for debris-flow prediction based on rainfall in an Alpine catchment. *Natural Hazards and Earth System Sciences, 21*(9), 2773–2789. https://doi.org/10.5194/nhess-21-2773-2021

Huang, J., Huang, R., Ju, N., Xu, Q., & He, C. (2015). 3D WebGIS-based platform for debris flow early warning: A case study. *Engineering Geology, 197*, 57–66. https://doi.org/10.1016/j.enggeo.2015.08.013

Hübl, J., & Fiebiger, G. (2005) Debris-flow mitigation measures. In: Debris-flow Hazards and Related Phenomena, Jakob M, Hungr O (eds). Praxis, Springer: Berlin/Heidelberg; 445–487.

Hungr, O., Morgan, G.C., Van Dine, D.F, & Lister, R. D. (1987). Debris flow defenses in British Columbia. In: J. E. Costa, & G.F. Wieczorek (Eds.), *Debris flows/avalanches: Process, recognition, and mitigation* (pp. 201–222). Geological Society of America, Reviews in Engineering Geology 7.

Hürlimann, M., Coviello, V., Bel, C., Guo, X., Berti, M., Graf, C., Hübl, J., Miyata, S., Smith, J. B., & Yin, H.-Y. (2019). Debris-flow monitoring and warning: Review and examples. *Earth Science Reviews, 199*, 102981. https://doi.org/10.1016/j.earscirev.2019.102981

Ikeya, H. (1989). Debris flow and its countermeasures in Japan. *Bull. Int. Assoc. Engn. Geol., 40*, 15–33.

Innes, J. L. (1983). Debris flows. *Progress in Physical Geography, 7*(4), 469–501.

Itakura, Y., Inaba, H., & Sawada, T. (2005). A debris-flow monitoring devices and methods bibliography. *Natural Hazards and Earth Systems Sciences, 5*, 971–977.

Jakob, M., & Hungr, O. (2005). *Debris-flow hazards and related phenomena* (739 p). Berlin-Heidelberg: Praxis-Springer.

Jakob, M., Owen, T., & Simpson, T. (2012). A regional real-time debris-flow warning system for the District of North Vancouver. *Canada. Landslides, 9*(2), 165–178. https://doi.org/10.1007/s10346-011-0282-8

Jiang, Z., Fan, X., Siva Subramanian, S., Yang, F., Tang, R., Xu, Q., & Huang, R. (2021) Probabilistic rainfall thresholds for debris flows occurred after the Wenchuan earthquake using a Bayesian technique. *Engineering Geology, 280*(105965). https://doi.org/10.1016/j.enggeo.2020.105965.

Johnson, J. B., & Palma, J. L. (2015). (2015) Lahar infrasound associated with Volcán Villarrica's 3 March 2015 eruption. *Geophysical Research Letters, 42*(15), 6324–6331. https://doi.org/10.1002/2015GL065024

Kean, J., Coe, J., Coviello, V., Smith, J., Mccoy, S. W., & Arattano, M. (2015). Estimating rates of debris flow entrainment from ground vibrations. *Geophysical*

Research Letters, 42(15), 6365–6372. https://doi.org/10.1002/2015GL064811

Krøgli, I. K., Devoli, G., Colleuille, H., Boje, S., Sund, M., & Engen, I. K. (2018). The Norwegian forecasting and warning service for rainfall- and snowmelt-induced landslides. *Natural Hazards and Earth Systems Sciences, 18*, 1427–1450. https://doi.org/10.5194/nhess-18-1427-2018

Kumagai, H., Palacios, P., Maeda, T., Castillo, D. B., & Nakano, M. (2009). Seismic tracking of lahars using tremor signals. *Journal of Volcanology and Geothermal Research, 183*(1–2), 112–121. https://doi.org/10.1016/j.jvolgeores.2009.03.010

LaHusen, R. (2005). Debris-flow instrumentation. In M. Jakob & O. Hungr (Eds.), *Debris-flow hazards and related phenomena* (pp. 291–304). Springer.

Marchetti, E., Walter, F., Barfucci, G., Genco, R., Wenner, M., Ripepe, M., McArdell, B., & Price, C. (2019). Infrasound array analysis of debris flow activity and implication for early warning. *Journal of Geophysical Research: Earth Surface*, 567–587. https://doi.org/10.1029/2018JF004785.

Marchi, L., Cazorzi, F., Arattano, M., Cucchiaro, S., Cavalli, M., & Crema, S. (2021). Debris flows recorded in the Moscardo catchment (Italian Alps) between 1990 and 2019. *Natural Hazards and Earth System Sciences, 21*, 87–97. https://doi.org/10.5194/nhess-21-87-2021.

Marra, F., Nikolopoulos, E. I., Creutin, J.-D., & Borga, M. (2014). Radar rainfall estimation for the identification of debris-flow occurrence thresholds. *Journal of Hydrology, 519*, 1607–1619. https://doi.org/10.1016/j.jhydrol.2014.09.039

Marra, F., Nikolopoulos, E. I., Creutin, J. D., & Borga, M. (2016). Space–time organization of debris flows-triggering rainfall and its effect on the identification of the rainfall threshold relationship. *Journal of Hydrology, 541*, 246–255. https://doi.org/10.1016/j.jhydrol.2015.10.010

Melillo, M., Brunetti, M. T., Peruccacci, S., Gariano, S. L., Roccati, A., & Guzzetti, F. (2018). A tool for the automatic calculation of rainfall thresholds for landslide occurrence. *Environmental Modelling and Software, 105*, 230–243. https://doi.org/10.1016/j.envsoft.2018.03.024

Mirus, B. B., Becker, R. E., Baum, R. L., & Smith, J. B. (2018). Integrating real-time subsurface hydrologic monitoring with empirical rainfall thresholds to improve landslide early warning. *Landslides, 15*(10), 1909–1919. https://doi.org/10.1007/s10346-018-0995-z

Nikolopoulos, E. I., Crema, S., Marchi, L., Marra, F., Guzzetti, F., & Borga, M. (2014). Impact of uncertainty in rainfall estimation on the identification of rainfall thresholds for debris flow occurrence. *Geomorphology, 221*, 286–297. https://doi.org/10.1016/j.geomorph.2014.06.015

Nikolopoulos, E. I., Borga, M., Creutin, J., & Marra, F. (2015a). Estimation of debris flow triggering rainfall: Influence of rain gauge density and interpolation methods. *Geomorphology, 243*, 40–50. https://doi.org/10.1016/j.geomorph.2015.04.028

Nikolopoulos, E. I., Borga, M., Marra, F., Crema, S., & Marchi, L. (2015b). Debris flows in the eastern Italian Alps: Seasonality and atmospheric circulation patterns. *Natural Hazards and Earth Systems Sciences, 15*, 647–656. https://doi.org/10.5194/nhess-15-647-2015

NOAA-USGS Debris Flow Task Force, 2005, NOAA-USGS debris-flow warning system - Final report: U.S. Geological Survey Circular 1283, 47 p.

Oorthuis, R., Hürlimann, M., Vaunat, J., Moya, J., & Lloret, A. (2023). Monitoring the role of soil hydrologic conditions and rainfall for the triggering of torrential flows in the Rebaixader catchment (Central Pyrenees, Spain). *Landslides, 20*, 249–269. https://doi.org/10.1007/s10346-022-01975-8

Palau, M. R., Berenguer, M., Hürlimann, M., & Sempere-Torres, D. (2022). Application of a fuzzy verification framework for the evaluation of a regional-scale landslide early warning system during the January 2020 Gloria storm in Catalonia (NE Spain). *Landslides, 19*, 1599–1616. https://doi.org/10.1007/s10346-022-01854-2

Palladino, M. R., Viero, A., Turconi, L., Brunetti, M. T., Peruccacci, S., Melillo, M., Luino, F., Deganutti, A. M., & Guzzetti, F. (2018). Rainfall thresholds for the activation of shallow landslides in the Italian Alps: The role of environmental conditioning factors. *Geomorphology, 303*, 53–67. https://doi.org/10.1016/j.geomorph.2017.11.009

Pavlova, I., Jomelli, V., Brunstein, D., Grancher, D., Martin, E., & Déqué, M. (2014). Debris flow activity related to recent climate conditions in the French Alps: A regional investigation. *Geomorphology, 219*, 248–259. https://doi.org/10.1016/j.geomorph.2014.04.025

Restrepo, P., Jorgensen, D. P., Cannon, S. H., Costa, J., Laber, J., Major, J., Martner, B., Purpura, J., & Werner, K. (2008). Joint NOAA/NWS/USGS prototype debris flow warning system for recently burned areas in Southern California. *Bulletin of the American Meteorological Society, 89*(12), 1845–1851.

Sättele, M., Bründl, M., & Straub, D. (2015). Reliability and effectiveness of early warning systems for natural hazards: Concept and application to debris flow warning. *Reliability Engineering and System Safety, 142*, 192–202. https://doi.org/10.1016/j.ress.2015.05.003

Schenato, L., & Pasuto, A. (2022). On the use of optical fiber sensors for debris flow monitoring: A review of recent achievements. In: H. H. Zhu, A. Garg, A. Zhussupbekov, & L. J. Su (Eds.), *Advances in geoengineering along the belt and road*. BRWSG 2021. Lecture notes in civil engineering (vol. 230). Singapore: Springer. https://doi.org/10.1007/978-981-16-9963-4_5

Schimmel, A., Hübl, J., McArdell, B. W., & Walter, F. (2018). Automatic identification of alpine mass movements by a combination of seismic and infrasound sensors. *Sensors (switzerland), 18*(5), 1–19. https://doi.org/10.3390/s18051658

Segoni, S., Piciullo, L., & Gariano, S. L. (2018). A review of the recent literature on rainfall thresholds for landslide occurrence. *Landslides, 15*, 1483–1501. https://doi.org/10.1007/s10346-018-0966-4

Thomas, M. A., Collins, B. D., & Mirus, B. B. (2019). Assessing the feasibility of satellite-based thresholds for hydrologically driven landsliding. *Water Resources Research, 55*, 9006–9023. https://doi.org/10.1029/2019WR025577

Tiranti, D., Cremonini, R., Marco, F., Gaeta, A. R., & Barbero, S. (2014). The DEFENSE (debris Flows triggEred by storms - nowcasting system): An early warning system for torrential processes by radar storm tracking using a Geographic Information System (GIS). *Computers and Geosciences, 70*, 96–109. https://doi.org/10.1016/j.cageo.2014.05.004

Toreti, A., Schneuwly-Bollschweiler, M., Stoffel, M., & Luterbacher, J. (2013). Atmospheric forcing of debris flows in the Southern Swiss Alps. *Journal of Applied Meteorology and Climatology, 52*(7), 1554–1560. https://doi.org/10.1175/JAMC-D-13-077.1

Walsh, B., Jolly, A. D., & Procter, J.N. (2016). Seismic analysis of the 13 October 2012 Te Maari, New Zealand, lake breakout lahar: Insights into flow dynamics and the implications on mass flow monitoring. *Journal of Volcanology and Geothermal Research, 324*(October 2012), 144–155. https://doi.org/10.1016/j.jvolgeores.2016.06.004.

Walter, F., Burtin, A., McArdell, B. W., Hovius, N., Weder, B., & Turowski, J. M. (2017). Testing seismic amplitude source location for fast debris-flow detection at Illgraben, Switzerland. *Natural Hazards and Earth Systems Sciences, 17*(6), 939–955. https://doi.org/10.5194/nhess-17-939-2017

Winter, M. G., Dent, J., Macgregor, F., Dempsey, P., Motion, A., & Shackman, L. (2010). Debris flow, rainfall and climate change in Scotland. *Quarterly Journal of Engineering Geology and Hydrogeology, 43*(4), 429–446.

Yin, H.-Y., Huang, C.-J., Chen, C.-Y., Fang, Y.-M., Lee, B.-J., Chou, T.-Y. (2011) The present development of debris flow monitoring technology in Taiwan – A case study presentation. International Conference on Debris-Flow Hazards Mitigation: Mechanics, Prediction, and Assessment, Proceedings, pp. 623–631. https://doi.org/10.4408/IJEGE.2011-03.B-068

Zhou, W., & Tang, C. (2014). Rainfall thresholds for debris flow initiation in the Wenchuan earthquake-stricken area, southwestern China. *Landslides, 11*(5), 877–887. https://doi.org/10.1007/s10346-013-0421-

Zimmermann, M. N. (2005) Analysis and management of debris-flow risk at Sörenberg (Switzerland). In: M. Jakob and O. Hungr (eds), Debris-flow Hazards and Related Phenomena, Praxis-Springer, Berlin-Heidelberg, pp. 615–634.

Land Use Planning on Debris-Flow Fans

19

Wendy Saunders, Richard Woods, Jeff Farrell and Craig Batchelar

Abstract

Land use planning is the most influential option for managing levels of natural hazard risk, as high-risk land uses can be avoided before an event occurs; and existing uses can be regulated to reduce risks. With international responsibilities for countries to address risk reduction for disasters—including land use planning—the Sendai Framework for Disaster Reduction and the United Nations Sustainable Development Goals provide two sources of mandates for countries to reduce risks through land use planning. Three key roles can be identified for planners in creating and maintaining resilient and sustainable communities: 1. 'Mainstreaming' resilience into planning—integrating resilience into land use and development considerations via the planning and implementation process; 2. Enhancing risk management processes—ensuring land use planning issues and opportunities are built into natural hazard management processes; and 3. Building back better—driving pre- and post-event recovery planning as it relates to land use and the built environment. To enable planning for debris flows, mapping the debris-flow extent is key, so that appropriate risk-based zoning and policies can be developed and implemented. Three case studies are presented to highlight the role of planning in managing the risk from debris flows, from increasing restrictions to the managed retreat of a high-risk area. The case studies present proactive and reactive planning responses to the debris-flow hazard and risk, examples of how the debris-flow extents can be mapped, and associated land use restrictions applied to reduce risk. While our perspective is predominantly from Aotearoa New Zealand, there are many themes and challenges raised that will be common across many countries. These include how to manage existing development on debris-flow fans.

W. Saunders (✉)
Toka Tū Ake EQC, Wellington, New Zealand
e-mail: WSaunders@eqc.govt.nz

R. Woods
GNS Science, Lower Hutt, New Zealand

J. Farrell
Whakatāne District Council, Whakatāne, New Zealand

C. Batchelar
Planning Consultant, Auckland, New Zealand

19.1 Introduction

Land use planning is a key 'risk treatment' option of the risk management process (see Strouth et al., this volume), and the planning process is a 'user' of risk assessments. Land use

M. Jakob et al. (eds.), *Advances in Debris-flow Science and Practice*, Geoenvironmental Disaster Reduction,
https://doi.org/10.1007/978-3-031-48691-3_19

planning provides a proactive approach to controlling risk, both now and in the future, and can prohibit, or at least severely restrict, development in high-risk areas. Applied at the local, site-specific level, it also attracts international attention for its ability to reduce risks through international treaty obligations, such as the Sendai Framework for Disaster Reduction and the United Nations Sustainable Development Goals.

In this chapter we outline the critical role land use planning plays in managing debris-flow risks. This is achieved by first explaining what planning is, and how planning is included in the aforementioned international treaties, which oblige countries to use planning as a risk reduction tool. We then discuss the relationship between risk management and spatial planning in Aotearoa New Zealand, and the resulting risk-based planning approach, where planning controls become more restrictive as the risk increases. Following, we outline the importance of mapping, zoning and risk-based land use planning policy responses to manage debris-flow risk; and then present three case studies showing examples of how planning has had a role in managing debris-flow risk in Squamish, Canada; and in Queenstown and Matatā, New Zealand.

19.2 What is Planning?

Land-use planning is known by various terms: town planning, town and country planning, urban planning, city planning, environmental planning, resource management planning, and many more. The terms are commonly used to cover the public policy intervention related to the ordering and regulation of land use in an efficient, sustainable, and ethical way. Numerous academic- and practice-based definitions exist of spatial planning, and these are, to some extent, context specific, varying between different countries and jurisdictions (Baker, 2014). Planning is defined by the UN Office of Disaster Risk Reduction as "The process undertaken by public authorities to identify, evaluate and decide on different options for the use of land, including consideration of long term economic, social and environmental objectives and the implications for different communities and interest groups, and the subsequent formulation and promulgation of plans that describe the permitted or acceptable uses" (UNDRR, 2020, p. 16). Planning makes deliberate decisions to achieve desired outcomes through five key activities (Fig. 19.1): empowering people, strategic foresight, providing public goods, investing in and operating infrastructure, and regulating negative effects of development on the environment.

Land use planning is one tool in the toolbox for managing natural hazards: building performance, engineering works, warning systems, evacuation planning, catchment management, insurance, and infrastructure resilience all contribute together to reducing and managing risks. Of all the tools, land use planning is the most influential option for managing levels of natural hazard risk, as high-risk land uses can be avoided before an event occurs; and existing uses can be regulated to reduce risks. But it is also a challenging policy area that must balance a range of (sometimes competing) priorities, in addition to the management of disaster risk. Using land use planning for effective disaster risk management does not necessarily imply that there should be no development in high-risk areas; rather, we should be planning smarter for land uses that can still be sustainable, although in certain circumstance avoidance or retreat will be the best option. As the level of risk increases, so can the level of planning restrictions and requirements.

19.3 International Obligations to Plan for Natural Hazards and Risks

There are international obligations for countries to address risk reduction for disasters including through land use planning, with the Sendai Framework for Disaster Reduction and the United Nations Sustainable Development Goals (SDGs) being two key sources.

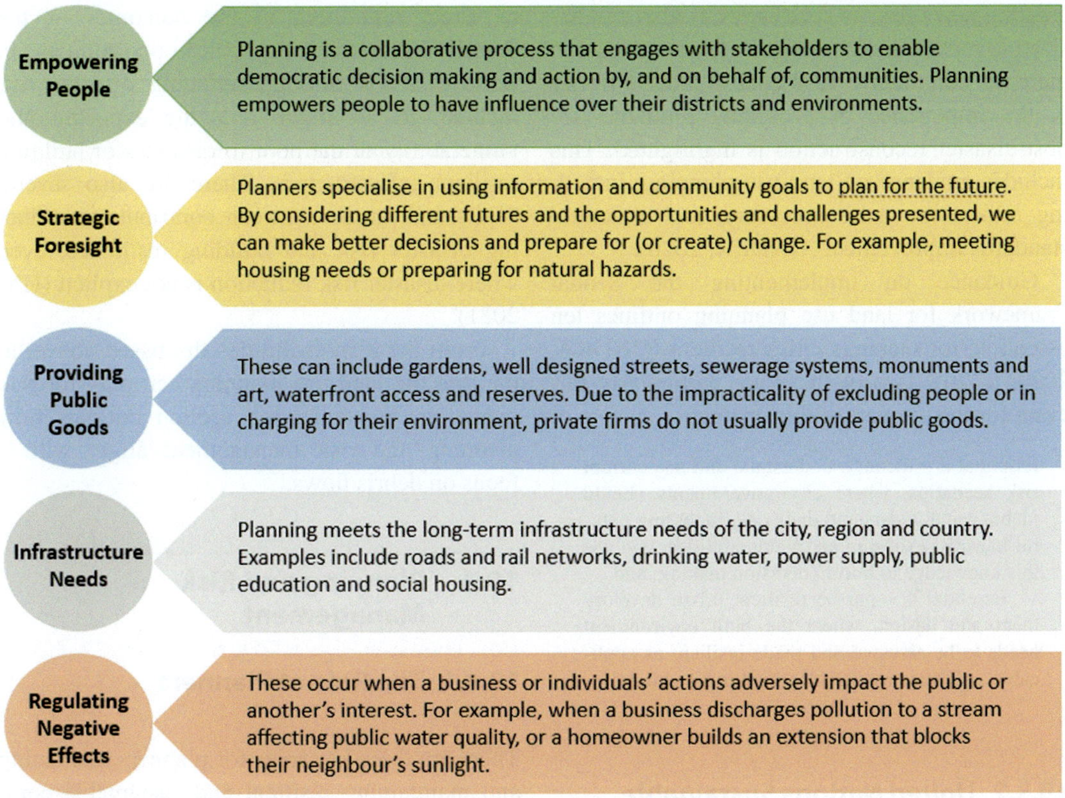

Fig. 19.1 Key activities of land use planning (NZPI, accessed 22 November 2021)

19.3.1 Sendai Framework for Disaster Risk Reduction

The Sendai Framework for Disaster Risk Reduction 2015–2030 (Sendai Framework) is a 15-year voluntary, non-binding agreement that sets a far-reaching, people-centered approach to disaster risk reduction (DRR). The Sendai Framework applies to risks that are "small-scale and large-scale, frequent and infrequent, sudden and slow-onset disasters, caused by natural or man-made hazards as well as related environmental, technological and biological hazards and risks. It aims to guide the multi-hazard management of disaster risk in development at all levels as well as within and across all sectors" (The Sendai Framework, 2015). It articulates the need to improve understanding of risk and focus on addressing underlying drivers that cause risks. This translates into reducing existing risk; preventing future risk; and enhancing resilience to risks (UNDRR, 2020).

Within the Sendai Framework, urban/ land use planning is highlighted as part of national and local actions to be taken under Priority 3 (Investing in disaster risk reduction for resilience) and, to a lesser degree, under Priority 4 (Enhancing disaster preparedness for effective response, and to "Build Back Better" in recovery, rehabilitation, and reconstruction). As part of the former, it is suggested that national and local-level actions could include mainstreaming of disaster risk assessments into land use policy development and implementation, including the areas of urban planning, land degradation and informal housing, as well as in mapping and management of rural development planning. Furthermore, the importance of getting building codes, standards and (re)construction practices right (whether by revising existing or creating new ones, as necessary), is also emphasized to reduce disaster impacts to structures and increase community resilience. Under Priority 4, the importance of providing guidance for post-disaster reconstruction is highlighted. This includes guidance on land use planning, including relocation where necessary, and structural standards improvement (UNDRR, 2020).

Guidance on implementing the Sendai Framework for land use planning outlines ten essentials for makings cities resilient (UNDRR, 2020, p. 20). Two of these are particularly relevant for the purpose of this chapter:

> Essential 2 – identify, understand and use current risk scenarios, where city governments should identify and understand their risk, including natural hazards, exposure and vulnerabilities, and use this knowledge to inform decision making; and
>
> Essential 4 – pursue resilient urban development and design, where the built environment needs to be assessed and made resilient as applicable, informed by risk identified in Essential 2.

19.3.2 United Nations Sustainable Development Goals

The SDGs are a collection of 17 interlinked global goals designed to be a "blueprint to achieve a better and more sustainable future for all" (United Nations, 2022). The SDGs were set up in 2015 by the United Nations General Assembly and are intended to be achieved by the year 2030. Goal 11 focuses on Sustainable Cities and Communities, with the aim of making cities and human settlements inclusive, safe, resilient and sustainable. An action associated with this goal was that by 2020, the number of cities and human settlements adopting and implementing integrated policies and plans that included resilience to disasters and adaptation to climate change, was increased. This action had a direct link to the Sendai Framework, to holistically manage disaster risk at all levels (The Global Goals, 2022).

The 2030 Agenda for Sustainable Development recognizes and reaffirms the urgent need to reduce disaster risk. In addition to direct references to the outcomes of the Sendai Framework, specific opportunities to achieve SDGs through reducing disaster risk include, for example, reducing exposure and vulnerability of the poor to disasters or building resilient infrastructure. There are also several SDGs and targets that can contribute to reducing disaster risk and building resilience, even where disaster risk reduction is not explicit (UN, 2021).

From the responsibilities discussed above, to the role the land use planning can play for risk reduction, the following section outlines how planning and risk management align, with a focus on debris flows.

19.4 Planning and Risk Management

19.4.1 The Role of Planners

There are three key roles for planners in creating and maintaining resilient and sustainable communities (PIA, 2015):

1. 'Mainstreaming' resilience into planning—integrating resilience into land use and development considerations via the plan making and implementation process.
2. Enhancing risk management processes—ensuring land use planning issues and opportunities are built into natural hazard management processes.
3. Building back better—driving pre- and post-event recovery planning as it relates to land use and the built environment.

Each of these roles can be achieved through integration with the two functions of land use planning: (1) the regulatory framework to control activities and manage risks involving plan development—the use of maps, zones, policies, and methods for policy implementation, which may include rules; and (2) applications for activities where the rules require an assessment against the plan's objectives and policies. This is where site-specific risk assessments (see Strouth et al., this volume) may be required.

Risk-based planning is an approach applicable to (1) and (2) above, where the level of planning restrictions increase as the risk increases. For example, Fig. 19.2 shows the Aotearoa New Zealand activity status categories, which become more restrictive, and require more detailed information in planning applications, as the risk increases.

Managing the risks from any natural hazard requires a combination of activities across these different options, with land use planning providing a proactive opportunity to reduce risks before an event occurs, while also providing for reactive planning to capture the opportunity to reduce risks after an event. Four key planning options exist to manage natural hazard risks: avoid, control (mitigate), transfer, and accept, as outlined in Table 19.1.

Within the risk management process planning options are often the 'risk treatment' part of a risk assessment for a development; and the planning process utilizes the risk assessment process to inform policy development and responses. This is exemplified further with a case study from Queenstown, Aotearoa New Zealand in a following section.

Risk assessment and management of that risk should be incorporated within the spatial planning process (Fig. 19.3) to achieve greater sustainability and resiliency of society's development (Saunders & Becker, 2015).

The framework in Fig. 19.3 shows how risk management and spatial planning are interconnected. The risk management process includes scientific studies to which spatial planning databases might contribute, particularly regarding vulnerability analysis, political decision-making processes related to risk, and implementation processes in which compliance with spatial plan specifications should also be considered. Similarly, the risk management process informs and contributes to the scientific analyses performed during plan making in which risk assessment data can be integrated with planning issues and problems (Sliuzas et al., 2015).

Greiving et al. (2006) identified four possible roles of spatial planning in risk management that are still relevant today. As they pertain to debris flows, these are:

1. Keeping areas free of future development that are; (a) hazard prone, particularly with history of occurrence of disaster events, (b) needed to lower the effects of hazardous event (e.g. debris catches), and c) needed to enhance effectiveness of disaster response (e.g. evacuation routes,)
2. Differentiated decisions on land use—allocating land for different uses based on hazard intensity, frequency or other hazard criteria. For instance, fan areas may be used for agriculture or forestry purposes, but prohibited for residential use or siting of critical buildings.
3. Regulating land use by legally binding status—for instance regulating building density or use such as prohibition of childcare facilities on debris-flow fans.
4. Hazard modification—spatial planning can contribute to the reduction of hazard potential from debris flows by influencing land use within the debris flow catchment.

Other important aspects of the Fleischhauer et al. (2006) approach are the importance of both horizontal and vertical interplays between

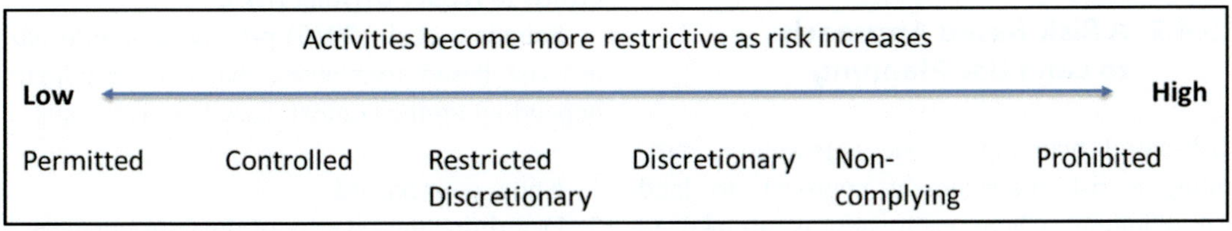

Fig. 19.2 Resource consent categories in Aotearoa New Zealand

Table 19.1 Options for managing risks through land use planning (adapted from Ministry for the Environment, 2022)

Option	Description	Planning response
Avoid	Avoidance of activities as the risk is too high (intolerable) e.g. land may be used for recreational activities like parks, gardens, or reserves	Avoid through zoning, locating development away from high hazard areas Managed retreat where development exits
Control	Control activities taking a risk-based approach, where activities become more restrictive as risk increases	Do not locate critical facilities in the high hazard zone Use of mitigation measures like engineered structures, warning systems, or density limitations
Protect/accommodate	Protection of property through engineering works and options	Use of protective structures and engineering mitigation measures, bunds, set backs
Transfer	Residual risk is managed through insurance	Continue to monitor and evaluate hazard and risk policies
Accept	Risks are low and accepted by society Risks should be monitored over time to ensure levels do not increase beyond acceptable levels	Enable development
Retreat	Relocating existing development away from high risk areas	Prohibited development, removal of existing use rights, property acquisition

organizations operating under different mandates. These interplays are a result of the multiple entities at multiple levels involved in both spatial planning and risk management, and the possible lack of an overriding authority that can provide consistency. This consistency is important between sectoral goals and plans; the position of spatial planning as an important end-user of risk assessment data; and ensuring that the most up to date information is exchanged from one realm to the other so that the consequences of risk changes in the real-world are reflected in plans and planning decisions as soon as possible. Furthermore, the implications of proposed planning decisions, which may have implications for future vulnerability levels, are to be assessed for their influence on risk.

19.4.2 A Risk-Based Approach to Land Use Planning

Risk-based planning is one approach to providing a risk management approach to land use planning, where each step is aligned to common risk management steps, as shown in Fig. 19.4. Risk-based planning—with a focus on consequences then likelihood, rather than solely focusing on the likelihood of the hazard occurring—provides the greatest opportunity to reduce and manage risks from natural hazards, including debris flows.

Key to risk-based planning is the ability to delineate between different levels of risk (such as acceptable, tolerable, intolerable), and link these to suitable land use policies. An acceptable level of risk needs to be based on measurable indicators that allow risk levels to be monitored over time, enabling towns and cities to undertake sustainable development that does not exceed acceptable levels of risk- and may even act to mitigate existing risk. Measurable indicators also make it possible to track the efficacy of efforts to reduce existing risks.

Saunders et al. (2015) provide a natural hazard risk-based framework that can be adapted depending on the context, based on five steps:

1. Know your hazard;
2. Determine the severity of the consequences;

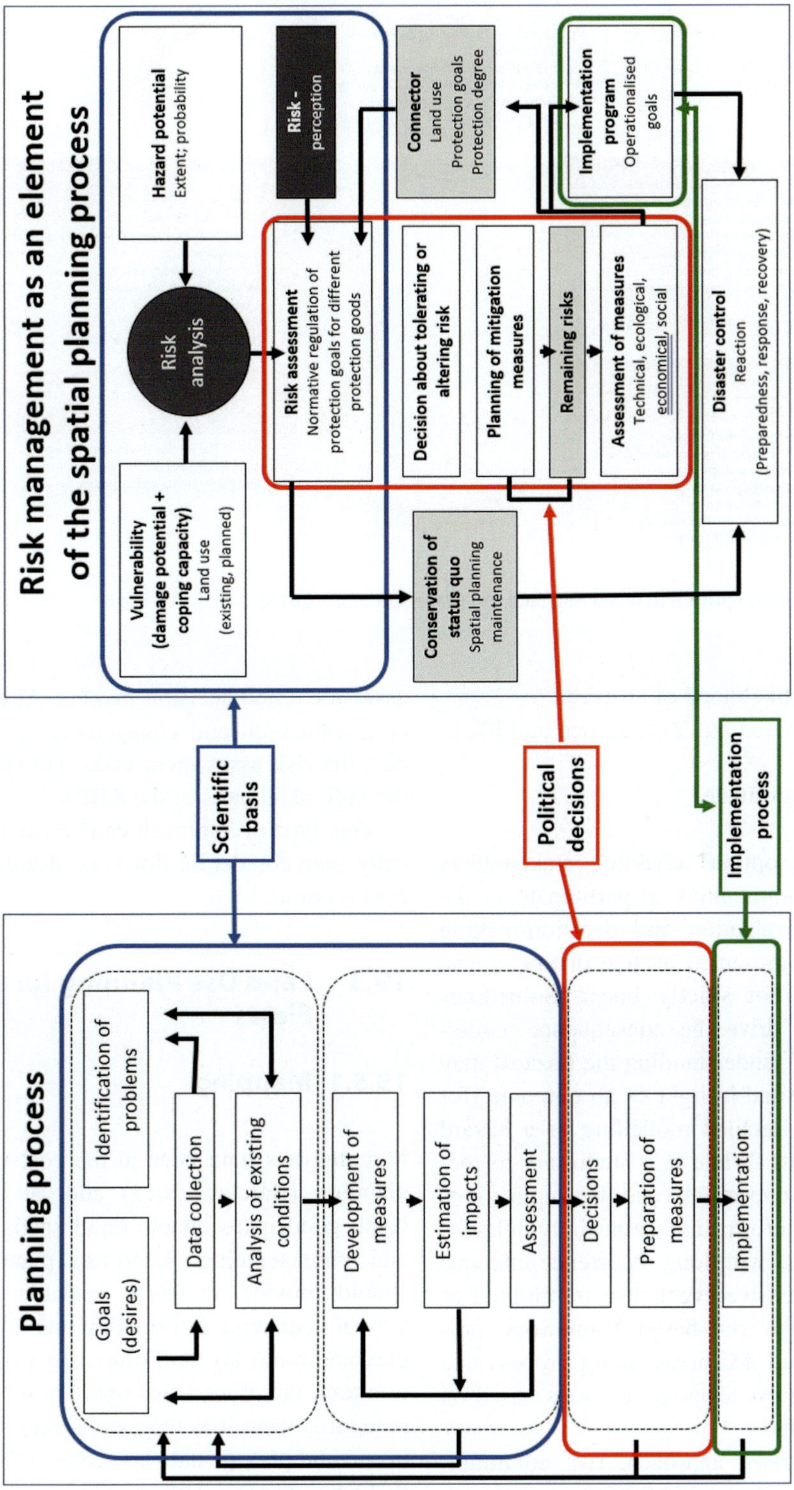

Fig. 19.3 Risk management as an element of the spatial planning process (Greiving & Fleischhauer, 2006, p. 116)

Fig. 19.4 Risk based approach to managing land use (Saunders et al., 2015)

3. Evaluate the likelihood of an event;
4. Assess risk based on consequence and likelihood; and
5. Monitor and evaluate.

Each step has options enabling stakeholders and affected communities to participate in the risk analysis, evaluation and decision-making process. While presented as five distinct steps, the process is not strictly linear. Sometimes likelihood may drive the consequence assessment, or Step 1 (understanding the hazard) may need to be revisited in light of an outcome (for example, more detailed modelling of a hazard may be required). While it is important to recognize areas of uncertainty and the implications of incomplete information (especially relevant for hazard events with long recurrence intervals where comprehensive event data is minimal or non-existent), the risk-based framework provides a structured decision-making process that can enable progress towards decisions based on current knowledge.

In the risk-based approach, risk communication and engagement are fundamental to risk assessment and decision-making. At each phase, communication and engagement activities support the risk assessment tasks and contribute to the individual parts of the RBPA.

This type of approach can be used to specifically plan for debris flows, as discussed in the next section.

19.5 Land Use Planning for Debris Flows

19.5.1 Mapping

Mapping has long been at the core of land-use planning (Barclay, 2013) and, at a practical level, good maps enable rapid spatial overview and understanding. Land-use planning maps should provide an easily readable and visual way of displaying information that is often complex and multi-layered. Maps are a communication tool; therefore, it is important for drafters to recognize what it is they are aiming to communicate and provide data and content that matches that need. From a land use planning perspective, a good map related to hazard or risk has zones that match regulatory thresholds and requirements (Kelly et al., 2020), as maps form the spatial layer of planning regulations.

Maps developed for planning purposes should be at a scale that reduces ambiguity for the property owner. Lines or boundaries on maps can often be contentious, particularly when planning policies restrict development opportunities based on these lines. Spatial planning maps showing hazard and/or risk overlays are often not assessed on a property-by-property level basis, and so there are inherent uncertainties in the placement of 'the line'. The post-earthquake Christchurch District Plan in New Zealand (https://ccc.govt.nz/the-council/plans-strategies-policies-and-bylaws/plans/christchurch-district-plan) provides an example of how this can be addressed for land instability hazards; while not specific to debris flows, it has been used for rockfall, cliff collapse, boulder roll, and mass movement, and could be applied to debris flows. In the Christchurch District Plan, if the named land instability risk line is on a property, the owner can apply to have the property level risk reassessed, but it must be in accordance with an agreed methodology that was used for the original assessment. If 'the line' moves (either increasing or decreasing spatially over the property), the planning maps are updated to show the new placement. Being able to contest the line in this robust and transparent manner provides a mechanism for line placement uncertainties to be addressed.

19.5.2 Zoning

Zoning is a long-used technique to divide areas of land or water into distinct areas to manage effects, activities or uses. Zoning has been described as allowing "the district plan to create bundles of activities considered generally appropriate in each zone or area, in recognizing the constraints of the environment and that some activities may not be appropriate in every location" (Quality Planning, 2017).

While the terms 'overlays' and 'coverages' are sometimes used in place of zoning, the principles and objectives of zoning are essentially the same, for example to manage or separate incompatible uses; for the avoidance or control of natural hazards; or for the identification and management of areas where similar outcomes are sought, or bundles of effects need to be managed.

The technique of zoning has been criticized by some as being too directive or inflexible, however in many cases analysis of the criticism of zoning shows that it is the policies and rules rather than the zoning itself that is at issue.

Figure 19.5 shows how a debris flow in the Bay of Plenty township of Matatā in New Zealand has been zoned using two methods: one as a high, medium or low (residential) debris risk zone; the other a Coastal Protection Zone (CPZ) over areas subject to high risk.

The CPZ is described as a strip of land adjacent to the Coastal Marine Area along the coastline where development or use may have a direct effect on the visual amenity and/or ecology of the coast. The Zone may also provide a level of protection against coastal hazard events. The CPZ is predominantly an open space zone that is not generally intended for development.

Once mapped, policy responses can then be applied to the zoned area, reflecting the level of risk that is identified (see case studies for examples).

19.5.3 Risk-Based Policy Responses

There are several planning options available to reduce and manage risk through land use planning policies, such as (Bay of Plenty Regional Policy Statement, 2016):

(a) Ensuring new subdivision and development avoids specific hazard locations (e.g. avoiding sensitive development such has residential land uses, while allowing certain recreational activities on debris fans);
(b) Replacement or modification of existing development over time to reduce potential

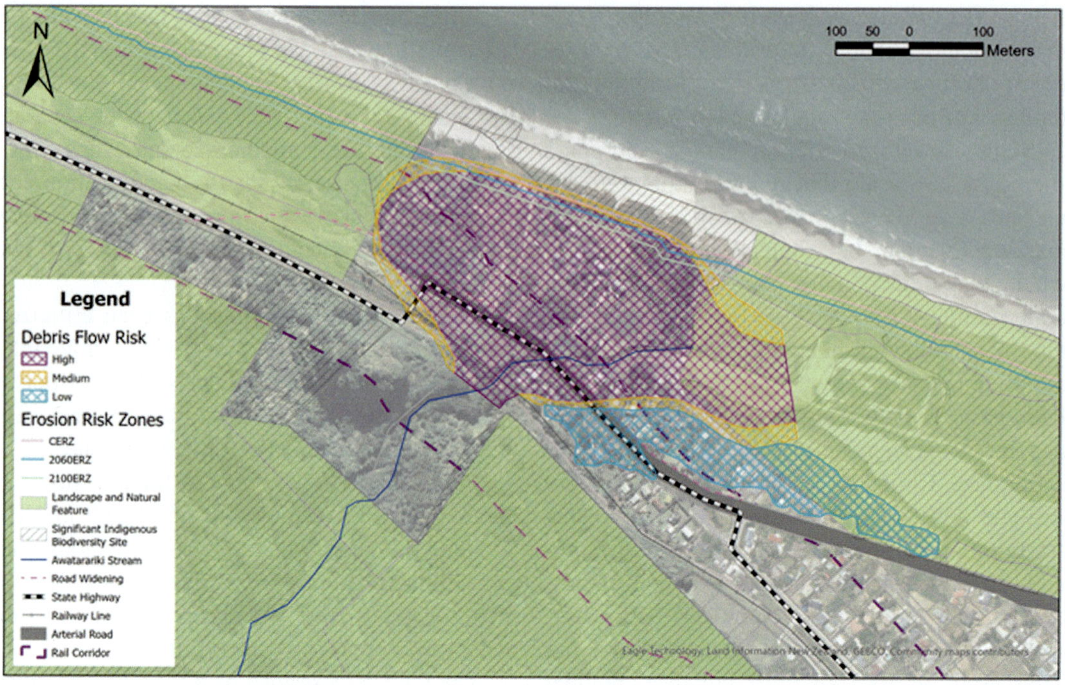

Fig. 19.5 Zoning of a debris flow in Matatā, New Zealand, showing high, medium and low risk overlay zones from debris flows and Coastal Protection Zone

consequences; this could include changes of use that reduce life safety risk (e.g. change a residential dwelling to a commercial use);

(c) Providing only for low intensity activities in specific locations, including activities with a minimum size that allow for the debris fan to be avoided (e.g. rural residential);

(d) Setbacks and undeveloped buffer land within areas of new subdivision and development;

(e) Restoration, retention or enhancement of natural defences against natural hazards as part of development proposals and promotion of the sustainable functioning of such natural defences to reduce the risk to existing development (e.g. trees and natural bunds to slow the debris flow);

(f) Property-specific works (e.g. debris nets and slope stability works) as part of development proposals;

(g) Urban and building design that takes into account measures to reduce the risks from debris flows (e.g. heights of building platforms, retention or reinstatement of stormwater overland flow paths, hazard resilient buildings and construction materials); and

(h) Ensuring new development anticipates possible hazard event emergencies and provides means to enable effective responses by people and communities including requiring:

(i) Hazard warning systems (e.g. rainfall thresholds for debris flow activation); and/or

(ii) Urban form and transport infrastructure (including for motor vehicles, cycles and pedestrians) that enables rapid and efficient evacuation; and/or

(iii) Provision for, and safeguarding of, safe and accessible evacuation routes and zones (including, where appropriate, vertical evacuation zones).

The spectrum of risk-based policies (which range from communicating the risk through to managed retreat) may be related to risk thresholds, for example intolerable, tolerable and acceptable (Strouth et al., this volume). If a policy requires risks to be managed to a tolerable level, the land use planning responses can require controls of certain activities to ensure the land use falls within the tolerable level (for example see Table 19.2).

19.6 Case Studies

19.6.1 Controlling Land Use, Controlling Risk: Squamish, British Columbia, Canada

The Cheekeye Fan is identified in the District of Squamish Official Community Plan (OCP) as a debris flow/flood hazard area. Schedule D1 of the OCP includes a map of the Cheekeye Fan hazard zones (see Fig. 19.6). Policies in the OCP state that there will be no permanent development in zones 1 and 2 and that any building development in zones 3 and 4 must be accompanied by a Debris Flow Management Plan and require implementation of appropriate mitigation measures (District of Squamish, 2017).

In addition, according to provincial legislation and the District's OCP, any development on hazard lands requires a report by a qualified professional (QP) certifying that the land is safe for the intended use. Any actions stipulated in the QP report are captured in a restrictive covenant in favor of the municipality (District of Squamish, 2017).

The OCP also states that the flood hazard zones identified in Schedule D1 are based on past geotechnical studies and may be reviewed when new natural hazard and risk management information becomes available.

In 2017 new information about the extent of the Cheekeye Fan debris-flow area emerged as a result of a proposed large development project on the Fan. Based on this new information, QP safety certification was also required for projects within zone C5 as identified on Schedule D-1 of the Official Community Plan. Any QP's are required to review a list of eight background information and study reports when completing hazard and risk assessments (District of Squamish, 2017).

19.6.2 Evidence Informed Planning Response: Queenstown, New Zealand

To manage the risk posed by future debris flows in Queenstown, New Zealand, the Queenstown Lakes District Council (QLDC) sought to understand the socio-economic consequences of various land-use planning policy options for their District Plan. RiskScape™ software (Paulik et al., 2022) was used to estimate the direct losses incurred under various debris-flow scenarios for each land-use policy option. These results were used to inform further socio-economic assessments for QLDC and the public to decide on a preferred policy option.

RiskScape™ estimated losses to buildings located on the urbanised Reavers Lane and Brewery Creek active alluvial fan areas. RiskScape™ is an open-source software with a flexible modelling engine for multi-hazard risk analysis centred on a well-established conceptual framework for risk quantification:

$$R = f_c(H_i, E, V_i) \qquad (19.1)$$

where risk (R) is a function (f_C) of the consequences from a hazard event (H) impacting an exposure (E) (i.e. element-at-risk). Consequences are determined from the exposure vulnerability (V) to an impact type and

Table 19.2 Debris flow land use planning policy responses and possible planning actions (adapted from Queensland Reconstruction Authority, 2012)

Land use responses and Possible Planning Actions

Planning evaluation risk category	Land use response & description	Possible land use transition strategies	Possible planning actions	Land use options (relative to urban/rural location)
Intolerable risk	Retreat from specific existing urban areas Expand into new areas suitable for urban development *The strongest land use response required to avoid risks to life or property. This would involve limiting land uses (e.g. 'back-zoning' in existing areas) and active measures to move people or property out of harms way*	• Actively limit future development in this area that may increase risk to life or property through strong zoning controls • Promote transition of at-risk existing uses & promote low-impact, non-urban uses • Discourage further intensification of existing uses • Consider how to maintain community connectivity in areas to be transitioned • Investigate complementary measures (e.g. voluntary purchase) to actively reduce existing at-risk people and property in this area • Investigate structural controls to further reduce risk to life and property	Strategic Framework: • Intents/outcomes limits development in these areas that would create unacceptable risk Zoning: • Limited Development • Environmental Management • Conservation • Sport & Recreation • Open Space • Rural • Risk-constrained zones as required (e.g. Residential Living—High Risk Zone)	Appropriate (subject to assessment): Market (temporary only) Outdoor Sport and Recreation Park Permanent Plantations Cropping

(continued)

Table 19.2 (continued)

Land use responses and Possible Planning Actions				
Tolerable risk (subject to ALARP)	Adapt existing areas Expand into new areas suitable for urban development Maintain agricultural and rural landscape values *A considered approach to land use and urban design is required where a greater range of land uses may be appropriate than in areas of highest risk, but others generally remain inappropriate*	• Discourage sensitive land uses but permit majority of land uses • Density increases may be appropriate in line with good planning principles (e.g. infill development)—where strong emergency management, evacuation routes & early warning systems are available	Strategic Framework: • Intents/Outcomes discourages incompatible land uses in these areas Zoning: • Risk-constrained zones as required (e.g. Residential Living—Medium Risk Zone)	Inappropriate: Child Care Centre Community Care Centre Community Residence Correctional Facility Educational Establishment Emergency Services High impact industry hospital Intensive animal husbandry intensive Horticulture Major Electricity Infrastructure Major Sport, Recreation and Entertainment Facility Medium Impact Industry Non-resident Workforce Accommodation Noxious and Hazardous Industry Relocatable Home Park Residential Care Facility Retirement Facility Substation Telecommunications Facility Tourist Park
				Appropriate: All other uses (subject to assessment)
Broadly acceptable risk	Adapt existing areas Expand into new areas suitable for urban development *Minimal land use changes required to respond to risk—urban design controls may be implemented to improve resilience*	• Broad consideration to be given to concern of debris flow—no specific strategy suggested • Land uses and density increases appropriate in line with good planning principles (e.g. infill development)—where strong emergency management, evacuation routes & early warning systems are available	Strategic Framework: • Intents/outcomes support appropriate development in these areas Zoning: • No changes based on debris flow concern	Appropriate: All uses subject to appropriate built form controls being achieved

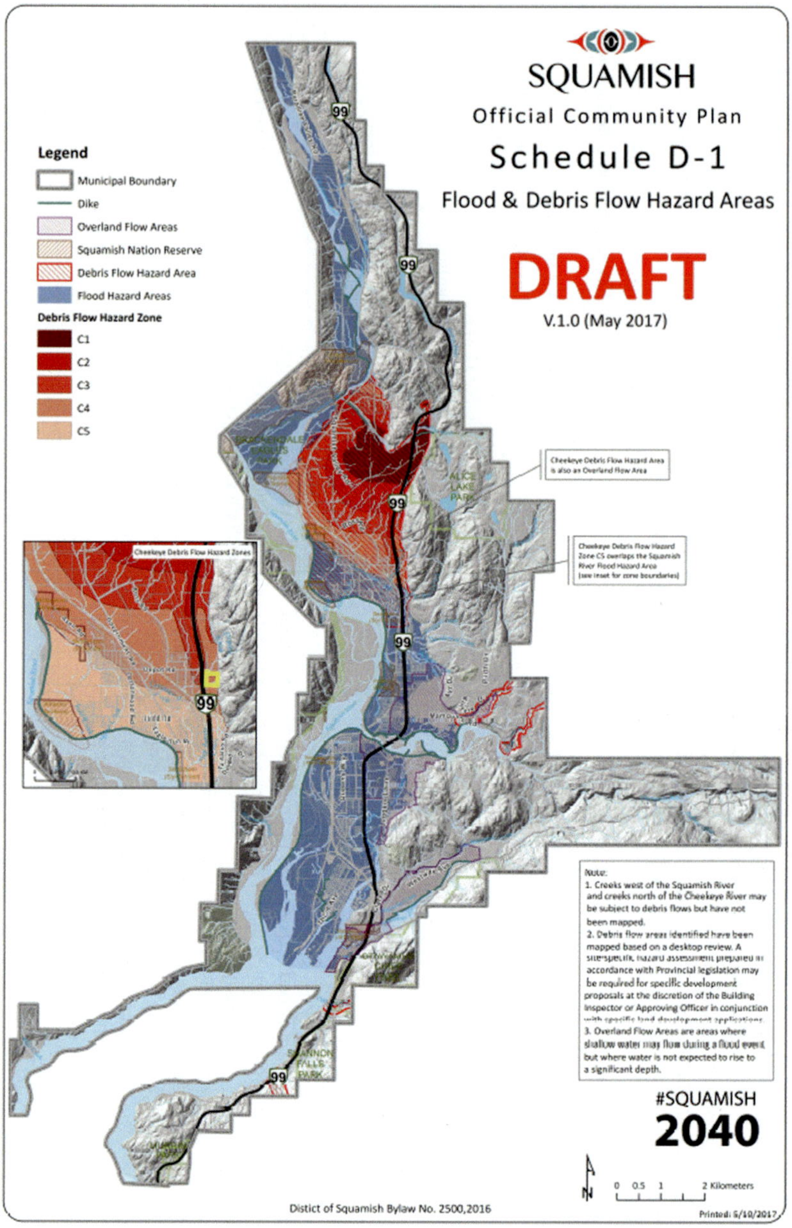

Fig. 19.6 Cheekeye debris flow hazard areas and zones (District of Squamish, 2017, p. 4)

magnitude in response to either single or multiple hazard events (*i*) (Paulik et al., 2022).

To estimate the risk (*R*) to the Reavers Lane and Brewery Creek alluvial fan study areas, RiskScape™ used debris-flow datasets (*H*) sourced from Beca (2020), building datasets (*E*) sourced from field visits (Woods et al., 2021) and vulnerability models (*V*) from Massey et al. (2019). Table 19.3 presents the range of debris-flow scenarios (Fig. 19.7) and their respective

Table 19.3 Hazard layers and their respective return periods created in Rapid Mass Movement System (RAMMS) debris-flow software by Beca (2020) and used to estimated losses by RiskScape™

Study area	Hazard layer	Return period range (years)
Reavers lane	Small debris flow	100–2,500
	Moderate debris flow	2,500–6,700
	Large debris flow	6,700–20,000
Brewery creek	Small debris flow	50–200
	Moderate debris flow	200–2,500
	Large debris flow	2,500–10,000

return period ranges, based on the geological characteristics of the catchments, as modelled by Beca (2020).

Using these datasets, RiskScape™ estimated both direct monetary loss and the associated building damage state, from 'Insignificant damage' to 'Destroyed', for each building (Woods et al., 2021) (Fig. 19.8). Due to the level of epistemic uncertainty in modelling the direct monetary losses, these are aggregated when presented in the final results (i.e. not per building).

After modelling all debris-flow scenarios at both study areas, RiskScape™ results represent the losses and damage that could result if each of these scenarios occurred in the present (2020). When assessing the impact of future policy options, these results are considered the *Baseline*.

Using the *Baseline*, four options for managing built form in both study areas were investigated. Each option involved a different set of modifications to the planning provisions, resulting in different 'future built environments.' Each of these future built environments are then applied in RiskScape™, using a modified exposure layer (*E*), (Fig. 19.9), to create losses expected for each of the respective hazard scenarios. When the same hazard event is applied to each of the four future built environments, it is possible to compare how effective each option is at reducing losses.

The four options for managing the built form are:

- Uncontrolled—maximise development allowed by the existing planning provisions on each site in the study areas out to 2120.
- Managed—manage (control) the development in the higher-risk parts of the study areas while allowing uncontrolled development to occur in areas of lower risk out to 2120.
- Reduce—removal of buildings in higher-risk areas while maintaining existing buildings in areas of lower risk out to 2120.
- Engineering—development continues without specific hazard restrictions, as for Uncontrolled, with engineering structures to manage debris flows out to 2120.

In consultation with QLDC land-use planners, new exposure datasets were created to represent each of these options and subsequently modelled in RiskScape™ using all debris flow scenarios. Results were tabulated and graphed showing the aggregated building losses and differences between each policy option as compared to the 2020 baseline (Table 19.4 and Fig. 19.10). To provide further insight to which hazard events would incur the greatest losses over time, the results were annualised (Fig. 19.11).

In summary, the study identified that the Uncontrolled policy option increases direct losses for all modelled scenarios due to the increase in urban form exposure. The Managed policy option reduces direct building losses for all events at both locations when compared to the 2120 Uncontrolled policy option. For the Reduced option, the reduction in exposed buildings in high hazard locations decreases direct building losses at both study locations for all hazard scenarios. The modelled Engineering scenarios (modelling the effect of a debris fence) reduced direct building losses for all

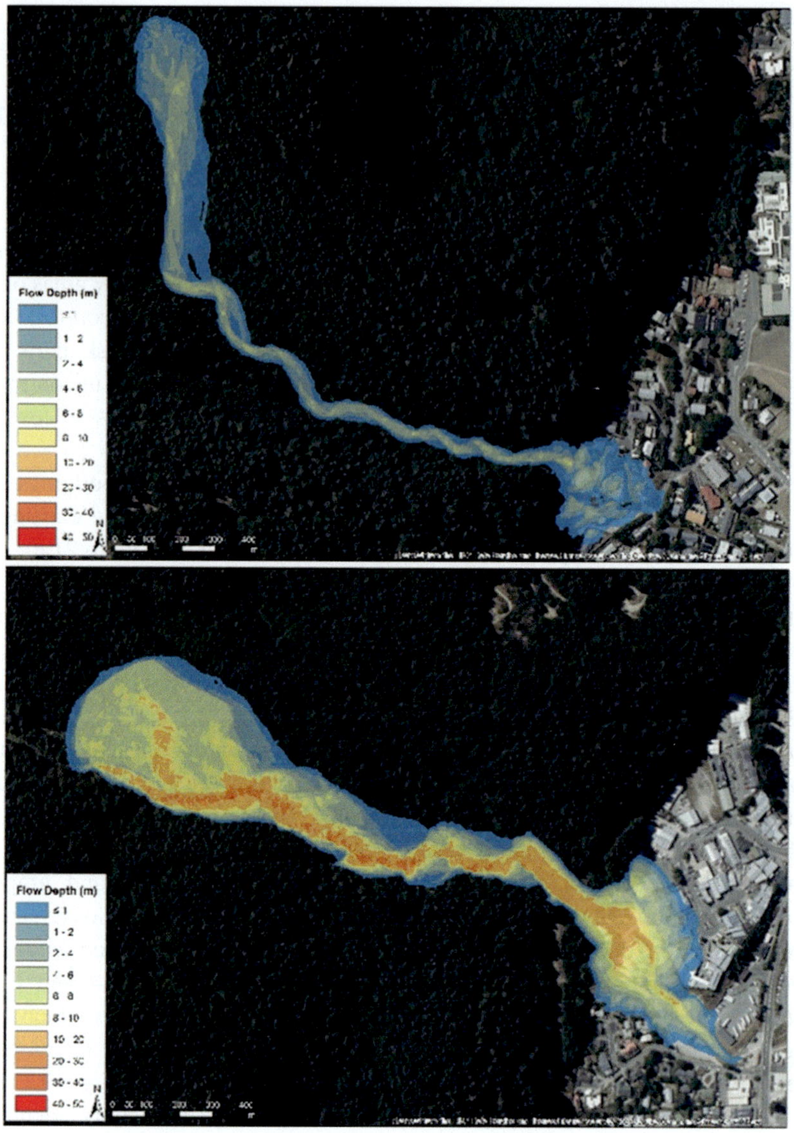

Fig. 19.7 Spatial representations of the hazard layers created by Beca (2020) used to estimated debris flow losses by RiskScape™, Reavers Lane (top, 2,500 year return period) and Brewery Creek (bottom, 2,500–10,000 year return period)

events at both locations when compared to the 2120 Uncontrolled policy option. Notably, while the Managed, Reduced and Engineering options result in reduced losses, this study did not assess the cost of implementing each policy option (cost versus benefit analysis). RiskScape™ results were used to inform a more comprehensive socio-economic assessment which included cost versus benefit analyses (Foy et al., 2021) for QLDC to seek public input to the preferred land use policy option.

While the preferred land-use policy option will reduce some of the risk from debris flows at the study areas, residual risk to life and property

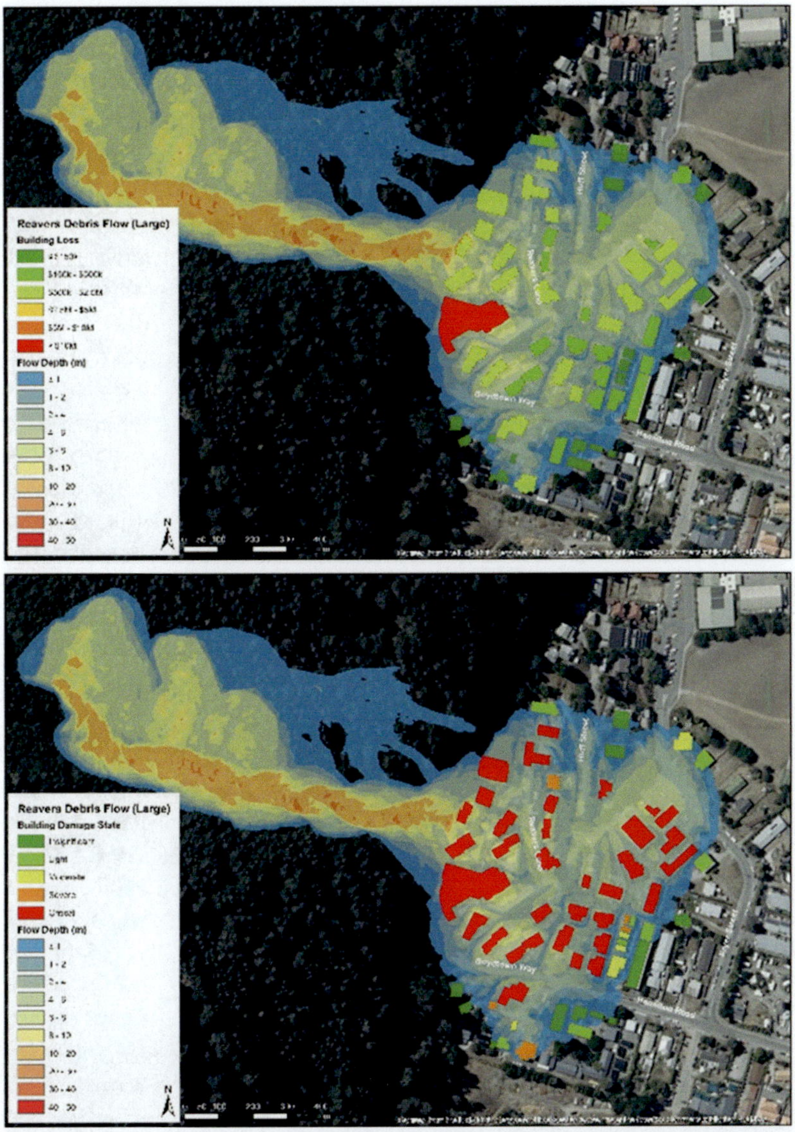

Fig. 19.8 Maps showing the modelled distribution of direct monetary losses (top) and associated building damage states (bottom) by RiskScape™ for a large (6,700–20,000 year return period) debris flow at the Reavers Lane study area

remains. As such, Woods et al. (2021) presents a range of options for local emergency management authorities to consider to further minimise the remaining residual risk.

Using the results of these assessments and options, public consultation and feedback is underway to decide the next land use steps to be decided by elected officials. Possible next steps include the decision to undertake further investigations into one or more preferred option, or notify a plan change for the preferred option.

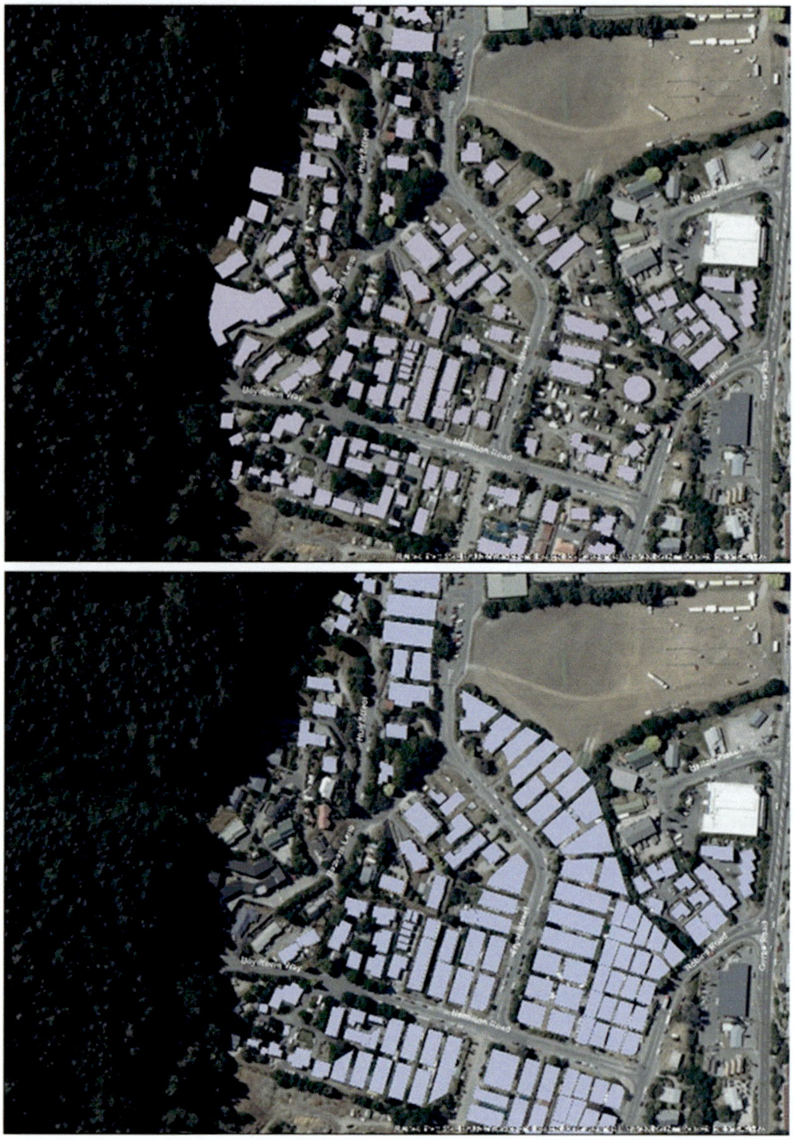

Fig. 19.9 Maps showing the baseline exposure dataset (top) compared to the exposure representation of one of the proposed policy options (bottom). Note an increase in urban form on the periphery of the alluvial fan and removal of buildings on the western fan extent (Woods et al., 2021)

19.6.3 Managed Retreat from Debris Flow Risk: Matatā, New Zealand

Matatā is a small coastal settlement of around 700 people located on the Bay of Plenty coast of the North Island. The township is positioned on a strip of land between the Pacific Ocean and steep hills formed from historic debris avalanches and debris flows from deeply incised stream catchments. On 18 May 2005, the township was hit by several debris flows triggered by heavy rain in four adjacent stream catchments. Twenty-seven houses were destroyed, and 87 properties damaged. Although there were no fatalities in the 2005 event, the scale and extent of boulder and tree deposition in that event was a widely recognised serious threat to

Table 19.4 An example of total aggregated building debris flow losses for the Reavers Lane study area (black text). Red text shows the actual and percentage differences between the uncontrolled and baseline scenarios. Green text shows the actual and percentage differences between managed, reduced and engineering scenarios compared to uncontrolled

	Return Period (Years)	2020 Baseline ($M)	2120 Uncon-trolled ($M)	2120 Man-aged ($M)	2120 Reduced ($M)	2120 Engi-neering ($M)
Total Losses	100–2500 (Small)	13.5	19.1 (5.5) (41%)	10.3 (8.8) (-46%)	0 (19.1) (-100%)	N/A
	2500–6700 (Moder-ate)	17.1	34.6 (17.4) (102%)	15.4 (19.2) (-55%)	1 (33.6) (-97%)	27.1 (7.5) (-22%)
	6700–20,000 (Large)	39.2	132.5 (93.3) (238%)	66 (66.5) (-50%)	20.9 (111.6) (-84%)	127.2 (5.3) (-4%)

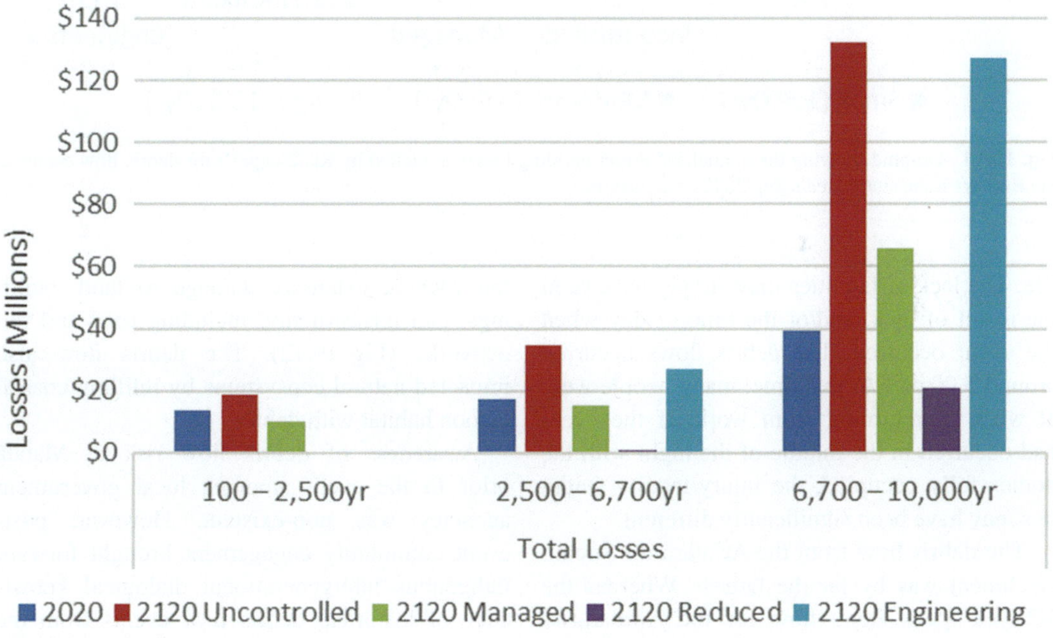

Fig. 19.10 Graphic showing the direct building losses modelled by RiskScape™ for debris flow events at the Reavers Lane study area using 2020 as a baseline

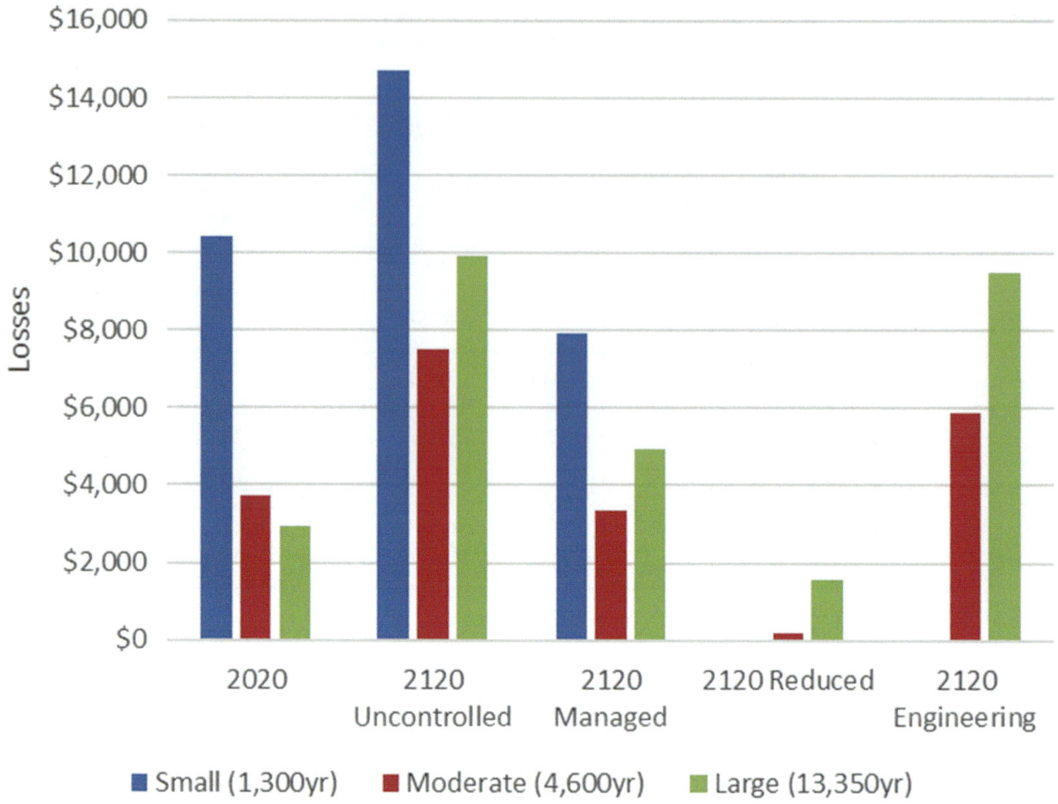

Fig. 19.11 Graphic showing the annualised direct building losses modelled by RiskScape™ for debris flow events at the Reavers Lane study area using 2020 as a baseline

life. The lack of fatalities may simply have been the result of luck, and/or the time of day when the event occurred. The debris flows occurred around 5.00 pm. At this time, many people were at work or returning from work. If the event had occurred in the middle of the night with the homes fully occupied, the injury/fatality statistics may have been significantly different.

The debris flow from the Awatarariki Stream catchment was by far the largest. Whereas the 100-year peak flood flow for the Awatarariki Stream is 60 m^3/s, the peak debris flow during the 2005 event was estimated at 700 m^3/s, a factor of 12 higher. Approximately 300,000 m^3 of debris was deposited on the Awatarariki debris fan causing extensive damage to land, buildings, and infrastructure including road and rail networks (Fig. 19.12). The debris flow also impacted natural ecosystems by infilling coastal lagoon habitat with debris.

Awareness of debris flow risk at Matatā prior to the event amongst local government agencies was non-existent. However, post-event community engagement brought forward indigenous intergenerational dialogical knowledge of historical debris flow events from the Awatarariki catchment (Shearer, 2005). The debris fan was considered tapu (sacred) by some Māori elders due it not only being the site of a famous battle, but importantly, also due to the

Fig. 19.12 Awatarariki Debris Fan (Matatā) after the May 2005 debris flow (Photo credit: GNS Science)

destructive nature of the wandering tail of the taniwha (large supernatural beings) that live in the Awatarariki stream.

An initial estimate of the rainfall return period that triggered the 2005 debris flows was 200 to 500 years (McSaveney et al., 2005). A recent updated meteorological report (Blackwood & Basset, 2019) which accounting for climate change projections, estimates that, by the turn of the century, a storm of the same magnitude as the 2005 event could have a return period of between 40 to 80 years. As the 4.5 km^2 steep deeply incised Awatarariki catchment provides an unlimited sediment supply, rainfall-initiated debris flows would therefore occur more frequently within the Awatarariki catchment in the future.

Initially, the Whakatāne District Council decided to construct engineered Debris-Flow Control Systems in all the affected catchments to protect the community from future debris-flow events up to the magnitude experienced in 2005. This was successfully implemented in three catchments, albeit with some community opposition over need and cost to District ratepayers.

However, this engineered approach proved not feasible for the Awatarariki Stream catchment. Various barrier and diversion options were considered. Due to uncertainties, the preferred method of partial containment with a flexible 'ring net' barrier to retain 100,000 m^3 of debris, a spillway, and raised building platforms was revised during the engineering peer review

Fig. 19.13 Awatarariki Debris Flow Risk Areas, Matatā, New Zealand. Pink = high risk area; orange = medium risk area; yellow = low risk area (Tonkin & Taylor Ltd, 2015)

process to a 'ring net' barrier only for containment of 300,000 m^3 of debris, an amount equivalent to the total amount of debris deposited on the fan during the May 2005 event.

Technical approval by the regulatory agency in this method was challenging due to uncertainties generated by the scale of the proposal with the ring net barrier being several orders of magnitude larger than any similar application of this technology elsewhere. When whole of life-cycle costs (including potential replacement every 50 years) and clean up and rebuild costs following an event were factored in, the elected members of the Whakatāne District Council resolved to abandon the engineering approach and decided to pursue land use planning-based options.

In 2013, the Whakatāne District Council completed a quantitative hazard assessment for landslides and debris flows at Matatā. A debris-flow risk assessment for the Awatarariki debris fan was completed in 2015. The risk assessment identified the levels of risk to life and property on parts of the fan as being high, i.e. greater than $10^{-4}a^{-1}$ (Fig. 19.13).

Management of significant risks from natural hazards is a matter of national importance under the Resource Management Act (RMA), the cornerstone land use planning statute in New Zealand. The Bay of Plenty Regional Policy Statement natural hazard policies takes a risk-based approach similar to that outlined earlier. The policies were promulgated under the RMA and required the Whakatāne District Council to reduce identified high risk natural hazard locations to a lower level of risk (medium, and lower if practicable). This means that property owners were not at liberty to tolerate "high" risks.

The Whakatāne District Council developed the Awatarariki Debris-Flow Risk Management Programme to reduce the high loss-of-life risk and thereby comply with the Resource Management Act and the Bay of Plenty Regional Policy Statement.

Risk reduction options considered included: engineering or structural interventions;

catchment management; and early warning and evacuation. These options were identified as having potential to achieve the required risk reduction but were excluded upon investigation because they did not meet the statutory tests of being "reasonably practicable" and were not aligned with a precautionary policy approach.

For example, a best practice multi-staged early warning system was developed but ultimately not implemented. Testing this system proved it would be unlikely to enable all people potentially present in the hazard zone to evacuate safely.

The catchment has a healthy indigenous forest cover. Investigation showed that a debris-flow triggering storm event would override any any catchment management/soil conservation measure such as vegetation enhancement, physical land contouring, or any other structural measure. The risk of future debris flows caused by extreme rainfall would therefore not be materially changed through catchment management initiatives.

Some individual property owners did not accept the results of either the risk assessment or the risk reduction investigations. The most effective measure to reduce risk was identified as managed retreat which included a proposal to enable relocation for owners of property within the high-risk area.

There is no statutory power under existing New Zealand Statute to force acquisition of land that is affected by high-risk natural hazards. A detailed business case to support funding of voluntary managed retreat was successfully negotiated through District, Regional and Central Government.

Owners of all the 34 affected private properties reached agreement for sale of their property to the local authority and all houses have were removed from the high-risk zone between 2019 and 2022. Keys to the success of the acquisition programme were the financial quantum of the acquisition offers and that no monetary contribution was required from property owners. Acquisition offers reflected the market value of the property at the time the offer was made with no discount applied for the presence of the debris-flow risk. This meant property owners received sufficient funds to relocate to an equivalent property.

Coupled with the voluntary managed retreat programme, the land use planning approach for managing debris-flow risks on the Awatarariki Fan area needed to be changed to appropriately address the significant risk from the debris flow hazard that had been identified. This provided a regulatory backstop if the voluntary managed retreat programme proved to be ineffective.

Changes to the Operative Whakatāne District Plan included identifying a Debris- Flow Policy Area on the planning maps, rezoning the high-risk area from "Residential" to a restrictive "Coastal Protection Zone" and prohibiting all activities in the high-risk area, other than those that relate to transitory recreational use of open space.

Because of legally enshrined existing use rights, changes to the local District Plan rules could only be effective in managing new development or redevelopment. However, because a Regional Plan is can apply to activities with existing use rights, a Regional Plan rule can be used to remove existing residential activities within the respective regional council's jurisdictional area that are subject to high natural hazard risk.

The process of obtaining commitment for funding resulted in the voluntary nature of the managed retreat programme largely being removed. Therefore, changes to the Operative Regional Natural Resources Plan were made including policies that state the intention to reduce the natural hazard risk on the Awatarariki Fan from high risk (i.e. greater than 10^{-4}) to at least a medium risk level (i.e. greater than 10^{-3} but less than 10^{-4}) and a rule prohibiting residential activities on residential sites within the high-risk area after 31 March 2021.

It is unlikely that these land use regulations will need to be applied as all the residential uses have or will cease, resulting in low risk.

Changes are currently being considered to New Zealand's land use planning legislation, including improved management of natural hazard risk. Central Government policy development has commenced on a new Climate Adaptation Act to specifically address the complex governance, policy and funding challenges associated with managed retreat.

While this Matatā case study has a climate change component (increased frequency of high intensity rainfall events), the primary driver was a public policy response to managing an identified high loss-of-life risk to a small community. As such, it has provided New Zealand policy makers with a valuable opportunity to better understand the challenges involved where managed retreat is identified as the preferred natural hazard risk reduction solution. It has highlighted the need for a clear policy framework on standardizing risk assessments definition of a tolerable risk threshold, retrospective interventions such as managed retreat, and how programmes of managed retreat should be accessed, funded, and delivered. The present gap in national guidance on these issues contributed to a protracted and challenging planning process at Matatā (more than a decade) which increased the cost and impacts on the affected community.

19.7 Discussion

Two approaches are required to manage the risks from debris flows through land use planning: one that addresses legacy planning issues, and one that addresses development before it occurs. Each of these case studies presents a different planning approach to the debris-flow hazard that the specific community faces. But there are also commonalities between the case studies, in particular: (1) debris-flow hazard can be mapped, and land use planning controls used to restrict or reduce development; (2) planning controls can be used to manage legacy development issues (i.e. if a debris fan is developed, the risk can still be managed and reduced).

The Squamish example provides a proactive approach to planning around a known debris fan, with land use planning restrictions linked to each of the five mapped debris-flow hazard zones, and flood hazard areas. These restrictions are based on the mapped areas, which can become out of date if there is no process for updating information. In this example, the Squamish plan allows for the mapped debris zones to be updated when new information becomes available, so that the restrictive development policies can apply to any extended mapped areas.

The Queenstown Lakes District Council example presents a reactive planning approach, where the development has already occurred in the debris-flow areas, and the risk now requires management. This is a common occurrence, where legacy planning issues and development have resulted in debris fans being developed, when the hazard information was not available and/or the debris fan features were not identified prior to development. Dollar losses can be modelled, and will aid decision makers and the community on the future planning response to manage the risk.

In Matatā, the risk from a future debris flow was deemed too high to allow development to continue in the mapped debris-flow risk area, so rather than proposed planning restrictions while allowing the land use to continue, the decision was made to instigate a managed retreat process to remove all development from the mapped debris-flow risk area. These is at the extreme end of the spectrum of reactive land use planning responses, but a very effective option to remove the risk.

In each case, risk assessments are key to understanding the hazard, understanding the risk, and identifying and evaluating the risk treatment options. With many communities around the world now experiencing the direct impacts of a changing climate, rainfall intensities are increasing, and new or reactivation of debris flows must be planned for, regardless of whether development has already occurred or is planned. With the ability to map the hazard, assess risk and administer planning controls, many communities will need to start using these tools to better prepare and reduce risks from debris flows.

19.8 Conclusions

Land use planning is a key risk treatment option for managing the risks from debris flows. A spectrum of policy responses is available for managing debris flow areas, from having permitted activities, such as forestry and agriculture, to the most restrictive, prohibiting development and relocating existing land uses, such as in the Matatā case study, where residential homes were purchased and removed from the high-risk debris-flow area.

A risk-based planning approach provides a robust and transparent method for controlling land use in debris-flow areas. While it is easiest to avoid inappropriate development on debris fans rather than manage the risk once developed, legacy planning issues (such as not being aware of the debris flow risk prior to development) have resulted in the management of existing risks being a more common response. Each of the case studies showed how the risks can be managed in existing developed areas.

There are many challenges associated with managing land use on debris fans, such as having robust information for assessments and for the public, understanding risk assessments and the options and implications they have, having the political support for restrictive land use controls, and gaining community acceptance of land use controls. With other tools, such as physical mitigation works, warning systems, and insurance, good land use planning can limit and reduce the risks communities face from debris flows.

References

Baker, M. (2014). Land-use planning. In A. C. Michalos (eds.), *Encyclopedia of quality of life and well-being research*. Springer. https://doi.org/10.1007/978-94-007-0753-5_1600. Accessed 22 Nov 2021. https://link.springer.com/referenceworkentry/10.1007%2F978-94-007-0753-5_1600#howtocite

Barclay, S. (2013). The meaning of cartography: Perspectives from Collins Bartholomew and the *Times Atlas*. *The Cartographic Journal, 50*(2), 121–127.

Bay of Penty Regional Council. (2016). *Appendix M—natural hazards risk reduction measures*. Bay of Penty Regional Policy Statement. https://atlas.boprc.govt.nz/api/v1/edms/document/A4439707/content

Beca. (2020). Natural hazards affecting Gorge Road, Queenstown. Christchurch (NZ): Beca. 74 p. + appendices. Prepared for Queenstown Lakes District Council.

Blackwood, P., & Bassett, T. (2019). Matatā flooding 18 May 23005: Meteorology update. Report for Whakatāne District Council.

District of Squamish. (2017). Memo to qualified professionals, prospective homeowners, builders, and developers in the Cheekeye Fan. Interim Development Strategies on the Cheekeye Fan. From District of Squamish, Development Services, 14 July 2017.

Foy, R., Yeoman, R., & Ashby, H. (2021). Gorge road hazards, social and economic impacts report. M.E Consulting, Auckland, New Zealand, 118p.

Greiving, S., & Fleischhauer, M. (2006). Spatial planning response towards natural and technological hazards. In P. Schmidt-Thomē (ed.), *Natural and technological hazards and risks affecting the spatial development of European regions* (pp. 109–123). Special Paper 42, Geological Survey of Finland.

Greiving, S., Fleischhauer, M., & Lückenkötter, J. (2006). A methodology for an integrated risk assessment of spatially relevant hazards. *Journal of Environmental Planning and Management, 49*(1), 1–19. https://doi.org/10.1080/09640560500372800

Kelly, S. D., Saunders, W. S. A., Payne, B., & Mathieson, J. (2020). Mapping natural hazards and risk for land-use planning for district plans. GNS Science report 2020/20, Lower Hutt (NZ): GNS Science. 52 p.

Massey, C. I., Thomas, K.-L., King, A. B., Singeisen, C., Taig, T., & Horspool, N. A. (2019). SLIDE (Wellington): vulnerability of dwellings to landslides (Project No. 16/SP740). Lower Hutt (NZ): GNS Science. 76 p. (GNS Science report; 2018/27).

McSaveney, M. J., Beetham, R. D., & Leonard, G. S. (2005). The 18 May 2005 debris flow disaster at Matatā: Causes and mitigation suggestions. Institute of Geological and Nuclear Sciences for Whakatāne District Council, Client report 2005/71.

Ministry for the Environment. (2022). Aotearoa New Zealand's first national adaptation plan. Ministry for the Environment, Wellington. https://environment.govt.nz/assets/publications/climate-change/MFE-AoG-20664-GF-National-Adaptation-Plan-2022-WEB.pdf

NZPI. (2021). What is planning. New Zealand planning institute. Accessed 22 Nov 2021. https://planning.org.nz/planning-in-nz/what-is-planning

Paulik, R., Horspool, N., Woods, R., Griffiths, N., Beale, T., Magill, C., Wild, A., Popovich, B., Walbran, G., & Garlick, R. (2022). RiskScape: A flexible multi-hazard risk modelling engine. *Natural Hazards*. https://doi.org/10.1007/s11069-022-05593-4PIA, 2015: National land use planning guidelines for disaster resilient communities. Planning Institute of Australia/Australian Government Attorney-General's Department, 99p.

Planning Institute of Australia (PIA). (2015). National land use planning guidelines for disaster resilient communities. https://www.planning.org.au/documents/item/7804

Quality Planning. (2017). Further information examples—zoning as a tool in plans. https://www.quality-planning.org.nz/node/621#zoning. Accessed 20 Feb 2022.

Queensland Reconstruction Authority. (2012). Planning for stronger more resilient floodplains. Queensland Reconstruction Authority, 91p.

Saunders, W. S. A., & Becker, J. (2015). A discussion of resilience and sustainability: Land use planning recovery from the Canterbury earthquake sequence, New Zealand. *International Journal of Disaster Risk Reduction, 9*, 10p.

Saunders, W. S. A., Beban, J. G., & Kilvington, M. (2015). Risk-based land use planning for natural hazard risk reduction. GNS Science Miscellaneous Series 67, GNS Science, Lower Hutt. https://www.gns.cri.nz/Home/RBP/Risk-based-planning/A-toolbox

Shearer, I. (2005). Draft report to Matata recovery projects Part D Historical Research.

Sliuza et al. (2015). https://www.cdema.org/virtuallibrary/index.php/charim-hbook/use-case-book/2-landuseplanning/2-1-national-land-use-plan

The Global Goals. (2022). 11—sustainable cities and communities. https://www.globalgoals.org/goals/11-sustainable-cities-and-communities/. Accessed 22 March 2022.

Tonkin & Taylor Ltd. (2015). *Supplementary risk assessment*. Debris Flow Hazard, Matatā, Bay of Plenty, Client Report for Whakatāne District Council.

UNDRR. (2020). Implementation guide for land use and urban planning—words into action. United Nations Office for Disaster Risk Reduction, Geneva, 45p. https://www.preventionweb.net/files/67430_landuseandurbanplanningforpublicrev.pdf. Accessed 23 Nov 2021.

United Nations. (2021). *Disaster risk reduction*. UN Department of Economic and Social Affairs: Sustainable Development. https://sdgs.un.org/topics/disaster-risk-reduction. Accessed 25 Nov 2021.

United Nations. (2022). Sustainable development goals. https://www.un.org/sustainabledevelopment/sustainable-development-goals/. Accessed 22 March 2022.

Woods, R. J., Thomas, K.-L., Scheele, F. R., Canessa, S., & Horspool, N. A. (2021). Queenstown debris flow and rockfall loss modelling for land-use planning policy options. Lower Hutt (NZ): GNS Science. 35 p. Consultancy Report 2021/07.

Correction to: Advances in Enbris-flow Science and Practice

Matthias Jakob, Scott McDougall and Paul Santi

Correction to:
M. Jakob et al. (eds.), *Advances in Debris-flow Science and Practice*, Geoenvironmental Disaster Reduction,
https://doi.org/10.1007/978-3-031-48691-3

This book published under the series "Geoenvironmental Disaster Reduction (Series ID: 16107)" was inadvertently published with the incorrect series ISSN number. This has now been amended in the book to the correct ISSN:

Print ISSN: **2946-5710**
Electronic ISSN: **2946-5729**

The updated version of the book can be found at
https://doi.org/10.1007/978-3-031-48691-3

M. Jakob et al. (eds.), *Advances in Debris-flow Science and Practice*, Geoenvironmental Disaster Reduction,
https://doi.org/10.1007/978-3-031-48691-3_20

GPSR Compliance

The European Union's (EU) General Product Safety Regulation (GPSR) is a set of rules that requires consumer products to be safe and our obligations to ensure this.

If you have any concerns about our products, you can contact us on ProductSafety@springernature.com

In case Publisher is established outside the EU, the EU authorized representative is:

Springer Nature Customer Service Center GmbH
Europaplatz 3
69115 Heidelberg, Germany

Batch number: 10731137

Printed by Printforce, the Netherlands